# ECOLOGY

## Global Insights & Investigations

**Peter Stiling**
University of South Florida-Tampa

ECOLOGY: GLOBAL INSIGHTS & INVESTIGATIONS, SECOND EDITION

Published by McGraw-Hill Education, 2 Penn Plaza, New York, NY 10121. 

Some ancillaries, including electronic and print components, may not be available to customers outside the United States.

This book is printed on acid-free paper.

1 2 3 4 5 6 7 8 9 0 RMN/RMN 1 0 9 8 7 6 5 4

ISBN 978-1-25-925231-0
MHID 1-25-925231-0

www.mhhe.com

# Brief Contents

# About the Author

**Peter Stiling** obtained his PhD from University College, Cardiff, Wales, in 1979. Subsequently, he became a postdoctoral fellow at Florida State University and later spent two years as a lecturer at the University of the West Indies, Trinidad. Dr. Stiling is currently a Professor of Biology at the University of South Florida at Tampa. He teaches graduate and undergraduate courses in ecology and environmental science as well as introductory biology. He has published over 120 scientific papers in journals such as *Ecology, Oecologia, Oikos, Global Change Biology, Biological Invasions,* and *Science,* and has received funding from the National Science Foundation, the National Institute for Global Environmental Change, and others. He is also a co-author of *Biology,* a text for biology majors also published by McGraw-Hill. Dr. Stiling's research interests include plant-insect relationships, parasite-host relationships, biological control, restoration ecology, coastal biology, and the effects of elevated carbon dioxide levels on plant-herbivore interactions. In 2012 he was elected a Fellow of the American Association for the Advancement of Science, and in 2013 he was awarded the Theodore and Venette Askounes-Ashford Distinguished Scholar Award for the University of South Florida.

**About the Cover** Gray wolves were exterminated from the western United States by the 1930s and from the western Great Lakes by the middle of the century. As a result, densities of elk, their prey, increased and densities of aspen and willow trees, eaten by the elk, declined. Furthermore, densities of beavers and songbirds, both of which utilized these tree species, were reduced. In 1995 the U.S. Fish and Wildlife Service introduced 66 wolves from Southwestern Canada into Yellowstone Park and Central Idaho. As explained in this book, the numbers of wolves increased rapidly, reaching about 6,100 in the northern Rocky Mountains and western Great Lakes area. The elk declined, tree density increased, and songbird numbers went up. In June 2013 the federal government moved to end endangered species protection for gray wolves in the lower 48 states, opening up the possibility of wolf management by states, once again upsetting the balance of nature in such areas.

# Preface

## Unique Approach

I have been an active researcher and teacher of ecology at the University of South Florida for 20 years. Most of my students are sophomores, juniors, or seniors who have completed their prerequisites in mathematics, chemistry, and basic biology classes. Many of these students want to connect what they learn in class to real-world problems. One of the biggest changes that we as a society face is global change, those alterations in the global environment that may alter the Earth's ability to sustain life. As defined by the U.S. Global Change Research Project, such alterations include changes in climate, land productivity, oceans or other water resources, atmospheric chemistry, and ecological services. Global change is effected by a variety of factors, including elevated atmospheric carbon dioxide, sulfur dioxide, and nitrogen dioxide; invasive species; habitat destruction; overharvesting; and water pollution. Thus, one of my goals in creating this ecology textbook is to show how ecological studies are vital in understanding global change. A focus on global change, however, is just one of the book's innovative areas of emphasis.

This book makes use of McGraw-Hill Education's unique Connect system, a powerful online learning assignment and assessment solution. For each chapter, there are multiple-choice, true/false, matching, or ranking questions, plus additional types of questions relating to art and photographs, art labeling, and fill-in-the-blank. I have found that the use of such material as homework results in an improvement in grades. Nearly all my students have responded positively to this approach, as it keeps their skills sharp and helps them prepare for exams. I hope your students will find it useful, too.

# Features

## Global Insight

The book's main aim is to teach the basic principles of ecology and to relate these principles to many of the Earth's ecological challenges, of which there are many. For example, climate change threatens to warm many areas of the globe and change precipitation patterns. The ranges of species as diverse as birds and butterflies have changed in response to global warming. Invasive species, from plants to pythons, outcompete many native species in all areas of the planet. Overhunting has caused numerous fish stocks to plummet. Pollution has impacted many aquatic animals. Most chapters include a "Global Insight" feature that highlights how such global change is important in all areas of ecology, from organismal to ecosystem ecology.

In the United States, the worst oil spill in the country's history occurred on April 20, 2010. The *Deepwater Horizon* semi-submersible offshore drilling platform burned and sank about 41 miles off the Louisiana coast. Oil was estimated to be flowing at 20,000–40,000 barrels (3.2–6.4 million liters) per day, causing a resultant oil slick of at least 2,500 square miles (6,500 $km^2$). It is widely thought that the spill has resulted in an environmental disaster with extensive and long-lasting impacts on marine and coastal habitats, especially salt marshes.

**Global Insight**

**Biological Control Agents May Have Strong Nontarget Effects**

In the 1970s, two gall flies, *Urophora affinis* and *U. quadrifaciata,* were introduced as biological control agents of diffuse knapweed, *Centaurea diffusa,* the weed discussed in the Feature Investigation (see Figure 1.A). The flies lay their eggs inside the flowerheads and create a tumor-like swelling called a gall, inside which the fly larvae feed. The hope was that the gall flies would reduce flowerhead production and, thus, the spread of the weed. Unfortunately, seed reductions have not been enough to control the plant. As a result, extensive knapweed and gall fly populations now coexist in many states (**Figure 1.B**).

Dean Pearson and Ragan Callaway (2006) showed that deer mice, *Peromyscus maniculatus,* feed on knapweed galls, and the increased gall production has provided a food subsidy, enabling deer mice populations to double in size. This is troubling because deer mice are a reservoir of Sin Nombre hantavirus, which causes the deadly hantavirus pulmonary syndrome in humans. Mice testing positive for hantavirus were over three times more abundant in the presence of gall-infested knapweed than in areas where it was absent.

This story has at least two lessons. First, unless biological control agents effectively reduce their target populations, releasing these agents may do more harm than good. Second, it is important to take into account the many complex interactions between species in nature. The chapters that follow provide more evidence of this process.

## Feature Investigation

Throughout the book, numerous studies have been provided to illustrate ecological hypotheses, but often, because of space constraints, experimental methods and results are presented in summary fashion. However, it is valuable to give students a firsthand look at how ecological studies are conducted. To this end, most chapters contain a "Feature Investigation" that outlines a hypothesis being tested, the methods researchers use to perform their studies, the data they collected, and the conclusions they reached. Some of these Feature Investigations are classic studies; others present cutting-edge research and techniques.

**Feature Investigation**

**Reto Zach Showed How Large Whelks Are the Optimum Prey Size for Crows**

At coastal habitats in western Canada, crows commonly feed on whelks. The crows are not strong enough to break open the whelks on their own, so they fly into the air and drop the whelks onto the rocky shore below, where they break open. Reto Zach (1979) made a series of interesting observations about northwestern crows, *Corvus caurinus,* foraging on whelks, *Thais lamellosa.* First, crows selected only large whelks, about 3.8–4.4 cm long. Second, they always seemed to fly about 5 m high before they dropped the whelk. Third, they performed this same 5-m-high flight many times until the whelk broke. A series of questions arose. Why do crows select only large whelks? Why do they fly to about 5 m high before dropping the whelk? Why do they repeat this operation even if the whelk did not break the first time? Why don't they fly higher?

Zach assumed that crows were foraging in an optimal manner, and he set out to test this assumption experimentally. First, he erected a 15-m tower on a rocky beach, complete with a movable platform the height of which could be adjusted. Next, he collected small, medium, and large whelks and dropped them from various heights. At low heights, the whelks would not break unless they were dropped scores of times. At about 5 m, whelks began to need fewer drops to break, but large whelks, being heavier, required the fewest drops of all (**Figure 4.A**). Increasing the height of the drop above 5 m did little to change the results. Furthermore, the chance of a whelk breaking was independent of the number of previous drops. The probability of breakage of large whelks at 5 m was about 1 in 4.

Finally, Zach calculated the average number of kilocalories the crows spent in ascending to 5 m and dropping the whelks, and obtained a value of 0.5 kilocalories. He determined that the calorific value of the large whelk was 2.0 kilocalories, so there was an average net gain of 1.5 kilocalories from foraging on large whelks if the whelk broke on the first drop. The net calorific gain from medium whelks, which require more drops to break, was actually a net loss of 0.3 kilocalories, and foraging on small whelks was even less profitable. Thus, the decision of crows to forage only on large

**HYPOTHESIS** Crows forage optimally on coastal whelks.

**STARTING LOCATION** Beaches of Mondarte Island, British Columbia, Canada.

| | Experimental level | Conceptual level |
| --- | --- | --- |
| 1 | Collect sample of 58 live whelks of the same size as those taken by crows. | Determine size of whelks eaten by crows. |
| 2 | Erect 15 m high pole on beach with a small platform from which whelks could be dropped from various heights. | Determine height required for breaking whelks. |
| 3 | Drop whelks of three size classes: small = 1.6–2.2 cm long, medium = 2.7–3.3 cm, and large = 3.8–4.4 cm. Whelks of each size class were dropped from 2, 3, 4, 5, 6, 7, 8, 10, and 15 m. | Determine breakage height of small, medium, and large whelks. |

4 THE DATA

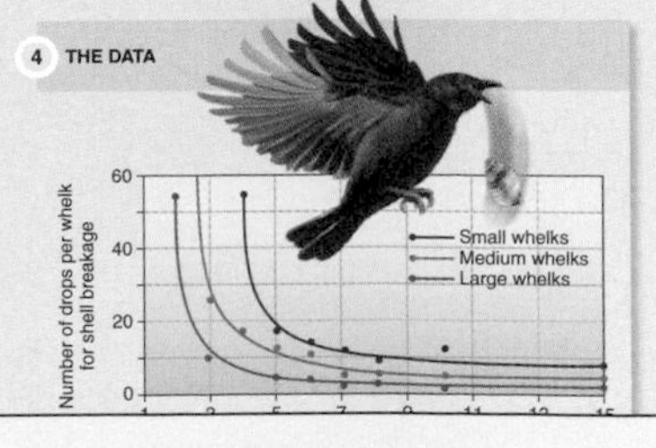

## Learning System

Each chapter begins with an outline that consists of the main section headings contained within each chapter. These headings are written in the form of an ecological statement that summarizes the material within the section. For this reason, reading the chapter outlines provides a summary of the entire chapter. Following the outline is a vignette, which provides a modern case history relevant to the chapter, followed by a one- or two-paragraph synopsis that introduces students to some key terms and summarizes what is to follow in the body of the chapter. Each example illustrates a hypothesis raised in the chapter.

# Features

## Annotated Art Program

Most of the chapters are built around the art. You cannot easily teach an ecological concept without reference to the data. Each piece of art has been carefully rendered or selected to illustrate a particular ecological concept. The art house, Laserwords, has provided some of the best renderings of line art that are available today. In addition, the photo researchers, Photo Affairs, Inc., have provided new and original photographs that increase the "wow" factor of the illustrations. Often the art is associated with an Ecological Inquiry question, which prompts students to think in more detail about the concept or example.

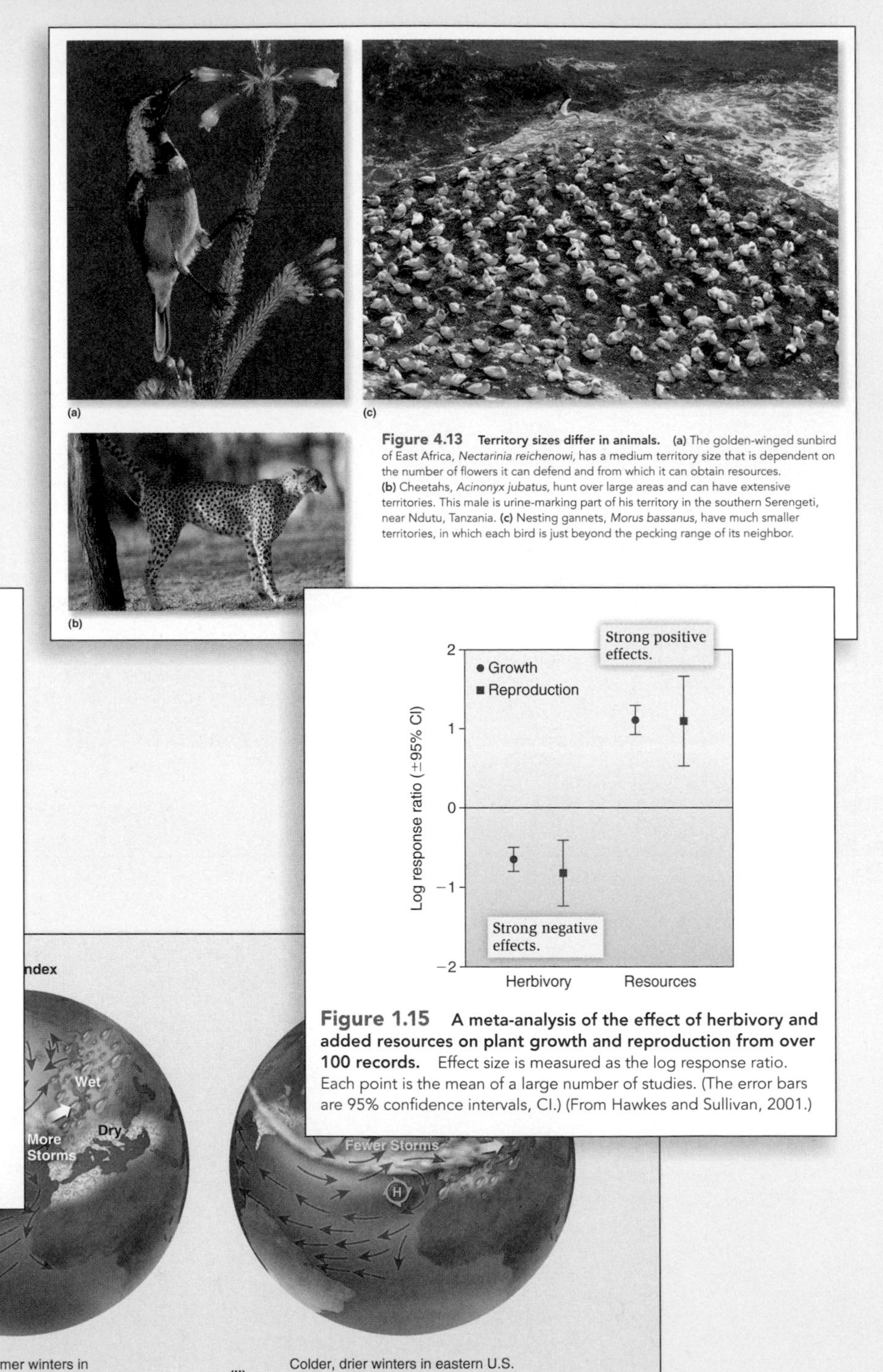


**Figure 4.13 Territory sizes differ in animals.** **(a)** The golden-winged sunbird of East Africa, *Nectarinia reichenowi*, has a medium territory size that is dependent on the number of flowers it can defend and from which it can obtain resources. **(b)** Cheetahs, *Acinonyx jubatus*, hunt over large areas and can have extensive territories. This male is urine-marking part of his territory in the southern Serengeti, near Ndutu, Tanzania. **(c)** Nesting gannets, *Morus bassanus*, have much smaller territories, in which each bird is just beyond the pecking range of its neighbor.

**Figure 1.15 A meta-analysis of the effect of herbivory and added resources on plant growth and reproduction from over 100 records.** Effect size is measured as the log response ratio. Each point is the mean of a large number of studies. (The error bars are 95% confidence intervals, CI.) (From Hawkes and Sullivan, 2001.)

**Figure 4.B The North Atlantic Oscillation.** **(i)** Positive phase: mild, wet North American and northern European winters. **(ii)** Negative phase: cold, drier, but snowy, North American and northern European winters.

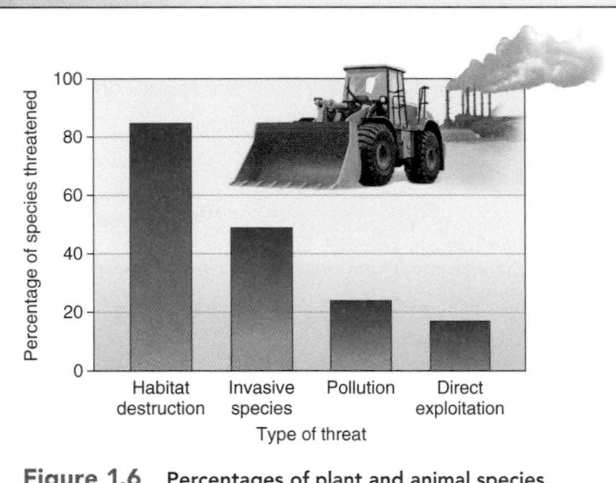


**Figure 1.6 Percentages of plant and animal species threatened by various causes.** Species can suffer from multiple threats, so categories do not sum to 100%. (From data in Wilcove et al., 1998.)

**ECOLOGICAL INQUIRY**

Habitat degradation is of paramount importance to all groups, but pollution and overexploitation affect some taxa more than others. Why is this?

## Sections and Chapters

The book is structured around seven sections that represent the core ecological disciplines:

- Organismal Ecology
- Physiological Ecology
- Population Ecology
- Species Interactions
- Community Ecology
- Biomes
- Ecosystems Ecology

The chapters within these sections are not equal in length, and while some may be taught in one class, others will need to be explored over a whole week. In each chapter, all important ecological terms are in boldface at their first mention in the text, or if not, in their appropriate chapter or subsection. In each case, terms are defined within the sentence in which they are used. Although the book contains an extensive glossary, most students should be able to understand these ecological terms at their first mention. Ecology is a rich field of hypothesis testing, and many of the ecological hypotheses that are being tested are boldfaced. The existence of multiple hypotheses to explain the same phenomena underscores what a dynamic and exciting field modern ecology is.

has been transformed by human action. Acid rain is carried from one country to another. Carbon dioxide pumped out by the industrial centers of developed nations has increased in the atmosphere worldwide, from the poles to the equator. More nitrogen is added to the land via fertilizers than by all natural sources combined. Pesticides have been detected in unintended targets, such as human breast milk and the tissues of penguins. Many species have been driven to extinction. Now, more than ever, there is a strong impetus to understand how natural systems work, how humans change those systems, and how in the future we can reverse these changes.

**Check Your Understanding**

**1.1** If an organism is limited in its distribution by cold temperatures, is it limited by biotic or abiotic factors?

**1.2 The Scale of Ecology: From Organisms to Ecosystems**

Ecology ranges in scale from the study of an individual organism through the study of populations to the study of communities and ecosystems (**Figure 1.3**). This section introduces each of the broad areas of organismal, population, community, and ecosystems ecology. This hierarchical scheme is an organizational framework for the succeeding chapters of the book.

**1.2.1 Organismal ecology investigates how individuals' adaptations and choices affect their reproduction and survival**

**Organismal ecology** can be divided into three main subdisciplines: evolutionary ecology, behavioral ecology, and physiological ecology. The first area, **evolutionary ecology**, considers how organisms have evolved to adapt to their environment through interactions with individuals, populations, and other species. The great biologist Theodosius Dobzhansky said, "Nothing in biology makes sense except in the light of evolution." This is as true in ecology as in any other biological discipline. For example, one may argue about what controls penguin numbers in the Southern Hemisphere, but a nagging question remains—why are there no penguins in the Northern Hemisphere? The answer is not insufficient food or too many predators. Penguins evolved in the Southern Hemisphere and have never been able to cross the Tropics to colonize northern waters. Evolutionary ecology addresses the genetic variation that exists within species and how this genetic variation is lessened by factors, such as habitat fragmentation, that can reduce the viability of populations.

**Behavioral ecology** focuses on how the behavior of an individual organism contributes to its survival and reproductive success, which in turn affects the abundance of a population. For example, forest tent caterpillars, *Malacosoma* species, are infamous for residing in silken tents and defoliating trees in deciduous forests (**Figure 1.4**). How can we explain this behavior? During the daytime this tent provides a safe refuge from predators. Group living enables the caterpillars to fabricate a large tent, which a single larva could not easily construct. At night the caterpillars leave the tent to forage. They must locate a tree whose branches have not been defoliated and remember where it is so that they can return to forage on subsequent nights. Here again group living becomes an advantage, because the caterpillars lay down silk trails between the tent and the leaves so that they can relocate the leaves. In addition, the silk trails are marked with pheromones to draw in colony mates. The denser the concentration of caterpillars, the stronger the trail. Group living is therefore advantageous in this species.

The third area, **physiological ecology**, investigates how organisms are physiologically adapted to their environment and how the environment impacts the distribution of species. Here we examine the effects of temperature, water, nutrient availability and other physical factors on the distribution and abundance of species.

**1.2.2 Population ecology describes how populations grow and interact with other species**

**Population ecology** focuses on populations—groups of interbreeding individuals that occur in the same place at the same time. A primary goal is to understand the factors that affect a population's growth and determine its size and density.

Although a population ecologist might focus on studying the population of a particular species, the relative abundance of that species is often influenced by its interactions with other species. Thus, population ecology includes the study of **species interactions**, such as competition, predation, mutualism, commensalism, herbivory, and parasitism. Much ecological theory is built upon the ecology of populations. Knowing what factors impact populations can help us reduce species endangerment, stop extinctions, and control invasive species.

**1.2.3 Community ecology addresses factors that influence the number of species in an area**

In a forest, we can find many populations of trees, herbs, shrubs, the herbivores that eat them, and the carnivores that in turn prey on the herbivores. This assemblage of many populations that live in the same place at the same time is known as a **community**. Communities occur on a wide variety of scales, from small pond communities to huge tropical rain forests. At the largest scales, these communities are known as biomes. Ecologists have examined different terrestrial, coastal, and aquatic biomes and determined the factors that limit the distributions of biomes such as tropical rain forests, temperate deciduous forests, or temperate grasslands. Biodiversity studies focus on why certain areas have high numbers of species (such areas are called species rich), whereas other areas have low numbers of species (these are called species poor).

6 CHAPTER 1 An Introduction to Ecology

## End-of-Chapter Material

Each chapter ends with a unique summary that specifically links each main concept with its corresponding piece(s) of art. There are 10 multiple-choice self-test questions with answers that can be accessed online at the book's website. There are also 3–5 broader conceptual questions that require essay answers of a paragraph or longer. All these questions refer to material that is explained in the textbook.

To highlight the importance of understanding graphs, images, or observations, most chapters also include a "Data Analysis" question that provides data sets and asks the student to analyze the data and provide an explanation or conclusion.

**SUMMARY**

1.1 **Ecology: The Study of Organismal Interactions**

- Many species are under threat from global change (Figure 1.1). Ecologists study the interactions among organisms and between organisms and their environments.
- The tools of ecologists have changed from collecting jars and nets to chemical autoanalyzers and computers (Fi
- Organismal ecology considers how individuals are adapted to their environment and how the behavior of an individual organism contributes to its survival and reproductive success and the population density of the species (Figure 1.4).
- Population ecology explores those factors that influence a population's growth, size, and density.

1.2 The S

- Th of eco

20

**TEST YOURSELF**

| Species | Community A | Community B |
|---|---|---|
| 1 | 4 | 8 |
| 2 | 3 | 6 |
| 3 | 0 | 4 |
| 4 | 11 | 2 |
| 5 | 2 | 0 |

Questions 1–8 refer to the above table.

1. What is the species richness of community B?
   a. 1.279
   b. 0.625
2. What is the Gini-Simpson index of diversity for community A?
   a. 1.279
   b. 0.625
   c. 0.6
   d. 3.60
   e. 4
3. What is the Shannon index for community B?
   a. 1.279
   b. 0.625
   c. 0.6
   d. 3.60
   e. 4

**CONCEPTUAL QUESTIONS**

1. Distinguish between ecology and environmental science.
2. Distinguish between organismal, population, community, and ecosystems ecology.
3. Outline the five-stage process of hypothesis testing.
4. Comment on the advantages and disadvantages of laboratory, field, and natural experiments.
5. What are some of the effects that introduced species can have on native species?

**DATA ANALYSIS**

It is thought that predators such as foxes, raccoons, and skunks can contribute to high death rates of nesting ducks. In one North Dakota study, trappers were paid to remove these three species from a number of different sites over the two-year period 1994–1996. During this time, more than 2,404 predators were removed. Investigators measured the nest success of numerous duck species, including blue-winged teal *(Anas discors)*, mallards *(A. platyrhynchos)*, gadwalls *(A. strepera)*, northern pintails *(A. acuta)*, and northern shovellers *(A. clypeata)*. The data shown give nest success in a number of sites in 1994–1996 at trapped and untrapped sites. Summarize the data and discuss whether you think predators affect nest success.

| Year | Site | Percent Nest Success, Trapped | Site | Percent Nest Success, Untrapped |
|---|---|---|---|---|
| 1994 | 1 | 45 | 2 | 14 |
| 1995 | 1 | 35 | 2 | 28 |
| 1995 | 5 | 56 | 3 | 29 |
| 1995 | 6 | 36 | 4 | 17 |
| 1995 | 8 | 52 | 7 | 24 |
| 1996 | 9 | 32 | 10 | 34 |
| 1996 | 12 | 19 | 11 | 19 |
| 1996 | 13 | 53 | 14 | 16 |
| 1996 | 15 | 34 | 16 | 13 |

## Updates and Additions

### Changes to the Second Edition

One of the main changes to ***Ecology: Global Insights & Investigations 2e***, has been to simplify some of the more difficult concepts that students found challenging, such as population growth and diversity indices. My overall goal has been to make the material more accessible to students. I have included more explanatory captions on the artwork and have added subsection headings to the summaries. Some of the existing material has been updated and new examples added. Some of the chapter-specific changes include:

- **Chapter 1:** Illustration of the scientific method using a real-life example, the oak winter moth in Canada, a pest of apple orchards. I return to this example, and that of the invasive Burmese python in Florida, several times throughout the book to reinforce basic principles.
- **Chapter 2:** A new discussion of Darwin's involvement in revegetating Ascension Island, and a new discussion of the genetic suppression of invasive species.
- **Chapter 3:** A new figure updates Wallace's zoogeographic regions of the world.
- **Chapter 4:** A new example of Hamilton's Rule using wild turkeys. The section on game theory has been simplified.
- **Chapter 5:** A new figure showing how skin-surface blood vessels react to warm and cool conditions.
- **Chapter 6:** A new section on the physical properties of water.
- **Chapter 7:** A simplified table on soil composition and a new figure illustrating ruminant digestive systems.
- **Chapter 10:** A new example of exponential growth for Burmese pythons in the Florida Everglades. A new example of logistic growth using the Seychelles warbler. The section on the effects of time lags on population growth has been greatly simplified.
- **Chapter 11:** The section reviewing studies of competition has been simplified by removing some meta-analyses and discussions of competition between biological control agents. An extra figure has been included to clarify the Lotka-Volterra competition model. The sections on resource partitioning and character displacement have also been simplified, and several complex figures have been removed.
- **Chapter 13:** There is an update on the efforts to control the invasive brown tree snake in Guam. Several of the examples showing the effects of predators on prey have been removed.
- **Chapter 14:** The material on beneficial herbivory has been removed, and the discussions of the effects of plants on herbivore densities has been moved to Chapter 16. There is a new Data Analysis question.
- **Chapter 17:** The section on diversity indices has been simplified, and the more complex material on weighted indices, regional diversity, and cluster analysis has been eliminated.
- **Chapter 20:** There is a new discussion of Pleistocene re-wilding and assisted migration.
- **Chapter 23:** The distribution map of the world's kelp forests has been updated.
- **Chapter 24:** New information is provided on the physical properties of rivers and the effects of humans on rivers.
- **Chapter 25:** The distinction between detritivores and decomposers has been made clearer.
- **Chapter 26:** New captions are provided to highlight the most important points of the biogeochemical cycles.

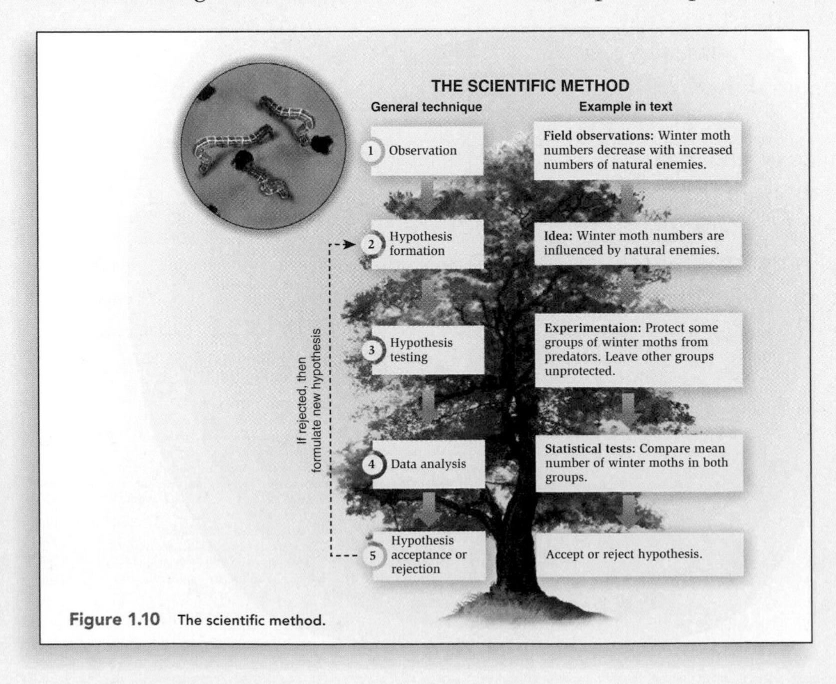


**Figure 1.10** The scientific method.

# Teaching and Learning Supplements

McGraw-Hill ConnectPlus™ Ecology is a web-based assignment and assessment platform that gives students the means to better connect with their coursework, with their instructors, and with the important concepts that they will need to know for success now and in the future.

With McGraw-Hill Connect™ Ecology, instructors can deliver assignments, quizzes, and tests online. Nearly all the questions from the text are presented in an autogradable format and tied to the text's learning objectives.

- Instructors can edit existing questions and author entirely new problems.
- Track individual student performance—by question, assignment, or in relation to the class overall—with detailed grade reports.
- Integrate grade reports easily with Learning Management Systems (LMS) such as WebCT and Blackboard. And much more.

By choosing Connect Ecology, instructors are providing their students with a powerful tool for improving academic performance and truly mastering course material. Connect Ecology allows students to practice important skills at their own pace and on their own schedule. Important, students' assessment results and instructors' feedback are all saved online—so students can continually review their progress and plot their course to success.

Some instructors may also choose ConnectPlus Ecology for their students. Like Connect Ecology, ConnectPlus Ecology provides students with online assignments and assessments, plus 24/7 online access to an eBook—an online edition of the text—to aid them in successfully completing their work, wherever and whenever they choose.

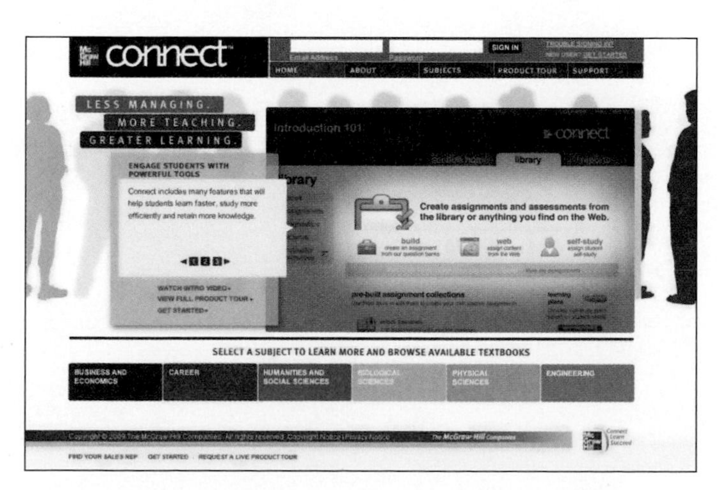


Craft your teaching resources to match the way you teach! With McGraw-Hill Create, **www.mcgrawhillcreate.com,** you can easily rearrange chapters, combine material from other content sources, and quickly upload content you have written—like your course syllabus or teaching notes. Find the content you need in Create by searching through thousands of leading McGraw-Hill textbooks. Arrange your book to fit your teaching style. Create even allows you to personalize your book's appearance by selecting the cover and adding your name, school, and course information. Order a Create book and you'll receive a complimentary print review copy in 3–5 business days or a complimentary electronic review copy (eComp) via email in minutes. Go to **www.mcgrawhill-create.com** today and register to experience how McGraw-Hill Create empowers you to teach *your* students *your* way.

## www.mhhe.com/stilingecology2

This text-specific website offers an extensive array of teaching tools. In addition to all of the student assets available, this site includes:

- Answers to review questions
- Class activities
- PowerPoint lecture® presentations
- Interactive world maps

## Presentation Tools

Accessed from the instructor side of your textbook's website, instructors will find the following digital assets for *Ecology: Global Insights & Investigations* at **www.mhhe.com/stilingecology2:**

**Color Art** Full-color digital files of ALL illustrations in the text can be readily incorporated into lecture presentations, exams, or custom-made classroom materials.

**Photos** Digital files of ALL photographs from the text can be reproduced for multiple classroom uses.

**Additional Photos** Full-color bonus photographs are available in a separate file.

**Tables** Every table that appears in the text is provided in electronic format.

**Animations** Full-color animations that illustrate many different concepts covered in the study of ecology are available for use in creating classroom lectures, testing materials, or online course communication. The visual impact of motion will enhance classroom presentations and increase comprehension.

**PowerPoint Lecture Outlines** Ready-made presentations written by Peter Stiling that combine art, photos, and lecture notes are provided for each of the 27 chapters of the text. These outlines can be used as they are, or tailored to reflect your preferred lecture topics and sequences.

**PowerPoint Slides** For instructors who prefer to create their lectures from scratch, all illustrations, photos, and tables are preinserted by chapter into blank PowerPoint slides for convenience.

## Test Bank

The test bank has been authored by Peter Stiling. Based on his years of education experience, he has put together a variety of questions. This computerized test bank that uses testing software to quickly create customized exams is available on the text website. The user-friendly program allows instructors to search for questions by topic or format, edit existing questions or add new ones, and scramble questions for multiple versions of the same test. Word files of the test bank questions are provided for those instructors who prefer to work outside the test-generator software.

### *Annual Editions: Environment 13/14*

by Eathorne

ISBN 0-07-351562-0/978-007-351562-5

The **Annual Editions** series is designed to provide convenient, inexpensive access to a wide range of current articles from some of the most respected magazines, newspapers, and journals published today. **Annual Editions** are updated on a regular basis through a continuous monitoring of over 300 periodical sources. The articles selected are authored by prominent scholars, researchers, and commentators writing for a general audience. **Annual Editions** volumes have a number of organizational features designed to make them especially valuable for classroom use: a general introduction; an annotated table of contents; a topic guide; an annotated listing of supporting World Wide Web sites; **Learning Outcomes** and a brief overview at the beginning of each unit; and a **Critical Thinking** section at the end of each article. Each volume also offers an online **Instructor's Resource Guide** with testing materials. *Using Annual Editions in the Classroom* is a general guide that provides a number of interesting and functional ideas for using **Annual Editions** readers in the classroom. Visit **www.mhhe.com/annualeditions** for more details.

### *Taking Sides: Clashing Views on Environmental Issues*

Expanded 15th edition © 2014

ISBN 0-07-351454-3 / 978-007-351454-3

**Taking Sides** volumes present current controversial issues in a debate-style format designed to stimulate student interest and develop critical thinking skills. Each issue is thoughtfully framed with *Learning Outcomes*, an *Issue Summary*, an *Introduction*, and an *Exploring the Issue* section featuring *Critical Thinking and Reflection, Is There Common Ground?*, and *Additional Resources*. **Taking Sides** readers also offer a *Topic Guide* and an annotated listing of *Internet References* for further consideration of the issues. An online **Instructor's Resource Guide** with testing material is available for each volume. *Using Taking Sides in the Classroom* is also an excellent instructor resource. Visit **www.mhhe.com/takingsides** for more details.

# Acknowledgments

In completing this book I am especially grateful to the many reviewers who graciously gave of their time in reading chapters. These folks are listed below. A great many other people also helped me immensely. Fran Simon and Lynn Breithaupt at McGraw-Hill provided help ranging from major advice on themes to minor help with housekeeping issues. Lisa Bruflodt, lead project manager, helped organize this team.

I would appreciate feedback on the text, be it art, narrative, or end-of-chapter material. If you know of a better example than the ones I have provided, please feel free to let me know. If there are good data sets to use in the end-of-chapter material or better multiple choice questions, I'd be pleased to hear about them. I will continue combing the latest literature to keep the book current.

## Reviewers

David Arieti, *Columbia College of Missouri*
Elizabeth Bancroft, *Southern Utah University*
David Bass, *University of Central Oklahoma*
Steve Blumenshine, *California State University–Fresno*
Robi Burks, *Southwestern University*
Chris Butler, *University of Central Oklahoma*
David Byres, *Florida State College at Jacksonville*
Hua Chen, *University of Illinois at Springfield*
Jay P. Clymer III, *Marywood University*
Mark Davis, *Macalester College*
Hudson DeYoe, *University of Texas Pan American*
Elisabeth Elder, *Louisiana State University at Alexandria*
Jarrod H. Fogarty, *Mississippi State University–Meridian*
Todd Fredericksen, *Ferrum College*
Jamie M. Kneitel, *California State University, Sacramento*
Ned Knight, *Linfield College*
Lynn A. Mahaffy, *University of Delaware*
Lauren Mathews, *Worcester Polytechnic Institute*
Paul Mickelson, *Central Lakes College*
Ana R. Otero, *Emmanuel College*
Craig Plante, *College of Charleston*
Karen Plucinski, *Missouri Southern State University*
Chris Romero, *Front Range Community College/Larimer Campus*
Raymond Russell, *University of the Incarnate Word*
Thomas W. Schoener, *University of California, Davis*
Stewart Skeate, *Lees-McRae College*
William Sluis, *Trine University*
L. Brooke Stabler, *University of Central Oklahoma*
Katie Stumpf, *Northland College*
Todd Tracy, *Northwestern College*
Edward Unangst, *US Air Force Academy*
Gary Villa, *Lynn University*
Lynn Westley, *Lake Forest College*
John B. Williams, *South Caroline State University*

# Contents

CHAPTER 9

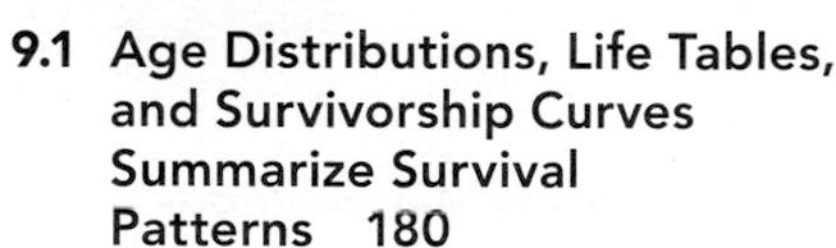

## Life Tables and Demography 179



CHAPTER 10

## Population Growth 195

## SECTION FOUR Species Interactions

## CHAPTER 14

## Herbivory 285



## CHAPTER 15

## Parasitism 303



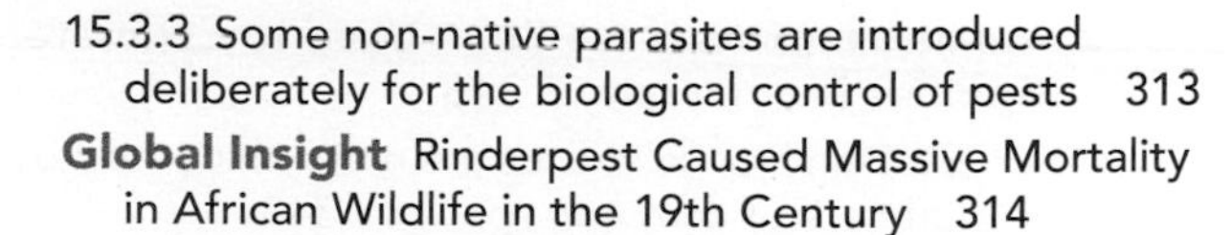



## CHAPTER 16

## Population Regulation 321

# SECTION FIVE Community Ecology

## CHAPTER 17

## Species Diversity 341

CHAPTER 20

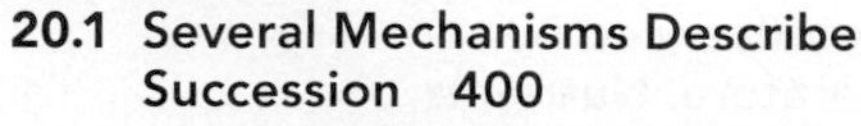

## Succession 399

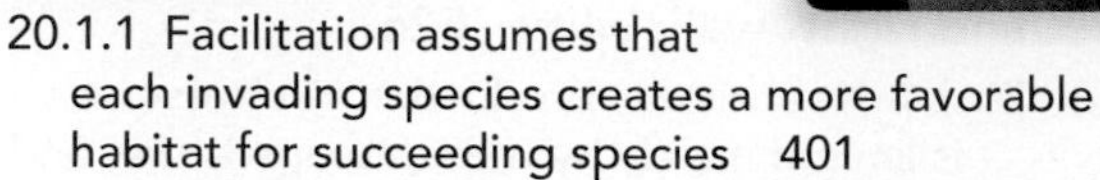



CHAPTER 21

## Island Biogeography 415



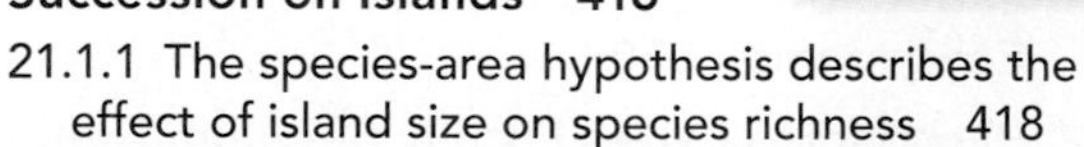



# SECTION SIX Biomes

CHAPTER 22

## Terrestrial Biomes 435

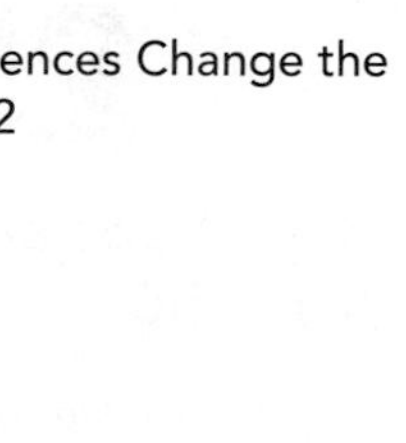

CHAPTER 23

## Marine Biomes 467

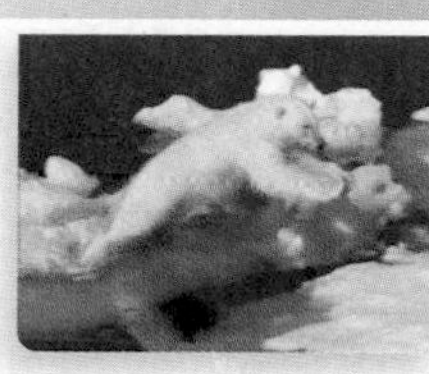

Population sizes of the Panamanian golden frog, *Atelopus zeteki*, have diminished greatly over the past 20 years, while populations of many other species of harlequin frogs have disappeared entirely. Ecologists are investigating the reasons for this decline.

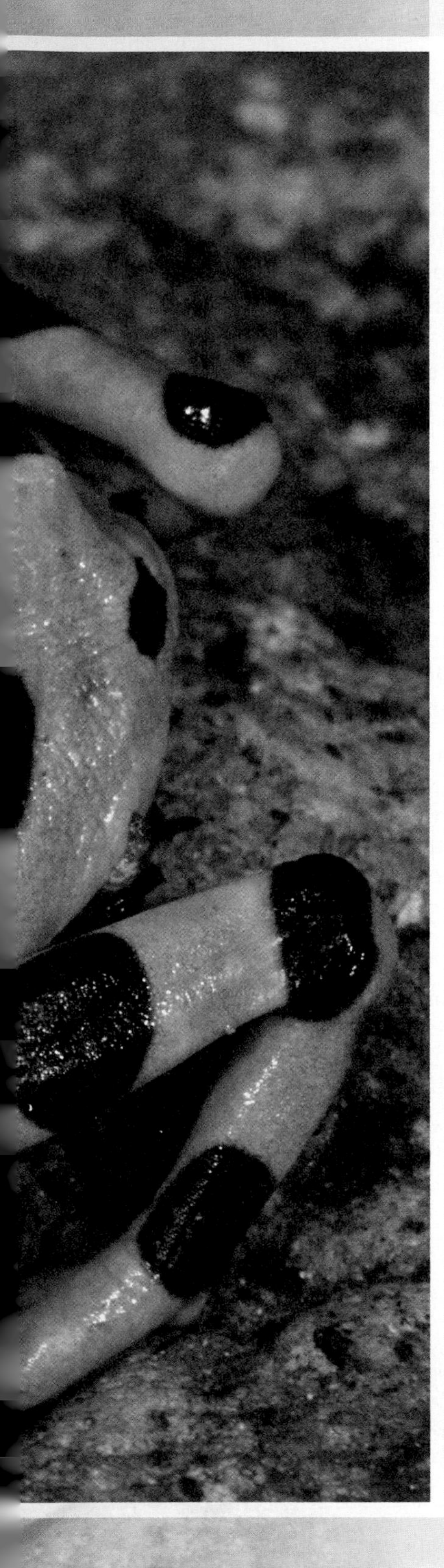

CHAPTER 1

# An Introduction to Ecology

## Outline and Concepts



From 1986 to 2006, two-thirds of the 110 species of harlequin frogs in mountainous areas of Central and South America became extinct. Researcher J. Alan Pounds and his colleagues noted that populations of species, such as the Panamanian golden frog, *Atelopus zeteki,* a type of harlequin frog, had been greatly reduced. The question was why.

Pounds's study identified the culprit as a disease-causing fungus, *Batrachochytrium dendrobatidis,* but implicated global warming as the agent causing the elevated fungal outbreaks. One effect of global warming is increased cloud cover that reduces daytime temperatures and raises nighttime temperatures. Researchers believe that this combination has created favorable conditions for the spread of *B. dendrobatidis,* which thrives in slightly cooler daytime temperatures. Pounds was quoted as saying, "Disease is the bullet killing frogs, but climate change is pulling the trigger." In 2011, Tina Cheng and her colleagues (Cheng et al., 2011) tested retroactively whether increased frequency of *B. dendrobatidis* was linked to declines in Central American frogs. They used a molecular sampling technique that was able to detect the presence of this fungal pathogen in formalin-preserved museum specimens. Frog specimens were examined before, during, and after amphibian declines in southern Mexico, Guatemala, and Costa Rica. At all locations, frogs collected pre-decline were free of *B. dendrobatidis,* whereas frogs collected during the decline exhibited the disease. The data suggested an initial emergence of this fungal pathogen in southern Mexico in the early 1970s, followed by a southward spread to Guatemala in the 1980s, to Costa Rica by 1987, and eventually to the Panama Canal in 2002 (**Figure 1.1**).

Global warming is having a profound effect on the distribution and abundance of many organisms, from bacteria, plants, and insects to fish, birds, and mammals. Global warming is one element of global change. The U.S. Global Change Research Act of 1990 defined global change as "changes in the global environment (including alterations in climate, land productivity, oceans or water resources, atmospheric chemistry, and ecological systems) that may alter the capacity of the Earth to sustain life." This act also mandated the U.S. Global Research Program to integrate federal research on changes in the global environment and their implications for society. Such change encompasses habitat destruction, the introduction of invasive species, direct exploitation, and the addition of pollutants to the environment. Habitat destruction reduces the abundance and genetic diversity of plants and animals. Invasive species can prey on, parasitize, or outcompete native species, and once in an area, are difficult to control. Direct exploitation of organisms includes hunting, harvesting, and collecting. Pollutants, from toxic chemicals to carbon dioxide, are released into the environment and affect organisms on scales local to global, respectively. One of the important themes in this book is to examine ecological concepts in light of the effects of global change on the distribution and abundance of life on Earth. So, for example, we are interested in how diseases spread and how a changing climate can exacerbate the effects of diseases, often with disastrous consequences, as in the case of *B. dendrobatidis* and the Central American frogs. However, the main purpose of this text is to understand the discipline of ecology: to determine the causes of the distributions and abundance of organisms.

## 1.1 Ecology: The Study of Organismal Interactions

**Ecology** is the study of interactions among organisms and between organisms and their environment. The interactions among living organisms are called **biotic** interactions, while those between organisms and their physical environment are termed **abiotic** interactions. These interactions in turn govern the population sizes of plants, animals, and other organisms, as well as the numbers of species in an area. The first part of this chapter introduces the four broad areas of ecology: organismal, population, community, and ecosystems ecology.

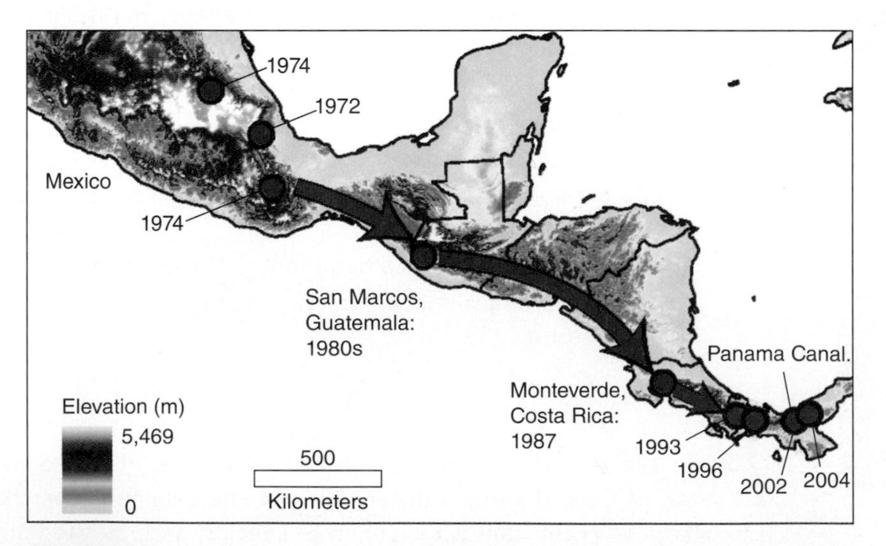


**Figure 1.1 The spread of *Batrachochytrium dendrobatidis,* the fungus causing disease of frogs throughout Central America, over time.** The disease was first found in Mexico and spread through lower Central America. (After Cheng et al., 2011.)

Next we explore the topic of global change and its major elements: habitat loss, invasive species, direct exploitation, and pollution. Finally, we discuss how ecologists approach and conduct their work.

Ecological studies have important implications in the real world, as will be illustrated with examples throughout the book. However, keep in mind that ecology is not the same as **environmental science**, the application of ecology to real-world problems. To use an analogy: ecology is to environmental science as physics is to engineering. Both physics and ecology provide the theoretical framework on which more applied studies can be pursued. Engineers rely on the principles of physics to build bridges. Environmental scientists rely on the principles of ecology to solve environmental problems, such as determining the effects of global warming on species distributions or controlling outbreaks of pests. Ecology provides the necessary framework for understanding how populations are affected by features of the physical environment, such as temperature and moisture, and by other organisms. For example, competitors displace one another, herbivores affect plant abundance, and predators and parasites impact prey populations.

Ecologists' tools of the trade have changed over the years to better enable them to answer more sophisticated questions. Before 1960 the field of ecology was dominated by taxonomy, natural history, and speculation about observed patterns. An ecologist's tools of the trade might have included sweep nets, quadrats (small, measured plots of land used to sample living things), and specimen jars (**Figure 1.2a**). Since that time, an explosion in both the scope of ecological studies and available technology has occurred, and ecologists have become active in investigating environmental change on regional and global scales. Ecologists have adapted concepts and methods derived from agriculture, physiology, biochemistry, genetics, physics, chemistry, and mathematics. Now an ecologist's equipment is just as likely to include portable computers, satellite-generated images, and chemical autoanalyzers (**Figure 1.2b**).

The research of ecologists is being used to an increasing extent to formulate solutions for the world's ills. These are many. Between one-third and one-half of Earth's land surface

(a)

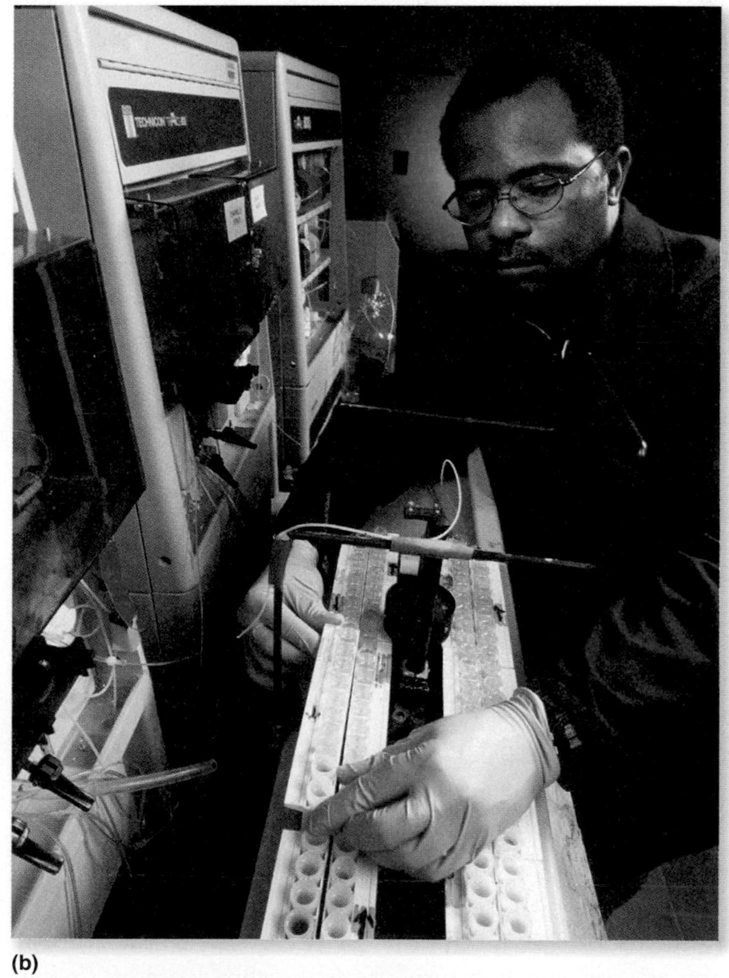

(b)

**Figure 1.2 Ecological equipment.** The "tools of the trade" have changed over the years. **(a)** Ernst Haeckel, sitting, with his friend Allers in Italy, 1862. Haeckel coined the term "ecology," and also the terms "phylum" and "kingdom protista." Allers is holding a sweep net, used for catching insects. **(b)** This ecologist is using a flow-through calorimetric autoanalyzer to determine the chemicals present in groundwater.

has been transformed by human action. Acid rain is carried from one country to another. Carbon dioxide pumped out by the industrial centers of developed nations has increased in the atmosphere worldwide, from the poles to the equator. More nitrogen is added to the land via fertilizers than by all natural sources combined. Pesticides have been detected in unintended targets, such as human breast milk and the tissues of penguins. Many species have been driven to extinction. Now, more than ever, there is a strong impetus to understand how natural systems work, how humans change those systems, and how in the future we can reverse these changes.

**Check Your Understanding**

**1.1** If an organism is limited in its distribution by cold temperatures, is it limited by biotic or abiotic factors?

## 1.2 The Scale of Ecology: From Organisms to Ecosystems

Ecology ranges in scale from the study of an individual organism through the study of populations to the study of communities and ecosystems (**Figure 1.3**). This section introduces each of the broad areas of organismal, population, community, and ecosystems ecology. This hierarchical scheme is an organizational framework for the succeeding chapters of the book.

### 1.2.1 Organismal ecology investigates how individuals' adaptations and choices affect their reproduction and survival

**Organismal ecology** can be divided into three main subdisciplines: evolutionary ecology, behavioral ecology, and physiological ecology. The first area, **evolutionary ecology**, considers how organisms have evolved to adapt to their environment through interactions with individuals, populations, and other species. The great biologist Theodosius Dobzhansky said, "Nothing in biology makes sense except in the light of evolution." This is as true in ecology as in any other biological discipline. For example, one may argue about what controls penguin numbers in the Southern Hemisphere, but a nagging question remains—why are there no penguins in the Northern Hemisphere? The answer is not insufficient food or too many predators. Penguins evolved in the Southern Hemisphere and have never been able to cross the Tropics to colonize northern waters. Evolutionary ecology addresses the genetic variation that exists within species and how this genetic variation is lessened by factors, such as habitat fragmentation, that can reduce the viability of populations.

**Behavioral ecology** focuses on how the behavior of an individual organism contributes to its survival and reproductive success, which in turn affects the abundance of a population. For example, forest tent caterpillars, *Malacosoma* species, are infamous for residing in silken tents and defoliating trees in deciduous forests (**Figure 1.4**). How can we explain this behavior? During the daytime this tent provides a safe refuge from predators. Group living enables the caterpillars to fabricate a large tent, which a single larva could not easily construct. At night the caterpillars leave the tent to forage. They must locate a tree whose branches have not been defoliated and remember where it is so that they can return to forage on subsequent nights. Here again group living becomes an advantage, because the caterpillars lay down silk trails between the tent and the leaves so that they can relocate the leaves. In addition, the silk trails are marked with pheromones to draw in colony mates. The denser the concentration of caterpillars, the stronger the trail. Group living is therefore advantageous in this species.

The third area, **physiological ecology**, investigates how organisms are physiologically adapted to their environment and how the environment impacts the distribution of species. Here we examine the effects of temperature, water, nutrient availability and other physical factors on the distribution and abundance of species.

### 1.2.2 Population ecology describes how populations grow and interact with other species

**Population ecology** focuses on populations—groups of interbreeding individuals that occur in the same place at the same time. A primary goal is to understand the factors that affect a population's growth and determine its size and density.

Although a population ecologist might focus on studying the population of a particular species, the relative abundance of that species is often influenced by its interactions with other species. Thus, population ecology includes the study of **species interactions**, such as competition, predation, mutualism, commensalism, herbivory, and parasitism. Much ecological theory is built upon the ecology of populations. Knowing what factors impact populations can help us reduce species endangerment, stop extinctions, and control invasive species.

### 1.2.3 Community ecology addresses factors that influence the number of species in an area

In a forest, we can find many populations of trees, herbs, shrubs, the herbivores that eat them, and the carnivores that in turn prey on the herbivores. This assemblage of many populations that live in the same place at the same time is known as a **community**. Communities occur on a wide variety of scales, from small pond communities to huge tropical rain forests. At the largest scales, these communities are known as biomes. Ecologists have examined different terrestrial, coastal, and aquatic biomes and determined the factors that limit the distributions of biomes such as tropical rain forests, temperate deciduous forests, or temperate grasslands. Biodiversity studies focus on why certain areas have high numbers of species (such areas are called species rich), whereas other areas have low numbers of species (these are called species poor).

**Figure 1.3** **The scales of ecology.** **(a)** Organismal ecology. What is the drought tolerance of this zebra? **(b)** Population ecology. How does drought influence the growth of zebra populations in Africa? **(c)** Community ecology. How does drought influence the number of species in African grassland communities? **(d)** Ecosystems ecology. How does water flow among plants, zebra, and other herbivores and carnivores in African grassland communities?

Although ecologists are interested in species richness for its own sake, a link also exists between species richness and community function, such as the ability to extract soil nutrients or to produce biomass. Ecologists generally believe that, for any given habitat, species-rich communities function better than species-poor communities. Community ecologists are also interested in community change. One hypothesis holds that more species make a community more stable, that is, more resistant to disturbances such as invasive species. We also know that species composition changes in a predictable way over time and in particular after a disturbance, such as fire or a flood. Ecologists call this process succession.

### 1.2.4 Ecosystems ecology describes the passage of energy and nutrients through communities

An **ecosystem** is a living, biotic community and its nonliving abiotic environment. Ecosystems ecology deals with the flow of energy and the cycling of nutrients among organisms within a community and between organisms and the environment. Understanding this flow of energy and nutrients requires knowledge of feeding relationships between species, called food chains.

**Figure 1.4** Tent caterpillars, *Malacosoma americana,* and their tent, Cape Cod, Massachusetts.

The second law of thermodynamics states that in every energy transformation, free energy is reduced because heat energy is lost from the ecosystem in the process. There is, therefore, a unidirectional flow of energy through an ecosystem, with energy dissipated at every step. An ecosystem needs a recurring input of energy from an external source—in most cases, the sun—to sustain itself. In contrast, chemicals such as nitrogen and phosphorus do not dissipate and constantly cycle between abiotic and biotic components of the environment, often becoming more concentrated in organisms higher in the food chain.

Global changes are affecting ecological processes at all these scales. In the next section, we discuss the main elements of change.

**Check Your Understanding**

**1.2** What might be a disadvantage to group living for many species?

## 1.3 Global Change Involves Habitat Destruction, Invasive Species, Direct Exploitation, and Pollution

Throughout the history of life on Earth, **extinction**—the process by which species die out—has been a natural phenomenon. The average life span of a species in the fossil record is around 4 million years. To calculate the current extinction rate, we could take the total number of species estimated to be alive on Earth at present, around 5 million, and divide it by 4 million, giving an average extinction rate of 1.25 species each year (or 1,250 species in 1,000 years). Thus, for the 4,000 species of living mammals, using the same average life span of around 4 million years, we would expect one species to go extinct every 1,000 years; this is termed the "background extinction rate."

However, it can be argued that the fossil record is heavily biased toward successful, often geographically wide-ranging species, which have a longer than average persistence time. The fossil record is also biased toward vertebrates and marine mollusks, both of which fossilize well because of their hard body parts. If background extinction rates were 10 times higher than the rates perceived from the fossil record, then extinctions among the 4,000 or so living mammals today would be expected to occur at a rate of one every 100 years. For birds, the background extinction rate would be two species every 100 years.

No one disputes, however, that the extinction rate for species in recent times has been far higher than this. In the past 100 years, approximately 20 species of mammals and over 40 species of birds have gone extinct (**Figure 1.5a**). The term **biodiversity crisis** is often used to describe this elevated loss of species. **Conservation biology** studies how to protect the biological diversity of life at all levels. Many scientists believe that the rate of species loss is higher now than during most of geological history. Growth of the human population is thought to have led to the increase in the number of extinctions of other species (**Figure 1.5b**). The reason is that humans are responsible for many elements of global change.

To understand the threats to life on Earth in more modern times, it is essential for ecologists to examine the role of human activities in the extinction of species. In this section we examine the factors that are currently threatening species with extinction. We do not know all the threats to life on Earth, but habitat destruction, introduced species, direct exploitation, and pollution have been major human-induced threats. E. O. Wilson (2002) has referred to these threats using the acronym HIPPO—habitat destruction, introduced species, pollution, population (human), and overharvesting—though in truth, overpopulation by humans drives the other four mechanisms. David Wilcove and colleagues (1998) categorized threats to 1,880 species of imperiled plants and animals in the United States (**Figure 1.6**). Habitat destruction was the most important threat. Second was invasive species, which threatened almost half the endangered species in the United States. Pollution was also important, especially for freshwater species such as fish, mussels, and amphibians. Overexploitation (hunting and collecting) was of considerable importance for mammals, birds, reptiles, and some plants.

### 1.3.1 Habitat destruction reduces available habitat for wildlife

Habitat destruction includes deforestation, conversion of habitat to agricultural land, urbanization, the draining of swamps, strip mining, quarrying, dam construction, river channelization, and many other forms of land modification.

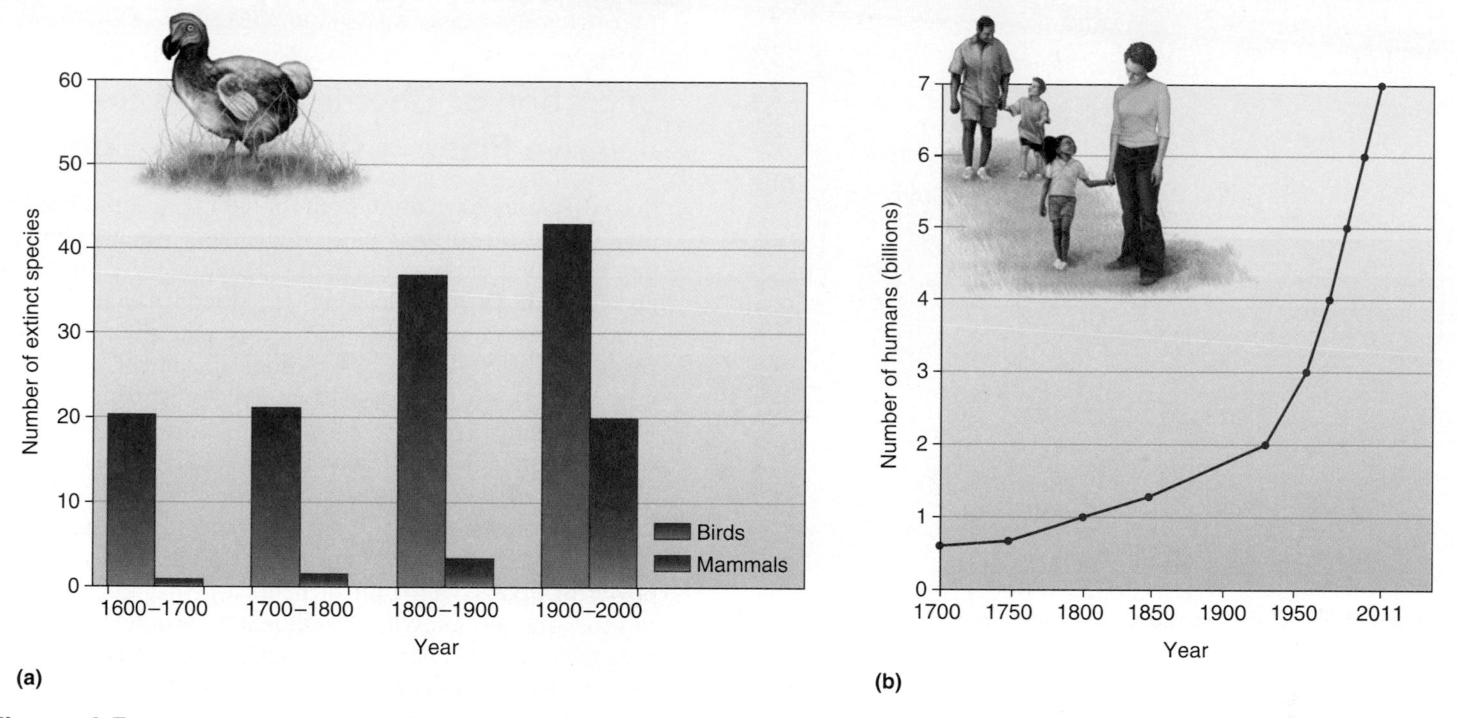


**Figure 1.5 Animal extinctions and human population growth.** **(a)** Increasing numbers of known extinctions in birds and mammals are concurrent with **(b)** an exponential increase in the global human population. These figures suggest that as human numbers increase, more and more species go extinct.

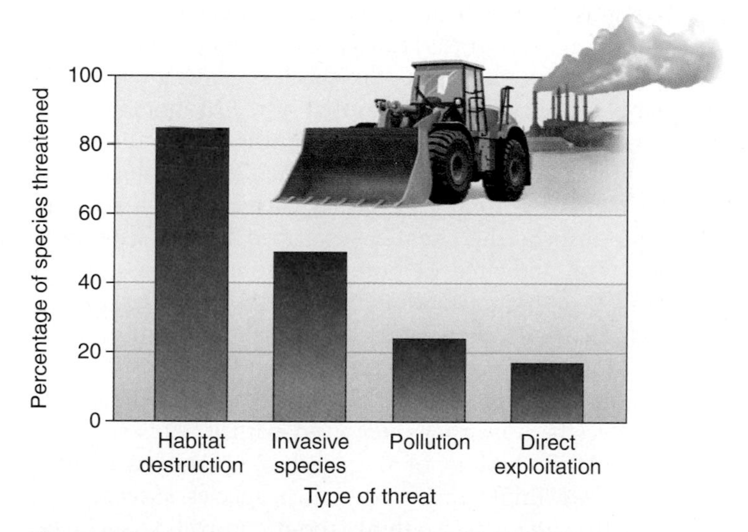


**Figure 1.6 Percentages of plant and animal species threatened by various causes.** Species can suffer from multiple threats, so categories do not sum to 100%. (From data in Wilcove et al., 1998.)

**ECOLOGICAL INQUIRY**

Habitat degradation is of paramount importance to all groups, but pollution and overexploitation affect some taxa more than others. Why is this?

**Deforestation**, the conversion of forested areas to non-forested land, is a prime cause of the extinction of species (**Figure 1.7**). About one-third of the world's land surface is covered with forests, and much of this area is at risk of deforestation. Although tropical forests are probably the most threatened, with rates of deforestation in Africa, South America, and Asia varying between 0.6% and 0.9% per year, the destruction of forests is a global phenomenon.

Among North American terrestrial wildlife, about one-quarter of the bird species (272 species) and more than 10% of mammal species (49 species) have an obligatory relationship with forest cover, meaning that they depend on trees for food and nesting sites. In terms of wildlife use, oaks are among the most valuable trees in North America. At least 100 species of birds and mammals include acorns in their diets, and for many species of wildlife, the annual acorn crop is a major determinant of their abundance. Most woodpeckers, as well as many other types of birds, nest in holes that they excavate in trees, and their food usually consists of insects collected on or in trees. The ivory-billed woodpecker, *Campephilus principalis,* the largest woodpecker in North America and an inhabitant of wetlands and forests of the southeastern United States, was widely assumed to have gone extinct in the 1950s due to destruction of its habitat by heavy logging (**Figure 1.8a**). In 2004 the woodpecker was supposedly sighted in the Big Woods area of eastern Arkansas by John Fitzpatrick and colleagues (2005). The sighting has yet to be confirmed, leading some to jokingly term this bird the "feathered Elvis."

Deforestation is not the only form of habitat destruction. More land has been converted to agriculture since 1945 than in the 18th and 19th centuries combined. The scouring of land to plant agricultural crops can create soil erosion, increased flooding, declining soil fertility, silting of rivers,

**Figure 1.7** **Deforestation.** Cascade mountains near Seattle, Washington, 1906.

and desertification. While the average area of land under cultivation worldwide averages about 12%, with an additional 26% given over to rangeland, this amount varies substantially between regions (look ahead to Table 12.C). Wetlands also have been drained for agricultural purposes and have been filled in for urban or industrial development. In the United States, as much as 90% of the freshwater marshes have disappeared in states such as Iowa and California, though the national average is approximately 53%. Urbanization, the development of cities on previously natural or agricultural areas, is the most human-dominated and fastest-growing type of land use worldwide, and it devastates the land more severely than practically any other form of habitat degradation.

### 1.3.2 Invasive species can cause extinctions of native species

Introduced species are those species moved by humans from a native location to another location. Most often the species are introduced for agricultural or landscaping purposes or as sources of timber, meat, or wool, and they need humans for their continued survival. Others, such as plants, insects, or marine organisms, are unintentionally transported via the movement of cargo by ships or planes. Regardless of the way they have been transported, some introduced species become **invasive species**, spreading naturally and outcompeting native species for space and resources (see **Feature Investigation**).

In the United States alone, there are over 4,500 invasive species, and 15% cause severe ecological or economic harm. Of introduced vertebrates, 142 species have self-sustaining populations in the wild. These include ring-necked pheasants, *Phasianus colchicus,* which were brought over by hunters, and Burmese pythons, *Python molorus,* which were introduced by pet owners. Of the 300 most invasive weeds in the United States, over half were brought in for gardening, horticulture, or landscape purposes. These include purple loosestrife, *Lythrum salicaria,* and Japanese honeysuckle, *Lonicera japonica,* in the Northeast; Kudzu, *Pueraria lobata,* in the Southeast; Chinese tallow, *Sapium sebiferum,* in the South; and leafy spurge, *Euphorbia esula,* in the Great Plains.

We can break down the interactions between introduced and native species into competition, predation, and

## Feature Investigation

### Secretion of Chemicals Gives Some Invasive Plants a Competitive Edge

Invasive plants have often been thought to succeed because they have escaped their natural enemies, primarily insects that remained in the country of origin and were not transported to the new locale. One way of controlling invasive species, therefore, has been to import the plant's natural enemies. This is known as **biological control.** However, new research on thc population ecology of diffuse knapweed, *Centaurea diffusa,* an invasive Eurasian plant that has established itself in many areas of North America, suggests a different reason for the success of invasive species.

Researchers Ragan Callaway and Erik Aschehoug (2000) hypothesized that the roots of *Centaurea* secrete powerful toxins, called **allelochemicals**, that kill the roots of other species, allowing *Centaurea* to proliferate. To test their hypothesis, Callaway and Aschehoug collected seeds of three native Montana grasses—Junegrass, *Koeleria cristata;* Idaho fescue, *Festuca idahoensis;* and Bluebunch wheatgrass, *Agropyron spicata*—and grew each of them both with and without the exotic *Centaurea* species (**Figure 1.A**). As hypothesized, *Centaurea* depressed the biomass of the native grasses. When the experiments were repeated with grasses native to Eurasia, *Koeleria laerssenii, Festuca ovina,* and *Agropyron cristatum,* the species' growth was inhibited, but significantly less than that of the Montana species.

As a further test of their hypothesis, not described in Figure 1.A, Callaway and Aschehoug modified their experiments by adding activated carbon to the soil; the carbon absorbs the chemical excreted by the *Centaurea* roots. With activated carbon, the Montana grass species increased in biomass compared to the previous experiments. The researchers concluded that *C. diffusa* outcompetes Montana grasses by secreting an allelochemical, and that Eurasian grasses are not as susceptible to the chemical's effect because they coevolved with it. This study on the population biology of an introduced plant has changed the way we think about why such species succeed, and could affect the way we think about controlling such species in the future, calling into question the effectiveness of biological control of some invasive plant species.

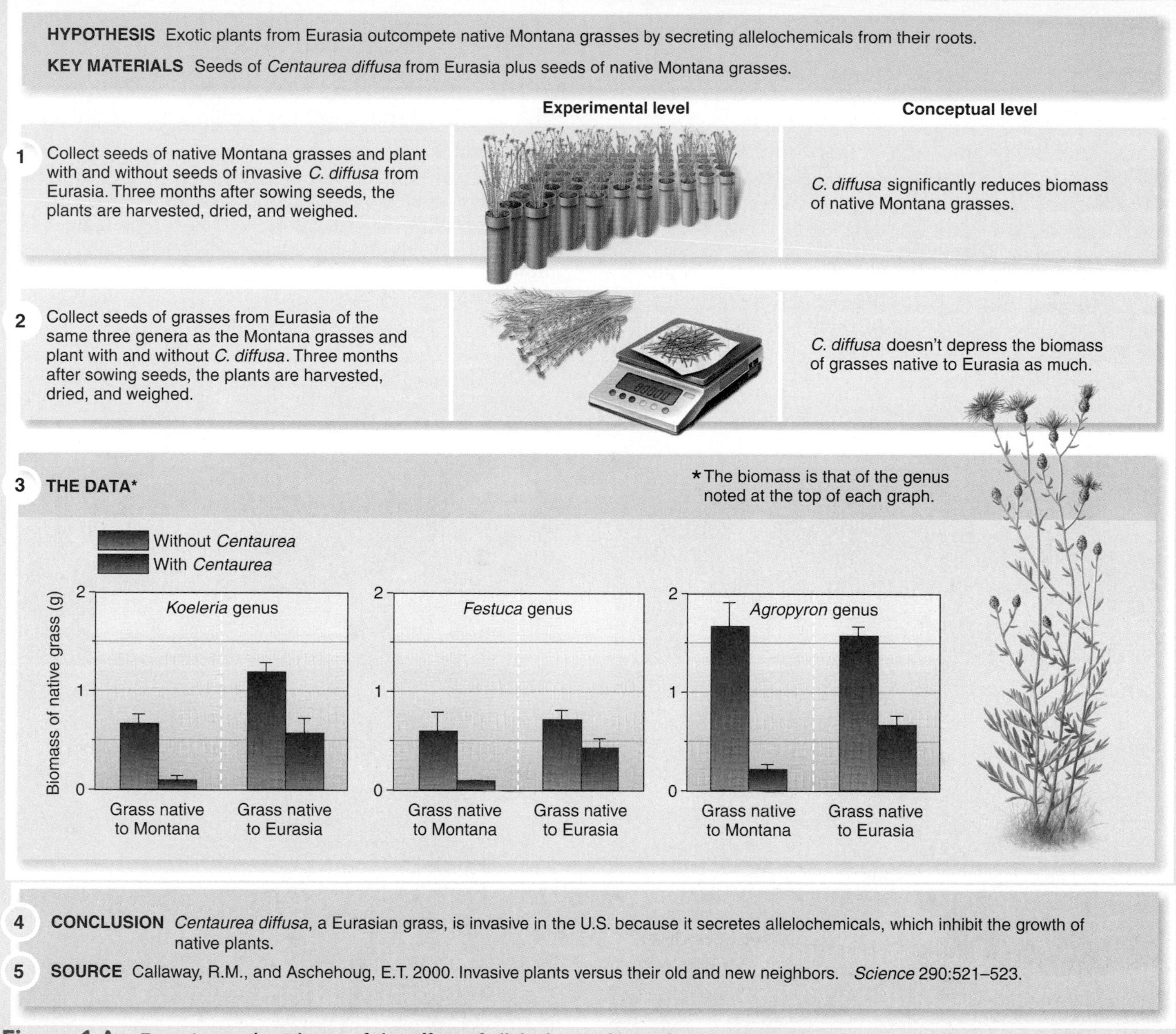

**HYPOTHESIS** Exotic plants from Eurasia outcompete native Montana grasses by secreting allelochemicals from their roots.

**KEY MATERIALS** Seeds of *Centaurea diffusa* from Eurasia plus seeds of native Montana grasses.

| | Experimental level | Conceptual level |
|---|---|---|
| 1 Collect seeds of native Montana grasses and plant with and without seeds of invasive *C. diffusa* from Eurasia. Three months after sowing seeds, the plants are harvested, dried, and weighed. | | *C. diffusa* significantly reduces biomass of native Montana grasses. |
| 2 Collect seeds of grasses from Eurasia of the same three genera as the Montana grasses and plant with and without *C. diffusa*. Three months after sowing seeds, the plants are harvested, dried, and weighed. | | *C. diffusa* doesn't depress the biomass of grasses native to Eurasia as much. |

3 **THE DATA***

*The biomass is that of the genus noted at the top of each graph.

Without *Centaurea*
With *Centaurea*

Biomass of native grass (g)

*Koeleria* genus — Grass native to Montana; Grass native to Eurasia

*Festuca* genus — Grass native to Montana; Grass native to Eurasia

*Agropyron* genus — Grass native to Montana; Grass native to Eurasia

4 **CONCLUSION** *Centaurea diffusa*, a Eurasian grass, is invasive in the U.S. because it secretes allelochemicals, which inhibit the growth of native plants.

5 **SOURCE** Callaway, R.M., and Aschehoug, E.T. 2000. Invasive plants versus their old and new neighbors. *Science* 290:521–523.

**Figure 1.A** **Experimental evidence of the effect of allelochemicals on plant production.**

parasitism (disease). For example, introduced Norway maple, *Acer platanoides,* tolerates very shady conditions and outcompetes many plant species in the central and northeastern United States. Predation by lighthouse keepers' cats has annihilated populations of ground-nesting birds on small islands around the world. The brown tree snake, *Boiga irregularis,* which was accidentally introduced onto the island of Guam, has decimated the country's native bird populations (look ahead to Figure 13.10). Parasitism and disease carried by introduced organisms have also been important in causing extinctions. Avian malaria in Hawaii, spread by introduced mosquito species, is believed to have contributed to the demise of up to 50% of native Hawaiian birds, such as honeycreepers (**Figure 1.8b**). Ecological communities are highly

## Global Insight

### Biological Control Agents May Have Strong Nontarget Effects

In the 1970s, two gall flies, *Urophora affinis* and *U. quadrifaciata,* were introduced as biological control agents of diffuse knapweed, *Centaurea diffusa,* the weed discussed in the Feature Investigation (see Figure 1.A). The flies lay their eggs inside the flowerheads and create a tumor-like swelling called a gall, inside which the fly larvae feed. The hope was that the gall flies would reduce flowerhead production and, thus, the spread of the weed. Unfortunately, seed reductions have not been enough to control the plant. As a result, extensive knapweed and gall fly populations now coexist in many states (**Figure 1.B**).

Dean Pearson and Ragan Callaway (2006) showed that deer mice, *Peromyscus maniculatus,* feed on knapweed galls, and the increased gall production has provided a food subsidy, enabling deer mice populations to double in size. This is troubling because deer mice are a reservoir of Sin Nombre hantavirus, which causes the deadly hantavirus pulmonary syndrome in humans. Mice testing positive for hantavirus were over three times more abundant in the presence of gall-infested knapweed than in areas where it was absent.

This story has at least two lessons. First, unless biological control agents effectively reduce their target populations, releasing these agents may do more harm than good. Second, it is important to take into account the many complex interactions between species in nature. The chapters that follow provide more evidence of this process.

**Figure 1.B Chain of effects following biological control attempts against invasive knapweed, *C. diffusa,* in Montana.** **(a)** Spotted knapweed invades a field near Missoula, Montana; **(b)** banded gall fly, *Urophora affinis,* is introduced to control it; **(c)** deer mouse, *Peromyscus maniculatus,* feeds on galls; **(d)** deer mice increase the incidence of hantavirus in the area. A worker from the University of New Mexico weighs a mouse caught in traps near Placitas, Nevada, during a study of hantavirus.

(a)

(b)

(c)

(d)

**Figure 1.8 Extinctions and threats to species in the past.** **(a)** The ivory-billed woodpecker, the third-largest woodpecker in the world, was long thought to be extinct in the southeastern U.S. because of habitat destruction, but it was supposedly rediscovered in 2004. This nestling was photographed in Louisiana in 1938. **(b)** Many Hawaiian honeycreepers were exterminated by avian malaria from introduced mosquito species. This 'Apapane, *Himatione sanguinea,* is one of the few remaining honeycreeper species. **(c)** The passenger pigeon, which may have once been the most abundant bird species on Earth, was hunted to extinction for its meat. **(d)** The shells of peregrine falcon eggs were thinned by of the accumulation of DDT in the parents' diet. This resulted in egg cracking and increased chick mortality. A normal egg, on the left, is darker and thicker than the DDT-affected egg, on the right.

interconnected, so that the introduction of one species can often have far-reaching effects. Pearson and Callaway showed how gall flies, introduced to control knapweed in the western United States, provide a food subsidy for deer mice, allowing populations to increase (see **Global Insight,** Figure 1.B).

### 1.3.3 Direct exploitation decreases the density of populations

Direct exploitation, particularly the hunting of animals, has been the cause of many extinctions in the past. Two remarkable species of North American birds, the passenger pigeon and the Carolina parakeet, were hunted to extinction by the early 20th century. The passenger pigeon, *Ectopistes migratorius,* was once the most common bird in North America, probably accounting for over 40% of the entire number of birds (**Figure 1.8c**). Flock sizes were estimated to be over 1 billion birds. It may seem improbable that the most common bird on the continent could be hunted to extinction for its meat, but that is just what happened. The flocking behavior of the birds made them relatively easy targets for hunters, who used special firearms to harvest the birds in quantity. In 1876, in Michigan alone, over 1.6 million birds were killed and sent to markets in the eastern United States. The Carolina parakeet, *Conuropsis carolinensis,* the only species of parrot native to the eastern United States, was similarly hunted to extinction by the early 1900s.

Many whale species were driven to the brink of extinction prior to a moratorium on commercial whaling issued in 1988 (look ahead to Figure 13.20). Steller's sea cow, *Hydrodamalis gigas,* a 9-meter-long manatee-like mammal, was hunted to extinction in the Bering Straits only 27 years after its discovery in 1740. A poignant example of human excess in hunting was the dodo, *Raphus cucullatus,* a flightless bird native only to the island of Mauritius that had no known predators. A combination of overexploitation and introduced species led to its extinction within 200 years of the arrival of humans. Sailors hunted it for its meat, and the rats and pigs they brought to the island, the latter as a food source, destroyed the dodos' eggs and chicks in their ground nests.

Many species of valuable plants have also been severely reduced for their human uses, including West Indian mahogany, *Swietenia mahogani,* in the Bahamas, and Lebanese Cedar, *Cedrus libani,* which in Lebanon has been reduced to a few scattered forest remnants. Rare cacti and orchids have also been threatened by collectors, who seek to own a rare organism or to profit from its sale.

### 1.3.4 Pollution may have strong direct effects on organisms locally and cause global change via climate alterations

Pollutants released into the environment come in many forms. Gaseous pollutants include carbon monoxide, carbon dioxide ($CO_2$), sulfur dioxide, and nitrogen oxides, most of which come from the burning of fossil fuels. Increasing $CO_2$ levels have already lowered the pH of the ocean by 0.1 pH units, and this increased acidification is likely to continue. At decreased pH, many calcifying organisms that produce shells or plates will be negatively impacted. Some models predict that by 2050, oceans may be too acidic for corals to calcify.

A variety of chemicals are applied to agricultural crops to kill pests, and many of these have nontarget effects on wildlife. DDT is a case in point (also look ahead to Figure 27.A). In the 1950s and 1960s, DDT was commonly used against a variety of agricultural pests and disease-carrying insects such as mosquitoes. Accumulation in food chains resulted in high levels of DDT in top predators such as birds of prey. Here DDT interfered with calcium deposition of eggs, resulting in cracked eggs, poor hatching, and lower population densities (**Figure 1.8d**).

Aquatic pollutants include numerous pesticides that run off into lakes and rivers from agricultural fields. In marine environments, oil spills have devastating effects along many of the world's coastlines. Freshwater aquatic systems are perhaps most at risk from pollution from both pesticides and fertilizers containing high levels of nitrogen and phosphorous. These nutrients can cause rapid increases in the growth of algae, which can kill many other forms of life (look ahead to Chapter 27).

Perhaps the single most important pollutant, however, is carbon dioxide because of its effect on global warming. Increasingly, global warming is viewed as a significant threat to species (look ahead to Chapter 5). As noted at the beginning of this chapter, global warming has been implicated in the dramatic decrease in the population sizes of frog species in Central and South America. Chris Thomas and colleagues (2004) employed computer models to simulate the movement of species' ranges in response to changing climate conditions in six biodiversity-rich regions, covering all of the different climatic regions of the world. The models predict that unless fossil fuel use is cut drastically, climate change will cause 15–37% of the species in those regions to become extinct by the year 2050.

What types of organisms are most threatened with extinction? According to data from the International Union for Conservation of Nature and Natural Resources, amphibians are now the most threatened group of organisms, with mammals a fairly close second (**Figure 1.9**). Amphibians' ability to fight infection depends strongly on environmental temperature, which is gradually changing. Large mammals, because of their large habitat requirements, are especially prone to extinction from habitat destruction, and many species of cats are threatened from overexploitation because of their fur.

### Check Your Understanding

**1.3** The introduction of gall flies to control invasive knapweed in Montana permitted an increase in the numbers of deer mice. Apart from an increase in hantavirus in the area, what other effects might this have on the local community?

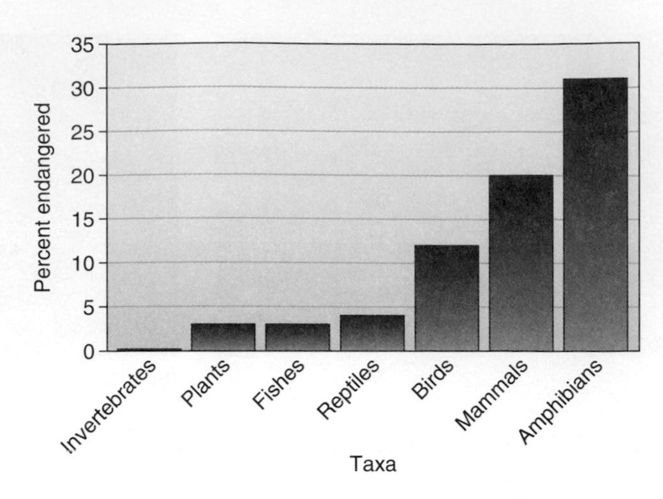


**Figure 1.9** **Levels of endangerment to various groups of organisms.**

**ECOLOGICAL INQUIRY**

Why are amphibians among the most endangered taxa on Earth?

## 1.4 Ecological Methods Include Observation, Experimentation, Analysis, and Mathematical Modeling

In biology, the scientific method involves a five-stage process:

1. observation,
2. hypothesis formation,
3. hypothesis testing,
4. data analysis,
5. acceptance or rejection of the hypothesis (**Figure 1.10**).

It is important to note that a hypothesis is never really proven. We may conduct further hypothesis testing and fail to disprove a hypothesis. After many such tests, biologists may accept that a hypothesis is true. Over time, we may use the term "theory" to explain a natural phenomenon that is supported by a large body of evidence. For example, Charles Darwin formulated the theory of evolution. In science, the term "theory" denotes an idea or set of ideas that explain a vast amount of data and are well supported by the evidence. This contrasts with the nonscientific or everyday usage of the word "theory," which connotes a more casual idea, a guess almost, that may explain an observation. For example, one student might propose a theory that the reason another student is missing from class is because he is at the beach.

Hypothesis testing often takes the form of the experimental or comparative method. In ecology, as in biology in general, much of our insights come from the experimental method, where we manipulate one variable but control other variables. For example, in late 16th-century Europe, it was generally thought that heavier objects fell faster than lighter objects. The Italian scientist Galileo tested this hypothesis with an experiment. He dropped objects of different mass from the top of the Leaning Tower of Pisa. Galileo found that the objects fell at the same rate, regardless of their mass, thus disproving the hypothesis. The comparative method requires collecting data on two groups that are then compared. A classic example is the effect of smoking on lung cancer in humans. We cannot experimentally expose humans to cigarette smoke on a daily basis over long periods, but we can compare groups of humans that have voluntarily smoked for long periods (>30 years) to those that have never smoked, and compare the incidence of cancer.

### 1.4.1 Observation includes taking data on natural systems

The oak winter moth, *Operophtera brumata,* is a small moth whose larvae feed on a variety of trees and shrubs, especially oaks, in Europe and Asia. Egg hatch often coincides with bud burst, so that young caterpillars can feed on soft new foliage. Often eggs hatch before leaves appear and, rather than starve, caterpillars "balloon" away on silken threads in the hope of locating trees whose leaves have already appeared. As this is a risky business, being able to feed on a variety of food plants maximizes chances of survival. Sometimes the caterpillars can reach great numbers, defoliating trees and reaching pest status. In the early 1900s, winter moths were accidentally introduced into North America, where larvae fed on many hardwood tree species such as oaks, elms, and maple. In Nova Scotia about 40% of red oak stands were killed in some areas. Winter moths also became a problem in apple orchards in British Columbia and the northwest United States, including Oregon.

How could these moths be controlled? A thorough understanding of the biology of the winter moth was needed. First, ecologists drew up a web of interactions between the winter moth and the factors that could impact populations (**Figure 1.11**). These interactions are many and varied, and they include interactions with the following factors:

- Abiotic factors such as temperature and rainfall: Warm temperatures can accelerate egg hatch, so that larvae appear before buds burst and leaves are available to feed on.
- Natural enemies, including bird predators of adults and caterpillars, insect parasites, bacterial parasites, and pupal predators. Bird predators consume relatively few adult moths and caterpillars. Bacterial parasites tend to

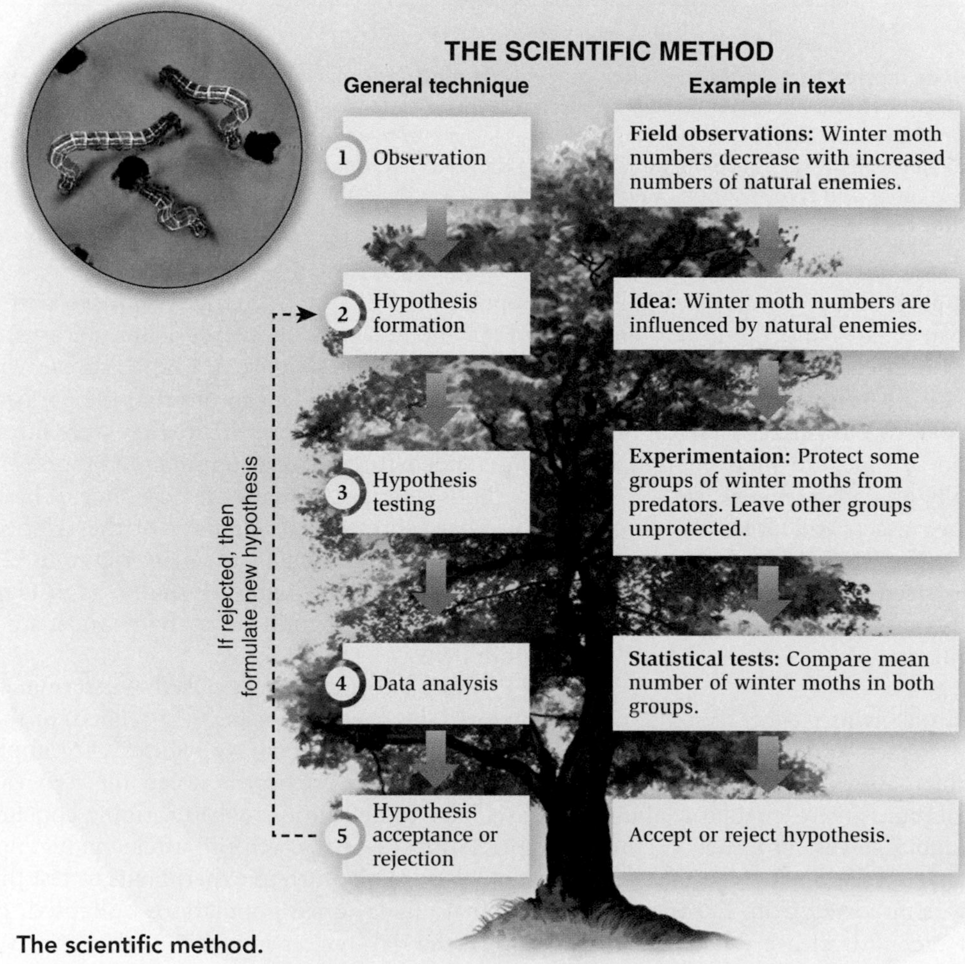


**Figure 1.10** The scientific method.

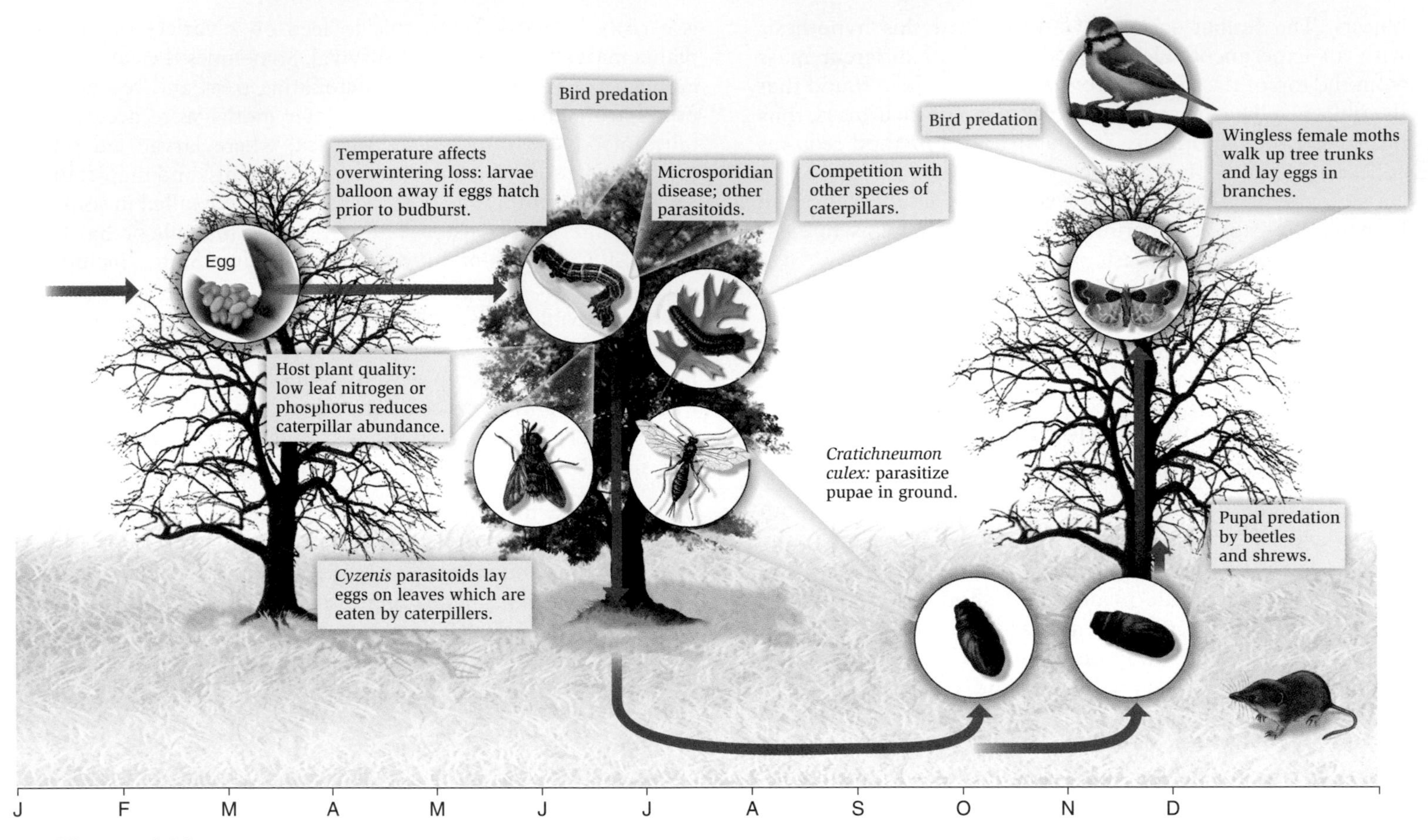


**Figure 1.11 Interaction web of factors that might influence oak winter moth populations.** Eggs hatch in early April, coincident with bud burst. If leaves are not available, caterpillars balloon away on silken threads to try to locate trees with leaves. Caterpillars feed on the leaves and may compete with other leaf feeders. Larvae of the parasite *Cyzenis*, which lay their eggs on the foliage, are unwittingly devoured by the caterpillars. Adult *Cratichneumon* directly parasitize pupae. Caterpillars drop to the ground to pupate in the soil and leaf litter. Pupal predation occurs from soil-dwelling beetles and shrews. Adult, wingless, female oak winter moths climb up trees to lay their eggs in November and December. (Redrawn from Varley, 1971.)

kill larvae before they pupate. Insect parasites usually lay their eggs inside the caterpillars and the developing parasite gradually eats the caterpillar from the inside out, emerging as an adult parasite the following summer when new caterpillars are available to parasitize. Predators of pupae in the leaf litter and soil include small mammals such as shrews but especially predatory beetles.

- Competitors, including other insects and larger vertebrate grazers that feed on leaves. The most common competitors are other species of leaf-feeding caterpillars.
- Host plants, including increases or decreases in either the quality or quantity of the plants.

In Canada, fluctuations in oak winter moth population sizes were monitored for many years to determine if population sizes were affected by fluctuations in abiotic or biotic interactions, such as levels of parasitism, predator abundance, or the quality of available leaves, but such factors caused relatively little mortality. In the 1950s the tachinid fly parasite *Cyzenis albicans* was imported from Europe and released. Within five years of tachinid release, winter moth populations collapsed. Ecologists noticed that in areas where winter months were more abundant, parasitism of larvae by *Cyzenis* increased (**Figure 1.12a**). In areas with few moths, parasitism was low. This meant that the parasites were killing more caterpillars in areas where they were more common and this tended to control caterpillar numbers. Using statistical methods ecologists usually create a "line of best fit," a straight line that represents a summary of the relationship between these two variables, as shown in Figure 1.12a. However, if the points were not highly clustered, as in **Figure 1.12b**, you would have little confidence that parasitism affects winter moth density.

Many statistical tests are used to determine whether or not two variables are significantly correlated. In this book, unless otherwise stated, graphs like Figure 1.12a imply that a meaningful relationship exists between the two variables. Ecologists have to be cautious when forming conclusions based on correlations. For this reason, after conducting observations, ecologists usually turn to experiments to test their hypotheses.

In Canada, once populations collapsed, continued control of the oak winter month was thought to arise not from

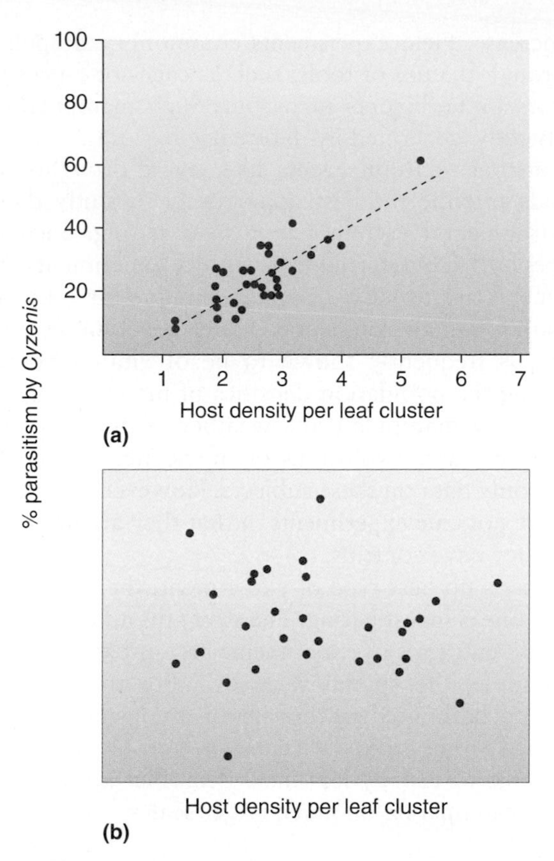


**Figure 1.12 The effects of larval parasitism on oak winter moth abundance.** **(a)** Percentage parasitism of winter moth larvae in Nova Scotia is dependent on the number of caterpillars per leaf cluster of 4–5 leaves. Data from apple orchard in Nova Scotia in 1983. (After Roland, 1986b.) In **(b)** there is no apparent relationship between the two variables.

> **ECOLOGICAL INQUIRY**
>
> What would it mean if the line of best fit sloped in the opposite direction?

larval parasitism but from pupal predation. By the late 1960s. Oak winter moth populations were now so low that generalist beetle and shrew predators already present could keep the winter moths at low densities.

## 1.4.2 Experimentation involves manipulating a system and comparing results to an unmanipulated control

Continuing with the oak winter moth example, an experiment to test the hypothesis that pupal predation controls winter moth abundance might involve removing pupal predators and examining subsequent winter moth pupal survival. Reduced predation might be achieved by putting out pitfall traps to catch predatory beetles and small mammals or applying a pesticide to the soil in July to kill the predators prior to winter moth pupation. Ecologists could then look at pupal survivorship over the course of a winter. If predators are having a significant effect, then removing them should cause winter moth pupal survival to increase. Thus, there would be two groups: a group of trees with predators denied access (the experimental group), and a group of trees with predators still present (the control group), with equal numbers of trees in both groups at the start of the experiment. Any differences in winter moth pupal survival in the future would be due solely to differences in predation.

Performing an experiment several times is called **replication.** Ecologists might replicate the experiment 5 times, 10 times, or even more. At the end of the replications, you would sum the total number of emerging winter moths, divide the sum by the number of replications, and calculate the mean.

In the experimental group, let's suppose that the surviving numbers of winter moth pupae per tree in each replicate are 5, 4, 7, 8, 12, 15, 13, 6, 8, and 10; the mean number of surviving pupae would be 8.8. In the control group, which still allows predator access, the numbers surviving might be 2, 4, 7, 5, 3, 6, 11, 4, 1, and 3, with a mean of 4.6. Without predators, the mean number of pupae surviving would therefore be almost double the mean number surviving with predators. Your data analysis would give you confidence that predators were indeed the cause of the changes in winter moth pupal abundance.

The result of such a predator removal experiment is illustrated graphically by a bar graph (**Figure 1.13**). Jens Roland applied the pesticide Diazinon to the soil prior to winter moth predation. This treatment killed beetle larvae and adults but not winter moth pupae, because the pesticide had degraded by the time winter moth larvae dropped down to the forest floor. Results showed that pupal predation was indeed important in influencing winter moth pupal densities. Ecologists use a variety of tests to determine whether the differences between treatments and controls are statistically significant. We won't look at the mechanics of these tests, but in this book, when experimental and control groups are presented as differing, these are considered to be statistically significant differences unless stated otherwise.

You might notice that in some graphs there are vertical lines around the mean. These lines represent a measure of the spread of the points around the mean. Two different measures are the standard error of the mean and the standard deviation. The standard error is the standard deviation divided by the square root of the number of observations. The smaller the bars, the tighter the replicate values are around the mean and, usually, the more significant the differences are between the treatment and control groups. Standard deviations and standard errors are calculated using the values of all the replicates, and the smaller the lines, the closer these replicate values tend to be. Large differences in replicates lead to large standard deviations and reduce our faith in the repeatability of the results. Studies of the oak winter moth, both in Canada and Europe, have revealed a great deal to ecologists about how natural systems function, and we will revisit these studies several times later in the book.

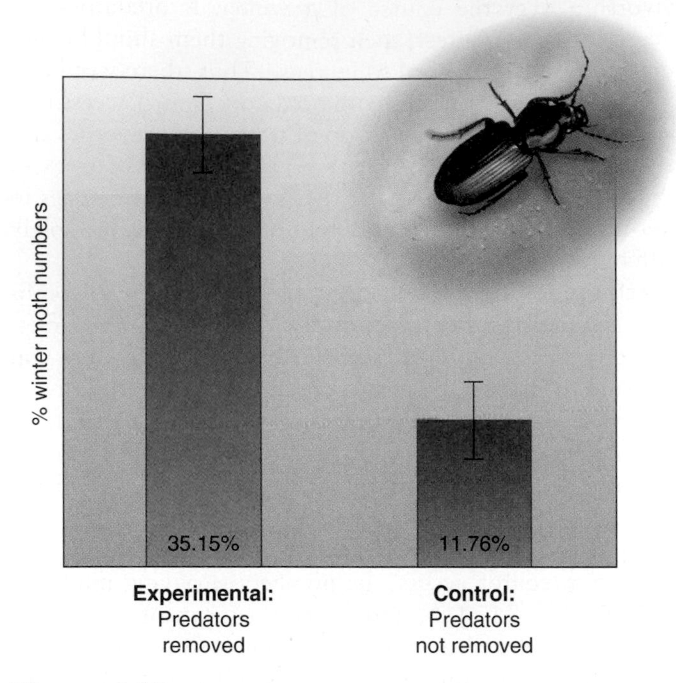


**Figure 1.13 Graphic display of results of a predator removal experiment.** The two bars represent the average percentage of winter moths preyed upon by beetles where beetle predators are removed (experimental) and where predators are not removed (control). Data are from Roland, 1986a. The vertical lines (standard deviations or standard errors) give an indication of how tightly the individual replicate results are clustered around the mean. The shorter the lines, the tighter the cluster of replicates and the more confidence we have in the result. Although these lines were originally absent, they have been added strictly for illustrative purposes.

Experiments can be classified into three main types: laboratory experiments, field experiments, and natural experiments. Laboratory experiments allow the most exact regulation of factors such as light, temperature, and moisture, while only the factor of interest is varied—such as increasing nitrogen availability to plants in pots by adding fertilizer. The biotic community represented in a laboratory experiment is simplified, however, so conclusions based on laboratory results are limited. Laboratory experiments are best used to study the physiological responses of individual organisms rather than the dynamics of reproducing populations.

Field experiments are conducted outdoors and have the advantage of operating on natural rather than artificially contrived populations or communities. The most commonly used manipulations include the local elimination or addition of competitors, predators, or herbivores. Density of the target species can then be monitored to see whether it increased or decreased with the treatment, relative to controls. Charles Darwin used a field experiment to demonstrate that the introduction of grazing animals increases the number of plant species on a lawn. The number of species is increased because grazers often eat the most common species, preventing these species from outcompeting other species, whose numbers then increase. Field experiments commonly manipulate species through the use of tools, such as cages or fences to keep predators or herbivores in or out. Such manipulations are unlikely to be generated by nature itself.

Sometimes natural events like severe droughts, freezes, or floods provide the best opportunity to study the effects of environmental extremes in a field setting. Such natural extremes are often referred to as natural experiments. Natural experiments are usually the sole technique for following the time path of an environmental change beyond a few decades. Weather is frequently shown to be of vital importance in influencing the population densities of many species, but we cannot easily manipulate the weather. Natural experiments involving volcanic explosions or hurricanes commonly provide the only data on these subjects. However, natural experiments are not true experiments in that they are not replicated nor do they have controls.

There is no best type of experiment; the choice depends on what one is investigating. The strengths and weaknesses of these different types of experiments are outlined in **Table 1.1**. For example, the spatial scale of laboratory experiments is likely to be limited to the size of a constant-temperature laboratory room, around 0.01 ha (hectare), and that of field experiments to usually less than 1 ha. Natural experiments, however, can operate on much larger scales.

### 1.4.3 Meta-analysis allows data from similar experiments to be combined

One problem with experiments is that they take a lot of time, money, and effort. These constraints frequently lead to low levels of replication and then to what is known as a type I error: the declaration that a hypothesis is false when it is actually true. The experiment may be said to have low statistical power.

**Table 1.1 The strengths and weaknesses of different types of experiments in ecology.** (Abbreviated from Diamond, 1986.)

| | Laboratory experiment | Field experiment | Natural experiment |
|---|---|---|---|
| 1. Regulation of independent variables | High | Medium | None |
| 2. Maximum temporal scale | Low | Low | High |
| 3. Maximum spatial scale | Low | Low | High |
| 4. Scope (range of manipulations) | Low | Medium | High |
| 5. Realism | Low | High | High |
| 6. Generality to other systems | Low | Medium | High |

For example, suppose that fertilization of plants increases herbivore densities, but such increases can be detected only by performing 10 replicates of an experiment in which we add fertilizer to plants. If we perform only 5 replicates, our standard errors or standard deviations may be larger than if we use 10 replicates. Larger standard errors reduce confidence in the results, and we cannot as easily conclude that they are meaningful or significant.

Now imagine that 100 of these fertilization experiments were reported in the literature: 90 with an insufficient number of replicates (perhaps 5), and 10 with sufficient numbers of replicates (10 or more). If we summarized the literature by reviewing it, we would say 90 studies failed to find a significant effect of fertilizer on plants, and 10 found a significant effect—even though these results were purely a reflection of sample size. If most of the studies have low statistical power, the failure to demonstrate the phenomenon will be perpetuated. The few studies with high statistical power that do demonstrate the phenomenon will be outweighed by the majority of experiments with low statistical power that fail to show it.

One technique for detecting the true strength of replicated experiments is meta-analysis, a method for combining the results from different experiments that weights the studies based primarily on their sample sizes. The method was pioneered in ecology by Jessica Gurrevitch and colleagues (1992). In meta-analysis the data are not re-analyzed, but instead the results from a number of different studies are examined to see whether together they demonstrate an effect that is significant. Meta-analysis starts by estimating the effect size of a treatment from every experiment, then pooling all the effects together to get one overall effect size, usually indicated by the variable $d$ (**Figure 1.14**). The individual effect sizes are weighted by the number of replicates performed for each experiment. An experiment with only a few replicates of a treatment would not be weighted as heavily as an experiment with 10 or 20 replicates.

As well as the effect sizes, some measure of the variation in experimental results is noted by drawing bars around the mean, indicating confidence intervals. These bars are usually called 95% confidence intervals, meaning that it is likely that this is the range within which the mean value will fall 95% of the time. This is analogous to using bars to represent the standard deviation around the mean. In meta-analysis, if these bars do not overlap each other on different treatments or between treatment and control, then the differences are probably statistically significant. If the bars overlap each other, the treatments generally are not significantly different. If the bars overlap the zero value on the y-axis, then the effect probably is not significant. Meta-analysis is being incorporated more frequently into ecology, and many different meta-analyses are mentioned in this book.

Some authors present the results of their meta-analyses data in terms of a different metric called the log response ratio, ln r. Let's consider a meta-analysis by Christine Hawkes and Jon Sullivan (2001) that examines studies comparing the effects of herbivores and resources on plant growth and

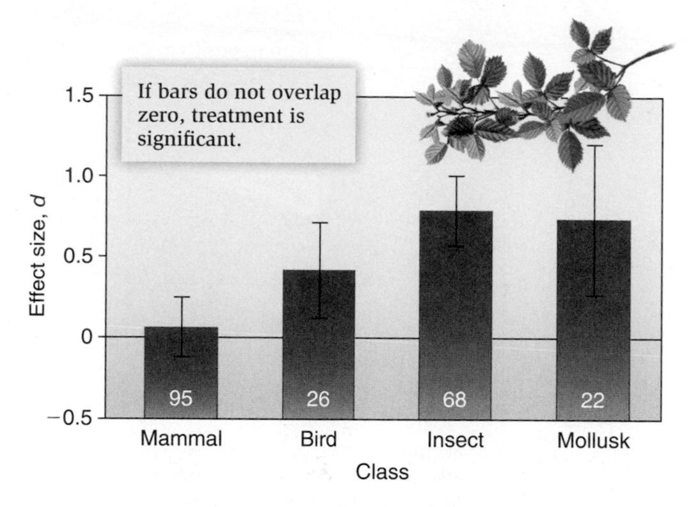


**Figure 1.14 A meta-analysis of the effect of herbivore type on plant biomass from 246 experiments.** Here, different types of herbivores were excluded from their host plants and the resultant increase in biomass was measured. Effect size is measured as the variable $d$. Each point is the mean of a given number of studies ($n$ is given in the bar).

**ECOLOGICAL INQUIRY**

Is the effect of insects on plants significantly stronger than that of mammals? (Modified from Bigger and Marvier, 1998.)

reproduction. The herbivory log response ratio = ln [plant growth with herbivores / plant growth without herbivores]. Thus, a negative value of this ratio suggests suppression by herbivores. Similarly, the resources log response ratio = ln [plant growth with added resources / plant growth without added resources]. A positive value of this ratio suggests increased growth by added resources. In this case, added resources were light, water, or nutrients. In total, 81 records from 45 studies were included in the growth analysis and 24 records from 14 studies in the reproduction analysis.

The advantage of this technique is that it estimates the effect as a proportional change resulting from experimental manipulations. In this meta-analysis, herbivory reduced plant growth and reproduction by about 0.6 and 0.75, or 60% and 75%, respectively, while increased resources increased growth and reproduction by over 100% (**Figure 1.15**). Here, the spread of the points around the mean are given by 95% confidence intervals.

### 1.4.4 Mathematical models can describe ecological phenomena and predict patterns

Sometimes it is very difficult or impossible to perform an experiment. In such cases we might turn to the use of mathematical models. Suppose we thought that disease was causing the demise of an endangered species such as the giant panda. We could not in good conscience experimentally expose giant

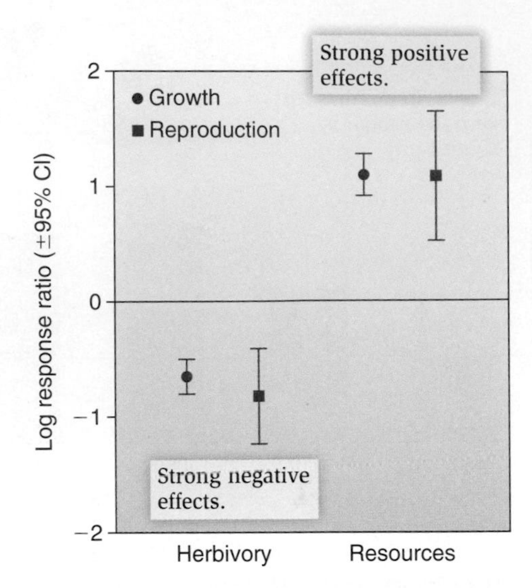


**Figure 1.15 A meta-analysis of the effect of herbivory and added resources on plant growth and reproduction from over 100 records.** Effect size is measured as the log response ratio. Each point is the mean of a large number of studies. (The error bars are 95% confidence intervals, CI.) (From Hawkes and Sullivan, 2001.)

pandas to the disease to see whether it decreased the population size of pandas. Instead, we might try to mathematically model what would happen. We could construct a model that incorporated the density of pandas, and the frequency and lethality of the disease, and field-test our model on other systems in which we are able to empirically examine the effects of diseases. For example, pandas suffer from SARS (severe acute respiratory syndrome) and bird flu, and the effects of both diseases can be modeled in other wildlife species. If the model worked well, then we might be able to predict how disease affects giant pandas. We would then have an explanation for the effects of disease in two systems and perhaps a general explanation for the effects of diseases in all systems. The collection of more empirical data allows such models to be further refined or rejected entirely. Models can give us valuable signposts as to how natural systems might work, what further data we need to collect to test our ideas, or what further observations we have to make.

This text uses all these ecological methods to address the principles of organismal, population, community, and ecosystems ecology by studying natural systems and systems that are undergoing or have undergone change. Change can come from natural events such as storms or fires, as well as from human-induced threats. Often synergistic (greater than additive) effects occur between agents of change. The record number of 28 named Atlantic basin tropical storms in 2005 was thought by some to be a result of global warming. Hurricanes are themselves agents of change as they destroy habitat. In south Florida, the spread of invasive plant species, such as Brazilian pepper, *Schinusterebinthifolius,* and Punk tree, *Melaleuca quinquenervia,* from Australia, is aided by hurricane-force winds. When hurricane Wilma crossed the Everglades late in 2005, it created a chaotic tumble of trees and light gaps of turned-over dirt that provided ideal germination sites for seeds of these invasive species. Many ecologists feel that another invasive pest, Old World climbing fern, *Lygodium microphyllum,* was spread around south Florida by hurricane Andrew in 1992. This illustrates a general ecological principle that many scenarios are context-dependent; that is, under certain conditions, one event in one place can lead to other events. Many agents of global change cause such chain reactions. This is a theme that we will explore throughout the book.

### Check Your Understanding

**1.4** Explain how you might set up a field experiment to determine whether insect pests are affecting crop yields. Be careful to include appropriate treatments and controls.

## SUMMARY

**1.1 Ecology: The Study of Organismal Interactions**

- Many species are under threat from global change (Figure 1.1). Ecologists study the interactions among organisms and between organisms and their environments.
- The tools of ecologists have changed from collecting jars and nets to chemical autoanalyzers and computers (Figure 1.2).

**1.2 The Scale of Ecology: From Organisms to Ecosystems**

- The field of ecology can be subdivided into broad areas of organismal, population, community, and ecosystems ecology (Figure 1.3).
- Organismal ecology considers how individuals are adapted to their environment and how the behavior of an individual organism contributes to its survival and reproductive success and the population density of the species (Figure 1.4).
- Population ecology explores those factors that influence a population's growth, size, and density.
- Community ecology studies how populations of species interact and form functional communities, and what factors influence the number of species in an area.
- Ecosystems ecology examines the flow of energy and the cycling of nutrients among organisms within

a community and between organisms and their environment.

**1.3 Global Change Involves Habitat Destruction, Invasive Species, Direct Exploitation, and Pollution**

- Organismal extinction has accompanied increases in human population growth (Figure 1.5).
- The main elements of global change are habitat destruction, introduced species, pollution, and direct exploitation (Figures 1.6–1.8, Figure 1.A).
- Amphibians and mammals are the most threatened groups of organisms on Earth (Figure 1.9).
- Population changes in one target species may have profound effects on other species that feed on or are fed on by the target species (Figure 1.B).

**1.4 Ecological Methods Include Observation, Experimentation, Analysis, and Mathematical Modeling**

- Ecological methods focus on observation, experimentation, data analysis, and mathematical modeling. A variety of graphical techniques exist that help determine whether two variables are related or whether experimentally altering one variable causes a significant change in the other (Figures 1.10–1.13).
- Types of experiments vary from those done in the laboratory to those done in the field (Table 1.1).
- Meta-analysis is a statistical technique that combines results from similar experiments to give more robust results (Figures 1.14, 1.15).

## TEST YOURSELF

1. Interactions among organisms are called:
   a. Abiotic interactions
   b. Biotic interactions
   c. Behavioral interactions
   d. Physiological interactions
   e. Evolutionary interactions
2. The organisms most threatened with extinction are believed to be:
   a. Birds
   b. Amphibians
   c. Mammals
   d. Fish
   e. Reptiles
3. Community ecology focuses on:
   a. The intersection of ecology and evolution
   b. How the behavior of an organism contributes to its survival and reproduction
   c. How organisms are physiologically adapted to their environment
   d. The factors that affect population growth
   e. The factors that influence the number of species in a given area
4. The small vertical bars around graphical means are usually called:
   a. Replicates
   b. Averages
   c. Standard deviations
   d. Standard errors
   e. Both c and d
5. Which is not an example of habitat loss?
   a. Deforestation
   b. Agriculture
   c. Urbanization
   d. Draining of wetlands
   e. Overharvesting
6. The main elements of global change are:
   a. Habitat destruction
   b. Introduced species
   c. Direct exploitation
   d. Pollution
   e. All of the above
7. The passenger pigeon, once the most common bird in North America, was driven to extinction by:
   a. Pollution
   b. Overharvesting
   c. Habitat destruction
   d. Invasive species
   e. All of the above
8. E. O. Wilson has referred to the threats to life on Earth with the acronym:
   a. RHINO
   b. FISH
   c. HIPPO
   d. FROG
   e. TOAD
9. Many frogs in Central America are threatened by extinction from:
   a. An invasive species
   b. A virus
   c. Bacteria
   d. A protist
   e. A fungus
10. Studies of the flow of energy and nutrients among organisms is the realm of which branch of ecology?
    a. Organismal
    b. Community
    c. Ecosystems
    d. Population

## CONCEPTUAL QUESTIONS

1. Distinguish between ecology and environmental science.
2. Distinguish between organismal, population, community, and ecosystems ecology.
3. Outline the five-stage process of hypothesis testing.
4. Comment on the advantages and disadvantages of laboratory, field, and natural experiments.
5. What are some of the effects that introduced species can have on native species?

## DATA ANALYSIS

It is thought that predators such as foxes, raccoons, and skunks can contribute to high death rates of nesting ducks. In one North Dakota study, trappers were paid to remove these three species from a number of different sites over the two-year period 1994–1996. During this time, more than 2,404 predators were removed. Investigators measured the nest success of numerous duck species, including blue-winged teal *(Anas discors)*, mallards *(A. platyrhynchos)*, gadwalls *(A. strepera)*, northern pintails *(A. acuta)*, and northern shovellers *(A. clypeata)*. The data shown give nest success in a number of sites in 1994–1996 at trapped and untrapped sites. Summarize the data and discuss whether you think predators affect nest success.

| Year | Site | Percent Nest Success, Trapped | Site | Percent Nest Success, Untrapped |
|---|---|---|---|---|
| 1994 | 1 | 45 | 2 | 14 |
| 1995 | 1 | 35 | 2 | 28 |
| 1995 | 5 | 56 | 3 | 29 |
| 1995 | 6 | 36 | 4 | 17 |
| 1995 | 8 | 52 | 7 | 24 |
| 1996 | 9 | 32 | 10 | 34 |
| 1996 | 12 | 19 | 11 | 19 |
| 1996 | 13 | 53 | 14 | 16 |
| 1996 | 15 | 34 | 16 | 13 |

SECTION ONE

# Organismal Ecology

Organismal ecology investigates how adaptations and behavioral choices by individuals affect their reproduction and survival and thus, ultimately, their distribution and abundance. As we mentioned in Chapter 1, organismal ecology can be divided into three subdisciplines.

The first subdiscipline, evolutionary ecology, focuses on how evolution is central to explanations of why certain organisms occur in certain areas on Earth. For example, the duck-billed platypus, *Ornithorhynchus anatinus,* and the echidnas, *Zaglossus* and *Tachyglossus* species, are egg-laying mammals, called monotremes, that occur only in Australia. Why are there no platypuses or echidnas in other areas of the globe? The answer is not insufficient food or too many predators; these animals once occurred in other areas of the globe but were outcompeted by placental mammals. However, by the time placental mammals evolved, Australia was a separate continent and terrestrial placental mammals could not cross into Australia. Monotremes and marsupials were safe from competition with placental mammals.

The second subdiscipline, behavioral ecology, concerns the survival value of behavior. It examines the relationship between behavior and ecology. For example, in determining the distribution and abundance of organisms we find that many animals are solitary while others are social and occur in dense herds or flocks. Being solitary or gregarious has large influences on overall densities. Group living promotes competition for food but may reduce the risk of predation. Group living also increases the likelihood of finding a mate but may increase the spread of disease. Behavioral ecology examines these strategies and also investigates why males of some species mate with multiple females yet males of other species have just one partner.

The third subdiscipline, physiological ecology, forms an interface with physiology and ecology. An understanding of the effects of the abiotic environment on plants and animals is vital if we are to understand the distribution and abundance of life on Earth. For example, in the Northern Hemisphere, the north-facing slopes of mountains are more shaded than the southern slopes (the reverse is true in the Southern Hemisphere). This influences temperature and soil moisture levels, with the result that different species may be present on adjoining mountain slopes of different aspects. Understanding the influences of abiotic conditions on organisms is of vital importance as we attempt to predict the future distribution patterns of life on Earth under conditions of global change.

Why does this Koala occur nowhere else on Earth but Australia? Characteristic faunas of different regions of the world point to a strong role of evolution in the distribution and abundance of life on Earth.

CHAPTER 2

# Evolution and Genetics

## Outline and Concepts



Because ecology is concerned with explaining the distribution and abundance of organisms around the world, genetics and evolutionary ecology are important parts of the discipline. For example, South America, Africa, and Australia all have similar climates, ranging from tropical to temperate, yet each continent has distinctive animals. South America is inhabited by sloths, anteaters, armadillos, and monkeys with prehensile tails. Africa possesses a wide variety of antelopes, zebras, giraffes, lions, baboons, the okapi, and the aardvark. Australia, which has no native placental mammals except bats, is home to a variety of marsupials such as kangaroos, koala bears, Tasmanian devils, and wombats, as well as the egg-laying monotremes, namely, the duck-billed platypus and four species of echidnas.

Most continents also have distinct species of plants; for example, *Eucalyptus* trees are native only in Australia. In South American deserts, succulent plants belong to the family *Cactaceae,* the cacti. In Africa, they belong to the genus *Euphorbia,* the spurges. In North America, the pines, *Pinus* spp., and firs, *Abies* spp., are common, but they do not occur south of the mountains of central Mexico. In contrast, palms are common in South America and do not generally occur north of the mountains of central Mexico, except for several genera in southern California and Florida.

A plausible explanation for these species distributions is that each region supports the fauna best adapted to it, but introductions have proved this explanation incorrect: European rabbits introduced into Australia proliferated rapidly, and *Eucalyptus* from Australia grows well in California. The best explanation is that different floras and faunas are the result of the independent evolution of separate, unconnected populations, which have generated different species in different places.

A knowledge of evolutionary ecology and of evolution itself is of paramount importance in understanding contemporary distributions of species. A thorough understanding of evolution must include some knowledge of genetics and how variation originates. Thus, in this chapter we begin with a brief history of the development of the theory of evolution and the mechanism of inheritance. This is followed by a discussion of how genetic variation is measured in organisms and how novel genotypes originate and are maintained in populations. The chapter ends with an important section that shows how present-day genetic diversity is vital to the continual existence of populations.

## 2.1 Evolution Concerns How Species Change over Time

A number of naturalists and philosophers, beginning with the ancient Greeks, suggested that many forms of life evolved from each other. The first to formalize and publish a theory of how species changed over time—"transformism" as he called it—was Jean-Baptiste Lamarck (1744–1829). The mechanism Lamarck proposed to explain how evolution works was based on the inheritance of acquired characteristics. He suggested that physiological events, such as use or disuse, determined whether traits were passed on to offspring. For example, someone who became strong through lifting weights would pass this trait on to their offspring. Lamarck used the long neck of the giraffe as an example. He proposed that giraffes, in their continual struggle to reach the highest foliage, stretched their necks by a few millimeters in the course of their lifetime. This increase in neck length was passed on to their offspring, which continued the process until the necks of giraffes reached their current proportions. Lamarck also proposed a "drive for complexity" such that living things evolved toward ever more-complex forms, ending in human "perfection." Though both ideas were eventually rejected, Lamarck was the first to develop a comprehensive evolutionary theory.

### 2.1.1 Charles Darwin proposed the theory of evolution by natural selection

Charles Robert Darwin (1809–1882) is considered the founder of modern evolutionary theory. Darwin was born into a wealthy family. Initially educated in medicine at Edinburgh, Darwin was nauseated by observing surgical operations because anesthetics had not yet been invented. He turned to theology at Cambridge University but maintained a strong interest in natural science. His family's wealth enabled Darwin to accept an unpaid job as a scientific observer on board HMS *Beagle,* which sailed on a 5-year world survey from 1831 to 1836, concentrating on South America (**Figure 2.1**). Darwin was in some ways "primed" to accept the theory of gradual biological change and evolution because he had read Charles Lyell's newly published *Principles of Geology* (1830). Lyell had taken the unprecedented step of describing the physical world as changing gradually through physical processes. Prior to this time, the prevailing view was that a few catastrophic events, such as the biblical flood, resulted in rapid change.

During the voyage of the *Beagle,* Darwin was able to view diverse tropical communities, some of the richest fossil beds in the world in Patagonia, and the Galápagos Islands, 600 miles west of Ecuador. The Galápagos contain a fauna different from that of mainland South America, with different animal forms on virtually every island. By the end the expedition, Darwin had amassed a wealth of data, described an astonishing array of animals, and built up a vast collection of specimens.

In March 1837, six months after Darwin's arrival back in England, the ornithologist John Gould pointed out that many of Darwin's specimens of mockingbirds from the Galápagos Islands were so different that they probably represented different species. Darwin also recalled that the island tortoise, *Geochelone elephantopus,* exhibited different growth forms on different islands. On larger, moister islands with abundant vegetation, such as Santa Cruz and Isabela, the tortoises have a domed shell and shorter necks, and they feed on grasses and low-lying shrubs (**Figure 2.2a**). On more arid islands, such as Espanola and Pinta, the vegetation grows above the ground and the tortoises have a saddle-back shape to their shell, which allows their neck greater upward movement so that they can access the higher vegetation (**Figure 2.2b**).

In 1838, two years after his return from the *Beagle* voyage, Darwin read a revolutionary book on human population growth by the English clergyman Thomas Malthus, who proposed that because the Earth was not overrun by humans, population growth must be limited by food shortage, disease, war, or conscious control. This idea became known as the **Malthusian theory of population.**

Darwin thought that the Malthusian theory of population could also apply to plant and animal populations. He made the brilliant deduction that these factors would act to the detriment of weaker, less well-adapted individuals, and that only the best adapted would survive and reproduce. Darwin had formulated his theory of **natural selection:** Better-adapted organisms would acquire more resources and leave more offspring. Nature "selects" individuals with traits that allow them to flourish and reproduce. This idea came to be known as survival of the fittest.

Over long periods of time, natural selection leads to **adaptation:** Given an evolutionary time span, a population's

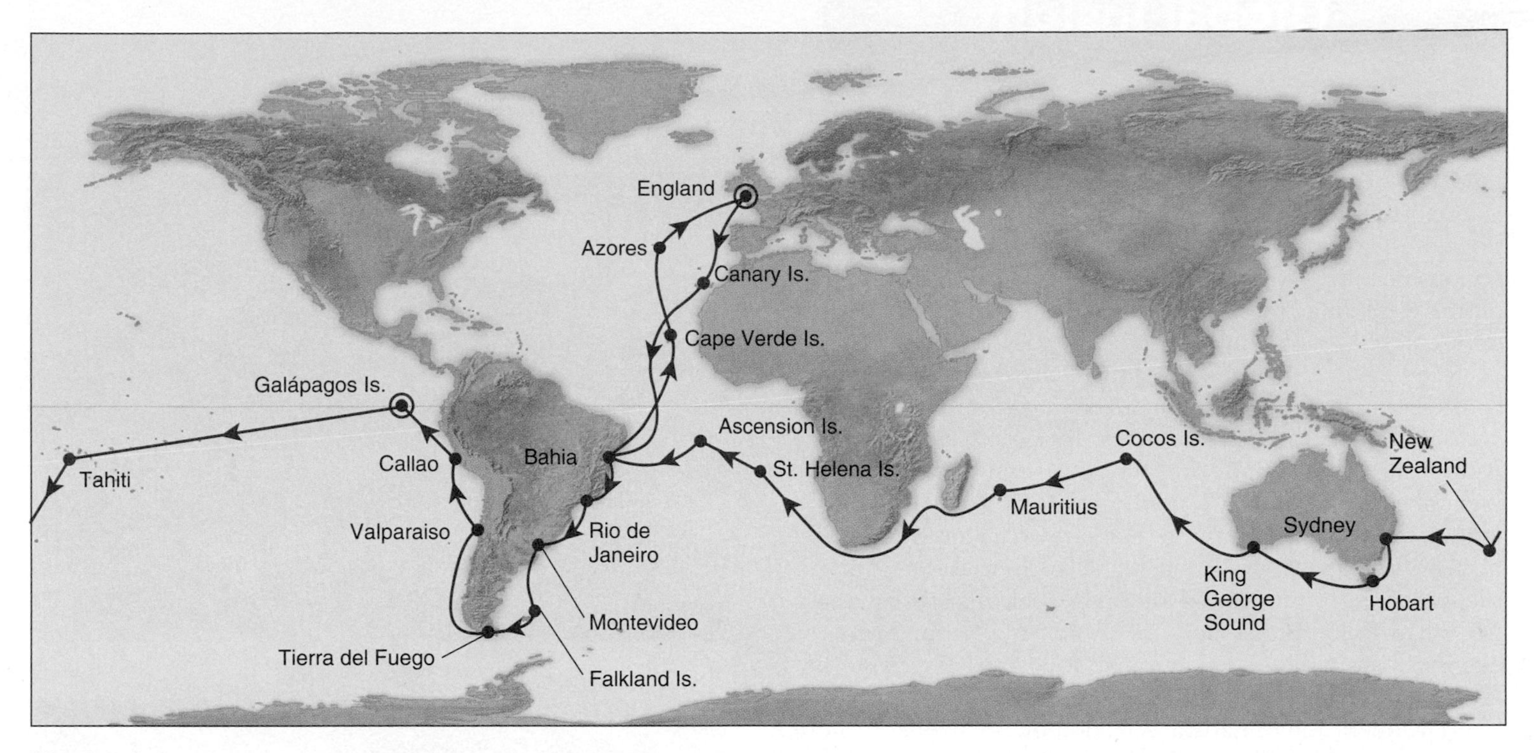


**Figure 2.1 The voyage of HMS *Beagle*, 1831–1836.** Most of Darwin's focus was on South America and the Galápagos Islands. However, toward the end of his voyage Darwin studied the vegetation on Ascension Island. This visit played an important role in the importation of many exotic plant species to Ascension.

**Figure 2.2 Two different forms of Galápagos Island tortoises.** **(a)** On moist islands, low-growing vegetation is present and tortoises have a dome-shaped shell. **(b)** On drier islands, the vegetation grows higher off the ground and tortoises with a saddle-back-shaped shell can lift their heads higher to feed on this vegetation.

## Global Insight

### Pollution Affects Color in the Peppered Moth, *Biston betularia*

One of the best-known examples of natural selection in action is the color change that has occurred in certain populations of the peppered moth, *Biston betularia,* in the industrial regions of Europe during the past 100 years. Originally these moths were uniformly pale gray or whitish in color. The first dark-colored, or melanic, individual was recorded in Manchester, England, in 1848. Over the next 100 years, the dark-colored forms came to dominate the populations of certain areas, especially those of extreme industrialization such as the Ruhr Valley of Germany and the Midlands of England. This phenomenon, an increase in the frequency of dark-colored mutants in polluted areas, became known as **industrial melanism.** A similar pattern occurred in North American forms of the peppered moth around the industrial areas of southern Michigan. Pollution did not directly affect mutation rates. For example, caterpillars feeding on soot-covered leaves did not give rise to dark-colored adults.

The operation of natural selection on the peppered moth was illustrated by H. B. D. Kettlewell (1955). He argued that normal pale gray/whitish forms are cryptic when resting on lichen-covered trees, whereas dark forms are conspicuous (**Figure 2.A**). In industrialized areas, lichens are killed off, tree bark becomes soot covered and darker, and the dark moths are more cryptic. Kettlewell demonstrated that birds were the selective force by releasing into rural and industrialized areas hundreds of pale forms and dark forms of moths marked with a small spot of paint. In the rural area of Dorset, England, he recaptured 13.7% of the pale morphs released but only 4.7% of the dark moths. In the industrial area of Birmingham, the situation was reversed; only 13.1% of pale morphs were recaptured as opposed to 27.5% of dark morphs. As a final test, Kettlewell and companions set up blinds and watched birds attack moths placed on tree trunks.

The white form of the peppered moth has made a strong comeback in Britain since the Clean Air Act was passed in

(i)

(ii)

**Figure 2.A The melanic and typical forms of the peppered moth, *Biston betularia.*** **(i)** On lichen-covered trees and **(ii)** dark trees, Sherwood Forest, Nottingham, England.

characteristics change to make its members better suited to their environment. For example, giraffes born by chance with longer necks would be better fed because they could reach more vegetation; as a consequence, they would be able to reproduce more successfully than shorter-necked giraffes. This trait would be passed on to their offspring, and over time, longer necks would become common.

Darwin reached two main conclusions about the origin of species. First, all organisms are descended with modification from common ancestors. All the prominent scientists of the day were convinced of this point within 20 years. Second, the mechanism for evolution was natural selection. This point was not fully accepted until the late 1920s, partly because of a widespread belief in blending inheritance, in which the traits of the parents were thought to be blended in the offspring, like the colors of two paints blending to produce an intermediate color. Natural selection would not work in such a system. For example, if a long-necked giraffe mated with a short-necked giraffe, the offspring would have a neck of medium length, and the advantage of a long neck would be lost.

Furthermore, Darwin knew nothing of the causes of hereditary variation and could not well answer questions on that subject. The evidence of genetics and Mendel's laws of heredity, first published in 1866, were available but had

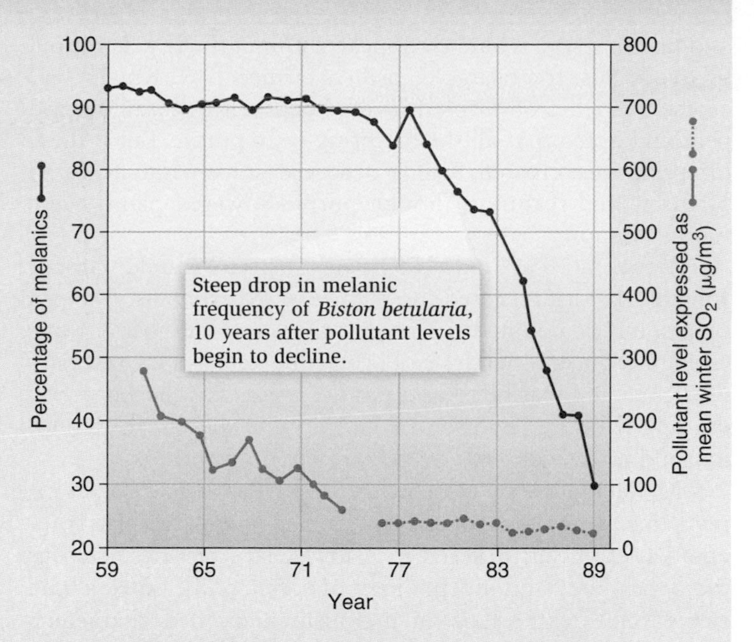


**Figure 2.B Decline in the proportion of melanic *Biston betularia* and pollutant levels in West Kirby, England.** Changes in moth numbers and sulfur dioxide concentrations over time. (After Clarke et al., 1990.)

**ECOLOGICAL INQUIRY**

If pollutant levels decreased to zero, would melanics become extinct?

1965. Sir Cyril Clarke trapped moths at his home on Merseyside, Liverpool, from 1959 to 1989. After 1975 there was a steep decline in dark-colored forms, and in 1989 only 29.6% of the moths caught were melanic (**Figure 2.B**). The mean concentration of sulfur dioxide pollution fell from about 300 $\mu g\ m^3$ in 1960 to less than 50 $\mu g\ m^3$ in 1975 and remained fairly constant until 1989. If the reappearance of the light-colored form of the moth continues at the same speed as the melanic form appeared in the last century, the melanic form will eventually be only an occasional mutant in the Liverpool area by the year 2020.

passed into obscurity, and they were not rediscovered until the early part of the 20th century. Today we can cite many examples of how natural selection operates in our changing world (see the **Global Insight** feature).

Darwin did not immediately publish his theory. Perhaps he was mindful of a hostile reception to an 1844 publication of Robert Chambers, which had discussed a divine plan whereby all living things had evolved from simple forms of life that had, in turn, arisen from nonliving matter. Darwin gathered more evidence in support of evolution for nearly 20 years, collecting data on a wide range of organisms. Eventually he was pushed into publication (Darwin, 1859) by the

**Figure 2.3 Ascension Island is a remote island in the Atlantic Ocean.** Its rich vegetation is almost completely exotic, having been introduced from many different areas of the world.

arrival of a manuscript by Alfred Russel Wallace, who had independently, and years later, arrived at the same conclusions.

It is also interesting to note that Darwin was involved in one of the largest-scale ecological experiments of all time. One of Darwin's last ports of call was Ascension Island, a barren, volcanic island located on the Mid-Atlantic Ridge (**Figure 2.3**). Ascension was a strategically important island for the Royal Navy, which established a base there to keep watch over the exiled emperor Napoleon on nearby St. Helena. A lack of fresh water limited the island's development. It is thought that Darwin, and his botanist friend Joseph Hooker, encouraged the navy to plant different vegetation on the island. Trees would capture moisture from the strong trade winds, reduce evaporation of existing sources of fresh water, and help create richer soils. By the late 1870s, after 20 years of regular shipments of plants from botanical gardens in Europe, South Africa, and South America, many different species had become common on the island, including eucalyptus, Norfolk Island pine, bamboo, and banana. The highest peak on Ascension, at 859 m (2,817 ft) was now called Green Mountain. The trees had created a damp forest where before there was only lava rock. The functioning forest, which is still increasing in extent today, is a mosaic of exotic species from all over the world. Some scientists believe that similar experiments could be used to help life establish a foothold on Mars.

### 2.1.2 Alfred Russel Wallace was codiscoverer of evolutionary theory

Alfred Russel Wallace (1823–1913) was born into poverty and was farmed out to an older brother in London at age 14, after only 6 years of schooling. He held down a succession of jobs until his brother died and left him some money. Wallace set sail immediately for the Amazon. A year later, a second brother joined him. Both men contracted yellow fever, which eventually killed the brother. After 4 years in the jungle, Wallace sailed for home with his precious collections of exotic plants, insects, and other organisms. En route the ship caught fire and sank. After 10 days on the open sea in a small boat, Wallace was saved, but 4 years of labor went down with the ship.

Back in England, Wallace began to prepare for a second voyage, this time to the Malay Archipelago as a professional collector and naturalist in the company of W. H. Bates (for whom Batesian mimicry is named; look ahead to Figure 13.4). It was there, during another bout of fever, that Wallace conceived the idea of natural selection. Wallace's one major advantage over Darwin was that he was persuaded before he left on his voyages that species evolve, and he was able to gather data with an eye to his evolutionary hypothesis.

Unfortunately, Wallace's earlier papers had been largely ignored by the scientific community and he was faced with the problem of lack of recognition. His solution was to send his manuscripts to Darwin, with whom he had previously corresponded. Darwin's higher standing in the scientific community made it more likely that he would be taken more seriously. Darwin sought the advice of friends, including geologist Lyell, and as a result of their suggestions, Darwin and Wallace had their theories presented jointly at a historic meeting of the Linnean Society of London on July 1, 1858. One year later, Darwin at last published *On the Origin of Species,* an abbreviated version of the manuscript based on his 20 years of work. Although Wallace deserves full credit as a codiscoverer of the chief mechanism of evolution, Darwin's subsequent work continued to explore the same ideas and principles inherent in the original work.

### 2.1.3 Gregor Mendel performed classic experiments on the inheritance of traits

In the 18th century, British farmers had crossed pea plants and noted that in crosses between two types, such as short and tall, one type would disappear within a single generation. In the 1790s, for example, British farmer T. A. Knight had crossed varieties of purple-flowered peas with white-flowered peas and noted that all the offspring were purple. But if these offspring were crossed, Knight observed, some white-flowered plants would reappear, though purple-flowered plants were more common.

Between 1856 and 1863, an Austrian monk, Gregor Johann Mendel, repeated these types of experiments with pea plants, but he counted the precise types and numbers of offspring produced. Mendel had been educated at the University of Vienna and was hired as a physics teacher at the Augustinian Abbey of St. Thomas in Brno (now in the Czech Republic), and it was here that he performed his experiments.

Mendel chose to work with the garden pea, *Pisum sativum,* for several reasons. First, it had many readily available varieties that differed in visible characteristics, such as the appearance and morphology of seeds, pods, flowers, and stems. Such features of an organism are called **characters.** Each of these characters in pea plants was found in two variants. For example, pea plant height had the variants known as tall and dwarf. Another was flower color, which had the variants purple and white.

A second important feature of garden peas is that they are normally self-fertilizing. In **self-fertilization,** a female gamete is fertilized by a male gamete from the same plant. Self-fertilization makes it easy to produce plants that breed true for a given trait, meaning that the trait does not vary from generation to generation. For example, if a pea plant with purple flowers breeds true for flower color, all the plants that grow from the seeds from these flowers will also produce purple flowers. A variety that continues to exhibit the same trait after several generations of self-fertilization is called a **true-breeding line.** Prior to conducting his crosses, Mendel had already established true-breeding lines in the strains of pea plants he had obtained. When two individuals with different characteristics are mated or crossed to each other, this is called a **cross-fertilization** or **hybridization** experiment, and the offspring are referred to as hybrids. For example, a hybridization experiment could involve a cross between a purple-flowered pea plant and a white-flowered pea plant.

A third reason for using garden pea plants was the ease of making crosses: The flowers are quite large and easy to manipulate. In some cases Mendel wanted his pea plants to self-fertilize, but in others he wanted to cross plants that differed with respect to some trait. In garden peas, cross-fertilization requires placing pollen from one plant on the stigma of another plant's flower. Mendel would pry open an immature flower and remove the stamens with scissors before they produced pollen, so that the flower could not self-fertilize (**Figure 2.4**). He then used a paintbrush to transfer pollen from another plant to the stigma of the flower whose stamens had been removed. In this way Mendel was able to cross-fertilize any two of his true-breeding pea plants and obtain any type of hybrid he wanted.

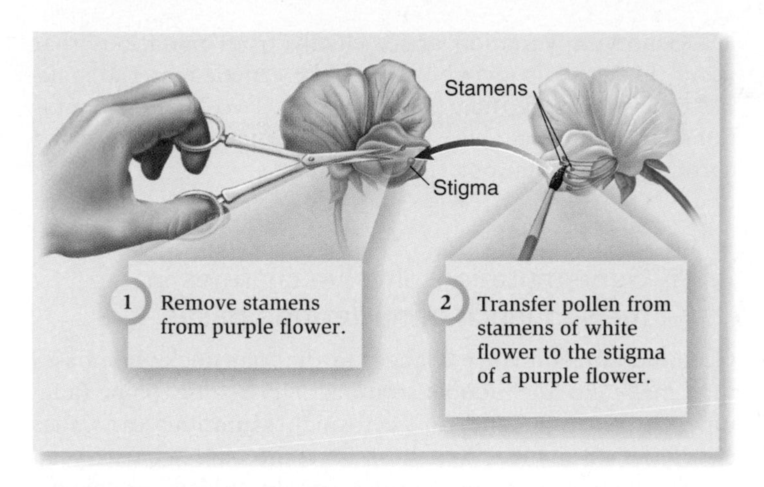


**Figure 2.4** **A procedure for cross-fertilizing pea plants.**

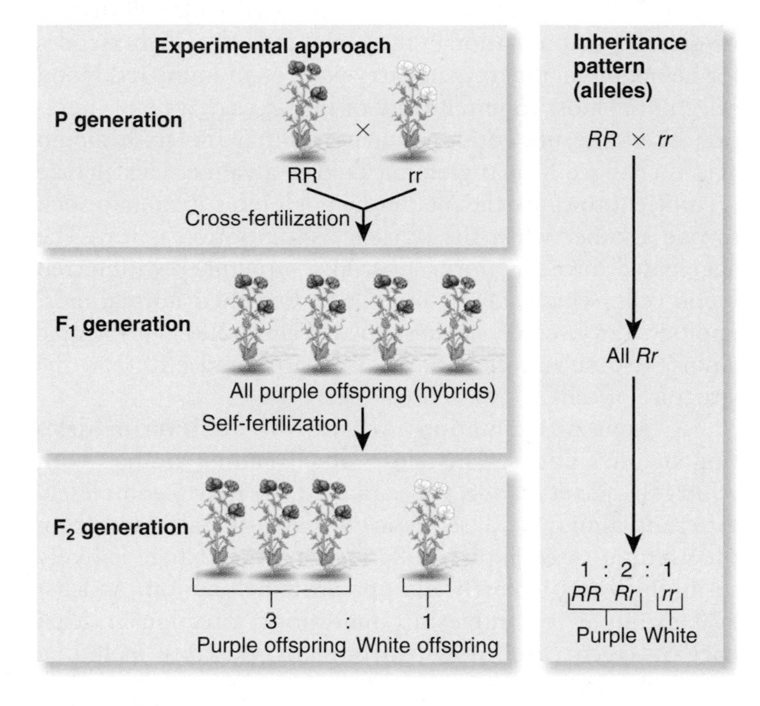


**Figure 2.5** **Mendel's crosses of pure-breeding purple- and white-flowered pea plants.** These crosses revealed that all offspring had purple flowers in the $F_1$ generation, but some offspring with white flowers would reappear in the $F_2$ generation.

Let's examine Mendel's crosses of pure-breeding pea lines with either purple or white flowers (**Figure 2.5**). The true-breeding parents are termed the **P generation** (parental generation), and a cross of these plants is called a P cross. The first generation offspring of a P cross are termed the **$F_1$ generation** (first filial generation, from the Latin *filius,* son). When the true-breeding parents differ with regard to a single trait, their $F_1$ offspring are called single-trait hybrids, or **monohybrids**. When Mendel crossed true-breeding purple- and white-flowered plants, he observed that all plants of the $F_1$ generation produced purple flowers.

Next, Mendel followed the transmission of this trait for a second generation. To do so, he allowed the $F_1$ monohybrids to self-fertilize, producing a generation called the **$F_2$ generation** (second filial generation). Although the white-flower trait reappeared in the $F_2$ offspring, three-quarters of the plants were purple and only one-quarter were white. Mendel's results were consistent with a particulate mechanism of inheritance, in which the determinants of traits are inherited as unchanging, discrete units. We term these alternative traits dominant and recessive. The term **dominant** denotes the displayed trait, while the term **recessive** denotes a trait that is masked by the presence of a dominant trait. Purple flowers are examples of a dominant trait; white flowers are examples of a recessive trait. We say that purple is dominant over white.

The genetic determinants of traits are called **genes** (from the Greek *genos,* birth). Every individual carries two genes for a given trait, and the gene for each trait has two variant forms, called **alleles**. For example, the gene controlling flower color in Mendel's pea plants occurs in two variants, called the purple allele and the white allele. The right-hand side of **Figure 2.6** shows Mendel's conclusions, using genetic symbols that were adopted later. The letters *R* and *r* represent the alleles of the gene for plant flower color. By convention, the uppercase letter represents the dominant allele (in this case, purple) and the lowercase letter represents the recessive allele (white).

When Mendel compared the numbers of $F_2$ offspring exhibiting dominant and recessive traits, he noticed a recurring pattern. Although he encountered some experimental variation, he observed approximately a 3:1 ratio between the dominant trait and the recessive trait. This quantitative observation allowed him to conclude that the two copies of a gene carried by an $F_1$ plant **segregate** (separate) from each other, so that each sperm or egg carries only one allele. The diagram in Figure 2.6 shows that segregation of the $F_1$ alleles results in equal numbers of gametes carrying the dominant allele (*R*) and the recessive allele (*r*). If these gametes combine with one another randomly at fertilization, as shown in the figure, this would account for the 3:1 ratio of purple to white-flowered plants in the $F_2$ generation. The idea that the two copies of a gene segregate from each other during transmission from parent to offspring is known today as Mendel's **law of segregation.**

Mendel's work showed that inheritance is generally particulate, that is, resulting from discrete factors, and that inherited factors could be passed down from ancient ancestors in the same form. Yet some variation must occur in populations if natural selection is to work. How does this variation originate? In the next section we discuss how novel genotypes arise.

### Check Your Understanding

**2.1** Darwin and Wallace conceived the idea of natural selection, but both knew little of the mechanisms by which traits are inherited. How was this dilemma resolved?

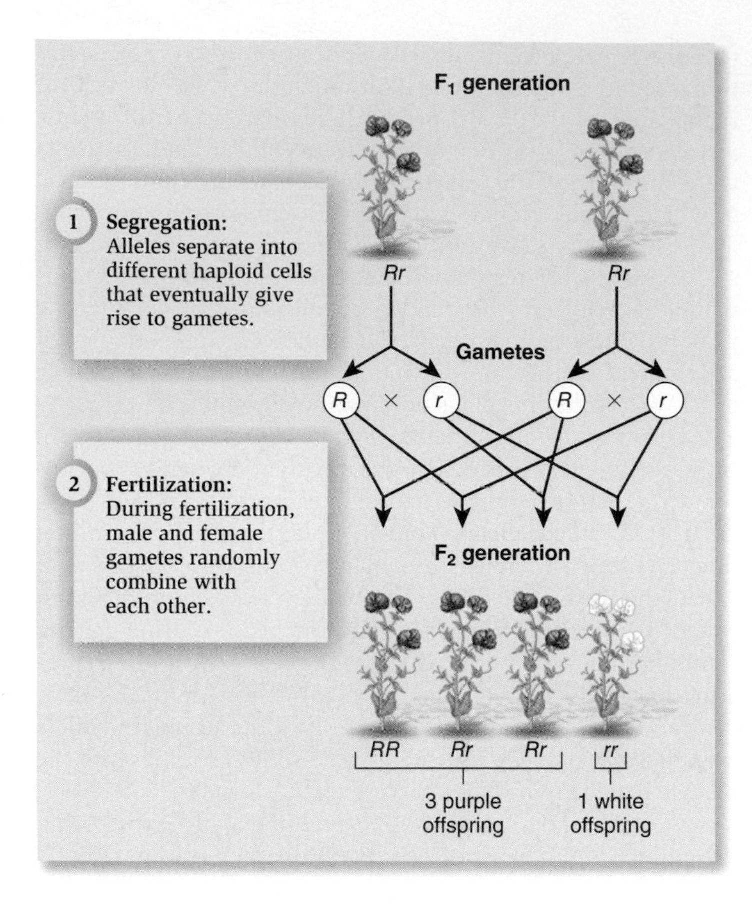


**Figure 2.6 Mendel's crosses of $F_1$-generation pea plants.** These crosses reveal a 3:1 ratio of purple : white flowered plants.

## 2.2 Gene and Chromosome Mutations Cause Novel Phenotypes

The term **genotype** refers to the genetic composition of an individual. In the pea plant example shown in Figure 2.6, *RR* and *rr* are the genotypes of the P generation and *Rr* is the genotype of the $F_1$ generation. An individual with two identical copies of a gene is said to be **homozygous** with respect to that gene. In the specific parental cross we are considering, the purple plant is homozygous for *R* and the white plant is homozygous for *r*. In contrast, a **heterozygous** individual carries two different alleles of the same gene. Plants of the $F_1$ generation are heterozygous, with the genotype *Rr*, because every individual carries one copy of the purple allele and one copy of the white allele. The $F_2$ generation includes both homozygous individuals (homozygotes) and heterozygous individuals (heterozygotes).

The term **phenotype** refers to the physical characteristics of an organism, which are the result of the expression of its genes. In the example in Figure 2.6, one of the parent plants is phenotypically purple for flowers and the other is phenotypically white for flowers. Although the $F_1$ offspring are heterozygous *(Rr)*, their phenotypes are purple because each of them has a copy of the dominant purple allele. In contrast, the $F_2$ plants display both phenotypes in a ratio of 3:1.

Genotypic variation arises chiefly from mutations that occur during the copying of DNA, the genetic material making up every living thing on the planet. Two kinds of mutations arise during DNA replication: gene mutations and chromosomal mutations.

### 2.2.1 Gene mutations involve changes in the sequence of nucleotide bases

Gene mutations involve changes in the four nucleotide bases that make up the double-stranded DNA base pairs (adenine, thymine, guanine, and cytosine). Mutations can cause two types of changes to genes. First, the base sequence can be changed; second, nucleotides can be added or deleted. **Point mutations** exchange a single nucleotide for another (**Figure 2.7**). The human disease known as sickle-cell disease involves a point mutation in the β-globin gene, which encodes for hemoglobin, the oxygen-carrying protein in the red blood cells. In the most common form of this disease, a point mutation alters the nucleotide sequence so that the sixth amino acid is changed from a glutamic acid to a valine. This change is sufficient to cause the mutant hemoglobin subunits to stick to one another when the oxygen concentration is low. The aggregated proteins form fiber-like structures within red blood cells, which causes the cells to lose their normal morphology and become sickle-shaped (**Figure 2.8**). This simple amino acid substitution thus has a profound effect on the structure of cells.

A **frameshift mutation** involves the addition or deletion of nucleotides. This shifts the "reading frame" with which the genetic code is deciphered, so that a completely different amino acid sequence occurs downstream from the mutation (see Figure 2.7). Such a large change is likely to inhibit or completely disrupt protein function. At least 470 examples of frameshift mutations affecting at least part of a gene are known in humans, possibly including Tay-Sachs disease.

| Mutation in the DNA | Effect on protein | Example |
|---|---|---|
| None | None | ATGGCCGGCCCGAAAGAGACC<br>Met Ala Gly Pro Lys Glu Thr |
| Base substitution | Point mutation changes one amino acid | ATGCCCGGCCCGAAAGAGACC<br>Met Pro Gly Pro Lys Glu Thr |
| Addition (or deletion) of single base | Frameshift—produces a different amino acid sequence | ATGGCCGGCACCGAAAGAGACC<br>Met Ala Gly Thr Glu Arg Asp |

**Figure 2.7 Gene mutations.** Point mutations involve changes in nucleotide bases at a single location. Frameshift mutations alter whole sequences and are often fatal.

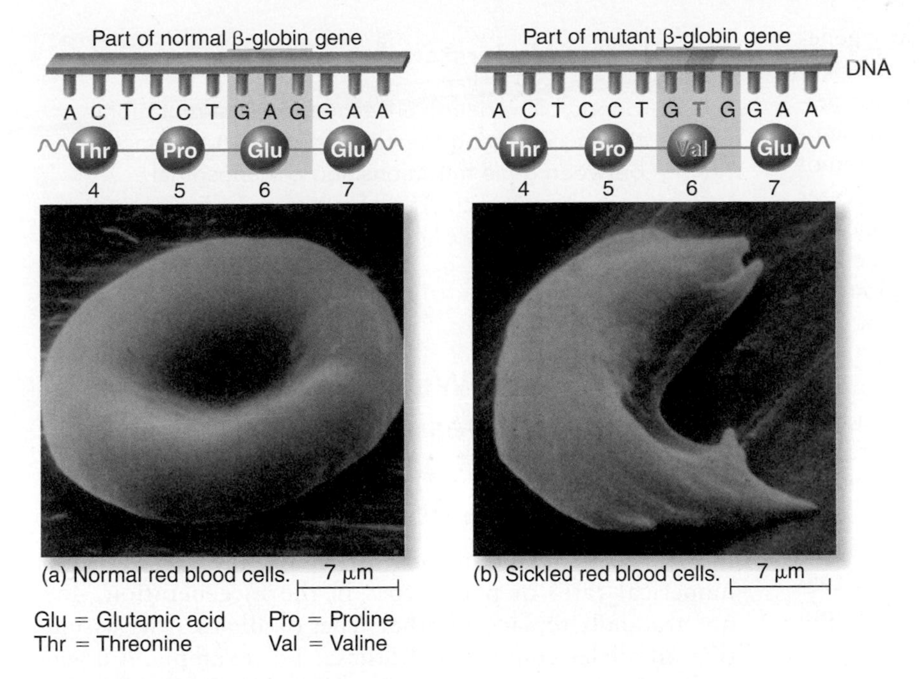


**Figure 2.8** **Point mutation changes the shape of red blood cells.**

The background rate of mutation is approximately one mutation for every 1 million genes. New mutations are more likely to produce proteins that have reduced rather than enhanced function. However, because so many mutations can occur, some can occasionally produce a protein that has a better ability to function. Although these favorable mutations are relatively rare, they may result in an organism with a greater likelihood to survive and reproduce. The favorable effect of such a mutation may cause it to increase in frequency in a population over the course of many generations.

### 2.2.2 Chromosome mutations alter the order of genes

Chromosomal mutations do not actually add to or subtract from the variability of the gene pool; they merely rearrange it, creating certain gene combinations. When the order of base pairs within the gene is unaffected, but the order of genes on a chromosome is altered (**Figure 2.9**), the chromosomes have undergone any of four types of changes: deletions, duplications, inversions, and translocations. These occur during meiosis when chromosomes are being duplicated.

A deletion is the simple loss of part of a chromosome and is the most common source of new mutations. A deletion is

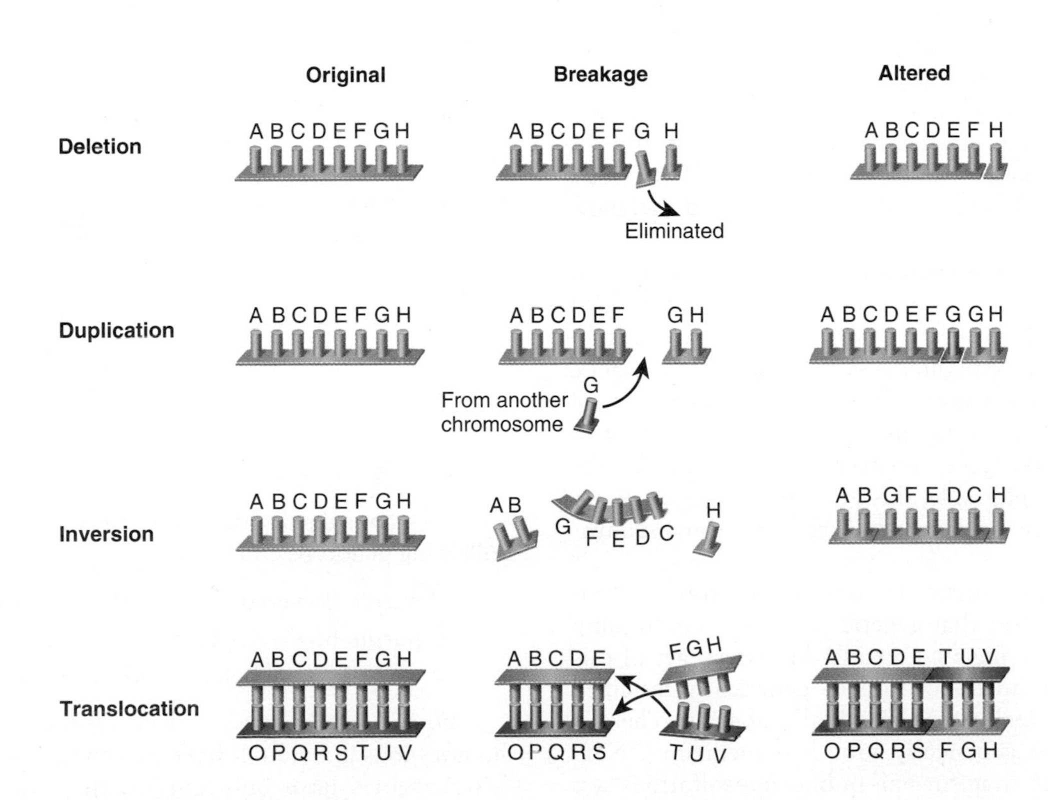


**Figure 2.9** **Chromosomal mutations.** Chromosome breakage and reunion can give rise to four principal structural changes: deletion, duplication, inversion, and translocation. Each letter represents a gene.

often lethal unless, as in some higher organisms, many genes have become duplicated.

Duplication occurs when two chromosomes are not perfectly aligned during crossing over; the result is one chromosome with a deficiency of genes and one with a duplication of genes. The duplication may be advantageous, in that greater amounts of enzymes may be produced from the duplicated genes. In yeasts, for example, an increase of the enzyme acid monophosphatase enables cells to more efficiently exploit low concentrations of phosphate in the medium in which the cells are growing.

An inversion occurs when a chromosome breaks in two places and the middle segment then turns around and refuses with the same pieces. Here the order of the chromosome's genes is reversed with respect to that on the unbroken chromosome. Such breaks probably occur during the prophase of cell division, when the chromosomes are long and slender and often bent into loops. In a translocation, two nonhomologous chromosomes break simultaneously and exchange segments.

### 2.2.3 Genetic suppression of invasive species

The release of genetically engineered organisms may prove to be a useful tool in the fight against invasive species. The mosquito *Aedes aegypti,* native to Africa, is a common invasive species in the Florida Keys. It is a known carrier of dengue fever, a flu-like virus that can cause intense joint pain in extreme cases. The disease had been eradicated in the U.S. until multiple cases were reported in Key West in 2009 and 2010. A British company plans to release genetically modified male mosquitoes in Key West as soon as it gains federal approval. The males, which don't bite, would mate with normal "wild type" females but the offspring would die and the overall population density of the mosquitoes would decrease. Trials in the Cayman Islands in 2009 and 2010 reduced mosquito densities by 80%.

Genetic methods of control are also being developed to control invasive vertebrates. Invasive carp in the rivers of Australia have been likened to rats with fins. Here researchers are developing a "daughterlesss" technology, a genetic construct that biases sex ratios toward males. Such control measures typically take a long time to act, usually more than 10 generations, as the gene construct spreads through the populations. On the plus side, this gives the biological community time to adjust to the gradual reduction of the invasive species.

In both projects scientists face opposition from the general public. There is fear that genetic constructs could jump to other species or be misused against humans. Part of this opinion stems from an opposition to genetically modified foods, though such feelings are gradually abating. There is broader support for genetically modified medicines. Nevertheless, it is clear that formal public consultations are needed as part of any genetic suppression program for invasive species.

**Check Your Understanding**

**2.2** Genotypic variation arises mainly from mutations that occur during the copying of DNA. Distinguish between gene mutations and chromosomal mutations. Which type of mutation is the most common source of genetic variation?

## 2.3 The Hardy-Weinberg Equation Describes Allele and Genotype Frequencies in an Equilibrium Population

How is it that a dominant allele, responsible for a 3:1 numerical ratio of phenotypes in the $F_2$ generation, does not gradually replace all other types of alleles, if we assume that all alleles confer equal fitness? For example, if a gene pool in one generation consists of 70% *R* alleles and 30% *r* alleles, what stops the proportion of *R* alleles from increasing dramatically? What will the proportion of alleles be in the next generation? This very question was posed and independently answered by a British mathematician, Godfrey Hardy, and a German physician, Wilhelm Weinberg, in 1908.

To understand Hardy and Weinberg's work, we first need to examine allele frequencies and **genotype frequencies** in a population. Allele and genotype frequencies are defined as:

$$\text{Allele frequency} = \frac{\text{Number of copies of a specific allele in a population}}{\text{Total number of all alleles for that gene in a population}}$$

$$\text{Genotype frequency} = \frac{\text{Number of individuals with a particular genotype in a population}}{\text{Total number of individuals in a population}}$$

Let's consider a population of 100 pea plants with the following genotypes:

49 purple-flowered plants with the genotype *RR*
42 purple-flowered plants with the genotype *Rr*
9 white-flowered plants with the genotype *rr*

When calculating an allele frequency, remember that homozygous individuals have two copies of an allele, whereas heterozygotes have only one. In this example, note, when totaling the number of copies of the *r* allele, that each of the 42 heterozygotes has one copy of the *r* allele, and each of the

9 white-flowered plants has two copies. Therefore, the allele frequency for $r$ is given as:

$$\text{Frequency of } r = \frac{(Rr) + 2(rr)}{2(RR) + 2(Rr) + 2(rr)}$$

$$\text{Frequency of } r = \frac{42 + (2)(9)}{(2)(49) + (2)(42) + (2)(9)}$$

$$= \frac{60}{200} = 0.3 \text{ or } 30\%$$

In other words, 30% of the alleles for this gene in the population are the $r$ allele.

The genotype frequency of $rr$ (white-flowered) plants is given by:

$$\text{Frequency of } rr = \frac{9}{49 + 42 + 9}$$

$$= \frac{9}{100} = 0.09, \text{ or } 9\%$$

Just 9% of the individuals in this population have white flowers.

Total allele and genotype frequencies are equal to 1, or 100%. In our pea plant example, the allele frequency of $r$ equals 0.3. Therefore, we can calculate the frequency of the other allele, $R$, as equal to $1.0 - 0.3 = 0.7$, because the frequencies of both must add up to 1.0.

To examine the relationship between allele frequencies and genotype frequencies in a population, let's consider the relationship between allele frequencies and the way gametes combine to produce genotypes (**Figure 2.10**). We can predict the outcome of a simple genetic cross between homozygotes or heterozygotes by making a Punnett square, named after the British geneticist Reginald Punnett. A Punnett square is a chart showing all possible allele combinations of a particular cross. If the allele frequency of $R$ equals 0.7, the frequency of a gamete carrying the $R$ allele, called $p$, also equals 0.7. If the allele frequency of $R$ is denoted by $p$, and the allele frequency $r$ is denoted by $q$, then $p + q = 1$. Therefore, if $p = 0.7$, then $q$ must be 0.3.

The frequency of producing an $RR$ homozygote, which produces purple flowers, $RR$, is $p^2$, which is $0.7 \times 0.7 = 0.49$, or 49%. The probability of offspring inheriting both $r$ alleles, which produces white flowers, is $qq$ or $q^2$, which is $0.3 \times 0.3 = 0.09$, or 9%.

In the Punnett square, two different gamete combinations can produce heterozygotes with purple flowers (see Figure 2.10). An offspring could inherit the $R$ allele from pollen and $r$ from the egg, or $R$ from the egg and $r$ from pollen. Therefore, the frequency of heterozygotes is $pq + pq$, which equals $2pq$. For the example, this would be $2(0.7)(0.3) = 0.42$, or 42%.

For a gene that exists in two alleles, the **Hardy-Weinberg equation** states:

$$(p + q)^2 = 1$$

or

$$p^2 + 2pq + q^2 = 1$$

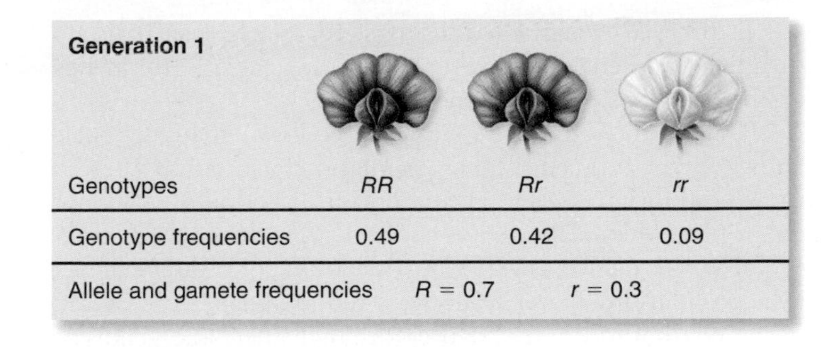


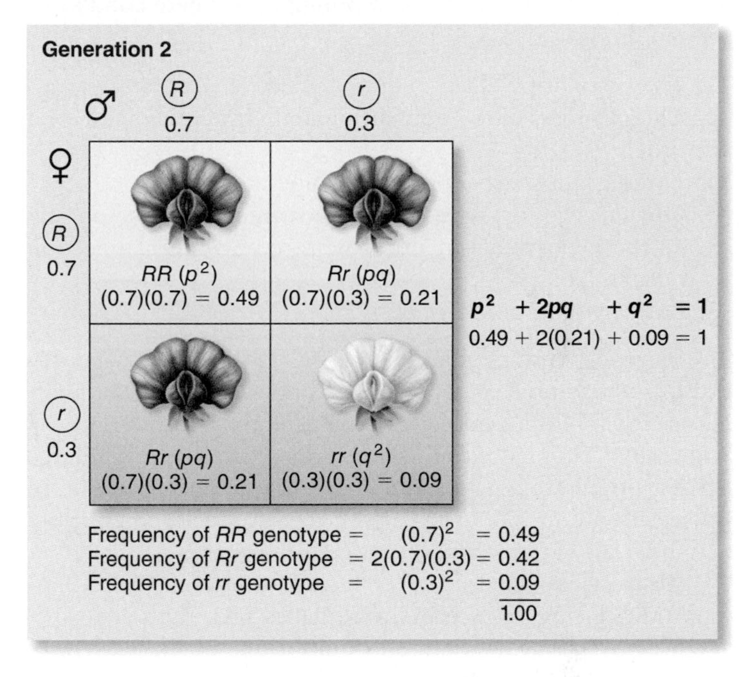


**Figure 2.10 Calculating allele and genotype frequencies in the next generation of a pea plant population.** Allele and genotype frequencies can be calculated using a Punnett square and the Hardy-Weinberg equation.

**ECOLOGICAL INQUIRY**

What is the frequency of white flowers in a population where the allele frequency of purple flowers, $R$, is 0.5? Assume that the population is in Hardy-Weinberg equilibrium and that $R$ and $r$ are the only two alleles.

If we apply this equation to our flower color gene, then

$$p^2 = \text{the genotype frequency of } RR$$
$$2pq = \text{the genotype frequency of } Rr$$
$$q^2 = \text{the genotype frequency of } rr$$

If $p = 0.7$ and $q = 0.3$, then

$$\text{Frequency of } RR = p^2 = (0.7)^2 = 0.49$$
$$\text{Frequency of } Rr = 2pq = 2(0.7)(0.3) = 0.42$$
$$\text{Frequency of } rr = q^2 = (0.3)^2 = 0.09$$

In other words, if the allele frequency of *R* is 70% and the allele frequency of *r* is 30%, then the genotype frequency of *RR* is 49%, *Rr* is 42%, and *rr* is 9%.

The Hardy-Weinberg equation predicts unchanging allele and genotype frequencies in a population, a situation referred to as equilibrium. When at equilibrium, a population is not adapting and evolution is not occurring. However, the Hardy-Weinberg prediction is valid only if certain conditions are met in a population. These conditions require that evolutionary mechanisms, those forces that can change allele and genotype frequencies, are not acting on a population. These conditions are as follows:

1. The population is large.
2. The members of the population mate randomly with each other.
3. No migration occurs between populations.
4. No survival or reproductive advantage exists for any of the genotypes—in other words, no natural selection occurs.
5. No new mutations occur.

In reality, no population satisfies the Hardy-Weinberg equilibrium completely. For example, in many organisms mating is not random and certain males contribute a relatively large number of genes to the gene pool. Nevertheless, in large natural populations with little migration and negligible natural selection and mutation, the Hardy-Weinberg equilibrium may be nearly approximated for certain genes.

Researchers often discover instead that allele and genotype frequencies for one or more genes in a particular species are not in Hardy-Weinberg equilibrium. In such cases, we would say that the population is in disequilibrium—in other words, evolutionary mechanisms are affecting the population. When this occurs, researchers may wish to identify the reason(s) why disequilibrium has occurred, because this may provide insight into factors impacting the future survival of the species.

**Check Your Understanding**

**2.3** Imagine a population of plants with three different flower colors: red, *RR;* pink, *Rr;* and white, *rr.* What are the frequencies of the red, pink, and white flowers where $R = 0.4$. Assume *R* and *r* are the only alleles and the population is at Hardy-Weinberg equilibrium.

## 2.4 Small Populations Cause the Loss of Genetic Diversity

So far, we have considered how genetic variation arises and how it is integral to the operation of natural selection. Genetic variation is also vital to the maintenance of healthy, modern-day populations. When population sizes become too small, individuals accumulate deleterious (harmful) mutations and survival of offspring is threatened. In species that are rare, such threats endanger the survival of the species.

### 2.4.1 Inbreeding is mating between closely related individuals

Suppose an individual is heterozygous for a rare, deleterious recessive allele. In most populations, population size is so great that when heterozygotes carrying such a rare, recessive allele mate, their partner likely would not carry the same allele. Half the offspring would be heterozygous, and half would be homozygous for the common form of the allele. None of the offspring would be homozygous for the deleterious allele. Now consider the case where an individual who is heterozygous for a deleterious allele mates with an individual also carrying the deleterious allele. In this case, one-quarter of the offspring would be homozygous for the deleterious allele and would suffer a loss of fitness. **Inbreeding,** or mating between closely related relatives, increases the chances of both parents carrying the same harmful alleles and thus of the production of homozygous offspring that exhibit the effects.

Inbreeding is more likely to take place in nature when population size becomes very small and the number of mates is limited. In many species, survivorship of offspring declines as populations become more inbred. This phenomenon was shown as long ago as the 19th century. At that time it was known that litter size in inbred laboratory rats declined by over 50% compared to noninbred lines, and the level of nonproductive matings, those in which no offspring were born, rose from 2% to 50%.

Generally, the more inbred the population, the more severe these types of problems become. This was shown by James Crow and Motoo Kimura (1970), who mathematically examined the loss in genetic variation with time for various types of inbreeding: self-fertilization (in plants), sibling matings, and first-cousin matings (**Figure 2.11**). Loss in genetic variation was highest with self-fertilization.

Zoos face huge problems because of inbreeding. Katherine Ralls and Jonathan Ballou (1983) examined the effects of inbreeding on juvenile mortality in captive populations of mammals, including ungulates, primates, and small mammals (**Figure 2.12**). In nearly all cases, species exhibited a higher mortality from inbred matings than from noninbred matings. For this reason, many zoos move animals between institutions to minimize inbreeding.

Theoretical calculations by Monroe Strickberger (1986) showed that the smaller the population size, *N*, the faster the genetic variation declines (**Figure 2.13**). This result has important consequences in the real world, where plant and animal populations are declining because of shrinking habitats (see **Feature Investigation**). As a result, conservation biology has become particularly concerned with the genetics of small populations. A rule of thumb is that a population of at least 50 individuals is necessary to prevent the deleterious effects of inbreeding for the immediate future.

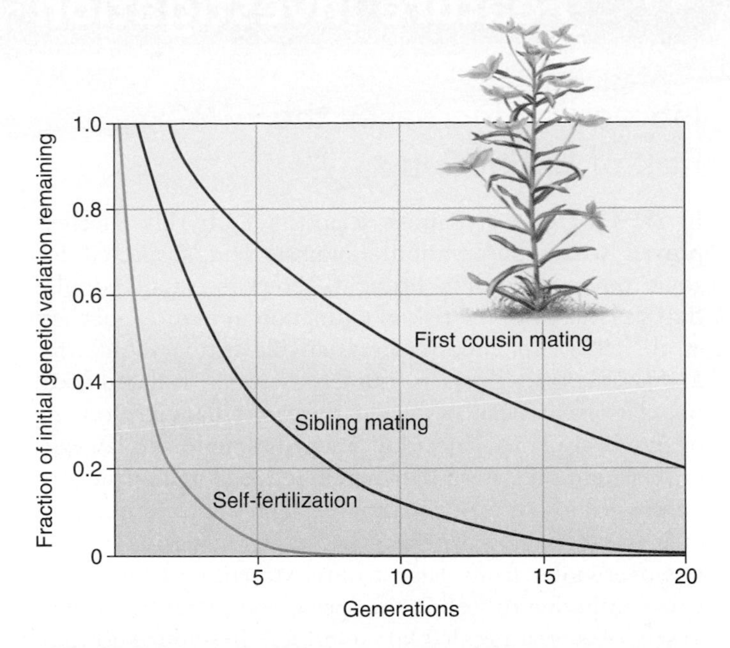


**Figure 2.11 Decrease in genetic variability is faster the greater the inbreeding.** Systems of mating are exclusive self-fertilization, sibling mating, and double first-cousin mating. (Redrawn from Crow and Kimura, 1970.)

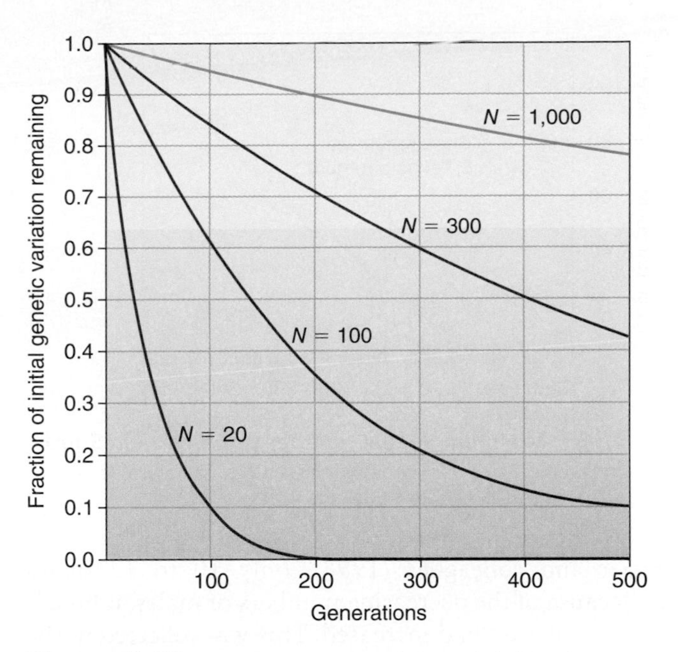


**Figure 2.13 Decrease in genetic variability due to finite population size.** The smaller the population size, $N$, the faster the decline in genetic variation in the population. (Modified from Strickberger, 1986.)

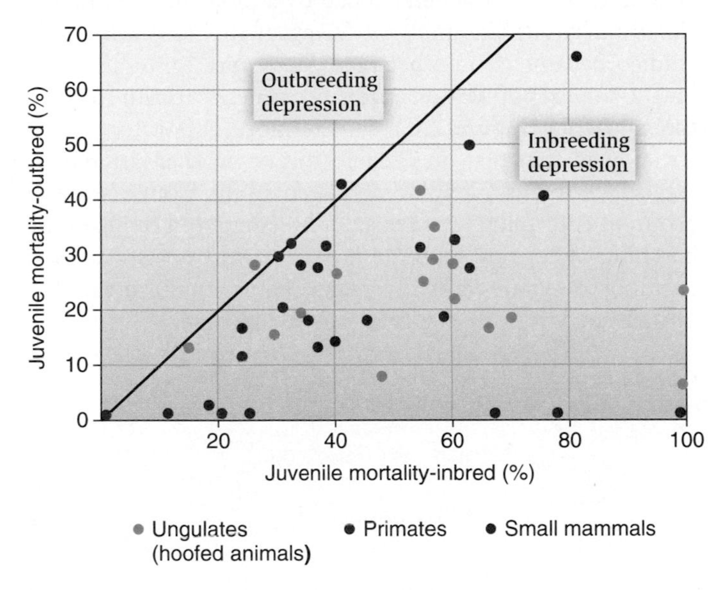


**Figure 2.12 The effects of inbreeding on juvenile mortality in captive populations of mammals.** Each point compares the percentage of juvenile mortality for offspring of inbred and noninbred matings. The line indicates equal levels of mortality under the two breeding schemes. Points above the line represent higher mortality from noninbred matings; nearly all points fall below the line, indicating higher mortality from inbred matings. The distance of a point below the line indicates the strength of the effect of level of inbreeding. (From data in Ralls and Ballou, 1983.)

Florida panthers, *Puma concolor*, have been isolated in South Florida since the early 1900s. A recent genetic analysis by Melanie Culver and colleagues (2008) compared DNA samples from museum specimens from the 1890s with samples from the 1980s and found that the genetic diversity of the 1980s cats was only a third that of the late 19th-century specimens. They suggested that at one point perhaps as few as six cats were alive. As a result, the Florida panther accumulated a series of unique traits, such as heart defects, deformed sperm, a kinked tail, and a cowlick of hair, that are rarely found in other panther populations. In addition, many adult males had one or no descended testes. Local officials recognized these indications of inbreeding, and in 1995 they released 8 females imported from the closest population, in Texas. The population increased from 25–30 individuals in 1995 to over 100 in 2007 (see **Figure 2.14**). Five of the 8 females produced litters. Five females died between 1995 and 2002, and the last 3 remaining Texas females in the wild were removed in 2002–2003 because they had produced about 20 kittens and were no longer breeding. A 2010 study by Warren Johnson and colleagues showed increased levels of genetic heterozygosity (having different versions of the same gene) in hybrids compared to the original Florida inbred stock. Hybrids also had fewer genetic abnormalities and were superior competitors, winning out in fights with inbred animals.

Another striking example of how the effects of inbreeding contributed to population decline involves the greater prairie chicken, *Tympanuchus cupido*. The male birds have a spectacular mating display that involves inflating the bright orange air sacs on their throat, stomping their feet, and spreading their tail feathers. The prairies of the Midwest were once home to millions of these birds, but as the prairies were converted to farmland, the range and population sizes of the bird shrank dramatically. The population of prairie chickens in Illinois decreased from 25,000 in 1933 to less than 50 in 1989. At that point, according to studies by Ronald

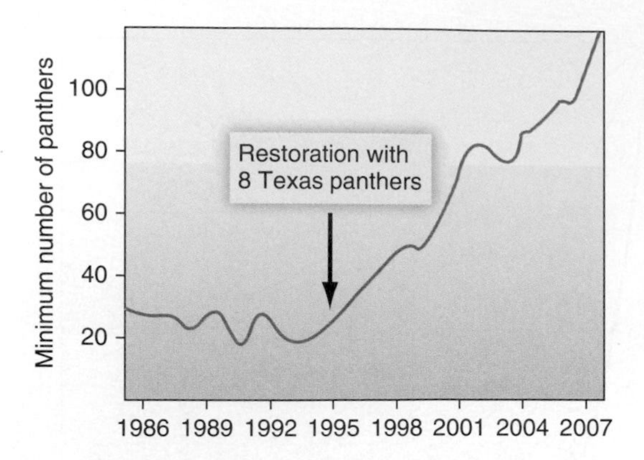


**Figure 2.14 Minimum annual panther population in South Florida.** Population size has steadily increased since eight Texas panthers were released in South Florida in 1995.

Westemeier and colleagues (1998), only 10 to 12 males existed. Because of the decreasing numbers of males, inbreeding in the population had increased. This was reflected in the steady reduction in the hatching success of eggs (**Figure 2.15**).

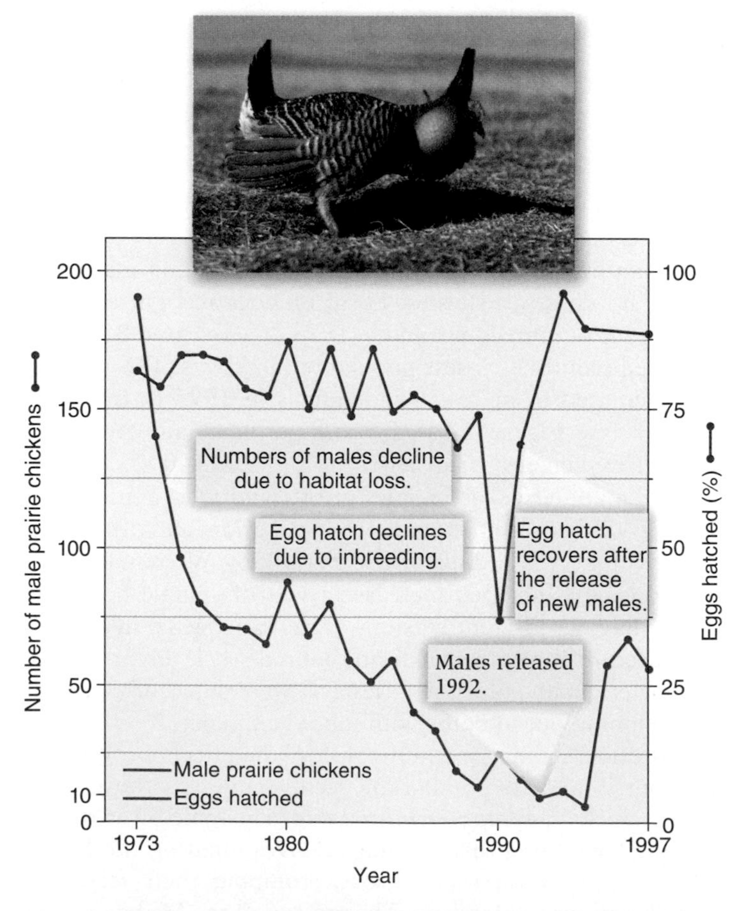


**Figure 2.15 Changes in the abundance and egg-hatching success rate of prairie chickens.** As the number of males decreased, inbreeding increased, resulting in a decrease in fertility, as indicated by a reduced egg-hatching rate. A translocation of males from Kansas and Nebraska in the early 1990s increased the egg-hatching success rate dramatically. (Modified from Westemeier et al., 1998.)

# Feature Investigation

## Inbreeding Increases the Risk of Extinction

In 1998 a group of Finnish scientists led by Ilik Saccheri proved what conservation biologists had suspected for some time: Inbreeding brought about by small population size increases the risk of extinction in nature (Saccheri et al., 1998). In Finland the Glanville fritillary butterfly, *Melitaea cinxia,* exists in numerous small, isolated local populations in meadows where the caterpillars feed on one or two host plants. The adult butterflies mate and lay eggs in June, and the caterpillars hatch and feed in conspicuous family groups of 50–250, making large-scale counting of their numbers in many meadows relatively easy. Caterpillars overwinter from August until March, with the survivors continuing to feed in the spring and pupating in May. Yearly censuses revealed larvae present in about 400 total meadows in an area of 3,500 $km^2$. Many populations were small, consisting of one group of caterpillars, the offspring of just one pair of butterflies. In 42 of the populations, the genetic variation was determined by a molecular technique called microsatellite analysis. Seven of the 42 populations studied became extinct between 1995 and 1996; all seven had a lower population size and genetic variation than the survivors (**Figure 2.C**). Furthermore, laboratory studies showed that just one generation of brother-sister mating, which might take place between adults from one small group of caterpillars, increased inbreeding and reduced the hatching of eggs by up to 46%. These studies also showed that inbred females laid fewer eggs than noninbred females, and the survival of the larvae was reduced.

**HYPOTHESIS** Inbreeding increases the likelihood of extinction.

**STARTING LOCATION** Åland, in southwestern Finland.

**Experimental level** | **Conceptual level**

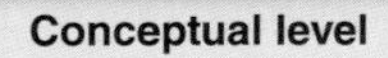

**1** Census an area (3,500 km$^2$) of suitable meadows in southwestern Finland. Meadows must contain one or both of the host plants *Plantago lanceolata* or *Veronica spicata*.

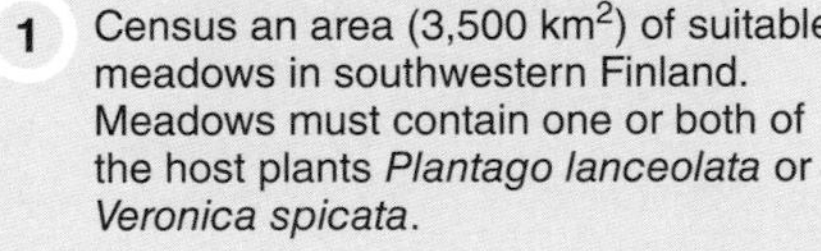

Locate a number (42) of local populations of Glanville fritillary butterfly in southwestern Finland. Populations should range from small (<5 larval groups) to large (>5 larval groups).

**2** Assess genetic variation at 8 loci by microsatellite DNA analysis.

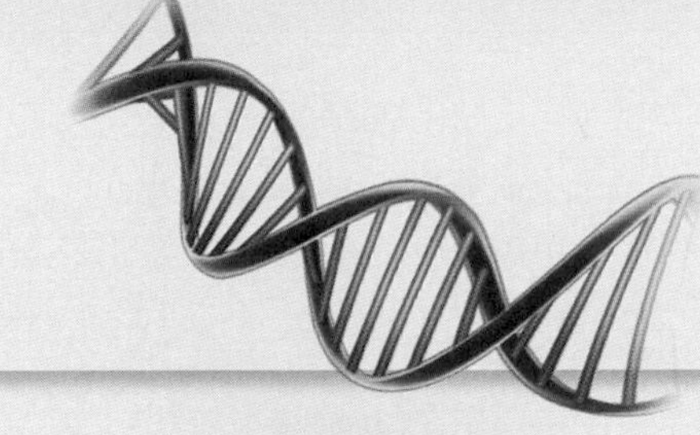

Identify genetic variability.

**3** Examine survival of all 42 populations between late summer 1995 and late summer 1996 by examining meadows for larval populations.

Monitor survival and extinction of all 42 populations.

**4** Model also includes other ecological variables such as patch size and butterfly density.

Construct a model linking likelihood of extinction with genetic variability.

**5 THE DATA** Probability of extinction is proportional to circle sizes. Extinct (purple) and surviving (open) populations are shown. The presence of more heterozygous loci increases the likelihood of survival.

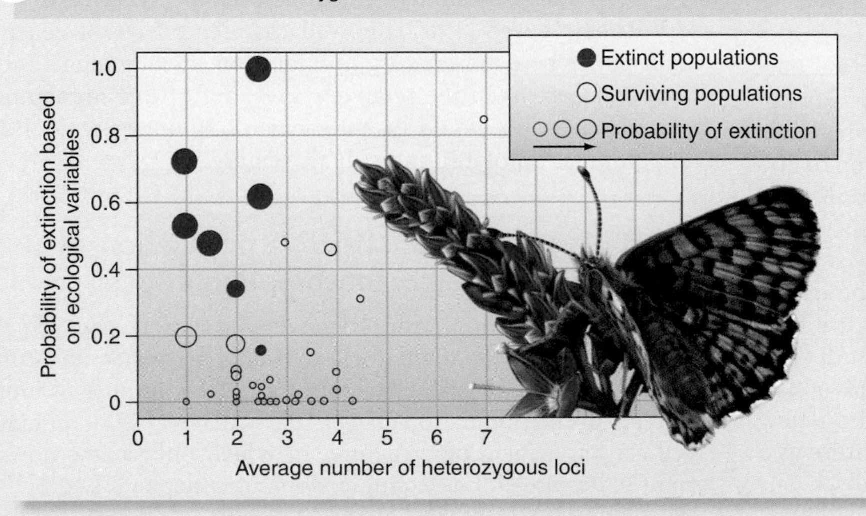

**Figure 2.C The effects of inbreeding on the Glanville fritillary butterfly.**

The prairie chicken population had entered a downward spiral toward extinction from which it could not naturally recover, a phenomenon called an extinction vortex. In 1992, conservation biologists began trapping prairie chickens in Kansas and Nebraska, where populations remained larger and more genetically diverse, and moved them to Illinois, bringing an infusion of new genetic material into the population. This translocation resulted in a rebounding of the egg-hatching success rate to over 90% by 1993. This increase in egg hatching and further release of males led to the appearance of 70 males of mixed origin by 1996.

## 2.4.2 Genetic drift refers to random changes in allele frequencies over longer periods of time

In small populations, there is a good chance that some individuals will fail to mate successfully purely by chance—for example, because of failure to find a mate—and this results in lowered birth rates. This is known as the Allee effect, after ecologist W. C. Allee, who first described it. If an individual that fails to mate possesses a rare gene, that genetic information will not be passed on to the next generation, resulting in a loss of genetic diversity from the population. **Genetic drift** refers to the random change in allele frequencies attributable to this type of chance.

Because the likelihood of an allele being represented in just one or a few individuals is higher in small populations as compared to large populations, small, isolated populations are particularly vulnerable to this type of reduction in genetic diversity. Such isolated populations will lose a percentage of their original diversity over time, approximately at the rate of $1/(2N)$ per generation, where $N$ = population size. This has a greater effect in smaller versus larger populations.

If $N = 500$, then $1/(2N) = 1/1{,}000 = 0.001$, or 0.1% genetic diversity lost per generation

If $N = 50$, then $1/(2N) = 1/100 = 0.01$, or 1% genetic diversity lost per generation

Due to genetic drift, a population of 500 will lose only 0.1% of its genetic diversity in a generation, whereas a population of 50 will lose 1%. Such losses become magnified over many generations. After 20 generations, the population of 500 will lose 2% of its original variation, but the population of 50 will lose 18%. For organisms that breed annually, this would mean a substantial loss in genetic variation over 20 years.

A rule of thumb for genetic drift is that a population size of at least 500 is necessary to decrease the drift effects over long time periods. This has been combined with the rule of thumb for small populations, that at least 50 individuals are needed to prevent inbreeding over relatively short time spans. Thus, the "50/500" rule has entered the literature as a "magic" number in conservation theory. Theoretically, if one guards against genetic drift by maintaining a population of 500 individuals, then inbreeding will not be a problem.

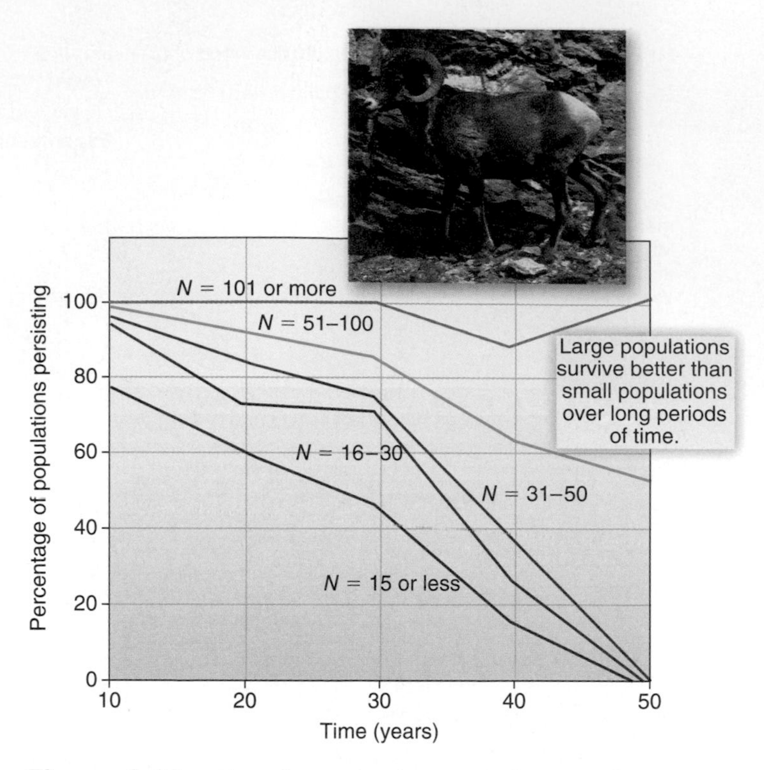


**Figure 2.16 The relationship between the size of a population of bighorn sheep and the percentage of populations that persist over time.** The numbers on the graph indicate population size (*N*); populations with more than 50 sheep almost all persisted beyond 50 years, while populations with fewer than 50 individuals died out within 50 years. (After Berger, 1990.)

Joel Berger's (1990) study of 120 bighorn sheep, *Ovis canadensis*, populations in the U.S. Southwest supported the idea that populations of at least 50 individuals survive better than smaller populations. He observed that 100% of the populations with fewer than 50 individuals became extinct within 50 years, while most of the populations with more than 50 individuals persisted for this time period (**Figure 2.16**).

Robert Lacey (1987) showed that the effects of genetic drift could be countered by immigration of individuals into a population. Even the relatively low rate of one immigrant every generation would be sufficient to counter genetic drift in a population of 120 individuals (**Figure 2.17**).

## 2.4.3 Knowledge of effective population sizes is vital to conservation efforts

In many populations, the effective population size, which is the number of individuals that contribute genes to future populations, may be smaller than the actual number of individuals in the population. This is particularly true in animals with a harem mating structure, in which only a few dominant males breed. For example, dominant elephant seal bulls, *Mirounga* spp., control harems of females, and a few males command all the matings. If a population consists of breeding

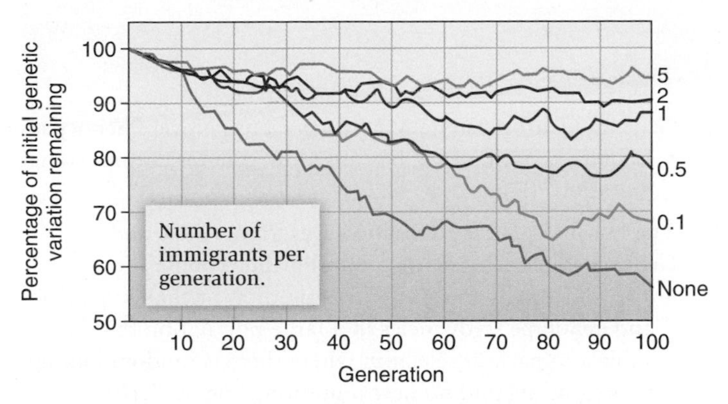


**Figure 2.17 The effect of immigration on genetic variability.** For a population of 120 individuals, even low rates of immigration (one immigrant per generation in a population of 120 individuals) can prevent the loss of heterozygosity from genetic drift. (After Lacey, 1987.)

males and breeding females, the effective population size is given by:

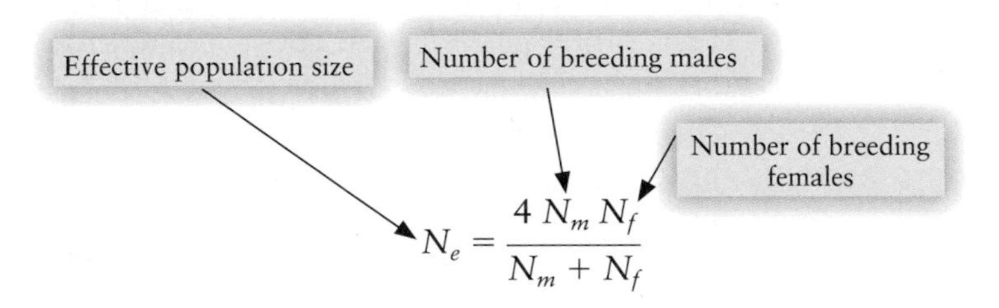


$$N_e = \frac{4\, N_m\, N_f}{N_m + N_f}$$

In a population of 500, with a 50:50 sex ratio and all individuals breeding, $N_e = (4 \times 250 \times 250)/(250 + 250) = 500$, or 100% of the actual population size. However, if 250 females breed with only 10 of 250 males, $N_e = (4 \times 10 \times 250) / (10 + 250) = 38.5$, or 8% of the actual population size.

Knowledge of effective population size is vital to ensuring the success of conservation projects. One notable project in the U.S. has involved planning the sizes of reserves designed to protect grizzly bear populations in the contiguous 48 states. The grizzly bear, *Ursus arctos,* has declined in numbers from an estimated 100,000 in 1800 to less than 1,000 at present. The range of the species is now less than 1% of its historical range and is restricted to six separate populations in four states (**Figure 2.18**). Research by biologist Fred Allendorf (1994) has indicated that the effective population size of grizzly populations is generally only about 25% of the actual population size because not all bears breed. Thus, even fairly large, isolated populations, such as the 200 bears in Yellowstone National Park, are vulnerable to the harmful effects of loss of genetic variation, because the effective population size may be as small as 50 individuals. Allendorf and his colleagues proposed that an exchange of grizzly bears between populations or zoo collections would help tremendously in promoting genetic variation. Even an exchange of two bears per generation between populations would greatly reduce the loss of genetic variation. L. Scott Mills and Allendorf (1996) concluded that, in the face of continuing habitat fragmentation and isolation, a minimum of 1 migrant and a maximum of 10 migrants per generation would be needed to minimize loss of genetic variation for populations of most endangered species, not only grizzly bears.

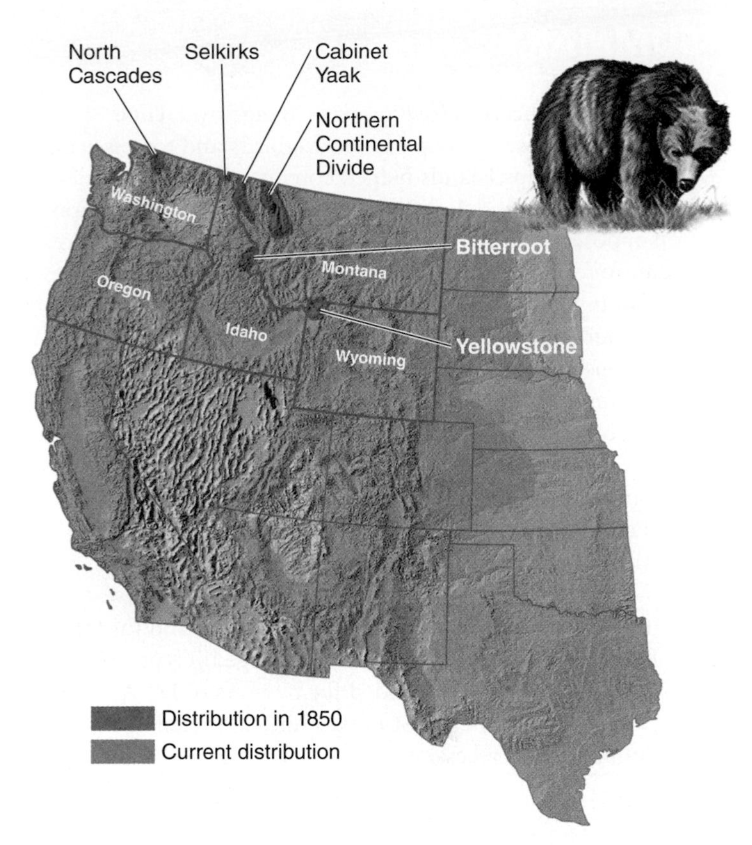


**Figure 2.18 The range of the grizzly bear is currently less than 1% of its historical range.** The range of the grizzly bear in the continental U.S. has contracted to just six populations in four states, as the population size has shrunk from 100,000 before the West was settled to about 1,000 today.

**ECOLOGICAL INQUIRY**

If only 500 males and 500 female grizzlies exist today, and only 25% of the males breed, what is the effective population size?

**Check Your Understanding**

**2.4** Why do managers go to the expense of moving males or females of large mammals between zoos to produce offspring?

## SUMMARY

**2.1 Evolution Concerns How Species Change over Time**

- Darwin's observations of mockingbirds and tortoises on the Galápagos Islands helped convince him that species evolve (Figures 2.1, 2.2). His theory of natural selection supposed that better-adapted individuals would acquire more resources and leave more offspring. Darwin was also involved in a large-scale ecological experiment, encouraging authorities to plant exotic plant species on Ascension Island, which at that time was relatively barren (Figure 2.3).
- There are many examples of natural selection in response to environmental changes. For example, peppered moths have changed color over time in response to pollution (Figures 2.A, 2.B).
- Mendel's crosses of pea plants provided a genetic mechanism for evolution. The crosses showed how individuals inherited factors from relatives and that these factors could be passed down unchanged (Figures 2.3–2.6). The factors were coded for by units of DNA called genes. Genes exist in two forms, or alleles, called dominant and recessive.

**2.2 Gene and Chromosome Mutations Cause Novel Phenotypes**

- Genetic variation may be caused by gene or chromosomal mutations (Figures 2.7–2.9).

**2.3 The Hardy-Weinberg Equation Describes Allele and Genotype Frequencies in an Equilibrium Population**

- The Hardy-Weinberg equation predicts unchanging allele and genotype frequencies in a large population that is not subject to natural selection where there is random mating, no migration, and no new mutations (Figure 2.10).

**2.4 Small Populations Cause the Loss of Genetic Diversity**

- Small populations are threatened by a loss of genetic variability. This loss may be caused by inbreeding, genetic drift, and limited mating.
- Small population size accelerates loss of genetic variation due to inbreeding and can increase rates of population extinction (Figures 2.11–2.15, Figure 2.C).
- Genetic drift can result in a loss of genetic diversity from a population by chance alone (Figure 2.16). The movement of a few individuals between populations can counteract the effects of genetic drift (Figure 2.17). Effective population size can be reduced by harem mating structures (Figure 2.18).

## TEST YOURSELF

1. A better-adapted organism acquires more resources and leaves more offspring than a less-well-adapted organism. This idea best conveys the theme of:
   a. Evolution
   b. Natural selection
   c. The Malthusian theory of population
   d. Phenotypic variation
   e. Genotypic variation
2. The theory of evolution by natural selection was proposed by:
   a. Jean-Baptiste Lamarck
   b. Charles Darwin and Alfred Wallace
   c. Charles Lyell and Thomas Malthus
   d. Gregor Mendel
   e. T. A. Knight and H. B. Kettlewell
3. The color change that has occurred in certain populations of the peppered moth, *Biston betularia,* in industrial areas of Europe is known as:
   a. Heterozygous phenotypes
   b. Frameshift mutation
   c. Industrial melanism
   d. Hardy-Weinberg equilibrium
   e. Inbreeding
4. The first generation offspring of true-breeding parents are termed the:
   a. P generation
   b. $F_1$ generation
   c. $F_2$ generation
   d. $F_3$ generation
   e. Monohybrids
5. The two variant forms of a gene are called:
   a. Alleles
   b. Loci
   c. Dominant
   d. Recessive
   e. Heterozygous
6. Which term refers to the genetic composition of an individual?
   a. Homozygote
   b. Heterozygote
   c. Genotype
   d. Phenotype
   e. Hybrid
7. When a chromosome breaks in two places and the middle segment turns around and re-fuses with the other two pieces, this is termed:
   a. Point mutation
   b. Duplication
   c. Deletion
   d. Inversion
   e. Translocation

8. What is the expected $F_2$ genotypic ratio of a monohybrid cross?
   a. 1:2:1
   b. 2:1
   c. 3:1
   d. 4:1
   e. 9:3:3:1
9. Approximately what percentage of genetic variation remains in a population of 25 individuals after three generations?
   a. 98
   b. 96
   c. 94
   d. 92
   e. 84
10. Small populations are threatened by the loss of genetic diversity from:
    a. Mutations
    b. Genetic drift
    c. Limited mating
    d. Outbreeding
    e. Cross fertilization

## CONCEPTUAL QUESTIONS

1. What evidence supports the theory of evolution?
2. What is the difference between genotype and phenotype?
3. How do novel genotypes arise in populations?
4. Define inbreeding, genetic drift, and limited mating, and discuss how they reduce the genetic variability of small populations.

*Neospiza wilkinsii* bunting, with a large bill, on the Tristan da Cunha archipelago. Tristan da Cunha is an isolated series of islands in the South Atlantic ocean.

CHAPTER 3

# Natural Selection, Speciation, and Extinction

## Outline and Concepts

In 2007 Peter Ryan and his South African colleagues described the parallel colonization of finches on two small islands in the Tristan da Cunha archipelago in the South Atlantic Ocean. Tristan is one of the world's most isolated island systems, lying midway between South America and the tip of South Africa. Of the three islands, two of them, Inaccessible and Nightingale, had been relatively untouched by humans and their associated pests, mice and rats. Both these islands had two species of *Neospiza* buntings that had evolved from finch tanagers blown there from South America across 3,000 km of ocean. Such long-distance dispersal is likely very rare. Ryan and colleagues discovered that each island had a small-billed seed generalist and a large-billed seed specialist, matching the availability of seeds on each island. Generalists eat a variety of seeds from different plant species. Specialists feed on seeds of just one plant species. Additional genetic evidence suggested that one small-billed and one large-billed species had evolved independently on each island. This mechanism was supported rather than the prevailing and simpler hypothesis that a large-billed species evolved on one island and a small-billed species evolved on another island, and subsequent dispersal of both species between islands formed different species of small-billed and large-billed *Neospiza* on each island. This work showed how the formation of species does not always occur via the simplest route but may involve more circuitous pathways.

As we discussed in Chapter 2, Charles Darwin and Alfred Wallace independently proposed the theory of evolution by natural selection. According to this theory, a struggle for existence results in the selective survival of individuals that have inherited genotypes that confer greater reproductive success. In this chapter we see how natural selection can follow four different patterns: directional, stabilizing, balancing, and disruptive. Given enough time, disruptive selection can lead to **speciation**, the formation of new species. There are many definitions of what constitutes a species. We will examine the species concept and the two main mechanisms of speciation, allopatric and sympatric speciation. New species arise from older species, but they may also go extinct after various lengths of time. In order to understand patterns of species origination and extinction, we will briefly examine the history of life on Earth and the pattern of species origination and extinction. Conservation biologists have a strong interest in determining where and how species are going extinct. In the last part of the chapter we address the current extinction crisis and the current factors endangering life on Earth.

## 3.1 Natural Selection Can Follow One of Four Different Pathways

At its simplest, natural selection will tend to lead to an increase in the frequency of the allele that confers the highest fitness in a given environment. In some cases, however, selection will act to maintain a number of alleles in a population, especially if the relative fitness of different alleles changes on a spatial or temporal scale. Here we describe four different patterns of natural selection: directional, stabilizing, balancing, and disruptive.

### 3.1.1 Directional selection favors phenotypes at one extreme

**Directional selection** favors individuals at one extreme of a phenotypic distribution that have greater reproductive success in a particular environment. One way in which directional selection may arise is that a new allele may be introduced into a population by mutation, and the new allele may confer a higher fitness in individuals that carry it (**Figure 3.1**).

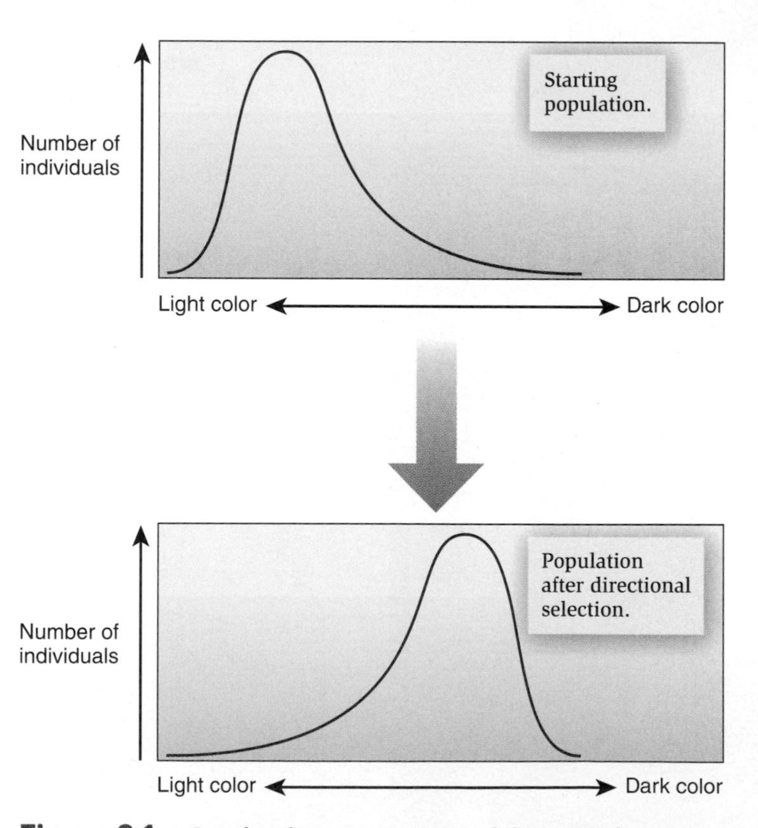


**Figure 3.1 Graphical representation of directional selection.** This pattern of selection selects for a darker phenotype that confers higher fitness in, for example, a polluted environment as with the peppered moth, *Biston betularia*.

In Chapter 2 we saw how darker color conferred greater fitness to peppered moths in polluted environments. If the homozygote carrying the favored allele has the highest fitness value, directional selection may cause this favored allele to eventually become predominant in the population.

(a) Daphne Major

(b) Medium ground finch

**Figure 3.2 The study site and study organism of the Grants' work on natural selection.** (a) Daphne Major, a small island in the Galápagos. (b) The medium ground finch, *Geospiza fortis.*

Another example of directional selection was provided by Peter and Rosemary Grant's study of natural selection in Galápagos finches. The Grants focused much of their work on one of the Galápagos Islands known as Daphne Major (**Figure 3.2a**). This small island (0.3 $km^2$) has a resident population of the medium ground finch, *Geospiza fortis* (**Figure 3.2b**). The medium ground finch has a relatively small crushing beak, allowing it to feed on small, tender seeds. The Grants quantified beak size among the medium ground finches of Daphne Major by carefully measuring beak depth, a measure of the beak from top to bottom. They compared the beak sizes of parents and offspring by examining broods over many years. The depth of the beak was inherited by offspring from parents, regardless of environmental conditions, indicating that differences in beak sizes are due to genetic differences in the population. This means that beak depth is a heritable trait. In the wet year of 1976, the plants of Daphne Major produced an abundance of small seeds that finches could easily eat. In 1977 a drought occurred and plants produced few of the smaller seeds and only larger drier seeds, which are harder to crush, were readily available. As a result, birds with larger beaks were more likely to survive, because they were better at breaking open the large seeds. In the year after the drought, the average beak depth of birds in the population increased by almost 10% (**Figure 3.3**).

### 3.1.2 Stabilizing selection favors intermediate phenotypes

**Stabilizing selection** favors the survival of individuals with intermediate phenotypes. The extreme values of a trait are selected against. An example of stabilizing selection involves

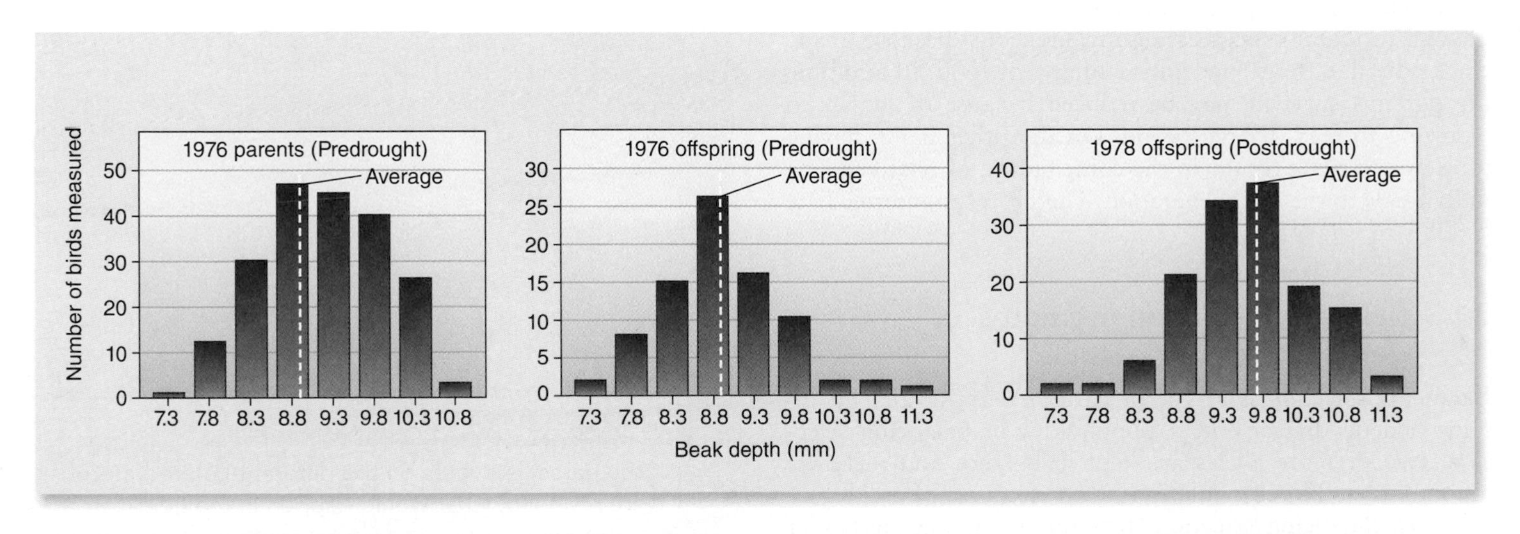


**Figure 3.3 Variation in the beak size of the medium ground finch, *G. fortis,* on Daphne Major in 1976 and 1978.** Beak size increased almost 10% the year following the drought.

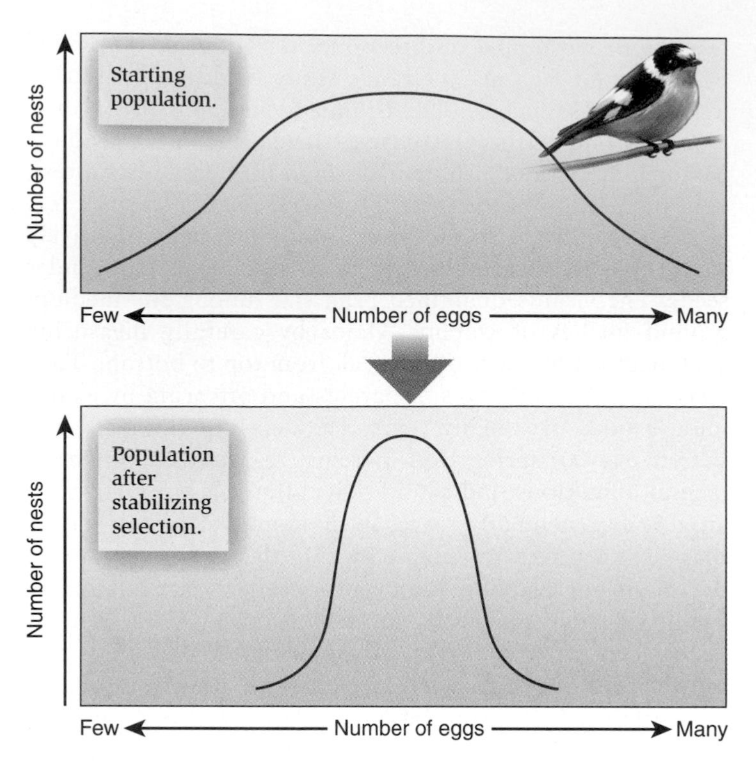


**Figure 3.4 Graphical representation of stabilizing selection.** Here the extremes of a phenotypic distribution are selected against while individuals with intermediate traits are favored. These graphs illustrate stabilizing selection in clutch size of birds.

clutch size (the number of eggs laid) in animals. British ornithologist David Lack suggested that birds that lay too many or too few eggs per nest have lower fitness values than do those that lay an intermediate number of eggs (**Figure 3.4**). Laying too many eggs is disadvantageous because many chicks die due to an inadequate supply of food. In addition, the parent's survival may be reduced because of the strain of trying to feed a large brood. On the other hand, having too few offspring results in the contribution of relatively few individuals to the next generation. Therefore, an intermediate clutch size is favored.

### 3.1.3 Balancing selection promotes genetic diversity

**Balancing selection** is a type of natural selection that maintains genetic diversity in a population. In balancing selection, two or more alleles are kept in balance and therefore are maintained in a population over the course of many generations. Balancing selection does not favor one particular

## Feature Investigation

### John Losey and Colleagues Demonstrated That Balancing Selection by Opposite Patterns of Parasitism and Predation Can Maintain Different-Colored Forms of Aphids

In frequency-dependent selection, the fitness of a genotype changes when its frequency changes. In other words, rare individuals have a different fitness from common individuals. John Losey and colleagues (1997) showed how the existence of both green and red color forms or morphs of the pea aphid, *Acyrthosiphon pisum,* were maintained by the action of natural enemies. This situation is known as a **balanced polymorphism**, where two or more alleles or morphs are maintained in a population. Aphids parasitized by the wasp *Aphidius ervi* become mummified; that is, they turn a golden brown and become immobile, stuck to their host plant. Aphids may also be eaten by ladybird beetles, *Coccinella septempunctata.* Green morphs suffered higher rates of parasitism than red morphs, whereas red morphs were more likely to be attacked by ladybird predators than were green morphs. Therefore, when parasitism rates were high relative to predation rates, the population of red morphs increased relative to green morphs, whereas the converse was true when predation rates were relatively high (**Figure 3.A**).

**ECOLOGICAL INQUIRY**

If the parasitism rates on the different-colored morphs were reversed, what would happen in the field?

**HYPOTHESIS** The action of parasitoids and predators maintains a color polymorphism in the pea aphid, *Acyrthosiphon pisum*.

**STARTING LOCATION** Alfalfa fields in south central Wisconsin, USA.

| | Conceptual level | Experimental level |
|---|---|---|
| 1 | Assess rate of predation of different aphid color morphs by ladybird beetles. | A single adult *C. septempunctata* beetle was released on caged alfalfa plants with 15 adults of each color morph for 4 hours. This experiment was replicated 27 times. The number of survivors was recorded. The predation rate was higher on red morphs, 0.91 aphids eaten per hour, than on green morphs, 0.73 aphids eaten per hour. |
| 2 | Determine whether parasitism is different on different color morphs | Sample 5 alfalfa fields on 5 days in the summer and dissect a total of 643 aphids for *Aphidius ervi* larvae. Parasitism rates were 53% on green morphs and 42% on red morphs. |
| 3 | Assess densities of aphids, parasitoids, and predators in the field. | 12 alfalfa fields were sampled roughly every 6 days throughout the summer. In each field, aphid density and color were recorded on 100 stems in 8 locations within the field. Twelve 3-minute walking scans were also made throughout each field to count parasitoids (*A. ervi*) and predators (*C. septempunctata*). |

4 **THE DATA** The data show that in the field, red aphid morphs are more common where parasitoids are abundant and green morphs predominate where predators are more common.

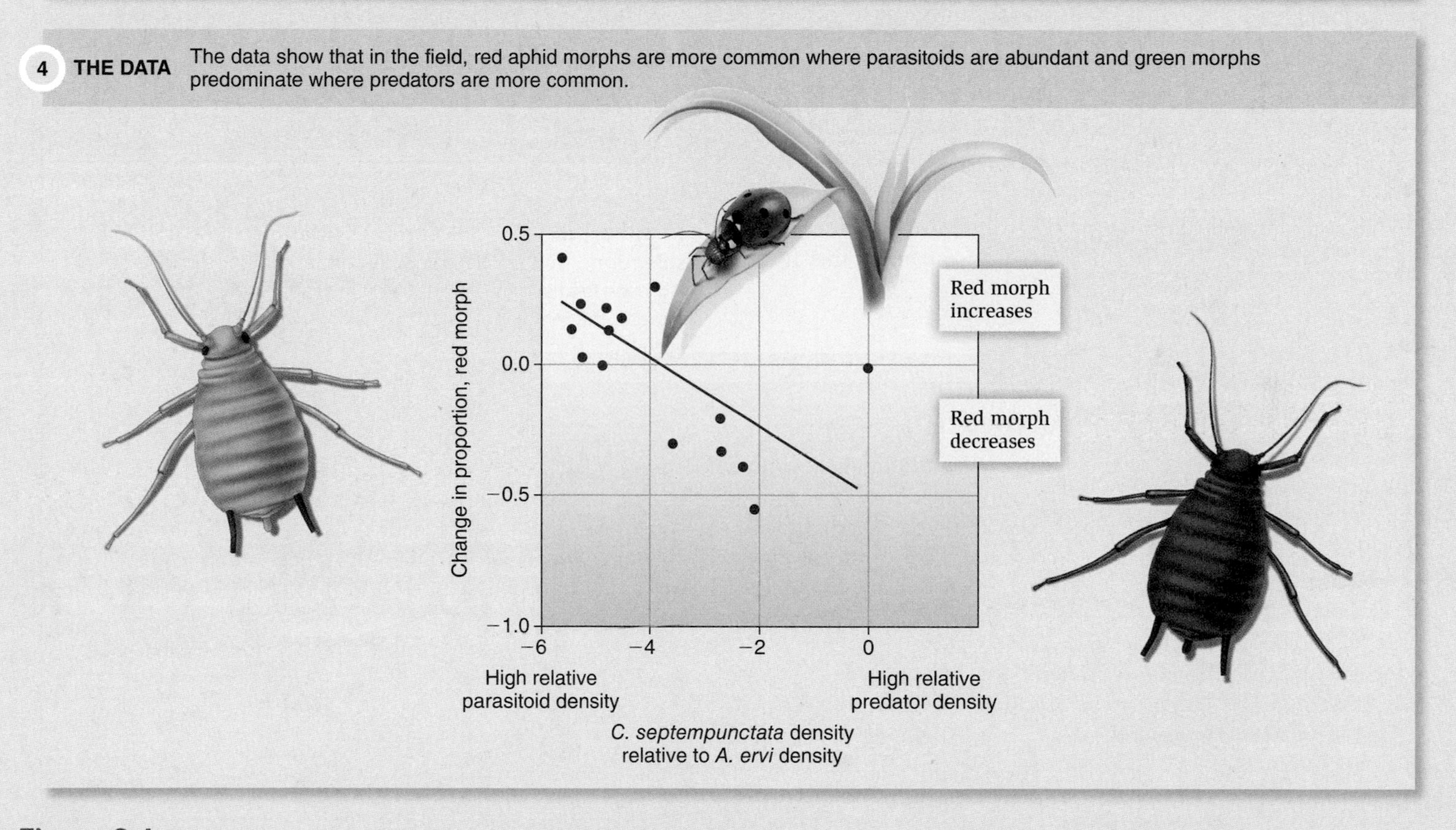


**Figure 3.A** The effects of natural enemies on aphid color morphs.

allele in the population. Population geneticists have identified two common pathways along which balancing selection can occur. The first is heterozygote advantage and the second is frequency-dependent selection.

A classic example of heterozygote advantage involves the sickle-cell allele of the human β-globin gene. A homozygous individual with two copies of this allele has sickle-cell disease, a hereditary disease that damages blood cells (refer back to Figure 2.8). The sickle-cell homozygote has a lower fitness than a homozygote with two copies of the normal and more common β-globin allele. However, in areas where malaria is endemic, the heterozygote has the highest level of fitness. Compared with normal homozygotes, heterozygotes have a 10–15% better chance of survival if infected by the malarial parasite, *Plasmodium falciparum*. Therefore, the sickle-cell allele is maintained in populations where malaria is prevalent, even though the allele is detrimental in the homozygous state.

Frequency-dependent selection is another mechanism that causes balancing selection (see **Feature Investigation**). In frequency-dependent selection, the fitness of one phenotype is dependent on its frequency relative to other phenotypes in the population. For example, many species of invertebrates exist as different-colored forms, identical in all respects except color. Visually searching predators often develop a search image for one color form, usually the more common form. The prey then proliferates in the rarer form until this form itself becomes more common.

### 3.1.4 Disruptive selection favors the survival of two phenotypes

**Disruptive selection** favors the survival of individuals at both extremes of a range, rather than the intermediate. It is similar but not identical to balancing selection, where individuals of average trait values are favored against those of extreme trait values. The fitness values of one genotype are higher in one environment, while the fitness values of the other genotype are higher in another environment. Janis Antonovics and Anthony Bradshaw (1970) provided an example of disruptive selection in colonial bentgrass, *Agrostis tenuis*. In certain locations where this grass is found, such as South Wales, isolated places contain high levels of heavy metals such as copper from mining. Such pollution has selected for mutant strains that show tolerance to copper. This genetic change enables the plants to grow on copper-contaminated soil but tends to inhibit growth on a normal, uncontaminated soil. This results in metal-resistant plants growing on contaminated sites that are close to normal plants growing on uncontaminated land (**Figure 3.5**).

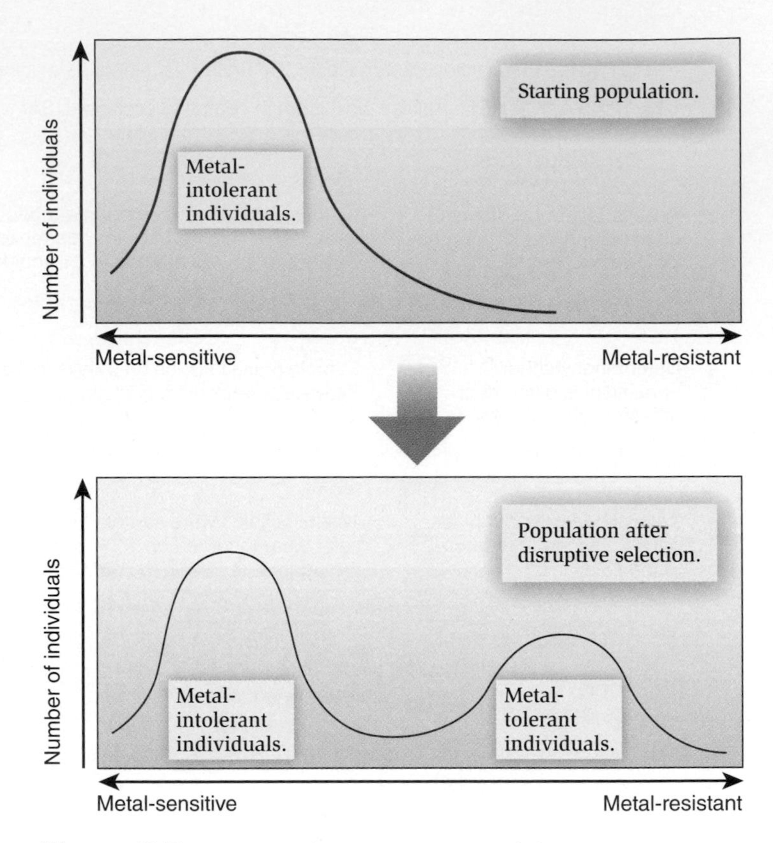


**Figure 3.5 Graphical representation of disruptive selection.** In this example, the normal wild-type colonial bentgrass, *Agrostis tenuis*, is intolerant to metals in the soil. A mutation creates a metal-tolerant variety, which can grow in soils contaminated with metals from mining operations.

**Check Your Understanding**

**3.1** What were the results of the Grants' study of *Geospiza fortis* finches in the Galápagos following the drought of 1977 and how did this work impact the theory of natural selection?

## 3.2 Speciation Occurs Where Genetically Distinct Groups Separate into Species

Over a long enough time span, disruptive selection can result in speciation. Here we describe two alternative mechanisms by which speciation occurs: allopatric speciation and sympatric speciation. But first we discuss the species concept. In any

**Table 3.1 Four Different Species Concepts.**

| Species Concept | Description |
|---|---|
| Biological | Species are separate if they are unable to interbreed and produce fertile offspring. |
| Phylogenetic | Differences in physical characteristics (morphology) or molecular characteristics are used to distinguish species. |
| Evolutionary | Phylogenetic trees and analyses of ancestry serve to differentiate species. |
| Ecological | Species separate based on their use of different ecological niches and their presence in different habitats and environments. |

discussion of speciation, it is valuable to have a good working concept of species. While one might think this is a simple matter, there are over 20 species concepts, each with its own advantages and disadvantages.

## 3.2.1 There are many definitions of what constitutes a species

There is considerable debate about what constitutes a species. We will consider four of the more widely accepted species concepts, the biological, phylogenetic, evolutionary, and ecological species concepts (**Table 3.1**).

### *Biological species concept*

Perhaps the best-known species concept is the **biological species concept** of Ernst Mayr (1942), who defined species as "groups of populations that can actually or potentially exchange genes with one another and that are reproductively isolated from other such groups." The biological species concept defines species in terms of interbreeding. It has been used to distinguish morphologically similar yet reproductively isolated species, such as the northern leopard frog, *Rana pipiens,* and the southern leopard frog, *R. utricularia* (**Figure 3.6**).

**(a)**

**(b)**

**Figure 3.6 The biological species concept.** The northern leopard frog, *Rana pipiens* **(a)**, and the southern leopard frog, *R. utricularia* **(b)**, appear very similar but are reproductively isolated from each other.

Despite its advantages, the biological species concept suffers from at least three disadvantages. First, for many species with widely separate ranges, we have no idea if the reproductive isolation is by distance only or whether there is some species-isolating mechanism. Second, especially in plants, individuals called hybrids often form when parents from two different species are crossed with each other and the resultant progeny develop. This greatly blurs species distinctions. Oak trees provide a particularly good example of confusion in species definitions. Oaks often form reproductively viable hybrid populations. That is, oaks from different species interbreed and their offspring are themselves viable, capable of reproducing with other oaks. For this reason, one might question whether the parental species should be called species at all. For example, *Quercus alba* and *Q. stellata* form natural hybrids with 11 other oak species in the eastern United States. It could be argued, humorously, that if these oaks cannot tell each other apart, why should biologists impose different names on them? Ecologists have noted many examples of viable hybrid formation when historically isolated species

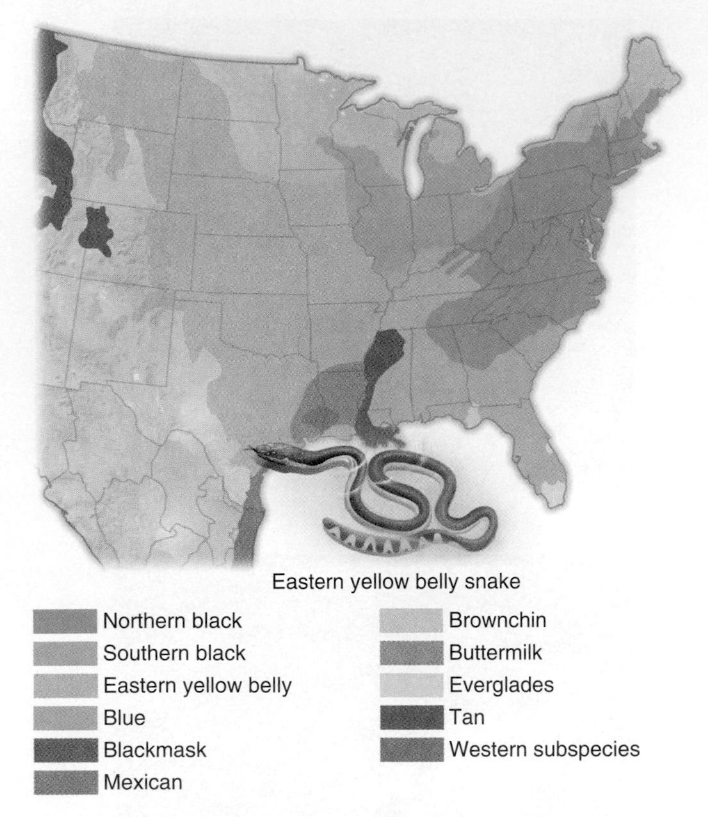


**Figure 3.7 Difficulties with the phylogenetic species concept.** The subspecies of the black racer, *Coluber constrictor,* appear different yet are members of the same species. (Modified from Conant, 1975.)

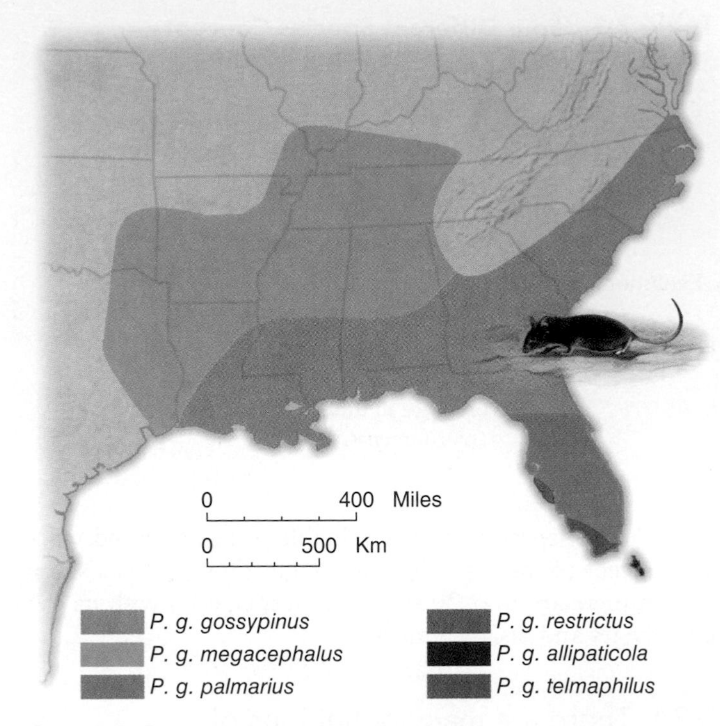


**Figure 3.8 Difficulties with the phylogenetic species concept.** The cotton mouse, *Peromyscus gossypinus,* exists as six subspecies. The subspecies possess slightly different coat colors and different lengths of tail and hind feet.

**ECOLOGICAL INQUIRY**

What would happen if you attempted to breed the blue and western yellow-bellied races of the black racer?

are brought into contact through climate change, landscape transformation, or transport beyond their historic boundaries (see **Global Insight**). Third, the biological species concept cannot be applied to asexually reproducing species, such as bacteria and some plants and fungi, or to extinct species.

### *Phylogenetic species concept*

Another popular definition of species is the **phylogenetic species concept**, which advocates that members of a single species are identified by a unique combination of characters. This definition incorporates the classic taxonomic view of species based on their morphological characters and, more recently, molecular features such as DNA sequences. A disadvantage of this concept is in determining how much difference between populations is enough to call them separate species. Many color differences in hair or feathers are controlled by a single gene. Conversely, many genetic changes have no discernible effect on the phenotype. Using this definition, many currently recognized subspecies or distinct populations would be elevated to species status. The black racer, *Coluber constrictor,* is a snake that exists as many different color forms or races throughout the United States (**Figure 3.7**). Some authorities argue that each race should be elevated to the status of a species. Similarly, the cotton mouse, *Peromyscus gossypinus,* exists as six formal subspecies in the southeastern U.S. (**Figure 3.8**).

### *Evolutionary species concept*

George Gaylord Simpson (1961) proposed the **evolutionary species concept**, whereby a species is distinct from other lineages if it has its own evolutionary tendencies and historical fate. For example, paleontologists have charted the course of species formation in the fossil record. One of the best examples documents the evolutionary changes that led to the development of many horse species, including modern horses. Some scientists believe this is both the best definition and, at the same time, the least operational, because lineages are difficult to examine and evaluate quantitatively. Incomplete fossil records and lack of transitional forms make lineages difficult to trace.

## Hybridization and Extinction

Introduced species can bring about a form of extinction of native flora and fauna by **hybridization**, breeding between individuals from different species. Purposeful or accidental introductions by humans or by habitat modification may bring previously isolated species together. For example, mallard ducks, *Anas platyrhynchos,* which occur throughout the Northern Hemisphere, have been introduced into many areas such as New Zealand, Hawaii, and Australia. The mallard has been implicated in the decline of the New Zealand gray duck, *A. superciliosa,* and the Hawaiian duck *A. wyvilliana* through hybridization (**Figure 3.B**). The northern American ruddy duck, *Oxyura jamaicensis,* similarly threatens Europe's rarest duck, the white-headed duck, *O. leucocephala,* which now exists only in Spain. The northern spotted owl, *Strix occidentalis,* is threatened in the Pacific Northwest by the recent invasion of the barred owl, *S. varia.* Hybrids and fertile offspring have been found.

Extinction from hybridization threatens mammals as well. Feral house cats, *Felix catus,* threaten the existence of the endangered wild cat, *F. silvestris.* In Scotland, 80% of wild cats have domestic cat traits, raising the question of at what point is an endangered species no longer a pure species. The Florida panther, *Puma concolor coryi,* a subspecies of the cougar, is listed as endangered by the U.S. Fish and Wildlife Service. Although over 100 Florida panthers now exist, most are hybrids between original Florida panthers and females captured in Texas and released in Florida in the 1990s. Some scientists question whether the Florida panther should be delisted. Hybrid plants also threaten natives with extinction. Many coastal temperate areas contain *Spartina* cordgrasses. In Britain, the native European cordgrass, *S. maritima,* hybridized with the introduced American smooth cordgrass, *S. alterniflora,* in about 1870 to form *S. anglica.* At first this hardy hybrid was seen as valuable in the fight against coastal erosion, and it was widely planted. With its dense root system and quick vegetative spread, it became invasive and displaced *S. maritima* from much of its native habitat.

**(i)** Mallard duck

**(ii)** New Zealand gray duck

**(iii)** Hybrid

**Figure 3.B Hybrid ducks.** **(i)** Mallard duck, *Anas platyrhynchos,* **(ii)** New Zealand gray duck, *A. superciliosa,* and **(iii)** the hybrid between them. Hybridization is threatening the New Zealand gray duck with extinction.

### *Ecological species concept*

American biologist Leigh Van Valen (1976) proposed the **ecological species concept**, in which each species occupies a distinct ecological niche, a unique set of habitat requirements. Competition between species is likely to result in each individual species occupying a unique niche. This species concept is useful in distinguishing asexually reproducing and morphologically similar species such as bacteria.

## 3.2.2 The main mechanisms of speciation are allopatric speciation and sympatric speciation

Two mechanisms have been proposed to explain the process of speciation. **Allopatric speciation** (from the Greek words *allo,* meaning different and *patra,* meaning fatherland) involves spatial separation of populations by a geographical barrier. For example, nonswimming populations separated by a river may gradually diverge because there is no gene flow between them. Alternatively, aquatic species separated by the emergence of land may undergo allopatric speciation. The emergence of the isthmus of Panama about 3.5 million years ago separated porkfish in the Caribbean Sea and the Pacific Ocean (**Figure 3.9**). Since this event, the two populations have been geographically separated and have evolved into a Caribbean porkfish species, *Anisostremus virginicus,* and the Panamic porkfish, *A. taeniatus.*

The alternative to allopatric speciation is **sympatric speciation** (from the Greek word *sym,* meaning alike), which occurs when members of a species that initially occupied the same habitat within the same range diverge into two or more different species. The metal-tolerant *Agrostis tenuis* plants in Wales (see Figure 3.5) are starting to show a change in their flowering season. Over time this population may evolve into a new species that cannot interbreed with the original metal-sensitive species. A common sympatric speciation mechanism in plants involves a change in chromosome number. Plants commonly exhibit polyploidy, meaning they contain three or more sets of chromosomes. At least 30–50% of all species of ferns and flowering plants are polyploid. Such changes can result in sympatric speciation. Polyploidy is much less common in animals, but some insects and about 30 species of reptiles and amphibians are polyploids. In many groups of herbivorous insects, individual species are restricted to individual host plant species; thus, as plants speciate, each has its own unique set of herbivores. Guy Bush (1994) and others have argued that sympatric speciation has occurred frequently among herbivorous insects. Sympatric speciation may also have been common in fish. In many isolated lakes, there has been a divergence of fish species. For example, cichlid fish have been isolated in the African rift valley lakes, Lakes Victoria, Malawi, and Tanganyika, for about 10 million years, and hundreds of species have arisen from a few founding lineages.

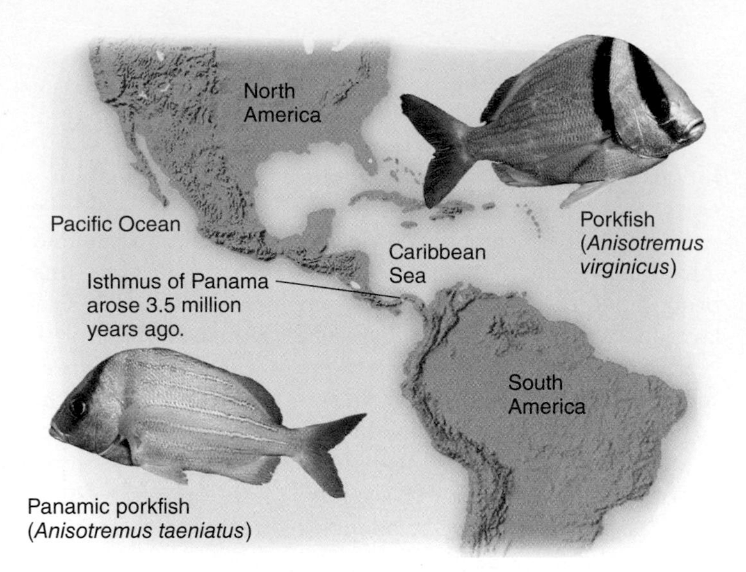


**Figure 3.9 Allopatric speciation in porkfish.** About 3.5 million years ago the isthmus of Panama arose, separating porkfish into two distinct populations with no opportunity for mixing. Since then, genetic changes in each population have led to the formation of two species, one in the Caribbean and one in the Pacific.

As we saw at the beginning of the chapter, recent evidence has suggested sympatric speciation among island birds in the South Atlantic as well.

**Check Your Understanding**

**3.2** Why is sympatric speciation more common in plants than in animals?

# 3.3 Evolution Has Accompanied Geologic Changes on Earth

The history of life on Earth and of the associated geological changes and formation of new taxa are summarized in **Table 3.2.** Following the appearance of eukaryotes around 1.2 billion years ago, most of our current taxa, from worms to tunicates, sprang into existence in the Cambrian explosion. The Earth's physical terrestrial environment, from climatic conditions to atmospheric oxygen, changed over the millennia as first the plants and invertebrates colonized the land, then the amphibians and reptiles, and finally the mammals appeared. In this section we review the history of life on Earth, together with associated geological changes, including drifting of the continents. We then discuss how continental drift and other factors have created disjunct distributions. Finally, we discuss the classification of modern biogeographic realms.

**Table 3.2** A brief history of life on Earth.

| Era | Period | Millions of Years from Beginning of Period | Major Geologic Changes | Major Evolutionary Events |
|---|---|---|---|---|
| **Cenozoic** | Quaternary | 2 | • Cold/dry climate<br>• Repeated glaciations in Northern Hemisphere | • Extinctions of large mammals<br>• Rise of civilization<br>• Evolution of *Homo* |
| | Tertiary | 65 | • Continents in approximately modern positions. India collides with Eurasia, Himalayas uplifted<br>• Atmospheric oxygen reaches today's level of 21%<br>• Drying and cooling trend in mid-Tertiary. Sea levels drop | • Radiation of mammals and birds<br>• Flourishing of insects and angiosperms |
| **Mesozoic** | Cretaceous | 144 | • Northern continents attached<br>• Gondwanaland drifting apart<br>• Sea levels rise<br>• Meteorite strikes Earth | • Mass extinctions of marine and terrestrial life, including last dinosaurs<br>• Angiosperms become dominant over gymnosperms |
| | Jurassic | 206 | • Two large continents form, Laurasia in the north and Gondwanaland in the south<br>• Climate warms<br>• Oxygen drops to 13% | • First birds and angiosperms appear<br>• Dinosaurs abundant<br>• Gymnosperms dominant |
| | Triassic | 251 | • Pangaea begins to drift apart<br>• Hot/wet climate<br>• Sea levels drop below current levels | • Mammals appear<br>• Mass extinction near end of period<br>• Increase of reptiles, first dinosaurs<br>• Gymnosperms become dominant |
| **Paleozoic** | Permian | 286 | • Continents aggregated into Pangaea, dry interior<br>• Large glaciers form<br>• Atmospheric oxygen reaches 30% | • Mass marine extinctions, including last trilobites<br>• Reptiles radiate, amphibians decline<br>• Metamorphic development in insects |
| | Carboniferous | 360 | • Climate cools<br>• Sea levels drop dramatically to present-day levels<br>• Oxygen levels increase dramatically | • Extensive forests of early vascular plants, especially ferns<br>• Amphibians diversify; first reptiles<br>• Sharks roam the seas<br>• Radiation of early insect orders |
| | Devonian | 409 | • Major glaciation occurs | • Seed plants appear<br>• Fishes and trilobites abundant<br>• First amphibians and insects<br>• Mass extinction late in period |
| | Silurian | 439 | • Two large continents form<br>• Warm/wet climate<br>• Sea levels rise | • Invasion of land by primitive land plants, arthropods<br>• Jawed fish appear |
| | Ordovician | 510 | • Mostly southern or equatorial land masses<br>• Gondwanaland moves over South Pole<br>• Climate cools resulting in glaciation and 50-m sea level drop | • Primitive plants and fungi colonize land<br>• Diversification of echinoderms, first jawless vertebrates<br>• Mass extinction at end of period |
| | Cambrian | 542 | • Ozone layer forms, blocking UV radiation and permitting colonization of land<br>• High sea levels | • Sudden appearance of most marine invertebrate phyla including crustaceans, mollusks, sponges, echinoderms, cnidarians, annelids, and tunicates |
| **Pre-cambrian** | | 1,200 | • Oxygen levels increase | • Origins of multicellular eukaryotes, sexual reproduction evolves, increasing the rate of evolution |
| | | 3,000 | • Moon is close to the Earth, causing larger and more frequent tides | • Photosynthesizing cyanobacteria evolve<br>• Oxygen is toxic for many bacteria |
| | | 3,900 | • No atmospheric oxygen | |

### 3.3.1 Early life caused changes in atmospheric oxygen and carbon dioxide

The original composition of Earth, formed by coalescence of material from the solar nebula 4.5 billion years ago, was largely a mixture of silicates, together with iron and sulfides. The planet was so hot that the iron melted and sunk to the center. Water was not present in a free form but was bound to hydrated minerals such as mica in the Earth's crust. Water released from rocks via volcanic explosions condensed to form the hydrosphere. We have a good idea about the composition of volcanic effluents by studying their emissions, which turn out to contain 50–60% water vapor, 24% carbon dioxide, 13% sulfur, and about 6% nitrogen. The atmosphere continued to be rich in carbon dioxide with little to no oxygen until about 2.5 billion years ago. We know this because of the absence of "red beds," sedimentary rocks stained red by iron oxide, in rocks older than 2.5 billion years. As a consequence, the climate would have been hot and steamy. Before life evolved, Earth had a reducing atmosphere, and only with the evolution of photosynthetic organisms, initially algae, about 3 billion years ago did an oxidizing atmosphere begin to form (**Figure 3.10**).

The essential step in the origin of life was the formation of replicating DNA or DNA-like molecules possessing the properties now found in genes. DNA became enclosed in membranes, which provided a stable physical and chemical environment and accelerated replication. For more than half a billion years there were no recorded living things on Earth. The earliest origins of life in the fossil record appeared about 3.5 billion years ago at the beginning of the Precambrian era. Unicellular prokaryotic life-forms, such as cyanobacteria, predominated. In these early days, atmospheric conditions were anaerobic, and fermentation provided most of the energy, but this process was inefficient and left most of the carbon compounds untapped. With the appearance of the first eukaryotes, about 2 billion years ago, chromosomes, meiosis, and sexual reproduction evolved. The long period needed for the action of prokaryotes and primitive eukaryotes to build up an oxygen layer through photosynthesis may explain the 2-billion-year gap between the origin of life and the appearance of multicelled aerobically respiring animals, metazoans. At the same time, the buildup of oxygen led to the formation of an ozone layer that shielded life from the harmful effects of radiation. The buildup of oxygen concomitantly led to the demise of many of the early anaerobic organisms. As we will see throughout this book, environmental conditions greatly affect the abundance and diversity of life.

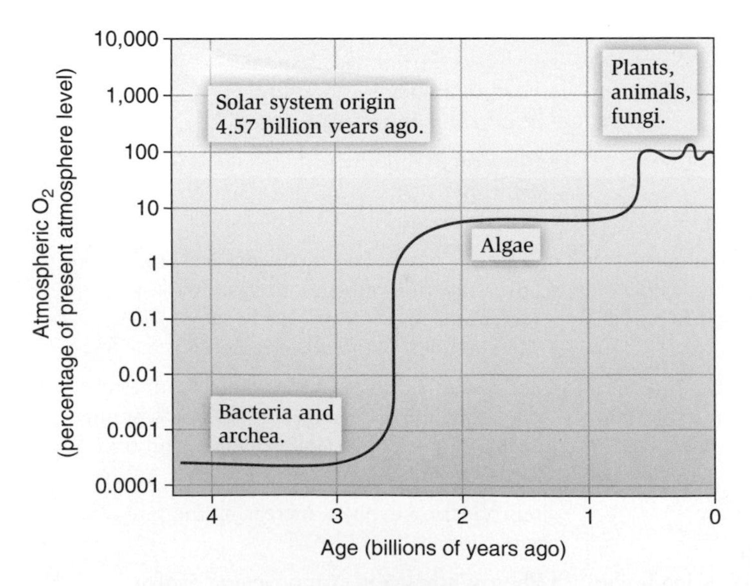


**Figure 3.10** **Changes in atmospheric oxygen over the Earth's history.** (After Kump, 2008.)

### 3.3.2 The evolution of multicellular organisms also accompanied atmospheric changes

We have a more detailed knowledge of atmospheric conditions and the history of life on Earth from about 600 million years ago (**Figure 3.11**). At about 530 million years ago, the

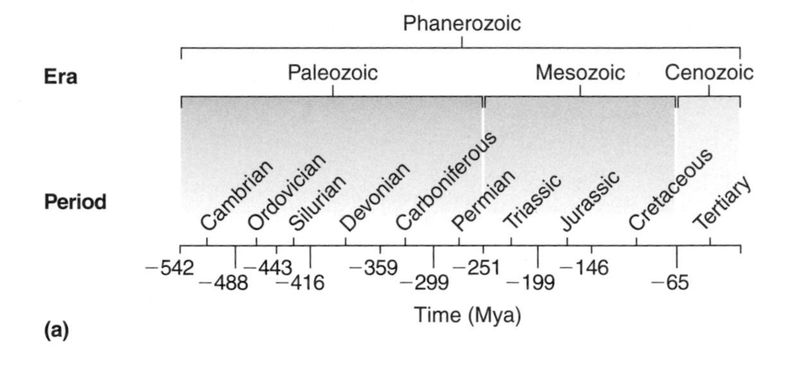


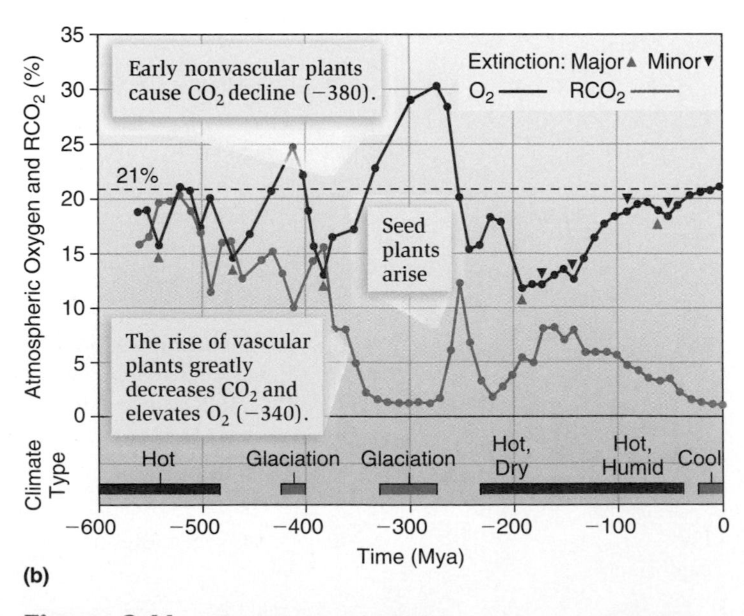


**Figure 3.11** **Changes in atmospheric oxygen and carbon dioxide since the Cambrian explosion.** $RCO_2$ is a multiplier for current atmospheric $CO_2$ levels. (After Ward, 2006.)

Cambrian explosion marked the appearance of most of our current marine invertebrate phyla—sponges, cnidarians, annelids, mollusks, crustaceans, echinoderms, and tunicates. Most organisms were soft-bodied and large. Without skeletons for muscle attachment, their movements would have been slow. They survived partly because no predators with jaws existed to prey on them. The appearance of skeletons permitted more diverse lifestyles and body forms. Among these early taxa are the now-extinct trilobites, whose closest living relative is the horseshoe crab, *Limulus polyphemus* (**Figure 3.12**).

During the Ordovician, the first chordates, jawless fish, were recorded. Also in the Ordovician, the first evidence of terrestrial life appeared as primitive plants and fungi colonized the land. As terrestrial vegetation formed and decayed, organic soils began to form. In the Silurian, the first jawed fish appeared, and arthropods and vascular plants invaded terrestrial habitats. In the Devonian, marine invertebrates, especially trilobites and corals, continued to diversify, and the first bony fishes appeared in the fossil record. The Devonian is sometimes known as the "age of fishes." Devonian fish were often heavily armored to defend against predation, in contrast with modern fish, which emphasize speed. The relatively high oxygen levels promoted large marine arthropod predators, the longest up to 10 feet long. Environment permitting, there is a tendency for animals to become larger over evolutionary time. This tendency is known as Cope's rule, after the 19th-century paleontologist Edward Cope, who suggested that large size protects against predation. Amphibians appeared at the end of the Devonian, as did the first insects, undoubtedly connected with the proliferation of land plants such as bryophytes. The amphibians would have been sluggish. Their salamander-like gait compressed the chest and lungs, making it difficult to breath and walk at the same time. Breaths were taken between steps, limiting periods of high activity. The proliferation of land plants caused carbon dioxide to decline and oxygen levels to increase.

**Figure 3.12 A living fossil.** The horseshoe crab, *Limulus polyphemus*, has existed unchanged for hundreds of millions of years. It is the nearest surviving relative of the trilobites. Most live in water off the coasts of eastern North America and Southeast Asia, but every spring they appear along the coasts to mate and lay eggs.

In the Carboniferous period, insects radiated. The reptiles arose, and the amphibians radiated briefly. Huge, lumbering 5-m amphibians appeared, quite unlike the small forms present today. The extensive forests of this period gave rise to today's rich coal beds and greatly increased atmospheric oxygen. Vascular plants were the first to use lignin for skeletal support. Carboniferous trees grew huge but had relatively shallow roots and fell over quite easily. However, bacteria that could decompose wood had not yet evolved, so the trees did not easily decompose. They lay on the ground and gradually became covered with sediment, and their reduced carbon would be buried. The lack of decomposition also allowed oxygen levels to rise. Some insects, such as dragonflies with 1-m wingspans, became very large, fueled by 30% oxygen levels. The lack of carbon dioxide cooled the planet and there were ice caps at each pole, with extensive glaciers reaching out from the mountains. Forest fires burned frequently and hot, but swampy conditions helped limit the fires' effects.

During the Permian period, the continents had aggregated into one central landmass, called Pangaea. Reptiles and insects underwent extensive radiation, and the amphibia suffered mass extinctions. Perhaps the most remarkable feature of this period, however, was the vast extinction of marine invertebrates, including the last of the trilobites and plankton, corals, and benthic invertebrates, on a scale that commonly implies some worldwide catastrophe. At the end of the Permian, oxygen levels dropped and carbon dioxide levels rose, increasing global temperatures and causing hot, dry conditions. Seed plants arose at this time, but plant life became scarce.

During the Mesozoic era, beginning 251 million years ago, Pangaea started to split up into a southern continent, Gondwanaland, and a northern one, Laurasia. By the end of the era Gondwanaland had formed South America, Africa, Australia, Antarctica, and India, which later drifted north, and Laurasia had begun to split into Eurasia and North America. The land now called the Sahara Desert was probably located near the South Pole 450 million years ago and has since passed through every major climatic zone. Now still drifting northward at 1–2 cm per year, the Sahara will move north 1° in the next 5–10 million years, and the climate and vegetation will change accordingly.

Following the Permian extinctions, marine invertebrates and reptiles began to diversify in the first Mesozoic period, the Triassic. Gymnosperms became large and dominant, and large herbivorous dinosaurs fed upon them. Early mammals also appeared. Oxygen levels dropped to 13%. By the Jurassic period, dinosaurs dominated the terrestrial vertebrate fossil records, and the first birds appeared. Peter Ward (2006) argues that both dinosaurs and birds developed extensive air sacs,

**Figure 3.13 Dilophosaurus.** Fossils of this predatory dinosaur, which was 6 m (20 ft) long, have been found in Arizona.

**ECOLOGICAL INQUIRY**

What caused the extinction of the dinosaurs?

extending the lungs and countercurrent blood flow. This highly efficient system was a wonderful adaptation for low atmospheric oxygen. Even today birds can fly over mountaintops, actively using flight muscles, despite low oxygen. The whole of the Mesozoic is generally known as the "age of reptiles." Turtles and crocodilians had appeared, and giant predatory dinosaurs stalked the Earth, preying on the herbivorous species (**Figure 3.13**). Advanced insect orders such as Diptera (flies) and Hymenoptera (wasps) were also evolving in conjunction with the first flowering plants, angiosperms. By the Cretaceous, dinosaurs had become extinct, as had many other animal groups, including, once again, much marine life, such as ammonites and planktonic Foraminifera. This extinction was, after the Permian, the second greatest extinction in the history of life. The explanation considered most likely is that a severe change took place, probably a cooling of the climate. What brought this climatic change about is the subject of much debate, with meteorite collisions featuring prominently. On land, all vertebrates larger than about 25 kg seem to have gone extinct, but extinction was virtually undetectable in fish.

In the early part of the Cenozoic era, most of the modern orders of birds and mammals arose, and the angiosperms and insects continued to diversify. By the middle of the Tertiary period, the world's forests were dominated by angiosperms. The continents arrived at their present positions early in the era but were connected and disconnected as the sea levels rose and fell. For example, during the early and late Cenozoic, Central America formed a series of islands between North and South America. Substantial numbers of vertebrate genera evolved in the Tertiary. During this time there existed *Paraceratherium*, an extinct rhinoceros. At 18 feet high at the shoulder, it is the largest land mammal known, weighing about 30 metric tons (**Figure 3.14**). Modern elephants rarely weigh more than 10 tons. Though the elephants evolved into a great diversity of forms, only three species survive today.

**Figure 3.14 Artist's rendition of *Paraceratherium*, a rhinoceros-like mammal.** At 5.5 m (18 ft) tall, it was the largest land mammal ever known, even larger than mammoths.

In the Quaternary period beginning about 2 million years ago, there were four ice ages, separated by warmer interglacial periods. During the ice ages, mammals adapted to cold conditions came southward, reindeer and arctic fox roamed in England, and musk ox ranged in the southern United States. Conversely, in the interglacial periods, species spread northward from the Tropics. Lions are known from northern England and the hippopotamus from the River Thames. During the glacial periods, some species that had moved south during the interglacial periods became restricted to isolated pockets of cool habitat, such as mountaintops, as the climate warmed. The most important extinctions at this time were of large mammals and ground birds, the so-called megafauna. For example, in North America, *Megatherium*, the giant 18-foot ground sloth, became extinct about 11,000 years ago. By 13,000 years ago, early humans had crossed the Bering land bridge into the New World, and these extinctions almost certainly were the first of many extinctions caused by human hunters.

### 3.3.3 Modern distribution patterns of plants and animals have been influenced by continental drift

The arrangement of the seas and landmasses on Earth has changed enormously over time as a result of **continental drift**, the slow movement of the Earth's surface plates. The present-day Earth consists of a molten mass overlain by a solid crust about 100 km thick. This crust is not a single continuous piece but is broken into about 14 irregular pieces, called tectonic plates (**Figure 3.15**). As the molten material below rises along

**Figure 3.15 Global tectonic plates.** There are about 14 major rigid slabs, called tectonic plates, that form the current surface of the Earth.

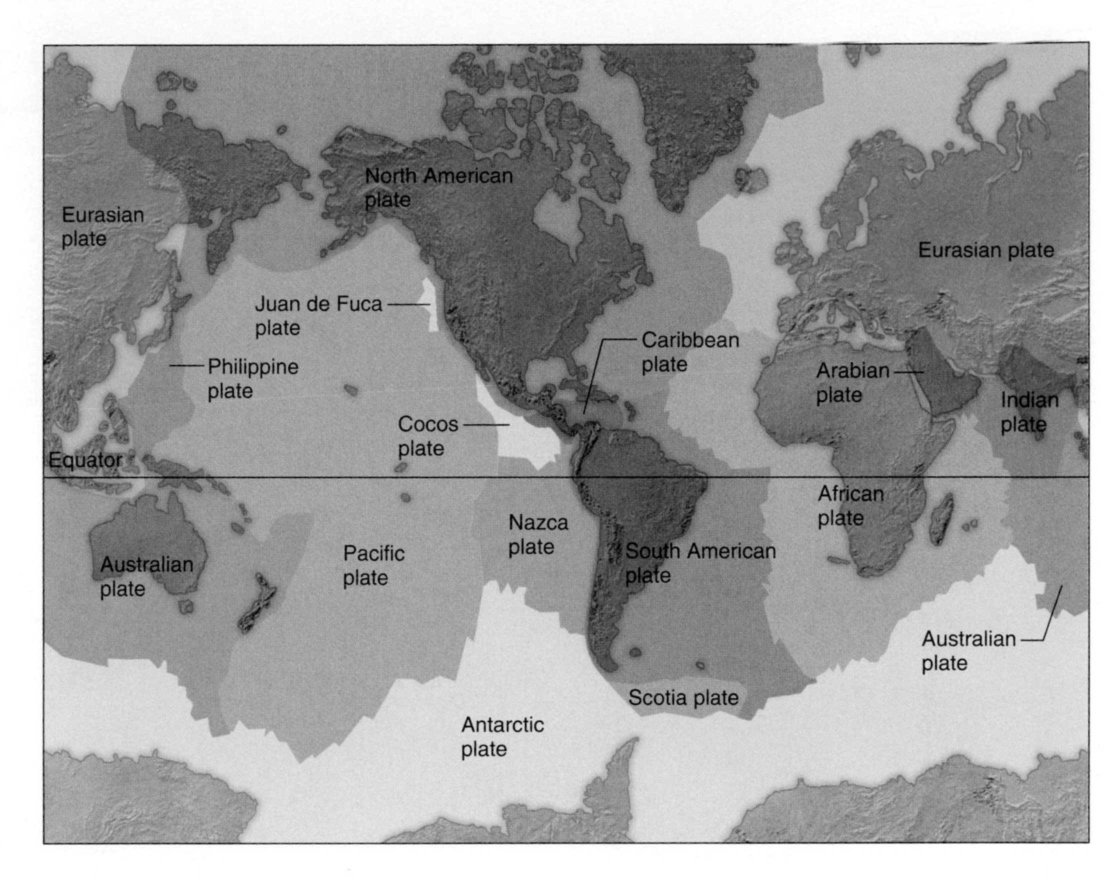


the cracks between the plates, it pushes them aside and cools to form new edges to the plates. The irregular, tumbled edges of these plates are the mid-oceanic ridges. As the plates are pushed aside, their opposite edges meet. Where they meet, one edge is forced under the other, a phenomenon called subduction (**Figure 3.16**). In subduction zones, mountain chains may be formed. Continental drift was first proposed by a German meteorologist, Alfred A. Wegener, in 1912 and has since been supported by a variety of geological and biological evidence.

The breakup of a supercontinental landmass into constituent continents and the eventual re-formation of a supercontinent is probably a cyclical event, with a distinct periodicity. As we noted earlier, the most recent of these supercontinents was Pangaea (meaning "all lands" in Greek), which subsequently broke up into Laurasia in the Northern Hemisphere and Gondwanaland in the Southern Hemisphere. These landmasses continued to break up into the present-day continents and drift farther apart (**Figure 3.17**).

Wegener's hypothesis about continental drift was based on several lines of evidence. The first was the remarkable fit of the South American and African continents, as part of Gondwanaland, shown in **Figure 3.18.** Furthermore, Wegener noted that the occurrence of matching plant and animal fossils in South America, Africa, India, Antarctica, and Australia was best explained by continental drift. Many of these fossils were of large land animals, such as the Triassic reptiles *Lystrosaurus* and *Cynognathus,* that could not have easily dispersed among continents, or of plants whose seeds were not likely to be dispersed far by wind, such as the fossil fern *Glossopteris.* Also, the discovery of abundant fossils in Antarctica was proof that this presently frozen land must have been situated much closer to temperate areas in earlier geological times.

The distribution of the essentially flightless bird family, the ratites, in the Southern Hemisphere is also the result of continental drift. The common ancestor of these birds

Subduction
Mid-ocean ridge
Slab pull
Plates
Subduction
Molten material
Outer core
Inner core

**Figure 3.16 Tectonic plates shift due to the movement of molten material.** A continent moves as the tectonic plate beneath it gradually shifts position.

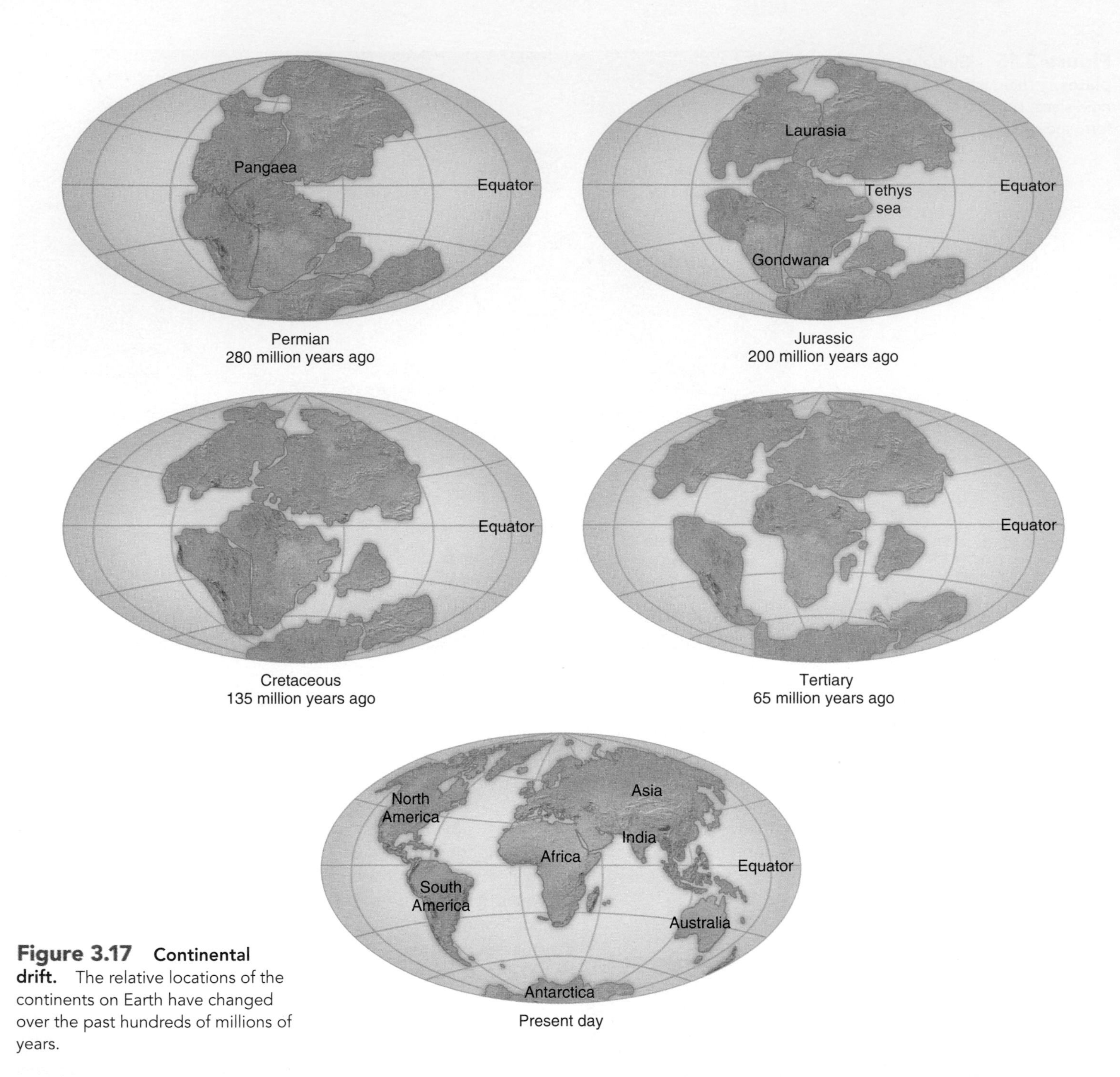


**Figure 3.17 Continental drift.** The relative locations of the continents on Earth have changed over the past hundreds of millions of years.

occurred in Gondwanaland. As Gondwanaland split apart, genera evolved separately in each continent, so that today we have ostriches in Africa, emus in Australia, and rheas in South America. Similarly, the southern beech trees of the genus *Nothofagus* have widely separate distributions in the Southern Hemisphere (**Figure 3.19**). The southern beech nuts have limited powers of dispersal and are not thought to be capable of germinating after long-distance ocean travel. However, trees produce abundant pollen that is found in many fossil records throughout Gondwanaland. Taken together, present-day and fossil records suggest a Gondwanaland ancestor.

Continental drift is not the only mechanism that creates disjunct distributions. The distributions of many present-day species are relics of once much broader distributions. For example, there are currently four living species of tapir, three in Central and South America and one in Malaysia (**Figure 3.20**). Fossil records reveal a much more widespread distribution over much of Europe, Asia, and North America. The oldest fossils come from Europe, making it likely that this was the center of origin of tapirs. Dispersal resulted in a more widespread distribution. Cooling climate resulted in the demise of tapirs in all areas except the tropical locations.

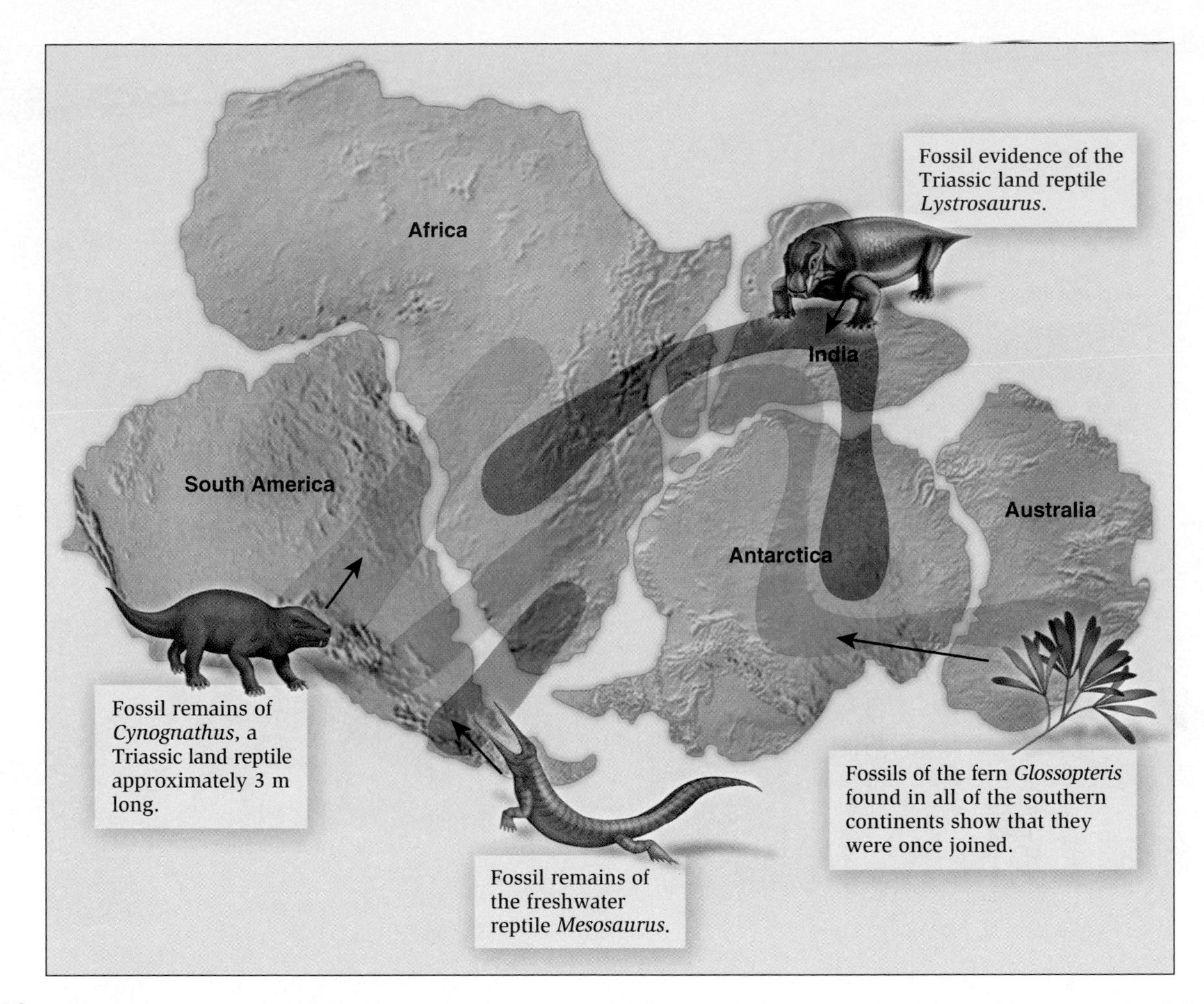


**Figure 3.18 The location of fossil plants and animals on present-day continents can be explained by continental drift.** The fossil remains of dinosaurs and ancient plants are spread across regions of South America, Africa, India, Antarctica, and Australia, which were once united as Gondwanaland.

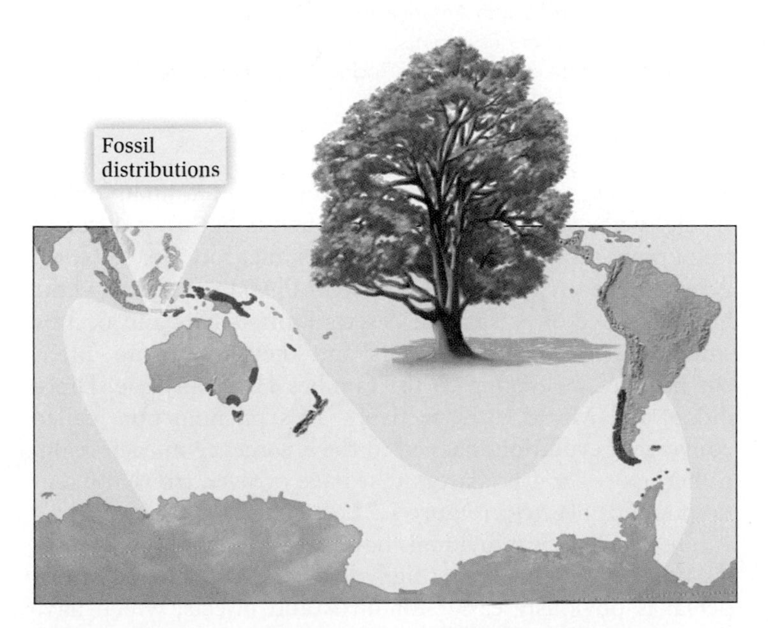


**Figure 3.19 The distribution of *Nothofagus* trees.** Present-day (dark) and fossil distributions (yellow shading) suggest a Gondwanaland ancestor. (After Heads, 2006.)

Another well-known example of a disjunct distribution is the restricted distribution of monotremes and marsupials. These animals were once plentiful all over North America and Europe. They spread into the rest of the world, including South America and Australia, at the end of the Cretaceous period when, although the continents were separated, land bridges existed between them. Later, placental mammals evolved in North America and displaced the marsupials there, apart from a few species such as the opossum. However, placental mammals could not invade Australia because the land bridge by then was broken.

The Elephantidae and Camelidae also have disjunct distributions. Elephants evolved in Africa and subsequently dispersed on foot through Eurasia and across the Bering land bridge from Siberia to North America, where many are found as fossils. They subsequently became extinct everywhere except Africa and India. Camels evolved in North America and made the reverse trek across the Bering bridge into Eurasia; they also crossed into South America via the Central

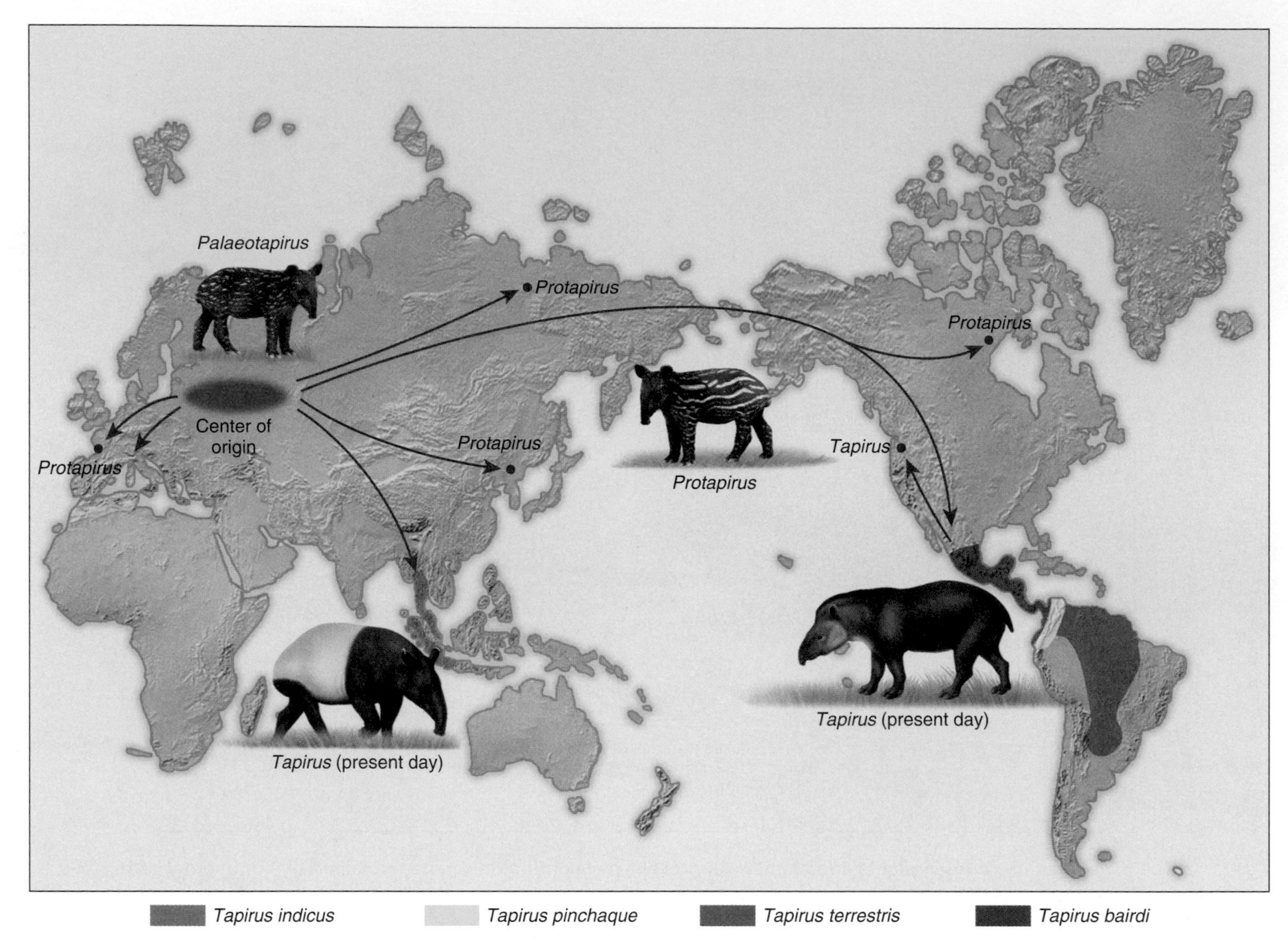


**Figure 3.20 Tapir distribution.** There are four living tapir species, three in Central and South America and one in Malaysia. Fossil evidence suggests a European origin of the ancestral *Paleotapirus* and a dispersal of later-evolving *Protapirus*. A more widespread distribution followed with tapirs dying out in other regions (marked with a red dot), possibly due to climate change. (After Rodriquez de la Fuente, 1975.)

American isthmus. They have since become extinct everywhere except Asia, North Africa, and South America.

Alfred Russel Wallace was one of the earliest scientists to realize that certain plant and animal taxa were restricted to certain geographic areas of the Earth, called biogeographic realms. For example, the distribution patterns of guinea pigs, anteaters, and many other groups are confined to Central and South America, from central Mexico southward. The whole area was distinct enough for Wallace to proclaim it the "neotropical realm." Wallace went on to divide the world's biota into six major realms, or zoogeographic regions. The most recent analyses divide the world into 11 major realms, with the Neotropical, Madagascan, and Australian being the most distinct (**Figure 3.21**).

Biogeographic realms correspond largely to continents but more exactly to areas bounded by major barriers to dispersal, like the Himalayas and the Sahara Desert. Within these realms, areas of similar climates are often inhabited by species with similar appearance and habits but from different taxonomic groups. For example, the kangaroo rats of North American deserts, the jerboas of central Asian deserts, and the hopping mice of Australian deserts look similar and occupy similar hot, arid environments, but they arise from different lineages, belonging to the families Heteromyidae, Dipodidae, and Muridae, respectively. This phenomenon, called **convergent evolution,** has led to the emergence in each realm of herbivores and predators that have evolved from different taxonomic ancestors (**Figure 3.22**).

In some cases individuals have been able to disperse from the area where the group originally evolved. This kind of dispersal is obviously easier for birds and insects, which have the power of flight, or for aquatic organisms, which can drift with the tide. Cattle egrets, *Bubulcus ibis,* crossed the South Atlantic from Africa to northern South America in the late

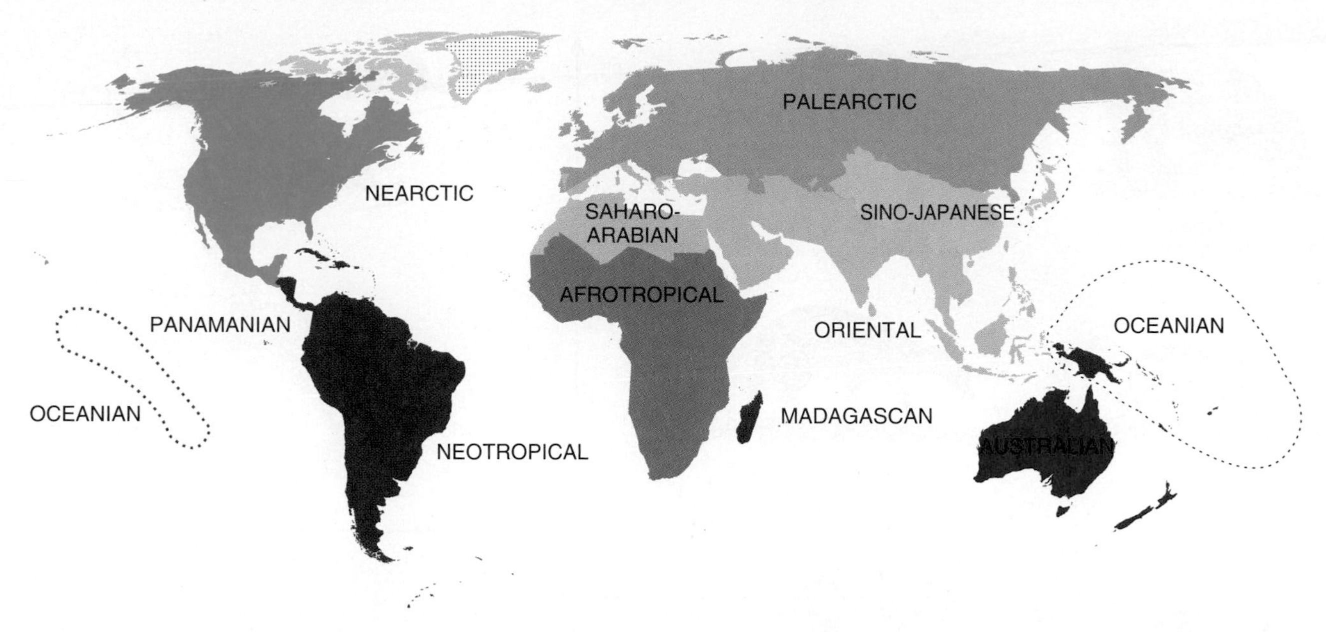


**Figure 3.21 The world's major zoogeographic realms.** Note that the borders do not always demarcate continents, but reflect major barriers to dispersal such as deserts, mountain ranges, or oceans. Eleven broad-scale realms are named. Color differences depict the relative phylogenetic differences between realms. The Panamanian and Neotropical realms are very similar, whereas the Australia and Madagascan realms are quite distinct. (After Holt et al., 2013)

1800s, possibly with the assistance of humans. Since then they have spread widely in South America and the United States. Humans, of course, have succeeded in transporting many species, such as rats, rabbits, sparrows, and starlings, outside their native ranges.

**Check Your Understanding**

**3.3** The giant anteater, *Myrmecophaga tridactyla*, of South America, and the Echidna, *Tachyglossus aculeatus*, of Australia look similar. Both have long snouts and eat ants, yet they are only distantly related. Explain how this is possible.

## 3.4 Many Patterns Exist in the Formation and Extinction of Species

Because the environment is constantly changing, new species evolve and others go extinct. Species become extinct when all individuals die without producing progeny. They disappear in a different sense when a species' lineage is transformed over evolutionary time or divides into two or more separate lineages. This is called pseudoextinction. The relative frequency of true extinction and pseudoextinction in evolutionary history is not yet known. Species extinction is a natural process. Fossil records show that the vast majority of species that have ever existed are now extinct. Leigh Van Valen (1976) described the evolutionary history of life as a continual race with no winners, only losers. He termed this view the **Red Queen hypothesis**, named for the red queen in Lewis Carroll's *Through the Looking Glass* who said to Alice, "It takes all the running you can do to keep in the same place." The analogy was that, in an ever-changing world, species must continually evolve and change in order not to go extinct. In this section we ask, What are the rates of formation of new species and the rates of extinction of old ones? Where are extinctions highest globally? We can possibly use this information to decrease rates of extinction.

### 3.4.1 Species formation may be gradual or sporadic

The rate of evolutionary change and species formation is not constant, though the tempo of evolutionary change is often debated. The concept of **gradualism** proposes that new species evolve continuously over long periods of time (**Figure 3.23a**). The idea of gradualism suggests that large phenotypic differences that produce new species are the result of long periods of small genetic changes that accumulate over time. Such gradual transitions are relatively rare in the fossil record. Instead, fossils of new species appear relatively rapidly and with few transitional types. Stephen Jay Gould and Niles Eldredge (1977) championed the idea of **punctuated equilibrium**, which suggests that the tempo of evolution is more sporadic (**Figure 3.23b**). According to this concept,

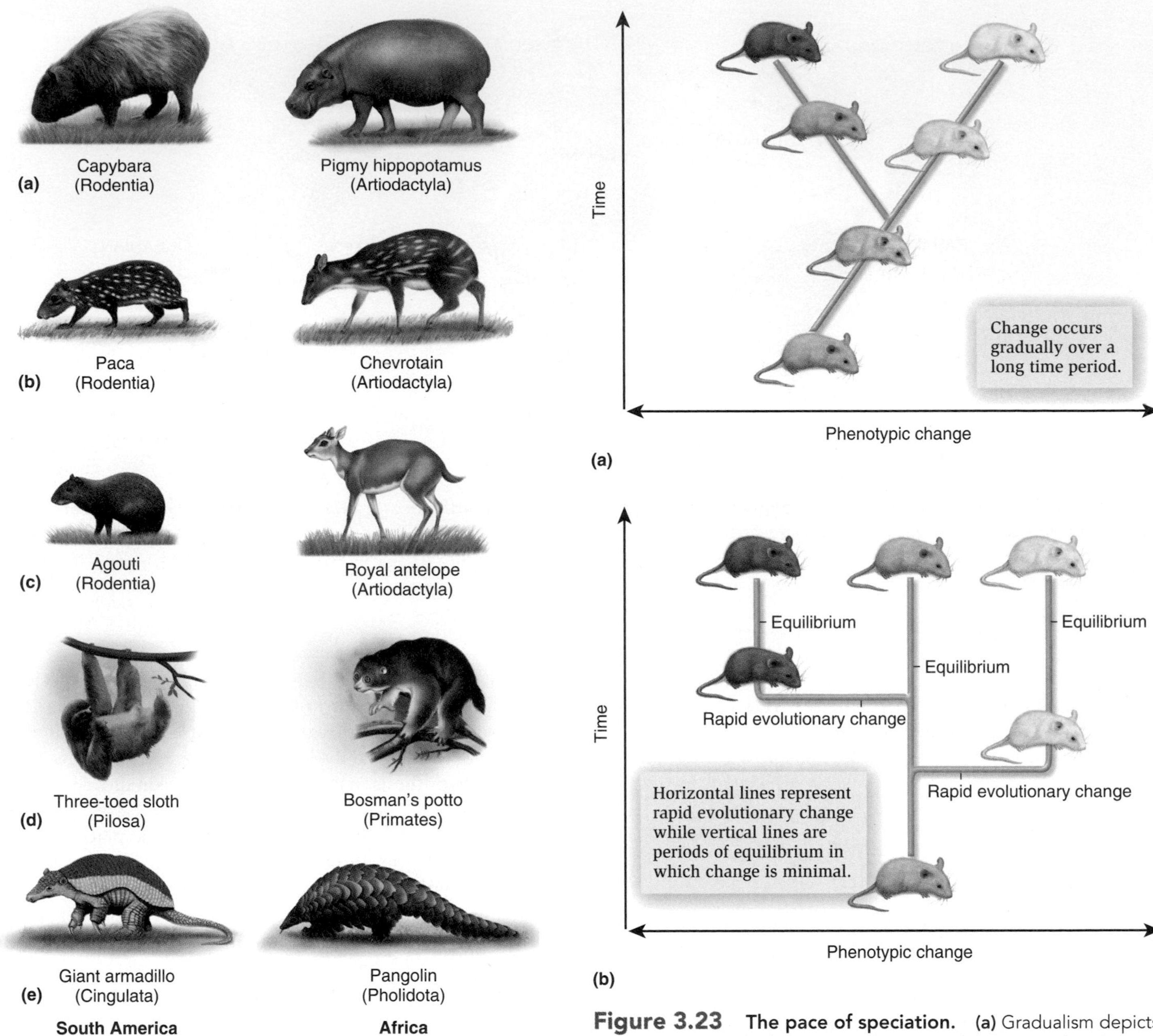


**Figure 3.22 Convergent evolution between tropical rain forest mammals of South America and Africa.** Common names and orders are shown, illustrating how similar-looking species have very different taxonomic affiliations.

**Figure 3.23 The pace of speciation.** **(a)** Gradualism depicts evolution as a gradual change in phenotype from accumulated small genetic changes. **(b)** Punctuated equilibrium depicts evolution as occurring during relatively long periods of stasis punctuated by periods of relatively rapid evolutionary change.

species remain relatively unchanged for long periods of time and are at equilibrium with their environment. These long periods of stasis are punctuated by relatively rapid periods of change. This concept is supported by the fact that fossils of new species usually appear suddenly, with few transitional types. Proponents of gradualism brush off this phenomenon as simply an inadequacy of the fossil record, but advocates of punctuated equilibrium argue that the fossil record accurately represents what happens in nature and that new species do appear suddenly. This process is too quick for the slow process of fossilization to reflect accurately. According to the idea of punctuated equilibrium, most evolution occurs not as "ladders," in which one species slowly turns into another and in turn into another, but as "bushes," where species arise relatively quickly and where only a few active tips survive in the long term.

How rapidly does speciation occur? The answer differs for different organisms. American and Eurasian sycamore trees, *Platanus occidentalis* and *P. orientalis,* have been isolated for at least 20 million years, yet still form fertile hybrids.

The selective forces on these two continents have obviously not been sufficient to cause reproductive isolation between these ecologically general species. However, several genera of mammals, including polar bears, *Ursus maritimus,* and *Microtus* voles, do appear to have originated relatively recently, roughly 200,000 years ago in the Quaternary period. Lake Nabugabo in Africa has been isolated from Lake Victoria for less than 4,000 years, yet it contains five species of fish that are found nowhere else on Earth. Again in Hawaii, at least five species of *Hedylepta* moth feed exclusively on bananas, which were introduced by the Polynesians only some 1,000 years ago.

### 3.4.2 Patterns of extinction are evident from the fossil record

For many taxa, five major mass extinction events appear in the geological record: one in each of the Ordovician, Devonian, Permian, Triassic, and Cretaceous periods. The causes of these extinctions have been much debated. The Ordovician extinction appears correlated with a huge global glaciation. The Permian was the largest recorded extinction for both fishes, 44% of families disappearing, and tetrapods, 58% of families disappearing. For the Permian extinction, geologically rapid changes in climate, continental drift, and volcanic activity are probably the most important causes, though a meteor strike has also been implicated. The causes of the Triassic and Devonian extinctions are not well known. Luis Alvarez and his colleagues (1980) at the University of California, Berkeley, suggested that the Cretaceous extinction may also have been associated with a single catastrophic event such as a meteor strike. The resultant dust cloud presumably blocked solar radiation, resulting in rapid global cooling and causing extinction of plants and animals alike. Groups that had little capacity for temperature regulation became extinct. Taxa with life history stages that were resistant to brief but intense cold, such as seed plants, insects, and endothermic birds and mammals, suffered fewer extinctions. The late Cretaceous extinction was far more significant for tetrapods than for other groups, with 75% of species in the fossil record disappearing at this time. Affected taxa were mainly confined to three major groups: the dinosaurs, plesiosaurs, and pterosaurs.

It is important to realize that extinction is the rule rather than the exception. Because the average species lives 5–10 million years in the fossil record, and the duration of the fossil record is 600 million years, the Earth's current number of plant and animal species represents about 1–2% of species that have ever lived. Leigh Van Valen (1973) suggested that over evolutionary time, the probability of the extinction of a genus or family is independent of the duration of its existence. Old lineages do not die out more readily than younger ones. On the other hand, past adaptations of species provide little preadaptation to extraordinary periodic conditions. There is some evidence that the survivors of mass extinctions tended to be the more ecologically generalist, having a broad diet and existing in a wide variety of habitats. Generalists also tend to have a greater breadth of geographic distribution, which appears to be important in enhancing survival.

### 3.4.3 Current patterns of extinction have been influenced by humans

As we noted in Chapter 1, it is indisputable that humans are the cause of accelerated extinction rates on Earth. Early anthropogenically caused extinctions were mostly due to hunting. As a result, the most widespread extinctions in the Quaternary period involved the so-called megafauna—mammals, birds, and reptiles over 44 kg in body size. The arrival of humans on previously isolated continents, around 40,000 years ago in the case of Australia and 15,000 years ago, or possibly earlier, for North and South America, coincides with large-scale extinctions in certain taxa. Australia lost nearly all its species of very large mammals, giant snakes, and reptiles, and nearly half its large flightless birds around this time. Similarly, North America lost 73% and South America 80% of their genera of large mammals around the time of the arrival of the first humans. The probable cause was hunting, but the fact that climate changed at around this same time leaves the door open for natural changes as a contributing cause of these extinctions. Many taxa were lost in Alaska and northern Asia, where human populations were never large, suggesting that in these cases climate change was the major cause.

However, the rates of extinctions on islands in the more recent past confirm the devastating effects of humans. The Polynesians, who colonized Hawaii in the 4th and 5th centuries C.E., appear to have been responsible for exterminating over 2,000 bird species, including around 50 of the 100 or so endemic species. Introduced predators, such as rats, aided in these extinctions. A similar impact probably was felt in New Zealand, which was colonized some 500 years later than Hawaii. There, an entire avian megafauna, consisting of 15 species of huge land birds, was exterminated by the end of the 18th century by the Maoris. This extinction was probably accomplished through a combination of direct hunting, large-scale habitat destruction through burning, and introduced dogs and rats. Only the smaller kiwis survived.

### 3.4.4 Extinction rates are higher on islands than on the mainland

Certain generalized patterns of extinction emerge on examination of the data. One of the strongest of these is the preponderance of extinctions on islands versus continental areas. While islands often have greater overall numbers of recorded extinctions (**Figure 3.24**), there are also lower numbers of species on islands than on continents, making the percentage of taxa extinct on islands even greater than on continents. The reason for high extinction rates on islands are many and varied. Many island species effectively consist of single

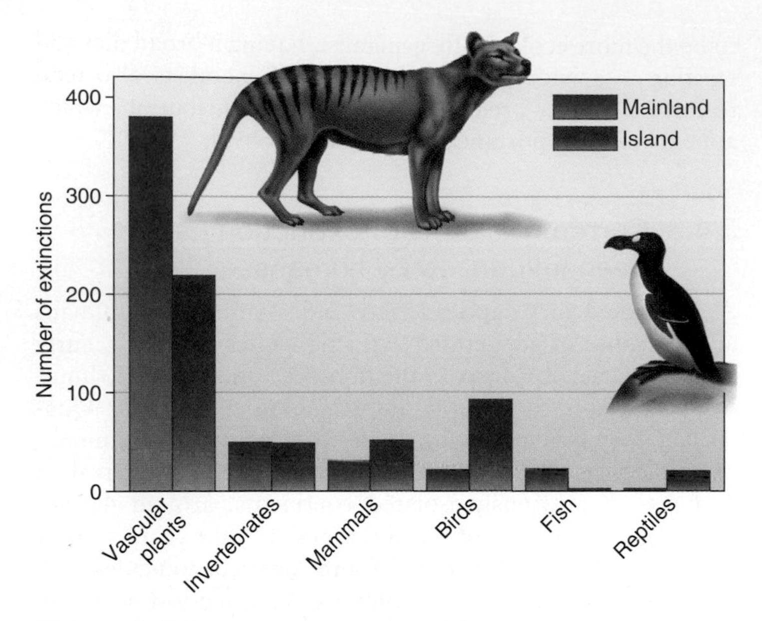


**Figure 3.24** **Recorded extinctions, from 1600 to the present, on continents and islands.**

**ECOLOGICAL INQUIRY**

Why do so many extinctions occur on islands?

populations. Adverse factors are thus likely to affect the entire species and bring about its extinction. In support of this idea, extinction rates of land birds of the Pacific islands tend to decrease with increasing island area (**Figure 3.25**). Also, species on islands may have evolved in the absence of terrestrial predators and may often be flightless. They may also have reduced reproductive rates. Finally, many island species were so tame and ecologically naive that when humans invaded,

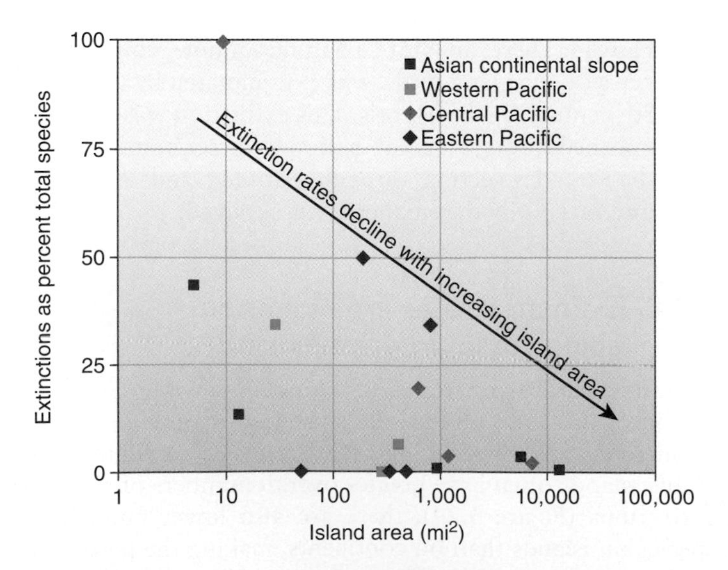


**Figure 3.25** **Extinction rates of Pacific island land birds.** Extinction rates tend to decrease with increasing land area. (After Greenway, 1967.)

they were easily killed. On Chiloe Island, off the coast of Chile, Darwin found the foxes so tame that he collected the species by hitting it over the head with his hammer. It was later named Darwin's fox (*Pseudalopex fulvipes*). Tameness, flightlessness, and reduced reproductive rates appear to be major contributory factors to species extinction, especially when novel predators are introduced.

### 3.4.5 Extinctions are most commonly caused by introduced species and habitat destruction

Introduced species and habitat destruction by humans have been the major factors involved in extinctions worldwide. In the United States these causes have been implicated in 38% and 36%, respectively, of known causes of extinctions (**Figure 3.26a**). Hunting and overcollecting also contribute significantly, causing 23% of extinctions. However, causes differ slightly for different taxa. Hunting and introduced species are much more important for mammals than for other animals (**Figure 3.26b**). For aquatic species, such as mollusks, habitat destruction, including pollution, has accounted for the majority of extinctions (**Figure 3.26c**).

The effects of introduced species can be assigned to competition, predation, or disease and parasitism. Competition may exterminate local populations, but it has less frequently been shown to extirpate entire populations of rare species. For predation there have been many recorded cases of extinction. James Brown (1989) noted that introduced predators such as rats, cats, and mongooses have accounted for at least 43.4% of recorded extinctions of birds on islands. Parasitism and disease by introduced organisms is also important in causing extinctions. As noted in Chapter 1, avian malaria in Hawaii, facilitated by the introduction of mosquitoes, is thought to have killed 50% of local Hawaiian bird species. Similarly, the American chestnut tree, *Castanea dentata,* and European and American elm trees, *Ulmus procera* and *U. americana* respectively, have been severely decreased in numbers by introduced plant diseases, though none of these has yet become extinct.

**Check Your Understanding**

**3.4** Why might length of evolutionary existence not guarantee future success to a lineage?

## 3.5 Degree of Endangerment Varies by Taxa, Geographic Location, and Species Characteristics

Knowing why species have gone extinct in the past helps us to recognize the problems that are likely to threaten species with extinction today. At least 20% of vertebrates are threatened with extinction, but the figure is much higher for amphibians (**Table 3.3**). Within the mammals some orders seem to have

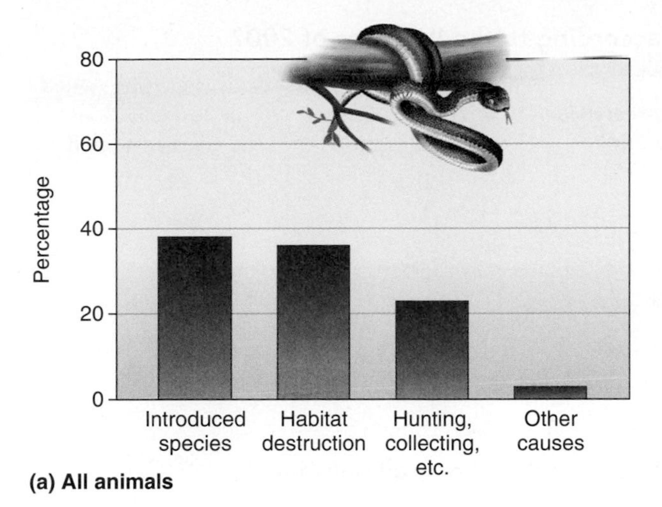


(a) All animals

(b) Mammals

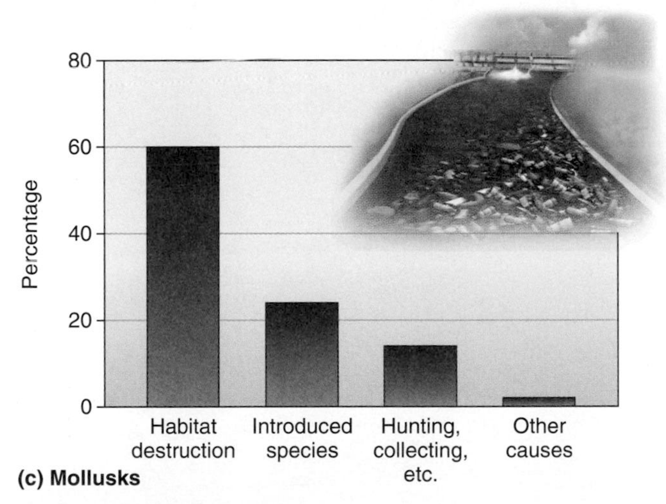


(c) Mollusks

**Figure 3.26 Causes of animal extinctions in the United States, 1600–1980.** **(a)** Causes of historical extinctions in all animals. **(b)** Causes of historical extinctions in mammals. **(c)** Causes of historical extinctions in mollusks.

**ECOLOGICAL INQUIRY**

Why is habitat destruction so much more important for mollusks than for other taxa?

a disproportionately high number of endangered species. For example, manatees and dugongs have four out of five species threatened. Members of the horses, primates, antelopes, and carnivores, are also highly threatened with 79%, 38%, 35%, and 28%, respectively, of their constituent species listed. Although these latter four orders combined contain only a little over 20% of the world's mammal species, they account for just over half of the endangered species. Vertebrates are probably more vulnerable to extinction than invertebrates, because they are much larger and require more resources and larger ranges. Most of the causal factors currently threatening species are anthropogenic in nature, and data from the United States show that current threats are similar to the causes of extinction, with habitat destruction, invasive species, pollution, and direct exploitation the most important (refer back to Figure 1.6).

## 3.5.1 Endangered species are not evenly distributed among geographical areas

Ecologists have determined which geopolitical areas contain the most endangered species (**Table 3.4**). Ecuador and the United States have the highest number of endangered species. This is in part due to relatively thorough biological inventories, and the fact that both countries are rich in species. The majority of threatened mammals occur in mainly tropical countries, with the highest numbers recorded from Indonesia (146), India (89), China (83), Brazil (73), and Mexico (72). Such countries may have large numbers of endangered mammals simply because they have more mammals in general; therefore, if the same percentage were threatened with extinction in each country, a country having a larger number of mammals to begin with would have a higher number endangered.

We can obtain some idea of the number of endangered species a country might be expected to have by graphically plotting the number of endangered mammal species against that country's area (**Figure 3.27**). Bigger countries should have more mammal species than smaller ones and so should have more endangered species. The line describes the relationships between area and number of endangered mammal species. Indonesia in particular has more endangered mammals in relation to country area than would be predicted statistically (points above the line), whereas the United States, for example, has fewer (points below the line). It may come as a surprise to find that the United States heads the list (Table 3.4) in terms of the numbers of endangered fishes and invertebrates, but this is probably because other countries have not inventoried such species as thoroughly.

In the United States the greatest numbers of endangered species occur in Hawaii, southern California, southern Appalachia, and the southeastern coastal states. Many endangered species are restricted to very small areas. For example, Andy Dobson and colleagues (1997) discovered that 48% of endangered plants and 40% of endangered arthropods are restricted to a single county.

**Table 3.3** Numbers of threatened vertebrate species, by taxon, according to the IUCN, as of 2007.

| | Number of described species | Number of species evaluated by 2007 | Number of threatened species in 2007 | Percentage threatened in 2007, as % of species described | Percentage threatened in 2007, as % of species evaluated |
|---|---|---|---|---|---|
| Mammals | 5,416 | 4,863 | 1,094 | 20% | 22% |
| Birds | 9,956 | 9,956 | 1,217 | 12% | 12% |
| Reptiles | 8,240 | 1,385 | 422 | 5% | 30% |
| Amphibians | 6,199 | 5,915 | 1,808 | 29% | 31% |
| Fishes | 30,000 | 3,119 | 1,201 | 4% | 39% |
| TOTAL | 59,811 | 25,238 | 5,742 | 10% | 23% |

**Table 3.4** Numbers of threatened species in twelve countries with the greatest overall numbers of threatened species.

| | Mammals | Birds | Reptiles | Amphibians | Fishes | Invertebrates | Plants | Total |
|---|---|---|---|---|---|---|---|---|
| Ecuador | 33 | 68 | 10 | 163 | 15 | 51 | 1,838 | 2,178 |
| United States | 41 | 74 | 32 | 53 | 166 | 571 | 242 | 1,179 |
| Malaysia | 50 | 40 | 21 | 46 | 47 | 21 | 686 | 911 |
| Indonesia | 146 | 116 | 27 | 33 | 111 | 31 | 386 | 850 |
| Mexico | 72 | 59 | 95 | 198 | 115 | 40 | 261 | 840 |
| China | 83 | 86 | 31 | 85 | 60 | 6 | 446 | 797 |
| Brazil | 73 | 122 | 22 | 25 | 66 | 35 | 382 | 725 |
| Australia | 64 | 50 | 38 | 47 | 87 | 282 | 52 | 623 |
| Colombia | 38 | 87 | 15 | 209 | 31 | 2 | 222 | 604 |
| India | 89 | 75 | 25 | 63 | 39 | 22 | 247 | 560 |
| Madagascar | 47 | 35 | 20 | 55 | 73 | 32 | 280 | 542 |
| Tanzania | 34 | 39 | 5 | 41 | 137 | 43 | 240 | 539 |

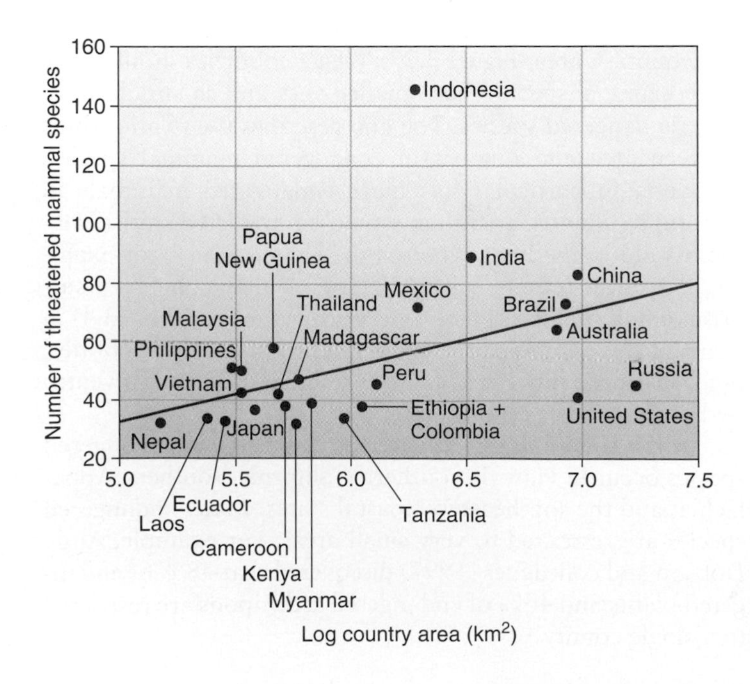


**Figure 3.27 The relationship between the number of threatened species in a country and the area of that country.** Data are for countries with greater than 30 species of threatened mammals. Countries with more threatened species than would be expected based on their area lie above the line, those with fewer threatened species fall below it.

### 3.5.2 Vulnerability to extinction can be linked to species characteristics

In order to further predict what types of human influences are critical in the extinction of wild species, it is useful to have some knowledge of the species traits that may be correlated with high levels of extinction. At least seven species characteristics have been proposed as factors affecting a species' sensitivity to extinction (**Figure 3.28**).

1. *Rarity.* Generally, rare species are more prone to extinction than common ones. This is not as intuitive as might first be thought. For example, a very common species might be very susceptible to even the slightest change in climate. A rare species, although it may exist at very low numbers, may be more resistant to climatic change and thus more persistent in evolutionary time as, for example, global warming proceeds. But what is rarity? Again, this is not as obvious as one might think. Deborah Rabinowitz (1981) showed that "rarity" itself may depend on different factors. A species is often termed rare if it is found only in one geographic area or one habitat type, regardless of its density there. A species that is widespread, but at very low density, can also be regarded as rare. Conservation by habitat management is much easier, and more likely to succeed, for species restricted to one area or habitat type than for widely distributed but rare species.
2. *Dispersal ability.* Species that are capable of migrating between fragments of habitat, such as between mainland areas and islands, may be more resistant to extinction. Even if a population goes extinct in one area, it may be "rescued" by immigrating individuals from another population.
3. *Degree of specialization.* It is often thought that organisms that are specialists are more likely to go extinct. For example, organisms that feed on only one type of plant, such as pandas, which feed on only a species of bamboo, are at a relatively high risk of extinction. Animals that have a broader diet may be able to switch from one food type to another in the event of habitat loss and are thus less prone to extinction. Plants that can live in only one soil type may be more prone to extinction.
4. *Population variability.* Species with relatively stable populations may be less prone to extinction than others. For example, some species, especially those in northern forests, show pronounced cycles. Lemmings reach very high numbers in some years, and the population crashes in others. It is thought that such species might be more likely to become extinct than others.
5. *Feeding level (animals only).* Animals feeding higher up the food chain usually have small populations (look ahead to Chapter 25). For example, birds of prey or predatory mammals number far fewer than their prey species, and, as noted earlier, rare species may be more vulnerable to extinction.
6. *Life span.* Species with naturally short life spans may be more likely to become extinct. Imagine two species of birds, one of which lives for 70 or 80 years and begins breeding at year 10, like a parrot, and the other, which is about the same size but begins breeding at age 2 or 3 and lives only to the age of 7 or 8 years, like an American robin. In the event of habitat degradation, the parrot, with its 80-year life span, may be able to "weather the storm" of a fragmented habitat for 10 years without breeding. The parrot can begin breeding again when the habitat becomes favorable. Species with naturally short life spans are not well able to wait out unfavorable periods.
7. *Reproductive ability.* Species that can reproduce and breed quickly may be more likely to recover after severe population declines than those that cannot. Thus, it is thought that those organisms with a high rate of increase, especially small organisms, like bacteria, insects, and rodents, are less likely to go extinct than larger species like elephants, whales, and redwood trees. For example, the passenger pigeon laid only one egg per year, and this low reproductive rate probably contributed to its demise.

We must decrease habitat destruction, reduce hunting, and prevent the release of exotic species in order to minimize threats to endangered species and increase their chance for survival. To identify species at risk for extinction before they reach endangered status, we must understand many aspects of their biology, including their abundance, how widely distributed they are, how specialized they are, and their reproductive rate and life span. The behavior of a species can also affect its reproductive ability and abundance, and therefore its susceptibility to extinction. In Chapter 4 we will explore the behavior of individuals and the life history strategies of organisms.

**Check Your Understanding**

**3.5** How is it that a rare species that is confined to a small geographic area may be less prone to extinction than a more widespread species?

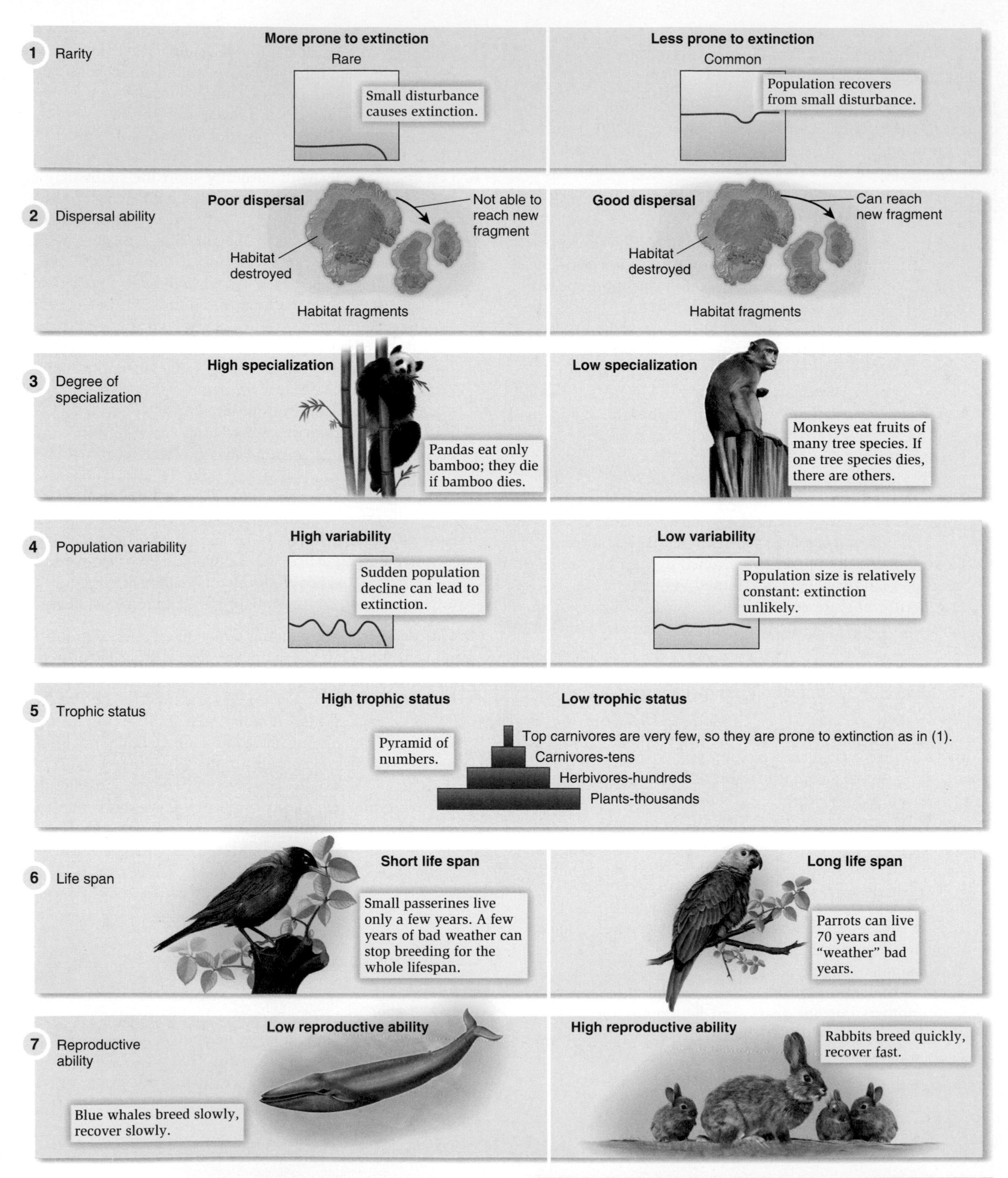


**Figure 3.28** Species characteristics that make them more or less prone to extinction.

**ECOLOGICAL INQUIRY**

What defines a species' rarity?

## SUMMARY

**3.1 Natural Selection Can Follow One of Four Different Pathways**

- Natural selection can follow one of four different paths: directional, stabilizing, balancing, and disruptive.
- Directional selection favors individuals at one end of a phenotypic distribution that have greater reproductive success in a particular environment (Figures 3.1–3.3).
- Stabilizing selection favors the survival of individuals with intermediate phenotypes, such as birds with intermediate clutch sizes (Figure 3.4).
- Balancing selection favors the maintenance of two or more morphs in a particular population (such as different colors of aphids), creating a situation known as a balanced polymorphism (Figure 3.A).
- Disruptive selection favors the existence of two or more genotypes that produce different phenotypes, as seen in different forms of plants that can survive in contaminated or uncontaminated soil (Figure 3.5).

**3.2 Speciation Occurs Where Genetically Distinct Groups Separate into Species**

- Among the four most important species concepts are the biological, phylogenetic, evolutionary, and ecological species concepts (Table 3.1).
- Biological species consist of groups of populations that are reproductively isolated from other such groups. This concept is useful for distinguishing morphologically similar species such as frogs (Figures 3.6).
- Phylogenetic species are identified by a unique combination of morphological or molecular features. Accepting the phylogenetic species concept would greatly increase the numbers of species on Earth, because many races of animals, such as black racers and deer mice, would be elevated to the status of species (Figures 3.7, 3.8).
- Evolutionary species have distinct historical lineages, and ecological species occupy different ecological niches.
- Hybridization between invasive and native species is threatening native species with extinction (Figure 3.B).
- Two main mechanisms explain the process of speciation, allopatric speciation and sympatric speciation.
- Allopatric speciation involves spatial separation of populations by a geographical barrier (Figure 3.9).
- Sympatric speciation involves separation of populations within the same habitat. In polyploidy, one species evolves into two species via changes in chromosome number. Insect populations may become reproductively isolated on different host plants, and other animals may become isolated in different habitats.

**3.3 Evolution Has Accompanied Geologic Changes on Earth**

- The history of life involves the appearance of unicellular organisms nearly 4 billion years ago and multicellular organisms about 1.2 billion years ago, both in the Precambrian era (Table 3.2). The evolution of life was accompanied by changes in atmospheric oxygen and carbon dioxide levels (Figures 3.10, 3.11).
- There was a tremendous increase in the number of invertebrate taxa on Earth in the Cambrian period, at the beginning of the Paleozoic era. Vertebrates appeared in abundance by the Ordovician period. Many millions of species have since gone extinct, while others, such as the horseshoe crab, appear relatively unchanged over hundreds of millions of years (Figure 3.12).
- The whole of the Mesozoic era is known as the age of reptiles, when giant dinosaurs roamed the Earth (Figure 3.13).
- Although mammals appeared early in the Mesozoic era, they did not flourish until the Cenozoic era, when huge species of elephant and rhinoceros were present (Figure 3.14).
- The surface of the Earth is not static but changes as underlying tectonic plates move (Figures 3.15, 3.16).
- The formation of a supercontinent and its breakup into constituent continents is probably a cyclical event; the latest supercontinent is known as Pangaea (Figure 3.17).
- Fossil records show that many species once occupied Gondwanaland, a southern supercontinent encompassing present-day South America, Africa, India, Australia, and Antarctica (Figure 3.18).
- Continental drift has affected the distribution patterns of life on Earth and has caused many disjunct distribution patterns (Figure 3.19). Range collapse has also caused disjunct distribution patterns (Figure 3.20).
- Distinct biogeographic realms with distinct flora and fauna are recognized in different parts of the world (Figure 3.21).
- Convergent evolution has led to the emergence of similar-looking floras and faunas in different realms where environmental conditions are similar (Figure 3.22).

**3.4 Many Patterns Exist in the Formation and Extinction of Species**

- The rate of evolutionary change may be gradual for some lineages, but for others it may show long periods of stasis punctuated by relatively quick periods of change (Figure 3.23).
- Over evolutionary time, many species have undergone periodic mass extinctions from causes as varied as global glaciations to meteor strikes.
- Humans have caused greatly accelerated extinction rates around the world. Some trends in extinction are evident.

- Extinction rates are higher on islands than the mainland (Figure 3.24) and higher on smaller islands than larger ones (Figure 3.25).
- The major causes of extinction in the past were habitat destruction, introduced species, and hunting or overcollecting (Figure 3.26).
- Most vertebrates have been evaluated as to their endangerment status, and amphibians are the most threatened taxa (Table 3.3).
- Most of the causal factors threatening species are anthropogenic in nature and are similar to the causes of extinction. The major threat is habitat loss and modification.

3.5 **Degree of Endangerment Varies by Taxa, Geographic Location, and Species Characteristics**
- Ecologists have determined which countries contain the most endangered species (Table 3.4).
- Some countries may have large numbers of endangered species simply because they have a large area and contain many species (Figure 3.27).
- At least seven species characteristics have been proposed to affect species sensitivity to extinction: rarity, dispersal ability, degree of specialization, population variability, feeding level, life span, and reproductive ability (Figure 3.28).

## TEST YOURSELF

1. The birth weights of human babies generally lie in the range of 3–4 kg (6–9 lb). Mothers cannot deliver much bigger babies, and small babies tend to have increased mortality. This is an example of:
   a. Genetic drift
   b. Disruptive selection
   c. Directional selection
   d. Stabilizing selection
   e. Sexual selection
2. The biological species concept is based on species':
   a. Unique morphological characteristics
   b. Reproductive isolation
   c. Unique morphological and molecular characteristics
   d. Evolutionary lineage
   e. Unique ecological niche
3. In what geological period did most phyla on Earth arise?
   a. Cambrian
   b. Ordovician
   c. Silurian
   d. Carboniferous
   e. Permian
4. The original supercontinental landmass present 280 million years ago was called:
   a. Gondwanaland
   b. Pangaea
   c. Laurasia
   d. Tethys
   e. MesoAmerica
5. The gradual movement of the Earth's landmasses across the surface of the world is known as:
   a. Glaciation
   b. Continental drift
   c. Biogeography
   d. Convergent evolution
   e. Range collapse
6. The biogeographic realm that incorporates South America is known as:
   a. Palearctic
   b. Nearctic
   c. Ethiopian
   d. Silurian
   e. Neotropical
7. Most recorded extinctions have been caused by:
   a. Introduced species
   b. Habitat destruction
   c. Hunting and overcollection
   d. a and b equally
   e. a, b, and c equally
8. Elimination of native species by introduced species has been commonly caused by:
   a. Competition
   b. Predation
   c. Parasitism and disease
   d. a and b
   e. b and c
9. As a percentage of prescribed species, which taxa is currently threatened most with extinction?
   a. Mammals
   b. Birds
   c. Reptiles
   d. Amphibians
   e. Fishes
10. Which of the following characteristics increase a species' sensitivity to extinction?
    a. Increased rarity
    b. Increased life span
    c. Increased reproductive ability
    d. Increased dispersal ability
    e. Decreased population variability

## CONCEPTUAL QUESTIONS

1. Explain the difference between directional, stabilizing, balancing, and disruptive selection.
2. Describe the advantages and disadvantages of the biological, phylogenetic, evolutionary, and ecological species concepts.
3. How can geographic isolation lead to speciation?
4. What factors can cause disjunct distribution patterns of plants and animals on Earth?
5. Explain the concepts of gradualism and punctuated equilibrium, which address the speed at which evolution occurs.

## DATA ANALYSIS

The frequency of the melanic form of *Biston betularia* near industrial centers of Michigan in 1959–1962 was 533 of 598 individuals. Clean air legislation was initiated in the United States in 1963. In 1994–1995, the frequency of melanics was 11 of 60 moths. Calculate the percentage of melanics in both instances and explain what happened.

When honeybees sting, they cannot easily remove the barbed stinger without fatally damaging themselves, so why do honeybees sting enemies when it means certain death to themselves?

CHAPTER 4

# Behavioral Ecology

## Outline and Concepts



Barry Richmond and colleagues (Liu et al., 2004) showed how the work ethic of monkeys is affected by a gene in a region of the brain called the rhinal cortex. Most primates—humans and monkeys included—tend to work harder when a deadline looms. Richmond's team trained four monkeys to release a lever at the exact moment a spot on a computer screen changed color from red to green. The monkeys had to complete this task three times and only on the third trial did they receive a food reward, regardless of how they performed on the first two trials. As an indication of how many trials were left, the monkeys could see a gray bar on the screen. As the bar became brighter the monkeys knew they were reaching the last trial and they worked more diligently for the reward. In the first two trials, the monkeys made more errors than in the last trial. Next, the team switched off the gene known to be involved in processing reward signals. To do this, the researchers injected a short strand of DNA into the monkey's brain. The effects lasted only 10–12 weeks, but during that time the monkeys were unable to determine how many trials were left before the reward was given and they worked vigilantly to receive the reward on every trial, making few errors even on trials one and two. Could such studies be performed with humans? Sufferers of obsessive-compulsive disorders and manic depressives also work for little personal reward.

**Behavior** is the observable response of organisms to external or internal stimuli. Defined as such, behavior includes a very broad range of activities. In this chapter we focus our attention on the field of **behavioral ecology**, the study of how behavior contributes to the differential survival and reproduction of organisms.

Behavioral ecology builds upon earlier work that focused primarily on how organisms behave. In the early 20th century, scientific studies of animal behavior, termed **ethology** (from the Greek *ethos,* habit, manner), focused on the specific genetic and physiological mechanisms of behavior called **proximate causes**. The founders of ethology, Karl von Frisch, Konrad Lorenz, and Niko Tinbergen, shared the 1973 Nobel Prize in Physiology or Medicine for their pioneering discoveries concerning the proximate causes of behavior. As an example of a proximate cause, we could hypothesize that an increase in serotonin, a chemical neurotransmitter, causes normally shy, solitary grasshoppers to morph into gregarious locusts. They become super social, change physically, becoming stronger and darker in color, and become more mobile. The swarms may contain millions of individuals.

However, we could also hypothesize that swarms appear after rains followed by a drought. The rains increase the growth of food plants, causing large numbers of grasshoppers, but the drought then reduces plant abundance, causing a food shortage. Only then is the swarming behavior of locusts triggered, causing the insects to gather in large swarms and fly off in search of more food. This hypothesis leads to a different answer than the hypothesis concerned with changes in serotonin levels. This answer focuses on the adaptive significance of swarming to the locusts, that is, on why a particular behavior evolved, in terms of its effect on reproductive success. These factors are called **ultimate causes** of behavior.

In this chapter we focus on the role of ultimate causes of behavior. We begin by exploring whether members of a single species cooperate with one another and behave in a mutually beneficial way, or tend to act selfishly. Some behavior seems altruistic; it appears to benefit the group at the risk of the individual. For example, a honeybee that stings a potential hive predator to discourage it pays with its life: the barbed stinger cannot be removed, and the associated abdominal tissues are ripped from the bee's abdomen when the bee attempts to leave. We also examine the advantages to some animals of being social and forming flocks or herds, compared to being solitary and territorial. Next, we consider how behavior shapes different mating systems, from monogamous relationships of one male and one female to polygamous systems where each individual mates with more than one partner. Last, we explore the role of sexual selection, a type of natural selection in which competition for mates drives the evolution of certain traits. You'll see that much of the chapter focuses on animal behavior, because the behavior of other organisms is more limited and less well understood. Much of behavioral ecology is based on the premises that animals should behave in ways that maximize benefits and minimize costs.

## 4.1 Altruism Is Behavior That Benefits Others at Personal Cost

A primary goal of an organism is to pass on its genes, yet we see many instances in which some individuals forego reproduction altogether, apparently to benefit the group. How do ecologists explain **altruism,** behavior that appears to benefit others at a cost to oneself? In this section we begin by discussing whether such behavior evolved for the good of the group or for the good of the individual. As you'll see, most altruistic acts serve to benefit the individual's close relatives, or kin. We conclude by examining reciprocal altruism as an attempt to explain the evolution of altruism among nonkin.

### 4.1.1 In nature, individual selfish behavior is more likely than altruism

One of the first attempts to explain the existence of altruism was called **group selection,** the premise that natural selection produces outcomes beneficial for the whole group or species. The British ecologist V. C. Wynne-Edwards (1962) argued that a group containing altruists, each willing to subordinate its interests for the good of the group, would have a survival advantage over a group composed of selfish individuals. In concept, the idea of group selection seemed straightforward and logical: A group that consisted of selfish individuals would overexploit its resources and die out, while the fitness of a group with altruists would be enhanced.

In the late 1960s the idea of group selection came under severe attack. Leading the charge was George C. Williams (1966), who argued that evolution acts through the individual—that is, that adaptive traits generally are selected for because they benefit the survival and reproduction of the individual rather than the group. Williams's main arguments against group selection were that mutation, immigration, and the lack of an ability to predict future resource availability would tend to prevent group selection. Let's see how each of these arguments works.

### Mutation

In a population where individuals limit their resource use, mutant individuals that readily use resources for themselves or their offspring will have an advantage. Consider a species of bird in which a pair lays only two eggs; that is, it has a clutch size of two and the resources are not overexploited for the good of the group. Two eggs would ensure a replacement of the parent birds but would prevent a population explosion. Imagine a bird arises that lays three eggs. If the population is not overexploiting its resources, sufficient food may be available for all three young to survive. If this happens, the three-egg genotype will eventually become more common than the two-egg genotype. This process would work for even larger brood sizes, such as four eggs or five eggs, and brood sizes would tend to increase until the parents could not provide for all their young. Field studies of great tits, *Parus major,* in Wytham Woods, England, show a median clutch size of eight to nine eggs, not because females couldn't incubate more, but because adult birds cannot reliably supply sufficient food for more than eight or nine chicks to survive without compromising their own chances for survival and breeding the next year.

### Immigration

Even in a population in which all pairs laid two eggs and no mutations occurred to increase clutch size, selfish individuals that laid more could still immigrate from other areas. In nature, populations are rarely sufficiently isolated to prevent immigration of selfish individuals from other populations.

### Resource prediction

Group selection assumes that individuals are able to assess and predict future food availability and population density within their own habitat. Little evidence has been found that they can. For example, it is difficult to imagine that songbirds would be able to predict the future supply of the caterpillars that they feed to their young and adjust their clutch size accordingly.

Most ecologists accept individual gain rather than group selection as a more plausible result of natural selection. Population size is more often controlled by competition, in which individuals strive to command as much of a resource as they can, than by cooperation. Such selfishness can cause some seemingly surprising behaviors. For example, male Hanuman langurs, *Semnopithecus entellus,* kill infants when they take over groups of females from other males (**Figure 4.1**). Investigators believe that the reason for the behavior is that when females are not nursing their young, they become sexually receptive much sooner, hastening the day when the male can father his own offspring. Infanticide ensures that a male will father more offspring, and the genes governing this tendency spread by natural selection. Males of carnivores such as lions and bears behave in the same way. A 1997 study by Jon Swenson and colleagues showed that the percentage of female brown bears, *Ursus arctos,* that became pregnant the year after losing a cub was 80%, compared to 0% of those whose cubs survived (**Figure 4.2a**). Where adult brown bears are hunted by humans and males are killed, new males move into a territory and seek to sire their own cubs. As a result, the number of infanticidal males increases and cub survival decreases (**Figure 4.2b**). In one area of Scandinavia where five male bears were killed, cub survival was only 72%; cub survival was 98% in an area where bears were not hunted. This effect only lasted for about 18 months, suggesting that after new male bears became residents, they did not kill cubs. Overall, reproductive output was reduced 30% by hunting, showing how disruptive the practice can be, beyond the mere removal of males.

**Figure 4.1 Infanticide as selfish behavior.** Male Hanuman langurs, *Semnopithecus entellus,* can act aggressively toward the young of another male, even killing them, hastening the day the females come into estrus and the time when the males can father their own offspring. Note that the mother is running with the infant.

If individual selfishness is more common than group selection, how then do we account for what appear to be examples of altruism in nature?

## 4.1.2 Altruistic behavior is often associated with kin selection

All offspring have copies of their parents' genes, so parents taking care of their young are helping to ensure survival of their own genes. Genes for altruism toward one's young are favored by natural selection, and the frequency of these genes will increase in subsequent generations because more surviving offspring will carry them.

In most organisms, the probability that any two individuals will share a copy of a particular gene by descent is a quantity, $r$, called the **coefficient of relatedness**. A mother and father are, on average, related to their offspring by an amount represented as $r = 0.5$, because half of an

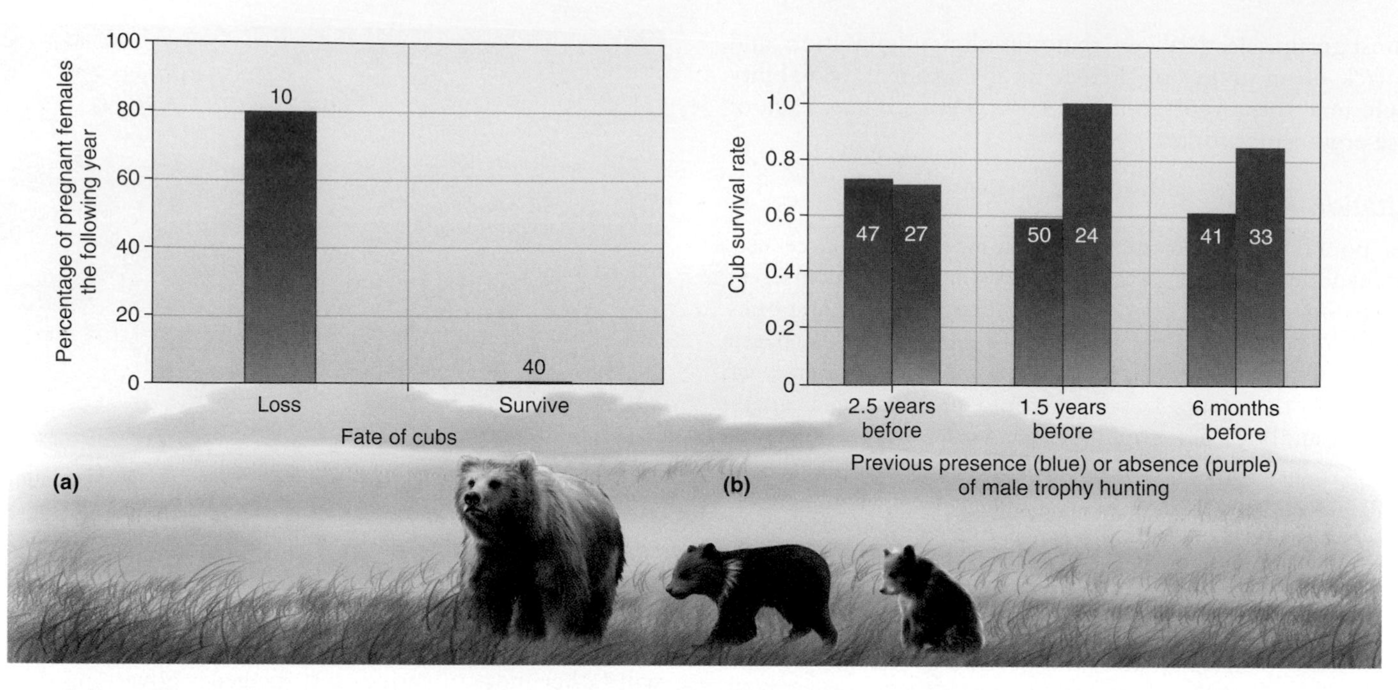


**Figure 4.2 Effect of infanticide on cub production and survival in brown bears.** **(a)** Percentage of females becoming pregnant with and without cub loss the previous year. **(b)** Cub survival rate in areas where adult males were killed or not, 2.5, 1.5, and 0.5 years prior. Sample sizes are given in or on bars. (After Swenson et al., 1997.)

offspring's genes come from its mother and half from its father. By similar reasoning, brothers or sisters are related by $r = 0.5$ (they share half their mother's genes and half their father's); grandchildren and grandparents, by 0.25; and cousins, by 0.125 (**Figure 4.3**). William D. Hamilton (1964) realized the implication of the coefficient of relatedness for the evolution of altruism. Not only can an organism pass on its genes through having offspring, but it also can pass them on through ensuring the survival of siblings, nieces, nephews, and cousins. This means an organism has a vested interest in protecting its brothers and sisters, and even their offspring.

The term **inclusive fitness** is used to designate the total number of copies of genes passed on through one's relatives, as well as one's own reproductive output. Selection for behavior that lowers an individual's own fitness but enhances the reproductive success of a relative is known as **kin selection.** Hamilton proposed that an altruistic gene will be favored by natural selection when

$$rB > C$$

where $r$ is the coefficient of relatedness of the donor (the altruist) to the recipient, $B$ is the benefit received by the recipient of the altruism, and $C$ is the cost incurred by the donor. This is known as **Hamilton's rule.**

Male wild turkeys, *Meleagris gallopavo,* form large display aggregations to attract females. Within these displays males form coalitions of two to four same-age birds that

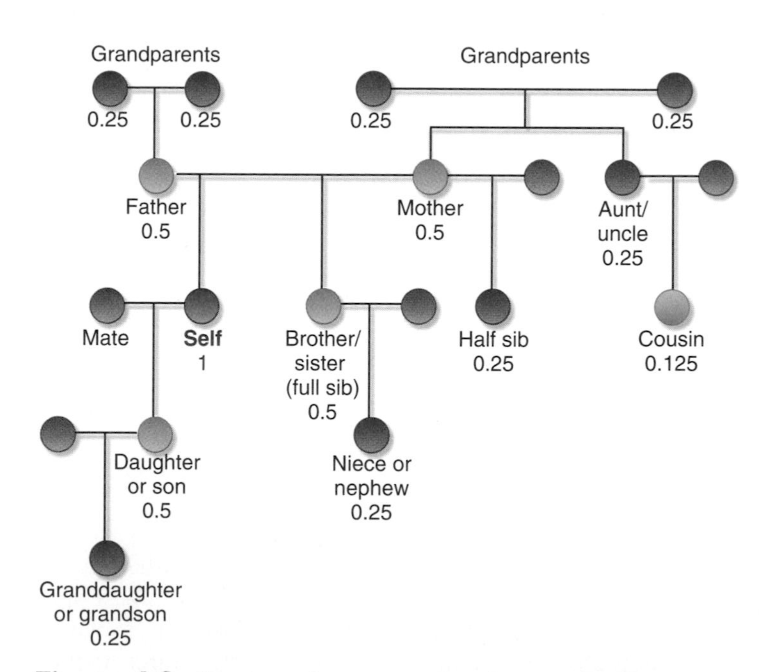


**Figure 4.3 Degree of genetic relatedness to "self" in a diploid organism.**

**ECOLOGICAL INQUIRY**

In theory, should you sacrifice your life to save seven cousins?

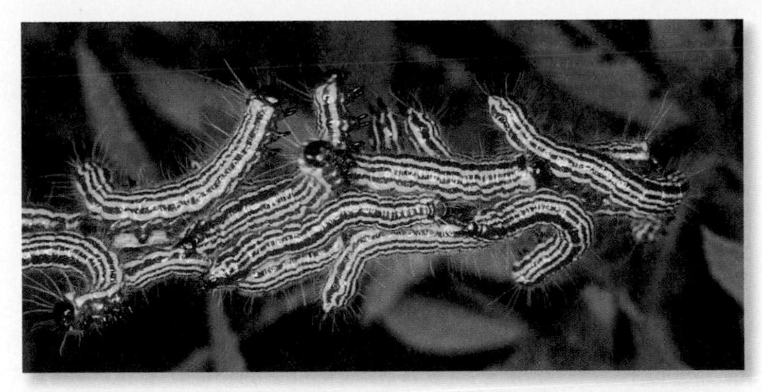

**Figure 4.4** **Kin selection explains clustering of some caterpillars.** *Datana* caterpillars exhibit a bright, striped warning pattern to advertise their bad taste to predators. All the larvae in the group are likely to be the progeny of one egg mass from one adult female moth. The death of the one caterpillar teaches a predator to avoid the pattern and benefits the caterpillar's close kin.

**Figure 4.5** **Kin selection helps explain the altruistic act of alarm calling.** This Belding's ground squirrel sentry is emitting an alarm call to warn other individuals, which are often close kin, of the presence of a predator. It is believed that by doing so, the sentry draws attention away from the others but becomes an easier target itself.

**ECOLOGICAL INQUIRY**

Do males or females tend to emit the most alarm calls?

court females and defend those females against other coalitions or solitary males. The dominant males in such coalitions perform all the matings with the females while the subordinate males perform none. Alan Krakauer (2005) showed how Hamilton's rule explains why subordinate males help dominant males in the absence of observable reproductive benefits for themselves.

First, coalitions are composed of relatives, mainly brothers, sometimes half brothers. The average coefficient of relatedness, $r$, between dominant males and subordinate males in coalitions of pairs was 0.42. Second, help provided by subordinate males increases the reproductive success of dominant males (average 7.0 offspring) compared to noncooperating solitary males (average 0.9 offspring). The benefit, $B$, to the dominant male is thus $7.0 - 0.9 = 6.1$. Subordinate males father no offspring, so the cost of helping to the subordinate is $0.9 - 0.0 = 0.9$. According to Hamilton's rule, kin selection should operate if $rB > C$. Here $(0.42 \times 6.1) = 2.6$ and $2.6 > 0.9$. There is a net benefit of $+1.7$ offspring where male turkey brothers form coalitions to attract females.

The minimum level of relatedness necessary to offset a subordinate's loss of reproductive opportunities can be calculated by setting $rB$ equal to $C$ and solving for $r$. As long as partners in the coalition are related by $r > 0.15$, cooperate behavior is favored, thus even half siblings form these partnerships.

Kin selection can also explain the group feeding behavior of distasteful caterpillars. Because caterpillars are soft-bodied creatures, they rely on bad taste or poison to deter predators, and this condition is advertised with bright warning colors. As one example, noxious *Datana* caterpillars, which feed on oaks and other trees, have bright red and yellow stripes and adopt a specific posture with head and tail ends upturned when threatened (**Figure 4.4**). Unless a predator is born with an innate avoidance of this prey type, it has to kill and eat one of the caterpillars in order to learn to avoid similar individuals in the future. It is of no personal use to the unlucky caterpillar to be killed. However, animals with warning colors often aggregate in kin groups because they hatch from the same egg mass and thus remain nearby. In this case, the death of one individual is likely to benefit its siblings, which will be less likely to be attacked in the future; thus its genes will be preserved. This explains how the genes for bright color and a warning posture are successfully passed on from generation to generation. In a case where $r = 0.5$, the number of brothers and sisters that survived might be 50, so $B = 0.5 \times 50 = 25$ and $C = 1$. Therefore, the benefit of 25 is much greater than cost, 1, so the genes for this behavior will be favored by natural selection.

A common example of altruism in social animals occurs when a sentry raises an alarm call in the presence of a predator. This behavior has been observed in Belding's ground squirrels, *Spermophilus beldingi* (**Figure 4.5**). The squirrels feed in groups, with certain individuals acting as sentries. As a predator approaches, the sentry typically gives an alarm call and the group members retreat into their burrows. In drawing attention to itself, the caller is at a higher risk of being attacked by the predator. However, in many groups those closest to the sentry are more likely to be offspring or brothers and sisters; thus, the altruistic act of alarm calling is thought to be favored by kin selection. Supporting this explanation is the fact that most alarm calling is done by females, because they are more likely to stay in the colony where they were born and thus have kin nearby, whereas the males are more apt to disperse far from the colony. The females are territorial, and territories tend to occur in matrilineal clusters and are cooperatively defended with female kin.

### 4.1.3 Altruism in social insects arises partly from genetics and partly from lifestyle

Perhaps the most extreme form of altruism is the evolution of sterile castes in social insects, in which the vast majority of females, known as workers, rarely reproduce themselves but instead help one reproductive female (the queen) to raise offspring, a phenomenon called **eusociality**. The explanation of eusociality lies partly in the particular genetics of most social insect reproduction. In the Hymenoptera (ants, bees, and wasps), females develop from the queen's fertilized eggs and are diploid—the product of fertilization of an egg by a sperm. Males develop from the queen's unfertilized eggs and are haploid—their sperm are genetically identical. Each daughter receives an identical set of genes from her father, and the other half of a female's genes come from her diploid mother, so the coefficient of relatedness ($r$) of sisters is 0.50 (from father) + 0.25 (from mother) = 0.75. Such a system of sex determination is called **haplodiploidy**. The result is that females are more related to their sisters (0.75) than they would be to their own offspring (0.50). This suggests that it is evolutionarily advantageous for females to stay in the nest or hive and care for other female offspring of the queen, which are their full sisters. It also explains why female honeybees are willing to die in defense of the hive, because the hive is full of genetically similar sisters.

Elegant though these types of explanations are, they do not provide the whole picture. Large eusocial colonies of termites exist, but termites are diploid, not haplodiploid. In this case, how do we account for the existence of eusociality?

Richard Alexander (1974) suggested that it was the particular lifestyle of these animals, rather than genetics, that promoted eusociality. He suggested that eusocial species exist when certain conditions are met, especially where some adults can prevent others from breeding. Such situations occur where nests or burrows are enclosed, escape is difficult, and a dominant individual, such as a queen, could prevent other individuals from reproducing.

In the 1970s a eusocial mammal was found that satisfied the predictions of Alexander's model: the naked mole rat, *Heterocephalus glaber.* Naked mole rats are diploid rodent species that live in arid areas of the Horn of Africa, including Ethiopia, Somalia, and Kenya. They range from 8 to 10 cm in length and from 30 to 35 g in weight. Their large protruding teeth are used for digging and their lips are sealed behind the teeth to keep out soil. Naked mole rats live in large underground colonies of 20–200 individuals. One colony may occupy an area of 20 football fields. The burrows are often deficient in oxygen but the mole rats have a low metabolic rate. They cannot regulate their body temperature, having no sweat glands or a layer of fat beneath skin. They sleep communally to keep warm, and some individuals will bask in burrows close to the surface during the day to get warm and later return to share their warmth with others. Only one female, the queen, and 1–3 males produce offspring (**Figure 4.6**). Litters typically consist of about 12 pups, and the queen may have 4–5 litters a year. The whole colony therefore consists of members of the same family. Mole rat colonies exhibit a division of labor, sometimes referred to as a caste system. Some individuals, both male and female, are soldiers who defend the colony from predators and mole rats from different colonies. Other individuals are primarily diggers and foragers. All individuals have a distinctive colony odor that they collect from rolling around in the toilet chamber. So dependent are members of the colony on one another that individual mole rats kept in zoos will die.

**Figure 4.6 A naked mole rat colony.** In this mammal species, most females do not reproduce and only the queen (shown resting on workers) has offspring.

A renewable mole rat food supply is present in the dry soils of Africa in the form of plant tubers. The dry climate forces many plants to have such underground reserves. Tubers of the plant *Pyrenacantha kaurabassana* weigh up to 50 kg and provide food for a whole colony, though the food would be insufficient if all the mole rats reproduced. Mole rats locate these more by luck than by judgment. One advantage of the large colony size is the large number of scouts for food.

Mole rats readily share food. Individuals would find it impossible to dig sufficient tunnels to find food. Most mole rats dig colonies rapidly in the brief rainy season when the soil softens. This partly explains why mole rats do well in large colonies. They often attack the center of a tuber and leave the outer skin intact, allowing the tuber to regenerate. Because the burrows are usually hard as cement, there are few ways to attack them, and a heroic effort by a mole rat blocking the entrance can effectively stop a predator, commonly a rufous beaked snake, *Rhamphiophis oxyrhynchu.* Though the attacked mole rat may die, its many siblings will live on. The queen manipulates the other colony members; she pushes and shoves them to perform their tasks. Such behavior inhibits nonbreeders from becoming reproductively active by suppressing hormone production. Queens can live 13–18 years. On occasion, large soldiers may leave the burrow and travel up to a mile away looking for a mate with which to establish a new colony. As Alexander argued, such lifestyle characteristics may provide an explanation for the evolution of eusociality in species such as termites and naked mole rats, in which both sexes are diploid.

## 4.1.4 Unrelated individuals may behave altruistically if reciprocation is likely

Kin selection can help explain instances of apparent altruism; however, cases of altruism are known to exist between unrelated individuals. What drives this phenomenon is a "You scratch my back, I'll scratch yours" behavior called reciprocal altruism. In **reciprocal altruism** the cost to the animal of behaving altruistically is offset by the likelihood of a return benefit. This occurs in nature, for example, when unrelated chimps groom each other.

Gerald Wilkinson (1990) noted that female vampire bats, *Desmodus rotundus,* exhibit reciprocal altruism via food sharing. Vampire bats that go 60 hours without a blood meal can die, because they can no longer maintain their body weight and thus their correct body temperature. Adult females will share their food with their young, with the young of other females, and with other unrelated females that have not fed. The females roost together in groups of 8–12 with their dependent young. A hungry female will solicit food from another female by approaching and grooming her. The female being groomed then regurgitates part of her blood meal for the other (**Figure 4.7**). The blood donor loses some weight and accelerates her time toward starvation. However, the weight gained by the recipient allows her to substantially delay the time to starvation, enabling her to forage for food on a subsequent night. The roles of blood donor and recipient are often reversed, and Wilkinson showed that unrelated females are more likely to share with those that had recently shared with them. The probability of a female getting a free lunch is reduced because the roost consists of individuals that remain associated with each other for long periods of time. The rationale of when individuals may be more likely to cooperate than not have been modeled by a branch of inquiry known as game theory.

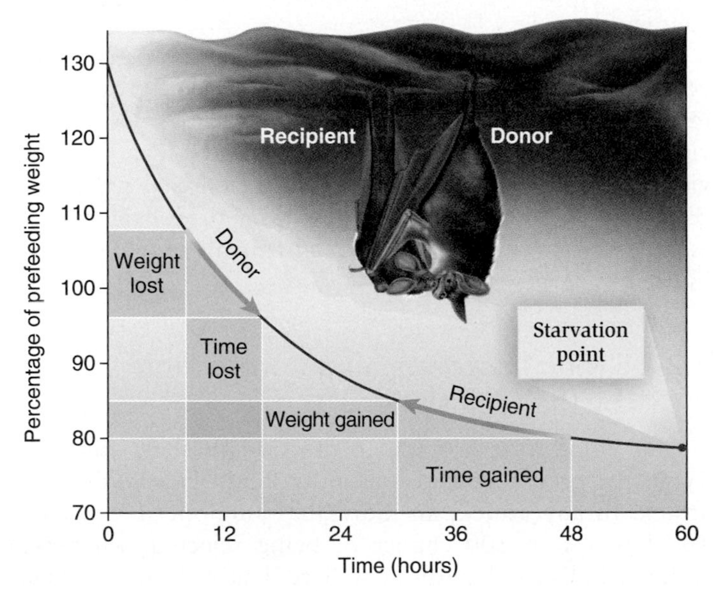


**Figure 4.7** **Reciprocal altruism.** A female vampire bat will regurgitate food to an unrelated female if needed, because at other times she may be the recipient of such a meal. Females return from feeding at about 30% above their prefeeding weight. During regurgitation, the weight loss by the donor may be about 15% and the weight gain by the recipient only 5%. However, the recipient delays starvation by about 18 hours whereas the donor accelerates time toward starvation by only about 6 hours. (From Wilkinson, 1990.)

In game theory, the rationale of reciprocal altruism is explained by the game called prisoner's dilemma. **Game theory** is a branch of mathematics and economics that studies interactions between agents. In behavioral ecology, the agents are individuals who are choosing different strategies in interactions with other individuals so as to maximize their benefits. For example, when is the best time to flee in a conflict and when is the best time to stay and fight? In game theory, an **evolutionarily stable strategy** (ESS) is a behavioral strategy that, if adopted by a population, cannot be invaded by any other strategy. This means that if all the members of a population adopt this strategy, no other strategy will yield a greater benefit to individuals over the long term. Imagine two suspects, A and B, who are arrested. The police have insufficient evidence for a conviction and they are relying on both prisoners to testify against the other. If the prisoners cooperate with one another and stay silent, the police cannot make their case and each prisoner gets a light 6-month sentence. However, both prisoners are in isolation and neither knows what the other will do. If prisoner A testifies against prisoner B, A will go free and B will serve 10 years, and vice versa. If both testify against the other, they each receive 5 years. The payoffs are summarized below:

**Prisoner's Dilemma**

| Prisoner A | Prisoner B | Payoffs |
|---|---|---|
| Stays silent | Stays silent | Each serves 6 months |
| Testifies | Stays silent | A goes free, B serves 10 years |
| Stays silent | Testifies | B goes free, A serves 10 years |
| Testifies | Testifies | Both A and B serve 5 years |

In this case, the best strategy for each prisoner is to testify, regardless of what the other chooses. If the other prisoner stays silent and you testify, you will walk free. If the other prisoner testifies and you testify, you receive a 5-year sentence, which is still better than 10 years. The relevance of the prisoner's dilemma to behavior ecology is that it shows how nature tends to discourage cooperation between individuals. However, in a social setting animals play the prisoner's dilemma repeatedly, and they have a memory of what co-contestants did on their last turn. When the game is played repeatedly, cheating, or failing to return a favor, is not an ESS, because contestants will remember an individual who cheats and they will not be as likely to cooperate with this individual in the future. A tit-for-tat strategy, where players do exactly as has been done unto them, is most stable. Reciprocal altruism is based on this strategy. If the recipient of a beneficial act does not reciprocate later on, it will be cut off in the future.

In 1980, Robert Axelrod and William Hamilton held a computerized "tournament" in which they invited participants to submit different "behavioral strategies" (computer programs) to determine which performed best when pitted against other strategies. The strategy of tit-for-tat won, illustrating that it is an ESS. However, one of the problems with tit-for-tat is that a single error in communication, which might be seen as cheating, can lead to a "death spiral" of tit-for-tats. In this situation "tit-for-two-tats" is a better, more forgiving strategy. Imagine a vampire bat that donates part of her blood meal to another female that does not return the favor. This failure of the second female to reciprocate may be because she did not have any meal to share. The first female may still donate to her a second time rather than fail to cooperate immediately.

**Check Your Understanding**

**4.1** Much research on ants shows that queens of some species mate with two or more males during their nuptial flights. How does this affect your viewpoint of the evolution of eusociality?

## 4.2 Group Living Has Advantages and Disadvantages

Much of animal behavior is directed at other animals, and some of the more complex behavior occurs when animals live together in groups such as flocks or herds. Although congregations promote competition for food and increased transmission of parasites and other diseases, benefits of group living can compensate for the costs involved. Many of these benefits relate to group defense against predators. Group living can reduce predator success in at least two ways: through increased vigilance and through protection in numbers. In turn, groups of predators may be able to kill more prey than solitary predators.

### 4.2.1 Living in groups may increase prey vigilance

For many predators, success depends on surprise. If an individual is alerted to an attack, the predator's chance of success is lowered. A wood pigeon, *Columba palumbus,* will take to the air when it spots a goshawk, *Accipiter gentilis*. Once one pigeon takes flight, the other members of the flock are alerted and follow suit. If each pigeon occasionally looks up to scan for a hawk, then the bigger the group, the more likely it is that one bird will spot the hawk early enough for the flock to take flight. This is referred to as the **many eyes hypothesis** (**Figure 4.8**). By living in groups, individuals may decrease the amount of time each spends scanning for predators and increase the time they have to feed. Of course, cheating is a possibility, because some birds might never look up, relying on others to keep watch while they keep feeding. However,

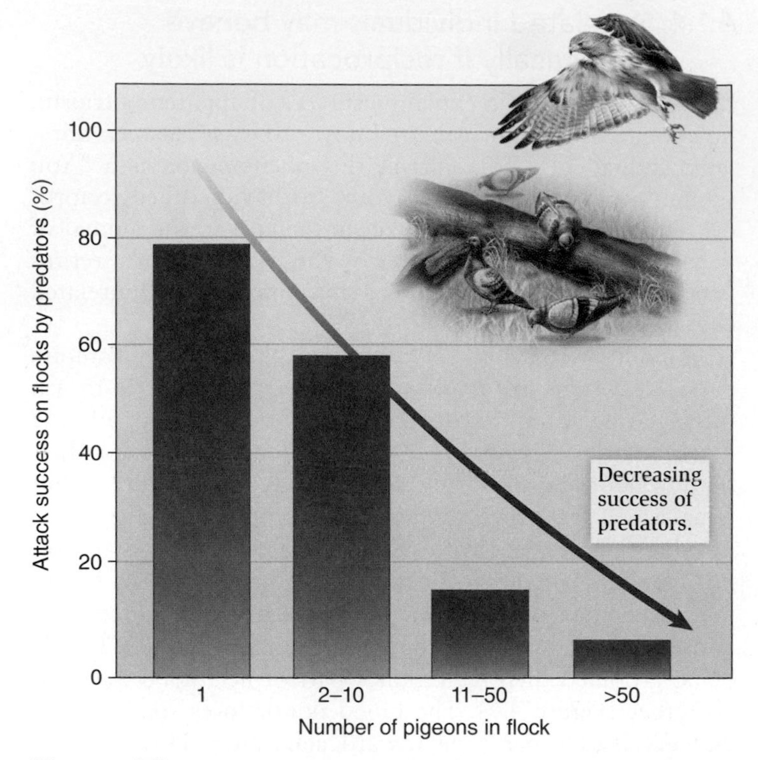


**Figure 4.8** **Living in groups and the many eyes hypothesis.** The larger the number of wood pigeons, the less likely it is that an attack will be successful. (After Kenward, 1978.)

the individual that happens to be scanning when a predator approaches is most likely to escape, a fact that tends to discourage cheating.

A recent meta-analysis by Guy Beauchamp (2008) generally supported the prediction that vigilance decreases with increased group size. Of 172 published relationships between vigilance and group size, most supported the idea that individual scan frequency decreases with increased group size (**Figure 4.9**). However, there was less support for the hypothesis that scan duration of individual animals should decrease as group size increased.

### 4.2.2 Living in groups offers protection by the "selfish herd"

Group living also provides protection in sheer numbers. Typically, predators take one prey item or individual per attack. In any attack, an individual antelope in a herd of 100 has a 1 in 100 chance of being selected, whereas a solitary individual without a herd has a 1 in 1 chance. Large herds may be attacked more frequently than a solitary individual, but a herd is unlikely to attract 100 times more attacks. Furthermore, large numbers of prey are able to defend themselves better than single individuals, which usually choose to flee.

Research has shown that within a group, each individual can minimize the danger to itself by choosing the location that is as close to the center of the group as possible. This was

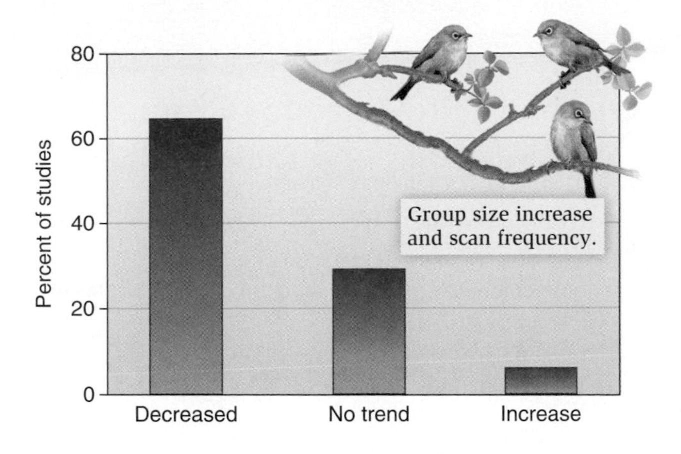


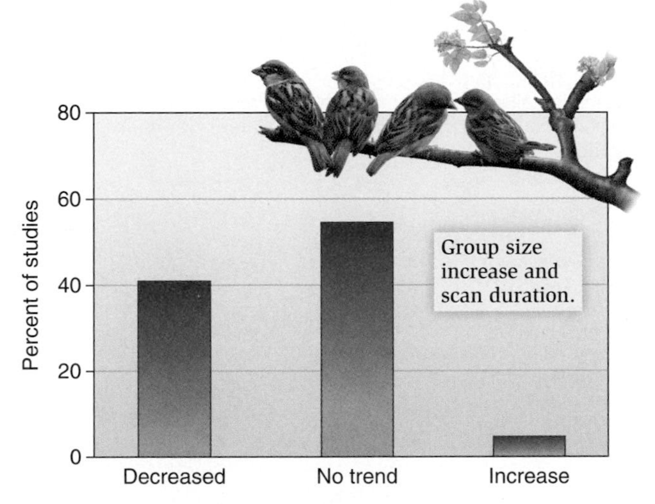


**Figure 4.9 The effects of increased group size in birds on vigilance.** Data reveal what percentage of studies showed decreased or increased behavior as group size increased. (After Beauchamp, 2008.)

the subject of a famous paper, "The Geometry for the Selfish Herd," by the British evolutionary biologist William Hamilton (1971). The explanation of this type of defense is that predators are likely to attack prey on the periphery because they are easier to isolate visually. Many animals in herds tend to bunch close together when they are under attack, making it physically difficult for the predator to get to the center of the herd (**Figure 4.10**).

## 4.2.3 Cooperative predators may perform better than solitary predators

Group living may allow predators such as lions, hyenas, and wolves the opportunity to take down prey of a disproportionately large size that a single predator would be unable to handle. Furthermore, when animals search for food within a group, each individual may be able to exploit the discoveries of others. This type of information sharing, be it deliberate or accidental, may result in a higher encounter with food items for individuals in groups than for solitary searching individuals.

**Figure 4.10 The "selfish herd."** Animals in the center of a herd may be more inaccessible to predators.

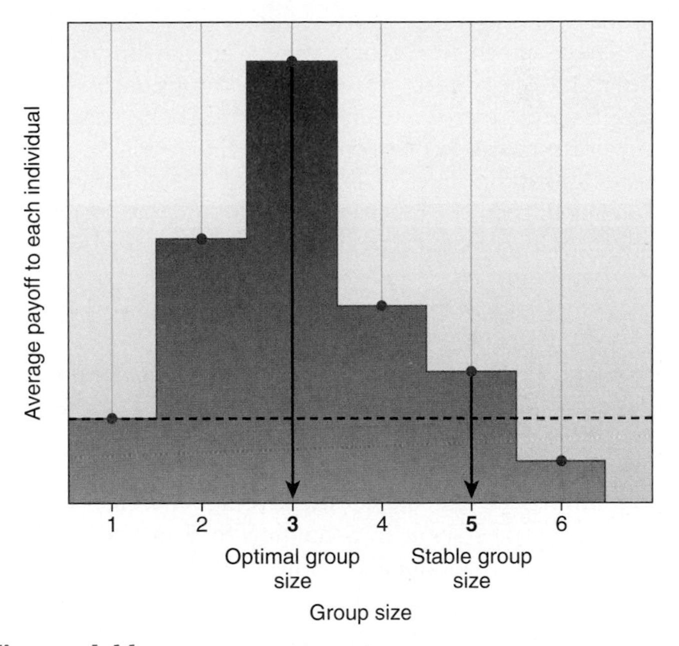


**Figure 4.11 Optimal and stable group size.** The payoff to a group of two or three predators is greater than for a solitary predator. The highest payoff is at a group size of three, which is the optimal size. However, other solitary individuals would benefit from joining the group, and group size increases to five individuals. In a group size of six, any one individual would do better on its own. The stable group size is then five. (After Giraldeau, 2008.)

Strangely, the actual size of predator groups may be different from the optimal size. Suppose that a solitary killer whale hunting for a seal has a given payoff for its efforts (**Figure 4.11**). If another killer whale joins it in the hunt, the seal is more easily caught and payoff for either individual increases. A third killer whale increases the payoff even more. Three whales is the optimal group size, where payoffs to each individual are maximized. When the group size becomes four or five, then the average payoff to each individual decreases, because even though the seal is more easily caught, the food

# Feature Investigation

## Reto Zach Showed How Large Whelks Are the Optimum Prey Size for Crows

At coastal habitats in western Canada, crows commonly feed on whelks. The crows are not strong enough to break open the whelks on their own, so they fly into the air and drop the whelks onto the rocky shore below, where they break open. Reto Zach (1979) made a series of interesting observations about northwestern crows, *Corvus caurinus,* foraging on whelks, *Thais lamellosa.* First, crows selected only large whelks, about 3.8–4.4 cm long. Second, they always seemed to fly about 5 m high before they dropped the whelk. Third, they performed this same 5-m-high flight many times until the whelk broke. A series of questions arose. Why do crows select only large whelks? Why do they fly to about 5 m high before dropping the whelk? Why do they repeat this operation even if the whelk did not break the first time? Why don't they fly higher?

Zach assumed that crows were foraging in an optimal manner, and he set out to test this assumption experimentally. First, he erected a 15-m tower on a rocky beach, complete with a movable platform the height of which could be adjusted. Next, he collected small, medium, and large whelks and dropped them from various heights. At low heights, the whelks would not break unless they were dropped scores of times. At about 5 m, whelks began to need fewer drops to break, but large whelks, being heavier, required the fewest drops of all (**Figure 4.A**). Increasing the height of the drop above 5 m did little to change the results. Furthermore, the chance of a whelk breaking was independent of the number of previous drops. The probability of breakage of large whelks at 5 m was about 1 in 4.

Finally, Zach calculated the average number of kilocalories the crows spent in ascending to 5 m and dropping the whelks, and obtained a value of 0.5 kilocalories. He determined that the calorific value of the large whelk was 2.0 kilocalories, so there was an average net gain of 1.5 kilocalories from foraging on large whelks if the whelk broke on the first drop. The net calorific gain from medium whelks, which require more drops to break, was actually a net loss of 0.3 kilocalories, and foraging on small whelks was even less profitable. Thus, the decision of crows to forage only on large whelks was logical, as was their habit of flying to a height of 5 m to drop them. More than that, Zach's research showed how theories relating to optimal foraging can be experimentally tested.

**HYPOTHESIS** Crows forage optimally on coastal whelks.

**STARTING LOCATION** Beaches of Mondarte Island, British Columbia, Canada.

| | Experimental level | Conceptual level |
|---|---|---|
| 1 | Collect sample of 58 live whelks of the same size as those taken by crows. | Determine size of whelks eaten by crows. |
| 2 | Erect 15 m high pole on beach with a small platform from which whelks could be dropped from various heights. | Determine height required for breaking whelks. |
| 3 | Drop whelks of three size classes: small = 1.6–2.2 cm long, medium = 2.7–3.3 cm, and large = 3.8–4.4 cm. Whelks of each size class were dropped from 2, 3, 4, 5, 6, 7, 8, 10, and 15 m. | Determine breakage height of small, medium, and large whelks. |

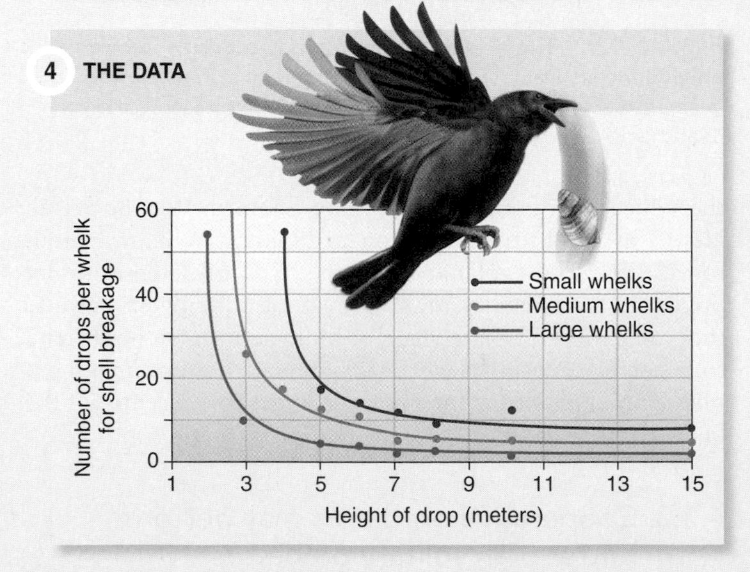


**Figure 4.A** Optimal foraging in coastal Canadian crows.

has to be shared among more individuals. However, living in a group of four or five is still a better option than hunting alone. When the group size is six or more, a whale should do better on its own than in the group. In this example, the stable group size is five individuals. Paradoxically, the optimal group size of three is not equivalent to the stable group size. Interestingly, when Robin Baird and Lawrence Dill (1996) examined group size in killer whales off the coast of British Columbia, they found that the actual group size in whales was three individuals, the same as the optimum group size and less than the stable group size of five. In this case, some group members may oppose the joining of others, though how this opposition becomes organized is unknown. Does one group member pay the costs of resisting invasion by others or is this a group behavior?

**Check Your Understanding**

**4.2** Guppies, *Poecilia reticulata*, are common aquarium fish. They were discovered by the naturalist Robert John Lechmere Guppy in Trinidad in 1866. In their native habitat, guppies live in tighter groups when they occur in streams in which predators are more common. Can you explain this phenomenon?

## 4.3 Foraging Behavior: The Search for Food

Food gathering, or foraging, often involves decisions about whether to remain at a resource patch and look for more food or look for a completely new patch. The analysis of these decisions is often performed in terms of **optimality modeling**, which predicts that an animal should behave in a way that maximizes the difference between the benefits of a behavior and its costs. In this case, the benefits are the nutritional or caloric value of the food items, and the costs are the energetic or caloric costs of movement. When the difference between the energetic benefits of food consumption and the energetic costs of food gathering is maximized, an individual organism is said to be optimizing its foraging behavior. Optimality modeling can also be used to investigate other behavioral issues such as how large a territory to defend. Too small a territory would contain insufficient food, and too large a territory would be too energetically costly to defend. Theoretically, then, each individual has an optimal territory size.

### 4.3.1 Optimal foraging maximizes the difference between the benefits and the costs of food gathering

The idea of optimal foraging proposes that in a given circumstance, an animal seeks to maximize its rate of energy gain (see **Feature Investigation**). The underlying assumption of optimal foraging is that natural selection favors animals that are maximally efficient at propagating their reproductive success and at performing all other functions that serve this function. In this model, the more net energy an individual gains in a limited time, the greater its reproductive success.

In some cases animals do not seem to be maximizing food intake. For example, animals seek also to minimize the risk of predation during food gathering. Therefore, some species may only dart out to take food from time to time. The risk of predation thus has an influence on foraging behavior. Recent studies of optimal foraging suggest that minimizing the risks of being preyed on is part of the cost of foraging. Newer models often take these risks into consideration.

Many species of ants are attacked by parasitic flies that lay eggs on the ant's head. These flies hover around ant colonies or at the nest entrances. Ants are relatively quick themselves but cannot easily avoid these fast-flying parasites. Once the parasite eggs hatch, the larvae bore their way into the ant's head, eventually killing it. In the Tropics, leafcutter ants, *Atta cephalotes*, are subject to attack by the fly *Neodohrniphora curvinervis*. Only the larger workers are attacked, because smaller workers' heads are too small to allow proper development of the fly. When Matthew Orr (1992) compared the size distributions of the leafcutter ants that foraged during the day to those that foraged at night, he discovered a large difference in the size distribution (**Figure 4.12**). Most foraging by larger workers is done at night, when it is too dark for the flies to see properly and their activity ceases. The lesser amount of foraging in the day is performed by smaller ants that are less susceptible to fly attack. Orr extended the activity of *Neodohrniphora* by positioning lights outside the leafcutter nests. In these cases, the foraging activities of the leafcutters were thrown into disarray. Ants were backed up around the nest, many rising on their hind legs to snap at the flies which, with the extended light, continued to try to parasitize the ants.

We learn two things from this study. First, foraging activity is influenced not only by energetic efficiency, but also by the threat of natural enemies. Second, changing the abiotic environment can have drastic effects on animal behavior and the interaction between predator and prey. Many ecologists feel that the increased temperatures of climate change will be accompanied by substantial changes in animal behavior. A case in point is how increased snowfall can subject moose to heavier predation by wolves, as both congregate in relatively snow-free areas (see **Global Insight**).

### 4.3.2 Defending territories has costs and benefits

Many animals, or groups of animals, such as a pride of lions, actively defend a **territory**, a fixed area in which an individual or group excludes other members of its own species, and sometimes other species, by aggressive behavior or territory marking. Territory owners tend to optimize territory size according to the costs and benefits involved. The primary benefit of a territory is that it provides exclusive access to a particular resource, whether it be food, mates, or sheltered nesting sites.

Global Insight

## The North Atlantic Oscillation Affects Snowpack, Wolf Behavior, and Moose Predation Rates

The North Atlantic Oscillation (NAO) is a cyclic variation of the storm track in the North Atlantic Ocean in response to changes in the atmospheric low pressure center off the coast of Iceland and the high pressure center off the Azores, which both shift north or south. A positive NAO index indicates a more northerly positioning of these pressure centers and a strong series of winter storms that push toward northern Europe (**Figure 4.B-i**). The result is warm, wet winters in Europe and the eastern United States. A negative NAO index indicates weaker, more southerly high and low pressure centers that push fewer and less powerful storms toward the Mediterranean. With a negative NAO index, northern Europe and the eastern U.S. are colder and the northern U.S. experiences snowier conditions (**Figure 4.B-ii**).

Eric Post and colleagues (1999) used 40 years of data from Isle Royale, an island in Lake Superior off the coast of Michigan, to show how changes in the NAO affect wolf behavior and moose populations. Heavy snowfall impedes wolf movement, encouraging wolves to travel in large packs. Negative values of the NAO index were correlated with both increased snowfall and increased wolf pack size (**Figure 4.C-i**). Wolves in larger packs killed more moose per day (**Figure 4.C-ii**). In times of heavy snowfall, both wolves and moose traveled more along the shorelines of Lake Superior, where snow was less deep, increasing the likelihood of encounters between the two species. Thus, moose winter mortality was dependent on wolf pack size, which was directly related to winter climate. The correlations did not stop there, however. The current year's density of moose was correlated to the previous year's pack size of wolves (**Figure 4.C-iii**). In turn, the growth of balsam fir and other fir trees was related to moose density in the previous year (**Figure 4.C-iv**). With less browsing by moose, growth in fir trees was noted. Thus, changes in the NAO affect snowfall, wolf behavior, moose density, and fir tree growth. Researchers have discovered that, over long time scales, global warming may strengthen the NAO index, thus lessening the U.S. snowfall and in turn reducing predation on moose and decreasing fir tree growth.

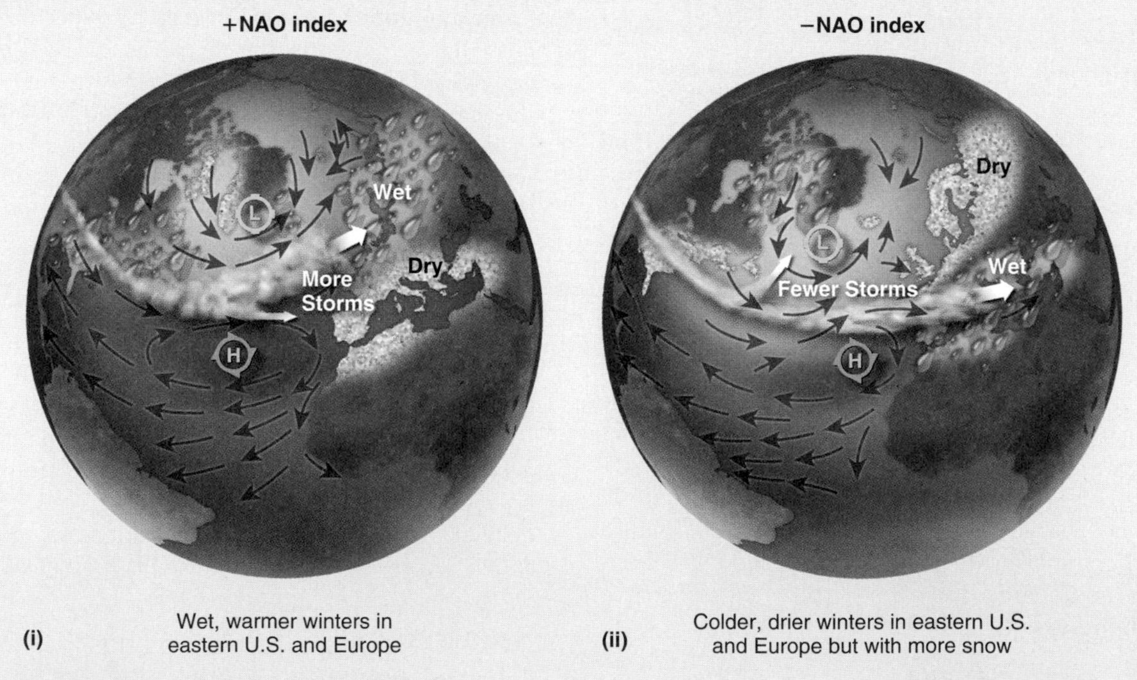


**Figure 4.B The North Atlantic Oscillation.** **(i)** Positive phase: mild, wet North American and northern European winters. **(ii)** Negative phase: cold, drier, but snowy, North American and northern European winters.

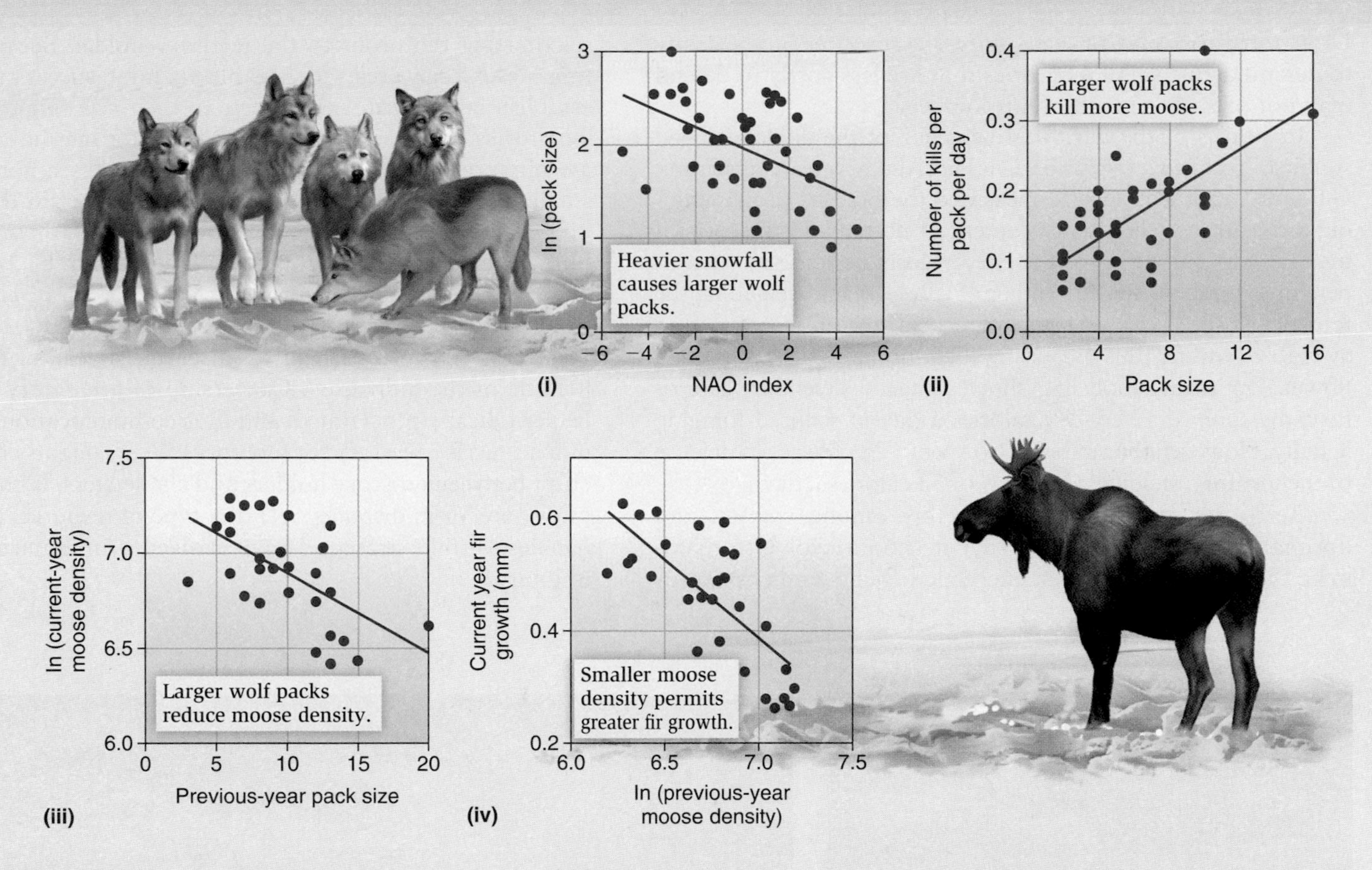


**Figure 4.C Progression of climatic influence on wolf behavior, moose density, and tree growth at Isle Royale, Michigan, 1959–1998.** Relationship of **(i)** wolf pack size with North Atlantic Oscillation (NAO), ln = natural logarithm; **(ii)** kills per day with pack size; **(iii)** current moose density with previous-year pack size; and **(iv)** fir tree growth with previous-year moose density. (After Post et al., 1999.)

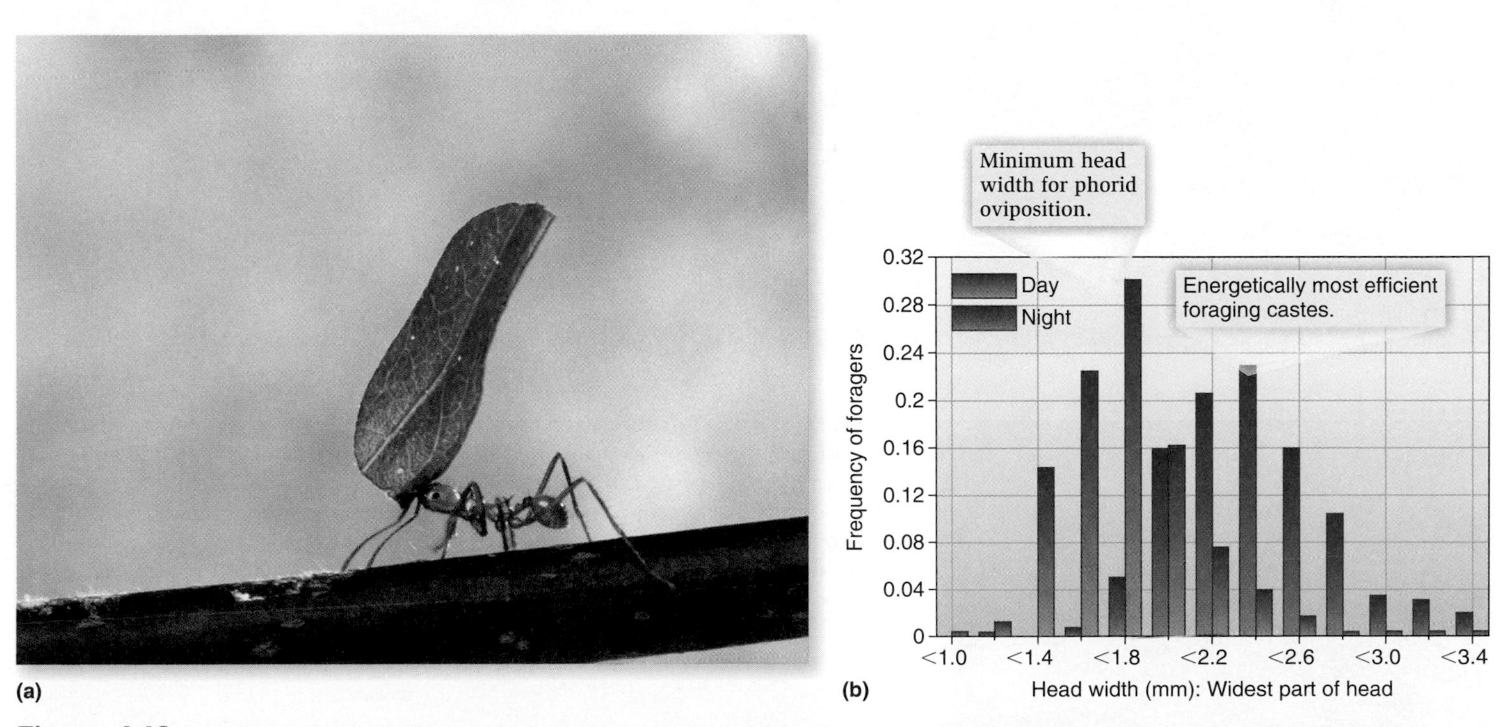


**Figure 4.12 Changes in foraging strategies in the presence of parasites.** **(a)** *Atta cephalotes.* **(b)** The most efficient leafcutter ants, *Atta cephalotes*, are larger individuals. However, larger individuals are restricted to foraging at night because of the activity of parasitic flies during the day. During the day only small ants forage, as these are not subject to the same levels of parasitism. (After Orr, 1992.)

Large territories may provide more of a resource but be costly to defend, while small territories that are less costly to defend may not provide enough of a resource.

In studies of the territorial behavior of the golden-winged sunbird, *Nectarinia reichenowi,* in East Africa, researchers Frank Gill and Larry Wolf (1975) measured the benefits of maintaining a territory as the energy content of nectar, and the costs of maintaining it in terms of the energy costs of activities such as perching, foraging, and defending (**Figure 4.13a**). Defending the territory ensured that other sunbirds did not take nectar from available flowers, thus increasing the amount of nectar in each flower. Optimality modeling showed that in defending a territory, the sunbird saved 780 calories a day in reduced foraging activity. However, the sunbird also spent 728 calories in defense of the territory, yielding a net gain of 52 calories a day.

Territory size differs considerably among species, and optimality modeling predicts that it should evolve to maximize the difference between energetic benefits and costs, thus maximizing the profit to the territory holder. Because cheetahs need large areas to be able to hunt successfully, they establish large territories relative to their size (**Figure 4.13b**). Territories set up solely to defend areas for mating or nesting are often relatively small. For example, male sea lions defend small areas of beach. The preferred areas contain the largest number of females and are controlled by the largest breeding bulls. The size of the territory of some nesting birds, such as gannets, is determined by how far the bird can reach to peck its neighbor without leaving its nest (**Figure 4.13c**).

Territories may be held for a season, a year, or the entire lifetime of the individual. Ownership of a territory needs to be periodically proclaimed, and thus communication between individuals is necessary for territory owners. Fights commonly erupt between territory holders and challengers. In these types of dispute, or in disputes over any type of resource, including females, certain strategies have evolved to minimize the risk of injury.

(a)

(c)

(b)

**Figure 4.13 Territory sizes differ in animals. (a)** The golden-winged sunbird of East Africa, *Nectarinia reichenowi,* has a medium territory size that is dependent on the number of flowers it can defend and from which it can obtain resources. **(b)** Cheetahs, *Acinonyx jubatus,* hunt over large areas and can have extensive territories. This male is urine-marking part of his territory in the southern Serengeti, near Ndutu, Tanzania. **(c)** Nesting gannets, *Morus bassanus,* have much smaller territories, in which each bird is just beyond the pecking range of its neighbor.

## 4.3.3 Game theory establishes whether individuals fight for resources or flee from opponents

John Maynard Smith (1976, 1982) investigated fighting behavior in animals by considering contests in which there are different sorts of strategies, modifying the prisoner's dilemma game. Imagine a game in which individuals of the same species engage in contests over resources. Some individuals, the "Hawks," always fight to win, risking injury to themselves in the process, while others, the "Doves," simply perform displays and always defer to Hawks in serious fights (**Figure 4.14**).

Maynard Smith arbitrarily gave the winner of a contest +50 and the loser 0. The cost of serious injury is −100 because the injured player may not be able to compete again for a long time, and the cost of spending time in a display is −10. It is assumed that Hawks and Doves reproduce in proportion to their scores. What happens when Hawks fight other Hawks or Doves, or when Doves fight Doves? We can analyze such a contest by constructing a two-by-two matrix with the average rewards for the four possible types of encounter (**Table 4.1**). Consider what happens if all individuals in the population behave as Hawks: half the population wins and gets a reward of 50, while half the population loses and gets −100. The average of +50 + (−100) is −25. In this situation, any individual playing Dove would do better, because when a Dove meets a Hawk it gets 0, which is not very good but still better than −25. Therefore, Hawk is not an evolutionarily stable strategy. As we mentioned earlier, an ESS is not necessarily the "best" strategy, it is merely the most resistant to invasion by individuals using alternative strategies or by mutants that arise within the population. In a population of all Doves, every contest is between a Dove and another Dove, and the score, on average is ½(50 − 10) + ½(−10) = +15. Even though the average score is positive, Dove is not an ESS either. In this population, any individual Hawk would do very well, and the Hawk strategy would soon spread, because when a Hawk meets a Dove it gets +50.

**Figure 4.14** **Hawk and Dove strategies.** In these dueling meerkats, the individual on the left is escalating the fight, behaving in a Hawk-like manner, while the individual on the right is not fighting and is in a submissive posture, a Dove-like strategy.

**Table 4.1 Average rewards to the attacker in the game between Hawk and Dove.**

| | Opponent | |
|---|---|---|
| **Attacker** | **Hawk** | **Dove** |
| Hawk | ½(50) + ½(−100) = −25 | +50 |
| Dove | 0 | ½(50 − 10) + ½(−10) = +15 |
| Costs and Rewards: Winner +50 | | Injury −100 |
| | Loser 0 | Display −10 |

While a population of pure Hawks or pure Doves is not stable, a mixture of Hawks and Doves could be stable. The stable equilibrium is the point at which the average reward for a Hawk is equal to the average reward for a Dove. For the rewards we have specified, the stable mixture can be calculated as follows. If $h$ is the proportion of Hawks in the population, the proportion of Doves is $(1 - h)$. The average reward, $H$, for a Hawk is the reward for each type of fight multiplied by the probability of meeting each type of contestant.

Therefore, the average reward for a Hawk will be:

$$H = -25h + 50(1 - h)$$

For Doves the average reward will be:

$$D = 0h + 15(1 - h)$$

At the stable equilibrium, $H$ is equal to $D$. When $H = D$, the proportion of Hawks $(h) = 7/12$, and the proportion of Doves $(1 - h) = 5/12$.

This stable equilibrium or ESS could be achieved in two ways. First, the population could consist of 7/12 individuals who always played Hawk and 5/12 who always played Dove. Alternatively, the population could consist of individuals who all adopted a mixed strategy, playing Hawk in 7/12 contests and Dove in the other 5/12.

Despite the simplicity of these models, some important conclusions can be drawn from them:

1. Fighting strategy is frequency dependent: It depends on what other animals are doing. A Hawk strategy is good for an individual in a population of Doves but not in one of Hawks.
2. The ESS is often a mixture of different strategy types, like Hawk and Dove. Such mixtures of displays and fighting are commonly observed in nature.
3. The ESS is dependent on the values of rewards. If the rewards are changed, the proportions of Hawks and Doves will change. However, as long as the cost of injury exceeds the benefit of winning, then all Hawk or all Dove will never be an ESS.

4. The frequency of Hawk behavior increases as the payoff increases. For males, one of the biggest payoffs is the opportunity to mate with a female, because failure to do so is equivalent to genetic death. Fights over females are therefore commonly severe, sometimes fatal, and Hawk-like strategies dominate.

In the real world, animals also adjust their strategies according to the vigor of their opponents. It would be pointless to adopt a Hawk strategy against a bigger opponent, even for a territory owner. Players may use a strategy that follows the rule "If larger, behave like a Hawk; if smaller, behave like a Dove." Age may also enter into the equation. Although a young animal and an old animal may be equal in size, young animals may give up sooner in a fight because fighting incurs risks, and a young animal risks a larger proportion of its future reproductive life. For an old male, each contest could represent his last chance to mate. One thing remains clear, male fights over females are common and severe. Do females ever fight over males and if not, why not? To answer these questions we turn to an examination of mating systems.

## Check Your Understanding

**4.3** Urban-dwelling American crows, *Corvus brachyrhynchos,* in California drop walnuts from the air to break them. Unlike northwestern crows dropping whelks, American crows sequentially reduce the height from which they drop walnuts, from 3 m on the first drop to about 1.5 m on the fifth drop. They also drop walnuts from lower heights in the presence of other crows. Why is their behavior different from that of the northwestern crows?

## 4.4 Mating Systems Range from Monogamous to Polygamous

In nature, males produce millions of sperm, while females produce far fewer eggs. It would seem that most males are superfluous, because one male could easily fertilize all the females in a local area. If one male can fertilize many females, why are there few species with a sex ratio of, say, 1 male : 20 females? The sex ratio is more often about 1:1, because of a phenomenon known as **Fisher's principle**, after the geneticist Ronald Fisher, who proposed it in 1930. If a population contained 20 females for every male, then a parent whose offspring were exclusively males could expect to have 20 times the number of $F_2$ progeny produced by a parent with the same number of female offspring. Under such conditions, natural selection would favor production of males, and males would become prevalent in the population. At such a point, females would be at a premium, and natural selection would favor their production. Such constraints operate on the numbers of both male and female offspring, keeping the sex ratio at about 1:1.

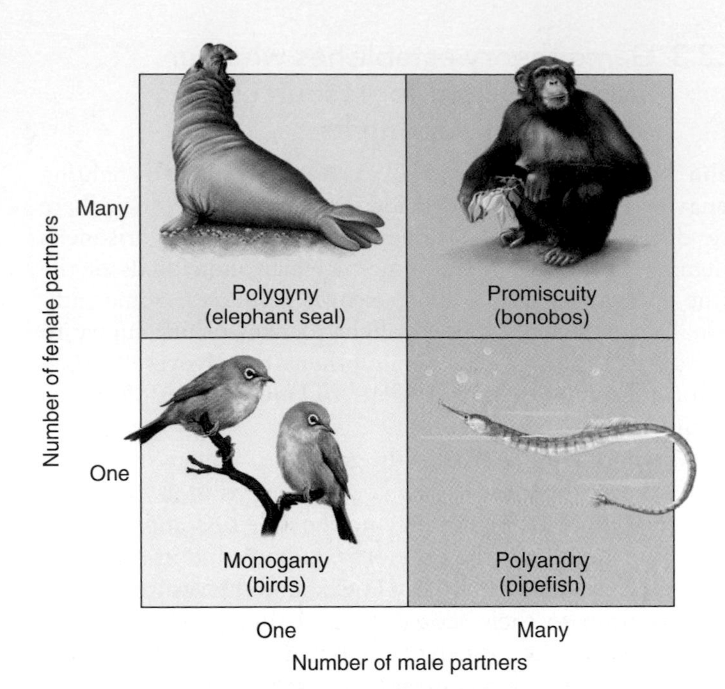


**Figure 4.15 The four different mating strategies of males and females.**

Even though the sex ratio is fairly even in most species, that doesn't mean that one female always mates with one male or vice versa. Four different types of mating systems occur (**Figure 4.15**). In some species, mating is **promiscuous**, with each female and each male mating with multiple partners within a breeding season. In **monogamy**, each individual mates exclusively with one partner over at least a single breeding cycle and sometimes for longer. In contrast, **polygamy** is a system in which either males or females mate with more than one partner in a breeding season. There are two types of polygamy. In **polygyny** (Greek for many females), one male mates with more than one female, but females mate only with one male. In **polyandry** (Greek for many males), one female mates with several males, but males mate with only one female.

In this section, we examine the characteristics of these mating systems and explore the role of sexual selection, a type of natural selection in which competition for mates drives the evolution of certain traits.

### 4.4.1 In promiscuous mating systems, each male or female may mate with multiple partners

Chimpanzees and bonobos are promiscuous; each male mates with many females and vice versa. In these cases sexual activity alleviates conflict. Intertidal and terrestrial mollusks are also usually promiscuous. Individuals copulate with several partners and eggs are fertilized with sperm from several different individuals. These mollusks are slow moving and they risk desiccation when searching for a mate. The risk of not finding a mate is believed to promote promiscuity. Similarly, humpback whales, *Megaptera novaeangliae,* do not live in stable groups and are widely scattered throughout an extensive breeding range in winter. Known mature female humpbacks have been seen to associate with multiple partners over a single winter, and genetic studies suggest a promiscuous mating system.

### 4.4.2 In monogamous mating systems, males and females are paired for at least one reproductive season

In monogamy, each individual mates exclusively with one partner over at least one breeding season and sometimes for longer. Monogamy occurs in fish, amphibians, reptiles, and mammals, but it is commonly observed only in birds, some rodents, and a few primate species. Males and females are generally similar in body size and appearance (**Figure 4.16a**). Several hypotheses explain the existence of monogamy. The first is the **mate-guarding hypothesis**, which theorizes that a male stays with a female to prevent her from being fertilized by other males. Such a strategy may be advantageous when receptive females are widely scattered and difficult to find. Monogamous male dunnocks, *Prunella modularis,* typically guard their female mates for 75% of the time they are sexually receptive. During this time other males have no access to the guarded female. When Nick Davies (1992) experimentally removed the dunnock-guarding males, intruder males quickly gained access to the females.

The **male assistance hypothesis** maintains that males remain with females to help them rear their offspring. Monogamy is common among birds, about 70% of which are monogamous; that is, the pairings remain intact during at least one breeding season. According to this hypothesis, monogamy is prevalent in birds because eggs and chicks take a considerable amount of parental care. Most eggs need to be incubated continuously if they are to hatch, and chicks require almost continual feeding. It is therefore in the male's best interest to help raise his young, because he would have few surviving offspring if he did not.

The **female-enforced monogamy hypothesis** suggests that females stop their male partners from being polygynous. Male and female burying beetles, *Nicrophorus defodiens,* work together to bury small, dead animals, which will provide a food resource for their developing offspring. Males will release pheromones to attract other females to the site. However, while an additional female might increase the male's fitness, the additional developing offspring might compete with the offspring of the first female, decreasing her fitness. As a result, on smelling these pheromones, the first female will physically interfere with the male's attempts at signaling, preserving the monogamous relationship.

### 4.4.3 In polygynous mating systems, one male mates with many females

In polygyny, one male mates with more than one female in a single breeding season, but females mate only with one male. Physiological constraints often dictate that female organisms must care for the young, at least in many organisms with internal fertilization such as mammals, birds, and some fish. Female parents are most often left "holding the baby"—the fetus develops inside the female's body, and the offspring are fed and cared for by her after birth. Because of these constraints, males are able to desert and mate with more females. Polygynous systems are therefore associated with uniparental care of young, with males contributing little. In species with male-male competition, males are often substantially larger than females, a phenomenon called sexual dimorphism (**Figure 4.16b**). Sexual maturity is often delayed in males that

(a) (b) (c)

**Figure 4.16 Sexual dimorphism in body size and mating system.** **(a)** In monogamous species, such as these Manchurian cranes, *Grus japonensis*, males and females appear very similar. **(b)** In polygynous species, like elk, *Cervus elaphus,* males are bigger than females and have large horns with which they engage in combat over females. **(c)** In polyandrous species, females are usually bigger, as with these golden silk spiders, *Nephila clavipes.*

fight because of the considerable time it takes to reach a sufficiently large size to compete for females.

Polygyny is influenced by the spatial or temporal distribution of breeding females. In cases where all females are sexually receptive within the same narrow period of time, little opportunity exists for a male to garner all the females. Where female reproductive receptivity is spread out over weeks or months, there is much more opportunity for males to mate with more than one female. For example, females of the common toad, *Bufo bufo,* all lay their eggs within a week, and males generally have time to mate with only one female. In contrast, female bullfrogs, *Rana catesbeiana,* have a breeding season of several weeks, and males may mate with as many as six females in a season. Several types of polygyny are known.

### Resource-based polygyny

Where some critical resource is patchily distributed and in short supply, certain males may dominate the resource and breed with more than one visiting female. In the lark bunting, *Calamospiza melanocorys,* which mates in North American grasslands, males arrive at the grasslands first, compete for territories, and then perform displays with special courtship flight patterns and songs to attract females. The major source of nestling death in this species is overheating from too much exposure to the sun. Prime territories are therefore those with abundant shade, and some males with shaded territories attract two females, even though the second female can expect no help from the male in the process of rearing young. Males in some exposed territories remain bachelors for the season. From the dominant male's point of view, resource-based polygyny is advantageous; from the female's point of view, there are costs. Although by choosing dominant males, a female may be gaining access to good resources, she may also have to share these resources with other females.

### Harem mating structures

Sometimes males defend a group of females without commanding a resource-based territory (**Figure 4.17**). This pattern is more common when females naturally congregate in groups or herds, perhaps to avoid predation, and is seen in equids (horses and zebras), some deer, lions, and seals. Usually the largest and strongest males command most of the matings, but being a harem master is usually so exhausting that males may only manage to remain the dominant male for a year or two. As we noted in the beginning of the chapter, in some monkeys the new harem master will kill existing young in order for the females to become sexually receptive sooner. Male lions also practice infanticide. Eighty percent of cubs generally die before reaching the age of 1 year old. In other species such as baboons and horses, males will harass females until they miscarry.

**Figure 4.17** Male elk defending harem of females.

### Communal courting

Polygynous mating can occur where neither resources nor harems are defended. In some instances, particularly in birds and mammals, males display in designated communal courting areas called **leks** (**Figure 4.18**). Females come to these areas specifically to find a mate, and they choose a prospective mate after the males have performed elaborate displays. Most females seek to mate with the best male, so a few of the flashiest males perform the vast majority of the matings. At a lek of the white-bearded manakin, *Manacus manacus,* of South America, one male accounted for 75 percent of the 438 matings where as many as 10 males were present. A second male mated 56 times (13 percent of matings), while six others mated only a total of 10 times.

## 4.4.4 In polyandrous mating systems, one female mates with many males

In most systems in which one individual mates with more than one individual of the opposite sex, the polygamous sex is the male. The opposite condition, polyandry, in which one female mates with several males, is more rare, and is practiced by some species of birds, fish, insects, and mammals. For example, honeybees are polyandrous because the queen often mates with multiple males. The males soon die off but the queen uses their stored sperm to fertilize her eggs. In polyandrous species, sexual dimorphism is present, with the females being the larger of the sexes (see **Figure 4.16c**). In the Arctic tundra, the summer season is short but very productive, providing a bonanza of insect food for two months. The productivity of the breeding grounds of the spotted sandpiper, *Actitis macularia,* is so high that the females can lay five clutches of 20 eggs in 40 days. Their reproductive success is limited not by food but by the number of males they can find to incubate

**Figure 4.18 Some male birds and mammals congregate in communal courting grounds called leks.** Black grouse, *Tetrao tetrix*, congregate on a moorland lek in Scotland in April. Females visit the leks, and males display to them.

**ECOLOGICAL INQUIRY**

On most leks, a few males command most of the matings, so why do many males congregate there instead of displaying on their own territories?

the eggs, and females compete for males, defending territories where the males sit.

Polyandry is also seen in some species where egg predation is high and males are needed to guard the nests. Male jacanas, a tropical bird, build a floating nest and guard and incubate the eggs. Female jacanas court males and confront other females to control large territories containing the nests of several males. Females are 50–60% heavier than males and have more developed spurs on their wings than males because the spurs are used in fighting. In the pipefish, *Syngnathus typhle,* males have brood pouches that shelter eggs and provide them with a supply of water rich in oxygen and nutrients. Females produce enough eggs to fill the brood pouches of two males, which are effectively in short supply, and may mate with more than one male.

## 4.4.5 Sexual selection involves mate choice and mate competition

**Sexual selection** is selection that promotes traits that will increase an organism's mating success. It can take two forms: intersexual selection and intrasexual selection. In intersexual selection, members of one sex, usually females, choose mates based on particular attractive characteristics such as color of plumage or sound of courtship song. In intrasexual selection, members of one sex, usually males, compete over partners, and the winner performs most of the matings. Let's explore each of these in a little more detail.

**Figure 4.19 Female choice of males based on nuptial gifts.** A male hangingfly presents a nuptial gift, a small moth, to a female.

### *Intersexual selection*

Females choose their prospective partners based on a variety of behaviors and traits. Female hangingflies, *Hylobittacus* spp., demand a nuptial gift of a food package, an insect prey item that the male has caught (**Figure 4.19**). Such a nutrient-rich gift may permit females to produce more eggs. Alternatively, the bigger the gift, the longer it takes the female to eat it and the longer the male can copulate with her and the more sperm he can transfer. Females will not mate with males that do not offer such a package. A review of nuptial gift-giving in insects by Karim Vahed (1998) found little evidence for the increased egg production hypothesis and more evidence for the increased sperm transfer hypothesis.

Males may also have parenting skills that females desire. Among 15-spined sticklebacks, *Spinachia spinachia,* males perform nest guarding, cleaning, and fanning of the offspring. Males display their parental skills through body shakes during courtship and females prefer to mate with males that shake their bodies the most energetically, apparently using this cue to assess the quality of the male as a potential father. Female gray partridges, *Perdix perdix,* choose mates that appear to be vigilant toward predators. This allows females more time to feed. Vigilance is indicated by a stereotypic position with raised head, stretched neck, and wings tucked in against the body. Some males can allocate up to 65% of their time in vigilance! Vigilant males also provide protection against intrusions from other males.

Often females choose mates without the offering of obvious material benefits or parenting skills, and instead make their choices based on plumage color or courtship display. In 1915 Ronald Fisher termed this phenomenon **runaway selection.**

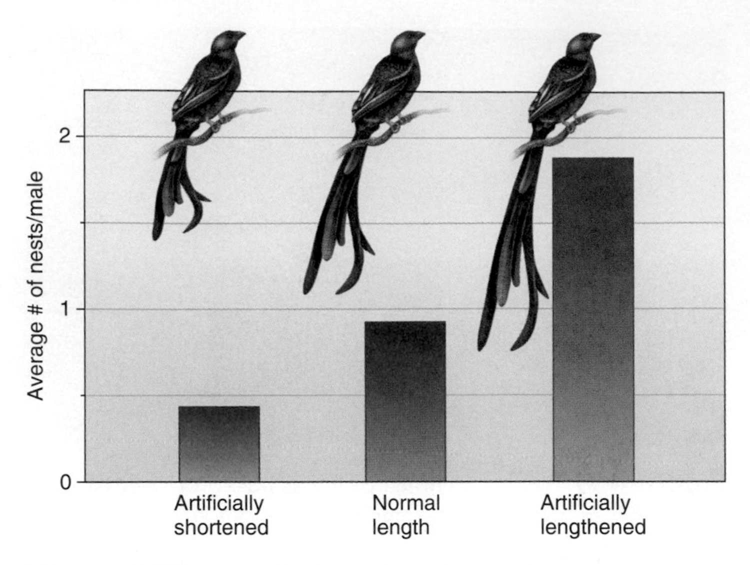


**Figure 4.20 Female choice based on male appearance.** Males with artificially lengthened tails mate with more females, and therefore have more nests, than males with a normal or an artificially shortened tail. (After Andersson, 1982.)

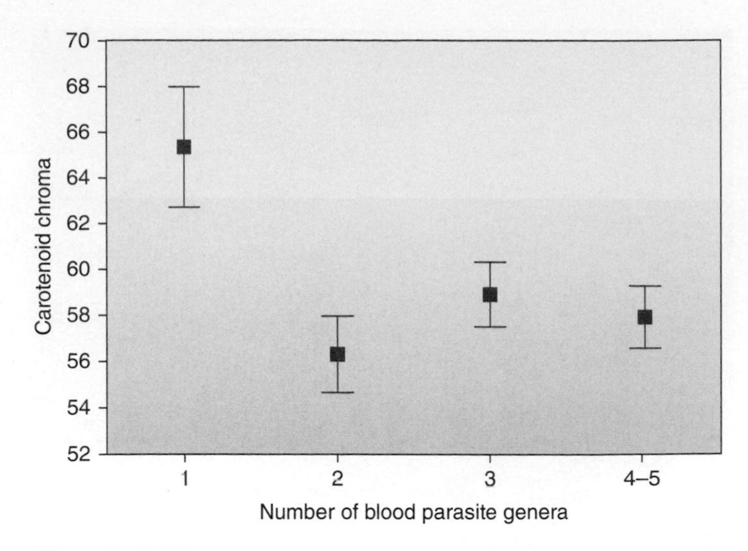


**Figure 4.21 Carotenoid chroma (plumage brightness) of blue tits in relation to blood parasite richness.** Bars indicate SE. (After del Cerro et al., 2010.)

A peahen may choose her peacock mate on the basis of his elaborate tail feathers. Her male offspring are likely to inherit this trait and be desirable mates themselves. The male African long-tailed widowbird, *Euplectes progne,* has long tail feathers that he displays to females via aerial flights. Researcher Malte Andersson (1982) experimentally shortened the tails of some birds by clipping their tail feathers, and he lengthened the tails of others by taking the clippings and sticking them onto other birds with superglue. Males with experimentally lengthened tails attracted four times as many females as males with shortened tails and they fathered more clutches of eggs (**Figure 4.20**). However, at some stage runaway selection must stop, otherwise females with extreme preferences would fail to find a mate.

Some researchers have suggested that ornaments such as excessively long tail feathers function as a sign of an individual's genetic quality, because the bearer must be able to afford this energetically costly trait. This hypothesis is called the **handicap principle.** However, other important benefits may be associated with plumage quality or quantity. Bright colors are often caused by pigments called carotenoids that help stimulate the immune system to fight diseases. Sara del Cerro and colleagues (2010) examined the brightness or chroma of the yellow color in the breast plumage of blue tits, *Cyanistes caerulens,* in Spain. More brightly colored birds were infected by just one genus of blood parasites, whereas duller colored birds had multiple genera of blood parasites, including *Trypanosoma* and *Plasmodium* (**Figure 4.21**).

In zebra finches and red junglefowl, colorful plumage has been associated with heightened resistance to disease, suggesting that females that choose such males are choosing genetically healthier mates. Marlene Zuk and colleagues (1990) infected a group of junglefowl, *Gallus gallus,* chicks by squirting saline containing worm eggs into their mouths. A control group received saline only. All chicks were allowed to grow and mature and their secondary sexual characteristics developed. These included size and color of male feathers and comb, and overall body size. Next, females were allowed to choose between a pair of tethered treatment and control males. The females preferred to mate with the control birds by a margin of 2:1. They were choosing based on feather and comb size and color, all of which were smaller or subdued in experimental birds.

### *Intrasexual selection*

In many species, females do not actively choose their preferred mate; instead, they mate with competitively superior males. In such cases, dominance is determined by fighting or by ritualized sparring. In the southern elephant seal, *Mirounga leonine,* females haul up onto isolated beaches to give birth and gain safe haven for their pups from marine and terrestrial predators. Following birth, they are ready to mate. In this situation, dominant males are able to command a substantial group and constantly lumber across the beach to fight other males and defend their harem. Over the course of many generations, such competition results in an increased body size and sexual dimorphism (**Figure 4.22**). Fewer than one-third of the males mate during the mating season. Large size also allows males to store more energy reserves, increasing their endurance and permitting them to be dominant over an entire season. On one beach, 8 males fertilized 348 females. Many other males died before having the opportunity to mate. If this scenario is taken one step further, we might expect the weaponry of males (horns, antlers, or teeth) to increase relative to body size in strongly polygynous species relative to monogamous ones.

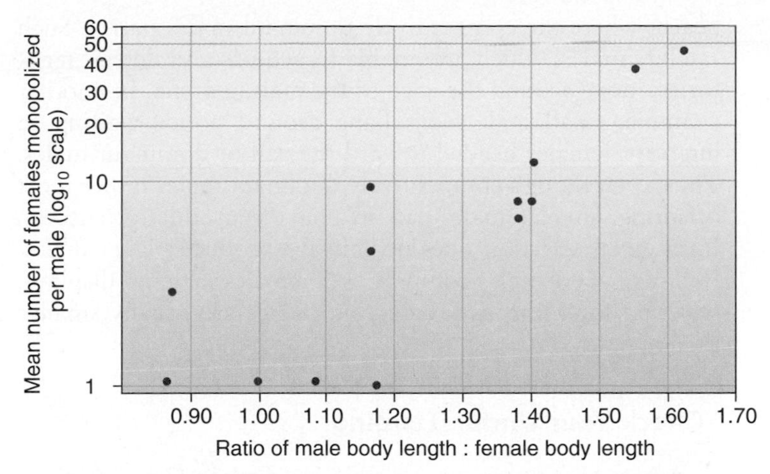

**Figure 4.22 Large male size correlates with success in mating.** **(a)** In southern elephant seals, the larger the male in relation to female size, the greater the number of females that can be monopolized and mated with. (After Alexander et al., 1979.) **(b)** These male elephant seals are fighting to maintain control of their female harems.

**ECOLOGICAL INQUIRY**

During fights between males, some pups are crushed as the males lumber across the beach. Why aren't the males careful to avoid this tragedy?

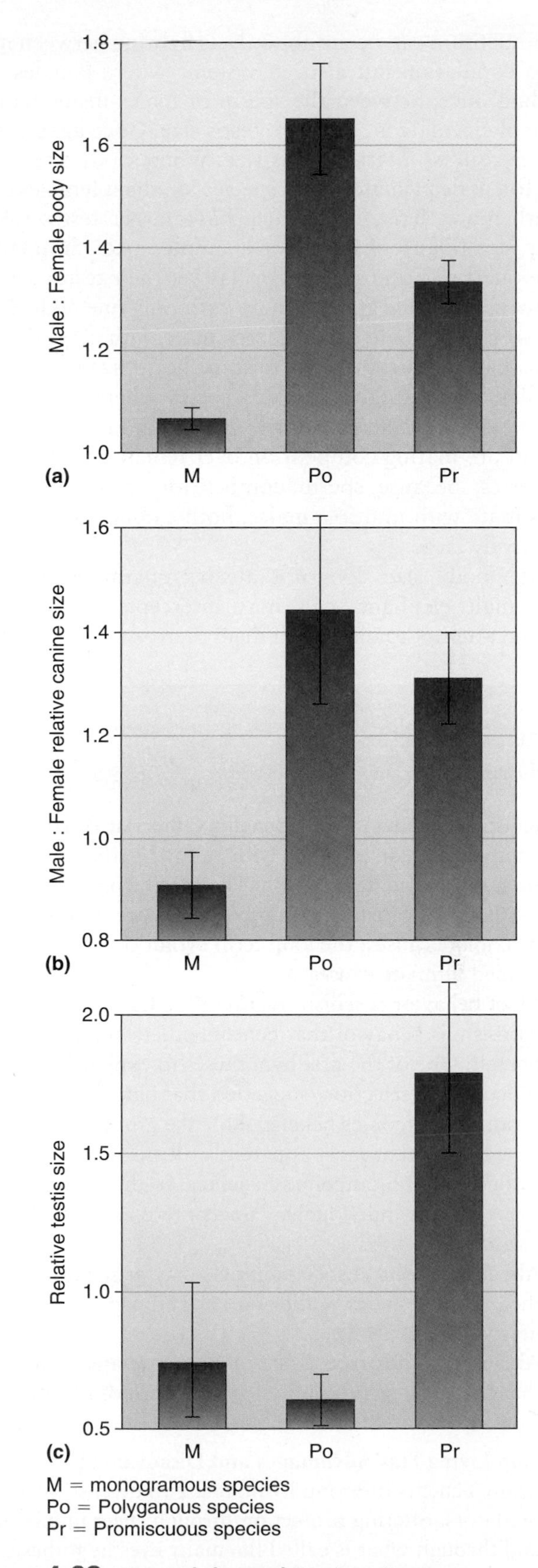


**Figure 4.23 Sexual dimorphism in primates.** **(a)** Body size dimorphism (adult male weight divided by adult female weight). **(b)** Relative canine size dimorphism. **(c)** Relative testis size dimorphism. Bars indicate standard errors. (From Harvey and Bradbury, 1991.)

Such is the case for primates where the primary weaponry is canine teeth. In primates, males are larger than females (**Figure 4.23a**), and larger body size is accompanied by increased canine size. Even when larger body size is taken into account, or controlled for, relative canine size is greatest in polygamous species where single males fight for control of females or in promiscuous species where females mate with multiple males (**Figure 4.23b**).

Competition can occur not only as fighting between males prior to copulation but also, in systems where females mate more than once, between the sperm of males inside females. Volume of ejaculate is related to testis size. Once again, larger body size results in larger testis size. Where body size is controlled for, male primates from species in which females copulate with more than one partner, have larger testes relative to body size (**Figure 4.23c**). For example, male chimpanzees are one-quarter of the size of gorillas yet their testes are four times heavier. Female gorillas mate with only one male during estrus, whereas female chimpanzees mate with several males. Humans have about the same ratio of body size to testis size as gorillas. Interestingly, primate species where single males mate with multiple females have relatively large canines to engage in pre-mating competition over females, but relatively small testes, because sperm competition is absent. Where females mate with multiple males, both canine and testis size are relatively large.

Large body size does not always guarantee paternity. Smaller male elephant seals may intercept females in the ocean and attempt to mate with them there, rather than on the beach, where the competitively dominant males patrol. Such satellite males, which are unable to acquire and defend territories, move around the edge of the mating arena. In another example, small male frogs hang around ponds waiting to intercept females headed toward the call of dominant males. Thus, even though competitively dominant males father most offspring, smaller males can still have reproductive success. Intrasexual selection does not always produce bigger males. In birds, where males compete over females in aerial displays, selection for agility is favored and males are usually smaller than females.

### Check Your Understanding

**4.4** In Malte Andersson's experiments (see Figure 4.20), the performance of male long-tailed widowbirds was compared between males with tails clipped, tails lengthened by adding tail feathers, and unmanipulated males. What do you think the best type of control would be?

## SUMMARY

**4.1 Altruism Is Behavior That Benefits Others at Personal Cost**

- Animals use complex behavior in food gathering or foraging and mate selection. The specific physiological mechanisms causing certain behaviors are termed proximate causes; the long-term evolutionary reasons are termed ultimate causes.
- Most behavior is selfish (Figures 4.1, 4.2).
- Altruism is behavior that benefits others at a cost to oneself. One of the first hypotheses to explain altruism, called group selection, suggested that natural selection produced outcomes beneficial for the group. Biologists now believe that most apparently altruistic acts are often associated with outcomes beneficial to those most closely related to the individual, a concept termed kin selection (Figures 4.3–4.5).
- Altruism among eusocial animals may arise partly from the unique genetics of the animals and partly from lifestyle (Figure 4.6).
- Altruism is known to exist among nonrelated individuals that live in close proximity for long periods of time, such as vampire bats (Figure 4.7).

**4.2 Group Living Has Advantages and Disadvantages**

- Many benefits of group living relate to defense against predators, offering protection through sheer numbers, and through what is called the many eyes hypothesis or the geometry of the selfish herd (Figures 4.8–4.11).

**4.3 Foraging Behavior: The Search for Food**

- The theory of optimal foraging assumes that animals modify their behavior to maximize the difference between their energy uptake and energy expenditure (Figures 4.A, 4.12).
- Animal behavior may change in response to global climate change (Figures 4.B, 4.C).
- The size of a territory, a fixed area in which an individual or group excludes other members of its own species, tends to be optimized according to the costs and benefits involved (Figure 4.13).
- Fighting strategies may be varied. For example, organisms may adopt Hawk or Dove strategies (Figure 4.14, Table 4.1).

**4.4 Mating Systems Range from Monogamous to Polygamous**

- There are four types of mating systems among animals: promiscuity, monogamy, polygyny, and polyandry (Figures 4.15, 4.16).
- The risk of not finding a mate is thought to promote promiscuity. Promiscuity is found in some mollusks and animals.
- Monogamous mating systems may be the result of mate guarding, male assistance, or female-enforced monogamy.
- Polygynous mating can occur where males dominate a resource, when males defend groups of females (harems),

or where males display in common courting areas called leks (Figure 4.17, 4.18).

- Polyandrous mating occurs where males provide most of the incubation or guarding of eggs or offspring, and some females compete for social rank.
- Sexual selection takes two forms: intersexual selection, in which the female chooses a mate based on particular characteristics, or intrasexual selection, in which males compete with one another for the opportunity to mate with a female (Figures 4.19–4.23).

## TEST YOURSELF

1. In a diploid organism, the coefficient of relatedness, *r*, between father and daughter is:
   - a. 0
   - b. 0.125
   - c. 0.25
   - d. 0.5
   - e. 0.75
2. In haplodiploid organisms, fathers are related to sons by:
   - a. 0
   - b. 0.125
   - c. 0.25
   - d. 0.5
   - e. 0.75
3. Which organisms do not have a haplodiploid mating system?
   - a. Ants
   - b. Termites
   - c. Bees
   - d. Wasps
4. In a polygynous mating system:
   - a. One male mates with one female
   - b. One female mates with many different males
   - c. One male mates with many different females
   - d. Many different females mate with many different males
5. Selection that lowers an individual's own fitness but enhances that of a relative is known as:
   - a. Kin selection
   - b. Inclusive fitness
   - c. Hamilton's rule
   - d. Altruism
   - e. Haplodiploidy
6. The many eyes hypothesis suggests prey flock because:
   - a. Predators have difficulty reaching prey in the center of the flock
   - b. Bigger flocks make it more likely that one individual will spot a predator
   - c. Individuals compete less in a social group
   - d. Individuals can move with the expenditure of less energy
   - e. Parasite transmission rates are decreased
7. Hamilton proposed that an altruistic gene will be favored by natural selection when:
   - a. $Br - C = 0$
   - b. $r = C/B$
   - c. $B = C/r$
   - d. $rB > C$
   - e. $C > Br$
8. In meiosis, any gene has what percentage chance of entering an egg?
   - a. 0
   - b. 25
   - c. 50
   - d. 75
   - e. 100
9. In naked mole rats, mothers are related to daughters by what percentage?
   - a. 0
   - b. 25
   - c. 50
   - d. 75
   - e. 100
10. Female peahens select male peacocks based on:
    - a. Mate competition between males
    - b. Suitability of males in guarding the eggs
    - c. Tail feather adornments
    - d. The size of the nuptial gift
    - e. Performance in a lek

## CONCEPTUAL QUESTIONS

1. Why is individual selection more likely than group selection?
2. What are the behavioral advantages and disadvantages of living in groups?
3. Why do some females accept a monogamous relationship with one male when they could mate with many different males?
4. What is the average reward to a Hawk in a Hawk versus Hawk confrontation, where the reward to the winner is +30 and all other rewards are the same as in Table 4.1?

## DATA ANALYSIS

1. Shore crabs, *Carcinus maenas,* eat mussels. The biggest mussels contain the most energy, yet *Carcinus* prefers eating medium-sized mussels as shown in the figure (after Elner and Hughes, 1978). Assume *Carcinus* are foraging optimally and discuss why they might prefer medium-sized mussels over small- or large-sized individuals.

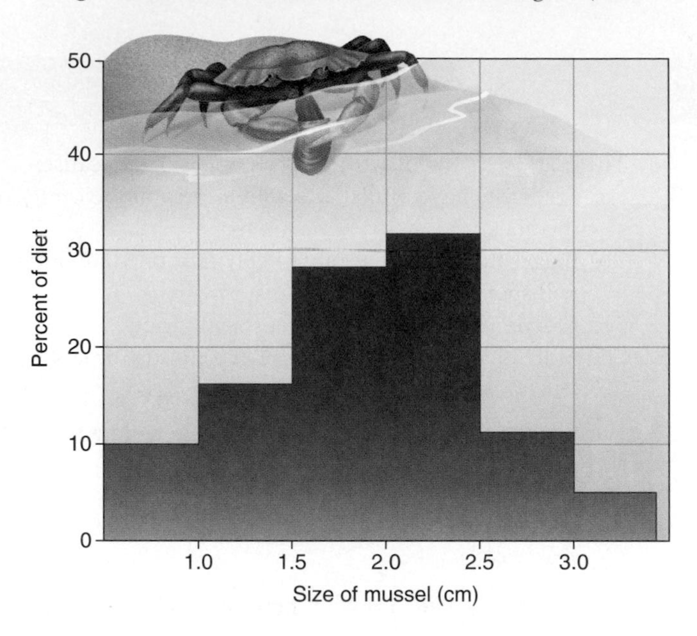


| | Numbers of species of caterpillars | |
|---|---|---|
| Dispersion Pattern | Warning Coloration | Camouflage |
| Solitary | 11 | 44 |
| Large family groups | 9 | 0 |

2. A study by Paul Harvey and colleagues (1983) examined the dispersion patterns of 64 different species of caterpillars (see table). Explain the results in terms of kin selection.
3. Female Belding's ground squirrels frequently deliver an alarm call when a predator approaches. When no relatives are present, alarm calls are produced less than 20% of the time when predators approach. When sisters or daughters are present, alarm calls are made in nearly 75% of the instances predators approach, and the frequency is about 70% when only daughters are present. Explain these results.

# Physiological Ecology

There exists a long history of bird watching in England and of collecting birds, banding, and recapturing them. During these captures and recaptures, birds are routinely weighed and measured. Such procedures have been going on since the 1960s. Incredibly, recent analyses show that the size of many British birds has been decreasing steadily at about 0.1% per year over the past 30 years. Why is this happening? Yoram Yom-Tov and colleagues (2006) analyzed bird size and weight data collected between 1968 and 2003 for 14 species of small passerine or perching birds. Species such as blue tits, *Cyanistes caeruleus,* great tits, *Parus major,* blackcaps, *Silvia atricapilla,* bullfinches, *Pyrrhula pyrrhula,* and dunnocks, *Prunella modularis,* showed steady decreases in weight, whereas others such as reed warblers, *Acrocephalus scirpaceus,* showed size decreases only over the past 5–10 years. Most of the affected species were insectivorous, but two were largely vegetarians. Some nested in the open, some nested in cavities. This variation in life histories lent support to the belief that temperature changes were behind the size changes. As we will see later, temperature affects body size. In animals, large volume : surface area ratios mean greater heat production capacity and less heat dissipating surface. Increased local temperatures would mean smaller birds could survive milder winters. In Britain, global warming of about 1.07°C occurred between 1968 and 2002. Similar effects have been seen in some German birds over the same time period, giving us confidence that this effect is occurring over a fairly wide area.

Physical features of the environment such as temperature, water, nutrient availability, light, salinity, and pH, which we term abiotic factors; limit both the distribution patterns of organisms and their abundance. Some species can tolerate a relatively wide range of physical conditions while others tolerate only a narrow range. Each species usually functions best over only a limited range of abiotic conditions known as a species' physiological optimum or **fundamental niche,** a term first coined by Joseph Grinnell in 1917 during his studies of the life histories of birds. Of course, some part of the fundamental niche may be already used by competitors. Therefore, the actual range of an organism in nature is termed the **realized niche** (**Figure S.1**). Other organisms, such as predators or parasites, so-called biotic factors, may also contribute to the formation of realized niches by limiting the areas inhabited by their prey or hosts. We will return to the concept of the realized niche in Chapter 11. In this section we will examine the features of the physical environment that determine the fundamental niches of organisms. The most important of these are temperature, the availability and chemistry of water, and the availability of nutrients.

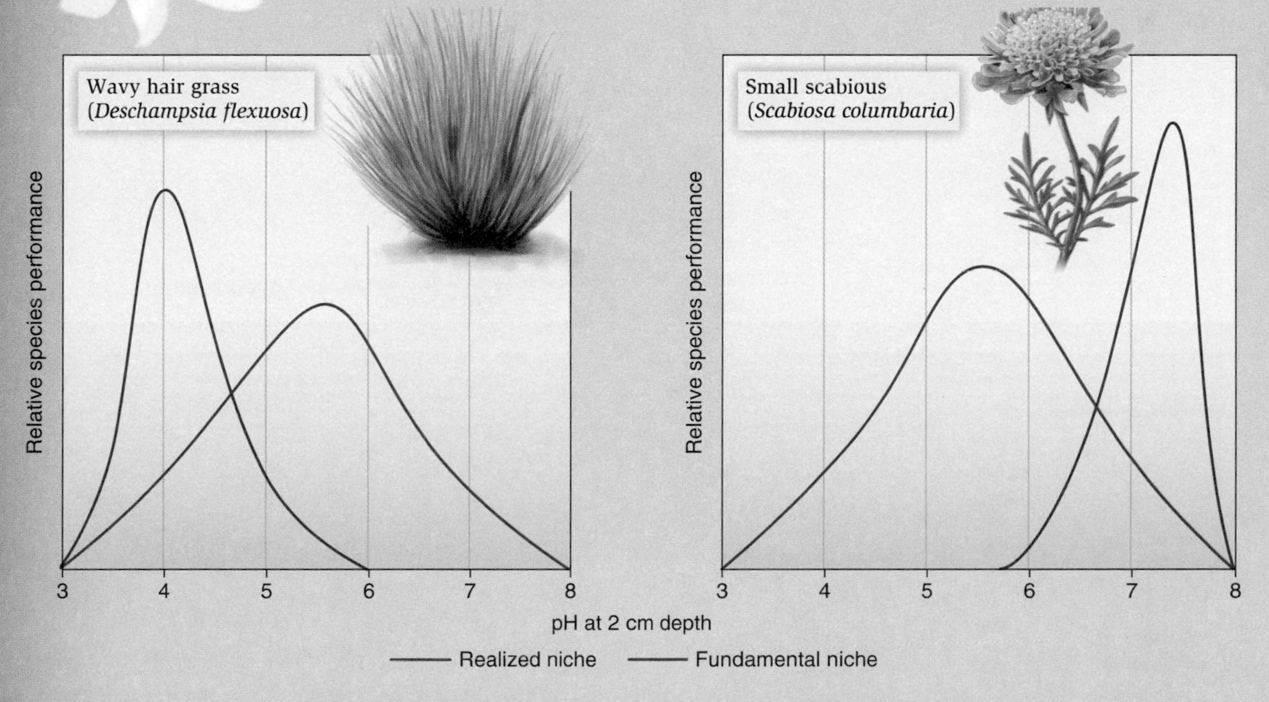


**Figure S.1 Fundamental and realized niches.** In Europe, wavy hair grass, *Deschampsia flexuosa,* and small scabius, *Scabiosa columbaria,* are two grasses that have broad fundamental niches and can grow across a wide range of pH in the laboratory. In the field, however, interactions with other plants limits their ranges to low pH and high pH respectively, so that their realized niches are much reduced. (After Rorison, 1969.)

Musk ox, *Ovibos moschatus,* can survive some of the coldest temperatures in the Northern Hemisphere.

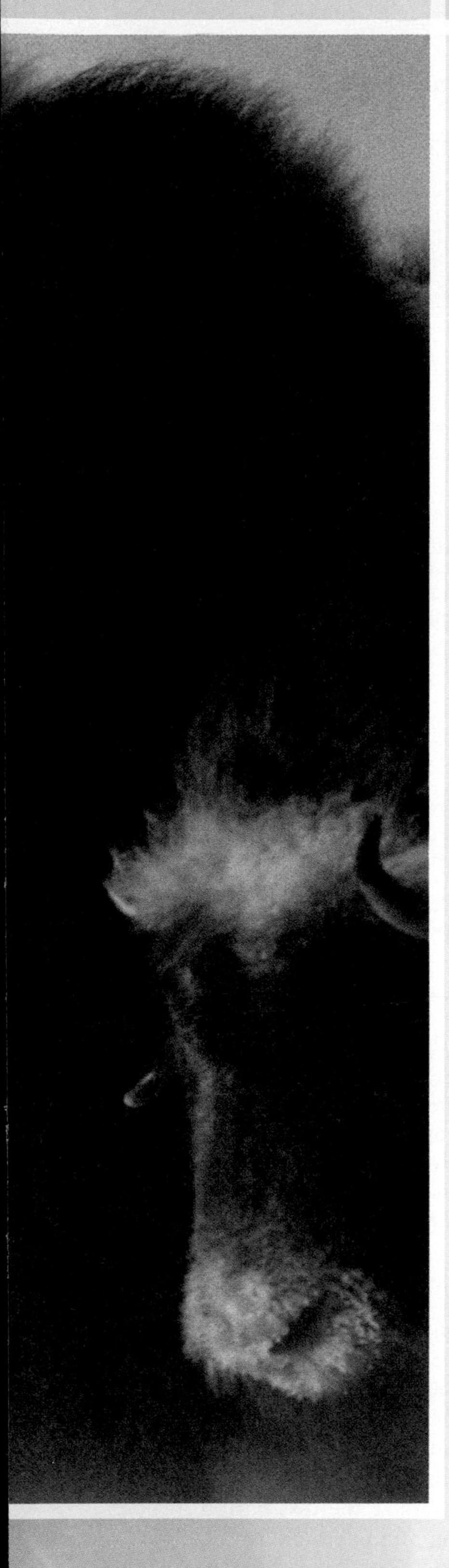

CHAPTER 5

# Temperature

## Outline and Concepts



Perhaps the most important factor in the distribution and abundance of organisms is temperature, because of its effect on biological processes, such as metabolic rates, and because of the inability of most organisms to regulate their body temperature precisely. For example, the organisms that form coral reefs secrete a calcium carbonate shell. Shell formation and coral deposition are accelerated at high temperatures but are suppressed in cold water. Coral reefs are therefore abundant only in warm water, and a close correspondence is observed between the 20°C isotherm for the coldest month of the year and the limits of the distribution of coral reefs (**Figure 5.1**). An isotherm is a line on a map connecting points of equal temperature. Coral reefs are located between the two 20°C isotherm lines that are formed above and below the equator.

Ecologists classify animals according to their source of heat and their ability to maintain their body temperature. **Ectotherms** rely on external heat sources to warm themselves, while **endotherms** generate their own internal heat. **Homeotherms** maintain an internal temperature within a narrow range, while the body temperature of **heterotherms** varies widely with environmental conditions (**Figure 5.2**). Most animals fall into two categories. Birds and mammals are generally endothermic and homeothermic; other vertebrates and invertebrates are ectothermic and heterothermic. However, there are exceptions to these classifications. During the winter, the body temperature of hibernating mammals drops as their metabolism slows. In effect, they are heterothermic at the transition period from fall to winter, and again from winter to spring, as they let their body temperature approach that of the ambient temperature. Similarly, deep-water fish behave like homeotherms, because the temperature of water, and therefore their body, is essentially constant.

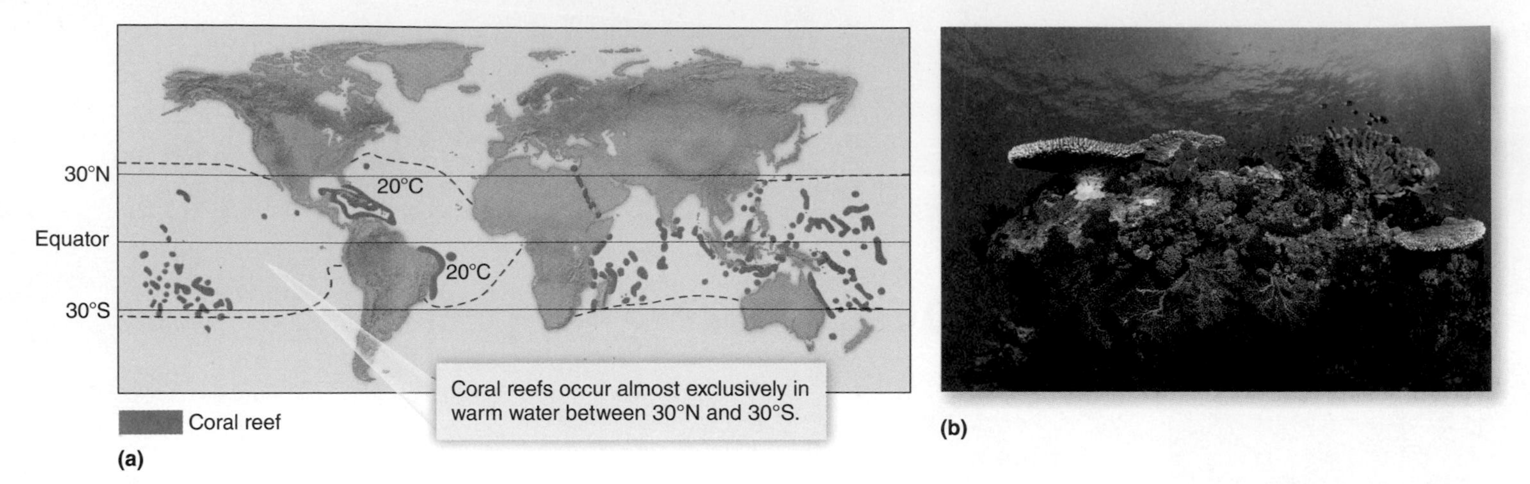


**Figure 5.1** **Coral reefs occur only in warm water.** **(a)** Coral reef formation is limited to waters bounded by the 20°C isotherm (dashed line), a line where the average daily temperature is 20°C during the coldest month of the year. **(b)** Coral reef from the Pacific Ocean.

**ECOLOGICAL INQUIRY**

Why are coral reefs absent on the west coasts of South America and Africa, until a little south of the equator?

In this chapter, we address how cold temperatures often limit the distribution of organisms, especially plants, many of which are highly susceptible to freezing. Next, we examine how hot temperatures may also limit the distribution of life on Earth. Finally, we discuss how the gradual warming of the Earth in recent times has affected species distribution patterns.

## 5.1 The Effects of Cold Temperatures, Especially Freezing, Are Severe

Animals use a variety of techniques to survive in cold temperatures. Some organisms change their behavior, curling up into a ball to reduce the exposed surface area. Others, such as penguins, may huddle together to conserve heat. Some species, notably bears, hibernate, spending the winter in a deep sleep in a secluded den where their pulse rate, breathing rate, and temperature are all reduced.

Arctic-dwelling vertebrates, including polar bears and seals, have multiple layers of fat, plumage, or fur to keep warm. Arctic birds often have thick layers of feathers, and the willow grouse or ptarmigans, *Lagopus lagopus,* even grow feathers on the soles of their feet. In a blizzard, these birds sometimes dive into a snow drift because the snow acts as good insulation. The diving strategy ensures that no tracks are left for predators to follow. Arctic musk ox, *Ovibos moschatus,* have two layers of protective fur; an outer layer of long hairs that protect them from wind, rain, and snow and a woolly inner layer that keeps them warm even at −40°C.

In some animals blood vessels near the surface of the skin can be adjusted in size to increase or decrease heat loss from

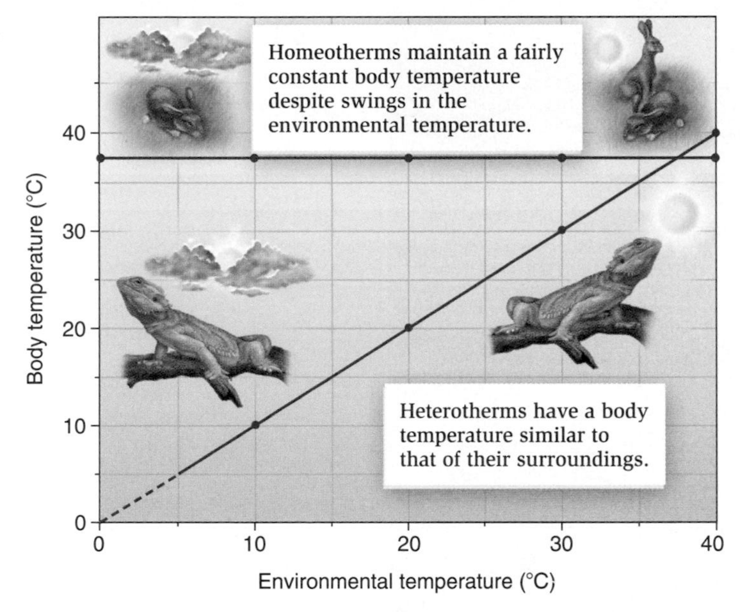


**Figure 5.2** **Body temperature of homeotherms and heterotherms in different environmental conditions.** Homeotherms maintain stable body temperatures across a range of environmental temperatures, whereas the body temperature of heterotherms varies with the external temperature.

**ECOLOGICAL INQUIRY**

Are homeotherms all endothermic, that is, do they generate their own heat?

the body (**Figure 5.3**). Skin surface blood vessels constrict on a cold day to retain warm blood deeper in the body, and they dilate on a hot day to maximize the dissipation of body heat to the environment. Diving animals such as ducks, seals, and whales reduce surface blood flow when they dive in cold waters so that they retain more of their body heat.

In cold conditions, many endotherms shiver, wherein the body undergoes rapid muscle contractions without any

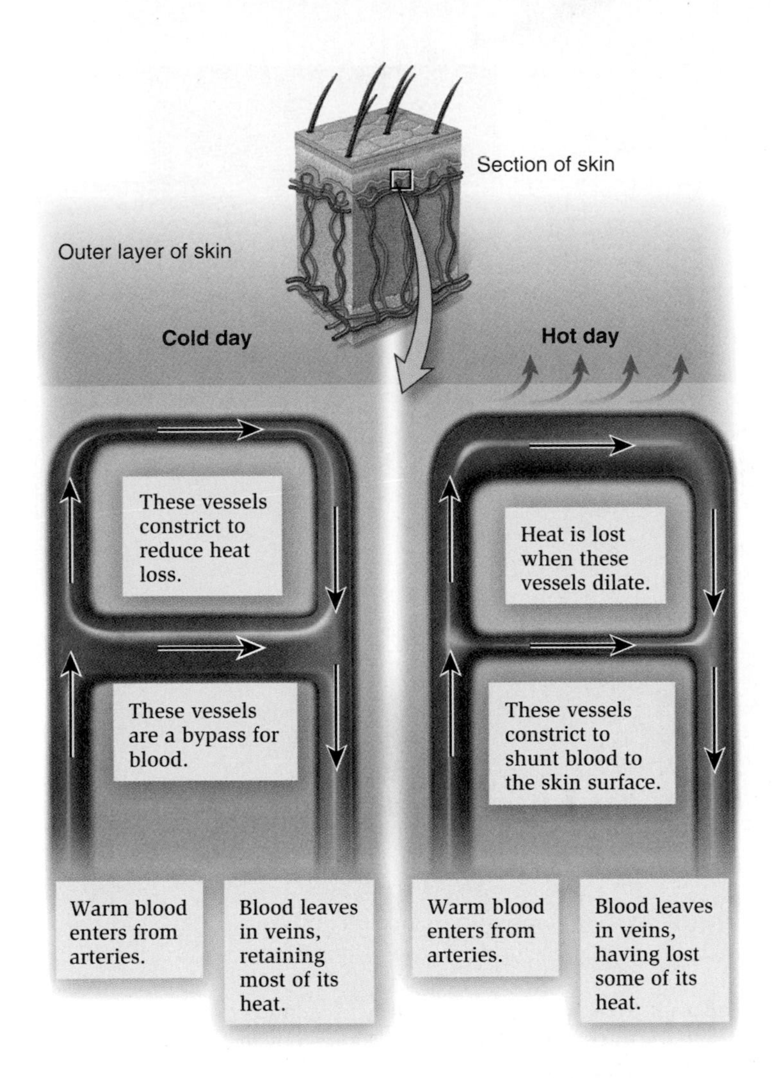


**Figure 5.3 Regulation of heat exchange in the skin.** Near the surface of some animals' skin the surface blood vessels constrict in cold temperatures, restricting blood flow and heat loss. These vessels dilate on hot days, increasing blood flow and heat loss. Arrows indicate the direction of blood flow.

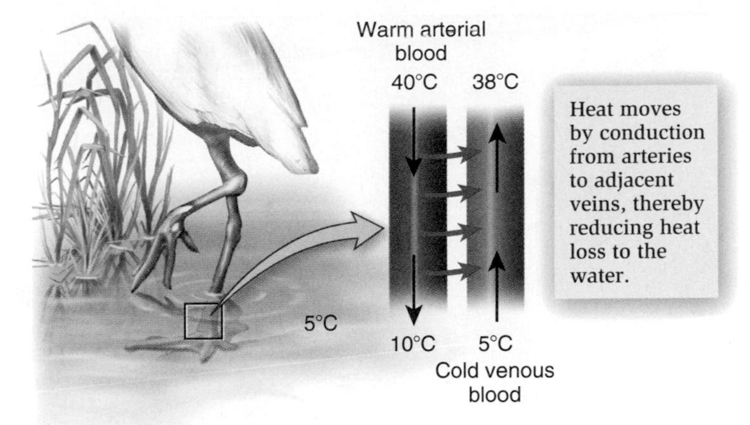


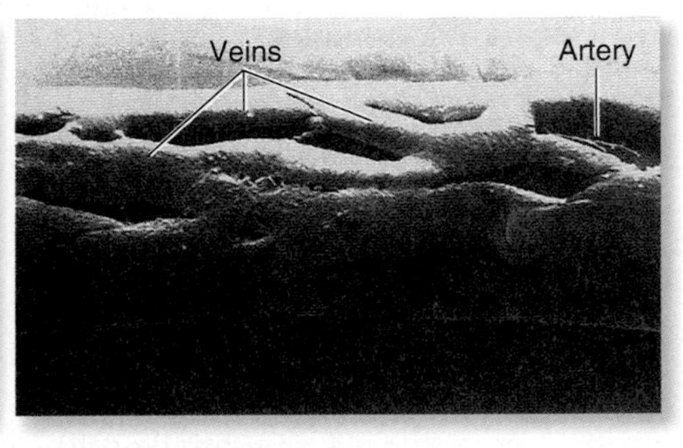


**Figure 5.4 Countercurrent exchange.** **(a)** Countercurrent heat exchange retains heat in the leg of an endotherm such as this bird. Black arrows in blood vessels indicate direction of blood flow **(b)** An SEM of the arrangement of veins surrounding an artery in a bird's leg. The artery is almost completely covered by overlying veins, allowing efficient heat exchange between the vessels.

locomotion. Virtually all of the energy liberated by the contracting muscles is used to produce internal heat. Many birds that remain in cold climates over winter shiver repeatedly. Arctic bees can raise their body temperatures by up to 15°C by shivering. Many small mammals of cold climates, such as hibernating bats or rodents, possess brown fat, a specialized heat-producing tissue. When the body temperature falls, brown fat cells increase their metabolic rate, and hence heat production.

Many organisms minimize heat loss to the environment via **countercurrent heat exchange,** a mechanism that conserves body heat by minimizing heat loss in the extremities and returning heat to the body core (**Figure 5.4**). Countercurrent heat exchange occurs primarily in the extremities, such as in dolphin flippers or bird legs. In cold temperatures, venal blood returning from the extremities is cool while arterial blood is warmer. Because artery and vein are positioned close together, in countercurrent heat exchange, heat moves by conduction from the artery to the adjacent vein. The temperature of arterial blood drops considerably, so that by the time it reaches the body surface, it has cooled, reducing heat loss to the environment and returning heat to the body's core. Many ectotherms, such as fish, use countercurrent heat exchange to warm their muscles. Swimming muscles generate heat, which warms the blood in the veins leaving the muscles. Without a countercurrent system, this heat would be lost as the warm blood enters the gills and is cooled by the adjacent water. However, heat from the warm veins is conducted into nearby arteries and returned to muscles rather than being lost from the gills. Because the muscle arteries are positioned close to the veins, heat is conducted to the arteries and heat loss is minimized. Warm temperatures increase muscle efficiency.

Some insects and a few species of amphibians release large amounts of glucose into the blood, which lowers the freezing point in much the same way as adding antifreeze lowers the freezing temperature of water in your car radiator. This ability to withstand freezing is called **supercooling.** Other species combine carbohydrates with proteins to produce glycoproteins. Glycoproteins can permit some invertebrates to withstand temperatures as low as −18°C without freezing. Antarctic fish also have glycoproteins in their blood,

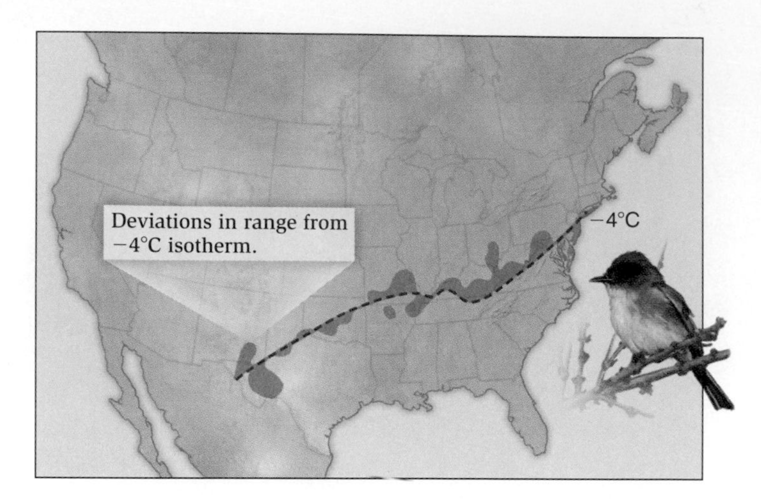


**Figure 5.5 Distribution map of the Eastern phoebe, *Sayornis phoebe*.** The dashed line marks the –4°C isotherm of average minimum January temperatures. The area of deviation between the range boundary of the Eastern phoebe, north or south of this isotherm, is shaded. (From Root, 1988.)

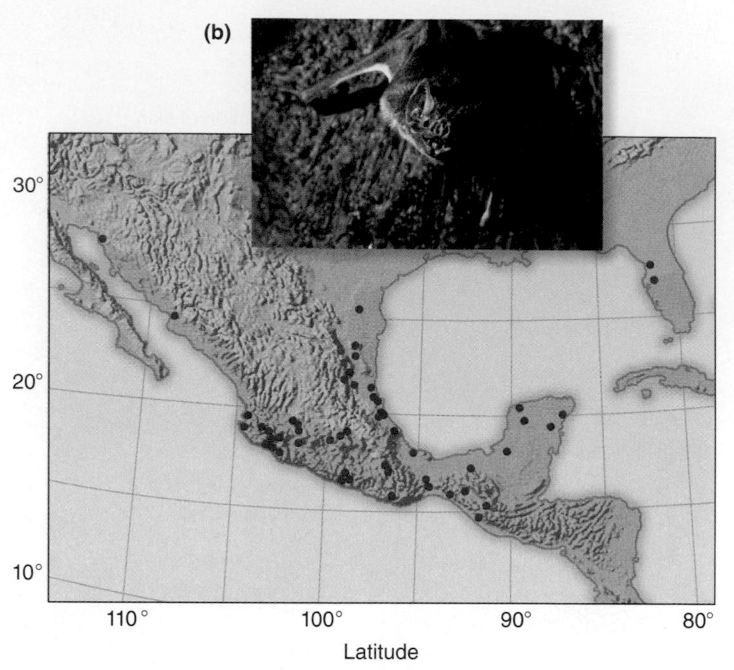


**Figure 5.6 Vampire bats occur south of the 10°C isotherm in January.** **(a)** The northern distribution of the vampire bat in Mexico is limited to areas warmer than the 10°C isotherm in January, because of the animal's poor capacity to thermoregulate. Its southern distribution in Argentina and Chile is limited by the same isotherm. (After McNab, 1973.) **(b)** The vampire bat, *Desmodus rotundus.*

and some species accumulate sodium, potassium, or chloride ions or urea, all of which lower the blood freezing point.

Other arctic species survive freezing temperatures using biochemical and behavioral mechanisms. The Greenland sulfur butterfly, *Colias hecla,* takes 2–3 years to complete its life cycle. Each year the caterpillars feed during the summer before freezing over the winter and rethawing the next spring. Some parasitic invertebrates rely on their hosts to keep them warm. The botfly parasites of vertebrates pass the winter inside their host. The nose botfly takes refuge inside the nose of a caribou, where it lays its eggs. The maggots crawl through the nose passages into the throat, where they feed. In the spring, the fully formed maggots are coughed up. Once on the ground they pupate and soon emerge as adults to begin the life cycle again.

Despite these adaptations, the geographic range limits of ectothermic and endothermic animals are affected by temperature. Many, especially birds, migrate to warmer climates. Arctic terns, *Sterna paradisaea,* create their own "endless summer" by migrating between the Arctic and the Antarctic, enjoying the summers of both. The Eastern phoebe, *Sayornis phoebe,* is a small bird with a northern winter range that coincides with an average minimum January temperature of above −4°C (**Figure 5.5**). Such limits are probably related to the energy demands associated with cold temperatures. Cold temperatures mean higher metabolic costs, which in turn are dependent on high feeding rates. Below −4°C, the Eastern phoebe cannot feed fast enough or, more likely, find enough food to keep warm. Other small endotherms with high basal metabolic rates, such as shrews, must eat almost continuously and will die if deprived of food for as little as a day. Similarly, cold temperatures limit the distribution of the vampire bat, *Desmodus rotundus.* The vampire bat is found in an area from central Mexico to northern Argentina. Its range in Mexico is limited to that area where the average minimum temperature in January is above 10°C. Because of the bat's poor capacity for thermal regulation, it cannot survive in areas below that temperature (**Figure 5.6**).

**ECOLOGICAL INQUIRY**

Should customs officials be concerned about the entry of vampire bats into the U.S.?

In plants, cold temperature can be lethal because cell membranes may start to leak. Even immediate chilling to 10°C injures tropical plants, because ions leak out of the cells and the photosynthetic and respiratory systems become inefficient. In cold climates many plants exist only below ground, as roots, or close to the ground, insulated by snow cover. They begin growth when spring warms temperatures. New buds may actually be formed at the end of summer and remain dormant over winter, ready to burst out in the spring. The northern boundary of wild madder, *Rubia peregrina,* in Europe corresponds to the 4.5°C January isotherm (**Figure 5.7**). Although this temperature is well above freezing, the shoots do not grow well in lower temperatures.

## 5.1.1 Freezing temperatures are lethal for many plant species

Freezing temperatures are probably the single most important factor limiting the geographic distribution of tropical and subtropical plants. Only low-elevation tropical areas are frost free (**Figure 5.8**). Freezing creates multiple problems for plants. First, cells may rupture if the water they contain freezes. Freeze-tolerant plants may add molecules to their cell water that lower the freezing temperature, which is another example of supercooling. Second, water availability decreases during freezes. At −4°C and below, the ground freezes solid, effectively stopping water uptake. Cell dehydration can occur, similar to what happens in a drought.

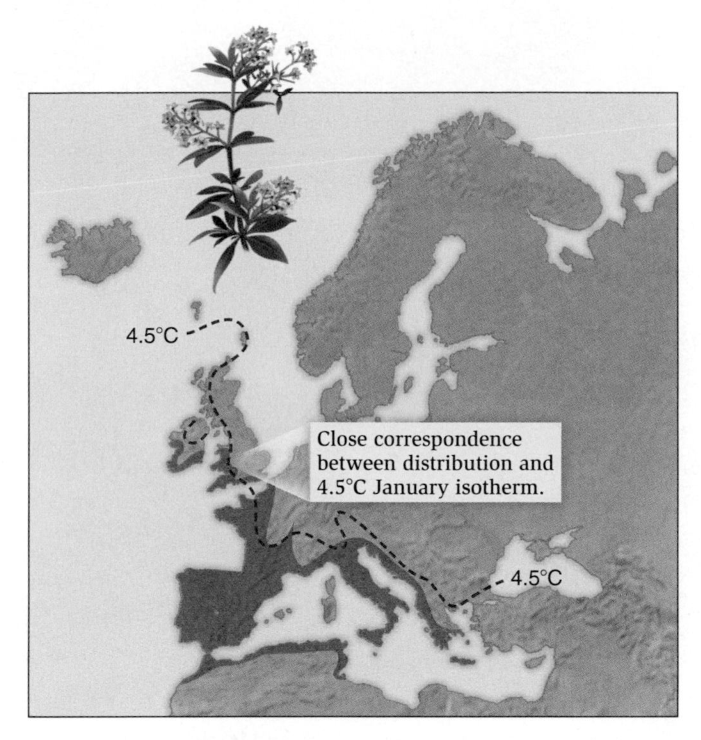


**Figure 5.7 The distribution of wild madder, *Rubia peregrina*, in Europe is limited by cold temperature.** Madder distribution (shaded) and the location of the January isotherm for 4.5°C are shown. (After Salisbury, 1926.)

Disruption of cells via freezing is especially lethal to plants that produce poisonous chemical defenses against herbivores. White clover, *Trifolium repens,* possess a cyanide defense against herbivores. However, frost injury is lethal, because when cell membranes are disrupted, toxins are released into the plant's other tissues. As a result, the cyanide-producing form is less common in colder regions of Europe (**Figure 5.9**).

## 5.1.2 Animal body size changes in different temperatures

A number of patterns relate changes in animal body size and extremity length over different geographic areas to different temperature regimes. The best-known patterns are Bergmann's rule and Allen's rule. In 1847 Carl Bergmann noted that among closely related mammals or birds, the largest species occurred at higher latitudes, where it is colder. This is known as **Bergmann's rule.** The logic behind this rule is that individuals with greater mass have a smaller surface area to volume ratio, which helps to conserve heat. For example, polar bears are the largest of all known bear species. However, many ecologists are reluctant to make these cross-species kinds of comparisons. Instead, they define Bergmann's rule as the tendency for the average body mass of animal populations within a species to increase with latitude. We can see this in the body size of moose, *Alces alces,* in Sweden (**Figure 5.10**). Support for the rule has been shown in mammals, birds, salamanders, turtles, flatworms, and ants.

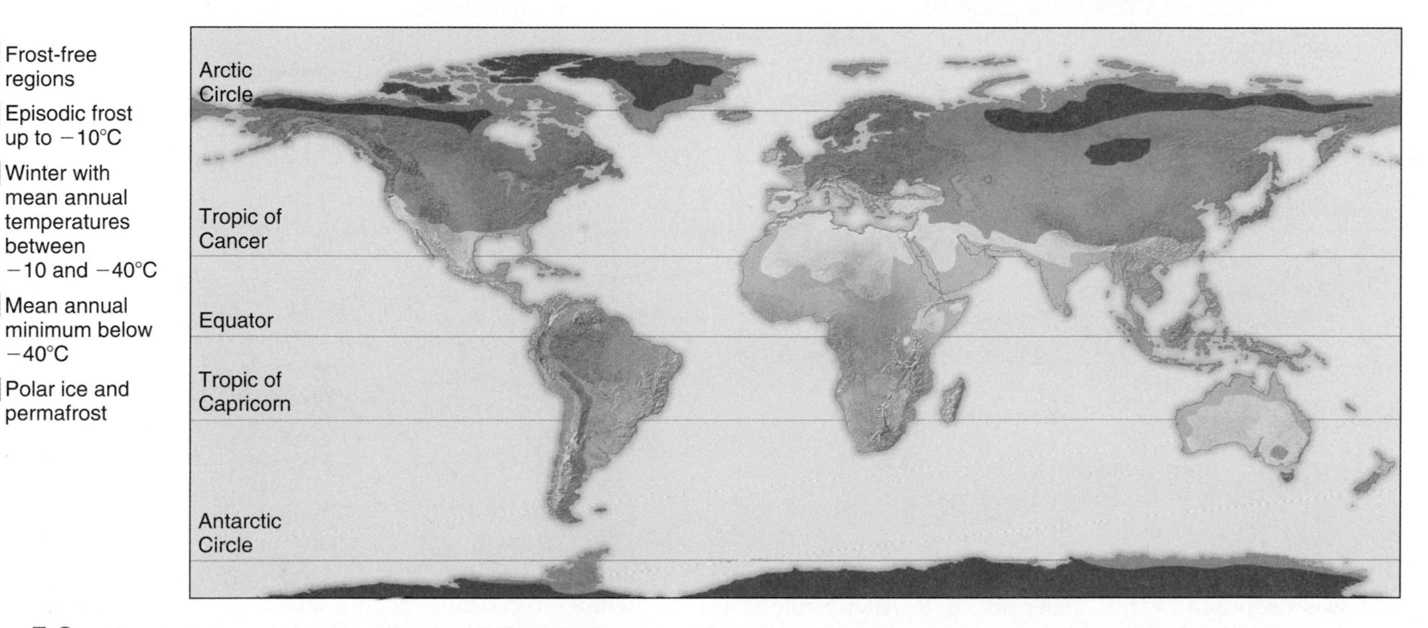


**Figure 5.8 The global occurrence of frost and frost-free zones.** Frost-free zones are confined to low-elevation tropical areas.

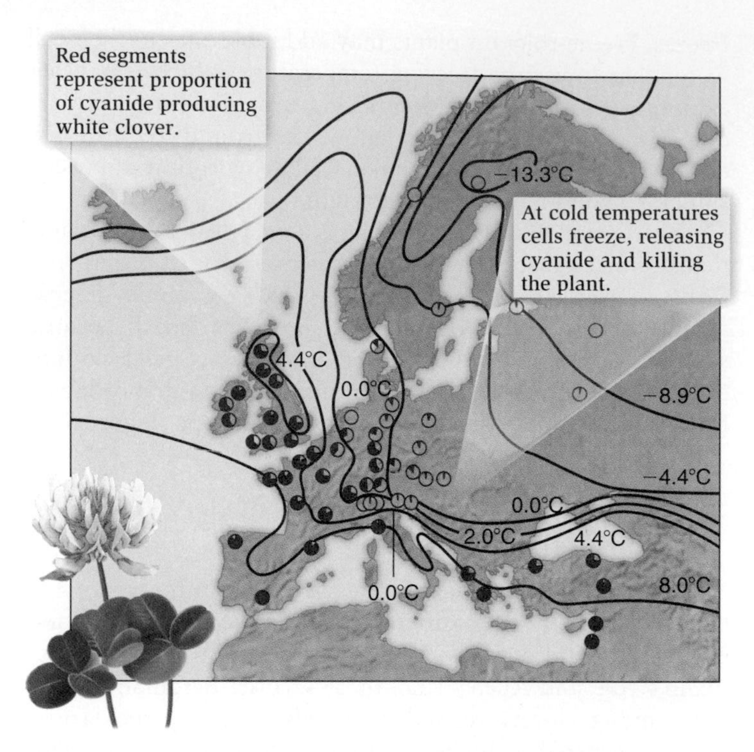


**Figure 5.9 Frequency of the cyanide-producing form in populations of white clover, *Trifolium repens*, is affected by temperature.** Cyanide-producing forms are represented by the red section of each circle. The cyanogenic form is more common in warmer regions. Lines are January isotherms. (After Daday, 1954.)

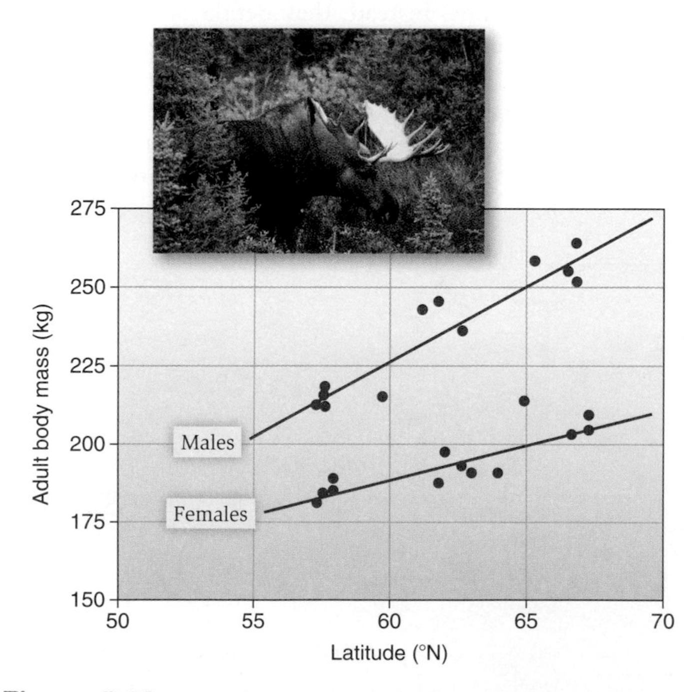


**Figure 5.10 Bergmann's rule.** Body size of moose, *Alces alces*, in Sweden increases with latitude. (After Sand et al., 1995.)

Another pattern relating to cold temperature adaptation was proposed by J. C. Allen in 1878. **Allen's rule** states that among closely related endothermic (warm-blooded) vertebrates, those living in colder environments tend to have shorter appendages than those living in warmer environments. A classic example is the ear size of hares and foxes. Those species in colder, arctic environments have shorter ears than those in temperate or hot desert environments (**Figure 5.11**). The greater the surface area, the greater the heat loss.

**Figure 5.11 Allen's rule.** Desert-dwelling species, such as **(a)** the Antelope jackrabbit, *Lepus alleni*, and **(c)** the Cape fox, *Vulpes chama*, have larger ears than species in Arctic climates, such as **(b)** the Arctic hare, *Lepus arcticus*, and **(d)** the Arctic fox, *Alopex lagopus*.

### Check Your Understanding

**5.1** Why should cold temperatures limit the distribution of endothermic animals?

## 5.2 Hot Temperatures Limit Many Species' Distributions

High temperatures are also limiting for many plants and animals, because relatively few species can survive internal temperatures more than a few degrees above their metabolic optimum. At the beginning of this chapter we saw that corals are sensitive to low temperature. They also are sensitive to high temperatures. When temperatures are too high, the symbiotic dinoflagellates that live within corals and contribute to their coloration die and are expelled. The coral tissue loses its color and turns a pale white, a phenomenon known as coral bleaching. In the winter of 1982–1983, an influx of warm water from the eastern Pacific raised temperatures just 2–3°C for 6 months on the coast of Panama, but this was enough to kill many of the reef-building corals, *Millepora* spp. By May 1983 just a few individuals of one species, *M. intricata,* were alive. Scientists are concerned that future increases in ocean temperatures may increase the frequency of coral bleaching.

Many organisms cope with heat stress by producing increased amounts of heat shock proteins (HSPs). At high temperatures, proteins may either unfold or bind to other proteins to form misfolded protein aggregations. HSPs act to prevent these types of events from taking place. In sick cells, HSPs can also help move irreparably damaged proteins from inside the cell to outside the cell, where the immune system recognizes the proteins and helps rid the organism of them. HSPs normally constitute only about 2% of the cell's soluble protein content, but this can increase to 20% when a cell is stressed by heat. HSPs are extremely common and are found in all organisms, from bacteria to plants and animals. In the Tropics, high temperatures can substantially decrease the growth rates and productivity of many crop species. There is now substantial interest in identifying crop strains with naturally high HSP levels and identifying thermally tolerant varieties for use in crop-breeding programs.

Some of the life-history stages most resistant to heat are the resting spores of fungi, cysts of nematodes, and seeds of plants. Dry wheat grains can withstand temperatures of 90°C for short periods of up to 10 minutes. Some bacteria are also incredibly heat resistant. Natural hot springs are home to *Thermus aquaticus,* which grows at temperatures of 67°C. DNA polymerase, called *Taq* polymerase, from this organism is commonly used in the process called polymerase chain reaction (PCR), a technique that can produce millions of copies of a specific DNA sequence. It can withstand the high temperatures needed to separate the strands of DNA. Some thermophilic bacteria collected from deep-sea vents have even been cultured at temperatures of 100°C, much higher than the temperatures that were originally thought to limit life.

High temperatures are particularly stressful when accompanied by lack of water, as in hot deserts. Plants have evolved several mechanisms to live in such environments. Xerophytes such as cacti are able to survive in environments with little or no water. They have few or no leaves and a large stem that can store water. They photosynthesize through their green stems. Phreatophytes are plants that have adapted to arid environments by growing very long roots, enabling them to acquire water at great depths. Annual plants grow quickly when the rains come, produce seeds, and die. The heat-resistant seeds remain dormant until the next season.

Plant leaf shape can also change with temperature. Simple leaves, with only one blade, are advantageous in cool, shady environments because they provide the largest light absorption surface (**Figure 5.12a**). However, large leaves can overheat in hot, dry environments and the leaves of some species form leaflets, which promote heat dissipation (**Figure 5.12b**).

Animals possess many behavioral adaptations to avoid heat stress, including radiation, conduction, convection, and evaporation (**Figure 5.13**). Radiation is the emission of electromagnetic waves by the surface of objects. The rate of

(a)

(b)

**Figure 5.12 Variation in leaf shape according to temperature.** **(a)** Large, simple leaves, such as this maple leaf, *Acer saccharum,* are photosynthetically efficient in shady conditions but can overheat in sunny environments. **(b)** Compound leaves with small leaves, or leaflets, like this dwarf date palm, *Phoenix canariensis,* are more common in hot environments because they encourage heat dissipation.

**Figure 5.13 Heat exchange.** The four methods of heat exchange in animals: radiation, conduction, convection, and evaporation.

**ECOLOGICAL INQUIRY**

When an elephant flaps its ears, is it losing heat by radiation, conduction, convection, or evaporation?

emission is governed by the relative differences of the radiating body surface and the environment. A positive difference results in heat loss, while a negative difference, in which the outside temperature is warmer than the body, results in heat gain. Thermal imaging cameras show radiated heat emanating from ectotherms as orange and red colors (**Figure 5.14**). In **conduction,** the body surface loses or gains heat through direct contact with cooler or warmer substances such as air or water. The greater the temperature difference, the faster the rate of conduction. Some animals, such as elephants, have enlarged ears to accelerate the rate of heat loss via conduction. Water has an even greater ability to store heat than air of the same temperature. Aquatic animals therefore lose large amounts of heat to their environment. Terrestrial animals can quickly cool themselves off in colder water. **Convection** is the transfer of heat by the movement of air or water next to the body. Convection is aided by air or water currents. When elephants flap their ears, they create wind currents that help them cool down even more. **Evaporation** occurs when organisms lose water from their surfaces, and heat from a plant's leaves or an animal's body is used to drive this process. Many animals are able to regulate heat loss via evaporation. Nerves

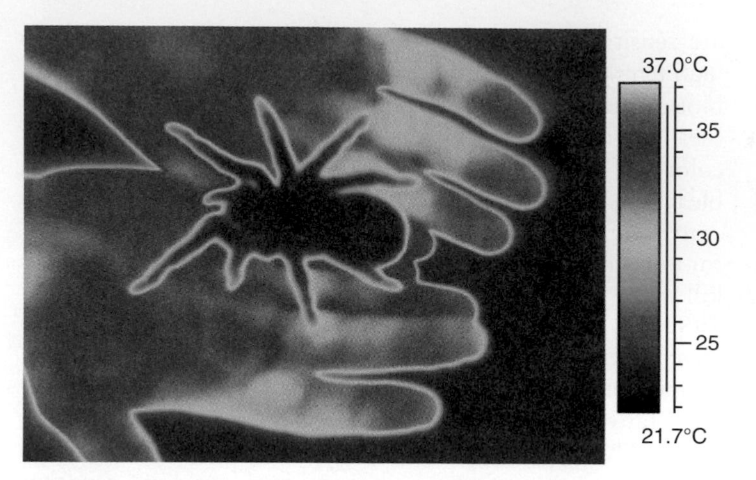


**Figure 5.14 Visualization of heat exchange in an ectotherm and an endotherm.** Thermal imaging cameras can detect heat radiated from an animal's body. Note the warm skin of the endotherm (the human) and the cold skin of the ectotherm (the tarantula), even though both animals are at the same environmental temperature.

to the sweat glands stimulate the production of sweat, a dilute aqueous solution containing sodium chloride. In endotherms that lack sweat glands, panting helps evaporate water from the tongue surface.

## 5.2.1 Some species depend on fire for their existence

The ultimate high temperatures that organisms face are brought about by fire. For example, before the arrival of Europeans in North America, fires started by lightning occurred frequently and regularly in the pine forests of what is now the southeastern United States. Here, there are 60–100 days per year with thunderstorms that generate lightning (**Figure 5.15**). These fires, because they were so frequent, consumed leaf litter, dead twigs and branches, and undergrowth before they accumulated in great quantities. As a result, no single fire burned hot enough or long enough in one place to damage large trees—each fire swept by quickly and at a relatively low temperature (**Figure 5.16a**). The dominant plant species of these areas came to depend both directly and indirectly on frequent, low-intensity fires for their existence. The long-leaf pine, *P. palustris,* has serotinous cones that remain sealed by resin until the heat of a fire melts them open and releases the seeds. Long-leaf pines therefore depend directly on fire for their reproductive success. The trees also depend on fires to thin out competing species that would otherwise eventually take over. The *Sequoia* or redwood trees of California also have serotinous cones.

Management practices that attempt to maintain forests in their natural state by completely preventing forest fires often have the opposite result. First, trees like long-leaf pine or jack pine simply stop reproducing in the absence of fire. Second, species like pines and wiregrass that depend on fire to suppress their competitors are replaced by species characteristic of other communities. Finally, when a fire does occur, fuel has

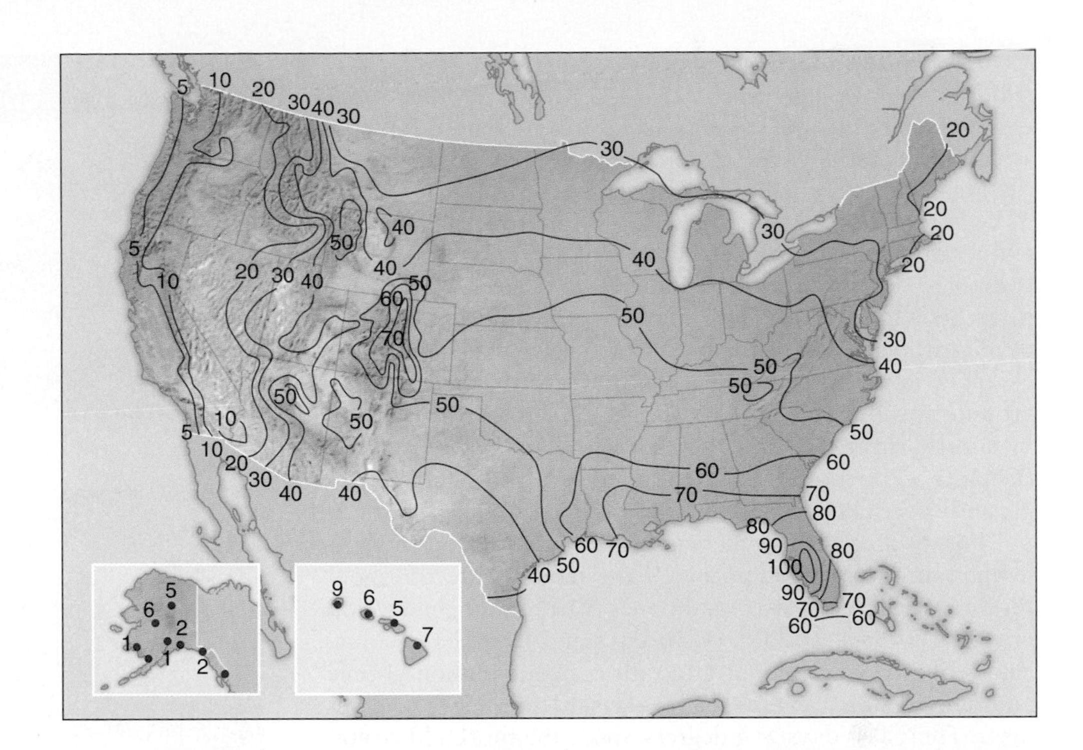


**Figure 5.15 Thunderstorm frequency.** The average number of days with thunderstorms across the U.S.

(a)

(b)

**Figure 5.16 Forest fires.** **(a)** Where fires are frequent, fuel loads are low and fires do not damage forest trees. **(b)** Where fires have been suppressed over long periods of time, when they do occur they often burn out of control, creating crown fires, which reach high into the tree canopy and kill most living organisms.

had a much longer period in which to accumulate on the forest floor and the result is a fire that is so large and burns so hot that it consumes seeds, seedlings, and adult trees, native and competitor alike (**Figure 5.16b**).

## 5.2.2 Temperature extremes may be more critical than temperature averages

A species' range usually is limited not by average temperatures but instead by the frequency of occasional extremes, such as hard freezes. Farmers know this only too well. The northern distribution of oranges in Florida and the southern distribution of coffee in Brazil are limited by the frequency and strength of periodic freezes, not by the average temperatures for the coldest months. One useful way to establish what factors control the natural limits of species is to experimentally move organisms outside their normal range and monitor survivorship. Of course, such movements should be done only in a carefully controlled manner and when there is no risk that the species will become invasive in their new habitats. Physical extremes, such as fires or severe freezes, may be apparent only in isolated years, so ecologists often have to wait many years to determine whether extremes are limiting the distribution and abundance of plants and animals in the field.

Despite the obvious relationships between species distributions and temperature, we need to be cautious about relating solely the two. The temperatures measured for constructing isotherm maps are not always the temperatures that the organisms experience. In nature an organism may choose to lie in the sun or hide in the shade, both of which affect the temperatures it experiences. Shade temperatures under vegetation in deserts can be 10–20°C lower than sun temperatures. Such local variations of the climate within a given area, or **microclimate,** can be important for a particular species. For example, in the spring, temperate forests turn green from the ground upward. Herbaceous species leaf out first, followed by shrubs, shorter trees, and taller trees. This occurs because the air is warmer closer to the ground. Dark soil helps warm ground cover first.

For most organisms there is a linear relationship between temperature and development. If the temperature threshold for the development of a grasshopper is 16°C, it might develop in, say, 100 days at 20°C or 50 days at 24°C. The absolute length of time is not important; rather, a combination of time and temperature, called **degree-days,** determines development. Thus, 100 days $\times$ 4 degrees above the threshold = 400 degree-days, and 50 days $\times$ 8 degrees = 400 degree-days also. Basking in the sun to raise temperatures accelerates grasshopper development. For plants, the threshold temperature can be the lowest temperature at which photosynthesis occurs. The geographic ranges of many crops and native species are well correlated with degree-day values.

Because so many species are limited in their distribution patterns by global temperatures, ecologists are concerned that if global temperatures rise, many species will be driven to extinction or their geographic ranges will shrink and the location of agricultural areas and forests will be altered.

### 5.2.3 Wind can amplify the effects of temperature

One way that wind is created is by temperature gradients. As air heats up, it becomes less dense and rises. As hot air rises, cooler air rushes in to take its place. For example, hot air rising in the Tropics is replaced by cooler air flowing in from more temperate regions, thereby creating northerly or southerly winds (look ahead to Figure 22.5). Wind amplifies the effects of temperature. It increases heat loss by evaporation and convection (the windchill factor). Wind also directly affects living organisms in a variety of ways. It contributes to water loss in organisms by increasing the rate of evaporation in animals and transpiration in plants. Winds can also intensify oceanic wave action, with resulting effects for organisms. On the ocean's rocky shore, seaweeds survive heavy surf by a combination of holdfasts and flexible structures. The animals of this zone have powerful organic glues and muscular feet to hold them in place (**Figure 5.17**). Wind can also be an important mortality factor in terrestrial systems. Hurricanes or gale-force winds can kill trees, breaking trunks or completely uprooting individuals.

(a)

(b)

**Figure 5.17 Animals and plants of the intertidal zone hold tight to their rocky surface.** **(a)** The brown alga *Laminaria digitata* has a holdfast that enables it to cling to the rock surface. **(b)** The mussel *Mytilus edulis* attaches to the surface of a rock by proteinaceous threads (byssal threads) that extend from the animal's muscular foot.

**ECOLOGICAL INQUIRY**

What other organisms have powerful glues or muscles to enable them to hold tight to the rock faces of the intertidal zone?

**Check Your Understanding**

**5.2** What is microclimate? List various ways organisms can alter their microclimate.

## 5.3 The Greenhouse Effect Causes the Earth's Temperature to Rise

The Earth is warmed by the **greenhouse effect.** In a greenhouse, sunlight penetrates the glass and heats the plants inside. The heat is reradiated but the glass acts to trap the heat inside. Similarly, solar radiation in the form of short-wave energy passes through the atmosphere to heat the surface of the Earth. This energy is then radiated from the Earth's warmed surface back into the atmosphere, but in the form of long-wave infrared radiation. Atmospheric gases absorb much of this infrared energy and radiate it back to the Earth's surface, causing its temperature to rise further (**Figure 5.18**). It should be noted that the greenhouse effect is important to life on Earth; without some type of greenhouse effect, global temperatures would be much lower than they are, perhaps averaging only −17°C compared with the existing average of +15°C.

The greenhouse effect is caused by a small group of gases, mainly water vapor, that together make up less than 1% of the total volume of the atmosphere. After water vapor, the four most significant greenhouse gases are carbon dioxide, methane, nitrous oxide, and chlorofluorocarbons (**Table 5.1**).

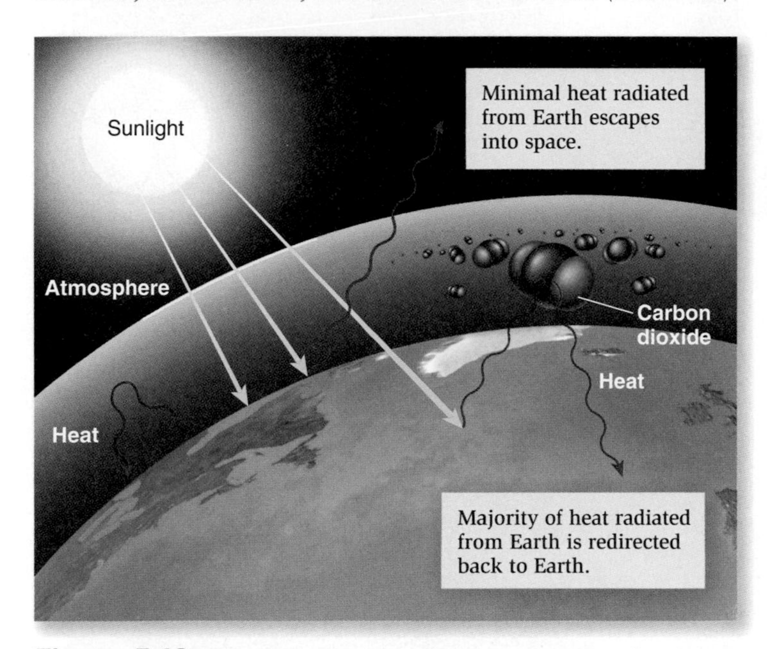


**Figure 5.18 The greenhouse effect is caused by the insulating effect of atmospheric carbon dioxide.** Solar radiation, in the form of short-wave energy, passes through the atmosphere to heat the Earth's surface. Long-wave infrared energy is radiated back into the atmosphere. Most infrared energy is reflected by atmospheric gases, including carbon dioxide molecules, back to Earth, causing global temperatures to rise.

Ecologists are concerned that human activities are increasing the greenhouse effect, resulting in the gradual elevation of the Earth's surface temperature, a process called

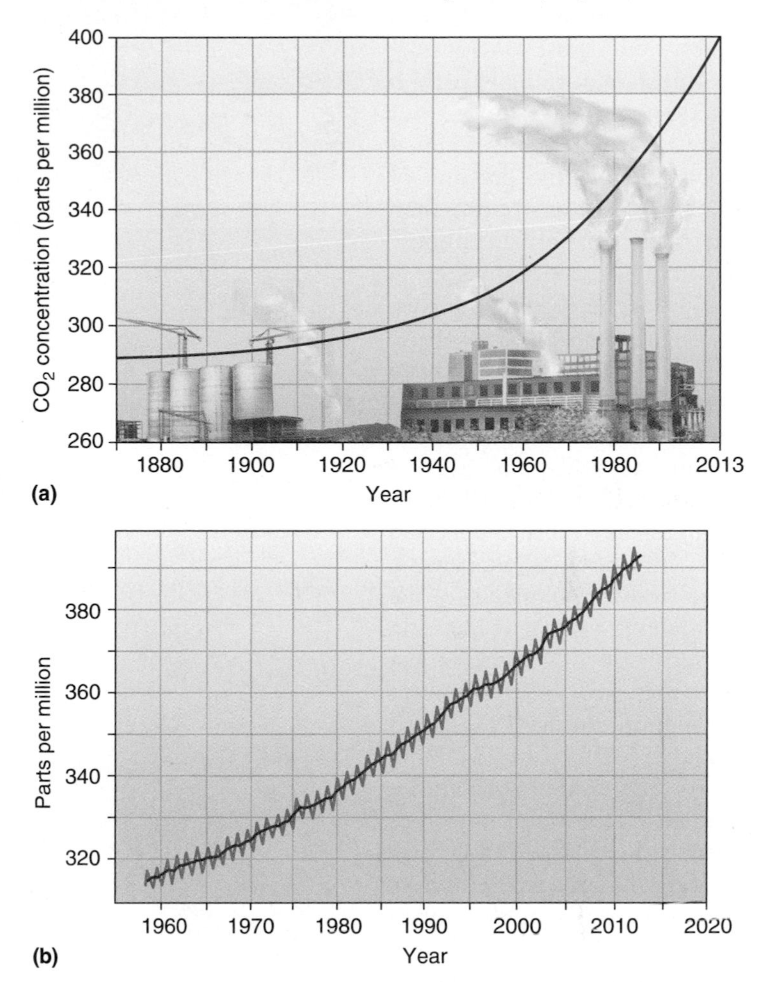


**Figure 5.19 The increase of atmospheric carbon dioxide.** **(a)** Increases in the burning of fossil fuels have caused a steady increase in atmospheric $CO_2$ since the late 19th century. **(b)** Since 1957, $CO_2$ levels have shown consistent increases in air taken directly from Mauna Loa, Hawaii.

**Table 5.1 The major greenhouse gases and their contribution to global warming.**

| | Carbon dioxide ($CO_2$) | Methane ($CH_4$) | Nitrous oxide ($N_2O$) | Chlorofluorocarbons (CFCs) |
|---|---|---|---|---|
| Relative absorption per ppm of increase[*] | 1 | 32 | 150 | >10,000 |
| Atmospheric concentration (ppm)[†] | 400 | 1.75 | 0.315 | 0.0005 |
| Contribution to global warming | 73% | 7% | 19% | 1% |
| Percent from natural sources; type of source | 20–30%; volcanoes | 70–90%; swamps, gas from termites and ruminants | 90–100%; soils | 0% |
| Major anthropogenic sources | Fossil fuel use, deforestation | Rice paddies, landfills, biomass burning, coal and gas exploitation | Cultivated soil, fossil-fuel use, automobiles, industry | Previously manufactured (for example, aerosol propellants) now banned in the U.S. and the EU |

[*]Relative absorption is the warming potential per unit of gas; water vapor is not included in this table.

[†]ppm = parts per million

**global warming.** Most greenhouse gases have increased in atmospheric concentration since the advent of industrial times. Of those increasing, the most important is carbon dioxide, $CO_2$. As Table 5.1 shows, $CO_2$ has a lower global warming potential per unit of gas (relative absorption) than any other major greenhouse gas, but its concentration in the atmosphere is much higher. Concentrations of atmospheric $CO_2$ have increased from about 280 ppm (parts per million) in the preindustrial 19th century to 400 ppm in 2013 (**Figure 5.19a**). Since 1957, air samples have been collected directly at Mauna Loa, Hawaii, a relatively unpolluted site. The data show a 25% increase in atmospheric $CO_2$ levels in just 57 years, from 313 ppm to 400 ppm during 1957–2013 (**Figure 5.19b**). In addition, there are seasonal oscillations. The Northern Hemisphere has greater land area and plant biomass than the Southern Hemisphere, so in the northern summer more $CO_2$ is used up by plants and atmospheric $CO_2$ level declines slightly. In the northern winter, less $CO_2$ is absorbed and atmospheric $CO_2$ levels increase.

To predict the effects of global warming, most scientists focus on a future point, about 2100, when the concentration of atmospheric $CO_2$ will have doubled, that is, increased to about 700 ppm compared with the late 20th-century level of 350 ppm. Ecologists argue that at that time, average global temperatures will be about 1–6°C (about 2–10°F) warmer than present and will increase an additional 0.5°C each decade. This increase in heat might not seem like much, but it is comparable to the warming that ended the last ice age. Assuming that this scenario of gradual global warming is accurate, scientists are beginning to consider what the consequences on natural and human-made ecosystems might be. Many species can adapt to slight changes in their environment within their own lifetime, a process called **acclimation.** For example, humans acclimate to different climates. Someone visiting a hot climate for the first time may overheat rapidly. After a few days to weeks, acclimation begins as the body starts to sweat sooner and a higher volume of sweat is produced, which is more effective at cooling the body. Long-term exposure to hot conditions causes blood vessels to dilate, increasing blood flow to the skin and dissipating heat by conduction.

However, the anticipated changes in global climate are expected to occur too rapidly to be compensated for by

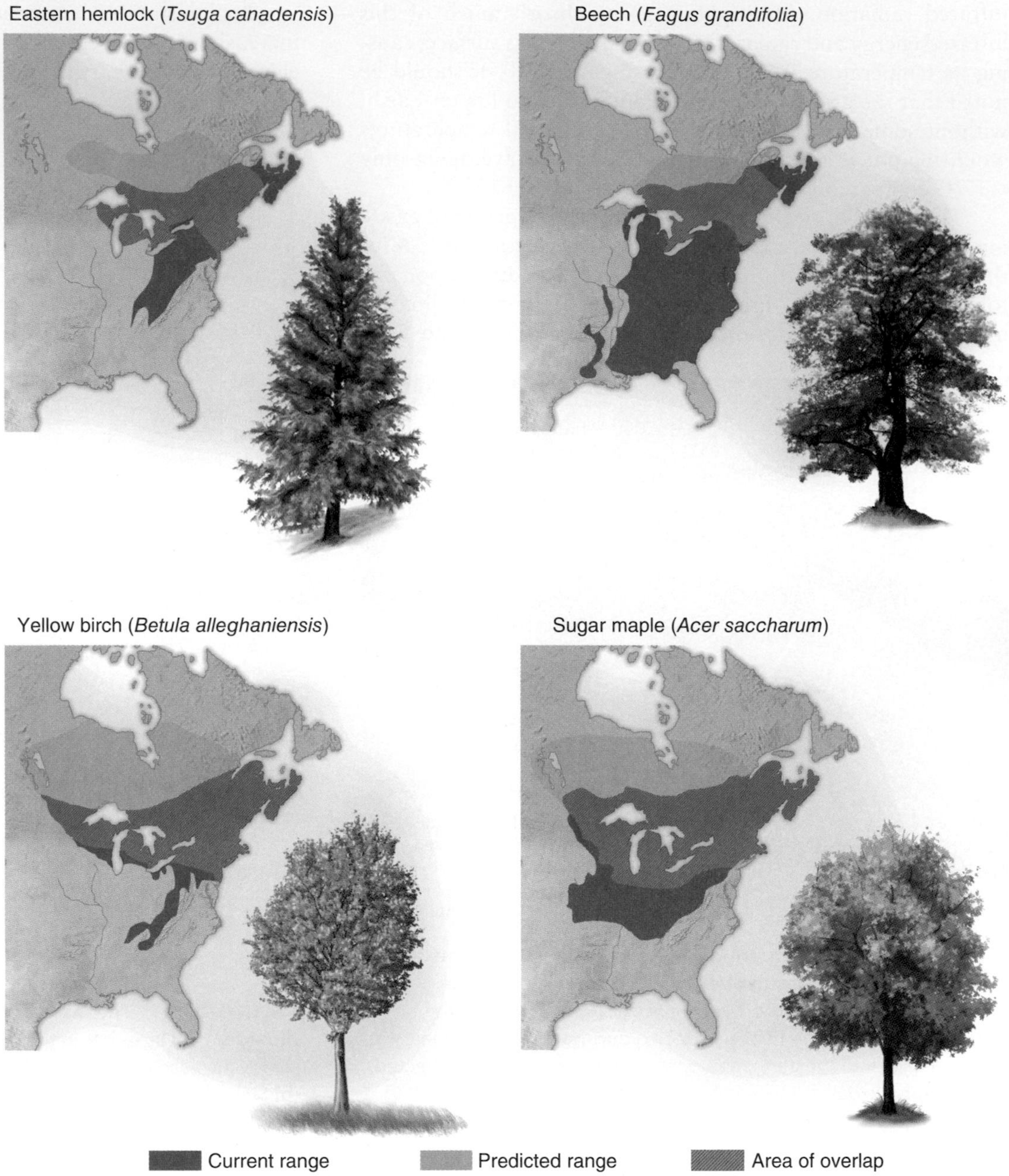

Eastern hemlock (*Tsuga canadensis*)
Beech (*Fagus grandifolia*)
Yellow birch (*Betula alleghaniensis*)
Sugar maple (*Acer saccharum*)
Current range
Predicted range
Area of overlap

**Figure 5.20 The range of North American trees could be reduced by global warming.** The present geographic ranges of four species (blue shading) and their potential ranges under doubled $CO_2$ levels (yellow shading) in North America. These new ranges are based on the temperature requirements of the trees. Green shading indicates the region of overlap, which is the only area where the species would be found before it spread into its new potential range. (After Gates, 1993.)

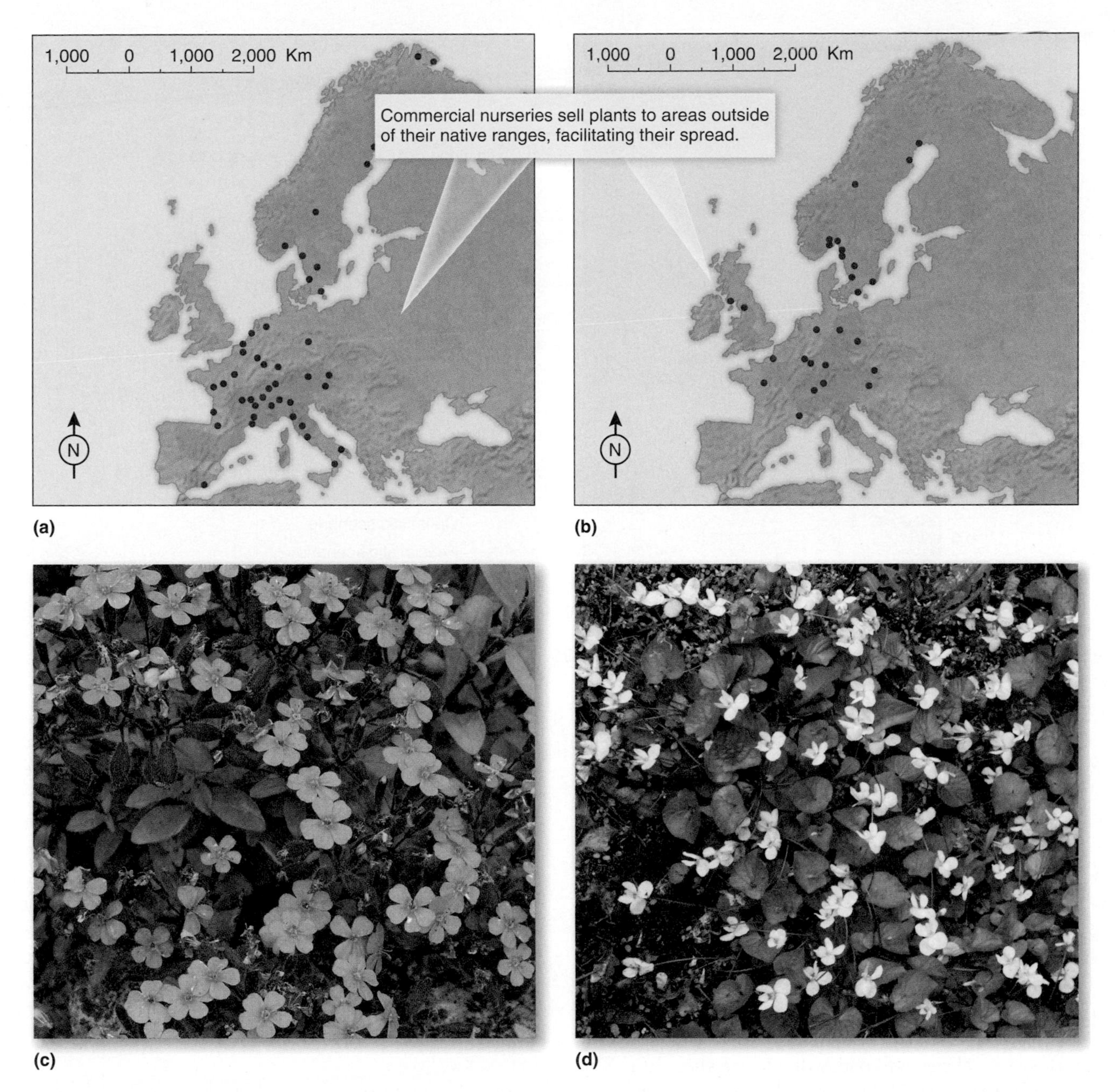


**Figure 5.21 Natural (shaded) and commercial ranges of two native European plant species** **(a and c)** Rock soapwort, *Saponaria ocymoides,* and **(b and d)** European wild ginger, *Asarum europaeum.* Commercial nurseries where these plants are sold are shown as red dots. In these cases the commercial range exceeds the native range by over 1,500 km. (After Van der Veken et al., 2008.)

adaptation. Plant species cannot simply disperse and move north or south into the newly created climatic regions that will be suitable for them. Many tree species take hundreds, even thousands, of years for seed dispersal. Paleobotanist Margaret Davis was among the first to investigate how species' ranges would be changed in a globally warmed world (Davis and Zabinski, 1992). She predicted that in the event of a $CO_2$ doubling, many tree species in North America would suffer range contractions because their southern ranges would retreat northward (**Figure 5.20**). Of course, this contraction in the trees' distribution could be offset by the creation of new favorable habitats in Canada.

Most scientists believe that the climatic zones would shift toward the poles faster than trees could either migrate via seed dispersal or evolve, hence extinctions would occur. However, as Sebastiaan Van der Veken and colleagues (2008) showed, some plants have already been given a head start on range shifts. Many commercial nurseries provide plants in areas well out of their natural range. They compared the natural ranges of 357 native European plant species, compared with the commercial ranges, based on 246 plant nurseries through Europe. For 73% of native species, the commercial range exceeded the natural range, with a mean difference of 1,000 km (**Figure 5.21**). For many native plants, commercial

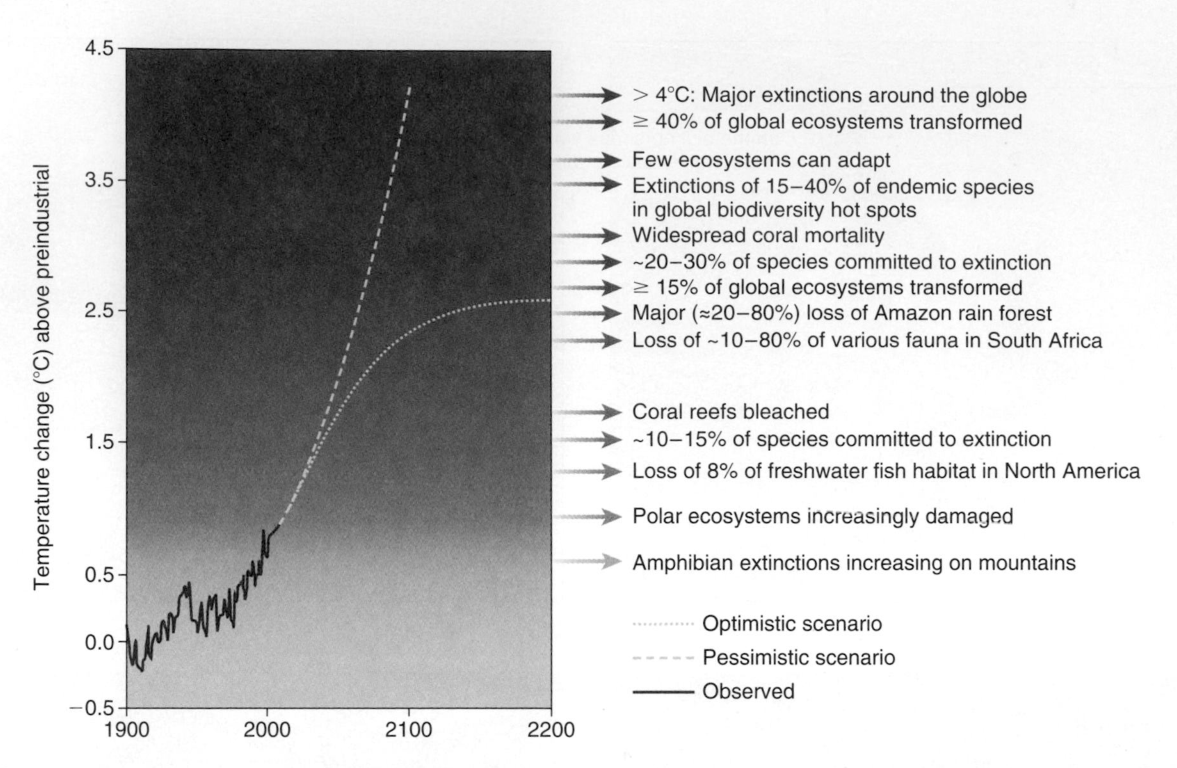


**Figure 5.22** **Changes in global temperature from 1900 to 2006 and beyond.** The range of ecological effects from various levels of global warming are shown on the right. (After Kerr, 2007.)

nurseries have in effect provided a head start on migration. Range expansions have also been noted in other taxa (see **Feature Investigation**).

The 2007 report of the Intergovernmental Panel on Climate Change (IPCC) suggested that a business-as-usual scenario with the same level of increasing greenhouse gas emissions would lead to a 2°C warming by about the middle of this century. Associated with this temperature rise are a variety of effects, including changes in species' phenologies (see **Global Insight**), increasing damage to polar ecosystems, increased coral bleaching, and perhaps the extinction of 10–15% of species (**Figure 5.22**). With a 2–3°C warming, several meters of sea-level rise will occur due to ice sheet loss. Sea level has already risen 10–25 cm over the past century, and seawater incursion into coastal forests has already killed trees in many areas. Part of the problem here is the long time lag between warming temperatures and time for ice to melt. Even if future temperature increases were prevented, ice would continue to melt for many years. For example, the huge thermal inertia of the ice sheets of Greenland mean they would take hundreds of years to melt, even though the increase in temperature needed to melt them would be reached in 50–100 years. You might envision this as being similar to an ice cube on a kitchen counter—it takes a while for the ice cube to melt, even though room temperature is considerably above freezing. This scenario means that even if no greenhouse gases were added to the atmosphere after 2010, we would already be committed to at least a quarter of the sea-level rise that could be expected in the century.

## Check Your Understanding

**5.3** What effects may global warming have on the phenology (timing) of natural phenomena?

Global Insight

## Global Warming Is Changing Species Phenologies

**Phenology,** the timing of life-cycle events such as flowering, egg laying, or migration will change with increased global warming. Terry Root and colleagues (2003) performed a meta-analysis on 61 studies, involving 694 species, that had examined shifts in spring phenologies over a time span of about 30 years. For example, the North American common murre, *Uria algae,* bred, on average, 24 days earlier per decade, and Fowler's toad, *Bufo fowleri,* bred 6.3 days earlier. Over an average decade, the estimated mean number of days changed in spring phenology was 5.1 days earlier. Most taxa, such as invertebrates, amphibians, birds, and nontree plants, show this type of change in phenology, whereas trees show lesser changes (**Figure 5.A**).

More critical than the change in phenology is the potential disruption of timing between interacting species such as herbivores and host plants, or predators and prey. For example, if plants flower early, before their pollinators take flight, then fruit production and seed set may be greatly reduced. All is well if the phenologies of all species are sped up by global warming. However, a review by Marcel Visser and Christian Both (2005) suggested this is the exception rather than the rule. These authors found only 11 studies that addressed the question of altered synchrony, in 9 predator–prey interactions and 2 insect–plant interactions. However, the results were discouraging. In 7 of the 11 cases, interacting species responded differently to temperature changes, putting them out of synchrony. For example, in the Colorado Rockies, yellow-bellied marmots, *Marmota flaviventris,* now emerge from hibernation 23 days earlier than they did in 1975, changing the relative phenology of the marmots with their food plants, which were not yet ready to be consumed.

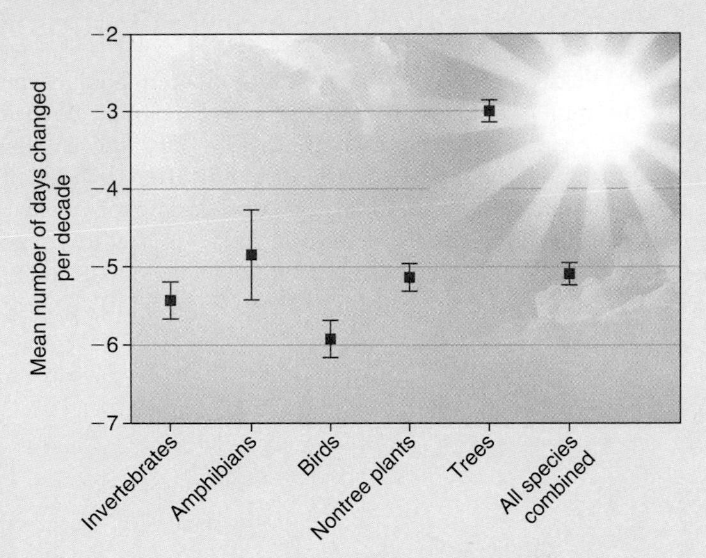


**Figure 5.A Phenological changes in response to global warming.** This meta-analysis summarizes temporal changes in animal behavior such as migration or hibernation times, and temporal changes in plant morphological events such as bud burst or the onset of flowering. (From Root et al., 2003.)

## Feature Investigation

### Rachel Hickling and Colleagues Showed That the Northerly Limits of a Wide Range of Taxonomic Groups Are Shifting Poleward

Rachel Hickling and colleagues (2006) documented range shifts in many different British taxonomic groups over a recent 25-year period between the 1960s and the 1990s. Their study was facilitated by the fact that after decades of natural history study, Britain has perhaps the most extensive time-scale and long-term distribution data for a wide range of taxa. For all taxa studied, except lizards and snakes, there was a consistent northward shift in range (**Figure 5.B**). Out of a total of 329 species studied, 275 shifted northward at their range margin, 52 shifted southward, and 2 did not change. The average northward shift across all species was between 30 and 60 km. For the three amphibian and reptile species included in the analysis, habitat loss from the northern edges of their ranges were thought to have driven them southward.

**HYPOTHESIS** The northerly limits of many animals have been shifting poleward in response to global warming.

**STARTING LOCATION** Great Britain

| | Conceptual level | Technique level |
|---|---|---|
| 1 | Examine northerly distribution of animal species in a variety of taxonomic groups whose northerly range limit is reached in Britain. | Examine distribution data from Biological Records Centre and British Trust for Ornithology. Introduced or migratory species are excluded. |
| 2 | Record northerly limits in 1960s and 1970s. | Each taxa was deemed present or absent in 10 km grid squares all over Britain. The northern most grid square was taken as the northerly limit. |
| 3 | Re-record northerly limits in 1980s and 1990s. | Repeat technique from step 2. This gives a distribution over two distinct time periods within the last 40 years for each taxonomic group. In most cases each recording period was 11 years long with a 14-year gap in between, corresponding to a 25-year period between the midpoint of the two recording periods. |

**4 THE DATA**

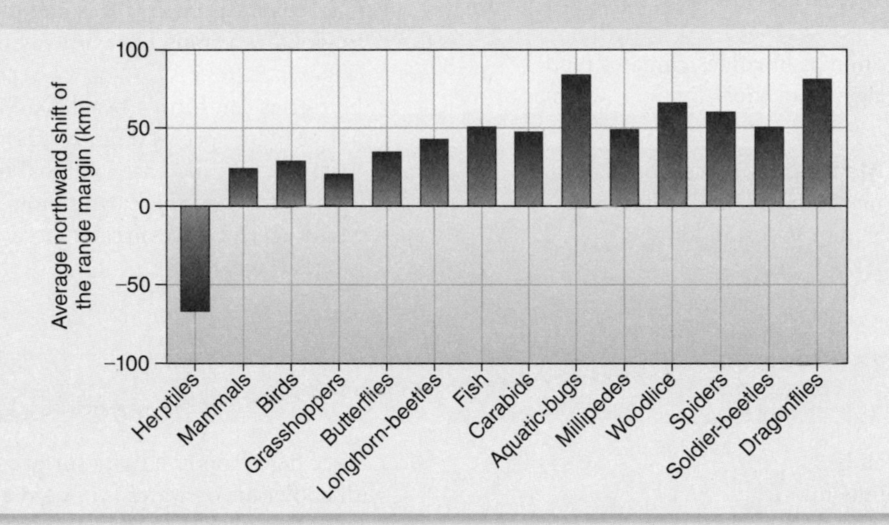


**Figure 5.B Rachel Hickling and colleagues have shown that many species have increased their northerly range in Britain in recent times.** (From Hickling et al., 2006.)

## SUMMARY

**5.1 The Effects of Cold Temperatures, Especially Freezing, Are Severe**

- Abiotic factors determine the fundamental niches of organisms (Figure S.1).
- Temperature exerts important effects on the distribution of organisms because of its effect on biological processes and the inability of most organisms to regulate their body temperature (Figure 5.1).
- The body temperature of heterotherms changes in different environmental conditions, whereas that of homeotherms remains constant (Figure 5.2).
- Animals use regulation of blood vessels near the skin and countercurrent heat exchange systems to maintain body heat in cold environments (Figure 5.3, 5.4).
- Despite their adaptations, the distribution of many homeotherms, especially smaller species such as birds and bats, is limited by cold temperatures (Figures 5.5, 5.6).
- Low temperatures, particularly freezing, limit the distribution of many plants (Figure 5.7).
- Frost is an important environmental factor that limits the distribution of many plant species whose cells burst upon freezing (Figures 5.8, 5.9).
- Bergmann's rule states that animals in colder climates are larger than similar species in warmer climates (Figure 5.10).
- Allen's rule states that animals in colder climates tend to have shorter appendages than those living in warmer climates (Figure 5.11).

**5.2 Hot Temperatures Limit Many Species' Distributions**

- Many organisms have physiological and behavioral mechanisms to allow them to withstand high temperatures (Figures 5.12–5.14).
- Fire initiated by lightning is a frequent threat to organisms, yet many species have special adaptations to allow them to survive low-temperature fires (Figures 5.15, 5.16).
- Wind can amplify the effects of temperature and can modify wave action (Figure 5.17).

**5.3 The Greenhouse Effect Causes the Earth's Temperature to Rise**

- In the greenhouse effect, short-wave solar radiation passes through the atmosphere to warm the Earth and is radiated back into the atmosphere as long-wave infrared radiation. Much of this long-wave radiation is absorbed by atmospheric gases and reradiated back to Earth's surface, causing Earth's temperature to rise (Figure 5.18).
- The major anthropogenic atmospheric gases causing the greenhouse effect are carbon dioxide, methane, nitrous oxide, and chlorofluorocarbons (Table 5.1).
- An increase in anthropogenic atmospheric gases is increasing the greenhouse effect, causing global warming, a gradual elevation of the Earth's surface temperature. Ecologists expect that global warming will have a large effect on the distribution of the world's organisms (Figure 5.19).
- Changes in species' ranges and phenologies in response to global warming have already been detected (Figures 5.20, 5.A).
- Nurseries can form a beachhead for range expansion of native plant species (Figure 5.21).
- Even a 2–3°C increase in global temperatures will have many effects on the Earth's biota (Figure 5.22).
- Many species have already increased their northerly ranges (Figure 5.B).

## TEST YOURSELF

1. In size, a realized niche can be:
   a. Greater than a fundamental niche
   b. Smaller than a fundamental niche
   c. The same size as a fundamental niche
   d. a and c
   e. b and c
2. The distribution of vampire bats is limited by:
   a. Rainfall
   b. Cold temperatures
   c. Warm temperatures
   d. pH
   e. Wind
3. The tendency for endothermic animals to be bigger in cold areas is known as:
   a. Cope's rule
   b. Allen's rule
   c. Gloger's rule
   d. Bergmann's rule
   e. Leibig's rule
4. Loss of heat from the body surface through direct contact with cooler air or water is known as:
   a. Radiation
   b. Conduction
   c. Convection
   d. Invection
   e. Evaporation
5. The discomfort you feel on a humid day is due to the failure of:
   a. Radiation
   b. Conduction
   c. Convection
   d. Evaporation
6. In which two states in the U.S. is lightning the most frequent?
   a. Georgia and Florida
   b. California and Hawaii
   c. Colorado and Florida
   d. Florida and Texas
   e. Louisiana and Texas

7. Which anthropogenic gas contributes most to global warming?
   a. Methane
   b. Chlorofluorocarbons
   c. Carbon dioxide
   d. Nitrogen oxides
   e. Sulfur dioxide
8. Solar radiation in the form of __________ energy heats the Earth's surface, whereas __________ infrared energy is radiated back to the atmosphere.
   a. short-wave, short-wave
   b. short-wave, long-wave
   c. long-wave, short-wave
   d. long-wave, long-wave
9. The data show that atmospheric carbon dioxide levels have increased by what percent over the past 46 years?
   a. 1
   b. 5
   c. 12
   d. 25
   e. 52
10. Sea-level rise due to global warming over the past century is thought to be in the range:
   a. 1–5 cm
   b. 5–10 cm
   c. 10–25 cm
   d. 25–50 cm
   e. 50–100 cm

## CONCEPTUAL QUESTIONS

1. Explain the greenhouse effect.
2. What effects is global warming having on the distribution of plants and animals?
3. Why are low temperatures in general, and frost in particular, such limiting factors for plants?
4. The U.S. Forest Service and state agencies now regularly manage many areas by burning them. Explain this seemingly contradictory behavior.

## DATA ANALYSIS

The figure details the distribution limits of loblolly pine, *Pinus taeda,* in the U.S. Southeast and the calculated limits based on a model of Hocker (1956). What factors do you think might set the limits of this distribution?

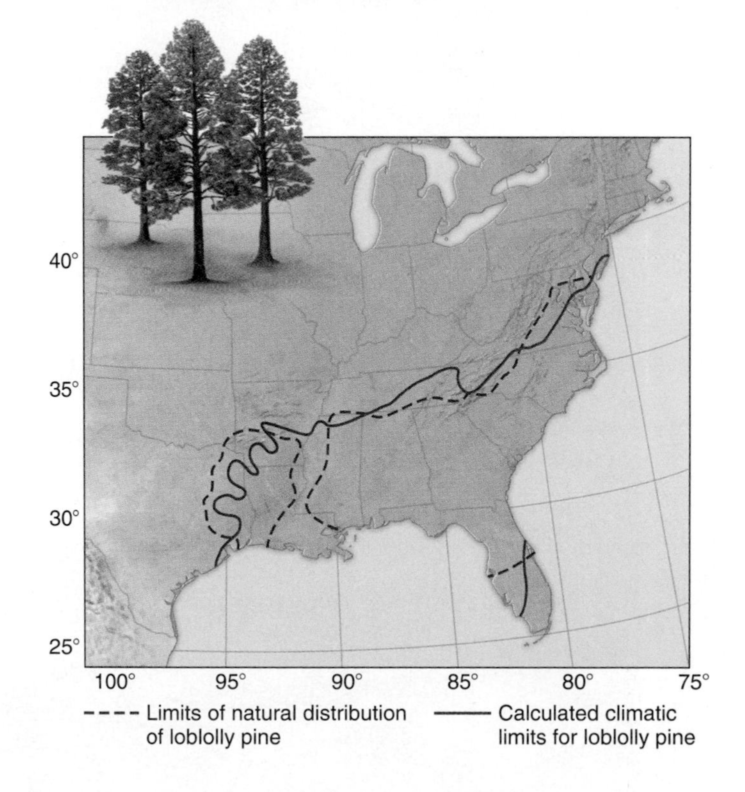

Lesser flamingoes, Kenya.

CHAPTER 6

# Water

## Outline and Concepts



High-salinity lakes can support the cyanobacterium *Spirulina,* which is a source of protein for flamingoes that feed on them. In the 1960s a lake in Kenya that was the traditional breeding ground of a large flamingo flock became flooded, and close to 2 million birds moved to Lake Magadi, a shallow lake fed by saltwater springs. The birds bred, but intense heat caused evaporation, and many thousands of baby flamingoes became salt encrusted. Fortunately a group of conservationists was visiting the area and chipped off the mineral shackles, saving about 27,000 young birds from death. The flamingo colony returned to its original lake the following year. The salinity of the flamingo's lake is critical. Too low and it would not support *Spirulina*, too high and it encrusts the chicks' legs.

Water has an important effect on the abundance of organisms. Cytoplasm is 85–90% water, and without moisture there can be no life. Water performs crucial functions in all living organisms. It acts as a solvent for chemical reactions, takes part in the reactions of hydrolysis, is the means by which animals eliminate wastes, and is used for support in plants and in some invertebrates. In this chapter we see not only how water availability is critical to many organisms, but also how the salinity and pH of water is of vital importance.

## 6.1 Water Availability Affects Organismal Abundance

Water is an abundant compound on Earth that exists in several different forms—ice, water, and water vapor. At most temperatures on Earth, water is found primarily as a liquid. In this section we will examine the properties of water and see how important water is to living organisms.

### 6.1.1 Water has unusual physical properties

Water has some amazing properties (**Figure 6.1**). The covalent bands linking the two hydrogen atoms to the oxygen atom are polar, and while oxygen has a slight negative

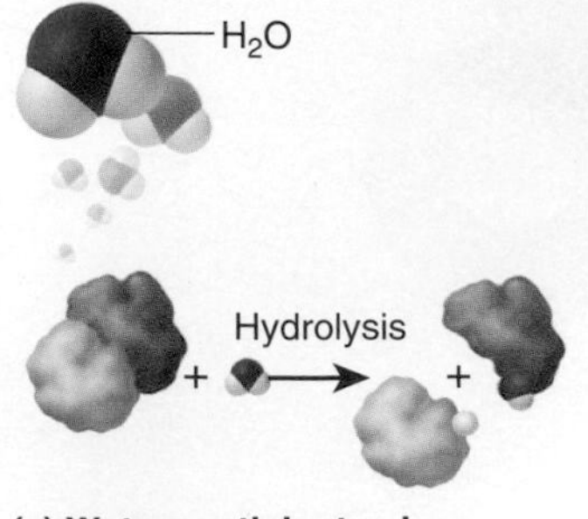


**(a) Water participates in chemical reactions.**

**(b) Water provides support.** The plant on the right is wilting due to lack of water.

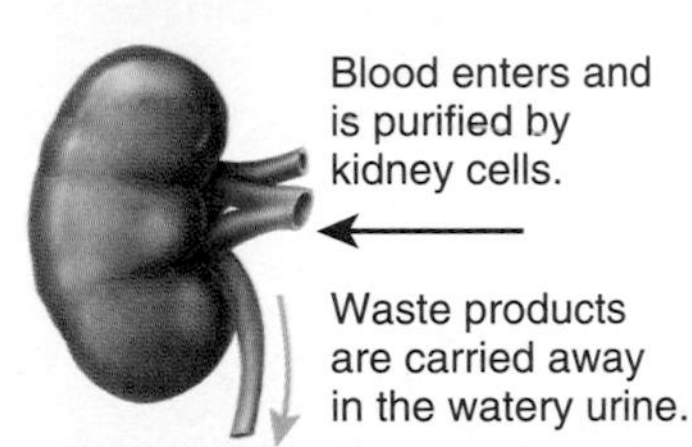


**(c) Water is used to eliminate soluble wastes.**

**(d) Evaporation helps some animals dissipate body heat.**

**(e) The cohesive force of water molecules aids in the movement of fluid through vessels in plants.**

**(f) Water in saliva serves as a lubricant during—or as shown here, in anticipation of—feeding.**

**(g) The surface tension of water explains why this water strider doesn't sink.**

**Figure 6.1** Some of the amazing roles of water in biology. In addition to acting as a solvent, water serves many crucial functions in nature.

charge, hydrogen has a slight positive charge. The hydrogen bonds of water cause molecules to be attracted to each other, a phenomenon known as cohesion. Surface tension is a measure of the attraction of molecules at the water surface. This surface tension allows some lightweight insects such as water striders to walk on the surface of a pond without sinking.

To dissolve in water, a substance must be able to separate with positive ions being attracted to the oxygen and negative ions to the hydrogen. For example, in water, table salt (NaCl) dissociates into $Na^+$, attracted to the negatively charged oxygen, and $Cl^-$, attracted to the positively charged hydrogen. Substances such as salt, which contain ionic bonds, or those with polar covalent bonds, dissolve in water and are said to be hydrophilic or water-loving. Molecules consisting primarily of carbon and hydrogen are not generally water-soluble, because the carbon-carbon or carbon-hydrogen bonds are nonpolar. Such molecules are said to be hydrophobic or water-fearing. Biological membranes, such as those that encase cells, are largely made up of nonpolar molecules.

Water exists in three states: solid (ice), liquid (water), and gas (water vapor). In the liquid state weak hydrogen bonds between molecules are continuously being formed and re-formed. At higher temperatures these bonds are broken more frequently and molecules escape into the air as water vapor. At lower temperatures, the rate of hydrogen bond breakage decreases. At 0°C the water molecules lie in an open arrangement with greater intermolecular distances (**Figure 6.2**). This makes ice less dense than water and allows it to float on the water surface. Pure water freezes at 0°C and vaporizes at a 100°C. However, adding solutes to water decreases its freezing point below 0°C and raises its boiling point above 100°C.

An important feature of water is that it is incompressible. This property is important because many organisms use water for support. For example, water provides turgidity (stiffness) for plants and supports the bodies of invertebrates such as worms. Water also has a high specific heat, defined as the amount of heat energy needed to raise 1 gram of a substance by 1°C. This is why large bodies of water have relatively stable temperatures compared to nearby landmasses. This is also why evaporative cooling via sweating is so effective, because converting liquid water to gaseous water vapor uses a lot of body heat.

## 6.1.2 Water has great importance to living organisms

The water content of plant cells depends on osmosis and turgor pressure. Osmosis is the movement of water across membranes to balance solute concentrations. Water diffuses from a solution that is hypotonic (lower solute concentration) into a solution that is hypertonic (higher solute concentration). Turgor pressure is the hydrostatic pressure that increases as water enters plant cells. Plant cell walls restrict the extent to which cells can expand, hence the pressure on the cell wall increases as more water enters the cell. A property known as relative water content (RWC) is a measure of turgidity and hence the water content of plants. To estimate RWC, three measurements are needed: fresh weight, turgid weight, and dry weight. First, a sample of freshly cut material is weighed to get fresh weight. Next, this material is completely hydrated

**Figure 6.2 Structure of liquid water and ice.** In its liquid form, the hydrogen bonds between water molecules continually form, break, and re-form, resulting in a changing arrangement of molecules from instant to instant. At temperatures at or below its freezing point, water forms a crystalline matrix called ice. In this solid form, hydrogen bonds are more stable. Ice has a hexagonally shaped crystal structure. The greater space between $H_2O$ molecules in this crystal structure causes ice to have a lower density compared to water. For this reason, ice floats on water.

in water in an enclosed, lighted chamber until a turgid weight is reached. Finally, the sample is dried to a constant weight, usually in a drying oven, to get the dry weight. These measurements are used to calculate RWC according to the equation

$$\mathrm{RWC} = \frac{(\text{fresh weight} - \text{dry weight})}{(\text{turgid weight} - \text{dry weight})} \times 100$$

In arid conditions, RWC values of less than 50% cause most plants to die, whereas values closer to 100% indicate that plants are not being affected by drought stress. Agriculturalists are very interested in developing new strains of crops that are more tolerant of dehydration, and RWC measurements evaluate the effectiveness of such crops at tolerating water stress. Molecular biologists have recently genetically engineered rice plants with a barley gene that confers dehydration tolerance.

Plants have a variety of strategies to deal with water stress. First, many plants undergo osmotic adjustment. Increased amounts of the amino acid proline or sugars such as glucose and fructose increase solute concentration, drawing water into the cells. In addition, sugars bind to phospholipids, helping stabilize cellular membranes, preventing them from becoming damaged during times of drought. Some plants of arid habitats close their stomata in the day and open them at night, when it is cooler, allowing carbon dioxide to enter the leaf without too much water loss. Water-stressed plants may also drop their leaves, a process called leaf abscission. Leaf abscission reduces the amount of root mass needed to supply water under arid conditions. For example, in North American deserts, the ocotillo, *Fouquieria splendens,* produces leaves after rains but abscises them when it becomes dry. This process may occur multiple times a year.

The distribution patterns of many plants are limited by available water. Some plants, such as the water tupelo tree, *Nyssa aquatica,* in the southeast U.S., do best when completely flooded and are thus found predominantly in swamps. In contrast, coastal plants that grow on sand dunes experience very little fresh water. Their roots penetrate deep into the sand to extract moisture. In cold climates, water can be present but locked up as permafrost and, therefore, unavailable; this is termed a frost-drought situation. Deciduous trees drop their leaves in fall because in frozen soil, roots would be unable to take up enough water to keep the leaves hydrated. Alpine trees can be affected by frost drought. The trees stop growing at a point on the mountainside where they cannot take up enough moisture to offset transpiration losses. This is the point where a combination of low temperatures and high winds makes transpiration exceed water uptake. This point, known as the **tree line** or timberline, is readily apparent on many mountainsides (**Figure 6.3**). Not surprisingly, the density of many desert plants is limited by the availability of water. For example, a significant correlation is observed between increased rainfall and increased creosote bush density in the Mojave Desert (**Figure 6.4**).

**Figure 6.3 The timberline.** On many mountains, such as this one in the Canadian Rockies, the upper limit of trees exists as a line well defined by drought stress.

**ECOLOGICAL INQUIRY**

Is the timberline set by low temperatures or drought?

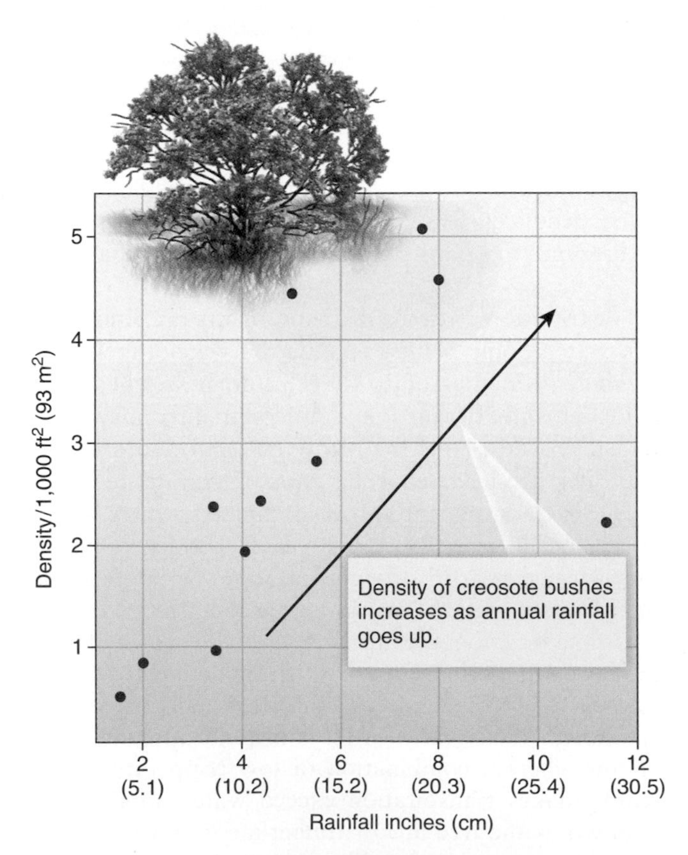


**Figure 6.4 The density of creosote bush, *Larrea divaricata*, in the U.S. Southwest is governed by rainfall.** Data are from 11 sites in the southwestern U.S. (After Woodell et al., 1969.)

Animal cells lack cell walls and do not experience turgor pressure. However, animals lose water as a consequence of respiration or the elimination of wastes. Proteins and RNA molecules exist for finite periods of time and are broken down into smaller components to be excreted. The nitrogenous wastes that result are often toxic at high concentrations and must be eliminated. Nitrogenous wastes usually exist in three forms: ammonia or ammonium ions, urea, or uric acid, depending on the type of animal involved and the environment in which it lives (**Figure 6.5**). Excretory organs such as flame cells, nephridia, gills, or kidneys excrete these wastes. Ammonia is so toxic that animals that produce it excrete it immediately. Most animals that excrete ammonia are therefore aquatic, where water is not limiting and wastes can be expelled repeatedly. Some terrestrial snails and crustaceans can excrete ammonia in gaseous form, reducing the need for water. All mammals, most amphibians, some marine fishes, some reptiles, and some terrestrial invertebrates convert ammonia to urea, which is then excreted. Urea is less toxic than ammonia and can be stored in urinary bladders. This reduces the need for constant excretion and thus the need for water. However, the production of urea requires the expenditure of energy. Birds, insects, and most reptiles produce uric acid or nitrogenous compounds called purines. These are also less toxic than ammonia but are even more energetically costly to produce. They are not very water-soluble and are excreted with salts and other waste products in a semiliquid paste, which conserves water even more. Some birds, such as vultures and storks, excrete urine on their legs, cooling themselves by evaporation. This behavior is termed **urohydrosis.**

The distribution patterns and population densities of animals are often strongly influenced by water availability. Because most animals depend ultimately on plants as food, their distribution is directly linked to those of their food sources. Such a phenomenon regulates the number of buffalo, *Syncerus caffer,* in the Serengeti area of Africa. In this area, grass productivity is related to the amount of rainfall in the previous month. Buffalo density is governed by food availability, so there is a significant correlation between buffalo density and rainfall (**Figure 6.6**). The only exception occurs in the vicinity of Lake Manyara, where groundwater promotes plant growth. Other African wildlife may respond more directly to rainfall. Large herds of wildebeest, *Connochaetes taurinus,* may turn in their tracks to head toward a rainstorm. Rainfall in African grasslands is localized but may be visible from up to 80 km away. By the time the wildebeest arrive, fresh grass may be about to appear.

In North America, populations of white-tailed deer in Texas are influenced by precipitation in the previous year (**Figure 6.7a**). Drought years are particularly hard on wildlife populations and reproduction is reduced, lowering the population density in the subsequent year. However, once rainfall approaches long-term mean rainfall levels, reproduction and hence population densities bounce back, and even above-normal rainfall may not increase densities further.

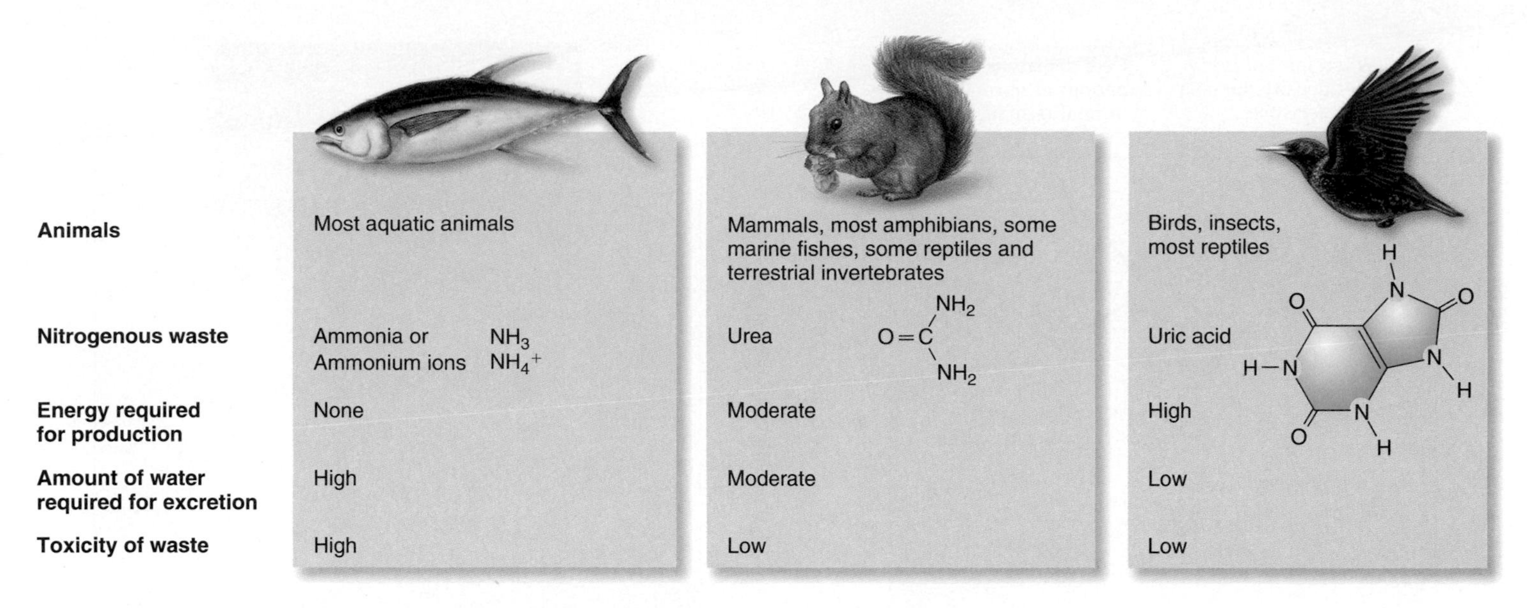


**Figure 6.5** The three main types of nitrogenous wastes produced by animals.

**ECOLOGICAL INQUIRY**

Why don't mammals produce ammonia, as it is cheap to produce?

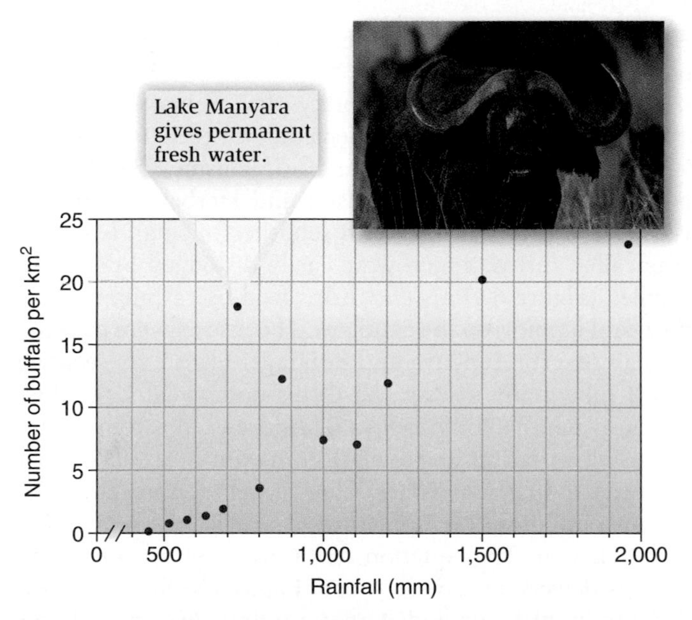


**Figure 6.6 Rainfall limits the density of buffalo.** In the Serengeti area of Africa, buffalo density is dependent on grass availability, which itself is dependent on annual rainfall. The main exception is where there is permanent water, such as Lake Manyara, where greater water availability leads to higher grass growth and buffalo densities. (After Sinclair, 1977.)

More than half the population of North American mallard ducks, *Anas platyrhynchos,* and pintail ducks, *A. acuta,* breed in small lakes of the northern plains. However, these small, shallow lakes and large ponds may dry out periodically in times of summer drought, reducing the habitat area available for waterfowl. A census known as the July Pond Index was developed to assess the availability of ponds and lakes for ducks. Over a period of many years, a strong relationship was shown between the number of these lakes that contain water and the breeding population of ducks the next year (**Figure 6.7b**).

Desert animals find shelter under the shade of plants or in burrows. Kangaroo rats, *Dipodomys* spp., seal their burrows to promote the recycling of moisture from their own breath (**Figure 6.8**). There are few large mammals in deserts, because it is hard for them to find such shelter and to obtain sufficient water. Many desert animals are nocturnal, inactive during the hot daylight hours and active at night when it is cooler. Some species, such as the kangaroo rats, can get all the water they need from their food and do not drink even when water is available. Each carbohydrate molecule, $C_6H_{12}O_6$, can potentially yield six $H_2O$ molecules. Kit foxes, *Vulpes macrotis,* also obtain some water via this reaction. The rest comes directly from their rodent food. More prey is consumed for water than for energy demands.

The importance of water in limiting animal population density was brought home by an extraordinary event in the Galápagos Islands caused by an El Niño Southern Oscillation (ENSO). In the tropical Pacific Ocean, strong trade winds normally keep warm water pushed against the coasts of New Guinea and Australia (**Figure 6.9a**). Cold, nutrient-rich water wells up along the coast of the Americas, supporting large fish populations. Heavy rains are concentrated in the western Pacific. The Galápagos Islands, located off the coast of Ecuador, in the eastern Pacific, are normally very dry. In El Niño conditions, however, weakened trade winds allow warm water to drift eastward toward the Americas. This stops the cooler nutrient-rich water from welling up. The heat of the ocean increases cloud formation and rainfall along the eastern Pacific,

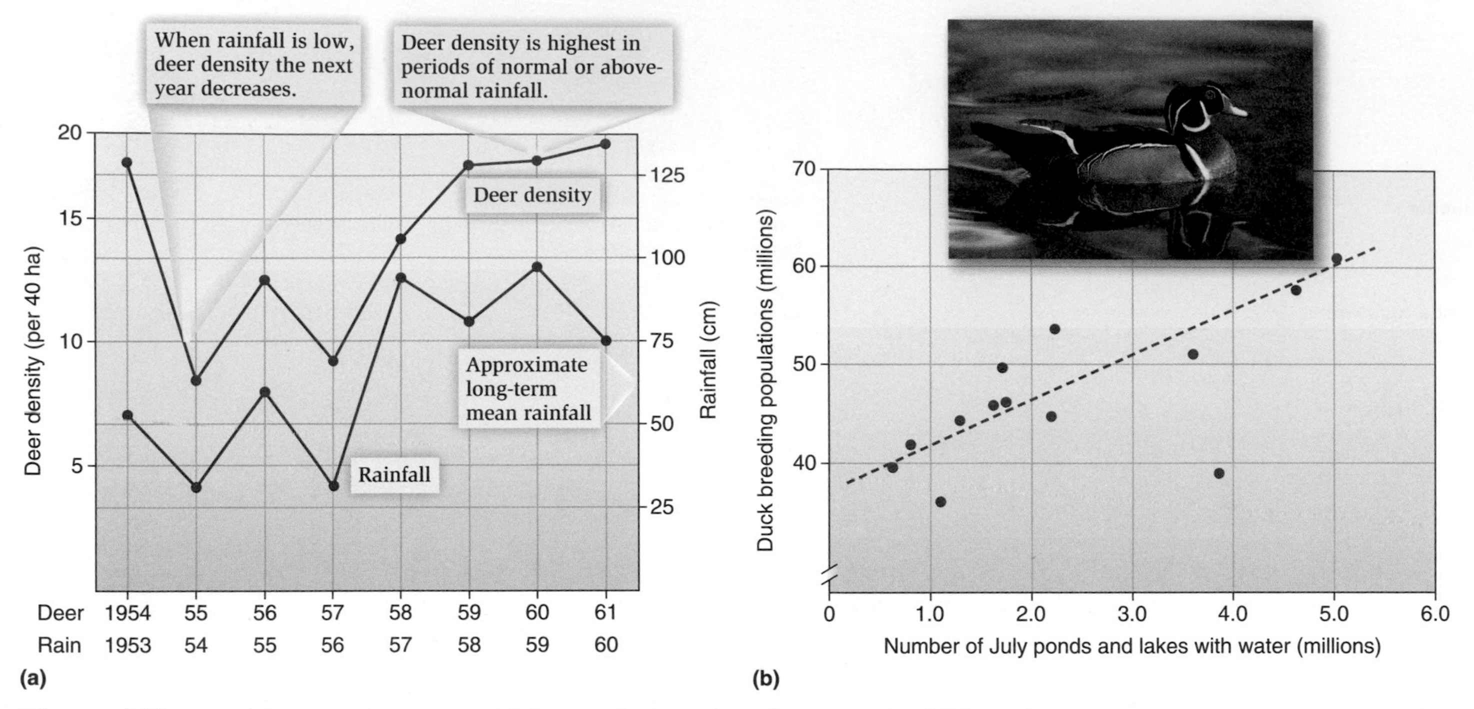


**Figure 6.7** **The influence of water availability on the breeding densities of wildlife in the U.S.** **(a)** White-tailed-deer density in Texas is influenced by precipitation in the preceding year. (After Teer et al., 1965.) **(b)** The number of breeding ducks in the northern plains is strongly limited by the number of water-containing ponds and lakes in the previous July. (After Crissey, 1969.)

**Figure 6.8** **Water conservation by desert animals.** This Ord's kangaroo rat, *Dipodomys ordii*, seals its burrow in the daytime to prevent water loss.

while decreasing rainfall in the western Pacific (**Figure 6.9b**). During the strong 1982–1983 El Niño, rainfall on Isla Genovesa in the Galápagos Islands increased from its normal 100–150 mm to 2,400 mm. Plants responded with prodigious growth, and certain finches, *Geospiza* spp., bred up to eight times rather than their normal maximum of three, probably because of increased abundance of fruits and seeds. Future changes in global precipitation patterns and the frequencies of El Niño events may therefore be critically important (see **Global Insight** feature).

Not only does water availability play an important indirect role in influencing animal densities via their food, it can directly affect the densities of small organisms. The Australian entomologists James Davidson and Herbert Andrewartha (1948) studied densities in rosebushes of small (<1 mm) insects called thrips, which were easily dislodged by rainfall and killed (**Figure 6.10a**). The thrips feed by rasping the leaf surface and lapping up the exudates. The insects were counted every day (except Sundays and holidays) for 81 consecutive months! Like many insects, the thrips underwent large fluctuations in density. Davidson and Andrewartha found that 78% of the variation in population maxima was accounted for by variations in weather. They developed an equation based on rainfall and temperature to predict the number of thrips each year. The equation did a great job of predicting the thrips density in any one year (**Figure 6.10b**). Davidson and Andrewartha concluded that rainfall and, to a lesser degree, temperature were the overriding influences on the density of thrips. Rainfall was important to these tiny insects, easily dislodging them from the leaves on which they fed and creating huge day-to-day fluctuations in their densities. Temperature affected reproductive rates. Additional variation was explained by the size of the population that overwintered. A large overwintering population gave rise to a large population the following growing season.

Similar correlations between abiotic variables and population density were found for other organisms. W. J. Francis

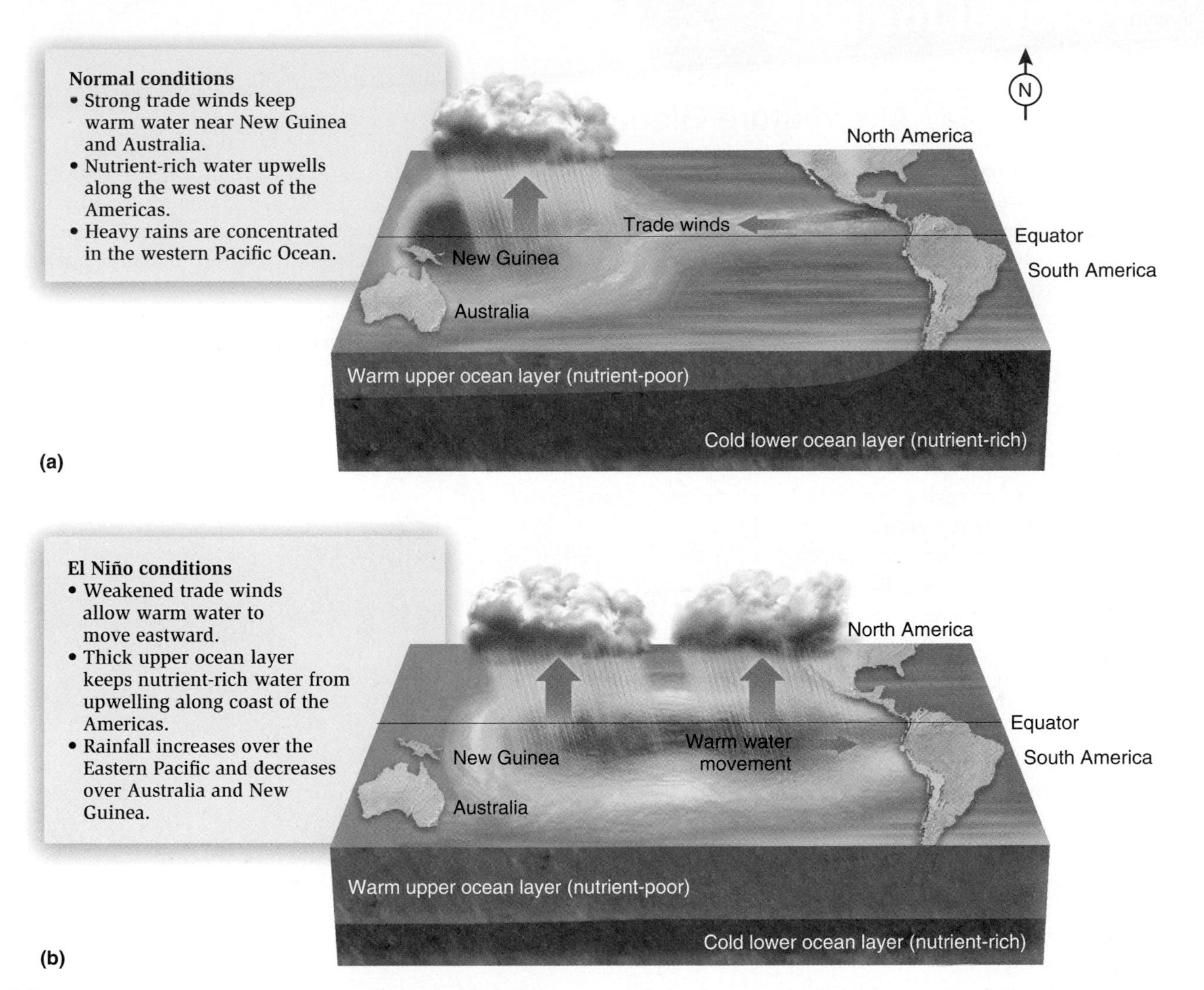


**Figure 6.9** **The El Niño Southern Oscillation, ENSO.** **(a)** In non–El Niño years, trade winds keep warm water piled up against Australia and New Guinea. **(b)** In an El Niño year, the trade winds decrease and warm water flows back across the Pacific Ocean toward the Americas.

**Figure 6.10** **The effects of rainfall on insect densities.** **(a)** Thrips insects, *Thrips angusticeps.* **(b)** Comparison of means of observed annual population densities of thrips, *Thrips imaginis,* and densities predicted by a model based on temperature, rainfall, and numbers of overwintering thrips. The means agree well, showing that thrips densities are well predicted by abiotic factors. (From data in Davidson and Andrewartha, 1948.)

Global Insight

## Global Warming May Alter Future Global Precipitation Patterns

An increase in global temperatures may alter global precipitation. Higher temperatures increase evaporation, and warmer air holds more water vapor than cooler air. This will generally lead to more frequent and heavier precipitation. Over the period 1986–2000, the average increase in precipitation over land was 3.5 mm per year. However, over normally dry areas there is no more moisture to evaporate, so the increased temperatures will likely lead to widespread droughts (**Figure 6.A**). We can make the analogy that the atmosphere is like a sponge; warm moist air allows the sponge to absorb more water, but in dry conditions nothing more can be rung out of the sponge. Northern and southern temperate areas have already shown precipitation increases, as have tropical and subtropical areas. However, desert areas such as the Sahel region south of the Sahara have become drier. In essence, the wet get wetter and the dry get drier.

Once again, changed precipitation patterns will alter the distribution patterns and densities of some organisms. When Margaret Davis mapped out the future distribution of the sugar maple, *Acer saccharum*, in a globally warmed world and took into account changed precipitation patterns, the predicted future area of distribution was considerably less than the prediction taking into account only temperature changes (**Figure 6.B**).

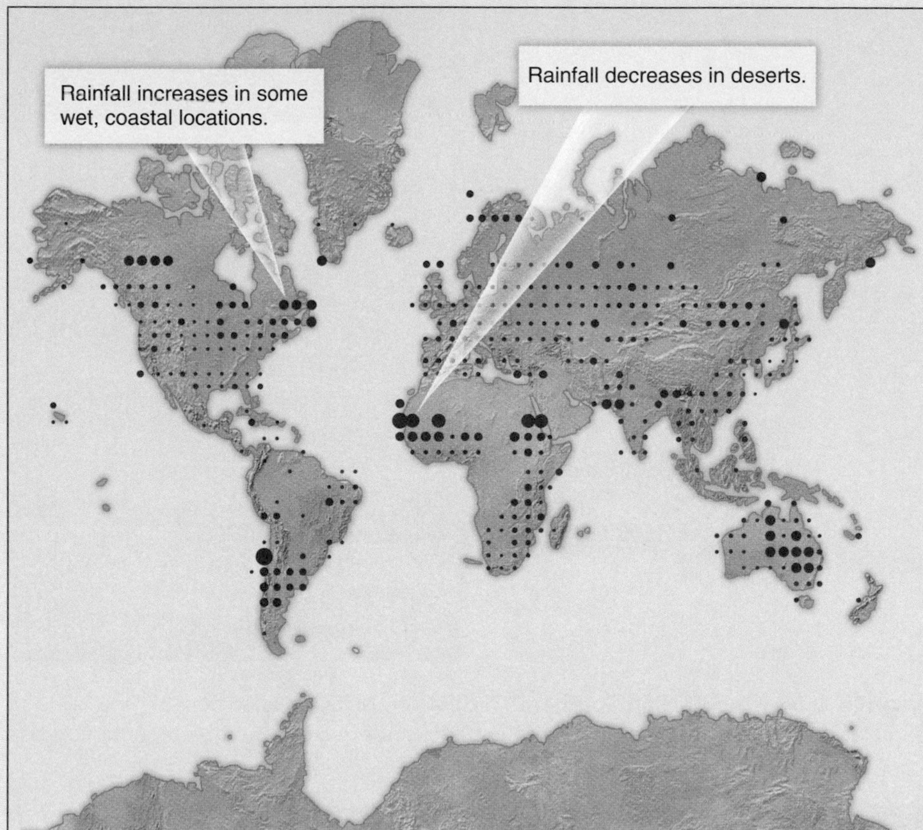


**Figure 6.A The effects of global warming on precipitation.** Reported trends, 1900–2000.

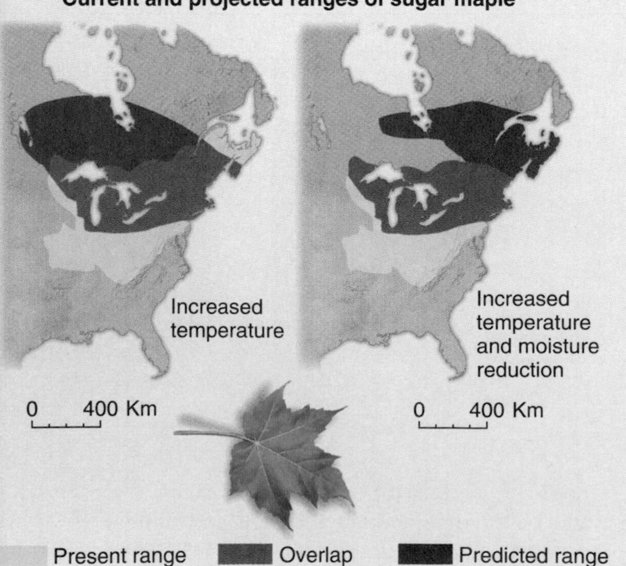


**Figure 6.B The effects of global warming and precipitation changes on sugar maple, *Acer saccharum*.** **(a)** Under temperature changes only; **(b)** under temperature and precipitation changes. (After Davis and Zabinski, 1992.)

(1970) developed an equation based on soil moisture and seasonal rainfall that explained 99% of the variation in the density of California quail, which feed on plant seeds and insects. In this case, soil moisture alone was most important, explaining 83.1% of the variation. It acted mainly by influencing the density of insects on which the quail fed. The remaining variation may be explained by the quantity and quality of plant seeds that the quail consume. In dry years there are fewer seeds, and high levels of phytoestrogens in the seeds inhibit quail reproduction. In wet years seeds are abundant and phytoestrogens are virtually absent.

**Check Your Understanding**

**6.1** Why do you think the geographic range of the huge redwood tree, *Sequoia sempervirens,* is limited to coastal areas of central and northern California and southern Oregon?

## 6.2 Salt Concentrations in Soil and Water Can Be Critical

Salt concentrations vary widely in aquatic environments and have a great impact on osmotic balance in animals. Oceans contain considerably more dissolved minerals than rivers do. This is because oceans continually receive the nutrient-rich waters of rivers, and the sun evaporates pure water from ocean surfaces, making concentrations of minerals such as salt even higher. However, closed inland watersheds can steadily accumulate salts to the point that very high salinities result. Such is the case in the Great Salt Lake in Utah, the Dead Sea bordering Israel, and many smaller lakes in Africa. As we saw at the beginning of the chapter, very high salt concentrations can cause problems for aquatic life.

The phenomenon of osmosis influences how aquatic organisms cope with different environments. Freshwater fish are hyperosmotic (having a higher concentration of ions) to their environment and tend to gain water by osmosis as it passes over the thin tissue of the gills and mouth. To counter this, the fish continually eliminate water in a dilute urine (**Figure 6.11**). Freshwater fish produce up to 30% of their body weight per day in urine. However, to avoid losing all dissolved ions, many ions are reabsorbed into the bloodstream at the kidneys. Other freshwater dwellers such as frogs absorb water through their skin, and they also produce copious dilute urine. Marine fish are hypo-osmotic (having a lower concentration of ions) to their environment and tend to gain salts and lose water across the mouth and gills. They drink more sea water to compensate for this loss, but the water contains a higher concentration of salts. This ingested salt is excreted at the gills and kidneys, producing a concentrated urine.

Other animals have different adaptations to the salinity of marine environments. Marine turtles and seabirds ingest seawater when consuming prey, or drink it, having no access to

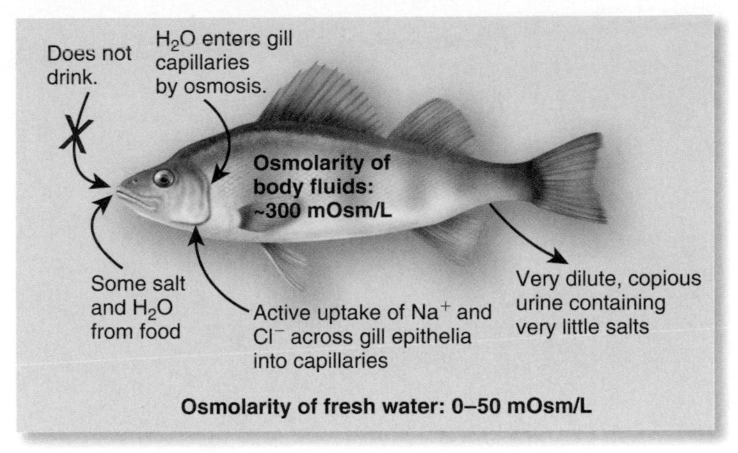


**(a) Freshwater fish**

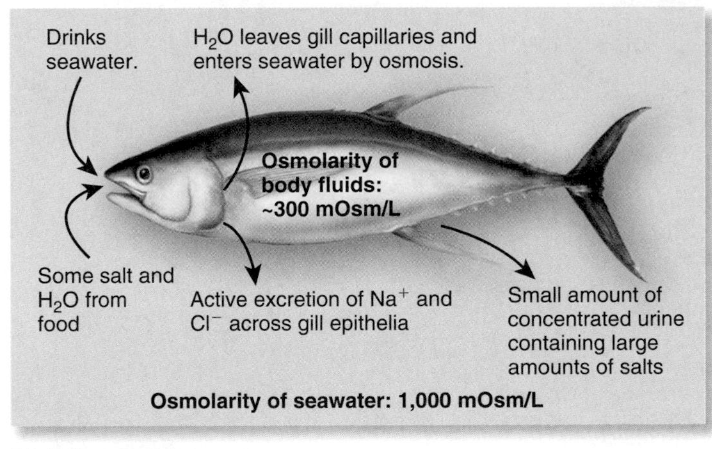


**(b) Saltwater fish**

**Figure 6.11 The effect of salt concentration on aquatic organisms.** The challenges of **(a)** freshwater and **(b)** saltwater living for fish are overcome by different strategies.

fresh water. These animals have specialized structures called salt glands located around the eyes and nostrils (**Figure 6.12**). Salt is transported from the blood to the interstitial fluid into the tubules of the gland. A highly concentrated, often viscous solution is secreted through a central duct to the outside. Female turtles appear to be crying when they lay their eggs on the beach, but they are merely excreting salt from their bodies.

Salt in the soil affects the growth of plants. In arid terrestrial regions, salt accumulates in soil where water settles and then evaporates. This can be of great significance in agriculture, where continued watering in arid environments, together with the addition of salt-based fertilizers, greatly increases salt concentration in soil and reduces crop yields. A very few terrestrial plants are adapted to live in saline soil along seacoasts. Here the vegetation consists largely of **halophytes,** species that can tolerate higher salt concentrations in their cell sap than regular plants. Some halophytes, such as glasswort, *Salicornia perennis,* accumulate inorganic salts in their vacuoles, balancing solute concentration inside and outside of the cell and preventing osmotic water loss. Species such as mangroves and *Spartina* grasses have salt glands that excrete salt to the surface

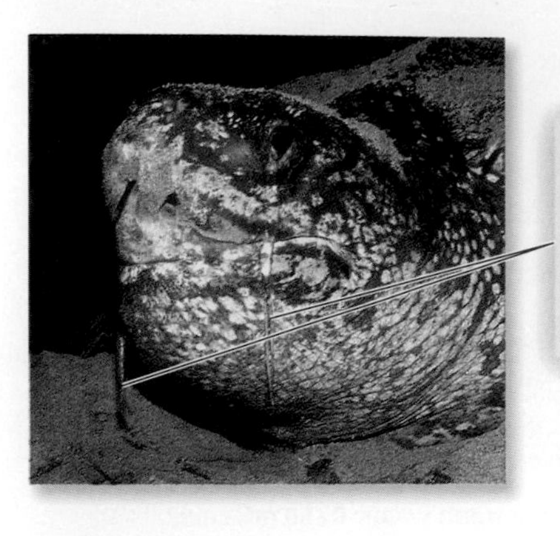


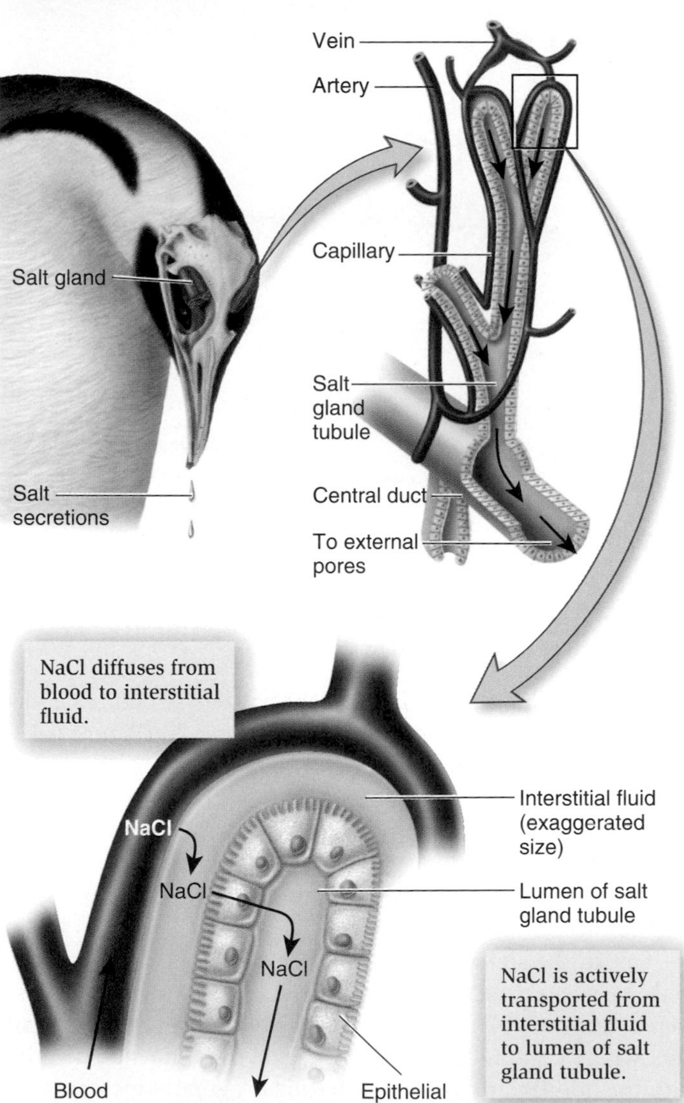


**Figure 6.12 Salt glands in turtles and sea birds are an adaptation to a marine lifestyle.** The viscous solution that moves through the central duct of the salt gland to the outside is high in salt. These glands occur near the nose and eyes, and the secretions make it appear as though the animal is crying.

**Figure 6.13 Some plant species have adaptations for salty conditions.** Special salt glands in *Spartina* leaves exude salt, enabling this grass to exist in saline intertidal conditions.

of the leaves, where it forms tiny white salt crystals (**Figure 6.13**). Windborne salt can affect the distribution of plants on sand dunes, too. On the Atlantic coast of the United States, sea oats, *Uniola paniculata,* can withstand high atmospheric salt and thus inhabit areas near the shoreline, while another dune grass, *Andropogon littoralis,* is less tolerant and occurs in protected areas back from the shoreline.

**Check Your Understanding**

**6.2** Are marine or freshwater bony fish hypo-osmotic to their environment and what does this mean for them?

## 6.3 Soil and Water pH Affect the Distribution of Organisms

Pure water has the ability to disassociate or ionize into hydrogen ions, $H^+$ ions, and hydroxide ions, $OH^-$ ions. In pure water the concentrations of $H^+$ and $OH^-$ in water are both $10^{-7}$ mols per liter, or 0.0000001 mols per liter, a relatively small amount. When certain substances are dissolved in water, they may release or absorb $H^+$ or $OH^-$ ions, thereby altering the relative concentrations of these ions. Molecules that release hydrogen ions in solution are called **acids.** Two examples are shown below:

$$HCl \rightarrow H^+ + Cl^-$$

(hydrochloric acid) (chloride)

$$H_2CO_3 \rightarrow H^+ + HCO_3^-$$

(carbonic acid) (bicarbonate)

Hydrochloric acid is called a **strong acid** because it completely dissociates into $H^+$ and $Cl^-$ ions when added to water. By comparison, carbonic acid is a **weak acid,** because some of it will remain in the $H_2CO_3$ state when dissolved in water.

Compared to an acid, a **base** has the opposite effect when dissolved in water—it lowers the $H^+$ concentration. This can occur in two ways. Some bases release $OH^-$ when dissolved in water:

$$NaOH \rightarrow Na^+ + OH^-$$
(sodium hydroxide) (sodium ion)

When a base such as NaOH raises the $OH^-$ concentration, some of the $H^+$ ions bind to these $OH^-$ ions to form water. Therefore, increasing the $OH^-$ concentration lowers the $H^+$ concentration. Alternatively, other bases, such as ammonia, react with water:

$$NH_3 + H_2O \rightarrow NH_4^+ + OH^-$$
(ammonia) (ammonium ion)

Both NaOH and ammonia have the same effect—they lower the concentration of $H^+$ ions. NaOH achieves this by increasing the $OH^-$ concentration, whereas $NH_3$ reacts with water to produce $OH^-$.

The addition of acids and bases to water can greatly change the $H^+$ and $OH^-$ concentrations over a very broad range. Therefore, chemists and biologists use a log scale to describe the concentrations of these ions. The hydrogen ion concentration is expressed as the solution's **pH,** which is defined as the negative logarithm to the base 10 of the hydrogen ion concentration:

$$pH = -\log10\,[H^+]$$

A solution where the pH is near 7.0 is said to be neutral because $[H^+]$ and $[OH^-]$ are nearly equal. An **acidic** solution has a pH that is below pH 7.0; an **alkaline** solution has a pH above 7.0. A solution at pH 6 is said to be more acidic, because the $H^+$ concentration is 10-fold higher than a solution at pH 7.0. Note that as the acidity increases, the pH decreases. Strong acids, such as sulfuric acid, $H_2SO_4$, can produce pH values of almost 0 that are 1 mol of $H^+$ per liter of water. **Figure 6.14** shows the pH scale, along with the pH values of some familiar fluids.

Why is pH of importance to ecologists? The answer is that $H^+$ ions are very reactive and can readily bind to many kinds of ions and molecules. For example, calcium carbonate, which makes up limestone rock, dissolves readily in acidic water:

$$H^+ + CaCO_3 \rightarrow Ca_2^+ + HCO_3^-$$

Water in limestone areas therefore contains calcium ions and is known as "hard" water. Calcium ions are essential to life, and species such as snails, which need especially high levels of calcium to produce calcium carbonate shells, prefer areas with hard water.

Variation in pH can have a major impact on the distribution of organisms. Normal rainwater has a pH of about 5.6, which is slightly acidic, because the absorption of atmospheric $CO_2$ into rain droplets forms carbonic acid. Thus, most organisms have evolved to live in slightly acidic water. However, most plants grow best at a soil water pH of about 6.5, a value at which soil nutrients are most readily available to plants. For plants, most roots are damaged below pH 3 or above pH 9. Furthermore, at a pH of 5.2 or less, nitrifying bacteria do not function properly, which prevents organic matter from decomposing. In general, alkaline soils, such as chalk and limestone, the so-called high-lime soils, carry a much richer flora, and associated fauna, than do acidic soils (**Figure 6.15**). Plants have been broadly classified as calciphobes, growing only in acid soils; calciphiles, growing only in basic soils; and neutrophiles, tolerant of either condition. Calciphobes, such as rhododendrons, *Rhododendron* spp., which include azaleas, can live in soils with a pH of 4.0 and less. Calciphiles, such as alfalfa, *Medicago sativa,* blazing star, *Chamaelirium luteum,* and southern red cedar,

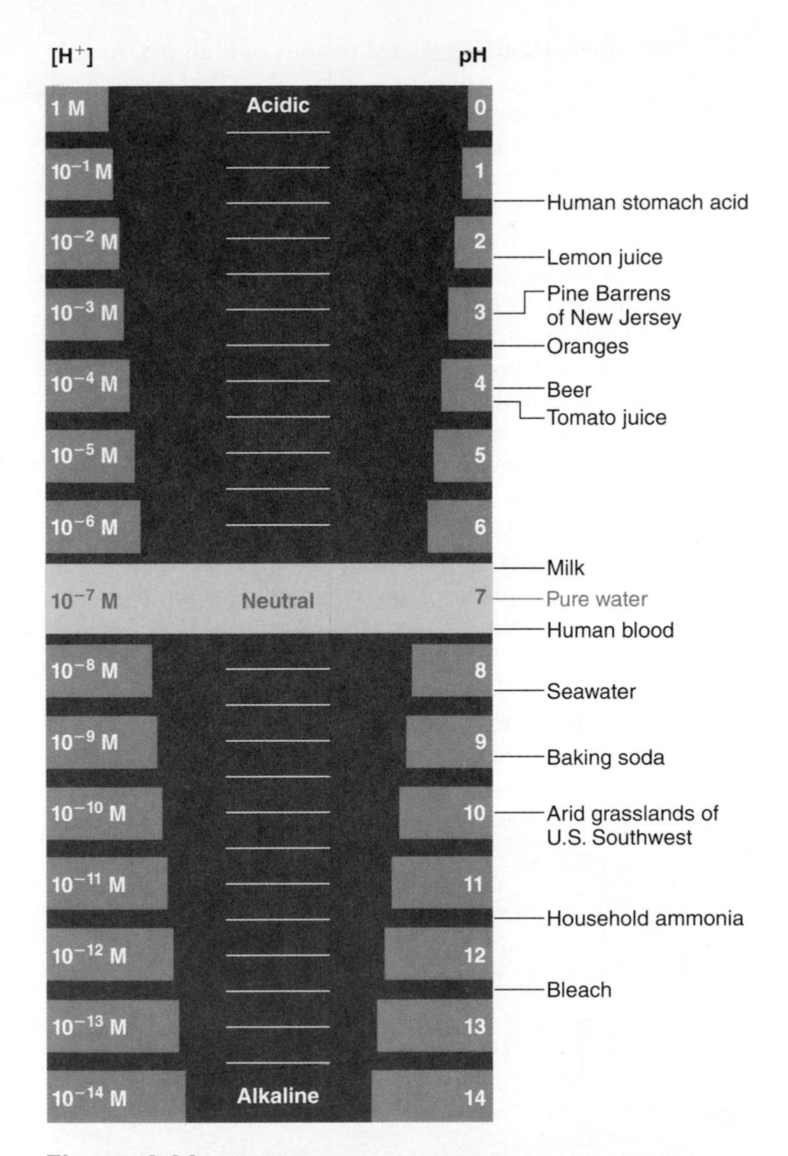


**Figure 6.14 The pH scale of hydrogen ion concentration and the relative acidities of some common substances.**

**ECOLOGICAL INQUIRY**

What is the pH of carbonated drinks?

*Juniperus silicicola,* are restricted to soils of high pH, mainly because they are susceptible to acidity.

Generally, the number of fish and other species decreases in acidic waters. In the 1970s, brown trout, *Salmo trutta,* disappeared from about one-third of the lakes in southern Norway because of lake acidification. The optimal pH for most freshwater fish and bottom-dwelling invertebrates, called benthic invertebrates, is between 6.0 and 9.0. Acidity in lakes increases the amount of toxic metals, such as mercury, aluminum, and lead, that leaches into the water from surrounding soil and rock. Both too much mercury and too much aluminum can interfere with gill function, causing fish to suffocate. Interestingly, Florida has a great number of acidic lakes, more than any other region in the U.S., but most of these are naturally acidic, and most fish found in these lakes are tolerant of the relatively low pH levels.

The susceptibility of both aquatic and terrestrial organisms to changes in pH explains why ecologists are so concerned about **acid rain,** precipitation with a pH of less than 5.6. Acid rain results from the burning of fossil fuels such as coal, natural gas, and oil, which releases sulfur dioxide and nitrogen oxide into the atmosphere. These gases react with water in the air to form sulfuric acid and nitric acid, which fall to the Earth's surface in rain or snow. When this precipitation drains into rivers and especially lakes, it can turn them very acidic and they lose their ability to sustain fish and other aquatic life. For example, fish such as walleye and smallmouth bass fail to reproduce, then disappear altogether, as lake pH decreases. Lake trout disappear from lakes in the eastern U.S. and Ontario, Canada, when the pH falls below about 5.2. Even lower pH can result in demineralization of fish skeletons, resulting in deformed fish. The influences of decreasing pH in an Ontario lake over a 10-year period in the 1960s and 1970s were staggering (**Figure 6.16**).

Lakes are not the only bodies of water that have been increasing in acidity. In oceans, despite their huge volume, the pH of seawater from all seas has decreased by 0.1 units over the 20th century. This decrease is due to an increase in atmospheric $CO_2$, which produces carbonic acid in the oceans. Because of the logarithmic nature of the pH scale, this is equivalent to a 30% increase in acidity. Ecologists predict that pH will decrease another 0.14–0.35 units by the end of the century. As a result, the rate of calcification of hard-bodied marine organisms such as corals is likely to decrease. A study by Ove Hoegh-Guldberg and colleagues (2007) showed that increasing acidity and ocean warming would severely impact coral reef formation by the end of the century and that reefs would cease to be dominated by corals. At atmospheric $CO_2$ levels of 480 ppm, which will be reached between 2050 and 2100, algal communities will become more common where coral reefs now

(a)

(b)

**Figure 6.15 Species-rich floras of chalk grassland compared to species-poor floras of acidic soils. (a)** In the lime-rich downs of Dorset County, England, there is a much greater variety of plant and animal species than at **(b)** a heathland site in England. Heathlands are characterized by acidic, nutrient-poor soils.

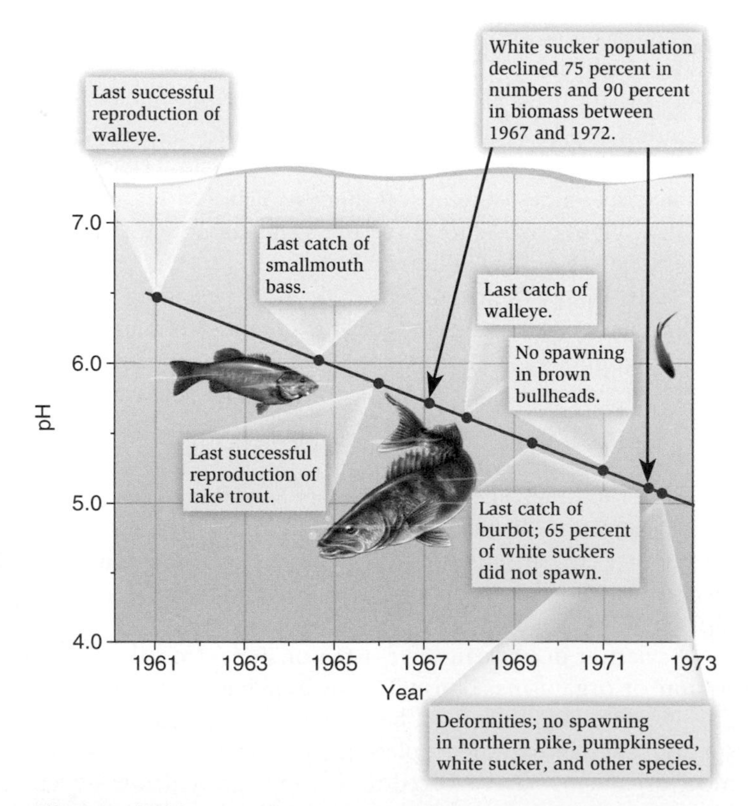


**Figure 6.16 Summary of the influences of acid rain on a variety of fish populations in George Lake, Ontario.** (From Beamish et al., 1975.)

**Figure 6.17** **Acid rain can damage terrestrial vegetation.** These trees on Clingmans Dome, in Great Smoky Mountains National Park, have been damaged by acid rain.

exist. Some Great Barrier Reef corals have already shown a 20% reduction in growth rate over the past 15–20 years. Interestingly, geological studies report an absence of calcified organisms during the early Triassic period, when $CO_2$ levels were five times as high as they were in 2000.

Acid rain is important in terrestrial systems, too (see **Feature Investigation**). For example, acid rain can directly affect forests by killing leaves or pine needles, as has happened on some of the higher mountaintops in the Great Smoky Mountains (**Figure 6.17**). It can also lower soil pH, which can result in a loss of essential nutrients such as calcium and nitrogen. Low calcium in the soil results in calcium deficiencies in plants, in the snails that consume the plants, and in the birds that eat the snails, ultimately causing weak eggshells that break before hatching. Decreased soil pH also kills certain soil microorganisms, preventing decomposition and resulting in reduced nitrogen levels in the soil. Decreases in soil calcium and nitrogen weaken trees and other plants, and may make them more susceptible to insect attack.

### Check Your Understanding

**6.3** Why do acidic soils generally support fewer plant species than alkaline soils?

## Feature Investigation

### Ralph Hames and Colleagues Showed How Acid Rain Has Affected the Distribution of the Wood Thrush in North America

The wood thrush, *Hylocichla mustelina,* is a widespread forest-dwelling bird that breeds most commonly along the slopes of the Appalachian Mountains, from northern New England to the Smoky Mountains (**Figure 6.C**). It winters from southeastern Mexico to Panama. In 2002, Ralph Hames and colleagues at the Cornell Laboratory of Ornithology showed a strong negative effect of acid rain on the probability of breeding by this species and suggested that acid rain continues to contribute to the decline of some breeding birds in the eastern United States.

Hames and colleagues collected data on the thrush's abundance from 650 study sites across its range. At each site volunteers used playback of wood thrush territorial calls to elicit a response from resident breeders. Other variables such as vegetation type and habitat fragmentation were taken at each site. Independent measures of wood thrush abundance were taken from the Breeding Bird Survey, an annual survey of bird populations during their breeding season conducted by volunteers under the supervision of the United States Geological Survey. Each survey consisted of a 39.4-km (20-mile) route with fifty 3-minute stops where the total number of birds seen and heard was recorded. The large number of volunteers covered 3,700 separate 39.4-km routes throughout the United States. These data were combined with measurements on long-term acid rain deposition from the National Atmospheric Deposition Project. The breeding probability of the wood thrushes was highly negatively affected by acid deposition, particularly in high-elevation zones with low-pH soils. The mechanism was thought to be calcium depletion in the soil that affected the quality and quantity of the prey, such as snails, isopods, millipedes, and earthworms. Growing nestlings have a particularly high calcium demand. As well as demonstrating the negative effects of acid rain on breeding birds, the project also showed the value of large-scale volunteer-based census data on answering important large-scale questions.

**HYPOTHESIS** Acid rain has negatively affected the abundance of the wood thrush in North America.

**STARTING LOCATION** The northeastern United States.

| | Conceptual level | Experimental level |
|---|---|---|
| 1 | Examine the breeding density of wood thrushes in North America. | (a) Volunteers in the Cornell Laboratory of Ornithology's Birds in Forested Landscapes (BFL) gather data on wood thrush abundance. At each of the many study sites, volunteers use a playback of wood thrush territorial calls to elicit a response. Each site is visited twice in the breeding season. (b) Independent measures of thrush abundance are obtained from the Breeding Bird Survey, an annual survey of bird abundance in the U.S. conducted by volunteers under the supervision of the U.S. Geological Survey. |
| 2 | Obtain data on acid deposition and soil pH. | The National Atmospheric Deposition Project (NADP) collects weekly samples of acid deposition at over 200 sites across the U.S. In addition, soil pH from the top 5 cm of the soil surface is collected through STATSGO, a nationwide database of soil properties compiled by the Natural Resources Conservation Service. |
| 3 | Analysis of the data from all soil pH levels and acid deposition were imported into a geographical system called ARCVIEW. | The response variable was the presence or absence of attempted breeding by the wood thrush. |

4 **THE DATA** The final model, based on data from 653 sites, showed a highly significant negative effect of acidic deposition on the predicted probability of wood thrush breeding.

**Figure 6.C** The effects of acid rain on the abundance of wood thrushes.

## SUMMARY

**6.1 Water Availability Affects Organismal Abundance**

- $H_2O$ is a polar molecule that exists in three states—liquid, ice, and water vapor (gas). Water molecules participate in many chemical reactions, and water serves many crucial functions in nature (Figures 6.1, 6.2).
- The availability of water has an important effect on the abundance of plants (Figures 6.3, 6.4).
- Animals need water for a variety of activities, including eliminating wastes (Figure 6.5).
- The abundance of many animals, from buffalo to ducks and deer, is limited directly and indirectly by rainfall (Figures 6.6, 6.7).
- Desert animals have many adaptations to reduce water loss (Figure 6.8).
- El Niño weather conditions can greatly affect global rainfall distribution and hence organismal distributions (Figure 6.9).
- Global warming is changing global rainfall patterns and threatens to alter organismal distribution patterns (Figures 6.A, 6.B).
- Rainfall can directly affect the survival of small insects such as thrips (Figure 6.10).

**6.2 Salt Concentrations in Soil and Water Can Be Critical**

- Many organisms have special adaptations to live in saline conditions (Figures 6.11–6.13).

**6.3 Soil and Water pH Affect the Distribution of Organisms**

- The pH of water, and other liquids, can vary over a wide scale (Figure 6.14).
- Soil pH can drastically alter the types of plants growing in a region (Figure 6.15).
- Acid rain has greatly affected the abundance of both aquatic and terrestrial organisms (Figures 6.16, 6.17, Figure 6.C).

## TEST YOURSELF

1. If a plant has a fresh weight of 100 g, a dry weight of 50 g, and a turgid weight of 130 g, what is its relative water content?
   a. 50
   b. 60
   c. 62.5
   d. 70
   e. 160
2. Which type of animals commonly produce ammonia as a waste product?
   a. Fish
   b. Reptiles
   c. Birds
   d. Mammals
   e. All of the above
3. In a globally warmed world, what will happen to global precipitation?
   a. All areas will become drier
   b. All areas will become wetter
   c. All areas will remain the same
   d. Dry areas will become wetter, and wet areas will become drier
   e. Dry areas will become drier, and wet areas will become wetter
4. Which type of plants are tolerant of high salt in their environment?
   a. Halophytes
   b. Xerophytes
   c. Phreatophytes
   d. Calciphobes
   e. Calciphiles
5. What is the approximate pH of seawater?
   a. 4
   b. 8
   c. 5
   d. 6
   e. 10
6. Which of the following is a strategy that plants use to cope with water stress?
   a. Increase proline concentration in sap
   b. Decrease sugar concentration in sap
   c. Open stomata during the day
   d. Reduce leaf abscission
   e. Increase urohydrosis
7. Which of the following is a naturally occurring weak acid?
   a. Hydrochloric acid
   b. Carbonic acid
   c. Sodium hydroxide
   d. Sulfuric acid
8. Hard water is generally rich in which ions?
   a. Calcium
   b. Magnesium
   c. Potassium
   d. Sodium
   e. Sulfur
9. Normal rainwater has a pH of about;
   a. 4.0
   b. 5.6
   c. 6.5
   d. 7.0
   e. 8.5
10. Which nitrogenous waste product is so toxic that animals need to excrete it immediately?
    a. Urea
    b. Ammonia
    c. Uric acid
    d. Purines
    e. Urohydrosis

## CONCEPTUAL QUESTIONS

1. Why are there three forms of nitrogenous waste that animals produce?
2. What are some of the adaptations that organisms use to live in saline environments?
3. Why do slightly alkaline soils support a richer flora than slightly acidic soils?

## DATA ANALYSIS

The Japanese ecologist Y. M. Park was interested in why only one species of grass, *Digitaria adscendens,* could grow on sand dunes, which are an extremely arid environment. In many other habitats *Digitaria* coexisted with another grass, *Eleusine indica.* Park collected seeds of both species and germinated them in moist sand and maintained the seedlings in glass tubes. Park watered all tubes with a nutrient solution every 10 days for 40 days. He then established two treatments for each species: continue watering for 19 days and leave unwatered. The roots from a sample of each treatment were measured for length, through the glass tubes, then removed and weighed. The data are shown in the figure. Explain the results.

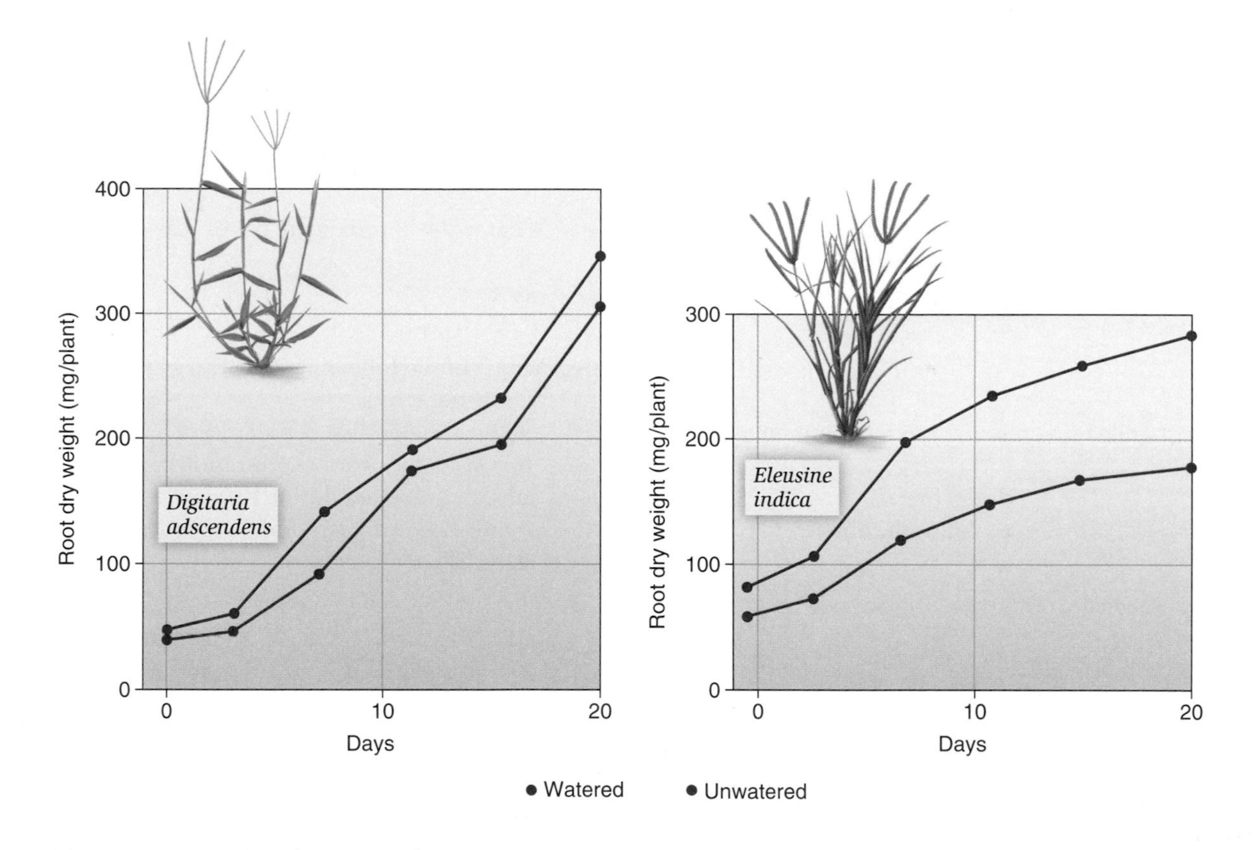

Scientists inspect a row of transgenic poplar trees capable of absorbing and breaking down soil contaminants.

CHAPTER 7

# Nutrients

## Outline and Concepts



Trichloroethylene, commonly known as TCE, is a widespread environmental contaminant. Forty percent of all abandoned hazardous waste sites are contaminated with TCE. Until recently TCE was widely used in a range of applications, including as a dry cleaning solvent and an anesthetic, and it continues to be used as a degreasing agent for metal parts. It is a human carcinogen and causes a range of neurological effects.

Scientists have been searching for safe ways to remove TCE and other toxic chemicals from hazardous waste sites. One approach, termed bioremediation, involves the use of plants that take up pollutants. After taking up the chemicals, the plants are harvested and the site is replanted until pollutants have been reduced to environmentally safe levels. Recently scientists have genetically engineered poplar trees to efficiently remove TCE from the environment. Products of mammalian cytochrome genes oxidize a range of compounds, including TCE. In a 2007 study, Sharon Doty and colleagues incorporated the gene that produces cytochrome in rabbit livers into poplar trees. These genetically engineered trees were found to metabolize TCE and other pollutants nearly 100 times faster than unaltered poplars. If used in the field, the trees would be cut down before flowering so there would be no chance of pollination with wild relatives.

Organisms require a wide variety of macronutrients and trace elements, in addition to the basic building blocks of hydrogen, carbon, and oxygen (**Table 7.1**). The elements required in the greatest amounts are nitrogen (N), phosphorous (P), potassium, sulfur, calcium, magnesium, iron, sodium, and chlorine. These are known as macronutrients. In addition, there is a whole host of trace elements needed in very small or trace amounts. Animals get most of their nutrients from plants or other animals. Plants acquire nutrients from water, usually in the soil. By dry weight, plants contain about 1.5% N and 0.2% P, and animals about 6.5% N and 1.0% P. The Earth's crust contains only about 0.03% N and 0.11% P, which means these elements are in short supply for many living things. A knowledge of soils is therefore critical for determining nutrient supply. The large carbon mass of both plants and animals, 45% and 57% by dry weight, respectively, is supplied by atmospheric $CO_2$. The Earth's crust contains only 0.18% C. In this chapter, we first see how soil development affects the availability of nutrients for life on Earth. Then, we detail how plants and microbes are affected by low and high nutrient levels, and later how nutrient levels also impact animal populations. Next, we discuss how light, carbon dioxide, and oxygen availability can also impact the distribution and abundance of different species. Finally, we discuss how species distributions are often limited by multiple abiotic factors.

**Table 7.1 Macronutrients and trace elements required by organisms and a partial list of their functions.**

| Nutrient | Function |
|---|---|
| **Macronutrients** | |
| Nitrogen (N) | Proteins and nucleic acids |
| Phosphorous (P) | Nucleic acids, phospholipids, and bone |
| Potassium (K) | Sugar formation in plants, protein synthesis in animals, solute in animal cells |
| Calcium (Ca) | Plant cell walls, regulation of cell permeability, bone, muscle contraction, blood clotting |
| Magnesium (Mg) | Chlorophyll and many enzyme activation systems |
| Iron (Fe) | Hemoglobin and enzymes |
| Sodium (Na) | Extracellular fluids of animals, osmotic balance, nerve transmission |
| **Trace Elements** | |
| Chlorine (Cl) | Chlorophyll, osmotic balance |
| Manganese (Mn) | Chlorophyll, fatty-acid synthesis |
| Zinc (Zn) | Auxin production in plants, enzyme systems, sperm development, tissue repair |
| Copper (Cu) | Chloroplasts, enzyme activation, oxygen-binding molecule in some invertebrates |
| Boron (B) | Vascular plants and algae |
| Molybdenum (Mo) | Enzyme activation systems |
| Aluminum (Al) | Nutrient for ferns |
| Silicon (Si) | Nutrient for diatoms |
| Selenium (Se) | Nutrient for planktonic algae, enzyme systems |
| Fluorine (F) | Bone and teeth formation |
| Chromium (Cr) | Glucose metabolism |
| Iodine (I) | Thyroid hormone |
| Sulfur (S) | Amino acids |

## 7.1 Soil Development Affects Nutrient Levels

Soil is the complex loose terrestrial surface material in which plants grow. It has three components: (1) weathered fragments of parent rock in various stages of breakdown, (2) water, and (3) minerals and organic compounds resulting from the decay of dead plants and animals. Soil is also teeming with life, much of it microscopic. When any organism dies, the process of decay occurs in the soil through the activities of soil organisms, especially bacteria and fungi (look ahead to Figure 25.2). **Humus,** finely ground organic matter, is produced, and eventually the minerals are absorbed by plant roots. Optimal plant growth occurs at around 8% organic matter, though some plants are adapted to humus-poor soils. Soils that contain less than 1% organic matter are said to be humus-poor, while those containing greater than 8% organic matter are humus-rich.

Because most soil-forming processes, such as litter fall, the fall of senescent plant leaves, tend to act from the top down, soil develops a vertical structure referred to as the **soil profile** (**Figure 7.1**). Soil scientists recognize four main layers, or horizons. The top layer is the O horizon, or organic layer, subdivided into Oi, the litter layer, and Oa, the humus layer. The leaf litter layer may fluctuate seasonally, whereas the humus layer, the zone of decomposition, does not. The general range of litter fall in temperate ecosystems is 250–450 $g/m^2$ per year, or 1–2 tons per acre, although ranges from almost 0 (deserts) to 72,000 $g/m^2$ (tropical forests) have been reported. The rate at which the litter decays varies from a half-life (that is, 50% decay) of 0.36 year in tropical forest to 1.1 years in temperate forest and 11.23 years in northern coniferous forest.

The next layer is the topsoil, or the A horizon, where most of the plant roots are located. Dead organic matter is added to this layer as it mixes with the soil below. Heavy rains can wash away or leach the nutrients from this layer of the soil and deposit them deeper down. Organic material is largely absent from the next layer, the B horizon, and small amounts of weathered mineral parent material are present. Materials leached from the A horizon may be deposited in the B layer. Fine layers of clay may also be deposited here, forming a clay pan, which may cause waterlogging. Below this is the C horizon, generally consisting of weathered parent material such as bedrock. Tree roots commonly penetrate to the C layer.

Soil texture is based on the sizes of mineral particles making up the soil, classified as gravel, $>2$ mm; sand, $<2$ mm; silt, $<0.05$ mm; and clay, $<0.002$ mm. Soils made up mainly of small particles like clay are called heavy soils. Water soaks into heavy soils slowly but is retained well; such soils have the potential for being very fertile. Light soils, such as sandy soils, are well aerated and allow free movement of roots and water but are relatively infertile because water and nutrients leach out of them.

The difference in potential fertility between light and heavy soils results from the way minerals are retained. Most elements that form positive ions, or cations, when dissolved, such as calcium and magnesium, are stored on the surface of particles. By contrast, anions are dissolved in soil water. Both are more prevalent in heavy and humus-rich soils. Plant roots remove cations and replace them with hydrogen ions. The potential fertility of a soil depends primarily on cation exchange capacity, a measure of the number of sites per unit of soil on which hydrogen can be exchanged for mineral cations. Clay soils may have a cation exchange capacity from 2 to 20 or more times that of sand. Acidity can greatly reduce the fertility of any soil type. Even in clay soils, if most of the sites on the particles are already filled up with hydrogen ions, as in acid conditions, the soil will be infertile.

Soil development is strongly influenced by five features: climate, parent material, age, topography, and living organisms. Climate includes rainfall, evaporation, and temperature. Many tropical soils are nutrient-poor, as nutrients have been washed out. Parent material is the bedrock, which may be igneous, sedimentary, or metamorphic. Igneous rocks are of volcanic origin. Sedimentary rocks result from the wearing away of rock by water or wind and its redeposition and recementation. Metamorphic rocks are igneous or sedimentary rocks that have been changed by high pressures and temperatures deep underground. Age is also important. Young soils, often less than 10,000 years old, occur in areas recently uncovered by glaciers, such as northern North America and Europe, or areas recently covered by sediments, such as river valleys. Older soils are generally tropical and may be 100,000 years old or more. The older the soil, the greater the chances it has been weathered by the elements. Topography can change the runoff rates of water. Vegetation can cause changes in soil properties and, in turn, is dependent upon them. For example, in the northern Midwest, coniferous forest, deciduous forest, and grassland grow within a few miles of one another, and the soil differences are entirely due to the differing effects of the vegetation. Coniferous forest produces needles that are slow to decompose and are strongly acidic. As a result, strongly acidic soils dominate, with a heavy layer of underdecomposed litter. Deciduous forest produces a thinner, less acidic litter, and this grades into the soil below. In grassland, the soil is less acidic still, and each year the entire aboveground part of the plant dies back and falls to the ground, allowing decomposing organisms access to the nutrients at or near the soil surface. Grassland soil is dark because of the addition of this organic material and because of organic material that is added to the soil from the death of many fine roots.

One of the clearest examples of soil effects on the distribution of plants is the serpentine soils that occur in scattered areas all over the world, including California. Serpentine rock is basically a magnesium iron silicate. Plants that grow there must be tolerant of high magnesium but low nitrogen and phosphorous. Such conditions are lethal to many plants, hence serpentine soils have little value for agriculture or forestry. However, many species have adapted to these conditions, forming stunted, endemic communities not found on other soils (**Figure 7.2**).

**Figure 7.1 Soil profile.** Ecologists recognize four soil horizons in a typical soil.

**Figure 7.2 The effects of soils on vegetation are dramatic.** The serpentine soil in the background supports only a stunted, endemic plant community. Compare this to the more luxurious vegetation in the foreground. Napa County, California, 70 miles north of San Francisco.

### Check Your Understanding

**7.1** In which soil layer are minerals commonly found?

## 7.2 Plant Growth Is Limited by a Variety of Nutrients

Soil consists mainly of oxygen and silicon (**Table 7.2**) with lesser amounts of aluminum, iron, carbon, and other elements. Trace elements needed by plants but at levels of less than 0.1% include zinc, nickel, copper, chlorine, cobalt, boron, and molybdenum. A comparison of the availability of elements in the soil and those accumulated annually by vegetation suggests that certain elements are likely to be a **limiting factor**, that is, most scarce in relation to need. For example, without renewal, soil nitrogen would be depleted in 40 years. Nitrogen is usually the most limiting factor for plants, because it is required in large amounts for the synthesis of amino acids and other cellular constituents. Although the Earth's atmosphere is 78% nitrogen gas ($N_2$), plants cannot use nitrogen in this form. Atmospheric nitrogen has a strong triple bond holding the nitrogen atoms together. To be useful to plants, soil nitrogen must occur in a combined form with another element. Such combined forms as ammonia ($NH_3$), ammonium ions ($NH_4^+$), or nitrate ($NO_3^-$) are also known as **fixed nitrogen**. In nature, nitrogen fixation is performed by photosynthetic cyanobacteria, which occur in aquatic systems and the soil surface, and by nonphotosynthetic actinobacteria in the genera *Clostridium, Klebsiella,* and *Azotobacter.* These prokaryotes possess the enzyme nitrogenase, which catalyzes the reduction of $N_2$ to $NH_4^+$. Because oxygen also binds to nitrogenase, nitrogen fixation occurs predominantly in low-oxygen habitats, such as the soil or inside special structures in plant roots. Cyanobacteria require a high-light environment to ensure that sufficient carbon is available for nitrogen fixation, primarily supplied to the bacteria as malate or succinate. Cyanobacteria accomplish nitrogen fixation in special thick-walled cells known as heterocycsts, which reduce the inward diffusion of oxygen. Soil prokaryotes excrete excess fixed nitrogen, and their death releases more nitrogen to the soil. Some plants actually contain nitrogen-fixing bacteria such as *Rhizobium* and *Frankia* in root nodules, where oxygen is low. *Rhizobium* cannot fix nitrogen outside of their host plants. The bacteria can sense chemicals secreted by the root and enter the root via a deformed root hair. Their infection triggers the formation of the nodule (**Figure 7.3**). Nitrogenase production has a high energy requirement but plants have an abundant supply of carbohydrate.

Some plants have adapted to low soil nitrogen levels by supplementing their diet with nitrogen-rich animals. About 600 species have modified leaves that enable them to capture mainly insects and some small vertebrates. Some species, so-called carnivorous plants, may obtain 80–90% of their nitrogen from animals. Carnivorous plants may have active or passive trapping mechanisms. The passive trappers wait for prey to fall or wander into a trap. For example, pitcher plants have modified leaves that are fused to form a pitcher that can hold rainwater (**Figure 7.4a**). The lip and sides of the pitcher are slippery, and small animals, including insects, lizards, and frogs, fall into the pitchers and cannot escape. They are digested by microbes in the water. Active trappers, such as Venus flytraps and sundews, have traps that are stimulated by touch. The Venus flytrap, *Dionaea muscipula,* has a series of active traps, each formed by two-lobed leaves that are edged with lance-shaped teeth (**Figure 7.4b**). When open, the inner

**Table 7.2** Typical concentrations of elements in the soil.

| Nutrient | Soil Content (% weight) |
|---|---|
| Oxygen (O) | 49 |
| Silicon (Si) | 33 |
| Aluminum (Al) | 7.1 |
| Iron (Fe) | 4.0 |
| Carbon (C) | 2.0 |
| Calcium (Ca) | 1.5 |
| Potassium (K) | 1.4 |
| Magnesium (Mg) | 0.5 |
| Sodium (Na) | 0.5 |
| Nitrogen (N) | 0.2 |
| Manganese (Mn) | 0.1 |
| Phosphorous (P) | 0.1 |
| Sulfur (S) | 0.1 |

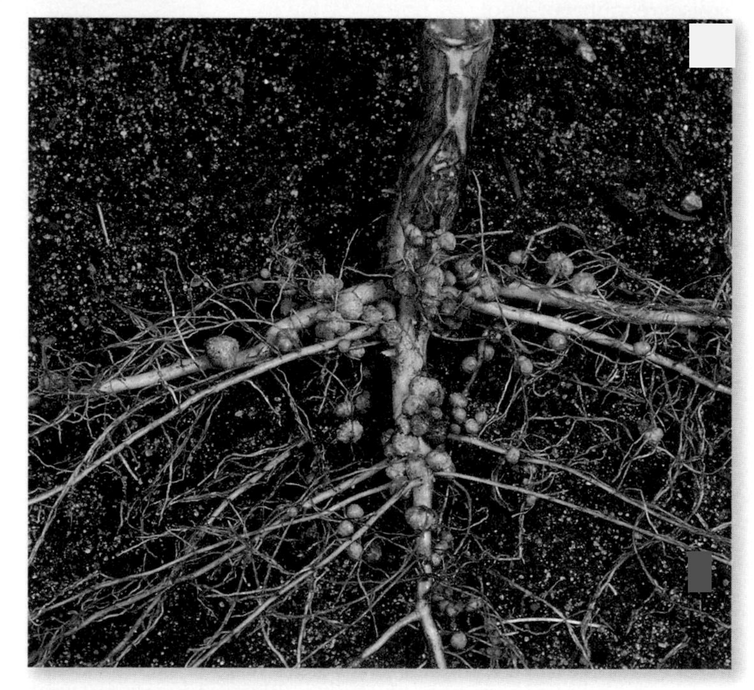

**Figure 7.3** Root nodules on soybean roots contain nitrogen-fixing *Rhizobium* bacteria.

**ECOLOGICAL INQUIRY**

Why do nitrogen-fixing bacteria often occur in special nodules on plant roots?

(a) (b)

**Figure 7.4** **Carnivorous plants** **(a)** Pitcher plants digest small animals that fall into the water-filled pitcher and cannot escape. **(b)** The Venus flytrap, *Dionaea muscipula,* has an active trap that is triggered by small animals touching sensory hairs.

leaf surfaces secrete carbohydrates that attract prey. Modified hairs inside the trap act as triggers. When more than one hair is touched, or one hair is touched twice within 20 seconds, the leaf lobes shut. Digestive enzymes are then secreted by the inner leaf surfaces. Traps typically digest their meal within 10 days and may go through three or four digestive cycles. Sundews, *Drosera* spp., which occur in most wet areas of the world, have glandular, sticky hairs on their leaves. Invertebrate prey become stuck to these hairs and, in their struggle to escape, become hopelessly entangled. The hairs also produce enzymes that digest the prey.

Although some soils are nutrient-deficient, soil nutrient levels in many areas are increasing. The continued combustion of fossil fuels and fertilizer use by modern agriculture both contribute to increased nitrogen levels. On average, these processes have increased nitrogen deposition from about 1–3 kg/ha/yr to 7 kg/ha/yr over the U.S. and to more than 17 kg/ha/yr in Europe. These increases are sufficient to change the plant species composition in many habitats. Some common species increase in abundance, outcompeting rarer species and driving them to extinction (see **Feature Investigation**). In addition, runoff from terrestrial systems may concentrate in rivers, estuaries, and wetlands, increasing aquatic nitrogen levels.

Anna Tyler and colleagues (2007) showed how increased nitrogen input in salt marshes favors the spread of invasive species. In Willapa Bay, Washington State, the smooth cordgrass, *Spartina alterniflora,* was introduced repeatedly between 1894 and 1920 from its native Atlantic and Gulf Coast habitats during commercial oyster transport. By 1997 about 6,000 ha of the 23,000 ha mudflat had been invaded. There were no native plants on the mudflat to slow its spread. The same species was introduced to San Francisco Bay by the U.S. Corps of Engineers in 1973 for marsh restoration, even though a native cordgrass, *Spartina foliosa,* was present. Here the two *Spartina* species hybridized, creating a hybrid that invaded historically unvegetated mudflats. In both locations the invasive *Spartina* prevented wading birds from feeding on the bare mudflats. Tyler and colleagues added fertilizer to five plots in the form of ammonium chloride salt wrapped in nylon tissue. The fertilizer was placed in 15-ml centrifuge tubes inserted into the sediment, to allow gradual diffusion of nitrogen into the pore water. Tubes were exchanged every 6 weeks in the 4-month growing season. Stem density of invasive *Spartina* plants was measured in each plot and compared to data from unfertilized control plots. Aboveground biomass and stem density were significantly increased in both Willapa Bay and San Francisco Bay. At unfertilized control sites at Willapa Bay there was a significant correlation between aboveground *Spartina alterniflora* biomass and sediment nitrogen (**Figure 7.5**). Tyler and her colleagues had shown that increased nitrogen deposition increased the success of invasive species and the strength of their effects on native habitats.

Phosphorous is another limiting nutrient for plants. Plants obtain their phosphorous from the ion known as phosphate, $HPO_4^{2-}$. Although often abundant in some soils, phosphate binds tightly with clay soils, iron and aluminum oxide, and calcium carbonate. Many plant roots form close

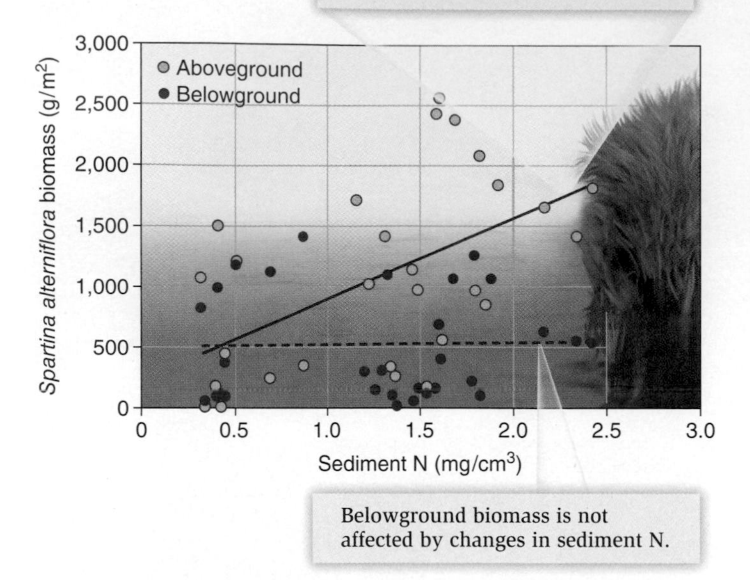


**Figure 7.5 Relationship between biomass of the invasive *Spartina alterniflora* on control plots at Willapa Bay, Washington State.** Significant increases were observed for aboveground biomass as sediment nitrogen increased, but not for belowground biomass. (After Tyler et al., 2007.)

**ECOLOGICAL INQUIRY**

Why does the aboveground portion of *Spartina* respond more to higher soil N than the belowground part?

associations with fungi that acquire phosphorous and pass it on to the roots. Other plants develop long roots and root hairs to increase their capacity to absorb phosphate. Of all nutrients, nitrogen and phosphorous are most often the limiting factors for plant growth.

Some soil minerals may be toxic to plants at high concentrations, such as aluminum, lead, chromium, and mercury, and the trace elements boron, copper, and zinc. At concentrations of 10 μg/g of plant tissue, boron is an important trace element needed for cell walls, but it is toxic at concentrations of 100 μg/g or higher. High aluminum concentrations inhibit mineral and water uptake. Aluminum toxicity is more common in acidic soils, which are frequently found in tropical and subtropical areas. Some plants can accumulate aluminum or other toxic metals in their tissues, often sequestering them in cell vacuoles. These plants are known as hyperaccumulators, an adaptation that allows them to grow on metal-rich soils. As we saw at the beginning of the chapter, hyperaccumulators can be planted to remove metals from soils, a process known as phytoremediation. Other examples of phytoremediation exist where plants or microbes are used to clean up polluted areas (see **Global Insight**).

## Feature Investigation

### Christopher Clark and David Tilman Showed How Low-Level Nitrogen Deposition Has Reduced the Number of Species in Midwest Prairies

Atmospheric deposition of nitrogen occurs at two to seven times preindustrial levels in many developed nations because of fossil-fuel combustion and agricultural fertilization. Both these phenomena are predicted to increase. This study investigates the impact of continued, low-level nitrogen on natural systems. David Tilman set up and managed a long-term experiment in Minnesota in which low levels of nitrogen were applied to grasslands over a 23-year period (**Figure 7.A**). Christopher Clark and David Tilman (2008) analyzed the effects of such nitrogen additions on the number of plant species present. Even at the lowest levels of nitrogen addition, long-term exposure reduced the number of plant species, relative to control plots. All types of plants were affected, but rarer species were affected more than common ones, as they were outcompeted by rapidly growing common species. When the addition of nitrogen stopped, the researchers noticed that the number of plant species recovered, indicating that some effects of nitrogen deposition are reversible.

**Check Your Understanding**

**7.2** Why does the Venus flytrap not close its traps if a single trigger is touched once?

**HYPOTHESIS** Low level nitrogen deposition can change grassland floras.

**STARTING LOCATION** Minnesota, U.S.A.

| | Conceptual level | Experimental level |
|---|---|---|
| 1 | Mimic continued low-level rates of atmospheric nitrogen addition over a long period. | Add pelletized $NH_4NO_3$, ammonium nitrate fertilizer. |
| 2 | Establish different levels of nitrogen addition to mimic natural rates. | Select 102 plots. Vary nitrogen deposition from 10, 34, or 95 kg/ha/yr, from 1982 to 2004. There were 11 replicates of each treatment plus 11 plots with no added fertilizer. All plots received P, K, Ca, Mg, and trace metals to ensure limitation was by N only. |
| 3 | Stop additional nitrogen addition halfway through the experiment and examine recovery. | From 1991 onward, all treatments were stopped on 6 of the replicates per treatment. |
| 4 | Plant species number and biomass can be affected by climatic variation and other factors, so experimenters controlled for climatic variation and other variables. | Each year plant species numbers were expressed as a proportion of the species present in the control plots. |

5 **THE DATA** Relative plant species number was reduced on all treatments relative to controls (Figure 7.A-i). The arrows on the year axis indicate at what time the treatments became different from controls. For example, it took 8 years, until 1990, for the low nitrogen addition to significantly reduce the number of plant species. Following the cessation of nitrogen addition on some plots, the number of species present recovered to those of control plots after 13 years, by 2004 (Figure 7.A-ii).

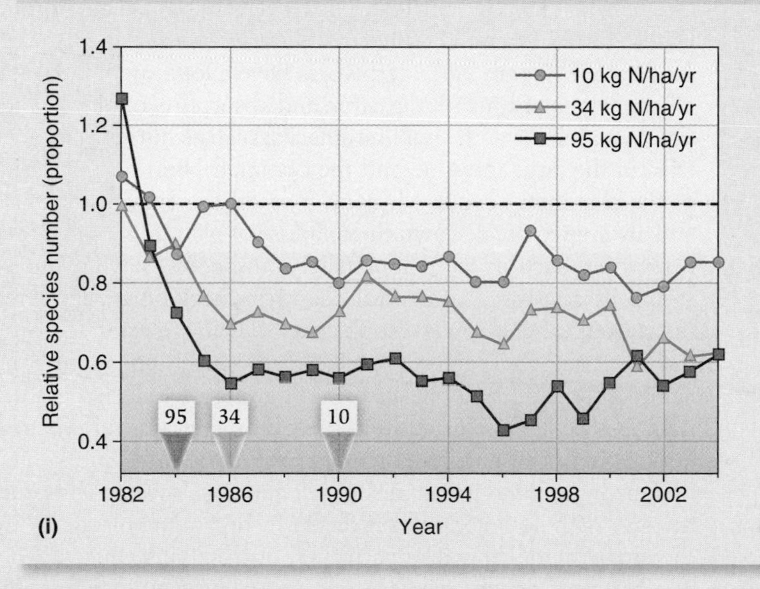


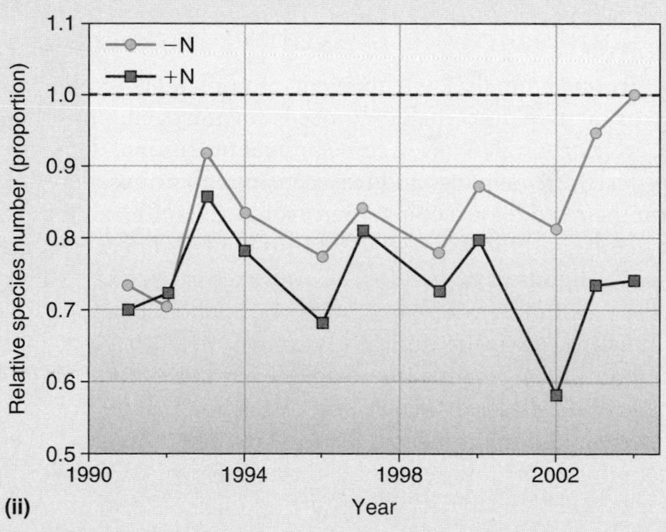


**Figure 7.A** The effects of low-level nitrogen addition on the number of species in a grassland.

## Global Insight

### Polluted Areas May Be Restored Using Living Organisms

Restoration of polluted areas can involve **bioremediation**, the use of living organisms, usually microbes or plants, to detoxify polluted habitats such as dump sites or oil spills. Some bacteria can detoxify contaminants, converting them to inert substances, while certain plants can accumulate toxins in their tissues and are then harvested, removing the pollutant from the system. Ecologists have used microbes in the degradation of sewage for over 100 years. More recently, the bacteria *Geobacter sulfurreducens* has been used to treat borrow pits contaminated with uranium. *Geobacter* precipitates uranium from contaminated water, thereby purifying it.

In 1975 a leak from a military fuel storage facility released 80,000 gallons of jet fuel into the sandy soil at Hanahan, South Carolina. Soon the groundwater contained toxic chemicals such as benzene. By the 1980s it was found that naturally occurring microorganisms in the soil were actively consuming many of these toxic compounds and converting them into carbon dioxide. In 1992, nutrients were delivered in pipes to the contaminated soils to speed up the action of the natural microbial community. By the end of 1993, contamination had been reduced by 75%. The increasing interest in bacterial **genomes**, the entire genetic complement of an individual, is providing opportunities for understanding the genetic and molecular bases of degradation of organic pollutants.

Heavy metals such as cadmium or lead are not readily absorbed by microorganisms. **Phytoremediation**, the treatment of environmental problems through the use of plants, is valuable in these cases. Plants absorb contaminants in the root system and store them in root biomass or transport them to stems and leaves. After harvest, a lower level of soil contamination will remain. Several growth/harvest cycles may be needed to achieve cleanup. Sunflower, *Helianthus annuus,* may be used to extract arsenic and uranium from soils. Pennycress, *Thalsphi caerulescens,* is an accumulator of zinc and cadmium, while lead may be removed by Indian mustard, *Brassica juncea,* and ragweed, *Ambrosia artemisifolia.* Floating aquatic plants have been used to help absorb heavy metals from polluted water bodies. Some polychlorinated biphenyls (PCBs) have been removed by transgenic plants containing genes for bacterial enzymes.

## 7.3 Herbivore Populations Are Limited by Plant Nutrient Levels

As a group, animals have nutrient requirements similar to those of many plants, including needs for nitrogen, phosphorous, and carbon. However, some animal nutritional demands are different from plants and between animal groups, depending on their mode of feeding. **Herbivores** are animals that eat only plants, and their digestive systems contain microbes that aid in the digestion of cellulose. Carnivores eat primarily animal flesh or fluids, whereas omnivores, such as bears, foxes, and humans, have the ability to survive on both plant and animal products. Twenty amino acids exist in the bodies of all animals, regardless of their mode of feeding. Many animals obtain their amino acids from the food they eat, though plants contain only small amounts of eight amino acids. Herbivores have evolved the capacity to synthesize these missing amino acids. However, carnivores and omnivores lack this ability and must obtain these eight amino acids—isoleucine, leucine, lysine, methionine, phenylalanine, threonine, tryptophan, and valine—solely from their meat diet. These amino acids are termed essential amino acids. They are no more essential to survival than other amino acids. The term "essential" refers only to the fact that they come from food. Humans who are vegetarians must find ways to obtain these essential amino acids. Animals also need certain essential fatty acids and vitamins. Among vertebrates, primates and guinea pigs cannot synthesize their own vitamin C and must therefore consume it in their diet.

The digestive system of animals depends on diet. Among vertebrates, the length of the small intestine varies greatly with food source (**Figure 7.6**). Herbivores have a long small intestine, allowing greater time for digestion and absorption of plant material. In mammals with simple stomachs, such as horses, microbes exist in the large intestine and the cecum, a blind outpocketing of the alimentary canal. Microbes possess the enzyme cellulase and are able to break down the cellulose of plant material. Other herbivores, such as ruminants (sheep and cows) have complex stomachs consisting of several chambers beginning with three outpockets of the lower esophagus collectively referred to as the forestomach (**Figure 7.7**). The forestomach consists of the rumen, reticulum, and omasum. In ruminants, cellulose-digesting microbes exist in the rumen and reticulum. The rumen may store large volumes of partially digested food, known as cud, which is sometimes regurgitated, rechewed, and swallowed again. Carnivores, with a higher protein diet, which is much more readily digested, generally have a relatively short small intestine.

Nutrient deficiencies are known to limit animal population sizes. In 1959, pronghorn antelopes, *Antilocapra americana,* were introduced on to the Hawaiian Island of Lanai for sport. The phosphate-deficient soils and low plant phosphorous levels led to brittle pronghorn bones that often broke in the hands of hunters. In addition, the pronghorns did not breed as expected and often wandered away from their grassland habitats, perhaps in search of richer grazing. In Minnesota, iodine-deficient soils led to small deer populations. Thyroid glands in these deer produced thyroxine levels only one-half to one-third of normal, resulting in smaller

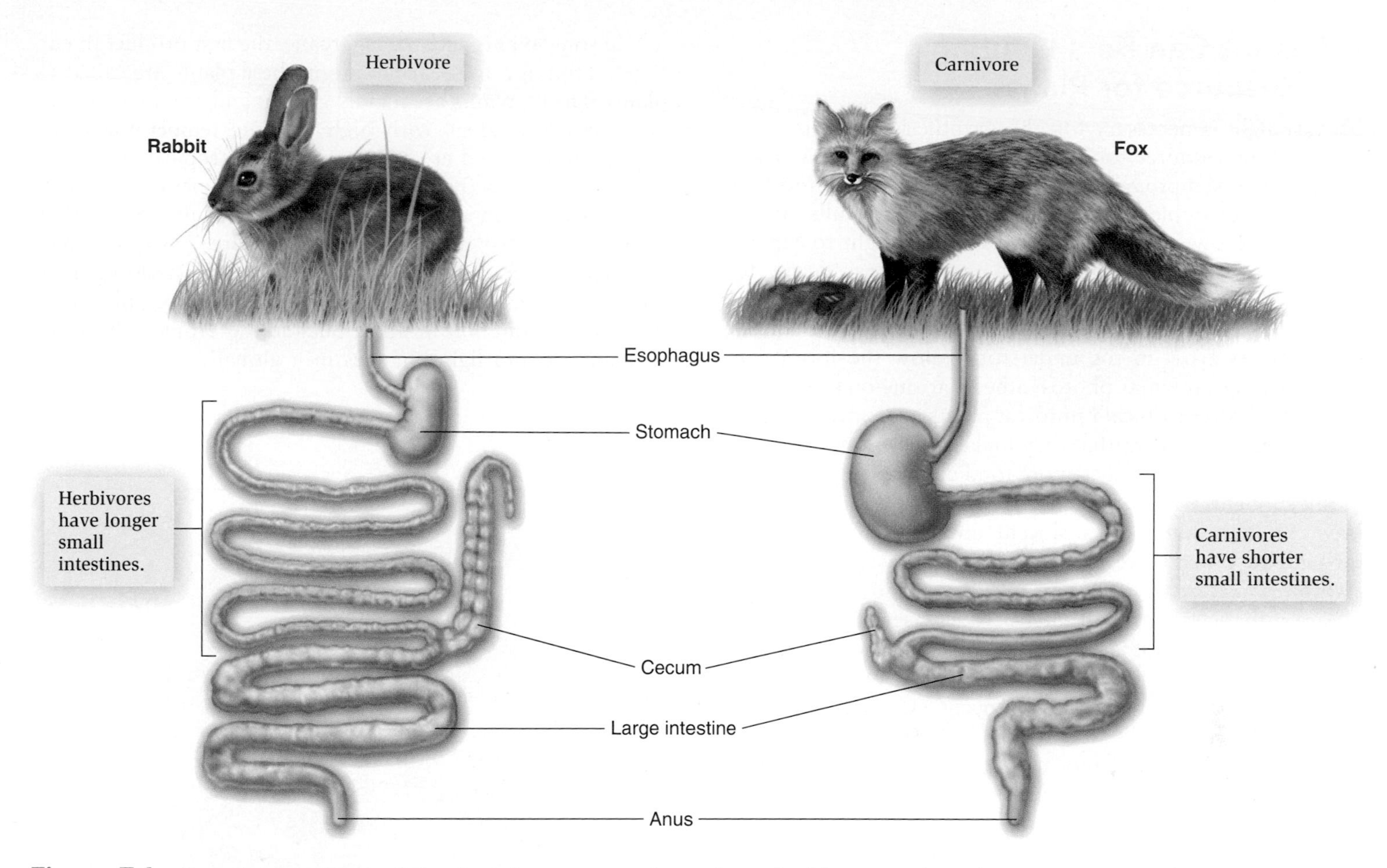


**Figure 7.6** **Length comparison of the digestion systems of vertebrate herbivores and carnivores.**

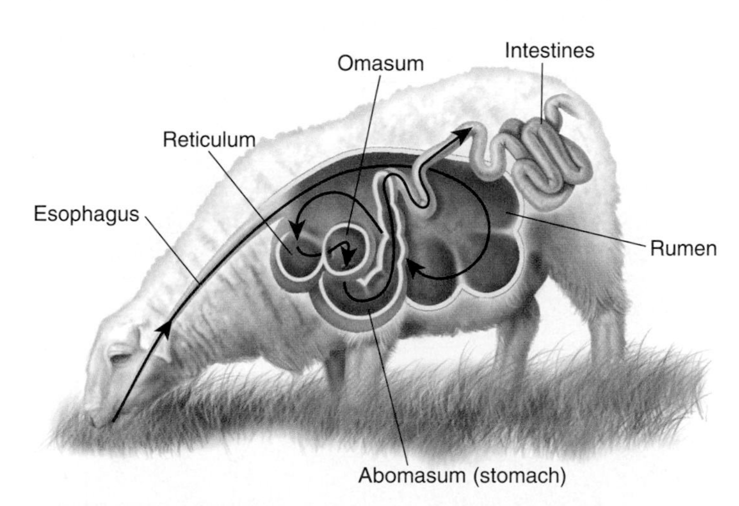


**Figure 7.7** **Digestive tract of a ruminant.** Ruminants have a complex arrangement of three modified outpockets, together called the forestomach, arising from the esophagus; these outpockets are called the rumen, reticulum, and omasum.

male testes and decreased reproduction. Iodine supplements increased thyroxine levels and testis size of males.

Because sodium is essential for many animal body functions, yet occurs only at low levels in most plants, herbivores seek out alternative sources such as so-called salt licks. Licks are natural formations of mineral-bearing soils, often associated with saline springs. Salt pellets applied to de-ice roads in Canada contribute to traffic accidents involving moose, who are attracted to roadside pools of water with salt residues. Sodium salts are preferred, but calcium and magnesium salts in bone also attract herbivores. Rodents gnaw on old bones, and field mice eat away at shed antlers.

While animals are influenced by soil nutrients, many species can themselves change the levels of soil nutrients and organic matter. Nesting birds can deposit large volumes of excrement, called guano, on vegetation, altering the soil characteristics underneath, especially by increasing soil nitrogen and phosphate. Large heron colonies in Alabama were shown to increase the soil phosphate concentration 60-fold and increase soil pH 1.2 units. Often the vegetation itself will die under such conditions, forcing the birds to change nesting sites. Wood rats construct dens of plant material, and the mixture of compost and droppings can greatly boost soil nitrogen.

### Check Your Understanding

**7.3** Though the flesh of dead animals is often eaten by decomposers, why don't we see their bones more frequently?

## 7.4 Light Can Be a Limiting Resource for Plants

Because light is necessary for photosynthesis to occur, it can be a limiting resource for plants. Insufficient light may result in deficiency symptoms as with plants that receive too little water. For example, insufficient light often results in leaf abscission. However, what may be sufficient light to support the growth of one plant species may be insufficient for another. Many plant species grow well in shady conditions, such as tropical or temperate forests. In temperate forests, oak saplings grow in the understory below the forest canopy, reaching maximal photosynthesis at one-quarter of full sunlight. Shorter tropical rainforest plants are so well adapted to growing in low light that they make excellent houseplants. Shade plants often produce more total chlorophyll than those in full sunlight, which maximizes their light absorption. Their leaves are also thin and translucent, allowing sunlight to pass through to other leaves. Other plants, like sugarcane, *Saccharum officinarum,* grow best in full sun and increase their photosynthetic rate as light intensity increases (**Figure 7.8**).

One reason photosynthetic rates vary among plants is that there are three different biochemical pathways by which the photosynthetic reaction can occur: $C_3$, $C_4$, or CAM. $C_3$ plants fix carbon dioxide by incorporating it into ribulose bisphosphate to produce molecules of phosphoglycerate, each of which contains three carbon molecules, hence the name $C_3$. The enzyme that catalyzes this reaction is called rubisco, and it is the most abundant protein in chloroplasts and perhaps even on Earth. Ninety percent of plants employ the $C_3$ pathway.

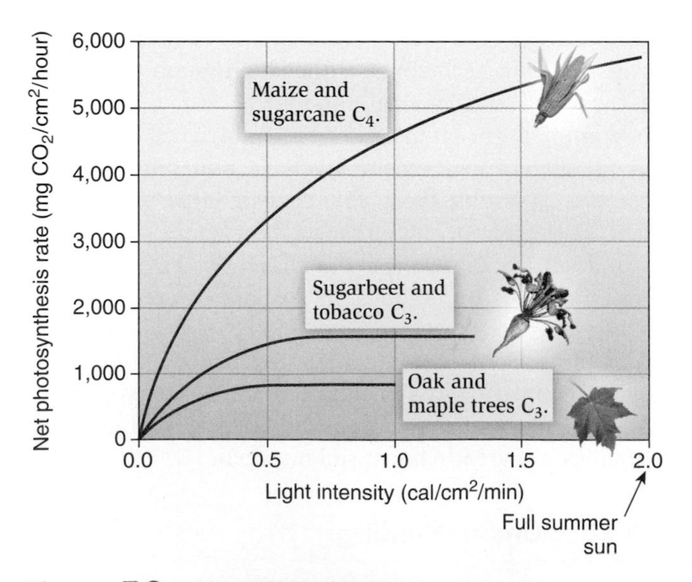


**Figure 7.8 The effect of light intensity on photosynthetic rate in different plants.** The arrow represents the light equivalent of full summer sun. (After Zelitch, 1971.)

In some plants such as sugarcane, the first product of carbon fixation is a four-carbon atom; these plants are called $C_4$ plants. The $C_4$ plants are mainly grasses and sedges. $C_4$ plants grow faster in areas with high daytime temperatures and intense sunlight and are more common in tropical areas than in temperate ones (**Figure 7.9a**). In North America, $C_4$ plants are more common in southerly areas of the United States than in more northerly areas of Canada (**Figure 7.9b**). In cooler, cloudier temperate areas, $C_3$ species can tolerate lower light and live in areas where $C_4$ plants cannot. Thus, $C_3$ plants are much more common in areas outside the Tropics. Projected temperature and light changes in a globally warmed world

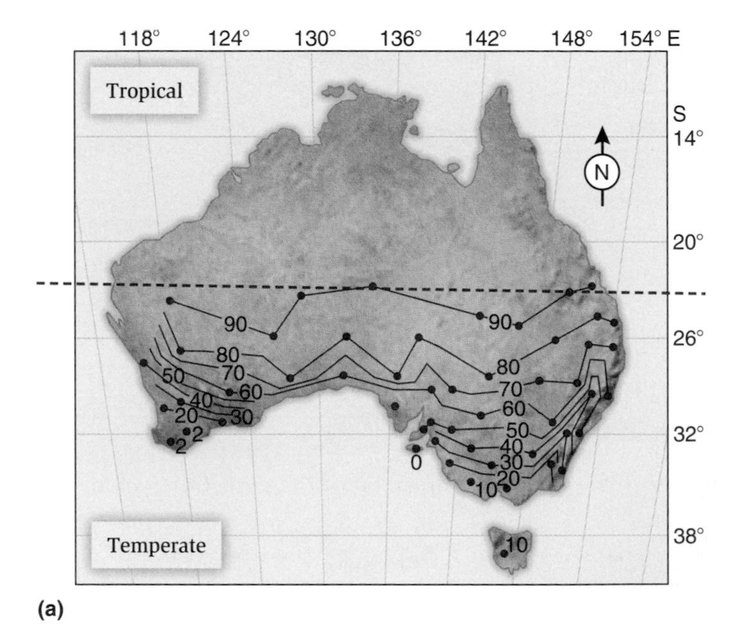


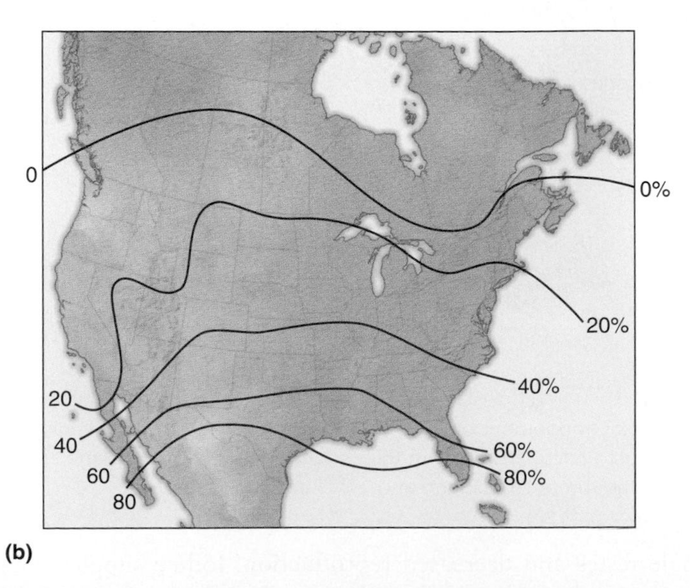


**Figure 7.9 Distribution maps of percentage of $C_4$ grasses in native grass floras of (a) Australia, (b) North America.** $C_4$ grasses become more common in warmer areas of both continents. (After Teeri and Stowe, 1976; and Hattersley, 1983.)

(a)

(b)

**Figure 7.10 Algae growing at different ocean depths.** **(a)** Many algae that grow in tidal pools are green, just like terrestrial plants. **(b)** In contrast, at 75-m depth, most seaweeds are pink and red because the pigments can utilize the blue-green light that occurs at such depths.

could change these distributions. Finally, some $C_4$ plants take up carbon dioxide at night by a process called crassulacean acid metabolism and are known as CAM plants. Such species generally live in hot, arid environments and include cacti. To avoid water loss, CAM plants keep their stomata closed in the day and open them at night when it is cooler and the relatively humidity is higher.

In aquatic environments, light may be an even more limiting factor because water absorbs light, preventing photosynthesis at depths greater than 100 m. Most aquatic plants are limited to a fairly narrow zone close to the surface, where light is sufficient to allow photosynthesis to exceed respiration. This zone is known as the **photic zone** (look ahead to Figure 23.9). In marine environments, seaweeds at greater depth have wider thalli (leaflike light-gathering structures) than those nearer the surface, because wide thalli collect more light. In addition, in aquatic environments, plant color changes with depth. At the surface, plants appear green, as they are in terrestrial conditions, because they absorb red and blue light, but not green. At greater depths, red light is absorbed by the water, leaving predominantly blue-green light. Red algae, which reflect red light when we see them at the surface, occur in deeper water, where there is no red light, because they possess pigments that enable them to utilize the available blue-green light efficiently (**Figure 7.10**).

### Check Your Understanding

**7.4** Why are marine algae red at 75-m depth if there is no red light?

## 7.5 Carbon Dioxide and Oxygen Availability Limit Organismal Growth and Distributions

Although light is a critically important resource for plants, most plant dry mass originates from carbon dioxide, $CO_2$. Atmospheric $CO_2$ concentrations are fairly low, about 400 ppm (parts per million), although this is gradually increasing as more fossil fuels are burned (refer back to Figure 5.19b). Most plants would require 1,000 ppm of atmospheric $CO_2$ to saturate photosynthesis. In greenhouses, farmers can enrich the $CO_2$ level and double the growth rate of tomatoes, cucumbers, and other vegetables. In temperate forests, increasing the level of $CO_2$ to around 700 ppm has increased growth by about 23%.

The efficiency with which plants absorb $CO_2$ from the atmosphere also influences their distribution patterns. $C_4$ plants can absorb $CO_2$ much more effectively than $C_3$ plants. Consequently, they do not have to open their leaf stomata as much, and so they lose less water than $C_3$ plants. $C_4$ plants are therefore more frequent in arid conditions. $C_3$ plants are favored in wetter parts of the world. Also, concentration of rubisco, the enzyme that converts $CO_2$ to either a $C_3$ or a $C_4$ molecule, is generally 60–80% lower in $C_4$ plants. Because rubisco makes up 25% of the leaf nitrogen, total leaf nitrogen content is generally lower in $C_4$ plants. As a consequence, $C_4$ plants are nutritionally less attractive to some herbivores. Thus, a change in $C_3/C_4$ abundance in a globally warmed world might affect herbivores too.

Organisms also differ in their need for oxygen. Eukaryotes are **obligate aerobes**, meaning they require oxygen to live. However, many prokaryotes live in oxygen-poor conditions. Soil-dwelling prokaryotes may live in a complete absence of oxygen, if they live deep underground, and may be obligate anaerobes. **Obligate anaerobes**, such as *Clostridium perfringens,* which causes gas gangrene, are poisoned by oxygen. Treatment of gas gangrene usually involves placing patients in a hyperbaric chamber where oxygen levels are raised, killing the bacteria. Some species may or may not use oxygen, depending on its availability, and are known as **facultative anaerobes**. Some intertidal worms are facultative anaerobes, depending on the conditions they live in. For example, *Nereis* polychaetes may live anaerobically in oxygen-deficient mud yet live aerobically where substrates are better oxygenated. Some fungi, such as yeasts, are also facultative anaerobes.

The availability of oxygen to terrestrial animals is dependent upon altitude. The traditional measure of atmospheric pressure is how high a column of mercury, Hg, can be forced upward in a device called a manometer. At sea level the atmospheric pressure is 760 mmHg. As elevation increases, the amount of atmosphere pressing down on organisms decreases and the atmospheric pressure decreases (**Figure 7.11**). In Denver, elevation 1,700 m, atmospheric pressure decreases 16%, to about 640 mmHg. Atmospheric pressure, P, is the sum of the pressures exerted by all gases in the air, mainly nitrogen and oxygen. Each gas has a partial pressure, designated by P with a subscript, depending on the gas. The partial pressure of oxygen, $P_{O_2}$ at sea level is about 160 mmHg (760 mm $\times$ 0.21% oxygen = 160 mm). The higher the altitude, the lower the partial pressure of oxygen. Pressure gradients drive the process of oxygen diffusion from air or water across respiratory surfaces. Thus, as elevation increases, the rate of diffusion of oxygen into the bloodstream decreases. At the top of Mount Everest there is only a third of the oxygen that is available at sea level.

Organisms that live at high altitudes have special adaptations that enable them to acquire sufficient oxygen. Llamas, *Lama glama,* live in the Andes mountains, at altitudes up to 4,800 m, where $P_{O_2}$ is only about 85 mmHg, or 53% that of sea level. Llama hearts and lungs are much bigger than for a comparable-sized mammal living at sea level, and llamas have higher numbers of red blood cells per unit volume of blood. These adaptations provide increased oxygen-carrying capacity and cardiac output needed to sustain life at such high altitudes. In addition, their hemoglobin has a different amino acid sequence from that of other animals, giving them a high affinity for binding oxygen even at low partial pressures.

The solubility of oxygen in water is affected by four factors: atmospheric gas pressure, temperature, the presence of other solutes, and depth. The partial pressure of oxygen in water is the same as that of the partial pressure of oxygen in the air that presses up against it. So, if the $P_{O_2}$ of air is 160 mmHg, the $P_{O_2}$ of water is also 160 mmHg. For this reason, high-altitude lakes have lower oxygen content than lakes at sea level. Gas solubility also decreases with increasing temperature, so that cold waters have higher oxygen contents than warmer tropical lakes. Other solutes, such as salts, reduce the amount of dissolved gas. For this reason, marine environments generally contain lower oxygen levels than freshwater ones. Finally, oxygen content generally decreases with depth, because there is less mixing with surface oxygen and fewer photosynthetic oxygen-producing organisms in deeper waters. The presence of rooted vegetation or phytoplankton can increase oxygen content. Oxygen is present in water at a concentration of only about 7 ml/L, so to obtain sufficient oxygen an aquatic animal must process about 30 times as much water as air-breathers. This results in active pumping of water across gill surfaces. The high dissolution of carbon dioxide enables water-breathers to easily rid themselves of excess carbon dioxide.

Diving animals such as birds and mammals are not limited by oxygen content of the water, because they only breathe at the surface. However, diving requires its own adaptations. Many diving animals have a high number of red blood cells, and some animals, such as seals, store extra blood cells in the spleen. At greater depths, the spleen contracts, ejecting the extra blood cells into circulation. The muscles of diving animals also contain large quantities of myoglobin and its associated oxygen. This permits blood to be routed to critical structures such as the brain and sense organs. Such adaptations permit some diving animals to remain submerged for over an hour.

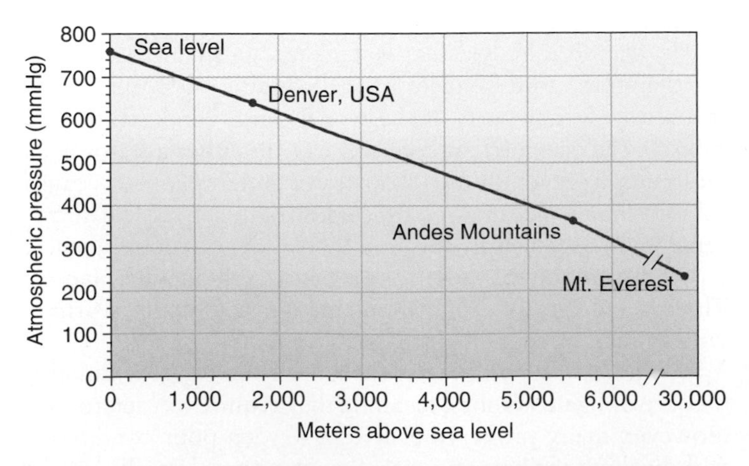


**Figure 7.11** **Atmospheric pressure decreases as altitude increases.**

**Check Your Understanding**

**7.5** In a world where $CO_2$ concentrations are ever increasing, which type of plant—$C_3$, $C_4$, or CAM—will be favored, and how might this affect herbivores?

## 7.6 Species Distributions Are Often Limited by Multiple Abiotic Factors

As we have learned so far in Section Two, species may be limited by temperature, water, nutrients, or other abiotic factors. Often, however, rather than just one limiting factor there are many interacting factors. Plant growth rates may be affected by the availability of both light and nitrogen. However, species that are tolerant of low light levels are not necessarily tolerant of low nitrogen availability. For example, tree species of eastern deciduous forests differ with respect to their tolerance to shade and available nitrogen (**Table 7.3**). The data in Table 7.3 indicate where each species might be common within a large forest region containing areas with varying light and nitrogen availability. For example, white ash requires intermediate light levels and high nitrogen, and it tends to occur in small disturbances on sites with rich soil. It may now be apparent why there are so many species of tree in a given forest region. Nine distinct types of environment can be delineated by their light and nitrogen availability alone.

For animal species too, a knowledge of all variables influencing range is of vital importance, because abundance may be affected by different factors in different parts of the range. A case in point is the Burmese python, *Python molorus,* an invasive species now well established in south Florida and spreading north. The Burmese python is a large snake up to 7 m (22 feet) long and 90 kg in weight, with a voracious appetite for species as large as white-tailed deer and alligators. As we will see later, there have been severe mammal declines coincident with the proliferation of Burmese pythons in south Florida (look ahead to Figure 13.12b). Burmese pythons, a subspecies of the Indian rock python, are regularly sold as pets in the U.S., where they have also often been released into the wild. Their large size and appetite probably contribute to pet snakes being released as they become too much trouble to handle. Most likely, a small number of adult snakes were released prior to 1985, but since then their numbers have increased dramatically (look ahead to Figure 10.5). Most people consider these snakes to be inhabitants of the tropical jungle, but in fact their native range includes parts of temperate China and the Himalayas. Gordon Rodda and colleagues (2009) matched the natural distribution limits of the Burmese python to monthly rainfall and temperature statistics from 149 locations near the edges of its range in Asia (**Figure 7.12a**). Much of the northern range limit was set by cold temperatures, but in the west aridity is likely to be a limiting factor. Burmese pythons can live in places with 2 months of mean temperatures as low as 2°C and zero rainfall. The southern limit ends on peninsular Malaysia south of

**Table 7.3 Separation of some deciduous tree species in the northeastern United States by light and nitrogen availability.**

| Nitrogen Availability | Light Availability: Shade | Light Availability: Intermediate | Light Availability: Sunny |
|---|---|---|---|
| *Poor* | Hickory | White oak, chestnut oak | Bigtooth aspen |
| *Intermediate* | Beech, red maple | Basswood | Trembling aspen |
| *High* | Sugar maple, black gum | White ash, northern red oak | Tulip poplar |

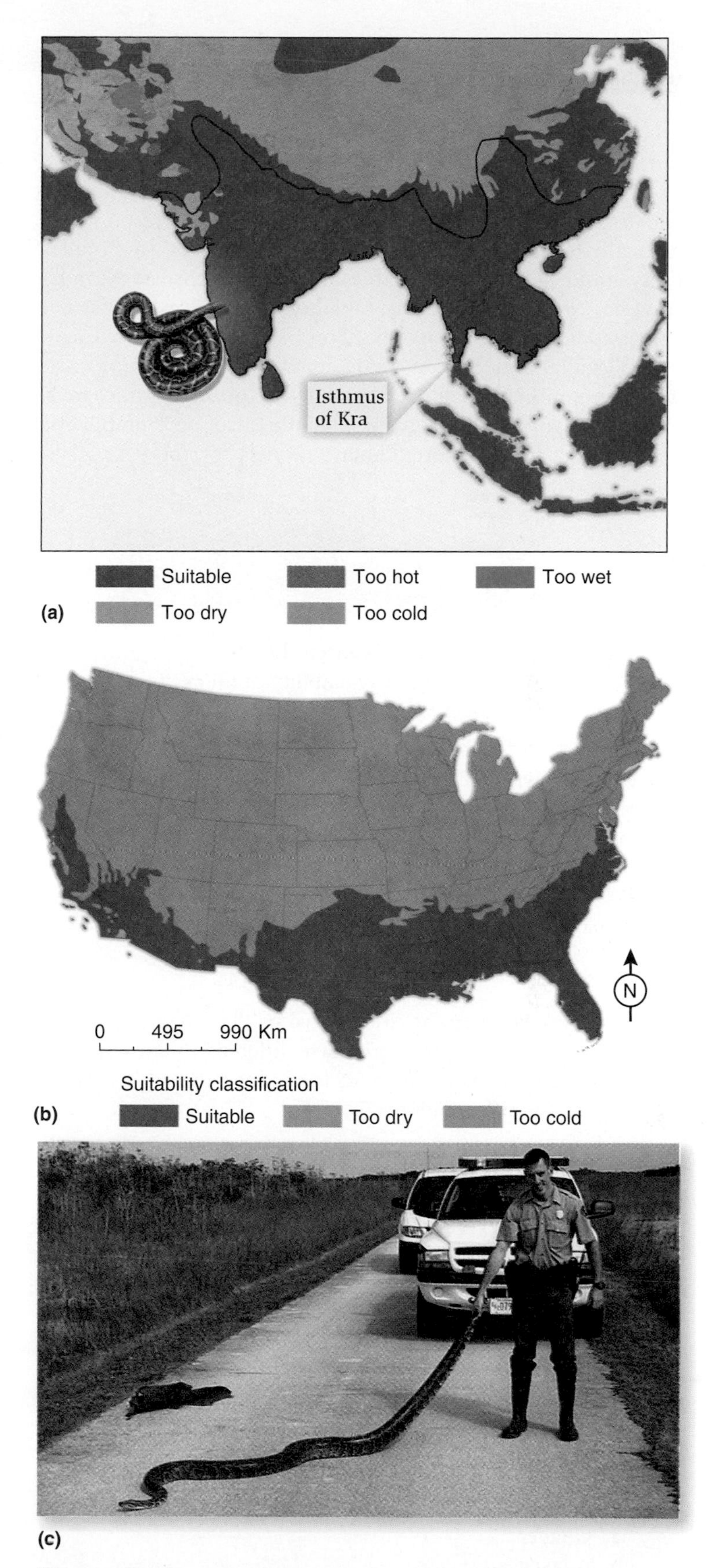


**Figure 7.12 Range limits of the Burmese python.** **(a)** Native range limits. **(b)** Projected range limits in the United States. (After Rodda et al., 2009.) **(c)** Large Burmese pythons are now regularly encountered in the Everglades.

the Isthmus of Kra, though areas south of this were deemed suitable and pythons do occur on Java, Sumba, and the southwestern arm of Sulawesi. It is possible that localized competition with the reticulated python, *Python reticulatus*, limits the Burmese python at the southern end of the range.

Although the range of the Burmese python straddles the equator, the more temperate parts of the range correspond climatically to most of the southern U.S. (Figure 7.12b). Populations could likely exist in large areas of California, Nevada, Arizona, New Mexico, Texas, Oklahoma, Arkansas, Louisiana, Mississippi, Alabama, Georgia, South Carolina, North Carolina, Virginia, and Florida. Only small areas of the Colorado desert in southern California would be too dry. Large areas of Mexico and the Neotropics would also be suitable for the pythons. Furthermore, climate models for the year 2100 suggest a possible northern expansion of range by 100–200 km. Nutrient availability, in the form of suitable food, may limit the spread of the pythons, but this is yet to be determined. On Tuesday, January 17, 2012, the U.S. Fish and Wildlife Service banned the importation of four species of large snakes into the United States and prohibited them from being transported across state lines: the Burmese python, yellow anaconda, northern and southern African pythons.

### Check Your Understanding

**7.6** Under which aquatic conditions would oxygen levels be greatest?

## SUMMARY

**7.1 Soil Development Affects Nutrient Levels**
- Soil conditions and the availability of nutrients can profoundly affect vegetation and the abundance of animals (Figures 7.1, 7.2; Tables 7.1, 7.2).

**7.2 Plant Growth Is Limited by a Variety of Nutrients**
- Many plants are limited by nutrients, and some supplement their intake of nitrogen by trapping and consuming small animals (Figures 7.3, 7.4).
- Low-level atmospheric nitrogen deposition can radically change the plant species present in an area and increase the likelihood of invasion by exotic species (Figures 7.A, 7.5).

**7.3 Herbivore Populations Are Limited by Plant Nutrient Levels**
- Herbivores and carnivores have different abilities to synthesize certain amino acids and their digestive systems may also differ (Figures 7.6, 7.7).

**7.4 Light Can Be a Limiting Resource for Plants**
- Light may influence the distribution of $C_3$ and $C_4$ plants. It is often a limiting resource for plants in both terrestrial and aquatic environments (Figures 7.9, 7.10).

**7.5 Carbon Dioxide and Oxygen Availability Limit Organismal Growth and Distributions**
- Atmospheric $O_2$ and $CO_2$ levels change with altitude (Figure 7.11) and can also affect the abundance of organisms.

**7.6 Species Distributions Are Often Limited by Multiple Abiotic Factors**
- Most organisms are simultaneously limited by the availability of multiple abiotic factors (Figure 7.12, Table 7.3).

## TEST YOURSELF

1. Which of the following are major nutrients for plants?
   a. Selenium and fluorine
   b. Manganese and zinc
   c. Copper and boron
   d. Aluminum and silicon
   e. Nitrogen and phosphorous
2. In a generalized profile of the soil, which layer contains the most partially decomposed organic matter?
   a. O layer
   b. A layer
   c. B layer
   d. C layer
   e. D layer
3. The large carbon mass of plants and animals is supplied by;
   a. Soil
   b. Fresh water
   c. Air
   d. Lightning strikes
   e. Salt water
4. In North America, the proportion of $C_4$ plants in the native grassland flora increases with decreasing latitude.
   a. True b. False
5. The term "essential amino acids" is used to denote amino acids that are essential to carnivores and omnivores.
   a. True b. False

6. Most desert succulents employ which type of photosynthesis?
   a. $C_3$
   b. $C_4$
   c. CAM
   d. a and b
   e. b and c
7. What color of light is predominant at greater depths of water?
   a. Red
   b. Yellow
   c. Blue-green
   d. Ultraviolet
   e. Infrared
8. In most aquatic environments, algae or plants are not present below which minimum depth?
   a. 10 m
   b. 50 m
   c. 100 m
   d. 500 m
   e. 1,000 m
9. Species that cannot live in the presence of oxygen are termed:
   a. Obligate aerobes
   b. Obligate anaerobes
   c. Facultative anaerobes
   d. $C_3$
   e. $C_4$
10. Carnivores have a relatively longer intestine than herbivores.
    a. True
    b. False

## CONCEPTUAL QUESTIONS

1. Explain why most soil-forming processes act from the top down.
2. Why would decreasing chronic atmospheric nitrogen deposition be valuable to natural systems?

## DATA ANALYSIS

Peter Reich and colleagues (1992) collected much data on the longevity of leaves, their nitrogen content, and other features from a wide array of plant species in different habitats. Some of their data are shown in the figure. What does this relationship tell you, and where would coniferous plants fit?

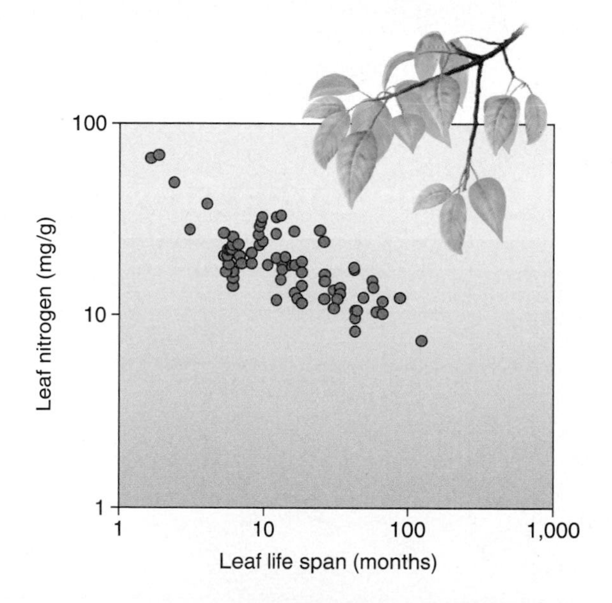

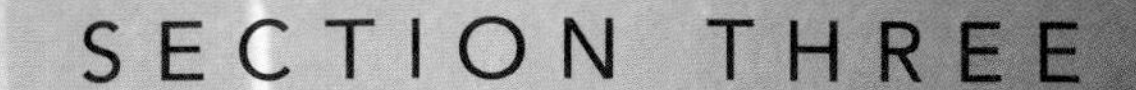

SECTION THREE

# Population Ecology

A population can be defined as a group of interbreeding individuals of the same species occupying the same area at the same time. In this way, we can think of a population of water lilies in a particular lake, the lion population in the Ngorongoro crater in Africa, or the human population of New York City. However, the boundaries of a population can often be difficult to define, though they may correspond to geographic features, such as the boundaries of a lake or forest, or be contained within a mountain valley or a certain island. Individuals may enter or leave a population, such as the human population of New York City. Thus, populations are often fluid entities defined by a particular area, with individuals moving into (immigrating to) or out of (emigrating away from) an area. For the purposes of simplicity, in most of our discussion of populations we will assume that immigration and emigration of populations occur at the same rate and so cancel each other out as factors.

This section explores **population ecology**, the study of how populations grow. To study populations, we need to employ some of the tools of **demography**, the study of birth rates, death rates, age distributions, and the sizes of populations. In Chapter 8 we begin our discussion by exploring characteristics of populations, including density and how it is quantified. In Chapter 9 we show how life tables and survivorship curves help summarize demographic information such as birth rates, death rates, and population growth per generation. In Chapter 10 we examine population growth by determining how many reproductive individuals are in the population and their fertility rate. The data are then used to construct simple mathematical models that allow us to analyze and predict population growth. We use these population concepts and models to explore the growth of human populations as well as populations of other organisms.

Population growth affects population size and population density, and knowledge of both can help us make decisions about the management of species. How long will it take for a population of an endangered species to recover to a healthy level if we protect it from its most serious threats? For example, in the northern Rockies, after being eliminated in the wild, gray wolves have been reintroduced, despite the protests of ranchers who fear livestock losses. A knowledge of wolf population growth rates and population densities would allow us to determine at what point in time populations might constitute a threat to livestock.

Cougars live in fragmented habitats in southern California and rely on habitat corridors to travel between habitat fragments.

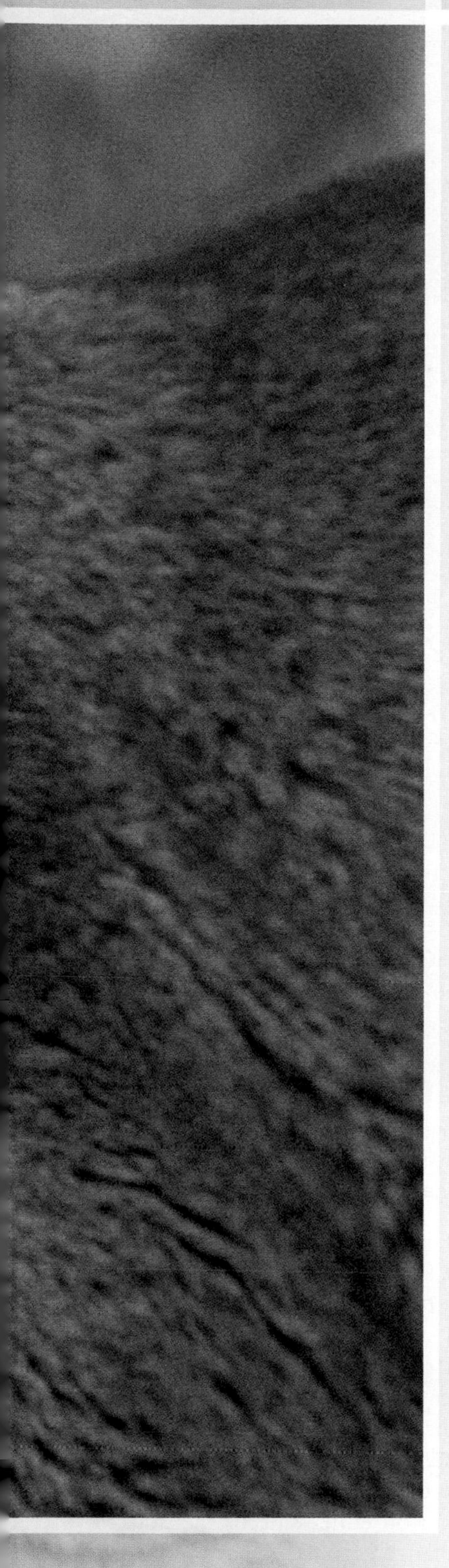

CHAPTER 8

# Demographic Techniques and Population Patterns

## Outline and Concepts



In southern California the cougar, *Puma concolor,* exists mainly as semi-isolated populations found mostly in small mountain ranges. These habitat areas are linked by riparian corridors, vegetated corridors that occur along the floodplains of watercourses such as rivers or streams. Radiotelemetry studies show that cougars often use these corridors to travel between habitat patches or colonize empty areas. Development often threatens to sever these links. In the 1990s one of these corridors was proposed for development by the city of Anaheim. This would have isolated a large 150-$km^2$ habitat patch. Cougars would eventually have become extinct in this isolated area, and the cougar population as a whole would have decreased. Ecologists recommended the development plan be modified to leave the corridor undeveloped. Maintaining connections between habitats has been shown to be important in sustaining wildlife populations in many other areas. For example, on a small scale, road underpasses allow key deer, *Odocoileus virginianus,* to move from the north of Big Pine Key in the Florida Keys to the south of the island. On a large scale, regional corridors could allow movement of black bears, *Ursus americanus,* between the Osceola and Ocala National Forests in Florida.

In this chapter we discuss characteristics of populations, beginning with population size and **population density**, the number of organisms in a given unit area or volume.

We examine a variety of techniques that are used to measure population sizes and densities, from simple counts of plants per square meter to more complex mark-recapture techniques for animals. We also discuss how individuals may be clumped together or spread out in their habitat. Loss of habitat has caused populations to exist in habitat fragments, especially in urban areas, which may or may not be connected by habitat corridors. We examine the effects of habitat fragmentation on populations and dispersal between fragments, and explore the characteristics of metapopulations, a series of small, separate populations that affect each other.

## 8.1 Various Techniques Are Used to Quantify Population Size and Density

The simplest method to measure population size is to visually count the number of organisms in a population in a given habitat. We can reasonably do this only if the area where the population exists is small and the organisms are relatively large. For example, we can determine the number of gumbo limbo trees, *Bursera simaruba,* on a small island in the Florida Keys. Sometimes, large conspicuous animals can be counted using binoculars. In this way we could count the number of butterflies or dragonflies over various ponds or patches of flowers. Small insects can be dislodged from vegetation by spraying soap solution, then counted as the number that fall off into trays laid beneath the plant. Other insects may be dislodged by beating the vegetation, such as a tree branch, with a stick and collecting the insects by placing a cloth sheet underneath. Counts of insects on different branches, of about the same size, can be compared. Sometimes a chemical knockdown agent, such as pyrethrum, is applied as a mist or aerosol to maximize the numbers of individuals dislodged.

Population ecologists calculate the density of plants or animals in a small area and use this data to estimate the total abundance over a larger area. For plants, algae, or other sessile organisms such as intertidal animals, it is fairly easy to count numbers of individuals per 0.25 $m^2$ or, for larger organisms such as trees, numbers per hectare. However, many plant individuals are clonal, that is, they grow in patches of genetically identical individuals, so that rather than count individuals we can also use the amount of ground covered by plants as an estimate of vegetation density.

### 8.1.1 Quadrats are used to quantify population densities of more sedentary species

Plant ecologists use a sampling device called a **quadrat**, a square frame often, but not always, measuring 50 × 50 cm and enclosing an area of 0.25 $m^2$ (**Figure 8.1**). They then count the numbers of plants of a given species inside a series of quadrats to obtain a density estimate per square meter. For example, if you counted densities of 20, 35, 30, and 15 plants in four quadrats, you could reliably say that the average density of this species was 100 individuals per square meter. For larger plants, such as trees, a quadrat would be ineffective. Many ecologists perform **line transects** to count such organisms. A line transect is essentially a long piece of string that is stretched out, and the number of trees along its length are counted. Belt transects include species within a short distance of the line. For example, to count tree species on islands in the Florida Keys, graduate student Mark Barrett laid out 100-m belt transects and counted all the trees within 1 m on either side of the line. In effect his quadrat shape had changed from a square to a long thin rectangle encompassing 200 $m^2$. By performing five such transects he obtained estimates of tree density per 1,000 $m^2$. Quadrats can also be used to count the number of some insects per 0.25 $m^2$ of grassland. Quadrats are initially placed in the field, then revisited later after the insects have resettled.

**Figure 8.1 The quadrat sampling technique.** This plant ecologist in Dartmoor National Park, England, is using a 50 cm × 50 cm quadrat to count the number of plants per 0.5 $m^2$.

**Figure 8.2 Sampling methods for insects.** (a) Suction traps in Iowa suck in flying insects. (b) D-Vac suction sampler to sample insects on vegetation.

## 8.1.2 Traps are used to quantify relative population densities of more mobile species

Many different trapping methods exist for quantifying animal densities (**Figure 8.2a**). Suction traps, like giant aerial vacuum cleaners, can suck flying insects from the sky. An electric fan pulls or drives air through a fine gauze cone that filters out the insects. Insect numbers per volume of air drawn in can be calculated. For small or hidden grassland species, sweep nets can be passed over vegetation to dislodge and capture the insects feeding there. By measuring the volume of the open mouth of the net and by counting the number of sweeps made, we can arrive at a rough estimate of the number of animals caught per area swept. Alternatively we can thoroughly sweep a habitat of known area. Some insects can be vacuumed off their plants using a Dietrick suction sampler, commonly referred to as a "D-Vac" (**Figure 8.2b**). These suction samplers have a hose with an opening of known area, so that, by placing the hose over different areas of the vegetation, known areas can be sampled and compared. Suction samplers collect a mixture of insects and loose plant debris into a collecting bag that must then be sorted. Such sorting is often done in the laboratory, under a microscope.

For soil-dwelling organisms, including those inhabiting leaf litter or those inhabiting mud from aquatic habitats, organisms cannot be easily counted in the field. Instead, samples of known volume are taken and returned to the laboratory, where they are placed under warming lights. The organisms move away from the heat and drop into collecting fluids underneath (**Figure 8.3**). This combination of warming lamps, sample-holding funnels, and containers with collecting fluid is

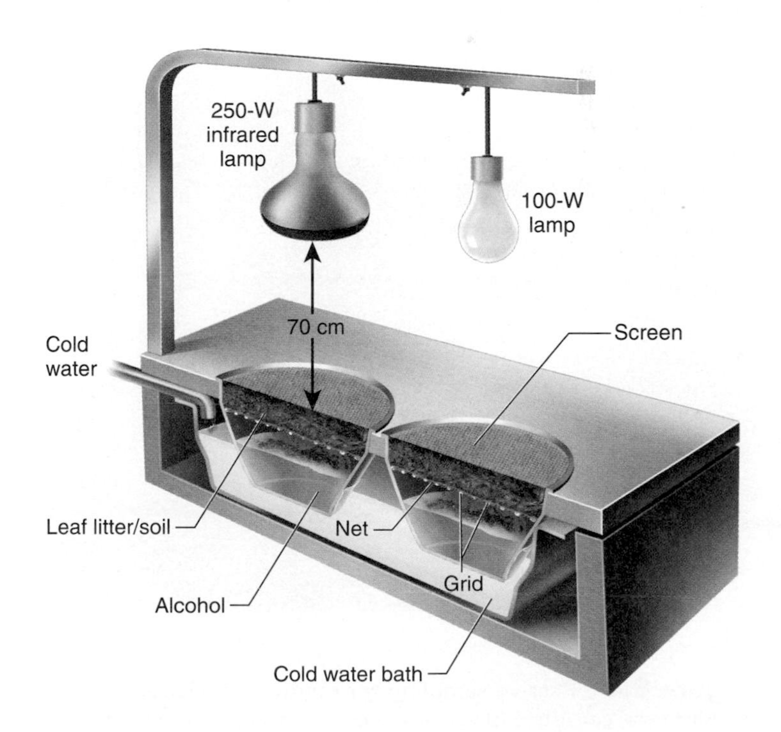


**Figure 8.3 Berlese funnel.** This apparatus heats soil or litter samples, causing the small organisms inside the sample to move away from the heat and fall into preservative underneath.

(a)

(b)

(c)

**Figure 8.4 Passive sampling for relative density estimates.** **(a)** Mist nets, consisting of very fine mesh, entangle birds. **(b)** Pitfall traps set into the ground catch wandering species such as beetles and spiders. These have rain covers to prevent the animals from drowning. **(c)** Baited live traps catch small mammals; here, foxes on Guam.

known as a Berlese funnel, after the Italian entomologist who designed it.

Apart from these absolute estimates of population density, ecologists use a number of methods that give relative densities. Many of these involve passive sampling. For example, mist nets, consisting of very fine netting spread between trees, can entangle flying birds and bats (**Figure 8.4a**). Pitfall traps set into the ground can catch species wandering over the surface, such as spiders, lizards, or beetles (**Figure 8.4b**). Baited live traps can catch small mammals (**Figure 8.4c**). Population density can thus be estimated as the number of animals caught per trap or per unit area where a given number of traps are set—for example, 10 traps per 100 $m^2$ of habitat.

### 8.1.3 Mark-recapture is used to quantify population sizes of mobile species

To estimate population size, population biologists may also capture animals and then tag and release them (**Figure 8.5**). The rationale behind the **mark-recapture technique** is that after the tagged animals are released, they mix freely with unmarked individuals and within a short time are randomly mixed within the population. The population is resampled and the numbers of marked and unmarked individuals are

**Figure 8.5 The mark-recapture technique is often used to estimate population size.** An ear tag identifies this Rocky Mountain goat, *Oreamnos americanus*, in Olympic National Park, Washington State. Recapture or recensusing of such marked animals permits accurate estimates of population size.

**ECOLOGICAL INQUIRY**

If we mark 110 Rocky Mountain goats and recapture 100 goats, 20 of which have ear tags, what is the estimated total population size?

recorded. We assume that the ratio of marked to unmarked individuals in the second sample is the same as the ratio of marked to unmarked individuals in the population. Thus:

$$\frac{\text{Number of individuals marked in first catch}}{\text{Total population size, N}} = \frac{\text{Number of marked recaptures in second catch}}{\text{Total number of second catch}}$$

Let's say we catch 50 largemouth bass in a lake, mark them with colored fin tags, and release them back into the lake. A week later we return to the lake and catch 40 fish, and 5 of them have been previously tagged. If we assume that no immigration or emigration has occurred, which is quite likely in a closed system like a lake, and we assume there have been no births or deaths of fish, then

$$\frac{50}{N} = \frac{5}{40}$$

The total population size is given by rearranging the equation:

Total population size,

$$N = \frac{\text{Number of marked individuals in first catch} \times \text{Total number of second catch}}{\text{Number of marked recaptures in second catch}}$$

Using our data:

$$N = \frac{50 \times 40}{5} = \frac{2{,}000}{5} = 400$$

From this equation, we estimate that the lake has a total population size of 400 largemouth bass. This could be useful information for game and fish personnel who wish to know the total size of a fish population in order to set catch limits. However, the mark-recapture technique can have drawbacks. Some animals that have been marked may learn to avoid the traps or lose their tags. Recapture rates will then be low, resulting in an overestimate of population size. Imagine that instead of 5 tagged fish out of 40 recaptured fish, we only get 2 tagged fish. Now our population size estimate is 2,000/2 = 1,000, a dramatic increase in our population size estimate. On the other hand, some animals can become trap-happy, particularly if the traps are baited with food. This would result in an underestimate of the population size. Tag types vary with study organism. Larger mammals may receive ear tags, while smaller mammals and birds receive bands or rings. Insects may be painted in various colors. Dusting with fluorescent pigments may permit detection of marked recaptures by the use of ultraviolet light.

Because of the limitations of the mark-recapture technique, ecologists also use other, more novel methods to estimate population size of large animals. For some larger terrestrial or marine species, captured animals can be fitted with a radio collar, a collar with a small attached radio transmitter that enables researchers to follow the animals telemetrically using an antennal tracking device. The species' home ranges can be determined and population estimates developed based on the area of available habitat. For many species with valuable pelts, we can track population densities through time by examining pelt records taken from trading stations. We can also estimate relative population size by examining catch per unit effort, which is especially valuable in commercial fisheries. Although we can't easily count the number of fish in an area of ocean, we can count the number caught, say, per 100 hours of trawling.

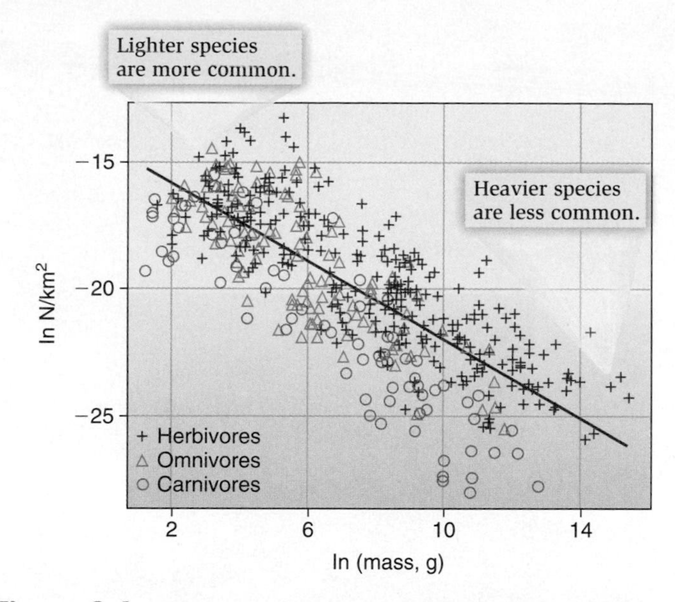


**Figure 8.6 The relationship of population density with individual species mass in terrestrial mammals.** The larger the mammal, the lower the population density. (After Brown et al., 2004.)

Generally there is a negative relationship between body mass and population density (**Figure 8.6**). This makes sense when one considers that, compared to smaller organisms, fewer large organisms can be found in the same area of habitat. Nevertheless, the relationship has much variation, due in part to the type of habitat where organisms live and their lifestyle. For example, fewer carnivores of a given weight occur in an area compared to the number of herbivores.

**Check Your Understanding**

**8.1** Which techniques might you use to determine the population densities or relative densities of the following organisms: (a) bats, (b) butterflies, and (c) ants?

## 8.2 Patterns of Spacing May Be Clumped, Uniform, or Random

Individuals within a population can show different patterns of spatial dispersion; that is, they can be clustered together or spread out to varying degrees. The three basic kinds of dispersion pattern are clumped, uniform, and random. We can visualize these patterns by imagining people in a meeting room.

(a) Clumped

(c) Random

(b) Uniform

**Figure 8.7 Three types of dispersion.** **(a)** A clumped distribution pattern, as in these plants clustered around an oasis in the Namib desert, Africa, often results from the uneven distribution of a resource—in this case, water. **(b)** A uniform distribution pattern, as in these nesting gannets, may be a result of competition or social interactions. **(c)** A random distribution pattern, as in these chinstrap penguins on an Antarctic hillside has no underlying cause.

**ECOLOGICAL INQUIRY**

Which of these three types of dispersion is most common in nature and why?

If some people know each other, they will get together in small groups, creating a clumped pattern. If people do not know each other, and are perhaps even wary of each other, they might maintain a certain minimum personal distance between one another to produce a uniform dispersion. If nobody thinks or cares about their position relative to anyone else, we would get a random dispersion.

The type of dispersion observed in nature can tell us a lot about what processes shape group structure. The most common dispersion pattern is **clumped**, because resources tend to be clustered in nature. For example, certain plants may do better in moist conditions, and moisture is greater in low-lying areas (**Figure 8.7a**). Social behavior between animals may also promote a clumped pattern. Many animals are clumped into flocks or herds.

On the other hand, competition may cause a **uniform** dispersion pattern between individuals, as between trees in a forest. At first the pattern of trees and seedlings may appear random, because seedlings develop from seeds dropped at random, but competition between roots may cause some trees to be outcompeted by others, causing a thinning out and resulting in a uniform distribution. Thus, the dispersion pattern starts out random but ends up uniform. As we saw earlier (refer back to Figure 4.13c), uniform dispersions may also result from social interactions, as between some nesting birds, which tend to keep an even distance from each other (**Figure 8.7b**). Territoriality between animals also causes uniform dispersion patterns.

Perhaps the rarest dispersion pattern is **random**, because resources in nature are rarely randomly spaced. Where resources are common and abundant, as in moist, fertile soil, the dispersion patterns of plants may be random and lacking a pattern. Some animals may also exhibit a random dispersion pattern (**Figure 8.7c**).

**Check Your Understanding**

**8.2** What is the dispersion pattern of students in a half-full classroom? How might this change at test time?

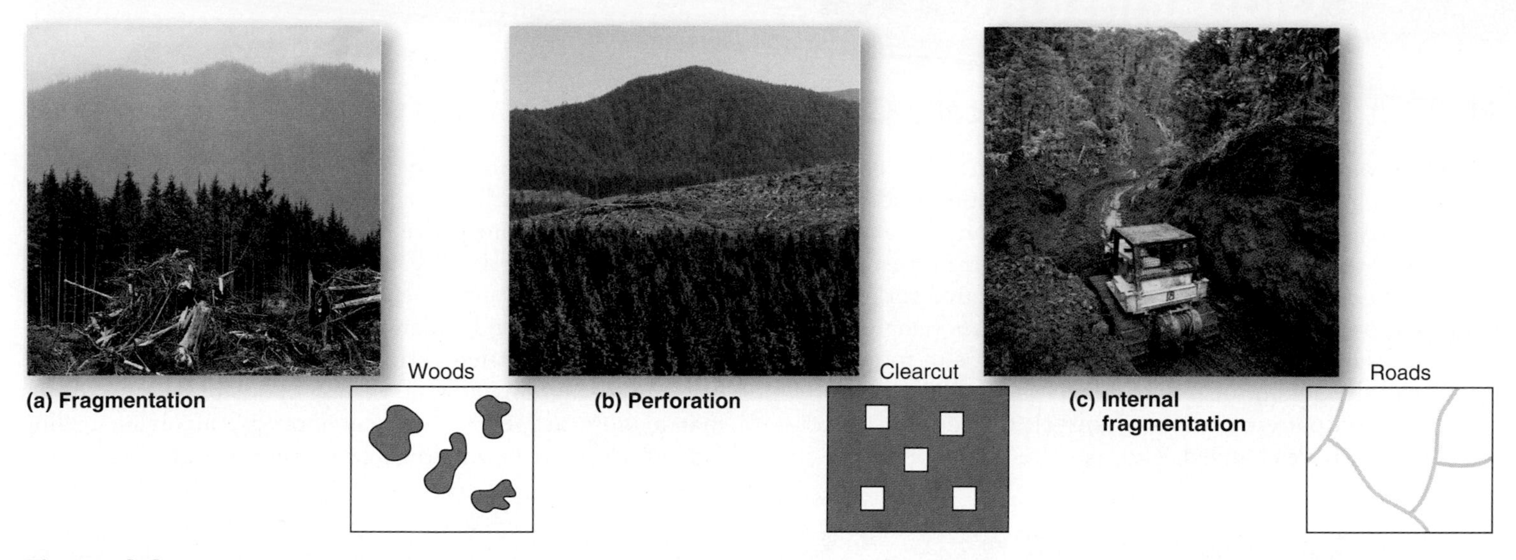


**Figure 8.8 Different types of habitat.** **(a)** Fragmented, with deforestation leaving habitat remnants. **(b)** Perforated, where habitat is largely intact but small openings are created. **(c)** Internal fragmentation, where roads or trails dissect habitat, effectively breaking it into smaller pieces.

**ECOLOGICAL INQUIRY**

What are some of the natural causes of fragmentation?

## 8.3 Fragmented Habitats Affect Spatial Dispersion

Many populations may be fragmented because of habitat destruction caused by deforestation, agriculture, urbanization, or the building of roads. These phenomena can affect population dispersion patterns. In the case of deforestation, sometimes only isolated patches of forest are left intact, and the habitat is said to be **fragmented** (**Figure 8.8a**). Alternatively, the forest might be left largely intact with only small clear-cut areas within it, creating what is called **perforated** habitat (**Figure 8.8b**). The erection of power lines across a forest or the building of roads can isolate patches of forest from one another and create what is known as **internal fragmentation** (**Figure 8.8c**). For example, in the Everglades, Interstate 75 (also known as Alligator Alley) bisects Florida panther habitat. A large chain-link fence stretches the entire 100-mile length of this highway and blocks the movement of panthers and other large vertebrates, including alligators, deer, and raccoons. However, for other species, such as birds, butterflies, and plants, such a barrier is easily crossed. The degree of connectivity between patches may be affected by the properties of the landscape, including the proximity and connectivity of fragments. Habitat destruction has profoundly changed the dispersion patterns of many organisms (see **Global Insight**).

One of the biggest problems with habitat fragmentation is that the fragments of habitat that remain may be smaller than the territory or home range of a single individual of some species. For example, in the western United States, cougars, *Puma concolor,* require habitats greater than 400 $km^2$ (1 $km^2$ = 100 hectares), and grizzly bears, *Ursus arctos,* require 900 $km^2$. Small fragments of habitat preserved in 1-$km^2$ parks are virtually useless for such species. They do not contain enough prey and they promote interactions with humans.

Even small birds might not be found in small forest fragments, despite the fact that the forest is many times larger than the size of their territory. Such birds are said to be area-sensitive, and they breed only in large forest areas. For example, some neotropical bird migrants that nest in the United States require relatively large areas of forest, greater than 10 ha, before there is a large probability that they will inhabit it (**Figure 8.9**). Even larger areas may be required for breeding. Such relationships are known for many other bird and mammal species. The result is that very large areas are needed to be certain that the full complement of species is present. Few parks in the United States and Canada are large enough to maintain their historic complement of mammals. As a result, the number of native mammal species in most parks has declined.

**Check Your Understanding**

**8.3** What are some of the human-made causes of habitat fragmentation?

Global Insight

## Habitat Destruction Has Radically Changed the Dispersion Patterns of Many Species

Over the course of time, humans have subverted much original habitat to agricultural or urban use. In the process they have greatly reduced the extent of some natural habitats to a fraction of their original extent. For example, in Europe and North America, forested areas have been cleared for agricultural purposes and only a few isolated patches remain. Over large scales of counties, the dispersion patterns of forest trees has changed from uniform to clumped (**Figure 8.A**). Similarly, the dispersion patterns of many animals that depend on forested habitat have changed. The last remaining population of wild reindeer, *Rangifer tarandus,* in southern Norway was virtually a continuous herd prior to 1900. Human settlements, roads, and power lines have split the population into 26 distinct herds (**Figure 8.B**). Because reindeer prefer to live at least 5 km distant from human settlements, increased development will likely reduce and isolate existing herds even more. Furthermore, these isolated herds do not migrate as much as the original population and tend to stay in one place. This means that a more active management strategy, involving culling individuals, must be employed to prevent overgrazing.

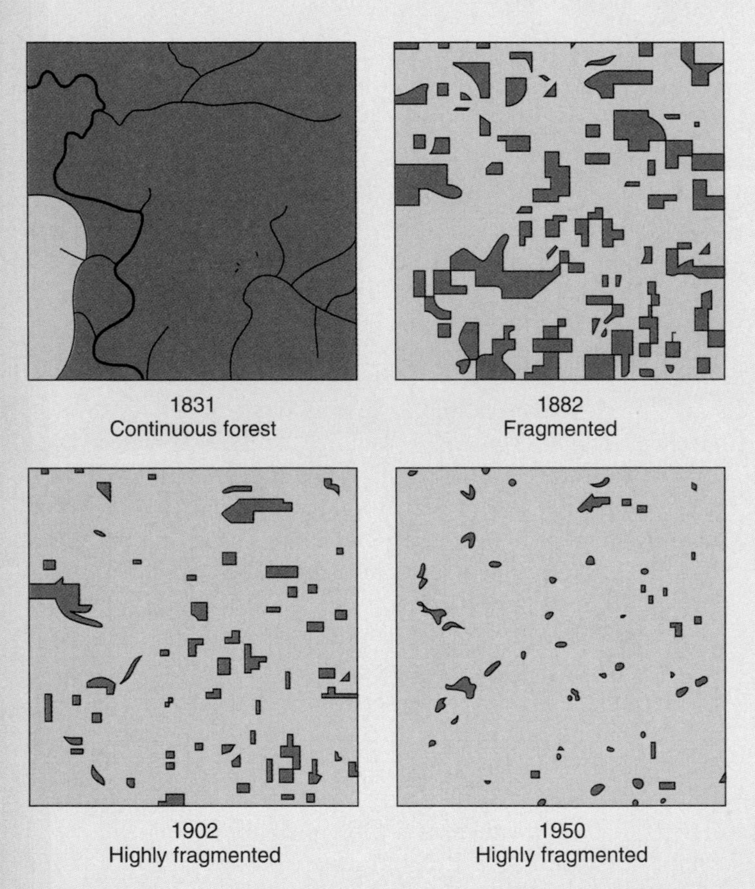


**Figure 8.A Changes in the wooded area of a Wisconsin county over time.** Data are from Cadiz Township, Green County, Wisconsin, following European settlement after 1831 until 1950. Shaded areas represented forested areas. (After Curtis, 1956.)

Atlantic Ocean
Sweden
Norway
Oslo
(a)
Atlantic Ocean
Sweden
Norway
Oslo
(b)
Range

**Figure 8.B Current and former range of reindeer in southern Norway.** **(a)** Prior to 1900, **(b)** as of 2000. Note how the formerly contiguous herd has been divided up into 26 separate herds, mainly by human activities. (After Nelleman et al., 2001.)

**ECOLOGICAL INQUIRY**

Does agriculture increase habitat fragmentation?

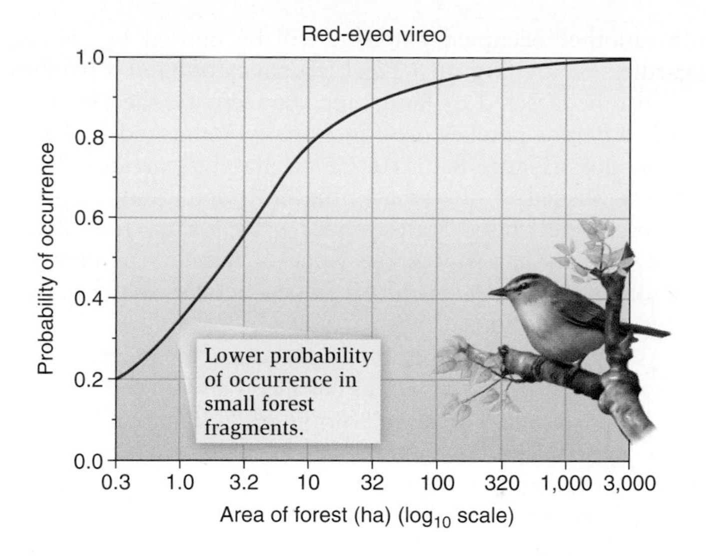


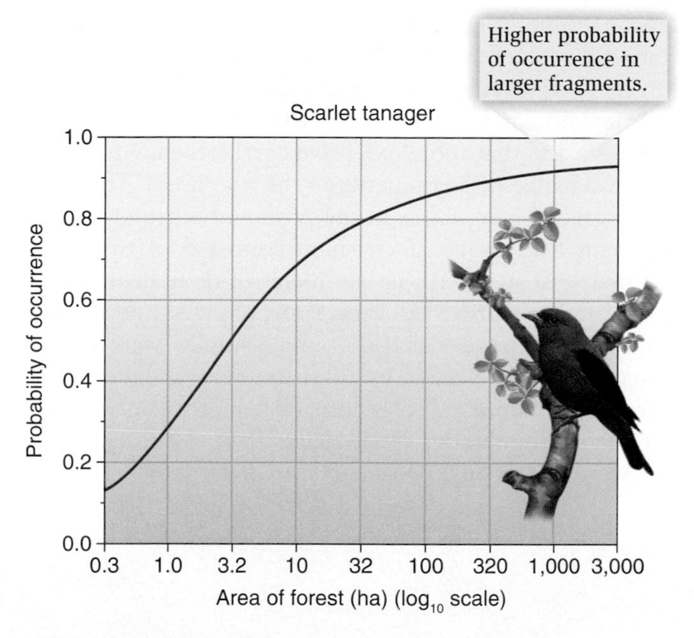


**Figure 8.9 Effect of forest area on occurrence of neotropical migrant birds in the United States.** Note the log scale. (From Robbins et al., 1989.)

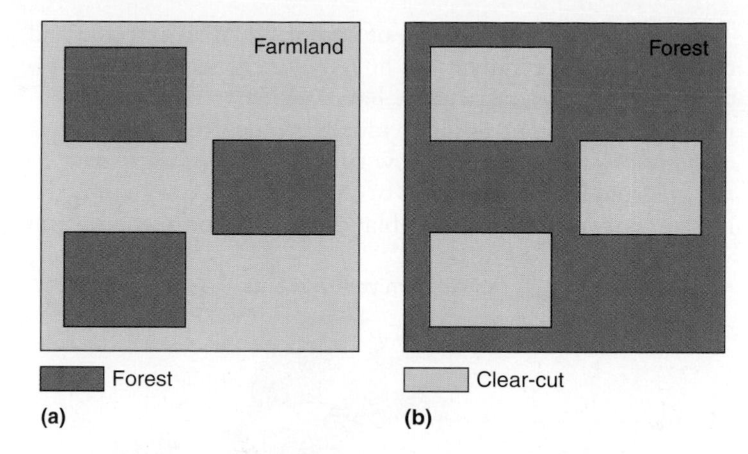


**Figure 8.10 The matrix is the most extensive element in an area.** **(a)** Agricultural matrix with embedded forest patches. **(b)** Forest matrix with embedded clear-cuts.

**Figure 8.11 The effect of the matrix on dunlin densities in Scotland.** Dunlin densities decline with distance from their feeding areas in pools, but densities are generally greater in a lower-biomass matrix (solid dot) than in a higher-biomass matrix (open dot). (After Lavers et al., 1996.)

## 8.4 Landscape Ecology Concerns the Spatial Arrangement of Habitats and Organisms

**Landscape ecology** is a subdiscipline of ecology that examines the spatial arrangement of elements in populations and communities. Such elements consist of **habitat** patches, which contain the food and environmental conditions necessary to support a particular organism, and nonhabitat patches where these same organisms cannot live. At the community level, spatial arrangements of habitats can influence the number of species present, as we will see later (look ahead to Section 21.2). At the population level, the spatial arrangement of habitats can influence population densities. In landscape ecology the **matrix** is defined as the most extensive element of an area. In agricultural areas the matrix is farmland that may have embedded patches of remnant forests (**Figure 8.10a**). In urban areas the matrix is built-up areas or suburbs with embedded parks. In forests the matrix is coniferous or deciduous trees with embedded clear-cuts (**Figure 8.10b**). The condition of the matrix may affect the degree of habitat occupancy. Dunlin, *Calidris alpine,* are European wading birds common on Scottish estuaries where feeding pools are present. As the distance from a cluster of pools increases, dunlin density falls. However, dunlin densities are greater around pools embedded in a low-biomass vegetation matrix of European heather, *Calluna vulgaris,* and deergrass, *Trichophorum cespitosum,* than they are where the matrix consists of higher-biomass *C. vulgaris* and purple moor grass, *Molinia caerulea* (**Figure 8.11**). Low-biomass vegetation may permit quicker detection of predators.

Many landscapes consist of completely disconnected and distinct groups of individuals linked by migrants. Bigger habitat patches, not surprisingly, have a greater chance of supporting bigger groups of individuals, but degree of isolation is also important. Western yellow robins, *Eopsaltria griseogularis,* are found only in patches of shrubland greater than 20 ha in size. However, if the shrubland patch is more than 2 km from another occupied patch, it will be unused by robins, regardless of size (**Figure 8.12**). Occupancy of habitat patches may also be affected by **landscape connectivity**, the extent to which different patches are functionally connected by habitat corridors (**Figure 8.13**). In China, habitat corridors have been established to link small, adjacent populations of giant pandas. In the United States, research by Joshua Tewksbury and colleagues has shown the value of habitat corridors in promoting the movement of organisms between patches (see **Feature Investigation**).

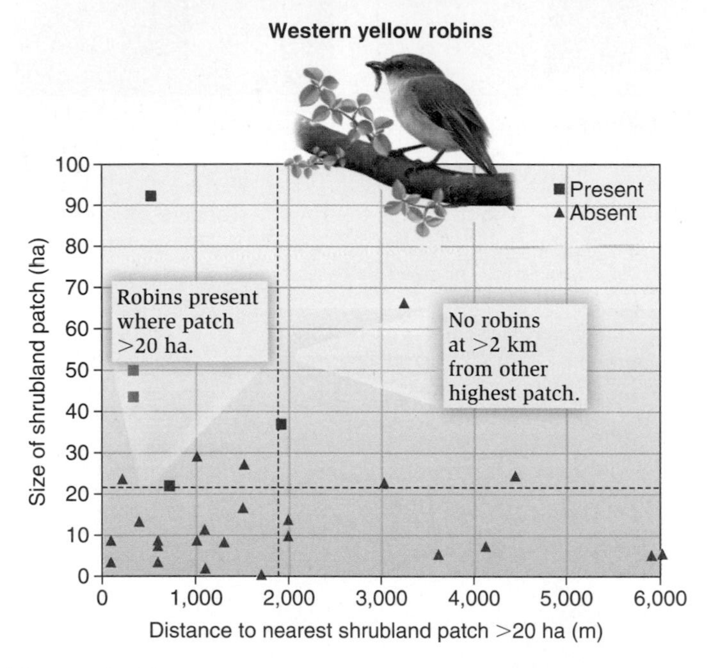


**Figure 8.12** **Landscapes and the effect of isolation.** Occupancy of shrubland patches by western yellow robins decreases with increasing distance from other patches. (After Lambeck and Hobbs, 2002.)

As mentioned at the beginning of the chapter, there is much interest in establishing a habitat corridor between the Osceola and Ocala National Forests in North Florida (**Figure 8.14**). The Ocala National Forest contains approximately 1,660 $km^2$ of black bear habitat, and the Osceola National Forest contains 638 $km^2$. In between is a patchwork of public conservation lands and private lands. Conservationists have suggested establishing a 90-km-long, 30-km-wide corridor to link the two forests and promote the interchange of animals. Would the corridor be used? Jeremy Dixon and colleagues (2006) suggested there may already be some movement of bears between these two forests and that more extensive corridors may foster even greater exchange of bears between the two areas. They established a grid of hair snares in the proposed corridor to examine current bear usage. Each snare consisted of two strands of barbed wire attached to a perimeter of three or more trees. The investigators provided attractants, pastries, and raspberry extract to encourage visitation. Hair samples were collected and genetically analyzed to establish bear sex and identity. The samples were compared to genotypes from those of the Ocala and Osceola populations. Hair snares from 31 different black bears were identified in the corridor area, showing that the

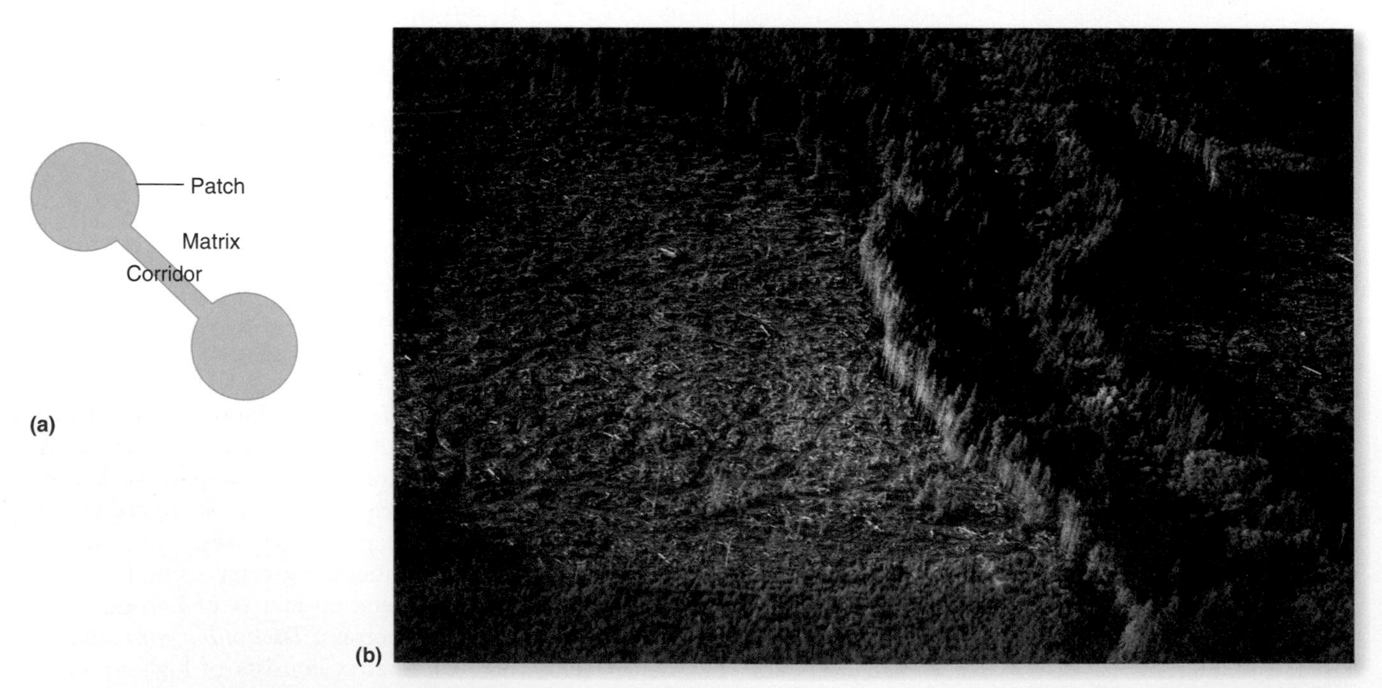


**Figure 8.13** **Landscape corridors.** **(a)** Corridors may link patches of habitat, facilitating movement of organisms. **(b)** A strip of trees may act as a corridor between patches of forest.

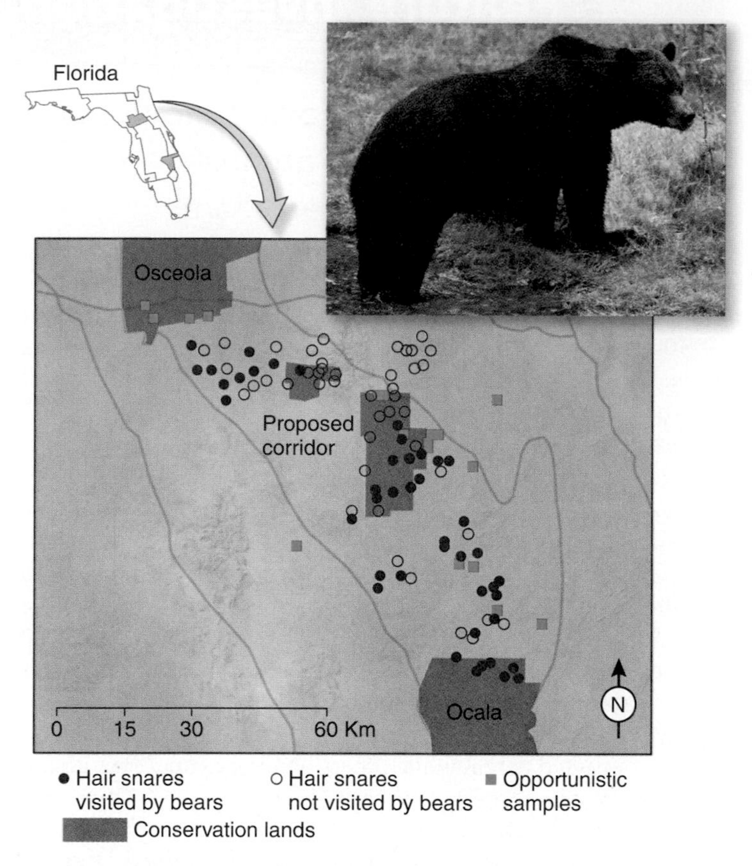


**Figure 8.14** **Locations of bear hairs on snares and proposed habitat corridor between Osceola and Ocala National Forests, Florida.** (After Dixon et al., 2006.)

corridor is functional. Most bears were related to the Ocala population, which is bigger than that of Osceola, but at least some were from Osceola.

### Check Your Understanding

**8.4** Can you think of any disadvantages of habitat corridors?

## 8.5 Metapopulations Are Separate Populations That Mutually Affect One Another via Dispersal

Fragmented landscapes create fragmented populations, many of which may exchange individuals via migration, creating one linked series of subpopulations known as a metapopulation. A **metapopulation** is a series of small, separate populations in individual habitat patches that mutually affect one another. Four conditions must be satisfied by a metapopulation:

1. Each habitat patch must be able to support a breeding population.
2. Any subpopulation must be prone to extinction.
3. Recolonization has to be possible.
4. Subpopulation growth and decline are asynchronous to avoid the extinction of the entire metapopulation.

In this scenario, if individuals in one patch go extinct, other individuals in other patches survive, and they supply individuals that disperse and recolonize the patches where the extinction occurred. Extinction may be caused by local catastrophes such as fires, floods, the effects of natural enemies, or even chance fluctuations in numbers of individuals. Because fragmented populations are often small, such effects may exterminate all individuals in a patch. Even if the individuals from the patch that supplied the colonists themselves all become extinct, this patch will be recolonized again later. Metapopulations are viewed as populations persisting in a balance between local extinction and colonization.

One of the first mathematical treatments of metapopulation dynamics was formulated by Richard Levins (1969), who predicted that the proportion of occupied patches will stabilize over time to $1 - (x/m)$, where $x$ is the extinction rate of populations in patches per unit time and $m$ is the rate of movement between patches. To get these data, ecologists monitor many patches over a long time period to determine the number of patches where populations go extinct. Movement rate is obtained by marking individuals in patches and recapturing them in the same or different patches. If the extinction rate per time period, say, a year, is 0.10, and movement rate, $m$, is 0.4, then the proportion of occupied patches when a population is stable is $1 - (0.1/0.4) = 0.75$. This model is a good first step in determining whether a metapopulation is stable or not. If the proportion of occupied patches is less than predicted necessary by Levins's model, then extinction of the whole metapopulation is likely.

Susan Harrison (1991) illustrated the classical description of a metapopulation and three related situations (**Figure 8.15**). In **core-satellite metapopulations**, sometimes called mainland-island metapopulations, persistence depends on the existence of one or more extinction-resistant populations, usually inhabiting large patches, which constantly supply colonists to small peripheral patches that often become extinct. The second alternative is **patchy populations**, in which dispersal between patches or populations is so high that colonists always "rescue" populations from extinction before it actually occurs. The system is then effectively a single extinction-resistant population. The last alternative is **nonequilibrium metapopulations**, in which local extinctions occur in the course of species' overall regional decline, with fragmentation of their habitat reducing their population density. The failure of populations to disperse effectively eliminates a true metapopulation scenario. Habitat destruction often creates nonequilibrium metapopulations.

One of the best-known studies of a core-satellite metapopulation in nature was that of the Bay checkerspot butterfly, *Euphydryas editha bayensis* (**Figure 8.16**), studied by Harrison and her colleagues (1988). It consisted, in 1987, of one population of 783 adult butterflies on a 2,000-ha habitat patch called Morgan Hill in Santa Clara County, California,

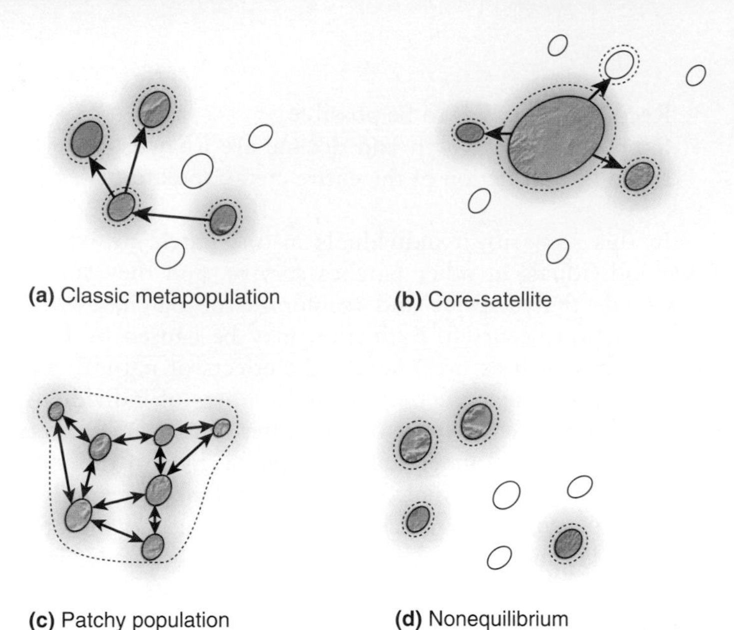


**Figure 8.15 Different kinds of metapopulations.** Closed circles represent habitat patches; filled = occupied; unfilled = vacant. Dashed lines indicate the boundaries of "populations." Arrows indicate migration (colonization). **(a)** Classic metapopulation. **(b)** Core-satellite metapopulation (common). **(c)** Patchy population. **(d)** Nonequilibrium metapopulation (differs from (a) in that there is no recolonization) often happens as part of a general regional decline. (Modified from Harrison, 1991.)

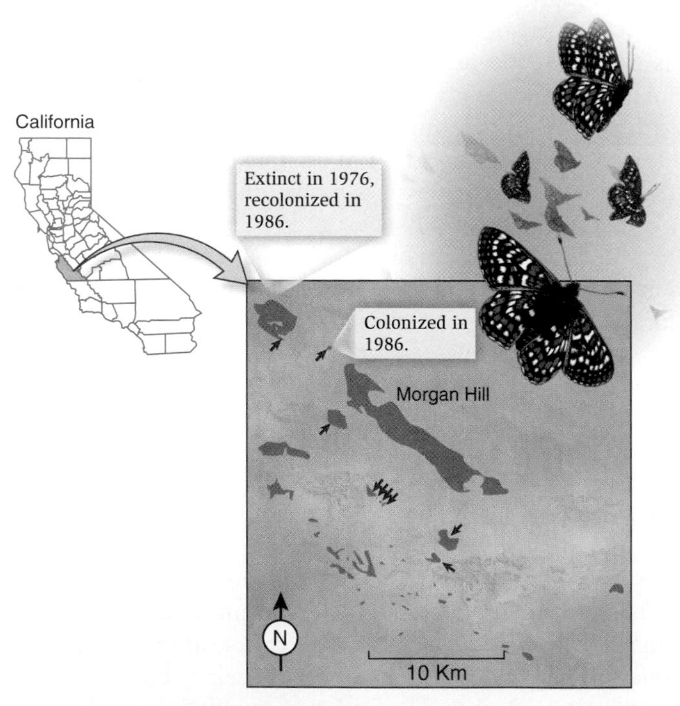


**Figure 8.16 Metapopulation of the Bay checkerspot butterfly, *Euphydryas editha*.** Occurrence of groups of individuals at Morgan Hill, near San Francisco, California. The green areas represent patches of the butterfly's serpentine grassland habitat. The 2,000-ha patch labeled Morgan Hill supported a population on the order of 783 adult butterflies in 1987 and acted as the source population. The nine smaller patches labeled with arrows supported populations on the order of 10–350 butterflies in that year. Many other small patches were found to be suitable but unoccupied. (Reproduced from Harrison, 1991.)



## Feature Investigation

### Joshua Tewksbury Showed How Connecting Habitat Patches via Corridors Facilitated Plant and Animal Movement

Joshua Tewksbury and his colleagues (2002) created eight 50-ha landscapes on the 1,240-km$^2$ Savannah River Site, a National Research Park in South Carolina (**Figure 8.C-i**). All landscapes comprised a matrix of mature, 40- to 50-year-old coniferous forest with a series of five clear-cuts that became vegetated by smaller herbaceous plants and grasses. These clear-cuts were the habitat focused on by the researchers. Movement rates of plants and animals were compared from a clear-cut square 1-ha patch, created at the center of all landscapes to four outlying patches, at least 150 m from the central patch. A 25-m-wide corridor connected the central patch to one of the outside patches, but all other patches were unconnected. Some of these patches had added 75-m$^2$ "wings" to compensate for the added area of the corridor on the connected patch, while others had the extra 75 m$^2$ added on to create a rectangular shape.

Butterflies were captured and marked in the central patch, and all peripheral patches were examined later for these marked butterflies. To measure pollen movement (fruits per flower) and fruit set (seeds per patch), researchers planted eight mature male deciduous holly trees, *Ilex verticillata,* in the central patch and three females in each peripheral patch. Other deciduous holly trees were not present in the area. Flowers and fruits were counted on each female tree. To examine seed movement, the researchers planted Yaupon holly, *Ilex vomitoria,* and wax myrtle, *Myrica cerifera,* in the central patches. All other *I. vomitoria* in the area, which occur in well-defined patches, were removed. Wax myrtle was too common to remove, so plants and their fruits in the central patch were dusted with a fluorescent powder. Bird perches were planted in peripheral patches to encourage seed-dispersing birds such as Eastern bluebirds (*Sialia sialis*), other thrushes, and warblers (*Dendroica* sp.) to perch. Bird fecal matter was captured in traps under the perches, where *Ilex* seeds or the presence of fluorescent powder could be detected.

Results showed conclusively that butterfly movement of two common species, the common buckeye, *Junonia coenia,* and the variegated fritillary, *Euptoieta claudia,* was increased in the corridor-connected peripheral patches over either of the other two types of peripheral patch, winged or rectangular (**Figure 8.C-ii,iii**). Similarly, pollen movement, fruit set, and seed movement (proportion of fruits with fluorescent dye) was increased by the presence of corridors (**Figure 8.C- iv,v,vi**). Tewksbury and coworkers had shown how corridors can increase plant and animal movement in a fragmented landscape.

**HYPOTHESIS** Connecting patches via corridors promotes the movement of organisms between patches.

**STARTING LOCATION** The Savannah River Site, South Carolina.

| | Conceptual level | Experimental level |
|---|---|---|
| 1 | Create both isolated and connected patches of habitat of equal area, where some patches are connected to central areas. | Create five clear cuts of equal area in a coniferous forest. One patch is central to four others, each of which is 150 m distant from it. One patch is connected to the central patch via a 25-m-wide corridor. Replicate design 8 times. (Figure 8.C-i) |
| 2 | Mark animals (butterflies) and plant pollen, seeds, and fruits in central patches. | After 1 year, herbaceous vegetation grew in the clear cuts. Two species of butterflies were caught in nets in the central patch, marked, and released. Pollen, fruit, and seed of common plants were identified and marked by various techniques including dusting with fluorescent powder. |
| 3 | Recapture animals and plant pollen, seed, and fruit in peripheral patches to study movement. | Butterflies were recaptured in peripheral patches using nets. Fruit and seed moved by birds was recovered using collecting baskets underneath plastic poles which acted as bird perches. |

4 **THE DATA** Results show movement of butterflies, plant pollen, fruit, and seeds between patches was facilitated by corridors. Figures 8.C-ii-vi

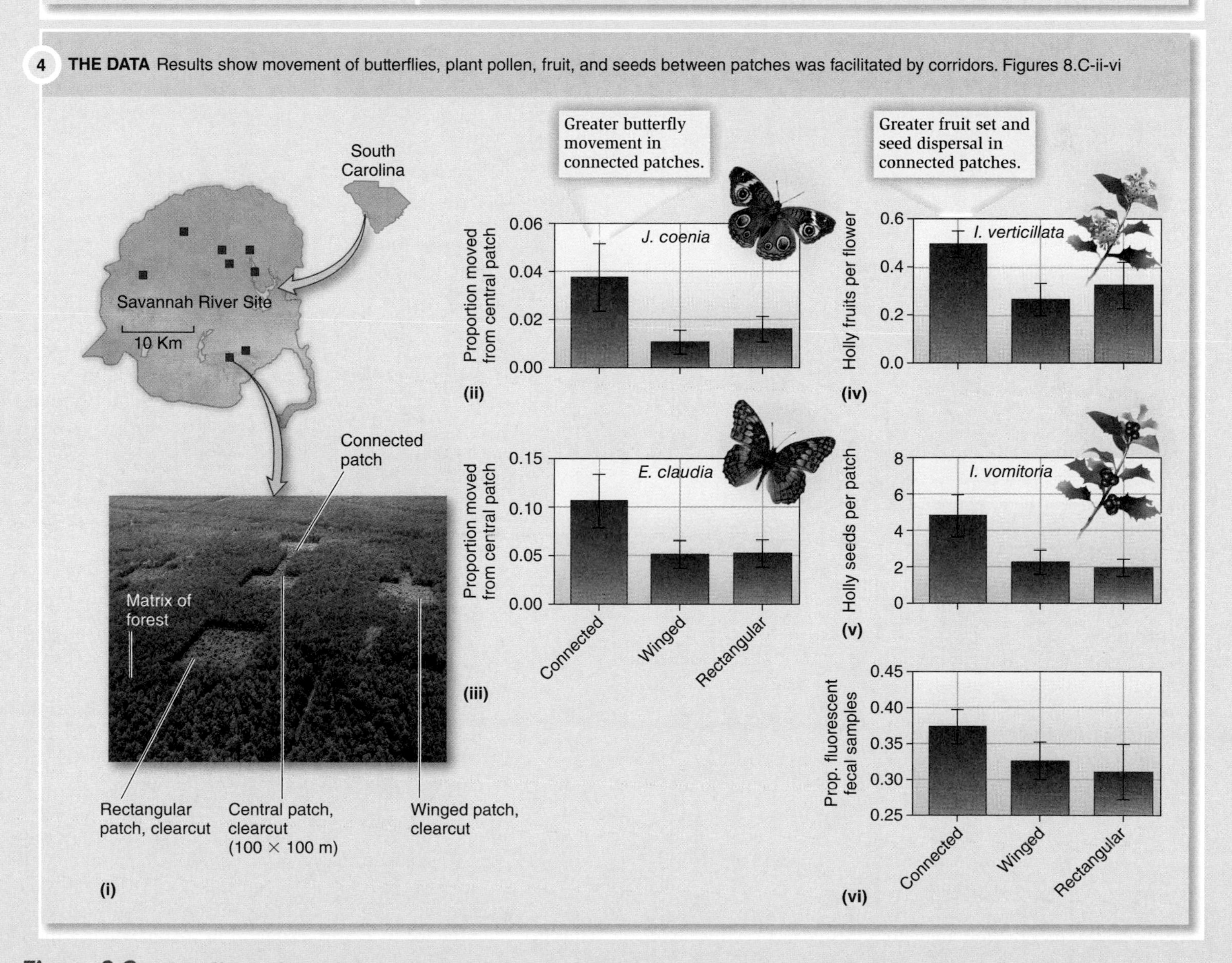


**Figure 8.C** The effects of corridors on plant and animal movement.

**Figure 8.17 The southern California spotted owl metapopulation.** Lines show possible dispersal routes. Numbers in parentheses are the estimated maximum numbers of owls that could be supported in each area. (After Lahaye et al., 1994.)

and nine populations of 10–350 adult butterflies on patches of 1–250 ha. Of 27 small patches of habitat in the region found to be suitable for populations to live, only those closest to the large patch were occupied. This pattern of occupancy could not be explained by differences in quality of the habitats. Instead, a distance effect appeared to indicate that the butterfly's capacity for dispersal was limited and that the large population acted as the dominant source of colonists to the small patches, supporting the core-satellite model.

Metapopulation scenarios have been invoked for many other systems, including moths on small islands off the Finnish coast, butterflies in alpine meadows of the Canadian Rockies, bush crickets on patches of habitat in Norway, marsh rabbits in the Florida Keys, pikas near Bodie, California, and spotted owls in southern California (**Figure 8.17**). In most cases the data show that core-satellite or nonequilibrium metapopulations were the most common scenario. Frequent extinctions of populations occurred as habitat loss, often caused by development, reduced the size of many patches.

As we noted earlier, the rate of movement of organisms affects patch occupancy rates. High occupancy rates reduce the rate of extinction, and many organisms have evolved strategies to maximize dispersal. In many plant species, seeds are wind-blown or encased in fruits that are dispersed by animals (look ahead to Figure 12.4). Most insect herbivores have to relocate to new host plants quickly, so the insects are winged to help them disperse over long distances. Many sessile marine organisms, such as algae, limpets, and barnacles, have larval stages that disperse widely. Even spiders and caterpillars can disperse for many kilometers when they spin long silken threads that catch the wind and the insects balloon away.

Ecologists have proposed various dispersal classifications. Prominent among these are passive and active dispersal. Passive dispersal is used by many species of plants. Their seeds fall close to the parent, while a few may be blown some distance away. Seed density usually declines quickly as distance from the parent plant increases. Passive dispersal may be facilitated by additional agents. For example, seeds that are eaten by animals often possess thick, hard seed coats. When defecated they are provided with a source of fertilizer. Some seeds also possess a mild laxative to speed up their passage through the animal gut and prevent digestion. A small percentage of such seeds may be transported for a long distance, especially by birds. However, in all cases, whether the dispersing seeds end up in a suitable location for germination is somewhat a matter of chance.

In active dispersal, organisms control their settlement. Most plants have passive dispersal, but many animals make decisions about where to settle. Aphids and other dispersing insects can decide whether a host plant is suitable for colonization; if they decide it is not, they may disperse further. Likewise, dispersing freshwater organisms that drift downstream can stop or keep drifting if the habitat is not suitable.

### Check Your Understanding

**8.5** Habitat corridors connecting patches of fragmented habitat may greatly facilitate the establishment of a metapopulation. However, continuous unfragmented habitat is still a better option. Why might this be so?

## SUMMARY

Population ecology studies how populations grow and what promotes and limits growth.

**8.1 Various Techniques Are Used to Quantify Population Size and Density**

- Ecologists measure population density, the numbers of organisms in a given unit area, in many ways, including quadrats, traps, and the mark-capture technique (Figures 8.1–8.5).
- Larger animals tend to have lower population densities (Figure 8.6).

**8.2 Patterns of Spacing May Be Clumped, Uniform, or Random**

- Individuals within populations show different patterns of dispersion, including clumped (the most common), uniform, and random (Figure 8.7).

**8.3 Fragmented Habitats Affect Spatial Dispersion**

- Populations may be fragmented because of many different types of disturbance, including deforestation and the construction of roads (Figure 8.8).
- Range contraction of many species has resulted from deforestation and other development (Figures 8.A, 8.B).
- Some species are “area sensitive,” that is, they do not use potentially suitable areas of habitat unless these areas are much bigger than the species’ territory (Figure 8.9).

**8.4 Landscape Ecology Concerns the Spatial Arrangement of Habitats and Organisms**

- Landscape ecology is a subdiscipline of ecology that examines the spatial arrangement of elements in communities and ecosystems and the effects on populations of organisms (Figures 8.10–8.14, Figure 8.C).

**8.5 Metapopulations Are Separate Populations That Mutually Affect One Another via Dispersal**

- Metapopulations are a series of small, separate groups of individuals that mutually affect one another (Figures 8.15–8.17).

## TEST YOURSELF

1. The number of organisms in a given unit area is termed population:
   a. Dispersion
   b. Dispersal
   c. Density
   d. Number
   e. Growth
2. Berlese funnels are normally used to sample organisms from:
   a. Seawater
   b. Soil
   c. Trees
   d. The air
   e. Fresh water
3. A student decides to conduct a mark-recapture experiment to estimate the population size of mosquito fish in a small pond near her home. In the first catch, she marked 45 individuals. Two weeks later she captured 62 individuals, of which 8 were marked. What is the estimated size of the population based on these data?
   a. 134
   b. 348
   c. 558
   d. 1,016
   e. 22,320
4. Plant competition from shading and root interactions usually results in which dispersion pattern?
   a. Clumped
   b. Uniform
   c. Random
   d. Fragmented
   e. Indirect dispersion

5. The most common habitat of a landscape is called the:
   a. Fragment
   b. Matrix
   c. Perforation
   d. Corridor
   e. Metapopulation
6. Mist nets are often used to sample:
   a. Butterflies
   b. Birds
   c. Bats
   d. a and b
   e. b and c
7. Species that wander along the ground are often counted using:
   a. Quadrats
   b. Pitfall traps
   c. Suction samples
   d. Sweep nets
   e. Funnels
8. Competition between nesting birds on a rock face would tend to promote which dispersion pattern?
   a. Clumped
   b. Uniform
   c. Random
   d. Metapopulation
   e. Fragmented
9. A metapopulation where species are in general decline in abundance in most patches is known as which type of metapopulation?
   a. Core-satellite
   b. Mainland-island
   c. Patchy population
   d. Nonequilibrium metapopulation
   e. Regional extinction
10. Small animals are usually more common than large animals.
    a. True
    b. False

## CONCEPTUAL QUESTIONS

1. Describe and list the assumptions of the mark-recapture technique.
2. Describe the three types of dispersion.
3. Is habitat destruction promoting the formation of metapopulations? Explain.

**Connect Ecology** helps you stay a step ahead in your studies with animations and videos that bring concepts to life and practice tests to assess your understanding of key ecological concepts. Your instructor may also recommend the interactive ebook.

Visit **www.mhhe.com/stilingecology** to learn more.

Dall mountain sheep, *Ovis dalli*, Denali National Park, Alaska.

CHAPTER 9

# Life Tables and Demography

## Outline and Concepts



The Dall mountain sheep, *Ovis dalli,* lives in mountainous regions, including the Arctic and sub-Arctic regions of Alaska. In the late 1930s the U.S. National Park Service was bombarded with public concerns that wolves were responsible for a sharp decline in the population of Dall mountain sheep in Denali National Park (then Mt. McKinley National Park). Shooting the wolves was advocated as a way of increasing the number of sheep. Because meaningful data on sheep mortality were nonexistent, the Park Service enlisted biologist Adolph Murie to collect relevant information. In addition to spending many hours observing interactions between wolves and sheep, Murie also collected thousands of sheep skulls and determined the sheep's age at death by counting annual growth rings on the horns. Murie's study was one of the first attempts to systematically collect data on all life history stages of an organism to determine when most individuals died and what may have killed them.

This chapter follows Murie's lead. For a variety of organisms, we examine how long individuals survive in a population and at what age they die. Such information is typically summarized in tables but can also be presented graphically. Later, we gather information about the reproductive rates of individuals of various ages in the population. When we know how long females survive and the reproductive rates of females of different ages, we can make predictions about how the population will grow. In the following discussions we assume that there is no immigration into or emigration out of the population and that any changes in population size are a result of births and deaths.

## 9.1 Age Distributions, Life Tables, and Survivorship Curves Summarize Survival Patterns

Population sizes are not always constant, and ecologists are often interested in how populations increase, and decrease, over time. One way to determine the survival and mortality of individuals in a population is to examine a cohort of individuals from birth to death. A **cohort** is a group of same-aged young that grow and survive at similar rates. A population is often made up of organisms from a variety of different cohorts, each of different ages, so we say the population consists of different age classes. An **age class** consists of individuals of a particular age, for example, three-year-olds. For most animals and plants, monitoring a cohort involves marking a group of individuals in a population as soon as they are born or germinate and following their fate through their lifetime. Researchers use this information to construct a life table. A **life table** provides data on the number of individuals alive in different age classes and the age-specific survival or mortality rates in these age classes. We can also present information from life tables graphically in the form of survivorship curves. A **survivorship curve** is a graphical representation of the numbers of individuals alive in a population at various ages.

Two types of life tables exist, a cohort life table and a static life table. The **cohort life table** follows a cohort of individuals from birth to death just as we have described. Cohort life tables can be used to estimate the age-specific probabilities of survival. A **static life table** accomplishes the same goal, but instead of following a cohort of individuals from life to death, data are gathered on the age structure of a given population at one point in time. For some long-lived organisms such as tortoises, elephants, or trees, following an entire cohort from birth to death is impractical, so a snapshot approach is used. This is the approach that Adolf Murie adopted in his studies of the Dall mountain sheep.

### 9.1.1 Age distributions reflect survival and mortality patterns

Numbers of individuals in different age classes can be calculated for any time period, but they often represent 1 year. Males are not often included in these calculations, because they are typically not the limiting factor in population growth. For example, even if there were only a few males in the population, they could probably fertilize all the females. However, if there were only a few females, then very few young would be born and population growth would be severely slowed. We expect that a population increasing in size should have a large number of young, because individuals are reproducing at a high rate. On the other hand, a decreasing population should have few young because of limited reproduction. An imbalance in age classes can have a profound influence on a population's future. For example, in an overexploited fish population, the bigger, older reproductive age classes are often removed and population growth is severely curtailed. Other populations may experience removal of younger age classes. Where populations of white-tailed deer are high, they overgraze the vegetation and eat many young trees, leaving only older trees, whose foliage is too high up for them to reach (**Figure 9.1**). This can have disastrous effects on the future population of trees, for while the forest might consist of healthy mature trees, when these die, there will be no replacements. Removal of deer predators such as panthers and wolves often allows deer numbers to skyrocket and survivorship of young trees in forests to plummet.

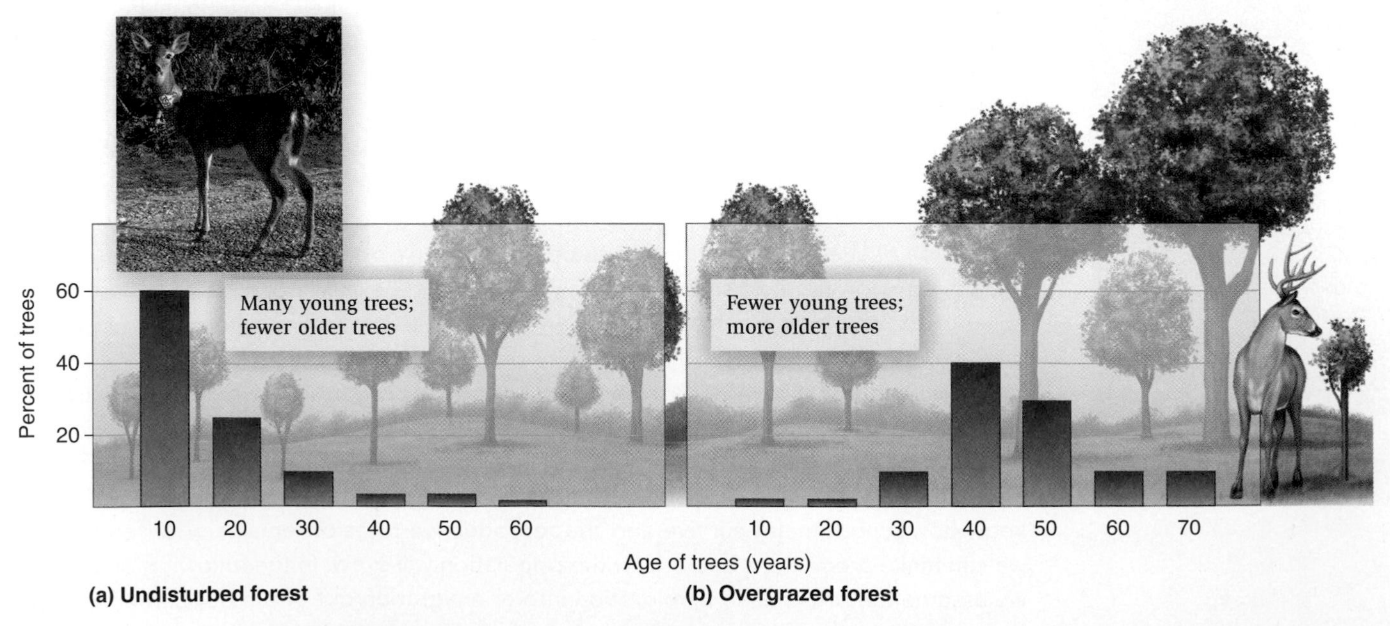


**Figure 9.1 Theoretical age distribution of two forest tree populations.** **(a)** Age distribution of an undisturbed forest with numerous young trees, many of which die as the trees age and compete with one another for resources, leaving relatively few big, older trees. **(b)** Age distribution of a forest where overgrazing has reduced the abundance of young trees, leaving mainly trees in the older age classes.

## 9.1.2 Static life tables provide a snapshot of a population's age structure from a sample at a given time

Let's examine a static life table for the North American beaver, *Castor canadensis*. Prized for their pelts, by the mid-19th century these animals had been hunted and trapped to near extinction. Beavers began to be protected by laws in the 20th century, and populations recovered in many areas, often growing to what some considered to be nuisance status. In Newfoundland, Canada, legislation supported trapping as a management technique, but was the beaver population actually increasing? From 1964 to 1971, trappers provided mandibles from which jaws were extracted for age classification, and government scientists could use this data to examine population growth. If many jaws were from, say, 1-year-old beavers, then such animals were probably common in the population. If the number of jaws from 2-year-old beavers was low, then we know there was high mortality for the 1-year-old age class. From the mandible data, researchers constructed a life table (**Table 9.1**). The number of individuals alive at the start of the time period, in this case a year, is referred to as $n_x$, where $n$ is the number and $x$ refers to the beginning of a particular age class. By subtracting the value of $n_x$ in any given year from the number alive at the start of the previous year, we can calculate the number dying in a given age class or year, $d_x$. Thus $d_x = n_x - n_{x+1}$. For example, in Table 9.1, 273 beavers were alive at the start of their sixth year ($n_5$) and only 205 were alive at the start of the seventh year ($n_6$); thus, 68 died during the sixth year: $d_6 = n_5 - n_6$, or $d_6 = 273 - 205 = 68$.

A little later on we'll return to the beaver story to see how the population fared. As mentioned at the beginning of the chapter, Adolf Murie collected detailed data on the survivorship of the Dall mountain sheep. In 1947 Edward Deevey put Murie's data in the form of a life table that listed each age class and the number of skulls in it (**Table 9.2**). Murie had collected 608 skulls, but Deevey expressed the data per 1,000 individuals to allow for comparison with other life tables.

**Table 9.1** Life table for the beaver, *Castor canadensis*, in Newfoundland, Canada.

| Age class, $x$ | Number alive at start of year, $n_x$ | Number dying during year, $d_x$ |
|---|---|---|
| 0–1 | 3,695 | 1,995 |
| 1–2 | 1,700 | 684 |
| 2–3 | 1,016 | 359 |
| 3–4 | 657 | 286 |
| 4–5 | 371 | 98 |
| 5–6 | 273 | 68 |
| 6–7 | 205 | 40 |
| 7–8 | 165 | 38 |
| 8–9 | 127 | 14 |
| 9–10 | 113 | 26 |
| 10–11 | 87 | 37 |
| 11–12 | 50 | 4 |
| 12–13 | 46 | 17 |
| 13–14 | 29 | 7 |
| 14+ | 22 | 22 |

From data in Payne (1984).

**Table 9.2** Static life table for the Dall mountain sheep, *Ovis dalli*, based on the known age at death of 608 sheep dying before 1937.*

| Age class | Number alive | Number dying | Proportion surviving | Mortality rate | Average no. alive in age class | Total years lived | Life Expectancy |
|---|---|---|---|---|---|---|---|
| $x$ | $n_x$ | $d_x = (n_x - n_{x+1})$ | $l_x = (n_x/n_0)$ | $q_x = (d_x/n_x)$ | $L_x = (n_x + n_{x+1})/2$ | $T_x = L_x$ | $e_x = (T_x/n_x)$ |
| 0–1 | 1000 | 199 | 1.000 | .199 | 900.5 | 7058 | 7.1 |
| 1–2 | 801 | 12 | 0.801 | .015 | 795 | 6158 | 7.7 |
| 2–3 | 789 | 13 | 0.789 | .016 | 782.5 | 5363 | 6.8 |
| 3–4 | 776 | 12 | 0.776 | .015 | 770 | 4581 | 5.9 |
| 4–5 | 764 | 30 | 0.764 | .039 | 749 | 3811 | 5.0 |
| 5–6 | 734 | 46 | 0.734 | .063 | 711 | 3062 | 4.2 |
| 6–7 | 688 | 48 | 0.688 | .070 | 664 | 2351 | 3.4 |
| 7–8 | 640 | 69 | 0.640 | .108 | 605.5 | 1687 | 2.6 |
| 8–9 | 571 | 132 | 0.571 | .231 | 505 | 1081.5 | 1.9 |
| 9–10 | 439 | 187 | 0.439 | .426 | 345.5 | 576.5 | 1.3 |
| 10–11 | 252 | 156 | 0.252 | .619 | 174 | 231 | 0.9 |
| 11–12 | 96 | 90 | 0.096 | .937 | 51 | 57 | 0.6 |
| 12–13 | 6 | 3 | 0.006 | .500 | 4.5 | 6 | 1.0 |
| 13–14 | 3 | 3 | 0.003 | 1.00 | 1.5 | 1.5 | 0.5 |

*Data are expressed per 1,000 individuals.

Deevey calculated several other valuable statistics from Murie's data. First, he was interested in survivorship. Second, he wanted to know the mortality rate within each age class. Third, he calculated the life expectancy of sheep at various ages.

Survivorship, or $l_x$, is calculated by dividing the number of individuals in an age class by the number in the original cohort; that is,

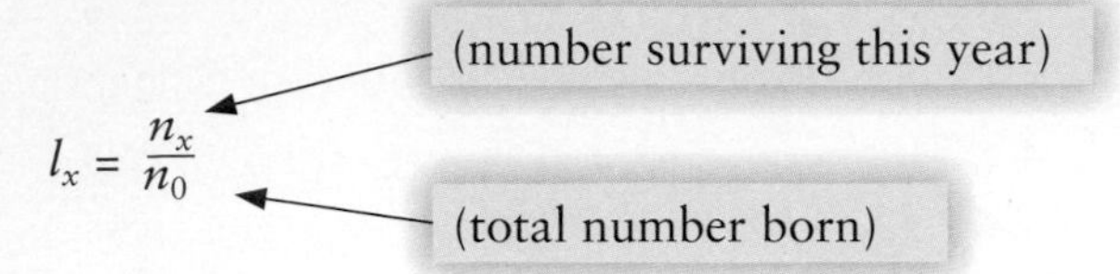


Survivorship to the sixth age class (5–6 years) is therefore 734/1,000, or 0.734. This value tells us that over three-quarters of those individuals born will survive to at least 5 years old.

The mortality rate within each age class is estimated by the variable $q_x$, which is calculated as

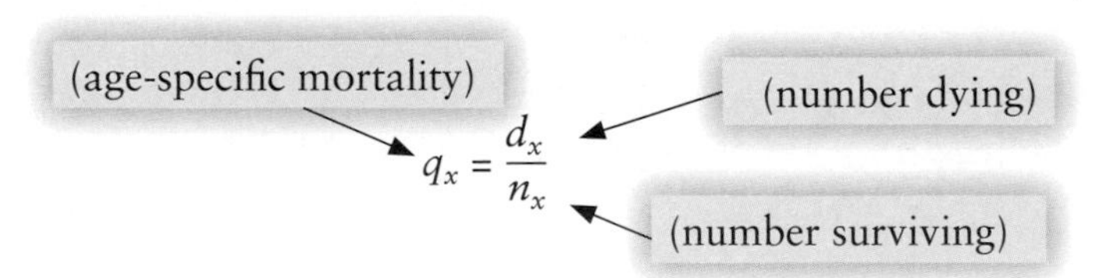


This equation expresses the number of deaths in an age class ($d_x$) as a proportion of the number of individuals that were alive at the beginning of that age class ($n_x$). It gives a valuable measure of proportional mortality within an age class. For example, in the sixth age class (5–6 years), there were 46 deaths out of 734 individuals, so $q_x = 46/734$, or 0.063. This mortality rate allows us to determine whether mortality increases with age, decreases with age, or is independent of age. In the Dall mountain sheep, $q_x$ is high in the first age class, meaning a lot of young sheep die. Then $q_x$ drops dramatically and remains low until about age 8, when it increases quickly. Murie suggested that the high mortality of both young and old sheep may have been due to wolves. He thought straying youngsters were easily picked off and older animals became arthritic and were easily chased down. Such predation would not be expected to dramatically reduce the sheep population, because it was not affecting reproductive females. The Park Service eventually ended a limited wolf control program that had been in effect since 1929. However, Graham Caughley (1966) compared the mortality rates of Dall mountain sheep with domestic sheep in New Zealand, where wolf predation is absent (**Figure 9.2**). It is interesting that the mortality rates between these two species are very similar in all ages except in lambs. This suggests that Murie was correct about wolf predation impacting young sheep, but it also suggests that if old Dall sheep were not preyed on by wolves, they would have died in some other way. The data also show that mortality rates for female Dall mountain sheep are greater than for males.

Another parameter that Deevey estimated was life expectancy, the average number of additional age classes an individual can expect to live at each age. This calculation requires two intermediate steps. First, the average number of individuals alive in each age class, $L_x$, must be calculated. Individuals may die at the beginning of an age class or at the end. Because $n_x$ is the average number of individuals alive at the beginning of each age class, and $n_{x+1}$ is the number alive at the beginning of the next age class, the number alive during an age class can be estimated by the relationship:

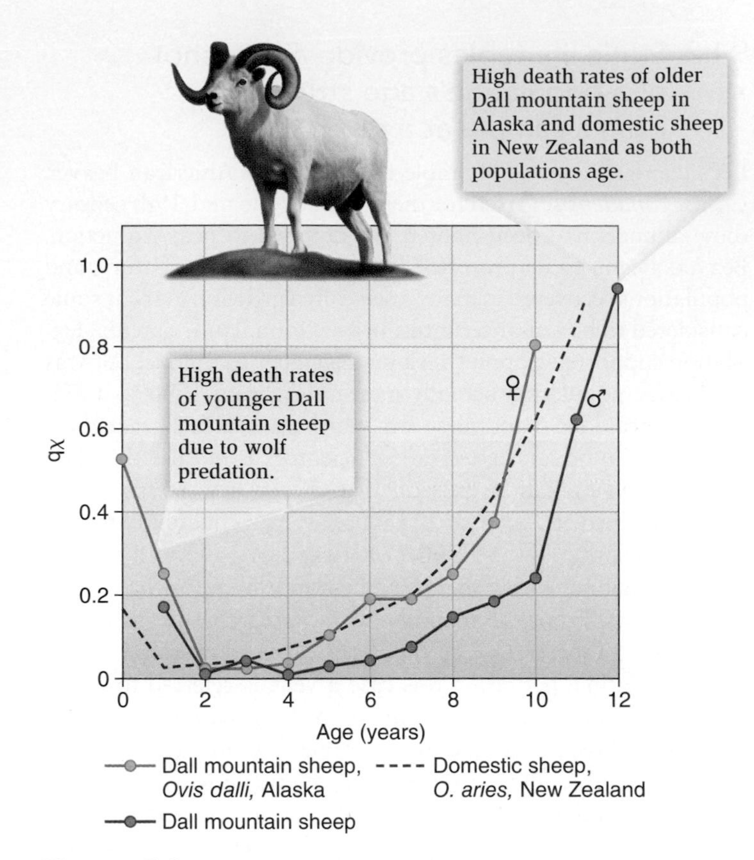


**Figure 9.2 Comparison of age-specific mortality, $q_x$, between Dall mountain sheep, *Ovis dalli*, in Alaska and domestic sheep, *Ovis aries*, in New Zealand.** Note that the data for males and females are provided for Dall mountain sheep for all years except those in their first year. The similarities between the curves later in life suggest wolf predation does not greatly affect older Dall sheep. (After Caughley, 1966.)

$$L_x = \frac{n_x + n_{x+1}}{2}$$

A numerical example, from the sixth and seventh rows of the data is

$$L_5 = \frac{n_5 + n_6}{2} = \frac{734 + 688}{2} = 711$$

On average, at any point in time, one could expect to find 711 individuals alive in the sixth age class. Of course, they could all die right at the beginning of the age class, or right at the end, but chances are good that about 711 would be alive at the midpoint in time of the age class. The second step in determining life expectancy is to calculate a quantity called total years lived, $T_x$. This is a purely intermediate step and, unlike the other columns in a life table, is without real biological meaning. $T_x$ is calculated by summing the values of $L_x$ in age class $x$ and all subsequent (older) age classes. Thus,

$$T_x = \sum_x^{\infty} L_x$$

The value of $T_x$ for the five-year-old sheep (sixth age class) is therefore:

$$T_6 = L_6 + L_7 + L_8 + L_9 + L_{10} + L_{11} + L_{12} + L_{13}$$
$$= 711 + 664 + 605.5 + 505 + 345.5 + 174 + 51 + 4.5 + 1.5 = 3{,}062$$

The sheep's life expectancy, can now be calculated as follows:

$$e_x = \frac{T_x}{n_x}$$

The value of $e_x$ indicates the average number of additional age classes an individual can expect to live at each age, so

$$e_5 = \frac{3{,}062}{734} = 4.2$$

Therefore, 5-year-old sheep can be expected to live, on average, for another 4.2 years.

Taken as a whole, these calculations allow important features of a population to be quantified: What is the life expectancy of individuals in this population? How does survivorship change with age? When in an individual's life is the mortality rate highest? The life expectancy of sheep actually increases in the second age class, indicating that newborn sheep struggle to survive.

Despite the value of static life tables, there are some assumptions that limit their accuracy. Paramount among these is the assumption that equal numbers of offspring are born each year. For example, if the rate of mortality of 2-year-old sheep were identical to the rate of 4-year-old sheep but there were more 2-year-old sheep born because of favorable climate in that particular year, then more skulls of 2-year-old sheep would be found later on and a higher rate of mortality of 2-year-olds would be assumed. There is often no independent method for estimating the birth rates of each age class. In addition, there are methodological difficulties in constructing static life tables. It is often difficult to accurately assess the age structure of population. In some cases, growth rings on trees, on fish scales, or on horns of ungulates make it possible to determine the age of an animal. Annual growth rings also occur in the genital plates of sea urchins and the shells of some mollusks, but this is relatively unusual. Perhaps for these reasons, the cohort life table is often reported.

## 9.1.3 Cohort life tables follow an entire cohort of individuals from birth to death

In cases where cohort life tables are constructed, population censuses must be conducted frequently but only for a limited time, usually less than a year for insects or annual plants. In these cases, the age classes may be weeks or months, not years. For other organisms, like many birds or small mammals, an annual or biannual census for up to 10 years will suffice (see **Feature Investigation**). A cohort life table for the American robin, *Turdus migratorius,* based on banded nestlings that were subsequently recensused, is shown in **Table 9.3.**

**Table 9.3 Life table for the American robin, *Turdus migratorius*.**

| Age (years) | $n_x$ | $d_x$ | $l_x$ | $q_x$ |
|---|---|---|---|---|
| 0–1 | 568 | 286 | 1.00 | 0.505 |
| 1–2 | 282 | 152 | 0.497 | 0.539 |
| 2–3 | 130 | 74 | 0.229 | 0.569 |
| 3–4 | 56 | 36 | 0.099 | 0.642 |
| 4–5 | 20 | 15 | 0.035 | 0.750 |
| 5–6 | 5 | 2 | 0.008 | 0.400 |
| 6–7 | 3 | 3 | 0.005 | 1.000 |

From data in Farner (1945). Based on returns of 568 birds banded as nestlings.

## 9.1.4 Survivorship curves present survival data graphically

Organismal survival, from both static or cohort life tables, can be quickly compared by representing the data graphically, creating what is known as a survivorship curve. For example, using the data from the beaver life table, we can plot the numbers of surviving individuals against age (**Figure 9.3**). The value of $n_x$, the number of individuals, is typically expressed on a logarithmic scale. Ecologists use a logarithmic scale to examine rates of change with time, not change in absolute

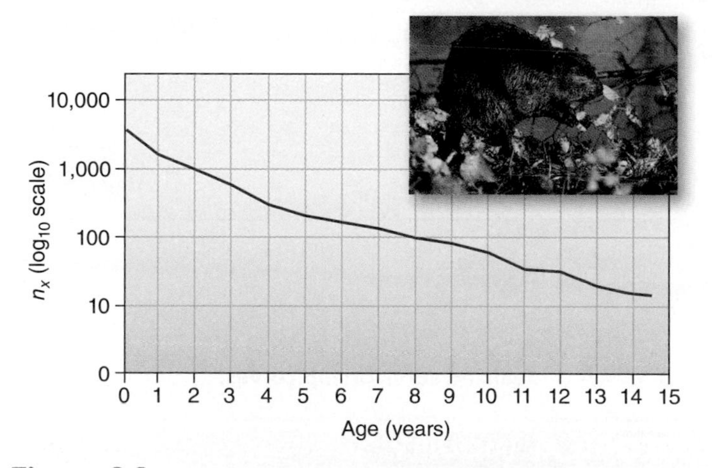


**Figure 9.3 Survivorship curve for the North American beaver.** The survivorship curve is generated by plotting the number of surviving beavers, $n_x$, from any given cohort of young, usually measured on a log scale, against age.

numbers. Also, the use of logs makes it easier to examine a wide range of population sizes. For example, if we start with 1,000 individuals and 500 are lost in year 1, the log of the decrease is

$$\log_{10}1{,}000 - \log_{10}500 = 3.0 - 2.7 = 0.3 \text{ per year.}$$

This does not mean a rate of loss of 0.3 beavers per year; rather, it is the difference between the starting and ending population sizes, where both are expressed on a log scale. If we start with 100 individuals and 50 are lost, the log of the decrease is similarly

$$\log_{10}100 - \log_{10}50 = 2.0 - 1.7 = 0.3 \text{ per year.}$$

In both cases the rates of change, 0.3 per year, are identical even though the absolute numbers are different. Plotting the $n_x$ data on a log scale ensures that regardless of the size of the starting population, the rate of change of one survivorship curve can more easily be compared to that of another species. On a log scale, the survivorship curve for the beaver is almost a diagonal line, meaning that it follows a fairly uniform rate of death over its life span.

Survivorship curves generally fall into one of three patterns (**Figure 9.4**). In a type I curve, the rate of loss for juveniles is relatively low, and most individuals are lost later in life, as they become older and more prone to sickness and predators. Organisms that exhibit type I survivorship have relatively few offspring but invest much time and resources in raising their young. Many large mammals, including humans, exhibit type I curves. Survivorship for the Dall mountain sheep in Denali National Park

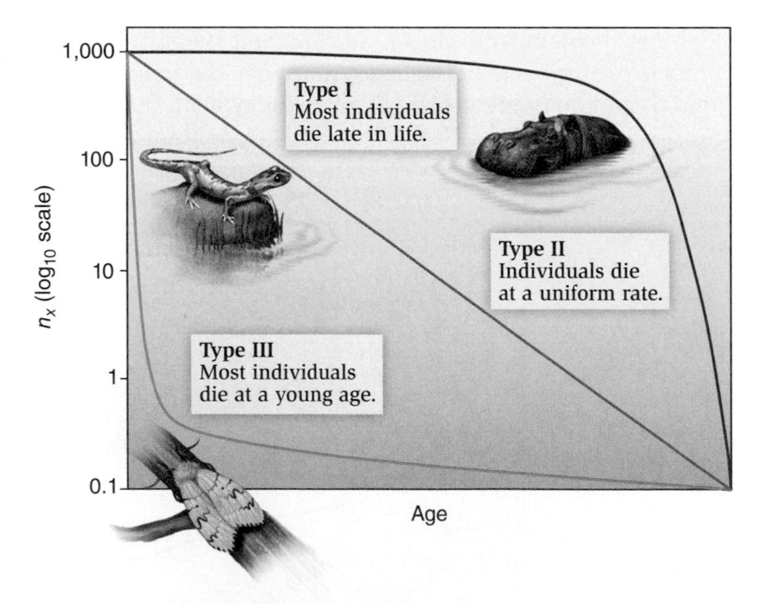


**Figure 9.4** **Idealized survivorship curves.**

**ECOLOGICAL INQUIRY**

Which type of survivorship curves would be typical of a butterfly, a turtle, and a human, respectively?

## Feature Investigation

### Frederick Barkalow Jr. and Colleagues Constructed a Cohort Life Table for the Eastern Gray Squirrel

The Eastern gray squirrel, *Sciurus carolinensis,* is one of the most important game animals in North America. Wildlife biologists were interested to evaluate the effects of artificial nest boxes on populations. Their hope was that nest boxes would increase squirrel survival and population densities. As a first step in understanding squirrel survival they constructed a cohort life table for an area without nest boxes. In this case the researchers based their conclusions on the survival of 493 male squirrels and 530 females marked at between 1 and 7 weeks of age. The survival of every marked squirrel was followed for all years after being marked. In this case, however, the marking occurred not all in 1 year, but was spread out over 8 years, between 1956 and 1963. The team did not have the resources to band all the squirrels at one time. The data from all years were compiled into a composite life table. Barkalow and colleagues marked a total of 1,023 squirrels, but, as Deevey had done with the Dall mountain sheep, they converted their data to survival of a cohort of 1,000 squirrels for ease of comparison with other organisms (**Figure 9.A**). Life expectancy of newborn squirrels was short, about 1 year, but improved to over 2 years once squirrels had survived their first year. This is because the mortality for baby squirrels is concentrated early in their first year and a different formula was used to calculate average number alive in year 1, which is 538.9 (lower than 623.6, calculated using the standard method—that is, [1,000 + 247.3]/2 = 623.6). Although the data are not shown here, nest boxes did increase squirrel survival by reducing the mortality associated with predation and bad weather that was experienced in natural cavities for both young and adult squirrels.

**HYPOTHESIS** Nest boxes impact the survivorship of American squirrels.

**STARTING LOCATION** North Carolina, U.S.A.

| | Conceptual level | Experimental level |
|---|---|---|
| 1 | Mark and recapture squirrels at yearly intervals. | Establish nest boxes and mark and sex nestlings at yearly intervals. Nestlings were toe-clipped and given ear tags when recaptured. |
| 2 | Record recaptures of tagged squirrel every year. Animals failing to be recaptured are counted as dead. | Table (i) |
| 3 | Establish a composite life table for the squirrel. This table uses data from all years to create one survivorship table. The known number of survivors for year 1 is the summed number of survivors for year 1 following each tagging event. For any year this is the diagonal sum of all the columns. For example, the survivors for year 1 is 8 for 1957, 60 for 1958, 61 for 1959, 58 for 1960, 19 for 1961, 4 for 1962, 18 for 1963, and 25 for 1964, giving a total of 253. | Table (ii) |
| 4 | Calculate mortality rates and life expectancy. | Table (iii) |

**5 THE DATA**

(i)

| Year Marked | Nestlings Marked | Recaptures 1957 | 1958 | 1959 | 1960 | 1961 | 1962 | 1963 | 1964 |
|---|---|---|---|---|---|---|---|---|---|
| 1956 | 40 | 8 | 4 | 3 | 2 | 0 | 0 | 0 | 0 |
| 1957 | 138 | | 60 | 30 | 28 | 13 | 9 | 4 | 3 |
| 1958 | 229 | | | 61 | 26 | 12 | 10 | 7 | 3 |
| 1959 | 193 | | | | 58 | 26 | 19 | 12 | 9 |
| 1960 | 162 | | | | | 19 | 13 | 8 | 6 |
| 1961 | 99 | | | | | | 4 | 1 | 1 |
| 1962 | 82 | | | | | | | 18 | 6 |
| 1963 | 80 | | | | | | | | 25 |

(ii)

| Age, $x$ | Total Known Alive | Maximum Available for Recapture | Known Alive Per 1,000 Available |
|---|---|---|---|
| 0–1 | 1,023 | 1,023 | 1,000 |
| 1–2 | 253 | 1,023 | 247.3 |
| 2–3 | 106 | 943 | 112.4 |
| 3–4 | 71 | 861 | 82.5 |
| 4–5 | 43 | 762 | 56.4 |
| 5–6 | 25 | 600 | 41.7 |
| 6–7 | 7 | 407 | 17.2 |
| 7–8 | 3 | 178 | 16.9 |

(iii)

| Age, $x$ | $n_x$ | $d_x$ | $q_x$ | $L_x$ | $T_x$ | $e_x$ |
|---|---|---|---|---|---|---|
| 0–1 | 1,000.0 | 752.7 | 0.753 | 538.9 | 989.6 | 0.99 |
| 1–2 | 247.3 | 134.9 | 0.545 | 179.9 | 450.7 | 1.82 |
| 2–3 | 112.4 | 29.9 | 0.266 | 97.4 | 270.8 | 2.41 |
| 3–4 | 82.5 | 26.1 | 0.316 | 69.5 | 173.4 | 2.10 |
| 4–5 | 56.4 | 14.7 | 0.261 | 49.0 | 103.9 | 1.84 |
| 5–6 | 41.7 | 24.5 | 0.588 | 29.4 | 54.9 | 1.32 |
| 6–7 | 17.2 | 0.3 | 0.017 | 17.1 | 25.5 | 1.48 |
| 7–8 | 16.9 | 16.9 | 1.000 | 8.4 | 8.4 | 0.50 |

**Figure 9.A** **Life table for the American Squirrel, *Sciurus carolinensis*, in North Carolina.**
(After Barkalow et al., 1970.)

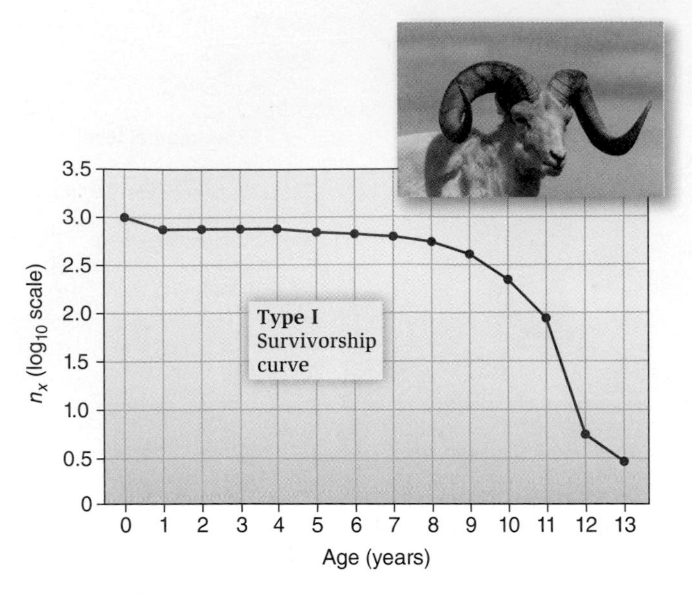


**Figure 9.5** Survivorship curve for the Dall mountain sheep, *Ovis dalli.*

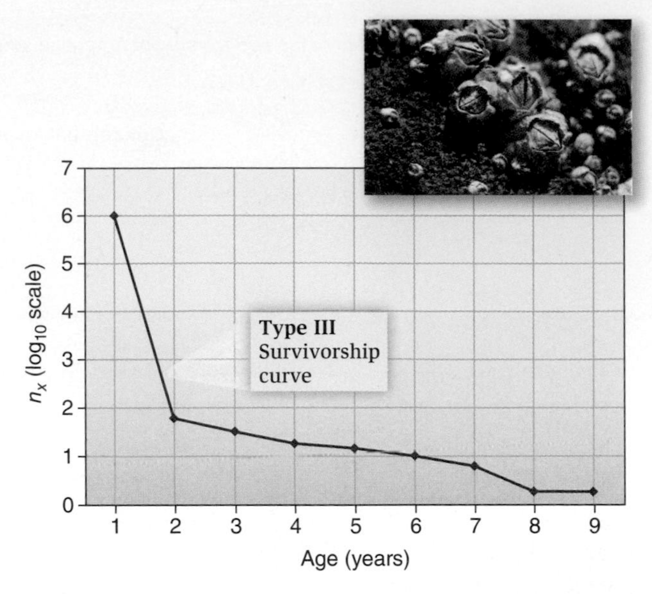


**Figure 9.6** Survivorship curve for the barnacle, *Balanus glandula.*

best fits a type I curve (**Figure 9.5**). There is a slight initial decline in survivorship as young lambs are lost, and then the survivorship curve flattens out, indicating that the sheep survive well through about age 7 or 8. Then the sheep decline rapidly in number as they age. These data underlined what Murie had previously observed, which was that death rates are greatest for the youngest and oldest members of the population.

At the other end of the scale is a type III curve, in which the rate of loss for juveniles is relatively high, and the survivorship curve flattens out for those organisms that have avoided early death. Many fish and marine invertebrates fit this pattern. Most of the juveniles die or are eaten, but a few reach a favorable habitat and thrive. For example, many juvenile barnacles are lost at sea, but once they find a suitable rock face on which to attach themselves, barnacles grow and survive very well. A survivorship curve for the barnacle *Balanus glandula* of Washington State shows huge mortality in the first year of life (**Figure 9.6**). Many insects and plants also fit the type III survivorship curve because they lay many eggs or release hundreds of seeds, respectively, and most of these die. Type II curves represent a middle ground, with fairly uniform death rates over time. Species with type II survivorship curves include many birds, small mammals, reptiles, and some annual plants. The beaver population we talked about earlier most closely resembles this survivorship curve. Keep in mind, however, that these are generalized curves and that few populations fit them exactly. Furthermore, humans can impact the survival of various species that they harvest, depending on what age classes are impacted (see **Global Insight**).

What's the difference in accuracy between a static and a cohort life table? It is very difficult to get data for both types of life table for most populations to compare the two techniques, but for humans it is possible because we can use birth and death records for people of all ages censused at a certain time and we can follow a cohort of individuals born at this same time. For example, let's examine the survivorship of some of the British peerage born in 1800 in England, for which there is good data, and follow them for their entire lives to obtain a cohort survivorship curve. We can also take a cross section of a hypothetical English population at 1812 to get a static survivorship curve (**Figure 9.7**). The 1812 static curve actually relates to people born before 1812, while the cohort table gives us information on death rates after 1800. In an 1812 static survivorship curve, 72-year-old people would have been born in 1740 and fewer would have survived to 1812, compared to 70-year-old people born in 1812 and still living in 1882, because medicine and diet had improved. Thus, static curves ignore environmental variation, because their predictions of how many 70-year-olds will be alive in 1882 is based on data collected from 70-years-olds in 1812, and it ignores medical and dietary improvements.

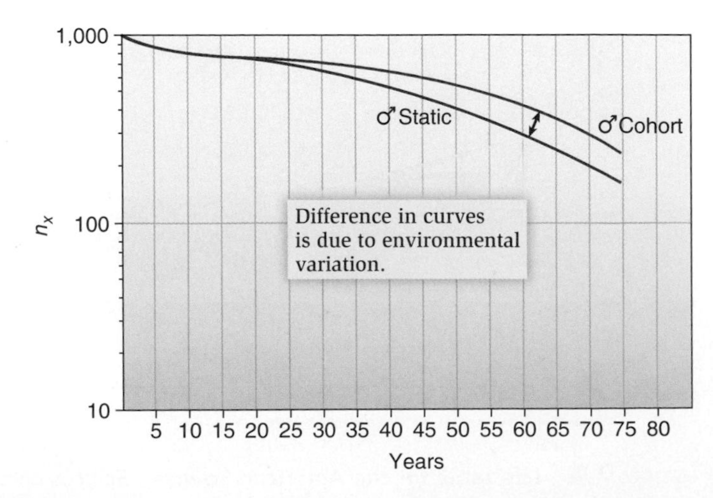


**Figure 9.7** Comparison of a cohort survivorship curve for humans born in 1800 with a static survivorship curve for 1812. (After Hutchinson, 1978.)

## Global Insight

## Hunting, Overcollecting, and Grazing Can Greatly Affect Survivorship Curves

Hunting can change the appearance of an animal's survivorship curve, just as grazing pressure from herbivores can change survivorship curves for plants. The removal of trophy males (deer with large antlers or antelopes with large horns) can impact mortality rates, as can the removal of large fish, especially because many large fish are females. In southern Arizona, grazing by cattle reduces the life expectancy of the grasses hairy grama, *Bouteloua hirsuta,* and common wolfstail, *Lycurus phleoides* (**Figure 9.B**). However, grazing appears to increase the life expectancy of cottontop, *Trichachne californica.* Presumably the preferred and competitively dominant grasses are eaten first, leaving the less preferred species to thrive in the absence of competition.

The effect of humans can be seen in an examination of survivorship curves of species considered to be pests in some urban areas. **Figure 9.C** shows survivorship curves for populations of foxes, *Vulpes vulpes,* in two cities in England: London in the southeast and the smaller city of Bristol in the southwest. Fox control is practiced in London via trapping, gassing, and digging out litters of cubs, but not in Bristol. Mortality of cubs is higher in London than in Bristol, as it is for mortality in most other age classes. As a result, the fox population was 20% lower in an area of London that was the same size as Bristol.

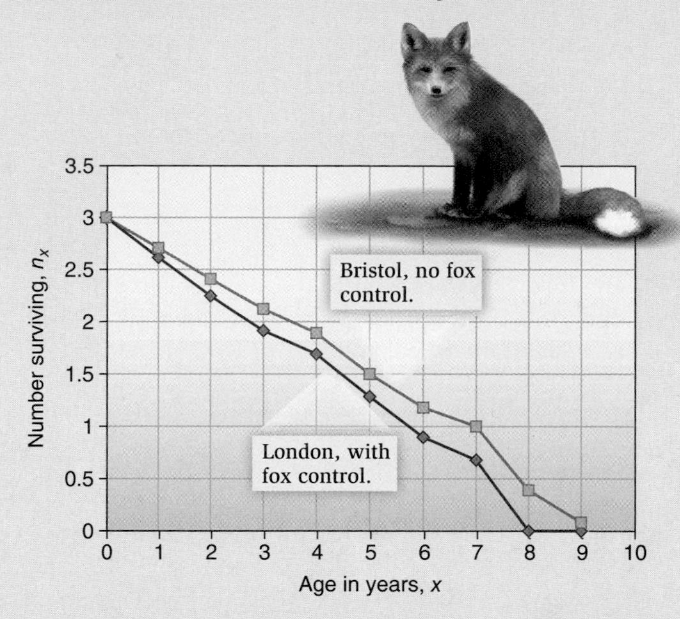


**Figure 9.C Survivorship curves for foxes, *Vulpes vulpes,* in London, where control is practiced, and Bristol, where no control occurs.** (From data in Harris and Smith, 1987.)

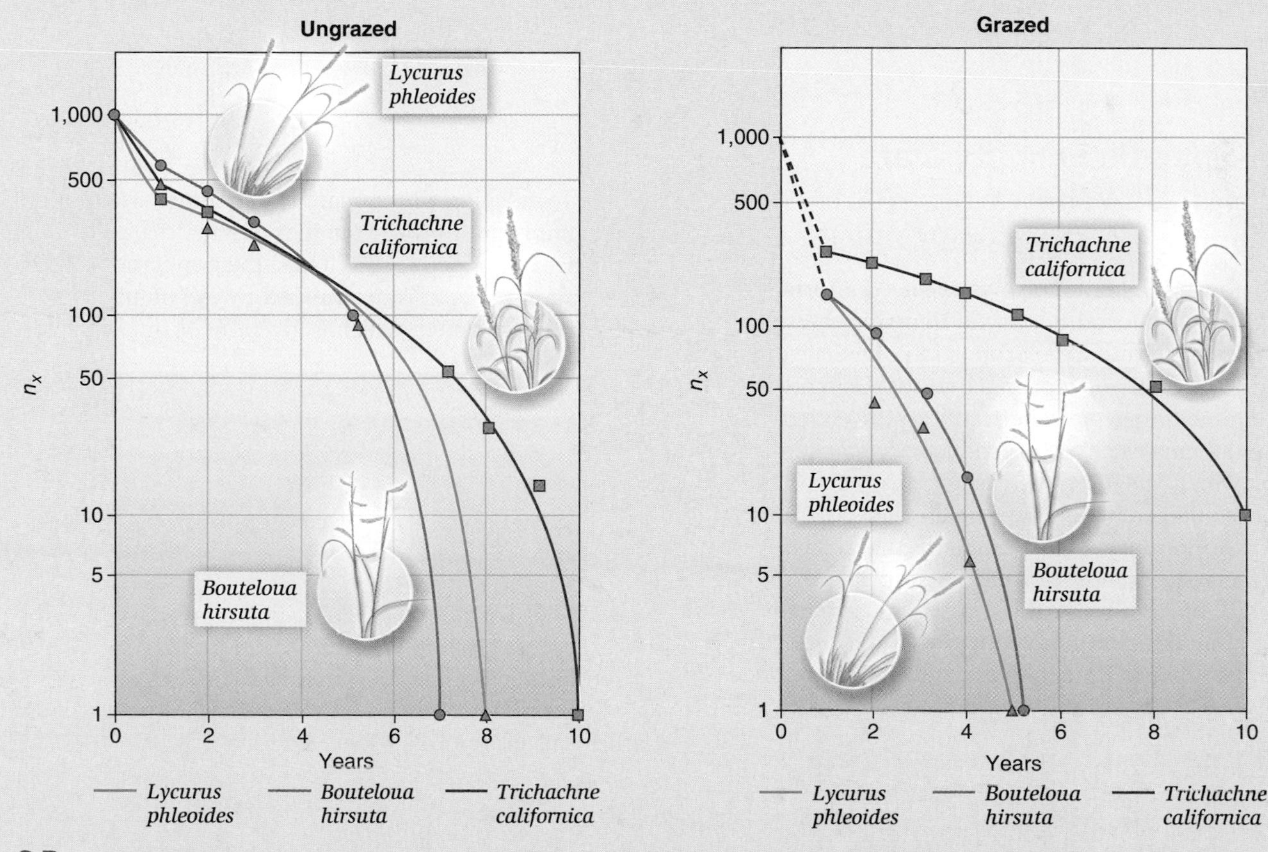


**Figure 9.B Survivorship curves for grazed and ungrazed grasses in Arizona.** (After Sarukhán and Harper, 1973.)

In the natural world, environmental variation includes years of good and bad climate or food supply. This means that in reality, the cohort life table is always more accurate than the static life table. Unfortunately, we do not always have the time and resources to follow populations of organisms from birth to death, particularly if they are long-lived. Static life tables are therefore a quicker and cheaper method to get a rough idea of how the individuals of a population are distributed among age classes.

### Check Your Understanding

**9.1** The following data refer to the survivorship of red deer on a Scottish island. Complete the table by calculating the $q_x$ column.

| Age (years) $x$ | Proportion of original cohort surviving to the beginning of age-class $l_x$ | Mortality rate $q_x$ |
|---|---|---|
| 1 | 1.000 | |
| 2 | 1.000 | |
| 3 | 0.939 | |
| 4 | 0.754 | |
| 5 | 0.505 | |
| 6 | 0.305 | |
| 7 | 0.186 | |
| 8 | 0.132 | |
| 9 | 0.025 | |

## 9.2 Age-Specific Fertility Data Can Tell Us When to Expect Population Growth to Occur

To calculate how a population grows, we need information on birth rates as well as mortality and survivorship rates. For any given age, we can determine the proportion of female offspring that are born to females of reproductive age. Using these data we can determine an **age-specific fertility** rate, called $m_x$. For example, if 100 females at age $x$ produce 75 female offspring, $m_x = 0.75$, which means that, on average, each female of age $x$ produces 0.75 female offspring in that year. With this additional information, we can calculate the growth rate of a population. First, we use the survivorship data to find the proportion of females alive at the start of any given age class. Recall that the survivorship rate, termed $l_x$, equals $n_x/n_0$, where $n_0$ is the number alive at time 0, the start of the study, and $n_x$ is the number alive at the beginning of age class $x$. Let's return to the beaver life table (**Table 9.4**). The proportion of the original beaver population still alive at the start of the sixth age class, $l_5$, equals $n_5/n_0 = 273/3{,}695$, or 0.074. This means that 7.4% of the original beaver population survived to age 5. Next we multiply the data in the two columns, $l_x$ and $m_x$, for each row, to give us a column $l_xm_x$, an average number of offspring per female. This column represents the contribution of each age class to the overall population growth rate. An examination of the beaver age-specific fertility rates illustrates a couple of general points. First, for this beaver population in particular, and for many organisms in general, there are no babies born to young females. Next, as females mature sexually, age-specific fertility goes up and remains fairly high until later in life, when females reach postreproductive age.

The number of offspring born to females of any given age class depends on two things: the number of females in that age class and their age-specific fertility rate. Thus, although fertility of young beavers is very low, there are so many females in the age class that $l_xm_x$ for 1-year-olds is quite high. Age-specific fertility for older beavers is much higher, but the relatively few females in these age classes cause $l_xm_x$ to be low. Maximum values of $l_xm_x$ occur for females of an intermediate age, 3–4 years old in the case of beaver. The overall growth rate per generation is the number of offspring born to all females of all ages, where a generation is defined as the mean period between birth of females and birth of their offspring. Thus, to calculate the generational growth rate, we sum all the values of $l_xm_x$, that is, $\Sigma l_xm_x$, where the symbol $\Sigma$ means "sum of." This summed value, $R_0$, is called the **net reproductive rate**, the average number of female offspring produced by all the females in a population over the course of a generation, where a generation constitutes the reproductive life of a female.

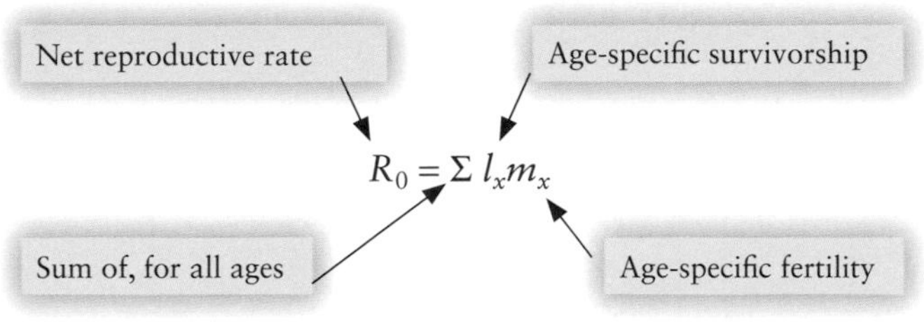


To calculate an estimate of the future size of a population, we simply multiply the number of females in the population by the net reproductive rate. Thus, the population size in the next generation, $N_{t+1}$, is determined by the number in the population now, at time $t$, which is given by $N_t$, multiplied by $R_0$.

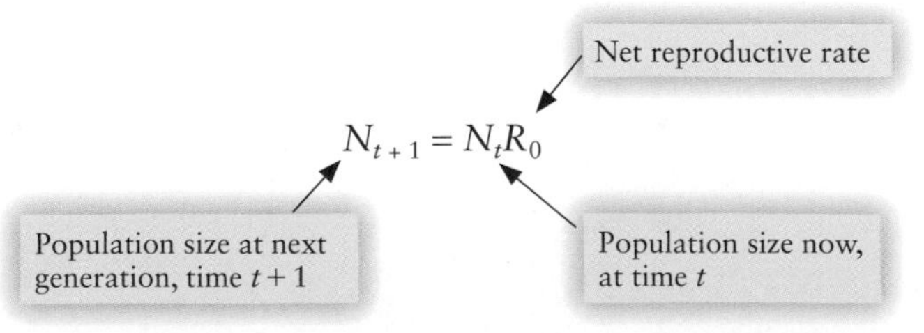


Let's consider a theoretical example in which the number of beavers alive now, $N_t$, is 1,000 and $R_0 = 1.1$. This means the beaver population is reproducing at a rate that is 10% greater than simply replacing itself. The size of the population next generation, $N_{t+1}$, is given by

$$N_{t+1} = N_tR_0$$
$$N_{t+1} = 1{,}000 \times 1.1$$
$$= 1{,}100$$

**Table 9.4** **Life table and age-specific fertility rates for the beaver, *Castor canadensis,* in Newfoundland, Canada.** This information allows us to calculate the net reproductive rate per generation, $R_0$.

| Age class, $x$ | Number alive at start of year, $n_x$ | Number dying during year, $d_x$ | Proportion alive at start of age interval, $l_x$ | Age-specific fertility, $m_x$ | Average number of offspring per age class $l_x m_x$ | $x\, l_x m_x$ |
|---|---|---|---|---|---|---|
| 0–1 | 3,695 | 1,995 | 1.000 | 0.000 | 0 | 0 |
| 1–2 | 1,700 | 684 | 0.460 | 0.315 | 0.145 | 0.145 |
| 2–3 | 1,016 | 359 | 0.275 | 0.400 | 0.110 | 0.220 |
| 3–4 | 657 | 286 | 0.178 | 0.895 | 0.159 | 0.477 |
| 4–5 | 371 | 98 | 0.100 | 1.244 | 0.124 | 0.496 |
| 5–6 | 273 | 68 | 0.074 | 1.440 | 0.107 | 0.535 |
| 6–7 | 205 | 40 | 0.055 | 1.282 | 0.071 | 0.426 |
| 7–8 | 165 | 38 | 0.045 | 1.280 | 0.058 | 0.406 |
| 8–9 | 127 | 14 | 0.034 | 1.387 | 0.047 | 0.376 |
| 9–10 | 113 | 26 | 0.031 | 1.080 | 0.033 | 0.297 |
| 10–11 | 87 | 37 | 0.024 | 1.800 | 0.043 | 0.430 |
| 11–12 | 50 | 4 | 0.014 | 1.080 | 0.015 | 0.165 |
| 12–13 | 46 | 17 | 0.012 | 1.440 | 0.017 | 0.204 |
| 13–14 | 29 | 7 | 0.007 | 0.720 | 0.005 | 0.065 |
| 14+* | 22 | 22 | 0.006 | 0.720 | 0.009 | 0.126 |
| | | | | Net reproductive rate, $\Sigma l_x m_x = 0.938$ | | **4.368** |

*Last row gives a summary of data for all beavers ages 14 and older.
From data in Payne (1984).

Thus, the number of beavers in the next generation is 1,100 and the population will have grown larger. How long is a beaver generation? Population ecologists generally refer to a generation time, $T$, as the average time for a female organism to grow from a fertilized egg and produce more fertilized eggs. We can calculate $T$ from our life table by multiplying the $l_x m_x$ column by $x$, the age in years, summing the column and dividing by $R_0$. Here we assume that all the babies born to mothers over their lifetimes were born to them at some age, $T$, instead. We are thus calculating a mean age of reproduction.

$$\text{Thus, } T = \frac{\Sigma\, x l_x m_x}{R_0}$$

$$\text{For our beaver data, } T = \frac{4.368}{0.938} = 4.657$$

The average generation time for beaver in Newfoundland is about 4.6 years. Generally, generation time increases as organismal size increases (**Figure 9.8**).

In estimating population growth, much depends on the value of $R_0$. If $R_0 > 1$, then the population will grow. If $R_0 < 1$, the population is in decline. If $R_0 = 1$, then the population size stays the same and we say it is at **equilibrium.** In the case of the beavers, Table 9.4 reveals that $R_0 = 0.938$, which is less than 1, and therefore the population is in decline. This is valuable information, because it tells us that at that time, the beaver population in Newfoundland needed more protection,

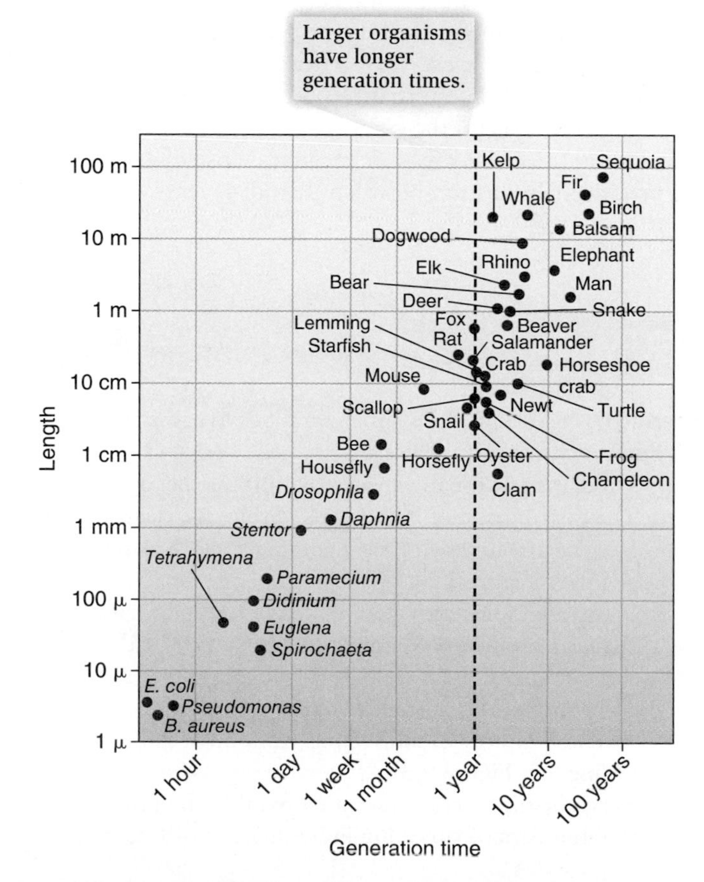

**Figure 9.8** **The relationship of length and generation time for a variety of organisms.** (After Bonner, 1965.)

**Table 9.5** A life table and age-specific fertility for the barnacle, *Balanus glandula*.

| Age (years) $x$ | Number of survivors $n_x$ | $l_x$ | $m_x$ | $l_x m_x$ | $x l_x m_x$ |
|---|---|---|---|---|---|
| 0 | 1,000,000 | 1.000 | 0 | 0 | |
| 1 | 62 | 0.0000620 | 4,600 | 0.285 | 0.285 |
| 2 | 34 | 0.0000340 | 8,700 | 0.296 | 0.592 |
| 3 | 20 | 0.0000200 | 11,600 | 0.232 | 0.696 |
| 4 | 15.5 | 0.0000155 | 12,700 | 0.197 | 0.788 |
| 5 | 11 | 0.0000110 | 12,700 | 0.140 | 0.700 |
| 6 | 6.5 | 0.0000065 | 12,700 | 0.082 | 0.492 |
| 7 | 2 | 0.0000020 | 12,700 | 0.025 | 0.175 |
| 8 | 2 | 0.0000020 | 12,700 | 0.025 | 0.200 |

From data in Connell (1970).

perhaps in the form of limits on trapping and hunting, in order to attain a population level at equilibrium.

In his study of the barnacles of Washington State, Joe Connell (1970) also provided age-specific fertility data (**Table 9.5**). Connell was interested in determining which factors were most important in influencing population sizes of barnacles, and to do this he gathered life table and fertility data. Despite huge mortality of young stages, barnacles 1 year old or older were very fertile, and Connell estimated each female could lay up to 12,700 eggs. With such high levels of egg production, the population had a net reproductive rate, $R_0$, of 1.282 and was increasing. Most of the offspring are produced by adults between 1 and 4 years old because survival of older age classes is poor. Generation time, $T$, was about 3.1 years. Predation by several species of snails of the genus *Thais* proved to be the most influential factor on population size, though the data are not shown here.

### Check Your Understanding

**9.2** The following is a fertility table for the wildebeest, *Connochaetes taurinus*, in Africa. Calculate $R_0$ and $T$, generation time.

| $x$ | $l_x$ | $m_x$ |
|---|---|---|
| 0 | 1.00 | 0 |
| 1 | 0.80 | 0 |
| 2 | 0.70 | 0.11 |
| 3 | 0.63 | 0.46 |
| 4 | 0.57 | 0.46 |
| 5 | 0.41 | 0.46 |
| 6 | 0.31 | 0.46 |
| 7 | 0.23 | 0.46 |
| 8 | 0.17 | 0.46 |
| 9 | 0.14 | 0.46 |
| 10 | 0.14 | 0.46 |
| 11 | 0.12 | 0.46 |
| 12 | 0.09 | 0.46 |
| 13 | 0.06 | 0.46 |
| 14 | 0.05 | 0.46 |
| 15 | 0.04 | 0.46 |
| 16 | 0.02 | 0.46 |

## SUMMARY

**9.1 Age Distributions, Life Tables, and Survivorship Curves Summarize Survival Patterns**

- Populations of many species exhibit distinct age classes. The distribution of individuals within age classes may later be affected by other phenomena, such as natural enemies (Figure 9.1).
- Static life tables provide a snapshot of a population's age structure and survivorship and mortality of individuals in different age classes (Tables 9.1, 9.2, Figure 9.2).
- Cohort life tables provide similar information but follow an entire cohort of individuals from birth to death (Table 9.3, Figure 9.3).
- Survivorship curves illustrate life tables by plotting the numbers of surviving individuals at different ages (Figure 9.A).
- Survivorship curves generally fall into one of three types: type III, with high juvenile mortality, type II with constant mortality throughout life, or type I, with high juvenile survival (Figures 9.4–9.6).
- Hunting and overgrazing can cause large changes in survivorship curves (Figures 9.B, 9.C).
- Survivorship curves generated from static life tables may be easier to construct, but they ignore environmental variation and may be slightly less accurate than survivorship curves from cohort life tables (Figure 9.7).

**9.2 Age-Specific Fertility Data Can Tell Us When to Expect Population Growth to Occur**

- Age-specific fertility and survivorship data help determine the overall growth rate per generation called the net reproductive rate ($R_0$) (Tables 9.4, 9.5).
- Generally, generation time increases as organismal size increases (Figure 9.8).

## TEST YOURSELF

1. Survivorship of which of the following organisms is most likely to be analyzed with a static life table?
   a. Butterfly
   b. Annual plant
   c. Elephant
   d. American robin
   e. Barnacle

2. Complete the following hypothetical life table for a bird and calculate the net reproductive rate, $R_0$:

| Age Class, $x$ | # Alive, $n_x$ | # Dying, $d_x$ | Age-Specific Survivorship, $l_x$ | Age-Specific Fertility, $m_x$ | $l_x m_x$ |
|---|---|---|---|---|---|
| 0–1 | 1,000 | 300 | | 0 | |
| 1–2 | 700 | | 0.7 | 0.5 | |
| 2–3 | | 200 | | 1.0 | |
| 3–4 | 300 | 150 | 0.3 | 1.0 | |
| 4–5 | | 150 | | 1.0 | |

   a. 0
   b. 0.3
   c. 0.35
   d. 1.3
   e. 1.5

3. _____ survivorship curves are usually associated with organisms that have high mortality rates in the early stages of life.
   a. Type I
   b. Type II
   c. Type III
   d. Types I and II
   e. Types II and III

4. If the net reproductive rate ($R_0$) is equal to 0.5, what assumptions can we make about the population?
   a. This population is essentially not changing in numbers
   b. This population is in decline
   c. This population is growing
   d. This population is in equilibrium
   e. None of the above

5. In a population of beavers, if $n_2 = 500$, $n_3 = 300$, and $n_4 = 200$, what is $d_3$?
   a. 200
   b. 300
   c. 2.5
   d. 0.66
   e. 100

6. If $n_0 = 100$ and $d_0 = 20$, what is $q_0$?
   a. 0.2
   b. 0.8
   c. 5.0
   d. 800
   e. 2

7. If $n_2 = 800$ and $d_2 = 200$, what is $n_1$?
   a. 0.25
   b. 4.00
   c. 600
   d. 1,000
   e. 160,000

8. If $n_x = 600$ and $n_{x+1} = 400$, what is $L_x$?
   a. 0.67
   b. 1.5
   c. 200
   d. 500
   e. 1,000

9. If $N_t = 100$ and $R_0 = 0.5$, what is the value of $N_{t+1}$?
   a. 50
   b. 500
   c. 200
   d. 0.005
   e. 100

10. Type I survivorship curves include:
    a. Weedy plants
    b. Marine invertebrates
    c. Large mammals
    d. Insects
    e. Fish

## CONCEPTUAL QUESTIONS

1. What are the main differences between static and cohort life tables, and which is more accurate?

2. Describe the differences between type I, type II, and type III survivorship curves and give examples.

## DATA ANALYSIS

1. The age structure for a gray squirrel population in North Carolina is given below. (a) Calculate $l_x$, the proportion alive at the start of each age interval. (b) If $m_0 = 0.05$, $m_1 = 1.28$, and $m_x$ for every other age group is 2.28, calculate $R_0$. Is the squirrel population increasing or decreasing?

| Age in Years | $n_x$ |
|---|---|
| 0 | 1000 |
| 1 | 253 |
| 2 | 116 |
| 3 | 89 |
| 4 | 58 |
| 5 | 39 |
| 6 | 25 |
| 7 | 22 |

2. Black rhinoceros, *Diceros bicornis,* skulls were collected and aged, based on mandible size from Tsavo National Park in Kenya, and a life table was constructed. The number of deaths in each age class has been expressed per 1,000 individuals born. Calculate $n_x$, $L_x$, and life expectancy, $e_x$. Comment on possible errors that may underestimate mortality of young rhinos. Hint: Think back to Chapter 7 and what can happen to young skulls.

| Age, $x$ | $d_x$ | Age, $x$ | $d_x$ |
|---|---|---|---|
| 0 | 160 | 20 | 12 |
| 1 | 141 | 21 | 11 |
| 2 | 93 | 22 | 11 |
| 3 | 68 | 23 | 10 |
| 4 | 50 | 24 | 10 |
| 5 | 33 | 25 | 9 |
| 6 | 31 | 26 | 7 |
| 7 | 31 | 27 | 6 |
| 8 | 31 | 28 | 6 |
| 9 | 25 | 29 | 6 |
| 10 | 25 | 30 | 6 |
| 11 | 26 | 31 | 6 |
| 12 | 25 | 32 | 5 |
| 13 | 25 | 33 | 4 |
| 14 | 25 | 34 | 4 |
| 15 | 25 | 35 | 2 |
| 16 | 21 | 36 | 2 |
| 17 | 17 | 37 | 1 |
| 18 | 16 | 38 | 0 |
| 19 | 14 | | |

3. The survivorship curves given below detail the survivorship of female (a) African and (b) Asian elephants in zoos and in the wild. Data are provided for zoo captive-born, zoo wild-born and moved to zoos, and wild-born with natural mortality. Based on what you have learned in this chapter and the previous chapters, discuss what is happening.

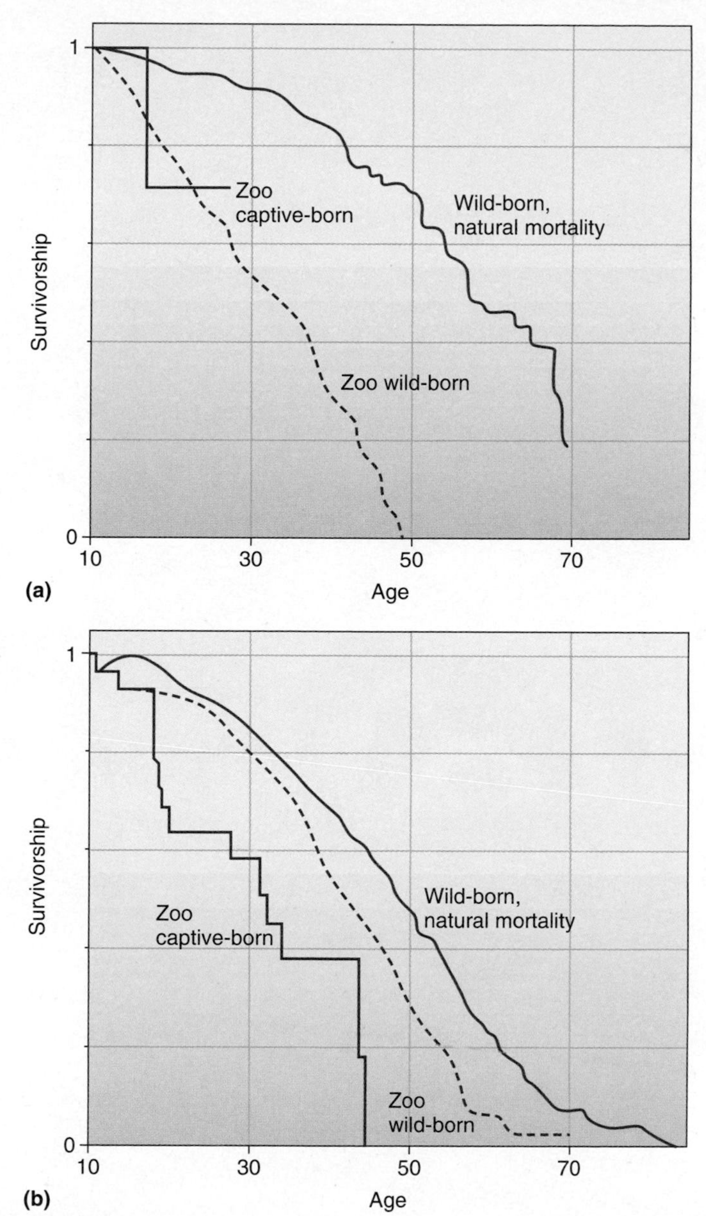

Black-footed ferrets were reintroduced into Shirley Basin, Wyoming, in the early 1990s. Population ecologists are currently monitoring their population growth.

CHAPTER 10

# Population Growth

## Outline and Concepts



The last known population of black-footed ferrets, *Mustela nigripes,* was discovered in 1981 near Meeteetse, Wyoming. In many other areas, prairie dogs, the primary prey of these animals, had been exterminated by cattle ranchers who mistakenly think that cattle break their legs by stumbling in prairie dog holes and worry that prairie dogs reduce forage quality for livestock. A dearth of prairie dogs has led to the demise of the ferrets. Shortly after 1981 all but 18 of the 100 known ferrets in Meeteetse died of distemper. The remainder were captured in 1988, inoculated against distemper, and bred in captivity. Seven females were used as genetic founders for the captive population, with the intent of reestablishment in the wild later on. Populations have now been established in Arizona, Colorado, Montana, South Dakota, Utah, Wyoming, and Chihuahua, Mexico.

In Wyoming, an area called Shirley Basin was one of those targeted for reintroductions of captive-born animals. During 1991–1994, Shirley Basin received 228 ferrets, but diseases again triggered a decline in the population size. By 1997 only 5 ferrets were found. Extinction seemed imminent. Monitoring efforts, which might disturb the animals, decreased. Surprisingly, 52 animals were found by 2003, and 223 were present by 2006. How did the population increase this fast?

In this chapter we will examine how variation in the net reproductive rate, $R_0$, can affect population growth rates. Other population growth models can also provide us with valuable insights into how populations grow over shorter time periods. The most simple of these assumes that populations grow if, for any time interval, the number of births is greater than the number of deaths. We will examine three models of population growth. The first two, geometric growth and exponential growth, assume resources are not limiting, and both models are similar in that they result in prodigious growth. The third, and perhaps more biologically realistic model, logistic growth, assumes resources are limiting, and it results in eventual stable population sizes. Later in the chapter we consider how other factors, such as natural enemies, might limit population growth and we discuss how different life history strategies influence population growth. Finally, we examine the special case of human population growth.

## 10.1 Unlimited Population Growth Leads to J-Shaped Population Growth Curves

Population growth often proceeds in a pattern similar to how savings accrue in an interest-bearing bank account. The greater the interest rate, the faster your savings grow. Similarly, the greater the population growth rate, the faster a population grows. We can imagine an interest that paid annually or a population that breeds once a year. In these cases we say the population grows geometrically. Other banks pay interest quarterly or monthly, and compound growth causes your savings to accrue quicker. In the same way, some species reproduce almost continuously and generations overlap. All ages of individuals, from babies to juveniles and reproductive adults are present at the same time. In these cases we say the population grows exponentially. Geometric growth and exponential growth are described by slightly different mathematical models, but both produce J-shaped population growth curves.

### 10.1.1 Geometric growth describes population growth for periodic breeders

Once we know a population's net reproductive rate, $R_0$, we can begin to predict how the population will grow. Let's consider the simplest type of growth, that of an annually breeding species with no limits to growth at the present, such as a butterfly that breeds once a year and has a life span of 1 year. We determine growth by knowing the population size now, at time $t$, and by knowing $R_0$. Because no females survive more than one year, population growth is described by the equation

$$N_{t+1} = R_0 N_t$$

where $N_t$ is the population size of females at generation $t$, $N_{t+1}$ is the population size of females at generation $t + 1$, and $R_0$ is the net reproductive rate or number of females produced per female per generation. If $R_0 = 1.5$ and $N_t = 100$, then $N_{t+1} = 1.5 \times 100 = 150$. At the second generation, $N_{t+2}$, the population size would be $150 \times 1.5 = 225$. Similarly, population size at $N_{t+3}$ would be 337.5; at $N_{t+4}$, 506.25; and after five generations, $N_{t+5} = 759.375$. Of course, we can't have a fraction of a butterfly in nature, so we would have to round these numbers up or down to the nearest whole number. We can write this relationship more generally as

$$N_t = N_0 \, R_0^{\,t}$$

where $t$ is the number of generations and $N_0$ is the number of females currently in the population. In our example

$$N_0 = 100, \; R_0^{\,t} = 1.5^5 = 7.59375,$$
$$\text{so } N_t = 759.375.$$

Clearly, much depends on the value of $R_0$. In the simplest models we assume that $R_0$ remains constant. When $R_0 < 1$, the population eventually declines to extinction; when $R_0 = 1$, the population size remains constant; and when $R_0 > 1$, the population increases. When $R_0 = 1$, the population is at equilibrium, where no changes in population size occur. Even if $R_0$ is only fractionally above 1, population increase is rapid (**Figure 10.1**) and a characteristic J-shaped curve results. We refer to this characteristic curve as **geometric growth.**

How do field data fit this simple model for geometric growth? Predator populations have been found to increase in geometric fashion when released into favorable sites. The last wild wolves in Montana, Wyoming, and Idaho were killed in the early 1900s as part of a government-funded wolf extermination project. At the time it was the policy of the federal government to exterminate wolves everywhere, even in national parks. The last wolves in Yellowstone National Park were killed in 1924, when two pups were killed at Soda Butte Creek in the northeast corner of the park. In the 1960s and 1970s, public attitudes changed, and wolves received legal protection under the Endangered Species Act of 1973. Wolves from Canada slowly began to filter into northern Montana in the 1980s. After a long, heated debate, wolf reintroductions were started in the continental United States a decade later. Between 1995 and 1996, 66 wolves were captured near Alberta, Canada, and released in Yellowstone Park, Wyoming, and central Idaho. The combined population of wolves in these states plus Montana has shown geometric growth ever since (**Figure 10.2a**). The black-footed ferrets discussed at the beginning of the chapter also showed geometric growth, despite the occurrence of plague and distemper diseases (**Figure 10.2b**). A litter size of

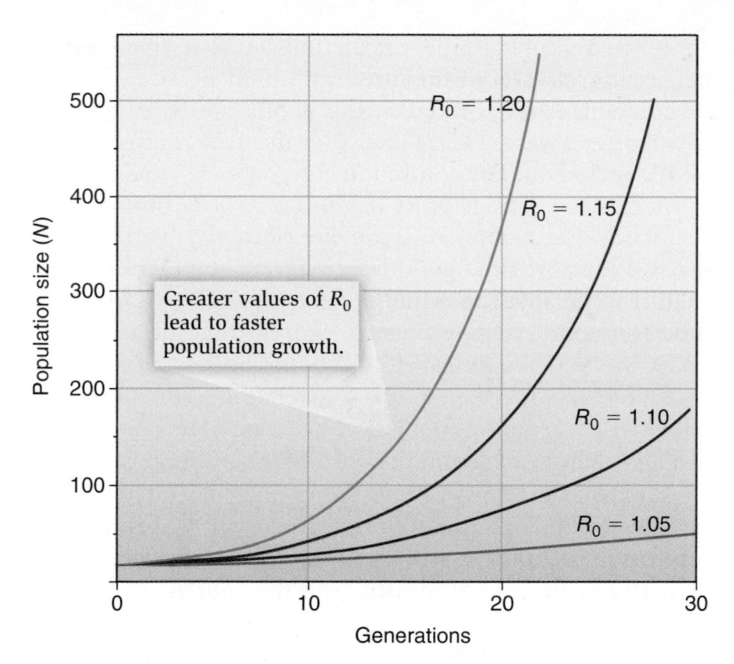


**Figure 10.1** **Geometric growth.** Four examples of geometric population growth where $N_0 = 10$.

What is the population size after five generations if $N_0 = 10$ and $R_0 = 1.4$?

3–5 young per year helped to fuel this growth. Remarkably, an average finite rate of increase of 59% per year was estimated from the minimum number alive for the period 2000–2006. Let's examine what this term means.

Because of the effort involved in calculating $R_0$, the net reproductive rate, ecologists sometimes use a shortcut to predict population growth. Imagine a bird species that breeds annually. To measure population growth, ecologists count the number of birds in the population, $N_0$. Let's say $N_0 = 100$. The next year ecologists count 105 birds in the same population, so $N_1 = 105$. The **finite rate of increase,** $\lambda$, is the ratio of the population size from one year to the next, calculated as

$$\lambda = N_1/N_0$$

In this case $\lambda = 1.05$. $\lambda$ is often given as percent annual growth, and $t$ is a number of years. Let's consider a population of birds growing at a rate of 5% per year. After 5 years, how many birds would there be? For the butterfly population growth example we talked about earlier, we wrote $N_t = N_0 R_0{}^t$, Similarly, here we calculate $N_t$ as $N_0\lambda^t$.

If

$$N_t = 100, \lambda = 1.05$$

and

$$t = 5$$

Then

$$N_{t+5} = 100\ (1.05)^5 = 127.6$$

(a) Wolves

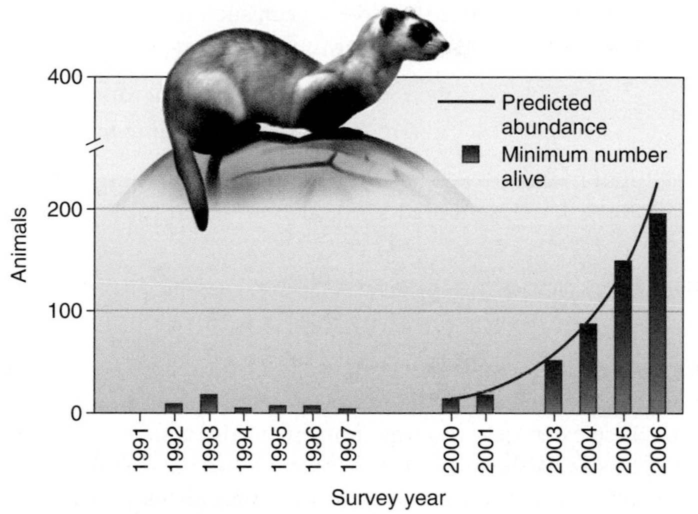


(b) Black-footed ferret

**Figure 10.2** **Geometric growth of predators.** **(a)** Wolves reintroduced in Montana, Idaho, and Wyoming. (From data in the Rocky Mountain Wolf Recovery 2005 Interagency Annual Report.) **(b)** Black-footed ferrets reintroduced to Shirley Basin, Wyoming. (After Grenier et al., 2007.)

We can rewrite this equation in many other ways to calculate other variables, such as cumulative finite rate of increase, average finite rate of increase, and time to reach a certain population size. To solve for cumulative finite rate of increase, we write:

$$(N_t - N_o/N_o) \times 100$$

In our bird example, cumulative finite rate of increase is given by:

$$(127.6 - 100/100) \times 100 = 27.6\%$$

Thus, the bird population has grown by 27.6% over 5 years. To get an average finite rate of increase, we rearrange the equation:

$$\text{Average finite rate of increase} = (N_t/N_0)^{(1/t)}$$
$$\text{For our birds, } \lambda = (127.6/100)^{(1/5)}$$
$$= 1.276^{.20}$$
$$= 1.05 \text{ or } 5\% \text{ per year}$$

How long will it take the bird population to reach 200? We rearrange our equation by taking logarithms of both sides.

$$\log N_t = \log N_0 + t(\log\lambda)$$
$$t = \log N_t - \log N_0/\log\lambda$$
$$= \log(N_t/N_0)/\log\lambda$$
$$\text{so } t = \log(200/100)/\log 1.05$$
$$= 14.2$$

The bird population of 100 would take 14.2 years to reach a population of 200.

What's the difference between $R_0$ and $\lambda$? $R_0$ represents the net reproductive rate per generation, and $\lambda$ represents the finite rate of increase over some time interval, often a year. Where species are annual breeders that live 1 year, such as annual plants, $R_0 = \lambda$. For species that breed for multiple years, $R_0 \neq \lambda$. Just as

$$N_t = N_0 R_0^t, \text{ where } t = \text{a number of generations}$$
$$\text{so } N_t = N_0 \lambda^t, \text{ where } t = \text{a number of time intervals}$$

Populations grow when $R_0$ or $\lambda > 1$, populations decline when $R_0$ or $\lambda < 1$, and are at equilibrium when $R_0$ or $\lambda = 1$.

### 10.1.2 Exponential growth describes population growth for continuous breeders

So far we have considered population growth of organisms that have discrete breeding periods, either once a year or perhaps once every few months. For most larger plants and animals, this is likely to be true. However, for bacteria, internal parasites, some insects, some tropical organisms or humans, reproduction can be continuous. To understand population growth under such conditions, a slightly different type of model is required. To understand the difference between periodic and continuous breeders, let's examine a situation where we have a population of 1,000 organisms increasing at a finite rate of 10% per year. The population size at the end of year 1 will be 1,000 + (10% of 1,000) = 1,100. The following table gives the corresponding values for subsequent years.

| Year | $\lambda$ = *10% per year* | $\lambda$ = *5% per every 6 months* |
|---|---|---|
| 0 | 1,000 | 1,000 |
| 1 | 1,100 | 1,102 |
| 2 | 1,210 | 1,215 |
| 3 | 1,331 | 1,340 |
| 4 | 1,464 | 1,477 |
| 5 | 1,610 | 1,629 |

If we repeated these calculations with a finite rate of increase of 5% every 6 months, our numbers would be a little different. After 6 months, the population size would be 1,050; after 1 year, 1,102; after 2 years, 1,215; after 3 years, 1,340; and so on. For continuous breeders we would have to repeat the calculations every day or every hour, so our values would be slightly different again. For example, dividing a year into 1,000 short time periods, we have a population size of 1,000.1 in the first thousandth of the year, 1,000.2 in the second thousandth, and so on. If we repeat this for 1,000 time intervals, we will end up with 1,105 organisms at the end of the first year. This population now contains five more individuals than a similar population breeding annually.

The change in population size over any time period can be written as the number of births per unit time interval minus the number of deaths per unit time interval. For example, in a population of 1,000 apartment-dwelling cockroaches, if there were 100 births and 50 deaths over the course of one week, then the population would grow in size to 1,050 the next week. We can write this formula mathematically as

$$\frac{\text{Change in numbers}}{\text{Change in time}} = \text{Births} - \text{Deaths}$$

Or $$\frac{\Delta N}{\Delta t} = B - D$$

The Greek letter $\Delta$ indicates change, so that $\Delta N$ is the change in number and $\Delta t$ is the change in time; $B$ is the number of births per time unit; and $D$ is the number of deaths per time unit.

The numbers of births and deaths can be expressed per individual in the population, so the birth of 100 cockroaches to a population of 1,000 would represent a birth rate, $b$, of 100/1,000, or 0.10 per individual per week. Similarly, the death of 50 cockroaches in a population of 1,000 would be a death rate, $d$, of 50/1,000, or 0.05 per individual per week. Now we can rewrite our equation giving the rate of change in a population.

$$\frac{\Delta N}{\Delta t} = bN - dN$$

In our cockroach example,

$$\frac{\Delta N}{\Delta t} = 0.10 \times 1{,}000 - 0.05 \times 1{,}000 = 50$$

If $\Delta t$ = 1 week, the cockroach population would increase by 50 individuals in a week. If $\Delta t$ = 1 day, there would likely be fewer cockroaches born, perhaps only 15, but the number dying would also be fewer, perhaps only 7. In this case $b = 0.015$, $d = 0.007$, and so

$$\frac{\Delta N}{\Delta t} = 0.015 \times 1{,}000 - 0.007 \times 1{,}000 = 8$$

The cockroach population would increase by only 8 individuals.

Ecologists often simplify this formula by representing $b - d$ as $r$, the **per capita rate of increase.** Thus, $bN - dN$ can be written as $rN$. To most accurately determine growth rates of continuous breeders, ecologists are interested in population growth rates that are shorter than whole generation times, so-called instantaneous growth rates. So, instead of writing

$$\frac{\Delta N}{\Delta t}$$

ecologists write

$$\frac{dN}{dt}$$

which is the notation of differential calculus. The equations essentially mean the same thing, except that $dN/dt$ reflects very short time intervals. Thus, for cockroach growth over a week,

$$\frac{dN}{dt} = rN = (0.10 - 0.05) \times 1{,}000 = 50$$

We can rewrite this equation to find population sizes at various times. The population size at time $t$ is given by

$$N_t = N_0 e^{rt}$$

where $N_0$ is the population size now, at time 0; $e$ is the base of natural logarithms, a constant of about 2.7; $r$ is the per capita rate of increase; and $t$ is the number of time intervals. For example, if $N_0 = 10$, $t = 2$, and $r = 0.1$,

$$N_{t+2} = 10e^{0.1\times 2} = 10e^{0.2} = 12.214$$

How is the per capita rate of increase, $r$, related to the net reproductive rate, $R_0$, we introduced at the beginning of the chapter? The two quantities are very similar. $R_0$ reflects a generational growth rate, while $r$ denotes an instantaneous rate. In the financial world $r$ is called the (APR) Annual Percentage Rate. $R$ is termed the Effective Annual Rate (EAR) expressed on an annual basis. An APR of 12.99% on a credit card, compounded monthly, would be equivalent to an EAR of 13.78% because of the compound nature of the interest. As a rough approximation, dividing the natural log of $R_0$ by the generation time ($T$) gives us $r$.

Thus,

$$r \approx \frac{\ln R_0}{T}$$

For the beaver population in Newfoundland that we discussed in Section 9.1.2, even though the beavers breed in discrete seasons, we can calculate a rough approximation for $r$. The net reproductive rate, $R_0$ was 0.943 and generation time was 4.638 years, so that

$$r = \frac{\ln 0.943}{4.638} = -0.0126$$

Generally, for a variety of organisms, the per capita rate of increase, $r$, is affected by organismal size and generation time. $r$ decreases as generation time increases (**Figure 10.3a**). When conditions are optimal for the population, $r$ is at its maximum rate and is called the **intrinsic rate of increase** (denoted $r_{max}$). Thus, the rate of population growth under optimal conditions is $dN/dt = r_{max} N$. The value for $r_{max}$ is always a positive number, whereas the measured value of $r$ may be positive or negative, depending on whether the population is increasing or decreasing. $r_{max}$ has been found to decrease with body weight (**Figure 10.3b**). For many ectotherms, $r_{max}$ is dependent upon

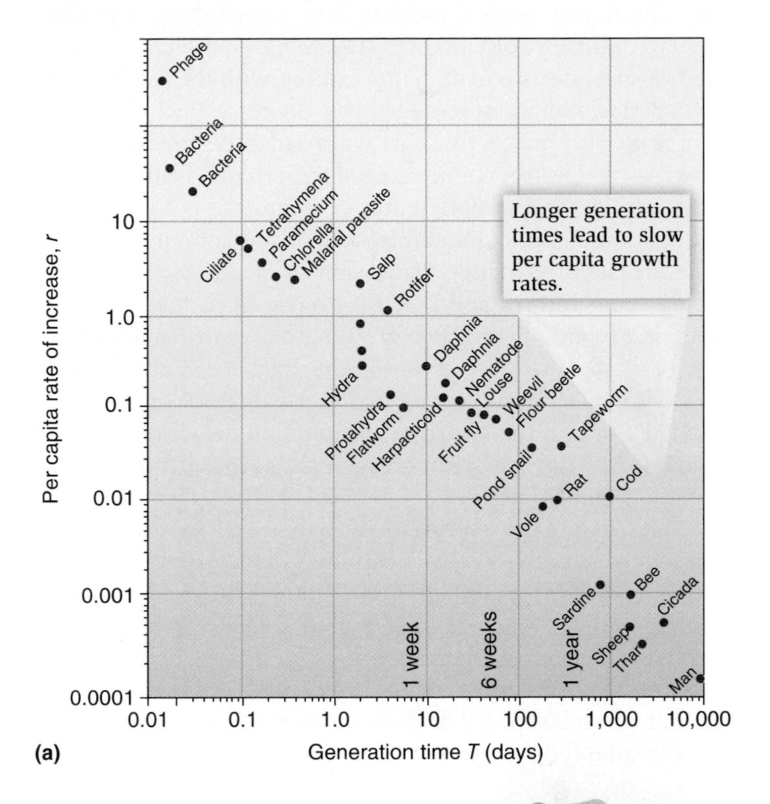


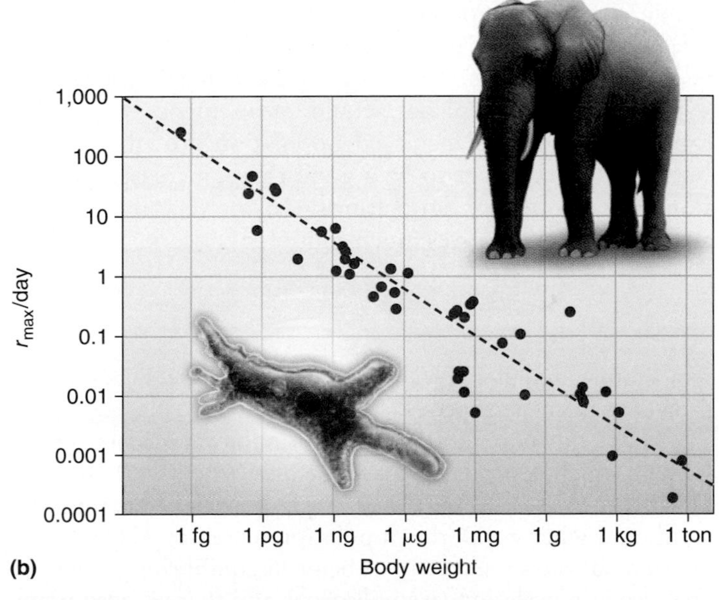


**Figure 10.3 Variation in the per capita growth rate *r* with generation time and organismal size.** **(a)** $r$ decreases with generation time. Data are from a variety of organisms. (After Heron, 1972.) **(b)** $r_{max}$ decreases with body weight. (After Blueweiss et al., 1978.)

## Global Insight

### Population Growth May Change in Response to Global Warming

Population growth of ectotherms, the predominant organisms on Earth, is greatly affected by body temperature. Many ectotherms are adapted to live at a particular optimal temperature, $T_0$, either in a cold arctic environment, temperate climate, or a hot tropical area. Some ecologists have argued that the particular adaptations of cold-adapted species have perfectly compensated for the effects of cold temperatures, with the result that in such species, temperature-dependent processes including population growth match those of warm-adapted species. This is known as the perfect compensation hypothesis (**Figure 10.A-i**). Alternatively, the thermodynamic constraint hypothesis argues that adaptations of cold-adapted organisms cannot fully overcome the thermodynamic depression of biochemical reactions in cold temperatures and that they have much lower rates of processes, such as population growth, than warm-adapted species. This is known as the thermodynamic constraint hypothesis, sometimes referred to as "warmer is better." In a globally warmed world, if the "warmer is better" model were true, then populations of many ectotherms, such as mosquitoes, would increase (**Figure 10.A-ii**).

Melanie Frazier and her colleagues (2006) tested these competing hypotheses by compiling data from 65 insect studies that had examined population growth at a variety of temperatures. They argued that if the perfect compensation hypothesis were operating for cold-adapted species, $r_{max}$ would be the same as for warm-adapted species of similar taxonomic affiliations. The data showed $r_{max}$ was much lower for cold-adapted species than for warm-adapted species, and that $r_{max}$ declined with a drop in $T_0$, supporting the thermodynamic constraint hypothesis. Frazier and colleagues summarized their results by saying the data supported the "warmer is better" hypothesis. Thus, for cold-adapted insect species, a globally warmed planet would mean higher population growth rates. While increased population growth rates for butterflies might seem to be a good thing, increased population growth rates for agricultural pests or vectors of disease such as mosquitoes gives much cause for concern.

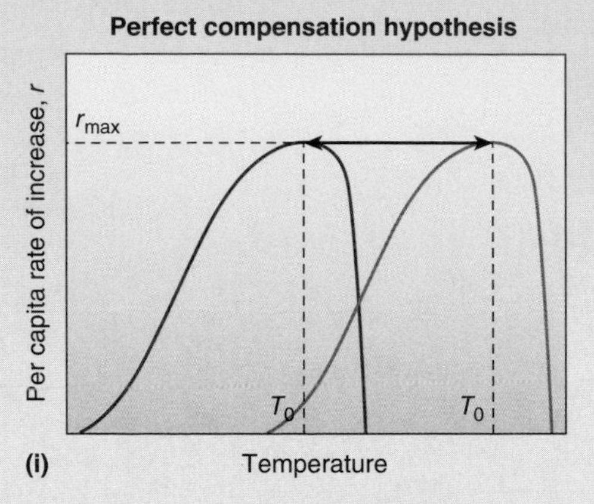


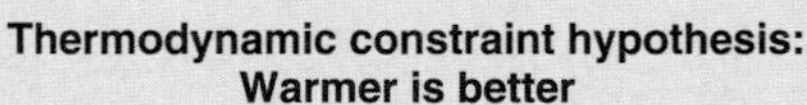


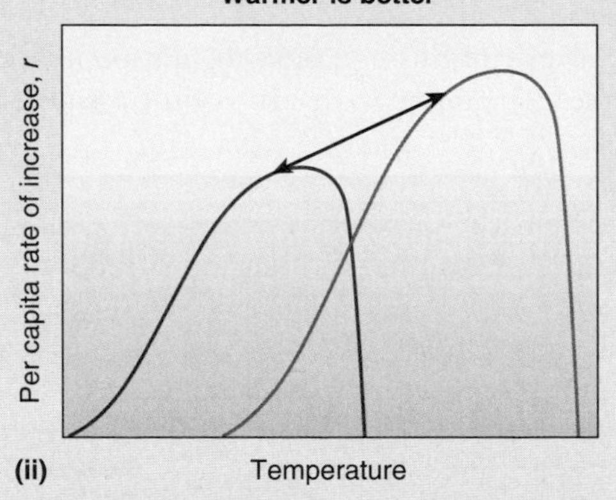


**Figure 10.A Two different hypotheses predicting how population growth rate of ectotherms responds to temperature adaptation.** **(i)** The perfect compensation hypothesis, which predicts biochemical adaptations of cold-adapted species (red line) permit the intrinsic rate of population increase, $r_{max}$, to be the same as in warm-adapted species (orange line). $T_0$ represents the optimal temperature. **(ii)** The thermodynamic constraint hypothesis (warmer is better), where species adapted to warmer temperatures (orange curve) have a higher $r_{max}$ than cold-adapted species (red line). (After Frazier et al., 2006.)

temperature, and global warming will likely increase population growth of many pest species (see **Global Insight**).

How do populations of continuous breeders grow? Clearly, much depends on the value of $r$. When $r < 0$, the population decreases; when $r = 0$, the population remains constant; and when $r > 0$, the population increases. When $r = 0$, the population is referred to as being at equilibrium, where no changes in population size will occur and there is **zero population growth.** For the Newfoundland beaver population the slight negative growth rate, $r = -0.0126$, indicates the population is in decline, though this is something we knew from our examination of the net reproductive rate, $R_0$.

Even if $r$ is only fractionally above 0, population increase is rapid, and when plotted graphically a characteristic J-shaped curve results (**Figure 10.4**). We refer to this type of population growth as **exponential growth.** Note that both exponential growth and geometric growth give very similar J-shaped growth curves. How do field data fit the simple model for exponential growth? The growth of some colonies of yeast and bacteria fit the classic J-shaped curve in much the same way as growth for the annual breeders, such as black-footed ferrets and wolves, fits the geometric pattern. The growth of the global human population over time also shows dramatic increases over short periods of time. Because of its

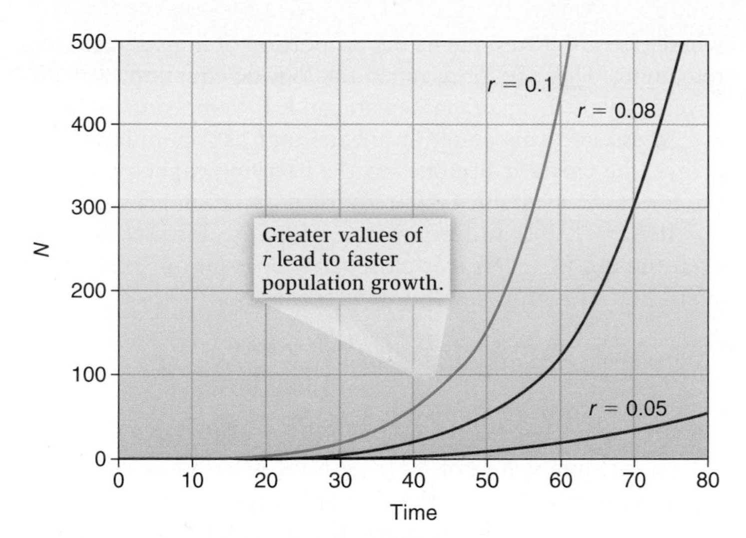


**Figure 10.4** **Exponential population growth.** In theory, a population with unlimited resources could grow indefinitely. Starting population, $N_0$, = 1. (After Case, 2000.)

large importance, we will examine human population growth separately later in the chapter.

Unfortunately, the growth of many invasive species also fits the pattern of exponential growth. As we mentioned in Chapter 7, Burmese pythons, *Python molurus,* native to Southeast Asia, are now well established in south Florida. John Wilson and colleagues (2011) fitted an exponential growth curve to python abundance based on capture numbers (**Figure 10.5**). From a relatively small number of pythons, likely released prior to 1985, population densities in the Everglades National Park were estimated conservatively at 30,000 individuals in 2007.

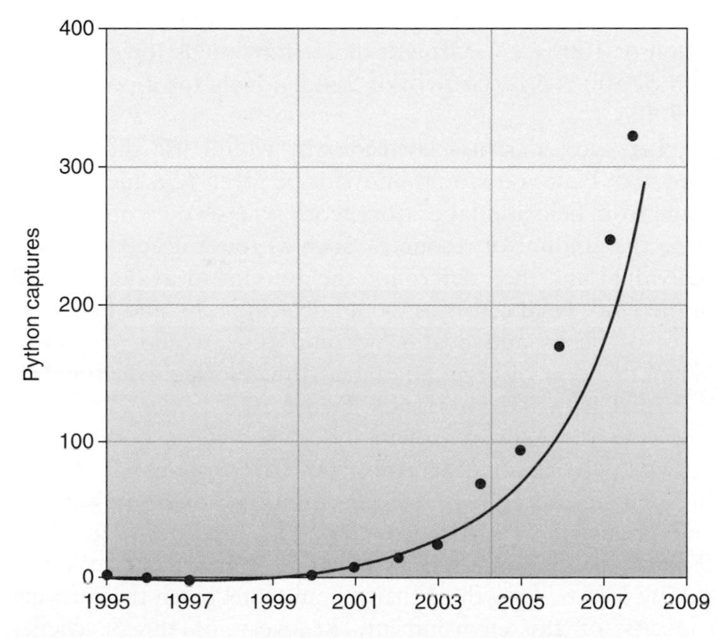


**Figure 10.5** **Exponential growth of invasive Burmese python, *Python molurus,* in the Everglades National Park, as evidenced by numbers captured.** (After Wilson et. al., 2011.)

Many organisms exhibit exponential growth for a short period of time when they are introduced into a new environment. Suppose you have a pond that is being overgrown by an aggressive weed that doubles in number every day. If unchecked, it would cover the pond in 30 days. Weed growth is very slow, almost negligible, to begin with, and you decide not to remove the weed until the pond is half covered. When will that be? On the twenty-ninth day. You will have one day to save your pond. There is an important message here. We cannot wait too long to control invasive species, such as insect or weed pests that are growing rapidly. We must act as soon as possible. Dan Simberloff (2003) has noted that in the fight to control invasive species, the most effective approach is to discover them early and attempt to eradicate or contain them before they spread. For example, the tropical alga *Caulerpa taxifolia* was discovered in a tiny area, a few square meters wide, of the Mediterranean sea around Monaco in 1984, and could have been eliminated by hand removal. By 1989 it had grown to 2 1/2 acres, and by 2001, researchers had found that the alga had invaded 50 square miles of seabed along nearly 120 miles of the coasts of Spain, France, Monaco, Italy, Croatia, and Turkey. *Caulerpa* infestations have shown up in the California coastline and the Gulf of Mexico. Contrast this with the discovery of the Caribbean black-striped mussel, *Mytilopsis sallei,* in Darwin Harbor, Australia, in 1999. Within 9 days the Australians treated the whole area with liquid bleach and copper sulfate. Eradication was achieved before the infestation got out of control.

How are the variables $r$ and $\lambda$ related? $r$ is the per capita rate of increase, which gives an instantaneous growth rate, while $\lambda$ is the finite rate of increase over a specific time period. The relationship between the two parameters is given by:

$$\lambda = e^r$$

Again, values of $r$ are generally less than rates of $\lambda$, because growth rates based on $r$ compound faster. However, at low values of $r$, the finite rate of increase is very similar. For example, if $r = 0.012$, then $\lambda = 1.012$, or 1.2% annual growth. For higher values of $r$, $\lambda$ tends to be a higher. For example, where $r = 0.47$, $\lambda = 1.599$ or 60% annual growth.

Many ecologists, particularly those interested in human population growth, are interested in how long it takes a population to double, that is, $\lambda = 2$. There are several steps to understanding how to calculate the doubling time. First, we know that the finite rate of increase over many years is given by $e^{rt}$, where $t$ is a number of years. We need to know how many years it takes the population to double, that is:

$$2 = e^{rd}$$

where $d$ is the time to double. We can take the natural logarithm of both sides,

$$\ln 2 = rd$$

or

$$0.693 = rd$$
$$d = 0.693/r$$

Therefore, if we know $r$, we can calculate the population doubling time. Many ecologists approximate this equation as:

$$d = 0.7/r$$

This relationship is called "the rate of 70": take 70 and divide it by the growth rate. Here, the growth rate is cast as the percentage increase in a year, because at low rates of $r$, per capita rate of increase is roughly equivalent to the finite rate of increase, $\lambda$, expressed as a percentage. For example, in 2006 the world's population grew by 1.23%. If that rate were to continue unchanged, the world's population would double in 57 years, because 70/1.23 = 57. You can make a similar calculation about accrual of money in a savings account. If your bank pays you 5% interest, your money will double in 70/5 = 14 years.

**Check Your Understanding**

**10.1** 1. If the world's population was 6.5 billion in 2006, and it grew at an annual rate of 1.23%, how long would it take to reach 10 billion?

2. What's the average finite rate of increase for a deer population starting with 1,000 individuals and ending with 1,800 after 5 years?

## 10.2 Limited Resources Lead to S-Shaped Population Growth Curves

Despite its applicability to rapidly growing populations, the exponential growth model is not appropriate in all situations. The model assumes unlimited resources, which is not always the case in the real world. For many species, resources become limiting as populations grow. We need a different type of mathematical model to describe how population growth occurs under conditions of limited resources.

### 10.2.1 Logistic growth results in an upper limit to population size

**Logistic growth** occurs where there are limited resources and thus an upper boundary for population size. The upper boundary for the population size is known as the **carrying capacity ($K$).** Although logistic growth can occur for both periodic and continuous breeders, we will illustrate the logistic model using continuous breeders, which have a per capita rate of increase, $r$. Logistic growth, which takes into account the amount of available resources, is given by

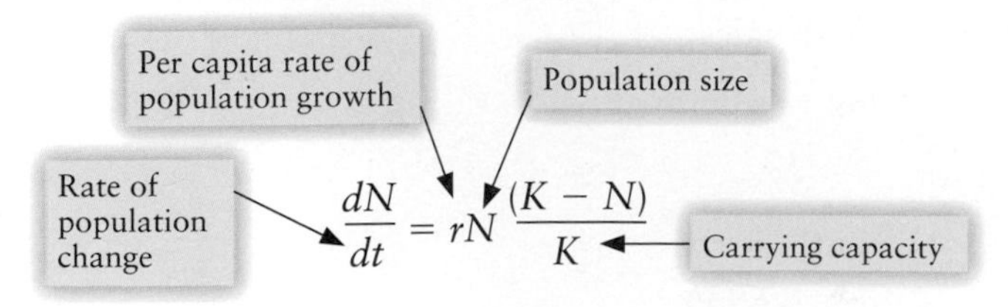


$$\frac{dN}{dt} = rN\frac{(K - N)}{K}$$

where $(K - N)/K$ represents the proportion of unused resources remaining. This equation, called the **logistic equation,** was discovered by the Belgian mathematician P. F. Verhulst in 1838.

In essence, this equation means that as a population, $N$, grows, the closer it becomes to the carrying capacity, $K$, and the fewer the available resources for population growth.

If $K$ = 1,000 and population sizes are low ($N$ = 100), even though $(K - N)/K$ is close to 1, population sizes are so small that growth is small. In this scenario

$$\frac{dN}{dt} = (0.1)\,(100) \times \frac{(1{,}000 - 100)}{1{,}000}$$
$$= 9$$

At medium values of $N$, $(K - N)/K$ is less close to a value of 1 but population growth is larger because there are a larger number of reproducing females. If $K$ = 1,000, $N$ = 500, and $r$ = 0.1, then

$$\frac{dN}{dt} = (0.1)\,(500) \times \frac{(1{,}000 - 500)}{1{,}000}$$
$$= 25$$

At larger values of $N$, $(K - N)/K$ becomes small, resources are close to being used up, and population growth is again small. If $K$ = 1,000, $N$ = 900, and $r$ = 0.1, then

$$\frac{dN}{dt} = (0.1)\,(900) \times \frac{(1{,}000 - 900)}{1{,}000}$$
$$= 9$$

By comparing these three examples, we see that absolute growth is small at low and high values of $N$ and is greatest at intermediate values of $N$. Absolute growth is greatest when $N = K/2$. Try some calculations to verify this for yourself. Proportionally, growth is greatest at low values of $N$ and decreases as $N$ increases. Growth of 9 individuals for a population of 100 is 9%. Growth of 25 individuals for a population of 500 is 5%. Growth of 9 individuals for a population of 900 is 1%.

Let's consider how an ecologist would use the logistic equation. First you would know or be given $K$, which would come from field and laboratory work where you would determine the amount of resources, such as food, needed by each individual and then determine the amount of available food in the wild. Field censuses would determine $N$, and field censuses of births and deaths per unit time would provide $r$. When this type of population growth is plotted over time, an S-shaped growth curve results (**Figure 10.6**).

Does the logistic growth model provide a better fit to growth patterns of organisms than the exponential model? In some instances, such as laboratory cultures of bacteria and yeasts, the logistic growth model provides a good fit (**Figure 10.7a**). Some species introduced into new areas also exhibit logistic growth as their populations reach the carrying capacity of the environment. Recovery of the Seychelles warbler, *Acrocephalus sechellensis,* on Cousin Island in 1960–1990 showed a logistic growth pattern (**Figure 10.7b**).

At one time the world population of this endemic warbler was reduced to just 26 individuals, but following long-term management by the International Council for Bird Preservation, numbers rebounded spectacularly until all available habitat on this 29-ha island was used up. However, for many other populations there is little support for the logistic model. In nature there are frequent variations in temperature, rainfall, or resources that in turn cause changes in population growth and carrying capacity and thus in population sizes. The uniform conditions of temperature and resource levels of the laboratory do not usually exist in nature. In addition, a time lag between availability of resources and time of reproduction can cause large overshoots and undershoots of population densities relative to the carrying capacity.

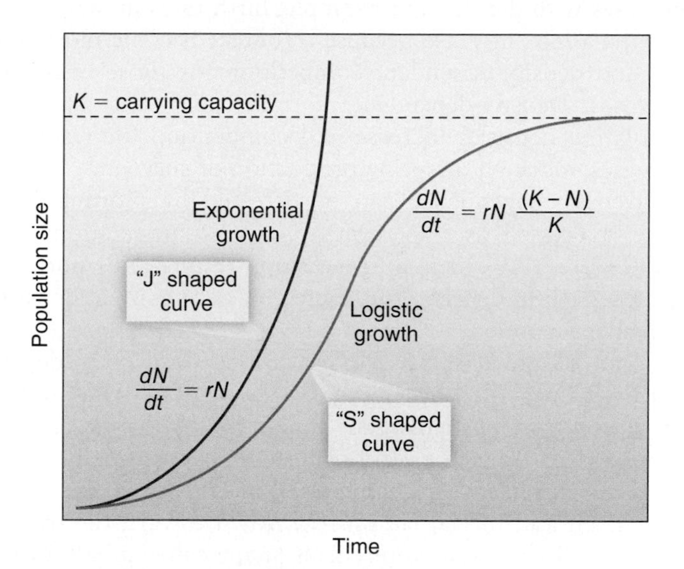


**Figure 10.6 Logistic growth yields an S-shaped population growth curve.** This contrasts to the J-shaped curve of exponential growth.

**ECOLOGICAL INQUIRY**

What is the population growth per unit of time when $r = 0.1$, $N = 200$, and $K = 500$?

## 10.2.2 Time lags can influence whether or not a population reaches an upper limit

The Australian physicist turned ecologist Robert May (1976) surprised ecologists by showing how a time lag could radically change the shape of population growth curves. A **time lag** is a delay in response to change. For example, when the young of most species are born they are small and consume less resources than the adults. There is a time lag in the effects of resources on population growth related to the developmental period of the offspring. We can incorporate this time lag into the logistic growth equation.

For example, in an unlagged population, if $r = 0.3$, $K = 1{,}000$ and $N = 900$, then

$$\frac{dN}{dt} = (0.3)\,(900) \times \frac{(1{,}000 - 900)}{1{,}000}$$
$$= 27$$

The new population size is $900 + 27 = 927$.

If there is a time lag such that at the time the population is 900, the effects of crowding are only being felt as though the population were 500, then

$$\frac{dN}{dt} = (0.3)\,(900) \times \frac{(1{,}000 - 500)}{1{,}000}$$
$$= 135$$

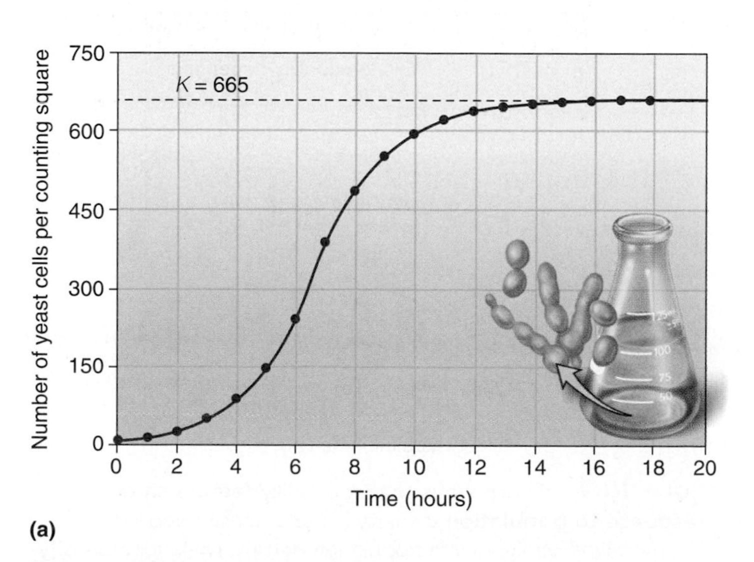


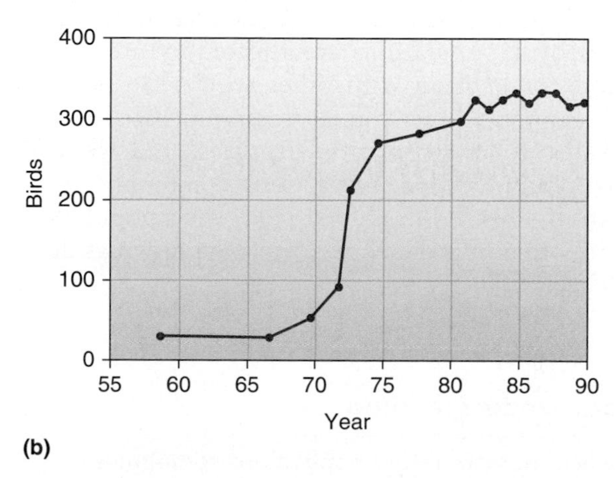


**Figure 10.7 Logistic growth is shown by some populations.** **(a)** Growth of yeast cells in laboratory cultures. (After Pearl, 1927.) **(b)** The growth of the Seychelles warbler on Cousin Island between 1960 and 1990 (After Komdeur, 1992). These populations show the typical S-shaped growth curve.

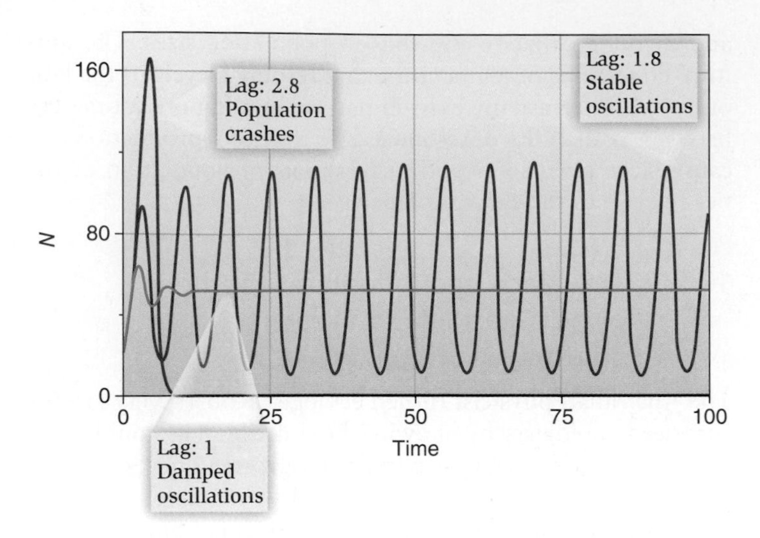


**Figure 10.8 The effects of varying time lags on population growth.** The data are from a population model and do not refer to any particular species. Here $r = 1.0$, $K = 50$, and the time lag is either 1.0, 1.8, or 2.8 time units. (After Case, 2000.)

so the new population size is $900 + 135 = 1{,}035$. The effect of the time lag is to increase, or decrease, the population growth rate from the ordinary logistic. It is also now possible for a population to overshoot its carrying capacity. With small time lags, the population undergoes oscillations that dampen with time until $K$ is reached (**Figure 10.8**). As the length of the time lag increases, the population fluctuations increase in size and the population enters into a stable oscillation called a limit cycle, rising and falling around $K$ but never converging on the equilibrium value. We can think of a population at just below the carrying capacity having enough food for every individual to reproduce. After reproduction, however, few of the new individuals will gain enough resources to reproduce; the population will fall below the carrying capacity again, and the population may fluctuate so widely in size that it crashes to extinction.

It is easy to see how time lags can completely disrupt a population from following a logistic growth curve. Also, as we will discover, populations are affected by predators, parasites, and competition with other species. In Section IV we will examine how competitors, mutualists, and natural enemies affect population densities, and we will explore situations in which interactions commonly influence population sizes. As described next, such population reductions are often influenced by a process known as density dependence.

### Check Your Understanding

**10.2** Why is it that laboratory populations sometimes show good fits to the logistic growth curve while field populations often do not?

## 10.3 Density-Dependent Factors May Limit Population Size

A **density-dependent factor** is a factor whose influence varies with the density of the population. Density-dependent factors affect a higher proportion of individuals when population densities are higher and a lower proportion when population densities are lower. Alternatively, density-dependent factors may affect all individuals, but the effect per individual increases with density. For example, birth rates may decrease as populations increase because resources become more limited and density-dependent competition for those resources increases. Density-dependent mortality may also occur as population densities increase and competition for resources increases, reducing offspring production or survival.

Density dependence can be detected by plotting mortality, expressed as a percentage, against population density (**Figure 10.9**). If a positive slope results and mortality increases with density, the factor is acting in a density-dependent manner.

One of the earliest demonstrations of density dependence was provided by George Varley (1941), who studied gall insects, in particular the small fly *Urophora jaceana* (**Figure 10.10**). Female *U. jaceana* lay eggs in small groups within flower heads of knapweed, *Centaurea nemoralis*. The eggs hatch and the larvae burrow down toward the ovules, which swell up into a woody flask-shape called a gall. Inside the gall, the fly larvae feed and develop. Larvae may be parasitized by a small wasp, *Eurytoma curta*, females of which penetrate the knapweed flower heads with their ovipositors and lay eggs on the young *U. jaceana* larvae. The wasp larvae completely consume the fly larvae, and wasps emerge from the gall instead of flies. Varley censused 20 different 1-$m^2$ patches of knapweed. He discovered that where patches of

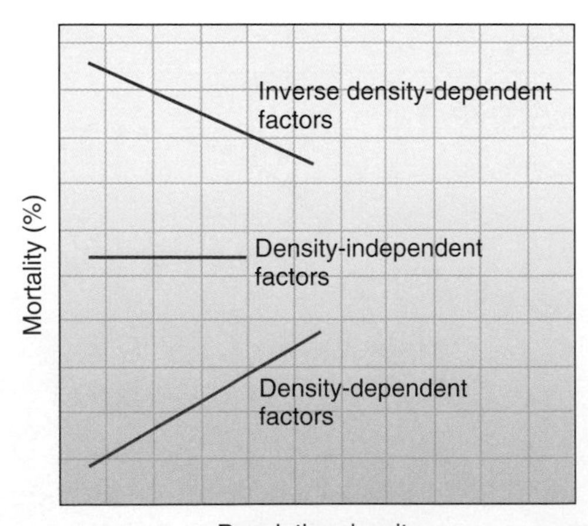


**Figure 10.9 Three ways that mortality factors change in response to population density.** For a density-dependent factor, mortality increases with population density, while for a density-independent factor, mortality remains unchanged. For an inverse density-dependent factor, mortality decreases as a population increases in size.

(a) (b)

**Figure 10.10 Insect galls.** **(a)** These galls on goldenrod, *Solidago canadensis,* caused by the fly *Eurosta solidaginis,* are a common sight in North America. **(b)** Cross section of gall caused by the fly *Urophora jaceana* on knapweed, *Centaurea,* showing fly larva (right) and pupa (left).

**Figure 10.11 Density dependence.** **(a)** Density-dependent attack of fly larvae inside galls on knapweed by parasitic wasps. (After Varley, 1941.) **(b)** Density-dependent attack of winter moth pupae in leaf litter by predatory beetles and shrews. Here, mortality is expressed on a log scale as a *k*-value (look ahead to Figure 16.11), and host density is also plotted on a log scale. (After Varley et al., 1973.)

flower heads contained a greater number of fly larvae hosts, a greater proportion of them were parasitized (**Figure 10.11a**).

Density dependence can also be detected by plotting other expressions of mortality against host density, and it can be detected over time as well as spatially. As we will see later, in Chapter 16, ecologists sometimes express mortality in terms of a killing power, where the killing power, or *k* value, is given by

$$k = \log N_{(t)} - \log N_{(t+1)}$$

where $N_{(t)}$ is the population size before it is subject to the mortality factor and $N_{(t+1)}$ is the population size afterward. Because *k* is expressed as a log value, it is plotted against log *N*, the log of the abundance of the population. As before, a positive slope indicates density dependence. After his gall fly studies, George Varley went on to conduct one of the most detailed ecological studies ever undertaken, when he followed populations of the oak winter moth from 1950 until 1962. We introduced this study in Chapter 1 and will discuss it in greater depth in Chapter 16, but it is worth noting here that Varley discovered that the killing power of pupal predators, beetles and shrews, acted in a density-dependent manner over time (**Figure 10.11b**). In the years where pupal densities of winter moth were high, pupal predation was high, but in the years where pupal density was low, predation was low. In this case, density dependence was acting over time, preventing large outbreaks of oak winter moths in any one year.

A **density-independent factor** is a mortality factor whose influence is not affected by changes in population size or density. Here a flat line results when mortality is plotted against density. In general, density-independent factors are physical factors such as weather, drought, freezes, floods, and disturbances such as fire. For example, in hard freezes a considerable percentage of organisms such as birds or plants are usually killed, no matter how large the population size, because most individuals are susceptible. However, even physical factors such as weather can sometimes act in a density-dependent manner. In an environment where there are many bears and a limited number of dens, some individuals will not have a den. In such a situation, a cold winter could kill a high percentage of bears. If, on the other hand, there are few bears, most would have a den to provide them with protection from a particularly harsh winter. In this case, the cold would kill a lower percentage. It is also worth noting that some biotic factors, like introduced diseases, to which native organisms are not adapted, can act in a density-independent manner, because they may kill virtually all their hosts, regardless of their density. Such was the case for chestnut blight, which killed nearly all American chestnut trees in the middle of the 20th century (look ahead to Figure 15.9).

Finally, a source of mortality that decreases with increasing population size is considered an **inverse density-dependent factor.** Here a negative slope results when mortality is plotted against density. For example, if territorial predators, such as a pride of lions, always ate the same number of wildebeest prey, regardless of wildebeest density, they would be acting in an inverse density-dependent manner, because they would be taking a smaller proportion of the population at higher density. Some mammalian predators, being highly territorial, act in this manner on herbivore densities.

Determining which factors act in a density-dependent fashion has practical implications. Foresters, game managers, and conservation biologists alike are interested in learning how to maintain populations at equilibrium levels. For example, if disease were to act in a density-dependent manner on white-tailed deer, there wouldn't be much point in game managers' attempting to kill off predators, such as mountain lions, to increase herd sizes for hunters, because proportionately more deer would be killed by disease.

Which factors tend to act in a density-dependent manner? Tony Sinclair (1989) reviewed many studies of density dependence in 51 populations of insects, 82 populations of large mammals, and 36 populations of small mammals and birds (**Figure 10.12**). Data for insects showed a wide variety of causes of density dependence, and no one density-dependent factor was of overriding importance. Food was more important for large mammals: being bigger, they need a large amount of food, and obtaining it is of critical importance. Space and social interactions were more important for smaller mammals and birds, because they are more territorial. Decreases in available space increase mortality in these groups. However, Sinclair noted that the effects of disease and parasitism had probably been grossly understudied for animals other than insects, because of the difficulty inherent in studying parasites in the field. These effects are therefore likely to be underrepresented in his data.

Not all mortalities result in density-dependent control. This might appear strange, until we realize that there may be specific biological reasons why some mortalities do not operate in a density-dependent fashion. For example, a searching insect parasite may not always lay her eggs in dense concentrations of knapweed flower heads; she may lay some of her eggs elsewhere, even in a seemingly suboptimal place. The reason may be that some local catastrophe could occur in the best area, wiping out all her progeny. For example, a herd of cattle could also be attracted to a dense congregation of knapweed and eat it all. If the wasp oviposits in a few solitary galls in out-of-the-way places, some progeny would survive such a catastrophe, and so this behavior would be selected for. This phenomenon has become known as spreading the

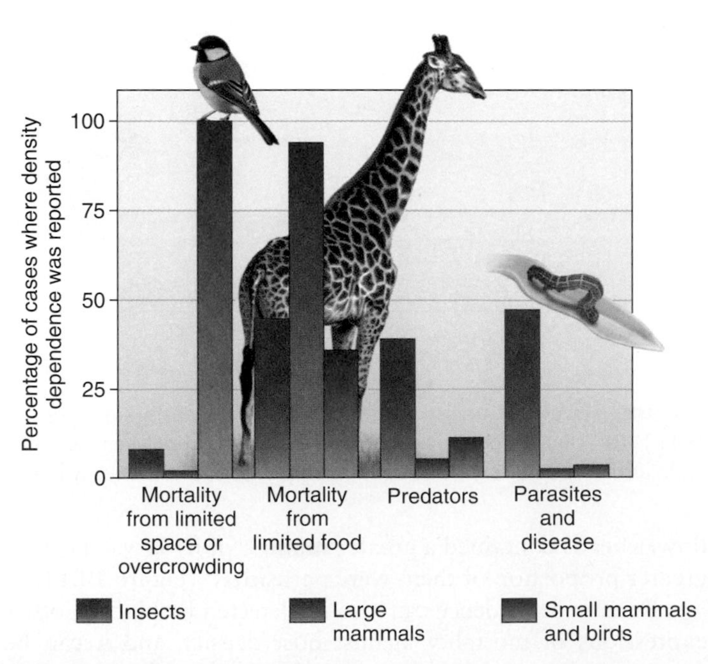


**Figure 10.12 The frequency of density-dependent mortalities in animal populations.** Percentage of reports of separate populations demonstrating density dependence for different animal groups. (After Sinclair, 1989.)

risk. The behavior of some organisms does seem to conform to a risk-spreading strategy.

### Check Your Understanding

**10.3** Biological control is the intentional release of natural enemies against pest organisms. *Mastrus ridibundus,* a parasitic insect from Asia, was released in the U.S. Pacific Northwest to attack codling moth larvae, *Cydia pomonella,* which attack fruits. The data below show the relationship between mean host density of codling moth larvae and percent parasitism by *M. ridibundus* in six different orchards. Assume that parasites can move between orchards. Plot percent parasitism against mean host density. What do the data tell you?

| Orchard | Mean host density per tree | Mean percentage parasitism |
|---|---|---|
| A | 2.9 | 40.3 |
| B | 3.0 | 36.9 |
| C | 3.5 | 30.9 |
| D | 4.4 | 11.6 |
| E | 6.8 | 8.6 |
| F | 7.7 | 7.4 |

## 10.4 Life History Strategies Incorporate Traits Relating to Survival and Competitive Ability

**Life history strategies** are sets of physiological and behavioral features that incorporate not only reproductive traits but also survivorship and length-of-life characteristics, preferred habitat type, and competitive ability. Life history strategies have important implications for how populations grow and indeed for the reproductive success of populations and species. Here we will discuss several different types of life history strategy, iteroparity versus semelparity, continuous versus seasonal iteroparity, $r$ and $K$ selection, and Grime's triangle.

### 10.4.1 Reproductive strategies include reproduction in a single event or continuous breeding

Organisms differ in the timing of reproduction, producing offspring in one single event or reproducing at intervals throughout their lifetime. Where all offspring are produced in a single reproductive event, we term this pattern **semelparity** (from the Latin *semel,* once, and *parere,* to bear). Semelparity is common in insects and other invertebrates, and also occurs in organisms such as salmon, bamboo grasses, and agave plants (**Figure 10.13a**). These individuals reproduce once only and die.

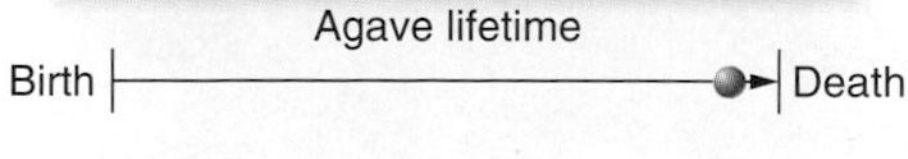


**(a) Semelparity**

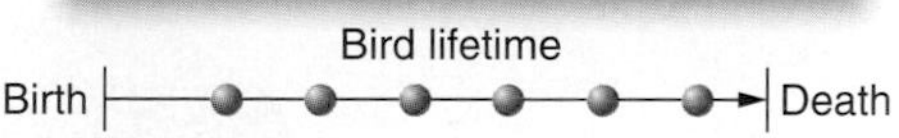


**(b) Iteroparous (seasonal): Bird**

Chimpanzee lifetime
Birth Death

**(c) Iteroparous (continuous): Chimpanzee**

**Figure 10.13 Differences in reproductive strategies.** Species such as **(a)** this Century plant, *Agave shawii,* in Baja, California, is semelparous, meaning that it breeds once in its lifetime and then dies. **(b)** This contrasts with the blue tit, *Parus caeruleus,* and **(c),** the chimpanzee, *Pan troglodytes,* which are iteroparous and breed more than once or repeatedly in their lifetime.

**ECOLOGICAL INQUIRY**

In an environmentally predictable environment, which strategy would be preferred: semelparity or iteroparity?

Semelparous organisms may live for many years before reproducing, like the agaves, or they may be **annual** plants that develop from seed, flower, and drop their seed within a year.

Other organisms reproduce in successive years or breeding seasons. The pattern of repeated reproduction at intervals throughout the life cycle is called **iteroparity** (from the Latin *itero,* meaning to repeat). Iteroparity is common in most vertebrates and perennial plants, such as trees. Among iteroparous organisms, much variation occurs in the number of reproductive events and in the number of offspring per event. Many species, such as birds or mammals, or temperate forest trees, have distinct breeding seasons (seasonal iteroparity) that lead to distinct groups of individuals all born at the same time (**Figure 10.13b**). For a few species, individuals reproduce repeatedly and at any time of the year. This is termed continuous iteroparity and is exhibited by some tropical species, many parasites, and some primates (**Figure 10.13c**). It is worth emphasizing that these different strategies are not the result of active choices by organisms but are sets of adaptive, evolved traits over which the organisms have very little control.

Why do species reproduce in a semelparous or iteroparous mode? The answer may lie in part in environmental uncertainty. If survival of juveniles is very poor and unpredictable, then selection favors repeated reproduction and long reproductive life to increase the chance that juveniles will survive in at least some years. We can again think of this as bet-hedging. If the environment is stable, then selection favors a single act of reproduction, because the organism can devote all its energy to making offspring, not to maintaining its own body. Under favorable circumstances, the same per unit biomass of annuals produces more seeds than the same per unit biomass of trees, which have to invest a lot of energy in maintenance. However, when the environment becomes stressful, annuals run the risk of their seeds not germinating. They must rely on some seeds successfully lying dormant and germinating after the environmental stress has ended. A good contrast is agave and yucca plants. Agaves look very much like yuccas but have shallow roots, are sensitive to yearly variation in rainfall, and are semelparous. Yuccas have deep roots, are less sensitive to variation in rainfall because they can tap into deep supplies of ground water, and are iteoparous.

### 10.4.2 *r* and *K* selection represent two different life history strategies

One of the first attempts to categorize life history strategies was made by Robert MacArthur and Edward Wilson (1967), who proposed the idea of *r* and *K* selection. According to McArthur and Wilson, such strategies can be considered a continuum. At one end are species, termed ***r*-selected species**, that have a high rate of per capita population growth, *r,* but poor competitive ability (**Figure 10.14a**). An example is a weed that produces huge numbers of tiny seeds and therefore has a high value of *r.* Weeds exist in disturbed habitats such as gaps in a forest canopy where trees have blown down, allowing light to penetrate to the forest floor. An *r*-selected species like a weed grows quickly and reaches reproductive age early, devoting much energy to producing a large number of seeds that disperse widely. These weed species are small in size, and individuals do not live long. Populations pass a few generations in the light gap before it closes.

At the other end of the continuum are species, termed ***K*-selected species,** that have more or less stable populations adapted to exist at or near the carrying capacity, *K,* of the environment (**Figure 10.14b**). An example is an oak tree that exists in an undisturbed, mature forest. Oak trees grow slowly and reach reproductive age late, having to devote

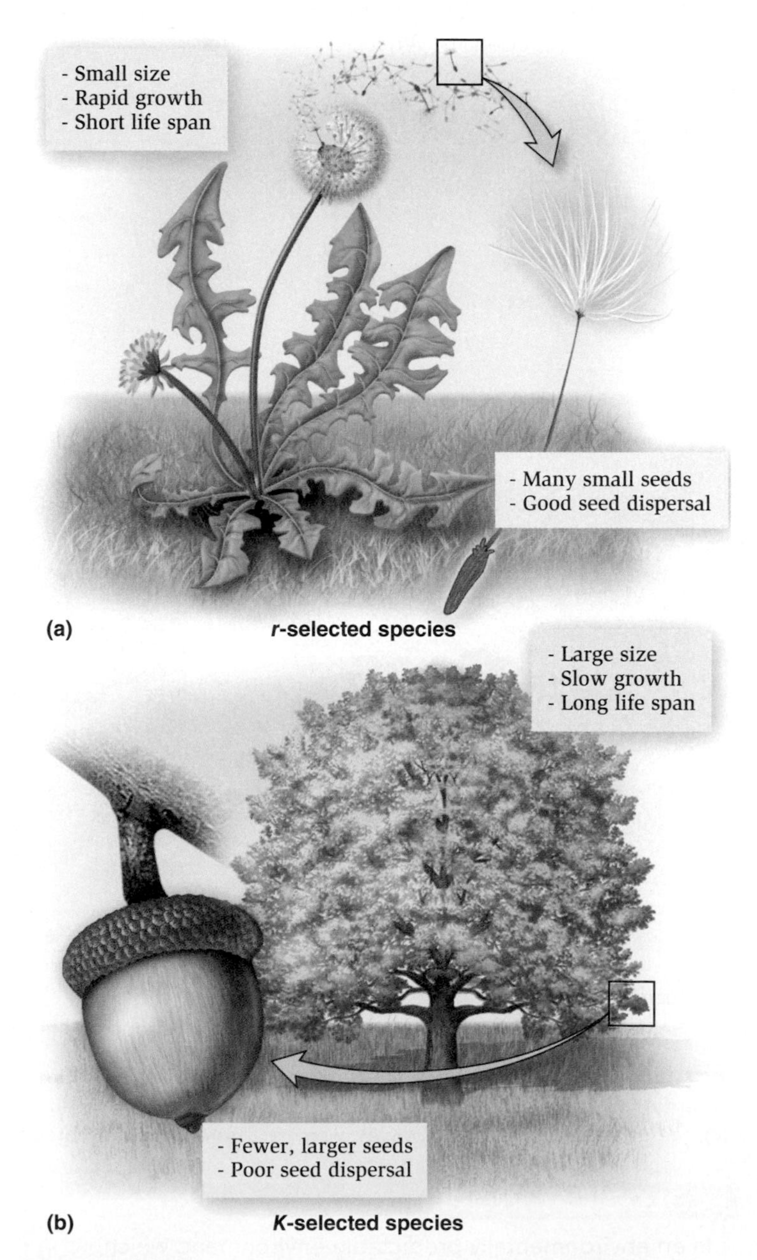


**Figure 10.14 Life history strategies.** Differences in traits of a dandelion **(a)** and an oak tree **(b)** illustrate some of the differences between *r*- and *K*-selected species.

much energy to growth and maintenance. A *K*-selected species like a tree grows large and shades out *r*-selected species like weeds, eventually outcompeting them. Such trees have a long life span and produce seeds repeatedly when mature. These seeds are bigger than those of *r*-selected species. Consider the acorns of oaks versus the seeds of dandelions: acorns contain a large food reserve that helps them grow, whereas dandelion seeds are small and must rely on whatever nutrients they can gather from the soil. Other plants, such as shrubs or herbs, have life histories that are intermediate on the *r*-*K* continuum.

While the weed–tree example is a useful way to think about the *r*- and *K*-selection continuum, other organisms can be *r*- or *K*-selected, too. For example, bacteria and insects can be considered small, *r*-selected species that produce many young and have short life cycles. Mammals, such as elephants, grow slowly, have few offspring, and reach large sizes typical of *K*-selected species. Eric Pianka (1970) noted various other life history features of *r*- and *K*- selected species including reproductive strategy, dispersal ability, growth rates, and population size (**Table 10.1**).

In a human-dominated world, almost every life history attribute of a *K*-selected species sets it at risk of extinction. First, *K*-selected species tend to be bigger, so they need more habitat in which to live. Second, *K*-selected species tend to have fewer offspring and so their populations cannot recover as fast from disturbances like fire or overhunting. California condors, for example, produce only a single chick every other year. Third, *K*-selected species breed at a later age, and their generation time is long. Gestation time in elephants is 22 months, and elephants take at least 7 years to become sexually mature. Large *K*-selected species such as the giant sequoia; terrestrial mammals like elephants, rhinoceros, and grizzly bears; and marine mammals like blue whales and sperm whales all run the risk of extinction.

What are the advantages to being a *K*-selected species? In a world not disturbed by humans, *K*-selected species would fare well. Being a *K*-selected species is as viable an option as being *r*-selected. However, in a human-dominated world, many *K*-selected species are selectively logged or hunted or their habitat is altered, and the resulting small population sizes make extinction a real possibility.

### 10.4.3 Grime's triangle is an alternative to *r* and *K* selection

Alternatives to the *r*- and *K*-selection continuum have been proposed as scientists have suggested that more than two strategies are needed to encompass the variety of life history strategies among different organisms. Plant ecologist Phillip Grime (1977, 1979) proposed a scenario of three strategies: ruderals, competitors, and stress tolerators. Ruderals are adapted to take advantage of habitat disturbance, and they include annual plants adapted to colonizing disturbed areas. Competitors are adapted to live in highly competitive but benign environments and include many tree species. Stress tolerators are adapted to cope with extreme environmental conditions such as high soil salt or temperatures that exist in salt marshes and deserts, respectively. Such species include mangroves at the coast and cacti in arid lands. The triangular representation of this scheme (**Figure 10.15**) implies a trade-off between the three strategies. One cannot be a good stress tolerator and a good competitor at the same time. However, many perennial herbs adopt a mixture of these strategies and tend to lie in the center of Grime's triangle. Animals can be

**Table 10.1 Characteristics of *r*- and *K*-selected species.**

| Life history feature | *r*-selected species | *K*-selected species |
|---|---|---|
| Intrinsic Rate of Increase, $r_{max}$ | High | Low |
| Development | Rapid | Slow |
| Reproductive rate | High | Low |
| Reproductive age | Early | Late |
| Body size | Small | Large |
| Length of life | Short | Long |
| Competitive ability | Weak | Strong |
| Survivorship | High mortality of young (Type III) | Low mortality of young (Type I) |
| Population size | Variable | Fairly constant |
| Dispersal ability | Good | Poor |
| Habitat type | Disturbed | Not disturbed |
| Example | Weedy plants, small fish, insects, bacteria | Canopy trees, large mammals, some parrots, large turtles |

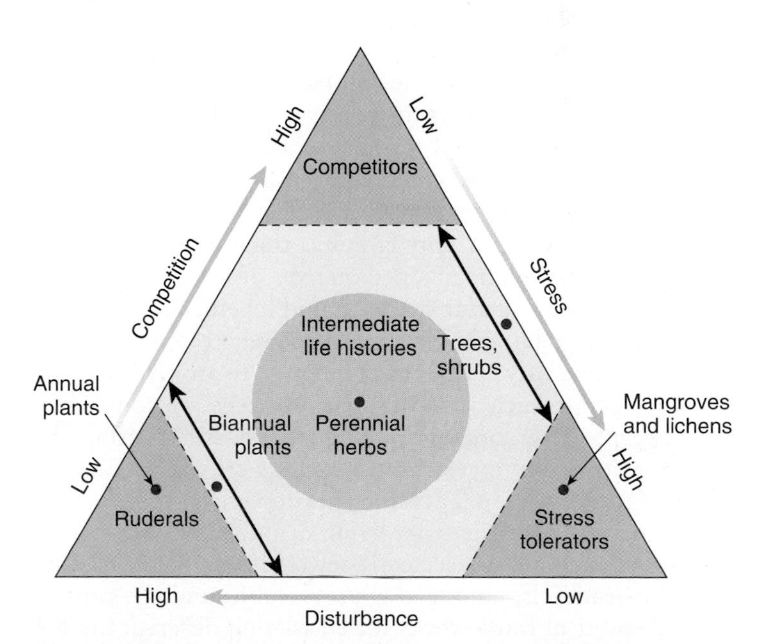


**Figure 10.15 Grime's triangle.** Plant life histories based on a model in which stress, disturbance, and competition are the important selective factors. (Based on data in Grime, 1979.)

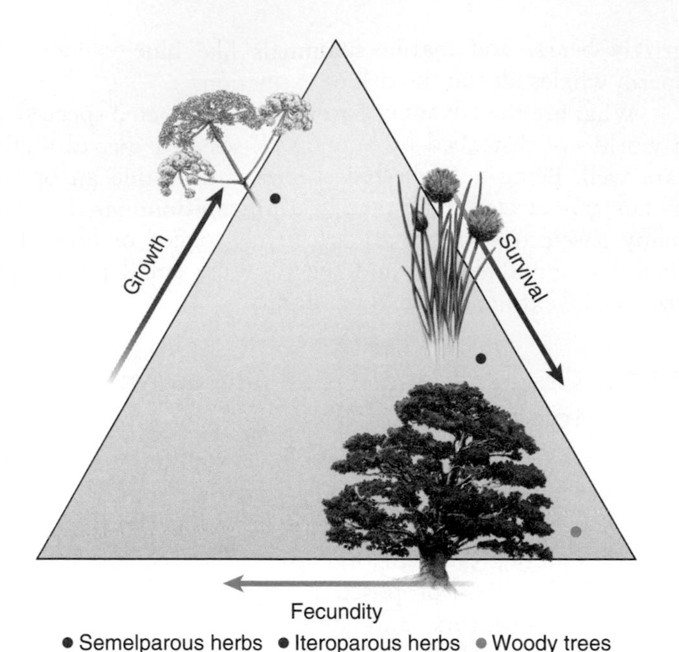


**Figure 10.16** **The growth-survival-fecundity triangle.** Here the distribution of species of perennial plants is mapped in the growth-survival-fecundity triangle.

classified according to this scheme as well with desert animals, as stress tolerators, having burrows to escape extreme conditions; insects that feed on annual plants as ruderals; and large vertebrate carnivores and herbivores in forests or grasslands as competitors.

Jonathan Silvertown and his colleagues (1993) provided a demographic interpretation of Grime's ideas (**Figure 10.16**). They argued that potential reproductive output, called **fecundity**, growth, and survival can be more accurately measured than the position of a species along a relatively ambiguous stress gradient. This allows the relative contributions of fecundity, growth, and survival to the overall reproductive rate, *r*, to be calculated. Species could succeed by producing a lot of seeds, growing large, or surviving a long time as adults, but not all three. Again a demographic trade-off was implied. Individuals can then be mapped on axes of fecundity, growth, and survival, with similar species found together. Semelparous herbs all have relatively high fecundity, whereas woody trees have good longevity and survival but with relatively low levels of reproduction each year. Iteroparous herbs are somewhere in between. Other organisms could be classified using this system, with many invertebrates such as crustaceans having a high fecundity, elephants and giraffes growing to a large size, and turtles and parrots living a long time. These classification systems—*r*-selected/K-selected, competitors/stress tolerators/ruderals, and growth/survival/fecundity—provide good theoretical frameworks for classifying different life history strategies.

### 10.4.4 Population viability analysis uses life history data to predict extinction probability

Based on what we have learned so far in this and other chapters, many factors can influence population size and thus viability, the likelihood that a population may go extinct. Small populations can suffer from inbreeding and effective population sizes can be reduced by harem mating structure. Environmental variation in temperature and rainfall may affect carrying capacities, and many species suffer population reduction from habitat loss or overexploitation. Population sizes at any given point in time are also affected by the age structure of the population and population growth rates. In 1981 Mark Schaffer proposed a technique called population viability analysis (PVA) that provides an estimate of the minimum population size needed to preserve a given species. It is traditionally defined as the process that determines the probability that a population will go extinct within a given number of years. The technique, which we will not detail here, takes into account the effects of chance events on populations, such as droughts and fires, the effects of disease and competitors, and genetic effects from inbreeding, and calculates the minimum number of individuals necessary for a population to persist over long periods of time in the face of these threats. Specifically, Schaffer proposed that a minimum viable population (MVP) for any given species in any habitat is the smallest population having a 99% chance of remaining intact for 1,000 years. Schaffer used computer simulation models to show that the population of grizzly bears in Yellowstone National Park in 1978 had a less than 95% chance of surviving only 100 years, let alone 1,000 years.

David Reed and colleagues (2003) used PVA, using a computer program called Vortex, to estimate the MVPs for 102 vertebrate species, based on actual life history data. The primary variable of interest was the mean population size required for a 99% probability of persistence for 40 generations. The 102 species included 1 fish, 2 amphibians, 18 reptiles, 28 birds, and 53 mammals. The mean MVP size for all species was 7,316, suggesting that, on average, a minimum habitat area capable of supporting approximately 7,000 sexually mature adults is necessary to maintain vertebrate populations in the wild.

#### Check Your Understanding

**10.4** In Grime's classification system, how would you classify the following types of plants: (a) slow growth rate, low level of seed production, dense leaf canopies; (b) small evergreen leaves, low level of seed production, long life spans, intertidal; and (c) relatively small size, rapid growth, existence in disturbed areas, annuals?

## 10.5 Human Populations Continue to Grow

In 2011 the world's population was estimated to be increasing at the rate of 145 people every minute: 2 in developed countries and 143 in less developed countries. The United Nations' 2010 projections pointed to a world population stabilizing at around 10 billion near the year 2150, as would happen with a logistic growth pattern. However, until now human population growth has better fit an exponential growth pattern than a logistic one. In this section, we examine human population growth trends in more detail and discuss how knowledge of a population's age structure can help predict its future growth. We then investigate the carrying capacity of the Earth for humans and explore how the concept of an ecological footprint, which measures human resource use, can help us determine this carrying capacity.

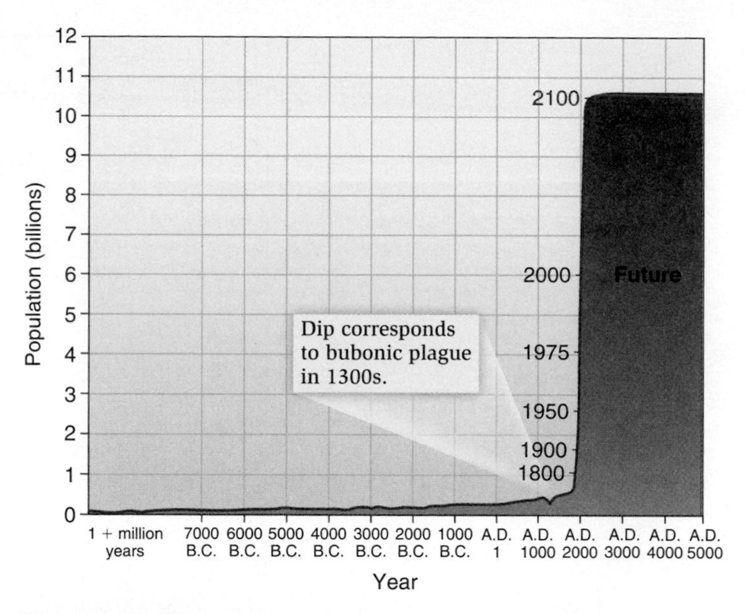


**Figure 10.17 The growth pattern of the world human population.** If and when human population growth will level off are issues of considerable debate. Current UN projections are for leveling off around 10 billion in the year 2100.

### 10.5.1 Human population growth shows extreme recent increases

Until the beginning of agriculture and the domestication of animals, about 10,000 B.C.E. (before the Common Era), the average rate of population growth was very low. With the establishment of agriculture, the world's population grew to about 300 million by 1 C.E. (Common Era) and to 800 million by the year 1750. Between 1750 and 1998, a relatively tiny period of human history, the world's human population surged from 800 million to 6 billion (**Figure 10.17**). In 2011 the number of humans was estimated at 7 billion. Considering this phenomenal increase in growth, the two biggest questions are, when will the human population level off and at what level?

Human populations can exist at equilibrium densities in one of two ways:

1. *High birth and high death rates.* Before 1750 this was often the case, with high birth rates offset by deaths from wars, famines, and epidemics.
2. *Low birth and low death rates.* In Europe beginning in the 18th century, better health and living conditions reduced the death rate. Eventually social changes such as increasing education for women and marriage at a later age reduced the birth rate.

The shift in birth and death rates with development is known as the **demographic transition** (**Figure 10.18**). In the first stage of the transition, birth and death rates are both high, and the population is in equilibrium. In the second stage of this transition, the death rate declines first, while the birth rate remains high. High rates of population growth result. In the third stage, the birth rates drop and death rates stabilize, so population growth rates become lower. In the fourth stage, both birth and death rates are low, and the population is again at equilibrium.

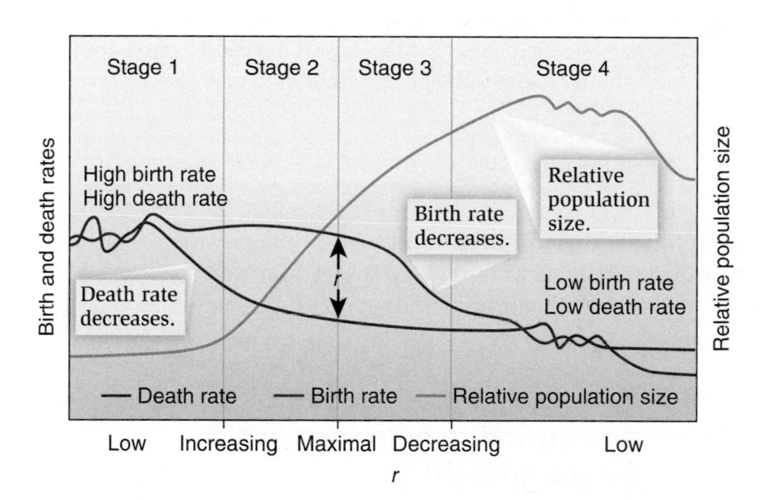


**Figure 10.18 The classic stages of the demographic transition.** The difference between the birth rate and the death rate determines the rate of natural increase or decrease.

The pace of the demographic transition between countries differs, depending on culture, economics, politics, and religion. This is illustrated by examining the demographic transition in Sweden and Mexico (**Figure 10.19**). In Mexico the demographic transition occurred more recently and was typified by a faster decline in the death rate, reflecting rapid improvements in public health. A relatively longer lag occurred between the decline in the death rate and the decline in the birth rate, however, with the result that Mexico's

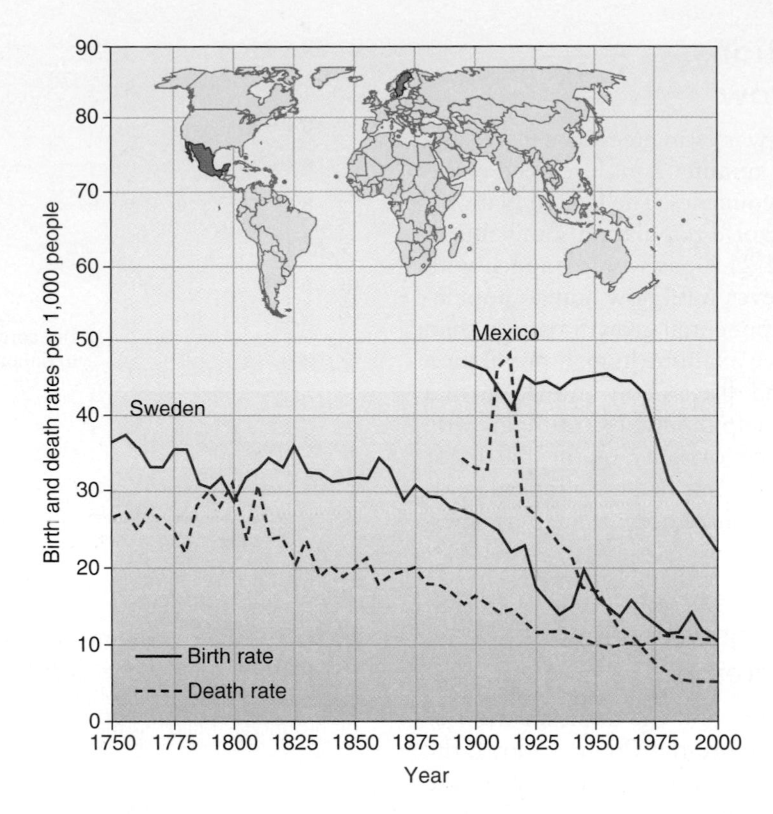


**Figure 10.19 The demographic transition in Sweden and Mexico.** While the transition began earlier in Sweden than it did in Mexico, the transition was more rapid in Mexico and the overall rate of population increase remains higher. (The spike in the death rate in Mexico prior to 1920 is attributed to the turbulence surrounding the Mexican revolution.)

population growth rate is still well above Sweden's, perhaps reflecting differences in culture or the fact that in Mexico the demographic transition is not yet complete.

### 10.5.2 Knowledge of a population's age structure helps predict future growth

Changes in the **age structure** of a population also characterize the demographic transition. In all populations, age structure refers to the relative numbers of individuals of each defined age class. This information is commonly displayed as a population pyramid (**Figure 10.20**). In West Africa, for example, children under the age of 15 make up nearly half of the population, creating a pyramid with a wide base and narrow top. Even if fertility rates decline, there will still be a huge increase in the population as these children move into childbearing age. The age structure of Western Europe is much more balanced. Even if the fertility rate of young women in Western Europe increased to a level higher than that of their mothers, the annual numbers of births would still be relatively low because of the low number of women of childbearing age.

### 10.5.3 Human population fertility rates vary worldwide

Most estimates propose that the human population will grow to around 10 billion people by the middle of the next century. These estimates depend on human **fertility**, the actual reproductive output of women. Global population growth can be examined by looking at **total fertility rates (TFR)**, the average number of live births a woman has during her lifetime if she were to live to the maximum age. It is based on age-specific fertility rates of all women in their childbearing years, 15–49, in a given year (**Figure 10.21**). Note that fertility is an actual reproductive rate and contrasts with fecundity, which is the potential reproductive output under ideal conditions. Thus, the use of fecundity would produce much larger population estimates over time. The global total fertility rate has been declining, from 4.47 in the 1970s to 2.59 in 2007. This is still greater than the average of 2.3 needed for zero population growth. Although one might think that total fertility rates of 2.0 would suffice for zero population growth, some mortality occurs before children reach reproductive age, so in reality an average TFR of 2.3 is needed to ensure population equilibrium. The total fertility rate differs considerably between geographic areas. In Africa, the total

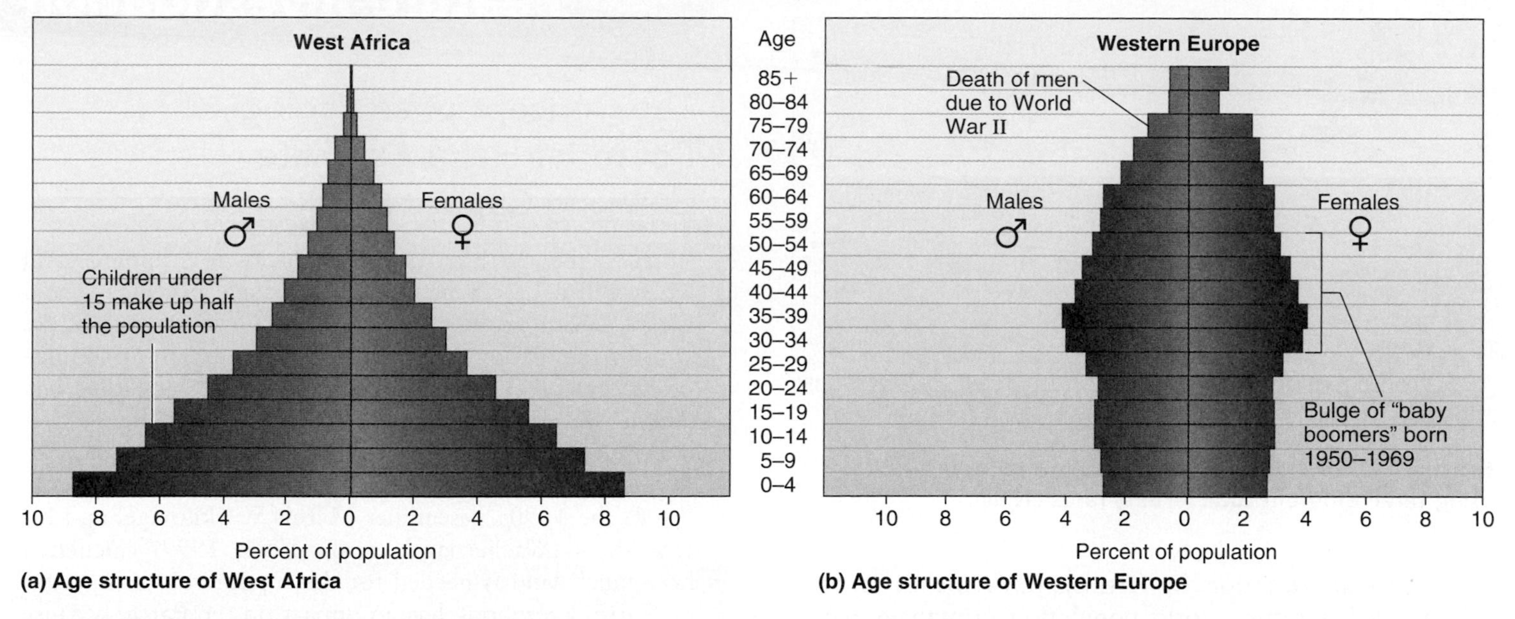


**Figure 10.20 The age structure of human populations in West Africa and Western Europe, as of 2000.** **(a)** In developing areas of the world, such as West Africa, there are far more children than any other age group. **(b)** In the developed countries of Western Europe, the age structure is more evenly distributed. The bulge represents those born in the post–World War II baby boom, when birth rates climbed due to stabilization of political and economic conditions.

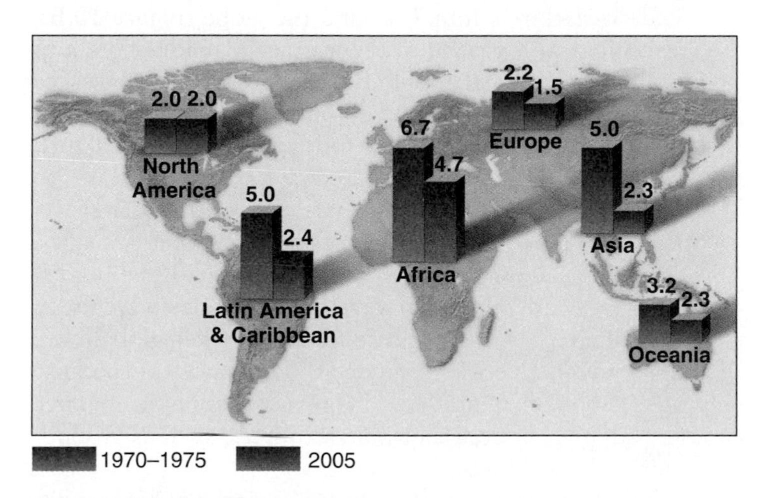


**Figure 10.21 Total fertility rates (TFRs) among major regions of the world.** Data refer to the average number of children born to a woman during her lifetime.

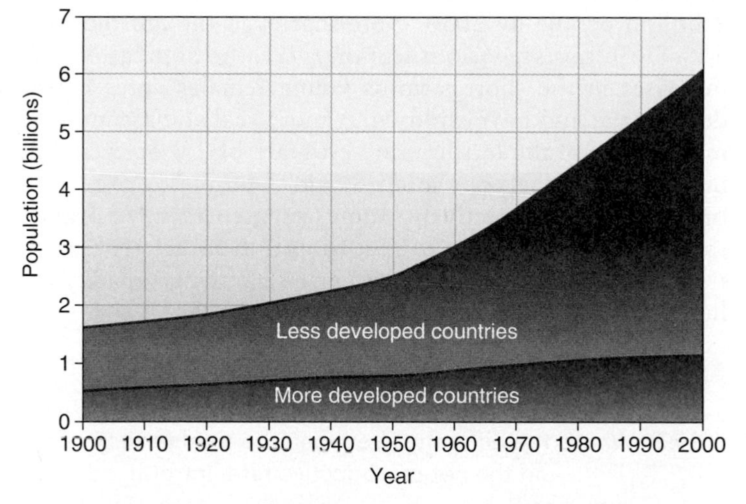


**Figure 10.22 A comparison of the population growth of developed and less developed countries.** Population growth in more developed countries (including North America, Europe, Japan, and Australia) has nearly stabilized at around 1.1 billion people, while that of developing countries continues to skyrocket.

fertility rate of 4.6 in 2010 has declined substantially since the 1970s, when it was around 6.7 children per woman. In Latin America and Southeast Asia, the rates have declined considerably from the 1970s and are now at around 2.3. Canada and most countries in Europe have a TFR of less than 2.0 (it's slightly above that in the U.S.). In Russia, fertility rates have dropped to 1.34. In China, while the TFR is only 1.7, the population there will still continue to increase until at least 2025 because of the large number of women of reproductive age.

In the developed countries, total fertility rates are about 2.1, and the population has nearly stabilized at a little over 1 billion people. In developing countries, population is still increasing dramatically (**Figure 10.22**).

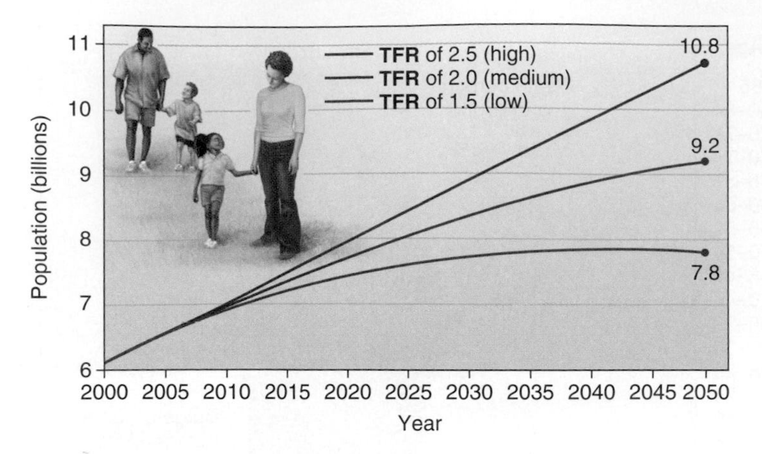


**Figure 10.23** **Population predictions for 2000–2050, using three different total fertility rates (TFRs).**

The wide variation in fertility rates makes it difficult to predict future world population growth. A recent United Nations report shows world population projections to the year 2050 for three different growth scenarios: low, medium, and high (**Figure 10.23**). The three scenarios are based on three different assumptions about fertility rate. Using a low fertility rate estimate of only 1.5 children per woman, the population would reach a maximum of about 8 billion people by 2050. Note that even though the lowest TFR in this scenario is less than 2.1, the population still increases in the short term as young females enter reproductive age and have children. A more realistic assumption may be to use the fertility rate estimate of 2.0 or even 2.5, in which case the population would continue to rise to 10 or 16 billion, respectively. Some researchers have argued that determining the ecological footprint of different countries may help us calculate the Earth's carrying capacity for humans (see **Feature Investigation**).

### Check Your Understanding

**10.5** What's the difference between the total fertility rate (TFR), and the net reproductive rate, introduced earlier as $R_0$?

## Feature Investigation

### The Concept of an Ecological Footprint Helps Estimate Carrying Capacity

What is the Earth's carrying capacity for humans, and when will it be reached? Estimates have been many and varied. Much of the speculation on the future size of the world's population centers on lifestyle. To use a simplistic example, if everyone on the planet ate meat extensively and drove large cars, then the carrying capacity would be a lot lower than if people were vegetarians and used bicycles as their main means of transportation.

In the 1990s, researcher Mathis Wackernagel and his coworkers (Wackernagel et al., 1996, 1999) calculated how much land is needed for the support of each person on Earth. Everybody has an impact on the Earth, because they consume the land's resources, including crops, wood, oil, and other supplies. Thus, each person has an **ecological footprint,** the aggregate total of land needed for survival in a sustainable world. The average footprint size for everyone on the planet is about 3 hectares (1 ha = 10,000 m$^2$), but a wide variation is found around the globe (**Figure 10.B**). The ecological footprint of the average Canadian is about 7.6 hectares versus about 10 hectares for the average American. In most developed countries, the largest component of land is for energy, followed by food and then forestry. Much of the land needed for energy serves to absorb the $CO_2$ emitted by the use of fossil fuels. If everyone required 10 hectares, as the average American does, we would need three Earths to provide us with the needed resources. Many people in developing countries are much more frugal in their use of resources. However, globally we are already in an ecological deficit. This is possible because many people currently live in an unsustainable manner, using supplies of nonrenewable resources, such as groundwater and fossil fuels.

What's your personal ecological footprint? Several different calculations are available on the Internet that you can use to find out. A rapidly growing human population combined with an increasingly large per capita ecological footprint makes many aspects of global change inevitable, especially habitat loss and pollution.

**Figure 10.B** **Calculating ecological footprints of each nation.**

**ECOLOGICAL INQUIRY**

What is your ecological footprint?

**HYPOTHESIS** To determine the ecological footprint of each nation.

| | Conceptual level | Experimental level |
|---|---|---|
| 1 | Determine how much land area is needed for each commodity for each country. | To keep things simple we will consider food, forest products (firewood, sawn wood, and paper), fossil energy, and urban areas. For Italy, for vegetable and fruit production, Consumption = Production + Imports – Exports. Consumption = 33,323,000 + 2,613,956 – 5,185,679 = 30,751,277 tons. |
| 2 | Determine the amount of land necessary to produce this food based on the global average. | In this case yield = 18t/ha. Land needed = Consumption / Average Yield/ha = 30,751,277 / 18 = 1,708,404 ha. |
| 3 | Determine the per capita footprint of vegetables and fruit by dividing by the size of the human population. | 1,708,404 / 57,127,000 = 0.0299 ha per person. |
| 4 | Repeat for all products. | Food = 2.1 ha per person, Forest = 0.3 ha per person, Fossil energy = 1.6 ha per person (amount of land needed to absorb all the $CO_2$ produced). Urban area = 0.2 ha per person. |
| 5 | Determine the total per capita footprint by summing the per capita footprints for all commodities. | 4.2 ha per person for Italy. |

**6 THE DATA**

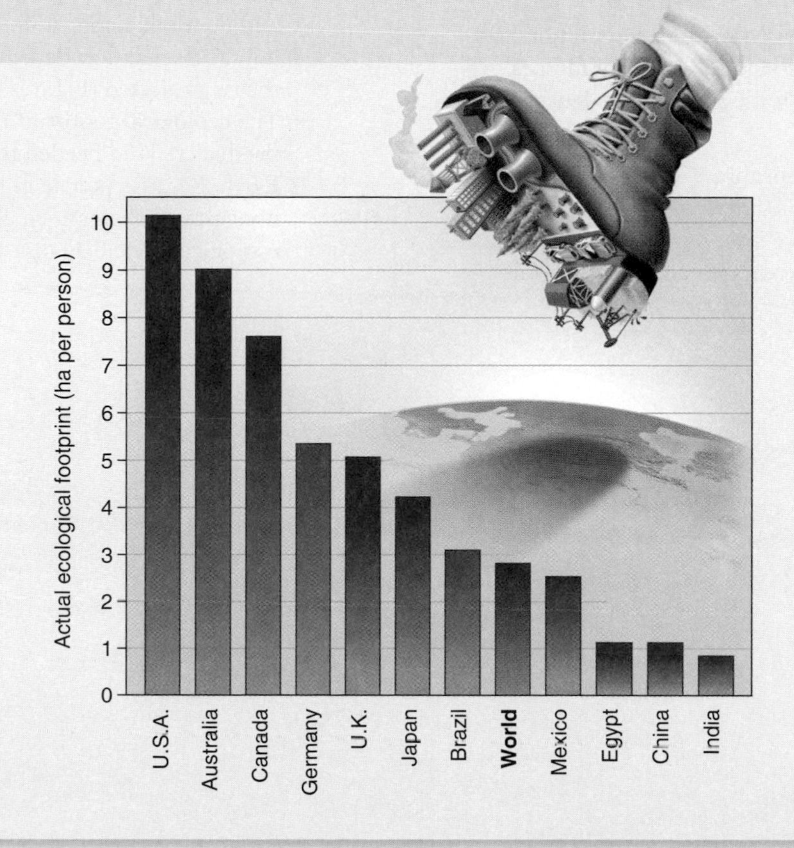

## SUMMARY

**10.1 Unlimited Population Growth Leads to J-Shaped Growth Curves**

- The net reproductive rate ($R_0$) determines how populations that breed at discrete time intervals grow over time. Geometric (J-shaped) growth occurs where $R_0 > 1$ (Figures 10.1, 10.2).
- The per capita growth rate ($r$) helps determine how continuously breeding populations grow over any time period. Per capita growth rates decrease as generation times increase (Figure 10.3).
- The population growth rate of many ectotherms will increase in a globally warmed world (Figure 10.A).
- When $r > 0$, exponential (J-shaped) growth occurs. Exponential growth can be observed in an environment where resources are not limited (Figures 10.4, 10.5).

**10.2 Limited Resources Lead to S-Shaped Population Growth Curves**

- Logistic (S-shaped) growth takes into account the upper boundary for a population, called carrying capacity, and occurs in an environment where resources are limited (Figures 10.6, 10.7).
- Variations in temperature, rainfall, or resource quantity or quality cause changes in carrying capacity and population growth patterns.
- A time lag in the logistic growth model can also cause large variations in population growth resulting in damped oscillations and stable limit cycles (Figure 10.8).

**10.3 Density-Dependent Factors May Limit Population Size**

- Density-dependent factors are mortality factors whose influence varies with population density (Figures 10.9–10.12).

**10.4 Life History Strategies Incorporate Traits Relating to Survival and Competitive Ability**

- Life history strategies are a set of features including reproductive traits, survivorship and length-of-life characteristics, habitat type, and competitive ability (Figure 10.13).
- Life history strategies can be viewed as a continuum, with *r*-selected species (those with a high rate of population growth but poor competitive ability) at one end and *K*-selected species (those with a lower rate of population growth but better competitive ability) at the other (Figure 10.14, Table 10.1).
- Other life history schemes (Grime's triangle and others) attempt to classify life history strategies as either ruderals, competitors, or stress tolerators or as species that tend to emphasize fecundity, growth, or survival (Figures 10.15, 10.16).

**10.5 Human Population Growth**

- Up to the present, human population growth has better fit an exponential growth pattern than a logistic one (Figure 10.17).
- Human populations in industrializing countries have been moving from states of high birth and death rates to low birth and death rates, a shift called the demographic transition (Figures 10.18, 10.19).
- Differences in the age structure of a population, the numbers of individuals in each age group, are also characteristic of the demographic transition (Figure 10.20).
- Though they have been declining worldwide, total fertility rates (TFRs) differ markedly in developing and developed countries. Predicting the growth of the human population depends on the total fertility rate that is projected (Figures 10.21–10.23).
- The ecological footprint refers to the amount of productive land needed to support each person on Earth. Because people in many countries live in a nonsustainable manner, globally we are already in an ecological deficit (Figure 10.B).

## TEST YOURSELF

1. If $R_0 = 1.1$ and $N_t = 1{,}000$, what is $N_{t+1}$?
   a. 1,100
   b. 1,210
   c. 1,331
   d. 1,464
   e. 1,610
2. The maximum number of individuals a certain area can sustain is known as:
   a. The intrinsic rate of growth
   b. The resource limit
   c. The carrying capacity
   d. The logistic equation
   e. The equilibrium size
3. Which of the following factors may change carrying capacity over time?
   a. Weather pattern changes
   b. Numbers of other species that are present in the habitat
   c. Deforestation
   d. Soil chemistry changes
   e. All of the above
4. The maximum rate of population growth for a continuously breeding population is known as:
   a. Exponential growth
   b. Logistic growth
   c. Geometric growth
   d. Time lag growth
   e. Intrinsic rate of increase
5. If $r = 0.1$, $K = 100$, and $N = 50$, what is $dN/dt$?
   a. 5
   b. 10
   c. 2.5
   d. 1
   e. 100
6. Age structure refers to:
   a. The number of fertile females in a population
   b. The relative number of individuals in each age group
   c. The relative number of females
   d. The carrying capacity of the Earth for humans
   e. The average age of women when they bear children
7. The current human population of the Earth is approximately:
   a. 7 million
   b. 70 million
   c. 1 billion
   d. 7 billion
   e. 100 billion
8. The amount of land necessary for survival for each person in a sustainable world is known as:
   a. The sustainability level
   b. The ecological impact
   c. The ecological footprint
   d. Survival needs
   e. All of the above
9. If $b = 0.55$, $d = 0.45$, and $N = 1{,}000$, then $dN/dt =$
   a. 1,222.2
   b. 100
   c. 5,500
   d. 4,500
   e. 1,000
10. What is the minimum condition needed to prevent a continuously breeding population from going extinct?
    a. $r < 0$
    b. $r = 0$
    c. $r > 0$
    d. $r = 1$
    e. $r < 1$

## CONCEPTUAL QUESTIONS

1. Define density-dependent and density-independent factors.
2. Differentiate between *r*- and *K*-selected organisms and between ruderals, competitors, and stress tolerators.
3. Discuss the differences and similarities between geometric and exponential growth.
4. Describe the effects of time lags on population growth.
5. Discuss the concept of demographic transition.

## DATA ANALYSIS

1. You survey an annually breeding butterfly population and census 1,000 females per hectare. The next year you census 1,200 females. What is the finite rate of increase, $\lambda$? What will the population be 5 years from now if the rate of increase is the same each year? When will the population reach 10,000?
2. American holly trees, *Ilex opaca,* are important landscaping trees but are attacked by leafmining flies, *Phytomyza ilicicola,* whose larvae form blotched-like mines in the leaves. *Opius striativentris* is the predominant parasite of the leafminer in Kentucky. Plot the percent leaves mined and the percent parasitism data from the 20 trees given in the table to see if there is a pattern of density-dependent parasitism.

| Tree number | Percent leaves mined | Percent parasitism |
|---|---|---|
| 1 | 84.7 | 14.1 |
| 2 | 94.3 | 25.5 |
| 3 | 62.5 | 8.2 |
| 4 | 64.9 | 8.8 |
| 5 | 90.9 | 29.8 |
| 6 | 34.0 | 7.4 |
| 7 | 42.0 | 4.1 |
| 8 | 27.6 | 40.7 |
| 9 | 19.1 | 40.5 |
| 10 | 28.4 | 26.6 |
| 11 | 13.7 | 28.6 |
| 12 | 75.8 | 5.1 |
| 13 | 90.9 | 2.6 |
| 14 | 92.6 | 54.7 |
| 15 | 2.3 | 57.9 |
| 16 | 49.0 | 8.3 |
| 17 | 45.5 | 37.5 |
| 18 | 68.5 | 42.3 |
| 19 | 92.6 | 4.6 |
| 20 | 82.0 | 5.4 |

3. At Anglesey, an island off the north coast of Wales, grazing by introduced rabbits prevented trees from colonizing many areas. When rabbits were reduced by the introduction of the myxoma virus in 1954, willow trees began to recolonize. The following table gives the approximate number of trees in a 4-ha plot over a 30-year period. Plot the data and determine what type of growth was exhibited.

| Year | Number of willow trees |
|---|---|
| 1966 | 0 |
| 1968 | 0 |
| 1970 | 10 |
| 1972 | 130 |
| 1974 | 250 |
| 1976 | 350 |
| 1978 | 425 |
| 1980 | 450 |
| 1982 | 520 |
| 1984 | 520 |

SECTION FOUR

# Species Interactions

In this section we turn from considering populations on their own to investigating how they interact with populations of other species that live in the same locality. As we can see from the table below, species interactions can take a variety of forms (**Table IV.1**). **Competition** is an interaction that affects both species negatively (−/−), as both species compete over food or other resources. Sometimes this interaction is quite one-sided, where it is detrimental to one species but not to the other (−/0), an interaction called **amensalism**. Natural enemies all have a positive effect on one species and a negative effect on the other (+/−). However, while predators always kill their prey, the hosts of parasites and herbivores often survive their attacks. **Facilitation** denotes species interactions that result in benefits to at least one of the species involved, and often to both species. Facilitation can be split into two categories. The first, **mutualism**, is an interaction in which both species benefit (+/+), while the second, **commensalism**, benefits one species and leaves the other unaffected (+/0). Last is the interaction, or rather lack of interaction, termed **neutralism**, when two species occur together but do not interact in any measurable way (0/0). Neutralism may be quite common, but few people have quantified its occurrence.

**Table IV.1** **Summary of the types of species interactions. (+ = positive effect; 0 = no effect; − = negative effect)**

| Nature of Interaction | Population of species 1 | Population of species 2 |
|---|---|---|
| Competition | − | − |
| Amensalism | − | 0 |
| Predation, Herbivory, Parasitism | + | − |
| Mutualism | + | + |
| Commensalism | + | 0 |
| Neutralism | 0 | 0 |

To illustrate how species interact in nature, let's consider a rabbit population in a woodland community (**Figure IV.1**). To determine what factors influence the size and density of the rabbit population, we need to understand the range of its possible species interactions. For example, the rabbit population could be limited by the quality of available food. It is also likely that other species, such as deer, use the same resource and thus compete with the rabbits for food. The rabbit population could be limited by predation from foxes or by parasitism in the form of the virus that causes the disease myxomatosis. It is also possible that other associations, such as mutualism or commensalism, may occur. This section examines each of these types of species interactions in turn, beginning with competition, the most studied interaction among species. We conclude with discussions of conceptual models of species interactions and analytical techniques that ecologists use when trying to determine which factors are most important in influencing the population densities within ecological systems.

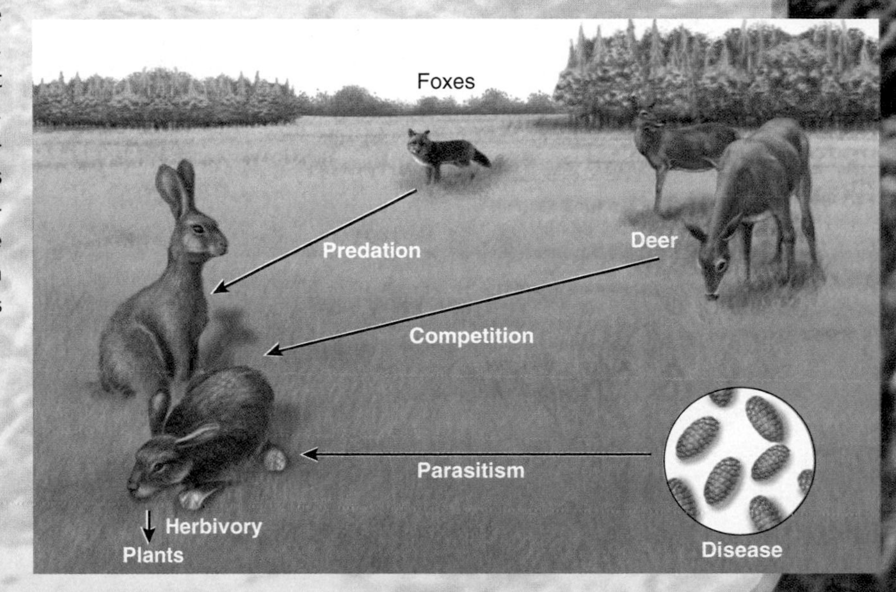


**Figure IV.1** **Species interactions.** These rabbits can interact with a variety of species, experiencing predation by foxes, competition with deer for food, and parasitism from various disease-causing organisms.

Red squirrel, *Sciurus vulgaris*.

CHAPTER 11

# Competition and Coexistence

## Outline and Concepts



The eastern gray squirrel, *Sciurus carolinensis*, is native to the eastern and midwestern United States and southern and eastern Canada. It has been transported to several different areas of the world, including the western United States, Britain, Ireland, Italy, and South Africa. In Britain and Ireland, populations of the native red squirrel, *S. vulgaris,* have declined dramatically as the gray squirrel has spread. Competition between the two squirrel species has been implicated as the causative reason. Gray squirrels are larger and stronger than red squirrels. However, there could be other reasons for the decline of the red squirrels. As in any field of scientific inquiry, it is valuable to examine all the possibilities.

First, red squirrels are less tolerant of habitat destruction and fragmentation than are gray squirrels, so recent development may have impacted red squirrels. However, an overall loss of woodland should not have caused gray squirrels to expand their range and increase their population densities. Second, squirrelpox virus is carried by British gray squirrels. Grays rarely die from the disease, but they may spread it to reds, whose mortality rate from this disease appears to be 100% within 4 to 5 days after being infected. However, there is no evidence of squirrelpox virus in Italy, where gray squirrels are also expanding their range and replacing red squirrels. There is evidence that gray squirrels are more efficient at exploiting acorns than red squirrels, because they are better able to neutralize the defensive chemicals. Gray squirrel populations increase with increased acorn crops, whereas reds do not. However, the replacement of reds by grays still occurs in conifer forests.

Ultimately, reduction of red squirrels' food reserves by grays may be to blame. Gray squirrels often pilfer red squirrels' food caches, causing a reduced energy intake in red squirrels, a decrease in body mass, and a decrease in breeding success. Red squirrels are now thinly distributed in Britain, usually in areas where grays are absent: the Isle of Wight, Scotland, and northern England. However, restoration may increase the population of red squirrels. The removal of gray squirrels on the Island of Anglesey has allowed the red squirrels to breed.

In this chapter we will see how ecologists have studied different types of competition and how they have shown that the competitive effects of one species on another can change as the environment changes or as different predators or parasites are present. Next, we discuss why competition is an important phenomenon to take into consideration when studying the effects of introduced species. Globally there are thousands of species of weeds that have originated from other countries and are outcompeting native vegetation or interfering with agricultural crops. Many animal species have also been introduced and compete with native fauna. By studying such interactions we gain insight into when and where competition is most likely to cause environmental problems. Later in the chapter we describe various mathematical models that have been used to predict the outcomes of competitive interactions. Finally, although species may compete, we will also learn how sufficient differences in lifestyle or morphology can reduce the overlap in their ecological niches, thus allowing them to coexist. Although we discussed the concept of the niche in Chapter 5 in terms of an optimal range of a species, Charles Elton (1927) used the word "niche" to denote an organism's ecological role in the community. Thus, he defined an organism's **niche** by its diet and physical distribution pattern. Elton (1958) later wrote, "When an ecologist says, 'There goes a badger,' he should include in his thoughts some definitive idea of the animal's place [niche] in the community to which it belongs, just as if he had said, 'There goes the vicar.'" (A vicar is the English equivalent of a clergyman.) Elton meant that we should know how an organism makes its living, as well as where it occurs, in order to predict which other organisms it may compete with. If two organisms live in the same habitat but don't overlap in resource use, meaning they eat different things, they are less likely to compete.

## 11.1 Several Different Types of Competition Occur in Nature

Several different types of competition are found in nature (**Figure 11.1**). Competition may be **intraspecific**, between individuals of the same species, or **interspecific**, between individuals of different species.

Competition can also be characterized as exploitation competition or as interference competition. In **exploitation competition**, organisms compete indirectly through the consumption of a limited resource, with each obtaining as much as it can, as when plants compete for water or nutrients. In animals, exploitation competition occurs in a similar fashion, such as when fly maggots compete for food in a mouse carcass—often not all the individuals can command enough of the resource to survive and only a few become adult flies. When herbivorous insects cooccur on a host plant, they may compete over resources such as leaves (see Figure 11.1). In **interference competition**, individuals interact directly with one another by physical force or intimidation. Often this force is ritualized into aggressive behavior associated with territoriality. In these cases, strong individuals survive and take the best territory, and weaker ones perish or at best survive under suboptimal conditions. Interference competition occurs most commonly in animals. Both intraspecific and interspecific competition may be caused by exploitation competition, interference competition, or both.

Competition between species is not always equal. In some cases, one species has a strong effect on another, but the reverse effect is negligible. Thus, some interactions involving competition are a −/0 relationship, called **amensalism**, rather than a −/− relationship. Such asymmetric competition can be observed between plants, where one species produces and secretes from its roots chemicals that inhibit the growth of

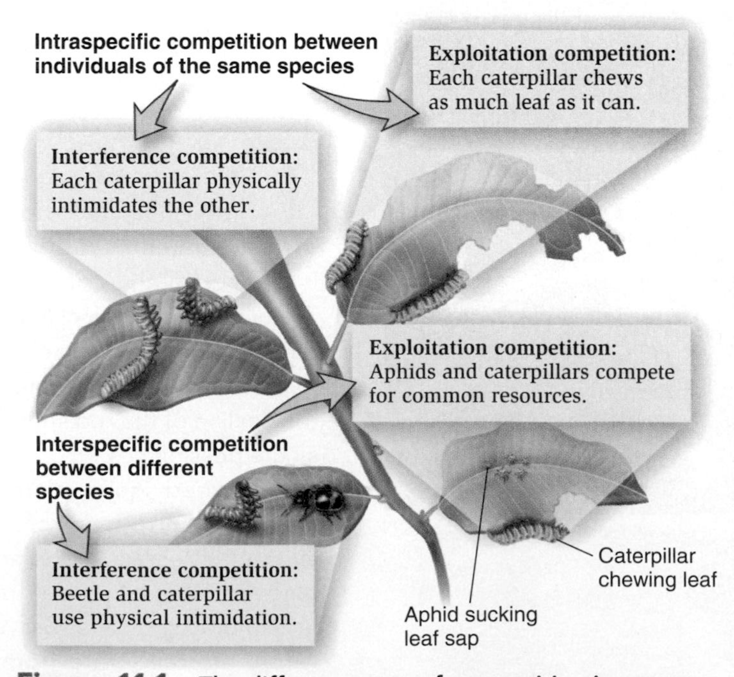


**Figure 11.1** The different types of competition in nature.

another species. For example, black walnut, *Juglans nigra,* produces from its roots a chemical called juglone, which kills roots of neighboring plants. This phenomenon is termed **allelopathy**, and the chemicals are known as allelochemicals. The action of penicillin, derived from a fungus, which directly inhibits the growth of bacteria and other fungi, is a classic case of allelopathy. As we noted in Chapter 1, some invasive plant species have been shown to secrete allelochemicals and this negatively impacts native plants and increases the spread of the invasives. In **apparent competition**, two species do not necessarily compete for the same resource but they do share at least one natural enemy. This often leads to the exclusion of the more susceptible of the species. We alluded to this phenomenon at the beginning of the chapter when we mentioned how red squirrels are very susceptible to squirrelpox virus, while gray squirrels are not. In North America, the usual host of the meningeal worm, *Parelaphostrongylus tenuis,* is the white-tailed deer, *Odocoileus virginanus,* which is tolerant to the infection. Moose, *Alces alces,* however, are potential hosts, and in this species the worm causes severe neurological damage, even when very small numbers of the nematode are present in the brain. The deleterious effects include direct mortality and reduced resistance to other disease. This differential pathogenicity of *P. tenuis* makes the white-tailed deer a potential competitor with moose. In Maine and Nova Scotia, moose are not usually found in areas frequented by white-tailed deer.

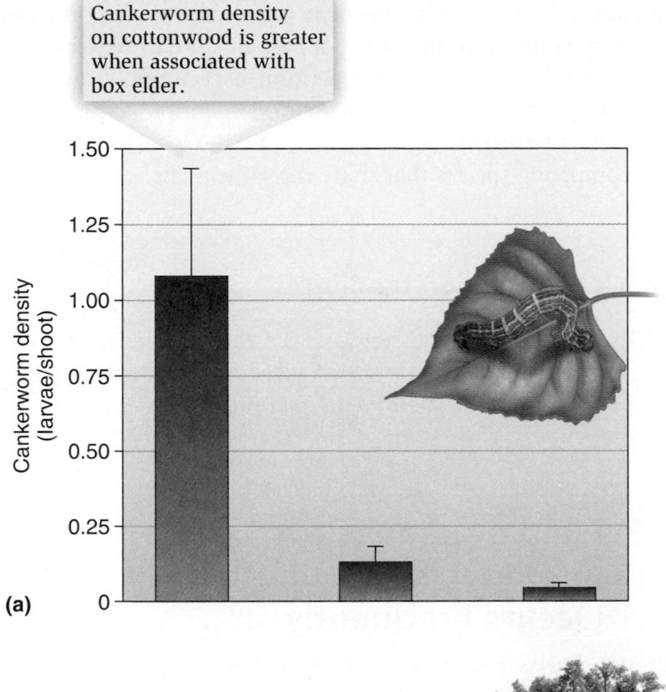


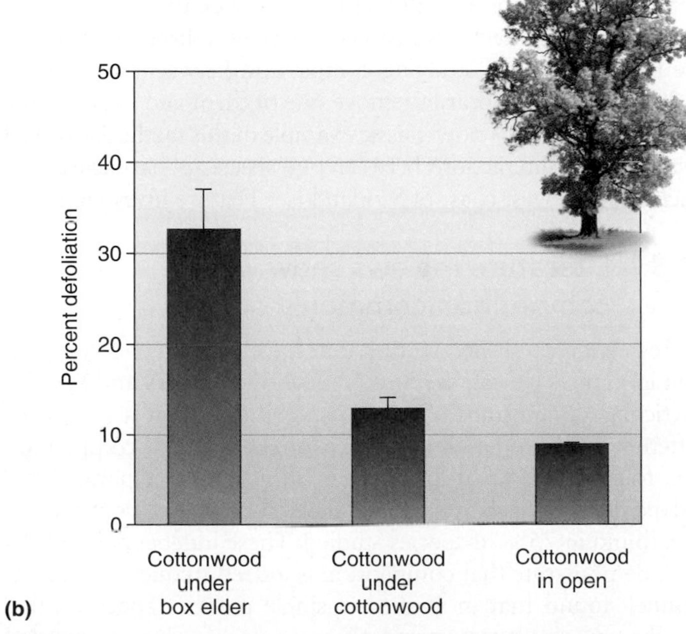


**Figure 11.2 Associational susceptibility.** The number of cankerworms on cottonwood plants **(a)** and the damage done to leaves **(b)** both increase when cottonwoods occur under box elder trees. (Modified from White and Whitham, 2000.)

Not all cases of apparent competition result in the complete elimination of one of the host species. **Associational susceptibility** occurs in plants where herbivores spill over from one species onto another. Jennifer White and Thomas Whitham (2000) demonstrated associational susceptibility between cottonwood, *Populus* sp., and box elder trees, *Acer negundo.* In northern Utah, the caterpillars of the moth *Alsophila pometaria,* known as fall cankerworms, prefer to feed on box elder trees and are rarely found on isolated cottonwood trees. However, when cottonwoods occur under box elder, the cankerworms spill over and cause increased defoliation of the cottonwoods, though without actually killing them (**Figure 11.2**). In the next section we will see how one researcher documented interspecific exploitative competition and apparent competition mediated by a parasite in the same laboratory system.

**Check Your Understanding**

**11.1** How should we classify the competition between the red and gray squirrels outlined at the beginning of the chapter?

## 11.2 The Outcome of Competition Can Vary with Changes in the Biotic and Abiotic Environments

Using experiments to temporarily add or remove individuals of one species and examining the results on the other species is often the most direct method to investigate the effects of competition. In the late 1940s, biologist Thomas Park (1948, 1954) began a series of experiments examining competition between two flour beetles, *Tribolium castaneum* and *Tribolium confusum,* in which he cultured both species together and systematically varied temperature and moisture. These beetles were well suited to study in the laboratory because large colonies could be grown in relatively small containers that contain finely sieved whole meal flour. Thus, many replicates of each experiment were possible to confirm that results were consistent.

Park conducted the experiments by putting the same number of beetles of both species into a container and counting the number of each type that were still alive after a given time interval. *T. confusum* usually won, but in initial experiments, unknown to Park, the beetle cultures were infested with a protozoan parasite called *Adelina triboli* that killed some beetles and preferentially killed *T. castaneum* individuals. In these experiments, *T. confusum* won in 66 out of 74 replicates (89%) because it was more resistant to the parasite. In subsequent experiments the parasite was removed and *T. castaneum* won in 12 out of 18 replicates (67%). Two things were evident from these experiments. First, the presence of a parasite (a biotic factor) was shown to alter the outcome of competition, an example of apparent competition. Second, with or without the parasite, there was no absolute victor. For example, even with the parasite present, sometimes *T. castaneum* won. Thus, some random variation, which we call stochasticity, was evident.

Park then began to vary temperature and moisture (abiotic factors) and found that competitive ability was greatly influenced by climate (**Figure 11.3**). Generally, *T. confusum* did better in dry conditions, and *T. castaneum* in wet conditions. However, *T. confusum* also won in cold wet conditions. Once again, some stochasticity occurred, and victory was not always absolute. Later it was found that the mechanism of competition was largely predation of eggs and pupae by larvae and adults, which, as they ate the flour medium, would also bite the stationary eggs and pupae, killing them. In general the species were mutually antagonistic, that is, they bit more eggs and pupae of the other species than they did of their own.

Park's important series of experiments illustrated that the results of competition could vary as a function of at least three factors: parasites, temperature, and moisture. The experiments also showed how much stochasticity occurred, even in controlled laboratory conditions. But what of systems in nature, where far more variability exists? Is competition a common occurrence or not? Experiments often show that competition is frequent in nature and a strong enough force to cause the local extinction of some competing species that share the same niche.

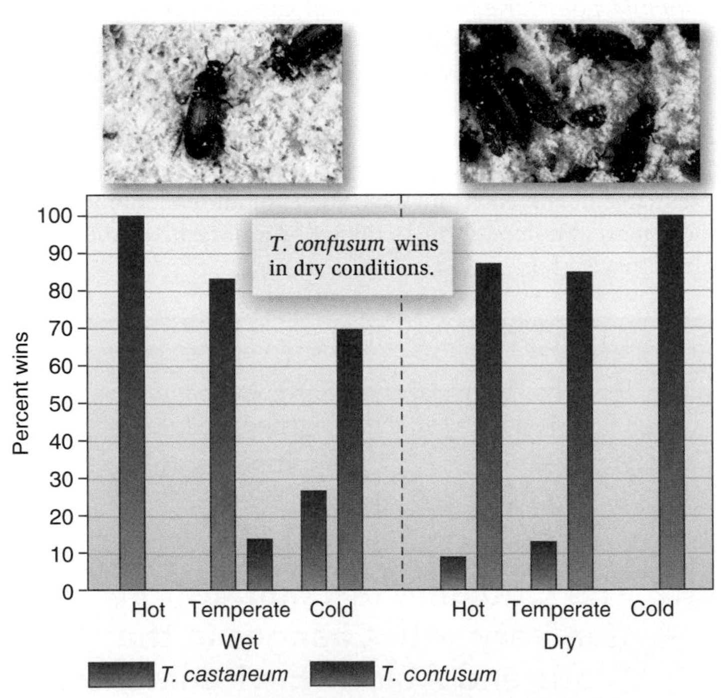


**Figure 11.3 Competitive ability can be influenced by the abiotic environment.** Results of competition between the flour beetles *Tribolium castaneum* and *Tribolium confusum* show that each species usually performs better in a given habitat; for example, *T. confusum* does better in dry conditions.

**ECOLOGICAL INQUIRY**

How would you classify competition between these two species?

**Check Your Understanding**

**11.2** What could cause some of the random variation that we see in the competition experiments between *Tribolium castaneum* and *T. confusum*?

## 11.3 Field Studies Show That Interspecific Competition Occurs Frequently

Although Park's experiments showed that competition was an important source of mortality in laboratory cultures, is it important in natural systems? Researchers have established that one of the best methods of studying competition between two species in nature is to temporarily remove one of them and examine the effect on the other. A now-classic example of this method involved a study of the interactions between two species of barnacles, conducted on the west coast of Scotland (see **Feature Investigation**).

### 11.3.1 Literature reviews show where competition commonly occurs

By reviewing published studies that have investigated competition in nature, we can see how frequently it occurs and in what particular circumstances it is most important. In a review of different field studies by Joseph Connell (1983), competition was found in 55% of 215 species surveyed. In a parallel but independent review by Thomas Schoener (1983), competition was found in 75% of species studied. These independent results both demonstrate that competition is indeed frequent in nature. Connell found that in studies of single pairs of species utilizing the same resource, competition is almost always reported (90%), whereas in studies involving more than two species, the frequency of competition drops to 50%. Why should this be the case? Imagine a resource such as a series of different-sized grains, with four species—ants, beetles, mice, and birds—feeding on them (**Figure 11.4a**). The ants feed on the smallest grain, the beetles and mice on the intermediate sizes, and the birds, on the largest. If only adjacent species compete with each other, competition would be expected only between the ant-beetle, beetle-mouse, and mouse-bird. Thus, competition would be found in only three out of the six possible species pairs, which means competition operates in 50% of the cases. Naturally, the percent of cases where competition is expected would vary according to the number of species on the resource axis. If there were only three species along the axis, we would expect competition in two of the three pairs (67%). If there were just two species utilizing the resource axis, we would expect competition in almost 100% of the cases (**Figure 11.4b**).

Some other general patterns were evident from Connell's and Schoener's reviews. Plants showed a high degree of competition, perhaps because they are rooted in the ground and cannot easily escape or perhaps because they are competing for the same set of limiting nutrients—water, light, and minerals. Marine organisms tended to compete more than terrestrial ones, perhaps because many of the species studied lived in the intertidal zone and were attached to the rock face, in a manner similar to that of plants. Because the area of the rock face is limited, competition for space is quite important.

Other ecologists have suggested that competition is more frequent in a **guild**, a group of species that feed on the same resource and in the same way. The term was originally coined by Dick Root (1967) in his studies of birds. Root recognized the leaf-gleaning guild, a group of birds that feed by removing insects from leaf surfaces, and other guilds that feed in different ways. For example, the ground-feeding guild includes species that forage for worms and other invertebrates on the ground. In an insect community feeding on a plant, we may have the leaf-chewing guild, the stem-feeding guild, and the root-feeding guild, as well as the flower-feeding guild. For example, the stem-boring guild of salt marsh cordgrass, *Spartina alterniflora,* consists of five insect species that feed by boring into the stems and eating stem tissue (**Figure 11.5**). Four species bore from the top of the plant downward and one feeds from the bottom upward. The narrow confines of the stem also means that these species frequently encounter each other. Competition would be expected in such circumstances. Peter Stiling and Donald Strong (1983) reported frequent competition in this guild as species slashed at each other with sharp mandibles during the course of feeding, often resulting in death.

Medical studies have also shown competition to occur frequently between bacteria, especially in the human intestinal system. The human body supports trillions of bacteria from about 300–1,000 different species in the digestive system. Here bacteria perform a multitude of functions, such as digestion of unutilized energy substrates, fermentation, and production of vitamins. They also competitively exclude harmful pathogenic bacteria such as *Clostridium difficile,* which can cause colitis, from adhering to the mucosal lining of the intestine. The process of fermentation, which produces lactic acid and lowers the colon pH, also helps reduce harmful species. These "barrier effects" may be reduced by antibiotic use if helpful gut "flora" are replaced by harmful species such as *C. difficile.* An emerging treatment is to restore beneficial species by fecal microbiota transplantation via donor feces.

### 11.3.2 Invasive species may outcompete native species

Worldwide, humans have succeeded in establishing thousands of species of non-native plants and invertebrates and hundreds of species of vertebrates. In some cases these were deliberate, well-intentioned introductions for economic gain. In others, introductions were accidental, with seeds or insects arriving in soil loaded onto ships for ballast. In the past 30 years, seawater has been

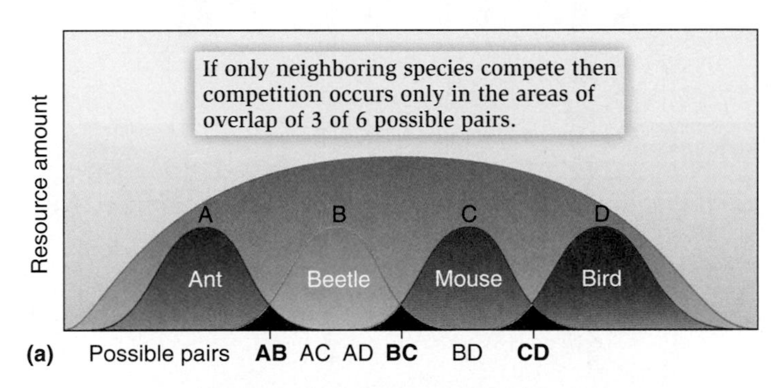


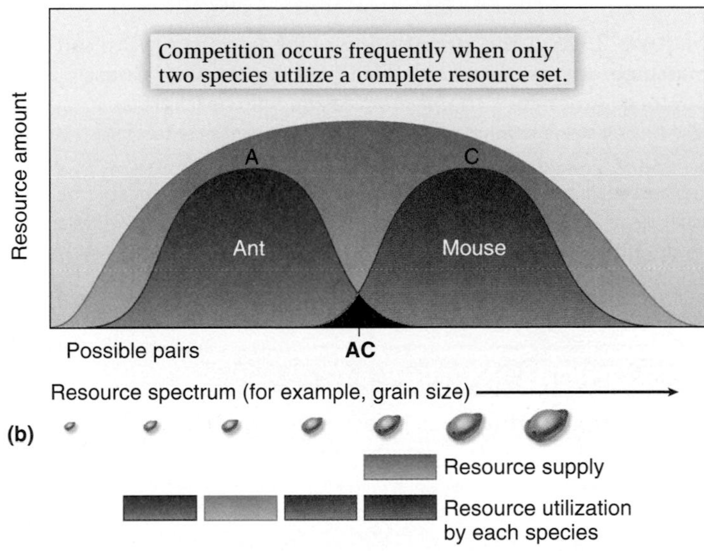


**Figure 11.4 The frequency of competition may vary according to the number of species involved.** **(a)** Resource supply and utilization curves of four species, A, B, C, and D, along the spectrum of a hypothetical resource such as grain size. If competition occurs only between species with adjacent resource utilization curves, competition would be expected between three of the six possible pairings: A and B; B and C; and C and D. **(b)** When only two species utilize a resource set, competition would nearly always be expected between them.

**ECOLOGICAL INQUIRY**

If five species utilized the resource set in part **(a)**, what would be the expected frequency of competition observed?

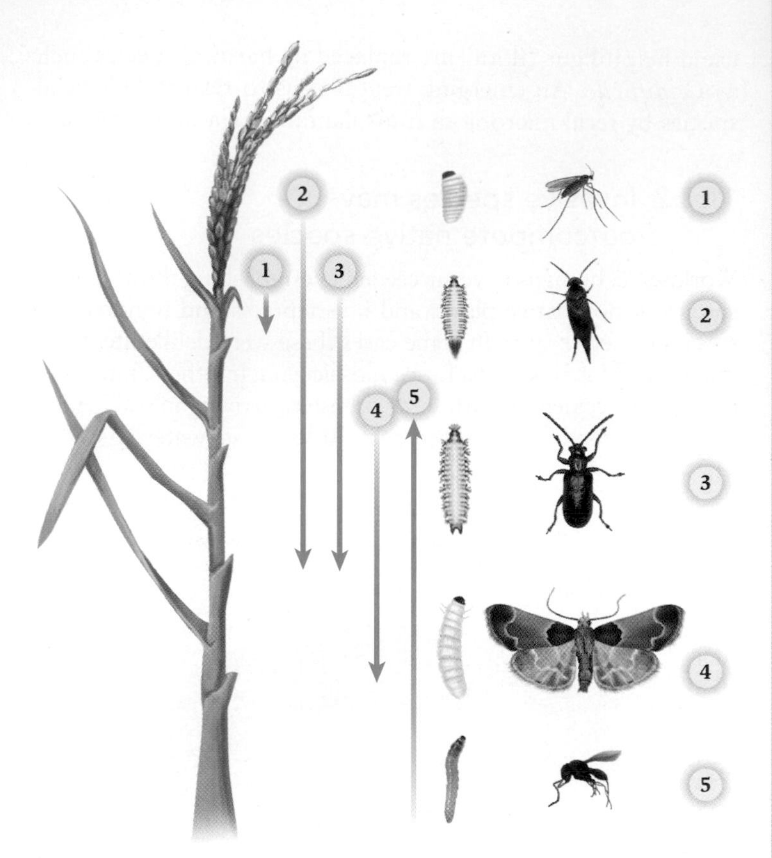


**Figure 11.5 The stem-boring guild associated with salt marsh cordgrass, *Spartina alterniflora*, on the Gulf Coast of North America.** (1) *Calamomyia alterniflorae* (Diptera). (2) *Mordellistena splendens* (Coleoptera). (3) *Languria taedata* (Coleoptera). (4) *Chilo plejadellus* (Lepidoptera). (5) *Thrypticus violaceus* (Diptera). Arrows indicate where in the stem the larva of each species is found and the direction it bores. *Calamomyia* does not move but feeds in the same position within the stem. Adults, shown at extreme right, are free living.

**ECOLOGICAL INQUIRY**

What other herbivorous insect-feeding guilds can you think of?

used as ballast and has been a major pathway for marine invasions. Regardless of the mechanism of introduction, many introduced species have spread, invading native habitats and assuming pest proportions. In many cases invasives compete with natives for resources, often displacing them. This has occurred for both vertebrate and invertebrate animals, and plants.

### *Vertebrates*

Steeve Côté and his students (Côté, 2005) documented the extinction of the black bear, *Ursus americanus,* on a Canadian island, caused by introduced white-tailed deer, *Odocoileus virginianus*. Anticosti Island is a large (8,000 km$^2$) island in the St. Lawrence Seaway off the coast of Quebec. Historical records reported black bears as extremely numerous.

## Feature Investigation

## Connell's Experiments with Barnacle Species Show That One Species Can Competitively Exclude Another in a Natural Setting

While the most direct method of assessing the effect of competition is to remove individuals of species A and measure the response of species B, such manipulations are often difficult to conduct outside the laboratory. If individuals of species A are removed, what is to stop them from migrating back into the area of removal?

In 1954 ecologist Joseph Connell conducted an experiment that overcame this problem. *Chthamalus stellatus* and *Semibalanus balanoides* (formerly known as *Balanus balanoides*) are two species of barnacles that dominate the Scottish coastline. Each organism's niche on the intertidal zone was well defined. *Chthamalus* occurs in the upper intertidal zone, and *Semibalanus* was restricted to the lower intertidal zone. Connell sought to determine what the range of *Chthamalus* adults might be in the absence of *Semibalanus* (**Figure 11.A**).

To do this, Connell obtained rocks from high on the rock face, just below the high-tide level, where only *Chthamalus* grew. These rocks already contained young and mature *Chthamalus*. He then moved the rocks into the *Semibalanus* zone, fastened them down with screws, and allowed *Semibalanus* to also colonize them. Once *Semibalanus* had colonized these rocks, he took the rocks out, removed all the *Semibalanus* organisms from one side of the rocks with a needle, and then returned the rocks to the lower intertidal zone, screwing them down once again. As seen in the data, the mortality of *Chthamalus* on rock halves with *Semibalanus* was fairly high. *Semibalanus* would undercut the *Chthamalus* as they grew, increasing the chances that *Chthamalus* would be washed away from the rock face. On the *Semibalanus*-free halves, however, *Chthamalus* survived well.

In other studies, Connell also monitored survival of natural patches of both barnacle species where both occurred on the intertidal zone at the upper margin of the *Semibalanus* distribution. In a period of unusually low tides and warm weather, when no water reached any barnacles for several days, desiccation became a real threat to the barnacles' survival. During this time, young *Semibalanus* suffered a 92% mortality rate, and older individuals a 51% mortality rate. At the same time, young *Chthamalus* experienced a 62% mortality rate compared with a rate of only 2% for more resistant older individuals. Clearly, *Semibalanus* is not as resistant to desiccation as *Chthamalus* and could not survive in the upper intertidal zone where *Chthamalus* occurs. *Chthamalus* is more resistant to desiccation than *Semibalanus* and can be found higher in the intertidal zone.

While the potential distribution (the fundamental niche, see Figure S.1) of *Chthamalus* extends over the entire intertidal zone, its actual distribution (the realized niche) is restricted to the upper zone. Connell's experiments were among the first to show that, in a natural environment, one species can actually outcompete another and affect its distribution within a habitat.

**HYPOTHESIS** Adult *Chthamalus stellatus* were being competitively excluded from the lower intertidal zone by the species *Semibalanus balanoides*.

**STARTING LOCATION** Two species of barnacles, *Chthamalus stellatus* and *Semibalanus balanoides*, grow in the intertidal zone of the rocky shores of the Scottish coast. In this habitat, they do not coexist.

**Experimental level**

**1** Remove rocks containing young *Chthamalus* from the upper intertidal zone to the lower intertidal zone, and fasten them down in the new location with screws.

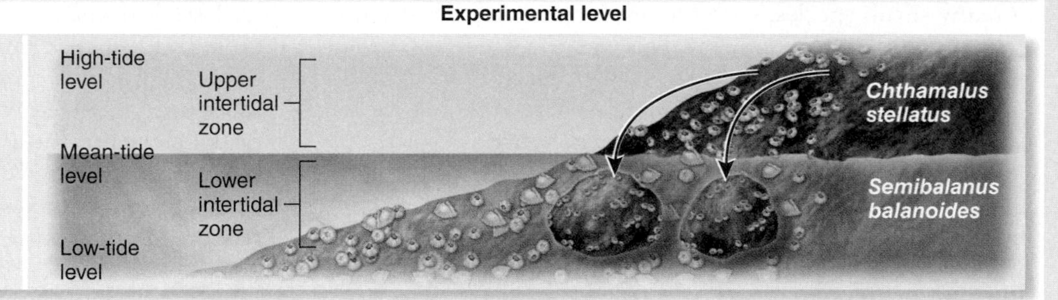


**2** Allow *Semibalanus* to colonize rocks.

**3** After colonization period is over, remove *Semibalanus* from half of each rock with a needle (leaving the other half undisturbed). Return the rocks to the lower intertidal zone and fasten them down once again. Monitor survival of *Chthamalus* on both sides.

**4** *Chthamalus* grows on the side where *Semibalanus* has been removed, indicating that *Semibalanus* may exclude *Chthamalus* from certain habitats.

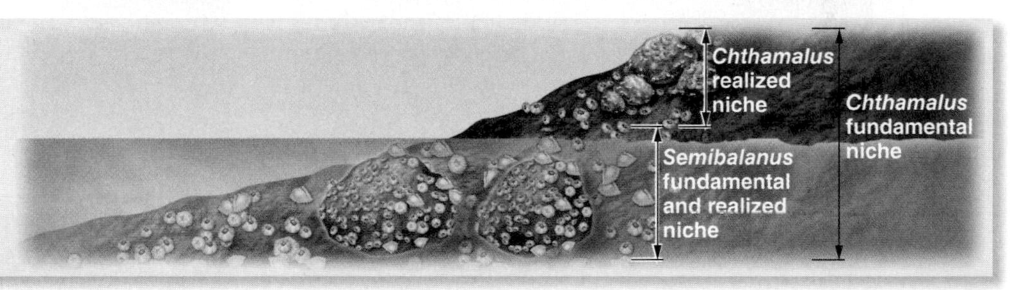


**5** **THE DATA**

| Rock No. | Treatment | Percent *Chthamalus* Mortality Over One Year | |
|---|---|---|---|
| | | Young Barnacles | Mature Barnacles |
| 13b | *Semibalanus* removed | 35 | 0 |
| | Undisturbed | 90 | 31 |
| 12a | *Semibalanus* removed | 44 | 37 |
| | Undisturbed | 95 | 71 |
| 14a | *Semibalanus* removed | 40 | 36 |
| | Undisturbed | 86 | 75 |

**Figure 11.A** Connell's experimental manipulation of species indicated the presence of competition.

In 1896, 220 white-tailed deer were introduced onto the island and by 1934, in the absence of predators, their numbers had increased to >50,000. The feeding pressure of these deer was sufficient to greatly reduce the abundance and sizes of many shrub species, with the result that berry density was reduced to only 0.28 berry/m$^2$. This was 235 times lower than the minimum 66 berries/m$^2$ necessary for black bears to maintain body mass. Bears need to build up energy reserves for the winter, and without the berries the bears became extinct.

Placental mammals have been found to be particularly serious competitors in Australia, where the introduced rabbit competes for burrows with an animal called the boodie, *Bettongia lesueur,* the only burrowing kangaroo, and with the rabbit bandicoot, *Macrotis lagotis*. Among the predators, dingoes, *Canis familiaris,* are thought to have excluded the thylacine, *Thylacine cynocephalus,* a native marsupial wolf-like animal, from mainland Australia (**Figure 11.6a,b**). Sheep introduced for the wool industry probably compete with a range of kangaroo species, especially the brush-tailed rock wallaby and larger species, such as the red kangaroo and the western gray kangaroo. Introduced vertebrates are a problem in aquatic environments too. In California, 48 of 137 species of freshwater fish are non-native. Of these, 26 have been well studied and 24 are known to have a negative impact on native fish. The introduction to Gatun Lake, Panama, of the cichlid fish, *Cichla ocellaris,* a native of the Amazon, is thought to have led within 5 years to the elimination of six of the eight previously common fish species. Similarly, after construction of the Welland Canal linking the Atlantic Ocean with the Great Lakes, many native fish populations were reduced in abundance through competition for food by the alewife, *Alosa pseudoharengus,* which invaded from the St. Lawrence Seaway.

### Invertebrates

In the continental U.S. alone, over 1,500 insect species have been introduced. Introduced insects may be useful in

**Figure 11.6 Introduced species often compete with native species.** **(a)** The thylacine in Australia may have been outcompeted by **(b)** feral dogs called dingoes. **(c)** Introduced plants can choke out native vegetation, as happens with purple loosestrife in the U.S. Northeast. **(d)** Barb goatgrass in California.

biological control, but some accidentally introduced species can quickly get out of hand. One of the worst cases involves the imported red fire ant, *Solenopsis invicta,* accidentally introduced in 1918 into Mobile, Alabama, in a ship's ballast containing soil from Argentina. Since 1918, fire ants have spread rapidly throughout the southeastern U.S. and have recently become established in California (**Figure 11.7a**). Though their northward spread is limited by their inability to overwinter in colder environments, they may yet spread northward along the western states. Fire ants outcompete many U.S. species of native ants and alter the appearance of old field habitats with their prominent nests (**Figure 11.7b**). Some of the most severe effects of competition between invertebrates occur in freshwater systems, where invasive mussels, crayfish, and other species outcompete natives (see **Global Insight**).

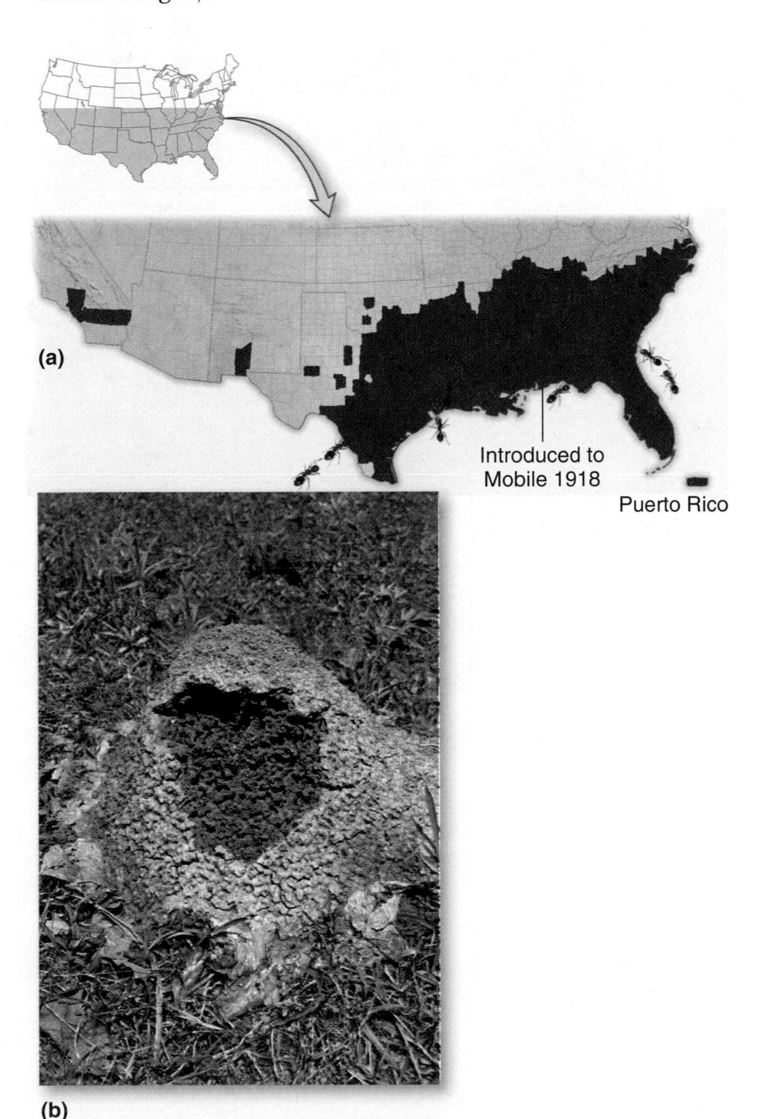


**Figure 11.7 The distribution of the fire ant, *Solenopsis invicta*, in the United States, 2005.** **(a)** Accidentally introduced into Mobile, Alabama, in 1918, it has since spread widely, outcompeting many native ant species. **(b)** The prominent nests alter the appearance of fields in the U.S. Southeast.

### *Plants*

Invasive plants are a problem throughout the world. In the U.S. Northeast, purple loosestrife is an intensive competitor, choking out species in native American wetlands (see **Figure 11.6c**). This, in turn, has endangered several species of ducks and a species of turtle that depend on the native plants for food and shelter. Chinese tallow trees, *Sapium sebiferum,* have produced near monocultures in the southern U.S., as has garlic mustard, *Alliaria petiolata,* in eastern and central U.S. forests. In California the introduced species barb goatgrass, *Aegilops triuncialis,* is found on serpentine soils (see **Figure 11.6d**). These soils have low nutrients and low moisture content, and are quite toxic, even to some native plants. Because barb goatgrass can tolerate these conditions, it can use the limited nutrients and water that are largely untapped by other plant species. The result is that barb goatgrass thrives and can form dense stands, choking out other species. In south Florida, three invasives—Brazilian pepper, *Schinus terebinthifolius,* Australian pine, *Casuarina equisetifolia,* and, especially punk-tree, *Melaleuca quinquenervia*—are suffocating native vegetation in the Everglades. In north Florida and southeastern states, kudzu vine, *Pueraria lobata,* imported from Japan to stabilize earthworks, smothers large areas of native habitat.

### *General trends*

Why are some invasive species such good competitors? First, invasive species do not enjoy unilateral competitive superiority. Many native species can coexist with invasives for long periods of time. However, in some instances invasive species quickly gain competitive superiority. There are at least 29 hypotheses currently being tested to account for the strong competitive ability of invasives. Four of the most common are listed here. The **enemy release hypothesis** suggests that invasives are released from their natural enemies, which do not accompany them during invasions, and can devote more resources to growth and thus competition. This leads to the evolution of increased competitive ability. The **superior competitor hypothesis** suggests invasives are more efficient users of natural resources than natives. The **propagule pressure hypothesis** suggests that invasives produce more progeny than some native species, and by sheer weight of numbers this permits them entry into natural communities and competitive superiority. Finally, the **ideal weed hypothesis** suggests preadaptation of some invaders to existing environmental conditions. With few limitations to many abiotic environments, these invaders can grow quickly and outcompete native species.

**Check Your Understanding**

**11.3** In south Florida, many invasive species are derived from plants sold at local nurseries. The longer the plant has been sold at nurseries, the more likely it is to have become invasive. What hypothesis does this support regarding the success of these invasive species?

Global Insight

## Invasive Rusty Crayfish Have Outcompeted Many Native North American Crayfish

Crayfish are a highly diverse group in North America, with 333 species comprising 75% of the global crayfish fauna. The rusty crayfish, *Orconectes rusticus,* is native to the Ohio River drainage. However, it has been spread by anglers, who use them as bait, to waters throughout Illinois, Michigan, Wisconsin, Minnesota, and parts of 11 other states. In these areas, rusty crayfish have outcompeted other species of crayfish, including native congeners, leading to declines in native species. The rusty crayfish have a relatively high metabolic rate and a large appetite. They greatly decrease the amount of plant biomass and consume abnormally high amounts of food. Their behavior has been argued to cause the underwater equivalent of clear-cutting forests. Julian Olden and colleagues (2006) documented a great decline in numbers of native crayfish species in Wisconsin compared to pre-invasion records (**Figure 11.B-i**). The rusty crayfish are replacing *O. propinquus* and *O. viralis* as the most dominant members of the crayfish fauna. Physiological studies have shown that rusty crayfish are absent from lakes with pH values lower than 5.5 and dissolved calcium concentrations less than 2.5 mg/L (**Figure 11.B-ii**). Unfortunately, a preliminary analysis of 527 lakes in Wisconsin shows 463, or 88%, to be suitable for rusty crayfish.

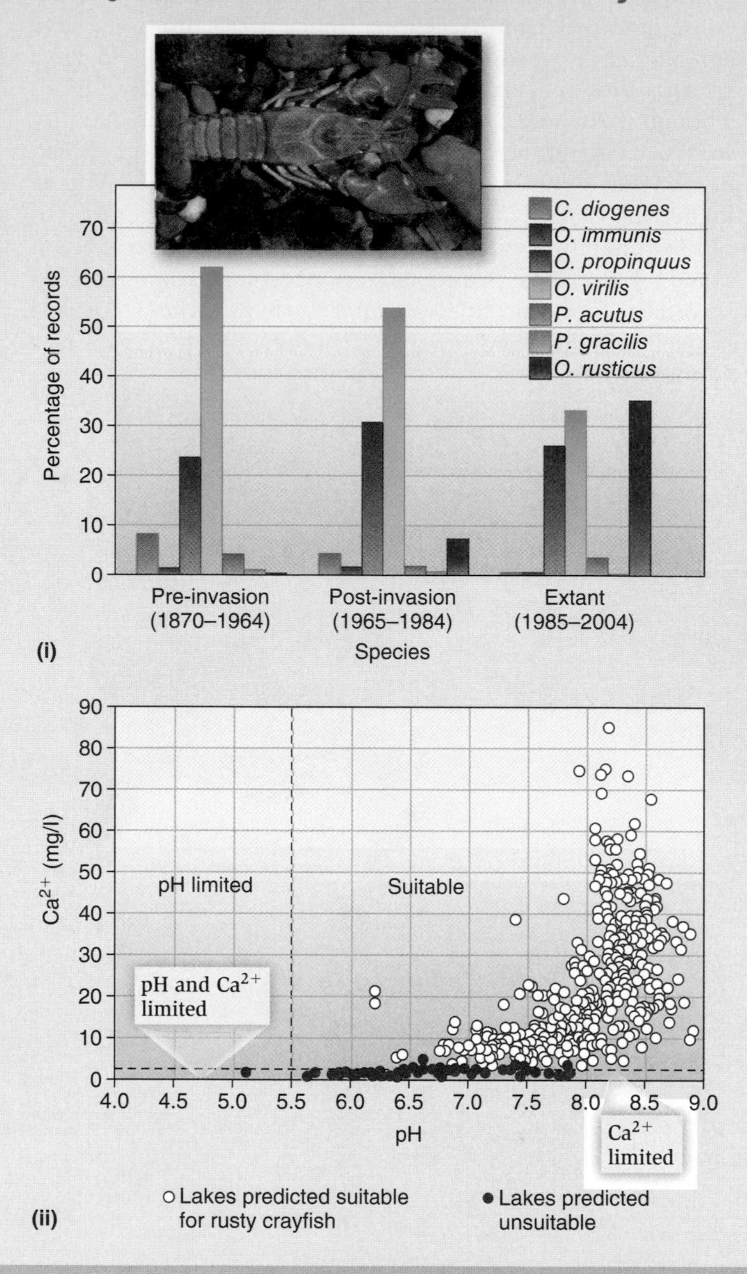


**Figure 11.B Competition from the rusty crayfish (red bar) has reduced native crayfish species abundance (other bars) in Wisconsin.** **(i)** Comparison of pre-invasion data (1870–1964) with post-invasion data (1965–1984) and current (extant) data (1985–2004). **(ii)** Physiological thresholds for rusty crayfish include low pH and low dissolved calcium. (From Olden et al., 2006.)

## 11.4 The Winners and Losers of Competitive Interactions May Be Predicted Using Mathematical Models

The first mathematical models used to describe the outcomes of competition were provided in the 1920s by two people working independently: Alfred J. Lotka (1925), an insurance actuary who worked for the Metropolitan Life Insurance Company in the United States, and Vito Volterra (1926), a professor of mathematical physics in Rome, Italy. Later, in the 1980s, David Tilman provided competition models based on patterns of resource use.

### 11.4.1 The lotka-volterra competition models are based on the logistic equation of population growth

The growth rate of populations of two species growing independently can be described using logistic growth equations. As before, $r$ is the per capita rate of population increase, $N$ is the population size, and $K$ is the carrying capacity. Here we introduce subscripts to refer to particular species, so $r_1$ is the per capita population growth rate of species 1, and $r_2$ is the per capita population growth rate of species 2.

For species 1:

$$\frac{dN_1}{dt} = r_1 N_1 \frac{(K_1 - N_1)}{K_1}$$

For species 2:

$$\frac{dN_2}{dt} = r_2 N_2 \frac{(K_2 - N_2)}{K_2}$$

In the single-species logistic equation, the growth of a population is affected by the population size relative to its carrying capacity. The quantity $K - N$ defines how far below carrying capacity a population is, and reflects the effects of intraspecific competition. This quantity can easily be modified to account for the presence of a second competing species. To do this, it is necessary to define a competition coefficient or conversion factor that quantifies the per capita competitive effect on species 1 of species 2 and vice versa. For example, if two individuals of species 2 take up the same amount of resources as one individual of species 1, then the per capita competitive effects on species 1 of species 2 is 0.5 (**Figure 11.8**). Likewise, the per capita competitive effects on species 2 of species 1 is 2.0. In this case, the competition coefficients are reciprocals of one another, but such is not always the case. These competition coefficients can be defined by the term $\alpha$ or $\beta$ in which

$\alpha$ = per capita competitive effect of species 2 of species 1

$\beta$ = per capita competitive effect of species 1 of species 2

The conversion factor allows the logistic equation to be modified to take into account the effects of both species on the growth rate of the other:

$$\frac{dN_1}{dt} = r_1 N_1 \frac{(K_1 - N_1 - \alpha N_2)}{K_1}$$

$$\frac{dN_2}{dt} = r_2 N_2 \frac{(K_2 - N_2 - \beta N_1)}{K_2}$$

The population size relative to carrying capacity is now defined by the abundance of both species, and the growth rate of each population is determined by per capita rate of population increase ($r$), carrying capacity for that species ($K$), and the population sizes of species 1 ($N_1$) and species 2 ($N_2$). These equations can be used to describe how the population sizes of both species changes together.

Such relationships can be expressed graphically (**Figure 11.9**). In the absence of species 2, population growth

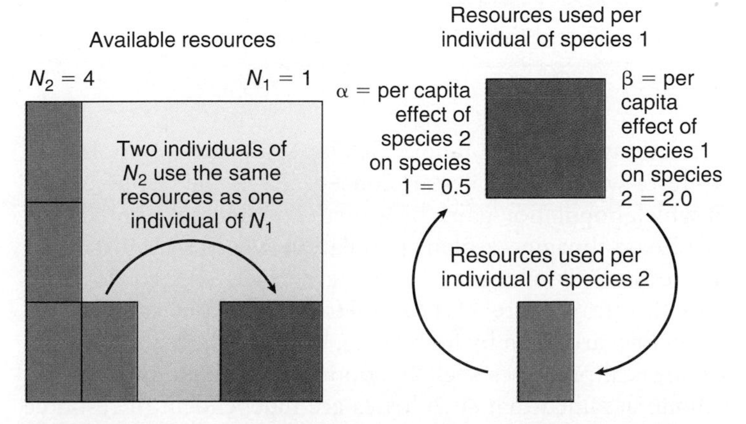


**Figure 11.8 Conceptualization of "conversion factors" $\alpha$ and $\beta$ before use in Lotka-Volterra competition equations.** $\alpha$ is the per capita competitive effect on species 1 of species 2, and $\beta$ is the per capita competitive effect on species 2 of species 1.

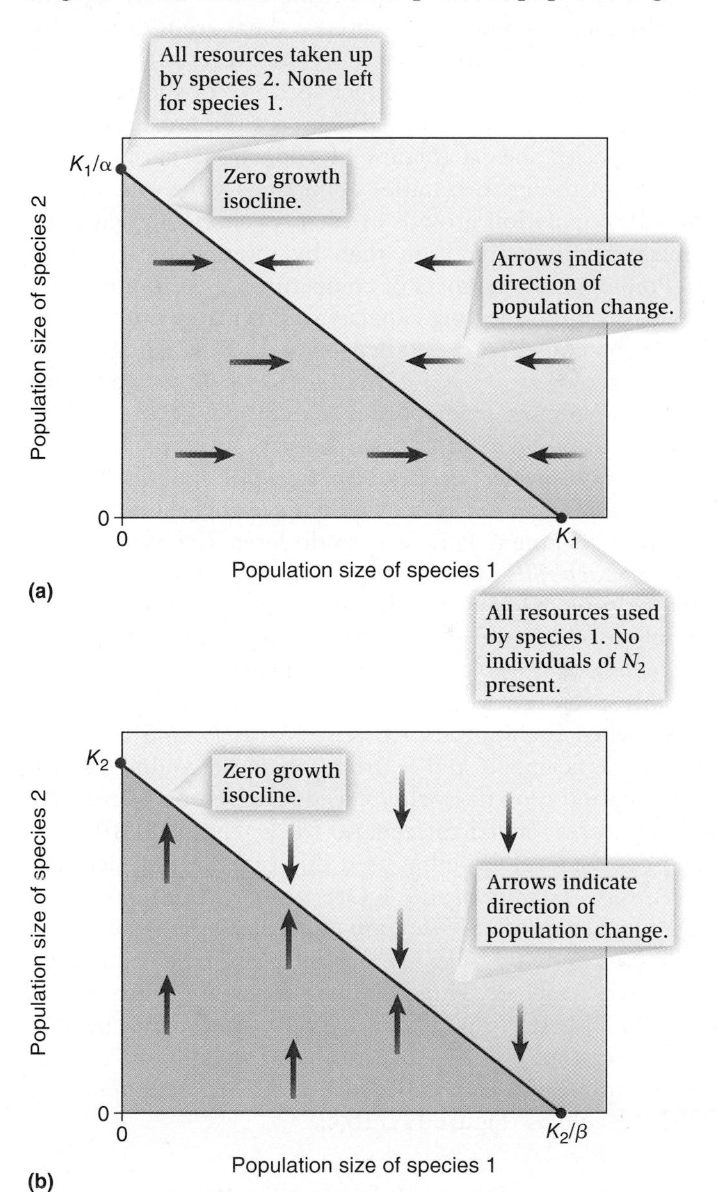


**Figure 11.9 Changes in the population sizes of competing species.** **(a)** changes in population size of species 1 when competing with species 2. Populations in the orange area will increase in size and will come to equilibrium at some point on the diagonal line. At any point on the diagonal there is no further population growth, that is, $dN/dt = 0$. Populations in the blue area will decrease in size until they reach the diagonal line. **(b)** Changes in population size of species 2 when competing with species 1.

of $N_1$ continues to the carrying capacity of the environment $K_1$ (**Figure 11.9a**). Alternatively, the whole environment may be filled with $N_2$ individuals, and no growth of $N_1$ is possible because $K_2$ has been reached. In this case the equivalent number of individuals of $K_2$ would be $K_1/\alpha$. These are the two extremes marked on the axes. Between these two extremes are many combinations of $N_2$ and $N_1$ at which no further growth of $N_1$ is possible. These points fall on the diagonal $dN_1/dt = 0$, which is often called the zero growth isocline. Population growth of $N_2$ can be represented by a similar diagram (**Figure 11.9b**). Combining the two figures and adding the arrows by vector addition illustrates what happens when the species co-occur. Essentially, there are four possible outcomes: species 1 goes extinct (**Figure 11.10a**), species 2 goes extinct (**Figure 11.10b**), either species 1 or species 2 goes extinct, depending on the initial densities (**Figure 11.10c**), or the two species coexist (**Figure 11.10d**). In scenario (c), the species with the greater initial abundance wins out. In scenario (d), population growth of each species is limited more by intraspecific competition than by interspecific competition. Predicting the winners of competition therefore requires knowledge of the carrying capacity of both competitors. This information comes from detailed study of their biology.

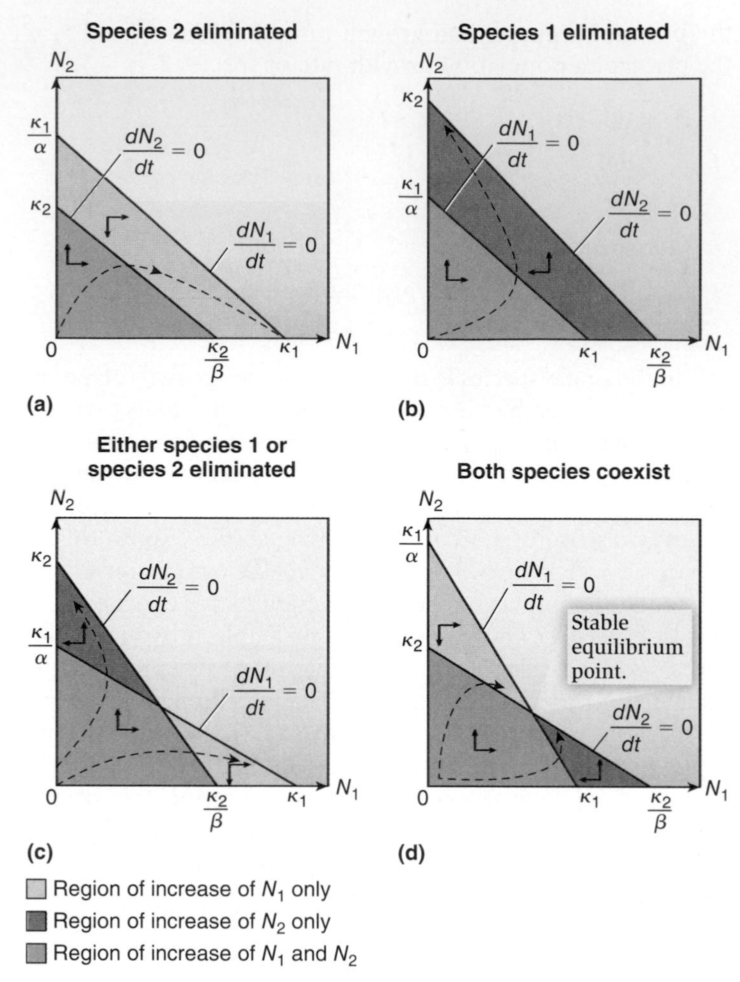


**Figure 11.10 The four consequences of the Lotka-Volterra competition equations.** Axes are population sizes of species 1 and species 2. Dotted lines represent the results of vector addition of the two vectors shown by the solid lines. There are four possible outcomes: **(a)** species 2 is eliminated; **(b)** species 1 is eliminated; **(c)** either species 1 or species 2 is eliminated, depending on who has fewer individuals when the competition begins; or **(d)** both species coexist at a stable equilibrium.

The ecologist A. C. Crombie (1946) showed how the Lotka-Volterra competition model could be applied to competition between grain-feeding beetles. Crombie allowed two species of beetle, *Tribolium confusum* and *Oryzaephilus surinamensis,* to compete over fine-grained flour, just as Thomas Park was to do later. *Tribolium* ate so many *Oryzaephilus* pupae that this latter species went extinct in the cultures. However, when both species were provided with whole wheat, the grains gave a measure of protection to *Oryzaephilus,* whose larvae pupated inside the wheat grains, and both species coexisted. Crombie was able to measure $K_1$ and $K_2$ when each species existed separately, and the competition coefficients $\alpha$ and $\beta$, from which he could calculate the population growth isoclines. The relative positions of the isoclines corresponded to scenario (d) in Figure 11.10, indicating coexistence. Crombie used different starting densities of beetles, 4 *Tribolium* and 4 *Oryzaephilus,* 100 *Tribolium* and 4 *Oryzaephilus,* 4 *Tribolium* and 100 *Oryzaephilus,* and 4 *Tribolium* and 400 *Oryzaephilus*. In each case a stable equilibrium was reached. Three of these cases are illustrated in **Figure 11.11a**. If we plot the numbers of each species separately over time for two of these experiments, we can see that the end result is the same, with some 375 *Tribolium* and 180 *Oryzaephilus* (**Figure 11.11b,c**).

**ECOLOGICAL INQUIRY**

If $K_1 > K_2/\beta$ and $K_1/\alpha < K_2$, which species wins?

## 11.4.2 Tilman's R* models predict the outcome of competition based on resource use

A drawback to the Lotka-Volterra competition model is that the mechanisms that drive the competitive process are not specified. David Tilman (1982, 1997) criticized this approach and emphasized that we need to know the mechanism by which competition occurs. Knowing the mechanism will enable better predictions of the outcome. Tilman's alternative to Lotka-Volterra was the R* (R star) concept. R* is the resource level at which population gains by growth equal losses to predators or disease. Imagine a plant population where the most limiting resource is nitrogen. As nitrogen levels rise, so do plant growth rates (**Figure 11.12a,b**). However, plants can also die when they are eaten by herbivores, killed by pathogens, or lost to physical processes such as floods or freezing temperatures. Tilman assumed that such losses are independent of resource level. At a certain resource level, commonly called R*, gains by growth equal losses to biotic and abiotic factors and so the plant population is at equilibrium. At resource levels beyond

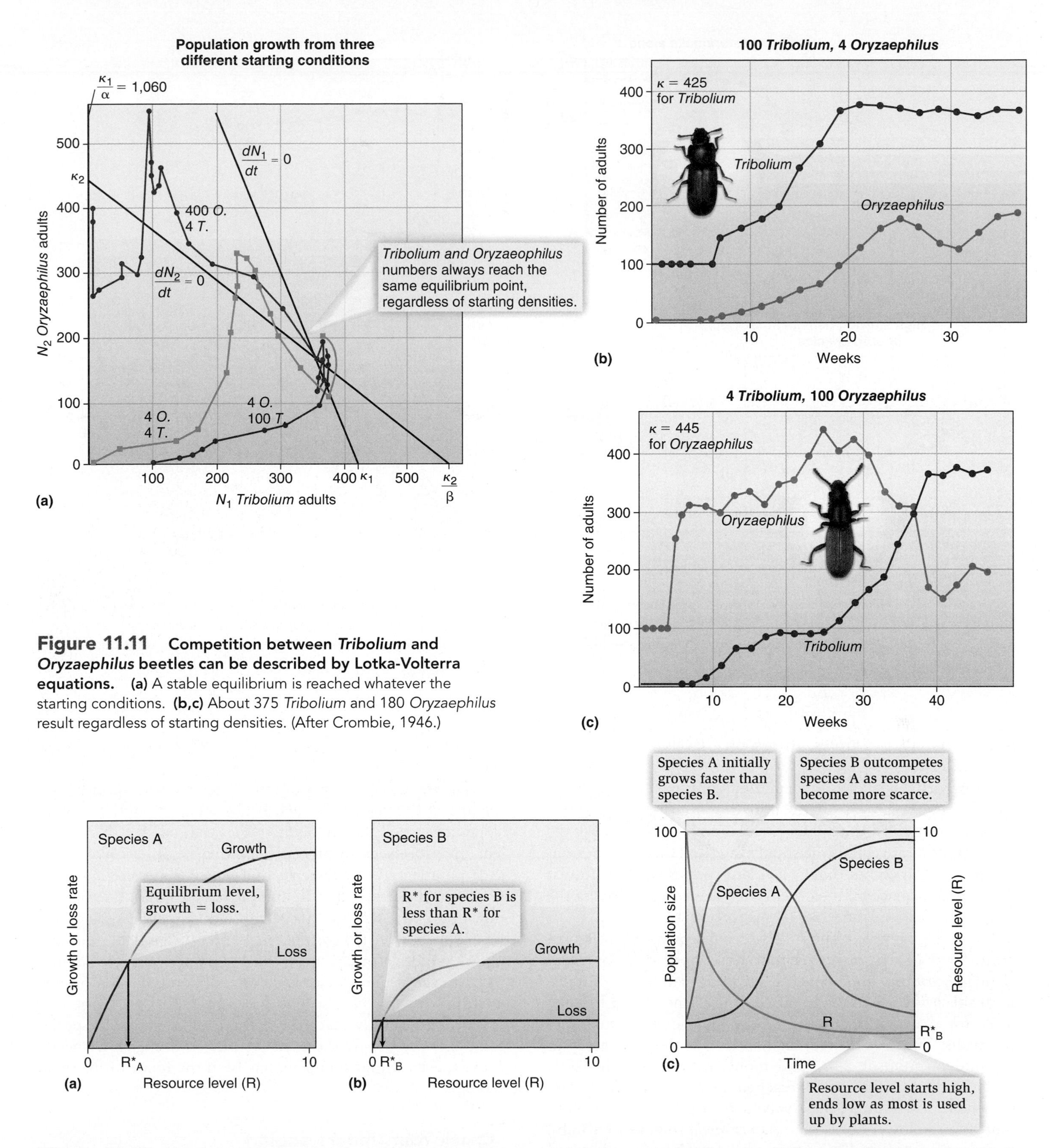


**Figure 11.11** **Competition between *Tribolium* and *Oryzaephilus* beetles can be described by Lotka-Volterra equations.** **(a)** A stable equilibrium is reached whatever the starting conditions. **(b,c)** About 375 *Tribolium* and 180 *Oryzaephilus* result regardless of starting densities. (After Crombie, 1946.)

**Figure 11.12** **Tilman's R* (R star concept) of competition between two species, A and B, based on their resource utilization curves** **(a)** Resource-dependent growth and loss for species A. The resource concentration at which growth equals loss is $R_A$*. **(b)** Resource-dependent growth and loss for species B and its $R_B$*. Growth is slower than for species A. **(c)** Dynamics of resource competition between species A and B. Note that species B, which has the lower R*, displaces species A and drives the resource concentration down to $R_B$*. (Reproduced from Tilman, 1997.)

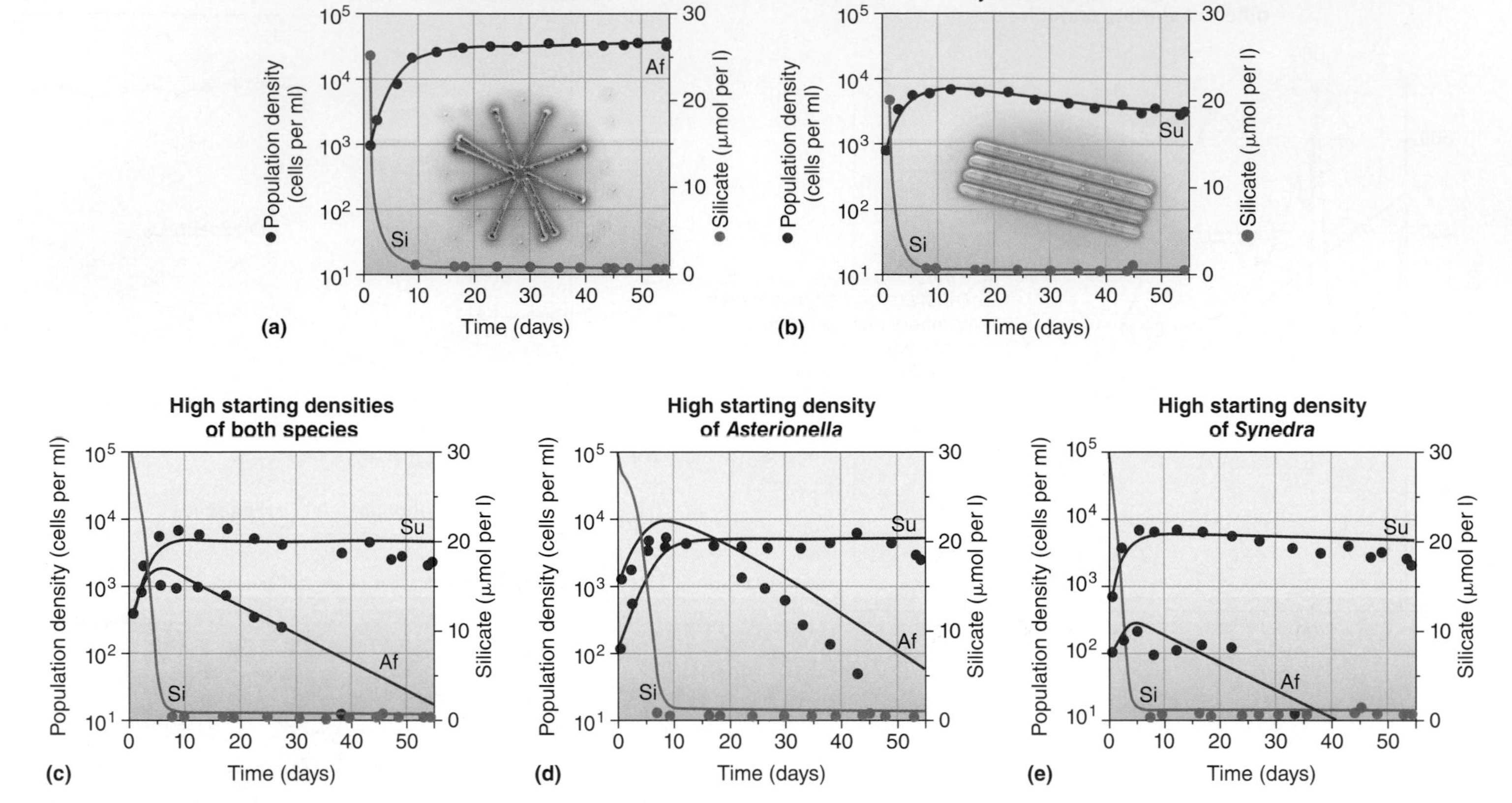


**Figure 11.13 Competition between diatoms over silicate.** **(a)** *Asterionella* maintains a stable population equilibrium in the laboratory and decreases silicate to a low level. **(b)** *Synedra* reduces silicate levels even more and therefore can exist at low levels of silicate in solution. **(c,d,e)** As a result, R* for *Synedra* is lower than that for *Asterionella*, and *Synedra* outcompete *Asterionella* regardless of starting densities. (After Tilman et al., 1981.)

R*, the population grows. At resource levels below R*, the population goes extinct. The level of the resource itself, in this case nitrogen, depends on the amount consumed by the plants and the supply rate, as provided by decomposition of organic matter. Eventually the plant population increases until the consumption rate equals the supply rate, and the resource concentration is at equilibrium. The plant species with the lowest R* should displace all the others from the habitat, driving the resource concentration to its R* (**Figure 11.12c**).

A good example of how the R* concept works was provided by Tilman's own study of freshwater diatoms *Asterionella formosa* and *Synedra ulna* (Tilman et al., 1981). Both species required many different nutrients, but a key element for diatoms is the availability of silicate to construct cell walls. Tilman cultured each species in the laboratory in a liquid medium at 24°C and monitored the silicate levels in the medium. Both species reached a steady carrying capacity when grown alone, but *Synedra* reduced the silicate concentration to lower levels than *Asterionella* (**Figure 11.13a,b**). *Synedra* was expected to outcompete *Asterionella* and indeed did so, even when its starting densities were much lower than those of *Asterionella* (**Figure 11.13c,d,e**). J. Bastow Wilson and colleagues (2007) found that virtually all of the 41 studies that have tested Tilman's R* concept have supported it.

Theoretically, for one limiting resource, the species with the lowest R* should displace all the others from the habitat. There are usually many plant species that coexist in an area. This may be because of many limiting resources like nitrogen, potassium, phosphorus, and other minerals, and also resources like water and light. If these resources fluctuate in time and place, no one species can outcompete all the others for all resources. Coexistence may also be possible because plants and other organisms live in a spatially variable habitat, so that even if one species is continually outcompeted in one area, it wins out in a second area, and seeds from the second area constantly reach the first area, allowing that species to apparently persist there. In fact, determining how species coexist, even if they compete, has been the focus of many an ecologist's career.

## Check Your Understanding

**11.4** In the Lotka-Volterra competition model, what is implied if $\alpha = \beta$?

## 11.5 Species May Coexist If They Do Not Occupy Identical Niches

In 1934 the Russian microbiologist Georgyi Gause began to study competition between three protist species, *Paramecium aurelia, Paramecium bursaria,* and *Paramecium caudatum,* all of which fed on bacteria and yeast, which in turn fed on an oatmeal medium in a culture tube in the laboratory. The bacteria occurred more in the oxygen-rich upper part of the culture tube, and the yeast in the oxygen-poor lower part of the tube. Because each *Paramecium* species was a slightly different size, Gause calculated population growth as a combination of numbers of individuals per milliliter of solution multiplied by their unit volume to give a population volume for each species. When grown separately, population volume of all three *Paramecium* species followed a logistic growth pattern (**Figure 11.14a**). When Gause cultured *P. caudatum* and *P. aurelia* together, *P. caudatum* went extinct (**Figure 11.14b**). Both species utilized bacteria as food, but *P. aurelia* grew at a rate six times faster than *P. caudatum* and was better able to convert the food into offspring.

However, when Gause cultured *P. caudatum* and *P. bursaria* together, neither went extinct. The population volumes of each were much less, compared to when they were grown alone, because some competition occurred between them. However, Gause discovered that *P. bursaria* was better able to utilize the yeast in the lower part of the culture tubes. *P. bursaria* have tiny green algae inside them that produce oxygen and allow *P. bursaria* to thrive in the lower oxygen levels at the bottom of the tubes. From these experiments Gause concluded that species with exactly the same requirements cannot live together in the same place and use the same resources, that is, occupy the same niche. His conclusion, later termed the **competitive exclusion principle** by Garrett Hardin (1960), essentially means that complete competitors cannot coexist. Species may avoid competitive exclusion by partitioning resources or being of different sizes, which permits them to feed on different-sized resources.

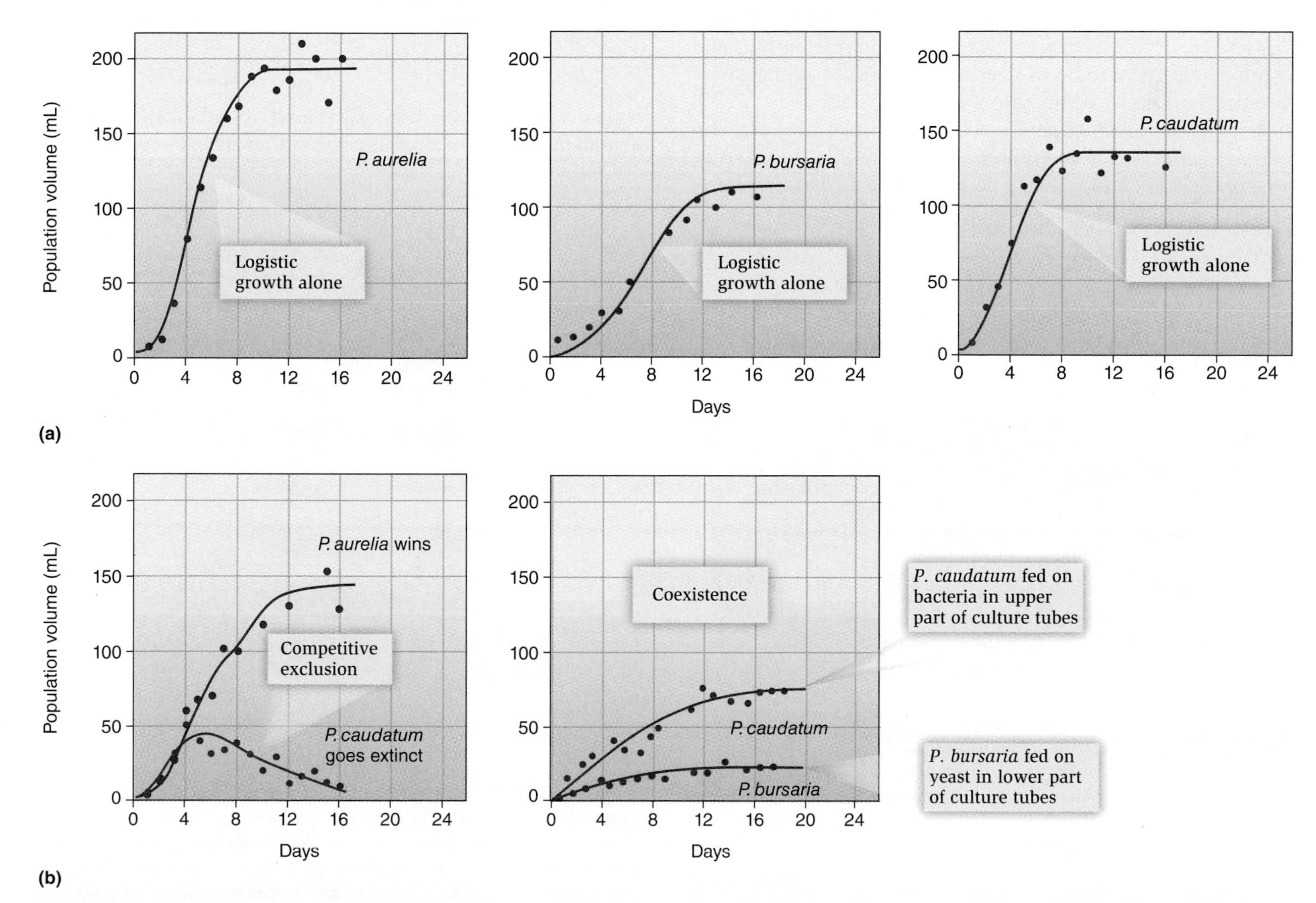


**Figure 11.14 Competition among *Paramecium* species.** **(a)** Each of three species, *Paramecium aurelia, Paramecium bursaria,* and *Paramecium caudatum,* grows according to the logistic model when grown alone. **(b)** When *P. aurelia* is grown with *P. caudatum,* the density of *P. aurelia* is lowered compared to when grown alone, but *P. caudatum* goes extinct. When *P. caudatum* is grown with *P. bursaria* (right), the population densities of both are lowered but they coexist.

## 11.5.1 Species may partition resources, promoting coexistence

If complete competitors drive one another to local extinction, how different do two species have to be to coexist and in what features do they usually differ? The term **resource partitioning** describes the differentiation of niches, both in space and time, that enables similar species to coexist in a community. In the previous example, *P. caudatum* and *P. bursaria* were partitioning resources. Just how much resource partitioning is necessary to permit coexistence? We will consider two approaches to this question. First, let's consider three species distributed on a resource set (**Figure 11.15**). Here, *K* is the resource availability or carrying capacity, such as the availability of different-sized seeds, and *d* is the distance between maximum values of resource utilization curves of different species. The term *w* represents one standard deviation, a statistical quantity that represents approximately 68% of the area on one side of the resource distribution curve of a species. Because *d* measures how far apart species resource utilization maxima are, and *w* measures how spread out species resource utilization curves are, the combination of *d* and *w* can define the amount of overlap in resource use and therefore the amount of competition. Richard Southwood (1978) and others have argued that if $d/w < 1$, species cannot coexist; if $1 < d/w < 3$, there will be competition yet coexistence between species; and if $d/w > 3$, species coexist harmoniously without competition.

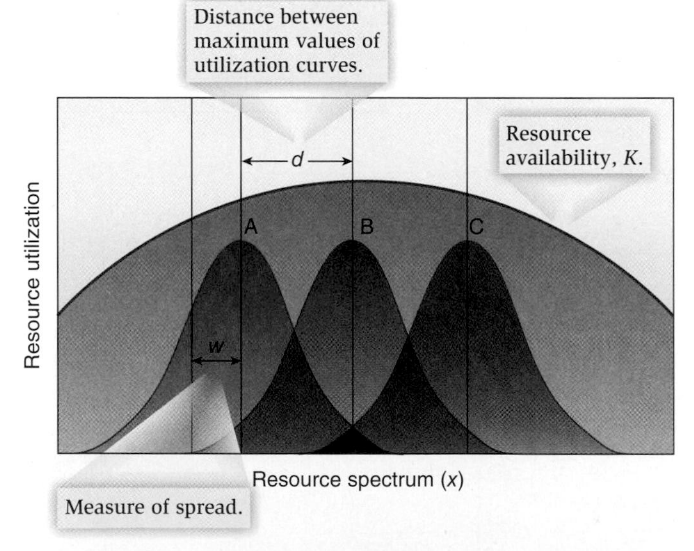


**Figure 11.15 Resource partitioning.** Three species, A, B, and C, with similar resource utilization curves, utilize a resource supply, *K*. *d* = distance between maximum values of resource utilization curves, *w* = standard deviation of resource utilization, and *d*/*w* = resource separation ratio.

**ECOLOGICAL INQUIRY**

According to the resource partitioning theory, if $d = 2.8$ and $w = 1.4$, could two species coexist?

The second approach to determining how much resource partitioning is necessary to permit coexistence is called **proportional similarity analysis.** Consider two insect species that feed on the leaves of the shrub shown in **Figure 11.16.** There are ten pairs of leaves on the shrub. The first species, 1, feeds toward the bottom of the plant, in the shade, on leaf pairs 4–8, while the other, species 2, prefers to feed on the sunny leaves toward the top of the plant, leaf pairs 1–6. In other words, the fundamental niche of species 1 is leaf pairs 4–8, while that of species 2, is leaf pairs 1–6. The two species overlap in resource use only on leaf pairs 4–6, called the area of **niche overlap.** We can represent this distribution diagrammatically in **Figure 11.17.** Is this overlap small enough to permit coexistence between the insects? The niche overlap between the two insect species that feed on the shrub can be measured by a quantity known as proportional similarity, *PS*. which is given by:

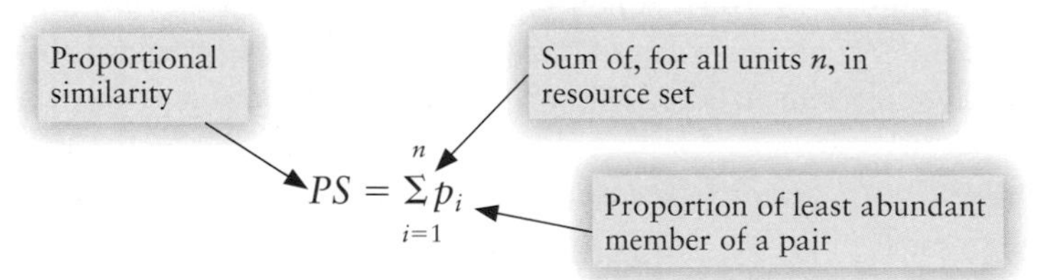


$$PS = \sum_{i=1}^{n} p_i$$

where $p_i$ is the proportion of the least abundant species of the pair in the *i*th unit of a resource set with *n* units. For example, on

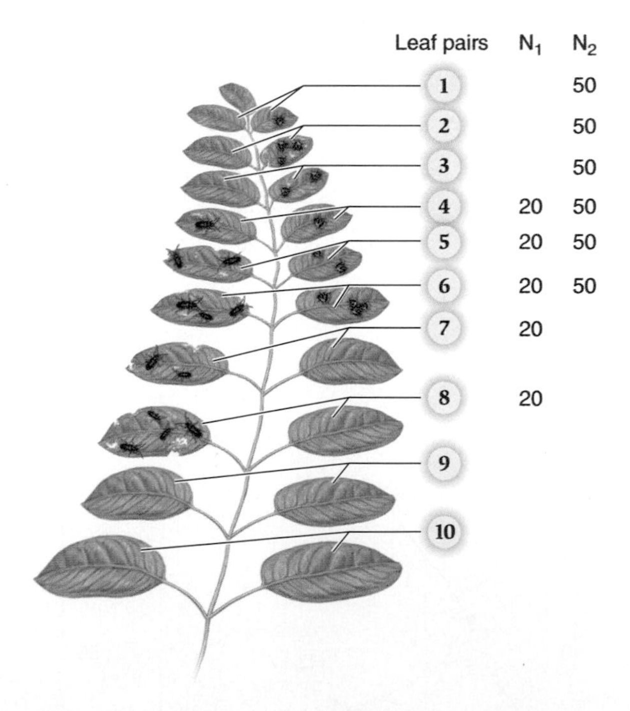


**Figure 11.16 Resource utilization relationships of two species.** Imagine two insect species which both feed on a discontinuously distributed resource like leaves on a shrub. There are ten pairs of leaves distributed from top to bottom on a plant. The first insect species has 100 individuals distributed evenly on leaf pairs 4–8, and the second insect species has 300 individuals distributed evenly on pairs 1–6. Thus, on each of leaf pairs 1–6 there are a sixth, 0.167, of the population of the second species, and on each of leaf pairs 4–8 there are a fifth of the population of the first species.

the first leaf pair, species 1 is absent so its proportion of occurrence is 0. Similarly, the proportion of abundance of species 2 is 0.167. The lowest of these two values is 0. In Figures 11.16 and 11.17, the proportional similarity for each of the ten units of resources is the sum of the proportional abundance of the lowest species. Going from leaf pair 1 through pair 10 we have $PS = 0 + 0 + 0 + 0.167 + 0.167 + 0.167 + 0 + 0 + 0 + 0 = 0.500$. The proportional similarity is then 0.500.

Proportional similarity values of less than 0.70 have been taken to indicate possible coexistence, while those greater than 0.70 indicate competitive exclusion, so in our insect example the species would be able to coexist despite some competition on leaf pairs 4, 5, and 6. Why 0.70? Similarities of 0.70 indicate differences of about 0.3 or 30%. Thirty percent differences have often been deemed necessary to promote coexistence; the empirical reasoning behind this is explained in the next section. However, species may differ not only along one resource axis but along many, such as leaf height and plant species. For two resource axes, proportional similarity indices can be combined—proportional similarity values of 0.8 and 0.6 on two axes multiply to give an overall *PS* of 0.48. Theoretically, coexistence along two resource axes would be permitted in cases where combined *PS* values = 0.49 or less, that is, $0.7 \times 0.7$.

### 11.5.2 Morphological differences may allow species to coexist

Yale biologist G. Evelyn Hutchinson (1959) examined the sizes of mouthparts or other body parts important in feeding, and compared their sizes between species when they were sympatric (occurring in the same geographic area) and allopatric (occurring in different geographic areas). Hutchinson's hypothesis was that when species were sympatric, each species tended to specialize on different types of food. This was reflected by differences in the size of body parts associated with feeding, also called feeding characters. The tendency for two species to diverge in morphology, and thus resource use, because of competition is called **character displacement.** In areas where species were allopatric, there was no need to specialize on a particular prey type, so the size of the feeding character did not evolve to become larger or smaller; rather it retained a "middle of the road" size that allowed species to exploit the largest range of prey size distribution.

In Hutchinson's time, one of the classic cases of character displacement involved a 1947 study, by David Lack, of several closely related species of finches that Charles Darwin discovered on the Galápagos Islands. When three species, *Geospiza fuliginosa, Geospiza fortis,* and *Geospiza magnirostris,* are sympatric on the islands of Pinta and Marchena, their bill depths are different. *Geospiza fuliginosa* has a smaller bill depth, which enables it to crack small seeds more efficiently. *Geospiza fortis* has a larger bill depth, which enables it to feed on bigger seeds. *Geospiza magnirostris* has an even larger bill size (**Figure 11.18**). The pattern for *G. fuliginosa* and *G. fortis* is repeated on two more islands, Floreana and San Cristobal. However, when these latter two species are allopatric, that is, existing on different islands, their bills are more similar in depth. Lack concluded that the bill depth differences evolved in ways that minimized competition.

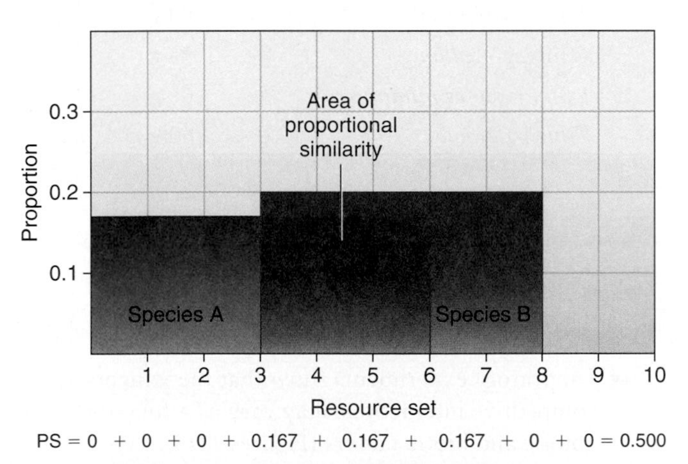


**Figure 11.17 Hypothetical distributions of species 1 and species 2 on a resource set subdivided into 10 resource units.** (Based on data in Figure 11.16.)

**ECOLOGICAL INQUIRY**

If species separated out along three resource axes, would they be able to coexist if the overall *PS* value was 0.6?

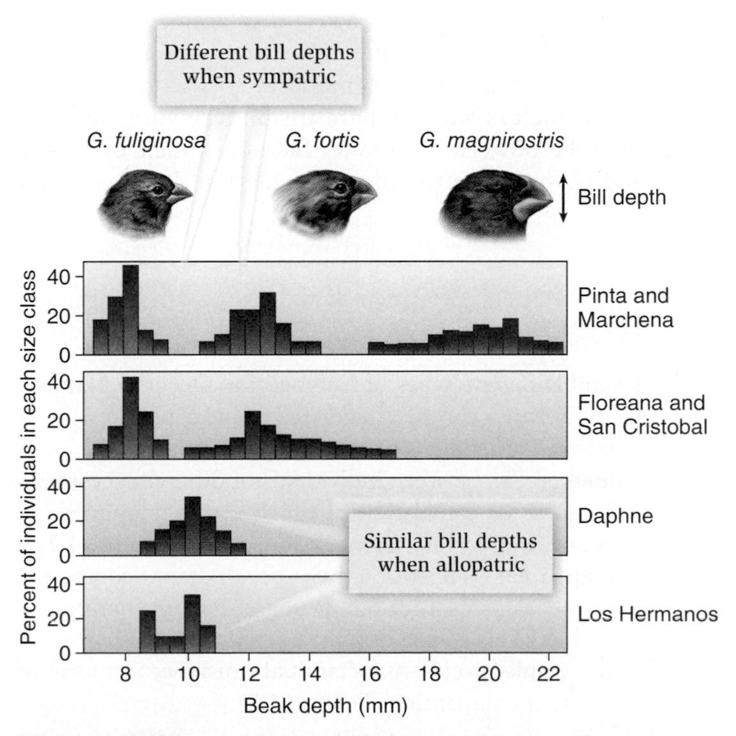


**Figure 11.18 Character displacement in beak size among Galápagos finches.** (After Lack, 1947.)

**Table 11.1 Comparison of feeding characters of sympatric and allopatric species.**

| | | Measurement (mm) when | | Ratio* when | |
|---|---|---|---|---|---|
| Animal (character) | Species | Sympatric | Allopatric | Sympatric | Allopatric |
| Weasels (skull) | *Mustela erminea* | 50.4 | 46.0 | 1.28 | 1.07 |
| | *Mustela nivalis* | 39.3 | 42.9 | | |
| Mice (skull) | *Apodemus flavicollis* | 27.0 | 26.7 | 1.09 | 1.04 |
| | *Apodemus sylvaticus* | 24.8 | 25.6 | | |
| Nuthatches (beak) | *Sitta tephronota* | 29.0 | 25.5 | 1.23 | 1.02 |
| | *Sitta neumayer* | 23.5 | 26.0 | | |
| Galápagos finches (beak) | *Geospiza fortis* | 12.0 | 10.5 | 1.43 | 1.13 |
| | *Geospiza fuliginosa* | 8.4 | 9.3 | Av = 1.26 | Av = 1.06 |

*Ratio of the larger to smaller character.

How great do differences between characters have to be in order to permit coexistence? Hutchinson surveyed the ecological literature of the time and noted that the ratio between feeding characters when species were sympatric, and thus competed, averaged about 1.3 (**Table 11.1**). In contrast, the ratio between feeding characters when species were allopatric, and did not compete, was closer to 1.0. Hutchinson proposed that the value of 1.3, a roughly 30% difference, could be used as an indication of the amount of difference necessary to permit two species to coexist.

In summary, competition occurs frequently in nature and involves several different competitive mechanisms. Competition over the same limited resource can lead to the extinction of one of the species occupying the same niche. However, species may coexist if their niche overlap is reduced through resource partitioning or character displacement. We also saw that the outcome of competitive interactions may vary according to both abiotic and biotic factors in the environment. Biotic factors such as parasitism, herbivory, and predation may be powerful mortality factors in their own right and could be sufficient to reduce population densities to the extent that competition does not occur. In the following sections we will examine the strength of these factors, beginning with predation.

### Check Your Understanding

**11.5** Assume that the size of some birds gives us a clue as to their diet. Bigger birds eat bigger prey items. According to a 1979 study of ground-feeding birds of Finnish peatlands, which of the following five species of coexisting birds would be most likely to go extinct?

| Species | Weight |
|---|---|
| *Numenius phaeopus* | 360 g |
| *Tringa nebularia* | 180 g |
| *Gallinago gallinago* | 95 g |
| *Lymnocryptes minimus* | 68 g |
| *Tringa glareola* | 60 g |

## SUMMARY

**11.1 Several Different Types of Competition Occur in Nature**

- Competition can be categorized as intraspecific, between individuals of the same species, or interspecific, between individuals of different species. Competition can also be categorized as exploitative competition or interference competition (Figure 11.1).
- In apparent competition, two species do not compete for the same resource but do share the same natural enemies. This can lead to the exclusion of the more susceptible species. Associational resistance is a form of apparent competition (Figure 11.2).

**11.2 The Outcome of Competition Can Vary with Changes in the Biotic and Abiotic Environments**

- Laboratory experiments show that the winners of competitive interactions may vary as a function of both abiotic and biotic factors (Figures 11.3).

**11.3 Field Studies Show That Interspecific Competition Occurs Frequently**

- Field experiments show that one species can exclude the other in a natural environment, affecting its distribution within a habitat (Figure 11.A).
- The frequency of competition may vary according to the number of species utilizing a resource (Figure 11.4).

- Competition may be particularly intense among members of a guild that feed in a similar way (Figure 11.5),
- Many invasive species are widely distributed and may outcompete native species (Figures 11.6, 11.7, Figure 11.B).

**11.4 The Winners and Losers of Competitive Interactions May Be Predicted using mathematical Models**

- Lotka-Volterra models of competition may be used to describe competitive interactions (Figures 11.8–11.11).
- Tilman's R* model may also be used to determine which species may outcompete others via resource competition (Figures 11.12, 11.13).

**11.5 Species May Coexist If They Do Not Occupy Identical Niches**

- Species may partition resources to minimize competition and promote coexistence (Figures 11.14–11.17).
- Morphological differences between species, especially between their feeding apparatus, may also permit species coexistence (Figure 11.18, Table 11.1).

## TEST YOURSELF

1. Two species of birds feed on similar types of insects, and nest in the same tree species. This is an example of:
   a. Intraspecific competition
   b. Interference competition
   c. Exploitation competition
   d. Territorial competition
   e. Allelopathy
2. The experiments conducted by Thomas Park using flour beetles provided evidence that the results of competition are influenced by:
   a. Moisture
   b. Temperature
   c. Parasitism
   d. Stochasticity
   e. All of the above
3. In Lotka-Volterra competition models, if $K_2/\beta < K_1$ and $K_1/\alpha > K_2$, then:
   a. $N_1$ is eliminated
   b. $N_2$ is eliminated
   c. Either species may be eliminated, depending on starting conditions
   d. Both species coexist
   e. None of the above
4. Divergence in morphology that is a result of competition is termed:
   a. Competitive exclusion
   b. Resource partitioning
   c. Character displacement
   d. Amensalism
   e. Proportional similarity
5. Two species of barnacle, *Semibalanus balanoides* and *Chthamalus stellatus*, occur in the intertidal regions of Scotland. Which of the following statements is correct?
   a. *Chthamalus* occurs lower down the rock face than *Semibalanus*
   b. *Semibalanus* is more resistant to desiccation than *Chthamalus*
   c. *Semibalanus* outcompete *Chthamalus* over most of the rock face
   d. *Chthamalus* larvae settle lower down the rock face than those of *Semibalanus*
   e. *Semibalanus* has a larger fundamental niche than *Chthamalus*
6. Two species of insect are found on a species of plant with six leaf pairs. Species A occurs with equal abundance on each of the bottom five pairs and Species B occurs with equal abundance on each of the top five pairs. What is the proportional similarity?
   a. 0.2
   b. 0.4
   c. 0.6
   d. 0.8
   e. 1.0
7. In *d*/*w* analysis, what is not true?
   a. $w$ = 68% of the area on one side of the resource use curve
   b. $K_x$ = carrying capacity
   c. $d/w < 1$ = no competition
   d. $1 < d/w > 3$ = competition occurs
   e. $d/w > 3$ = coexistence possible
8. What makes invasive plant species good competitors?
   a. Loss of natural enemies
   b. Native species are very susceptible to allelochemicals secreted by invasives
   c. Invasives are more efficient utilizers of resources
   d. All of the above
   e. None of the above
9. A group of species that feed on the same resources, in the same way, is known as a:
   a. Group
   b. Guild
   c. Niche
   d. Fundamental Niche
   e. Realized niche
10. In a group of related bird species, larger species eat larger food items while smaller species eat smaller items. This is an example of:
    a. Competitive exclusion
    b. Interference competition
    c. Intraspecific competition
    d. Resource partitioning
    e. The ghost of competition past

## CONCEPTUAL QUESTIONS

1. Discuss the different types of competition found in nature.
2. Explain how the Lotka-Volterra models describe interspecific competition.
3. Define the competitive exclusion principle.
4. Discuss two different ways in which ecologists measure niche partitioning among species.
5. Explain how character displacement reduces interspecific competition.

## DATA ANALYSIS

*Desmodium* are small herbaceous plants that live in the oak woodlands of the midwestern United States. An ecologist performed an experiment whereby he planted each of two species 3 m from any other *Desmodium* (no competition), 10 cm from a large individual of a different species (interspecific competition), or 10 cm from a large individual of the same species (intraspecific competition). He then measured the increase in length of all leaves as a measure of growth. The results are shown below. What can you conclude?

| | *Transplant species* | |
|---|---|---|
| | ***D. nudiflorum*** | ***D. glutinosum*** |
| Grown alone | 25.0 | 130.0 |
| Grown next to *D. nudiflorum* | 9.6 | 41.5 |
| Grown next to *D. glutinsoum* | 2.6 | 25.7 |

A European buff-tailed bumblebee.

CHAPTER 12

# Facilitation

## Outline and Concepts



Approximately 85.5% of flowering plants depend on insects to transfer pollen from one individual to another. The rest depend on wind. In the 1990s several studies suggested that some insect pollinators were declining, but there was little evidence of a wide-scale problem. That changed in 2006, when Jacobus Biesmeijer and colleagues compiled 1 million observations of native bee and hoverfly observations in Britain and the Netherlands and compared the numbers of species recorded in all areas before and after 1980. The study was conducted as part of a broad European Union biodiversity research program called ALARM (Assessing LArge scale Risks for biodiversity with tested Methods). The data came from a wide variety of sources, including modern scientific studies and records from as far back as the 19th century. Many areas showed declines in the numbers of bee species since 1980, and some also showed declines in numbers of hoverfly species (**Figure 12.1a**). Such changes suggest possible shifts in pollination services, because bees and hoverflies facilitate the transfer of pollen. The distribution of both British insect-pollinated and Netherlands bee-pollinated plants declined over a similar period (**Figure 12.1b**). At the same time, there were no decreases in the distributions of wind-pollinated or self-pollinating plants; in fact, the area of distribution of wind-pollinated species increased over this same time period. Such results are also cause for alarm because a loss of pollinators could also have an economic impact for agricultural plants.

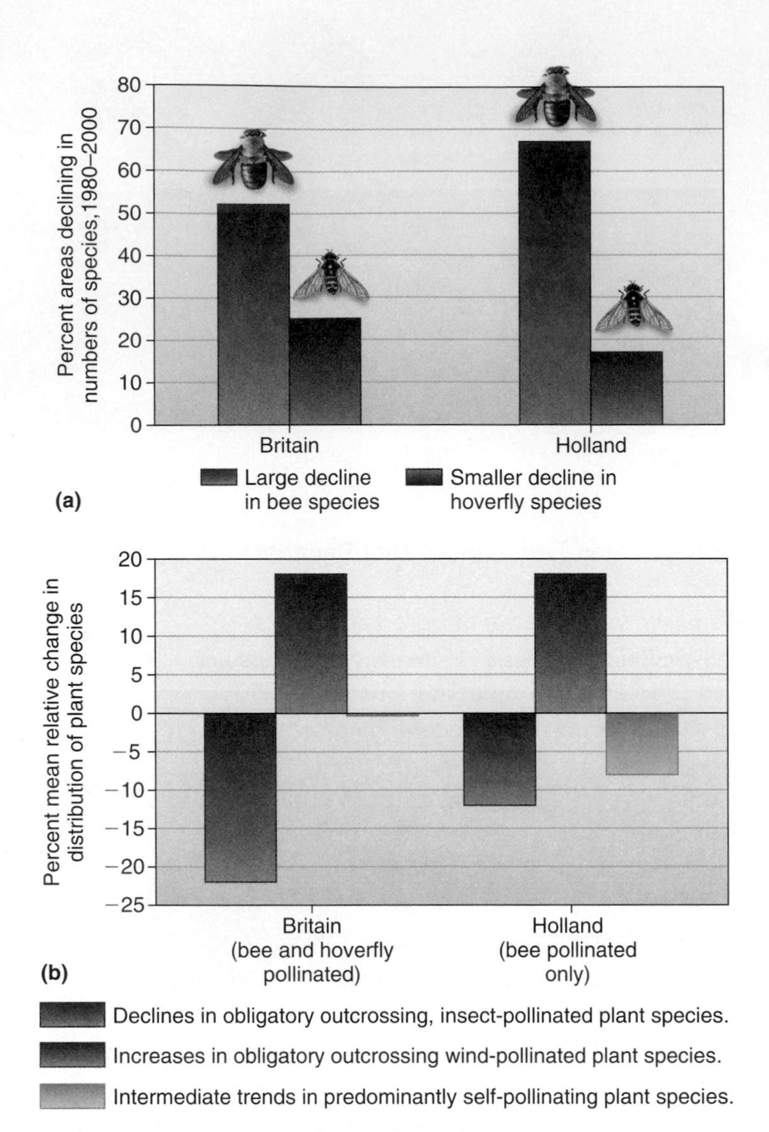


**Figure 12.1** **Pollinator decline in Europe.** **(a)** The percentage of areas where pollinators, mainly bees, declined in Britain and Holland from 1980 to 2000. **(b)** The decline in insect-pollinated plants over the same period.
(From data in Biesmeijer et al., 2006.)

Documenting pollinator changes in other areas and uncovering the causes of the decline remains a top research priority. Penelope Whitehorn and her colleagues (2012) showed that at least part of the decline is linked to widespread use of neonicotinoid pesticides, widely applied as a seed dressing on most major crops, including corn, sunflower, sugar beet, and cereals. Being systemic, these pesticides spread throughout the plant, even reaching the pollen, and deter insect herbivores. These researchers found that bee colonies treated with field-realistic levels of neonicotinoids showed reduced growth and suffered an 85% reduction in the production of new queens.

Facilitation describes species interactions that are beneficial to at least one of the species involved while the other species is unaffected or benefits as well. There are two types of facilitation: mutualism and commensalism, both of which we will examine in this chapter. In mutualism, both species benefit from the interaction. For example, in mutualistic pollination systems, both plant and pollinator benefit, the former by the transfer of pollen and the latter typically by a nectar meal. There are several different types of mutualism: dispersive, defensive, and resource-based. The dispersal of plant pollen or seeds is a dispersive mutualism. In defensive mutualism one species defends the other species in return for a reward, such as ants defending some plants against herbivory in return for a nectar reward. Resource-based mutualism involves two species which can survive and grow better when together than when alone. We will discuss numerous examples of these types of mutualism, and we will see that such interactions are not easily modeled mathematically because the models tend to produce unstable predictions. In commensalism, one species benefits and the other remains unaffected. For example, in some forms of seed dispersal, barbed seeds are transported to new germination sites in the fur of mammals. The seeds benefit, but the mammals are generally unaffected. As we will see, mutualism and commensalism tend to be more common in stressful environments, where it is hard to survive alone, than in more benign environments, where it is easier to prosper.

## 12.1 Mutualism Is an Association Between Two Species That Benefits Both Species

Many close associations are known between species in which both species benefit. For example, many deep-sea fish harbor bioluminescent bacteria in specialized organs. The organs provide safe haven for the bacteria while the fish take advantage of the light-emitting properties of the bacteria to communicate with other fish. In deep-sea thermal vent communities, sulfur-oxidizing bacteria live within the bodies of worms and mussels. The bacteria derive energy from the oxidation of sulfide-rich water and are able to form dense colonies inside the worms and mussels. In turn, the simple sugars the bacteria produce are used by their hosts. In the case of the worms, *Riftia pachytila,* there is no mouth, gut, or anus, only highly vascularized areas packed with these sulfur-oxidizing bacteria. On coral reefs, "cleaner" fish nibble parasites and dead skin from their client fish at specific cleaning stations. On reefs cleared of their cleaner fish, client fish develop skin diseases and their populations decline.

It is interesting to note that humans have entered into mutualistic relationships with many species. For example, the mutualistic association of humans with crops and livestock has resulted in some of the most far-reaching ecological changes on Earth (see **Global Insight**). This is because humans have planted huge areas of the Earth with crops, allowing these plant populations to reach densities that they would never reach on their own. In return, the crops have led to expanded human populations because of the increased amounts of food they provide. An extensive array of microbial species live in the human digestive tract, enabling humans to assimilate the otherwise indigestible cellulose in plant cell walls. These bacteria and protists break down the cellulose into monosaccharides that can be absorbed with other by-products of microbial digestion, including fatty acids and some vitamins.

Different types of mutualism occur in nature. Where neither species can live without the other, the mutualism is called **obligate mutualism**. For example, lichens are an inseparable mix of fungi and algae. This contrasts with **facultative mutualism**, in which the interaction is beneficial but not essential to the survival and reproduction of either species. For example, many ant species exist in mutualistic relationships with aphids, which provide them with a nutritious fluid called honeydew, but the ants can exist without them. Mutualisms can also be subdivided according to the services provided, regardless of whether the participants are obligate or facultative mutualists. **Dispersive mutualism** includes plants and the pollinators that disperse their pollen, and plants and the fruit eaters that disperse the plant's seeds. **Defensive mutualism** often involves an animal defending a plant or an herbivore. Finally, **resource-based mutualism** involves the increased acquisition of resources for both species.

### 12.1.1 Dispersive mutualism involves dispersal of pollen and seeds

Many examples of plant–animal mutualism involve pollination and seed dispersal. From the plant's perspective, an ideal pollinator would be a specialist, moving quickly among individuals but retaining a high fidelity to a plant species; otherwise, its pollen could be wasted. Two ways that plant species in an area promote the pollinator's fidelity are by specific flowering at certain times during the year and synchronized flowering of all individuals within a species. The plant should provide just enough nectar to attract a pollinator's visit. From the pollinator's perspective, it would be best to be a generalist and obtain nectar and pollen from as many flowers as possible in a small area, thus minimizing the energy spent on flight between patches. This suggests that although mutualisms are beneficial to both species, the species' needs are quite different.

#### *Pollination*

Although many flowers are pollinated by a wide array of animals, others are pollinated only by a specific species of pollinator. In turn, while some pollinators visit a wide variety of flowers, others visit only flowers of one particular plant species. These specializations, which result from coevolution, are known as **pollination syndromes**. For example, flowers pollinated by hummingbirds have bright colors to attract the birds, but little odor, because birds have a poor sense of smell. They have to produce copious nectar because hummingbirds have a high nutrient requirement. Other examples of pollination syndromes are given in **Table 12.1**.

**Table 12.1 Pollination syndromes.**

| Animal features | Coevolved flower features |
|---|---|
| **Bees** | |
| • Color vision includes UV, not red | • Often blue, purple, yellow, white (not red colors) |
| • Good sense of smell | • Fragrant |
| • Require nectar and pollen | • Provide nectar and abundant pollen |
| **Butterflies** | |
| • Good color vision | • Blue, purple, deep pink, orange, red colors |
| • Sense odors with feet | • Light floral scent |
| • Need landing place | • Provide landing place |
| • Feed with long tubular tongue | • Nectar in deep, narrow flora tubes |
| **Birds** | |
| • Color vision, includes red | • Often colored red |
| • Often require perch | • Strong, damage-resistant structure |
| • Poor sense of smell | • No fragrance |
| • Feed in daytime | • Open in daytime |
| • High nectar requirement | • Copious nectar in floral tubes |
| **Bats** | |
| • Color blind | • White or light, reflective colors |
| • Good sense of smell | • Strong odors |
| • Active at night | • Open at night |
| • High food requirements | • Copious nectar and pollen provided |
| • Often require perch | • Pendulous or borne on tree trunks |

After Brooker et al., 2008.

There are cases in which flower or pollinator try to cheat, gaining an advantage over the other. In the bogs of Maine, the grass pink orchid, *Calopogon pulchellus* (**Figure 12.2a**), produces no nectar, but it mimics the nectar-producing rose pogonia, *Pogonia ophioglossoides* (**Figure 12.2b**) and is therefore still visited by bees, *Bombus* species. Bee orchids, *Ophrys apifera,* mimic female bees; in trying to copulate with the flowers, male bees pick up and transfer pollen to the stigma of another flower (**Figure 12.2c**). So effective are the stimuli of flowers of this orchid that male bees prefer to mate with them even in the presence of real female bees! Conversely, some *Bombus* species cheat by biting through the petals at the base of flowers and robbing the plants of their nectar without entering through the tunnel of the corolla and picking up pollen (**Figure 12.2d**). In these cases, only one species of the pair is benefiting from the relationship, and what may have started out as a mutualism has turned into a different type of relationship. In the case of the grass pink orchid the

## The Mutualistic Relationships of Humans with Crops and Livestock Have Produced Dramatic Environmental Changes

The mutualisms that have produced perhaps the most far-reaching ecological effects on Earth involve those associated with human agriculture (**Figure 12.A**). Most introduced crops will always remain dependent on people for their survival, requiring that water and fertilizer be added or competing weeds and insect herbivores be removed. Most of the world's crops and livestock, whose origins are shown in **Figure 12.B**, have been moved around the globe relatively recently in ecological time. The population size of humans is increased by the presence of these crops and domesticated animals, and similarly, populations of crops and domesticated animals are increased in the presence of humans. Excluding Antarctica, a staggering 37.3% of the world's land area is given over to crops and grazing pastures for animals such as cattle and sheep (**Table 12.A**), and the number of livestock in the world runs into the billions (**Table 12.B**). No other human endeavor on Earth requires as much land as agriculture. Unfortunately, the ramifications for natural systems don't stop simply with the acreage of land used. They also include:

1. *Pollution of water bodies.* Runoff water from agricultural land often contains high levels of nitrogen and phosphorus from fertilizers as well as residual pesticides and manure, all of which contaminate streams. These can all affect communities of aquatic plants and animals.
2. *Loss of topsoil.* Some 4 billion tons of topsoil are washed into U.S. waterways each year, severely impacting the organisms that live there. Topsoil can also be whipped away by wind.

(i)

(ii)

**Figure 12.A Modern agriculture illustrates one of the most far-reaching kinds of mutualism.** **(i)** These crops could not exist without human help, and at the same time, crops sustain human populations. Such mutualisms between humans and crops and **(ii)** humans and livestock have radically altered the ecological landscape.

**Table 12.A Agricultural and grazing lands, relative to total area.**

| Continental Area | % Cropland | % Pastures | Total |
|---|---|---|---|
| World | 11.5 | 25.8 | 37.3 |
| Asia | 20.0 | 33.0 | 53.0 |
| Central America and the Caribbean | 15.6 | 36.7 | 52.3 |
| Europe | 13.1 | 7.9 | 21.0 |
| Middle East and North Africa | 7.6 | 28.1 | 35.6 |
| North America | 11.4 | 12.5 | 23.9 |
| Oceania | 6.2 | 48.2 | 54.4 |
| South America | 6.8 | 28.8 | 35.6 |
| Sub-Sahara and Africa | 8.1 | 33.9 | 42.0 |

**Table 12.B Numbers of livestock in the world in 2005.**

| Livestock | Millions |
|---|---|
| Chickens | 16,740 |
| Cattle | 1,355 |
| Sheep | 1,081 |
| Ducks | 1,046 |
| Pigs | 960 |
| Goats | 807 |
| Geese | 301 |
| Turkeys | 279 |
| Buffaloes | 174 |
| Horses | 55 |
| Asses | 40 |
| Camels | 19 |
| Mules | 12 |

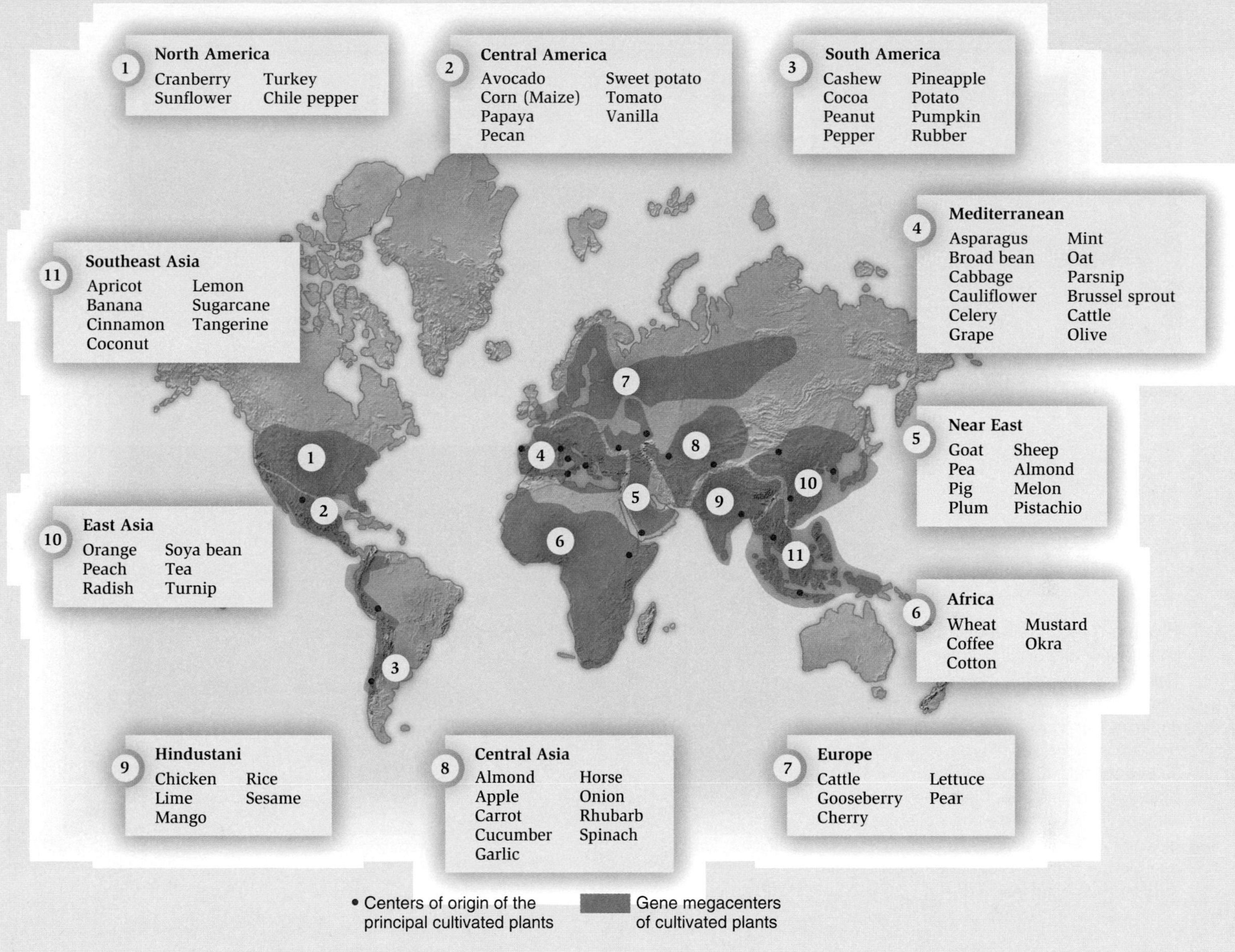


**Figure 12.B** Genetic origins of crops and livestock.

3. *Depletion of water supplies.* About 70% of the fresh water used by humans is expended on irrigation. Most of this water comes from underground aquifers at rates much in excess of natural recharge rates. In Africa, the Middle East, and China, groundwater levels are falling by as much as 1 meter per year. Many seasonal ponds and lakes are dramatically reduced in size, even dried up, by such activity, with dire consequences for amphibians, insects, fish, and other organisms that need standing water.
4. *Salinization.* Excess salt can result from continued irrigation for agriculture. Even "fresh" water contains minute quantities of dissolved salts, and as water evaporates or is used by plants, the salt is left behind and accumulates in the soil, killing crops and native vegetation alike.
5. *Desertification.* Reduced vegetative cover can result from severe overgrazing by livestock. The cattle population in the Sahel, the region immediately south of the Sahara desert in Africa, increased fivefold in the period 1940–1968, following the installation of wells. This ended the nomadic existence of local people and increased the likelihood of desertification in this region.
6. *Deforestation.* Agriculture and pasture lands not only reduce the amount of grasslands, they also cause deforestation and drainage of swamps as other habitats are converted to cropland or rangeland.

Crop and domestic animal mutualisms with humans are vital to sustaining human populations, but the costs to natural ecosystems and to other organisms are significant.

**Figure 12.2** **Mutualistic cheating.** **(a)** In Maine, the grass pink orchid, *Calopogon pulchellus*, produces no nectar but receives pollinators because it mimics **(b)** the rose pogonia, *Pogonia ophioglossoides*, which is a nectar producer. **(c)** Bee orchids, *Ophrys apifera*, mimic the shape of a female bee. Male bees copulate with the flowers, transferring pollen but getting no nectar reward. **(d)** *Bombus* bumblebee biting through the flower corolla to pilfer the nectar inside.

relationship is commensalism, and in the case of the bee species it is herbivory.

While some areas have witnessed declines in native pollinators, as described in the chapter opening, others have seen influxes of invasive pollinators. In south Florida, Hong Liu and Robert Pemberton (2009) studied the biology of *Euglossa viridissima*, a solitary orchid bee that has invaded from Mexico and Central America. This species is a buzz pollinator, meaning that it vibrates its indirect flight muscles to sonicate the pollen from the anthers. Some invasive plant species in south Florida, such as *Solanum* and *Senna* spp., are buzz-pollinated by native carpenter bees, bumblebees, and halticid bees. Liu and Pemberton tested the possibility that the addition of a new species of pollinator could increase fruit set. They installed small and large mesh cages around the flowers of invasive *Solanum torvum*. The small mesh cages let in only native halticid bees, which were smaller than the invasive orchid bees. The large mesh cages let in both species of pollinators. Fruit set was elevated in the large mesh cages, compared to the small mesh cages, indicating that the invasive orchid bee greatly increased seed set in the invasive *Solanum* (**Figure 12.3**). Liu and Pemberton noted that such "invasive mutualisms" could promote population growth of the weed and pose management problems.

### *Seed dispersal*

Mutualistic interactions are also highly prevalent in the seed-dispersal systems of plants. Fruits provide a valuable source of carbohydrates and vitamins. In return for this juicy meal, animals unwittingly disperse the enclosed seeds, which often pass through the digestive tract unharmed. Many plants signal fruit ripeness by changing color from green to red, orange, blue, or black. Fruits taken by birds and mammals often have attractive colors (**Figure 12.4a**); those dispersed by nocturnal bats are not always brightly colored but instead give off a pungent odor that attracts the bats (**Figure 12.4b**). In the floodplains of the Amazon, fruit- and seed-eating fish have evolved that disperse seeds (**Figure 12.4c**).

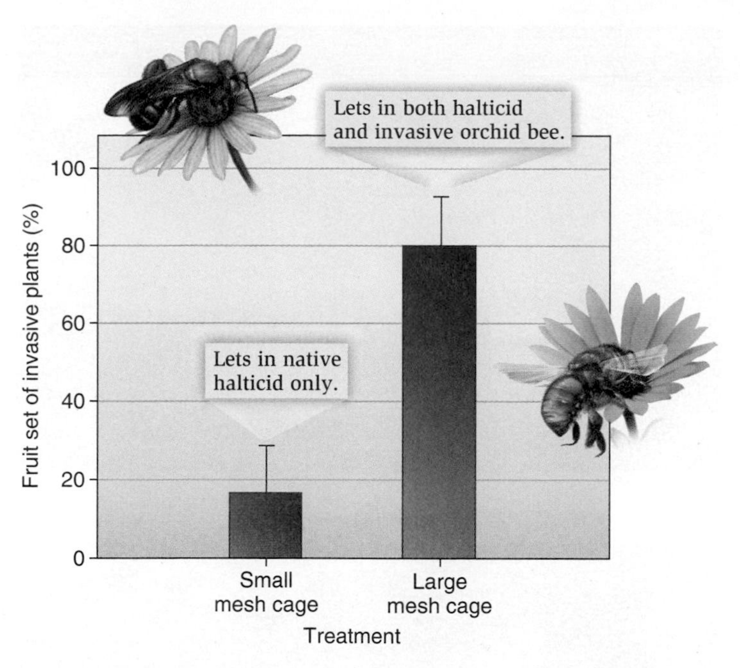


**Figure 12.3 Effect of native and invasive pollinators on fruit set in an invasive plant, *Solanum torvum*, in south Florida.** Flowers were protected by small mesh cages that let in halticids, or large mesh cages that let in all pollinators. (After Liu and Pemberton, 2009.)

**ECOLOGICAL INQUIRY**

Why did these researchers have a large mesh cage treatment? Why not have no cage at all?

A number of hypotheses have been suggested to explain why seed dispersal is so advantageous to plants. The **competition avoidance hypothesis** suggests that dispersal avoids competition with the parent plant, but on the other hand there is no guarantee that the seed will fall into an optimal habitat. The **predator escape hypothesis** suggests seed predators congregate under parent trees to feed on the bonanza of fallen seeds, so well-dispersed seeds suffer less predation. The **colonization hypothesis** suggests that constantly shifting environmental conditions for seed germination mean that parental location is not always a good predictor of seedling success. Many forest trees blanket the ground with propagules because light gaps are required for seedlings to grow to reproductive maturity, and gap formation is unpredictable. The **directed dispersal hypothesis** suggests some dispersers distribute seeds into optimal sites. These hypotheses are not mutually exclusive. Directed dispersal may be more widespread than originally thought. Two examples illustrate this. Mistletoes are obligate stem parasites whose seeds are dispersed by passerine birds. The seeds must be deposited directly on tree twigs of the correct diameter. Large twigs cannot be penetrated by the germinating mistletoe, and small twigs often die after infections. The medium-sized twigs are the ones most often perched on by birds.

(a)

(b)

(c)

**Figure 12.4 Dispersive mutualism.** **(a)** This blackbird, *Turdus merula,* is an effective seed disperser of brightly colored seeds. **(b)** Fruit bats, such as this flying fox, *Pteropus conspicillatus,* in Australia, disperse fruits with strong odors and weak colors. **(c)** This berry-eating *Piranha* fish is an important seed disperser in the Amazonia floodplain.

Furthermore, the seeds contain a sticky substance called viscin, which causes the seeds to clump together, even after passage through the bird's gut. They often stick to branches, or the birds, and are directly wiped onto the branch by the bird. Seeds defecated onto the ground die. Our second example concerns the tropical and temperate plants that produce seeds with lipid-rich attachments called elaiosomes or arils. Ants carry the seeds to their nests, eat the nutrient-rich elaiosome, and discard the seed. The nutrient-rich seed pile provides an excellent germination site for the seeds.

Cheating can also occur in seed-dispersal systems. Microbes appear to be good at cheating plants, for they readily attack the fruit without dispersing it. Daniel Janzen (1979) suggested that, over time, microbes have gained a selective advantage by causing fruit to "rot," rendering it objectionable to larger animals and reserving it for themselves.

## 12.1.2 Defensive mutualism involves one species defending another in return for a reward

One of the most commonly observed defensive mutualisms occurs between ants and aphids. Aphids are fairly defenseless creatures and easy prey for most predators. The aphids feed on plant sap and have to process a significant amount of this dilute food medium to get their required nutrients. Like most organisms, the aphid's digestive tract is not 100% efficient, and some of the sugars still remain in the excreted fluid, which is called honeydew. The ants drink the honeydew and, in return, protect the aphids from an array of predators by driving the aphid's predators away (**Figure 12.5a**).

In other cases, ants enter into a mutualistic relationship with a plant itself. The large thorns of acacia trees in Central America provide food and nesting sites for ants. In return, the ants bite and discourage both insect and vertebrate herbivores from feeding on the trees. Ants also trim away foliage from competing plants and kill neighboring plant shoots, ensuring more light, water, and nutrient supplies for the acacias. Among the most famous cases is bull's horn acacia, *Acacia cornigera,* which has large hornlike thorns filled with a soft pithy interior that is easy for the ants to hollow out (**Figure 12.5b**). The ants, generally *Pseudomyrmex ferruginea,* make nests inside the thorns. In addition, the *Acacia* provide two forms of food to the ants, Beltian bodies and extrafloral nectaries.

Beltian bodies are tiny protein-rich food nodules at the end of the leaflets. They are named after the early tropical naturalist Thomas Belt (1874), who first suggested that ants may indeed protect *Acacia* plants. The extrafloral nectaries are carbohydrate-rich glands located at the base of the leaves. In this particular example, the mutualism is obligate, though this is not always the case with ants and acacias. *Acacia* species that are not defended by ants have toxic compounds in their foliage to protect against herbivores, whereas ant-defended species do not. Daniel Janzen (1966) showed that removing the ants from bull's horn acacia allowed the number of insect herbivores to increase nearly eightfold. In turn, shoot growth was reduced sevenfold. In patrolling the vegetation the ants were active 24 hours per day—an unusual feature for ants. In return the *Acacia* retained its leaves year round, to provide food for the ants, whereas non-ant-defended *Acacia* normally

(a)

(b)

**Figure 12.5 Defensive mutualism involves species that receive food or shelter in return for providing protection.** **(a)** This red carpenter ant, *Camponotus pennsylvanicus,* tends aphids feeding on a twig. The ants receive sugar-rich honeydew produced by the aphids and in return protect the aphids from predators. **(b)** *Pseudomyrmex ferruginea* ants make nests inside the large, hornlike thorns of the bull's horn acacia and defend the plant against insects and mammals. In return, the acacia, *Acacia corrigera,* provide two forms of food to the ants: protein-rich granules called Beltian bodies and nectar from extrafloral nectaries, nectar-producing glands that are physically apart from the flower.

**ECOLOGICAL INQUIRY**

Is the relationship between red carpenter ants and aphids an example of facultative or obligatory mutualism?

drop their leaves in the dry season. Such ant–plant protection mutualisms have been found frequently, especially in the Tropics.

Felix Rosumek and colleagues (2009) conducted a meta-analysis of the effects of ant removal on plant herbivory, herbivore and predator abundance, and plant reproduction. They examined data from 81 studies. Ant removal increased herbivory by 50%, reduced plant biomass by 23%, and reduced plant reproduction by 24%. These results highlighted the strong positive effects of ants on plants. The effects were usually stronger in tropical environments than in temperate ones. The authors suggested that this effect was caused by higher ant diversity and pugnaciousness in tropical versus temperate communities.

Another defensive mutualism involves fungi that live inside the leaves of grasses and produce toxic chemicals that deter herbivores from feeding on the grasses. A study by Dawn Bazely and her associates (1998) showed how a defensive mutualism may be responsible for population fluctuations of Soay sheep, *Ovis aries,* on Hirta, one of many Scottish islands in the St. Kilda archipelago (**Figure 12.6**). The population crashes every 3–5 years, with up to 60% mortality. The main forage of the sheep is the grass *Festuca rubra,* which harbors in its blades an endophytic fungus, *Acremonium* sp. The fungus synthesizes toxic chemicals that function as an antiherbivore defense; in return, the fungus gains access to a food source within the plants' leaves. The fungi are in highest concentrations in the basal regions of the plants. As sheep numbers increase and pasture height decreases through grazing, the sheep encounter higher concentrations of the chemicals, which reduce the sheep numbers. When herbivory is reduced, grass grows longer and concentration of chemicals in the tips of the grass is less, permitting sheep population density to increase. Such studies are important to an understanding of the widespread losses to the livestock industry due to fungal-infected fescue grass, especially in the U.S. An understanding of such mutualisms may help minimize livestock losses.

### 12.1.3 Resource-based mutualism involves species that can obtain resources better together than alone

Many mutualisms are based on the fact that both partners can improve the supply of essential resources. About 90% of seed plants have mutualistic associations with fungi that live on or in the root tissue. These associations are called mycorrhizae (from the Latin *myco,* referring to fungi, and *rhiza,* meaning foot) (**Figure 12.7a**). The fungi require soluble carbohydrates from their host as a carbon source, and they supply mineral resources and water, which they are able to extract more efficiently from the soil than the host. Most plant species are susceptible to colonization by mycorrhizal fungi, though some, like tropical species, rely heavily upon them. In tropical systems the soil is nutrient-poor, nutrients having been washed away by heavy rains. Nutrients are locked up in living or recently dead and decaying organisms. Mycorrhizae are efficient at rapidly extracting remaining nutrients from decaying material.

Leaf-cutting ants of the tribe Attini, of which there are about 210 species, enter into a resource-based mutualistic relationship with a fungus (**Figure 12.7b,c**). A typical colony of about 9 million ants has the collective biomass of a cow and harvests the equivalent of a cow's daily requirement of fresh vegetation. Instead of consuming them directly, the ants chew the leaves into a pulp, which they store underground as a substrate on which the fungus grows. The ants shelter and tend the fungus, protecting it from competing fungi and helping it reproduce and grow. In turn, the fungus produces specialized structures known as gongylidia,

(a)

(b)

**Figure 12.6 Defensive mutualism.** **(a)** Soay sheep, *Ovis aries,* are less well able to digest grasses where these are defended by mutualistic fungi which live inside the leaves. **(b)** Boreray Island in the St. Kilda archipelago, Scotland.

**Figure 12.7 Resource-based mutualism.** **(a)** Mycorrhizae, like these on soybean, are mutualistic associations of fungi and plant roots. **(b)** Leaf-cutting ants, *Atta cephalotes,* in South America, cut leaves and chew them to a pulp underground. **(c)** The pulp supports a fungal colony on which the ants feed.

which serve as food for the ants. In this way, the ants circumvent the chemical defenses of the leaves, which are digested by the fungus, and the fungi exists as a healthy colony underground. Scientists discovered that these interactions are more complicated than originally thought when it was realized that these gardens were also infected with a second species of fungus, *Escovopsis,* which feeds on the first fungus species. In 2003 it was discovered that *Escovopsis* in turn was kept in check by a fourth species, an Actinomycete bacteria housed and nourished in pits on the ants' bodies, which produced chemicals to keep *Escovopsis* from depleting the fungus garden.

Another example of resource-based mutualism involves corals and the dinoflagellates that live within them. Coral contains unicellular algae, dinoflagellates of the genus *Symbiodinium,* which photosynthesize and provide oxygen and nutrients to their hosts, as well as give the coral its coloration. In return, the coral provides carbon dioxide and waste products used by the dinoflagellates. During conditions that place the coral under stress, such as temperatures above 30°C, dinoflagellate photosynthetic pathways are impaired and poisonous by-products accumulate instead of sugars. To save itself, the coral expels the dinoflagellates. Without the coloration of the dinoflagellates, the coral has a light or white appearance, hence the term "coral bleaching." Bleaching may be adaptive, because corals may become repopulated with different dinoflagellates that may be more resistant to stress. If corals regain their mutualistic partners within a few months, they can recover. If not, they usually die.

Lichens are combinations of algae, which provide the photosynthate, and fungi, which provide nutrients and a safe habitat for the algae. The lichenized fungi include a thin layer of algal cells, which form 3–10% of the weight of the lichen body. Of the 50,000 or so species of fungi, 25% are lichenized. Most nonlichenized fungi are parasites of plants or animals or involved in decomposition. Lichenized fungi occur in deserts and alpine regions, and across a wide range of habitats.

Vertebrate species can also enter into resource-based mutualisms. In African savannas, honeyguides are a group of birds that feed on honey and bee larvae, and can also digest beeswax. They have even been reported to eat candle wax inside churches. Honeyguides are unable to open bees' nests for themselves. For this, they enlist the help of the honey badger, *Mellivora capensis.* Having located a swarm of bees, honeyguides attract the attention of the honey badger by a distinctive call and by flying from perch to perch with tail feathers fanned out. The honey badgers follow the honeyguides to the bees' nest and rip it open and consume the contents, leaving the scraps for the honeyguides to feed on.

Just as some dispersive mutualisms are open to cheating, not all seemingly resource-based mutualisms are what they appear to be. Oxpeckers, *Buphagus* spp., were also thought to enter into a mutualistic relationship with African mammals, especially large ungulates on which they perch and remove ticks and earwax. However, the oxpeckers also feed on blood, and they often delay the scarring of wounds. Paul Weeks (2000) studied the relationship between oxpeckers and domestic cattle on a Zimbabwe farm. He managed to prevent the oxpeckers from gaining access to part of the herd, and then counted the number of ticks and scars on the two groups of cattle at the end of four weeks. After four weeks, the number of ticks was somewhat reduced when the oxpeckers were allowed to feed (**Figure 12.8a**), but there was also an increase in the number of wounds (**Figure 12.8b**). Presumably the oxpeckers would behave in a similar way on wild ungulates, and they are known to exploit wounds on hippopotami. Thus, the oxpeckers behave as much as vampires as honest cleaners, calling into question their overall benefit to ungulates.

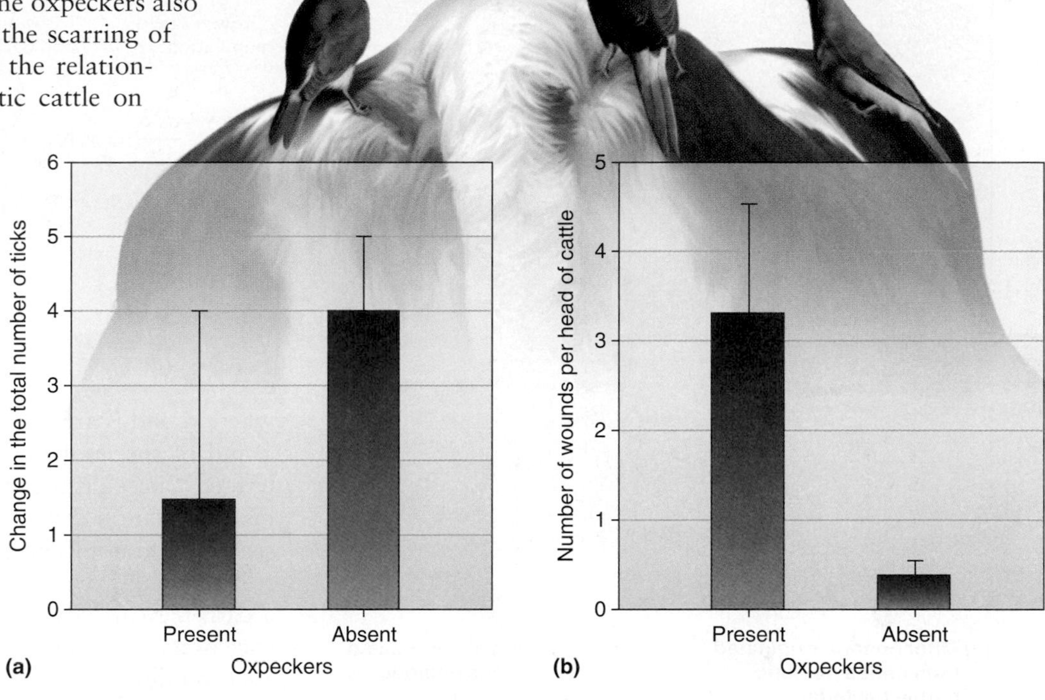


**Figure 12.8 The effects of oxpeckers on tick loads and wound frequency in Zimbabwe cattle over a four-week period.** Oxpeckers **(a)** reduce tick loads but **(b)** increase wound frequency, indicating they act as vampires as well as tickbirds. (After Weeks, 2000.)

## 12.1.4 Some mutualisms may be endosymbiotic, where one species lives in the body of another

Symbiosis is a relationship whereby two species live in close association with one another. Thus, parasitism can be a symbiotic relationship, as can some obligatory mutualisms such as that between ants and bull's horn acacia, or leaf-cutting ants and their fungus gardens. **Endosymbiosis** refers to a close association where one organism lives inside the body of another. Thus, some parasites such as tapeworms are endosymbionts, as are the bioluminescent bacteria in fish, the fungi in grasses, the fungi inside roots, the algae inside corals, and the algae inside lichens. The **endosymbiosis theory** has come to be associated with the idea that mitochondria of eukaryotes evolved from purple aerobic bacteria called α-proteobacteria and that chloroplasts evolved from cyanobacteria. According to this theory, billions of years ago an anaerobic cell ingested an aerobic bacterium but failed to digest it. The bacterium flourished in a food-rich cell, and its excess ATP also benefited the cell. Similarly, a cell must have captured a photosynthetic cyanobacterium and failed to digest it. The cyanobacterium continued to photosynthesize, and some of its ATP also benefited the cell. Eventually both bacteria and cyanobacteria lost the ability to live outside the cell, and the mutualism between eukaryotic cells and prokaryotic cells became obligate. The bacteria became known as mitochondria, and the cyanobacteria as chloroplasts (**Figure 12.9**).

There are several lines of evidence to support this theory. First, mitochondria and chloroplasts contain their own DNA, which may be left over from when they were independent organisms. Second, their DNA sequences indicate that mitochondria are very similar to a group of bacteria called α-proteobacteria and that chloroplasts originated from cyanobacteria. While this hypothesis enjoys wide acceptance, it suggests that the notion of evolution fueled solely by competition is incomplete, because cooperation and mutualism have obviously greatly influenced the evolution of life on Earth.

## 12.1.5 Mutualisms are not easily modeled mathematically

Unlike other species interactions, mutualisms are not easily modeled mathematically. We can show this by trying to model mutualisms with equations similar to the Lotka-Volterra competition equations (**Figure 12.10**). For facultative mutualisms, we could incorporate the positive effect of one species

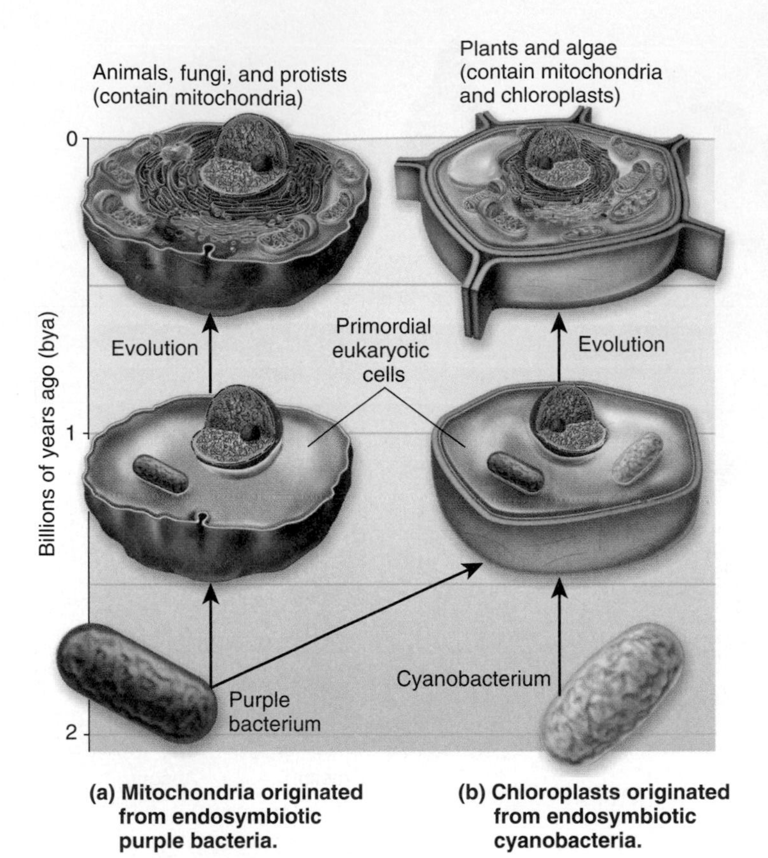


**Figure 12.9 The endosymbiosis theory.** **(a)** Modern mitochondria were derived from purple bacteria. **(b)** Chloroplasts were derived from cyanobacteria. (After Brooker et al., 2008.)

on the other by changing the signs of the competition coefficients such that

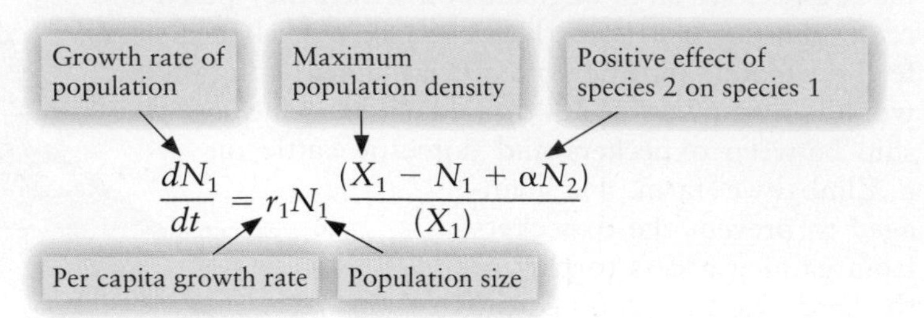


$$\frac{dN_1}{dt} = r_1 N_1 \frac{(X_1 - N_1 + \alpha N_2)}{(X_1)}$$

and

$$\frac{dN_2}{dt} = r_2 N_2 \frac{(X_2 - N_2 + \beta N_1)}{(X_2)}$$

where α and β are the positive effects of species 2 on species 1 and of species 1 on species 2, respectively, and are better termed "mutualism coefficients." In this situation, the maximal population density when each species is alone is represented by $X$, not $K$, because such densities no longer represent real carrying capacities. The reason is that although there is a carrying capacity, $K$, while a mutualist is alone, the presence of its fellow mutualist can increase the carrying capacity. Such modifications can lead to unstable situations in which both populations increase to unlimited size (**Figure 12.10a**). Reducing the coefficients α and β as the populations grow can stabilize the populations and cause a stable equilibrium to be reached (**Figure 12.10b**).

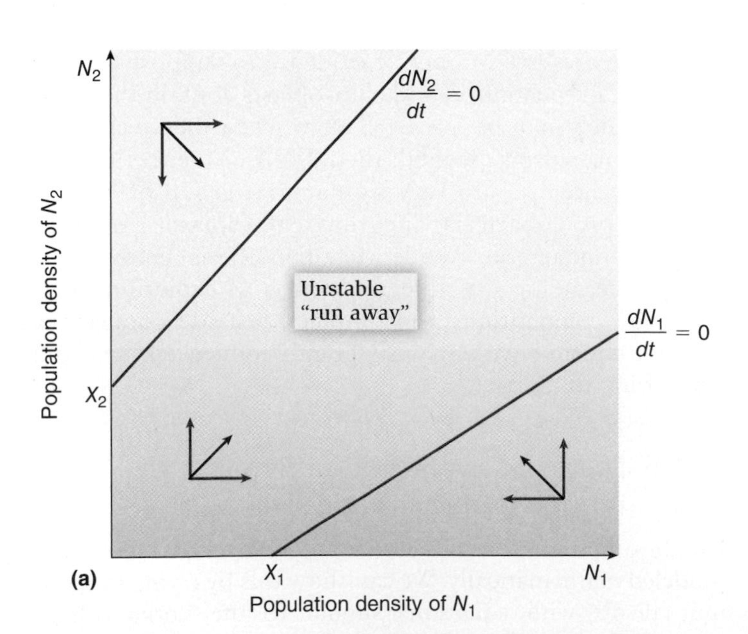


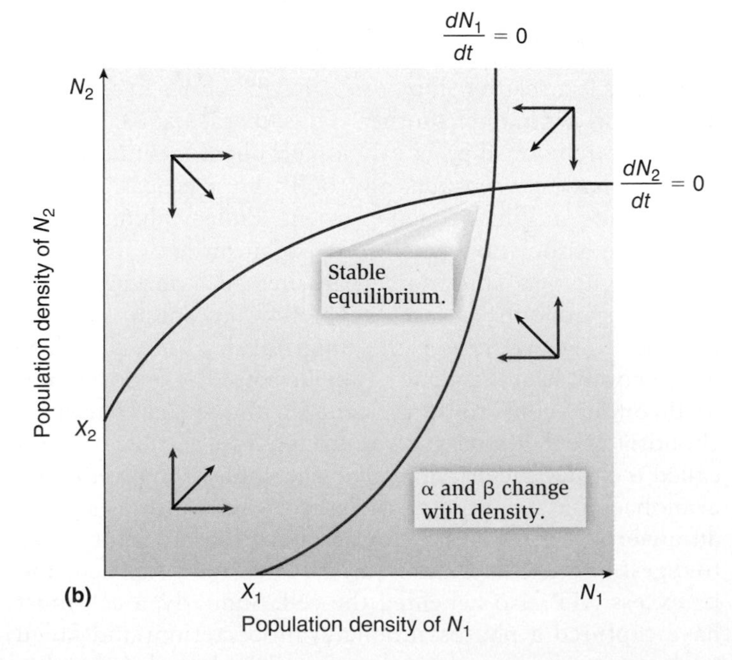


**Figure 12.10 Graphical models of facultative mutualism.** $X$ represents maximum carrying capacities in the absence of the mutualist. If the isoclines fail to cross **(a)**, then unstable dynamics result. Modifications to the Lotka-Volterra equations such that α and β change with density **(b)** can result in a stable equilibrium.

**Check Your Understanding**

**12.1** (a) Why is it difficult to understand mutualisms through mathematical models? (b) According to Janzen, why does fruit "rot" and not simply dry up and decompose?

## 12.2 Commensal Relationships Are Those in Which One Partner Receives a Benefit While the Other Is Unaffected

In commensalism, one member derives a benefit, while the other neither benefits nor is harmed. As with mutualisms, the possible benefits to commensalism include increased access to food, protection from enemies, or increased dispersal ability. Cattle egrets feed in pastures and fields among cattle, whose movements stir up insect prey for the birds. This creates more food for the cattle egret in fewer steps (**Figure 12.11**). The egrets benefit from the association, but the cattle are unaffected. Ecologists have adopted four additional terms to describe different types of commensalism. **Inquilinism** occurs when one species uses a second species for housing. For example, orchids grow in forks of tropical trees. Such plants are known as **epiphytes,** plants that use other plants for support but gain water and nutrients from moist air or from runoff. The tree is unaffected, but the orchid gains support and increased exposure to sunlight and rain. **Phoresy** occurs when one organism uses a second organism for transportation. Flower-inhabiting mites travel between flowers in the nostrils (nares) of hummingbirds. The flowers the mites inhabit live only a short while before dying, so the mite relocates to distant flowers by scuttling into the nares of visiting hummingbirds and hitching a ride to the next flower. Presumably the hummingbirds are unaffected. Many plants have developed seeds with barbs or hooks to lodge in animals' fur or feathers (**Figure 12.12**). In these cases, the plants receive free seed dispersal, and the animals receive nothing, except perhaps minor annoyance. This type of relationship is fairly common; most hikers and dogs have at some time gathered spiny or sticky seeds as they wandered through woods or fields. In **metabiosis**, an organism uses something produced by the first, usually after its death. For example, hermit crabs use snail shells for protection. Commensalism is very common in nature (**Table 12.2**).

Palatable plants can gain protection against herbivores through an association with unpalatable neighbors, a phenomenon known as **associational resistance**. This was shown by Peter Hamback and his colleagues (2000) in Sweden, where the chrysomelid beetle, *Galerucella calmariensis,* feeds on purple loosestrife, *Lythrum salicaria*. Sometimes purple loosestrife grows on its own, and sometimes it grows in thickets of *Myrica gale,* an aromatic shrub. *Myrica* secretes, from its roots and leaves, volatile chemicals that deter insects from feeding on it. The chemicals also interfere with *Galerucella* beetles searching for purple loosestrife, decreasing both the number of beetle larvae and, consequently, damage to the leaf when the plant grows in *Myrica* thickets (**Figure 12.13**). This has important implications for the biological control of invasive species. Purple loosestrife is an important invader of native habitats in the U.S. Northeast. This research shows that biological control would likely be less effective in *Myrica* thickets, because pockets of purple loosestrife would remain. These plants could reinfect areas already cleared by biological control agents.

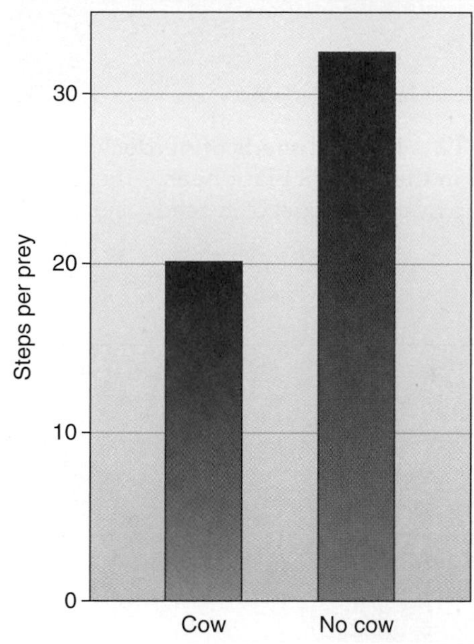


**Figure 12.11 Commensalism between egrets and cattle.** The feeding rates of cattle egrets are different when feeding alone or when associated with cattle. Associations with cattle lead to more feedings per minute and fewer steps per prey item. (After data in Heatwole, 1965.)

Commensal interactions are increasingly found to be common in plant communities. Ragan Callaway (1995) documented 169 cases of commensal interactions among plants. There were many mechanisms responsible. Prominent among these mechanisms were resource modifications where one species positively affected availability of resources such as light, water, or nutrients. Another mechanism was substrate modification, whereby one species might accumulate soil around its roots that the other could use. Callaway (1998) went on to

**Figure 12.12 Hooked seeds of burdock, *Arctium minus*, have lodged in the fur of a black bear.** The plant benefits from the relationship by the dispersal of its seeds, and the bear is not affected.

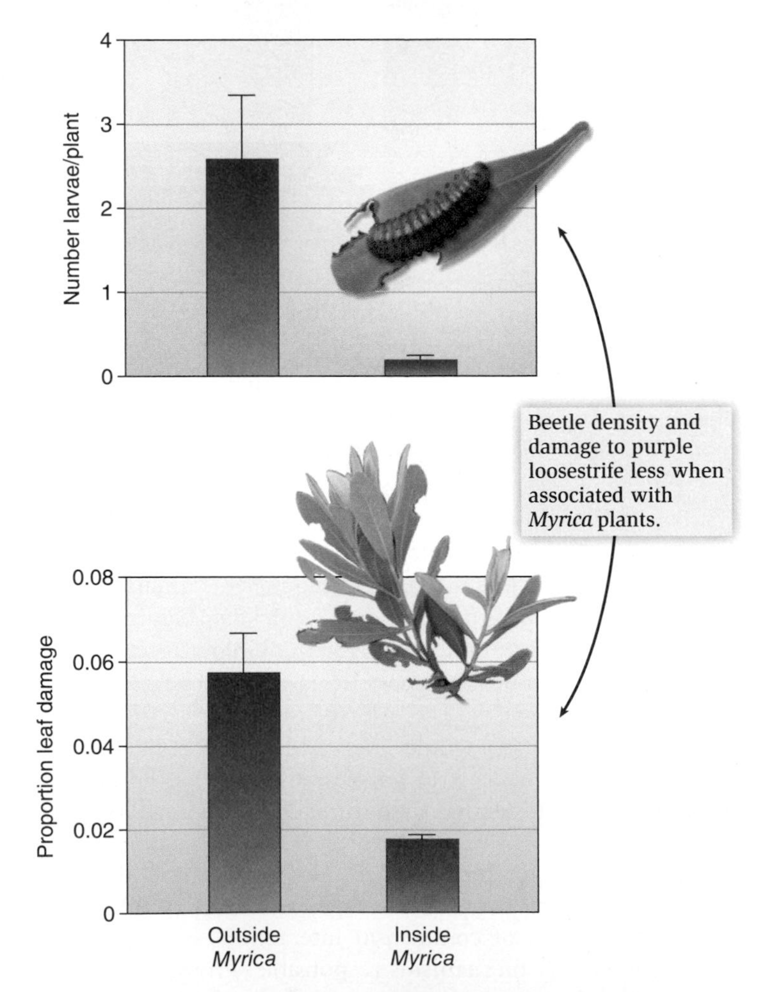


**Figure 12.13 Associational resistance.** The number of beetle larvae per plant and the proportion of leaves damaged on purple loosestrife plants is less when associated with *Myrica* plants. The *Myrica* produce volatile chemicals which interfere with the beetle's host location abilities. (Modified from Hamback et al., 2000.)

**Table 12.2 Some examples of commensalism.**

| Provider | Resource provided and mechanism | Beneficiary |
|---|---|---|
| Pitcher plant | Habitat of water pitcher | Mosquitoes |
| Yellow-bellied sapsucker | Drills holes to collect sap | Hummingbirds and other feeders |
| Shark | Food scraps, transport | Remora fish |
| Marine burrowing worms | Habitat (burrow) | Crabs |
| Cattle | Stir up insect food | Cattle egrets |
| Trees | Habitat | Orchids or other air plants (epiphytes) |
| Sloth | Habitat in hair | Green algae |
| Turtles | Carapace habitat | Baiscladia green algae |
| Whales | Habitat, body surface | Barnacles |
| Sea cucumber | Habitat, cloacal cavity | Pearl fish |
| Portuguese man-o'-war | Protection | Certain fish |
| Polar bears | Food scraps | Arctic fox |
| Anemones | Protection | Clownfish |
| Sea urchins | Protection | Clingfish |
| Insects | Transport | Mites or pseudoscorpions |

**ECOLOGICAL INQUIRY**

Classify these commensalisms as Phoresy, P; Inquilinism, I; Metabiosis, M; or other, O.

suggest that most of these facilitations were species-specific. In other words, the positive effects of one plant on another were not simply due to physical effects that could be caused by inanimate objects, like rocks, which provide shade. Beneficiaries were often found associated with particular benefactor species. This has huge implications for plant communities, because it suggests that many plant communities are not random groups of species but are highly interdependent assemblages.

### Check Your Understanding

**12.2** Figure 10.10 shows *Urophora jaceana* galls on knapweed, *Centaurea*. The fly larvae inside the galls may be parasitized by parasitic wasps. The galls themselves may also support other insect larvae that feed on the gall tissue but not the fly larvae. What is the correct term for such a phenomenon?

## 12.3 Facilitation May Be More Common Under Conditions of Environmental Stress

One of the oldest ideas about facilitation is that it is more common under unfavorable environmental conditions than in favorable ones (see **Feature Investigation**). For example, seed germination and seedling survival often increase under so-called nurse plants, plants of various species that alleviate physical stress by providing increased shade and soil moisture. In the salt marshes of the northeast United States, smooth cordgrass, *Spartina alterniflora,* buffers wave action and permits the growth of a group of perennial plants that, on their own, would not grow in a habitat with such high disturbance. In Spain, a strong facultative mutualism exists between two plants, the leguminous shrub *Retama sphaerocarpa* and understory plant *Marrubium vulgare,* in a semiarid region in the southeast. In these almost desert-like conditions, the soil moisture and nitrogen content were much improved where both species co-occurred. *Retama* shades *Marrubium,* providing a favorable microclimate, while *Marrubium* enhances the availability of water to *Retama.* Nutrient cycling is probably increased due to litter accumulation. As a result, both species had greater leaf area, leaf mass, shoot mass, more flowers, and higher leaf nitrogen content when growing together than when growing alone (**Figure 12.14**).

### Check Your Understanding

**12.3** If facilitation is more common under conditions of environmental stress, think of some habitats where facilitation might be common and some habitats where it might not be.

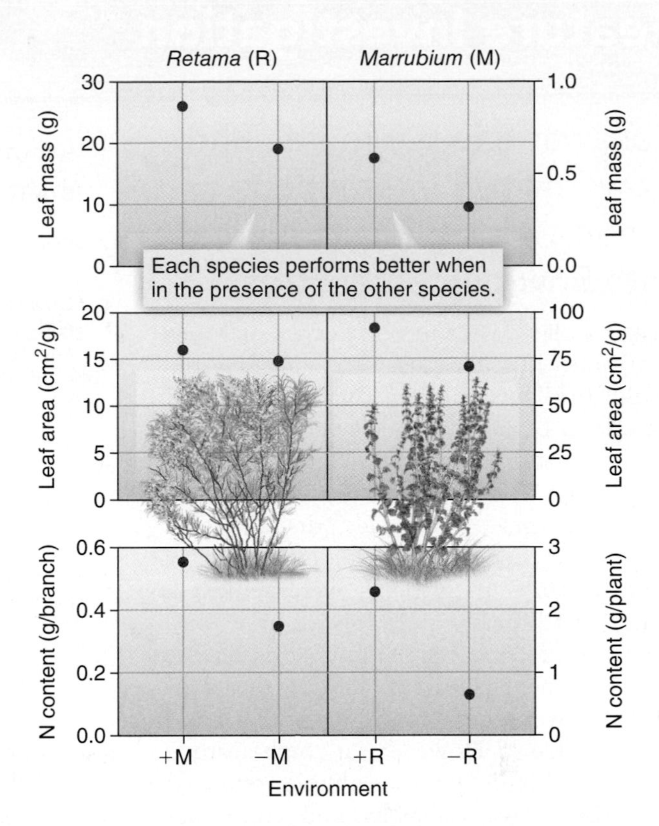


**Figure 12.14 Mutualism between two plant species in southeastern Spain.** Dry mass of leaves per branch, leaf area (LA), and nitrogen content in *Retama sphaerocarpa* (R) and *Marrubium vulgare* (M) growing in association (+R, +M) or alone (–R, –M) at Rambla Honda, Spain. (After Pugnaire et al., 1996.)

## Feature Investigation

### Callaway's Experiments Show How Positive Interactions Among Alpine Plants Increase with Environmental Stress

Ragan Callaway, of the University of Montana, and 12 colleagues from around the world (Callaway et al., 2002), measured the effects of thinning out neighboring vegetation on the growth of small herbaceous plants on 11 mountain ranges from Alaska to Europe. They tested the idea that all plants would do better when their competitors for light, water, and nutrients were removed. Indeed, at nonstressful, low-altitude sites where resources weren't limiting, this is what the experiments showed. Plants with neighbors removed grew, on average, 22% more than controls. However, at high-altitude locations, plants with neighbors removed grew 25% less. Callaway and colleagues reasoned that at stressful high altitudes the presence of neighbors improved microclimate, sheltered plants from wind, and stabilized the soil (**Figure 12.C**).

**HYPOTHESIS** Environmental stress affects plant mutualisms

**STARTING LOCATION** Worldwide alpine locations

| | Conceptual level | Experimental level |
|---|---|---|
| 1 | Choose stressful and nonstressful environmental locations. | At 11 alpine locations around the world, one site was situated in nonstressful subalpine vegetation, and one site was situated 300–1200 m higher in stressful alpine conditions.  |
| 2 | Perform removal of plant neighbors. | At each site, 3–10 target plant species were chosen and all neighboring species within 10 cm of target individuals were removed (neighbor removal). Vegetation around other target species was not disturbed (control). |
| 3 | Measure effects of experiments. | At least one growing season after the manipulations were performed, the number of leaves, flowers, fruit, and plant survival were measured at all sites. |

4 **THE DATA**

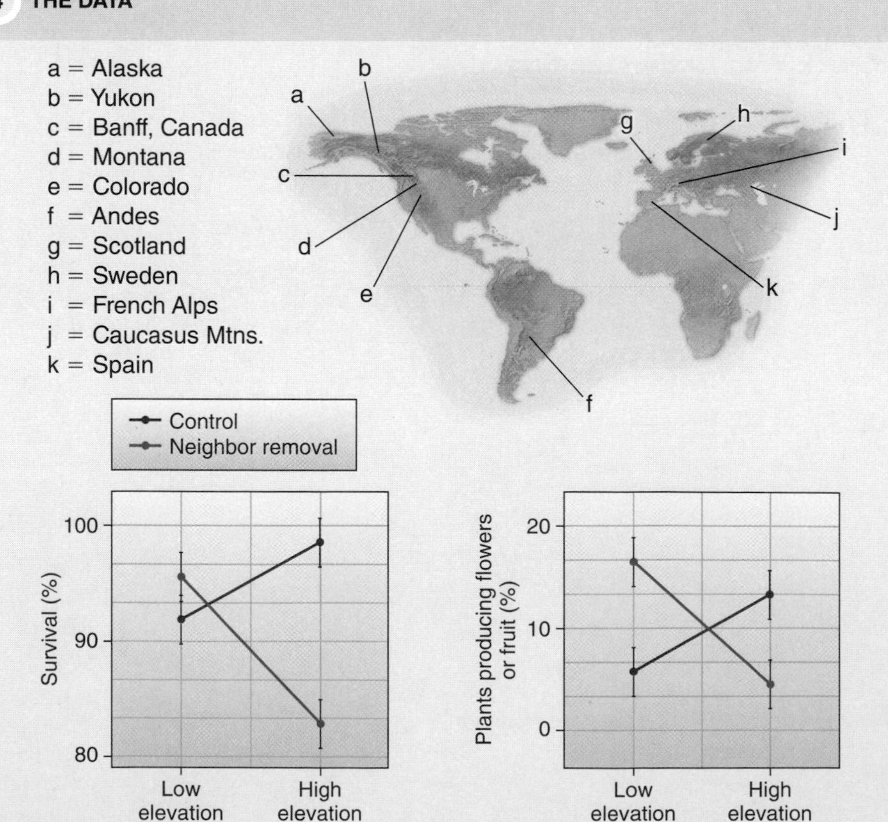


**Figure 12.C** **Determining the effects of environmental stress on plant mutualisms.**

## SUMMARY

**12.1 Mutualism Is an Association Between Two Species That Benefits Both Species**

- Mutualism is an association between two species that benefits both (chapter opener photo).
- Facilitation is an association between two species that benefits at least one of the species; two types are mutualism and commensalism.
- Pollination of flowers and dispersal of fruits are two good examples of mutualism. In pollination, both the plant and the pollinator benefit, the latter by a nectar meal. In seed dispersal, the disperser often gains a fruit meal. Pollinators, and the plants they pollinate, may be suffering a decline in parts of Europe (Figure 12.1).
- Mutualism involving humans and crops/livestock have caused some of the most drastic environmental changes on Earth (Figures 12.A, 12.B, Tables 12.A, 12.B).
- Sometimes one member of a mutualistic association will try to cheat the other member of its rewards (Figure 12.2).
- Dispersive mutualism involves plants and pollinators that disperse their pollen, and plants and fruit eaters that disperse the plant's seeds. Some invasive pollinators have facilitated the spread of invasive plants (Figures 12.3, 12.4, Table 12.1).
- Defensive mutualism typically involves an animal defending either a plant or an herbivore (Figures 12.5, 12.6).
- Resource-based mutualism involves the mutually beneficial increase of resources to both species (Figures 12.7, 12.8).
- In endosymbiotic mutualism, one species lives inside the body of another (Figure 12.9).
- Mutualisms cannot easily be modeled mathematically (Figure 12.10).

**12.2 Commensal Relationships Are Those in Which One Partner Receiv-es a Benefit While the Other Is Unaffected**

- In commensal relationships, one partner receives a benefit while the other is not affected (Figures 12.11–12.13, Table 12.2).

**12.3 Facilitation May Be More Common Under Conditions of Environmental Stress**

- Facilitation may be more prevalent in environmentally stressful conditions (Figures 12.C, 12.14).

## TEST YOURSELF

1. **Which of the following is not a mutualism?**
   a. The association between deep-sea fish and luminescent bacteria
   b. The association between fish and cleaner fish
   c. The association between tapeworms and pigs
   d. The association between cellulose-digesting microbes and herbivores
   e. The association between humans and crops
2. Fruits taken by birds tend to:
   a. Have bright colors
   b. Give off pungent odors
   c. Be unavailable to mammals
   d. All of the above
   e. None of the above
3. The colonization hypothesis suggests seed dispersal is important:
   a. To avoid seed predators
   b. To allow seeds to disperse to optimal sites
   c. Because the parental location is not always a good site for seeds
   d. To avoid competition with the parents
4. The mutualism of fungi that live in leaves and grasses can be classified as:
   a. Defensive mutualism
   b. Resource-based mutualism
   c. Dispersive mutualism
   d. Predator escape mutualism
5. When one species cannot live without the other, we use the term:
   a. Facultative mutualism
   b. Dispersive mutualism
   c. Defensive mutualism
   d. Resource-based mutualism
   e. Obligate mutualism
6. Which is the odd species out in the list below:
   a. Leaf cutter ant
   b. Underground fungus garden
   c. Gongylidia
   d. Symbiodinium
   e. *Escovopsis*

7. Lichens are combinations of:
   a. Plants and algae
   b. Algae and fungae
   c. Fungae and mycorrhizae
   d. Bacteria and plants
   e. Nematodes and algae
8. Commensal relationships can be represented by which of the following costs/benefits for the species involved?
   a. +/ +
   b. +/ −
   c. −/ +
   d. −/0
   e. +/0
9. Which of the following relationships is not a type of commensalism?
   a. Phoresy
   b. Parasitism
   c. Inquilinism
   d. Metabiosis
10. Phoresy involves:
   a. One species using a second species for housing
   b. One species using a second species for transport
   c. One species using something produced by the first
   d. Mutualism between two species

## CONCEPTUAL QUESTIONS

1. Distinguish between facultative and obligate mutualism.
2. What are the main differences between dispersive, defensive, and resource-based mutualisms?
3. What are the distinguishing features of a commensal relationship?
4. Why are facilitations more common under environmentally stressful conditions?
5. Explain the mechanism of associational resistance.

## DATA ANALYSIS

Because mammals are effective seed dispersers, a decrease in the abundance of mammals can affect plant densities. The data in the figure, parts (a) and (b), show the densities of two species of monkeys, white-faced and howler, in eight areas of Panama that vary in intensity of poaching. Data for other seed eaters show similar trends. The proportion of seeds dispersed and seedling density of palm trees, *Astrocaryum standleyanum,* are shown in parts (c) and (d). Explain what is happening here.

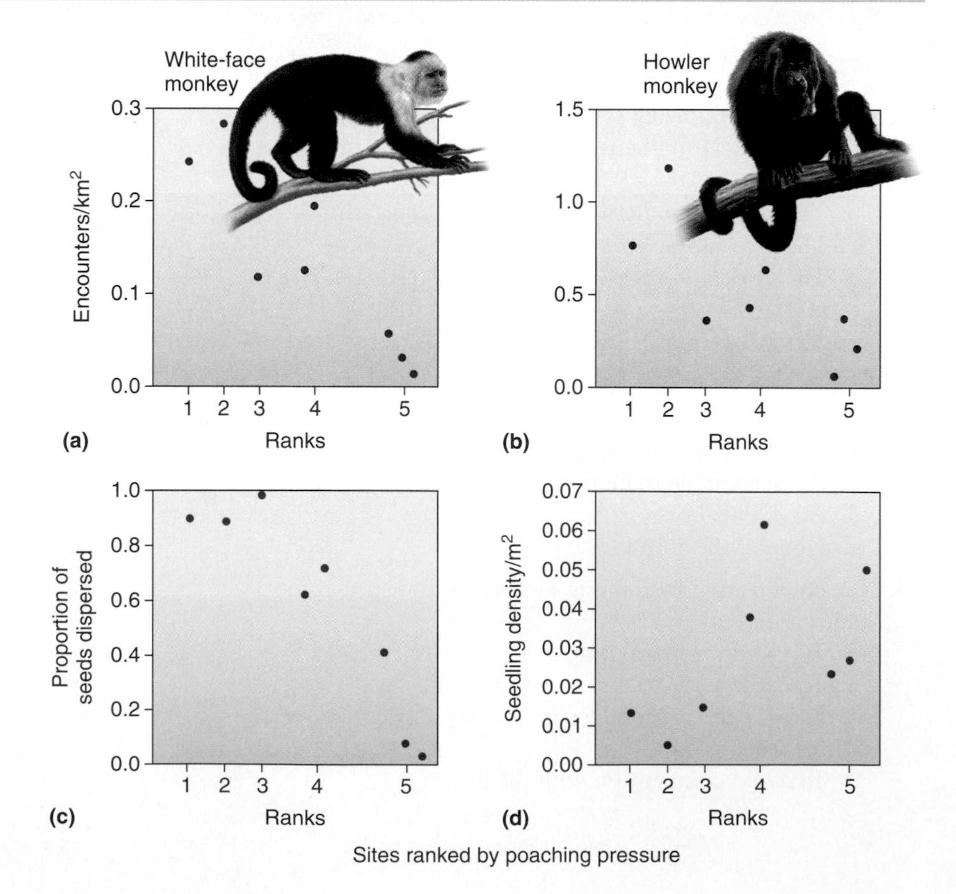

The horned lizard, a common lizard of Texas, has a variety of defenses against predators.

CHAPTER 13

# Predation

## Outline and Concepts



The Texas horned lizard, *Phrynosoma cornutum,* is the largest of eight species of horned lizards in the United States. It is the state reptile of Texas and Wyoming, and is the mascot of Texas Christian University, where it is commonly referred to as a horned toad. Horned lizards have several lines of defense against predators. First, their coloration acts as camouflage, allowing them to resemble the coloration of the habitat in which they live. If this doesn't work, and they are spotted, the lizards will run quickly and stop abruptly, challenging the predator's visual acuity. Next, they will puff themselves up, appearing to be bigger. The spines on their backs, made from modified scales, and the horns on their head make them difficult to ingest. Finally, if all else fails, the lizards can squirt a stream of blood at their attackers for a distance of up to 1 m. They do this by restricting the blood vessels leaving the head, increasing blood pressure around the eyes. The blood accumulates in sacks around the eyes, eventually rupturing the blood vessels. In addition to the surprise factor, the blood is distasteful to canine and feline predators. The occurrence of so many defense mechanisms in one species suggests that predation is an important selective pressure in nature.

In this chapter we examine the phenomenon of predation. Predation, herbivory, and parasitism are all interactions that have a positive effect for one species and a negative effect for the other. These categories of species interactions can be classified according to how lethal they are for the prey and the length of association between the consumer and prey (**Figure 13.1**). Each has particular characteristics that set it apart. Herbivory usually involves nonlethal predation on plants, whereas predation generally results in the death of the prey. Parasitism, like herbivory, is typically nonlethal and differs from predation in that the adult parasite typically lives and reproduces for long periods in or on the living host. Parasitoids, insects that lay eggs in living hosts, have features in common with both predators and parasites. They always kill their prey, as predators do, but unlike predators, which immediately kill their prey, parasitoids kill the host more slowly. Parasitoids are common in the insect world and include parasitic wasps and flies that feed on many other insects.

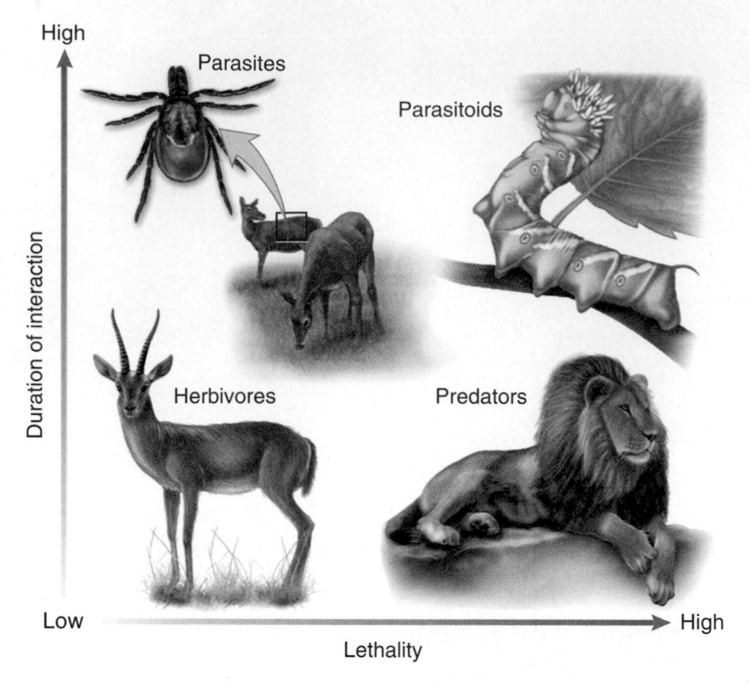


**Figure 13.1 Possible interactions between populations.** Lethality represents the probability that an interaction results in the death of the prey. Duration represents the length of the interaction between the consumer and the prey. (After Pollard, 1992.)

Predators feed on other living organisms known as prey, killing their prey either before they eat them or during the act of eating. We will survey the wide range of antipredator strategies that prey exhibit, from chemical defenses to warning coloration, camouflage, and mimicry. We will use predator-prey models to show how predator and prey populations can interact to produce cyclic oscillations. Finally, we will show how native and introduced predators impact native prey populations. In some human endeavors, such as whaling and fishing, humans act as the ultimate predators, severely impacting their prey populations.

## 13.1 Animals Have Evolved Many Antipredator Adaptations

Animals have evolved many different antipredator adaptations over time to reduce their chances of being eaten. These adaptations include chemical defenses, with or without warning coloration; cryptic coloration; mimicry; physical defenses; intimidation displays; and predator satiation (**Table 13.1**).

A great many species have evolved chemical defenses against predation. One of the classic examples of a chemical defense involves the bombardier beetle, *Stenaptinus insignis*, which was studied by Tom Eisner and coworkers (Eisner and Aneshansley, 1982). These beetles possess reservoirs of hydroquinone and hydrogen peroxide in their abdomen. When threatened, they eject the chemicals into an "explosion chamber," where the subsequent release of oxygen causes the whole mixture to be violently ejected as a hot spray that can be directed at the beetle's attackers (**Figure 13.2a**). Many other

**Table 13.1 Main types of defense employed by prey against predators.**

| Type of defense | Examples |
|---|---|
| 1. Chemical with aposematic coloration | Bombardier beetles, ladybird beetles, many butterflies |
| 2. Cryptic coloration | Grasshoppers, seahorses |
| 3. Batesian mimicry | Hoverflies and wasps |
| 4. Müllerian mimicry | Butterflies |
| 5. Physical defenses | Turtle shells, beetle exoskeletons, crab claws, scorpion stingers |
| 6. Intimidation displays | Frilled lizards, porcupine fish |
| 7. Predator satiation | 13-year and 17-year cicadas |

arthropods, such as millipedes, also utilize chemical sprays to deter predation. In *Neocapritermes taracua*, a tropical termite species, some workers rupture and release a sticky fluid when attacked. The mandibles of these workers wear out with time, making them less suitable for other jobs. At the same time, they build up their backpacks of sticky fluid, making them more suitable for suicidal fighting. This phenomenon has led scientists to note that while humans send their young men to war, social insects send their old ladies.

Often associated with a chemical defense is an **aposematic coloration**, or warning coloration, which advertises an organism's unpalatable taste. Ladybird beetles' bright red colors warn of the toxic defensive chemicals they exude when threatened, and many tropical frogs have bright warning coloration that calls attention to their skin's lethality (**Figure 13.2b**). Some animals synthesize toxins using their own metabolic processes; other animal poisons are acquired from plants. For example, monarch butterfly caterpillars, *Danaus plexippus*, accumulate emetic chemicals called cardiac glycosides from milkweed plants. Emetic chemicals cause predators to vomit. The caterpillars advertise their distastefulness with bold colors (**Figure 13.2c**). The chemicals are also present in the adult butterflies, which advertise their distastefulness with bright orange and black colors (**Figure 13.2d**).

**Cryptic coloration** is an aspect of camouflage, the blending of an organism with the background of its habitat. Cryptic coloration is a common method of avoiding detection by predators. For example, many grasshoppers are green and blend in with the foliage on which they feed. Leaf wing butterflies also mimic leaves. Even the veins of leaves can be mimicked on the butterfly's wings (**Figure 13.3a,b**). Stick insects mimic branches and twigs with their long, slender bodies. In most cases, these animals stay perfectly still when threatened, because movement alerts a predator. Cryptic coloration is prevalent in the vertebrate world, too. Many sea horses adopt a body shape and color pattern that mimics the habitats in which they are found (**Figure 13.3c**).

**Mimicry**, the resemblance of a species (the mimic) to another species (the model), also secures protection from predators. There are two major types of mimicry. **Batesian mimicry**, named after the English naturalist Henry Bates, is the mimicry of an

**Figure 13.2 Chemical defense.** (a) As it is held by a tether attached to its back, this bombardier beetle, *Stenaptinus insignis*, directs its hot, stinging spray at a forceps "attacker" squeezing its leg. **(b)** Aposematic coloration advertises the poisonous nature of this blue poison arrow frog, *Dendrobates azureus*, from South America. **(c)** Monarch butterfly caterpillars sequester poisonous cardiac glycosides from their milkweed host plants and advertise their toxicity via striking colors. **(d)** The adult monarch butterflies maintain the toxicity imparted to them from their caterpillar stages and are also warningly colored.

**ECOLOGICAL INQUIRY**

Do any mammals have chemical defenses?

unpalatable species (the model) by a palatable one (the mimic). The nonvenomous scarlet king snake, *Lampropeltis elapsoides,* mimics the venomous eastern coral snake, *Micrurus fulvius,* thereby gaining protection from would-be predators (**Figure 13.4a,b**). Other examples involve flies, especially syrphid flies, which are striped black and yellow to resemble stinging bees and wasps but are themselves harmless. In **Müllerian mimicry**, named after the German biologist Fritz Müller, many noxious species converge to look the same, thus reinforcing the basic distasteful design. One example is the black and yellow striped bands found on many different species of bees and wasps. Another example is the strong resemblance of the monarch, *Danaus plexippus,* queen, *D. gilippus,* and viceroy, *Limenitis archippus,* butterflies in North America (**Figure 13.4c–e**). Müllerian mimicry is also found among noxious Amazonian butterflies.

Physical defenses of prey against predators are varied and numerous. The shells of tortoises and freshwater turtles are a strong means of defense against most predators (**Figure 13.5a**). Many beetles have a tough exoskeleton that protects them from attack from other arthropod predators such as spiders. Other animals have developed horns and antlers, which, although primarily used in competition over mates, can be used in defense against predators. Many invertebrate species have powerful claws, pincers, or, in the case of scorpions, venomous stingers that can be used in defense as well as offense. Some groups of insects, such as grasshoppers, have a powerful jumping ability to escape the clutches of predators. Many frogs are prodigious jumpers, and flying fish can glide above the water to escape their pursuers.

Some animals put on displays of intimidation in an attempt to discourage predators. For example, a toad swallows air to make itself appear larger, frilled lizards extend their collars when frightened to create this same effect, and porcupine fish inflate themselves to large proportions when threatened (**Figure 13.5b**). All of these animals use displays to deceive potential predators about the ease with which they can be eaten.

**Predator satiation** is the synchronous production of many progeny by all individuals in a population to satiate predators and thereby allow some progeny to survive. Thirteen- and 17-year periodical cicadas, *Magicicada* spp., are so termed because the emergence of adults is highly synchronized to occur

**Figure 13.3 Cryptic coloration or camouflage.** **(a, b)** This Indian leaf butterfly, *Kallima paralekta,* mimics the leaves on which it rests. **(c)** A Pygmy seahorse, *Hippocampus bargibanti,* from Bali, blends in with its background.

**Figure 13.4** **Mimicry.** Batesian mimicry: An innocuous scarlet king snake, *Lampropeltis elapsoides* **(a),** mimics the venomous coral snake, *Micrurus fulvius* **(b)**. Müllerian mimicry: These viceroy **(c),** queen **(d),** and monarch **(e)** butterflies have the same basic warning coloration, and all three are toxic to birds.

**ECOLOGICAL INQUIRY**

Is the syrphid fly's mimicry of a yellow jacket (wasp) Müllerian or Batesian?

**Figure 13.5** **Intimidation and armor.** **(a)** This juvenile ornate box turtle, *Terrapene ornata,* has a strong shell that can deter predators much larger than itself. **(b)** In a display of intimidation, this porcupine fish, *Diodon holocanthus,* puffs itself up to look threatening and to prevent ingestion by gape-limited predators.

**Table 13.2 Additional prey defenses used against predators.**

| Prey | Defense |
|---|---|
| 1. Chameleons, octopuses | Change color for camouflage |
| 2. Decorator crabs, caterpillars | Cover body with debris for camouflage |
| 3. Sea cucumbers | Evisceration. Excrete portions of digestive tract, which contain toxic chemicals |
| 4. *Camponotus saundersi* ants | Malaysian ants which can self-destruct by squeezing abdominal muscles, causing glands to explode and spraying poison in all directions |
| 5. Gulls | Mob predators that approach colonial nest sites |
| 6. Honey bees | Mob Asian hornet scouts, vibrate their flight muscles to raise the temperature to lethal levels; obviates the need to use stingers |
| 7. Swallowtail butterfly caterpillars | Resemble bird droppings on leaf |
| 8. Ants, termites | Secrete alarm pheromone when threatened, causing neighbors to help attack the predator |
| 9. Some moths | Perform elaborate evasive maneuvers in response to bat sonar signals |
| 10. Porcupines | Quills can be ejected into predator |

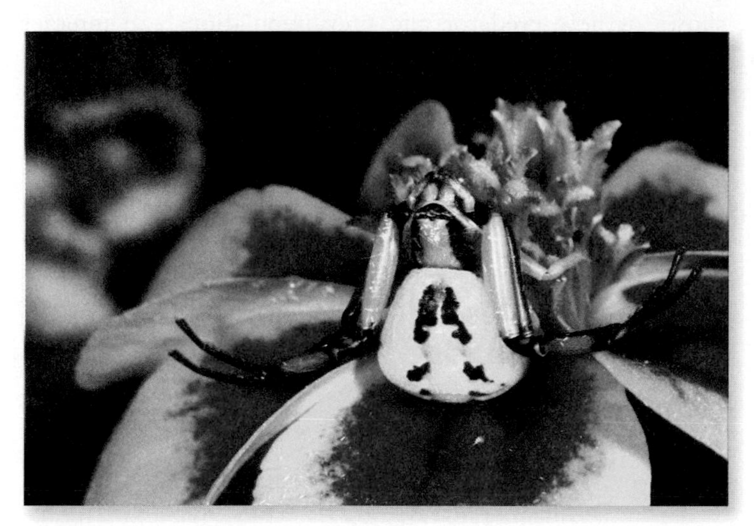

**Figure 13.6 Aggressive mimicry.** Crab spiders, such as the North American species, *Misumenoides formosipes,* often mimic the vegetation or flowers where they wait to ambush their prey.

once every 13 or 17 years. Adult cicadas live for only a few weeks, during which time females mate and deposit eggs on the twigs of trees. The eggs hatch 6–10 weeks later, and the nymphs drop to the ground and begin a long subterranean development, feeding on the contents of the xylem of roots. Because the xylem is low in nutrients, it takes many years for nymphs to develop, though there appears to be no physiological reason why some cicadas couldn't emerge after, say, 12 years of feeding, and others after 14. Their synchrony of emergence is thought to maximize predator satiation. Worth noting in this context is the fact that both 13 and 17 are prime numbers, and thus predators on a shorter multiannual cycle cannot repeatedly utilize this resource. For example, a predator that breeds every 3 years could not rely on cicadas being present as a food supply every few generations.

How common is each of these defense types? No one has done an extensive survey over the entire animal kingdom. However, Brian Witz (1989) surveyed studies that documented antipredator mechanisms in arthropods, mainly insects. By far the most common antipredator mechanisms were chemical defenses and associated aposematic coloration, noted in 51% of the examples. Other, less common, types of prey defense mechanisms are noted in **Table 13.2**. The existence of such a wide array of defenses against predators is a testament to the strong selective pressure applied by predators to their prey.

In an interesting turnaround, predators themselves have evolved adaptations to help them catch prey. Prominent among these is **aggressive mimicry**, where predators mimic a harmless model, allowing them to get close to prey. For example, crab spiders sometimes mimic the color of flowers and sit in the flower's center, waiting to catch a passing pollinator, such as a bee (**Figure 13.6**). Praying mantises also mimic flowers and leaves to get close to their prey. The humpback anglerfish, *Melanocetus johnsonii,* uses a modified dorsal spine as a bioluminescent fishing rod to attract prey close to their mouth. The alligator snapping turtle, *Macrochelys temminckii,* wriggles its tongue, which resembles a little pink worm, to attract small fish. Female bolas spiders of the genus *Mastophora* attract male moths by producing analogs of the female moth's pheromones. Female fireflies of the genus *Photuris* mimic the light-flashing patterns of females of the genus *Photinus* to attract—and then consume—*Photinus* males. The saber-toothed blenny, *Aspidontus taeniatus,* mimics the blue streak cleaner wrasse, *Labroides dimidiatus.* The saber-tooth mimics the cleaner's dance and is allowed to approach close to its prey, whereupon it bites off a piece of fin and flees. In flight, the zone-tailed hawk, *Buteo albonotatus,* resembles the turkey vulture, *Cathartes aura,* and flies among them. It peels off from the flock to ambush its prey.

### Check Your Understanding

**13.1** Many predators are visual searchers. One species of prey can sometimes exist as two completely different color forms or morphs. This is known as color polymorphism. How could this be a defensive strategy?

## 13.2 Predator-Prey Interactions May Be Modeled by Lotka-Volterra Equations

A good starting point for examining predator-prey interactions is with the models of Lotka and Volterra, who had already investigated population growth and the effects of interspecific competition (refer back to Chapters 10 and 11). From the early models of Lotka and Volterra we know that prey populations can increase exponentially according to the formula:

$$\frac{dN}{dt} = rN$$

How do predators affect such prey growth rates? First, we need to know the searching efficiency of the predators, which we will call *s*. This is an inherent behavior of the predators. Some species of predators are diligent searchers of prey, others merely sit and wait for passing prey. The actual number of prey attacked will also depend on the number of prey present, *N*, and the numbers of predators present, which we will designate as *P*. The consumption of prey is thus equivalent to the combination of these three features, *sPN*. We can work out the growth rate of the prey in the presence of predators as:

$$\frac{dN}{dt} = rN - sPN$$ ← Prey removal by predators

How do the predator populations grow? Predator birth rate depends on the value *sPN* and on how efficient predators are at turning food into new predators, a value we can term *e*. Some predators, such as hyenas, are very efficient in that they consume even the bones of their prey. In the absence of death, the predator population grows according to the formula:

$$\frac{dP}{dt} = esPN$$

However, we know also that there is predator mortality. Notwithstanding the effects of prey, the predator death rate is thus dependent upon a mortality rate, *m*, and the number of predators, *P*. The overall growth rate of the predator population is now:

$$\frac{dP}{dt} = esPN - mP$$ (Predator death: $mP$; Predator growth rate: $esPN$)

We can investigate the outcome of this model graphically, by drawing zero isoclines of prey and predators, in much the same way as we did in Chapter 11 when we investigated Lotka-Volterra models of competition. Remember, along a zero isocline, prey population growth is 0; thus when

$$\frac{dN}{dt} = 0$$

then $rN = sPN$

and $P = r/s$

Because *r* and *s* are constants, the prey isocline is a line for which *P* itself is constant (**Figure 13.7a**). Below this line of set predator abundance, prey increase, because predator pressure is insufficient to stop the prey population from growing. Above the line, prey decrease, because of high predator pressure. Similarly, for the predators, when:

$$\frac{dP}{dt} = 0$$

then $esPN = mP$

and $N = m/es$

Once again, because *m*, *e*, and *s* are all constants, the predator zero isocline is a line along which *N* is a constant (**Figure 13.7b**). To the left of this line of constant prey, predator abundance decreases, because there is not enough prey to support the predators. To the right of this line, predators increase, because sufficient prey exists to support them.

As with our investigations of the Lotka-Volterra model of interspecific competition, combining the two isoclines shows us how predator and prey populations can interact (**Figure 13.7c,d**). Initially, when the interaction begins, we

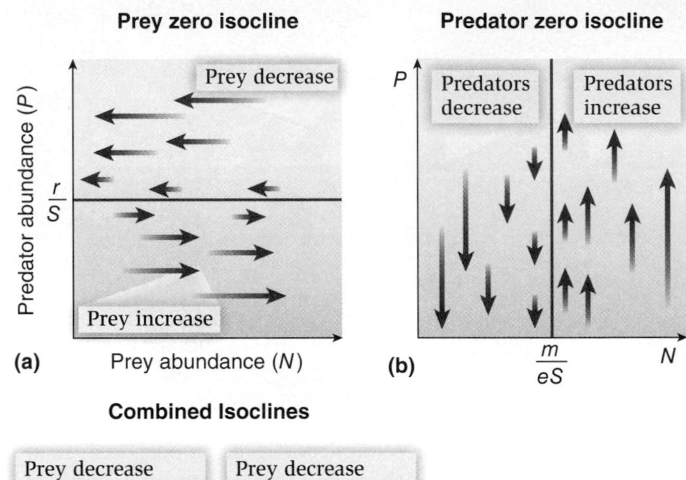


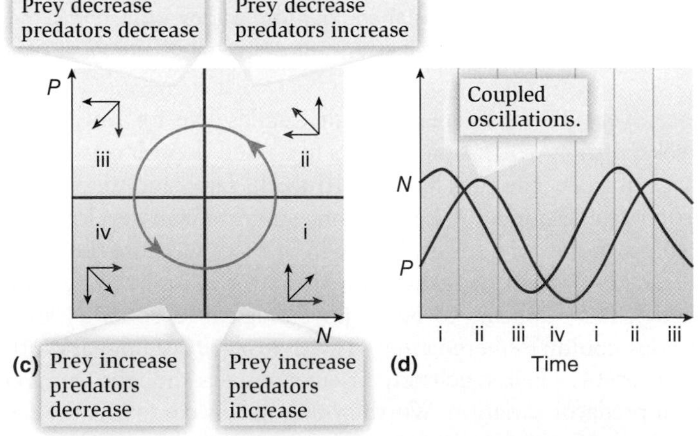


**Figure 13.7 The basic Lotka-Volterra predator-prey model.** **(a)** The prey zero isocline: The number of prey, *N*, increases at low predator abundance, *P*, and decreases at high predator abundance. **(b)** The predator zero isocline: The number of predators, *P*, increases when prey numbers are high and decreases when prey numbers, *N*, are low. **(c)** When prey and predator isoclines are combined, coupled oscillations result, which are expressed in **(d)** as numbers of prey and predators through time. (Modified from Begon et al., 1996.)

can assume moderate numbers of both predators and prey, and both populations increase (i). Next, predator pressure is so intense it starts to drive the prey population down (ii). As prey numbers decrease, there are insufficient prey to support the predators and the predator population decreases (iii). As predator pressure is relieved, the prey population rebounds (iv). Finally, in response to increased prey, predator numbers recover (back to i). The Lotka-Volterra model thus predicts an endless cycle of predator and prey, with a delay in the response of predators to prey driving the cycle.

Many authors have refined the Lotka-Volterra model because they have found it too simplistic. One of the more intuitive refinements involves a bending of both predator and prey isoclines (**Figure 13.8**). Let's consider the prey zero isocline first (**Figure 13.8a**). At low prey abundance the isocline is almost horizontal, as in the Lotka-Volterra model. However, at higher prey abundance, prey compete, and once they reach the carrying capacity, their abundance cannot increase; hence, the isocline curves downward. For the predators, we will vary predator efficiency. To begin with, the predator isocline slants to the right because more predators require more prey to sustain them (**Figure 13.8b**, curve i). Mutual interference between predators, as with fighting, lessens efficiency and would bend the isocline even more; because predators would spend more time fighting and less time hunting, they would need even higher prey densities to sustain them (curve ii). At some point the predator isocline may level off, because even a huge number of prey would not further increase predator numbers (curve iii). This is true, for example, if predators are territorial.

The dynamics of predator-prey interactions can now be determined by combining the modified predator and prey isoclines. Once again, exactly one stable point emerges for each combination of predator and prey isoclines: the intersection of the lines (**Figure 13.8c**). Two general conclusions emerge. The first is that predator-prey oscillations may still occur, but the oscillations can be damped and less severe through time (curve i). The second conclusion is that the greater the degree of mutual interference or territoriality between predators, the less the oscillations (curve ii in Figure 13.8c). In fact, at some point the oscillations may disappear altogether (curve iii). Thus, a relatively simple interaction of predator and prey can produce a variety of results, from coupled

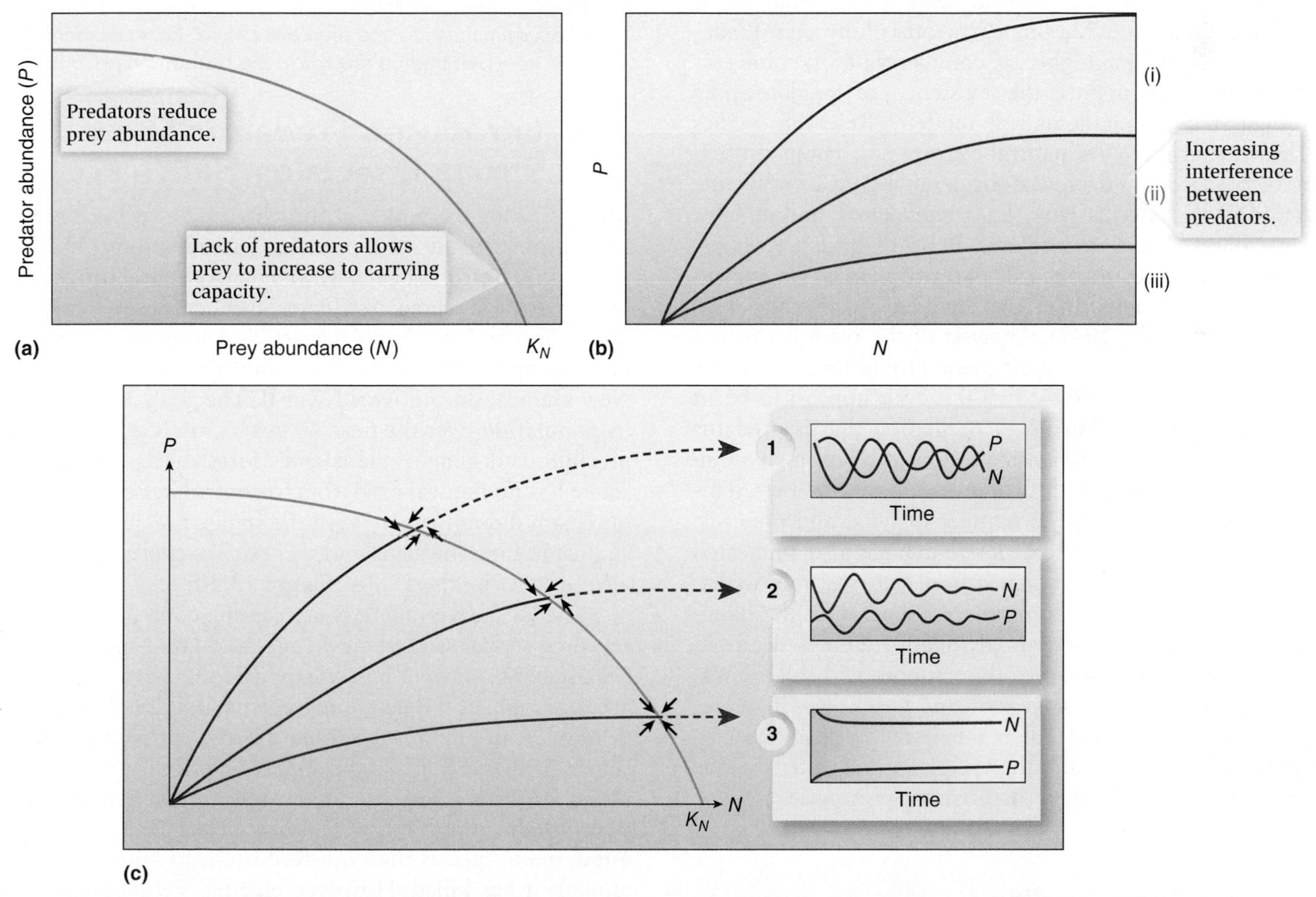


**Figure 13.8 A modified Lotka-Volterra model of predator-prey interactions.** **(a)** The prey zero isocline curves downward and intercepts the prey abundance axis at the carrying capacity of the prey, $K_N$. **(b)** The predator zero isoclines bend toward the right as successively higher levels of fighting and interference between predators require even higher prey densities for them to feed on (curves i and ii). For some species, higher prey densities will not increase predator densities further, as in curve iii. **(c)** Combining the modified predator and prey isoclines can result in damped oscillations of predator and prey (curves i and ii) or no oscillation at all (curve iii). (Modified from Begon et al., 1996.)

oscillations to damped oscillations to no oscillations at all, depending on predator behavior.

Our modification of the Lotka-Volterra model of predator-prey interactions has produced vastly different outcomes based on tweaking just one parameter—predator efficiency. Including more components in the model, such as refuges for prey and predators that feed on a variety of prey species, can further alter the outcomes of the model. For example, a predator that can sustain itself on a variety of prey species could depress the population of a rare prey well below the prey carrying capacity and drive the prey to extinction. Imagine that lions ate both wildebeest, which are common, and sable antelope, which are rare. Lions could sustain themselves on wildebeest yet still constitute a threat to sable antelope even though sable antelopes alone could not support the lion population. This type of subsidy of predators by common prey may explain how predators can impact densities of rare prey. In the U.S. Midwest, predators such as coyotes, *Canis latrans,* swift foxes, *Vulpes velox,* and badgers, *Taxidea taxus,* are attracted to high concentrations of prairie dogs, *Cynomys leucurus,* but then also attack ground-nesting birds in the same areas. Bruce Baker and colleagues (1999) placed 722 artificial nests on 74 white-tailed prairie dog colonies and 722 on nearby off-colony sites. Predation rates averaged 14% higher on colonies than off colonies.

Do field data support the existence of predator-prey cycles suggested by mathematical models? Research studies have shown that there are natural examples of coupled oscillations of predators and prey. Considerable data exist on the interaction of the Canada lynx, *Lynx canadensis,* and its prey, snowshoe hares, *Lepus americanus.* In 1942 British ecologist Charles Elton analyzed the records of furs traded by trappers to the Hudson's Bay Company in Canada over a 100-year period (Elton and Nicholson, 1942). Analysis of the records showed that a dramatic 9- to 11-year cycle existed for as long as records had been kept (**Figure 13.9**). At first this cycle appears to be an example of a stable predator-prey oscillation that is predator driven: The hares go up in density, followed by an increase in density of the lynx, which then depresses hare numbers. This is followed by a decline in the number of lynx, and the cycle begins again. However, more recent research has also implicated food availability as an additional causal factor of the oscillations. As the density of hares increases, they reduce the quantity and quality of their food supply. Plant numbers decrease, and remaining grazed plants produce shoots with high levels of chemical defenses. Hare starvation and weight loss increases susceptibility to predators. After the hare numbers decline, plant populations recover and the cycle repeats again. A plant-herbivore cycle thus influences the predator-prey cycle.

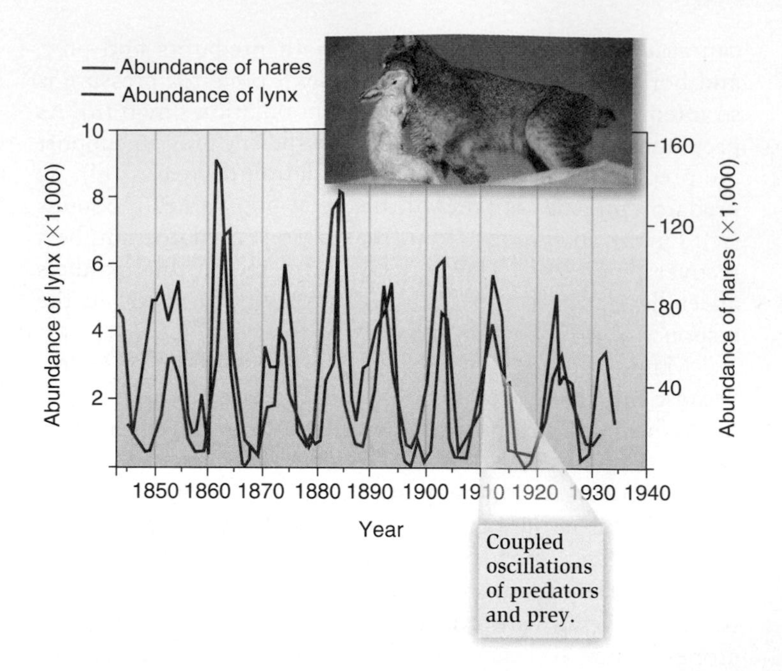


**Figure 13.9 Coupled predator-prey oscillations.** The 9- to 11-year coupled oscillations of the abundance of the snowshoe hare, *Lepus americanus*, and the Canada lynx, *Lynx canadensis*, were revealed from pelt trading records of the Hudson's Bay Company.

## 13.3 Introduced Predators Show Strong Effects on Native Prey

Many striking examples of the powerful effects of predators have been provided by predator introductions. The brown tree snake, *Boiga irregularis,* was inadvertently introduced by humans to the island of Guam, in Micronesia, shortly after World War II. It probably arrived as a stowaway in U.S. military transport ships from the Admiralty Islands, near Papua New Guinea, during World War II. The growth and spread of its population over the next 40 years closely coincided with a precipitous decline in the island's forest birds. On Guam, the snake has no natural predators to control it. Because the birds on Guam did not evolve with the snake, they had no defenses against it. Eight of the island's 11 native species of forest birds went extinct by the 1980s (**Figure 13.10**).

The U.S. government is attempting control measures, but trapping snakes is extremely time- and labor-intensive. However, scientists have a new weapon in their arsenal, tablets of acetaminophen, the active ingredient in Tylenol, packed into dead mice. In humans acetaminophen gives pain relief, but in snakes it disrupts the oxygen-carrying ability of the snake's blood proteins. Only 80 milligrams, equivalent to a child's dose, is lethal. Furthermore, the brown tree snake is one of only a few snake species that will feed on dead animals as well as animals it has killed. However, placing acetaminophen-laden dead mice on the forest floor could impact other wildlife. A solution was to attach long 1-2 m (4 ft) paper streamers to the mice and airdrop them on the forest canopy, where they lodge in the foliage and only the snakes can find them. A small subset of the airdropped mice were fitted with radio

### Check Your Understanding

**13.2** Many burrowing animals can escape their predators by going underground. What effect might such prey refuges have on predator-prey relationships?

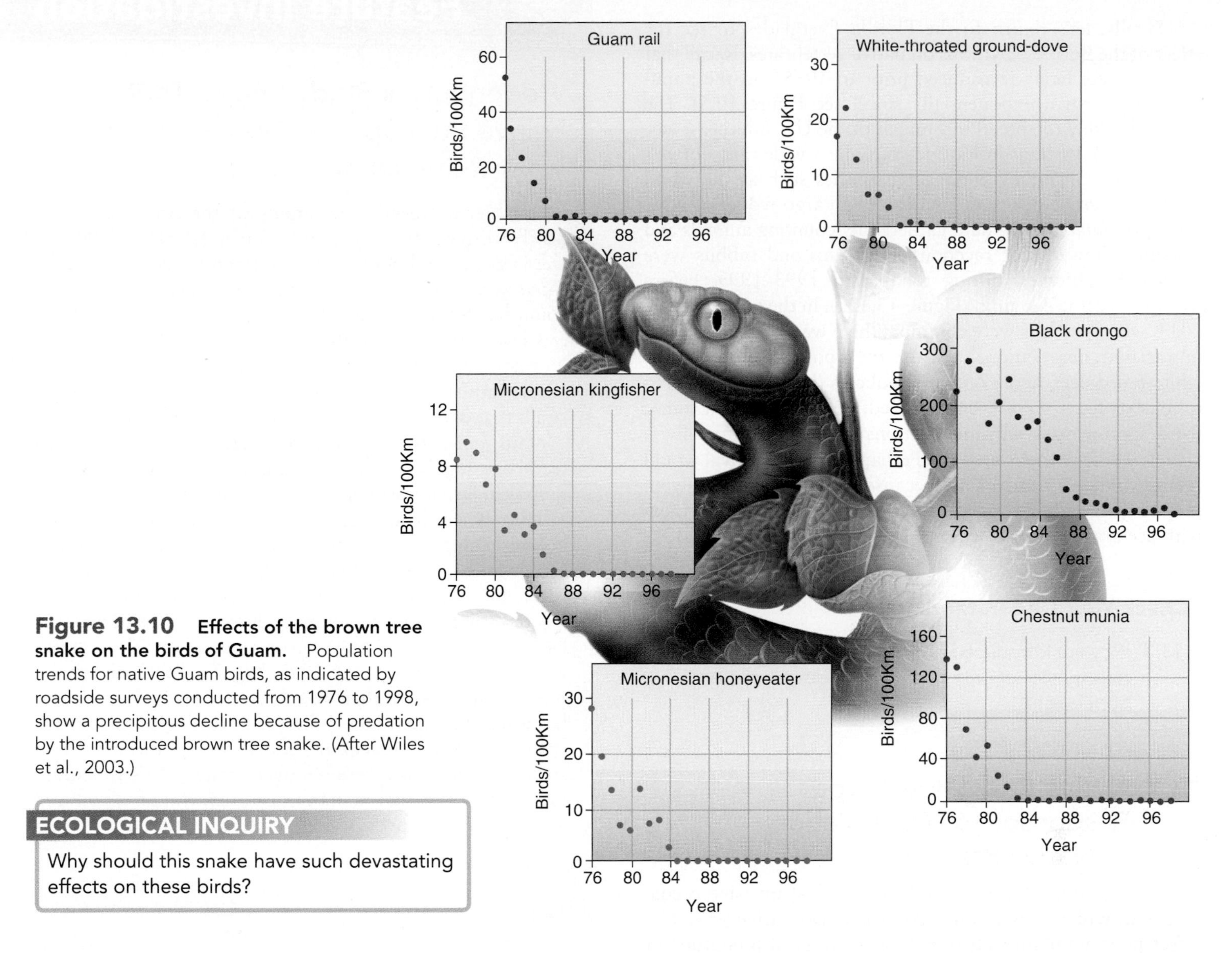


**Figure 13.10 Effects of the brown tree snake on the birds of Guam.** Population trends for native Guam birds, as indicated by roadside surveys conducted from 1976 to 1998, show a precipitous decline because of predation by the introduced brown tree snake. (After Wiles et al., 2003.)

**ECOLOGICAL INQUIRY**

Why should this snake have such devastating effects on these birds?

transmitters. If a signal is found to move, it is because the mouse has very likely been eaten by a snake. The first airdrop was made in 2010, and it is too early to see how effectively this control measure is working. However, in some other systems, removal of exotics has resulted in a rebound of native prey (see **Feature Investigation**).

Sea lampreys, *Petromyzon marinus,* are primitive jawless fish that feed by rasping holes in the sides of larger fish species and feeding on their blood (**Figure 13.11a**). They spawn in freshwater streams, and the juveniles return to salt water (or one of the Great Lakes as a substitute) to develop. Such anadromous fish have special modifications of their kidneys to allow them to exist in fresh water or salt water. Sea lampreys found their way into Lake Ontario in the mid-1800s by way of the Erie Canal (**Figure 13.11b**). Improvements to the Welland Canal allowed lamprey to bypass Niagara Falls and colonize the rest of the Great Lakes. In 1921 lampreys were found in Lake Erie and by 1938 they had been found all the way up into Lake Superior. Lake trout, *Salvelinus namaycush,* were the preferred prey in the Great Lakes. The lake trout fishing industry had declined somewhat in the 1940s due to overfishing, but the decline was hastened after the arrival of lampreys (**Figure 13.11c**). By the 1960s, lake trout catches had been reduced by over 90% of their historic averages, and the fishery collapsed.

In the late 1940s the State of Michigan began looking at control measures for lamprey. Weirs were erected across the mouths of streams to prevent the return of spawning adults, but the weirs were difficult to maintain and were not 100% effective. Attention turned to chemical control of juvenile lampreys, which spend 3–5 years buried in stream sediments, filter feeding. The chemical TFM (trifluoromethyl nitrophenol) proved effective in controlling juvenile lampreys in streams and was first applied in 1958 to streams feeding into Lake Superior. Treatments reduced lamprey densities by 90%, and the population of lake trout increased in the 1980s, aided by an active program of restocking. To maintain lamprey control, streams are treated every 3–5 years. Removal of exotics is time-consuming and costly but may lead to recovery of native prey.

Finally, let's return to the Florida Everglades to see the effect of the Burmese pythons on native vertebrates. Recall that pythons were likely introduced prior to 1985 and the population has grown exponentially since (see Figure 10.5). The species is likely to spread into much of the U.S. Southeast (see Figure 7.12). Pythons in Florida consume a wide range of vertebrate prey, including endangered species such as the wood stork, *Mycteria americana,* and the Key Largo woodrat, *Neotoma floridana.* Survey data gathered by counting animals and roadkills showed that raccoons, opossums and rabbits were commonly observed during the period 1993–1999, before pythons became common (**Figure 13.12a**). In the period 2003–2011, when pythons were common, there was a 99.3% decline in raccoon observations, 98.9% for opossums, 94.1% for white-tailed deer, and 87.5% for bobcats (**Figure 13.12b**). No rabbits or foxes were even observed. In South Florida today, these species exist only in locations peripheral to the Everglades, where pythons are rare. Perhaps python removal would permit wildlife to recover, but because constrictors feed only on prey they have killed, it would not be feasible to control them with acetaminophen, as was done with the brown tree snake.

### Check Your Understanding

**13.3** Why are introduced predators able to threaten rare species of prey that cannot sustain predator populations?

## 13.4 Native Prey Show Large Responses to Manipulations of Native Predators

Native prey often have few defenses against invasive predators with which they did not co-evolve. Do native predators affect prey populations as much as introduced predators, or do they take only individuals that would die anyway from disease or starvation? In actuality, native predators have also been shown to impact their native prey in many areas. In North America, humans have for many years exterminated wolves, bobcats, coyotes, and pumas under the assumption that, because predators kill to survive, they must depress the numbers of natural prey as well as numbers of cattle or sheep. While it is true that deer herds appear to increase dramatically after the removal of a predator, we have to acknowledge the effect of environmental change, such as clearing forests and providing rich agricultural fields, which has gone on concurrently and which would tend to boost deer numbers.

Much of the best data on the effects of native predators on native prey comes from studies of game animals, such as pheasants or ducks, where predators are removed by game managers to increase flock sizes for hunters. In the prairie pothole region of North Dakota, shallow lakes serve as breeding grounds for a variety of duck species, including blue-winged teal, *Anas discors,* mallards, *A. platyrhynchos,* gadwalls, *A. strepera,* northern

## Feature Investigation

### Vredenburg's Study Shows That Native Prey Species May Recover After Removal of Exotics

In the Sierra Nevada mountains of the American West, nearly all the lakes and ponds above 2,100 m (7,000 ft) were fishless, and populations of amphibians thrived, with *Rana muscosa,* the yellow-legged frog, perhaps the most common. Since the mid-1800s, a variety of trout species have been released in these lakes to support a fishing industry. Supply by airplanes ensured even the most remote lakes were stocked with trout on a regular basis, with a result that over 80% of naturally fishless lakes in the Sierra contain non-native trout. Unfortunately, these introduced trout are highly effective predators of native frogs and tadpoles. Densities of *R. muscosa* were hugely depressed in lakes where trout were released (**Figure 13.A-i**). This is not surprising, because the species had evolved in a fishless environment. The good news is that such effects might be reversible. Beginning in 1997 Vance Vredenburg and others began to use gill nets to remove introduced fish from five lakes. Because of the huge effort required in fish removal, removal in lake 1 began in 1997, lake 2 in 1998, and so on. Results have shown that frog densities have rebounded to levels seen in nearby lakes where fish were never introduced (**Figure 13.A-ii**) (Vredenburg, 2004). Recoveries have been strongest where fish were removed earliest. This encouraging result means that in many cases removal of introduced predators may allow systems to return to normal.

**HYPOTHESIS** Native *Rana muscosa* frogs in lakes of the Californian Sierra Nevada recover where exotic trout are removed.

**STARTING LOCATION** California, Sierra Nevada

| | Conceptual level | Experimental level |
|---|---|---|
| 1 | Determine densities of native *R. muscosa* tadpole and frog densities in lakes where trout may or may not be present. | Examine densities of introduced trout in 50 lakes in Sierra Nevada, California, at Kings Canyon National Park (map). Use gill nets to catch trout and visual examinations of the shoreline to count tadpoles and frogs. Some lakes had introduced trout and others did not. |
| 2 | Eliminate exotic predator or reduce it to very low population levels. | Use gill nets to remove trout from five lakes. Lakes had no upstream trout populations and a downstream barrier (that is; waterfall with no jump pool) to prevent recolonization. Removal began in 1997 in Lake 1, 1998 in Lake 2, and so on. |
| 3 | Monitor densities of native *R. muscosa* to see if it recovers. | Examine densities of *R. muscosa* in five removal lakes, eight control lakes with trout (fish controls), and eight control lakes which never had fish introduced (fishless controls). Frog density was estimated by visual counts along shoreline from 1997 to 2003. |

4 **THE DATA** Lakes 1–5 are the trout removal lakes; * = fishless control lakes, used in the study and ** = fish control lakes, used in the study.

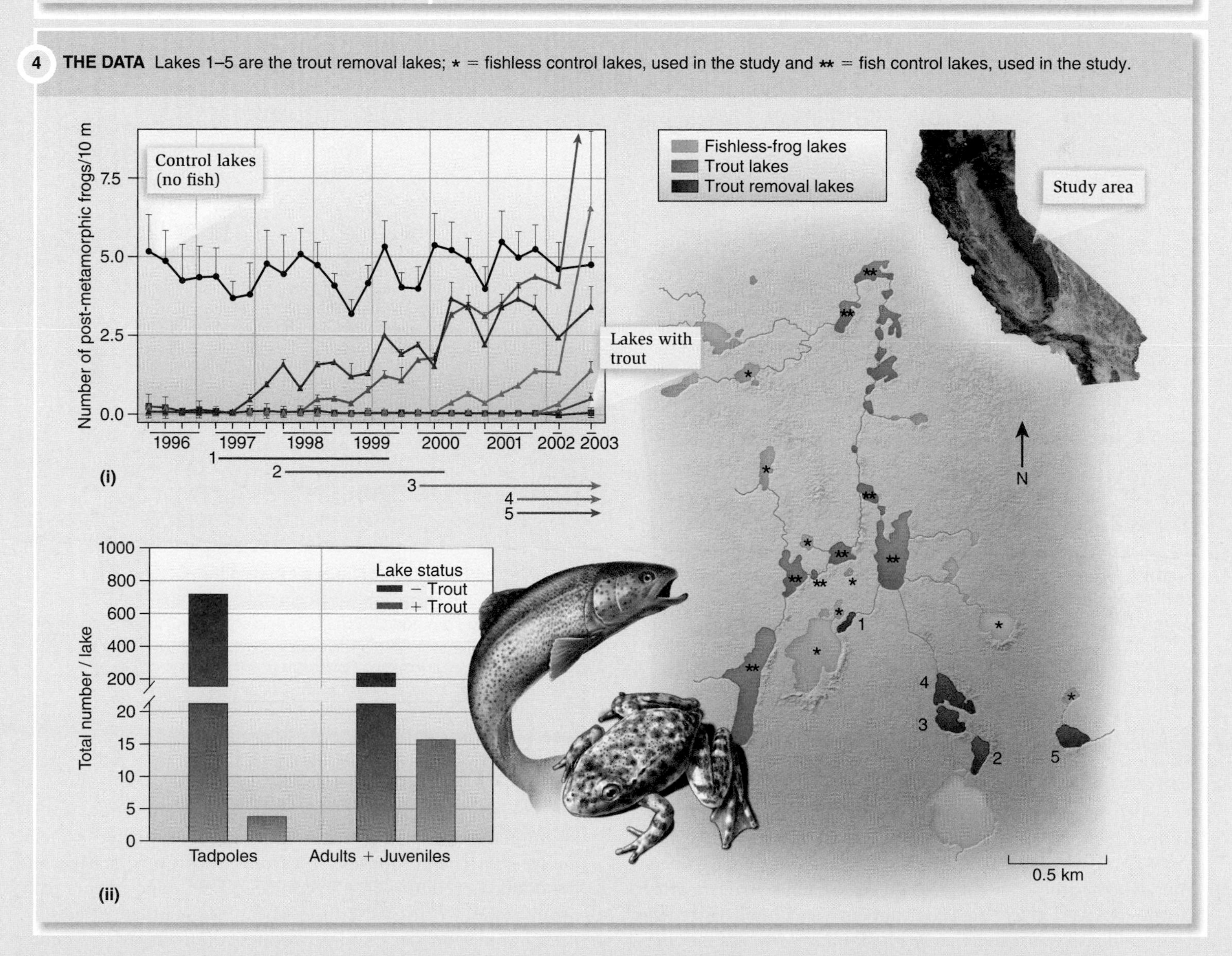


**Figure 13.A Effect of exotic trout removal on native *Rana muscosa* frog populations in lakes of the Californian Sierra Nevada.** **(i)** Recovery of frogs in lakes following trout removal. **(ii)** Effect of trout on densities of frog tadpoles and adults plus juveniles.

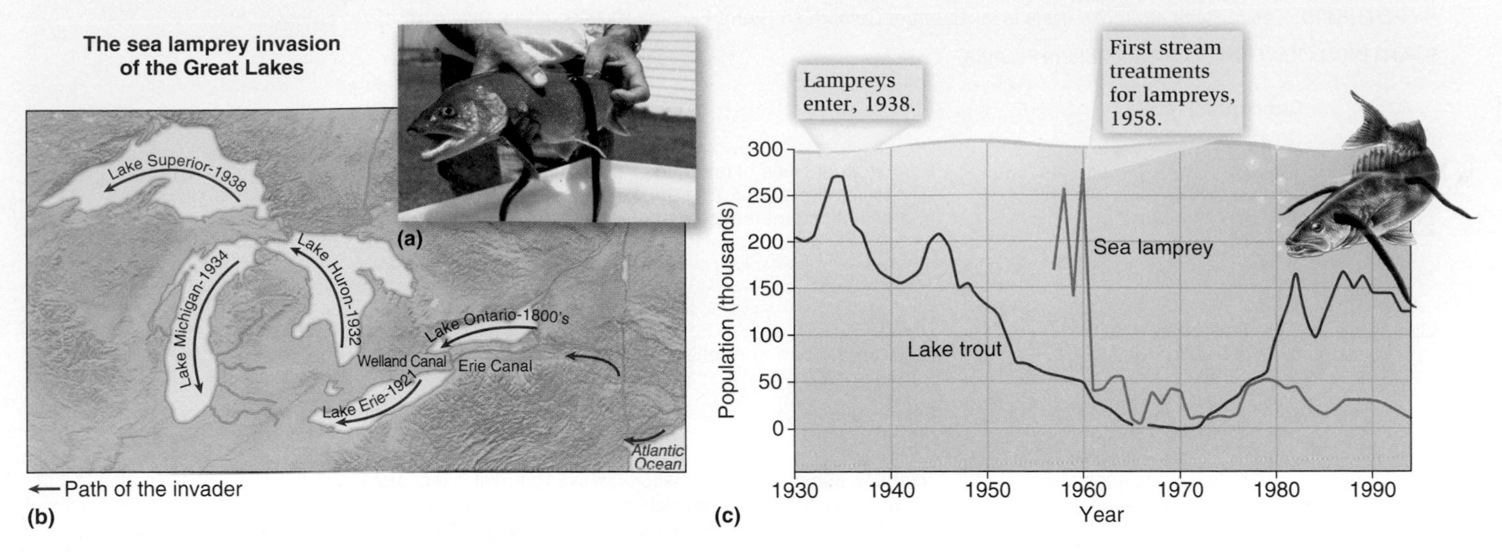


**Figure 13.11** **Effects of invasive sea lampreys on the lake trout population in the Great Lakes.** **(a)** Sea lamprey attached to a lake trout. **(b)** Historical passage of lampreys into the Great Lakes from the Atlantic. **(c)** Effects of lampreys on the trout population in Lake Superior. Note that detailed records of sea lamprey populations were not available before 1956.

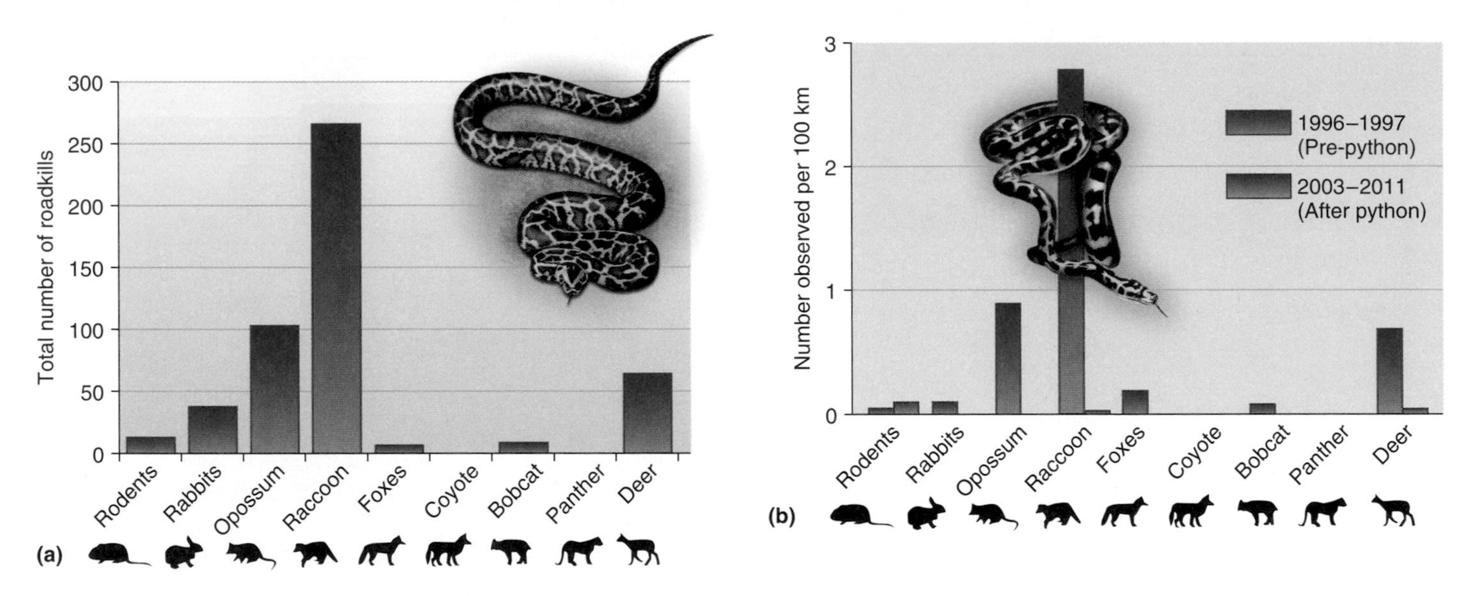


**Figure 13.12** **Effect of invasive Burmese pythons on wildlife abundance in the Everglades National Park, Florida.** **(a)** Numbers of roadkills recorded by park staff in 1993–1999, before pythons became common. Raccoons and opossums were common at this time. **(b)** Decline in mammal abundance per 100 km of road survey (live and roadkill) prior to pythons becoming common (1996–1997) and after they became common (2003–2011). Numbers below the bar represent the change in numbers of observations/100 km for each species or group. (After Dorcas et al., 2012.)

pintails, *A. acuta,* and northern shovelers, *A. clypeata.* Medium-sized predators such as red fox, *Vulpes vulpes,* and raccoon, *Procyon lotor,* have increased due to eradication of large predators such as wolves and coyotes, which used to prey on and compete with them. Pamela Garrettson and Frank Rohwer (2001) embarked on a predator removal experiment to determine whether these native medium-sized predators were affecting nest success. Experienced trappers were hired to remove foxes and raccoons, together with skunks, *Mephitis mephitis,* and mink, *Mustela vison.* Over 2,400 mammalian predators were removed from a series of sites over a 2-year period. Nest success almost doubled in areas where predators were removed, compared to control areas (**Figure 13.13**).

Do native predators have as strong effects on prey as introduced predators? Pälvi Salo and colleagues (2007) attempted to answer this question. They conducted a meta-analysis of the responses of vertebrate prey to native and introduced predators in 45 replicated field experiments from around the world. Response variables included effects on prey population sizes and reproductive effects. Most studies involved predatory mammals (24 studies), predatory mammals and birds combined (17), and predatory birds alone (4).

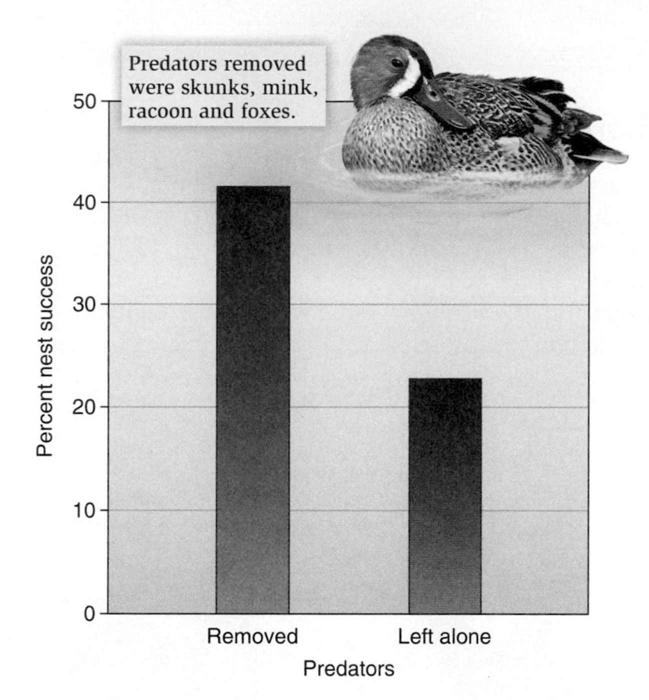


**Figure 13.13 Effects of native predator removal on nest success of ducks in North Dakota.** (After data in Garrettson and Rohwer, 2001.)

The effect of the manipulations was scored on mammalian prey, mainly mice and voles (27 studies), or birds, mainly waterfowl (18). Most studies were removals or reductions (41 studies) rather than additions (4), possibly because it is easier to fence or cage areas and keep predators out than it is to add them. The results showed that introduced predators had an impact double that of native predators (**Figure 13.14**). These data suggest that although predators greatly impact their prey, introduced predators are a greater threat to native prey than are native predators. Introduced predators are not the only new threat to native prey. Climate change also threatens to disturb the balance between native prey and their predators (see **Global Insight** feature).

### Check Your Understanding

**13.4** Why can't we be certain that removal of predators such as wolves or cougars causes deer populations to skyrocket?

## 13.5 Humans, as Predators, Can Greatly Impact Animal Populations

The ultimate predators are humans, who have succeeded in hunting a multitude of species to extinction. Here we consider the effects of human predators on fish and whale populations. Can people harvest part of a fish population without causing long-term changes in its equilibrium numbers? Can commercial fisheries exist without catastrophic collapses of fish populations? Good ecological and economic models exist that, if followed, might

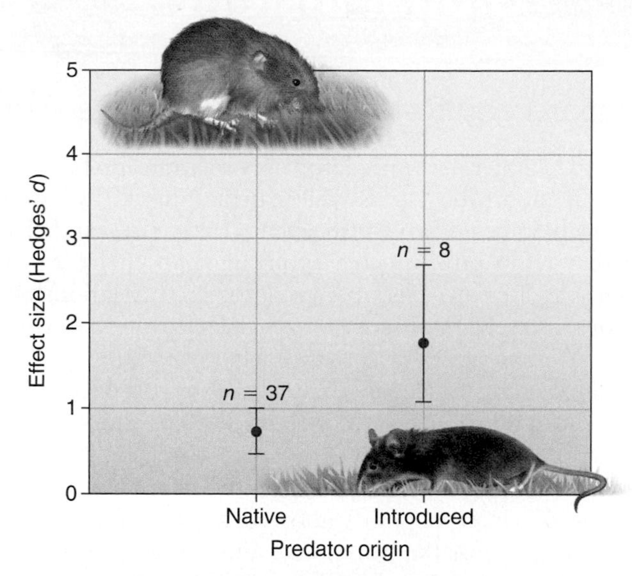


**Figure 13.14 Effects of native and introduced predators on native prey.** (After Salo et al., 2007.)

### ECOLOGICAL INQUIRY

In a separate analysis, investigators showed a much greater effect of introduced species in Australia than in other parts of the world. Why might the effects of introduced predators be so strong in Australia?

prevent overfishing. However, as we shall see, there also seem to be plenty of fisheries that don't follow these sound principles.

For a harvested population, the important measurement is the yield, expressed in terms of either weight or numbers. **Maximum sustainable yield** (MSY) is the largest number of individuals that can be removed without causing long-term changes in the population. This is like skimming the interest from a bank account but leaving the principal intact to ensure the same future interest payment in subsequent years. The maximum sustainable yield of a population is the point of the maximum population increase. According to the logistic model, population increase, $dN/dt$, occurs following the equation

$$\frac{dN}{dt} = rN\frac{(K-N)}{K}$$

Greatest population increase, and thus maximum yield, occurs at the midpoint of the logistic curve, at $K/2$ (**Figure 13.15**). Thus, maximum yield is obtained from populations at less than maximum density, when they are maximally expanding.

MSY can be estimated as

$$\text{MSY} = \frac{rK}{2}$$

where $r$ is the per capita growth rate and $K$ is the carrying capacity of organisms throughout the year. Thus, if $K/2 = 10{,}000$ and $r = 0.14$ then MSY $= 1400$.

## Global Insight

### Predator-Prey Relationships May Be Altered by Long-Term Climate Changes

As noted in Chapter 4, the predator-prey relationship between wolves and moose on Isle Royale can be altered by heavy snow, which impedes wolf movement. Recent research has shown that temperature increases brought about by global warming threaten to disrupt predator-prey relationships even more severely. Moose colonized Isle Royale by swimming from Canada, 20 miles away, early in the 20th century. Moose numbers increased, and by the winter of 1949, reports of moose starvation had surfaced and the government was faced with the issue of how to reduce moose numbers and alleviate starvation. Coincidentally, nature intervened.

The cold winter of 1949 allowed a pair of Canadian wolves to walk across the frozen lake and colonize the island. The wolves reproduced and wolf numbers began to increase, presumably reducing the numbers of moose. In 1958 wildlife biologist Durwood Allen of Purdue University began studying the wolf and moose interactions and providing precise data on wolf and moose abundance. Student Rolf Peterson took over in the 1970s, and the 50-year investigation is now the longest running predator-prey study on record. Allen noticed a gradual increase in wolf numbers and a concomitant decrease in moose in the 1970s (**Figure 13.B**). By 1980 a record 50 wolves had reduced moose to only 750 animals. The balance lasted until 1981, when a strict rule barring pets from the island was broken. A Chicago man had brought his sick dog to the island on a July 4th fishing trip in 1981, and canine parvovirus, a disease that ravages wolves as well as dogs, swept the island. By 1982 only 14 wolves remained. The moose numbers skyrocketed to 2,400 in 1995.

Once again, nature changed the game. A severe winter in 1995, with temperatures as low as −43°F, chilled the island. Heavy snow made foraging difficult. Snow melt in the spring was delayed, and by 1997 only 500 moose were left. Moose and wolf numbers slowly began to rebound. More recently, John Vucetich, a population biologist at Michigan Tech, has noticed a slide in moose numbers, brought about, he says, by climate change. Five of the summers in 2002–2008 were the warmest on record. Warm summers triggered devastating tick infestations, which reduced moose numbers, but the wolves were not affected. In summary, the wolf-moose interaction on Isle Royale has been upset, directly and indirectly, by climate change.

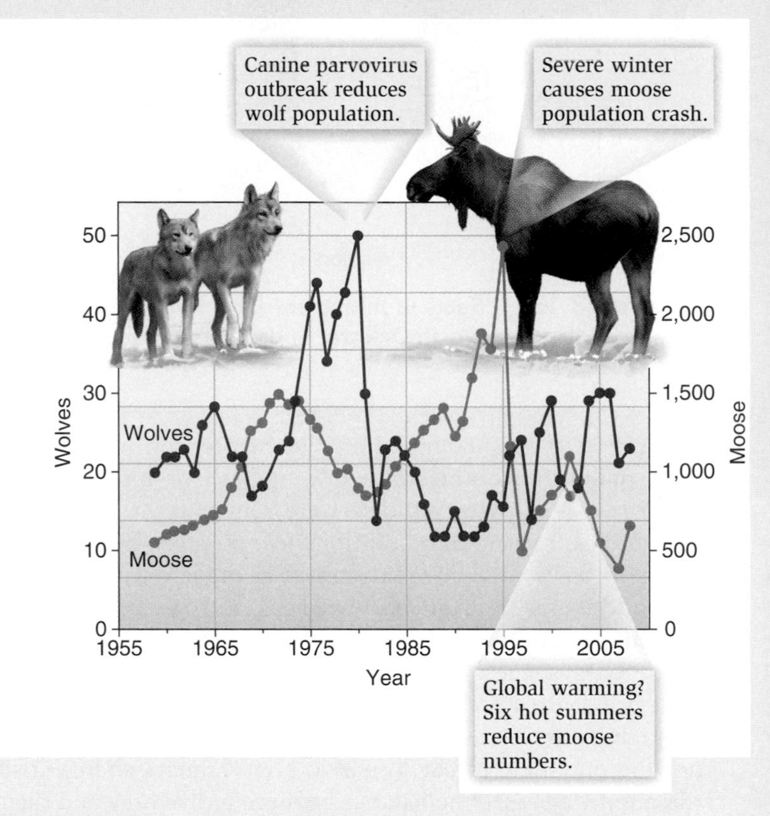


**Figure 13.B** Moose-wolf interactions on Isle Royale over the past 50 years.

Despite the simplicity of this model, it is usually difficult to harvest fish in a sustainable manner because biologists cannot easily go underwater and count the numbers of available fish. Often they turn to simple economic fisheries models, which predict that fisheries should be operated at well below MSY, thus protecting them from overexploitation (**Figure 13.16**). Let's consider that total economic costs increase linearly as fishing effort increases. Total revenue also increases with fishing effort, up to a point. After MSY is reached, any increase in effort would result in a decrease in revenue as the fish populations become overfished. The maximum profit is equal to the biggest difference between the cost and revenue curves, and as Figure 13.16 shows, this occurs below the effort needed to reach the maximum sustainable yield. And yet, many species of fish are harvested to the degree that the rate of removal exceeds the rate of reproduction, the fish populations crash, and the fishery collapses. In the case of the Canadian cod fishery, overfishing and collapse came in the early 1990s after hundreds of years fishing (**Figure 13.17**). Since that time, the species has not been economically fished.

How do fisheries become overfished? Most estimates of sustainable harvest are nearly always too high, because of overestimates of population size. Furthermore, incremental improvements are made to fishing gear over the years and the same level of effort results in increased catches. Finally, as the stock declines, the prices inch upward, making it more

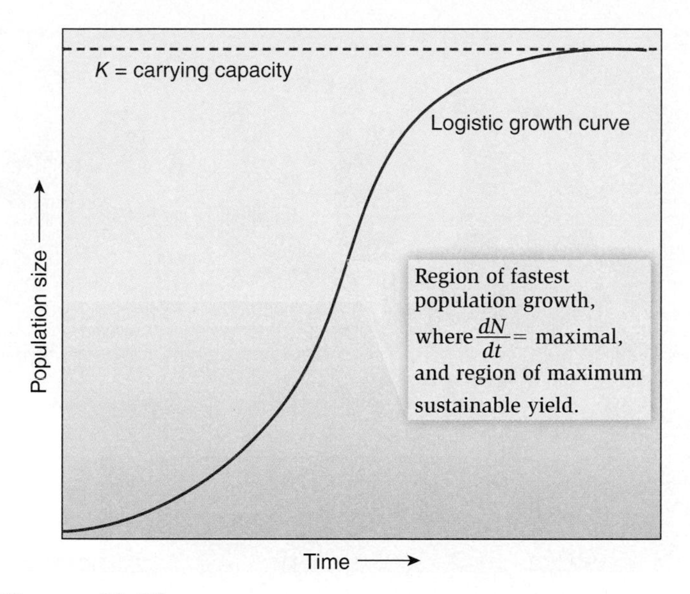


**Figure 13.15 Maximum sustainable yield.** Theoretically, maximum sustainable yield occurs at the midpoint of the logistic curve.

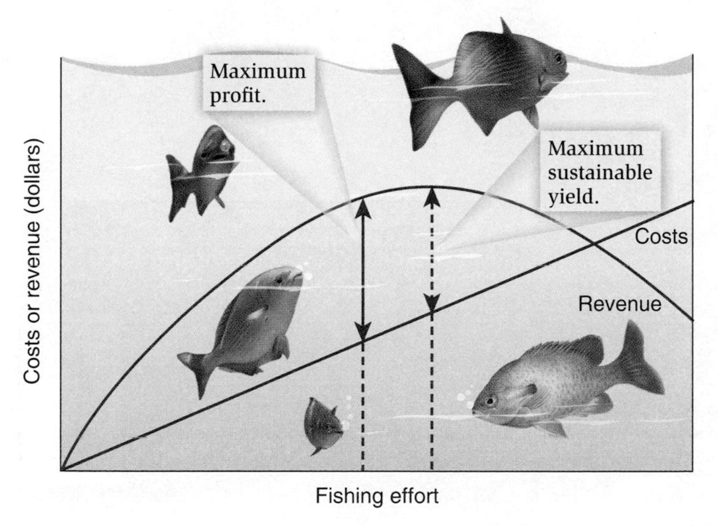


**Figure 13.16 Economic fisheries model.** Here it is assumed that costs increase with effort but revenue reaches a maximum at maximum sustainable yield. Maximum profit, however, occurs at a lower level than this.

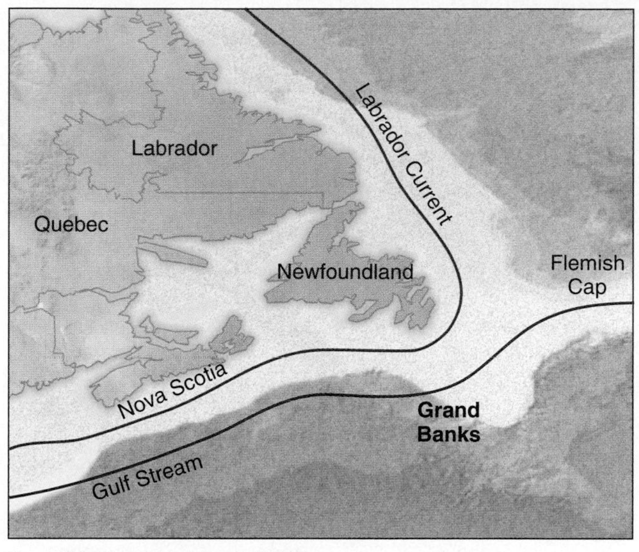


(a) Map showing the Grand Banks

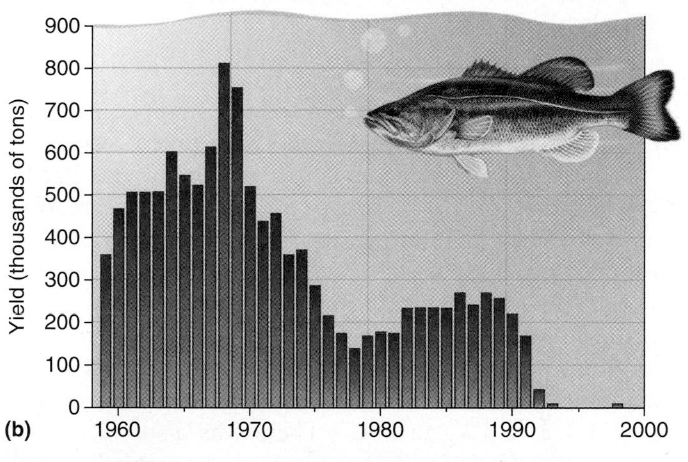


(b)

**Figure 13.17 Commercial cod fisheries. (a)** The Grand Banks area off the coast Newfoundland used to be a fertile fishing ground for cod. **(b)** This cod fishery collapsed between 1989 and 1992, when it was closed due to overfishing.

likely for commercial fishing to keep going. In the case of the cod industry, the introduction of radar and sonar allowed crews to pursue fish over huge areas. The crews also caught enormous amounts of noncommercial fish, which were discarded. Among these were important prey species for cod, such as capelin. Finally, although undersized cod, which cannot spawn, were returned to the ocean, such discards do not always survive. As the cod population went into a tailspin, increased effort resulted in more undersized fish being caught and more discarded.

One of the most famous examples of overfishing is the Peruvian anchovy industry, which collapsed in the 1970s. This was the largest fishery in the world until 1972, when it collapsed due to overfishing and a severe El Niño event. In the El Niño the coastal waters warmed, stopping cold, nutrient-rich water from coming to the surface. Incredibly, after 16 years of suspended fishing, from 1972 until 1988, the anchovy population began to rebound, but it took another 5 years to fully recover (**Figure 13.18**). There is hope that the Atlantic cod fishery on the Grand Banks will eventually recover, but the recovery will be slow because the fish do not spawn until age 7. A 2010 study showed that stocks near Newfoundland and Labrador were still only 10% of their original size.

Felicia Coleman and colleagues (2004) showed that even recreational fishing can severely impact some fish species. They conducted detailed censuses of the recreational and commercial fish landings in the United States in 2002. Among populations of fish species for which overfishing is a concern, recreational landings accounted for 23% of the total nationwide, rising to 38% in the south Atlantic and 64% in the

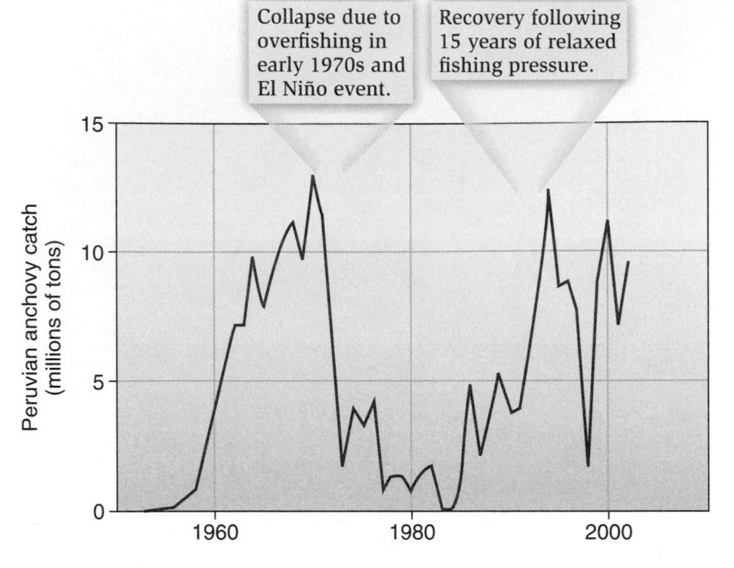


**Figure 13.18 The Peruvian anchovy industry.** Overfishing caused a collapse in the 1970s. A 16-year moratorium allowed the fisheries to rebound in the late 1980s.

Gulf of Mexico. Fifty-nine percent of red snapper, *Lutjanus campechanus,* in the Gulf of Mexico were taken recreationally, far outstripping the commercial catch. In the Pacific, 87% of bocaccio, *Sebastis paucispinis,* were taken recreationally, and in the south Atlantic, the figure was an astounding 93% for red drum, *Sciaenops ocellatus.*

Loren McClenachan (2009) used historical photographs to document the reduction of large trophy fish taken in the Florida Keys over the period 1956–2007 (**Figure 13.19**). The mean fish size declined from an estimated 19.9 ± 1.5 kg in 1956 to 2.3 ± 0.3 kg in 2007. There was also shift in the species caught. In 1956, large groupers, *Epinephellus* spp., were commonly landed together with large sharks >2 m. In 2007, small snappers, *Lutjanus campechanus,* and *Ocyurus chrysurus,* average length 34.4 cm, and sharks <1 m were landed. McClenachan used these data to argue that unfished reef communities support larger-bodied fish than fished reef communities.

The question as to whether whales should be exploited has been the subject of vigorous and worldwide debate since at least the 1960s. The level of popular interest in this question probably exceeds that concerning any other group of exploited animals. The history of whaling in general and Antarctic whaling in particular and has been characterized by a progression from more valuable or more easily caught species to less attractive ones, as stocks of the original targets were depleted (**Figure 13.20**). In the Antarctic, blue whales, *Balaenoptera musculus,* dominated the catches through the 1930s, but by the middle 1950s few were being taken, although the species was not legally protected until 1965. As the stocks of blue whales diminished, attention was turned to the fin whale, *B. physalus,* which was originally the most abundant of all whales in the southern ocean. By the 1960s, numbers of this

(a)

(b)

(c)

**Figure 13.19 Change in size of trophy fish caught on Key West charter boats.** **(a)** 1957, **(b)** early 1980s, and **(c)** 2007. (After McClenachan, 2009.)

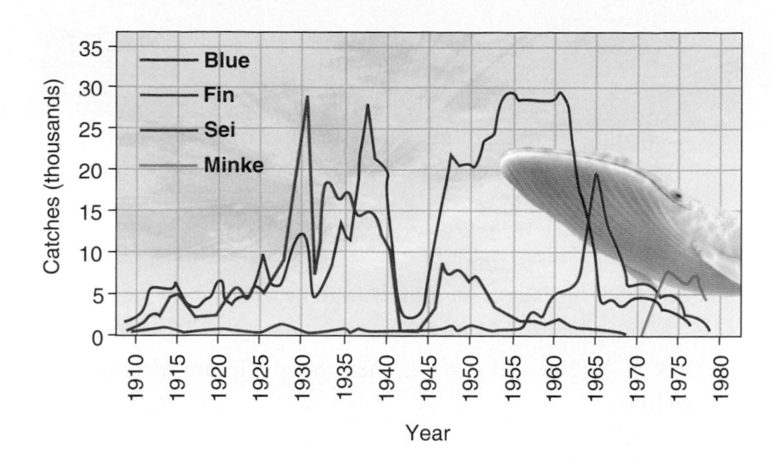


**Figure 13.20** Sequential decline of whale catches in the Antarctic, showing the strong effect of human predators.

species had collapsed rapidly. Humpback whales, *Megaptera novaeangliae,* though never very numerous, were attractive because of their high oil yield and the ease with which they could be caught. Catches were never very great, but the stocks in most areas collapsed dramatically in the early 1960s. Sei whales, *B. borealis,* were almost ignored by whalers until the bigger species were no longer available. They were hardly taken at all until about 1958, but then catches increased rapidly and reached a peak of about 20,000 in 1964–1965. Catches declined rapidly thereafter, this time due to the introduction of catch limits. Then the relatively small minke whales, *B. acutorostrata,* which were ignored in the southern ocean until 1971–1972, began to be taken. Since that time, minkes have been the largest component of the southern baleen whale catch.

The story is similar in the Northern Hemisphere, although the stocks were much smaller, about 20% of that in Antarctic waters. In 1982, perhaps because it could see the writing on the wall, the International Whaling Commission (IWC) voted for a moratorium on all commercial whaling. It was proposed that commercial whaling be ended in 1985–1986, a proposal that did not actually take effect until 1988. The good news is that following the moratorium, the populations of some whales have increased. Blue whales are thought to have quadrupled their numbers off the California coast during the 1980s, showing the impact that protection from a predator can have. A report out of Steve Palumbi's lab (Roman and Palumbi, 2003) estimated past population sizes of North Atlantic whales by examining mitochondrial DNA sequence variation. Population size estimates for fin and humpback whales suggested that prewhaling populations were 6–20 times higher than present-day population estimates and that full recovery of whale populations would take another 70–100 years.

### Check Your Understanding

**13.5** Why does maximum sustainable yield occur at the midpoint of the logistic curve and not where the population is at carrying capacity?

## SUMMARY

**13.1 Animals Have Evolved Many Antipredator Adaptations**

- Predation, herbivory, and parasitism all result in a positive effect for one species and a negative effect for the other, but these interactions differ in their lethality and duration (Figure 13.1).
- The many antipredator strategies that animals have evolved, including chemical defense, aposematic coloration, cryptic coloration, Batesian and Müllerian mimicry, physical defenses, and predator satiation, suggest that predation is a strong selective force (Figures 13.2–13.5, Tables 13.1, 13.2).
- Predators mimic various abiotic and biotic features of their environment to get close to their prey (Figure 13.6).

**13.2 Predator-Prey Interactions May Be Modeled by Lotka-Volterra Equations**

- Mathematical models suggest that the interaction between prey and their natural enemies can result in coupled oscillations. As the efficiency of predators increases, the oscillations become more damped (Figures 13.7–13.9).

**13.3 Introduced Predators Show Strong Effects on Native Prey**

- Invasive predators can have large effects on densities of native prey (Figure 13.A, Figures 13.10–13.12).

**13.4 Native Prey Show Large Responses to Manipulations of Native Predators**

- Native predators also greatly impact populations of native prey, though the effects are often not as great as those from introduced predators (Figures 13.13, 13.14, Figure 13.B).

**13.5 Humans, As Predators, Can Greatly Impact Animal Populations**

- Humans can greatly impact the densities of harvested species. The maximum yield of a prey population is when harvesting occurs at the midpoint of the logistic curve, at $K/2$ (Figures 13.15, 13.16).
- Overharvesting and overfishing can lead to the collapse of many harvested populations (Figure 13.17–13.20).

## TEST YOURSELF

1. Which of the following has low lethality but high intimacy?
   a. Parasitoids
   b. Herbivores
   c. Predators
   d. Parasites
2. Ladybird beetles employ which type of antipredator strategy?
   a. Aposematic coloration
   b. Cryptic coloration
   c. Batesian mimicry
   d. Müllerian mimicry
   e. Predator satiation
3. Syrphid flies employ which type of antipredator strategy?
   a. Aposematic coloration
   b. Cryptic coloration
   c. Batesian mimicry
   d. Müllerian mimicry
   e. Predator satiation
4. Mutual interference between predators tends to:
   a. Bend the predator isocline to the right
   b. Bend the predator isocline to the left
   c. Bend the prey isocline to the right
   d. Bend the prey isocline to the left
   e. Bend both predator and prey isocline
5. Which of the following are native predators?
   a. Brown tree snakes in Guam
   b. Sea lampreys in Lake Superior
   c. Foxes in America
   d. Rainbow trout in California
   e. Pythons in Florida
6. According to a recent meta-analysis in England, which type of predator removal had the strongest effects on prey populations?
   a. Predatory birds
   b. Predatory mammals
   c. Both birds and mammals together
   d. Both had the same effects
7. If $K = 20{,}000$ and $r = 0.2$, then the maximum sustainable yield is:
   a. 4,000
   b. 100,000
   c. 2,000
   e. 40,000
8. The maximum yield occurs at which point of the logistic curve?
   a. Beginning
   b. Middle
   c. Upper end
9. Which was the first group of whales to be overharvested in the southern ocean?
   a. Blue
   b. Fin
   c. Minke
   d. Sei
   e. Humpback
10. Cryptically colored prey often remain motionless in the presence of predators:
    a. True
    b. False

## CONCEPTUAL QUESTIONS

1. Discuss three different types of antipredator strategies that animals have evolved.
2. Explain the Lotka-Volterra predation models and describe what they tell us about predator-prey relationships.
3. Why do invasive predators have stronger effects than native predators?

## DATA ANALYSIS

A researcher in Wisconsin used a trained red-tailed hawk, *Buteo jamaicensis,* to capture three types of prey, eastern chipmunk, cottontail rabbit and gray squirrel, in field conditions. Chipmunks were small and easy to kill, rabbits were bigger and escaped more frequently, and squirrels were very wary and escaped into trees. The researcher retrieved the intact carcasses from the hawk and scored them as to their condition. Individuals with fractured bones, defective eyes, wounds, or low loin fat were classified as substandard. The hawk frequently failed in its attacks. Attack rate success against each of the prey species, and the percent classified as substandard, are given in the table. Explain the results.

| Prey Species | Ease of capture | Percent of attacks failed | Percent prey that were substandard |
|---|---|---|---|
| Chipmunk | Easy | 72 | 8 |
| Rabbit | Moderate | 82 | 21 |
| Squirrel | Difficult | 88 | 33 |

Bison within the Wind Cave National Park, South Dakota. Note the lack of small trees due to heavy herbivore pressure. Herbivore densities have increased in the absence of predation by wolves.

CHAPTER 14

# Herbivory

## Outline and Concepts



Herbivory involves predation of plants or algae. There is a staggering variety of herbivorous species that feed in a variety of different ways (**Figure 14.1**). Many different vertebrate species, such as deer, goats, sheep, cows, horses, antelopes, rhinoceros, elephants, and humans, feed on plants. In addition, there are over 360,000 species of herbivorous insects, primarily beetles, butterflies and moths, flies, sucking bugs, and grasshoppers. Beetles, grasshoppers, and lepidopteran caterpillars often chew plant leaves, and many hemiptera, the sucking bugs, pierce the plant's vascular system and tap into the phloem or, less commonly, the xylem. In addition, there are a variety of more specialized feeding strategies. Gall insects, which are most commonly flies, aphids, or wasps, lay their eggs on stems or leaf buds, and induce a tumor-like growth. The larvae feed safely inside the developing gall (**Figure 14.2a**). Leaf miners lay eggs on leaves and the larvae feed between the leaf surfaces, creating blister-like mines on leaves (**Figure 14.2b**). Some herbivore species even live deep underground, feeding on roots. Bark beetles feed just under the bark, while other beetles feed deep within the heart wood. We can classify many of these herbivores by what they eat. Folivores specialize in leaf feeding, while frugivores, such as some parrots, eat mainly fruit (**Figure 14.2c**). Nectarivores feed only on nectar, and granivores only on seeds. Many large vertebrate herbivores eat nearly all parts of the plant. **Monophagous** herbivores, most of which are insects, feed on one plant species or just a few closely related species. **Polyphagous** species, which are mainly mammals, feed on many different host species, often from more than one family. There are, however, exceptions. Pandas are monophagous because they feed only on bamboo,

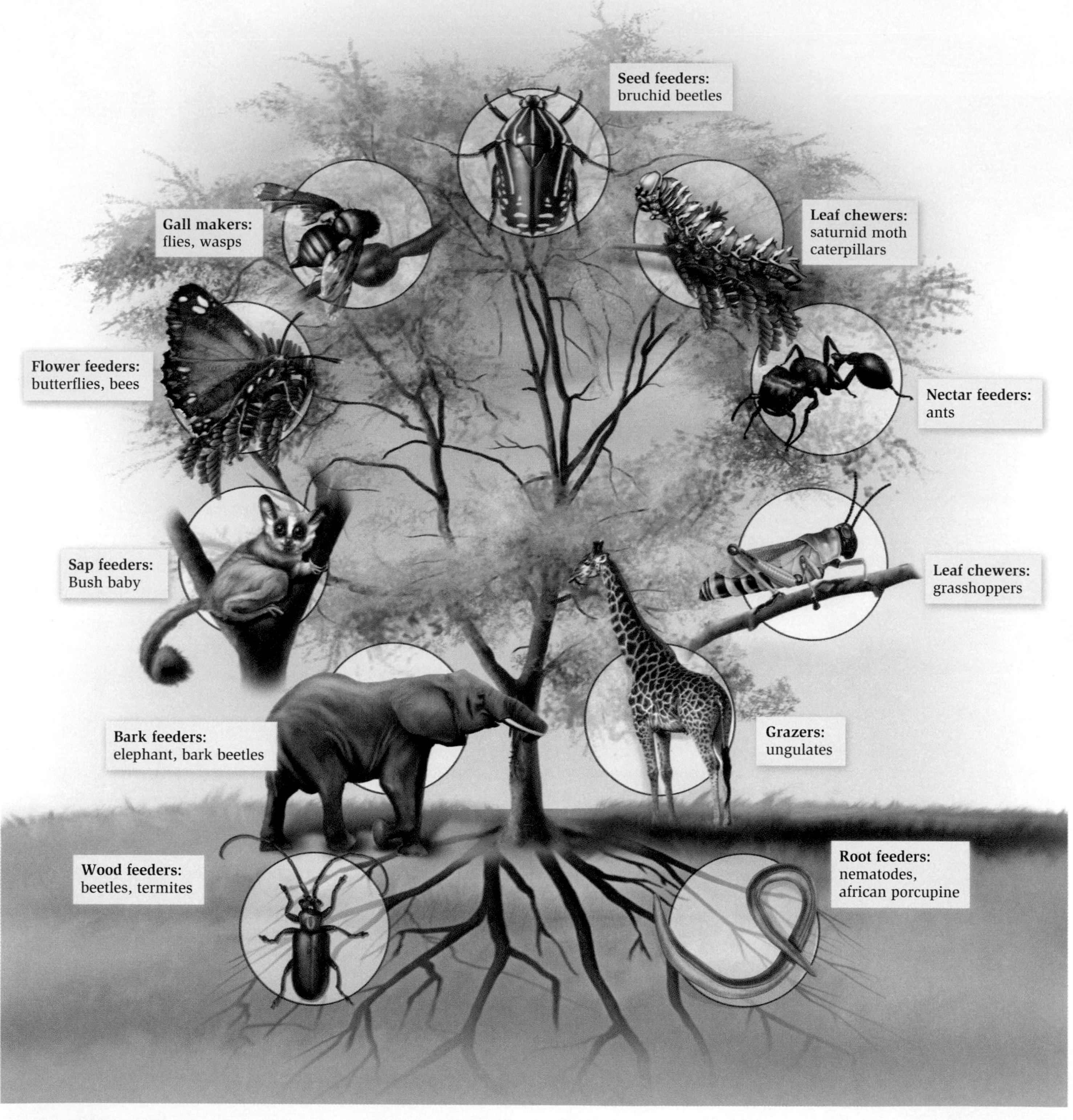


**Figure 14.1** **Herbivory.** A wide range of herbivores feed on plants, as on this African acacia tree.

and koalas specialize on eucalyptus trees. On the other hand, grasshoppers are polyphagous because they feed on a wide variety of plant species, including important crops.

With so many species of herbivores, it is not surprising to find that herbivory can be lethal to plants, especially for small species. However, many plant species defend themselves against herbivores, and some species, particularly larger ones, can regrow. In this chapter we take a closer look at plant defenses against herbivores and the ways that herbivores attempt to overcome them. Later in the chapter we discuss the effects of native and introduced herbivores on plant populations and the use of herbivores to control pest plants.

**Figure 14.2 Specialized herbivores.** **(a)** A gall insect on an oak leaf caused by a cynipid wasp. **(b)** A *Stigmella* leaf miner on an oak leaf in Wales. **(c)** A rose-ringed parakeet.

## 14.1 Plants Have a Variety of Defenses Against Herbivores

Plants present a luscious green world of food to any organism versatile enough to use it, so why don't we see more plant material being eaten by herbivores? After all, unlike most animals, plants cannot move to escape being eaten. Three hypotheses have been proposed to answer the question of why more plant material is not eaten. First, herbivores may have evolved mechanisms of self-regulation to prevent the destruction of host plants, ensuring a future supply of food. However, as we saw in Chapter 4, such a group selection type of explanation is unlikely to be true. Second, predators and parasites might keep herbivore numbers low, thus sparing the plants. There are many examples of predation of herbivores that support this view (look ahead to Chapter 16). Third, the plant world is not as helpless as it appears. The sea of green is armed with defensive spines, tough cuticles, and noxious chemicals (**Table 14.1**). The second and third hypotheses are not mutually exclusive, but we will focus here on the defensive properties of plants. **Constitutive defenses** are always present in the plant. **Induced defenses** are switched on only when herbivores attack. This means that plants do not always have to use valuable resources in the continued production of defenses, which allows resources to be diverted to other functions.

### 14.1.1 Mechanical defenses include spines and sticky hairs

A plant's first line of defense is mechanical. Thorns and spines deter vertebrate herbivores, though not invertebrates that are small enough to chew around the thorns. However, many

**Table 14.1 Strategies of plants to avoid being eaten.**

| Strategy | Example |
|---|---|
| 1. Mechanical defenses | Thorns, sticky hairs, silica |
| 2. Chemical defenses | Alkaloids, phenolics, terpenoids |
| 3. Mutualisms with defensive agents | Bull's horn acacia and ants |
| 4. Associational resistance | Purple loosestrife growing next to *Myrica* |
| 5. Mimic semiochemicals | Ecdysteroids mimic insect-molting hormone |

leaves also exhibit sticky hairs or trichomes on the leaves that trap and kill small insects. Other plants produce sticky resins that gum up the mouthparts of insects when they puncture the leaf veins. Low-lying plants are generally more spinose than larger plants, which may grow above the browse line of vertebrates. Plants with one or a few apical meristems, such as palms, often protect them with spines. In temperate locations, evergreen species such as holly, *Ilex* spp., are often the only browse available in winter and are spiny. African acacias have dense thorns on their outsides, where herbivory is common, but few on their insides, which are relatively protected from large herbivores such as giraffes (**Figure 14.3**). Young acacias are also very thorny. Recent experiments have confirmed that in some systems thorns are induced by the presence of herbivores (see **Feature Investigation**). In addition, tough fibers, which are common in plant bark, seed coats, and the outside of nuts, discourage herbivore feeding. Similarly, grasses and palms sequester silica, which makes them difficult to chew. Both fibers and silica, which are indigestible to animals, grind down mammalian teeth and insect mandibles, reducing feeding efficiency.

## 14.1.2 Chemical defenses include alkaloids, phenolics, and terpenoids

An array of unusual and powerful chemicals provide a second line of defense against herbivores. Such compounds are not part of the primary metabolic pathway that plants use to obtain energy. They are therefore referred to as **secondary metabolites**, or secondary chemicals. Most of these chemicals smell bad or are bitter tasting or toxic, and they deter herbivores from feeding and plant pathogens from colonizing and establishing. In some cases, as with flowers, the smell may be pleasant and serve to attract animals such as pollinators. Many fruits are also chemically defended, and such defenses may change as fruits ripen. Each plant tends to produce an array of secondary metabolites, and many plants share the same metabolites. Because of the sheer numbers of different plant species, many different types of secondary metabolites exist. Furthermore, production of many of these secondary metabolites is induced by herbivory. In an interesting twist, many of these compounds have medicinal properties that are

# Feature Investigation

## José Gómez and Regino Zamora Showed That Thorns Can Be Induced by Herbivory

Many ecologists have suggested that mechanical or chemical defenses are costly to produce and their production requires the diversion of valuable plant resources. Defenses should be switched on, or induced, only by the presence of herbivores. This hypothesis suggests that spiny plants should be more fit than non-spiny ones when herbivores are present, but less fit in herbivore-free environments. Induced responses are expected to be more frequent in long-lived plants, which may experience considerable changes in herbivory levels over their lifetime than short-lived plants. José Gómez and Regino Zamora provided elegant field tests of these ideas using a long-lived thorny shrub, growing in the mountains of Spain, which is eaten by the Spanish ibex, *Capra pyrenaica,* and domestic sheep. They showed that herbivory reduced fruit set (the ratio of fruit to flowers produced) and that thorn density decreased over the course of 2 years when the plants were protected from herbivores. They also showed that thorns use a considerable portion of the plant's resources and that, in the absence of herbivores, thorny plants produce less fruits (**Figure 14.A**).

**HYPOTHESIS** Thorns that protect plants from herbivores may be induced by herbivory.

**STARTING LOCATION** Sierra Nevada, southwest Spain

| | Conceptual level | Experimental level |
|---|---|---|
| 1 | Ungulates reduce fruit set. | Mark 50 shrubs in 1998. Exclude herbivores from 20 shrubs by use of a large meshed fence. The fence let pollinators in but kept herbivores out. Count thorn numbers in four 25 $cm^2$ quadrats located randomly on the shrubs. Select ten inflorescences on each shrub. Count flowers produced and number ripening to fruit. Where ungulates were excluded, fruit set was higher than where ungulates were present (14A-i). This illustrated the negative effects of herbivores on fruit set. |
| 2 | Thorns are a costly defense to produce. | Mark 40 shrubs protected from ungulates. In 20 shrubs, remove thorns from half the surface area using scissors, just as they are beginning to form. Leave thorns untouched on 20 other shrubs. Label 10 inflorescences on each of 40 shrubs. Count percent of inflorescences yielding fruit. Fruit set was higher in thorn removed plants (57.6%) than controls (33.7%) because resources had not been diverted into thorn production. This showed that producing thorns was expensive (14A-ii). |
| 3 | Relax ungulate feeding pressure and examine thorn production over time. | Thorn production on the ungulate excluded plants was monitored in 1998 and for 2 subsequent years. By 2000, thorn density had decreased almost 50% (14A-iii). |

**4 THE DATA** Herbivores reduce fruit set in shrubs. Thorns protect plants from herbivory. However, thorns are costly to produce and are produced more in the presence of high herbivore densities.

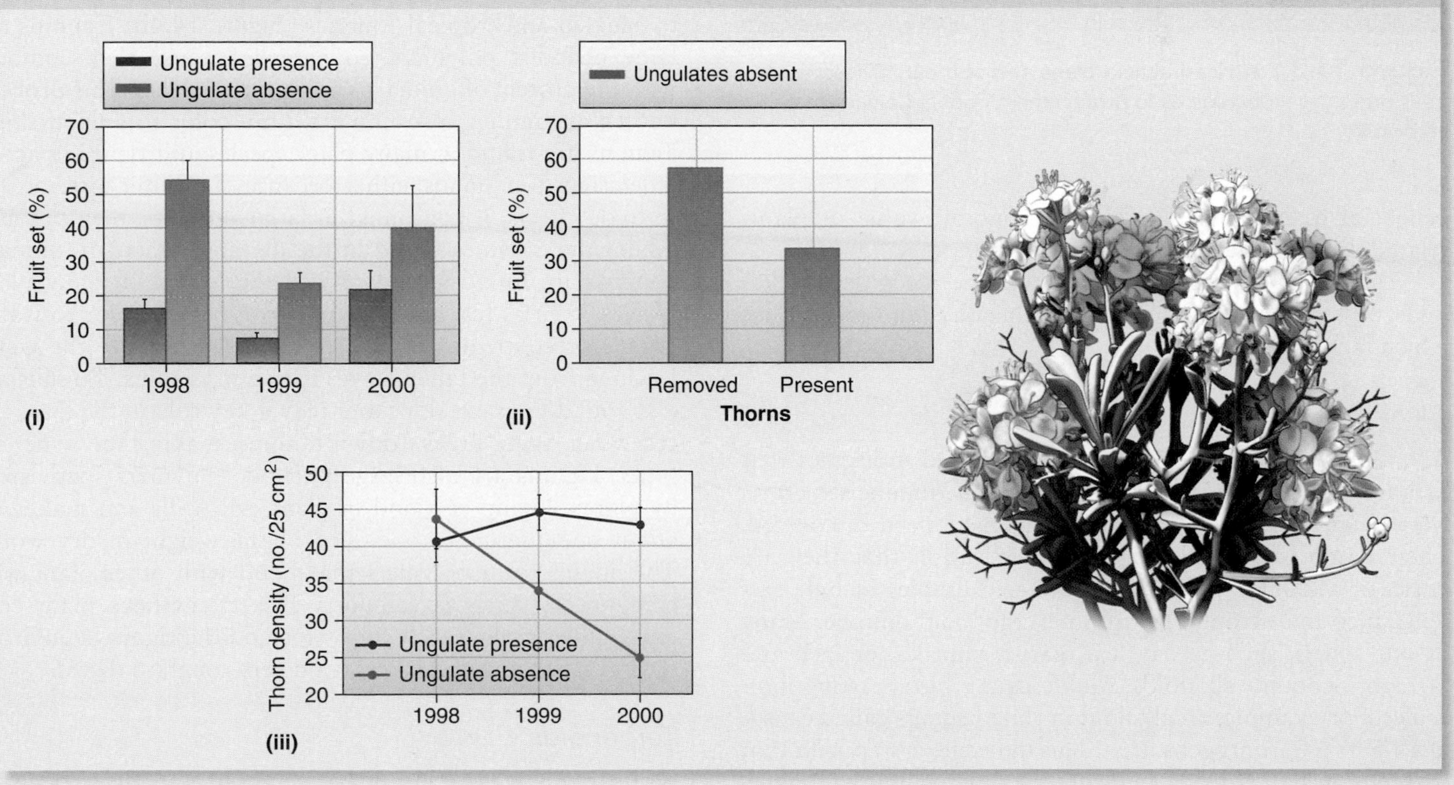


**Figure 14.A** **José Gómez and Regino Zamora showed that thorns are induced by herbivory.** (After Gomes and Zamora, 2002.)

**Figure 14.3** **African acacia trees and spines.** These trees often have long spines to protect them from large vertebrate herbivores.

beneficial to humans. The global economic value of plant-related drugs is over $1 trillion annually.

Three general chemical classes can be recognized: alkaloids, which contain nitrogen; and phenolics and terpenoids, which lack nitrogen.

### *Alkaloids*

Alkaloids are a group of structurally related molecules that contain nitrogen and usually have a cyclic, ringlike structure. More than 12,000 different alkaloids have been discovered. Their name is derived from the observation that they are basic, or alkaline, molecules. Familiar examples include caffeine, nicotine, atropine, morphine, ergot, and quinine. Many of our spices, such as bay leaf, garlic, paprika, pepper, and tarragon, contain alkaloids. Alkaloids are bitter-tasting molecules. For example, an alkaloid in chile peppers called capsaicin elicits a burning sensation. This molecule is so potent that one-millionth of a drop can be detected by the human tongue. Capsaicin may serve to discourage mammals from eating the peppers. Interestingly, birds, which serve to disperse seeds, do not experience the burning sensation of capsaicin. Many alkaloids also have an unpleasant odor.

Other alkaloids are poisonous, like the alkaloid atropine, a very potent toxin derived from deadly nightshade plants (**Figure 14.4a**). Animals that eat this plant and consequently ingest atropine become very sick and may die. It is unlikely that an animal that eats deadly nightshade and survives would choose to eat it a second time. Atropine acts by interfering with nerve transmission. In humans, for example, atropine causes the heart to speed up to dangerous rates because the nerve inputs that normally keep a check on heart rate are blocked by atropine. Other alkaloids are not necessarily toxic, but can cause an animal that eats them to become overstimulated (caffeine), understimulated (any of the opium alkaloids, like morphine), or simply nauseated because the compound interferes with nerves required for proper functioning of the gastrointestinal system.

### *Phenolics*

Phenolic compounds all contain within their structure a cyclic ring of carbon with three double bonds, known as a benzene ring. Approximately 5,000 phenolic compounds are known. Common categories of phenolics are the flavonoids, tannins, and lignins. Flavonoids are produced by many plant species and create a variety of flavors and smells. These can play a role as deterrents to eating a plant or as attractants that promote pollination. The flavors of chocolate and vanilla largely come from a mixture of flavonoid molecules. Vanilla is produced by several species of perennial vines of the genus *Vanilla* native to Mexico and tropical America (**Figure 14.4b**). Tannins are large phenolic polymers, so named because they combine with the protein of animal skins to form leather. This process, known as tanning, also imparts a tan color to animal skins. Tannins are found in many plant species and typically act as a deterrent to animals, either because of a bitter taste or due to toxic effects. If consumed in large amounts, they can also inhibit the enzymes found in the digestive tracts of animals. Tannins are found abundantly in the leaves of many plant species, such as tea, and they impart a brown color to water. Many forest streams have a coffee color because the water leaching from the fallen leaves is rich in tannins. Tannins are also found in grape skins and play a key role in the flavor of red wine. Aging breaks down tannins, making the wine less bitter. Lignins are also large phenolic polymers synthesized by plants. Lignin is found in plant cell walls and makes up about one-quarter to one-third of the weight of dry wood. The lignins form polymers that bond with other plant wall components, such as cellulose. This strengthens plant cells and enables a plant to better withstand the rigors of environmental stress. It also makes wood very tough to digest.

### *Terpenoids*

A third major class of secondary metabolites are the **terpenoids**, of which over 25,000 have been identified, more than any other family of naturally occurring products. Terpenoids are synthesized from five-carbon isoprene units. Terpenoids have a wide array of functions in plants. Notably, because many terpenoids are volatile, they are responsible for the odors emitted by many types of plants, such as menthol produced by mint (**Figure 14.4c**). The odors of terpenoids may attract pollinators or repel animals that eat plants.

(a)

(b)

(c)

**Figure 14.4 The three major classes of plant chemical defense.** **(a)** Alkaloids, as found in deadly nightshade, *Atropa belladonna.* **(b)** Phenolic compounds, as found in the vanilla plant, *Vanilla* sp. **(c)** Terpenoids, as found in these mint plants.

**ECOLOGICAL INQUIRY**

Are the alkaloids in deadly nightshade a qualitative or quantitative defense?

In addition, terpenoids often impart an intense flavor to plant tissues. Many of the spices used in cooking are rich in different types of terpenoids. Examples include cinnamon, cloves, and ginger. Additional terpenoids are menthol, camphor, and cannabinoids found in the cannabis plant. Terpenoids are found in many traditional herbal remedies and are under medical investigation for potential pharmaceutical effects. Other terpenoids, like the carotenoids, are responsible for the coloration of many species. An example is β-carotene, which gives carrots their orange color. Carotenoids are also found in leaves of some species, but their color is masked by chlorophyll, which is green. In the autumn, when chlorophyll breaks down, the color of the carotenoids becomes evident.

### 14.1.3 Chemical defense strategies change according to plant type and environmental conditions

Plant chemical defenses can be classified as quantitative or as qualitative, depending on the volume of defense present in the plant. **Quantitative defenses** are substances that are ingested in large amounts by the herbivore as it eats and that prevent energy gain from the digestion of food. Examples are tannins and resins in leaves, which may constitute 60% of the dry weight of the leaf. These compounds are not toxic in small doses, but they have cumulative effects. The more leaf the herbivores ingest, the more difficult it is for them to digest it. Paul Feeny (1970) was the first to document such a defense by testing oaks against externally feeding caterpillars, including the oak winter moth we discussed in Chapter 1. Feeny reared larvae on an artificial diet in the laboratory, where he was able to vary the amounts of tannin extracted from September oak leaves. As little as 1% tannin in the diet reduced larval growth rate and pupal weight. In the field he found that caterpillars of all species were more common on oaks in the spring, when the leaf concentrations of tannins are the lowest (**Figure 14.5**). As the season progresses and the leaf toughens, tannin concentrations increase and caterpillar feeding decreases. Many lepidopteran species are spring feeders that try to complete their feeding stages as caterpillars before higher concentrations of tannins kick in, just as the oak winter moth does.

**Qualitative defenses** are toxic substances that are effective in very small doses. These compounds are present in leaves at low concentrations, usually less than 1% of dry weight. Examples include cyanogenic compounds in leaves. Atropine, the toxin produced by deadly nightshade, is a most potent poison. The plant stores these poisons in discrete glands or vacuoles in order not to poison itself. An infusion of juice from berries of the European deadly nightshade, *Atropa belladonna,* was formerly dropped in women's eyes, causing dilation of the pupils to produce a "wide-eyed" look, hence the name "belladonna," beautiful lady. Once again, birds are immune to the poison and can safely disperse the seeds.

These two defense strategies, qualitative and quantitative, are correlated with plant apparency. **Apparent plants** are

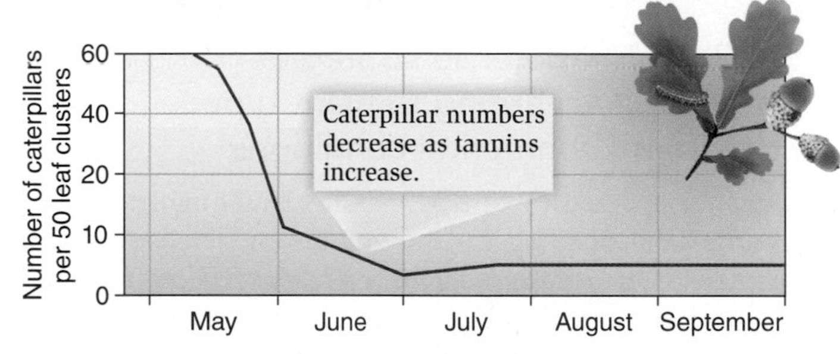


**Figure 14.5** **The effects of tannins.** The number of butterfly and moth caterpillars per 50 clusters of oak leaves in Britain is highest in the spring, when leaf tannin concentrations are lowest. Caterpillar abundance decreases in the summer as tannin concentrations increase. (Modified from Feeny, 1970.)

long-lived, large, and always apparent to the herbivores (for example, oak trees). This means that their herbivores can find them easily. Their defenses are thought to be mainly quantitative, effective against both monophagous and polyphagous herbivores, with a long history of association with these *K*-selected plants. **Unapparent plants** are small, ephemeral, difficult to find, and unavailable to herbivores for long periods (for example, weeds). Their defenses are thought to be mainly qualitative, guarding against polyphagous enemies like deer or other vertebrates, which would find them only by chance. Quantitative poisons would not work well for unapparent plants because by the time the herbivore had acquired a sufficient dose of poison, the plant may have already been consumed. Thus, trees nearly all contain digestibility-reducing compounds, and weeds contain toxins. **Table 14.2** summarizes some of the phenomena associated with apparent and unapparent plants.

Several other hypotheses have been proposed to account for the allocation of plant chemical defenses. We will briefly consider two of these. The **carbon-nitrogen balance hypothesis** is an attempt to explain how the types of defenses produced are influenced by the environmental conditions. It suggests that carbon and nitrogen are allocated to production of secondary metabolites only after requirements for growth are met first. Where soil nitrogen levels are low, plants tend to use nitrogen for growth and to produce more carbon-based defenses, mostly digestibility-reducing phenolic compounds such as tannins. Where low-carbon conditions exist, such as shady

**Table 14.2 Some features of apparent and unapparent plants.**

| Apparent plants | Unapparent plants |
|---|---|
| Often in monocultures | Often in polycultures (mixed species patches) |
| Large | Small |
| Long-lived | Short-lived |
| Chemical defenses against polyphagous and monophagous herbivores | Chemical defenses against polyphagous herbivores that may happen upon the plant |
| Quantitative defenses, such as tannins, often >1% of leaf fresh weight | Qualitative defenses, toxins, often <1% of leaf fresh weight |
| Example: trees | Example: weeds, annuals |

environments where light is limited and photosynthesis is constrained, carbon is used for growth and plants are more likely to produce nitrogen-rich toxins. The **optimal defense hypothesis** suggests that certain plant parts, such as flowers or seeds, are not so easily replaced as others, such as leaves or twigs. Such parts contain a higher proportion of a plant's defenses. Seeds of many edible fruits contain cyanogenic glycosides.

### 14.1.4 Plants may use additional defenses apart from spines and toxins

Plants may also engage in additional strategies to defend themselves. First, they sometimes enter into a mutualistic relationship with insects that defend them, such as with bull's horn acacia and ants (refer back to Figure 12.5b). Second, they can grow next to unpalatable plants that tend to deter herbivores, a phenomenon known as associational resistance (see Figure 12.13). Third, they may selectively abscise leaves heavily infested by sessile insects such as leaf miners or aphids, causing the leaves to fall to the ground and preventing the insect from completing its life cycle. Fourth, some specialized species-specific plant defenses exist. For example, some tropical vines of the genus *Passiflora* produce physical structures that mimic eggs of the *Heliconius* butterflies, whose larvae feed on them. Because females are less likely to lay eggs where other eggs are present, oviposition is discouraged.

Finally, the behavior of many insect herbivores is altered by chemical messengers known as **semiochemicals**. In some cases these are **pheromones**, which act as sex attractants between males and females. In others they are **allelochemicals** that affect behavior between different species. In many cases, plants produce analogs of these chemicals, which serve to disrupt herbivore behavior. Disruptions to behavior ultimately reduce reproductive success and hence future population densities. Examples of these phenomena follow.

Insect molting is controlled by molting hormone, which promotes molting, and by juvenile hormone, which prevents it. Some plants have been found to produce ecdysteroid-like compounds, which mimic the action of molting hormone and cause the insect to molt prematurely, often killing it. The common house plant flossflower, *Ageratum houstonianum,* produces precocenes, which prevent production of juvenile hormones. The lack of juvenile hormone causes juvenile herbivores to molt prematurely into adults, which are often sterile, reducing the likelihood of reproduction.

Some damaged plants produce herbivore-specific volatile chemicals that attract the natural enemies of the herbivores. Consuelo De Moraes and colleagues (1998) showed that the frequency of visits by the parasitic wasp *Cardiochiles nigriceps* to tobacco or cotton plants was increased in plants that had been damaged by the wasp's host, tobacco budworm, *Heliothis virescens,* compared to undamaged plants (**Figure 14.6a**). What's more, the plants produced distinct volatile compounds when attacked by *H. virescens* compared to attack by the maize earworm, *H. zea,* which is not the host of *C. nigriceps*. As a result, the percentage of visits by *C. nigriceps* to *H. zea*–damaged plants was much less than the number of visits to *H. virescens*–damaged plants. The frequency of visits was maintained when all the larvae and damaged leaves were removed from the plants, showing that the parasitic wasps were not responding to the smell of the caterpillar larvae themselves (**Figure 14.6b**).

**Figure 14.6 Percent of total visits by the parasitic wasp, *C. nigriceps,* to tobacco plants damaged by *H. virescens* (tobacco budworm) or *H. zea* (maize earworm), or undamaged (control).** **(a)** Caterpillars actively feeding on the leaves. **(b)** All caterpillars and damaged leaves removed. (After De Moraes et al., 1998.)

**Check Your Understanding**

**14.1** Do you think roots and leaves are equally chemically defended? What would the optimal defense hypothesis predict?

## 14.2 Herbivores May Overcome Plant Defenses and Impact Plant Populations

Herbivores can overcome plant defenses in five ways (**Table 14.3**). First, they use mechanical adaptations. Many vertebrate herbivores have large, flat molars and well-developed jaw muscles that help them grind up plant parts into more easily digestible pieces. Some herbivores, the ruminants, have evolved a complex stomach, consisting of four separate chambers, to help digestion (see Figure 7.7).

Second, many herbivores possess behavioral adaptations. Some species snip vegetation and delay consumption for a few days to allow secondary metabolites to decrease in concentration. Some small herbivores, such as caterpillars or beetles, chew through the vascular tissue that supplies certain defenses such as resins. In this way they create a leaf patch free of a defense and they then feed on this patch (**Figure 14.7**). Others, as we have seen, feed at a time of the year when secondary metabolite concentrations are at their lowest. Hemipterans feed on plant sap, thus avoiding many secondary metabolites that are present in root or leaf tissue.

Third, herbivores produce a wide array of digestive enzymes that help negate the effects of secondary metabolites. Many herbivores can produce tannin-binding salivary proteins, inactivating the toxic effects of tannins. Herbivore guts may possess protective molecular mechanisms that decrease secondary metabolite absorption or pump foreign substances back into the gut lumen. Herbivores can also biotransform

**Table 14.3 Strategies of herbivores to overcome plant defenses.**

| Strategy | Example |
|---|---|
| 1. Mechanical adaptation | Flat grinding teeth |
| 2. Behavioral adaptation | Feed when plant defenses are lowest |
| 3. Digestive enzymes | Mixed function oxidases |
| 4. Microbial symbionts | Microbes in stomachs |
| 5. Host manipulation | Gall-inducing insects |

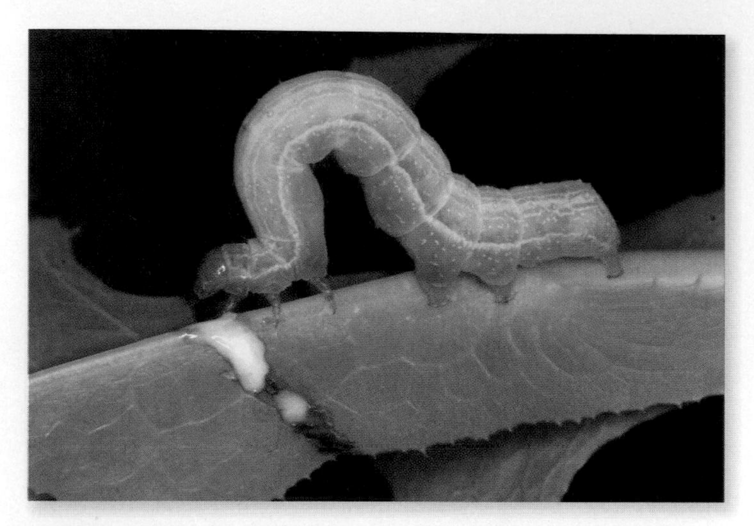

**Figure 14.7** A cabbage looper caterpillar, *Trichoplusia ni,* trenching a *Lobelia cardinalis* leaf before feeding on the distal tip (to the left) and circumventing the chemical defenses.

secondary metabolites into more polar metabolites readily excreted in urine or bile. The most important biotransformation is oxidation, which occurs in mammal livers and in insect midguts. Oxidation involves catalysis of the secondary metabolite to a corresponding alcohol by a group of enzymes known as mixed-function oxidases (MFOs). Conjugation, often the next step in detoxification, occurs by the uniting of the harmful element, or the compound resulting from the oxidation, with another molecule to create an inactive and readily excreted product. Conjugation enzymes are commonly produced in the liver or kidney.

Fourth, many herbivores have developed a mutualistic relationship with microbes that live in their digestive tracts and digest cellulose into monosaccharides, which can then be absorbed. In animals with simple stomachs, such as horses, these microbes live in the large intestine. In ruminants, they live in two of the four chambers of the stomach.

Lastly, some herbivores can manipulate the host plants to get food. For example, galling insects, including some flies, wasps, and aphids, can lay their eggs on flowers, buds, leaves, stems, or roots and induce a tumor-like growth. The most important galling insects are cynipid wasps, often called gall wasps. Many galls contain large amounts of tannins and have a bitter taste, hence the name "gall."

In addition to these adaptations, certain chemicals that are toxic to polyphagous herbivores actually increase the growth rates of adapted monophagous species, which can circumvent the defense or put the chemicals to good use in their own metabolic pathways. The Brassicaceae, the plant family that includes mustard and cabbage, contains acrid-smelling secondary metabolites called glucosinolates, the most important one of which is sinigrin. Large white butterflies, *Pieris brassicae,* preferentially feed on cabbage over other plants. If newly hatched larvae are fed an artificial diet, they perform much better when sinigrin is added to it. When larvae are fed cabbage leaves on hatching from eggs and are later switched to an artificial diet without sinigrin, they die rather than eat. In this case, the secondary metabolite has become an essential part of the diet.

Paul Ehrlich and Peter Raven (1964) suggested that this specialism of some herbivores on otherwise toxic plants supports the notion of an evolutionary arms race between plants and herbivores (**Figure 14.8**). In this scenario, the development of a plant defense is followed by the development of a detoxifying mechanism by herbivores, which leads to development of another plant defense, and so on. The result is a profusion of many specialized herbivores on plants. May Berenbaum (1981) tested this idea by examining the diversity of chemical defenses and the number of insect species feeding on members of the carrot family, *Umbelliferae*. These plants are defended by furanocoumarins, a type of phenolic, some of which are photoactive, that is, their toxicity is enhanced by ultraviolet radiation. Humans who come in contact with

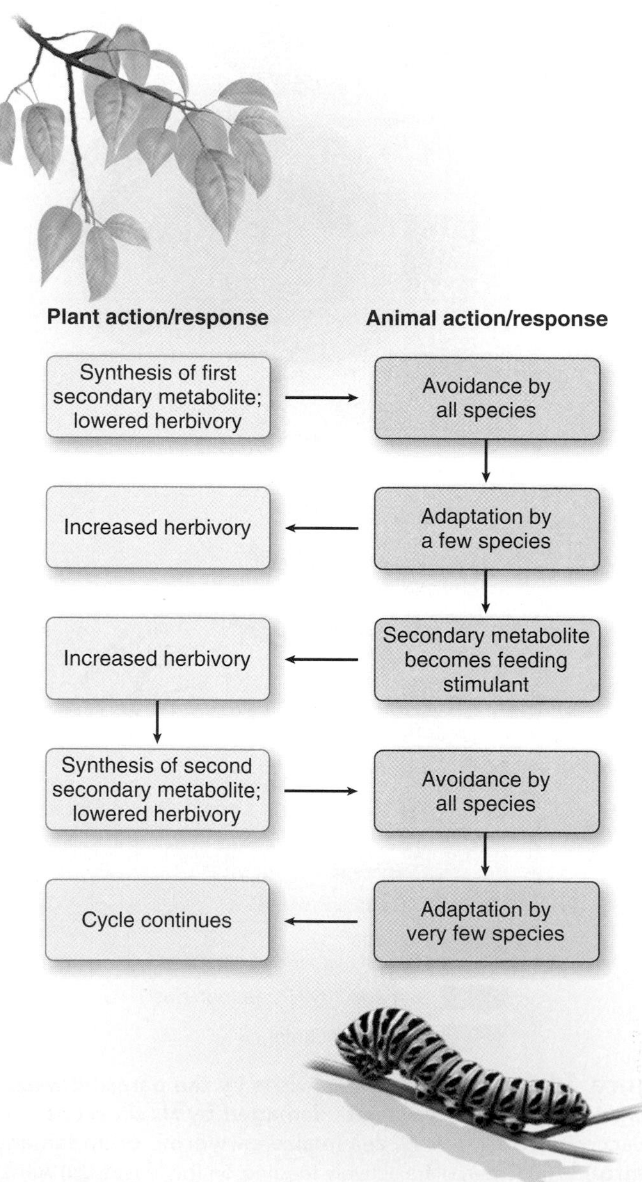


**Figure 14.8** The evolutionary arms race between plants and herbivores.

these compounds contract blisters, as in "celery picker's itch." The chemically simplest umbellifers were defended only by the phenolic compound hydroxycoumarin, the next group was defended by more complex linear furanocoumarins, and the most sophisticated group had complex angular furanocoumarins. In support of the evolutionary arms race hypothesis of plants and herbivores, the most specialized insects, feeding only on 1–3 genera of umbellifers, were all found on those with complex linear and angular furanocoumarins. The umbellifers without any furanocoumarins were fed upon by more polyphagous species (**Figure 14.9**).

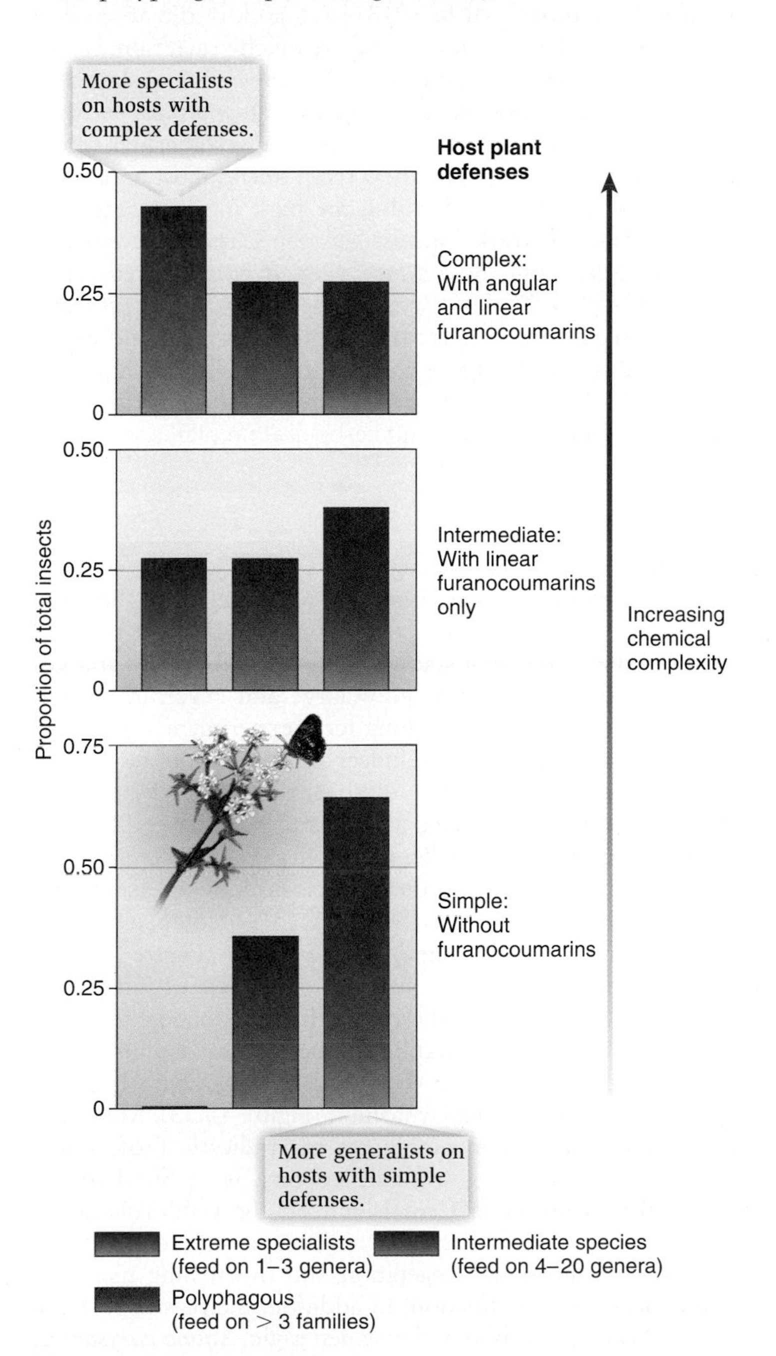


**Figure 14.9 Herbivory and host defenses.** Degree of host specialization of insects feeding on umbellifers, according to their complexity of secondary defenses. (From data in Berenbaum, 1981.)

**Check Your Understanding**

**14.2** Provide another example where a host plant chemical that is toxic to many species actually benefits one of its herbivores.

## 14.3 How Much Plant Material Do Herbivores Consume?

Some herbivores, such as tent caterpillars, can completely defoliate their host plants. While such high levels of defoliation are unusual, virtually all plants are eaten by some herbivores. Native herbivores can be especially damaging to vegetation where their predators have been removed (see **Global Insight**). In an extensive literature review, Hélène Cyr and Michael Pace (1993) showed that herbivory was tied to the growth rate of the plants being fed upon (**Figure 14.10**). In both aquatic and terrestrial systems, the higher the growth rate of the plants, the more the herbivores ate. Mean values for amount of plant material removed by herbivores were 18% for terrestrial plants, 30% for aquatic plants, and 79% for aquatic algae. Thus, herbivory can account for the removal of large amounts of plant and algal material. The reasons for the higher values in aquatic algae are that they lack some of the chemically sophisticated defenses of terrestrial or aquatic plants.

D. S. Bigger and Michelle Marvier (1998) examined which types of herbivores have the strongest effects on plants. We presented this data when we introduced the concept of meta-analysis in Chapter 1 (refer back to Figure 1.15). Invertebrate herbivores, such as insects, have stronger effects than vertebrate herbivores, such as mammals, at least in terrestrial systems. While one might consider large grazers like bison in North America, or antelopes in Africa, to be of considerable importance, the chances are that grasshoppers are of more importance in grasslands. In forests, invertebrate grazers such as caterpillars have greater access to many canopy leaves than vertebrates do and are likely to have greater effects.

### *Invasive herbivores*

As with predation, the strength of herbivory as a selective pressure on natural vegetation can also be illustrated where herbivores have been introduced into new environments and the native vegetation has few coevolved defenses. The list of such examples is long. Introduced insects such as the gypsy moth can defoliate trees completely. Introduced rabbits have devastated native vegetation in Australia and most other areas where they have been introduced. Goats introduced to Santa Catalina Island off the coast of California drove the local population of California sagebrush, *Artemisia californica,* to extinction. Indian elephants introduced to the Andaman Islands have eliminated cane, *Calamus* spp., and screwpine, *Pandanus* spp., except in steep and rocky slopes that the elephants cannot reach. More recently, exotic gall-forming

Global Insight

## Predator Removal in the United States Has Resulted in Overgrazing by Native Herbivores

In the summer of 1874, two years before the battle of the Little Bighorn, George Armstrong Custer led a military and scientific expedition to the Black Hills area of South Dakota. Custer's staff included a small posse of scientists, because science had been used to help justify the expedition's purpose. Skeptics had claimed the true reason for the expedition was to invade Sioux lands in search of gold. The scientists recorded the presence of elk, mule deer, white-tailed deer, pronghorn antelope, beaver, grizzly and black bears, cougars, and wolves. There were limited signs of coyotes and bison. Wolves were commonly seen and heard, especially howling at night. The expedition's botanist, Aris Donaldson, recorded an abundance of berry-producing shrubs. Flowers were so abundant that the soldiers plucked them without dismounting from the saddle, creating bouquets to decorate the head gear of the horses. Gold was also found, triggering a rush of miners to the area, sparking the last major Indian war of the Great Plains. A treaty with the Sioux in 1877 opened the Black Hills up to American settlement and livestock. Rewards were offered to settlers to kill wolves and other carnivores. By the end of the 1880s, open ranching was replaced by fenced pasture ranching. Elk, bison, and pronghorn were extirpated. In 1903, Wind Cave National Park was established along the southeastern foothills of the Black Hills. Elk, bison, mule deer, white-tailed deer, and pronghorn were reintroduced in 1913–1916 and, in the absence of wolves, reached high densities and ate much vegetation. Densities of young plains cottonwoods, *Populus deltoides,* lanceleaf cottonwood, *P. acuminata,* and bur oaks, *Quercus macrocarpa,* declined dramatically, and only large trees established before 1900 remain today. Without new recruitment, most hardwood species will be extirpated within the park in areas accessible to ungulates. The stark contrast between Custer's records and today's observations show a cascading effect from predator to herbivore to tree. Without wolves, elk and bison have thrived, reducing the abundance of berry-producing shrubs and young hardwood trees. Managing natural areas, such as Wind Cave National Park, requires a thorough understanding of the effects of predators on herbivores and herbivores on plants.

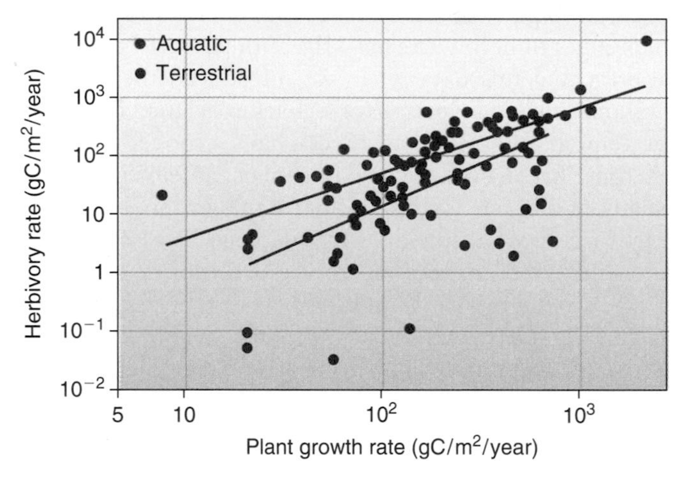


**Figure 14.10 Levels of herbivory.** Rate of herbivory, measured as carbon removed, increases with plant biomass in both aquatic systems (blue circles) and terrestrial systems (red circles). (Modified from Cyr and Pace, 1993.)

wasps, *Quadrastichus erythrinae,* from Southeast Asia, appeared in Hawaii in 2005, attacking the cherished native coral trees, *Erythrina variegata,* commonly known as wiliwili, which provide flowers for leis and wood for canoes. Within a year the wasps had spread to every island, killing and disfiguring trees as they created characteristic galls on the leaves (**Figure 14.11**).

On the sub-Antarctic island of South Georgia, reindeer, *Rangifer tarandus,* from Norway were imported between 1909 and 1925 to provide a source of fresh meat for the people working at whaling stations. Though only a few animals were introduced, the herds grew large and began to destroy local vegetation. In 1973 a long-term experiment was set up to examine the effects of reindeer on vegetation, and large, 10 × 10 m exclosures were established to exclude the reindeer. By 1984 the preferred winter and summer food plants of the reindeer, *Poa flabellata* and *Acaena magellanica,* had recovered substantially inside the exclosures, whereas nonpreferred food plants decreased as they were outcompeted by the now thriving *P. flabellata* and *A. magellanica* (**Figure 14.12**).

The good news is that some of the effects of introduced vertebrate herbivores can be reversed if the political and economic will is provided. Rabbit eradication on Phillip Island, Australia, has transformed the island as the native vegetation has recovered in spectacular fashion (**Figure 14.13**). Mammals are perhaps the easiest herbivores to eradicate. Professional hunters, often shooting from helicopters, were hired to kill the feral pigs on Santa Cruz Island off the California coast. The pigs, which were introduced as farm animals in the 1850s, were rooting up native vegetation and threatening nine rare plant species with extinction. In addition, the pigs attracted a new predator to the island, the golden eagle, *Aquila chrysaetos,* which preyed on piglets and also hunted the endangered island fox to low numbers. Another target was the 150,000 goats that ate their way through the vegetation of Isabela Island in

**Figure 14.11 The effect of introduced gall-forming wasps on Hawaii's coral trees.** **(a)** Infested tree with wasp galls, **(b)** cross section of galls showing wasp pupae, and **(c)** uninfested, healthy tree.

**ECOLOGICAL INQUIRY**

What are some of the other types of insects that initiate galls on plants?

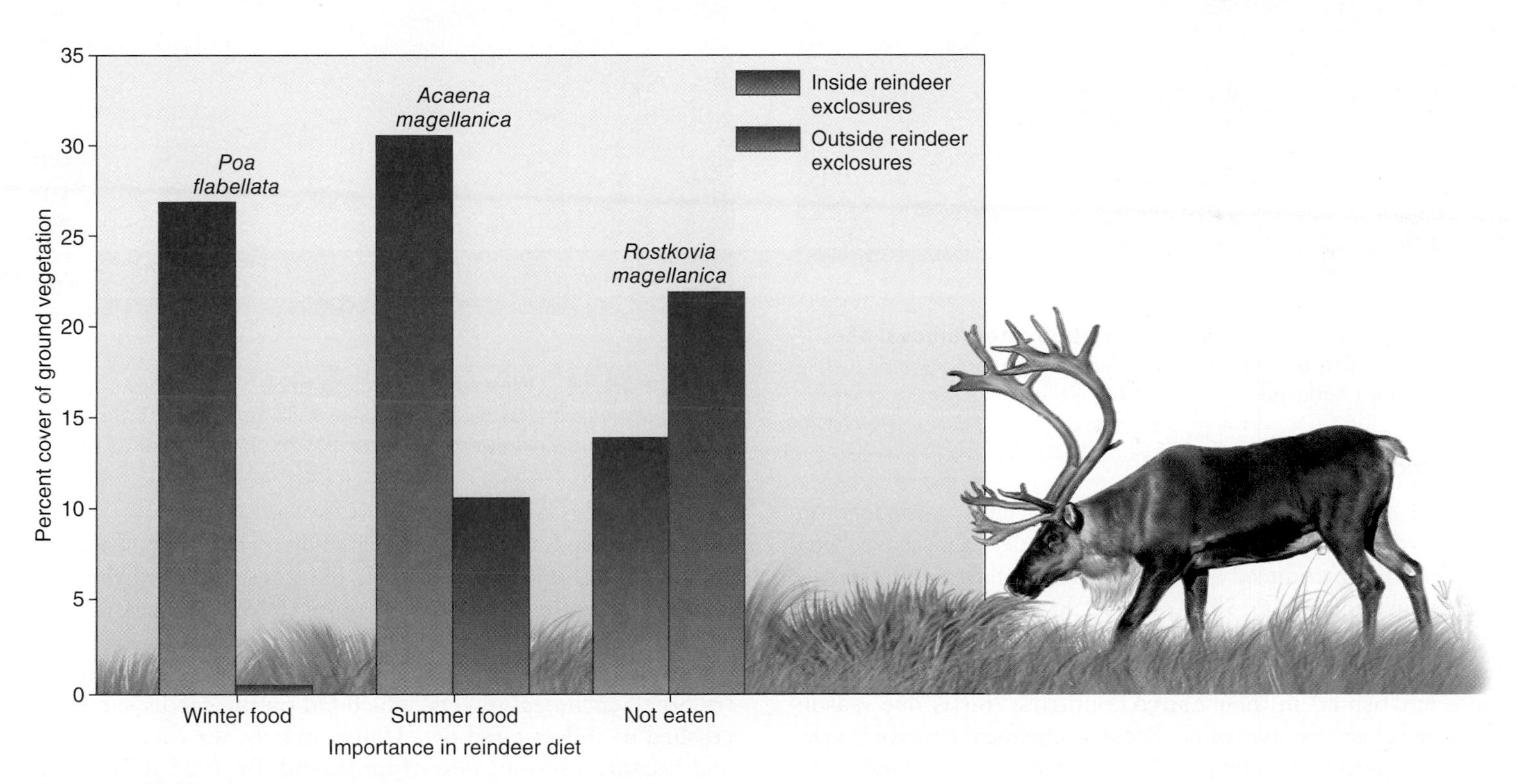


**Figure 14.12 Habitat restoration.** Recovery of native vegetation on South Georgia Island after exclusion of introduced reindeer. (After data in Vogel et al., 1984.)

the Galápagos. The difficult part was to kill the last few goats, which hid at the sound of the helicopter. The project employed "Judas goats," 600 radio-collared, companion-seeking individuals that led their helicopter-borne shooters to holdouts. Fitted with global positioning systems, their every movement was tracked. If this wasn't enough, "Super Judas" nannies were sterilized and implanted with hormones to attract the billies. The last feral goat was shot in March 2006.

(a)

(b)

**Figure 14.13 Recovery of vegetation after removal of introduced rabbits.** Phillip Island, Australia, **(a)** before and **(b)** after rabbit removal.

(a)

(b)

**Figure 14.14 Weed biological control.** The prickly pear cactus, *Opuntia stricta,* in Chinchilla, Australia, **(a)** before control by the cactus moth, *Cactoblastis cactorum*, 1928, and **(b)** after control, 1929.

### *Biological control*

Not all introduced herbivores have negative impacts on native vegetation. Many weeds are invaders that have been accidentally introduced as seeds in ship ballast or in agricultural shipments. Over 50% of the 190 major U.S. weeds are invaders from outside the United States. Many of these weeds have become separated from their native herbivores, which were left behind in their native countries; this is one reason the weeds become so prolific. Because chemical control is very expensive and damaging to the environment, many land managers have used biological control, where the native herbivore is reunited with the weed in its new country. There have been many successes in the biological control of weeds by introduced herbivores. Klamath weed, *Hypericum perforatum,* a pest of pastureland in California, was controlled by two French beetles. Floating fern, *Salvinia molesta,* choked a lake in New Guinea and was controlled by the weevil *Cyrtobagus salvinae,* introduced from Brazil, where the fern is native. Alligatorweed was controlled in Florida's rivers by the alligatorweed beetle, *Agasicles hygrophila,* from South America.

In 1839 the prickly pear cactus, *Opuntia stricta,* was imported from the Americas to Australia to start a dye industry. Small cochineal insects, which fed on the cactus, could be crushed to collect a red dye. Unfortunately, the cactus spread and became a serious pest of rangeland. By 1925 it occupied 240,000 km$^2$, which was rendered useless to sheep and cattle. In 1925 control measures were initiated by introducing the cactus moth, *Cactoblastis cactorum,* from South America. By 1932 the original stands of prickly pear had been destroyed (**Figure 14.14**). Thankful ranchers erected a statue in honor of the architect of this successful project, entomologist Alan Dodd. The success story was repeated in the Caribbean and

Hawaii in the 1960s. Unfortunately, *C. cactorum* has since spread from the Caribbean to Florida (1989) and the southeast U.S. (1998), where it is destroying native cacti. In 2006 the moth was discovered in the Yucatán Peninsula of Mexico, where it threatens huge areas of native cacti that are used to produce vegetables and cattle feed. Even with biological control, great care must be taken to minimize nontarget effects. Once "out of the bottle," biological control agents can themselves become invasive species.

### Check Your Understanding

**14.3** Refer back to the Data Analysis figure in Chapter 7. What might you infer about herbivory levels and levels of secondary metabolites on leaves of different life spans?

## SUMMARY

**14.1 Plants Have a Variety of Defenses Against Herbivores**

- Plants are subject to attack by many different types of herbivores, including specialist species (Figures 14.1, 14.2).
- Most plants have evolved an array of defenses against herbivores, including mechanical defenses such as thorns and spines and chemical defenses such as phenolics, alkaloids, and terpenoids (Figures 14.3, 14.4, Table 14.1).
- Both mechanical and chemical defenses may be induced by herbivores (Figure 14.A).
- Plant chemical defenses can be classified as quantitative or qualitative, depending on the volume of defense present in the plant (Figure 14.5).
- Apparent plants have mainly quantitative defenses, whereas unapparent plants have mainly qualitative defenses (Table 14.2).
- The behavior of some insect herbivores, and their natural enemies, is altered by chemical messengers emitted from plants (Figure 14.6).

**14.2 Herbivores May Overcome Plant Defenses and Impact Plant Populations**

- Herbivores can overcome plant defenses in numerous ways, such as by circumventing or detoxifying their chemical defenses (Figure 14.7).
- An evolutionary arms race exists between plants and the herbivores that feed on them, fueling the development of increasingly sophisticated defenses (Figures 14.8, 14.9).

**14.3 How Much Plant Material Do Herbivores Consume?**

- Herbivores can consume a considerable amount of plant biomass. The rate of herbivory generally increases as the growth of plant biomass increases (Figure 14.10). Invertebrates generally have stronger effects on their host plants than vertebrates do.
- Introduced herbivores have had devastating effects on many plant species (Figures 14.11–14.13).
- Some exotic herbivores are deliberately introduced as biological control agents of invasive weeds (Figure 14.14).

## TEST YOURSELF

1. A "specialist" herbivore that feeds on one or two closely related plant species is called:
   a. Monophagous
   b. Polyphagous
   c. Endophagous
   d. Ectophagous
   e. Holoparasitic

2. Which type of defenses are turned on by herbivores?
   a. Constitutive
   b. Induced
   c. Semiochemical
   d. Allelochemical
   e. Qualitative

3. Which of the following substances are alkaloids?
   a. Caffeine
   b. Nicotine
   c. Flavonoids
   d. a and b
   e. All of the above

4. Which is not a qualitative plant defense?
   a. Atropine
   b. Capsaicin
   c. Morphine
   d. Tannin
   e. Nicotine

5. Which is an attribute of an unapparent plant?
   a. Often grows in polycultures
   b. Large
   c. Often has quantitative defenses
   d. Short-lived
   e. Often has defenses against specialist herbivores

6. The flavor of wine comes from:
   a. Alkaloids
   b. Tannins
   c. Terpenoids
   d. a and b
   e. b and c

7. Many forest streams have a coffee color because the water leaching from the fallen leaves is rich in:
   a. Tannins
   b. Alkaloids
   c. Carbohydrates
   d. Terpenoids
   e. Sugars

8. Secondary metabolites synthesized from five-carbon isoprene units and common in spices are likely to be:
   a. Terpenoids
   b. Alkaloids
   c. Phenolics
   d. Tannins
   e. Sugars

9. Oak trees generally contain quantitative defenses:
   a. True
   b. False

10. Chemical defenses are found only in plants:
    a. True
    b. False

## CONCEPTUAL QUESTIONS

1. Describe some plant defenses against herbivory.
2. In what ways do herbivores overcome plant defenses?
3. What lines of evidence suggest that there is a coevolutionary arms race between plants and herbivores?

## DATA ANALYSIS

Toth and colleagues (2005) measured tannin levels in the brown seaweed *Ascophyllum nodosum* in plants exposed or not exposed to the herbivorous snail *Littorina obtusata* for 2 weeks (see accompanying figure). What do the data show?

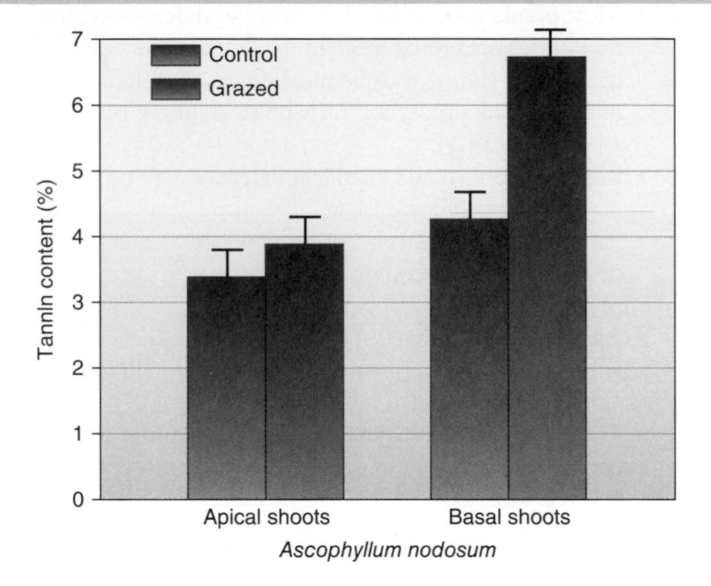

Red Colobus monkeys, *Piliocolobus tephrosceles*, Uganda.

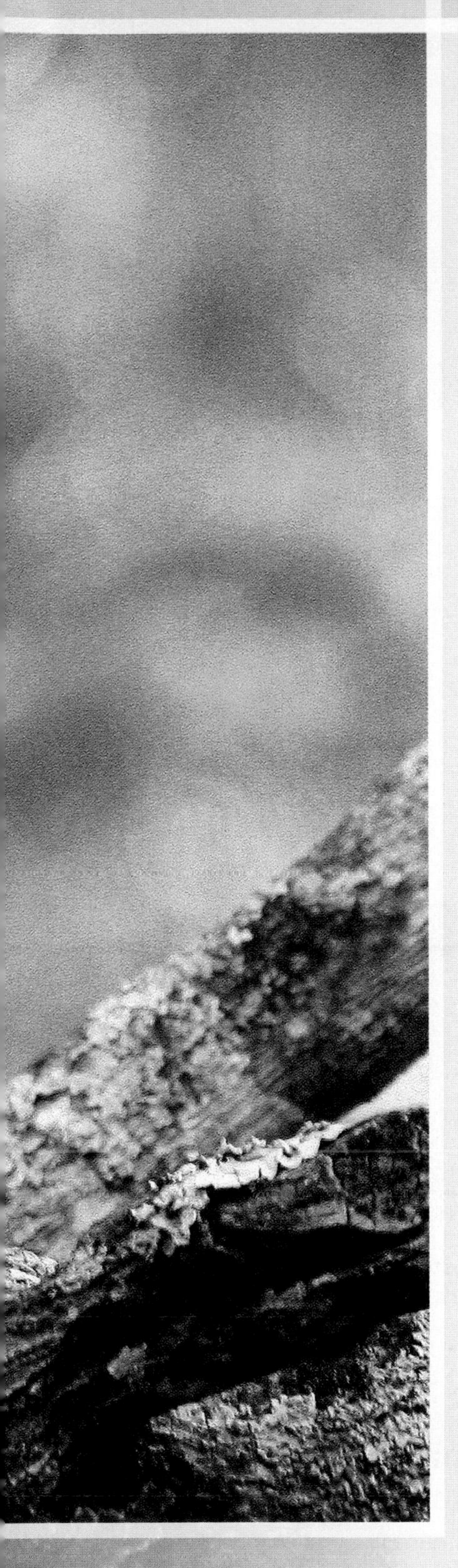

CHAPTER 15

# Parasitism

## Outline and Concepts



As we saw in Chapter 8, forest fragmentation can have severe impacts on forest-dwelling species. Thomas Gillespie and Colin Chapman (2006) conducted a study of red colobus monkeys, *Piliocolobus tephrosceles,* in fragmented forests in western Uganda. Here, colobus monkeys live in groups of 5 to 300 individuals, often forming mixed-species associations with other primates. The monkeys have a complex digestive tract that allows them to eat leaves. Gillespie and Chapman studied the utilization of nine forest fragments, ranging in area from 1.2 to 8.7 ha. They collected 536 fecal samples to determine the prevalence of infection with nematode parasites. The level of degradation and human use of each forest patch was assessed by counting the number and species of trees and of tree stumps remaining after harvest by local people. Infection risk was highest in the fragments where stump density was greatest. In these areas, it is believed, colobus monkeys are more likely to experience human contact and hence exposure to novel pathogens. In this case, human pathogens can spread to wildlife.

When one organism feeds on another, but does not normally kill it outright, the organism is termed a **parasite** and the prey is called a **host.** Some parasites remain attached to their hosts for most of their lives. For example, tapeworms spend their entire adult life inside the host's alimentary canal. Some, such as ticks and leeches, drop off their hosts

(a)

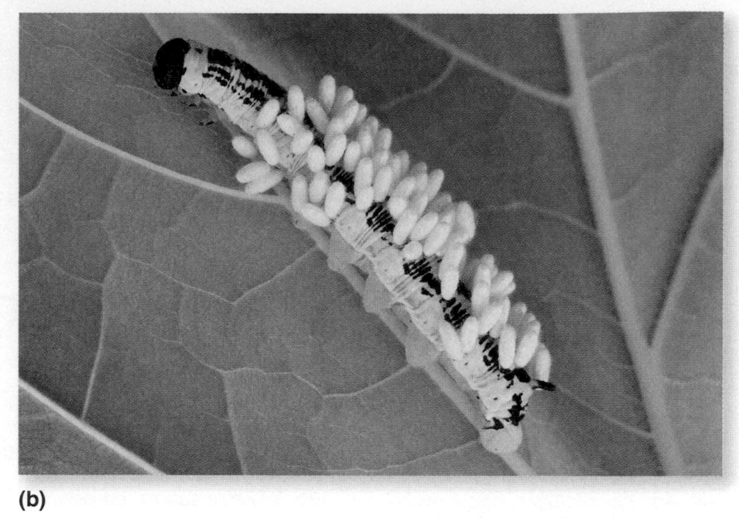

(b)

**Figure 15.1 Parasites and parasitoids.** **(a)** These parasitic South African bont ticks, *Amblyomma herbraeum*, are feeding on a white rhinoceros, but do not kill it. **(b)** These braconid wasp parasitoids, *Apanteles congregatus*, kill their hosts, in this case a Catalpa sphinx caterpillar, *Catalpa catalpae*.

**ECOLOGICAL INQUIRY**

Can you classify these two types of parasites? (Refer to section 15.1.1)

after prolonged periods of feeding (**Figure 15.1a**). Others, like mosquitoes, remain attached for relatively short periods.

There are vast numbers of parasites species, including traditionally known parasites such as viruses, bacteria, fungi, protozoa, flatworms (flukes and tapeworms), nematodes, and various arthropods (ticks, mites, and fleas). To some ecologists, herbivorous insects can be thought of as plant parasites because they feed on plants but generally do not kill them. In turn, most insect species are host to several other species of insect parasites that do not immediately kill their hosts and are termed parasitoids (**Figure 15.1b**). It is therefore clear that parasitism is a common way of life. A free-living organism that does not harbor parasitic individuals of a number of species is a rarity. Andy Dobson and colleagues (2008) noted that in the best-studied host and parasite taxa, mammals and their helminth parasites (internal wormlike parasites), each mammal hosts two cestode species, two trematodes, four nematodes, and one acanthocephalan per host species. Data for other vertebrate taxa such as birds showed slightly higher numbers of parasites, and for fish, amphibians, and reptiles, slightly less (**Figure 15.2**).

In this chapter we will examine some of the many different attributes and lifestyles exhibited by parasites, and the strength of their effects on their hosts. We note that the presence of intermediate hosts often makes modeling host-parasite interactions difficult, but we introduce some simple mathematical models. We also discuss the devastating effects some introduced parasites have had on native populations and how parasitism rates may be increased by climate change.

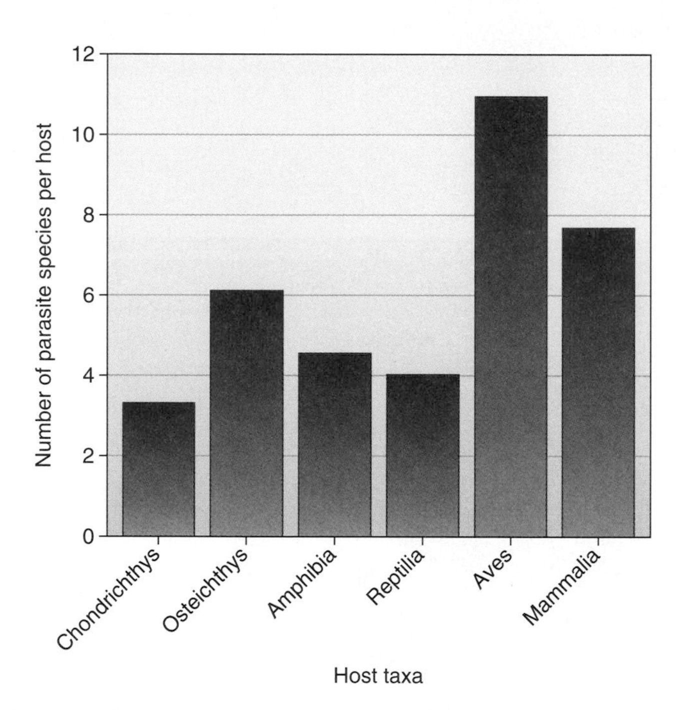


**Figure 15.2 Vertebrate hosts and the typical number of helminth parasites they support.** (From data in Dobson et al., 2008.)

## 15.1 Parasites Exhibit a Wide Range of Attributes and Lifestyles

Parasites are derived from various taxa and exhibit many different types of life history attributes and life cycles. In this section we begin by defining various types of parasites. Later we examine the lifestyles of parasites that utilize multiple host species and we investigate kleptoparasitism, a type of parasitism where one species steals resources from another.

### 15.1.1 Parasites can be classified in many different ways

We can distinguish various types of parasites based on their lifestyle. **Microparasites,** such as pathogenic bacteria and viruses, multiply directly within their hosts, usually inside the host's cells. They are microscopic in size and the term "disease" is usually used to describe these infections. Disease-causing organisms reproduce rapidly. Transmission between hosts may occur directly or be facilitated by other organisms such as blood-feeding insects. The virulence of a disease-causing organism is a measure of the degree of mortality it causes to its hosts. Usually the host has a strong immunological response to microparasitic infections. **Macroparasites,** such as schistosomes, live in the host but release infective juvenile stages outside the host's body. Such species are usually visible to the naked eye and have a relatively long generation time. Most require more than one host species to complete their life cycle. For macroparasitic infections, the immunological response is short-lived, the infections tend to be persistent, and the hosts are subject to continual reinfection. Insects are often host to other species of parasitic insects termed parasitoids. **Parasitoids** develop as internal parasites of insect immature stages, such as larvae or pupae, of hosts. The adult parasitoid lays its eggs in or on the host, and the larvae usually develop internally. Eventually the adult parasitoids emerge, ultimately killing the host organism. In this respect the interaction is similar to predation. Some parasitoids themselves serve as hosts for other parasitoids, which are known as hyperparasitoids.

Some of the differences between microparasites and macroparasites, and between parasitoids and predators, are outlined in **Table 15.1**.

**Ectoparasites,** such as ticks and fleas, live on the outside of the host's body, while **endoparasites,** such as pathogenic bacteria and tapeworms, live inside the host's body. Problems of definition arise with regard to plant parasites, which seem to straddle both camps. For example, some parasitic plants, such as mistletoe, exist partly outside of the host's body and partly inside. Outgrowths called haustoria penetrate inside the host plant to tap into nutrient supplies. Being endoparasitic on a host seems to require greater specialization compared to being ectoparasitic, perhaps because more-specialized mechanisms are needed to thwart the host's defenses. Thus, ectoparasitic animals like leeches tend to be polyphagous, having many host species, while internal parasites such as liver flukes are more often monophagous, having only one host.

Parasitic flowering plants may be classified as holoparasites or hemiparasites. **Holoparasites** lack chlorophyll and are totally dependent on the host plant for their water and nutrients. One famous holoparasite is the tropical *Rafflesia arnoldii,* which lives most of its life within the body of its host, a vine that grows only in undisturbed rain forests in Southeast Asia (**Figure 15.3a**). Only the flower develops externally, and

(a)

(b)

**Figure 15.3 Parasitic plants.** **(a)** *Rafflesia arnoldii,* the world's biggest flower, lives as a holoparasite in Indonesia. **(b)** Mistletoe, *Viscum album,* lives as a hemiparasite on trees in northern temperate climates, here at Richmond, England.

**Table 15.1 Variations in some general life history traits of parasitic and predatory organisms.**

| Trait | Microparasite | Macroparasite | Parasitoid | Predator |
|---|---|---|---|---|
| Body size | Microscopic | Smaller than hosts | Similar in size: wasp and moth | Often larger than prey unless social hunters |
| Intrinsic rate of population growth | Rapid | Faster than hosts | Comparable to hosts | Comparable but often slightly slower than prey |
| Number per host | One host usually supports populations of several different species | One host supports a few to many individuals of different species | One host usually supports one but can support many individuals | Each predator eats many prey individuals |
| Fatality for host | Mild to severe | Usually mild | Eventually fatal | Immediately fatal |

it is a massive flower, 1 m in diameter and the largest known in the world. **Hemiparasites** generally photosynthesize, but they lack a root system to draw water and thus depend on their hosts for that function. Mistletoe, *Viscum album*, is a hemiparasite (**Figure 15.3b**). Hemiparasites usually have a broader range of hosts than do holoparasites, which may be confined to a single or a few host species.

## 15.1.2 Many parasite life cycles involve more than one species of host

Many species of macroparasites cannot complete their life cycles without the presence of at least two different host species. Such parasites are said to have complex life cycles. The **definitive host** is the host in which parasites exhibit sexual reproduction.**Intermediate hosts** are species containing nonreproducing forms of the parasite. A case in point is the parasitic worm *Plagiorhynchus cylindraceus,* a parasitic worm that lives in the intestines of songbirds, such as starlings, and was studied by Janice Moore (1984) (**Figure 15.4**). This parasite is a member of the acanthocephala, or thorny-headed worms, whose gutless adults absorb nutrients directly. They produce eggs that exit with the host feces. Once on the ground, the feces are eaten by foraging arthropods such as pillbugs. Here the eggs hatch into a juvenile infective stage. Eventually the infected pillbugs are eaten by birds and the life cycle continues. Having two host species facilitates the transmission of the parasite from one bird host to another. In this case, the juvenile parasites cause a change in behavior of the pillbugs. Normally pillbugs seek out areas of high humidity, such as the cover of leaf litter, and they are concealed from predators. However, the behavior of parasite-infected individuals is changed and they wander more into open areas where they are more easily spotted by birds. The term "enslaver parasites" has been coined to describe parasites that induce changes in the behavior or color of one host, making that host more susceptible to being eaten by a second host.

Some parasites have a complex life cycle involving transmission between several hosts. For example, adult lancet flukes, *Dicrocoelium dendritium,* are about 1 cm long and live in cows. They spread their eggs in cow manure. Snails eat the eggs and the immature parasites hatch and infect their intestines. The snail coughs up these offspring in balls of slime, and ants eat these slime balls loaded with flukes. The flukes migrate to the ants' heads and make them climb to the tops of grass blades, where they are eaten by cows, and the cycle starts over again (**Figure 15.5**).

Numerous other examples of enslaver parasites exist. For example, the freshwater amphipod *Gammarus pulex* is regularly infected by two acanthocephalan species, *Pomphorhynchus laevis* and *Polymorphus minutus. Gammarus* is an intermediate host for both species. *P. laevis* can mature and reproduce only in fish species, and *P. minutus* can mature and reproduce only in birds. Frank Cézilly and colleagues (2000) showed that uninfected *Gammarus* avoid light and stay close to the bottom of the water column. Those infected by *P. laevis* are attracted to light, which presumably influences them to swim away from shelters and makes them more prone to fish predation. Those *Gammarus* infected by *P. minutus* swim near the water surface, where they are subject to predation by birds.

Other parasites can manipulate the feeding behavior of their vectors. The malaria parasite, a single-celled species in the genus *Plasmodium,* has a complex life cycle involving two hosts, mosquitoes and vertebrates that are the definitive

**Figure 15.4** **The starling-isopod life cycle of the acanthocephalan *Plagiorhynchus cylindraceus.*** (After Moore, 1984.)

**Figure 15.5 Parasite life cycles can be complex.** The life cycle of the lancet fluke, *Dicrocoelium dendriticum,* involves behavioral changes in ants, one of its three host species that increase its transmission rate.

hosts. *Plasmodium* interferes with the ability of the mosquito to draw up blood from its hosts. This increases the number of attacks the infected mosquitoes make in order to try to obtain enough blood. Increased attack rates maximize the transmission rates of the *Plasmodium* itself. Jacob Koella and colleagues (1998) showed that most uninfected mosquitoes generally fed on just one human host at night, and only 10% bit more than one person. This multiple biting of different hosts increased to 22% in malaria-infected mosquitoes. The saliva of the infected mosquitoes was changed, making the host's blood flow less freely into the mouthparts. Similar behavior is exhibited by leishmaniasis parasites in sand flies and bubonic plague parasites in fleas.

### 15.1.3 Kleptoparasitism involves one species stealing resources from another

**Kleptoparasitism,** meaning parasitism by theft, is a form of feeding where one animal takes food that another has caught. These interactions may be intraspecific or interspecific. For example, some small spider species live as kleptoparasites in the webs of larger species, feeding on prey remains of the larger spider. Some gulls and frigatebirds steal food from other bird species that dive into the ocean to catch it.

Brood parasitism is a specific kind of kleptoparasitism in which individuals manipulate others to feed and look after their young. For example, avian brood parasites include cuckoos, cowbirds, indigobirds, whydahs, and honeyguides. Cuckoos utilize a variety of host species, but individual females specialize on certain species. This means that some female cuckoos might always parasitize robins, whereas others might always parasitize blackbirds. Thus, some individual female cuckoos prefer certain host species while cuckoos as a whole utilize many different host species. Females are thought to imprint on the host species that raised them. Genes influencing egg color are passed down the maternal line, allowing females to mimic the egg color of their hosts. Male cuckoos mate with females of all lines, ensuring gene mixing. Why do hosts permit their nests to be parasitized? Parents can often recognize foreign eggs and eject them from the nest. The **mafia hypothesis** proposes that cuckoos or cowbirds repeatedly check their host's nests and destroy all eggs if their owner is not present. This causes the hosts to lay an additional clutch of eggs, giving the cuckoos or cowbirds an opportunity for parasitism. If the parasite egg has been removed, they again destroy the host's nest and kill or injure the nestlings. A study by Jeffrey Hoover and Scott Robinson (2007) investigated nest destruction by the brown-headed cowbird, *Molothrus ater,* and its host, the prothonotary warbler, *Pronotaria citrea*. They found that 56% of nests where cowbird eggs were rejected were destroyed compared to only 6% of nonrejected nests.

#### Check Your Understanding

**15.1** *Toxoplasma gondii* is an intracellular protozoan with a complex life cycle, able to infect all mammals. However, it can reproduce only in cats and its eggs are found only in cat feces. Why might rats infected with the protozoan parasite, *Toxoplasma gondii,* lose their fear of cats?

## 15.2 Hosts Have Evolved Many Different Types of Defenses Against Parasites

Just as with attack by predators or herbivores, attack by parasites elicits a plethora of defensive reactions. In vertebrates, many cells and organs contribute to fighting off parasite attack and are collectively known as the immune system. Two broad types of immunity are known, innate immunity and acquired immunity. Innate immunity is constituted by the body's defenses present at birth. For example, the skin and mucous membranes protect the body from invasion by microparasites. In addition, the innate immune system includes a set of chemical and cellular defense reactions that operate against parasites that breach the body's line of defense. For instance, macrophages engulf bacteria that penetrate the body. In phagocytosis, a cell engulfs a parasite and uses enzymes to destroy it. A variety of leukocytes, or white blood cells, are present to engulf viruses and bacteria. Infected areas often become inflamed. Inflammation causes increased blood flow to an area and increases the delivery of beneficial food, oxygen, and leukocytes. All organisms have an innate immune system. For example, insects have specific cellular defense reactions against parasitoids. The eggs of the parasitoid are rendered inviable by encapsulating them in a tough case. Acquired immunity develops after the body is exposed to a parasite. Acquired immunity can persist for long periods of time, may years in humans, and can cause the production of antibodies. In a way, these different types of immunity are similar to the constitutive and induced defenses in plants, which we mentioned at the beginning of Chapter 14.

As we mentioned in section 14.2, when discussing plants and their herbivores, an evolutionary arms race exists between parasite and host. Theoretically, parasites should have an advantage in the evolutionary arms race with their hosts because of their more rapid generation time. Hosts reproduce more slowly and have less time to adapt their defenses against parasites.

**Check Your Understanding**

**15.2** Can grooming of one mammal by another be viewed in the context of selfish behavior?

## 15.3 Parasites Can Cause High Mortality in Host Populations

The existence of a wide variety of host defenses against parasitism suggests that parasites must exert a strong selective pressure on host populations. It is very difficult to accurately assay levels of protozoa or bacteria in native hosts, but reliable data on infection rates from other parasites come from game animals and insects (**Table 15.2**). Infection rates of ducks, deer, and mice by helminths can be assayed from harvested animals, and parasitism rates may be 50% or more. Similarly, insect parasitism levels from native parasitoids can often be 50–100%. Plants too can be affected by parasitic plants, of which there are over 5,000 species worldwide. Marsh dodder, *Cuscuta salina* (**Figure 15.6**), is a common and widespread plant parasite in salt marshes of North America. In this section we describe a series of parasite-removal studies designed to determine the effect of native parasites on their hosts. Later, we see the devastating effects invasive parasites

**Table 15.2 Parasitism levels of some native vertebrates and invertebrates mentioned in the text.**

| Host | Parasite | Location | Parasitism level |
|---|---|---|---|
| Bighorn sheep | Lungworms | Rocky Mountains | 91% |
| Squirrels | Botflies | Mississippi | 50% |
| Moose | Tapeworms | Isle Royale | 68% |
| Snowshoe hare | Tapeworms | Canada | 64–72% |
| Ducks and Geese | Avian malaria | Northeast U.S. | 71% |
| Blue tits | Blowflies | Corsica | 90% |
| Gall flies | Hymenopterans | Florida | 90–100% |

**Figure 15.6 Dodder, *Cuscuta* spp., is a common plant parasite of many species.** Dodder parasitizing pickleweed, *Salicornia virginica*, in Carpentaria marsh, California.

have on native hosts, but we also see how some invasive parasites are used to control crop pests.

### 15.3.1 Parasite-removal studies show how native parasites strongly affect native host populations

As with studies of other species interactions, a direct method to determine the effect of parasites on their host populations is via parasite-removal studies where investigators remove the parasites and to reexamine the densities of their host populations. However, parasite-removal studies are difficult to perform, primarily because of the small size and unusual life histories of many parasites, which makes them difficult to remove from hosts. In addition, many parasites impair the health of their hosts rather than kill them directly, making the effects of parasites on host populations even more difficult to gauge. For example, weakened or unhealthy hosts may fall prey to predators more often than healthy, unparasitized hosts.

Some parasite removal studies have been performed using birds, which are relatively easy to catch on their nests. The common eider, *Somateria mollissima,* is a long-lived sea duck. Females do not feed during egg incubation and lose 40% of their body mass during egg laying and incubation. Low-body-weight ducks may desert their nests and return next year to breed again. Low body weight may be a result of a large burden of intestinal parasites that absorb much of the duck's ingested food. Svein Hanssen and colleagues (2003) treated some Norwegian eiders with an antiparasitic drug administered orally. Control ducks received only sterile water. Ducks were banded during the treatment phase, and researchers could monitor their return rates for the next 2 years. Treated ducks had a much higher return rate than control ducks, 69% to 18%, suggesting high mortality in eider ducks with heavy parasitc loads. In a different parasite-removal experiment, nest parasites were removed by microwaving bird nests and returning chicks to the parasite-free nests (see **Feature Investigation**).

Malarial parasites are thought to be particularly devastating to birds. Alfonso Marzal and colleagues (2005) tested this idea by inoculating house martins, *Delichon urbica,* with primaquine, designed to reduce infestation by the blood parasite *Haemoproteus prognei.* Individuals were captured at dawn while in their nests. Fifty-two adults from 26 nests were treated 9 days before egg laying. Forty-six adults from 23 nests received a saline placebo (controls). Subsequent bad weather caused some nest desertion, but about 30 treated adults and the same number of controls were available to assess treatment effects. When nestlings were 12 days old, a blood sample was taken from the parents to test for disease prevalence. Primaquine significantly lowered infection prevalence and intensity of infection. Clutch size, brood size at hatching, and brood size at fledging were all increased by parasite reduction, though body mass of offspring was not affected (**Figure 15.7**). This demonstrates strong effects of parasites on clutch size and other variables. What is the mechanism?

## Feature Investigation

### Sylvia Hurtrez-Boussès and Colleagues Microwaved Bird Nests to Eliminate Nest Parasites

The nests of birds such as blue tits, *Parus caerulus,* are often infested with parasitic blowfly larvae that feed on the blood of nestlings. Blowflies feed only on chicks, and adult flies lay their eggs in old nests after the young have fledged. Sylvie Hurtrez-Boussès (1997) and colleagues experimentally reduced the numbers of blowfly larval parasites in blue tit nests in Corsica, where there is one of the highest blowfly larvae infestations ever recorded, with more than 90% of nests affected. Parasite loads can reach up to 100 larvae per nest, and uptake can be up to 55% of the blood volume of the chick, causing anemia. Parasite removal was cleverly achieved by taking the nests from 145 nest boxes, removing the parasites from the young birds, microwaving the nests to kill the parasites, and then returning the nests and chicks to the wild. The success of chicks in microwaved nests was compared to that in non-microwaved (control) nests that were parasite infected. Many parasite-infected nests failed to fledge any young at all, possibly because the adult birds deserted. This could have been due to excessively high feeding demands of anemic chicks. Alternatively, the higher calling rates of hungry chicks begging for food from their parents could attract more predators. The parasite-free blue tit chicks also had greater body mass at fledging, the time when feathers first grow (**Figure 15.A**).

**HYPOTHESIS** Ectoparasites reduce the breeding success of birds.

**STARTING LOCATION** Corsica, France

| Conceptual level | Experimental level |
|---|---|
| **1** Examine infestation rates of nestlings by *Protocalliphora azura* and *P. falcozi* blowfly larvae. | Establish 145 nest boxes for blue tits, *Parus caeruleus*. On average, 65–70 blue tits use these nest boxes each year. Of 29 control broods, only one was free of parasites. |
| **2** Remove blowfly larvae from some nests, keep others as controls. | Choose 23 nests, remove nestlings, replace nests with disinfected nest of same size. Microwave old nest at 850W for 1 minute. Repeat at approximately 3 day intervals for 15 days. |
| **3** Determine effects of parasites on breeding success. | On day 2 after hatching, each chick in a microwaved or control nest was marked. At approximately 3 day intervals chicks were weighed to the nearest 0.1 g. Survival and weight were recorded over time. |

**4 THE DATA**

**Figure 15.A** Determining the effects of parasitic blowflies on bird chicks.

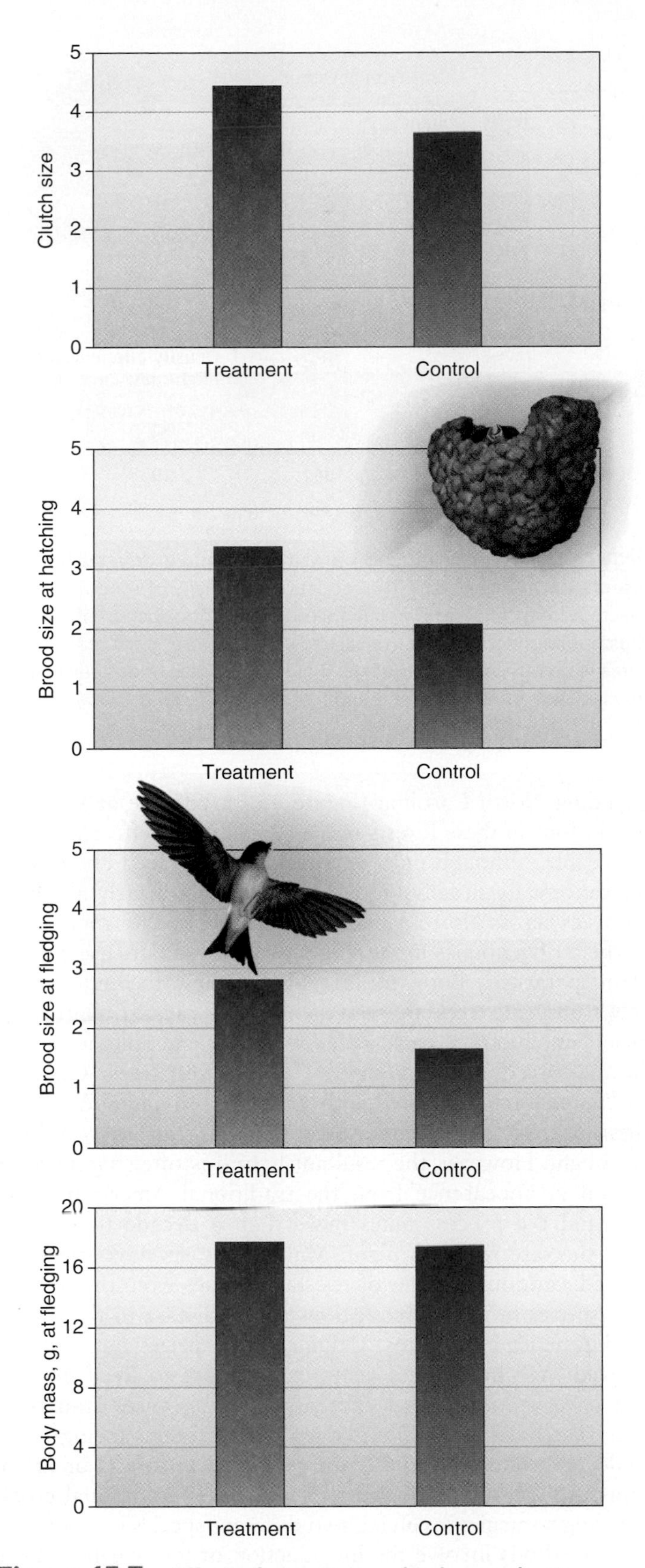


**Figure 15.7** **Effect of parasites on bird reproduction.** Clutch size, brood size at hatching, brood size at fledging, and body mass at fledging of house martins, with and without blood parasites.

Malarial parasites possibly reduce the ability of adults to forage successfully and acquire resources for egg production and subsequent fledging of chicks. Second, the functioning of the immune system may take away resources that could be used for reproduction and other functions.

Peter Stiling and Anthony Rossi (1997) were able to manipulate parasitism levels of a gall-making fly, *Asphondylia borrichiae*, on a coastal plant, *Borrichia frutescens*, on isolated islands in Florida. Gall flies are parasitized by four species of parasitoid wasps and may suffer parasitism levels of 90–100% by the end of the year. Both flies and galls can complete one generation in about 6–8 weeks in warm weather, and multiple generations are completed in a year. To create gall fly populations with low rates of parasitism, they allowed ungalled potted plants on one island to be colonized first by gall flies. They then removed these plants before the parasitoids could find them (low parasitism treatment). To produce gall fly populations with high rates of parasitism, they left some plants longer on the island to allow the parasitoids to colonize the galls. Using these plants, replicates of both high and low parasitism treatments were established on other islands, less likely to be visited by flies or wasps from other areas. Where parasitism of gall flies was high, numbers of new galls were significantly lower than where levels of parasitism were low (**Figure 15.8**). Toward the end of the year, in cooler weather, the development of the flies slowed, allowing parasitoids more time to locate the galls and parasitize the fly larvae. Parasitism on all islands increased, and gall densities decreased.

### 15.3.2 Invasive parasites may have even more devastating effects than native parasites

A common belief among parasitologists is that the degree of mortality caused by the parasite to the host depends on its evolutionary age of association. Older associations are thought to cause less harm and may even evolve toward commensalism or mutualism. For example, the myxoma virus produces a mild, nonlethal disease in its native hosts, the South American cottontail rabbit, *Sylvilagus floridanus*. In the 1950s this virus was used as a biocontrol agent of European rabbits, *Orytolagus cuniculus*, in Australia and England, where both are invasive pests, and killed more than 99% of the rabbits. Rabbits were introduced into England from France at around the time of William the Conqueror in the 11th century and possibly earlier. Other examples of the devastating effects of invasive parasites abound.

Chestnut blight, *Cryphonectria parasitica*, is a fungus from Asia that was accidentally introduced to New York around 1904 from imported Asian chestnut trees, *Castanea crenata*. At that time, the American chestnut tree, *Castanea dentata*, was one of the most common trees in the eastern United States. It was said that a squirrel could jump from one chestnut to another all the way from Maine to Georgia without touching the ground. The densest populations of chestnuts occurred in

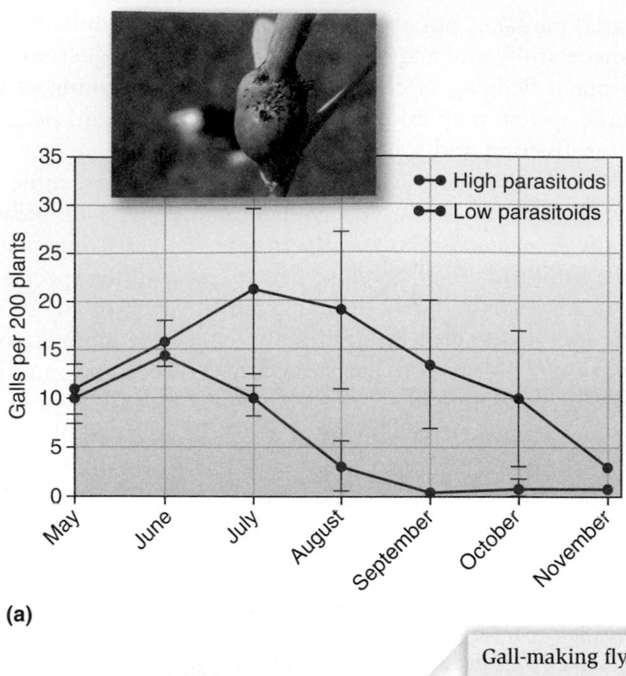


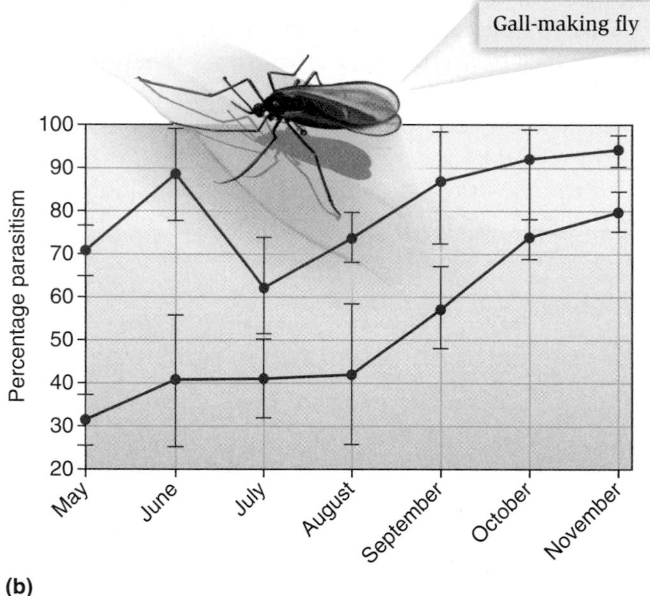


**Figure 15.8 Parasitoid reduction increases gall fly abundance.** **(a)** Numbers of new galls on islands with high and low levels of parasitoids. **(b)** Corresponding parasitism levels on the same islands. Means and standard errors are shown. (After Stiling and Rossi, 1997.)

## ECOLOGICAL INQUIRY

Why do gall numbers decrease in both treatments at the end of the year?

the Appalachian mountains. Here, humans and wildlife alike feasted on the bountiful supply of nuts. The tree was ingrained in American culture, as in "chestnuts roasting on an open fire." By the 1950s the disease had significantly reduced the density of American chestnut trees in all areas of the United States,

**Figure 15.9 Introduced parasites can have devastating effects on their hosts.** The reduction in density of American chestnut trees in North Carolina following the introduction of chestnut blight disease from Asia shows the severe effect that parasites can have on their hosts. By the 1950s this once-prevalent species was virtually eliminated. (From data in Nelson, 1955.)

including North Carolina (**Figure 15.9**). Because at least one tree in four in these forests was a chestnut, the effect was very noticeable, although oaks and hickories replaced chestnut in the canopy. Eventually the fungus eliminated nearly all chestnut trees across North America. Scientists hope for much from the field of genomics in their fight against such disease-causing plant parasites. Both traditional breeding techniques and new transgenic trees are seen as possible ways to reintroduce American chestnuts back to the wild. For example, using traditional breeding technology, Asian chestnut trees (*Castanea mollissima*) are being bred with some few remaining American chestnut trees to reduce the susceptibility of the latter to chestnut blight. However, the resultant hybrid is often significantly altered in appearance from the traditional American chestnut, and the process takes more than a decade to produce trees that are ready to plant. Many scientists have suggested limited, cautious transfer of resistance genes from the original host species in the source regions of the disease to the threatened American chestnuts. Original host species have usually evolved over millions of years of exposure to local diseases and have acquired genes that provide resistance. Transgenic trees that have received pathogen resistance-enhancing genes could be produced and then be replanted in forests or urban areas. An advantage of this technique over traditional crossbreeding strategies involving two different species is that transgenic methods involve the introduction of fewer genes to the native species. Also, fewer tree generations would be required to develop resistance. American scientists are currently working to enhance the American chestnut's resistance by inserting a gene taken from wheat. The gene, which encodes an enzyme called oxalate oxidase, destroys a toxin produced by the fungus

**Figure 15.10 Restoration of American chestnut trees in Tennessee.** In a double-barreled approach to restoration, Appalachian mountains scarred by strip-mining are being planted with American chestnut trees.

that causes chestnut blight. The whole process takes up to 2 years to produce plants ready to be transplanted back to the wild. The first two transgenic chestnuts were planted in New York in 2006, and 17 more followed in 2007. Five hundred blight-resistant trees were planted in Virginia, Tennessee, and North Carolina in 2009, but it will take many years to determine how successful these transplants will be (**Figure 15.10**).

Sudden oak death, first reported in central coastal California in 1995, is a recently recognized disease that is killing tens of thousands of oak trees and other plant species in California. The geographic range of the disease in natural ecosystems stretches over 750 km, from northern California to Oregon. Satellite imagery suggested that by 2009 about 20% of the existing forest trees in the Big Sur region had been killed. The country of origin of sudden oak death is unknown, though it is thought likely to have been introduced on rhododendrons by the commercial plant trade. The symptoms vary between species but include leaf spots, oozing of a dark sap through the bark, and twig dieback (**Figure 15.11**). Although sudden oak death is a forest disease, the organism causing this disease is known to infect many woody ornamental plants that are commonly sold by nurseries. In March 2004 a California nursery was found to have unknowingly shipped plants infected with sudden oak death to all 50 states. Following this discovery, California nurseries halted shipments of trees to other states in an attempt to stop the spread of the disease. In 2004 scientists mapped out the genome sequence of the disease-carrying fungus, *Phytophthora ramorum*. They are hoping that identifying the genes and their proteins will help develop specific diagnostic tests to quickly detect the presence of sudden oak death in trees, which is currently impossible to detect until a year or more after the tree is infected.

Examples of the effects of introduced disease on animals abound (see **Global Insight**). In the early 1980s, a distemper virus caused a greater than 70% drop in the remaining

**Figure 15.11 The effects of sudden oak death.** Sudden oak death is a recent and severe disease of many plant species in California that causes leaf spots, oozing of dark sap, and twig dieback on many plant species.

population of black-footed ferrets in Wyoming (see Chapter 10), and a few years later another outbreak exploded among seals and dolphins. Particularly worrying for conservationists is that many endangered animals are threatened by disease from domestic animals. In 1994, a thousand lions, one-third of the resident Serengeti National Park population, were wiped out by canine distemper virus, probably transmitted from domestic dogs. Some scientists have suggested that the demise of the thylacine, a marsupial wolf, in Tasmania was because of a distemper-like disease brought about by close association with dogs. To prevent the threat of disease, some populations of endangered species have been vaccinated against diseases; for example, mountain gorillas have been vaccinated against measles and African hunting dogs against rabies. Yet it would be very difficult to vaccinate more than a few species of wild animals against more than one or two diseases. Even the vaccinated animals may succumb to new diseases.

### 15.3.3 Some non-native parasites are introduced deliberately for the biological control of pests

As we noted earlier, when discussing the introduction of the myxoma virus against rabbits, not all introduced parasites are seen as detrimental by humans. Many are used as an effective line of defense against pests, especially insect pests

## Global Insight

### Rinderpest Caused Massive Mortality in African Wildlife in the 19th Century

*Rinderpest,* meaning cattle plague in German, belongs to a family known as Morbilliviruses, a family of single-stranded RNA viruses that includes measles and distemper. It causes fever, diarrhea, and mortality in many bovids, particularly in young animals. Rinderpest is spread by food or water contaminated by the feces of sick animals. Wildlife contract the disease from cattle. The disease is usually fatal in buffalo, wildebeest, eland, kudu, giraffe, and warthog. Other species, such as zebra, impala, gazelle, and hippopotamus, appear to suffer little. A major epidemic swept through Africa in the 1890s, probably from cattle imported into Somalia from India by an Italian military expedition. In India the disease caused only mild infections, but in Africa more than 80% of cattle died over the entire continent. The disease traveled 5,000 km in 8 years, arriving in North Africa in 1889 and reaching the Cape of Good Hope in 1897. Without grazing animals, grasslands developed thickets of vegetation that provided breeding grounds for tsetse flies, and human sleeping sickness increased.

In the Serengeti of Tanzania, the two dominant bovids are the wildebeest, *Connochoetes taurinus,* and the buffalo, *Syncerus cafter;* both were susceptible to rinderpest, and their numbers were reduced to about 200,000 and 30,000, respectively, following the rinderpest outbreak. Following the control of rinderpest in the 1960s, the population of both species increased until the 1980s when populations were impacted by poaching (**Figure 15.B**). The number of zebra, *Equus burchelli,* was also about 200,000, but it was not susceptible to rinderpest so its numbers changed little over this time span. Some parasitologists think that epidemics have a more severe impact on wildlife populations than does predation. This is a disconcerting finding for conservation biologists, because it means that even protected populations on small reserves could be wiped out by disease. Luckily, Dr. Walter Plowright developed an effective vaccine against rinderpest in the 1960s and the disease was brought under control, but not eliminated. The disease returned and spread in the 1970s from two small areas of Africa. In 1993 the Global Rinderpest Eradication Program (GREP), an initiative of the Food and Agriculture Organization (FAO) of the United Nations, began with a goal of vaccinating all cattle worldwide. In 2011 the GREP declared that rinderpest had been eradicated. Rinderpest is only the second disease ever eradicated, following smallpox.

(i)

(ii)

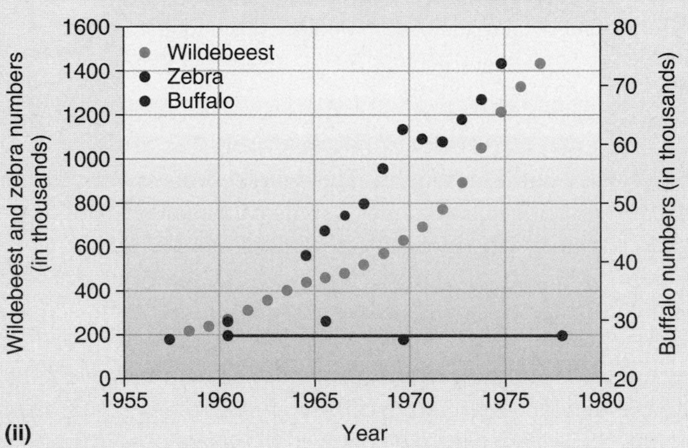


**Figure 15.B Rinderpest.** **(i)** Cattle herds were decimated by rinderpest in the late 19th century. **(ii)** Populations of wildebeest and buffalo, which are both susceptible to the disease, increased geometrically following rinderprest elimination. Numbers of zebra, which are not affected, remained steady. (After Grange et al., 1984.)

**ECOLOGICAL INQUIRY**

What type of population growth was exhibited by wildebeest and buffalo from 1958 to 1975, following the reduction of rinderpest?

of crops. The use of natural enemies, including parasitoids and predators, is known as **biological control.** This usage has spurred the scientific community into a search for more-effective biological control agents.

Only about 16% of classical biological control introductions attempted so far qualify as economic successes. Early campaigns used a hit-or-miss technique: release a variety of parasitoids and predators, and hope that one of them does the job. Others have recommended new approaches, such as the establishment of novel parasite-host associations. In these situations, new species of enemies, never before experienced by pests, are collected from around the world and released. The logic is that hosts have not had the opportunity to evolve defenses against these parasites from foreign lands.

Other authors have tried to find what attributes make a successful biocontrol agent. Carl Huffaker and Charles Kennett (1969) suggested five necessary attributes of a good agent of biological control:

1. General adaptability to the environment.
2. High searching capacity for hosts.
3. High rate of increase relative to the host's.
4. Good dispersal ability.
5. Minimal time lag effects in responding to changes in host numbers.

Later, Peter Stiling (1990) reviewed the factors affecting success in biological control. The factor of greatest importance was a climatic match between the control agent's locality of origin and the region where it was to be released. This finding shows that climatic variation is of vital importance in affecting biotic relations, and supported Huffaker and Kennett's idea that adaptability to the environment is important. The importance of climatic variation was underscored by Stiling's later (1993) analysis of reasons for biological control failures. He found that reasons for failure related to climatic mismatching of the enemy's preferred optimal abiotic range and the range where it was being released were more common (34.5%) than any other type of reason, including competition or parasitism by native insects.

It is important to note that biological control is not the risk-free alternative to chemical control it is often touted to be, because of effects on nontargeted species (see also Section 14.3). As early as 1983 Frank Howarth, a longtime entomologist in Hawaii, lamented the reduction in numbers of native Hawaiian butterflies and moths, which he thought was partly due to parasitoid wasp species introduced for biological control of insect crop pests, particularly pests of sugarcane. The wasps had attacked not only the target invasive pest but also nontarget native species. Howarth called for a more narrowly focused release effort rather than a hit-or-miss campaign—the opposite of what many earlier biocontrol campaigns had espoused.

**Check Your Understanding**

**15.3** How might you go about reducing the threat of canine distemper virus to Serengeti lions?

## 15.4 Host-Parasite Models Are Different from Predator-Prey Models

Because parasite-removal studies are difficult to do, ecologists have also examined the strength of parasitism as a mortality factor by modeling host-parasite relationships. Models of the effect of parasites on host populations differ from predator-prey models we discussed earlier, in section 13.2, for two reasons. First, the life cycle of many parasites involves intermediate hosts. Modeling parasite transmission between the various types of hosts can be extremely complex, involving the densities of many hosts and parasites at several different stages. Second, with respect to diseases, models of parasite population dynamics generally describe the parasite's population growth rate by the average number of new cases of a disease that arise from each infected host, $R_p$, rather than by $R_0$, the net reproductive rate of the parasite. The transmission threshold, which must be exceeded if a disease is to spread, is given by $R_p > 1$. If $R_p < 1$, a disease dies out. For microparasites that are transmitted directly from host to host, $R_p$ is influenced by

- $N_s$, the number of susceptible hosts in the population.
- $B$, the transmission rate of the disease, a quantity correlated with frequency of host contact and infectiousness of the disease.
- $L$, the average period of time over which the infected host remains infectious.

The value $R_p$ is related to these factors by the equation

$$R_p = N_s BL$$

Three generalizations can be made. First, as $L$, the period during which the host is infectious, increases, $R_p$ increases. An efficient parasite therefore keeps its host alive for a long time. In the extreme, some hosts remain infectious even after they are dead. For example, fungal parasites leave a residue of resting spores on the dead host. Second, if diseases are highly infectious, that is, $B$ is large, $R_p$ increases. Third, large populations of susceptible hosts, $N_s$, promote the spread of disease. By rearranging the equation we can obtain the critical threshold density, $N_T$, where $R_p = 1$. $N_T$ is an estimate of the number of susceptible hosts needed to maintain the parasite population at a constant size:

$$N_T = \frac{1}{BL}$$

If $B$ or $L$ is large, $N_T$ is small. For example, if $B = 0.1$ and $L = 0.01$, then $N_T = 1/0.1 \times 0.01 = 1/0.001 = 1{,}000$. Conversely, if $B$ or $L$ is small, the disease can persist only in a large population of infected hosts. For instance, if $B = 0.1$ and $L = 0.0001$, then $N_T = 100{,}000$.

Measles, rubella, smallpox, mumps, cholera, and chicken pox, for example, probably did not exist in ancient times, because the hunter-gatherer populations were small bands of 200–300 persons at most. In a small population there would be no infections like measles, which spreads rapidly and immunizes a majority of the population in one epidemic. Measles occurs endemically only in human populations larger than 300,000. In ancient times, typhoid, amoebic dysentery, and leprosy, diseases whereby the host remained infective for long periods, were probably the most common afflictions.

Malaria and schistosomiasis would also have been quite prevalent, because of the presence of outside vectors to serve as additional reservoirs. Human hookworm and roundworm can produce vast numbers of eggs per day, 15,000 and 200,000, respectively, so these too existed in very low-density human populations. Paradoxically, civilization has increased the kind and frequency of diseases suffered by humans by enlarging the human host pools and by domesticating certain animals. Many modern diseases have arisen because of humans' intimate association with animals and their viruses. Smallpox, for example, is similar to the cowpox virus, measles belongs to the group containing dog distemper and cattle rinderpest, and human influenza viruses are closely related to those found in hogs (swine flu) and birds (bird flu). Human immunodeficiency virus (HIV) is similar to a virus found in chimpanzees in Africa.

**Check Your Understanding**

**15.4** The myxoma virus, which was released against rabbits in Australia, is bloodborne and was originally transmitted by mosquitoes. The disease was lethal in a short time. Why was this lethality a disadvantage?

## 15.5 Parasitism May Be Increased by Climate Change

Particularly worrisome is the prospect that global warming will hasten disease development and transmission rates. In 2008 the Wildlife Conservation Society, based in the Bronx Zoo in New York, predicted a "deadly dozen" diseases, ranging from avian flu to yellow fever, which they said were likely to spread because of climate change. It is known that many plant pathogens cause greater damage at higher temperatures than lower ones. Many vector-borne diseases such as malaria, African trypanonosomiasis, Lyme disease, yellow fever, and dengue fever have increased in incidence or range in recent years, particularly at higher altitudes. Mercedes Pascual and Menno Bouma (2009) have studied the relationship between malaria endemnicity and elevation in Ethiopia. Here endemnicity is expressed as the percentage of the population with an enlarged spleen, enlarged because it is swollen as the body fights malaria. Enlarged spleens can be felt by trained experts with the subject lying in a recumbent position. Such techniques were used as an indirect marker for malaria long before plasmodia were known as the cause. Human population densities in eastern Africa increase with elevation, partly to escape malaria. Seventy-five percent of Ethiopia's population of 77.1 million people live between 1,500 m and 2,400 m in elevation. A 1°C increase in temperature would essentially shift the regression line relating malaria and altitude up 150 m (**Figure 15.12**). On average, 6 million people inhabit each 100 m of altitudinal range, so a 1°C increase in temperature would put an extra 9 million people at risk. Because adults are likely to have developed some degree of malaria immunity, the most serious threat is to children. Approximately 2.8 million additional children would be exposed to malaria with a rise of 1°C, and 410,000 extra people would become infected. The official mortality for 2003 during the malaria season was between 45,000 and 114,000 people, though this is likely only a quarter of unofficial estimates. Although Europe and North America would also be further threatened with malaria, most scientists suggest that the quality of human health care there will prevent increased outbreaks. However, such is not the case for diseases of other organisms. Avian malaria, *Plasmodium relictum,* an introduced disease in Hawaii, causes marked declines of forest birds in mid-elevation forests, and its effects are less at higher elevations where mosquitoes are limited by lower temperatures. Global warming could expose higher-elevation birds to this deadly parasite. As we mentioned in Chapter 1, climate change has altered patterns of fungal infection in tropical harlequin frogs, leading to the extinction of two-thirds of the known species in Central America. In Canada's Northwest Territories, increased soil temperatures have sped up the life cycle of a nematode worm parasite of musk oxen, *Ovibos moschatus.* This may be one factor behind a 50% decline in numbers of musk oxen since 1980 in heavily infested areas. Drew Harvell and colleagues (2009) have shown

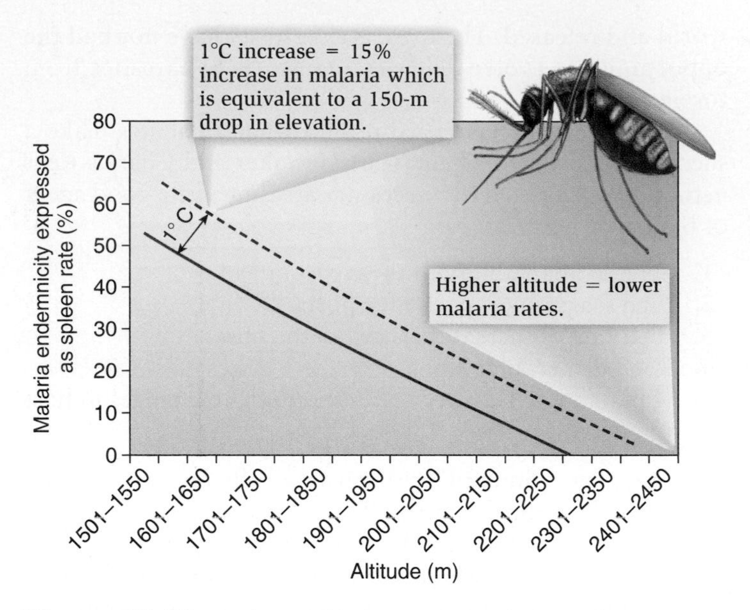


**Figure 15.12** **Relationship between malaria endemnicity, expressed as percentage enlargement rates in the spleen, and elevation in Ethiopia.** (After Pascual and Bouma, 2009.)

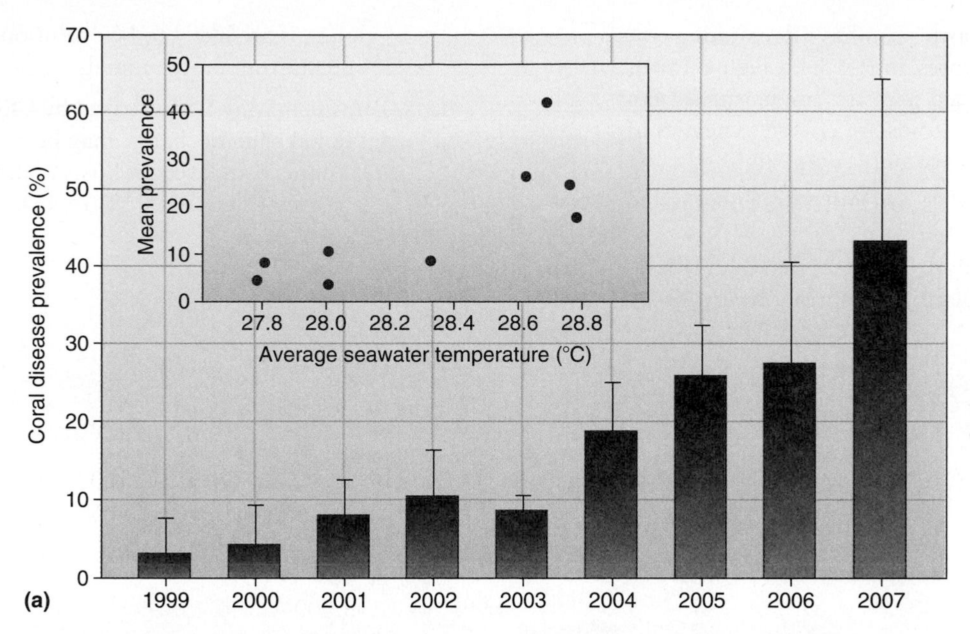


**Figure 15.13** **Increase in the prevalence of Caribbean yellow banding disease, CYBD, in Montastraea coral off the southwest coast of Puerto Rico at La Parguera, 1999–2007.** The inset shows the relationship between increased average water temperature and mean prevalence of the disease. (After Harvell et al., 2009.)

that the incidence of Caribbean yellow banding disease (CYBD) infecting coral reefs has increased significantly from the late 1990s to the mid-2000s (**Figure 15.13**). This is the most damaging disease of all four coral species of the genus *Monstrastaea,* the most important reef-building species in the area. The incidence of the disease increased as temperatures have increased. Even a 1°C increase over 8 years (from 1996 to 2007) was sufficient to increase the frequency of this disease more than sixfold. Warmer average winter water temperatures favor CYBD growth and pathogenicity.

### Check Your Understanding

**15.5** Hendra virus was discovered in Brisbane, Australia, in 1994, first killing horses and then people. Its origins were traced back to fruit bats. Thinking back to what we know about habitat fragmentation from the chapter introduction, construct a scenario for the spread of the disease from bats to people.

## SUMMARY

**15.1 Parasites Exhibit a Wide Range of Attributes and Lifestyles**

- Hosts may be parasitized by a variety of parasites that may be classified as microparasites, macroparasites, or parasitoids (Figures 15.1, 15.2, Table 15.1).
- Parasitism is a common lifestyle, and some parasites have complex life cycles involving multiple hosts (Figures 15.3, 15.4).
- Many parasites change the behavior of their hosts to facilitate parasite transmission (Figure 15.5).

**15.2 Hosts Have Evolved Many Different Types of Defenses Against Parasites**

- Hosts have developed complex mechanisms, including immunity, as defenses against parasites.

**15.3 Parasites Can Cause High Mortality in Host Populations**

- Attack rates of hosts by parasites may sometimes reach 50–100% (Table 15.2).
- Evidence from experimental removal of parasites confirms that parasites can greatly reduce host densities (Figures 15.6–15.8, Figure 15.A).

- Introduced parasites can have devastating effects on native prey (Figures 15.9–15.11, Figure 15.B).
- Not all introduced parasites are harmful. Many introduced parasitoids are effective enemies of economically important crop pests.

**15.4 Host-Parasite Models Are Different from Predator-Prey Models**

- Host-parasite models suggest that many diseases need large host populations to spread. Many diseases of humans are likely to be evolutionarily recent, having spread from other animals.

**15.5 Parasitism May Be Increased by Climate Change**

- Global climate change may be changing the frequency of parasitism levels of hosts and, by implication, host densities (Figures 15.12, 15.13).

## TEST YOURSELF

1. Which of the following has low lethality but high intimacy?
   a. Parasitoids
   b. Herbivores
   c. Predators
   d. Parasites
2. Which of the following natural enemies is eventually, but not immediately, fatal to its host/prey?
   a. Microparasites
   b. Macroparasites
   c. Parasitoids
   d. Predators
   e. Herbivores
3. Some flies steal food transported by ants and can be found beside their foraging trails. Such flies may best be termed:
   a. Hemiparasite
   b. Holoparasite
   c. Endoparasite
   d. Ectoparasite
   e. Kleptoparasite
4. Which type of host immunity to parasites is, by definition, present at birth?
   a. Innate
   b. Acquired
   c. Phagocytosis
   d. Social immunity
   e. All of the above
5. Which of the following is a native parasite?
   a. Dutch elm disease in the United States
   b. Rinderpest in Africa
   c. Chestnut blight in the United States
   d. Lungworms in bighorn sheep in the United States
   e. All of the above
6. If the transmission rate, $B$, is 0.01 and the period of time a host is infectious is 0.02, what is $N_T$, the critical threshold density?
   a. 50
   b. 500
   c. 5,000
   d. 50,000
   e. 100,000
7. What diseases likely existed when human densities were low and people existed as small bands of hunter-gatherers?
   a. Smallpox
   b. Measles
   c. AIDS
   d. Leprosy
   e. None of the above
8. Which plant disease recently appeared in California in the 1990s?
   a. Chestnut blight
   b. Dutch elm disease
   c. Sudden oak death
   d. Myxomatosis
   e. b and c
9. What is a suggested necessary attribute of a successful biological control agent?
   a. Low searching capacity
   b. Good dispersal ability
   c. High rate of increase relative to host's
   d. a and b
   e. b and c
10. Innate immunity against parasites is constituted by the body's defenses present at birth:
    a. True
    b. False

## CONCEPTUAL QUESTIONS

1. Describe some of the defenses that hosts use to reduce infection by parasites.
2. Why are introduced parasites often more deadly than native ones?
3. What are the attributes of a successful biological control agent?

## DATA ANALYSIS

Brucellosis, caused by a bacterium, *Brucella abortus,* is a highly contagious disease of ungulates, especially cattle, which causes females to abort their calves. It was probably brought to North America via imported cattle. Elk and bison also contract the disease, and there are about 4,000 bison in Yellowstone National Park. Some ranchers are strong advocates of culling the herd to reduce the chances of transmission to their cattle. The data show that the seroprevalence, the frequency of individuals in the population that tested positive for the disease, changed with herd size. Explain the data and why the researchers drew a vertical line at 200.

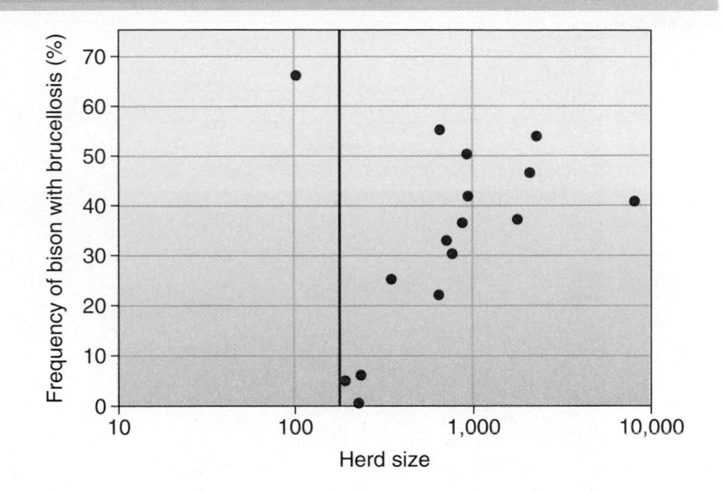

Islands in the Caroni Valley, Venezuela, were created when the river was dammed.

CHAPTER 16

# Population Regulation

## Outline and Concepts



In 1986 in the state of Bolivar, Venezuela, a hydroelectric dam created an impoundment in the Caroni Valley. Rising waters isolated a series of islands, from 200 m to 4.9 km from the mainland. In 1993 and 1994, John Terborgh and colleagues (2001) censused a series of these islands, varying in size from small islands, 0.25–0.9 ha, to medium-sized islands, 4–13 ha, to large islands, >150 ha, for vertebrates and selected invertebrates. On small- and medium-sized islands, large vertebrate predators, such as jaguars, were absent, because the islands were too small to support them. Animals inhabiting these islands were herbivores, including howler monkeys, iguana lizards, and leafcutter ants; seed predators, especially rodents; and insect predators such as spiders, lizards, and birds. Densities of howler monkeys, iguanas, leafcutter ants, and rodents were 10 to 100 times greater on small- and medium-sized islands than on large islands or the mainland. As a result, the densities of plant seedlings and saplings were severely reduced. Terborgh and colleagues argued that the absence of predators permitted herbivore numbers to explode, which in turn depressed plant abundance. The study was a test of two different viewpoints of population regulation. The top-down model holds that predators directly control the population densities of herbivores and herbivores directly control the density of plants. Therefore, predators indirectly control plant abundance. The bottom-up model proposes that the availability and quality of plants directly controls depredation by herbivores and herbivores influence predator abundance. Terborgh's data supported the top-down model.

In Chapters 11 to 15, we saw that competition, facilitation, predation, herbivory, and parasitism are all important in nature. How can we determine which of these factors, along with food availability and abiotic factors, is the most important in regulating populations? The question is one asked by academics and applied biologists, such as foresters, agriculturalists, marine biologists, and conservation biologists. Many different conceptual models have been proposed to describe which mortality factors are the most significant in influencing population densities. Some models stress the importance of so-called bottom-up factors, such as plant quality and abundance, acting from the bottom of the food chain in influencing the densities of herbivores or predators that feed on them. Others stress the importance of top-down factors, such as predators and parasites, acting from the top of the food chain on their animal or plant prey (**Figure 16.1**). Still others incorporate components of both these models.

In this section we will briefly discuss some of the evidence for the existence of top-down versus bottom-up control and some of the models that suggest when one might be more common or important than the other. We will also examine two empirical techniques, key factor analysis and indispensable mortality, that can help determine which factors influence populations.

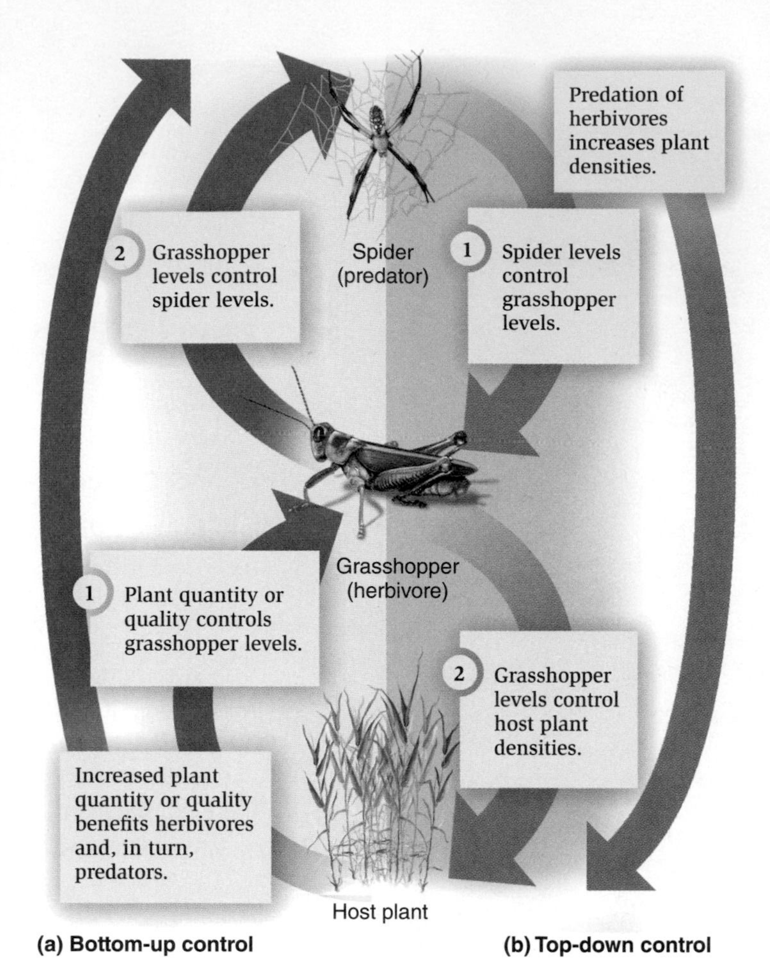


**Figure 16.1 Bottom-up and top-down control. (a)** Bottom-up control proposes that host plant quantity or quality limits the density of herbivores, which in turn sets limits on the abundance of predators. Taken together, this means that high levels of host plants would result in increased numbers of predators because of higher densities of the herbivores on which they prey. **(b)** Top-down control proposes that plant densities are limited by herbivores and that herbivores are limited by predators. Taken together, this means that high levels of predation would result in high densities of host plants because there would be fewer herbivores.

**ECOLOGICAL INQUIRY**

Is the effect of natural enemies on host plants a direct or indirect effect? (see Section 16.1.2)

## 16.1 Both Bottom-Up and Top-Down Effects Are Important in Natural Systems

### 16.1.1 Bottom-up models suggest prey quantity or quality limits enemy abundance

At least two lines of evidence suggest that bottom-up effects are important in limiting population densities of herbivores and carnivores. First, there is a progressive lessening of available energy passing from plants through herbivores to carnivores and to the carnivores that eat carnivores. This line of evidence, based on the thermodynamic properties of energy transfer, suggests that plants should regulate the population densities of all other species that rely on them. The thermodynamic properties of energy transfer are dealt with in more detail in Chapter 25.

Second, much evidence supports the **nitrogen-limitation hypothesis** that organisms select food in terms of the nitrogen content of the tissue. This is largely due to the different proportions of nitrogen in plants and animals (refer back to Chapter 7). Animal tissue generally contains about 4–5 times as much nitrogen as plant tissue; thus, animals favor high-nitrogen plants (**Figure 16.2**). In turn, the nitrogen content of the soil is often 0.1–0.5%, so many plants favor high-nitrogen soils. Water hyacinth, *Eichhornia crassipes*, is a floating aquatic plant native to South America that has become invasive in many tropical and subtropical regions, including Florida. Two specialist beetles, *Neochetina eichhorniae* and *N. bruchi*, were introduced from South America to Florida to control this weed. Ted Center and F. Allen Dray Jr. (2010) showed how water nutrient levels affected tissue nitrogen levels in the plant, which in turn affected beetle abundance and reproduction. In a series of 60 tanks, 15-9-12 N:P:K fertilizer was applied at 20 different rates, with each rate being replicated three times. As the fertilizer rate increased, the tissue nitrogen of the plants went up (**Figure 16.3a,b**). The treatments were assigned to three pools, low, intermediate, and high fertilizer categories. In each tank 12 weevils were introduced and allowed to feed and reproduce. After one generation the number of female weevils was counted and females were dissected to determine how many were reproductively active, that is, having healthy, functioning ovaries with oocytes. Numbers of females were highest in the high-fertilizer grouping (**Figure 16.3c**) as was the percent of reproductive females (**Figure 16.3d**). This suggests beetle numbers could be highest in regions of high water nutrients.

Gwen Waring and Neil Cobb (1992) reviewed the effects of nutrient addition on arthropod herbivore performance and abundance in a large number of studies. Plant fertilization, specifically nitrogen enhancement, had strong positive effects on the population sizes, survivorship, growth, and fecundity of almost all herbivores, including chewing organisms, sucking insects, galling insects, and phytophagous mites (**Figure 16.4**). Nearly 60% of 186 studies reported positive

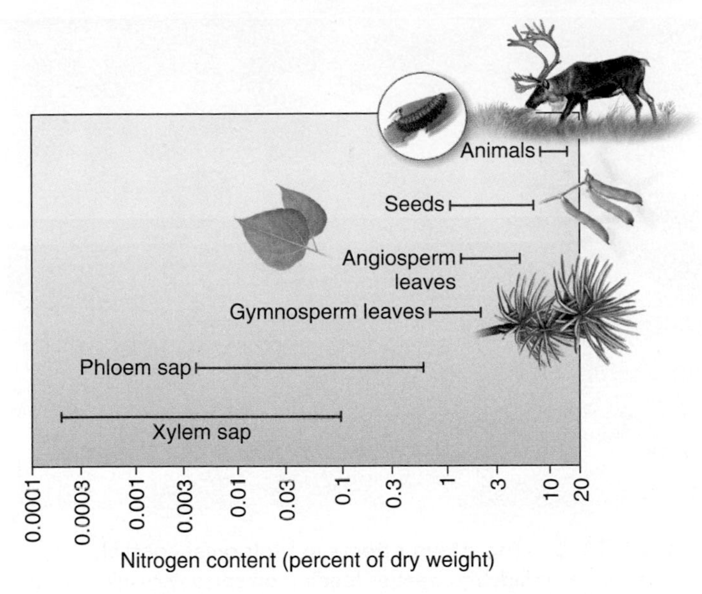


**Figure 16.2** **Nitrogen content of plants and animals.**

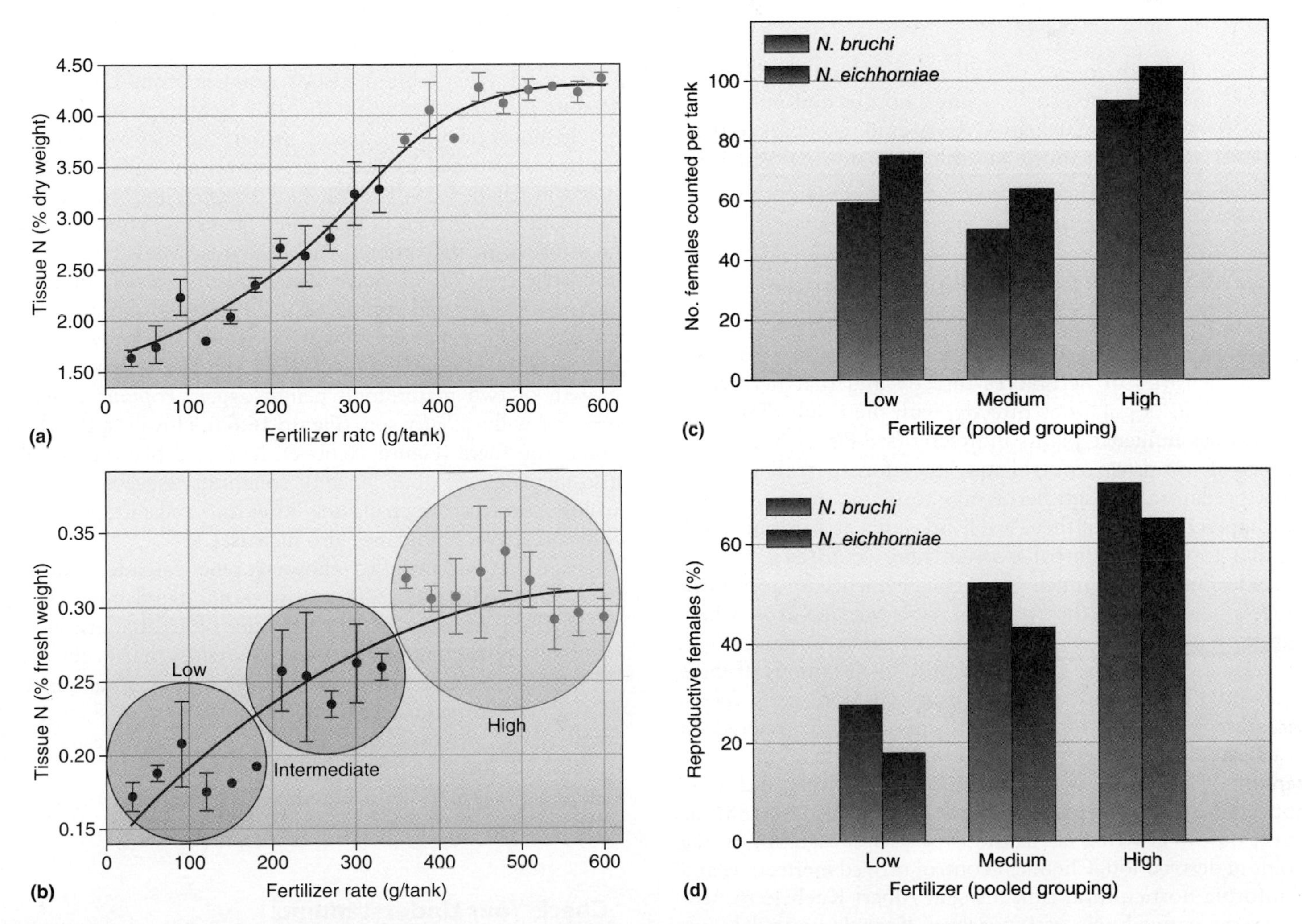


**Figure 16.3** **Bottom-up control of water hyacinth and its herbivores in Florida.** **(a, b)** As fertilizer rate of N, P, and K increases, % tissue N as dry weight or wet weight increases. Fertilizer rates were pooled as low, intermediate, and high. **(c)** Under high fertilizer rates, the highest number of female weevils was produced after one generation of growth. **(d)** The percent of reproductive female weevils was also highest under the highest fertilizer group. (After Center and Dray, 2010.)

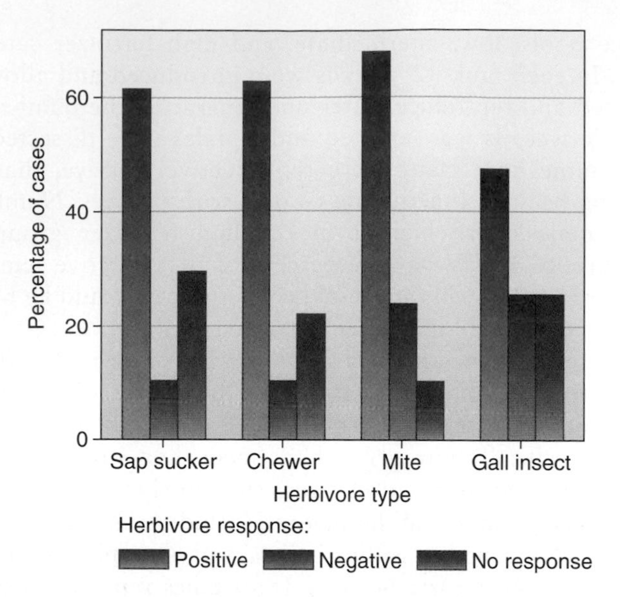


**Figure 16.4** **Plant-herbivore responses to fertilization.** Response of different types of herbivores to nitrogen fertilization. (After Waring and Cobb, 1992.)

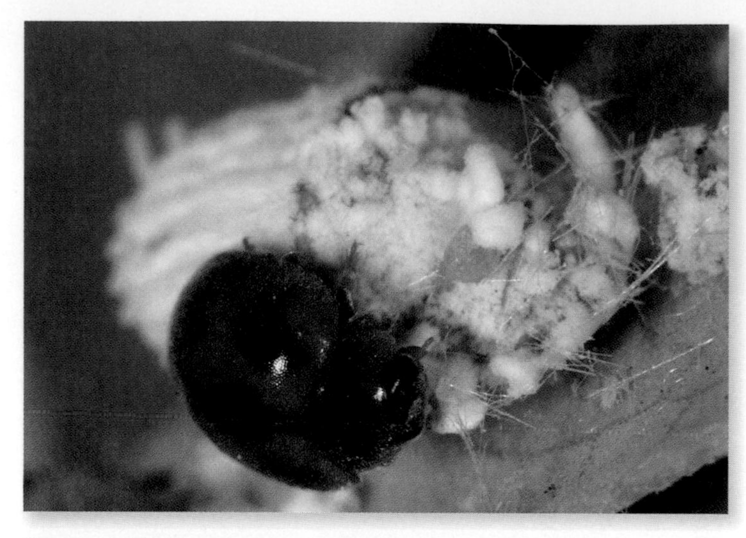

**Figure 16.5** **Top-down effects in biological control.** *Vedalia cardinalis* ladybird beetles feeding on cottony cushion scale, *Icerya purchasi,* which themselves feed on citrus. In the 1880s these beetles were released in California and almost single-handedly saved the citrus industry.

responses by herbivores to fertilization. Although the addition of other nutrients, such as phosphorous and potassium, can increase herbivore densities, the overall responses to these nutrients were much more variable, and positive responses were generally not as common as no responses or negative responses.

### 16.1.2 Top-down models suggest natural enemies affect prey abundance

Top-down models suggest that natural enemies control population densities of herbivores directly and that herbivores control plant populations directly, with the result that natural enemies influence plants indirectly (see Figure 16.1). The effects of top-down control may therefore percolate down from predators through herbivores to plants. As we will see in Chapter 25, each of these levels is known as a trophic level, so that top-down control is often referred to as a **trophic cascade**. Once again, much supporting evidence for top-down models comes from the world of biological control, where natural enemies are released to control pests of agriculture, increasing crop yields. The earliest and most famous of these cases involved control of the cottony cushion scale, *Icerya purchasi,* a pest of citrus in California. First discovered on acacia trees in northern California in 1868, this pest spread rapidly. It fed on a wide variety of host plants and soon appeared on citrus trees in southern California. By 1886 its effect on the growing citrus industry was devastating to the point of destruction. Chemical control proved ineffective, and California horticultural officials sent Albert Koebele to Australia, the native home of *I. purchasi.* Koebele located *I. purchasi*'s natural enemy, the ladybird beetle, *Vedalia cardinalis* (**Figure 16.5**), which he sent back to California in November 1888. By 1890 all infestations in the state were gone. In just one year, shipments of oranges from Los Angeles County increased from 700 to 2,000 freight car loads.

In nonagricultural systems, strong support for top-down control is provided by predator removal and addition. As we noted in Chapter 10, wolves, *Canus lupus,* are being reintroduced into many areas of the United States and populations are growing. In Banff National Park, Canada, Mark Hebblewhite and colleagues (2005) have watched colonization of certain areas by wolves and compared these areas to those closer to human settlements, which wolves avoid. The wolves reduce the densities of elk, *Cervus elaphus,* which in turn promotes growth of two major food plants, aspen, *Populus tremuloides,* and willow, *Salix* spp. (**Figure 16.6a**). However, the effects don't stop there (**Figure 16.6b–e**). Increased plant availability also increases the abundance of songbirds, especially obligate willow specialists such as the American redstart, *Setophaga ruticilla.* Beaver abundance also increases.

Aquatic systems also show trophic cascades. Ransom Myers and colleagues (2007) showed that overfishing severely depleted the densities of all 11 species of great shark including blacktip sharks, *Carcharhinus limbatus,* that occur along the eastern seaboard of the United States. This has resulted in a large increase in their main prey items, rays, skates, and small sharks. Myers and colleagues further suggested such increases in rays have resulted in a decrease in the density of their bivalve prey items, including bay scallops, *Argopecten irradians* (**Figure 16.7**).

**Check Your Understanding**

**16.1** Why might bottom-up control have logical primacy over top-down control?

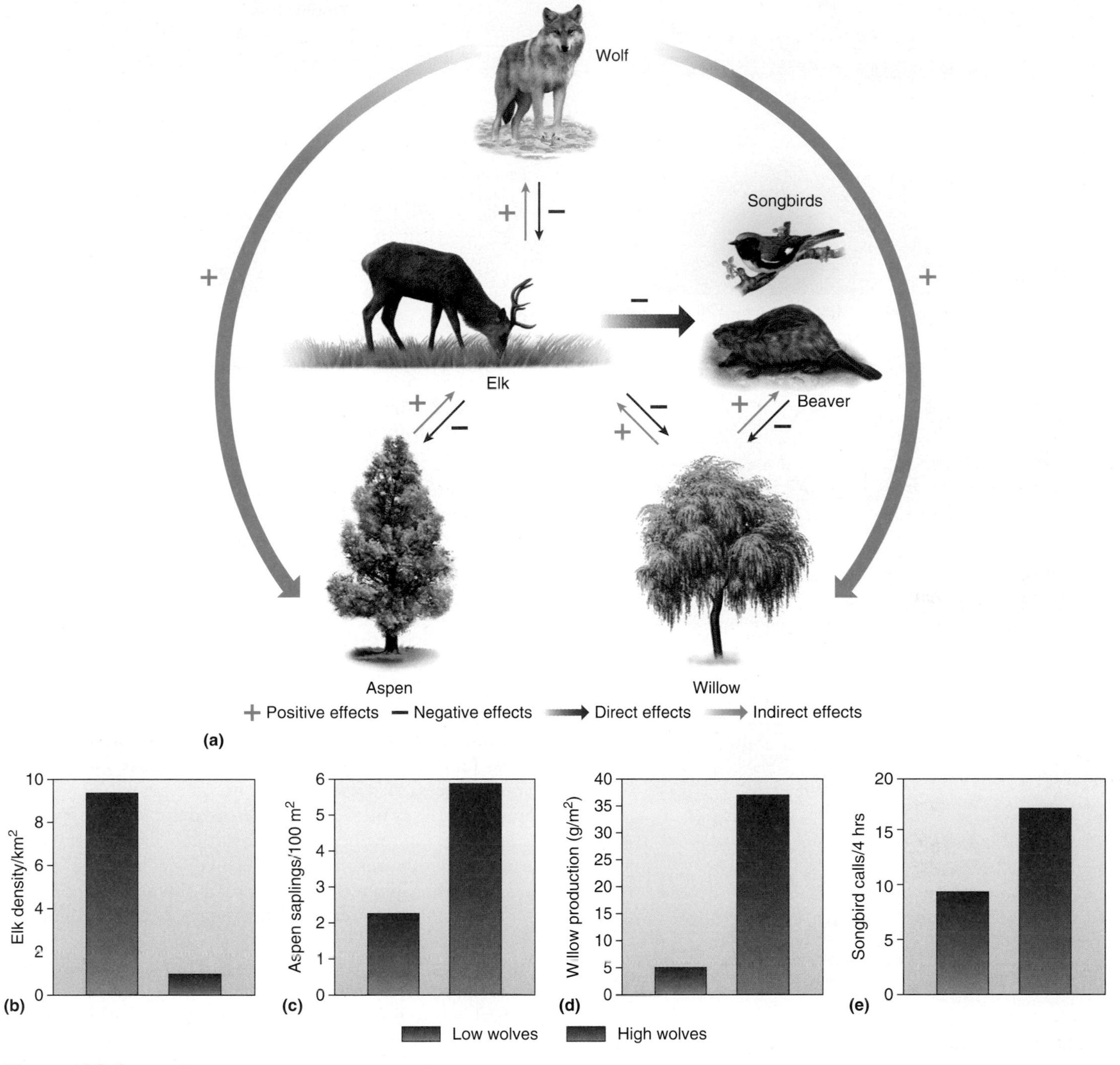


**Figure 16.6 Wolf addition studies provide support for the top-down model.** **(a)** The simplified trophic interactions of wolves, elk, plants, and other organisms at Banff National Park, Canada. Increased wolf abundance caused **(b)** decreased elk abundance, **(c)** increased abundance of aspen trees and **(d)** willow trees, and **(e)** increased songbird abundance. (From data in Hebblewhite et al., 2005.)

## 16.2 Conceptual Models Suggest That Top-Down and Bottom-Up Effects Vary in Importance in Different Environments

Several different models have been proposed to determine when top-down and bottom-up effects are likely to be important and when they are not. Among the first was the **green Earth hypothesis** of Nelson Hairston, Frederick Smith, and Larry Slobodkin (1960), which stated that because the Earth appears "green," and plants are common, herbivores must have little impact on plant abundance. They suggested that herbivores must therefore be limited by their predators. The implication was that because plants are limited not by herbivores, they must be limited by competition for resources, a bottom-up effect. Herbivores, suffering high rates of mortality

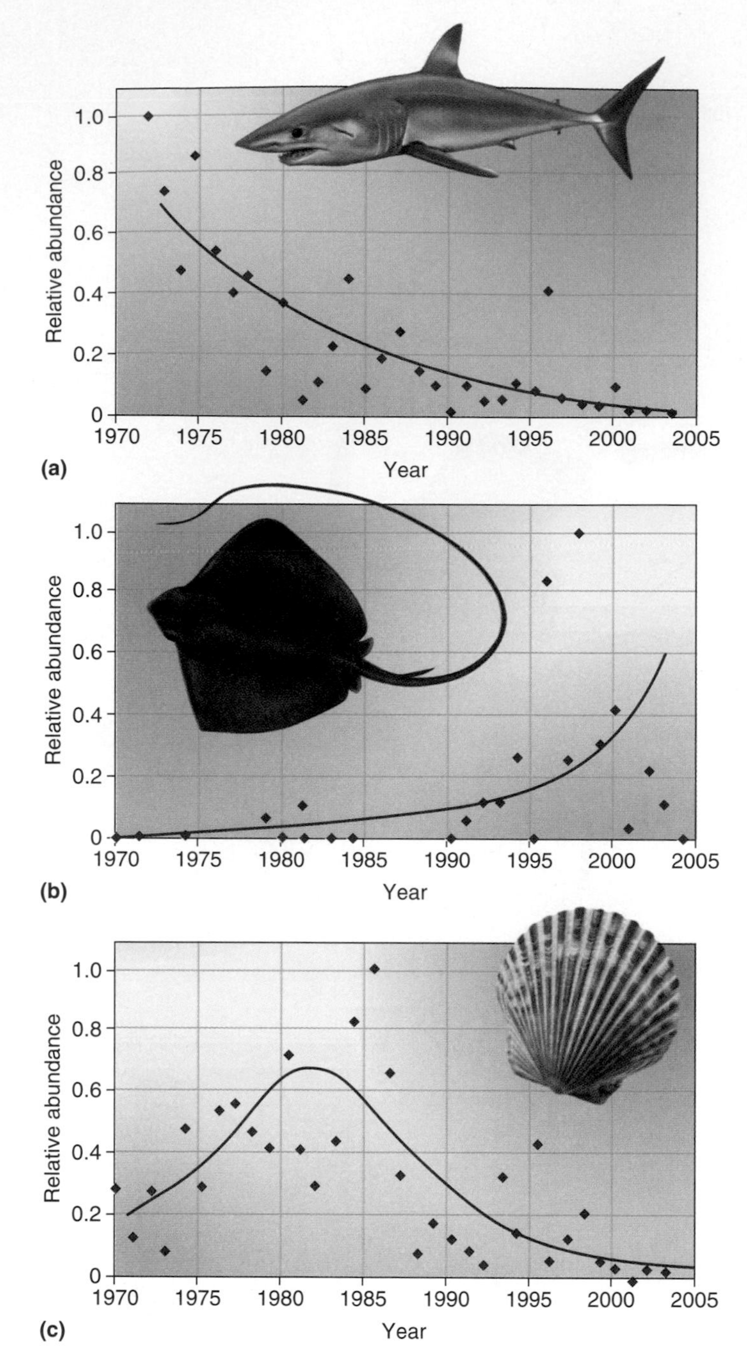


**Figure 16.7 Studies of the effects of shark removal support the top-down model.** **(a)** The removal of blacktip sharks, and others, causes **(b)** increases in rays and skates, their prey, and **(c)** a decrease in bay scallops, the prey of rays and skates. (After Heithaus et al., 2008.)

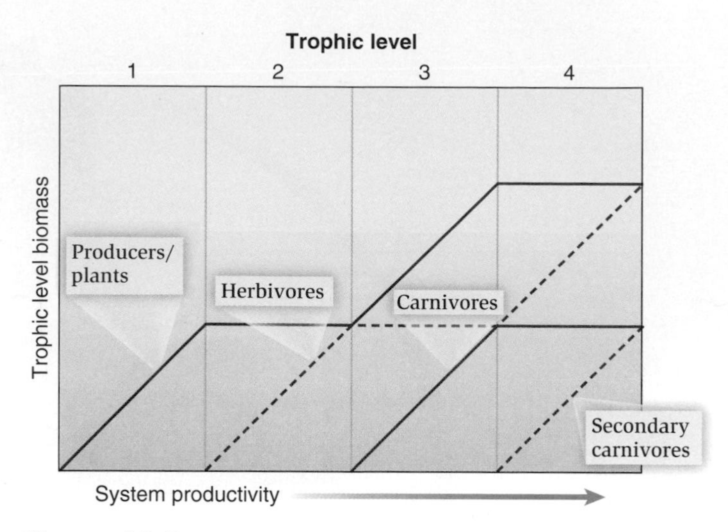


**Figure 16.8 Trophic biomass accrues with increased system productivity across four trophic levels.**

from natural enemies, exist at low densities and do not compete. Natural enemies themselves, being limited only by the availability of their prey, also compete.

More sophisticated models have built upon the green Earth hypothesis. Laurie Oksanen and coworkers (1981) suggested that the strength of mortality factors varies with plant productivity, the rate of production of plant biomass, a model they termed the **ecosystem exploitation hypothesis** (**Figure 16.8, Table 16.1**). Thus, for very simple systems, where few plants exist, not enough plant material is available to support herbivores, and plants must be resource limited, that is, limited by competition. Plant biomass increases as more resources, such as nutrients, become available. As plant biomass increases further, some herbivores can be supported, but there are too few herbivores to support carnivores. In the absence of carnivores, levels of herbivory can be quite high. Plant biomass is limited primarily by herbivory, not competition. As resources increase in the system, herbivore biomass increases but plant biomass does not. In these two-level systems, the abundance of herbivores is limited by competition for plant resources. As plant biomass increases still further and herbivores become more common, carnivores become abundant and reduce the number of herbivores, which in turn increases plant abundance. This is the green Earth scenario. Finally, as productivity increases still further, secondary carnivores can be supported, and these in turn depress the numbers

**Table 16.1 Effects of number of trophic levels on major mortality factors in natural systems.**

| Taxa | Plants only | Plants and herbivores | Plants, herbivores, and carnivores | Plants, herbivores, carnivores, and secondary carnivores |
|---|---|---|---|---|
| Plants | Competition | Herbivory | Competition | Herbivory |
| Herbivores | | Competition | Predation | Competition |
| Carnivores | | | Competition | Predation |
| Secondary carnivores | | | | Competition |

## Urbanization Can Change the Relative Effects of Top-Down and Bottom-Up Factors

Urbanization has been a continuing trend for most of recorded history. In 2007 a threshold was reached in that more people lived in urban than in rural areas. Urbanization fragments habitat, introduces exotic species, and has effects far beyond the city boundaries. Stan Faeth and his colleagues (2005) compared species interactions in Phoenix, Arizona, and in the nearby rural Sonoran Desert area. In the Sonoran Desert, plant biomass is limited primarily by water availability. Associated herbivores, are, in turn, limited mainly by plant quality and quantity (**Figure 16.A-i**). Natural enemies, such as avian predators, have relatively little effect on herbivore densities. This is a bottom-up driven system.

In urban Phoenix, plant biomass is much increased by watering. Even native Sonoran plants are generally more productive due to increased water usage, especially in dry periods when desert plants typically senesce. A higher and more consistently available plant biomass results from this increase in water resources. In turn, herbivores and their avian predators are more abundant. The increase in bird densities in suburban Phoenix translates into stronger top-down effects on herbivores. A long-term experiment in both Phoenix and the Sonoran Desert excluded birds from 40 brittlebush plants, *Encelia farinosa*, via the use of netting. The experiments were repeated in deserts, in desert remnants within the city, and in suburban yards. In the Sonoran Desert, bird exclusion had no effect on herbivore densities, which are limited by plant availability and water supply. Within the city, insect herbivores are more abundant and bird exclusion allowed herbivores to increase even more (**Figure 16.A-ii**). With increased productivity, top-down factors had increased in importance. These results show how global change via urbanization can affect species interactions. These interactions are important because many direct and indirect services are provided by living organisms in urban environments, such as pollination of home garden plants by insects, control of pests by predators and parasites, and the recycling of nutrients.

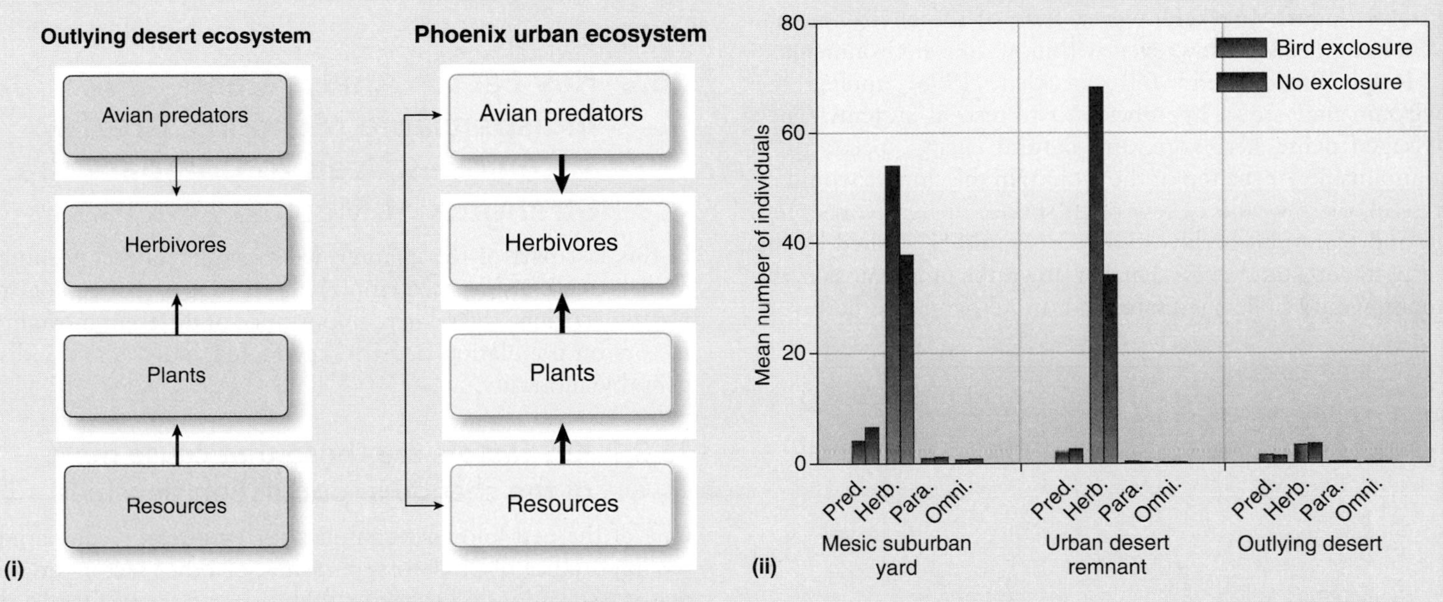


**Figure 16.A The effects of urbanization on top-down and bottom-up factors.** **(i)** In outlying, rural systems, desert plants are influenced by resources (water availability), which in turn affect other species (left). In urban systems, water subsidies increase plant growth and bird densities, and both impact herbivore densities (right). The size of the typeface in the box reflects differences in biomass, and the thickness of the arrows indicate relative importance of species interactions. **(ii)** Cage exclosures for birds had little effect in outlying desert areas. However, in mesic suburban yards and desert remnants within the city, bird-free plants supported more herbivores, but not predatory, parasitic, or omnivorous insects. (After Faeth et al., 2005.)

of carnivores, which in turn increases levels of herbivory and lessens competition between plants. Ecologists are currently examining the degree to which such models hold true in nature. Urbanization may affect the relative strengths of top-down and bottom-up factors in systems (see **Global Insight**).

The **environmental stress hypothesis** , originated by Bruce Menge and John Sutherland (1976), posits that in stressful habitats, higher trophic levels have little effect because they are rare or absent, and lower trophic levels are affected mainly by environmental stress. Being marine ecologists, Menge and Sutherland couched their arguments in terms of intertidal organisms. For example, on exposed rocky shores with severe abiotic pressures such as rough seas and high tides, predators and herbivores would be absent in the intertidal, and plants and mollusks would be limited by their ability to attach to the rock face. In environments of moderate stress there would be a little predation and herbivory, but not enough to affect plant densities. Here plant and mollusk densities would be higher and affected by competition for space on the rock face. In still more benign environments there would be many natural enemies. In low-stress environments, plant and mollusk abundance would be controlled by herbivory and predation, not competition or environmental stress. The strength of top-down factors increases as the environment becomes less stressful (**Figure 16.9**). However, tests of the environmental stress hypothesis are few, even within marine environments.

Ralph Preszler and Bill Boecklen (1996) applied the environmental stress hypothesis to terrestrial systems. They envisaged plant, herbivore, and natural enemy species on a mountainside. At the top of the mountain the climate would be stressful, there would be few plant species or herbivores present, and plant and herbivore density would be limited by the ability to withstand stress. Farther down the mountainside, the climate would be less stressful, and more plant and herbivore species would occur, increasing competition. Lower still, many polyphagous natural enemies would occur, and predation and parasitism of herbivores would be common. To test their ideas, Preszler and Boecklen investigated the different mortality factors operating on leaf miners, common plant herbivores, on oaks. Their data supported the environmental stress hypothesis and showed that parasitism of leaf miners was lowest at highest elevations, where parasites were rare, whereas host plant–induced mortality increased with elevation, consistent with the idea that plant quality decreased with environmental stress.

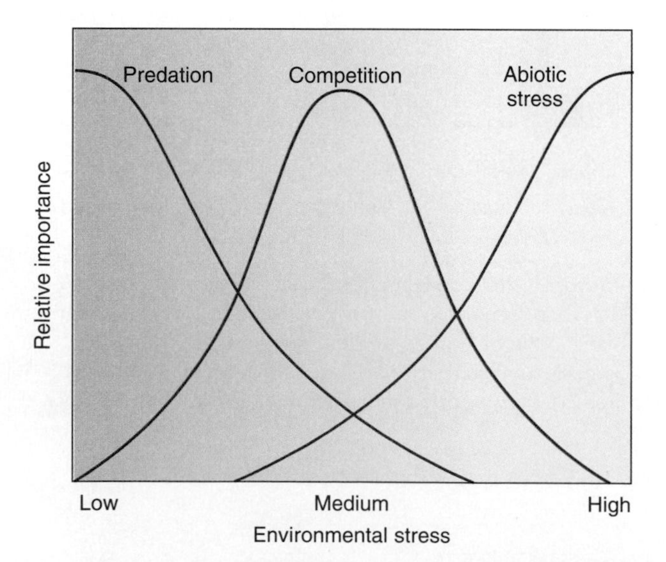


**Figure 16.9 The environmental stress hypothesis of Menge and Sutherland (1976).** The relative importance of predation, competition, and abiotic stress varies with environmental stress. (Modified from Bruno et al., 2003.)

In theory, the ecosystem exploitation hypothesis and environmental stress hypothesis offer plausible conceptual models to predict under which circumstances certain mortality factors are most important, but neither hypothesis has been widely tested. Next we will examine two empirical techniques that have been used to examine the strength of mortality factors in nature.

**Check Your Understanding**

**16.2** Researchers Juraj Halaj and David Wise (2001) found that removing top predators in four-trophic-level crop and woodland systems actually increased herbivory and reduced plant biomass. How is this possible?

## 16.3 Key Factor Analysis and Indispensable Mortality Are Two Techniques Used to Compare the Strengths of Mortality Factors

In this last part of the chapter, we move beyond conceptual models and consider two empirical techniques that ecologists use to determine the relative importance of different mortality factors on population density: key factor analysis and indispensable mortality.

### 16.3.1 Key factors are those that cause most of the change in population densities

One of the best-known techniques for empirically comparing the importance of predators, parasites, or other factors on the size of populations is key factor analysis, a technique first developed by R. F. Morris (1957) and refined by the British entomologists George Varley, G. R. Gradwell, and Michael Hassell (1973). Key factor analysis requires detailed information on the fate of a cohort of individuals: total mortality of a generation or cohort ($K$) is subdivided into its various causes, and the relative importance of these is compared. Here we will consider key factor analysis on the oak winter moth, the subject of Varley, Gradwell, and Hassell's work that we introduced in Chapter 1 and a species probably subject to the most comprehensive key factor analysis ever performed (**Table 16.2, Figure 16.10**).

Varley, Gradwell, and Hassell made detailed censuses of all stages of the oak winter moth on five oak trees at Wytham Wood in Berkshire, England. The adult moths emerge from

**Table 16.2** Life table for the oak winter month, 1955–1956.

| Life history stage/ mortality | Number alive per $m^2$ | Log number alive per $m^2$ | $k$-value |
|---|---|---|---|
| **Adult females, 1955** | 4.39 | | |
| Eggs (= females × 150) | 658 | 2.82 | |
| Larvae (after overwintering loss) | 96.4 | 1.98 | $0.84 = k_1$ |
| Killed by *Cyzenis* | 90.2 | 1.95 | $0.03 = k_2$ |
| Killed by other parasites | 87.6 | 1.94 | $0.01 = k_3$ |
| Killed by microsporidian disease | 83.0 | 1.92 | $0.02 = k_4$ |
| Pupae | 83.0 | 1.92 | |
| Killed by predators | 28.4 | 1.45 | $0.47 = k_5$ |
| Killed by *Cratichneumon* | 15.0 | 1.18 | $0.27 = k_6$ |
| **Adult females, 1956** (= half of the pupae) | 7.5 | | |

After Varley, Gradwell, and Hassell, 1973.

pupae in the soil in December. As the females crawl up the trunks of oak trees, the males mate with them. Thereafter the females lay eggs in crevices in the bark of branches. Varley and Gradwell counted the number of females per tree by using traps of fabric fastened around the tree trunk. Exactly one quarter of the tree was covered in this way, so the trap catch multiplied by four equaled the total numbers of females per tree. Next, the canopy area of the tree was calculated using the shadow of the tree area at noon when the sun is directly overhead, and the number of moths caught was divided by the canopy area to get a number per square meter. Varley and Gradwell dissected the females they caught in the traps to get the average number of eggs per female moth—about 150. They therefore knew the number of females and eggs per tree and per square meter of canopy.

Young winter moth larvae emerge from the eggs in the late winter/early spring so that they can feed on the newly developing foliage, which at this time has the lowest levels of antiherbivore chemicals, such as tannins (see Figure 14.5). The rewards of feeding on tannin-free leaves are high, but often larvae emerge so early that they appear prior to bud

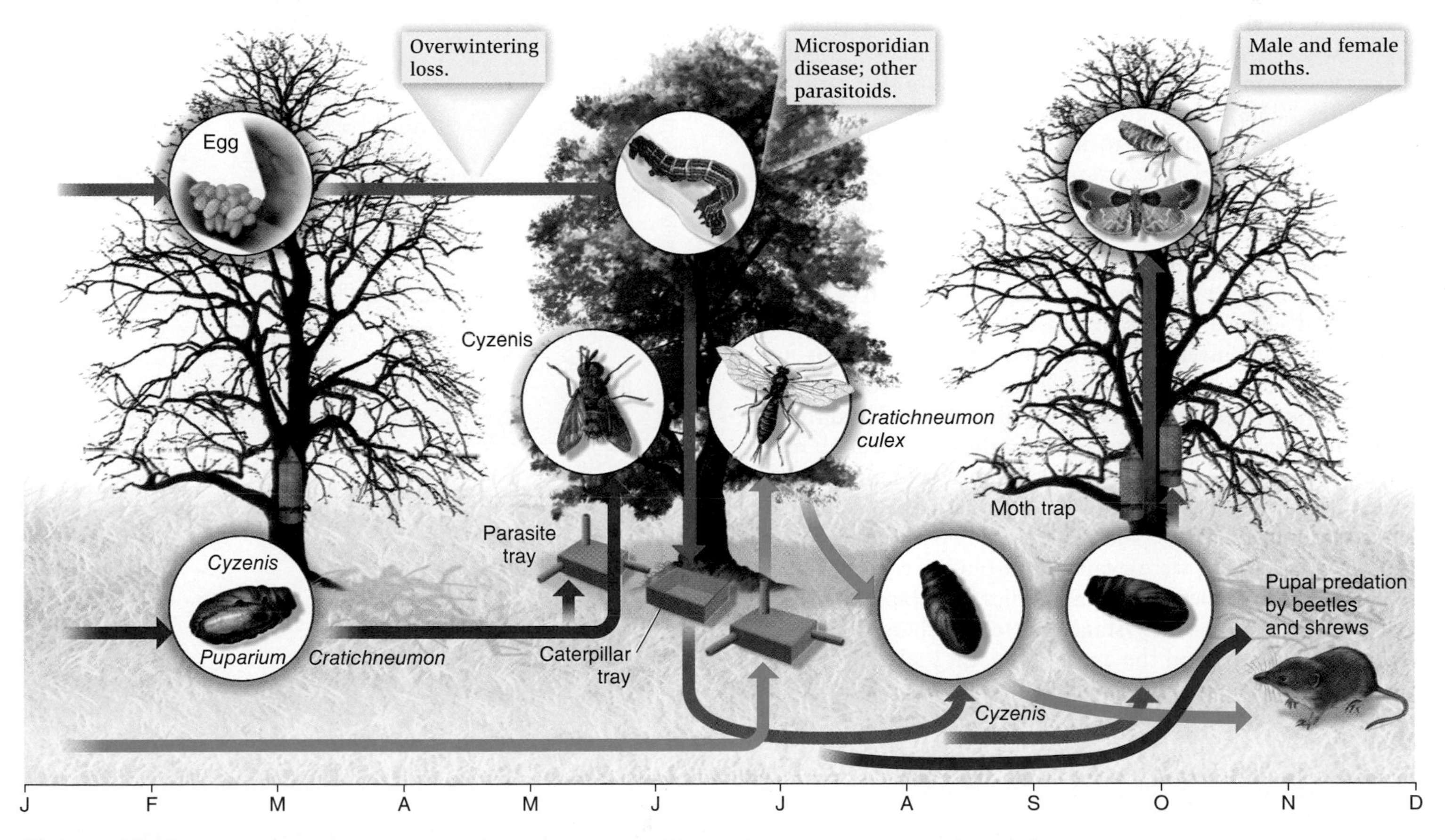


**Figure 16.10 Life cycle of the oak winter moth and sampling methods.** Adult, wingless, female winter moths, which climb up trees to lay their eggs, were counted in moth traps on the tree trunks and their eggs were dissected. Caterpillars, which occur on the leaves, were counted in the trays when they dropped to the ground to pupate. Larvae of the parasite *Cyzenis,* which lay their eggs on the foliage, and which are unwittingly devoured by the caterpillars, were counted by dissection of the fallen caterpillars. Adult *Cratichneumon,* which directly parasitize pupae, were counted upon emergence from the soil into the parasite traps. Pupal predation by soil-dwelling beetles and shrews was also estimated from these techniques. The different-colored arrows reflect the life cycles of the different organisms involved. (Redrawn from Varley, 1971.)

burst and face starvation. If there are no leaves present on the tree, the caterpillars disperse by ballooning away on silken threads, with a possibility of landing on trees where the leaves have already flushed. This is obviously a chancy business, and many larvae are lost; hence the mortality from dispersal is high. This loss was known as overwintering loss.

To count the caterpillars, the researchers relied on passive sampling to get estimates of caterpillar numbers. Fully grown caterpillars leave the foliage and spin down on silken threads to pupate in the soil. By putting two metal traps on the ground under each tree, each of 0.5 $m^2$, they could estimate the number of caterpillars per square meter. The traps filled with rainwater and the caterpillars were drowned, to be picked up and examined later. Caterpillars were dissected for parasites so that the number of healthy and sick caterpillars per square meter of canopy was determined. There is one main type of larval parasitoid: a tachnid fly called *Cyzenis*. The fly lays its eggs on oak foliage and the eggs are ingested by feeding caterpillars, in whose body they hatch. On average, only about 5–10% of caterpillars were infected with *Cyzenis*. Other parasites, mainly wasps, take a more active role in searching out caterpillars and inserting their ovipositors to lay eggs inside. However, their level of parasitism was even smaller, about 2–3%. A microspordian disease also infects some larvae but at a low level, about 4–5%. All this information was obtained from dissections of the drowned caterpillars in the traps.

After falling to the ground, the caterpillars pupate in the soil, where they are subject to additional mortality. First, a parasitoid of the pupae, *Cratichneumon*, can cause 40–50% mortality. Second, predators such as shrews and beetles eat 60–70% of all pupae, including parasitized ones. Thus, between the two, they cause 76–85% mortality at this stage. Later, Varley and colleagues inverted the trays they had used to catch caterpillars and poked a hole in the top to which they attached a small tube. They placed the inverted traps on the ground where any moth or wasp hatching from a pupa in the ground would emerge, move toward the light, and be caught in the tube. The difference in the numbers of caterpillars entering the soil and adult moths or parasitoids emerging was due to pupal predation. *Cyzenis* parasitism could also be estimated by counting the adult wasps emerging into the traps, and thus, for this parasitoid two estimates of parasitism were available—from dissections of caterpillars and from emergence of adult flies into traps.

At each life history stage, the researchers tracked the number of deaths and the cause of death, such as parasitism or predation. The importance of each mortality factor ($k$) was estimated by calculating the logarithm of the amount that the factor, $k$, reduced population size:

$$k = \log N_t - \log N_{(t+1)}$$

where $N_t$ is the density of the population before it is subjected to the mortality factor and $N_{(t+1)}$ is the density afterward.

So, for example, from Table 16.2, the killing power of the pupal parasitoid, *Cratichneumon*, in 1955–1956 is:

$$k_6 = \log 28.4 - \log 15.0 = 1.45 - 1.18 = 0.27$$

For larvae of the oak winter moth, mortality from overwintering loss is the largest:

$$k_1 = \log 658 - \log 96.4 = 2.82 - 1.98 = 0.84$$

Total generational mortality, $K$, can then be defined as the sum of the individual mortality factors $k_1$ through $k_n$:

$$K = k_1 + k_2 + k_3 + k_4 + \ldots + k_n$$

If this analysis is repeated for multiple generations, then graphs can be constructed of the total generation mortality ($K$) and the individual sources of mortality. This is done by plotting generational mortality $K$, and all its submortalities $k_1$ through $k_n$, for as many years as the study was done. The source of mortality that most closely mirrors overall generation mortality ($K$), by visual comparison, is then termed the **key factor**.

For oak winter moths, overwintering loss is the key factor (**Figure 16.11a**). For more precision, individual sources

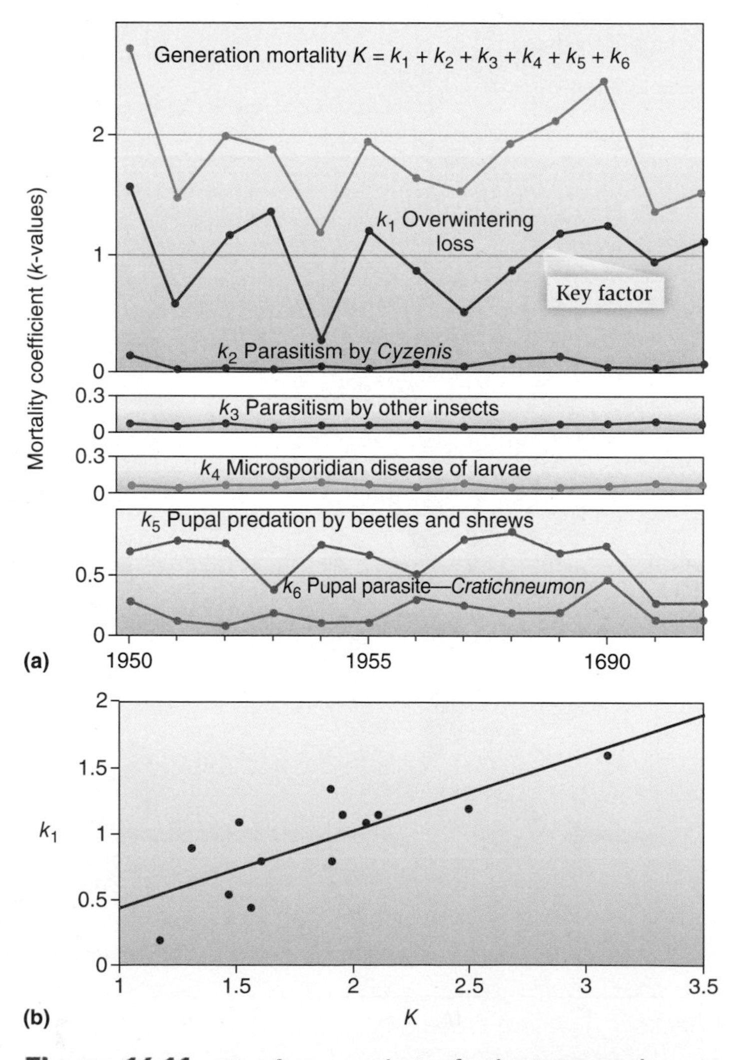


**Figure 16.11 Key factor analysis of oak winter moth populations.** **(a)** Changes in individual mortalities, expressed as $k$ values, show that the biggest contribution to changing the generation mortality $K$ comes from changes in winter disappearance. (Redrawn from Varley et al., 1973.) **(b)** Correlation of $k_1$, overwintering loss, with $K$ yields a straight-line relationship indicative of a key factor.

of mortality or $k$ values can be plotted on the y-axis against total mortality, $K$, on the x-axis, and the key factor is then the source of mortality with the biggest correlation, or line of best fit, with $K$ (**Figure 16.11b**). However, visual comparison is often preferred so that one can exclude a $k$ that is correlated with $K$ but makes only a small contribution to its variation.

Key factor analysis has been used to determine whether imported biological control agents are impacting host populations (see **Feature investigation**). For other animals, key factors are many and varied, ranging from overwintering loss to juvenile starvation. Unfortunately, few generalizations can be made as to which type of key factors operate on which types of population, because there are many different key factors operating on different life stages of organisms (**Table 16.3**).

Although key factor analysis has been an important tool for ecologists, it is not without problems. Tomo Royama (1996) provided several criticisms. First, key factors cannot always be precisely linked to specific mortality agents. Thus, overwintering loss in the winter moth may actually be due to egg death, as well as death by ballooning. The contribution of each of these to $k_1$, overwintering loss, is unknown, but it is possible that such a division could reduce the correlation between $k_1$ and $K$ so much that it would disappear. Often, census data are only good enough to determine key factor phases in life cycles, such as juvenile mortality or adult mortality. Second, there may be interactions between natural enemies, and such effects fail to show up in key factor analysis. For example, if *Cyzenis* had been the key factor in the winter moth life cycle, its own abundance may have been influenced by a bacterial parasite, but the importance of such an agent would not have been detected. Third, populations can be hugely influenced by migration—for example, by dispersal of egg-bearing females into or out of a population. Such a factor does not show up in key factor analysis.

**Table 16.3 Key factors for a variety of animals.**

| Organisms | Key factor |
|---|---|
| Oak winter month *Operopherata brumata* | **Overwintering Loss** |
| Tawny owl *Strix aluco* | Loss of birds outside the breeding season |
| African buffalo *Syncerus caffer* | Adult mortality |
| Partridge *Perdix perdix* | Chick mortality |
| Great tit *Parus major* | Variation in egg clutch size |
| White admiral butterfly *Ladoga camilla* | Larval predation |
| Small white butterfly *Pieris rapae* | Larval disease |
| Meadow spittlebug *Philaenus spumarius* | Weather |
| Wildebeest *Connochaetes taurinus* | Newborn mortality due to malnutrition of mother |
| Fir engraver beetle *Scolytus ventralis* | Larval starvation |

# Feature Investigation

## Munir and Sailer Used Key Factor Analysis to Examine the Success of an Imported Biological Control Agent

The tea scale, *Fiorinia theae,* is a sap-sucking pest of tea plants in Asia. The scale has been introduced into many warm areas of the world, including Florida, where it is a pest of ornamental plants, especially camelias, *Camelia japonica.* In 1976 an attempt was made to control tea scale in Florida by importing an insect parasitoid, *Aphytis theae,* from India. Badar Munir and Reece Sailer (1985) conducted a key factor analysis on multiple generations in 1977 and 1978 to determine whether the parasitoid was acting as a key factor. To do this they needed a thorough understanding of the biology of the scale.

The tea scale sits motionless under a protective armor-like covering and appears like a miniature scale on the leaf. Males are gnatlike insects with one pair of wings, whose sole purpose is to find and fertilize females. Females remain motionless and mature their eggs under their body armor, which is sealed to the leaf. When the eggs hatch, the armor is lifted off at the posterior end and the young nymphs, called crawlers, disperse, often only a few centimeters, to find new areas of the leaf, or other leaves, where they settle and develop. Settlers are soft bodied and fairly easy prey for other insects. Later they develop their armor coating and are known as nymphs. The nymphs feed for a relatively long time before turning into male or female scales known as prereproductives. In warm countries, many generations are passed in a year. Mortality may be caused by death of crawlers, $k_1$; mortality of settlers due to predators, $k_2$; mortality of nymphs due to parasitoids, such as *A. theae*, $k_3$; and mortality of prereproductive males and females due to parasitoids feeding on and killing them, $k_4$ (**Figure 16.B**).

**HYPOTHESIS** An imported biological control agent, *Aphytis theae,* will act as a key factor in the life cycle of the tea scale.

**STARTING LOCATION** Gainesville, Florida, USA

| | Conceptual level | Experimental level |
|---|---|---|
| 1 | Count number of tea scales on camelia leaves. | Ten infested camellia leaves were picked each month. Under a microscope in the laboratory, a 3 $cm^2$ area was marked and examined for all life stages of the scale, from crawler through settler, nymph, and adult. |
| 2 | Determine mortality of crawlers $k_1$. | Count the number of eggs under each female and the number of crawlers, the difference is mortality of crawlers from wind and rain. |
| 3 | Determine mortality of settlers $k_2$. | Count number of settlers and the difference between the number of crawlers and settlers is the number of settlers killed by predatory mites and thrips. |
| 4 | Determine mortality of nymphs $k_3$. | Count numbers of nymphs parasitized by introduced *A. theae* and by local parasitoids. Parasitized scale appear dark, in contrast to lighter living nymphs. |
| 5 | Determine mortality of prereproductives $k_4$. | Count the number of dead prereproductives killed by predation from feeding parasitoids. |

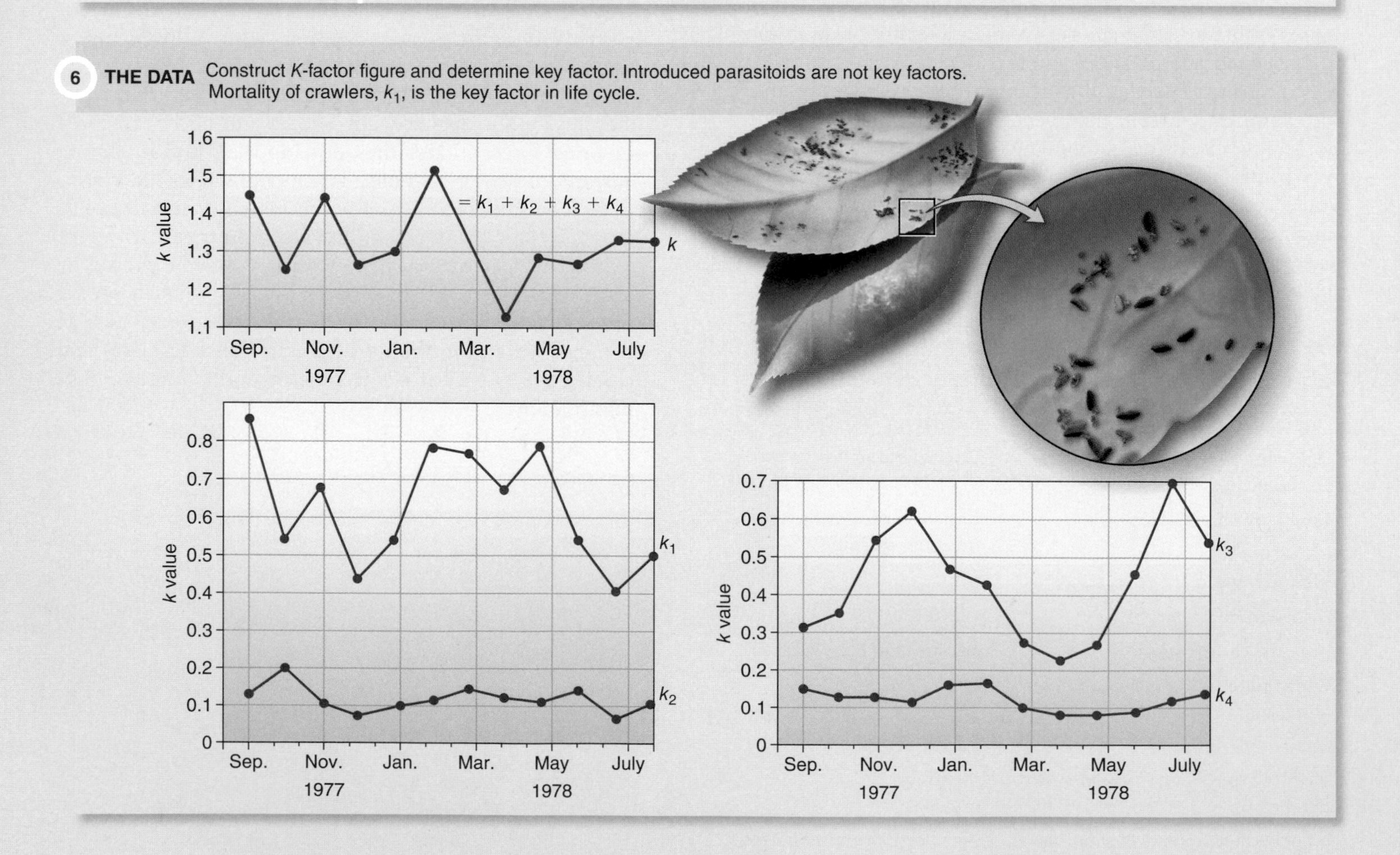


**Figure 16.B** Does a biological control agent act as a key factor?

### 16.3.2 Indispensable mortality is the amount of mortality from one factor that cannot be replaced by mortality from another factor

Sir Richard Southwood (1978) outlined other ways to look at the strength of mortality factors. One of the most important of these is indispensable mortality. Indispensable mortality is that part of the generation mortality that would not occur should the mortality factor in question be removed from the life system, after allowance is made for the action of subsequent mortality factors.

Let's examine the value of protecting sea turtle nests as a conservation measure (**Table 16.4**). If there are 1,000 eggs, let's assume 500 are normally killed by predators, such as raccoons that dig up eggs or birds that intercept hatchlings on their way to the ocean. This equates to 50% mortality. We'll assume that once the hatchlings are in the ocean, a further 400 are eaten by fish predators before they can develop into juveniles. This is equivalent to 80% mortality at this stage. Let's say of the remaining 100 juveniles, 90 are lost to fishing nets before they become reproductive adults. This is 90% mortality of juveniles. Ten turtles survive out of the original 1,000 eggs, for a total mortality of 99%. How would the population fare if we protected all the nests with cages to stop the raccoons and transported the hatchings to the ocean? Then 1,000 young turtles would enter the sea, where they would be reduced 80% by fish to 200. A further 90% would be lost to nets, leaving 20. This is two times the original number surviving but perhaps not what you'd expect from protecting so many eggs initially. Now total mortality is 98%, so indispensable egg mortality is 99 − 98 = 1%. Nearly all of the mortality that raccoons and birds would have inflicted has been made up by other factors. We would do better by protecting the juvenile turtles from fishing nets, because then 100 turtles would survive for a total mortality of only 90%. Indispensable mortality at the juvenile stage would then be 9% (**Figure 16.12**). This is why there is a U.S. federal law that requires the inclusion of turtle exclusion devices, trap doors that turtles can push open and escape through, in fishing nets. Deborah Crouse and her colleagues (1987) showed that protecting sea turtles in their adult reproductive condition yields better results than protecting the very young, especially if there are high mortalities at later stages, as in the turtle example. For conservation biologists, it means that conservation should focus on saving reproductive adults. For pest biologists, it means control techniques should focus on killing reproductive-age individuals. Robert Peterson and colleagues (2009) calculated the indispensable mortality from life tables of 28 insect species, 11 of which were non-native. Many of these 28 species were pest species against which biological control agents, such as parasitoids, predators, or pathogens had been released. The indispensable mortality from pathogens, predators, and parasitoids was only 18.6%, much less than that from nonnatural enemies, which was 35.1% (**Figure 16.13**). The nonnatural enemies category consisted of mostly abiotic factors, such as unfavorable temperatures, but also host plant quality factors, such as death from poor-quality host plants. The indispensable mortality from introduced natural enemies was 5.2%. Peterson and colleagues suggested that such low rates contribute to the low rates of success in biological control. As noted in Chapter 15, only 16% of biological control measures attempted so far qualify as economic successes.

**Table 16.4 Indispensable mortality for sea turtles.**

| | Stage | | | |
|---|---|---|---|---|
| Measure | Eggs | Hatchlings | Juveniles | Adults |
| $n_x$ (number alive) | 1,000 | 500 | 100 | 10 |
| $d_x$ (number dying) | 500 | 400 | 90 | |
| % mortality in stage | 50 | 80 | 90 | |
| % indispensable mortality | 1 | ? | 9 | |

Determining indispensable mortality has additional applications. Imagine that hunters take 20% of a deer population. Is this too high a percentage? It depends on whether this mortality is indispensable or not. If a population of deer is close to the carrying capacity and a cold winter would likely lead to starvation and death for 20% of the population, hunters might argue they were only taking the 20% of the population destined to starve anyway. In this scenario, hunting acts as a **compensatory mortality**. An alternative scenario is that 20% of the population dies from winter starvation no matter what the population size. In other words, hunting would add to the total mortality. In this scenario, hunting is acting as an **additive mortality**. Weather often acts as an additive mortality because it kills a fixed proportion of individuals, no matter how many have been killed earlier by natural enemies and disease. This is why weather can be so devastating. There has not been sufficient work on determining whether mortalities are generally additive or compensatory, although in many cases hunting acts as a compensatory mortality. This is because surviving individuals are faced with less competition for resources, and they often produce more offspring.

The overall conclusion from conceptual models of top-down and bottom-up effects, key factor analysis, and indispensable mortality is that the question of what influences population density requires much work for each species in question. There are no easy generalizations that ecologists can make before collecting the data. Each species, and even each population, requires a detailed study to determine what is influencing its density.

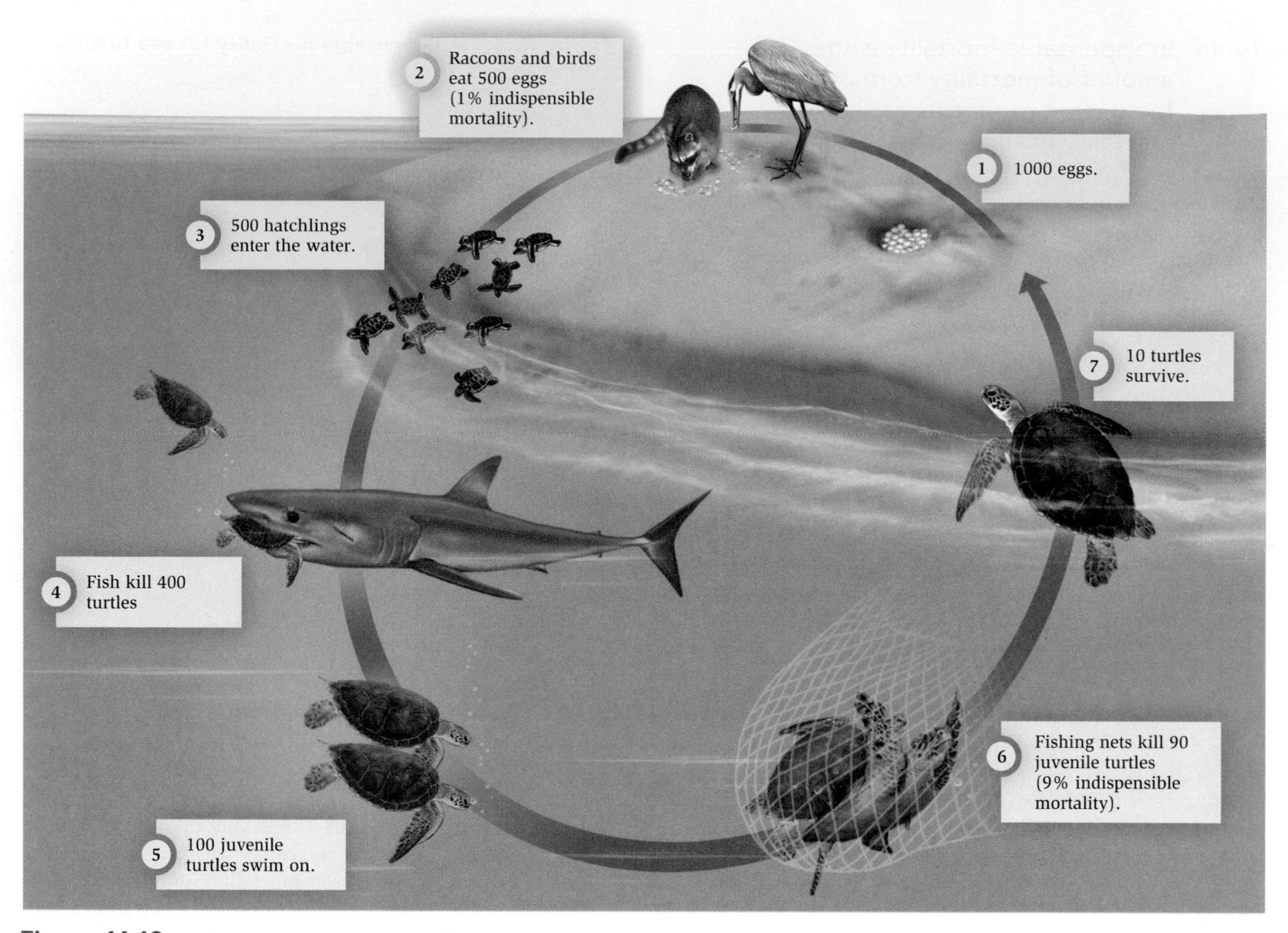


**Figure 16.12 Indispensable mortality in the life cycle of sea turtles.** Protecting eggs from egg predators may not do as much good as protecting juveniles and adults from fishing nets.

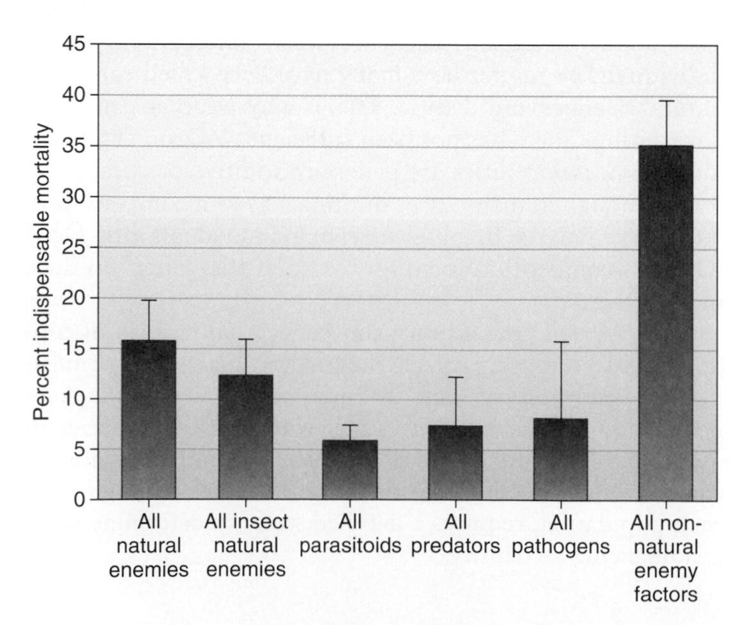


**Figure 16.13 Indispensable mortalities in insects by mortality types.** (After Peterson et al., 2009.)

## Check Your Understanding

**16.3** Let's return to the sea turtle example and imagine the same scenario as outlined in Table 16.4, except that we protect the hatchlings from fish predators by use of nets strung across open bays. This permits the turtles to grow into juveniles. What is the indispensable mortality in the hatchling stage?

## SUMMARY

**16.1 Both Bottom-Up and Top-Down Effects Are Important in Natural Systems**

- Conceptual models of species interactions underscore the importance of both bottom-up and top-down factors (Figure 16.1).
- Bottom-up models propose that plant quality or quantity regulates the densities of herbivore and predator species. Populations of most species are nitrogen limited (Figures 16.2–16.4).
- Top-down models propose that the densities of many species are suppressed by their natural enemies, and these effects cascade down through various trophic levels (Figures 16.5–16.7).

**16.2 Conceptual Models Suggest That Top-Down and Bottom-Up Effects Vary in Importance in Different Environments**

- The ecosystem exploitation hypothesis proposes that mortality varies according to system productivity (Table 16.1, Figure 16.8).
- The effects of humans can change the relative strengths of top-down and bottom-up factors—for example, through urbanization (Figure 16.A).
- The environmental stress hypothesis suggests environmental stress levels influence the strength of both bottom-up and top-down effects (Figure 16.9).

**16.3 Key Factor Analysis and Indispensable Mortality Are Two Techniques Used to Compare the Strengths of Mortality Factors**

- Key factor analysis and indispensable mortality are two empirical techniques we can use to compare the relative strengths of top-down and bottom-up effects on populations (Tables 16.2–16.4, Figures 16.10–16.13, Figure 16.B).

## TEST YOURSELF

1. Which type of tissue contains the highest concentrations of nitrogen?
   a. Xylem
   b. Phloem
   c. Leaves
   d. Seeds
   e. Animal
2. Wolf predation at Banff National Park, Canada, _____ songbird densities and _____ beaver densities.
   a. decreases, decreases
   b. decreases, increases
   c. increases, decreases
   d. increases, increases
3. According to the ecosystem exploitation hypothesis (EEH), in a four-trophic-level system, _____ would be the most important mortality for plants and _____ would be most important for herbivores.
   a. competition, competition
   b. competition, predation
   c. herbivory, competition
   d. herbivory, predation
4. According to the original environmental stress hypothesis of Menge and Sutherland, competition is most important at which level of environmental stress?
   a. Low
   b. Medium
   c. High
   d. a and b
   e. b and c
5. A trophic cascade implies the presence of:
   a. Bottom-up control
   b. Top-down control
   c. Strong competition
   d. Nutrient limitation
   e. No natural enemies
6. Cod feed on small fish and crabs, which in turn feed on zooplankton. Zooplankton feed on phytoplankton, which themselves use up nitrates in the water. If cod were overfished, which things should decrease?
   a. Zooplankton
   b. Small fish and crabs
   c. Phytoplankton
   d. a and d
   e. b and c
7. If $k_1 = 0.2$, $k_2 = 0.1$, $k_3 = 0.8$, and $k_4 = 0.4$ What is $K$?
   a. 0.1
   b. 0.5
   c. 1.0
   d. 1.5
   e. 1.7
8. What is the key factor in oak winter moth populations?
   a. Parasitism by *Cyzenis*
   b. Parasitism by other insects
   c. Microsporidian disease of larvae
   d. Pupal predation by beetles and shrews
   e. Overwintering loss
9. In an insect pest, if there are 5,000 eggs, 2,000 larvae, 1,000 pupae, and 100 adults, what is the percentage indispensable mortality at the larval stage?
   a. 1%
   b. 2%
   c. 3%
   d. 5%
   e. 10%
10. Adding fertilizer to plants often increases the numbers of insect herbivores. This supports the top-down view of population control.
    a. True
    b. False

## CONCEPTUAL QUESTIONS

1. Distinguish between top-down and bottom-up effects on populations.
2. What is key factor analysis, and how does it work?
3. Contrast a key factor with a density-dependent factor (discussed in Chapter 10).

## DATA ANALYSIS

1. The following data are from a key factor analysis of an insect over a 16-year period: $k_{oa}$ is mortality caused from reduction in the number of eggs laid, caused by predation of adults before egg laying and reduction in eggs laid per female. $k_{ob}$ is egg infertility and egg predation. $k_1$ through $k_4$ are mortalities of larval stages 1, 2, 3, and 4, from predation and larvae dropping off the plant. $K$ is the total generational mortality. Plot the data and determine the key factor.
2. Daniel Moon and Peter Stiling (2002) compared the strength of top-down and bottom-up control on two sap-sucking insects that fed on saltmarsh plants: *Prokelisia marginata,* which fed on saltmarsh cordgrass, *Spartina alterniflora,* and *Pissonutus quadripustulatus,* which fed on *Borrichia frutescens.* Both species lay their eggs in the plant stems, where they are attacked by a tiny parasitoid, *Anagrus* spp., a so-called fairy fly, which develops entirely within the eggs. Egg parasitism is by far the greatest source of mortality. Fairy flies are attracted to the color yellow; yellow cards coated with a sticky adhesive, like that on flypaper, can capture these parasitoids, which decreases parasitism rates. Moon and Stiling manipulated both these systems using the four same treatments: (1) fertilize the plants, (2) employ many sticky traps to reduce egg parasitism, (3) fertilize and use sticky traps together, and (4) leave some unmanipulated control plots. They counted the numbers of sap-sucking insects every month until midsummer and determined the foliar nitrogen content of the plants on all types of plots (see Figure). What did they find?

| Year | $k_{0a}$ | $k_{0b}$ | $k_1$ | $k_2$ | $k_3$ | $k_4$ | $K$ |
|---|---|---|---|---|---|---|---|
| 1963/64 | 1.3749 | 0.0079 | 0.1551 | 0.0953 | 0.0672 | 0.1902 | 1.8906 |
| 1964/65 | 0.9645 | 0.0022 | 0.2278 | 0.0032 | 0.0481 | 0.1214 | 1.3672 |
| 1965/66 | 0.3240 | 0.0287 | 0.6103 | 0.0115 | 0.0924 | 0.1574 | 1.2243 |
| 1966/67 | 0.8548 | 0.0097 | 0.3950 | 0.0012 | 0.0257 | 0.5673 | 1.8537 |
| 1967/68 | 0.5935 | 0.0097 | 0.4653 | 0.0015 | 0.0101 | 0.1597 | 1.2398 |
| 1968/69 | 1.2221 | 0.0043 | 0.3343 | 0.0103 | 0.0500 | 0.2801 | 1.9011 |
| 1969/70 | 0.9195 | 0.0128 | 0.1343 | 0.1135 | 0.0684 | 0.2323 | 1.4808 |
| 1970/71 | 1.0385 | 0.0075 | 0.1548 | 0.0593 | 0.0554 | 0.2332 | 1.5487 |
| 1971/72 | 0.9072 | 0.0141 | 0.1592 | 0.0611 | 0.0479 | 0.2420 | 1.4316 |
| 1972/73 | 0.8964 | 0.0071 | 0.2045 | 0.0417 | 0.0857 | 0.2666 | 1.5020 |
| 1973/74 | 1.0059 | 0.0066 | 0.2831 | 0.0682 | 0.0689 | 0.1481 | 1.5808 |
| 1974/75 | 1.2556 | 0.0072 | 0.3805 | 0.0442 | 0.0775 | 0.1707 | 1.9357 |
| 1975/76 | 1.0122 | 0.0086 | 0.1559 | 0.0324 | 0.0397 | 0.2273 | 1.4761 |
| 1976/77 | 1.2357 | 0.0085 | 0.1361 | 0.0371 | 0.0754 | 0.2194 | 1.7122 |
| 1977/78 | 1.1218 | 0.0085 | 0.2109 | 0.0143 | 0.0107 | 0.2868 | 1.6530 |
| 1978/79 | 1.4355 | 0.0085 | 0.2109 | 0.0172 | 0.0565 | 0.3098 | 2.0384 |

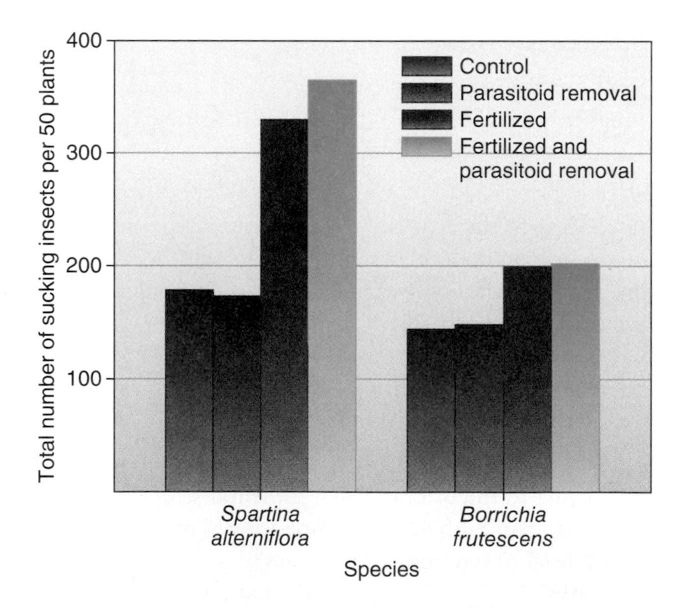

SECTION FIVE

# Community Ecology

So far, we have examined ecology in terms of the behavior of individual organisms, the growth of populations, and interactions between species. Most populations, however, exist not on their own but together with populations of many other species. This assemblage of many populations that live in the same place at the same time is known as a **community**. For example, a forest community consists of not only the tree species and other vegetation, but also pollinating insects, herbivores that feed on the vegetation, and predators and parasites of the herbivores. Communities can occur on a wide range of scales and can be nested. The forest community also encompasses smaller communities, such as an individual fallen and rotting tree, which forms a complete habitat for various species of insects, fungi, and bacteria that decompose the wood. Both of these entities, the forest and the rotting tree, are viable communities, depending on one's frame of reference with regard to scale.

**Community ecology** focuses on the factors that influence the number of species in a community. Ecologists explore the distribution of species, such as why, on a global scale, the number of species is usually greatest in the Tropics and declines toward the poles. Many conservation organizations, such as the World Wildlife Fund, strive to protect areas rich in plant and animal species. Knowing what processes affect the numbers of species in an area is therefore vital to the conservation process, but conservationists are also interested in the preservation of areas where there are sizeable populations of different species, not just a few individuals of a high number of species. In this section we compare a number of diversity indices designed to measure community diversity using data on the numbers of species and the distribution of individuals between species. Biologically diverse communities have a large number of species and a relatively equitable distribution of individuals among species.

Species richness is also an important concept, because it is thought that species-rich communities function more efficiently than those that are species-poor. For example, species-rich communities may better resist invasion by introduced species. In this section we also examine the link between species richness and community function. However, ecologists recognize that communities may change. For example, after a physical disturbance such as a fire, recovery often occurs in a predictable way that ecologists have termed succession. In certain situations, such as on islands recovering from physical disturbance, succession has been postulated to occur via waves of colonization of species from neighboring landmasses, followed by extinctions of some species. In this situation, the theory of island biogeography holds that island size and distance from the mainland govern the process of succession. While the model was developed to describe the factors affecting the species richness of island communities, it is now used to describe processes in any community surrounded by an unlike community, such as parks in an urban area. In the last two chapters of this section, Chapters 20 and 21, we examine the concepts of succession and island biogeography.

A waterhole community at Etosha National Park, Namibia, Africa. Species include greater kudu, giraffe, zebra, and various birds.

CHAPTER 17

# Species Diversity

## Outline and Concepts



There is currently much interest in preserving the Earth's biological diversity or **biodiversity**. Biodiversity can be considered on several levels, including genetic diversity, the genetic variation within species (refer back to Chapter 2); community diversity, the variation among communities (also look ahead to Chapters 22–24); and species diversity, the diversity among species. In 1710, tin ore was discovered by Dutch colonists on the Indonesian island of Bangka. Over the centuries the mining industry engaged in deforestation on a massive scale both to allow access to the tin ore and to use the trees as a source of fuel for the smelters. Deforestation has left only about 3% of the island's primary forest intact. Relatively recently, some companies have attempted minimal restoration of mining land by simply revegetating with seedlings planted into a small hole, filled with natural topsoil collected from nearby. Ecologists are interested in knowing whether the restored areas accrue a higher community diversity than unrestored sites. This involves counting individuals, species, and seeing how individuals are spread out among species. At first glance, researchers noted the same biodiversity in restored as unrestored sites. For example, both had nine recorded bird species. However, restored sites had more individual birds and these individuals were spread more evenly among species. Restoration did appear to increase biodiversity, at least for birds.

In this chapter we will discuss the nature of communities and how to quantify the diversity they contain. We explore some techniques used to measure species diversity, including diversity indices, rank abundance diagrams, and similarity coefficients. **Diversity indices** are a measure of the number of species in an area and the relative distribution of individuals among those species. **Rank abundance diagrams** are graphical plots of numbers of individuals per species against rank of species commonness in the community. **Similarity indices** compare how areas may be similar in biodiversity, in terms of the numbers of species they hold in common. Each of these biodiversity measures can be important in different circumstances, depending on the question of interest.

## 17.1 The Nature of Communities Has Been Debated by Ecologists

Ecologists have long held differing views on the nature of communities and how they are structured and function. Some of the initial work in the field of community ecology considered a community to be a collection of species equivalent to a superorganism, in much the same way that the body of an animal is more than just a collection of organs. In this view, termed the **organismic model**, individuals, populations, and communities have a relationship to each other that resembles the associations found between cells, tissues, and organs. American botanist Frederic Clements (1905), the champion of this viewpoint, suggested that ecology was to the study of communities what physiology was to the study of individual organisms. Clements's view of the community envisaged predictable and distinct associations of species, separate communities, separated by sharp boundaries. Clements was well aware that communities sometimes changed in response to physical disturbances, but he argued that this change occurred in a predictable way and that the original community always returned. In this view, competition between plant species fueled these changes.

Clements's ideas were challenged by botanist Henry Allan Gleason (1926). Gleason proposed an **individualistic model**, which viewed a community as an assemblage of species coexisting primarily because of similarities in their physiological requirements and tolerances. While acknowledging that some assemblages of species were fairly uniform and stable over a given region, Gleason suggested that distinctly structured ecological communities usually do not exist. Instead, species are distributed independently along an environmental gradient. Viewed in this way, communities do not necessarily have sharp boundaries, and associations of species are much less predictable and integrated than in Clements's organismic model.

By the 1950s many ecologists began to closely examine Clement's and Gleason's contrasting views about communities. Paramount among these studies were those of Robert Whittaker, who asserted the **principle of species individuality**, which stated that each species is distributed according to its physiological needs, that most communities intergrade continuously, and that competition does not create distinct vegetational zones. For example, let's consider an environmental gradient such as a moisture gradient on an uninterrupted slope of a mountain. Whittaker proposed that four hypotheses could explain the distribution patterns of plants and animals on the gradient (**Figure 17.1**):

1. Competing species, including dominant plants, exclude one another along sharp boundaries. Other species evolve toward a close, perhaps mutually beneficial association

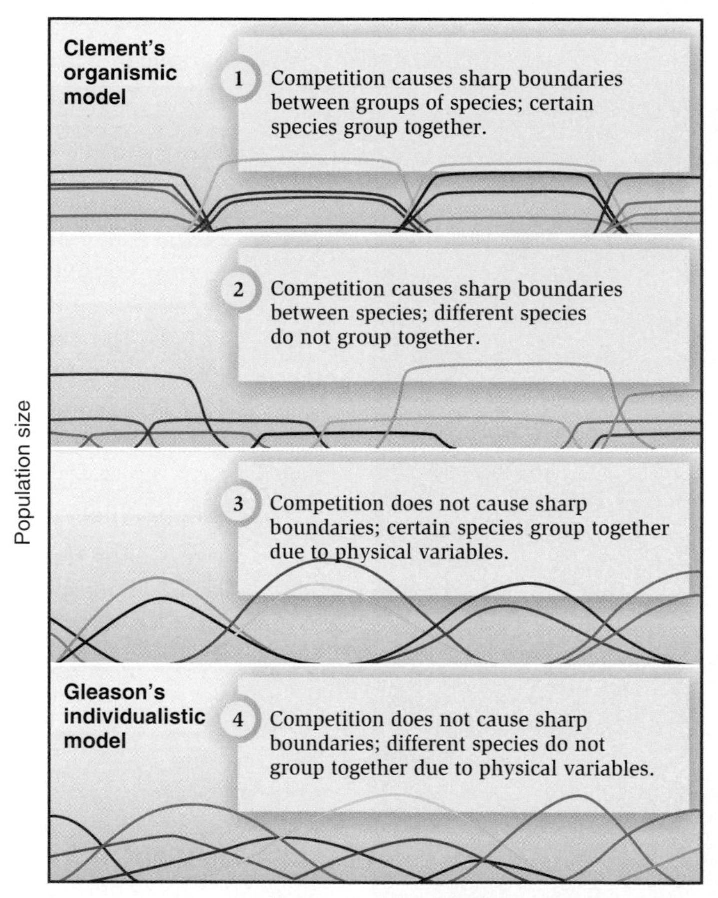


**Figure 17.1 Four hypotheses on how populations might relate to one another along an environmental gradient.** Each curve in each part of the figure represents one species and the way its population might be distributed along an environmental gradient such as a mountain slope.

**ECOLOGICAL INQUIRY**

Which scenario might best explain community development in areas with alternating serpentine and nonserpentine soil? (Refer back to Chapter 7 for the effects of serpentine soil on plant communities.)

with the dominant species. Communities thus develop along the gradient, each zone containing its own group of interacting species giving way at a sharp boundary to another assemblage of species. This corresponds to Clements's organismic model.
2. Competing species exclude one another along sharp boundaries but do not become organized into groups of species with parallel distributions.
3. Competition does not usually result in sharp boundaries between species. However, the adaptation of species to similar physical variables will result in the appearance of groups of species with similar distributions.
4. Competition does not usually produce sharp boundaries between species, and the adaptation of species to similar physical variables does not produce well-defined groups of species with similar distributions. The centers and boundaries of species populations are scattered along the environmental gradient. This corresponds to Gleason's individualistic model.

To test these possibilities, Whittaker (1970) examined the vegetation on various mountain ranges in the western U.S. He sampled plant populations along an elevation gradient from the tops of the mountains to the bases and collected data on physical variables, such as soil moisture.

The results supported the fourth hypothesis, that competition does not produce sharp boundaries between species and that adaptation to physical variables does not result in defined groups of species. Whittaker concluded that his observations agreed with Gleason's predictions that (1) each species is distributed in its own way, according to its genetic, physiological, and life cycle characteristics; and (2) the broad overlap and scattered centers of species populations along a gradient implies that most communities grade into each other continuously rather than form distinct, clearly separated groups. Whittaker's observations showed that the composition of species at any one point in an environmental gradient was largely determined by factors such as temperature, water, light, pH, and salt concentrations (factors discussed in Chapters 5–7). Yet it is true that the community at the top of a mountain is often much different from that at the bottom, so at broad scales, distinguishing associations of species can be useful. Such broad associations are very dependent on climate. Furthermore, there are sometimes abrupt changes from one community to another. These mostly correspond to abrupt abiotic changes, such as changes in soil conditions (refer back to Figure 7.2). We can compare the numbers of species in these communities and the distribution of individuals between species, giving us different values of biodiversity.

**Check Your Understanding**

**17.1** What is Whittaker's principle of species individuality?

## 17.2 Various Indices Have Been Used to Estimate Species Biodiversity

The simplest measure of biodiversity is **species richness**, which is a count of the number of species. Although many analyses of community diversity are performed this way, one major problem is that this approach does not take into account species frequency of occurrence or **relative abundance**. For example, imagine two hypothetical communities, A and B, both with two species and 100 individuals, as shown below.

| | Community A | Community B |
|---|---|---|
| Number of individuals of species 1 | 99 | 50 |
| Number of individuals of species 2 | 1 | 50 |

The species richness of community B equals that of community A, because both communities contain two species. However, community B is considered more diverse because the distribution of individuals between species is more even. One would be much more likely to encounter both species in community B than in community A, where one species dominates.

Species richness is very susceptible to sample size—the greater the number of individuals sampled, the higher the number of species recorded. Let's consider another example. You have been charged to find the most species-rich forest in the Appalachians, in which to establish a regional park. In one forest you count 100 trees and find 10 individuals of each of 10 species (community 1 in **Table 17.1**). In a second forest you also find 10 species in 100 trees counted but here the distribution is less equitable. Most of the trees are of one species (species j in Table 17.1). Community 1 is therefore more diverse than community 2. To measure biodiversity one must incorporate both species richness and the distribution of individuals among species in diversity indices. A great many of these indices exist, but we can divide them into two broad categories: dominance indices and information statistic indices.

### 17.2.1 Dominance indices are more influenced by the numbers of common species

Dominance indices are weighted toward the abundance of the commonest or dominant species. A widely used dominance index is Edward Simpson's diversity index (1949), which gives the probability of any two individuals drawn at random from a community belonging to different species. For example, the probability that two trees picked at random from a tropical forest will be the same species would be low, because there are many different tree species in any given tropical forest. In a boreal forest in Canada, where there are

**Table 17.1 Species richness and species abundance.**

The total number of species, called the species richness, and the total number of individuals, called species abundance, in two hypothetical communities, 1 and 2. Species richness and abundance in both communities are identical—both have 10 species and 100 individuals. However, individuals are more evenly distributed in community 1, so we say that community 1 is more diverse.

| | Individuals per species | |
|---|---|---|
| Species | Community 1 | Community 2 |
| a | 10 | 5 |
| b | 10 | 5 |
| c | 10 | 5 |
| d | 10 | 5 |
| e | 10 | 5 |
| f | 10 | 5 |
| g | 10 | 5 |
| h | 10 | 5 |
| i | 10 | 5 |
| j | 10 | 55 |
| Total individuals | 100 | 100 |
| Total number of species | 10 | 10 |

**ECOLOGICAL INQUIRY**

What is the Shannon index for community 2 (see page 345)?

only one or two species of tree in the forest, the chances of both individuals being of the same species would be relatively high. We calculate Simpson's diversity index, $D_S$, using the following formula:

$$D_S = \sum_{i=1}^{S} p_i^2$$

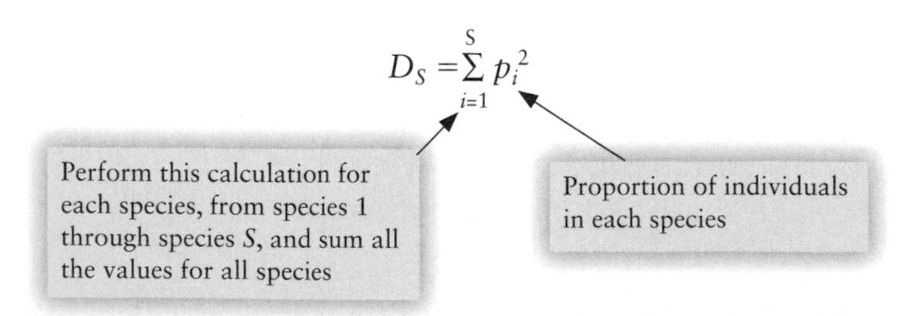


where $p_i$ is the proportion of individuals in the $i$th species. For example, let's imagine a community of five tree species in an area where there are 200 individual trees. For species 1, the number of individuals, $n = 100$, and the number of individuals of all tree species in the forest, $N = 200$. The proportion of species 1 in the community is then $0.5^2 = 0.25$. Two worked examples with communities of five tree species follows:

| Tree species | Number of individuals in Community A | $p_i$ | $p_i^2$ |
|---|---|---|---|
| 1 | 100 | 0.5 | 0.25 |
| 2 | 50 | 0.25 | 0.0625 |
| 3 | 30 | 0.15 | 0.0225 |
| 4 | 19 | 0.095 | 0.009025 |
| 5 | 1 | 0.005 | 0.000025 |
| Total | 200 | | |
| $D_S$ | | | 0.34405 |
| **Tree species** | **Number of individuals in Community B** | $p_i$ | $p_i^2$ |
| 1 | 100 | 0.5 | 0.25 |
| 2 | 100 | 0.5 | 0.25 |
| 3 | 0 | 0 | 0 |
| 4 | 0 | 0 | 0 |
| 5 | 0 | 0 | 0 |
| Total | 200 | | |
| $D_S$ | | | 0.50 |

The problem with the Simpson index is that more-diverse communities have a lower value of $D_S$ than less-diverse communities. To express greater diversity with a numerically greater value, we can use a complement form of the index by subtracting the value from 1 to obtain the Simpson complement index of diversity. This value is called the *Gini-Simpson index of diversity* ($D_{GS}$), where $D_{GS} = 1 - D_S$. Now the more-diverse community, A, actually does have a higher index of diversity than the less-diverse community, B. In our examples, the Gini-Simpson index for community A = 1 − 0.34405 = 0.65595, and for community B = 1 − 0.50 = 0.50. Values of the Gini-Simpson index range from 0 to 1. The disadvantage of the Gini-Simpson index is that it is again heavily weighted toward the most abundant species. The addition of many rare species of trees with one individual each will scarcely change the index. This is obvious from examining the contribution of tree species 5 in community A of the preceding table to the overall value of the index—it is 0.000025 or 0.007%. As a result, the Gini-Simpson index is of limited value in conservation biology if there are many rare species.

C. V. Chittibabu and N. Parthasarathy (2000) examined tree diversity in disturbed and human-impacted tropical forest in the Eastern Ghats, India. They examined diversity using the Simpson index on a series of 2-ha sites, and the results are converted here to the Gini-Simpson index. Tree diversity was greatest at site PS, an undisturbed site, but declined at more-disturbed sites where tree harvest, cattle grazing, and firewood collection were permitted (**Figure 17.2**). At the most-disturbed sites, cultivation of horticultural crops such as pineapple, banana, and others, together with soil excavation for cement factories, reduced tree diversity further and

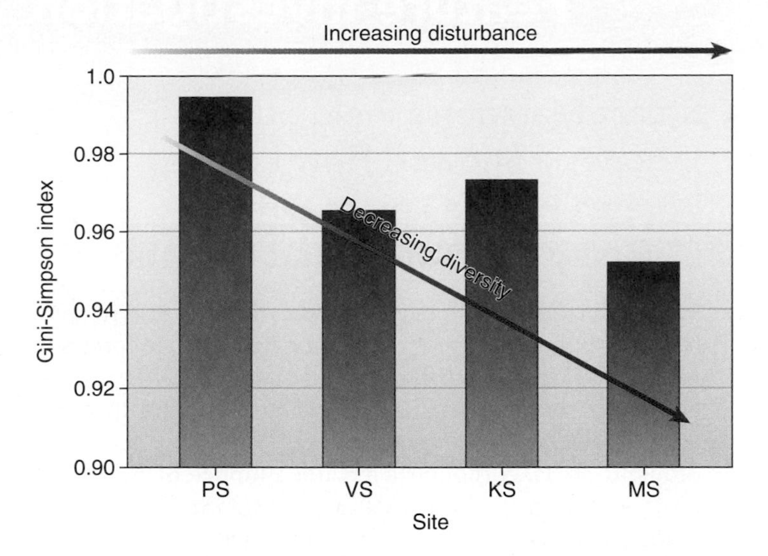


**Figure 17.2 Effects of disturbance on diversity of tree species in Indian tropical forest, as measured by the Gini-Simpson index.** (After data in Chittibabu and Parthasarathy, 2000.)

sites came to be dominated by a few tree species that could withstand these disturbances.

## 17.2.2 Information statistic indices are more influenced by the numbers of rare species

Information statistic indices are based on the rationale that diversity in a natural system can be measured in a way that is similar to measuring the information in a code. By analogy, if we know how to calculate the uncertainty of the next letter in a coded message, then we can then use the same technique to know the uncertainty of the next species to be found in a community. A message consisting of bbbbb has a low uncertainty because the next letter is virtually certain to be a b. A message consisting of bhwzj has a high uncertainty and hence a high index value. The higher the index value, the higher the uncertainty of being able to tell the next letter in the sequence, or the next species in a community. This means the community is diverse. The most commonly used information statistic index is the Shannon index.

The Shannon index, $H_S$, is given by

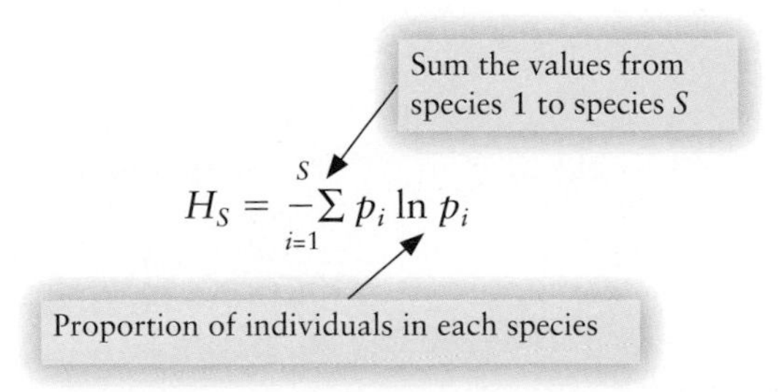


$$H_S = -\sum_{i=1}^{S} p_i \ln p_i$$

where $p_i$ is the proportion of individuals found in the $i$th species, ln is the natural logarithm, and $\Sigma$ is a summation sign. For example, for a species in which there are 50 individuals out of a total of 100 in the community, $p_i$ is 50/100, or 0.5. The natural log of 0.5 is −0.693. For this species, $p_i \ln p_i$ is then $0.5 \times -0.693 = -0.347$. For a hypothetical community with 5 species and 100 total individuals, the Shannon diversity index could be calculated as follows:

| Species | Abundance | $p_i$ | $\ln p_i$ | $p_i \ln p_i$ |
|---|---|---|---|---|
| 1 | 50 | 0.5 | −0.693 | −.347 |
| 2 | 30 | 0.3 | −1.204 | −.361 |
| 3 | 10 | 0.1 | −2.303 | −.230 |
| 4 | 9 | 0.09 | −2.408 | −.217 |
| 5 | 1 | 0.01 | −4.605 | −.046 |
| Total 5 | 100 | 1.00 | | −1.201 |

In this example, even the rarest species, species 5, contributes some value to the index. If a community had many such rare species, their contributions would accumulate. This makes the Shannon diversity index very valuable to conservation biologists, who are often interested in the number of rare species. Remember too that the negative sign in front of the summation changes the summed value to a positive number, so the index actually becomes 1.201, not −1.201. A positive number is more appealing than an index with a negative number. Values of the Shannon diversity index for real communities often fall between 0.5 and 3.5, with the higher the value, the greater the diversity. For example, Lisando Heil and colleagues (2007) documented Shannon diversity indices of between 0.5 and 1.2 for bird communities in the Andes. In addition, they noted that the presence of tourists in the high Andes scared many birds away, reducing diversity (**Figure 17.3**). Diversity of birds close to trails used by humans was much lower than diversity away from the trails.

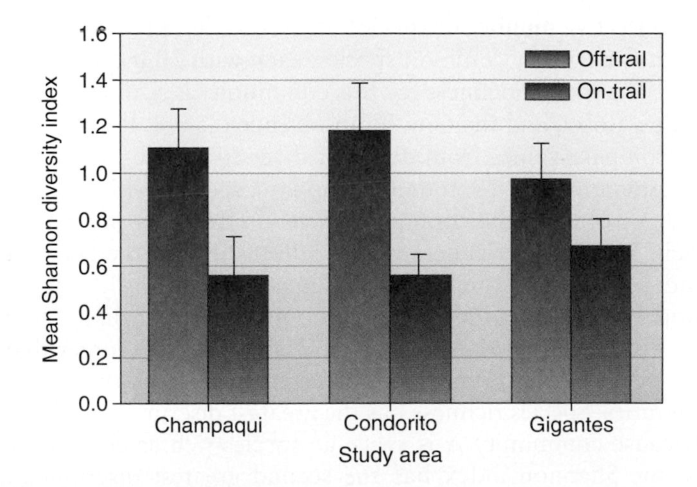


**Figure 17.3 Bird diversity in the Andes.** Here diversity is measured by the Shannon diversity index in two habitat types: undisturbed (off trail, no humans) and disturbed (on trail, with humans). (After Heil et al., 2007.)

Stuart Marsden used the Shannon index to show that the diversity of birds was significantly decreased in logged forests compared to unlogged forests in Indonesia (see **Feature Investigation**).

### 17.2.3 The effective number of species is a conceptually appealing measure of diversity

Robert MacArthur (1965) and more recently Lou Jost (2006) argued that traditional diversity indices are not easy to interpret or compare, and they championed another measure called

| Community A | | Community B | |
|---|---|---|---|
| Species | Species Abundance | Species | Species Abundance |
| A | 20 | A | 20 |
| B | 20 | B | 20 |
| C | 20 | C | 20 |
| D | 20 | D | 20 |
| E | 20 | E | 20 |
| Species richness | 5 | F | 20 |
| $H_S$ | 1.609 | G | 20 |
| $D_{GS}$ | 0.8 | H | 20 |
| | | I | 20 |
| | | J | 20 |
| | | Species richness | 10 |
| | | $H_S$ | 2.305 |
| | | $D_{GS}$ | 0.9 |

the **effective number of species**, $D_E$. Consider a community A with five equally common species, each with 20 individuals:

The species richness for this community is 5, the Shannon index 1.609, and the Gini-Simpson index is 0.8. It is difficult to compare values from different diversity indices. If we had a community of 10 equally abundant species with 20 individuals of each, the community would effectively be twice as rich. The species richness would reflect this, but the Shannon index, of 2.305, and the Gini-Simpson index, of 0.9, would not. The Shannon index of 2.305 is 43.2% larger than 1.609, and the Gini-Simpson index of 0.9 is only 12.5% larger than 0.8. Each index discriminates differently between these communities. Species richness has the greatest discriminant ability, because community A is twice as species-rich as community B, the Shannon index has the second greatest discriminant ability, with a 43.2% difference between communities, and the Gini-Simpson is a distant third. However, all indices can be converted to an effective number of species, an intuitively appealing species equivalent. Each index can be converted to an effective number of species using a different formula.

## Feature Investigation

### Stuart Marsden's Field Studies in Indonesia Showed How Logged Forests Have a Lower Bird Diversity Than Unlogged Forests

Stuart Marsden (1998) compared the diversity of bird communities in logged forests and unlogged, pristine forests in Indonesia (**Figure 17.A-i, ii**). To document diversity, Marsden established census stations in the two types of forest and recorded the identity and density of all bird species in different areas. Although a greater number of individual birds was seen in the logged areas (2,358) compared to unlogged areas (1,824), a high percentage of the individuals in the logged areas, 38.6%, belonged to just one species, *Nectarinia jugularis* (**Figure 17.A-iii**). While only one more bird species was found in the unlogged area than in the logged area, calculation of the Shannon diversity index showed an increased diversity of birds in the unlogged area, 2.284 versus 2.033 (**Table 17.A-iv**).

**HYPOTHESIS** Bird diversity is different in logged forests and unlogged forests in Indonesia.

**STARTING LOCATION** Seram, Indonesia

| | Conceptual level | Technique level |
|---|---|---|
| 1 | Locate areas to count birds | Establish 115 census stations along paths and water courses in logged and unlogged forests. |
| 2 | Census numbers of bird species and individuals at census stations. | Count perched or stationary birds at each census station for 10 minutes. Birds in flight were not counted due to the possibility of misidentification. |
| 3 | Compare bird diversity using Shannon diversity index. | Combine data from all census stations in logged or pristine forests for use in calculations. |

**4 THE DATA** See Table

| | Unlogged | | | Logged | | |
|---|---|---|---|---|---|---|
| **Species** | $N$ | $p_i$ | $p_i \ln p_i$ | $N$ | $p_i$ | $p_i \ln p_i$ |
| *Nectarinia jugularis*, Olive-backed sunbird | 410 | .225 | −.336 | 910 | .386 | −.367 |
| *Ducula bicolor*, Pied imperial pigeon | 230 | .126 | −.261 | 220 | .093 | −.221 |
| *Philemon subcorniculatus*, Grey-necked friarbird | 210 | .115 | −.249 | 240 | .102 | −.233 |
| *Nectarinia aspasia*, Black sunbird | 190 | .104 | −.235 | 120 | .051 | −.152 |
| *Dicaeum vulneratum*, ashy flowerpecker | 185 | .101 | −.232 | 280 | .119 | −.253 |
| *Ducula perspicillata*, White-eyed imperial pigeon | 170 | .093 | −.221 | 180 | .076 | −.196 |
| *Phylloscopus borealis*, Arctic warbler | 160 | .088 | −.214 | 140 | .059 | −.167 |
| *Eos bornea*, Red lory | 88 | .048 | −.146 | 73 | .031 | −.108 |
| *Ixos affinis*, Golden bulbul | 76 | .042 | −.133 | 31 | .013 | −.056 |
| *Geoffroyus geoffroyi*, Red-cheeked parrot | 44 | .024 | −.089 | 54 | .023 | −.087 |
| *Rhyticeros plicatus*, Papuan hornbill | 24 | .013 | −.056 | 27 | .011 | −.050 |
| *Cacatua moluccensis*, Moluccan cockatoo | 12 | .007 | −.035 | 1 | .001 | −.007 |
| *Tanygnathus megalorynchos*, Great-billed parrot | 9 | .005 | −.026 | 11 | .005 | −.026 |
| *Electus roratus*, Electus parrot | 7 | .004 | −.022 | 0 | 0 | 0 |
| *Macropygia amboinensis*, Brown cuckoo-dove | 6 | .003 | −.017 | 7 | .003 | −.017 |
| *Cacomantis sepulcralis*, Ruby-breasted cuckoo | 3 | .002 | −.012 | 0 | 0 | 0 |
| *Trichoglossus haematodus*, Rainbow lorikeet | 0 | 0 | 0 | 64 | .027 | −.097 |
| Total | 1,824 | 1.0 | | 2,358 | 1.0 | |
| Shannon Diversity Index | | | 2.284 | | | 2.037 |

**(iv)**

**5 CONCLUSION** Bird diversity is higher in unlogged forests than in logged forests.

**Figure 17.A Logged and unlogged forests in Indonesia.** **(i)** Logged forest. **(ii)** Unlogged forest. **(iii)** *Nectarinia jugularis*, the most common bird of logged forests, **(iv)** Shannon diversity indices for unlogged and logged forest.

**Table 17.2** Effective numbers of species using the Shannon and Gini-Simpson indices.

In communities 1 and 2 there are only two species present.

| | Relative abundance of species | | | | Diversity index | | Effective number of species | |
|---|---|---|---|---|---|---|---|---|
| Community | Species 1 | Species 2 | Species 3 | Species richness | $H_S$ | $D_{GS}$ | $H_S$ | $D_{GS}$ |
| 1 | 90 | 10 | – | 2 | 0.325 | 0.180 | 1.384 | 1.219 |
| 2 | 50 | 50 | – | 2 | 0.693 | 0.50 | 2.000 | 2.000 |
| 3 | 80 | 10 | 10 | 3 | 0.638 | 0.340 | 1.893 | 1.515 |
| 4 | 33.3 | 33.3 | 33.3 | 3 | 1.099 | 0.667 | 3.000 | 3.000 |

We will examine the effective number of species for the Gini-Simpson and the Shannon index. For the Shannon index we take the exponential and for the Gini-Simpson index we subtract it from unity and invert it. Using our example for community A, if we take the exponential of the Shannon index of 1.609 we get 5.00. If we subtract the Gini-Simpson index from unity and invert it, $1/1 - 0.8$, we also get 5.00. In this example, the effective number of species is the same number as the actual number of species, but it is not always so. The value of the effective number of species concept becomes more obvious when we consider communities with uneven species frequencies, as in the community illustrated in **Table 17.2.** Here, it is easier to compare communities using the effective number of species. We see that community 3 is 36.8% more diverse than community 1 when using the Shannon effective number of species. Where one species dominates a community, the Gini-Simpson effective number of species gives a lower value than the Shannon effective number of species. This is because the Gini-Simpson index is a dominance index and it more heavily weights the effects of the dominant species. Any index that involves the sums of squares of the frequencies, sometimes called an order 2 index, is therefore less satisfactory than an order 1 index such as the Shannon effective number of species. In turn, species richness is an order 0 index, because it does not take into account frequencies and is a function of the sum of the zeroth power of the frequency. However, we have already seen how species richness is very susceptible to sample size. For these reasons, Jost (2007) has argued that the Shannon effective number of species is the best measure of diversity. Unfortunately, ecologists lack data on abundance of individuals on large areas such as whole countries so, as we shall see in the next chapter, global patterns of biodiversity usually rely on species richness data. Shannon indices are usually available only for specific taxa in well-defined areas, such as Stuart Marsden's study of logged and unlogged forest patches in Indonesia.

### 17.2.4 Evenness is a measure of how diverse a community is relative to the maximum possible diversity

All diversity indices are affected by the number of species, the number of individuals, and their equitability or evenness. A higher number of species and a more even distribution of individuals among species both increase diversity. For any diversity index, the maximum diversity of a community is found when all species are equally abundant. We can then compute evenness, *E*, an index which compares the actual diversity value to the maximum possible diversity. For example, using the Shannon index:

$$E = H_S/H_{\max}$$

*E* is constrained between 0 and 1.0. In Table 17.2, $H_S$ for community 2 is maximal as the individuals are equally distributed among the species, $H_S = 0.693$, $H_{\max} = 0.693$, and $E = 1.0$. In community 1, $H_S = 0.325$ so $E = 0.325/0.69 = 0.469$. Adding species or increasing evenness both increase species diversity (Table 17.2). Therefore, a community with two species can be more diverse than a community with three species if the distribution of individuals between the two species is more even.

**Check Your Understanding**

**17.2** What is the effective number of species in the logged and unlogged areas in Table 17.2

## 17.3 Rank Abundance Diagrams Visually Describe the Distribution of Individuals Among Species in Communities

Descriptions of whole communities by diversity indices run the risk of losing information. A more complete picture of the distribution of species in a community is gained by plotting the proportional abundance (usually on a log scale) against rank of abundance. The abundance of the most common species appears on the extreme left and the rarest species on the extreme right. A rank abundance diagram can be drawn for the number of individuals in a community, biomass of individuals, ground area covered by plants, and other variables all plotted against rank abundance. There are many theoretical forms of rank abundance diagrams, but we will first consider perhaps the best known, the lognormal, then three more proposed by the Japanese ecologist Mutsunori Tokeshi.

## 17.3.1 The lognormal distribution is based on statistical properties of data

The lognormal distribution owes its place in ecology to Frank Preston (1948) who obtained a normal, or bell-shaped, curve when plotting number of species (y-axis) against log species abundance (x-axis) when abundance was grouped into various classes. Carrington Williams (1964) showed the existence of lognormal distributions for a variety of communities, such as the abundance of birds in Britain (**Figure 17.4**). Species abundance may be expressed in a variety of abundance classes, such as $\log_2$, $\log_3$, natural logs, or $\log_{10}$; for example, 1–10, 11–100, 101–1000, 1001–10,000, and so on, along the x-axis. The biological meaning of this distribution is that there are a few species that are very common or are very rare, and a lot of species that have an intermediate number of individuals. We can translate the lognormal distribution into a lognormal rank abundance plot by graphing the abundance of species versus their rank abundance, and obtain a sigmoidal-shaped curve. Any rank abundance plot with this shape is often attributed to a lognormal distribution (**Figure 17.5**).

There are at least two problems with the lognormal rank abundance plot. The first is that failure to sample all species in a community may result in a truncated lognormal distribution. The truncation is explained as being due to rare species that are present in the community but not in the sample. Undersampling would therefore result in a different-shaped rank abundance plot as the rare species, #35–36 and possibly #29–34 in Figure 17.5, would be absent. Williams (1964) illustrated this point with reference to moth samples taken at light traps in England. With few individuals sampled (**Figure 17.6a**), few rare species are captured. As larger samples are taken, more species would be obtained and the curve would move to the right (**Figure 17.6b,c**). Therefore, when

**Figure 17.4 The lognormal distribution.** Species abundance plotted against $\log_{10}$ abundance classes for British birds. (Data from Williams, 1964.)

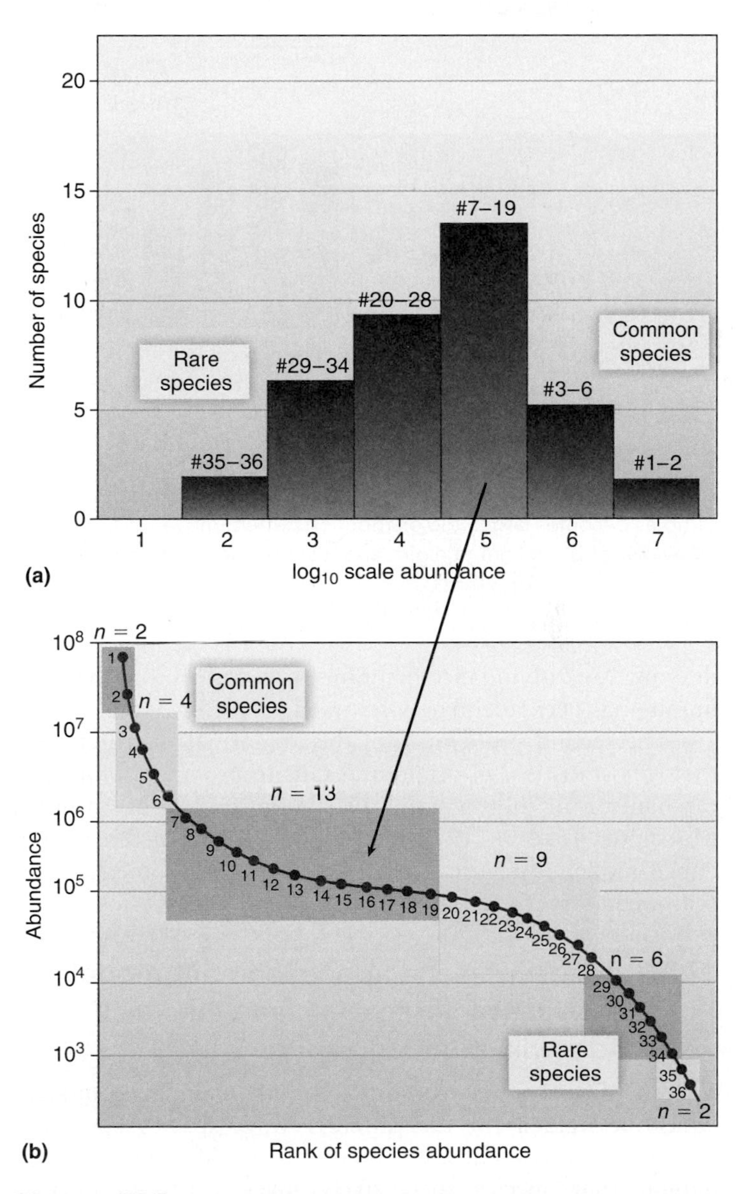


**Figure 17.5 The lognormal rank abundance plot.** **(a)** Each species is given a number. We then group the species in abundance classes of $\log_{10}$. The result is a normal distribution: some species with few individuals (#29–36), some species with many individuals (#1–6), but most with an intermediate number of species (#7–28). **(b)** In the rank abundance plot, the abundance of each species is plotted on a logarithmic scale against the species' rank, in order from the most abundant to least abundant species.

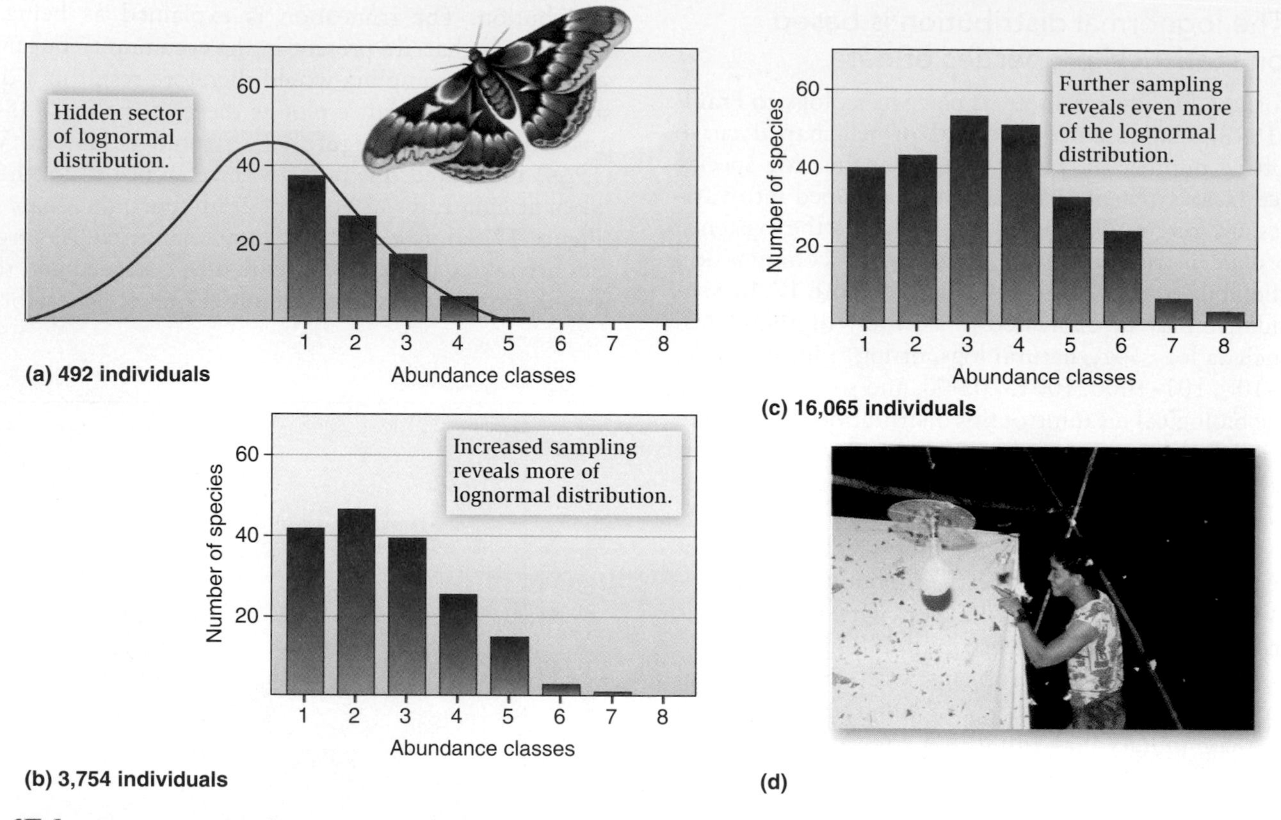


**Figure 17.6 The truncated lognormal distribution.** The distribution of abundance of moths captured by light traps in England varies with the number of samples taken. In the top figure **(a)**, the true distribution of abundance is hidden behind the y-axis. **(b, c)** As the number of samples becomes larger, the distribution pattern moves to the right to reveal the distribution of rarer species. **(d)** Adult moths are drawn to an ultraviolet or black light at night, and rest on a nearby sheet, where researchers can collect or identify them. (Reproduced from Williams, 1964.)

drawing rank abundance diagrams, it is critical that the community has been adequately sampled.

The second problem with the lognormal distribution is that it is statistical in origin and fails to provide a biological mechanism for the observed distribution. With this in mind, Mutsunori Tokeshi (1999) developed a series of rank abundance models that are more likely to provide mechanisms that might structure communities.

### 17.3.2 Tokeshi's niche apportionment models provide biological explanations for rank abundance plots

Tokeshi described many different rank abundance models, which he named for the different ways species apportion resources. We will consider three of these. Imagine an empty habitat that species successively invade. The dominance preemption model supposes that the first species to invade the community preempts a large fraction, >50%, of the available resources (**Figure 17.7**). The next species takes the same fraction of the remainder, and so on. Species abundance is proportional to resources taken. This model produces the least equitable species distribution (**Figure 17.8a**). The model fits communities of relatively few species where a single

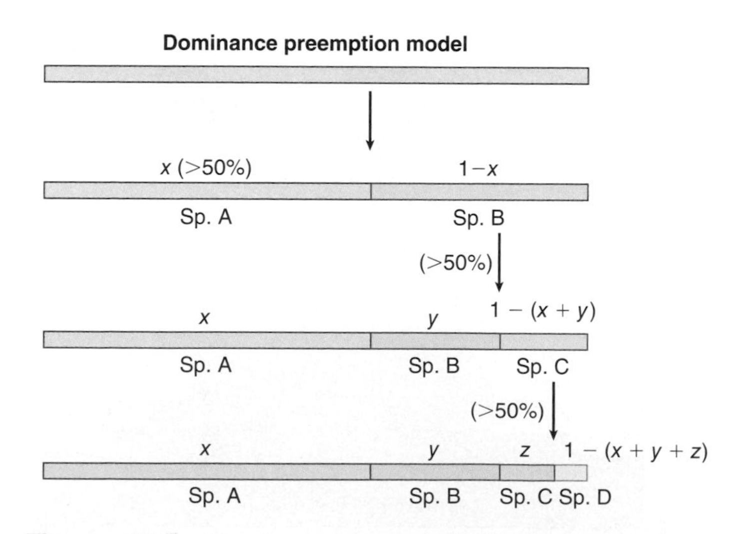


**Figure 17.7 The dominance preemption model.** This model assumes the same large fraction, 50% or more, of unused resources is taken by each successive species.

environmental factor is of dominating importance. Only a few species are fit to survive in such a habitat, so they are numerically dominant. Both cold-temperature communities and polluted communities fit this profile, so we are not

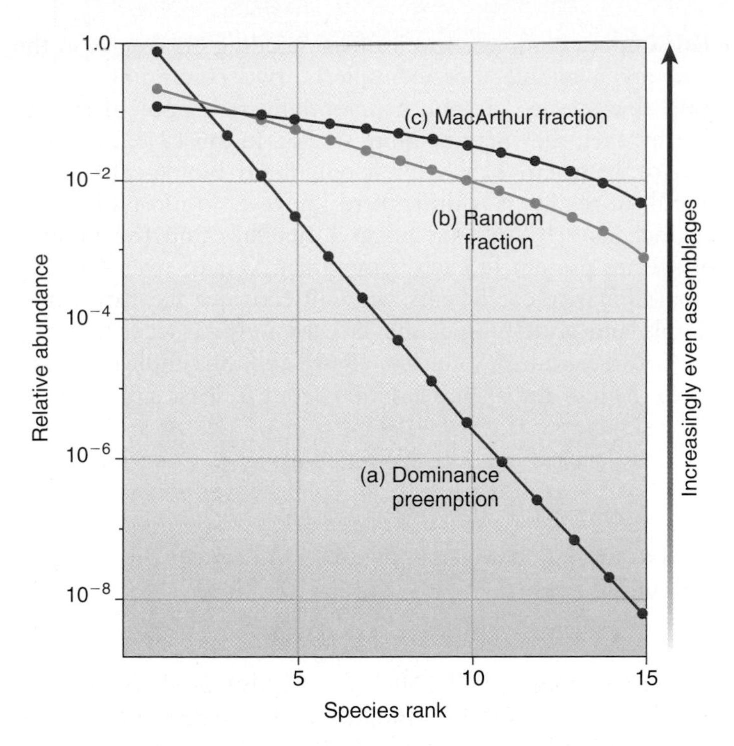


**Figure 17.8 Rank abundance plots based on Tokeshi's niche apportionment models.** (From Tokeshi, 1999.)

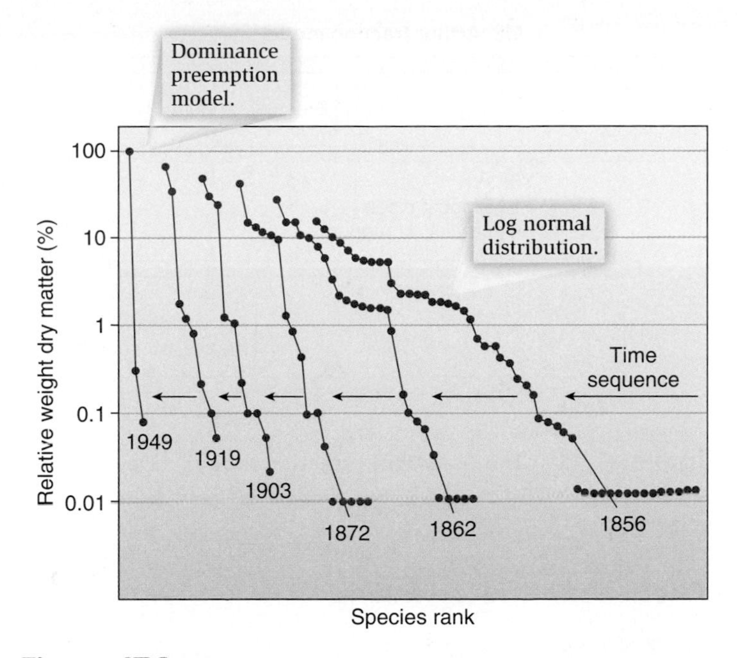


**Figure 17.9 Changes in relative abundances of grassland species at Rothamsted Experimental Station, UK, 1856–1949.** Nitrogen fertilizer has been added continuously since 1856. (After Kempton, 1979.)

**ECOLOGICAL INQUIRY**

If a community whose species abundance followed a lognormal distribution was not sufficiently sampled, which of these patterns might describe the data?

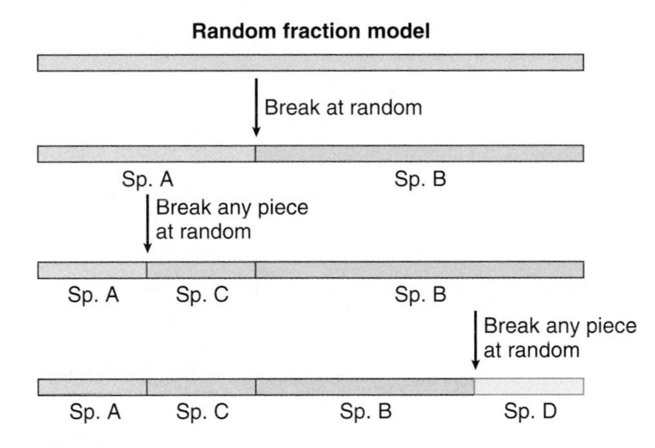


**Figure 17.10 The random fraction model.** This model assumes that any piece of the resource, used or unused, may be sequestered by successive species.

surprised to see limited species richness in such situations. Fits of data to the dominance preemption model might therefore be used as an indicator of polluted communities. Fertilization can cause pollution of communities when the application is chronic and sustained. The Park Grass experiment at Rothamsted, UK, has added nitrogen fertilizer continuously to some grassland plots since 1856. Here, researchers were interested whether organic manures or inorganic fertilizers provided the most increased yield. The resulting changes in relative abundance of grasses showed how the shape of the species abundance plots in these experiments has changed from lognormal to a dominance preemption model as one or two species have come to dominate these nitrogen-polluted plots (**Figure 17.9**).

The random fraction model supposes that each new species to invade the community commandeers a random fraction of the available resources, but that this fraction may be drawn from unused resources or resources already used by a competitor (**Figures 17.10, 17.8b**). It therefore envisions competition as a structuring mechanism of the community. This model gives a distribution close to the lognormal. The MacArthur fraction model is similar to the random fraction model, but supposes that it is usually the species with the largest niche that is likely to undergo competition from new invaders, and this results in a more equitable distribution than the random fraction model (**Figures 17.11, 17.8c**). It is named in deference to the ecologist Robert MacArthur, who had much earlier envisaged a model in which resources were apportioned randomly between species, in much the same way as a stick might break into multiple random pieces if dropped from the top of a tree. Currently, ecologists are determining how best to accurately fit their data to these types of models, and no consensus has emerged over which type of community best fits each model.

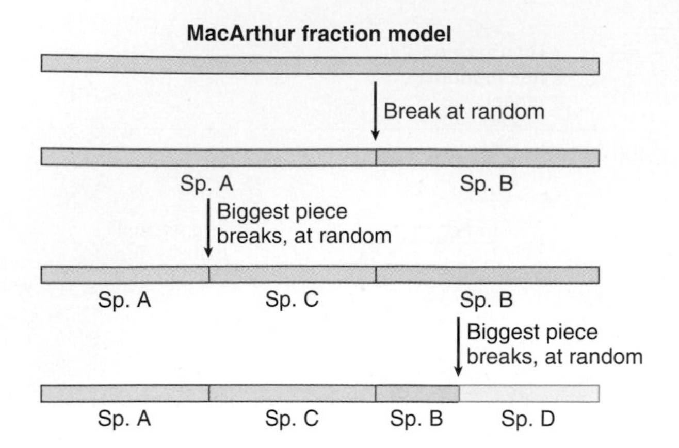


**Figure 17.11 The MacArthur fraction model.** Here, Tokeshi assumes that usually the largest portion of available resources is subdivided.

**Check Your Understanding**

**17.3** Why might polluted communities fit the dominance preemption rank abundance model?

## 17.4 Community Similarity Is a Measure of How Many Species Are Common Between Communities

To compare diversity between areas, one could simply compare diversity indices. Another method is to use indices called **similarity indices**, which compare diversity of sites directly. These indices compare the numbers of species common to all communities. Let's consider a simple matrix of presence/absence for two communities, A and B:

| | | Community A | Community A |
|---|---|---|---|
| | | No. of species present | No. of species absent |
| Community B | No. of species present | $a$ | $b$ |
| | No. of species absent | $c$ | $d$ |

where

$a$ = the number of species common to both communities
$b$ = the number of species in community B, but not A
$c$ = the number of species in community A, but not B
$d$ = the number of species absent in both communities

The number of species absent in both communities, $d$, is biologically meaningful only if the potential number of species that could be present in both communities is well known. This implies complete species lists detailing all species in the area are available. For example, if two communities both have few species, despite a potentially rich pool of species nearby, then they may be quite similar. In this case, $d$, a measure of the negative matches, is potentially biologically meaningful. In reality, it is difficult to know $d$, so most similarity coefficients rely only on positive matches and the number of species present in each community. Out of over 40 common similarity coefficients, we will consider the three most simple and widely used: the Jaccard index (Jaccard, 1912), the Sorensen index (Sorensen, 1948), and the simple matching index. Values for all indices vary from 0, least similar, to 1, most similar.

The Jaccard index is calculated using the equation

$$C_J = a / (a + b + c)$$

For the data outlined in Figure 17.A between the unlogged and logged sites:

$$C_J = 14 / (14 + 1 + 2) = 0.82$$

Monik Oprea and colleagues (2009) used the Jaccard index to investigate the similarity of bat species in different urban landscapes in southeastern Brazil. They were interested in whether wooded streets provided connectivity between urban parks, promoting the movement of bats. They compared the similarity of bat species captured in urban parks, wooded streets, and nonwooded streets. Unfortunately, wooded streets were more similar to nonwooded streets ($C_J = 0.500$) than to urban parks ($C_J = 0.273$), suggesting that wooded streets did not provide much connectivity for bats in urban landscapes.

The Sorensen index is similar to the Jaccard index and uses identical variables. However, the Sorensen index weights matches in species composition between the two samples $a$, which we know, more heavily than mismatches $b$ and $c$, which we are less sure of, because of the possibility of undersampling. Whether this type of weighting is valuable or not has yet to be resolved.

$$C_S = 2a / (2a + b + c)$$

Again, comparing the unlogged and logged sites (Figure 17.A),

$$C_S = 28 / (28 + 1 + 2) = 0.90$$

The simple matching index, $C_{sm}$, makes use of the number of species absent in both areas, $d$.

$$C_{SM} = \frac{a + d}{a + b + c + d}$$

Again, using the data in Figure 17.A and knowing $d = 2$,

$$C_{SM} = \frac{14 + 2}{14 + 1 + 2 + 2} = 0.84$$

Despite the differences in measurement of diversity using different indices, some large-scale patterns emerge. Paramount among these is that there is a pronounced latitudinal gradient in organismal diversity with relatively few species in high-latitude realms and many more species in equatorial regions. In Chapter 18, we will examine how species richness varies in different areas of the world and what drives this variation.

### Check Your Understanding

**17.4** How do the Jaccard and Sorensen indices differ from one another?

## SUMMARY

**17.1 The Nature of Communities Has Been Debated by Ecologists**

- Ecologists have held different views of communities and how they are structured and function, from tightly knit groups to loose associations of species (Figure 17.1).

**17.2 Various Indices Have Been Used to Estimate Species Biodiversity**

- A variety of indices has been used to estimate biodiversity, among the most common of which are species richness, the Gini-Simpson index, and the Shannon index,.
- Species richness is a count of the number of species present in a community (Table 17.1).
- Dominance indices such as the Gini-Simpson index are influenced strongly by common or dominant species (Figures 17.2).
- Information statistic indices such as the Shannon index take more account of rare species (Figures 17.3).
- Stuart Marsden used the Shannon index to show how bird diversity of logged forests in Indonesia was lower than that of unlogged forests (Figure 17.A).
- Converting diversity indices to an effective number of species makes comparisons between communities easier (Table 17.2). While there is no single best diversity index, some ecologist have suggested that the effective number of species based on the Shannon index is the most useful.
- Evenness indices measure actual diversity relative to maximum diversity.

**17.3 Rank Abundance Diagrams Visually Describe the Distribution of Individuals Among Species in Communities**

- Rank abundance diagrams provide a more complete summary of the distribution of individuals among species.
- The lognormal rank abundance diagram is based on the lognormal statistical distribution (Figures 17.5, 17.6).
- Tokeshi's niche apportionment models provide biological explanations for rank abundance diagrams (Figures 17.7–17.11).

**17.4 Community Similarity Is a Measure of How Many Species Are Common Between Communities**

- Community similarity indices such as the Jaccard index, the Sorensen index, and the simple matching index provide a direct measure of the similarity of species composition between sites.

## TEST YOURSELF

| Species | Community A | Community B |
|---|---|---|
| 1 | 4 | 8 |
| 2 | 3 | 6 |
| 3 | 0 | 4 |
| 4 | 11 | 2 |
| 5 | 2 | 0 |

Questions 1–8 refer to the above table.

1. What is the species richness of community B?
   a. 1.279
   b. 0.625
   c. 0.6
   d. 3.60
   e. 4

2. What is the Gini-Simpson index of diversity for community A?
   a. 1.279
   b. 0.625
   c. 0.6
   d. 3.60
   e. 4

3. What is the Shannon index for community B?
   a. 1.279
   b. 0.625
   c. 0.6
   d. 3.60
   e. 4

4. What is the Shannon effective number of species for community B?
   a. 1.279
   b. 0.625
   c. 0.6
   d. 3.60
   e. 4
5. Which measure of species richness has the best discriminant ability?
   a. Species richness
   b. Gini-Simpson
   c. Shannon
   d. lognormal
   e. Jaccard
6. What is the Jaccard index between the two communities?
   a. 0.4
   b. 0.6
   c. 0.75
   d. 0.8
   e. 1.0
7. What is the Sorensen index between the two communities?
   a. 0.4
   b. 0.6
   c. 0.75
   d. 0.8
   e. 1.0
8. What is the simple matching index between the two communities, assuming there are only these five species in the area?
   a. 0.4
   b. 0.6
   c. 0.75
   d. 0.8
   e. 1.0
9. Which rank abundance model has the least equitable distribution of species?
   a. Lognormal
   b. Dominance preemption
   c. Random fraction
   d. MacArthur fraction
10. There are five species in a community. Species 1 has 10 individuals and Species 2–5 each have 5. What is the evenness of the community?
   a. 0.33
   b. 0.20
   c. 1.5607
   d. 1.609
   e. 0.970

## CONCEPTUAL QUESTIONS

1. Why are information statistic diversity indices preferable to dominance indices for conservation biologists?
2. What is community evenness and how do we measure it?
3. Describe how you would construct a rank abundance diagram.

## DATA ANALYSIS

What are the Shannon indices and effective number of species for each of communities 1 and 2 in Table 17.1?

**Connect Ecology** helps you stay a step ahead in your studies with animations and videos that bring concepts to life and practice tests to assess your understanding of key ecological concepts. Your instructor may also recommend the interactive ebook.

Visit **www.mhhe.com/stilingecology** to learn more.

Coquerel's sifaka, *Propithecus coquereli*, an endangered lemur of Madagascar.

CHAPTER 18

# Species Richness Patterns

## Outline and Concepts



Biodiversity is threatened by human activities. Human-induced extinction rates are at least 100 times the geological background rate and will increase. Many ecologists are struggling to pinpoint areas of the globe where special conservation efforts are needed to prevent extinctions and maintain biodiversity. For example, the island of Madagascar separated from mainland Africa 65 million years ago and contains many unique species that have evolved independently. Its lowland rain forests and dry deciduous forests are home to over 11,000 species of plants found nowhere else on Earth, and hundreds of bird and mammal species, including lemurs, a group of primates unique to Madagascar. To prevent the loss of species diversity, we must understand not only how to measure diversity, a topic we considered in Chapter 17, but also what influences patterns of species richness. With this knowledge, conservation biologists can begin to identify and prioritize which areas to protect and conserve.

The number of species of most taxa generally increases from polar to temperate to tropical areas, a phenomenon known as the **latitudinal species richness gradient.** For example, the species richness of North American birds increases from Arctic Canada to Panama (**Figure 18.1a**). A similar pattern exists for mammals (**Figure 18.1b**), freshwater and marine organisms, and most other taxa. Exceptions to this pattern occur, however. The richness of penguin species does not increase toward the Tropics. Species richness is also increased by topographical variation. More mountains mean more hilltops, valleys, and differing habitats; thus, there is an increased number of species of birds and mammals in the mountainous western U.S. However, species richness is reduced by the peninsular effect, in which diversity decreases as a function of distance from the main body of land, and thus we see a decrease in the number of species in Florida and in Baja California.

At least 28 hypotheses for the polar-equatorial gradient of species richness have been advanced. In this chapter we will consider three of the most important, which propose that species richness increases with increased time, area, and productivity, respectively. We will also see how, on a more local scale, species richness can be affected by disturbance and natural enemies. Although they are treated separately here, these hypotheses are not mutually exclusive. All five can contribute to patterns of species richness. In addition, we will examine global patterns of community similarity between habitats. Finally, we address some of the strategies used to conserve species richness around the world.

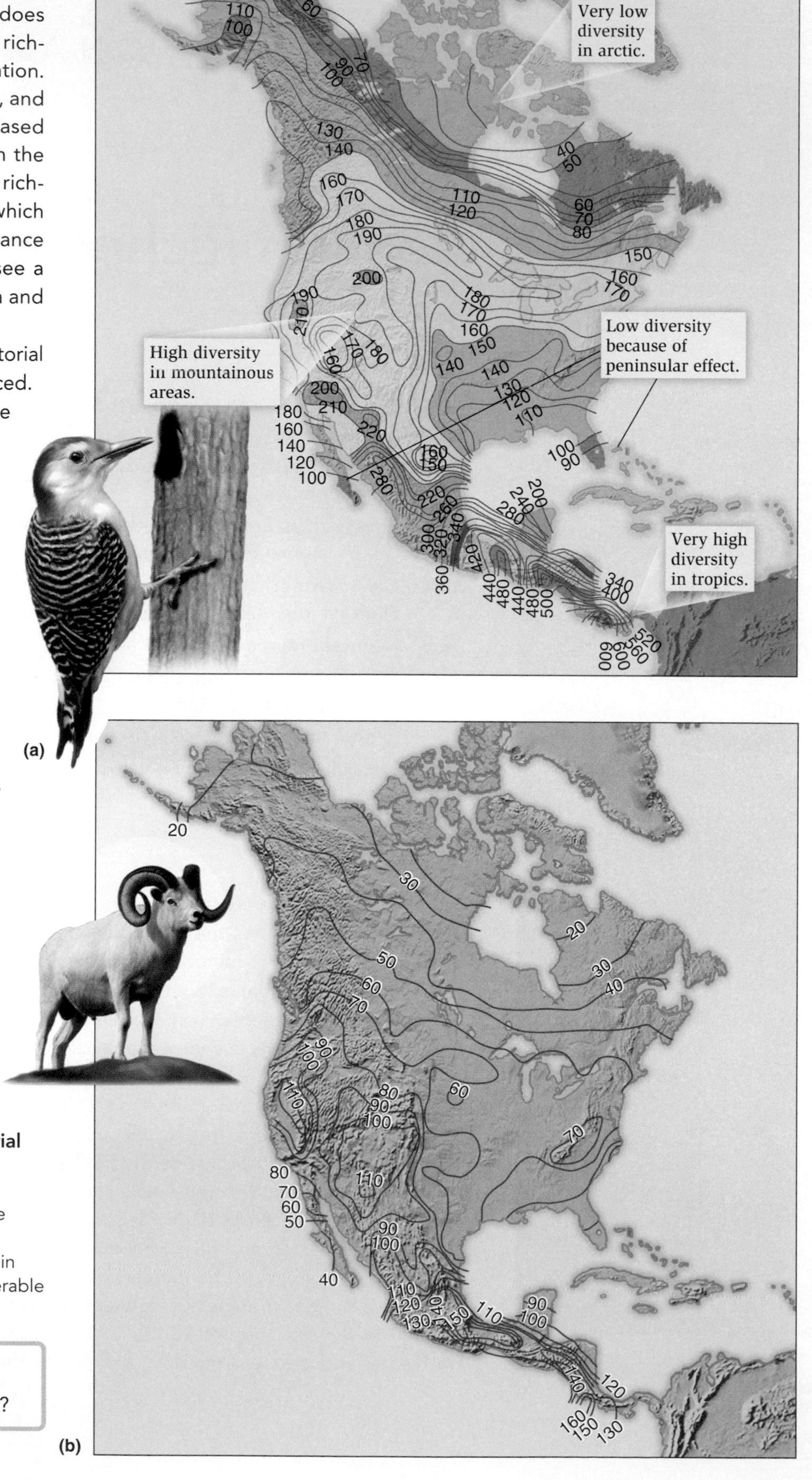


**Figure 18.1 Species richness in North American vertebrates follows a polar-equatorial gradient.** Contour lines show equal numbers of species. **(a)** Birds. (Redrawn from Cook, 1969.) **(b)** Mammals. (Redrawn from Simpson, 1964.) Note the pronounced latitudinal gradient heading south toward the Tropics and the high mammal diversity in California and northern Mexico, regions of considerable topographic relief and habitat diversity.

**ECOLOGICAL INQUIRY**

Why is bird diversity high in mountain areas?

## 18.1 The Species-Time Hypothesis Suggests That Communities Diversify with Age

The **species-time hypothesis** suggests that communities diversify, that is, gain species, with time and that temperate regions are less species-rich than tropical regions because they are younger and have only more recently recovered from glaciations and severe climatic disruption (**Figure 18.2a**). At the time of the last glacial maximum, 18,000 years ago, ice sheets covered much of North America and northern Europe, killing or displacing many temperate species not only from ice-covered areas but also from nearby areas where temperatures were greatly cooled (**Figure 18.3**). Tropical species were not affected.

The species-time hypothesis has several variations. One variation proposes that resident species in tropical zones have not yet evolved new forms to exploit vacant niches in temperate zones. Certain traits associated with temperate zones, such as freezing tolerance, present significant barriers to overcome. More than half the families of flowering plants have no temperate representatives. The second variation proposes that tropical species that could possibly live in temperate regions have not migrated back from the unglaciated areas into which the ice ages drove them. The third variation, the evolutionary speed hypothesis, proposes that evolution is faster in the Tropics due to high ambient temperatures, short generation times, and increased mutation rates, and this has produced more species (**Figure 18.2b**). Thus, tropical habitats are both a cradle for the generation of new taxa and a museum for the preservation of existing species. The time hypothesis and its variations contrast with other hypotheses that suggest that the accumulation of species is similar in temperate and tropical realms and that ecological factors ultimately affect species richness (**Figure 18.2c**). Such ecological factors include temperature and water availability and are considered under the species-energy hypothesis.

In support of the species-time hypothesis, H. John Birks (1980) found a significant correlation between the numbers of species of insects on various British trees and the evolutionary ages of these trees in Britain (**Figure 18.4**). Many of the tree species in Britain are relatively recent colonists, having appeared only in the last 18,000 years, following the departure of the glaciers that covered most of the islands. Pollen records from lake bottoms give us a good idea of how long a species has been present in Britain. Pollen present in deep lake sediments indicates a long presence, because much more recent sediment has overlaid the pollen. The presence of pollen solely in shallow sediments indicates a more recent arrival of a tree species. Birks used radiocarbon dating of pollen to more accurately estimate the length of time a tree had been present in Britain. He found that no species of tree had been present in Britain for longer than 13,000 years. He then gathered information on numbers of insect species present on trees from lists provided by other experts who had been examining

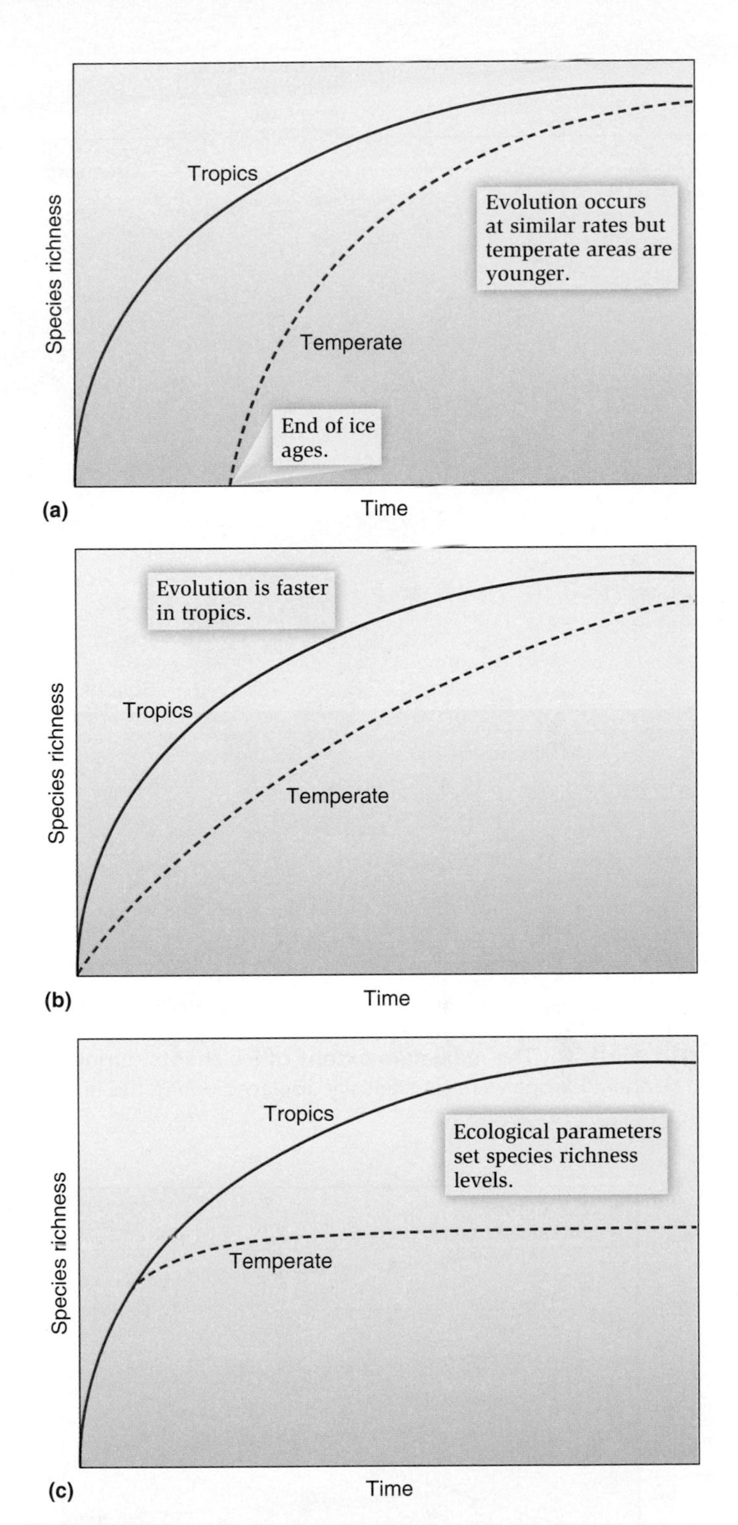


**Figure 18.2 The accumulation of species in tropical and temperate areas has been attributed to** **(a)** a longer time for speciation in the tropical areas following climate change in temperate areas, for example, the end of the ice ages, **(b)** higher speciation and/or extinction rates in the tropical areas, and **(c)** different ecological factors in each realm. (After Mittelbach et al., 2007.)

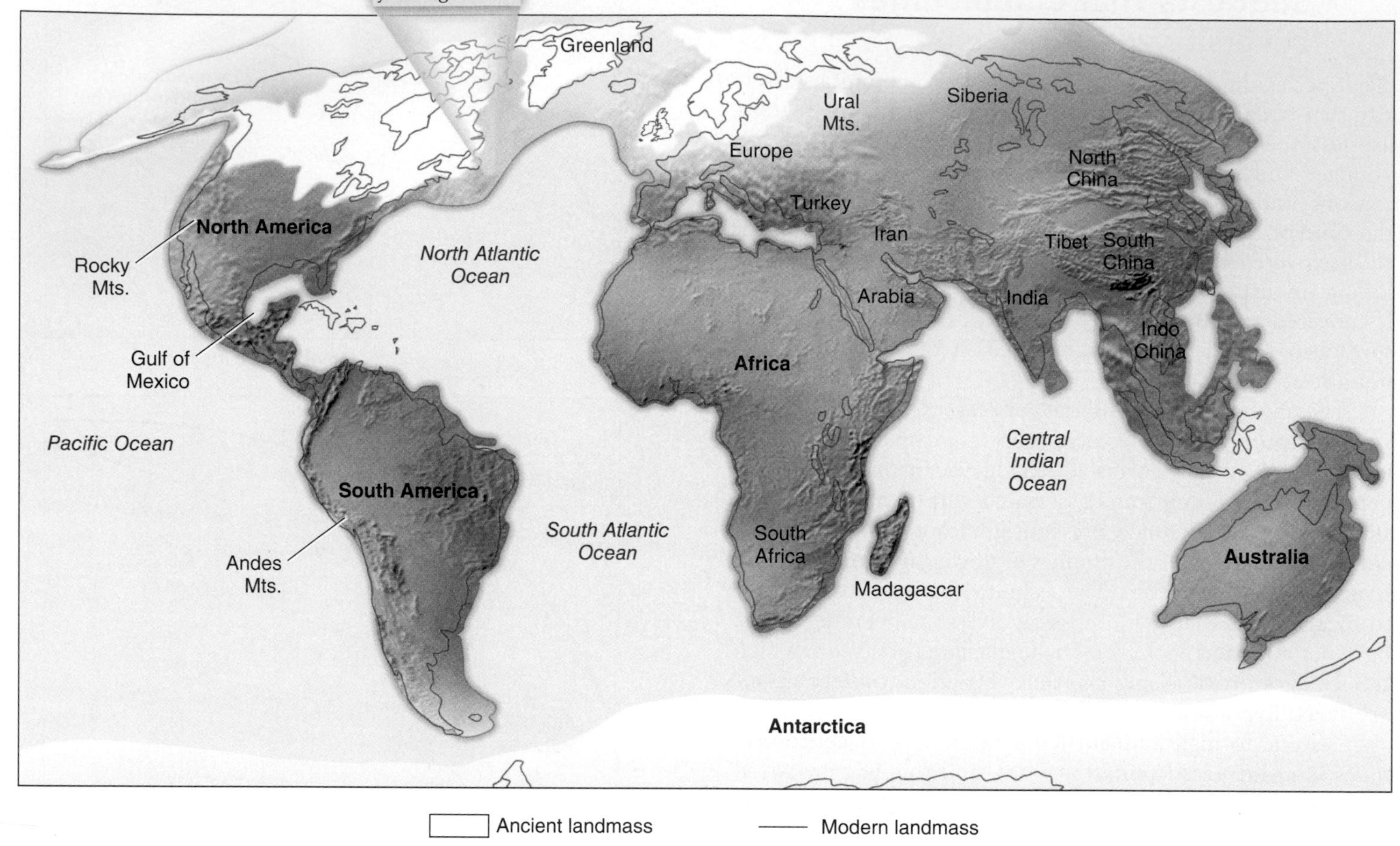


**Figure 18.3 The maximum extent of ice sheets during the last ice age 18,000 years ago.** Temperate species in North America and northern Europe were substantially displaced during this time period.

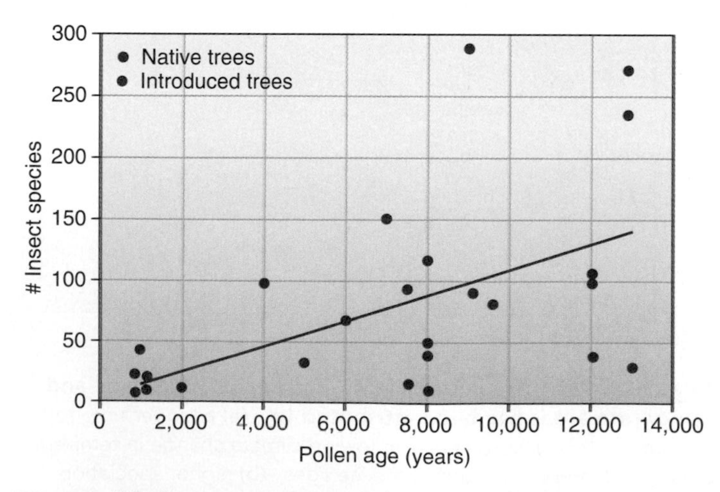


**Figure 18.4 Insect species richness on British trees shows the influence of evolutionary time.** (After Birks, 1980.)

the insect fauna of trees in Britain for many years. The significant relationship between pollen age and the total number of insects indicated to Birks that older tree species support more species of insects. This suggests that time is important in influencing species richness.

In a different comparison, ecologists compared the species richness of bottom-dwelling invertebrates, such as worms, in historically glaciated and unglaciated lakes in the Northern Hemisphere that occur at similar latitudes. Lake Baikal in Siberia is an ancient unglaciated temperate lake and contains a very diverse fauna; for example, 580 species of invertebrates are found in the bottom zone. Great Slave Lake, a lake of comparable size in northern Canada, where the ice has only relatively recently melted, contains only 4 species in the same zone. This comparison supports the species time hypothesis.

However, ecologists recognize drawbacks to the utility of the species-time hypothesis. First, other data sets do not support the species-time hypothesis. For example, Albert Bush and colleagues (1990) examined the helminth parasite richness of vertebrates. They suggested that, according to the species-time hypothesis, the oldest vertebrate lineages, the fish, should have more parasite species than the youngest lineages, birds and mammals, with lizards and snakes in between. Instead the data showed that habitat mattered most. Aquatic vertebrates, regardless of taxon, had more helminth species than terrestrial ones. Taxon age did not appear important. Among aquatic vertebrates, fish had fewer parasite species than mammals or birds. Second, even if the species-time hypothesis may help explain variations in the species richness of terrestrial organisms, it has limited applicability to marine organisms. We might not expect terrestrial species to redistribute themselves quickly following a glaciation, especially if there is a barrier like the English Channel to overcome. However, there seems to be no reason marine organisms couldn't relatively easily shift their distribution patterns during glaciations. Despite this, the polar-equatorial gradient of species richness still exists in marine habitats. Third, the evolutionary speed hypothesis applies only to ectotherms, whose generation times are directly impacted by temperature. However, endotherms also exhibit pronounced latitudinal species richness gradients.

Fourth, comparisons on speciation rates between tropical and temperate areas have yielded conflicting results. Shane Wright and colleagues (2006) compared the rate of molecular evolution in 45 congeneric or conspecific rainforest plant species, one member of which occurred in the Tropics and one in temperate realms. Tropical plant species had more than twice the rate of molecular evolution than their closely related temperate congeners. Jason Weir and Dolph Schluter (2007) compared the speciation and extinction rates of 309 sister pairs of New World birds and mammals and estimated their ages from molecular data. The highest speciation and extinction rates occurred at high latitudes and declined toward the Tropics. This conflicts with the idea that high tropical diversity is linked to high speciation rates.

**Check Your Understanding**

**18.1** Why did Weir and Schluter examine extinction rates of New World birds as well as speciation rates?

## 18.2 The Species-Area Hypothesis Suggests That Large Areas Support More Species

The **species-area hypothesis** proposes that large areas contain more species than small areas, because they can support larger populations and a greater range of habitats. Larger populations are more resistant to extinctions. Greater ranges of habitats promote increased speciation rates. There is much evidence to support the species-area hypothesis. Donald Strong (1974) showed that insect species richness on tree species in Britain was correlated with the area over which a tree species could be found (**Figure 18.5**). Strong divided the country up into 100-$km^2$ grids and totaled up all the grids where a tree species was found. The larger the number of grids where

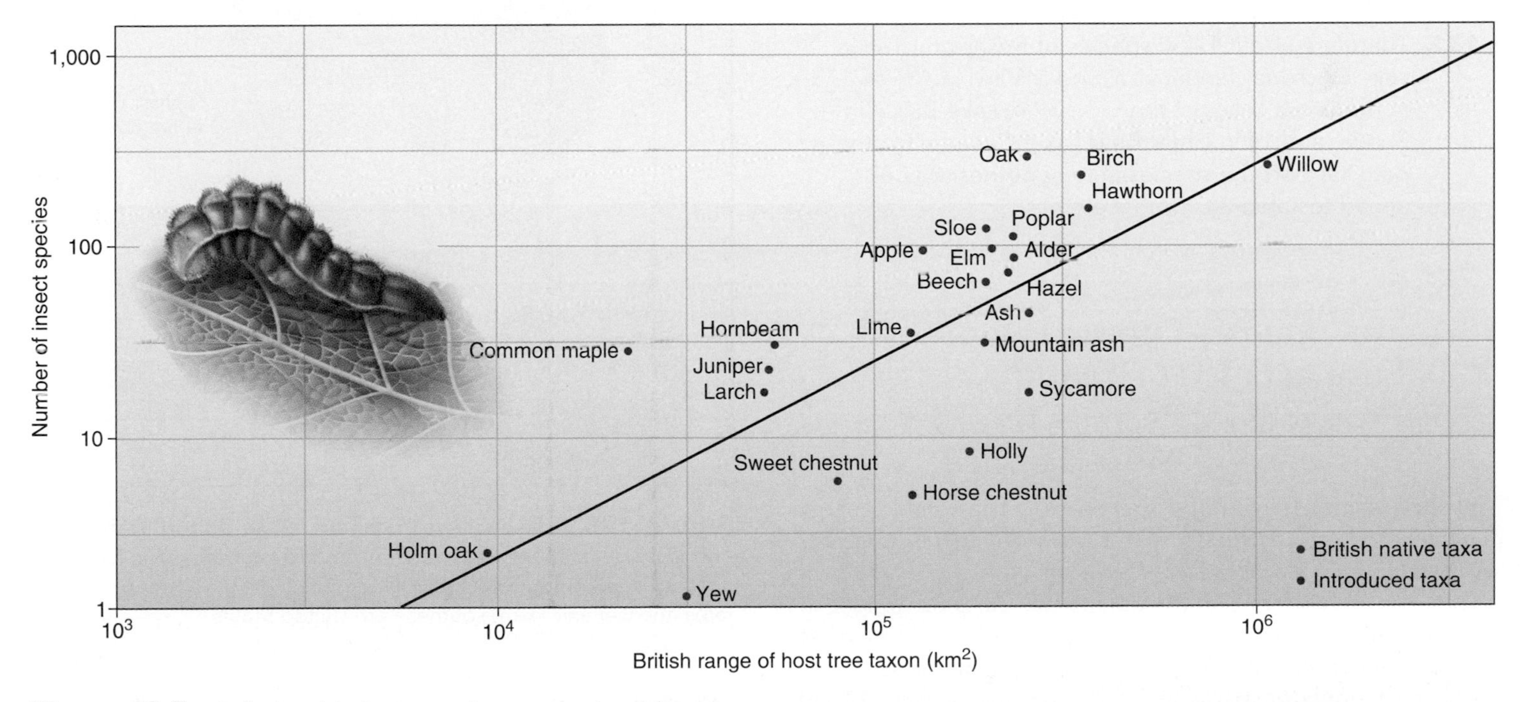


**Figure 18.5 Relationship between insect species richness on British trees and area.** A positive correlation is found between insect species richness and the host tree's present range (in $km^2$). Note the log scales. Range represents an area in Britain known to contain trees of each species. Red dots represent introduced species. (After Strong, 1974.)

**ECOLOGICAL INQUIRY**

How does this relationship relate to what we learned about the influence of secondary metabolites in Chapter 14?

a species was located, the greater the range a species occupied. Some introduced tree species, such as apple, were relatively new to Britain, having been introduced for agricultural purposes. Such species occupied relatively large areas and supported many different insect species, a fact Strong argued did not support the species-time hypothesis. Although Birks had argued that the evolutionary age of tree species influenced insect species richness, Strong suggested that area was more important, as the points clustered more tightly around the line of best fit. Data from other crop species planted across the world, such as sugarcane, also show that species richness of pests is better correlated with the area over which the crop was planted than the time since it was first planted.

A symmetry of climates exists between polar regions and temperate areas in the Northern and Southern Hemispheres. Symmetrically similar climates are adjacent only in the Tropics, however, which create one large area (**Figure 18.6**). This has been proposed as a reason why the Tropics have greater species richness than temperate or polar areas. However, the species-area hypothesis seems unable to explain why, if increased richness is linked to increased area, more species are not found in certain regions such as the vast contiguous landmass of Asia. Furthermore, tundra is a large land biome that has very low species richness. Finally, the largest marine system, the open ocean, which has the greatest volume of any habitat, has fewer species than tropical surface waters, which have a relatively small volume.

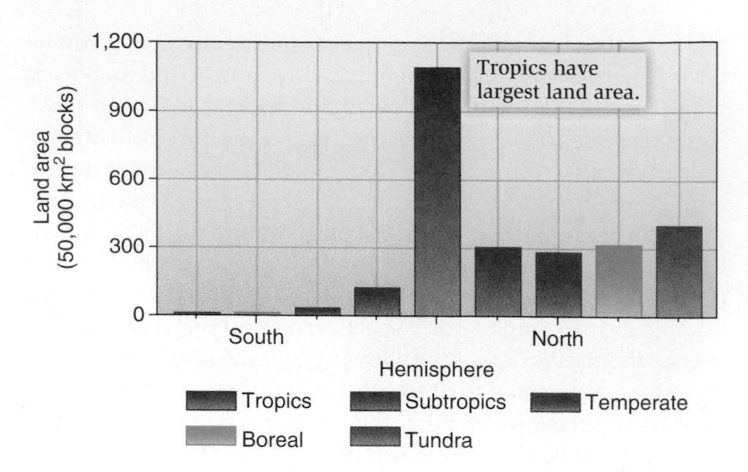


**Figure 18.6 Land areas in five broad zones of the Earth.** The Tropics are half an order of magnitude larger than other zones. (Redrawn from Rosenzweig, 1992.)

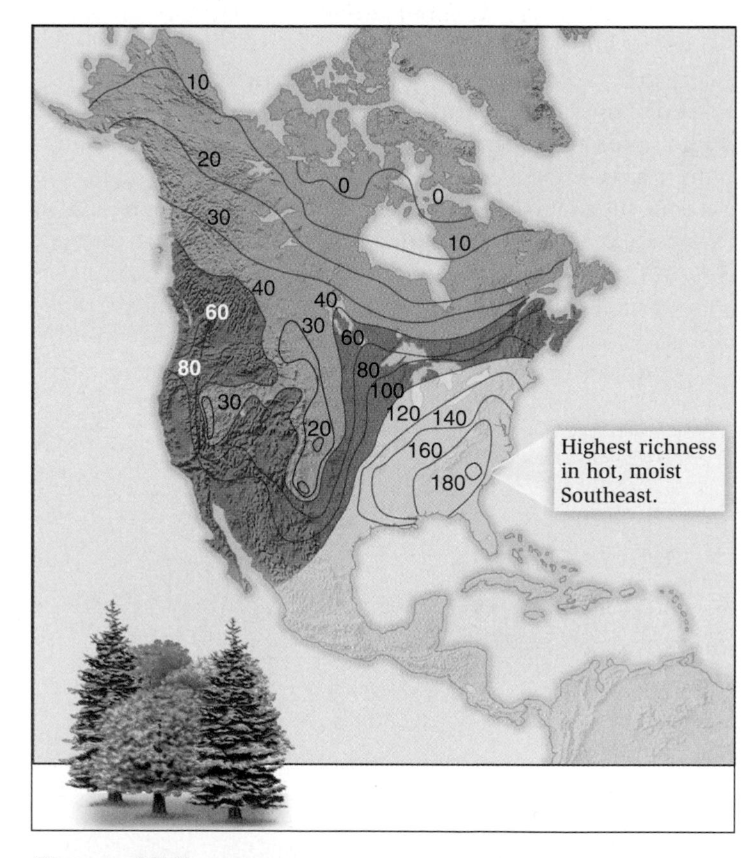


**Figure 18.7 Tree species richness in the United States and Canada is influenced by climate and rainfall.** The highest richness is associated with areas with the highest evapotranspiration rates, the wet and warm southeastern United States.

### Check Your Understanding

**18.2** There are about 1,450 species of freshwater fish in Lakes Victoria, Tanganyika, and Malawi in Africa; 173 in the Great Lakes of the U.S.; 39 in Lake Baikal in Russia; and 14 in Great Bear Lake, Canada. Do these data support the species-area hypothesis? Look ahead to Table 24.1 for lake areas.

## 18.3 The Species-Energy Hypothesis Suggests That Greater Productivity Permits the Existence of More Species

The **species-energy hypothesis** proposes that available energy determines species richness. Increased solar energy, together with an abundance of water, results in high productivity of plants. An increase in plant productivity, the total weight of plant material produced in a given time, leads to an increase in the number of herbivores and hence an increase in the number of predator, parasite, and scavenger species. David Currie and V. Paquin (1987) showed that the species richness of trees in North America is best predicted by the **evapotranspiration** rate, the rate at which water moves into the atmosphere through the processes of evaporation from the soil and transpiration of plants (**Figure 18.7**). The evapotranspiration rate is in turn influenced by the amount of solar energy and the amount of water in the ground. Hence, evapotranspiration rates, and tree richness, are highest in the hot, moist, southeast United States. Tree richness is low in the desert Southwest, because rainfall and water availability are low, even though temperatures are high.

Once again there are exceptions to this rule. Some tropical seas, such as the southeast Pacific off of Colombia and Ecuador, have low productivity but high species richness. On the other hand, the sub-Antarctic Ocean has a high productivity but low species richness. Estuarine areas, where rivers empty into the sea, are similarly very productive yet low in species, presumably because they are stressful environments for many organisms, being alternately inundated by fresh water and salt water. Some lakes that are polluted with fertilizers also have high productivity but low species richness. Finally, the productivity of temperate and tropical forests are very similar yet despite this, species richness is much higher in tropical forests than temperate ones. Also, it is not clear why increased productivity should promote increased species richness instead of high population densities of just a few species.

Robert Latham and Robert Ricklefs (1993) showed that while patterns of tree richness in North America support the species-energy hypothesis, the pattern does not hold for broad comparisons between continents. For example, the temperate forests of eastern Asia support substantially higher numbers of tree species, 729, than do climatically similar temperate forests of North America, 253, or Europe, 124. These three areas have different evolutionary histories and different neighboring areas from which species might have invaded. While the species-time, species-area, and species-energy hypotheses have been proposed as explanations for the pole-equator gradient in species richness, ecologists recognize that at more local scales, richness can be impacted by other phenomena. Next, we examine two of these phenomena, disturbance and natural enemies.

**Check Your Understanding**

**18.3** There are 222 ant species in Brazil, 134 in Trinidad, 101 in Cuba, 63 in Utah, and 7 in Alaska. Do these data support the species-energy hypothesis?

## 18.4 The Intermediate Disturbance Hypothesis Suggests That Species Richness Is Highest in Areas of Intermediate Levels of Disturbance

Joe Connell (1978) argued that local species richness was influenced by the frequency of disturbances such as fires and floods (**Figure 18.8**). He reasoned that, at low rates of disturbance, competitively dominant species would outcompete all other species, and only a few *K*-selected species would persist, yielding low species richness. *K*-selected species are those that exist near the carrying capacity of the environment, *K*, and that are good competitors. At high levels of disturbance, species would be continually driven extinct and only good colonists, which would be *r*-selected species, would persist, moving in from adjoining, but undisturbed, habitats. *r*-selected species have a high rate of per capita population growth, are often small, and can disperse well. Low species richness would result if only *r*-selected species were present. In this **intermediate disturbance hypothesis,** Connell argued the most species-rich communities would occur where the frequency of disturbance was intermediate, permitting both colonists and competitors to coexist. A variation of this hypothesis is that species richness is greatest in areas where disturbances are neither too large nor too small but are of intermediate size, sufficient to create patches of open habitat but not to obliterate all species. It also follows that richness would be relatively low soon after a disturbance and greatest at intermediate time intervals after a disturbance, when both colonists and competitors co-occur. At long time intervals after a disturbance, the colonists would likely have been replaced by superior competitors.

Wayne Sousa (1979) provided an elegant experimental test of the intermediate disturbance hypothesis in a marine intertidal situation. By examining the algal and animal species richness on intertidal boulders, he found that small boulders that were easily disturbed by waves carried a mean of 1.7 species, mainly *r*-selected species. Most species were crushed on these frequently moving boulders. Large boulders, which were rarely moved by waves, had a mean of 2.5 species, mainly *K*-selected species. On such boulders competitive dominants supplanted many other species. Intermediate-sized boulders, which had an intermediate frequency of disturbance, had the most species, an average of 3.7 per boulder, because they contained a mix of *r*- and *K*-selected species.

Other tests of the intermediate disturbance hypothesis provide additional supporting data. For example, Tsutomu Hiura (1995) examined the relationship between plant species

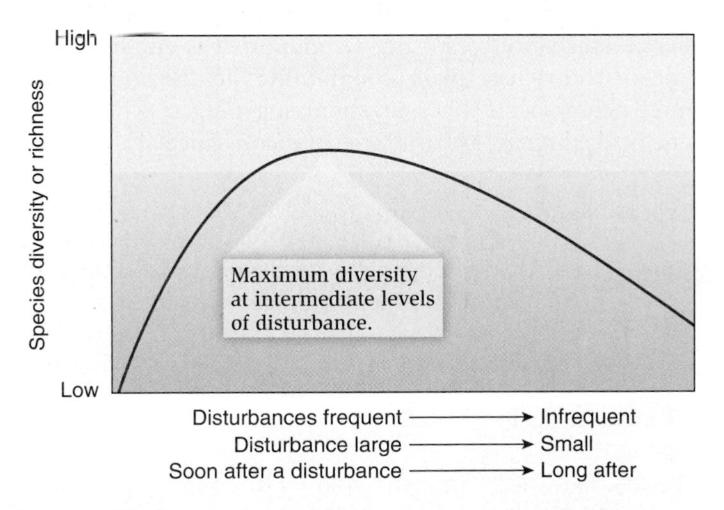


**Figure 18.8 The intermediate disturbance hypothesis.** According to this hypothesis, the species richness of a community is never in a state of equilibrium. High species diversity or richness is maintained by intermediate levels of disturbances, such as fairly frequent fires or windstorms. (Redrawn from Connell, 1978.)

(a)

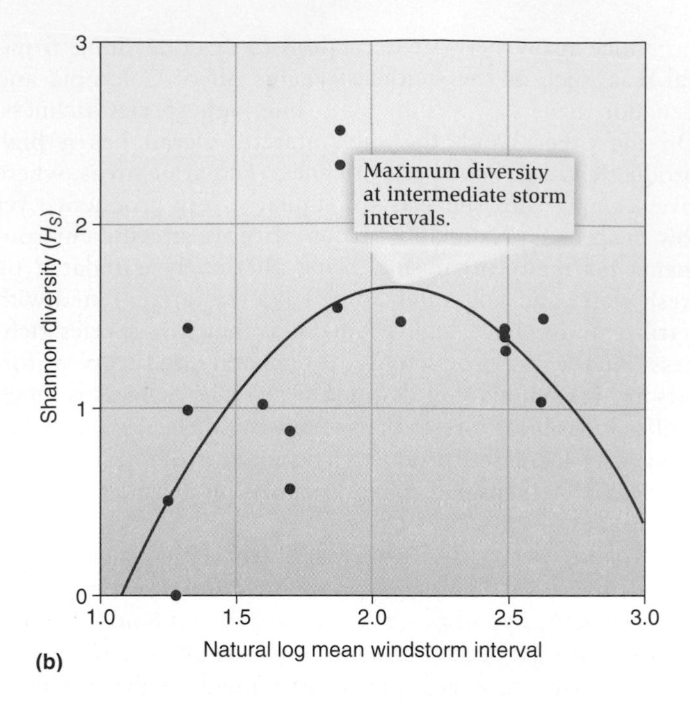


(b)

**Figure 18.9 The relationship between species diversity and disturbance interval.** **(a)** Japanese beech forests may be subject to gale-force winds during storms, Tsuta-numa, Aomori prefecture. **(b)** Tree diversity is greatest in areas of more frequent disturbance. Disturbance interval was measured as log mean windstorm interval. (After Hiura, 1995.)

diversity and storm frequency in beech forests in Japan over a scale of 10° latitude (**Figure 18.9a**). Locations that sustained an intermediate frequency of disturbance in terms of the mean windstorm interval had the highest species diversity (**Figure 18.9b**). Windstorms knock down trees, creating light gaps and allowing other plant species to thrive. Without windstorms only the strongest competitors would survive. However, very frequent windstorms kill many tree species and then only good colonists are present. Maximum diversity occurred in areas of intermediate storm intervals.

Other data also support the intermediate disturbance hypothesis. Coral reefs maintain high diversity in areas disturbed by hurricanes, and the richest tropical forests occur where disturbance by storms is common. It is interesting that some of the richest plant communities in the southeastern United States occur on army bombing ranges, which suffer some of the highest known rates of disturbance.

### Check Your Understanding

**18.4** In the tide pools along the rocky intertidal shoreline of Massachusetts, the periwinkle snail, *Littorina littorea*, grazes algae. The number of species of algae are shown against the density of snails below:

| Number of algal species | 3 | 5 | 9 | 12 | 3 | 6 | 1 |
|---|---|---|---|---|---|---|---|
| Density of snails per $m^2$ | 10 | 50 | 100 | 150 | 200 | 230 | 250 |

Plot the data and determine if they support the intermediate disturbance hypothesis.

## 18.5 Natural Enemies Promote Increased Species Richness at Local Levels

The natural enemies hypothesis supposes that the action of natural enemies reduces prey density and decreases prey competition, thus permitting the coexistence of more prey species. Robert Paine (1966) provided evidence to support this hypothesis from studies on the intertidal communities of the U.S. northwest coast at two wave-exposed sites: Mukkaw Bay and Tatoosh Island in Washington State, where the starfish *Pisaster* is a common predator (**Figure 18.10**).

*Pisaster* preys primarily on the mussel *Mytilus californicus*, but also on a predatory whelk, *Thais*, and on chitons, limpets, acorn barnacles, and bivalves. After removal of *Pisaster* from a section of the shore, species richness decreased from 15 to 8 species. The mussel *M. californicus* and a gooseneck barnacle, *Pollicipes polymerus*, increased, crowding out many other species. In unmanipulated sections of shore, *Pisaster* removed *Mytilus* and *Pollicipes*, preventing them from monopolizing space. Removal of any other single species from the system did not affect species diversity so drastically as *Pisaster* removal. For this reason, Paine termed *Pisaster* a keystone species, by analogy with the keystone that holds all the other stones in an arch in place (look ahead to Chapter 25). Such a phenomenon, wherein natural enemies permit the coexistence of many prey species, was noted even by Darwin, who observed more grass species coexisting in areas grazed by sheep or rabbits than in ungrazed areas.

When trying to explain spatial patterns in species richness, it is likely that different processes occur over different scales (**Figure 18.11**). On a continental scale, evolutionary speed and species-area effects may be important. On a regional scale, we know that the time since the last glaciation has the potential to change patterns of species richness. On a landscape scale, disturbance frequency may be important. At small scales, biotic interactions between species, such as the action of natural enemies, may also influence species richness. At any given point on the globe, species richness may be affected by the simultaneous interaction of these factors. Furthermore, global climate change is likely to drastically change species richness at local scales as temperature and precipitation regimes change (see **Global Insight**).

**Check Your Understanding**

**18.5** Why is tree species richness in the United States greatest in the Southeast but bird and mammal species richness is greatest in the Southwest?

**Figure 18.10 Natural enemies may promote increased species richness.** Predation by *Pisaster* increases the diversity of prey species on the western coasts of the United States and Canada.

**Continent**
Species area effects: Evolutionary speed

**Region**
Dispersal of species: Evolutionary time: Ecological time

**Landscape**
Disturbance frequency

**Local population**
Predation and other biotic interactions

**Figure 18.11 Processes affecting species richness at different spatial scales.** Each level includes all lower ones and nests within all higher ones.

## 18.6 Communities in Climatically Similar Habitats May Themselves Be Similar in Species Richness

Besides the generally recognized progression in species richness from poles to the equator, communities from different parts of the globe may show other patterns. For example, plant biologists have long noticed the similarity of vegetation in climatically similar areas around the globe: Cacti-like plants occur in deserts of all types. Do such areas show a similarity in degree of species richness?

The evidence is equivocal. Dolph Schluter and Robert Ricklefs (1993) provided a list of many examples of similar species richness in similar habitats around the globe (**Table 18.1**). In the same article, however, was a nearly equal-sized list of dissimilar species richness in similar habitats

**Table 18.1 Examples of nearly equal species richness in similar habitats around the globe.**

| Organism | Habitat | Area 1 (# species) | Area 2 (# species) |
|---|---|---|---|
| Plants | Desert | Arizona (250) | Argentina (250) |
| Plants | Semiarid | N. America (70) | Australia (65) |
| Sea anemones | Rocky shore | Washington (11) | S. Africa (11) |
| Ants | Desert | Arizona (25) | Argentina (25) |
| Lizards | Mediterranean scrub | California (9) | Chile (8) |
| Birds | Desert | Arizona (57) | Argentina (61) |
| Birds | Mediterranean scrub | California (30) | S. Africa (28) |
| Small mammals | Shrubland | California (7) | S. Africa (6) |

*Source:* After Schluter and Ricklefs, 1993.

GLOBAL INSIGHT

## Species Richness Could Be Reduced by Changing Climate

Much of the California coast is chaparral, a grassland habitat with isolated trees. It enjoys a Mediterranean climate with mild, wet winters and long, dry summers. Models have predicted a future climate of increased winter rains and a prolonged rainy period that stretches into the spring. Kenwynn Suttle and colleagues (2007) performed a long-term large-scale rainfall manipulation in a northern California grassland, beginning in 2001 and extending through 2005. In each of 18 large circular 70-m$^2$ plots, one of three treatments was applied: (1) additional winter rain from a sprinkler system, a 20% increase of natural rain over the normal rainy season; (2) additional spring rain from a sprinkler system, a 20% increase added from April through June; and (3) unmanipulated controls. Initially the watered communities responded with huge increases in biomass of nitrogen-fixing herbs and annual grasses, especially when the length of the rainy season was extended into spring. Plant species richness also increased (**Figure 18.A-i**), as did the richness of the herbivores that fed upon the plants and the predators of the herbivores, as reflected in the total number of invertebrate families (**Figure 18.A-ii**). Over time, these large increases did not occur where additional water was added in winter, the normal period of the rainy season. However, in the spring-watered plots, the growth of the annual grasses accelerated as the nitrogen-fixing plants died and decomposed, providing a rich source of nutrients. Eventually the annual grass litter suppressed germination of nitrogen-fixing and nonfixing herbs. Plant species richness decreased, as did the richness of the associated herbivores and predators. This trend continued into 2005, a year of exceptionally late natural rainfall that mirrored the spring water treatment. Changes in the timing of rainfall in California, as would occur under a scenario of global warming, will have dramatic effects on species richness on many levels.

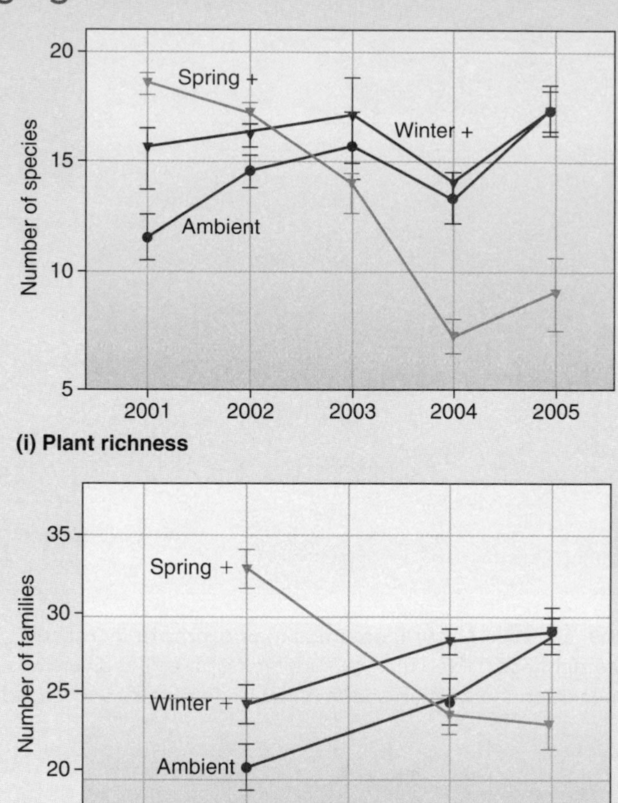


**Figure 18.A Species richness can be affected by the timing of rain.** Means and standard errors are shown. In 2005, naturally occurring spring rains mirrored the watering treatment. **(i)** Plant species richness. **(ii)** Invertebrate species richness. (After Suttle et al., 2007.)

around the globe (**Table 18.2**). They concluded that near-identical richness does not always result from similar environmental conditions.

John Lawton and colleagues (1993) examined community similarity between habitats in a different way, by examining convergence in guilds, the way species utilize a common resource. Bracken fern, *Pteridium aquilinum*, is a widespread and common native member of the flora of all the nonpolar continents (**Figure 18.12**). In many places it is regarded as a serious weed, and its herbivorous insect fauna has been thoroughly studied, sometimes with a view to controlling the plant by releasing biological control agents. Over a period of 20 years, surveys of insects were conducted in Hawaii, New Mexico, Great Britain, South Africa, Brazil, New Guinea, and Australia. Each species was identified taxonomically and by its feeding position on the plant and its feeding technique. Taxa include Collembola, soil feeders; Orthoptera, grasshoppers and crickets; Heteroptera and Homoptera, both types of sucking bugs; Thysanoptera, thrips that rasp plant surfaces; Diptera, flies; Lepidoptera, moths and butterflies; Coleoptera, beetles; Hymenoptera, wasps and ants; and Acarina, mites. The guild varied remarkably, giving no evidence of taxonomic similarity in the fauna (**Figure 18.13**). There are, for instance, no Coleoptera on bracken in Britain but many beetle species in New Guinea; no Diptera on the plants in South Africa but many fly species in Great Britain; no Hymenoptera in Australia but many in Great Britain. Insects appear to have independently colonized bracken in different parts of its range over evolutionary time. Hawaii has only one or two confirmed bracken herbivores, while New Guinea has the

**Table 18.2 Examples of highly dissimilar species richness in similar habitats around the globe.**

| Organism | Habitat | Area 1 (# species) | Area 2 (# species) |
|---|---|---|---|
| Mangroves | Mangal | Malaysia (40) | W. Africa (3) |
| Algae | Rocky shore | Washington (17) | S. Africa (3) |
| Chitons | Rocky shore | Washington (10) | S. Africa (3) |
| Insects | Streams | Australia (60) | N. America (26) |
| Bees | Desert | Argentina (188) | Arizona (116) |
| Bees | Mediterranean scrub | California (171) | Chile (116) |
| Ants | Desert | Australia (37) | N. America (16) |
| Lizards | Desert | Australia (27) | N. America (7) |
| Birds | Peatlands | Finland (33) | Minnesota (18) |

*Source:* After Schulter and Ricklefs, 1993.

richest fauna on bracken, with about 30 species. The variation in the total number of insect species exploiting bracken is partly a function of how common and widespread the plant is in each geographic region (look ahead to Figure 21.8).

Lawton and colleagues also asked if there was convergence in the ways in which bracken is partitioned among herbivores, with guilds including chewers, sap-suckers (that

**Figure 18.12** **Bracken fern, *Pteridium aquilinum*, in recently burned piney woods, central Louisiana.**

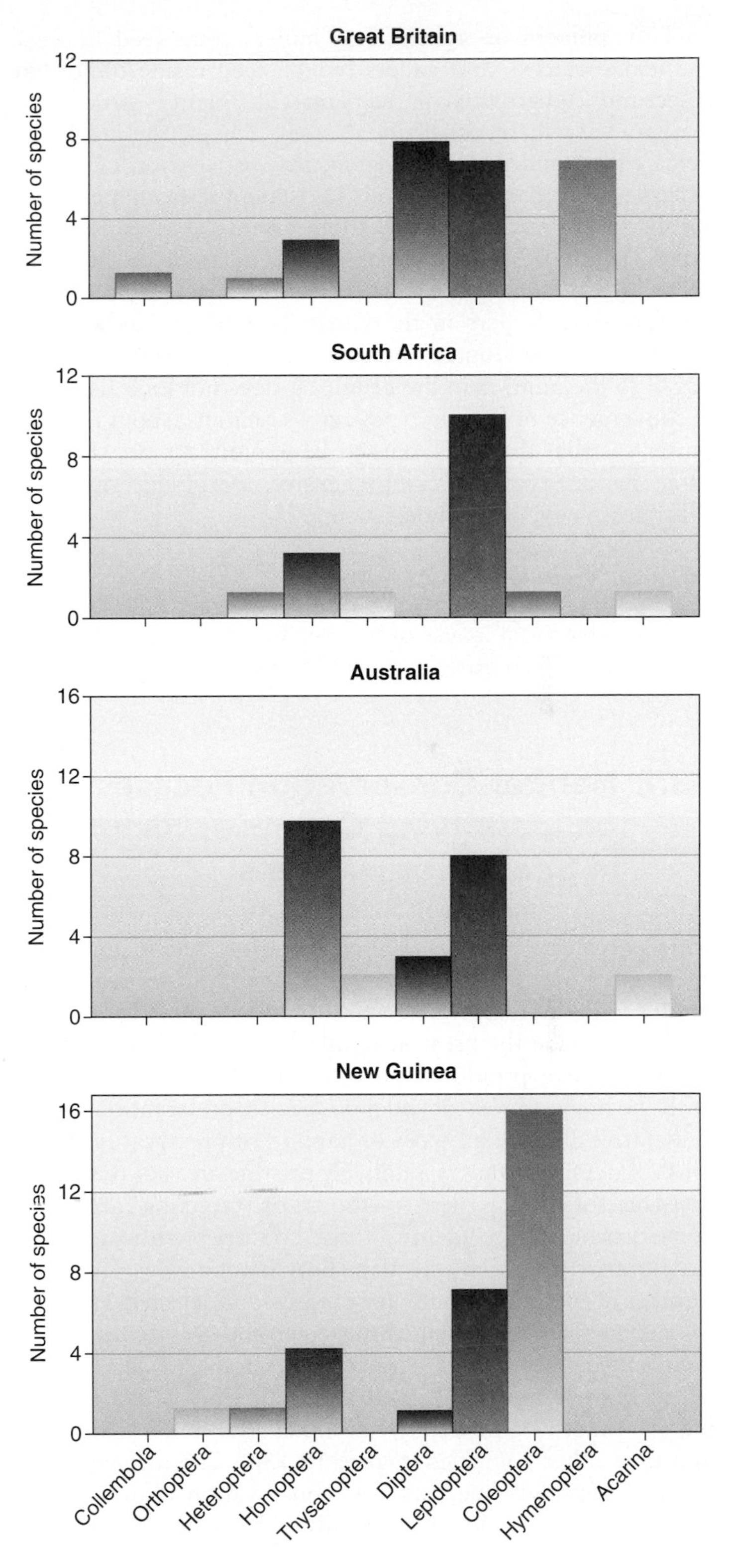


**Figure 18.13** **The taxonomic composition, by order, of the arthropod community on bracken fern, in different parts of the world.** The assemblage differs widely, showing little evidence of community similarity. (After Lawton et al., 1993.)

feed on phloem or xylem), leaf miners (that feed between the leaf surfaces), and gallers (which feed inside tumor-like insect-induced growths on the plant). The plant is structurally similar everywhere, consisting of stems, leaves, and major leaf veins called midribs. Once again, the distribution of species across resources on the plant is idiosyncratic from locality to locality, with numerous niches left vacant by certain feeding types (**Figure 18.14**). In other words, parts of the plant go unutilized in certain areas of the world. For example, there are few stem feeders in the Great Britain compared to the number in New Guinea and few chewers in Australia compared to the number in the Britain. It does not look like there is convergence of feeding types across regions. About the only pattern is that the leaf seems to be exploited more than the other parts. Few other comprehensive, worldwide studies of this nature have been undertaken.

**Check Your Understanding**

**18.6** Why might leaves of bracken fern be exploited more than other parts of the plant?

## 18.7 Habitat Conservation Focuses on Identifying Countries Rich in Species or Habitats

Conservation biologists often must make decisions regarding which areas and habitats should be protected to maximize species richness. Many conservation efforts have focused on saving habitats in so-called megadiversity countries because they often have the greatest number of species. More recent strategies have promoted preservation of certain key areas with the highest levels of endemic species, preservation of representative areas of all types of habitat, or preservation of the "last of the wild," that is, relatively pristine areas of the world.

One of the earliest methods of targeting areas for conservation was to identify those countries with the greatest numbers of species, the **megadiversity** countries. Using the number of species of plants, vertebrates, and selected groups of insects as criteria, Russell Mittermeier and colleagues (1997) determined that just 17 countries are home to nearly 80% of all known species. Australia, Brazil, China, and Colombia top the list, followed by Zaire, Ecuador, India, Indonesia, and nine other countries (**Figure 18.15a**). The megadiversity country approach suggests that conservation efforts should be focused on the most biologically rich countries. These are generally large countries, and large areas generally have the greatest number of species. Perhaps the greatest drawback of the megadiversity approach is that although megadiversity areas may contain the most species, they do not necessarily contain the most unique species. The mammal species list for Peru is 344 and for Ecuador, 271; of these, however, 208 species are common to both.

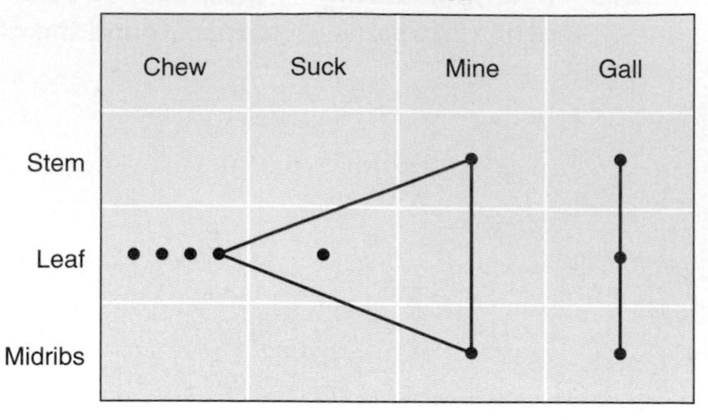


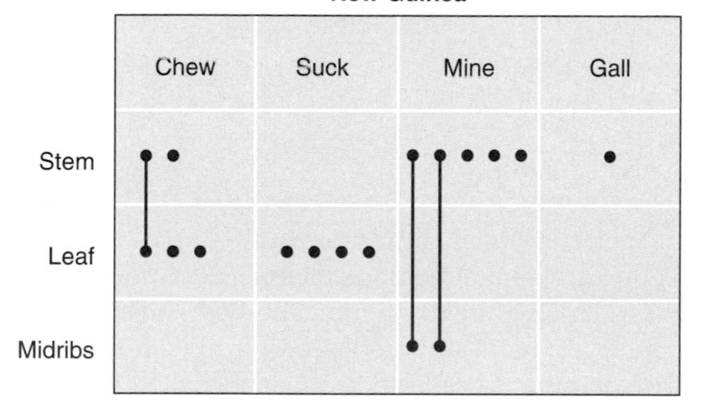


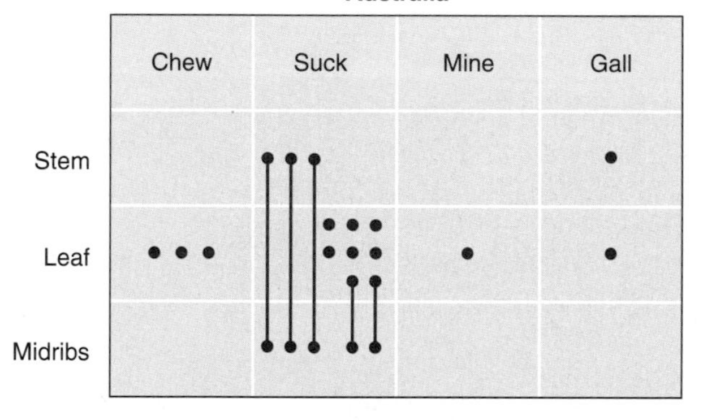


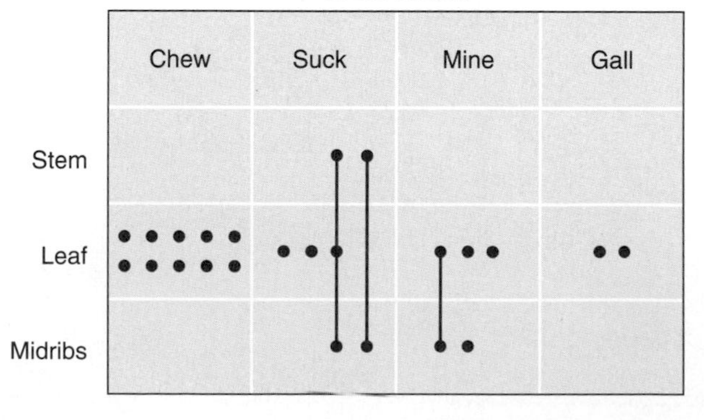


**Figure 18.14 Feeding guilds on bracken fern in different parts of the world.** Each dot represents one species. Feeding sites of species exploiting more than one part of the plant are joined by lines. Many parts of the plant remain underutilized by herbivores. (After Lawton et al., 1993.)

The method of setting conservation priorities adopted by the organization Conservation International takes into account the number of species that are **endemic,** found only in a particular place or region and nowhere else. This approach suggests that conservationists focus their efforts on **biodiversity hot spots,** those areas with the greatest number of endemic species. To qualify as a hot spot, a region must meet two criteria: it must contain at least 1,500 species of vascular plants as endemics, and it must have lost at least 70% of its original habitat. Vascular plants were chosen as the primary group of organisms to determine whether or not an area qualifies as a hot spot, mainly because most other terrestrial organisms depend on them to some extent.

Mittermeier and colleagues (2004) identified 34 hot spots that together occupy a mere 2.3% of the Earth's surface but contain 150,000 endemic plant species, or 50% of the world's total (**Figure 18.15b**). Of these areas, the tropical Andes and Sundaland, the region including Malaysia, Indonesia, and surrounding islands, have the most endemic plant species (**Table 18.3**). This approach posits that protecting geographic hot spots will prevent the extinction of a larger number of endemic species than would protecting areas of a similar size elsewhere. The main argument against using hot spots as the criterion for targeting conservation efforts is that the areas richest in endemics—tropical rain forests—would receive the majority of attention and funding, perhaps at the expense of other areas.

In a third approach to prioritizing areas for conservation, ecologists have recently argued that we need to conserve representatives of all major habitats that are under severe threat. The Pampas region of South America is arguably the most threatened habitat on the continent because of conversion of its grasslands to agriculture, but it does not compare well in richness or endemics to the tropical rain forests. Nevertheless, it is a unique area that without preservation could disappear. By selecting habitats that are severely threatened, many areas that are not biologically rich in species may be preserved in addition to the less immediately threatened, but richer, tropical forests. Jonathon Hoekstra and colleagues (2005) identified the biomes at greatest risk of habitat loss because of conversion to other uses such as agriculture, pasture lands, or plantation forests (**Figure 18.15c**). Within these areas, some regions were seen as more critical than others. Habitat loss has been most extensive in temperate grasslands, and tropical and temperate deciduous forest, where, in each case, over 40% of the land has been converted to other uses. At the same time, very little, less than 10%, of these agriculturally rich areas has been protected. Such habitats are thus termed "crisis ecoregions."

The final approach to conservation that we will discuss involves preservation of regions of the world relatively untouched by humans. Eric Sanderson and colleagues (2002) mapped out the extent of the human footprint on the globe. The areas of the Earth that fell within the lowest 10% of the human affected areas were termed the "10 percent wildest areas" or the "last of the wild" (**Figure 18.15d**). Such areas, because they are relatively pristine, offer a great opportunity for conservationists because of their relatively intact communities. Such areas include tundra and boreal forests of Russia and Canada as well as some desert biomes and tropical forests.

T. M. Brooks and colleagues (2006) have contrasted these four conservation strategies (**Figure 18.16**). They suggest that many conservation strategies weight the value of endemic species, taxonomically unique species, or unique habitats in terms of an attribute they called irreplaceability. The megadiversity country and biodiversity hot spot approaches rate high on the attribute of irreplaceability. Other strategies place greater value on the actual vulnerability of biomes that are currently being threatened. This includes the crisis ecoregion and biodiversity hot spot approaches. They also ranked the strategies in terms of being proactive or reactive in nature.

**Table 18.3 Global numbers of endemic species in the top 10 biodiversity hot spots of the world. Areas are ranked by the numbers of species of endemic plants.**

| Rank | Hot spot | Plants | Birds | Mammals | Reptiles | Amphibians | Freshwater fish |
|---|---|---|---|---|---|---|---|
| 1 | Tropical Andes | 15,000 | 584 | 75 | 275 | 664 | 131 |
| 2 | Sundaland | 15,000 | 146 | 173 | 244 | 172 | 350 |
| 3 | Mediterranean Basin | 11,700 | 32 | 25 | 77 | 27 | 63 |
| 4 | Madagascar | 11,600 | 183 | 144 | 367 | 226 | 97 |
| 5 | Brazil's Atlantic Forest | 8,000 | 148 | 71 | 94 | 286 | 133 |
| 6 | Indo-Burma | 7,000 | 73 | 73 | 204 | 139 | 553 |
| 7 | Caribbean | 6,550 | 167 | 41 | 468 | 164 | 65 |
| 8 | Cape Floristic Province | 6,210 | 6 | 4 | 22 | 16 | 14 |
| 9 | Philippines | 6,091 | 185 | 102 | 160 | 74 | 67 |
| 10 | Brazil's Cerrado | 4,400 | 16 | 14 | 33 | 26 | 200 |

*Source:* From Mittermeier et al., 2004.

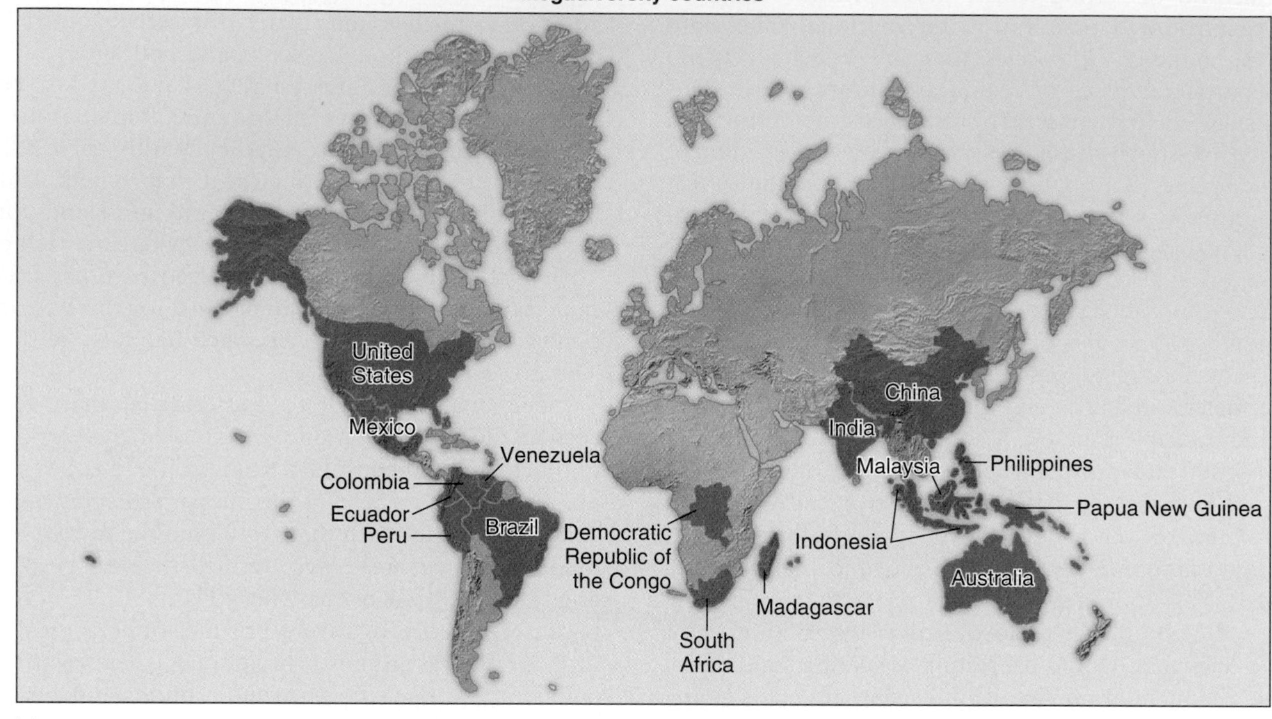


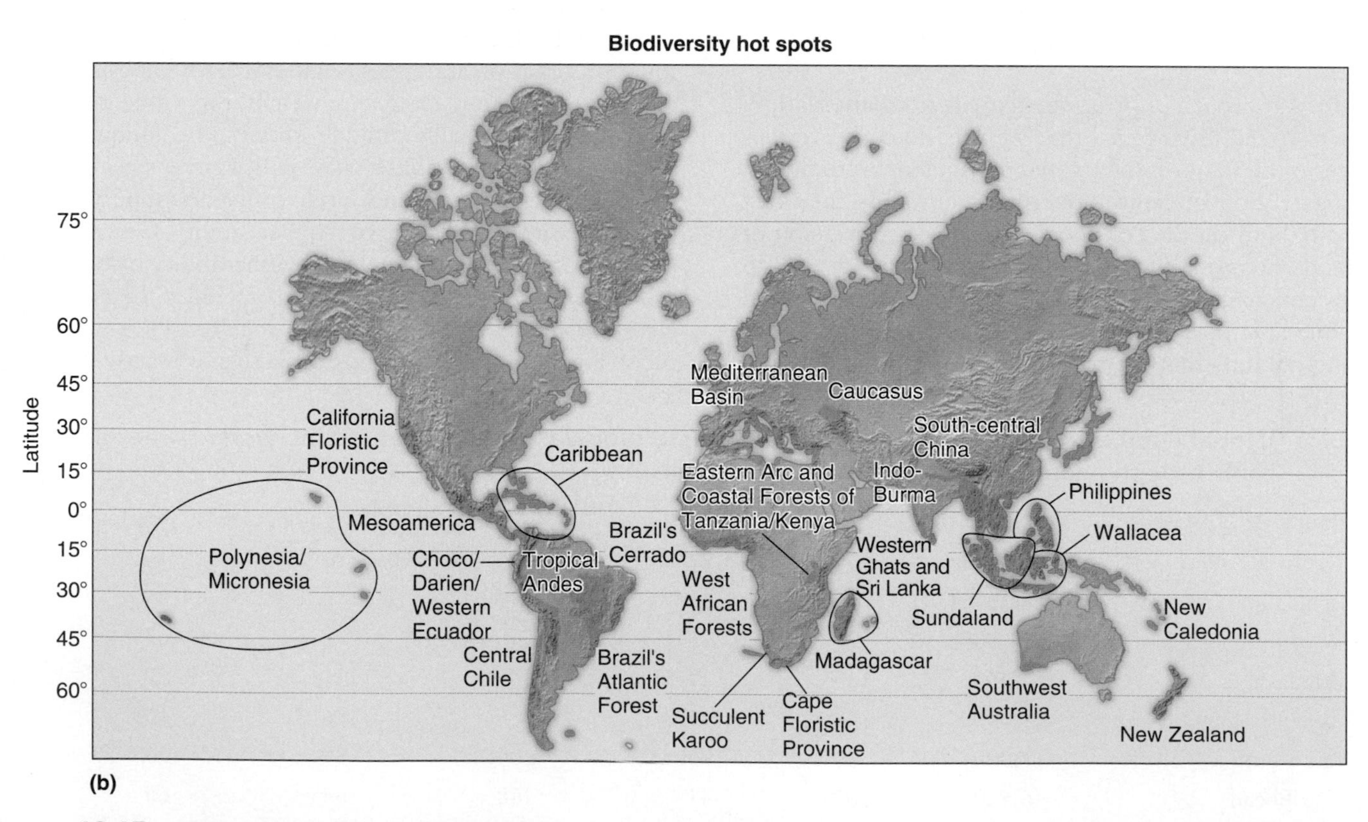


**Figure 18.15** **Maps of four global biodiversity conservation priority approaches.** **(a)** Megadiversity countries. (After Mittermeier et al., 1997.) **(b)** Biodiversity hot spots. (After Mittermeier et al., 2004.) **(c)** Crisis ecoregions. (After Hoekstra et al., 2005.) **(d)** "Last of the wild." (After Sanderson et al., 2002.)

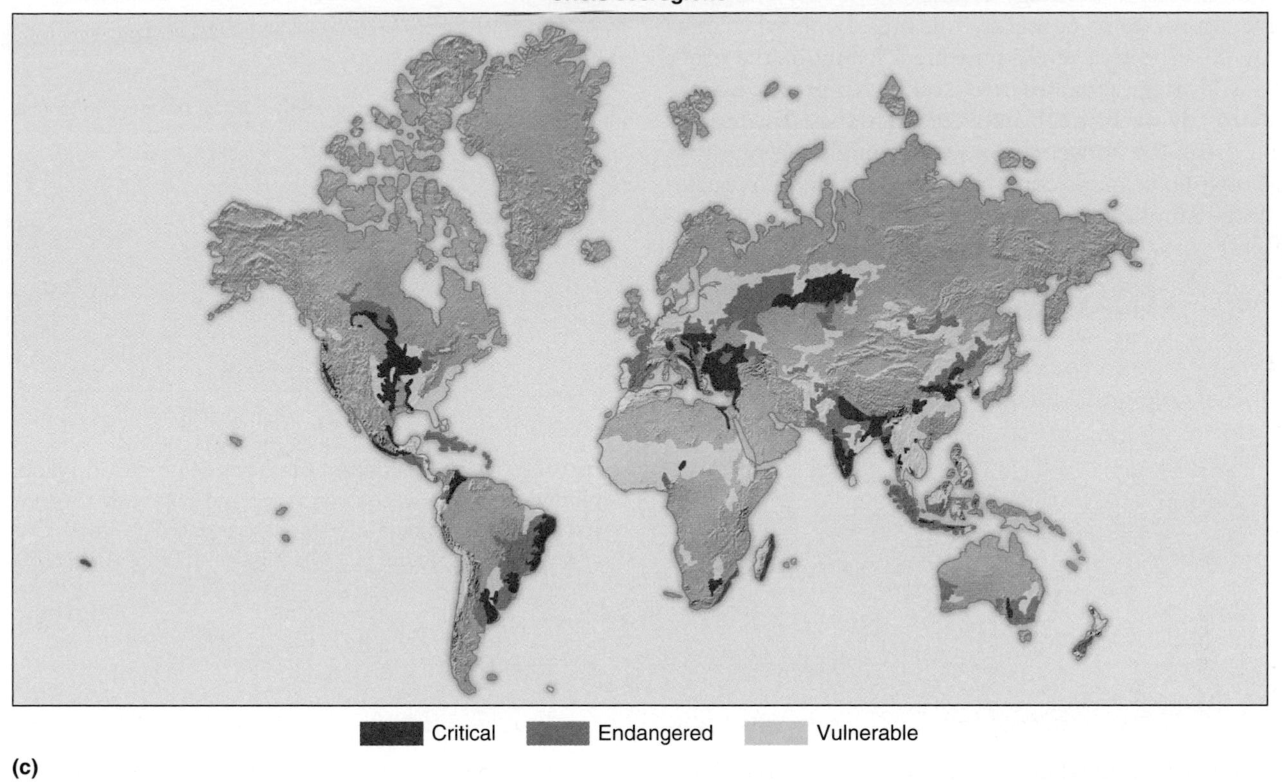
Crisis ecoregions
Critical
Endangered
Vulnerable

(c)

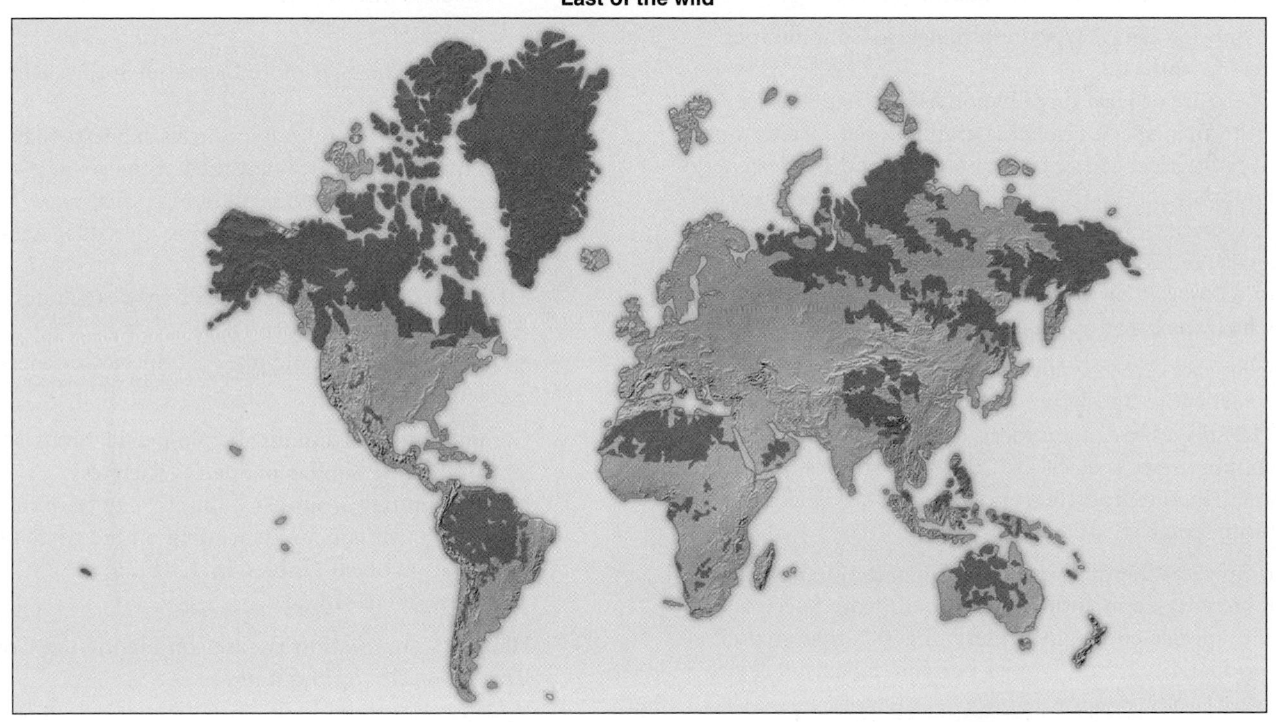
"Last of the wild"

(d)

The "last of the wild" strategy is one of the few that take a proactive approach to conservation, preserving wild areas before they have been severely impacted. Many of the other conservation strategies are reactive, seeking to preserve areas that have already been much impacted. The best strategy of identifying areas for conservation efforts might be one that creates a portfolio of areas containing those with high species richness, large numbers of endemic species, various representative habitat types, and relatively untouched areas.

### Check Your Understanding

**18.7** If we were to use most of our resources to protect crisis ecoregions, what might we be giving up?

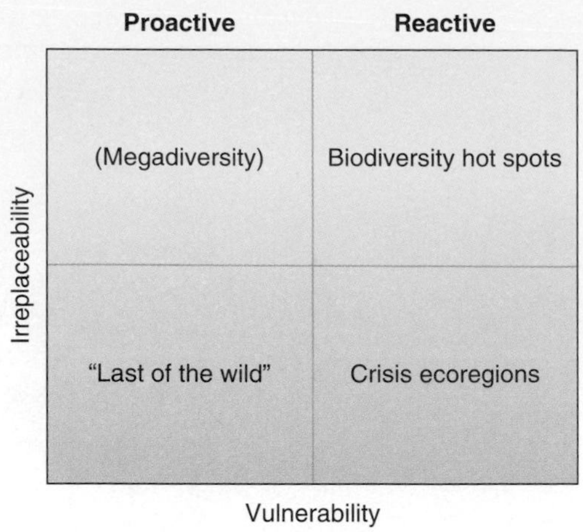

**Figure 18.16 Different methods of preserving global biodiversity.** The megadiversity country approach does not incorporate vulnerability as a criteria and emphasizes only high irreplaceability, hence its position in parentheses. (Modified from Brooks et al., 2006.)

## SUMMARY

- A strong latitudinal gradient in species richness exists for many taxa, with fewer species toward the poles and more species toward the equator (Figure 18.1).

**18.1 The Species-Time Hypothesis Suggests Communities Diversify with Age**

- There are at least three hypotheses to explain the latitudinal species richness gradient: the species-time hypothesis, species-area hypothesis, and species-energy hypothesis.
- The species-time hypothesis suggests that temperate areas are less species-rich because they have had less time to recover after the ice ages than have tropical areas, which were relatively unaffected (Figures 18.2–18.4).

**18.2 The Species-Area Hypothesis Suggests That Large Areas Support More Species**

- The species-area hypothesis suggests that larger areas contain more species and, since the Tropics have larger areas than temperate or polar biomes, they contain more species (Figures 18.5, 18.6).

**18.3 The Species-Energy Hypothesis Suggests That Greater Productivity Permits the Existence of More Species**

- The species-energy hypothesis suggests that greater productivity in the Tropics permits the existence of more species (Figure 18.7).

**18.4 The Intermediate Disturbance Hypothesis Suggests That Species Richness Is Highest in Areas of Intermediate Levels of Disturbance**

- Species richness at more local scales may also be affected by the frequency of disturbance and the action of natural enemies.
- The intermediate disturbance hypothesis suggests that species richness is highest in areas where disturbances, such as storms or fires, are of intermediate frequency (Figures 18.8, 18.9).

**18.5 Natural Enemies Promote Increased Species Richness at Local Levels**

- The natural enemies hypothesis suggests that natural enemies may reduce densities of the strongest prey competitors, thus allowing other, poorer competitors to exist, increasing overall species richness (Figure 18.10).
- Different processes may affect species richness at different scales (Figure 18.11).
- Climate change could alter the species richness of areas (Figure 18.A).

**18.6 Communities in Climatically Similar Habitats May Themselves Be Similar in Species Richness**

- Communities in similar habitats may be quite similar or dissimilar, depending on the particular habitat involved (Tables 18.1, 18.2; Figures 18.12–18.14).

**18.7 Habitat Conservation Focuses on Identifying Countries Rich in Species or Habitats**

- Conservation biologists use many different strategies to preserve species richness, focusing on megadiversity countries, biodiversity hot spots, crisis ecoregions, and the "last of the wild" (Figure 18.15).
- Conservation strategies may focus on regions of high vulnerability or high irreplaceability, and may be proactive or reactive (Figure 18.16).

## TEST YOURSELF

1. Which is not a variation of the species-time hypothesis?
   a. Resident species in the Tropics have not yet evolved forms to exploit similar temperate habitats
   b. Tropical species have not migrated back to temperate areas following glaciation
   c. Evolution is faster in the Tropics than in temperate areas
   d. Rates of disturbance are lower in temperate than in tropical areas
2. There are 580 species of benthic invertebrates in unglaciated Lake Baikal, in Russia, compared to only 4 species in glaciated Great Slave Lake in Canada. This supports the _____ hypothesis.
   a. Species-time
   b. Species-area
   c. Species-energy
   d. Natural enemies
   e. Intermediate disturbance
3. The number of insect herbivore species supported by introduced trees in Great Britain is similar to the number supported by native trees occupying the same geographic range. This supports the _____ hypothesis.
   a. Species-time
   b. Species-area
   c. Species-energy
   d. Natural enemies
   e. Intermediate disturbance
4. Which of the following explanations for the latitudinal species richness gradient explains the high numbers of tree species in the southeastern U.S. compared to the rest of the country?
   a. The intermediate disturbance hypothesis
   b. The species-area hypothesis
   c. The species-energy hypothesis
   d. The natural enemies hypothesis
   e. The species-time hypothesis
5. Which temperate area of the world has the highest number of tree species?
   a. North America
   b. Europe
   c. China
   d. Australia
   e. England
6. According to the intermediate disturbance hypothesis, which size of boulders supports the highest number of sessile species?
   a. Large
   b. Medium
   c. Small
   d. All of the above have equal numbers
7. The insect herbivores of bracken fern in different areas of the world show great taxonomic similarities.
   a. True
   b. False
8. Biodiversity hot spots contain:
   a. High numbers of species
   b. High numbers of endemic species
   c. Many different types of habitats
   d. High numbers of wilderness areas
9. Which is not a megadiversity country?
   a. Australia
   b. Brazil
   c. China
   d. Ecuador
   e. Russia
10. Climatically similar areas in different parts of the world always have the same number of species.
   a. True
   b. False

## CONCEPTUAL QUESTIONS

1. Distinguish between the following three hypotheses to explain the latitudinal species richness gradient: (a) the species-time hypothesis, (b) the species-area hypothesis, and (c) the species-energy hypothesis.
2. Distinguish between the following strategies used to conserve species richness: (a) megadiversity countries, (b) biodiversity hot spots, (c) crisis ecoregions, (d) "last of the wild."

## DATA ANALYSIS

Provide an explanation for global differences in coral species richness, shown as isobars in the accompanying figure. Assume that coral biomass is highest in areas with the highest numbers of species.

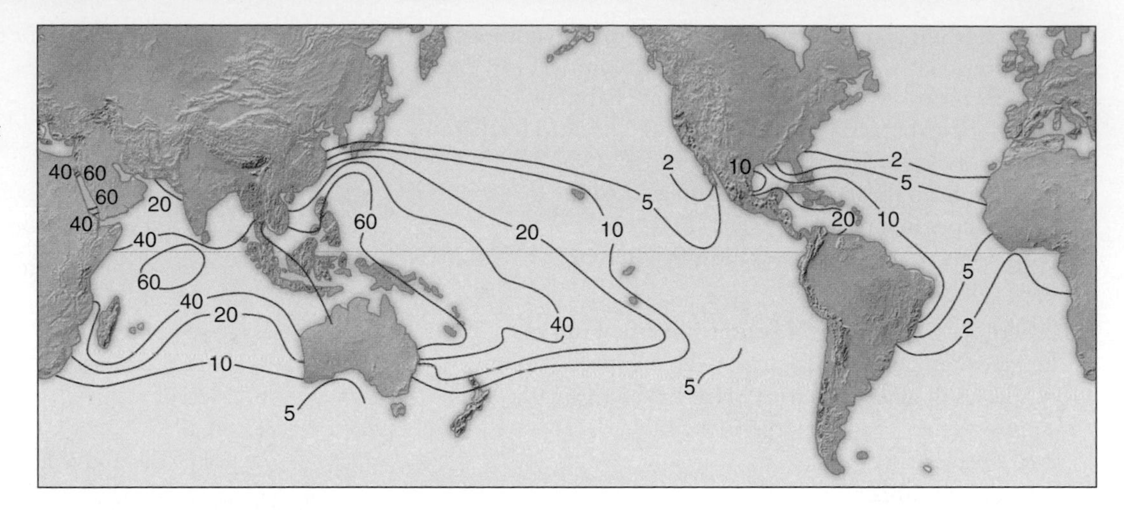

West Indian manatee, *Trichechus manatus*, in a clear Florida spring.

CHAPTER 19

# Species Richness and Community Services

## Outline and Concepts



In 2009 Jeff Corwin, an American conservationist and host for programs on *Animal Planet* and other television networks, published a book entitled *100 Heartbeats: The Race to Save the Earth's Most Endangered Species*. The Hundred Heartbeat Club had been created earlier by biologist E. O. Wilson to highlight the plight of animal species, such as Spix's macaw (*Cyanopsitta spixii*) in Brazil, the Chinese river dolphin (*Lipotes vexillifer*), and the Philippine eagle (*Pithecophaga jefferyi*), that have 100 or fewer individuals left alive (that is, species that have "fewer than 100 hearts beating on our planet"). Corwin's book was a result of 2 years' work traveling and researching such issues as the amphibian-killing chytrid fungus (refer back to the chapter-opening photo of Chapter 1), the killing of elephants for their tusks, and the destruction of Indonesia's rain forests. Sadly, many of the species in the Hundred Heartbeat Club are still headed toward extinction or have recently become extinct.

(a)

(b)

**Figure 19.1 Rare species may have traits useful to humans.** **(a)** Café marron, a wild relative of the coffee plant, here being propagated in London's Kew Gardens, is more tolerant of colder temperatures than most other coffee species. **(b)** The gastric brooding frog of Australia may have been able to turn off acid production in its stomach.

There is concern that loss of plant and animal species is undesirable, for several reasons. First, loss of species would impair future development of important products and processes in agriculture, medicine, and industry. For example, the Café marron, a wild relative of the coffee plant, native to a tiny island off the coast of Mauritius, might contain genes that would allow coffee to be grown in a cooler range of climates (**Figure 19.1a**). However, only one plant is left in the wild, so it is virtually impossible to experiment using crosses with wild coffee. Luckily the tree is now being cultivated in London's Kew Gardens, and more individuals should be available to use for restoration projects in its native habitat in the years to come. Taken further, the pharmaceutical industry is heavily dependent on information that is stored in plants. About 25% of the prescription drugs in the U.S. alone are derived from plants, and the 2009 U.S. market value of such drugs was estimated to be $300 billion. Many medicines come from plants found only in tropical rain forests; these include quinine, a drug from the bark of the Cinchona tree, *Cinchona officinalis,* which is used for malaria, and vincristine, derived from rosy periwinkle, *Catharanthus roseus,* which is a treatment for leukemia and Hodgkin's disease. The continued destruction of rain forests could mean the loss of billions of dollars in potential plant-derived pharmaceutical products. Endangered animals too could be valuable to medicine. The gastric brooding frog of Australia incubated its young in its stomach and gave birth through its mouth (**Figure 19.1b**). In the brooding period, female frogs were thought to be able to switch off acid production. Scientists were interested in how this trick was accomplished because it could have helped in ulcer treatment in humans. Sadly, the frog became extinct in the 1980s.

Second, it has been argued that humans have no right to destroy other species and the environment around us. There may be an instinctive bond between humans and other species. E. O. Wilson (1984) proposed a concept known as **biophilia,** which posits that humans have a love of life or living systems. In this hypothesis, Wilson suggests that we have a deep attachment to natural habitats and species because of our close association over millions of years.

Third, humans benefit from the services of natural communities (**Table 19.1**). Forests soak up carbon dioxide, maintain soil fertility, and retain water, preventing floods. Estuaries provide water filtration and protect coastal shores from excessive erosion. Plants and algae remove excess nutrients from rivers and streams, providing a buffer against nutrient pollution and providing us with clean drinking water. Other community services include the maintenance of predator populations to regulate pest outbreaks and reservoirs of pollinators to pollinate crops and other plants.

Economist Robert Costanza and colleagues (1997) attempted to calculate the monetary value of these services to various economies. They came to the conclusion that, at the time, community services were worth more than $33 trillion a year, nearly twice the gross national product of the world's economies combined ($19 trillion) (**Table 19.2**). If we were to include the value of these services in the cost of goods, most goods would cost vastly more than they currently do. While the open ocean has the greatest total global value of all ecosystems, because of its sheer size, perhaps a more meaningful statistic is the per hectare value of different communities. This statistic reveals that shallow aquatic communities, such as estuaries and swamps, are extremely

**Table 19.1 Examples of the world's community services.**

| Service | Example |
|---|---|
| Atmospheric gas supply | Regulation of carbon dioxide, ozone, and oxygen levels |
| Climate regulation | Regulation of carbon dioxide, nitrogen dioxide, and methane levels |
| Water supply | Irrigation, water for industry |
| Pollination | Pollination of crops |
| Biological control | Pest population regulation |
| Wilderness and refuges | Habitat for wildlife |
| Food production | Crops, livestock |
| Raw materials | Fossil fuels, timber |
| Genetic resources | Medicines, genes for plant resistance |
| Recreation | Ecotourism |
| Cultural | Aesthetic and educational value |
| Disturbance regulation | Storm protection, flood control |
| Waste treatment | Sewage purification |
| Soil erosion control | Retention of topsoil, reduction of accumulation of sediments in lakes |
| Nutrient recycling | Nitrogen, phosphorus, carbon, and sulfur cycles |

valuable because of their role in nutrient cycling, water supply, and disturbance regulation. They also serve as nurseries for aquatic life. These communities, once thought of as wastelands, are among the most endangered by pollution and development.

In this chapter we will examine the link between species richness and community services. Does the loss of species impair community services, and are species-rich communities more stable than species-poor communities? Is a species-rich community more able to resist invasion by introduced species? First, we will survey the different theories that have been proposed to link community services with species richness. We provide several lines of evidence to suggest that community services decrease with reduced levels of species richness. Next, we examine the effect of species richness on the ability of a community to resist invasion by non-native species. We show that a community's ability to resist invasion by non-native species varies according to scale. At small scales, increased species richness of native species reduces community invasibility by non-native species. However, at larger scales, high invasive species richness is correlated with high native species richness. We will explore and resolve this seeming paradox.

**Table 19.2 Valuation of the world's community services and value per hectare.**

| Biome | Total global value (US$ billion per year) | Total value (per ha) (US$) | Main services provided |
|---|---|---|---|
| Open ocean | 8,381 | 252 | Nutrient cycling |
| Estuaries | 4,100 | 22,832 | Nutrient cycling |
| Seagrass/algal beds | 3,801 | 19,004 | Nutrient cycling |
| Coral reefs | 375 | 6,075 | Recreational/disturbance regulation |
| Coastal shelf | 4,283 | 1,610 | Nutrient cycling |
| Tropical forest | 3,813 | 2,007 | Nutrient cycling/raw materials |
| Temperate forest | 894 | 302 | Climate regulation/waste treatment |
| Grasslands | 906 | 232 | Waste treatment/food production |
| Tidal marsh | 1,648 | 9,990 | Waste treatment/disturbance regulation |
| Swamps | 3,231 | 19,580 | Water supply/disturbance regulation |
| Lakes, rivers | 1,700 | 8,498 | Water regulation |
| Desert | 0 | 0 | |
| Tundra | 0 | 0 | |
| Ice, rock | 0 | 0 | |
| Cropland | 128 | 92 | Food production |
| Urban | 0 | 0 | |
| Total | 33,268 | | |

*Source:* After Costanza et al., 1997.

## 19.1 Four Hypotheses Explain How Species Richness Affects Community Services

Ecologists have debated the question of how the number of species affects the function of communities. In the 1950s Charles Elton (1958) proposed the **diversity-stability hypothesis,** which stated that the more species that are present, the more stable the community. The concept of stability is addressed later in the chapter. If we use stability as a surrogate for community services, Elton's hypothesis suggests that there is a linear correlation between species richness and community services (**Figure 19.2a**).

Brian Walker (1992) proposed an alternative to this idea, termed the **redundancy hypothesis** (**Figure 19.2b**). According to this hypothesis, ecosystems function well even at extremely low levels of diversity, so most additional species are functionally redundant. Species that play the same roles in a community can compensate for each other if some are lost under particular conditions. Two other alternative hypotheses relating species richness and community services have been proposed. The **keystone hypothesis** (**Figure 19.2c**) supposes that community services plummet as soon as biodiversity declines from its natural levels. The **idiosyncratic hypothesis** (**Figure 19.2d**) addresses the possibility that community services change as the number of species increases or decreases, but that the direction of change is not predictable. Determining which model best describes natural communities is very important, as our understanding of the effect of species loss on community services can greatly affect the way we manage our environment.

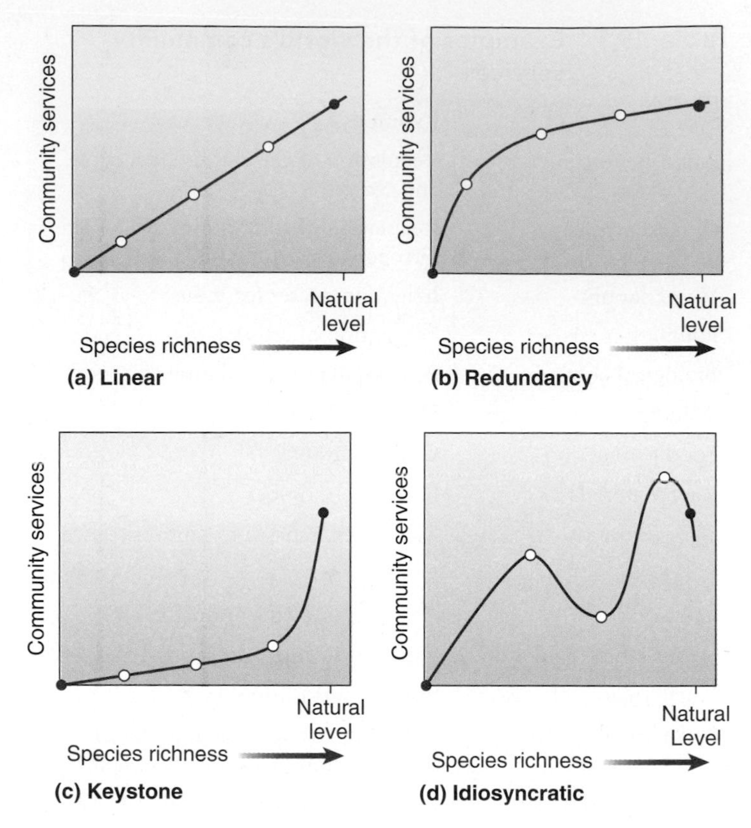


**Figure 19.2 Graphical representations of the possible relationships between species richness and community services.** The two solid dots represent the endpoints of a continuum of species richness. The first dot is at the origin, where there are no species and no community services. The second dot represents natural levels of species richness. In understanding these different hypotheses, we start at the far right-hand side of the figures, where there are natural levels of species richness, and we move toward the left, losing species and seeing what happens to community services. (Modified from Naeem et al., 2002.)

**ECOLOGICAL INQUIRY**

Which relationship between species richness and community function is best supported by the evidence?

### 19.1.1 Recent studies have investigated the relationship between species richness and community function

Experimental studies investigating the link between species richness and community services only began in the 1990s. Before this time, there was no clear consensus as to which hypothesis was supported by the available data. Would a rain forest function any differently with 30 species per hectare than with 300 species?

The earliest and most influential study on the link between species richness and community services was that conducted by Shahid Naeem and his colleagues in laboratories in England (see **Feature Investigation**). The equivalent of nearly $2 million was spent on constructing a facility where abiotic variables could be controlled and the number of species manipulated. The experiment showed that a loss in species richness led to a decline in ecosystem services. The implication was that, in natural systems, species extinction could result in an irreversible loss of community services. Later in the decade, field studies verified the link between species richness and community services.

In the mid-1990s David Tilman and colleagues (1996) performed field experiments at Cedar Creek, Minnesota, to determine the influence of species richness on community services. In these experiments Tilman's group sowed plots, each 3 × 3 m and on comparable soils, with seeds of 1, 2, 4, 6, 8, 12, or 24 species of prairie plants. Exactly which species were sown into each plot was determined randomly from a pool of 24 native species. Thus, particular combinations of species that may have interacted in particularly favorable ways were avoided. The treatments were replicated 21 times, for a total of 147 plots. The results showed that species-rich plots had an increased percentage of cover relative to species-poor

Feature Investigation

## Shahid Naeem's Ecotron Experiments Showed a Relationship Between Species Richness and Community Services

In the early 1990s Shahid Naeem and colleagues (1994) used a series of 14 environmental chambers in a facility termed the Ecotron, at Silwood Park, England, to replicate terrestrial communities that differed only in their level of species richness (**Figure 19.A**). The number of species in each chamber was manipulated to create high, intermediate, and low species-rich communities, each with four feeding levels: plants; herbivores such as insects, snails, and slugs; natural enemies such as parasitoids that fed on the herbivores; and decomposers such as earthworms. The experiment ran for just over 6 months, and species were only added after the feeding level below them was established. For example, herbivores were not added until plants were abundant.

Researchers monitored and analyzed a range of community services, including plant growth, nutrient retention rates, community respiration, and decomposition. One of the results was that plant growth, expressed as percent change in the amount of ground covered by leaves of plants, increased as species richness increased. The researchers suggested that this occurred because there was a greater variety of plant growth forms that could utilize light at different levels of the plant canopy. For example, some plant species grew close to the soil, while others grew taller. Species acted in a complementary way. A larger ground cover also meant a larger plant biomass, greater community respiration, and increased nutrient uptake rates and decomposition (not shown in **Figure 19.A**). For the first time, ecologists had provided an experimental demonstration that reduced species richness could alter or impair the functioning of a community.

**HYPOTHESIS** Reduced species richness can lead to reduced community functioning.

**STARTING LOCATION** Ecotron, a controlled environment facility at the Natural Environment Research Council (NERC) Centre for Population Biology in Silwood Park, England.

**Conceptual level** | **Experimental level**

1. Construct 14 identical experimental chambers.

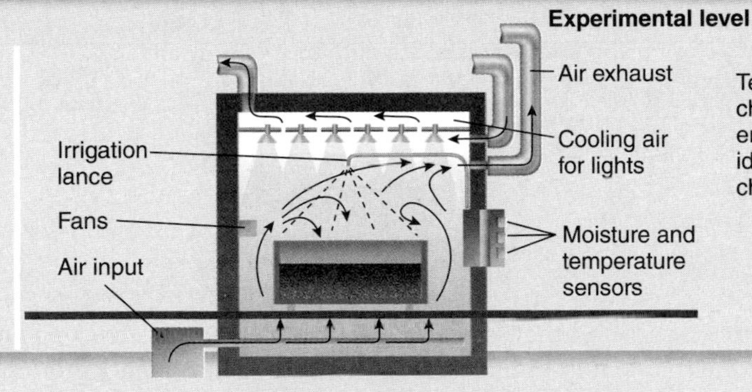


Temperature- and humidity-controlled chambers are used to control environmental conditions and allow identical starting conditions in all chambers.

2. Add different combinations of species to the 14 chambers. The species added were based on 3 types of model communities each with 4 trophic levels but with varying degrees of species richness.

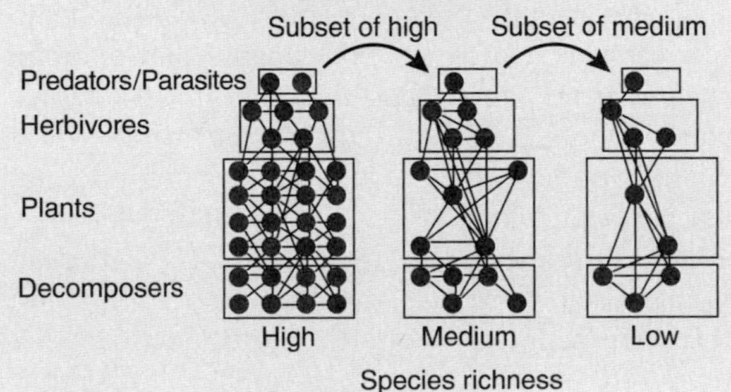


Circles represent species, and lines connecting them represent biotic interactions among the species. Note that the lower richness community is a subset of its higher richness counterpart and that all community types have 4 trophic levels.

3 Measure and analyze a range of processes, including vegetation cover and nutrient uptake.

Measurements help determine how each different type of community functions.

4 THE DATA

Data reveals that species-poor communities are less productive than species-rich communities.

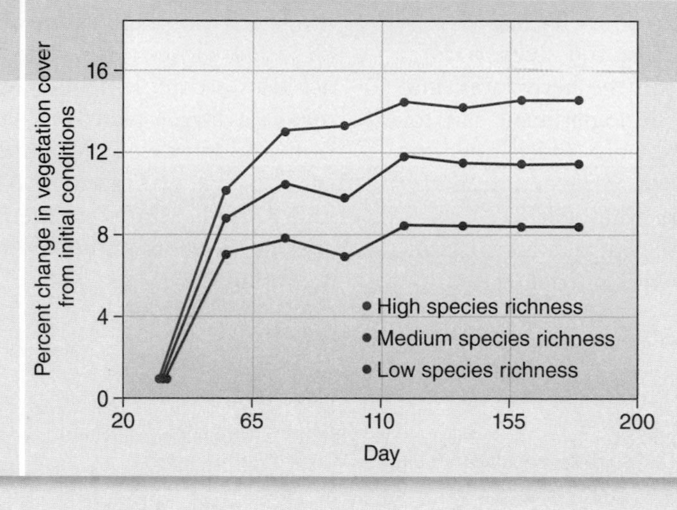


Plant productivity is linked to community richness as measured by the percent change in vegetation cover from initial conditions.

Data reveal that low-diversity communities have lower vegetation cover and are less productive than high-diversity communities.

**Figure 19.A** **Ecotron experiments comparing species richness and ecosystem function.**

plots (**Figure 19.3a**). This was the area of ground that is covered by plant biomass and is a good surrogate for community services. The mechanism proposed to explain the correlation between species richness and community services was again **species complementarity.** While some plants were tall and had deep roots, others were short with shallow roots. This allowed complementary resource use of light, water, and nutrients. Interspecific differences in soil use between plant species allowed fuller use of nitrogen, the main limiting factor in plant growth. So, as species richness increased, more soil nitrogen was used up (**Figure 19.3b**). The experiment had shown that species richness did influence community services in a field setting, as indicated by changes in percent cover and nitrogen removal.

Tilman's experiments supported the redundancy hypothesis, because the results showed that there is a point at which community service is maximized, beyond which additional species appear to have little to no impact. For example, uptake of nitrogen remains relatively unchanged as the number of species increased beyond 6. Since then, additional studies have been conducted to investigate the relationship between species richness and community function. Mark Schwartz and colleagues (2000) reviewed data from field and laboratory experiments that examined the influence of species richness on community function. Of the 13 studies where sufficient data existed to determine the shape of the relationships, all supported the redundancy hypothesis.

Anecdotal evidence also suggests that the redundancy hypothesis has the most support. Eastern American forest communities are still functional despite the loss of many species, such as the American chestnut tree, the passenger pigeon, and the Carolina parakeet. Whether such apparent functionality will be maintained over long time periods remains to be seen. On a larger scale, it is noteworthy that temperate forests of the Northern Hemisphere have virtually identical productivity (amount of plant biomass produced per unit time), despite huge differences in tree species richness: East Asia has far more species (729) than North America (253) or Europe (124). One rule of thumb that has been proposed is that average annual forest productivity tops out in the range of 10–40 tree species.

**Figure 19.3 Species richness affects community services.** The relationships between species richness and **(a)** percent cover of the ground by plant biomass and **(b)** uptake of nitrogen in experimental plots in Tilman's biodiversity experiments. With increasing species richness, percent cover (biomass) increases and more nitrogen is used up, leaving less nitrogen in the soil. (After Tilman et al., 1996.)

**ECOLOGICAL INQUIRY**

What is one of the dangers in interpreting this kind of graph, where community function appears to increase with species richness?

## 19.1.2 Plant species richness affects herbivore and predator species richness

In a separate set of experiments at Cedar Creek, Minnesota, Johannes Knops and colleagues (1999) examined the effects of plant species richness on insect herbivore and predator species richness. Their experiment involved 342 12 × 12 m plots that were planted with 1, 2, 4, 8, or 16 species from a pool of 18 grassland species. Because of the relatively large size of these plots, the investigators could sample them for organisms such as insect herbivores and natural enemies of the herbivores by passing over the plots with a sweep net and counting insect species caught in the net (**Figure 19.4a**). However, the large areas of the plots precluded rigorous weeding and maintenance. Thus, some nonplanted plant species invaded the plots, and some planted species died. Plant species diversity therefore varied between 1 and 16 species, and the variation was greater than the original categories of the 5 different groups of species. Results showed that insect herbivore species richness increased with plant species richness (**Figure 19.4b**). This can be explained by the fact that each plant species has some specialist herbivores that are specifically adapted to that particular host plant. Higher plant species richness therefore leads to higher herbivore species richness. Perhaps not surprisingly, the richness of predatory arthropods, such as spiders and insects that fed on the insect herbivores, was also greater on higher-richness plots where there was a greater variety of herbivores to feed on (**Figure 19.4c**). These experiments showed that increased plant species richness increased herbivore and predator richness, affecting community function at several trophic levels.

## 19.1.3 Increased natural enemy species richness increases herbivore suppression

A more recent study by William Snyder and colleagues (2006) showed that increased predator richness results in greater herbivore suppression—that is, reduction in herbivore abundance. In their experiments the researchers grew green peach aphids, *Myzus persicae,* and cabbage aphids, *Brevicoryne brassicae,* on collard plants, *Brassica oleracea,* with and without enemies. The predators were a diverse array of species and feeding types, including sit-and-wait *Nabis* bugs, active predatory bugs, *Geocoris pallens,* active predatory ladybird beetles, *Coccinella septempunctata* and *Hippodamia convergens,* and the parasitoid wasp *Diaeretiella rapae.* Although these natural enemies fed in a somewhat complementary way, there was still intraguild predation between them: the bugs ate parasitized aphids, both ladybird beetles fed on one another's eggs and the larvae fed on one another, and the larger *Nabis* and *Geocoris* bug stages fed on the smaller ones. The experiment included low-richness treatments where the aphids and one enemy species were added, and five high-richness treatments where four enemy species were added. One enemy species was not included in each of the five unique combinations. There were three replicates of each combination. Aphid densities in each treatment were compared to controls of aphids only, with no enemies (**Figure 19.5**). After 28 days, aphid densities were much lower in the high enemy richness treatments, showing that high enemy richness reduced herbivore abundance. There was no evidence that intraguild predation between predator species, or interspecific competition, caused high predator richness treatments to be less efficient than low-richness treatments; quite the contrary. The researchers also found that lower aphid densities allowed plants to grow bigger. High enemy richness thus allowed greater plant growth to occur. This is an example of the indirect effect discussed in Chapter 16, in which high levels of natural enemies are seen to benefit plants by reducing herbivory.

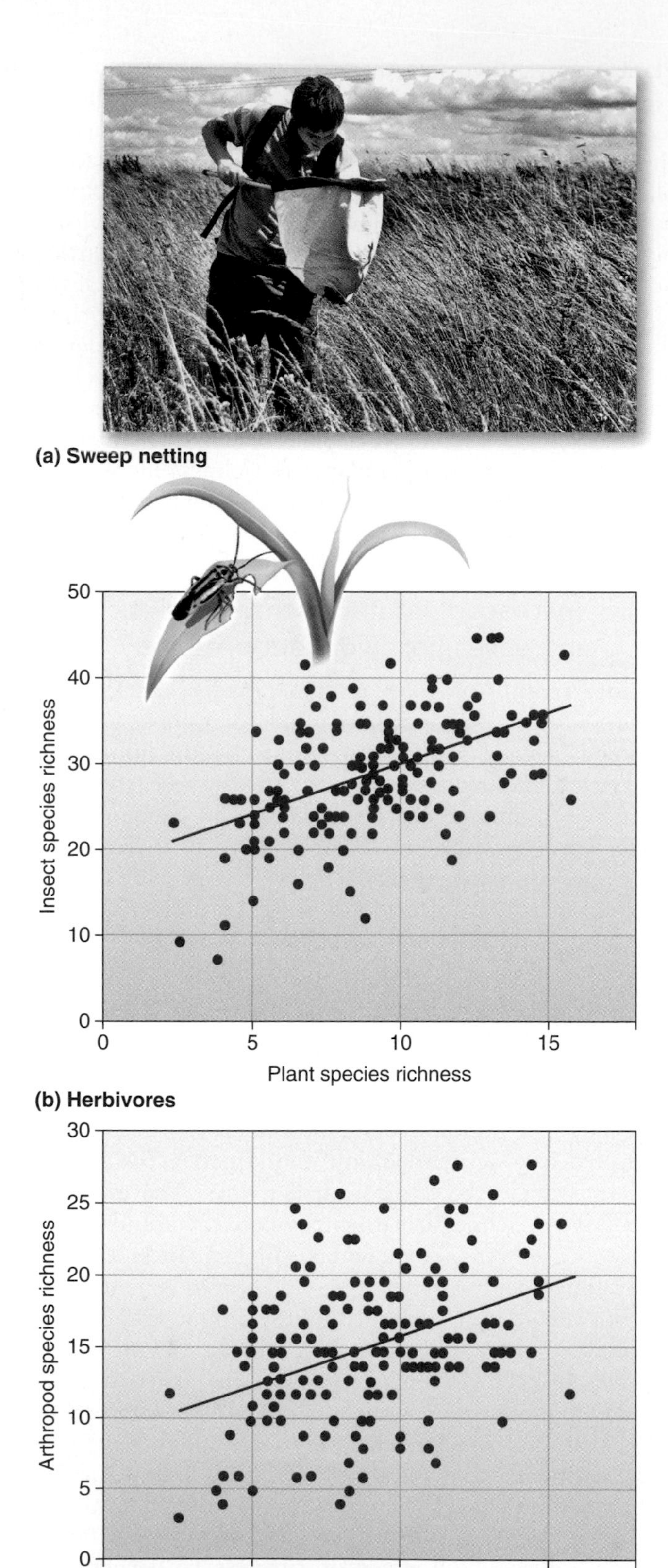


**Figure 19.4 Plant species richness affects herbivore and parasite community richness.** **(a)** Insect herbivore and predator species richness were sampled with a sweep net. **(b,c)** Richness of both groups increased with increasing plant species richness. (From Knops et al., 1999.)

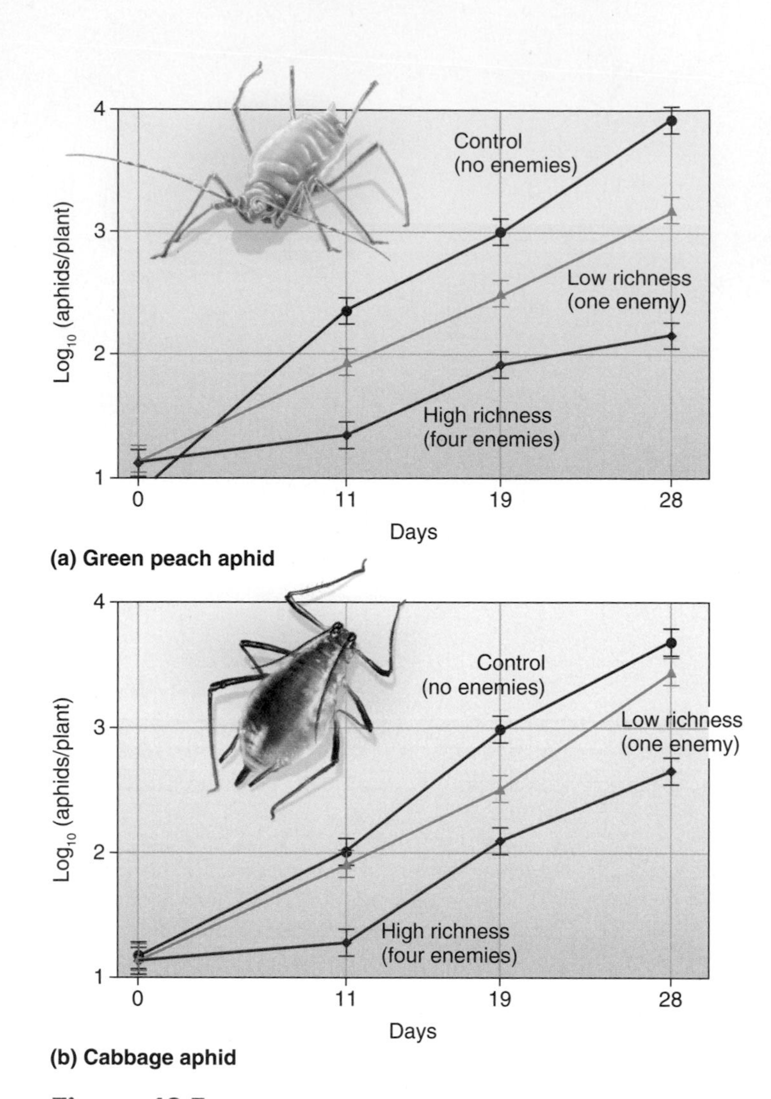


**Figure 19.5 Species richness affects community function at higher trophic levels.** Densities of **(a)** green peach aphids, *Myzus persicae*, and **(b)** cabbage aphids, *Brevicoryne brassicae*, through time for three enemy species-richness treatments: no enemies (control), low richness (one enemy species), or high enemy richness (four enemy species). (After Snyder et al., 2006.)

More recently, community diversity has also been linked to the prevalence of diseases in marine systems. Overfishing in tropical coral reef communities has allowed some coral-feeding fish species, which are not targeted by fishers, to become common, and these fish species spread disease as they feed (see **Global Insight**).

### 19.1.4 The sampling effect is the likely cause of increased performance in species-rich communities

While early experiments, such as Naeem's and Tilman's proposed that species complementarity was the likely cause of increased performance by species-rich communities, other explanations were proposed to account for the increase in

Global Insight

## Overfishing Reduces Fish Species Richness and Increases the Prevalence of Coral Disease

Coral reefs are among the most diverse systems on Earth, with many different coral species and the most species-rich fish communities in the world. They are under threat from poor water quality and destructive fishing practices. Laurie Raymundo and colleagues (2009) showed how overfishing in coral reefs of the Philippines decreased fish diversity and increased the prevalence of coral disease. At seven sites that examined the prevalence of coral disease on marine protected areas and on adjacent fished reefs, disease prevalence was highest at the fished sites (**Figure 19.B**). The mechanism that led to increased disease was competitive release of coral-feeding chaetodontid butterfly fishes. Butterfly fish are not targeted by fishers and tend to increase in abundance when competing fish are harvested. As these species feed on coral, they spread coral disease. In a separate analysis of a data set from Australia's Great Barrier Reef, the authors showed that the frequency of coral disease at different sites increased as butterfly fish abundance increased. The authors also showed that, among fished reefs, those with greater numbers of fish species were less diseased. The implications were that even a moderate reduction in fishing would reduce the frequency of disease on coral reefs.

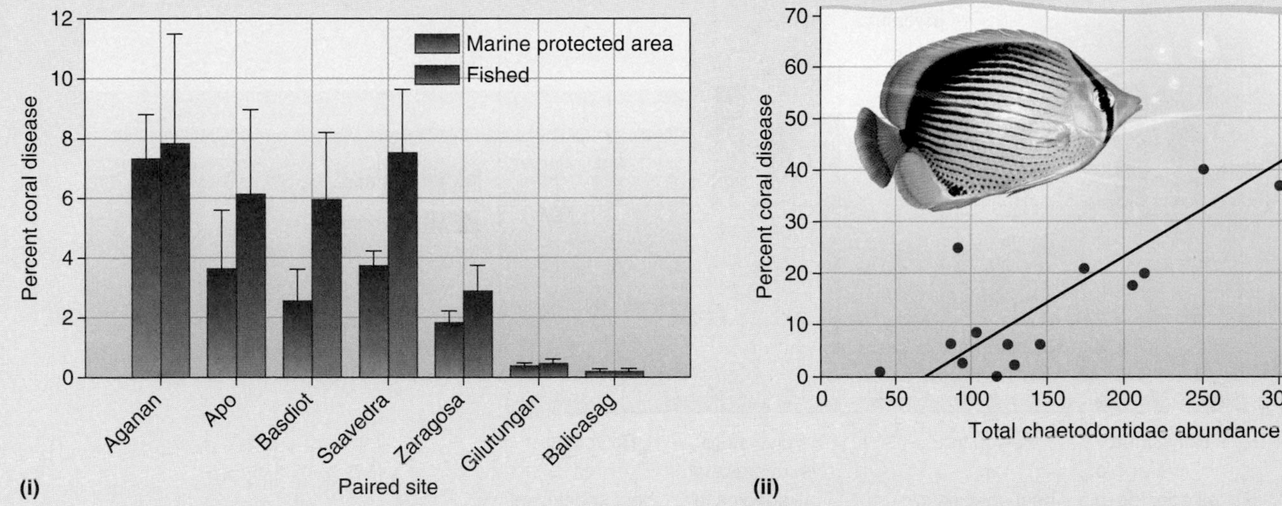


**Figure 19.B The effect of fish species richness on the prevalence of coral disease.** **(i)** Disease prevalence on paired protected and fished coral reefs in the Philippines. **(ii)** The frequency of coral disease increases with butterfly fish abundance on Australia's Great Barrier Reef. (After Raymundo et al., 2009.)

community function. Michael Huston (1997) suggested that species-rich communities also had a greater chance of containing a "superspecies," a species that performed a variety of functions exceptionally well. This became known as the **sampling effect.** That is, due to increased sample size, communities with many species are more likely to contain a large species that produces much biomass, boosting community productivity. Species-poor communities are less likely to contain such a species.

How do we tell which is the cause of increased performance: species complementarity or the sampling effect? One method is to compare the performance of a species-rich group, called a polyculture, with the average performance of all the same species grown on their own in monocultures. An increase in performance would prove that high species richness increases performance. Next, we compare the species-rich group with the performance of the best single species in monoculture. An increase in performance of the species-rich group is evidence of species complementarity. If the high species-rich group performs only as well as the best monoculture, then the sampling effect is causing the species-rich group to perform well.

As we mentioned earlier, many biodiversity/community services experiments have now been performed since Naeem's and Tilman's pioneering work. Bradley Cardinale and colleagues (2006) presented a meta-analysis of 111 such field, greenhouse, and laboratory experiments. They reviewed data

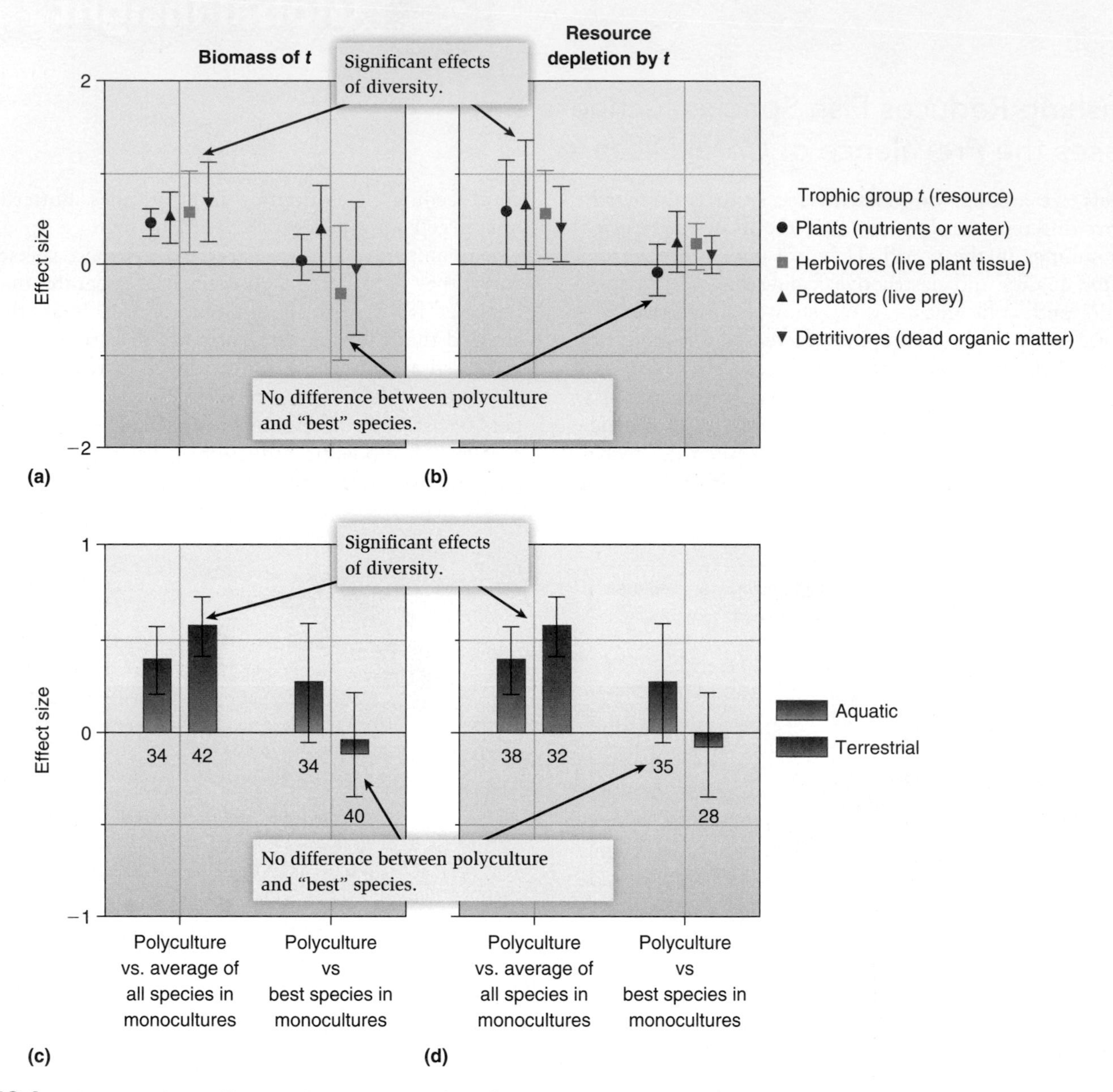


**Figure 19.6 The sampling effect explains increased performance in species-rich communities.** Data show mean proportional differences in standing stocks **(a,c)** or resource depletion **(b,d)** between polycultures and the average of all monocultures, or the best performing monoculture. Data are shown for many different trophic groups **(a,b)** and for different biome types **(c,d).** Means and 95% confidence intervals are shown. (From Cardinale et al., 2006.)

from a variety of groups, including herbivores, predators, detritivores, and plants, and in aquatic as well as terrestrial systems. There were significant effects of species richness on the biomass of all taxa when compared to the average of all monocultures (**Figure 19.6a**). Furthermore, this species richness effect carried over into resource depletion, the amount of resources, such as soil or water nutrients, used by the community (**Figure 19.6b**). These effects held true in both aquatic and terrestrial systems (**Figures 19.6c,d**). However, the sampling effect, not species complementarity, was the likely cause. Both biomass and resource depletion of high species-rich groups were not different from the best monocultures. Regardless of the mechanism, high species richness usually equates to increased community function.

### Check Your Understanding

**19.1** What are two possible drawbacks to Shahid Naeem's Ecotron experiments relating species richness to community function?

## 19.2 Species-Rich Communities Are More Stable Than Species-Poor Communities

As we mentioned previously, the English ecologist Charles Elton proposed a link between species richness and community stability in the 1950s. Elton (1958) suggested that a disturbance in a diverse, or species-rich, community would be cushioned by large numbers of interacting species and would not produce as drastic an effect as it would in a less rich community. For example, an introduced predator or parasite could cause extinctions in a species-poor system but would be less likely to have such effects in a species-rich–community. In this section, we investigate the link between species richness and stability. First, we need a good idea of what stability means.

### 19.2.1 A stable community changes little in species richness over time

A community is often seen as stable when little change can be detected in the number of species over a given time period. The frame of reference for detecting change may be a few years or, preferably, several decades. For example, long-term data from Eastern Wood, Bookham Common, in the county of Surrey, England, revealed that the number of bird species remained stable for nearly 30 years, from 1950 to 1979 (**Figure 19.7**). It is also important to realize that just recording the number of species in an area may obscure potential changes in the community, such as the fact that some species go extinct or emigrate and others immigrate.

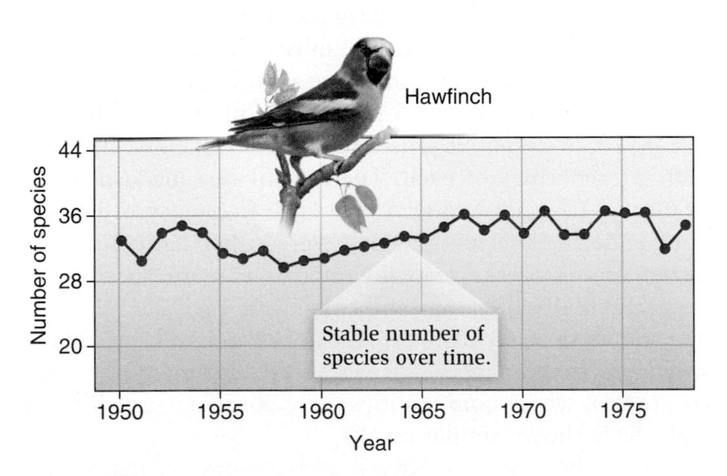


**Figure 19.7 Time plot of the total number of breeding bird species of Bookham Common, England.** The relative lack of change is taken to represent stability in the community. (Modified from Williamson, 1987.)

**ECOLOGICAL INQUIRY**

Are these data a watertight example of stability in nature over time?

Community stability is an important consideration in conservation biology. A decrease in the stability of a community over time may alert us to a possible problem. In the 1960s the composition of the raptor community changed as populations of many bird of prey species, including peregrine falcons, bald eagles, and osprey, declined precipitously. Eventually the decline was traced to the pesticide DDT (dichlorodiphenyltrichloroethane), which caused the birds' eggshells to become thin and break before the offspring could hatch (look ahead to Chapter 25). After DDT was banned later in the 1970s, raptor species began to recover.

There are two different ways of thinking about the concept of stability. **Resistance** refers to how tolerant a community is toward a disturbance before it changes. For example, some communities such as northern coniferous forest are very resistant to temperature changes and will exist at conditions of –20°C and 20°C. On the other hand, a light freeze kills much vegetation in subtropical Florida. **Resilience** refers to how quickly a community can return to equilibrium after a disturbance. Florida mangroves regrow quite quickly following leaf losses to freezes.

The concepts of resistance and resilience are sometimes correlated and sometimes not. Lakes are weakly resistant, because they can concentrate pollutants from a variety of sources, and weakly resilient because there is low through flow of water to cleanse the system. Rivers are not particularly resistant to pollutants but may be resilient because the fast-flowing water often cleanses them quickly. Tundra, with widespread vegetation but slow regeneration rates, is both weakly resistant and weakly resilient and is easily changed by disturbances such as off-road vehicles.

### 19.2.2 The diversity-stability hypothesis states that species-rich communities are more stable than species-poor communities

As we noted earlier, Charles Elton (1958) suggested that species-rich communities are more stable than species-poor communities. For example, Elton suggested that species-poor island communities are much more vulnerable to invading species than are species-rich continental communities. In support of this idea, Elton noted that introduced mosquito species have caused many extinctions of native birds on the islands of Hawaii, which are extremely susceptible to the avian malaria spread by the insects. Elton also argued that outbreaks of pests are often found on cultivated land or land disturbed by humans, both of which are species-poor communities with few naturally occurring species, especially natural enemies. His arguments collectively became known as the **diversity-stability hypothesis.**

Later, other ecologists began to challenge Elton's association of species richness with stability. They pointed out that many examples of introduced species have assumed pest proportions on continents, not just islands, including rabbits in Australia and pigs in North America. They noted that disturbed or cultivated land may suffer from pest outbreaks not because of its simple nature but because its individual components, including introduced and native species, often have no evolutionary history with one another, in contrast to the long associations evident in natural biomes. For example, in Europe, coevolved predators such as foxes prevent rabbit populations from increasing to pest proportions. Field experiments were needed to determine if a link exists between species richness and stability. Such a link was demonstrated in the 1990s.

David Tilman (1996) examined the relationship between species richness and stability in an 11-year study of 207 natural, not planted, grassland plots in Minnesota that varied in their species richness. He measured the biomass of every species of plant, in each plot, at the end of every year and obtained the average species biomass. He totaled the biomass of all species to obtain the community biomass. He then calculated how much this community biomass varied from year to year, using a statistical measure called the coefficient of variation. Year-to-year variation in plant community biomass was significantly lower in plots with greater plant species richness (**Figure 19.8**).

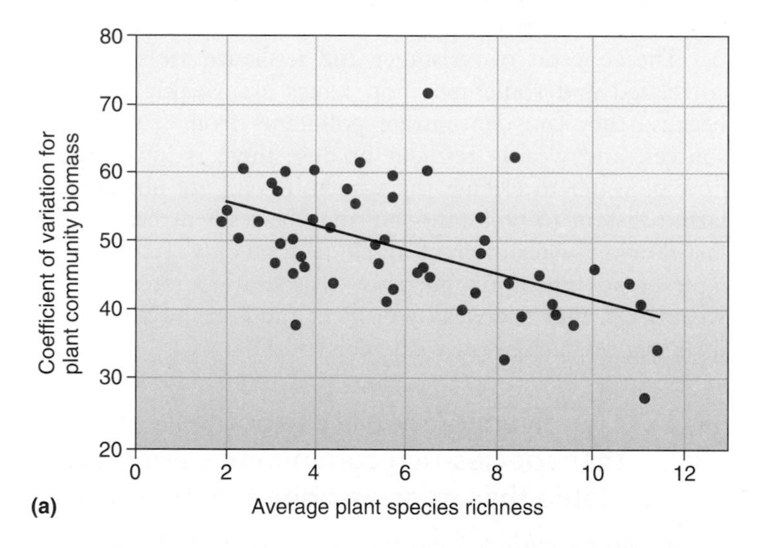

(b)

**Figure 19.8 Variation in community biomass decreases with increased species richness.** Tilman's 11-year study of grassland plots in Minnesota revealed that **(a)** year-to-year variability in community biomass was lower in species-rich plots. Only the data from one field are shown. **(b)** This aerial photograph shows the grassland plots.

Tilman suggested that increased richness stabilizes communities because species-rich plots were more likely to contain disturbance-resistant species that, in the event of a disturbance, could grow and compensate for the loss of disturbance-sensitive species. Thus, when a disturbance such as drought harmed drought-susceptible species that thrived in normal conditions, drought-resistant species increased in biomass and replaced them. Such decreases in disturbance-susceptible species and compensatory increases in other species acted to stabilize total community biomass. As in Tilman's earlier experiments, species complementarity was seen as an important mechanism underlying the correlation between richness and stability.

### 19.2.3 Species richness affects community resistance to invasion by introduced species

As part of a link between stability and diversity, Elton (1958) had suggested that species-rich communities would be more resistant to invasion by introduced species. This is known as the **biotic resistance hypothesis.** The logic was that species-rich communities would more fully utilize resources such as soil, nitrogen, and light, leaving little left over for invaders to use. More recently some authors have pointed out that species-rich communities may also be likely to contain super-species that particularly limit invasive species. Though these ideas sound logical, it is also possible that species-rich communities may be more likely to contain key facilitators, species that encourage invasion, especially if they interact with invaders in a mutualistic way. Some recent experiments have tested the link between community richness and invasibility.

Johannes Knops and colleagues (1999) tested the effects of species richness on community invasibility by using Tilman's field experiments in Minnesota. The experiment consisted of 147 plots of 3 × 3-m size. The plots had either 1, 2, 4, 6, 8, 12, or 24 native grassland species seeded within them, with 21 replicates of each. Their small size made detection of invading species relatively easy. The frequency of invaders not planted in the plots, which were either native or introduced species, decreased with increased plant species richness (**Figure 19.9**). Knops and coworkers suggested that they had shown that, 40 years previously, Elton had been right: as species richness increases, fewer species invade. Other similar studies, where communities were constructed on a small scale, have shown similar results.

Since the study by Knops and coworkers, many other experiments have addressed the role of biotic resistance in limiting invasions of introduced plants at small spatial scales. Researchers have examined whether the mechanism is via competition from native plants, suppression by native herbivores, or the effects of the soil fungal community, either mycorrhizae or pathogens, that prevent invasion. These studies were subject to a review and meta-analysis by Jonathan Levine and colleagues (2004). Levine and coworkers also distinguished

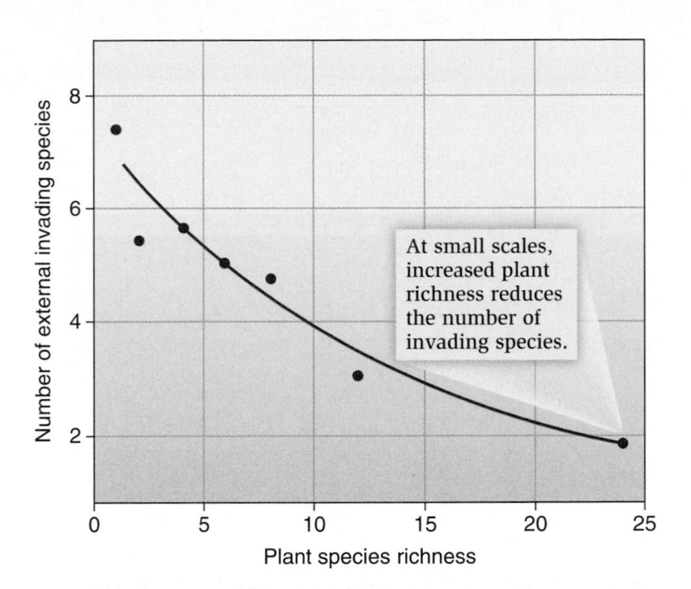


**Figure 19.9** **At small scales, increased native plant species richness reduces the number of invasive plant species.** (Reproduced from Knops et al., 1999; and Kennedy et al., 2002.)

whether such factors acted as a barrier to the invasion of introduced species or to regulate their impacts. There are several stages in invasion (**Figure 19.10a**). First, a species has to be introduced and then become established. Next, a species may spread and eventually impact native species. Biotic resistance may also regulate spread and impact. Results showed that species richness affected both invader establishment and invader performance (spread and impact), and that competition and suppression by native herbivores were prominent in both (**Figure 19.10b,c**). Soil fungal communities, which were not examined in the context of establishment, had no effect on performance.

Large-scale observational studies reveal a different pattern than the small-scale experimental studies, however. In large-scale studies, researchers have examined how native and invasive plant species richness varies over large spatial scales, such as between different counties. In Alabama, counties that had more native plant species also tended to have higher numbers of non-native species (**Figure 19.11a**). On a national scale, states with more native plant species also contain more

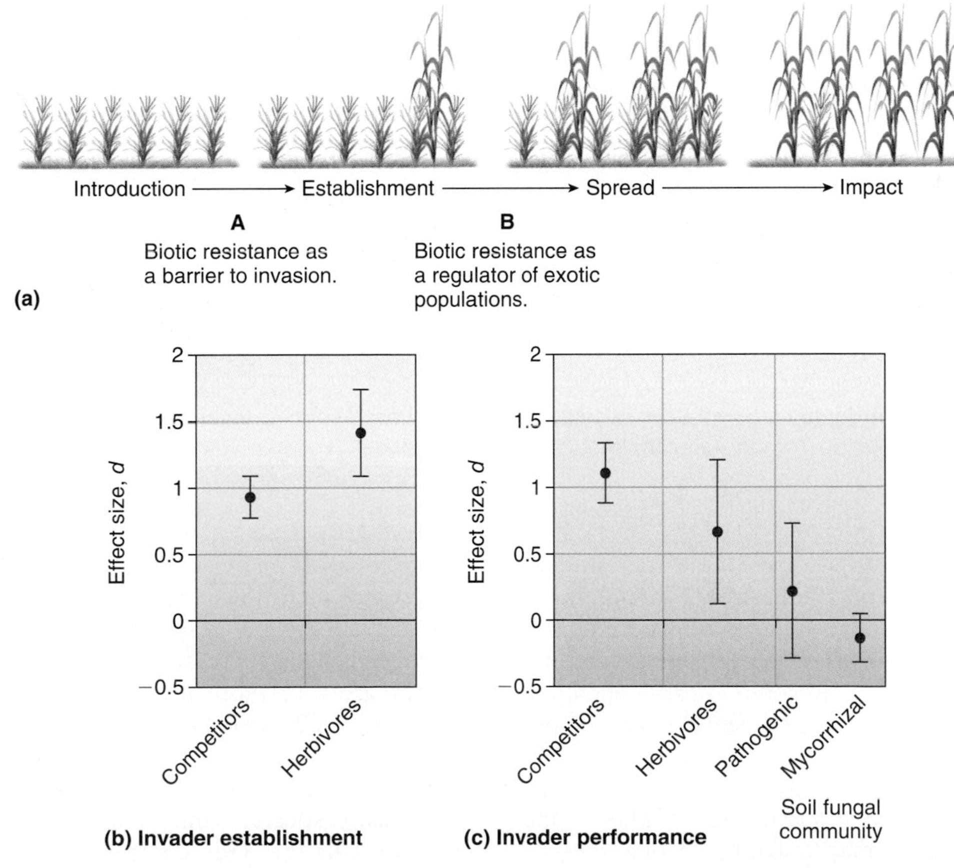


**Figure 19.10** **Effects of resident competitors, herbivores, and soil fungal communities on plant invaders.** **(a)** Concept of the invasion process. Effects of biotic factors on invader **(b)** establishment and **(c)** performance (spread and impact). (Reproduced from Levine et al., 2004.)

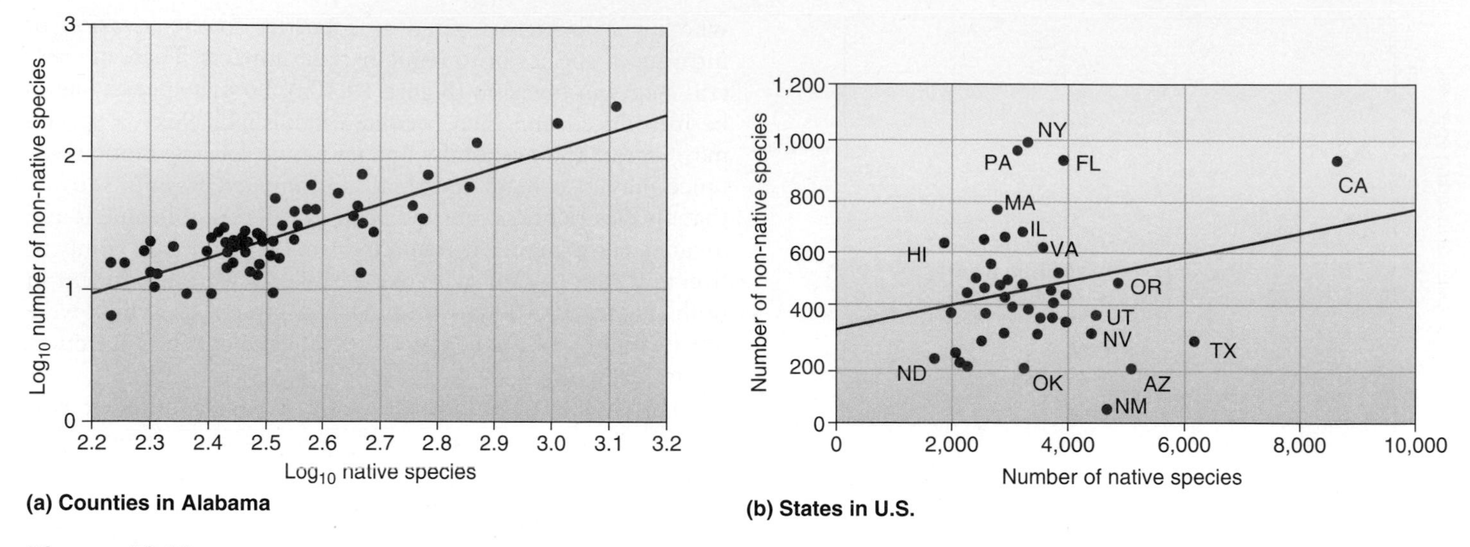


**Figure 19.11 At large scales, increased native plant species richness does not reduce the number of invasive plant species.** **(a)** Counties in Alabama. **(b)** States in the United States. (Redrawn from Stohlgren et al., 2003.)

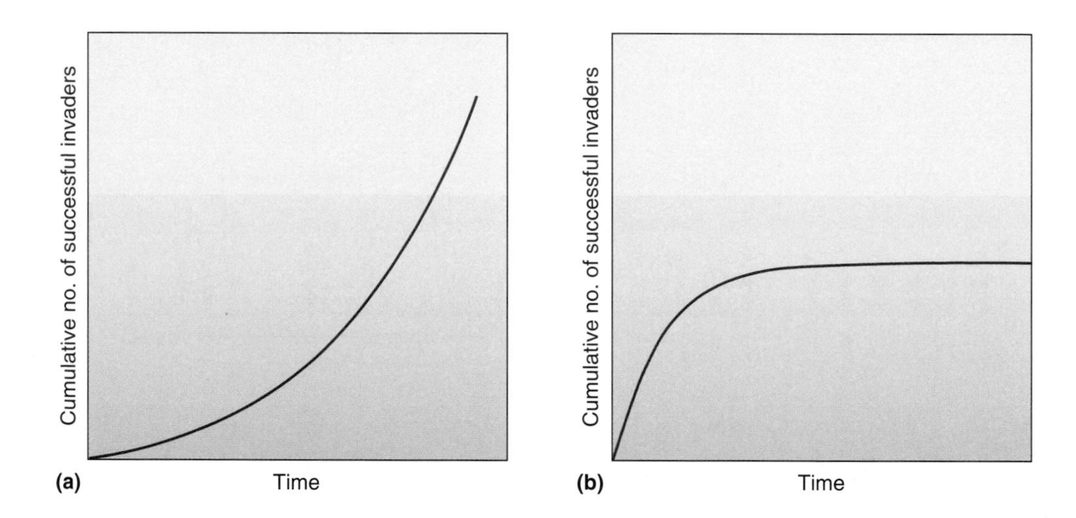


**Figure 19.12 Invasional meltdown leads to a runaway pattern of increased species invasion.** Temporal changes of exotic species consistent with **(a)** the invasional meltdown model and **(b)** the biotic resistance model.

non-native species (**Figure 19.11b**). In these large-scale studies, favorable abiotic or biotic conditions that result in high numbers of native plant species, such as high soil moisture or soil nutrients, also appear to favor invasions by introduced species.

Jonathan Levine and Carla D'Antonio (1999) concluded that neighborhood or small-scale experimental studies show that species richness does limit invasion by exotic species, but that this effect can be swamped out by other factors such as high nutrient availability at large spatial scales. They make the analogy that competitors may outcompete one another at small spatial scales but they often co-occur at large spatial scales due to similar habit requirements.

Some authors have suggested that once a community has been invaded by introduced species, these invaders facilitate invasion by other species, a runaway process termed **invasional meltdown** by Daniel Simberloff and Betsy Von Holle (1999). For example, some invasive plants can fix nitrogen, which increases soil nitrogen and could facilitate the invasion of more nitrogen-limited invasive species. As we saw in Chapter 12, introduced pollinators may facilitate seed set of plant species and, by implication, population growth. Under the invasional meltdown model, the number of invasive species would increase exponentially over time (**Figure 19.12a**). This contrasts with the biotic resistance model where the number of invasive species would tend to level off over time as competition or predation from previous invaders tends to suppress invasion by new species (**Figure 19.12b**).

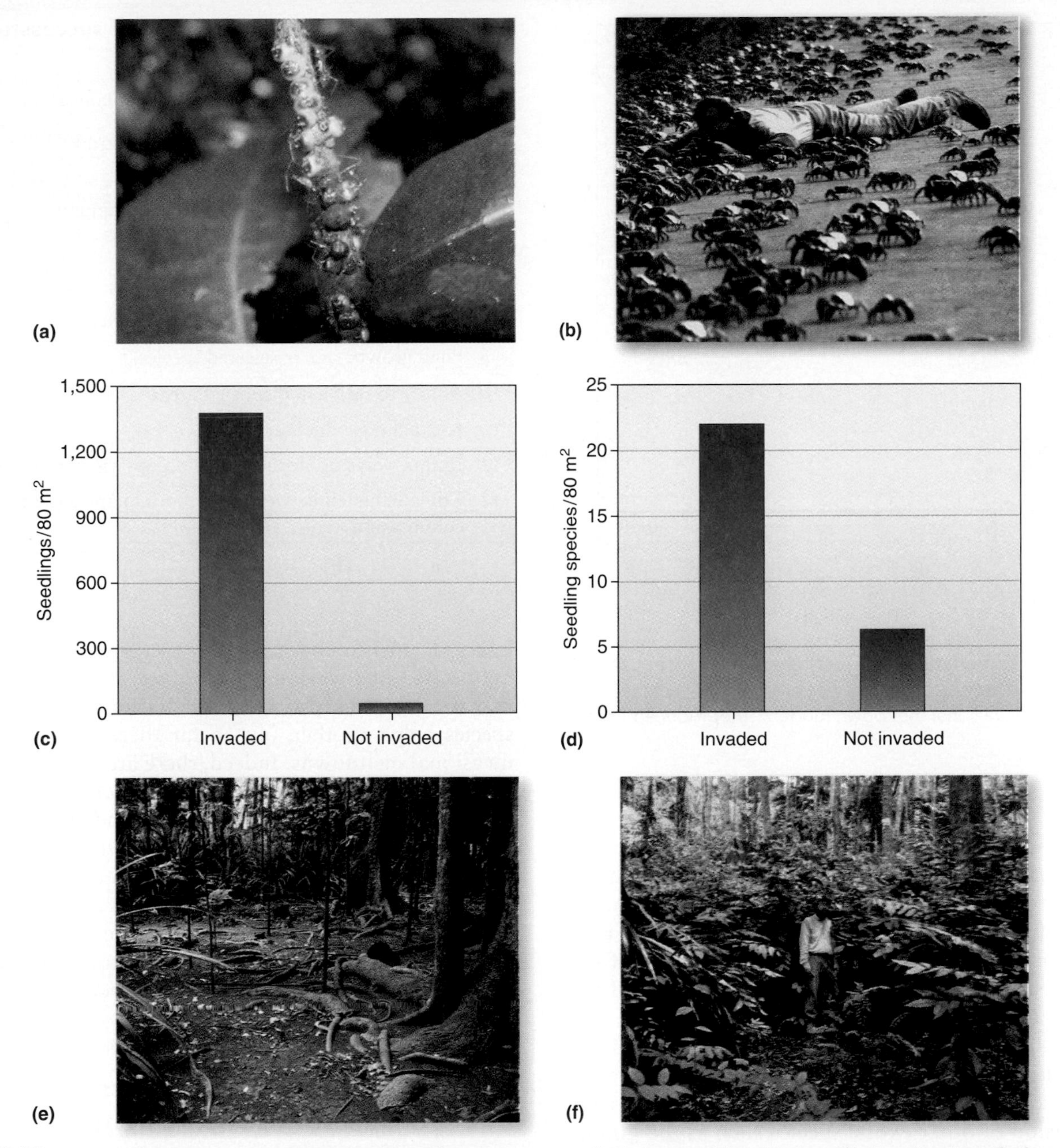


**Figure 19.13** **Impacts of the invasive crazy ant on the community of Christmas Island in invaded and uninvaded areas.** **(a)** Crazy ants, *Anoplolepis gracilipes,* tending to exotic scale insects, with incipient sooty mold on leaves, Christmas island. **(b)** Dennis O'Dowd photographing the red crab migration into the forest on Christmas island. Densities of **(c)** plant seedlings, and **(d)** seedling species per 80 $m^2$ in areas invaded and not invaded by ants. Appearance of forest understory **(e)** before and **(f)** after invasion by crazy ants. (Data and photos from O'Dowd et al., 2003.)

Dennis O'Dowd and colleagues (2003) described how an invasive ant modified the native community on Christmas Island, an isolated oceanic island in the northeastern Indian Ocean. The yellow crazy ant, *Anoplolepis gracilipes,* invaded Christmas Island in the 1930s but remained at low densities until the 1990s, when it formed high-density infestations called supercolonies. In areas where supercolonies are common on the island, this ant has entered into a mutualistic relationship with invasive sap-feeding insects called scale insects, *Coccus celatus* (**Figure 19.13a**). The scale insects are generalists, feeding on 10–20 tree species on the island. This ant–scale insect mutualism has resulted in increased scale and ant densities. In addition, the crazy ants kill the red land crab, *Geocarcoidea natalis,* a common herbivore (**Figure 19.13b**). It is estimated that 10–15 million red land crabs, one-quarter to one-third of the entire population, have been killed by crazy ants. Without the common seedling-feeding land crabs, the number of seedlings has

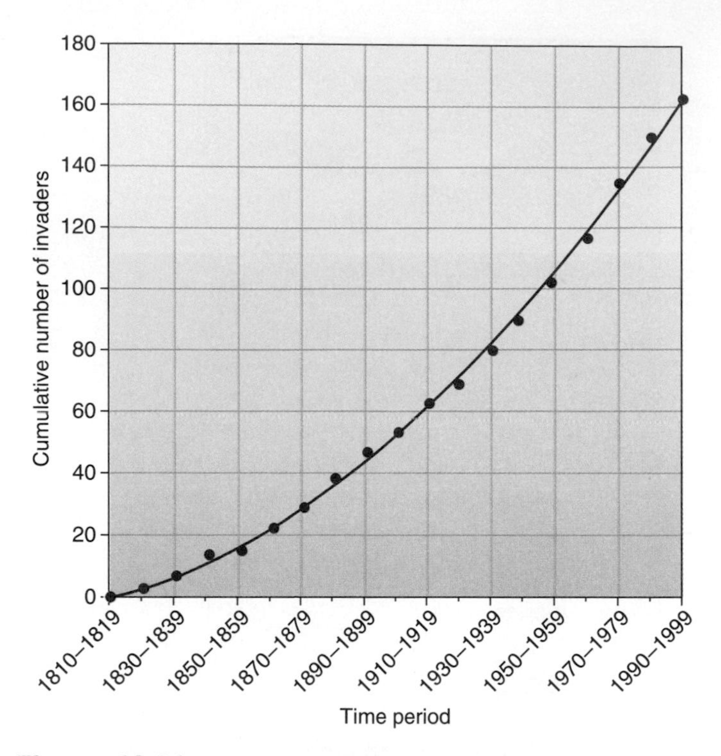


**Figure 19.14 Cumulative temporal changes in Great Lakes aquatic species.** The increase in species number over time is consistent with the invasional meltdown model. (Reproduced from Ricciardi, 2001.)

dramatically increased. In addition, seedling species richness has increased as many more species are able to grow in the absence of the crabs (**Figure 19.13c,d**). Thus, a mutualism between an invasive ant and sap-feeding insects has promoted huge increases in seedling growth due to lack of the dominant herbivore, altering the vegetative structure of the island (**Figure 19.13e,f**).

Anthony Ricciardi (2001) suggested that invasions of the Great Lakes freshwater community by exotic species such as the zebra mussel, *Dreissena polymorpha,* facilitates the invasion of other exotic species such as the Ponto-Caspian hydroid, *Cordylophora caspia,* which feeds on the zebra mussel larvae and uses the shells as a substrate. Similarly, the parasitic lamprey, *Petromyzon marinus,* which we discussed in Chapter 13, has suppressed the abundance of native salmon and facilitated entry of invading species such as the alewife, *Alosa pseudoharengus*. The data for the Great Lakes suggest that the number of invasive species has increased over time, consistent with the invasional meltdown model (**Figure 19.14**). However, it is hard to separate out the fact that, with increased global travel, the number of propagules of invasive species introduced into the Great Lakes might also be increasing, which may greatly contribute to the invasional meltdown pattern shown in Figure 19.14.

Despite the appeal of the invasional meltdown model, and the evidence from Christmas Island and the Great Lakes, Daniel Simberloff (2006) suggested that few watertight cases of invasional meltdown are known. He noted that there are many cases of facilitation, where one invasive species helps another, but so far there is little evidence of invasional meltdowns. Indeed, there are some cases of biotic resistance where invasive species impact one another negatively. For example, Blaine Griffen and colleagues (2008) showed that the Asian shore crab, *Hemigrapsus sanguineus,* introduced to the U.S. Atlantic coast in the 1980s, is replacing populations of the European green crab, *Carcinus maenas,* which was introduced in the mid-1800s. In the presence of *H. saguineus, C. maenas* consumes fewer mussels, its preferred prey, leading to lower growth rates and reduced population sizes.

**Table 19.3 Proposed attributes of successfully invading species.**

| | |
|---|---|
| 1. | Shedding of natural enemies during colonization |
| 2. | Absence of natural enemies in recipient community |
| 3. | Better competitor than native species |
| 4. | Presence of mutualists in recipient community |
| 5. | High dispersal ability |
| 6. | Association with humans |
| 7. | High reproductive rate (*r*-selected species) |
| 8. | High growth rate (*r*-selected species) |
| 9. | Ability to thrive in disturbed areas |
| 10. | Asexual reproduction (plants) |
| 11. | Unique ways of life; ability to occupy vacant niches |
| 12. | Climatic match between site of origin and site of colonization |

### 19.2.4 Invasive species may possess a variety of special life-history traits

As we noted in Chapter 1, invasive species are a major component of global change. As such, research has focused on what makes species invasive. Many traits have been linked to invasiveness (**Table 19.3**). Prominent among these has been a lack of natural enemies. The **enemy release hypothesis** states that invasive species are successful because they left their co-evolved natural enemies behind. This can arise as species are introduced without their natural enemies and because there are few natural enemies in the new habitat. The **evolution of increased competitive ability hypothesis** (EICA) suggests that a loss of natural enemies allows invasive species to devote more resources to competitive functions and less to defensive mechanisms. For example, in the absence of specialist insect herbivores, invasive plants may invest less in energetically costly chemical defenses and more to biomass production and reproduction.

In general, predicting which species will become invasive has not been very successful. A whole suite of traits related to growth and reproduction has been deemed important, yet no consensus as to which of these traits is most valuable has emerged. The house sparrow, *Passer domesticus,* spread throughout the entire United States within 50 years of its introduction in around 1850. The closely related tree sparrow, *P. montanus,* has largely been confined to the St. Louis area, where it was introduced in 1870. Norway rats, *Rattus norvegicus,* and black rats, *Rattus rattus,* are found throughout the world, but 54 other species of *Rattus* have not become worldwide pests.

Much of the research on what makes species invasive relies on post hoc evaluations of different communities and introduced species. It would be foolhardy, and illegal, to conduct experimental introductions of new species that may have a propensity toward invasiveness. Some of the best data on the success or failure of introduced species relates to biological control, where insect predators and parasitoids are released against pest insects of crops, orchards, and forests in the hope that the species will ultimately control their target pests. Peter Stiling (1993) reviewed the reasons for failure in 148 biological control campaigns. The most frequently cited reason for failure was improper climate in the area of introduction. Summers were too hot, winters too cold, or climate too dry for released enemies. This underscores the importance of climatic matching between site of origin of the invasive species and site of colonization and supports the **adaptation hypothesis,** which suggests that successful invaders are already pre-adapted to environmental conditions. More recently, Keith Hayes and Simon Barry (2008) conducted a similar but taxonomically broader study summarizing the results of 49 studies that examined 115 invasion characteristics in seven biological groups: birds, fin fish, insects, mammals, plants, reptiles/amphibians, and shellfish. Climate/habitat match was the only characteristic that was consistently significantly associated with invasive behavior. Other authors have shown that introduction effort, a measure of the number of individuals released into an area, is important; this idea is known as the **propagule pressure hypothesis.** For example, horticulture is an important source of non-native plant species. Plant nurseries sell hundreds of species of exotic plants over long time periods. Bob Pemberton and Hong Liu (2009) analyzed sales by the Royal Palm Nursery, in south Florida, from 1887 to 1930, and compared them to establishment and invasion patterns. Of the 1,903 non-native plant species sold by the nursery, 15% became established as invasives in south Florida. Pemberton and Liu showed that the proportion of plants that became established was related to the number of years sold (**Figure 19.15a**). The establishment of plants sold for 30 years or more was 70%! Similarly, the proportion of plants that became invasive also increased with the number of years sold (**Figure 19.15b**). The average number of years that species were sold from the nursery was 19.6 years for invasive species, 14.8 years for established species, and only 6.8 years for nonestablished plants (**Figure 19.15c**).

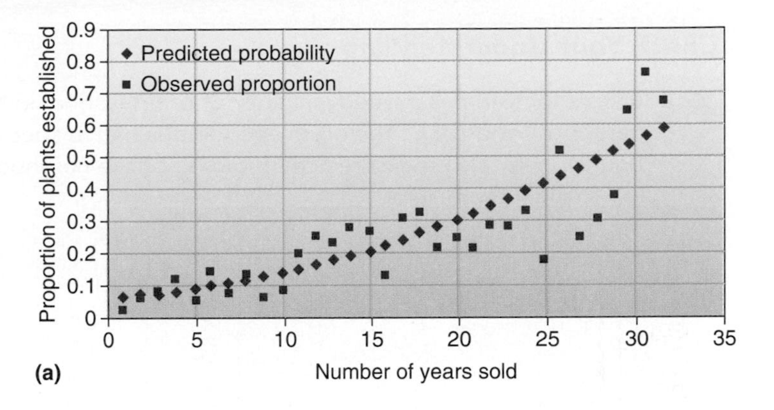


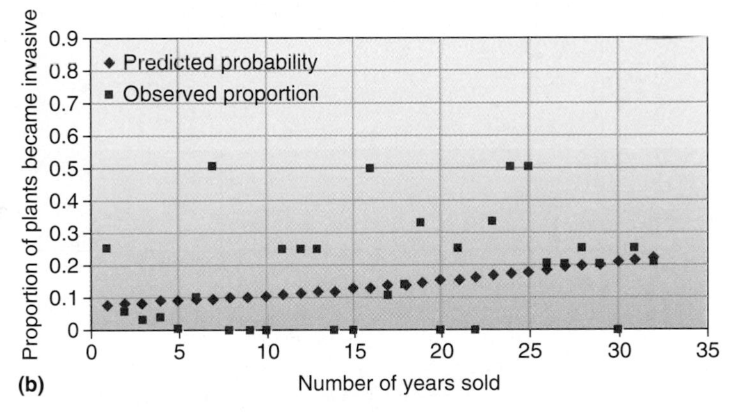


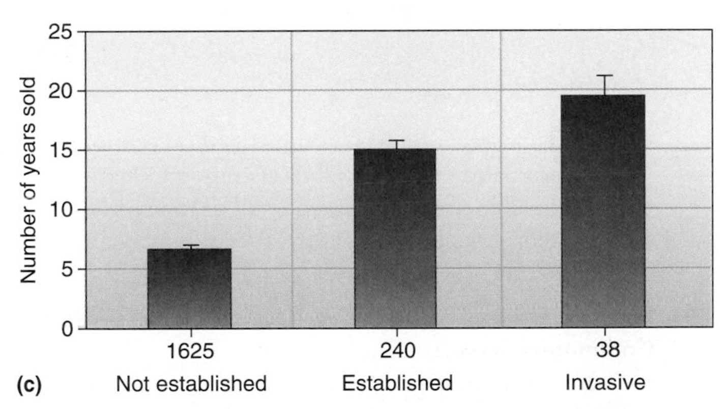


**Figure 19.15 The relationship between the number of years plants were sold in the Royal Palm Nursery and (a) establishment and (b) invasiveness in south Florida.** The predicted probability was based on a statistical property called a logistic regression model, **(c)** mean + standard error of years sold among nonestablished, established, and invasive species. (After Pemberton and Liu, 2009.)

## Check Your Understanding

**19.2** *Crepis tectorum* is a weedy invader of prairies. Shahid Naeem and colleagues (2000) created experimental neighborhoods of *C. tectorum* and various native species and measured *C. tectorum* biomass (see part **(a)** on the figure). They also measured the biomass of the neighboring plant roots **(b)** and the percentage light transmittance **(c)**. Explain the results.

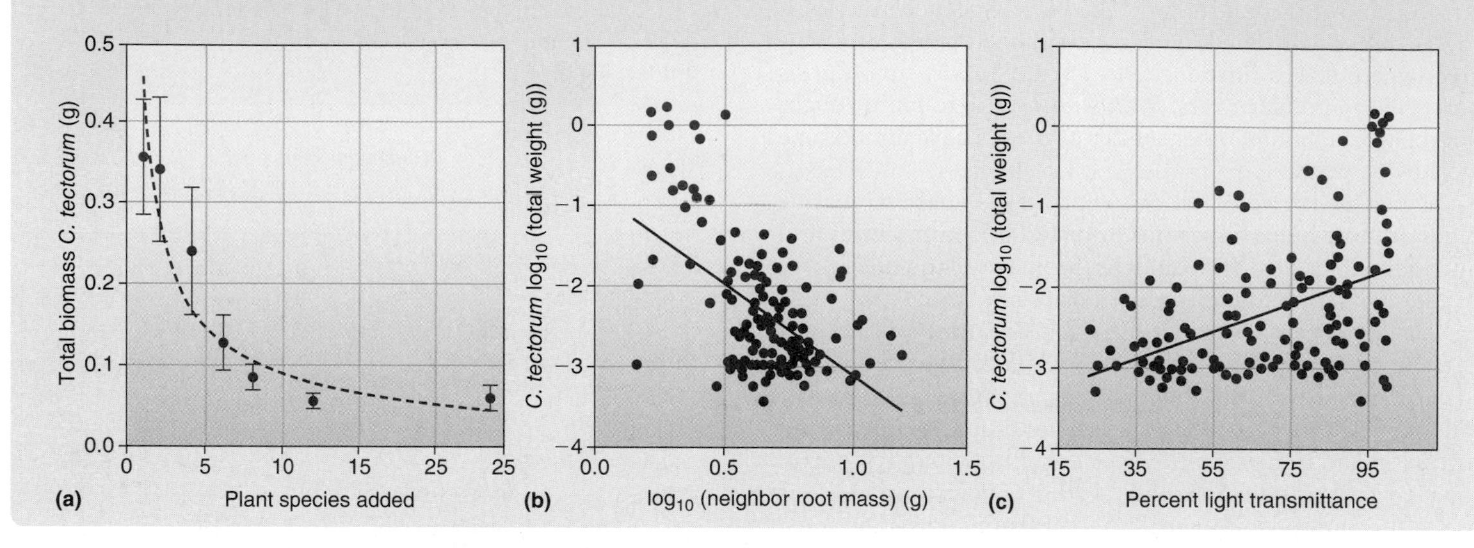


# SUMMARY

- Species conservation has been justified on economic, ecological, and ethical grounds (Figure 19.1).
- Functional communities provide all of the world's ecosystem services and have great economic value (Tables 19.1, 19.2).

### 19.1 Four Hypotheses Explain How Species Richness Affects Community Services

- There are at least four hypotheses concerning the effect of species richness on community function: linear, redundancy, keystone, and idiosyncratic (Figure 19.2).
- Experimental and field studies support the idea that species-rich communities perform better than species-poor communities. These studies also support the redundancy hypothesis, in which community function is impaired only when species richness is greatly reduced (Figures 19.A, 19.3).
- Experimental studies suggest species richness affects community function at higher trophic levels as well as lower trophic levels (Figures 19.4, 19.5).
- Increased plant species richness increases herbivore species richness and, in turn, predator species richness. High predator species richness is more effective than low species richness in reducing herbivore abundance and increasing plant growth.
- On coral reefs, overfishing causes the density of coral-feeding butterfly fishes to increase, by competitive release, and increased butterfly-fish feeding spreads coral disease (Figure 19.B).
- Reviews suggest that species-rich communities perform better than species-poor communities because of the sampling effect, not species complementarity (Figure 19.6).

### 19.2 Species-Rich Communities Are More Stable Than Species-Poor Communities

- Communities are stable when they change little in species richness over time (Figure 19.7).
- Tilman's field experiments established a link between plant species richness and community biomass stability (Figure 19.8).
- At small scales, field experiments show that high species richness limits invasibility by exotics (Figure 19.9).
- Biotic resistance to plant invaders is driven mainly by competition from native plant species and the effects of native herbivores (Figure 19.10).
- At large scales, observations suggest areas rich in native species also contain high numbers of exotic species (Figure 19.11).

- Invasional meltdown, the facilitation of exotic species by other exotics, has been noted in some communities but is not common (Figures 19.12–19.14).
- Predicting which species will become invasive is difficult, although many attributes associated with invasiveness have been proposed (Table 19.3).
- The probability of introduced plants becoming established and invasive increases with the number of years the plants are sold in nearby nurseries (Figure 19.15).

## TEST YOURSELF

1. Which of the following hypotheses proposes a quick decline in community function with decreasing species richness?
   a. Linear
   b. Keystone
   c. Community
   d. Idiosyncratic
   e. Redundancy

2. Brian Walker proposed the following hypothesis about the link between species richness and community function:
   a. Linear
   b. Keystone
   c. Community
   d. Idiosyncratic
   e. Redundancy

3. The presence of many more species than is needed for communities to function properly has been termed the _____ hypothesis.
   a. Linear
   b. Keystone
   c. Community
   d. Idiosyncratic
   e. Redundancy

4. Recent reviews have suggested that increased community function is linked to the increased species richness because of the _____ effect.
   a. Species complementarity
   b. Sampling
   c. Redundancy
   d. Idiosyncratic
   e. Keystone

5. The idea that humans have an innate love of life, coined by E. O. Wilson, is known as:
   a. Biodiversity
   b. Biophilia
   c. The call of the wild
   d. Biotheology
   e. The last of the wild

6. In the 1990s, economists calculated that the value of the world's ecological services was worth what percent of the world's gross national product?
   a. 10%
   b. 20%
   c. 50%
   d. 100%
   e. 200%

7. Rivers are seen as _____ to change.
   a. Resistant
   b. Resilient
   c. Resistant and resilient

8. Which mechanism most often prevents invasive plant species from invading native communities?
   a. Competition from native plants
   b. Suppression from native herbivores
   c. Effects of the soil fungal community
   d. a and b
   e. b and c

9. Large-scale studies have found that states with higher numbers of native plant species have fewer numbers of invasive plant species.
   a. True
   b. False

10. In South Florida, many plant species have become invasive because they are sold at many local nurseries.
    a. True
    b. False

## CONCEPTUAL QUESTIONS

1. What are some of the reasons why preservation of species richness is seen as important?
2. Distinguish between the following hypotheses linking species richness with community function: linear, redundancy, keystone, and idiosyncratic.
3. Distinguish between the two mechanisms that were proposed to explain the link between species richness and community function in Naeem's and Tilman's experiments.
4. Discuss two different ways of thinking about stability.
5. Why do small-scale experiments show that invasive species are less likely to invade species-rich areas, while large-scale observations show invasive species are more likely to be found in areas of high native-species richness?
6. What is invasional meltdown?
7. What are some of the proposed characteristics of invasive species?

## DATA ANALYSIS

The following graph was obtained by David Tilman in 1999 from his Cedar Creek, Minnesota, field site. It shows how aboveground plant biomass varies with the number of plant species. The dots represent individual replicates. The horizontal line is the biomass of the best monoculture. What can you conclude?

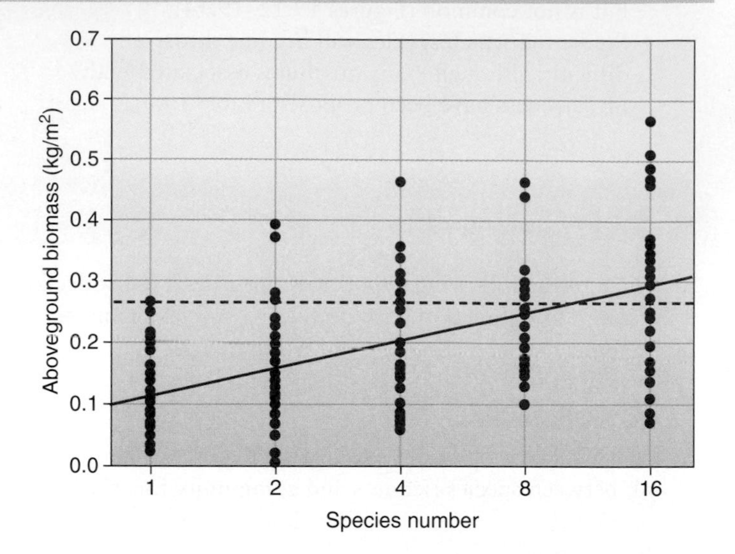

**Connect Ecology** helps you stay a step ahead in your studies with animations and videos that bring concepts to life and practice tests to assess your understanding of key ecological concepts. Your instructor may also recommend the interactive ebook.

Visit **www.mhhe.com/stilingecology** to learn more.

Mount St. Helens erupted in 1980. Since then, ecologists have been monitoring the return of plants and animals to its slopes.

CHAPTER 20

# Succession

## Outline and Concepts



At 8:32 a.m. on May 18, 1980, Mount St. Helens, a previously little-studied peak in the Washington Cascades, erupted. The blast felled trees over a 600-km$^2$ area, and the landslide that followed destroyed everything in its path, killing nearly 60 people. Tens of thousands of mammals, tens of millions of fish in lakes, and countless plants died as well. However, more than 30 years later, much of the area has experienced a relatively rapid recovery of plant and animal communities (**Figure 20.1**). First, millions of spiders ballooned in on silken threads. Plant seeds blown in by the wind germinated. Lupines, plants that can fix their own nitrogen, were the first plants to grow. Stands of red alder, willow, and cottonwood grew up around the area lakes. Elk returned to browse, bringing new plant species by defecating seeds. Pocket gopher tunnels helped mix the soil, and their mounds provided habitat for additional plants to grow. Their tunnels also provided refuges for returning frogs, newts, and salamanders.

The term **succession** denotes the gradual and continuous change in species composition and community structure over time following a disturbance such as a volcanic eruption. Other disturbances include earthquakes, landslides, hurricanes, tornadoes, floods, drought, fires, and a myriad of human activities. In this chapter we outline several different models that describe how communities change following disturbances. We then examine how species interactions such as competition and herbivory can deflect the path of succession, and we see how a community's species richness often increases during succession. Finally, we see how the theory of succession can be useful in restoration ecology, which seeks to return disturbed communities to a more natural condition.

(a)

(b)

**Figure 20.1** **Succession on Mount St. Helens.** **(a)** The initial blast occurred on May 18, 1980. **(b)** By 2001, only 21 years later, much of the areas initially covered by ash and mudflows had developed low-lying vegetation, and new trees sprouted up between the old dead tree trunks.

## 20.1 Several Mechanisms Describe Succession

Ecologists have developed several terms to describe community recovery in more detail. **Primary succession** refers to succession on a newly exposed site that was not previously occupied by soil and vegetation, such as bare ground caused by a volcanic eruption or the rubble created by the retreat of glaciers. In the case of volcanic eruptions, even though life occurred in the area previously, the deposition of many meters of ash has long since obliterated all living organisms. In primary succession on land, the plants must often help build up the soil, and thus a long time, possibly hundreds of years, may be required for the process. Only a tiny proportion of the Earth's surface is currently undergoing primary succession. This includes land around Mount St. Helens and volcanoes in Hawaii and off the coast of Iceland, and land behind retreating glaciers in Alaska, Canada, and Europe.

**Secondary succession** refers to succession on a site that has already supported life but that has undergone a disturbance, such as a fire, tornado, or flood, that has not killed all the native species. Clearing a natural forest and farming the land for several years is an example of a severe disturbance that does not kill all native species. Some plants and many soil bacteria, nematodes, and insects are still present. Cessation of farming may lead to a distinct secondary succession. The secondary succession in abandoned farmlands, also called old fields, can lead to a pattern of vegetation quite different from one that develops after primary succession following glacial retreat. For example, the plowing and added fertilizers, herbicides, and pesticides may have caused substantial changes in the soil of an old field, allowing species that require a lot of nitrogen to colonize. These species would not be present for many years in newly created volcanic or glacial soils.

Frederic Clements is often viewed as the founder of succession theory. His work in the early 20th century emphasized succession as proceeding to a distinct end point or **climax community.** Each phase of succession was called a sere or **seral stage.** The initial sere was known as the pioneer seral stage. Although disturbance could return a community from a later seral stage to an earlier seral stage, generally the community headed toward climax (**Figure 20.2**). Frequent disturbances

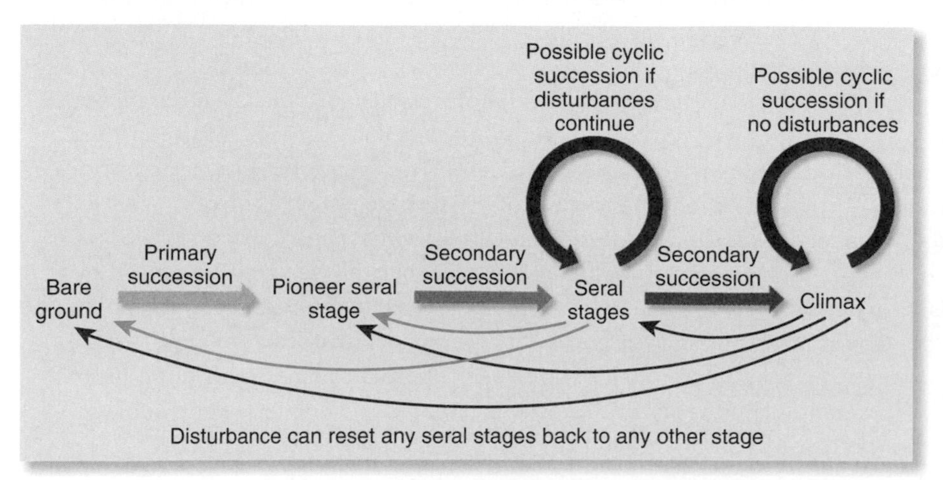


**Figure 20.2** **Diagrammatic representation of the different stages in succession.** Secondary succession can occur from any stage after the pioneer seral stage.

**ECOLOGICAL INQUIRY**

What types of disturbances could promote cyclic succession in midseral stages?

could lead to failure of a community to reach a climax community as one seral stage keeps replacing itself, a condition known as cyclic succession. In the absence of disturbances, a community would reach climax and undergo cyclic succession. For example, in a climax community, while forest trees may die, they are replaced by the same species of trees, which regrow in their place.

One of Clements's key assumptions was that each colonizing species made the environment a little different—for example, a little richer in soil nitrogen—so that it became more suitable for other species, which then invaded and outcompeted the earlier residents. This process, known as **facilitation,** supposedly continued until the most competitively dominant species had colonized, when the community was at climax. The composition of the climax community for any given region was thought to be determined by climate and soil conditions. While Clements's ideas on succession focused on the mechanism of facilitation, two other mechanisms affecting succession—inhibition and tolerance—have since been proposed by Joe Connell and Ralph Slatyer. In the next sections, we'll examine the evidence for each of them.

### 20.1.1 Facilitation assumes that each invading species creates a more favorable habitat for succeeding species

Succession following the gradual retreat of Alaskan glaciers is often used as a good example of facilitation. Over the past 200 years the glaciers in Glacier Bay have undergone a dramatic retreat of nearly 100 km (**Figure 20.3**). This is one of the few instances where we can trace the chronology of physical change in an area over time. In 1794 Captain George Vancouver visited the inlet and made notes on the positions of the glaciers. Since then, other explorers and scientists have visited the area and similarly left detailed notes on the positions of the glaciers. For example, John Muir, who was one of the conservationists instrumental in preserving Yosemite, visited in 1879. William Cooper, a professor at the University of Minnesota, inspired by Muir's writings, set up permanent study plots there in 1916 to study succession. By visiting different areas of Glacier Bay we can examine plant communities that have existed for different lengths of time along the edges of the bay. Ring counts of the oldest living trees also give approximations of when forests formed. Communities closest to the glaciers have been accumulating species for relatively short periods of time and are ecologically young. Communities at the mouth of Glacier Bay are relatively old, having been accumulating species since 1794, the time of Vancouver's visit. Sequential changes in community structure over physical distance, primarily influenced by time, constitute a **chronosequence.**

Succession in Glacier Bay follows a distinct pattern of vegetation. As glaciers retreat, they leave moraines—deposits of stones, pulverized rock, and debris that serve as parent material for soil—along the edges of the bay. In Alaska the bare soil has a low nitrogen content and scant organic matter.

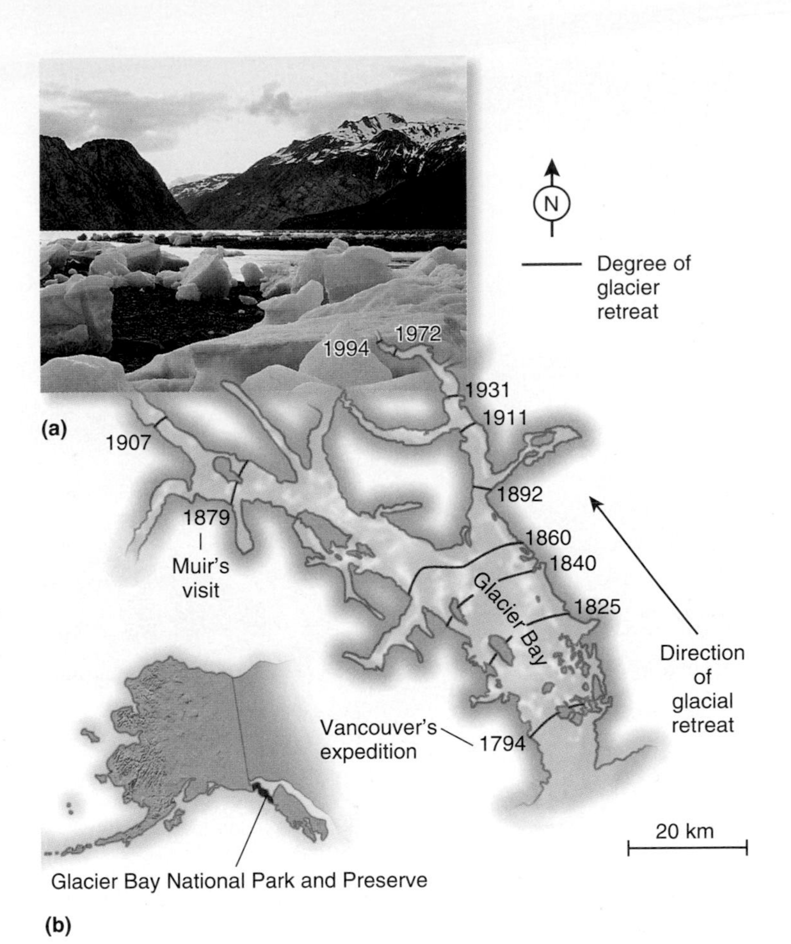


**Figure 20.3 The degree of glacier retreat at Glacier Bay, Alaska, since 1794.** **(a)** Primary succession begins on the bare rock and substrate evident at the edges of the retreating glacier. **(b)** The lines reflect the position of the glacier in 1794 and its subsequent retreat.

**ECOLOGICAL INQUIRY**

How do we know the position of the glacier in Glacier Bay 100 years ago?

In the pioneer seral stage, the soil is first colonized by a black crust of cyanobacteria, mosses, lichens, horsetails *(Equisetum variegatum),* and the occasional flowering plant called river beauty, *Chamerion latifolium* (**Figure 20.4**). Because the cyanobacteria are nitrogen fixers, the soil nitrogen increases a little, but soil depth and litterfall—fallen leaves, twigs, and other plant material—are still minimal. At this stage there may be a few seeds and seedlings of dwarf shrubs of the rose family, *Dryas drummondii,* alders, and spruce, but they are rare in the community. After about 40 years, *D. drummondii,* a nitrogen-fixing plant that contains bacteria, *Frankia* spp., dominates the landscape. Soil nitrogen increases, as does soil depth and litterfall, and alder trees begin to invade.

At about 60 years, alders, *Alnus sinuata,* form dense, close thickets. Alders also have nitrogen-fixing *Frankia* bacteria that live mutualistically in their roots and convert nitrogen from the air into a biologically useful form. The excess

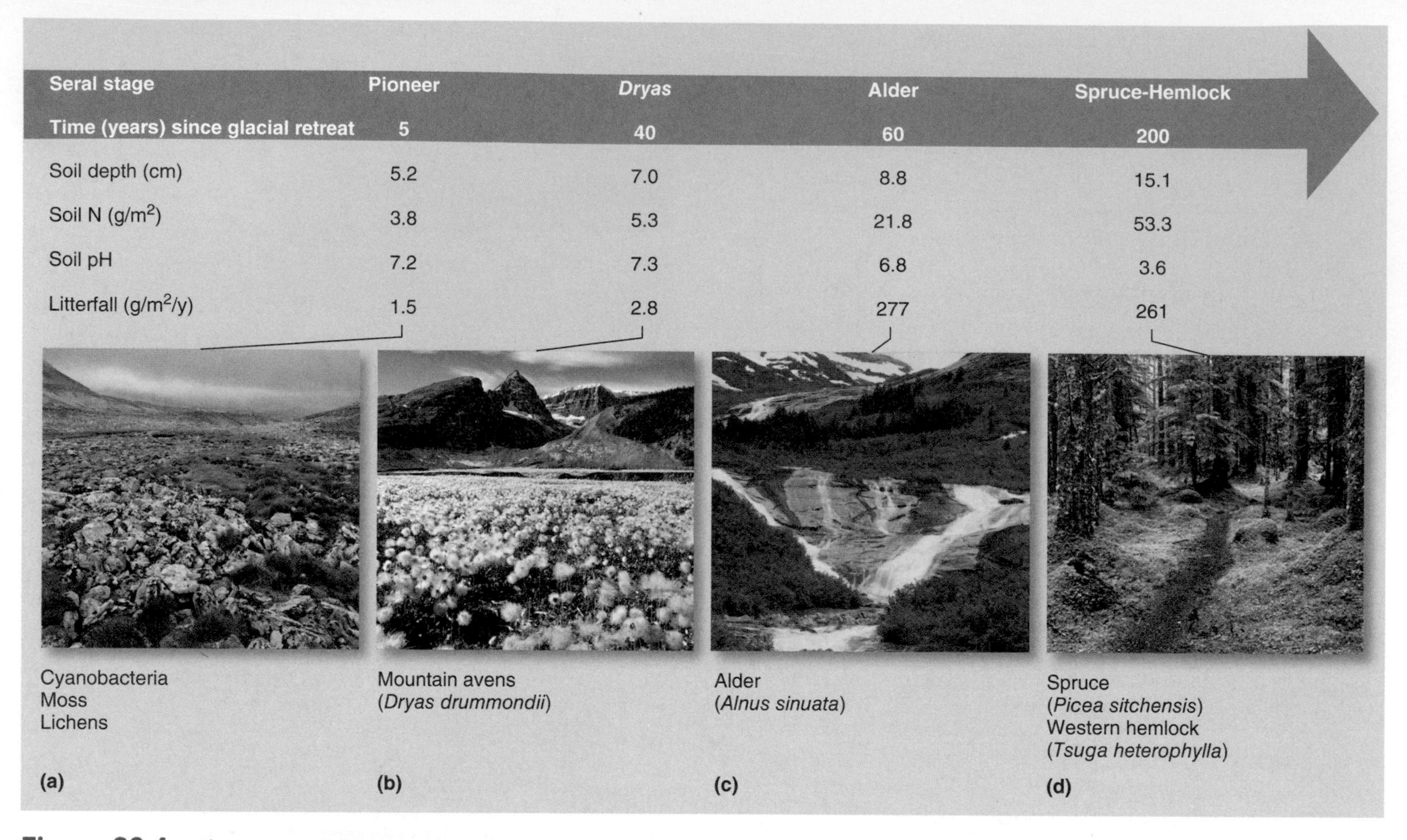


**Figure 20.4 The pattern of primary succession at Glacier Bay, Alaska. (a)** The first species to colonize the bare earth following retreat of the glaciers are small species such as cyanobacteria, moss, and lichens. **(b)** Mountain avens, *Dryas drummondii,* is a flower common in the *Dryas* seral stage. **(c)** Soil nitrogen and litterfall increase rapidly as alder, *Alnus sinuata,* invade. Note also the appearance of a few spruce trees higher up the valley. **(d)** Spruce, *Picea sitchensis,* and hemlock, *Tsuga* spp., trees in a climax spruce-hemlock forest at Glacier Bay, with moss carpeting the ground. Two hundred years ago, glaciers occupied this spot.

nitrogen fixed by these bacteria accumulates in the soil. Soil nitrogen dramatically increases, as does litterfall. Spruce trees, *Picea sitchensis,* begin to invade at about this time.

After about 75–100 years, the spruce trees begin to overtop the alders, shading them out. The litterfall is still high and the large volume of needles turns the soil acidic. The shade causes competitive exclusion of many of the original understory species, including alder, and only mosses carpet the ground. At this stage, seedlings of western hemlock, *Tsuga heterophylla,* and mountain hemlock, *Tsuga mertensiana,* may also occur, and after 200 years, a mixed spruce-hemlock climax forest results. Facilitation, in the form of increased soil nitrogen and depth, fuels succession in this community, but competition also plays a large role as later invaders outcompete earlier colonists.

What other evidence is there of facilitation? Succession in intertidal communities often supports the facilitation model. In New England salt marshes, *Spartina* grass facilitates the establishment of beach plant communities by stabilizing the rocky substrate and reducing water velocity, which enables seedlings of other species to emerge. Succession on sand dunes also supports the facilitation model. As long ago as the 19th century, Henry Chandler Cowles (1899) conducted research on the sand dune community along the southern shore of Lake Michigan (**Figure 20.5**). Cowles showed that

**Figure 20.5 Indiana Sand Dunes State Park.** Henry Chandler Cowles studied succession on these sand dunes in the late 19th century.

## Peter Vitousek and Colleagues Showed How Invasion by an Exotic Tree Changed Soil Nitrogen Levels in Hawaii

Fayatree, *Myrica faya,* is a small, nitrogen-fixing tree native to the Azores and Canary Islands. It was brought to Hawaii by Portuguese immigrants in the late 19th century. Because of its nitrogen-fixing capability, it became invasive on the nitrogen-deficient soils of Hawaii Volcanoes National Park on or before 1961. After 1961 *Myrica* continued to expand its coverage, occupying 12,200 ha in the park by 1985 and 34,365 ha in the whole of Hawaii. Peter Vitousek and his colleagues expected that the appearance of *Myrica* in the Hawaii Volcanoes National Park could significantly alter primary succession. They noted that primary succession generally involves a substantial increase in available nitrogen. The volcanic soil contained most nutrients plants need, except for nitrogen. The ability of *Myrica* to increase soil nitrogen levels would have the potential to alter the whole community that exists in the park, not just one or two species. While ecological studies on succession are still ongoing in this system, Vitousek and colleagues documented large effects of *Myrica* on soil nitrogen (**Figure 20.A**). Other nitrogen-fixing plants were present in the Hawaiian flora, but no other nitrogen fixers were able to colonize young volcanic soils.

**HYPOTHESIS** Invasive *Myrica* species can change soil conditions and the growth of native vegetation.

**STARTING LOCATION** Hawaii Volcanoes National Park

| | Conceptual level | Experimental level |
|---|---|---|
| 1 | Determine if nitrogen limits plant growth at Hawaiian sites. Use three different types of plot: young (following 1959 volcanic eruption), intermediate (following 1790 eruption) and old (1,000 – 2,000 years without eruption). No *Myrica* was present on any plots. | Fertilize plots containing the native tree *Metrosideros* with 12.5 g/m$^2$ of N, 6.2 g/m$^2$ of P, or a combined fertilizer with all macro and micro nutrients except N and P. There were 40 12.6-m$^2$ plots on the young sites, 32 225-m$^2$ plots on the intermediate-aged sites, and 12 200-m$^2$ plots on the old sites. Tree growth was measured on all plots. N fertilization increased tree growth on all plots, but relatively less on old plots (see Data part i) and only significantly on young and intermediate plots. |
| 2 | Determine if nitrogen fixation by the invasive species, *Myrica*, increases nitrogen availability. | Measure N inputs on young plots with and without *Myrica* (following its removal) (see Data part ii). |

**3 THE DATA** The invasive species *Myrica* changed soil conditions in Hawaii by increasing nitrogen availability. Native trees respond to increased nitrogen, especially on soils of young age following volcanic eruption. Therefore, the succession of native plant communities on volcanic soil could be sped up by the presence of nitrogen-fixing invasive species.

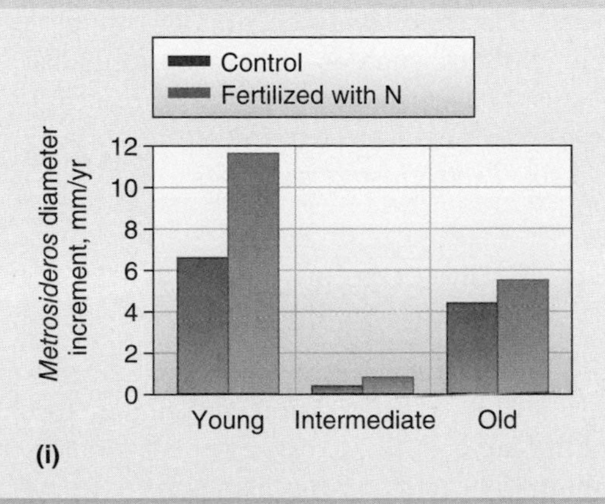

**Figure 20.A Peter Vitousek and colleagues demonstrated how invasive species affect soil properties in Hawaii.** (From data in Vitousek et al., 1987.)

these dunes were initially colonized and stabilized with salt-tolerant plants with extensive root systems such as beach grass, *Ammophila breviligulata*. Following dune stabilization, larger species such as dune willow, *Salix glaucophylloides,* and cottonwoods, *Populus* spp., appeared inland, followed eventually by trees such as basswood, *Tilia americana*. Thus, the sand dunes are dynamic changing entities, beginning with bare sand and beach grass, then, as they move inland as the sand blows, they are colonized by larger trees. Nitrogen-fixing invasive species can greatly increase soil nitrogen and have the ability to change whole communities (see **Feature Investigation**).

Facilitation also occurs in aquatic communities. Although soils do not develop in marine environments, facilitation may still occur when one species enhances the quality of settling and establishment sites for another species. The critical stage in the life cycle of many marine organisms occurs when planktonic larvae find suitable sites for attachment, where they develop into adult organisms. When T. A. Dean and Larry Hurd (1980) used experimental test plates to measure settling rates of marine organisms in the Delaware Bay, they discovered that hydroids enhanced the settlement of tunicates, and both facilitated the settlement of mussels, which were the dominant species in the community. In this experiment the smooth surface of the test plates prevented many species from colonizing, but once the surface became rougher, because of the presence of the hydroids, many other species were able to colonize.

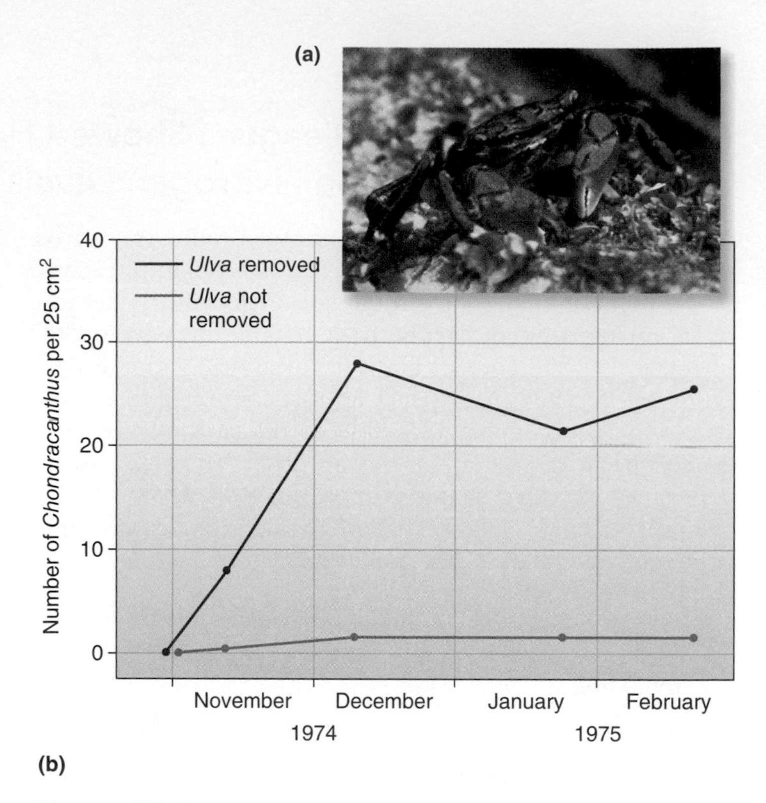


**Figure 20.6 Inhibition is a primary method of succession in the marine intertidal zone.** **(a)** Crab eating *Ulva* on rock face. **(b)** Removing *Ulva* from intertidal rock faces allowed colonization by *Chondracanthus*.

## 20.1.2 Inhibition implies that early colonists prevent later arrivals from replacing them

Although data on succession in some communities fit the facilitation model, researchers have proposed alternative hypotheses of how succession operates. Another view is that possession of space is all-important and that the colonists that get there first determine subsequent community structure. In this process, known as inhibition, early colonists may exclude subsequent colonists. For example, removing the litter of giant foxtail, *Setaria faberi,* an early successional plant species in New Jersey old fields, causes an increase in the biomass of a later species, eastern daisy fleabane, *Erigeron annuus*. The release of toxic compounds from decomposing *Setaria* litter or physical obstruction by the litter itself contributes to the inhibition of *Erigeron*. Without the litter present, *Erigeron* dominates and reduces the biomass of *Setaria*. Plant species that grow in dense thickets, such as some grasses, ferns, vines, pine trees, and bamboo, can inhibit succession, as can many introduced plant species. Invasive *Tamarix* trees in the southwestern U.S. and Australia increase soil salts, inhibiting succession. Similarly, invasive pine trees, *Pinus* spp., in South Africa produce nutrient-poor litter that lowers soil nutrient availability, also inhibiting succession.

Inhibition has been seen as the primary method of succession in the marine intertidal zone, where space is limited. In this habitat, early successional species are at a great advantage in maintaining possession of valuable space. Wayne Sousa (1979) created an environment for testing how succession works in the intertidal zone by scraping rock faces clean of all algae or putting out fresh boulders or concrete blocks. The first colonists of these areas were green algae, *Ulva* spp. Much later these were replaced by several species of red algae, including the large alga *Chondracanthus canaliculatus*. By removing *Ulva* from the substrate, Sousa showed that *Chondracanthus* was able to colonize more quickly (**Figure 20.6**). The results of Sousa's study indicate that early colonists can inhibit rather than facilitate the invasion of subsequent colonists. Succession may eventually occur because early colonizing species, such as *Ulva,* are more susceptible than later successional species, such as *Chondracanthus,* to the rigors of the physical environment and to attacks by herbivores, such as crabs, *Pachygraspus crassipes*.

## 20.1.3 Tolerance suggests that early colonists neither facilitate nor inhibit later colonists

The huge differences between facilitation and inhibition as mechanisms of succession prompted Joseph Connell and Ralph Slatyer (1977) to view the two as extremes on a continuum. The two researchers proposed a third mechanism of succession, which they termed **tolerance.** In this process, any species can start the succession, but the eventual climax community is reached in a somewhat orderly fashion.

Early species neither facilitate nor inhibit subsequent colonists. Connell and Slatyer found the best evidence for the tolerance model in Frank Egler's earlier work on floral succession. In Connecticut, Egler (1954) showed that secondary succession in plant communities is determined largely by species that already exist in the ground as buried seeds or old roots. Whichever species happened by chance to germinate or regenerate from roots closest in time to the occurrence of a disturbance—for example, a treefall that created a light gap—initiated the succession sequence. However, eventually larger, slower-growing species would outcompete the smaller pioneer species, and the forest trees would return.

The key distinction between the three models is in the manner in which succession proceeds. In the facilitation model, species are facilitated by previous colonists; in the inhibition model, they are inhibited by the action of previous colonists; and in the tolerance model, they are unaffected by previous colonists (**Figure 20.7**, **Table 20.1**).

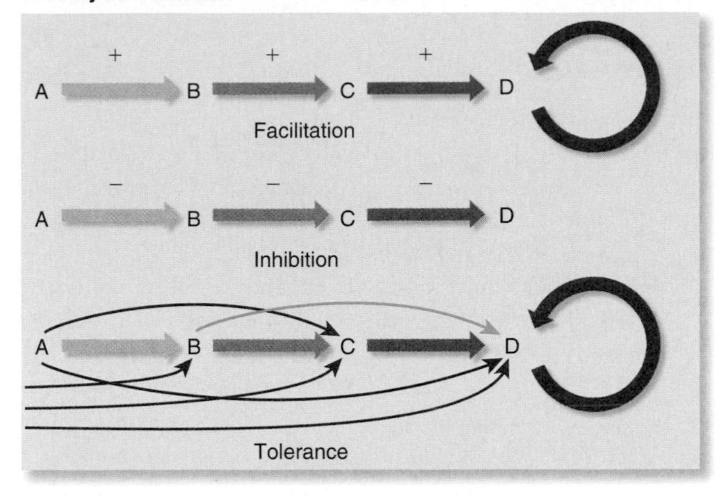


**Figure 20.7 Three models of succession.** A, B, C, and D represent four stages or seres. D represents the climax community. An arrow indicates "is replaced by," + = facilitation and, – = inhibition. The facilitation model is the classic model of succession. In the inhibition model, much depends on which species gets there first. In the tolerance model species are unaffected by previous colonists.

## 20.1.4 Facilitation and inhibition may both occur in the same community during succession

Sometimes the relative importance of facilitation and inhibition varies in different seral stages within the same community. Thus, facilitation may be important in early seral stages and inhibition important in later ones or vice versa. As a good example of this, let's consider another classic study, that of secondary succession in the Piedmont plateau of North Carolina (**Figure 20.8**). In the Piedmont region, much old-growth forest was felled for tobacco farms, then the

**ECOLOGICAL INQUIRY**

The inhibition model of succession implies competition between species, with early-arriving species tending to outcompete later arrivals, at least for a while. How does competition feature in the facilitation model of succession?

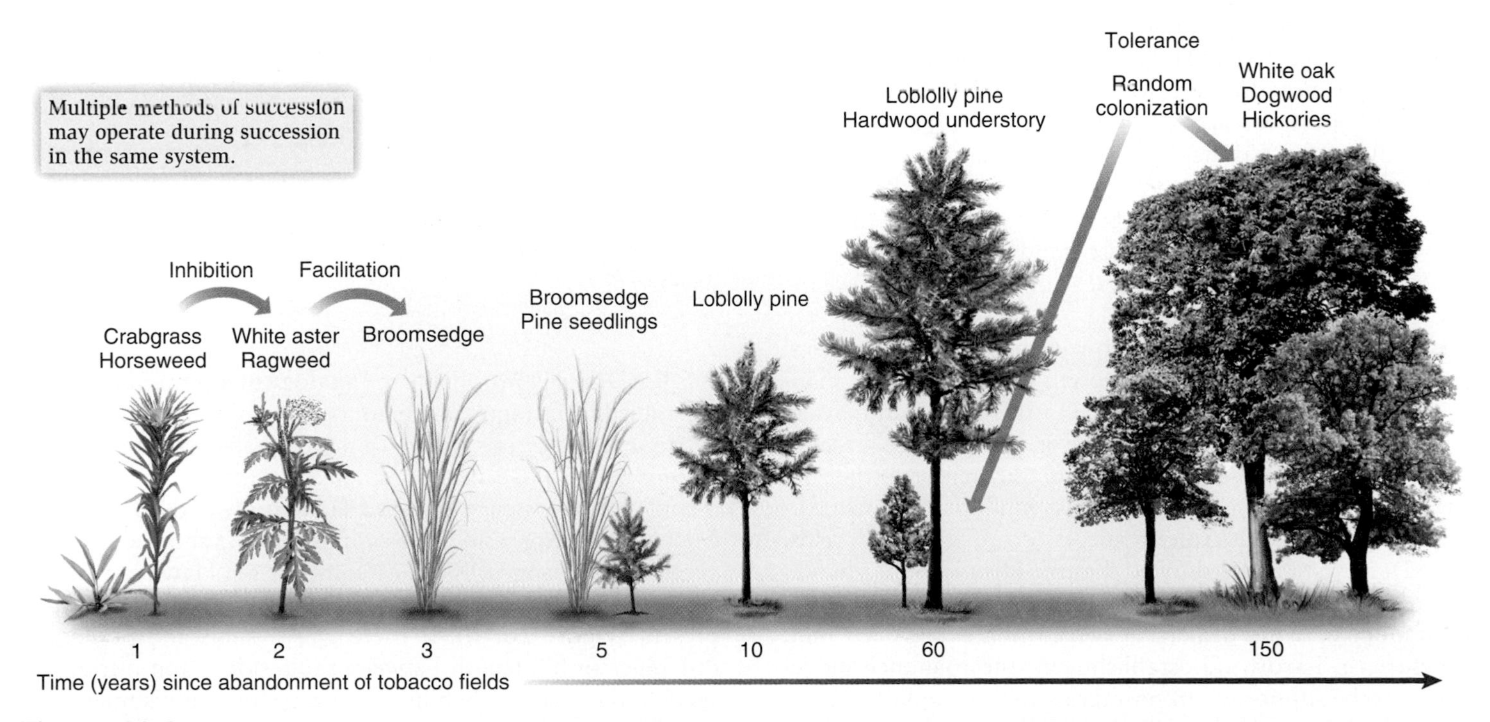


**Figure 20.8 Secondary succession in the Piedmont region of North Carolina following abandonment of tobacco fields.** Inhibition, facilitation, and tolerance are all important at various stages.

**Table 20.1** The main features of facilitation, tolerance, and inhibition, three models of succession.

| | Successional Model | | |
|---|---|---|---|
| Feature | Facilitation | Tolerance | Inhibition |
| 1. Type of species that become established first | Early successional | Any that can survive | Any that can survive |
| 2. Effect of early colonists on environment | Make less suitable for early-colonizing species, more suitable for later-colonizing species | Make less suitable for early-colonizing species, no effect on later-colonizing species | Makes less suitable for all colonists |
| 3. Effect of later colonists on early colonists | Elimination of early colonists by competition | Elimination of early colonists by competition | Later species unable to colonize if early colonists persist |
| 4. Time sequence | Continues until resident species no longer facilitate the invasion of other species | Continues until no species exist that can invade and displace resident | Continues unless early colonists are disturbed/removed |

fields were abandoned because soil nutrients were depleted and new forest was felled for new tobacco fields in different areas. This process operated repeatedly, so that there was a continuum of ages in the abandoned fields of the region on which to study succession. One year after the fields were abandoned, two species dominated: crabgrass, *Digitaria sanguinalis,* and horseweed, *Erigeron canadensis.* These species were regarded as the pioneer species. In the second year, these two species were joined by white aster, *Aster ericoides,* and ragweed, *Ambrosia artemisifolia.* Both of these latter species were absent or were of minor importance in year one but were dominant in year two. Twenty-six other plant species were less commonly found in the fields in year two. In the third year, species richness declined dramatically because yet a third species, broomsedge, *Andropogon virginicus,* a grass, became dominant and remained so for several years. During this time, seeds of loblolly pine, *Pinus taeda,* and some hardwood trees arrived via wind dispersal. The pines became established in year five and formed a closed canopy by year ten. Hardwood seeds arrived by wind dispersal, and because pine seedlings have difficulty germinating under a pine canopy, the hardwood understory thrived. Whichever hardwood species happened to germinate became common. By the time 100 years had passed, most forest stands had as many hardwoods, such as oaks and hickories, as pines, and by 200 years, only scattered pines remained.

Catherine Keever (1950) showed that in the early seral stages, horseweed inhibits the growth of asters, which manage to gain only a small foothold. On the other hand, asters stimulate the growth of their successor, broomsedge, because aster roots increase the amount of organic matter in the soil. Within the space of a few years, first inhibition, then facilitation plays a primary role in succession of these old fields. Diane DeSteven (1991) showed that the identity of hardwood species that became established under the pine depended on the amount of wind-dispersed seeds, often called **seed rain.** Therefore, in hardwood establishment, the tolerance model appears to be supported. In postagriculture secondary succession in the North Carolina Piedmont, each seral stage had a different mechanism or path to succession.

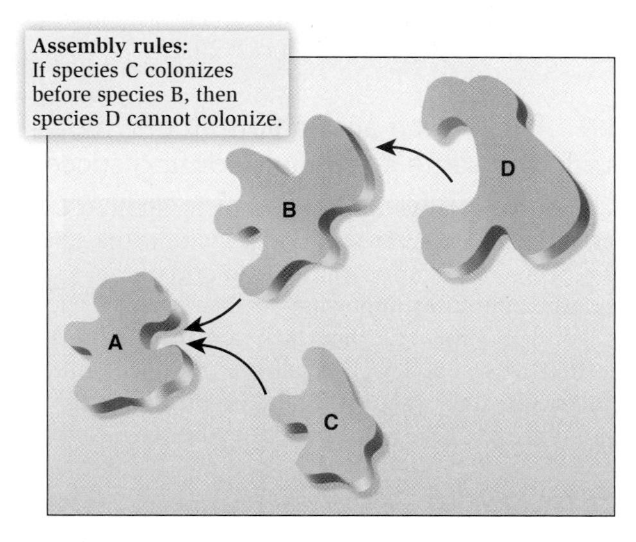


**Figure 20.9** **Assembly rules: Each piece of the puzzle represents a species in the community.** The first species to invade the community influences which other species can subsequently invade. (After Drake, 1990.)

**ECOLOGICAL INQUIRY**

How appropriate is this jigsaw puzzle analogy to natural systems?

Jim Drake (1990) made the analogy of a community being assembled in fashion similar to a jigsaw puzzle (**Figure 20.9**). If species A colonizes an area first, then species B or C could colonize, but if species C colonizes, then D cannot. This assumes the existence of both facilitation and competition. Michael Moulton and Stuart Pimm (1986) showed how the success of birds introduced to Tahiti or the Hawaiian Islands was dependent on the morphology of the species already present (**Figure 20.10**). If there was a small morphological difference in bill length between the resident and introduced species, then the probability of successful establishment was low. This builds on the concept introduced in Chapter 11 that morphological differences may allow species to coexist.

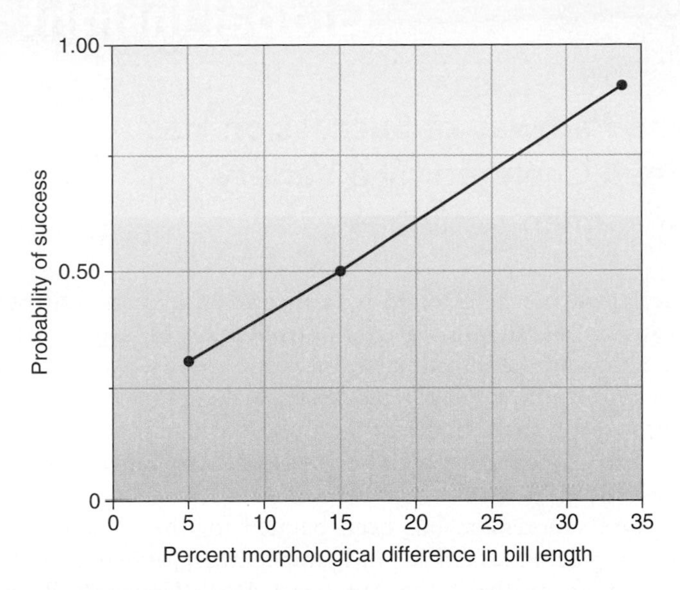


**Figure 20.10 Morphological differences and the probability of invasion success.** Invasion success for birds entering Hawaii increases as morphological differences in bill length increase. (From Moulton and Pimm, 1986.)

Where morphological differences between species are large, the probability of successful establishment increases.

### Check Your Understanding

**20.1** In the heathland soils of Europe, scotch heather, *Calluna vulgaris*, and cross-leaved heath, *Erica tetralix*, are gradually replaced by variegated purple moor grass, *Molinia caerulea*, and wavy hair grass, *Deschampsia flexuosa*. Adding *Calluna* litter or nitrogen fertilizer speeds up this process. Explain this phenomenon, and explain which mechanism of succession is supported.

## 20.2 Species Richness Often Increases During Succession

Species richness of both plants and animals is low in the early stages of succession but often rises quickly in later seral stages. Richard Inouye and colleagues (1987), who studied plant species richness of old fields in Minnesota, showed that the older the field, the greater the number of species. This was because there had been more time for species to colonize (**Figure 20.11**). Of course, such an increase in species richness can't go on forever. Fakhri Bazzaz (1975) also noted an increase in species richness of plants with age on abandoned fields in southern Illinois. He constructed rank abundance diagrams (refer back to Section 17.3) for fields between 1 and 40 years after abandonment (**Figure 20.12**). Newly abandoned fields were dominated by relatively few species of herbs. Shrubs and trees colonized older plots, species richness increased, and the shape of the rank abundance plot changed from a dominance-preemption model more toward lognormal. This means that the number of individuals is spread across more species in the latter years of succession as compared to the earlier years where a few species dominate the community.

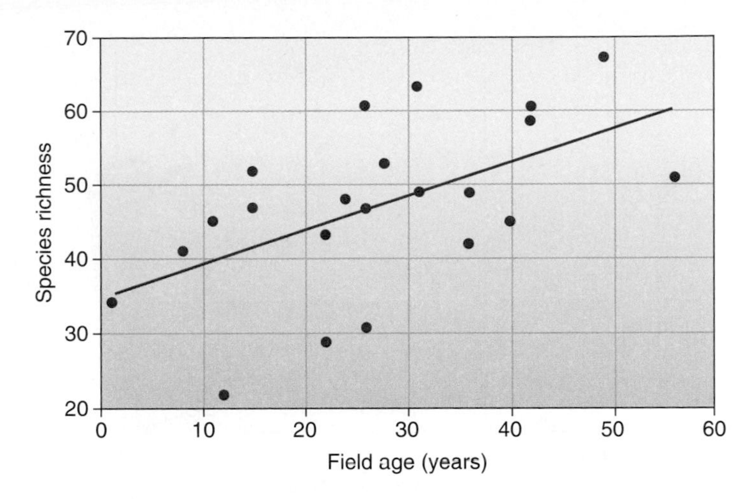


**Figure 20.11 Increase in plant species richness with field age in Minnesota old fields.** (After Inouye et al., 1987.)

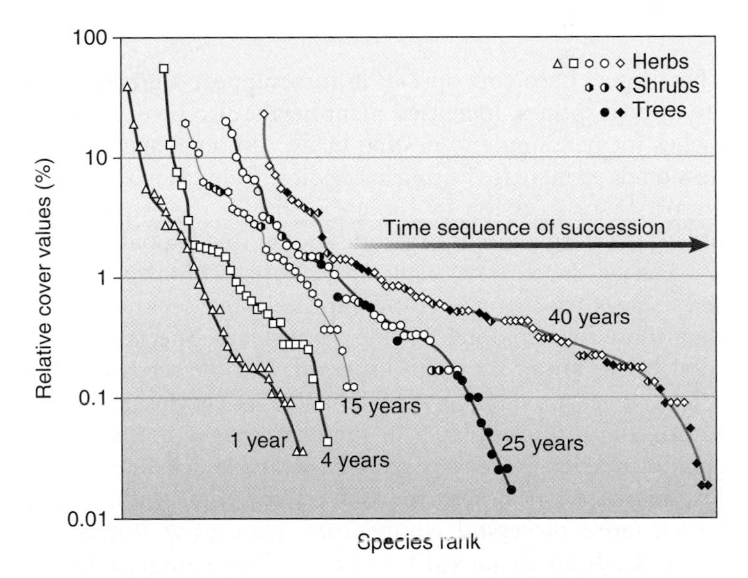


**Figure 20.12 The change in species abundance of plants with time in southern Illinois old fields.** Abundance is expressed as the percentage that a given species contributes to the total area covered by all species. (After Bazzaz, 1975.)

Animal species richness also increases as succession proceeds. In the North Carolina Piedmont plateau, David Johnston and Eugene Odum (1956) examined the richness of bird communities from grassland to mature oak-hickory forest in the same old tobacco fields where Catherine Keever worked (**Figure 20.13**). The number of bird species increased from 2 species in 1-year-old grasslands to about 20 species in the mature 150-year-old climax oak-hickory forests. This increase in bird species occurs because as more plants invade an area, the number of insect herbivores that feed on them increases.

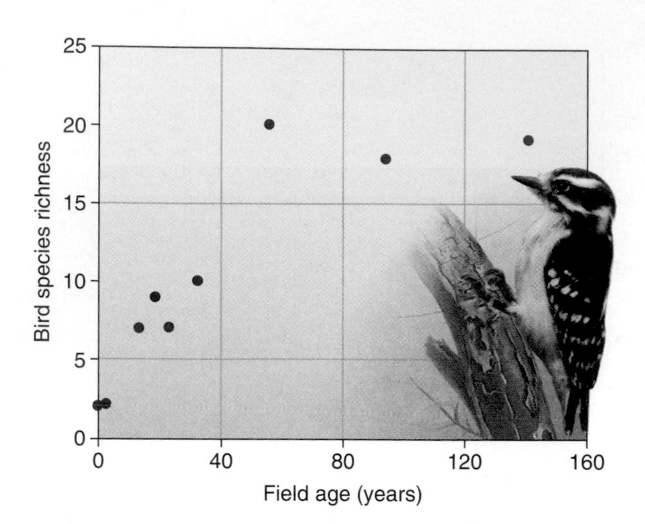


**Figure 20.13** **Increases in bird species richness on different-aged old fields in Georgia during the nesting season.** (Data from Johnston and Odum, 1956.)

These insect herbivore species in turn support a greater variety of bird species. Increased plant height also provides more niches for perching and nesting birds. Although species richness tends to increase during succession, humans may accelerate or slow succession by the use of fire or the introduction of large grazing animals or exotic species (see **Global Insight**).

During succession, plant species that colonize in early seral stages tend to have different life-history characteristics than those that colonize in later seral stages. Species in early seral stages are often wind-dispersed, and the seeds can live a long time, maximizing their chances of successful colonization. In addition, plants in the early stages of succession acquire nutrients quickly, grow fast, and reach maturity at low biomass. Species in the latter stages of succession are larger, more successful competitors, who grow slowly and whose seeds are dispersed by animals. They can grow in low light levels characteristic of closed forests. Species in later seral stage communities are relatively resistant to disturbance. **Table 20.2** summarizes many of these differences in life-history characteristics.

**Check Your Understanding**

**20.2** **In what ways can humans alter the trend of increasing species richness during succession?**

## 20.3 Restoration Ecology Is Guided by the Theory of Succession

**Restoration ecology** is the full or partial repair or replacement of biological habitats and/or their populations that have been degraded or destroyed. Following open-pit mining for coal or phosphate, huge tracts of disturbed land must be replenished with topsoil and a large number of species such as grasses,

**Global Insight**

### The Pathway of Succession Has Been Changed by a Variety of Human Activities

Succession can be affected by a myriad of human activities, including fire suppression, the introduction of large grazing animals, and the introduction of exotic species. As an example, consider how these processes have altered the community landscape in the U.S. desert Southwest, where much of what was once open grassland has succeeded to pinyon-juniper woodlands.

Fire suppression has been blamed for the expansion of pinyon pine, *Pinus edulis,* and juniper trees, *Juniperus monosperma,* in much of the American West (**Figure 20.B-i,ii**). In unmanaged areas, frequent fires kill low-growing shrubs

(i)

(ii)

**Figure 20.B** **The effects of fire suppression on succession.** Landscapes can be changed by the absence of fire. View from Enchanted Mesa, 100 km west of Albuquerque, in 1899 **(i)** and 1977 **(ii)**, after many years of fire suppression, showing juniper expansion.

and trees, but grasses resprout from their roots. Grasslands normally dominate fire-prone landscapes. Where fire is suppressed, the shrubs and trees survive and eventually outcompete the grasses. Subsequent shrub growth is difficult to reverse because the accumulation of nutrient-rich litter underneath trees and bushes occurs quickly. Nutrient enrichment promotes shrub growth but discourages grassland recovery because of competition. An aggressive program of cutting and burning pinyon pine and juniper woodlands in the 1960s and 1970s attempted to restore grazing lands.

Cattle and sheep brought to the American Southwest in the late 1500s only began to have a serious effect on the landscape in the late 1800s, when hundreds of thousands had been transported into the area by rail. The once grass-rich ponderosa pine forests were drastically altered, and recovery following this disturbance was hindered. The grazing habits of livestock provided the very conditions needed to stimulate juniper expansion, because juniper seeds germinate only after they have passed through the digestive tract of an animal. The effects of livestock on local vegetation are well illustrated by comparing ungrazed and grazed areas. Near Hatch, New Mexico, grazed areas bordering rivers are denuded of most of their vegetation compared to ungrazed areas (**Figure 20.C**).

Humans can also influence succession by the introduction of new plants. Cheyenne and Arapaho Indians carried a variety of tree nuts in a "trail mix" to sustain them. An isolated stand of pinyon pine, *Pinus* spp., exists at Owl Canyon, north of Fort Collins, Colorado, 250 km north of the nearest source population near Colorado Springs. From photographs taken in 1950 and 1989, we can see this 5-km$^2$ stand has shown a rapid increase in cover, characteristic of a recent and expanding population (**Figure 20.D-i,ii**). The nearest stands are too far distant to have provided the seeds to start this population. Instead, Native Americans may well have established this stand during a trek to the front range of the Rockies. Changes similar to these are likely to occur in other regions of the United States, but the data are still being accumulated.

Around the world, invasive species have changed communities, but it is not always clear when a species should be considered native or invasive. Some introductions occurred so long ago that they are now accepted as part of the natural fauna. Rabbits were probably introduced to Britain from France by William the Conqueror sometime after 1066. C. D. Preston and colleagues (2004) showed that 157 plant species had been introduced to Britain by humans, intentionally or accidentally, from the start of the Neolithic period, about 4,000 years ago, to 500 years ago. Some of these exotic species have been considered part of the British flora for generations, and some are threatened with extinction. Attaching the label "exotic" or "invasive" to plants may exclude them from some conservation schemes.

**Figure 20.C The effects of grazing on succession.** Grazed (left) and ungrazed (right) areas in Socorro County, New Mexico, are separated by a fence.

**(i)**

**(ii)**

**Figure 20.D The effects of seed introductions on succession.** Expansion of pinyon pine, *Pinus edulis*, at Owl Canyon, north of Fort Collins, Colorado, following seed introduction, possibly by Native Americans, in **(i)** 1950 and **(ii)** 1989.

**Table 20.2** Some characteristics of early and late seral stages in plant succession.

| Attribute | Early stages | Late stages |
|---|---|---|
| Seed dispersal | Good | Poor |
| Plant efficiency at low light | Low | High |
| Resource acquisition | Fast | Slow |
| Biomass | Small | Large |
| Species richness | Low | High |
| Species life-history | *r* | *K* |
| Seed dispersal vector | Wind | Animals |
| Seed longevity | Long | Short |

shrubs, and trees must be replanted. Aquatic habitats may be restored by reducing human impacts and replanting vegetation. In Florida, seagrass beds damaged by boat propellers are closed off to motorboats and the area is replanted.

The three basic approaches to habitat restoration are complete restoration, rehabilitation, and ecosystem replacement. In complete restoration, conservationists attempt to return a habitat to its condition prior to the disturbance. Under the leadership of ecologist Aldo Leopold, the University of Wisconsin pioneered the restoration of prairie habitats as early as 1935, converting agricultural land back to species-rich prairies (**Figure 20.14a**). The second approach aims to return the habitat to something similar to, but a little less than, full restoration, a goal called rehabilitation. In Florida, phosphate mining involves removing a layer of topsoil or "overburden," mining the phosphate-rich layers, returning the overburden, and replanting the area. Exotic species such as cogongrass, *Imperata cylindrica,* an invasive southeast Asian species, often invade these disturbed areas, and the biodiversity of the restored habitat is usually not comparable to that of unmined areas (**Figure 20.14b**). The third approach, termed replacement, makes no attempt to restore what was originally present but instead replaces the original community with a different one. The replacement could be a community that is simpler but more productive, as when deciduous forest is replaced after mining by grassland to be used for public recreation. Community replacement is particularly sensible for land that has been significantly damaged by past activities. It would be nearly impossible to recreate the original landscape of an area that was mined for stone or gravel. In these situations, wetlands or lakes may be created in the open pits (**Figure 20.14c**).

The hope is that restoration ecology can be guided by succession theory. In many restoration projects the aim is that the restored community will be an exact replica of the original. However, as we have seen, some species inhibit the arrival of others. Many invasive species have been present for long periods of time and prevent the colonization of natives. It is therefore sometimes difficult to restore a community without removing these species.

Robert Hilderbrand and colleagues (2005) have noted a series of myths associated with restoration, the first of which they termed the "field of dreams": build it and they will come. This is an ecological engineering view of restoration. For example, in wetlands restoration much emphasis is given to creating the correct elevational slope or grade so that hydrological conditions are restored. Less emphasis is placed on replanting because it is thought that restored areas will eventually revegetate on their own. While revegetation will occur naturally, the ensuing vegetation may not be of the correct type. Many wetland species have limited dispersal abilities and need to be planted. Many mined areas are restored by adding topsoil that has been stored following its initial removal. Woodland trees are planted in the hope that a native forest will return. However, many of these trees are late successional species that grow best in the shade, not full sunlight, so it is preferable to first plant early successional species.

Another myth is termed "fast forwarding." This myth states that we can greatly accelerate the path of succession toward our end goal. In some cases this might be true. For example, we can plant nitrogen-fixing legumes to boost soil nitrogen levels. In other cases, such as the previous forest

**(a) Complete restoration**

**(b) Rehabilitation**

**(c) Ecosystem replacement**

**Figure 20.14 Habitat restoration.** **(a)** The University of Wisconsin pioneered the practice of complete restoration of agricultural land to native prairies. **(b)** In Florida, phosphate mines are so degraded that complete restoration is not possible. After topsoil is replaced, some exotic species such as cogongrass often invade, allowing only habitat rehabilitation. **(c)** These old open-pit mines in Middlesex, England, have been converted to valuable freshwater habitats, replacing the wooded area that was originally present.

(a)

(b)

**Figure 20.15 Restoration ecology at Bluewater Creek, northern New Mexico.** Cows were removed from the area, non-native plant species were removed, and native tree species were planted. **(a)** A degraded (2008) and **(b)** a successfully restored (2009) community.

example, such shortcuts are difficult, if not impossible. The myth of the "cookbook" is that similar restoration techniques can be applied to many different areas. In reality, many communities differ substantially from one another. For example, some North American tall-grass prairie soils are, surprisingly, nutrient-poor, and the species that grow there are adapted to these conditions. Farming practices often increase soil nitrogen via fertilization, and returning these areas to native prairies requires reverse fertilization. How do we remove soil nitrogen? Nitrogen can be lowered by adding large quantities of carbon-containing organic material to the soil, such as wood shavings. This promotes microbial activity, which depletes soil nitrogen. Without such procedures, many prairie restoration projects face the continual problem of invasion by nitrogen-loving exotic species. Despite problems, restoration is a worthy goal, and when done properly, the end result is often a stunning improvement on the original degraded habitat (**Figure 20.15**). Furthermore, degraded but restored habitats often support more species than degraded habitats, and their ecosystem services are also improved. Jose Rey Benayas and colleagues (2009) conducted a meta-analysis of 89 restoration projects in a wide range of habitats across the globe. Degradation had been caused by logging, damming of rivers, mining, pollution, overgrazing, invasive species, and other factors. Restoration involved a variety of techniques, from simple cessation of the degrading action to invasive removal, replanting, or landscape engineering. Species richness, biodiversity, and a variety of ecosystem services were examined on degraded and restored habitats and compared to pristine, nondegraded habitats (**Figure 20.16**). Although restored habitats did not function as well as pristine areas, nor contain as many species, ecosystem services and biodiversity were much improved over degraded systems.

The last problem associated with restoration is that it can sometimes be difficult to determine the appropriate appearance of the original community. Do we go back

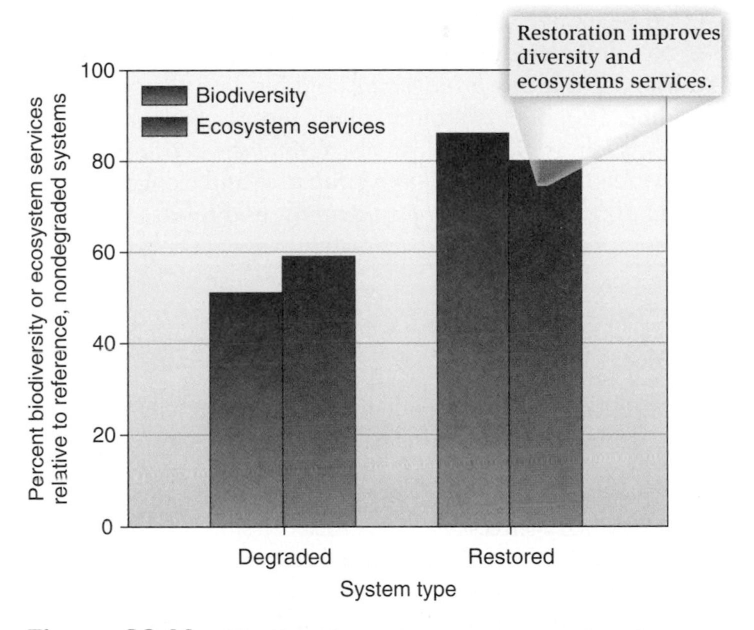


**Figure 20.16 Biodiversity and ecosystem services in restored and degraded habitats.** Restoration improves both biodiversity and ecosystem services over degraded habitats, but still does not reach 100%, the value for pristine systems.

100 or 200 years ago, or, in the case of many communities, before the arrival of Europeans or even native people? For example, dingoes of aborigines in Australia have profoundly affected native Australian communities. The arrival of the dingo is likely to have contributed to the extinction of the thylacine, *Thylacinus cynocephalus,* and the Tasmanian devil, *Sarcophilus harrisii,* on the Australian mainland. There is a strong movement in North America for so-called **Pleistocene re-wilding,** which would introduce large mammals from Africa and other continents, that would stand as proxies for North American species that went extinct in the Pleistocene,

**Figure 20.17** **The Iberian desman, *Galemys pyrenaicus*.** The habitat of this Pyrenean species is shrinking in Spain because of global warming. Will assisted migration—human dispersal of this species to new mountainous areas—save it from extinction?

13,000 years ago, at the hands of native people. For example, North American camels, *Camelops* spp., equids, mammoths, *Mammuthus primigenius,* and big cats went extinct at this time. Proponents would collect morphologically similar species, such as Bactrian camels, *Camelus bactrianus,* lions, *Panthera leo,* cheetahs, *Acinonyx jubatus,* and elephants, *Loxodonta africana* and *Elephas maximus,* and release them into huge game parks or let them range free. There is an equally strong movement against the idea.

Some ecologists believe that moving threatened species to new habitats, outside their native ranges, a concept known as **assisted migration,** is vital to their continued survival, especially in the face of global warming. Habitat of the Iberian desman, *Galemys pyrenaicus,* an amphibious, insect-eating mammal from the Pyrenees (**Figure 20.17**), is likely to vanish as the world warms. Species like the desman and some native possums in Australia are good candidates for moving to new areas which they could never reach on their own. Orchids in small mountainous areas of China are pushed higher and higher up mountain sides as the temperature increases and could easily go extinct unless moved to new habitats. Critics point out that the ecological impacts of assisted migration are unknown. For example, many invasive species are rare in their native habitats but assume pest proportions when moved.

### Check Your Understanding

**20.3** David Tilman suggested that plants appearing in the early stages of succession would tend to have relatively larger roots than those appearing later on, but that those appearing later would grow relatively taller than those appearing early. Try to explain this phenomenon in terms of biotic interactions.

## SUMMARY

- Succession is the gradual and continuous change in community structure over time.

**20.1 Several Mechanisms Describe Succession**

- Primary succession refers to succession on a newly exposed site not previously occupied by living organisms; secondary succession refers to succession on a site that has already supported life but has undergone a disturbance (Figures 20.1, 20.2).
- Succession following retreating glaciers and along sand dunes typically supports the facilitation model (Figures 20.3–20.5).
- Invasive species may also change the path of succession by increasing soil nitrogen (Figure 20.A).
- Succession on intertidal areas often supports the inhibition model (Figure 20.6).
- Three mechanisms have been proposed for succession: In facilitation, each species facilitates or makes the environment more suitable for subsequent species. In inhibition, initial species inhibit later colonists. In tolerance, any species can start the succession, and species replacement is unaffected by previous colonists (Figure 20.7, Table 20.1).
- Floral succession supports the tolerance model.
- Facilitation, inhibition, and tolerance may each occur during the same successional sequence (Figure 20.8), which has led some ecologists to describe community succession as being analogous to the assembly of a jigsaw puzzle (Figures 20.9, 20.10).

**20.2 Species Richness Often Increases During Succession**

- Species richness of both plants and animals, often, but not always, increases as succession proceeds (Table 20.2, Figures 20.11–20.13).
- Human activities such as fire suppression, introduced species, and the effects of livestock can alter the landscape and influence succession (Figures 20.B–20.D).

**20.3 Restoration Ecology Is Guided by the Theory of Succession**

- Restoration ecology can be guided by succession theory (Figures 20.14, 20.15).
- Many restored habitats have greater biodiversity and ecosystem services than degraded habitats (Figure 20.16).
- Determining the appropriate goal for restoration projects can be challenging. What is our target time period? Do we move species threatened by global warming to new suitable habitats, a process termed assisted migration? (Figure 20.17)

## TEST YOURSELF

1. Succession following a volcanic eruption is referred to as:
   a. Secondary succession
   b. Climax succession
   c. Seral stages
   d. Primary succession
   e. Facilitation
2. According to Clements, the end point of succession is known as:
   a. A climax community
   b. A seral stage
   c. A chronosequence
   d. Primary succession
   e. Secondary succession
3. Much of the coast of the Indonesian island of Sumatra was devastated by a 2004 tsunami. The recovery of coastal habitats in that area is an example of:
   a. Primary succession
   b. Secondary succession
   c. Tertiary succession
   d. The redundancy hypothesis
   e. The Keystone hypothesis
4. Each state of succession is called a:
   a. Series
   b. Set point
   c. Sere
   d. Climax
   e. Chronosequence
5. In which model of succession do existing species make it harder for new species to colonize?
   a. Inhibition
   b. Tolerance
   c. Assembly rules
   d. Facilitation
   e. Random colonization
6. What is considered to be the primary method of succession in the marine intertidal zone?
   a. Facilitation
   b. Tolerance
   c. Secondary
   d. Climax
   e. Inhibition
7. Which is a characteristic of species during the later seral stages of succession?
   a. Good seed dispersal
   b. Low plant efficiency at low light
   c. Small biomass
   d. Low species richness
   e. Short seed longevity
8. Which plants are dominant in the second seral stage of succession at Glacier Bay, Alaska?
   a. Mountain avens, *Dryas drummondii*
   b. Alder, *Alnus sinuata*
   c. Spruce trees, *Picea sitchensis*
   d. Cyanobacteria, lichens, moss
9. Facilitation and inhibition cannot happen in different seral stages in the same habitat.
   a. True
   **b. False**
10. Species richness usually decreases in the later stages of succession.
    a. True
    **b. False**

## CONCEPTUAL QUESTIONS

1. Distinguish between the three following models of succession: facilitation, inhibition, and tolerance.
2. What are the main approaches to habitat restoration?
3. List some of the advantages and disadvantages of Pleistocene re-wilding.

## DATA ANALYSIS

1. Look ahead to Figure 21.1, which details the numbers of plant species colonizing an Indonesian island following recovery after a volcanic explosion. How does succession help to explain the results?

The island of Rakata, Indonesia.

CHAPTER 21

# Island Biogeography

## Outline and Concepts



A massive volcanic explosion in 1883 on the island of Krakatau in Indonesia destroyed two-thirds of the island, originally 11 km long and covered in tropical rain forest. Life on the remainder, an island known as Rakata, and the two neighboring islands, Panjang and Sertung, was eradicated, suffocated by volcanic ash tens of meters deep. Recolonization on these islands has been studied by a series of scientists. Nine months after the 1883 eruption, the first reported colonist of Rakata was a spider spinning its web. Plant colonization of Rakata was greatly affected by how well plants were able to disperse. Wind- or sea-dispersed plants were able to colonize the island much more readily than plants dispersed by animals, such as birds (**Figure 21.1**). The early plant community was dominated largely by grasses. By 1929, over 40 years after the original eruption, animal-dispersed plants became as common as sea-dispersed ones, probably because the birds had become more abundant on the island. In 1930 a submarine eruption produced a fourth island, Anak Krakatau.

In Chapter 20 we explored the phenomenon called succession, the gradual and continuous change in species composition and community structure over time following a disturbance. On islands especially, succession is affected by the distance that dispersing organisms have to travel from undisturbed mainland areas. If a source pool of potential dispersing species is close by, then succession occurs quite rapidly. If there are long distances separating potential new colonists from areas to be colonized, then succession may be slower and some species may never colonize. Also, if an island to be colonized is large, it may accrue more species than smaller areas. In this chapter we investigate how the size of the area of habitat to be colonized, and the distance of the source pool of colonists, affect the special case of succession on islands. We also examine species turnover where certain immigrants to islands may be replaced by other species over time. Finally, we link the theory of island biogeography to the design of nature reserves, islands of natural habitat in a virtual sea of urban or developed land.

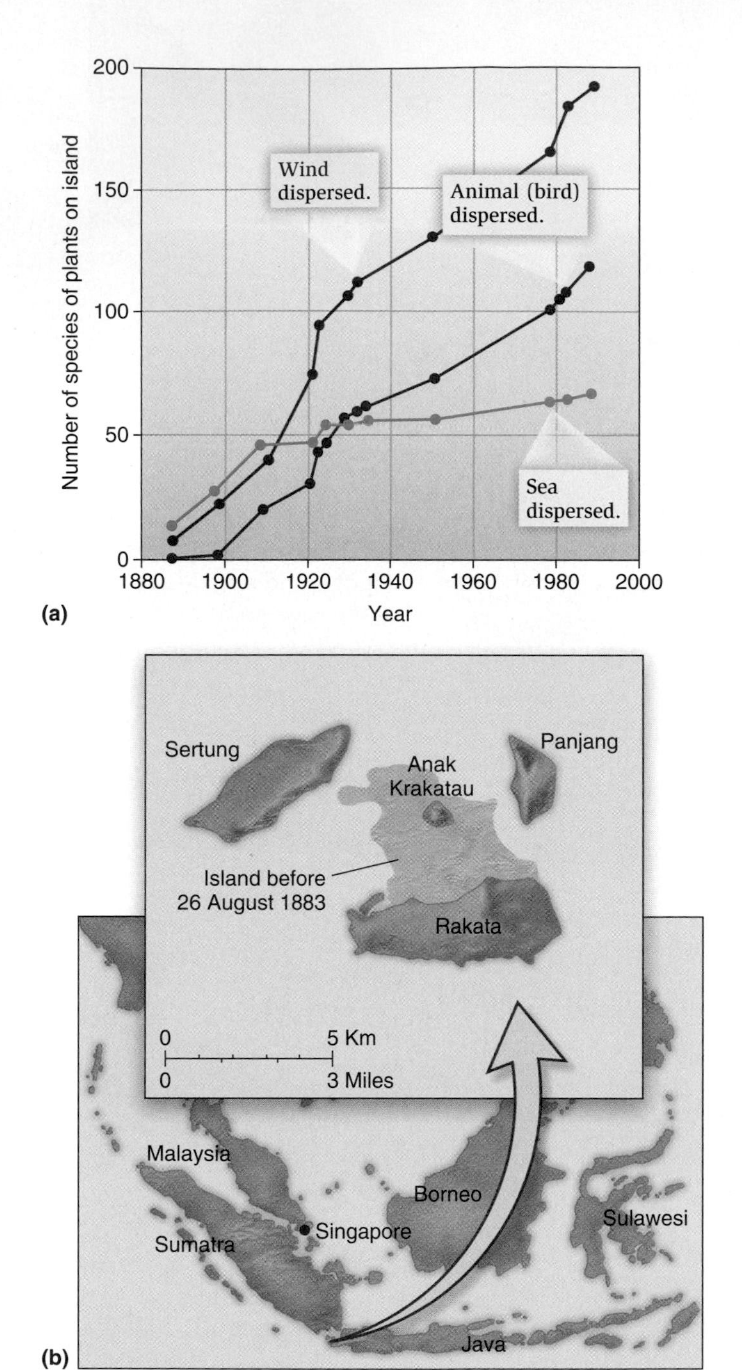


**Figure 21.1 Plant colonization on Rakata following volcanic eruption.** **(a)** Initially wind- and sea-dispersed plants, especially grasses, dominated the flora. **(b)** Map showing location of Rakata in southeast Asia. (From data in Whittaker et al., 1992.)

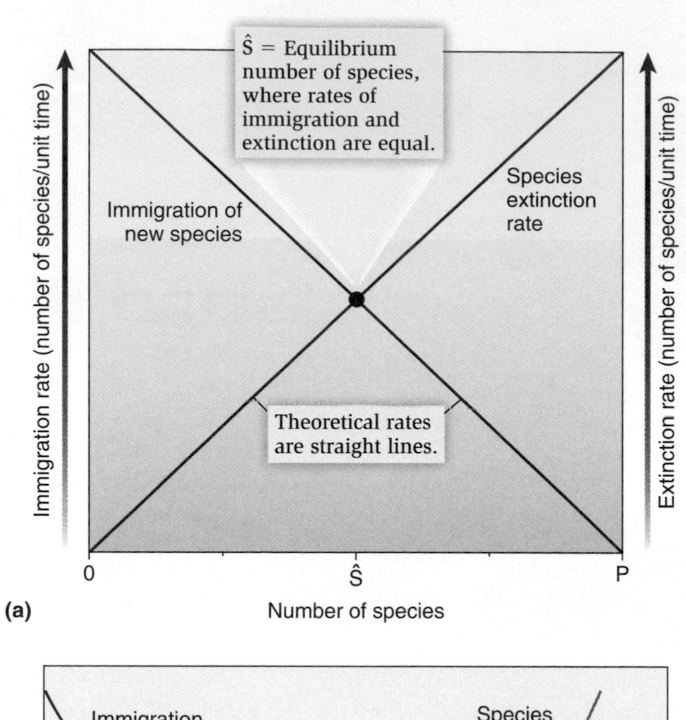


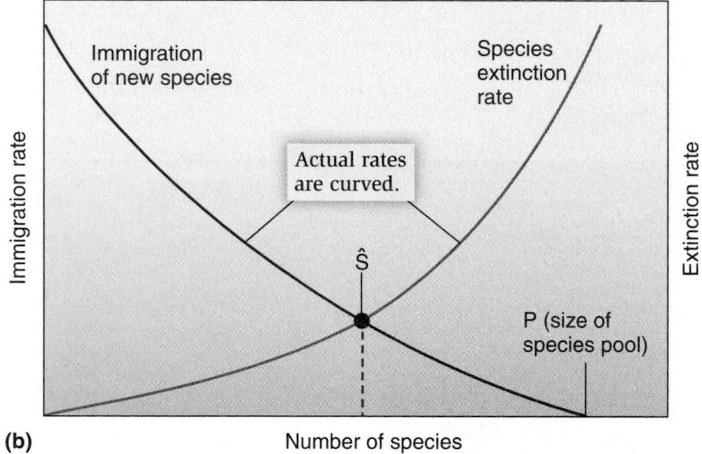


**Figure 21.2 The theory of island biogeography.** **(a)** The interaction of immigration rate and extinction rate produces an equilibrium number of species on an island, $\hat{S}$. $\hat{S}$ varies from 0 species to *P* species, the total number in the species pool of colonists. **(b)** In practice, the rate curves bend because some species immigrate more readily than others and because some species are better competitors than others.

## 21.1 The Theory of Island Biogeography Considers Succession on Islands

Robert MacArthur and E. O. Wilson (1963, 1967) developed a comprehensive model to explain the process of succession on newly formed islands, where a gradual buildup of species proceeds from a sterile beginning. Their ideas are termed the **theory of island biogeography**. In this section, we explore island biogeography and how well the theory's predictions are supported by data, particularly that provided by classic experiments in the Florida Keys.

MacArthur and Wilson's theory of island biogeography suggests that species repeatedly arrive on an island and either thrive or become extinct. The number of species tends toward an equilibrium number that reflects a balance between the rate of immigration and the rate of extinction (**Figure 21.2a**). The rate of immigration of new species is highest when no species are present on the island, so that each species that invades the island is a new species. As species accumulate, subsequent immigrants no longer represent new species, so the rate of immigration of new species drops. The rate of extinction is low at the time of first colonization, because few species are present and most have large population sizes. With the addition of new species, the population sizes of some species may happen to be smaller, so the probability of extinction of these species by chance alone increases. Species may continue to arrive and go extinct, but the number of species on the island remains the same.

MacArthur and Wilson reasoned that when plotted graphically, the immigration and extinction lines would both be curved, for several reasons (**Figure 21.2b**). First, species arrive on islands at different rates. Some organisms, including plants with seed-dispersal mechanisms and winged animals, are more mobile than others and will arrive quickly. Other organisms will arrive more slowly. This pattern causes the immigration curve to start off steep but get progressively shallower. On the other hand, extinctions start off slowly and rise at accelerating rates, because as later species arrive, competition increases and more species are likely to go extinct. Earlier-arriving species tend to be *r*-selected species, which are better dispersers, whereas later-arriving species are generally *K*-selected species, which are better competitors. Later-arriving species usually outcompete earlier-arriving ones, causing an increase in extinctions.

Since MacArthur and Wilson's work, their concept of island biogeography has been applied to mainland areas, where patches of particular habitat can be viewed as "islands" in a sea of other, unsuitable habitat. For instance, patches of grassland in the Craters of the Moon National Monument in Idaho are surrounded by extensive lava flows (**Figure 21.3**). For the animals that inhabit the grassland, the lava flows may as well be a real ocean, because the animals cannot survive in these inhospitable areas. Similarly, freshwater species that live in lakes behave as if they were in islands surrounded by a "sea" of dry land.

Dan Janzen (1968) extended the habitats-as-islands concept by proposing that individual host-plant species could be islands to their associated herbivore fauna, which was adapted to feed only on that particular type of vegetation. For example, many insects have sophisticated biochemical machinery that allows them to detoxify certain plant tissues. However, the machinery is so specialized that it will only work for one or two species of host plant. All other plants are essentially inedible to the insect. An example is the monarch butterfly caterpillar, which can feed only on milkweed plants. For the monarchs, milkweed patches are effectively islands surrounded by other vegetation that might as well be open ocean.

The strength of the island biogeography model was that it generated several falsifiable predictions:

1. The number of species should increase with increasing island size. This is also known as the **species-area hypothesis**. Extinction rates would be less on larger islands because population sizes would be larger and less susceptible to extinction (**Figure 21.4**).

**Figure 21.3 Habitat islands.** Idaho's Craters of the Moon National Monument contains patches of grassland habitat surrounded by inhospitable lava.

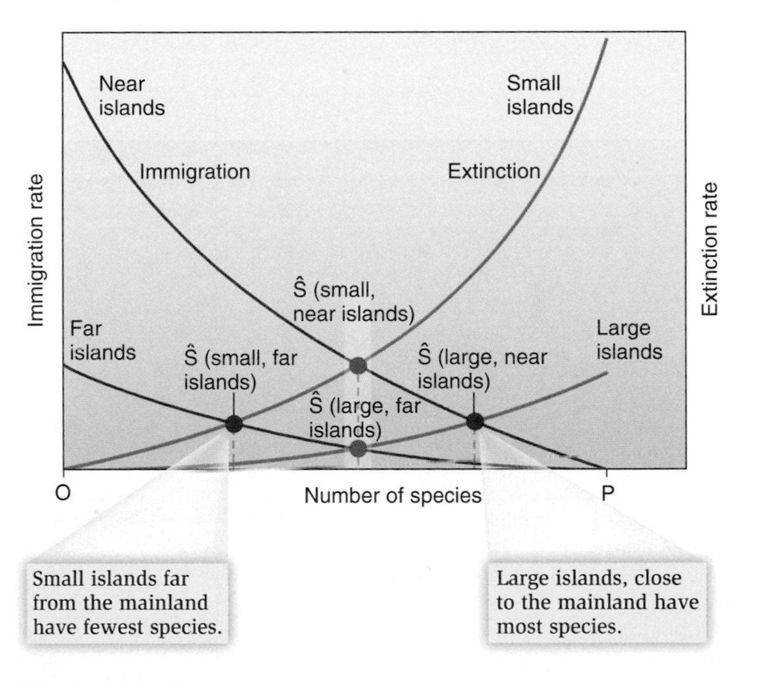


**Figure 21.4 Island biota size varies with distance from the source pool and island size.** An increase in distance, from near to far, lowers the immigration rate; an increase in island area, from small to large, lowers the extinction rate. The intersection of the immigration and extinction curves yields the equilibrium number of species. (After MacArthur and Wilson, 1963.)

2. The number of species should decrease with increasing distance of the island from the mainland, or the **source pool**, the pool of species that is available to colonize the island. This is also known as the **species-distance hypothesis.** Immigration rates would be greater on islands near the source pool because species do not have as far to travel (Figure 21.4).

3. The number of species on an island might remain the same, but the composition of the species should change continuously due to **species turnover**, as new species colonize the island and others become extinct. Species turnover should be considerable.

Let's examine these predictions one by one and see how well the data support them.

### 21.1.1 The species-area hypothesis describes the effect of island size on species richness

The West Indies has traditionally been a key location for ecologists studying island biogeography. This is because the physical geography and the plant and animal life of the islands are well known. Furthermore, the Lesser Antilles, from Anguilla in the north to Grenada in the south, enjoy a similar climate and are surrounded by deep water (**Figure 21.5a**). Robert Ricklefs and Irby Lovette (1999) summarized the available data on the species richness of four groups of animals—birds, bats, reptiles and amphibians, and butterflies—on 19 islands that varied in area over two orders of magnitude, 13–1,510 km$^2$. In each case, there was a significant relationship between area and species richness (**Figure 21.5b**). As area increased, so did species richness.

Note that these relationships are traditionally plotted on a double logarithmic scale, a so-called log-log plot, in which the horizontal axis is the logarithm to the base 10 of the area and the vertical axis is the logarithm to the base 10 of the number of species. A linear plot of the area versus the number of species would be difficult to produce, because of the wide range of area and richness of species involved. Logarithmic scales condense this variation to manageable limits and also produce a linear relationship instead of a curvilinear one.

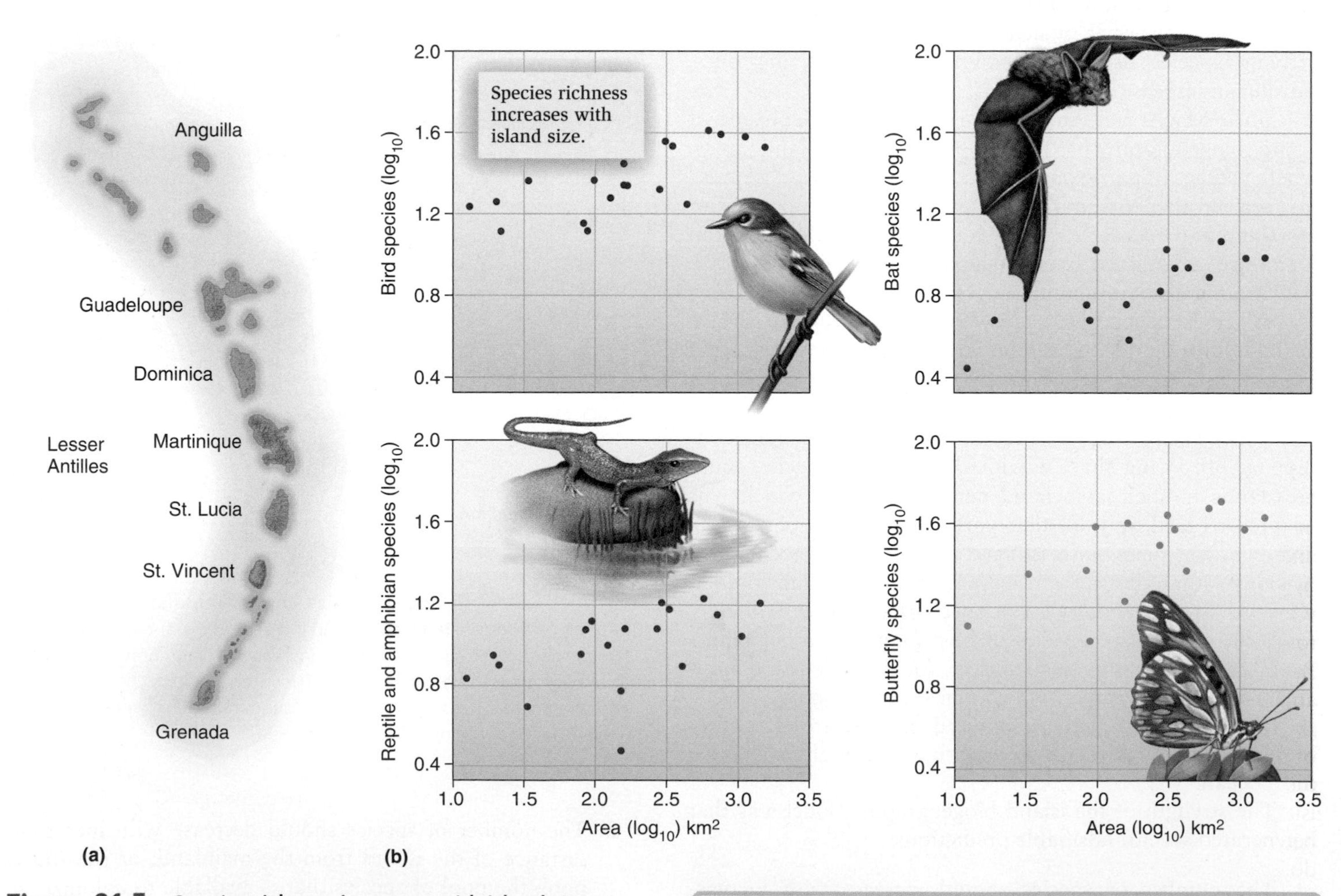


**Figure 21.5 Species richness increases with island size.** **(a)** The Lesser Antilles extend from Anguilla in the north to Grenada in the south. **(b)** The number of reptile and amphibian, bird, butterfly, and bat species increases with the area of an island. Note the logarithmic scales of both axes.

**ECOLOGICAL INQUIRY**

Approximately how large is the change in bird species richness across islands in the Lesser Antilles?

We can represent the relationship of species richness to area with the equation:

$$S = cA^z$$

or in logarithmic form,

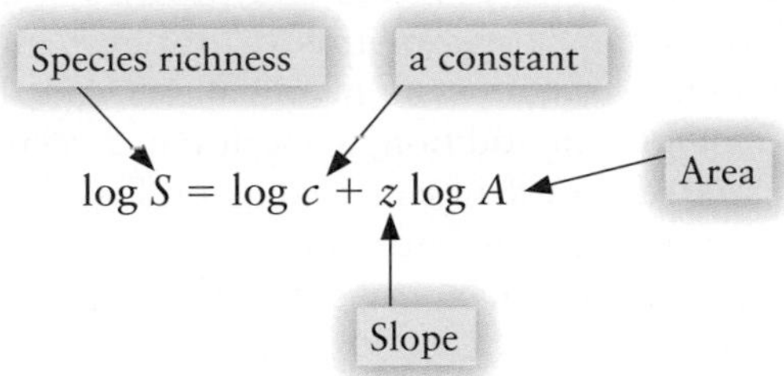


where $S$ = number of species, and $c$ and $z$ are both constants.

In this relationship, $z$ values represent the slope of the relationship between species richness and area. One of the early topics of discussion in island biogeography was the significance of variations in the value of $z$. A high value of $z$ indicates steep increases in species number as island size increases, whereas low values indicate much smaller differences in numbers of species between islands. Values of $z$ obtained from habitat islands, patches of suitable habitat on mainland areas, often vary from 0.15 to 0.25, lower than those from truly insular situations, where $z$ is often 0.20 to 0.40. This means that as larger areas are sampled, fewer new species are added on habitat islands than on true islands. The biological explanation for this phenomenon is that habitat islands contain some transient species from adjacent habitats, because on habitat islands, some species are simply resting while passing through an area. There are proportionately more transients passing through smaller areas, raising the apparent number of species in those areas, so the slope of the species-area curve is shallow for habitat islands. Islands, by contrast, are actual isolates with reduced migration rates, because species cannot rest up in the ocean between the islands, so the number of transients on a true island is reduced.

Differences in $z$ values may also result from differences in dispersal ability of different taxa. Jim Brown (1978) studied the distribution of forest-dwelling mammals and birds on mountaintops in the isolated ranges of the Great Basin area of Nevada in Utah. The mountain ranges are essentially isolated from one another, and the mammalian fauna is a relict community of a bygone age when rainfall was higher and this type of forest habitat was contiguous. Each mountaintop is a "sky island," essentially a forest remnant atop a mountain in a sea of desert (**Figure 21.6a**). Brown found a significant relationship between species richness and area for both birds and mammals. The species-area relationship for birds on these mountaintops (**Figure 21.6b**) had a slope of 0.165; that for mammals, 0.326 (**Figure 21.6c**). The slope of the line for mammals was thus more like that found on true islands. The reason is that there is little mammalian migration between mountaintops because mammals would have to walk down the mountain, across the valley, and up the next mountain. In this situation, "sky islands" behave more like true islands. In contrast, birds disperse more readily than mammals because they can fly between mountaintops, and the $z$ value in their case is more consistent with other habitat islands.

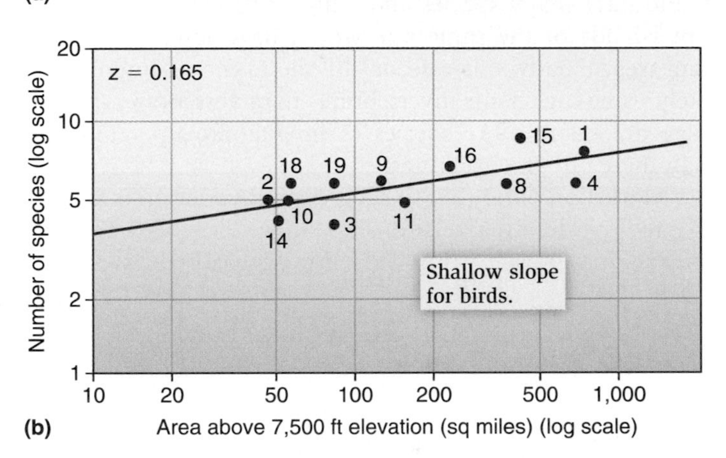


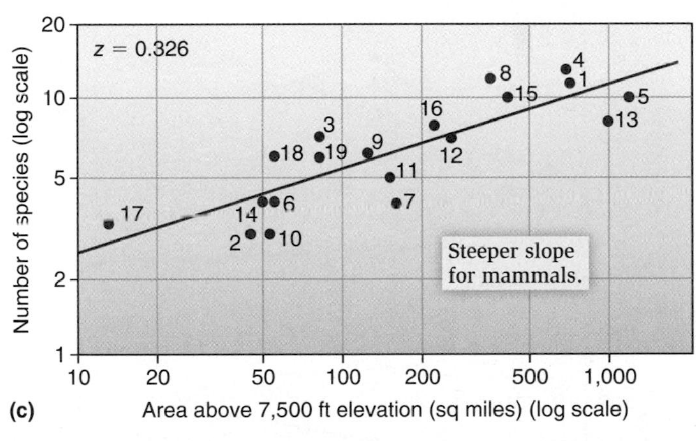


**Figure 21.6 The equilibrium theory of island biogeography applied to "sky islands."** **(a)** Map of the Great Basin region of the western United States showing the isolated mountain ranges between the Rocky Mountains on the east and the Sierra Nevada on the west. **(b)** The species-area relationship for the resident boreal birds of the mountaintops in the Great Basin shows a shallow slope. **(c)** The species-area relationship for the boreal mammal species shows a steeper slope. Numbers refer to sample areas on the map. (After Brown, 1978.)

## ECOLOGICAL INQUIRY

Why is the slope of the line steeper for mammals than for birds?

The constant $c$ is often thought of as the number of species per unit area—for example, per hectare of forest or grassland. It is affected by many factors and is likely to be higher for productive habitats. Tropical islands have greater $c$ values than temperate ones. For example, the islands of the Malayan Archipelago, a tropical location, and of the Shetland Islands, off the coast of Scotland, both have $z$ values of 0.31 for land birds. However, the $c$ value for Malayan Islands is 10.19 and for the Shetland Islands 6.90, indicating the Malayan Islands have 1.48 times as many bird species as Shetland Islands of the same area. This is not surprising, given that the tropical forest of Malaya is rich in species as compared to the more scantily vegetated Scottish islands. Distance to a rich source pool also affects $c$. A series of islands close to the mainland would have many species and a high $c$ value. A series of distant islands of the same size would have fewer species per unit area. Finally, $c$ is affected by the taxon of interest. It is likely to be higher for invertebrates than vertebrates, because there are many more species of invertebrates per unit area than there are of vertebrates.

Mark Lomolino (1989) showed that variation in $c$ values affected species-area relationships more than variation in $z$ values. First, most $z$ values vary little, typically between 0.15 and 0.35, which is not sufficient to change species number (**Figure 21.7a**). On the other hand, $c$ values vary much more, from less than 4 to over 16. Such variation is sufficient to quite dramatically affect species number (**Figure 21.7b**).

Apart from Ricklefs and Lovette's and Brown's studies, species-area relationships exist for birds of the East Indies, beetles on West Indian islands, ants in Melanesia, and land plants of the Galápagos. In addition, a species-area relationship was demonstrated for the insects feeding on British trees (refer back to Figure 18.5) and for the insects feeding on bracken fern (**Figure 21.8**). This provides strong support for this prediction of the theory of island biogeography.

Experimental support for the effect of area on species richness was provided by Daniel Simberloff (1978), a student of E. O. Wilson who studied islands of pure mangroves in the Florida Keys. He painstakingly crawled on his hands and knees to collect every species that fed on the islands, which were mainly very small insect species. Simberloff then created his own reduced-area islands by taking a chainsaw and felling trees to make the island smaller. The felled material was hauled away in a barge. (One could probably not get a permit to do this experiment today, as mangroves now are legally protected.)

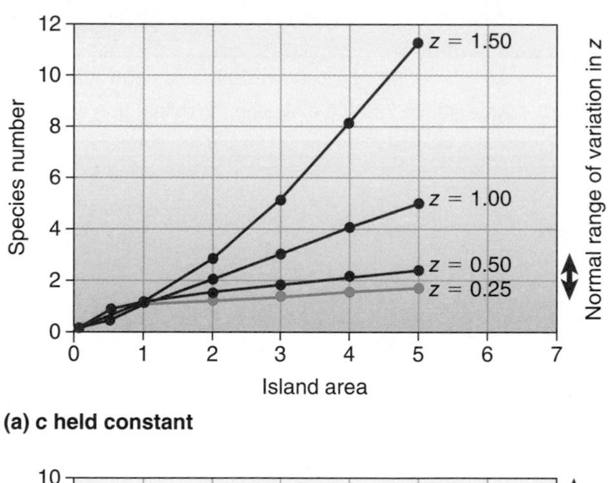


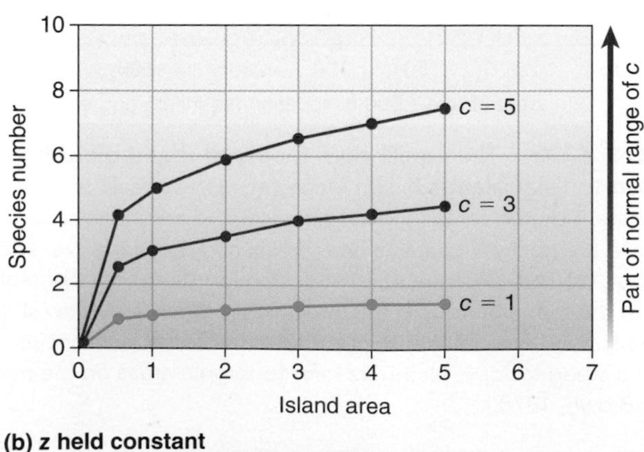


**Figure 21.7 The effects of varying values of c and z on the species-area relationship.** **(a)** Effects of varying $z$; $c$ is held constant at 1. **(b)** Effects of varying $c$; $z$ is held constant at 0.25. (After Lomolino, 1989.)

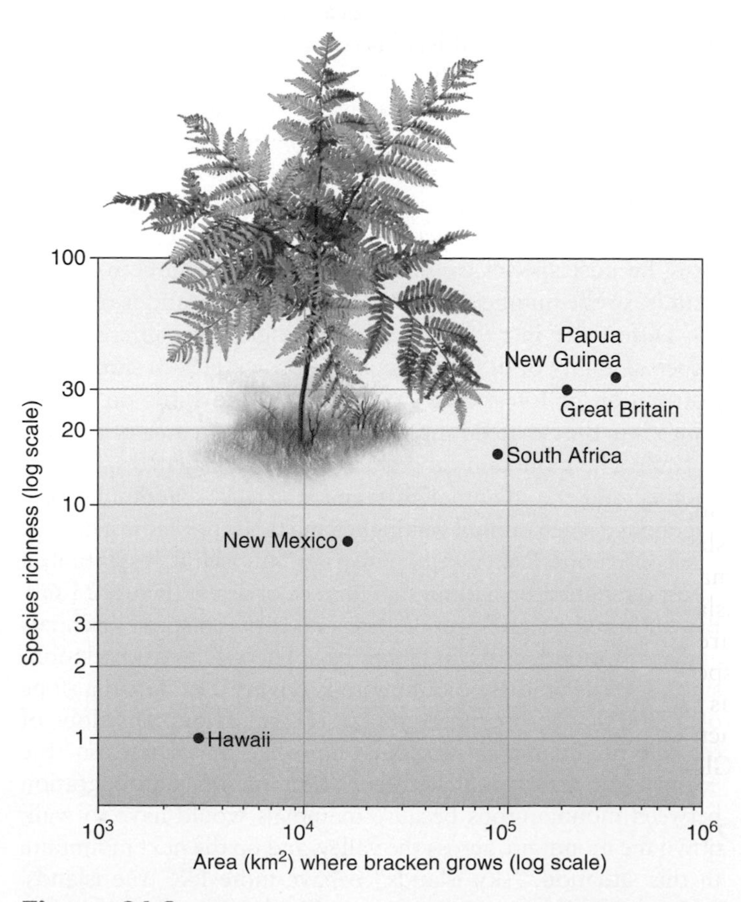


**Figure 21.8 Species-area relationship.** For the herbivores feeding on bracken fern in different areas of the world. (After Compton et al., 1989.)

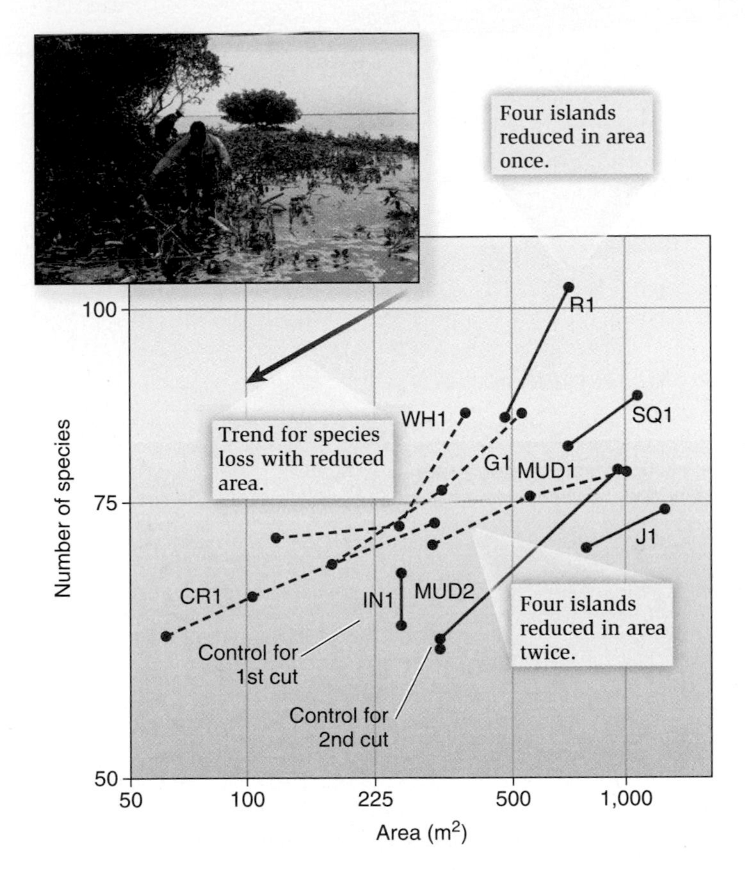


**Figure 21.9 Reduction in species numbers on mangrove islands with area reduced by chainsaw.** Solid lines and blue dots represent four islands where area was experimentally reduced once. Dotted lines and red dots represent four islands where area was reduced twice. (After Simberloff, 1978.)

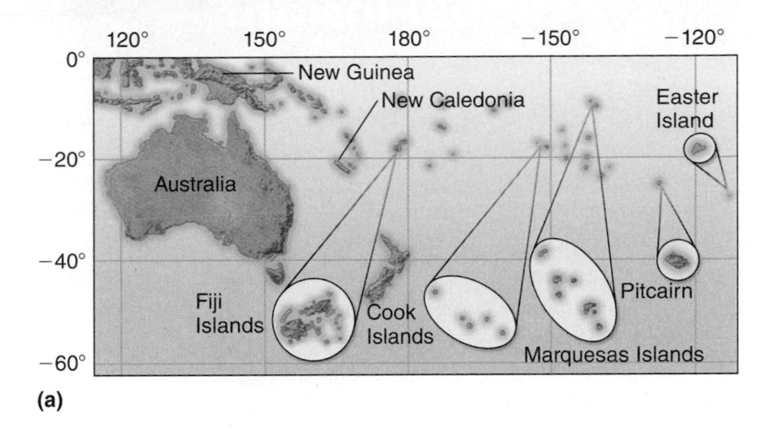


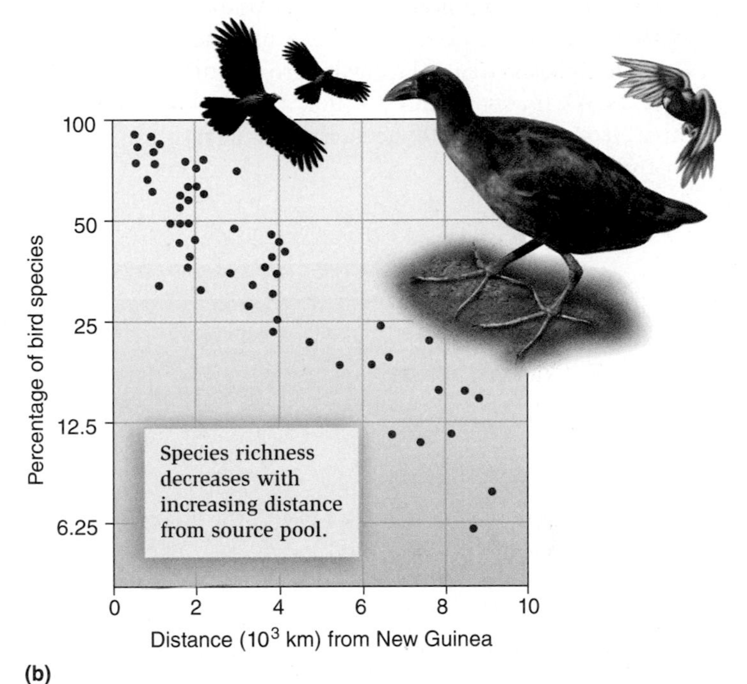


**Figure 21.10 Species richness decreases with distance from the source pool.** **(a)** Map of Australia, New Guinea, and these Polynesian Islands: New Caledonia, Fiji Islands, Cook Islands, Marquesas Islands, Pitcairn, and Easter Island. **(b)** The numbers of bird species on islands decreases with increasing distance from the source pool, New Guinea. Species richness is expressed as the percentage of that on New Guinea.

Simberloff reduced the area of eight islands and left one as a control (IN1). Seven months later, after enough time had passed for insects to become reestablished following this disturbance, he censused again. Insect densities had dropped on all eight islands but increased slightly on the control (**Figure 21.9**). He then further reduced the areas of four of the islands (WH1, CR1, MUD1, and G1) and left one island (MUD2) as a control. A year later a final census was made, and species richness had again dropped on the four islands but not on MUD2. The results clearly indicated that area affects species number. Some ecologists have used the species-area relationship to predict species extinction rates as wildlife habitat continues to shrink due to the human activities of deforestation, agriculture, and urbanization (see **Global Insight**).

## 21.1.2 The species-distance hypothesis describes the effect of island distance on species richness

MacArthur and Wilson also marshaled evidence for the effect of the distance of an island from a source pool of colonists, usually the mainland. In studies of the numbers of lowland forest bird species in Polynesia, they found that the number of species decreased with the distance from New Guinea, the source pool in this case (**Figure 21.10**). They expressed the richness of bird species on the islands as a percentage of the number of bird species found on New Guinea. There was a significant decline in this percentage with increasing distance, and more-distant islands contained lower numbers of species than nearer islands. This research substantiated the prediction that species richness declines with increasing distance from the source pool.

## Global Insight

### Deforestation and the Loss of Species

If species richness on habitat islands increases with area, extinction should result as area declines. Just how many species are lost with habitat shrinkage is the critical question. To some extent this depends on the value of $z$, the slope of the relationship between species richness and area. We can measure deforestation, or habitat area lost, fairly accurately given satellite images, but just how many species become extinct for each unit of habitat that is destroyed? We know that $z$ values are commonly found to be between 0.15 and 0.35 for a variety of taxa, so we can use these types of values in the species-area relationship to estimate species loss (**Figure 21.A**). If $c = 10$ and $z = 0.2$, a 90% decrease in habitat area, from 1,000 to 100 hectares, would result in a 36.9% loss of species:

$$S = cA^z$$
$$S = 10 \times 1{,}000^{0.2}$$
$$S = 39.81$$

If the area is reduced by 90%:

$$S = 10 \times 100^{0.2}$$
$$S = 25.11$$

This is a loss of 14.7 species, or 36.9%.
If $z = 0.35$, then the loss would be 55%.

$$S = cA^z$$
$$S = 10 \times 1{,}000^{0.35}$$
$$S = 112.20$$

If the area is reduced by 90%:

$$S = 10 \times 100^{0.35}$$
$$S = 50.11$$

This is a loss of 62.09 species, or 55%.

There are two conclusions from these examples. First, greater $z$ values, steeper slopes, result in greater species loss as habitat is lost. Second, because of the logarithmic nature of the species-area relationship, a 90% loss in forest habitat doesn't translate into a 90% loss in species.

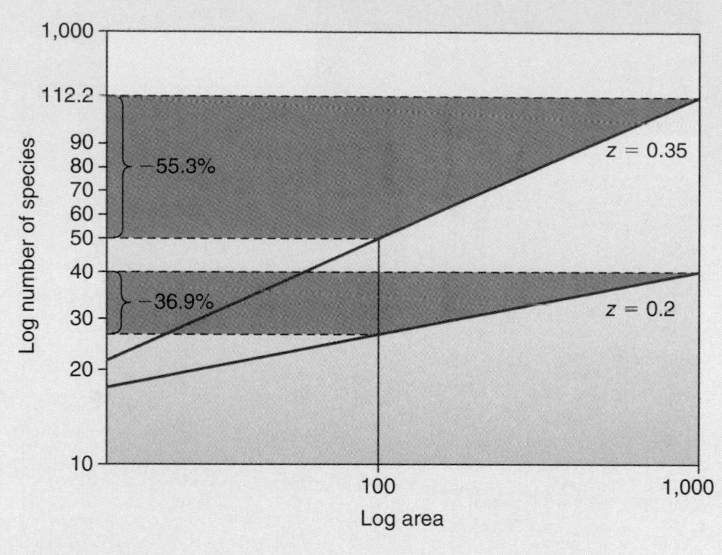


**Figure 21.A The effects of habitat loss on species abundance based on the species-area relationship, $S = cA^z$.** Here $c = 10$ and $z = 0.2$ or 0.35. Where $z = 0.35$, a 90% loss in habitat results in a 55% loss in species. Where $z = 0.2$, a 90% loss in habitat results in a 36.9% loss in species. (After Groom et al., 2006.)

Roger del Moral (2000) investigated the colonization by plants of four plots on Mount St. Helens following the eruption in 1980. Two "near" plots were less than 50 m from the intact forest, and two "distant" plots were over 200 m distant. Species accumulated quicker on near plots than on distant plots (**Figure 21.11a**). By the year 2000, the area covered by plants in the isolated plots was only 50% that of the near plots (**Figure 21.11b**). In part this was because fruit-eating birds that disperse seed seldom disperse more than 50 m from the forest.

Finally, Manuel Nores (1995) documented the number of bird species on montane forest islands on mountaintops near the Andes in Venezuela, Colombia, and Ecuador. The Andes acted as the source pool of the birds, and species richness on these sky islands decreased the farther they were from the Andes (**Figure 21.12**).

### 21.1.3 Species turnover on islands is generally low

Studies involving species turnover on islands are difficult to perform, because detailed and complete species lists are needed over long periods of time, usually many years, often decades, before and after disturbances. Many of the lists that do exist from before disturbances are compiled in a casual way and are not suitable for comparison with more modern data. For example, Jared Diamond (1969) studied the birds of the California Channel Islands in 1968 and compared his species lists to those of an earlier study published in 1917 by A. B. Howell. Diamond reported that many birds, about 5–10 species per island, reported from Howell's survey were no longer present but just as many species not reported by Howell had apparently colonized. The logic was that the species richness

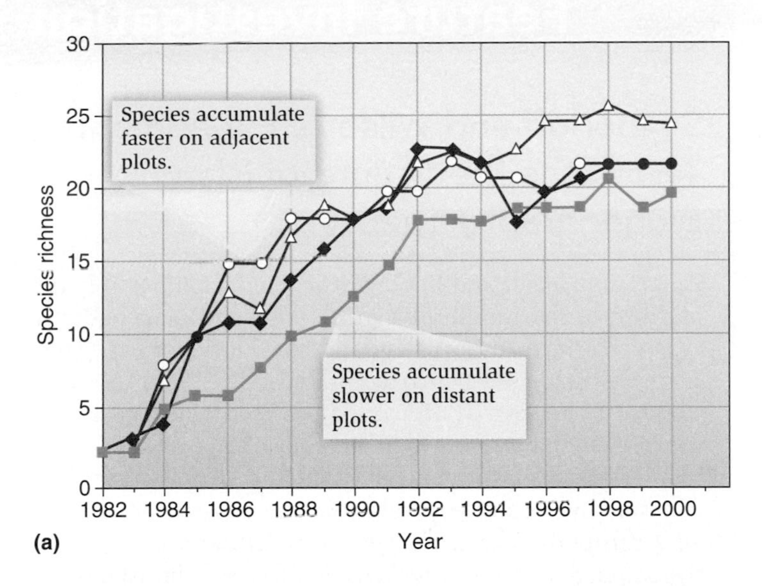


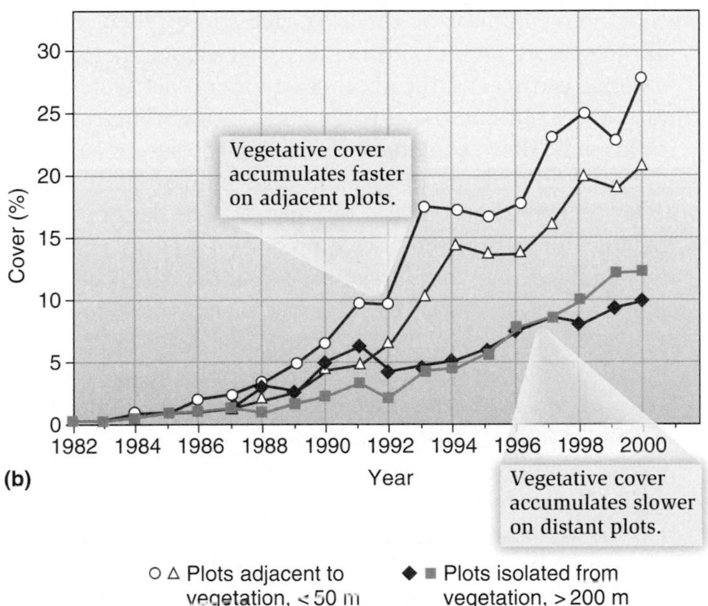


**Figure 21.11 Effects of distance on plant recolonization on Mount St. Helens, following the 1980 eruption.** **(a)** Species richness and **(b)** percent cover both recover faster on plots closer to intact vegetation (<50 m) than plots that were more isolated (>200 m). (After del Moral, 2000.)

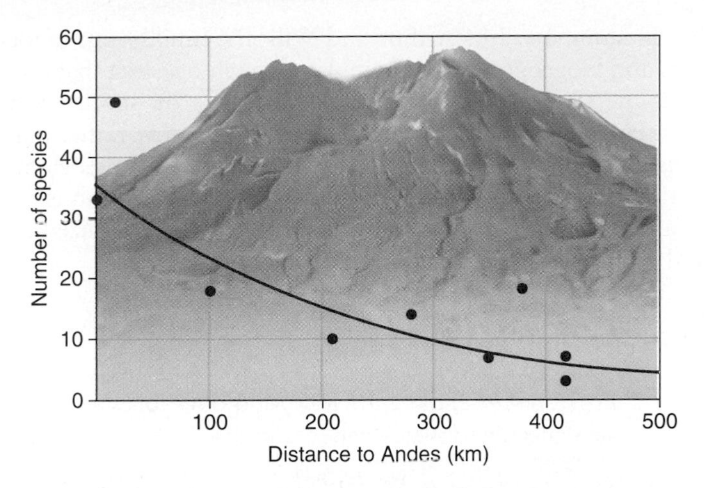


**Figure 21.12 Species richness decline of montane forest birds with increasing distance from the Andes mountains.** (After Nores, 1995.)

of islands was the same in 1969 as it was in 1917, but identities had changed, indicating species turnover. This was just what MacArthur and Wilson would predict. However, other authors were quick to point out that Howell's list was not exhaustive and was just a summary of all known breeding records, some from as far back as 1860–1870.

Diamond went on to carry out similar comparisons of bird species over time on islands near New Guinea, but once again the earlier surveys included no extensive field notes and Diamond was forced to compare his species richness lists of islands to collections of species made earlier. If a species had been seen but not collected, then the comparisons would be invalid. Francis Gilbert (1980) reviewed 25 investigations carried out to demonstrate turnover and found a lack of this type of rigor in nearly all of them. Furthermore, most of the observed turnover in these studies, usually less than 1% per year, or less than one species per year, appeared to be due to immigrants that never became established, not due to the extinction of well-established species. More-recent studies have revealed similar findings. The take-home message from most studies is that recorded rates of turnover are low, giving little conclusive support to this prediction of the theory of island biogeography. Only one study, carried out by Simberloff and Wilson (1969, 1970), was done with sufficient rigor that it was a good test of whether species turnover occurred, and even this study showed very low levels of turnover (see **Feature Investigation**).

More-recent studies have also shown low rates of species turnover. Jorge Rey (1983) defaunated islands of pure saltmarsh cordgrass, *Spartina alterniflora,* in north Florida and estimated turnover at only 0.13 species per island per week. Ian Abbott and Robert Black (1980) measured the known extinctions and immigrations of plants on 40 islands off the coast of western Australia. The islands were censused sporadically (one to four times between 1956 and 1978). Extinctions and immigrations tended to match each other, but again most turnover was low, with most islands showing turnover rates of between 0 and 2 species per census (or <0.1 species per year). Lloyd Morrison (1997) found low turnover of species on 77 cays in the Bahamas, which he surveyed annually over a 4-year period. Most of the observed turnover (usually <1% per year or less than 1 species per year) was due to immigrants that never became established. These studies indicate that recorded rates of turnover are low, which gives little support to the third prediction of the MacArthur-Wilson theory.

In summary, MacArthur and Wilson's equilibrium model of island biogeography has stimulated much research that confirms the strong effects of area and distance on species richness. However, species turnover appears to be low rather than considerable, which suggests that succession on most islands is a fairly orderly process. This means that colonization is not a random process, that the same species seem to colonize first, and that species gradually reappear in the same order.

**Check Your Understanding**

**21.1** Preston (1962) gave the following data for island size and bird species richness in the East Indies.

| Island | No. of bird species | Area (mi²) |
|---|---|---|
| New Guinea | 540 | 312,000 |
| Borneo | 420 | 290,000 |
| Philippines | 368 | 144,000 |
| Celebes | 220 | 70,000 |
| Java | 337 | 48,000 |
| Ceylon (now Sri Lanka) | 232 | 25,000 |
| Palawan | 111 | 4,500 |
| Flores | 143 | 8,870 |
| Timor | 137 | 18,000 |
| Sumba | 108 | 4,600 |

Construct a species area curve for these data. Which islands are richer in bird species than they should be, based on area alone? Why might this be?

## 21.2 Nature Reserve Designs Incorporate Principles of Island Biogeography and Landscape Ecology

In our exploration of the theory of island biogeography we noted that it could be applied not only to a body of land surrounded by water but also to isolated fragments of habitat. Seen this way, wildlife reserves and sanctuaries are in essence islands in a sea of human-altered land, either agricultural fields or urbanized areas. Conservationists have therefore utilized island biogeography modeling in the concept of nature preserve design. One question for conservationists is how large a protected area should be (**Figure 21.13a**). According to island biogeography, the number of species should increase with increasing area (the species-area effect); thus, the larger the area, the greater the number of species would be protected. In addition, there are other benefits of larger parks. For example, they are beneficial for organisms that require large spaces, including migrating species and species with extensive territories, such as lions and tigers.

# Feature Investigation

## Simberloff and Wilson's Experiments Tested the Predictions of Island Biogeography Theory

Daniel Simberloff and E. O. Wilson conducted possibly the best test of the equilibrium model of island biogeography ever performed, using islands in the Florida Keys. They surveyed small red mangrove islands, 11–25 m in diameter, for all terrestrial arthropods. With the help of a pest control company from Miami, they then enclosed each island with a plastic tent and had the islands fumigated with methyl bromide, a short-acting insecticide, to kill all arthropods. The tents were removed, and periodically thereafter Simberloff and Wilson surveyed the islands to examine recolonization rates. At each survey, they counted all the species present, noting any species not there at the previous census and the absence of others that were previously there but had presumably gone extinct (**Figure 21.B**). In this way, they estimated turnover of species on islands.

After 280 days, the islands had similar numbers of arthropod species as before fumigation. The data indicated that recolonization rates were higher on islands nearer to the mainland than on far islands, just as the island biogeography model predicts. However, the data, which consisted of lists of species on islands before and after extinctions, provided little support for the prediction of high turnover. Rates of turnover were low, only 1.5 extinctions per year, compared to the 15–40 species found on the islands within a year. Simberloff (1976) later concluded that turnover probably involves only a small subset of transient species that use islands only as a resting place as they are passing through an area, while the MacArthur-Wilson theory concerns resident species that feed and reproduce on the island.

Simberloff and Wilson also found that most of the species that returned to the islands were the species present before fumigation. This indicates the existence of biological processes that shape final community structure in the same way every time an island is recolonized, a finding that is contrary to the theory of island biogeography, which treats the dynamics of different colonizing species as equivalent, with community properties essentially unimportant.

**HYPOTHESIS** Island biogeography model predicts substantial turnover of species on islands.

**STARTING LOCATION** Mangrove islands in the Florida Keys.

**Conceptual level** | **Experimental level**

1 Each mangrove island is isolated.

Distant

Very near

Mainland

Take initial census of all terrestrial arthropods on 4 mangrove islands. Erect framework over each mangrove island.

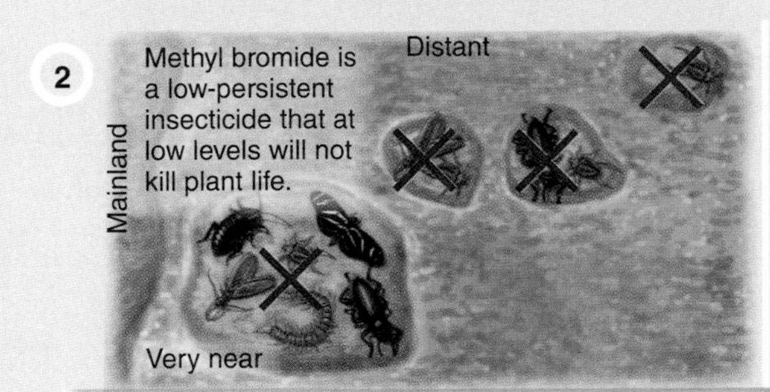


2 Methyl bromide is a low-persistent insecticide that at low levels will not kill plant life.

Cover the framework with tent and fumigate with methyl bromide to defaunate island.

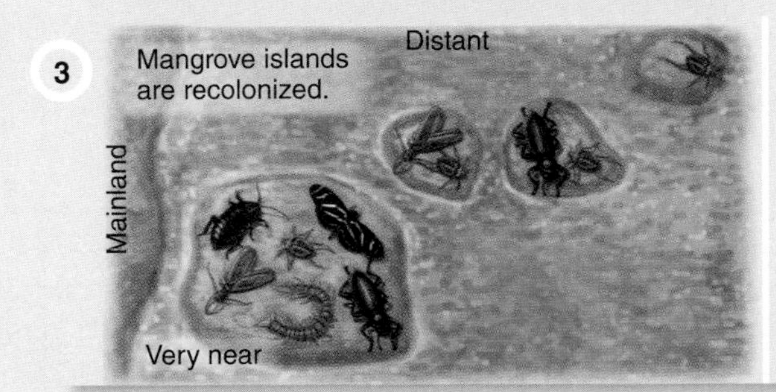


3 Mangrove islands are recolonized.

Remove the tents and conduct censuses every month to monitor recolonization.

4 **THE DATA** Island E2 was closest to the mainland and supported the highest number of species. E3 and ST2 were at an intermediate distance from the mainland, and E1 was the most distant.

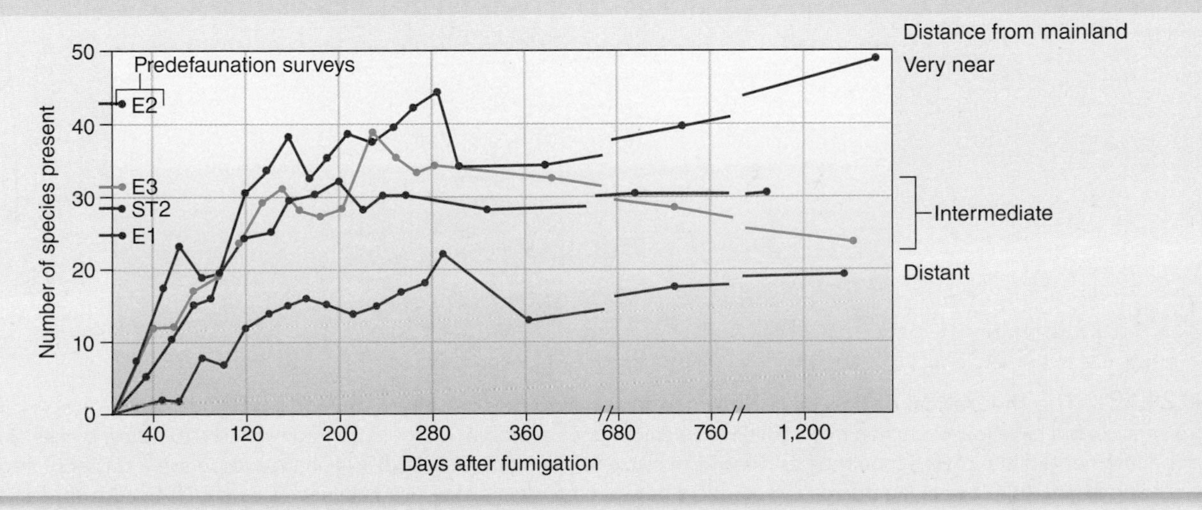


**Figure 21.B** Simberloff and Wilson's experiments on species turnover on islands.

Another question is whether it is preferable to protect one single, large reserve or several smaller ones (**Figure 21.13b**). This is called the **SLOSS debate** (for single large or several small). Proponents of the single, large reserve claim that a larger reserve is better able to preserve more and larger populations, because of lower extinction rates, than an equal area divided into small reserves. According to island biogeography, a larger block of equivalent habitat should support more species than all the smaller blocks combined. However, many empirical studies suggest that multiple small sites of equivalent area will contain more species, because a series of small sites is more likely to contain a broader variety of habitats than one large site. Looking at a variety of sites, researchers Jim Quinn and Susan Harrison (1988) concluded that animal life was richer in collections of small parks than in larger parks. In their study, having more habitat types outweighed the effect of area on species richness. In addition, there may be other benefits of a series of smaller parks, such as a reduction of extinction risk by a single event such as a wildfire or the eruption of disease.

As we learned in Chapter 8, **landscape ecology** is a subdiscipline of ecology that examines the spatial arrangement of elements in communities and ecosystems. In the design of nature reserves, one question that needs to be addressed is how close to one another reserves should be situated—for instance, whether to have three or four small reserves close to each other or farther apart or whether to have a linear or cluster arrangement of small reserves. Island biogeography implies that if an area must be fragmented, the sites should be as close as possible to permit dispersal (**Figure 21.13c,d**). In practice, however, having small sites far apart may preserve more species than having them close together, because once again, distant sites are likely to incorporate slightly different habitats and thus support a greater mix and higher number of species. Sufficient empirical data to test those ideas have yet to be collected.

Landscape ecologists have also suggested that small reserves should be linked together by **movement corridors**, thin strips of land that may permit the movement of species between patches (**Figure 21.13e**). Such corridors may facilitate

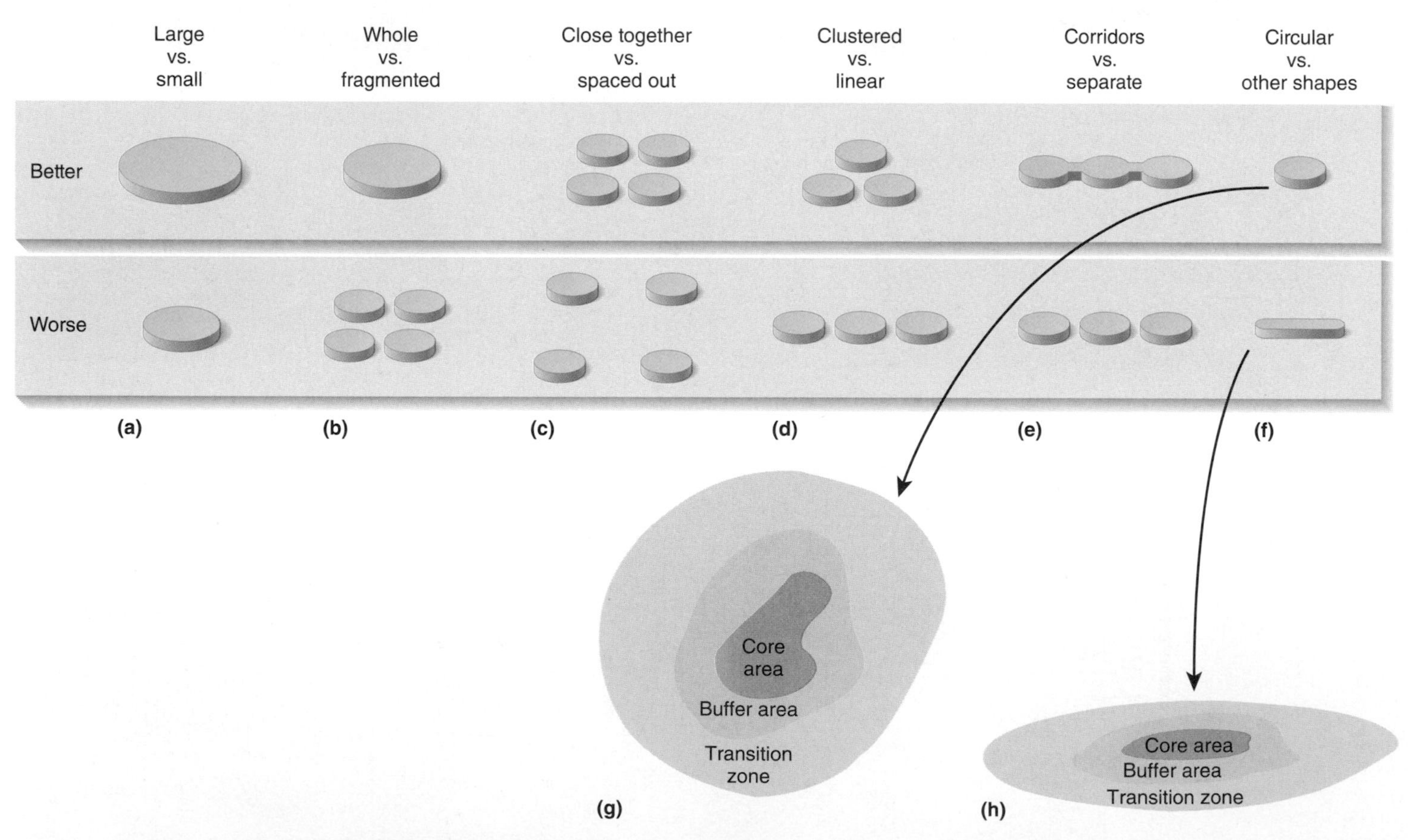


**Figure 21.13 The theoretical design of nature reserves, based on the tenets of island biogeography.** **(a)** A larger reserve will hold more species and have low extinction rates. **(b)** Given a certain area available, it should be fragmented into as few pieces as possible. **(c)** If an area must be fragmented, the pieces should be as close as possible to permit dispersal. **(d)** To enhance dispersal, a cluster of fragments is preferable to a linear arrangement. **(e)** Maintaining or creating corridors between fragments may also enhance dispersal. **(f)** Circular-shaped areas will minimize the amount of undesirable edge habitat. **(g, h)** Edge habitat may itself be delineated as transition zone or buffer area, and is relatively less in circular areas than in oval areas.

movements of organisms that are vulnerable to predation outside of their natural habitat or that have poor powers of dispersal between habitat patches. In this way, if a disaster befalls a population in one small reserve, immigrants from neighboring populations can more easily recolonize it. This avoids the need for humans to physically move new plants or animals into an area. However, there are disadvantages associated with corridors. First, corridors also can facilitate the spread of disease, invasive species, and fire between small reserves. Second, it is not yet clear if many species would actually use such corridors.

Ellen Damschen and colleagues (2006) used habitat patches in longleaf pine forest, connected and unconnected by corridors, to examine the effects of corridors on plant species richness. The patches had a rich herbaceous understory with over 100 plant species, in contrast to the surrounding dense pine forest. Each landscape consisted of a central 100 × 100-m patch surrounded by four other 100 × 100-m patches situated 150 m away. One of these outlying patches was connected to the central patch by a narrow 150-m corridor. This was the same experiment used by Joshua Tewksbury to examine the effects of corridors on plant and butterfly population densities, which we described in Chapter 8 (refer back to Figure 8.C). The habitat patches were created in 1999, and surveys revealed the plant community had recovered from the disturbance by 2001. Censuses of all plant species were made from 2001 to 2005, except in 2004, when the U.S. Fish and Wildlife Service burned the site as part of restoration management. Habitat patches connected by corridors retained more plant species than isolated, unconnected patches, and this difference increased over time (**Figure 21.14a**). In addition, the corridors did not promote invasion by exotic species (**Figure 21.14b**). Damschen and colleagues attributed the increase in richness in the presence of corridors to increased pollen movement by pollinators and increased seed deposition by seed predators. Both of these processes are mediated by animals, and corridors promote animal movement.

Parks are often designed to minimize **edge effects**, the special physical conditions that exist at the boundaries or "edges" of habitats (see **Figure 21.13f**). Habitat edges, particularly those between natural habitats such as forests and developed land, are often different in physical characteristics from the habitat core. For example, the center of a forest is shaded by trees and has less wind and light than the forest edge, which is unprotected. Many forest-adapted species thus shy away from forest edges and prefer forest centers. In many parks, habitat edges may themselves be broken down into transition zones and buffer areas. In transition zones, some land use, compatible with conservation within the reserve, such as grazing of cattle, is permitted. Between the transition zone and the core area is a buffer zone, where such activity is not permitted but full use by wildlife does not occur. Circular-shaped parks are preferred over long, skinny parks because the circular shape maximizes the amount of core area (see **Figure 21.13g,h**). Although this was not part of the original MacArthur-Wilson theory of island biogeography, it is a prominent part of park design. Some conservationists have combined the idea of core areas and corridors by suggesting that corridors should link core areas and consist of core areas themselves.

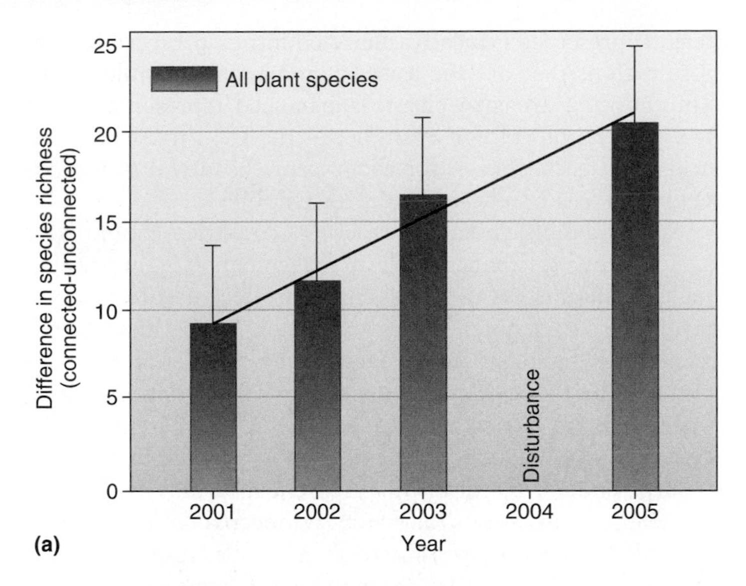


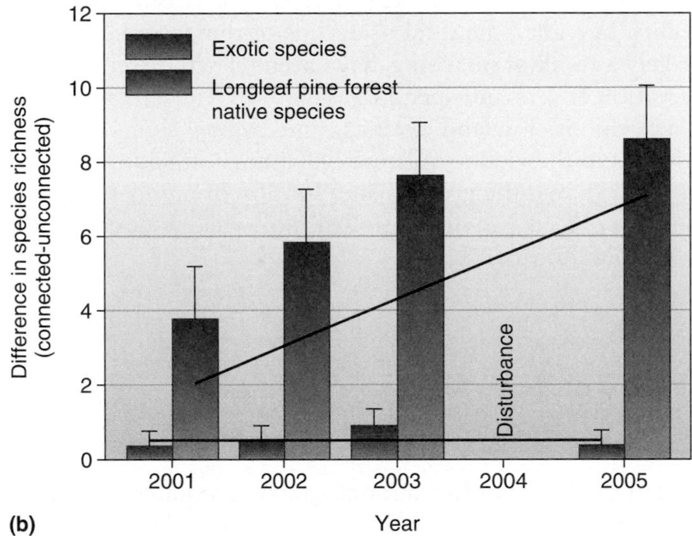


**Figure 21.14 The effects of corridors on plant species richness.** **(a)** The difference in plant species richness in patches connected and unconnected by corridors. **(b)** Differences in species richness in native and exotic species in connected and unconnected patches. (After Damschen et al., 2006.)

While the principles of island biogeography theory and landscape ecology are useful in illuminating conservation issues, in reality there is often little choice as to the size, shape, and location of nature reserves. Management practicalities, costs of acquisition and management, and politics often override ecological considerations, especially in developing countries where costs for large reserves may be relatively high. Economic considerations often enter into the choice of which areas to preserve. Typically, many countries protect areas in those regions that are the least economically valuable rather than choosing areas to ensure a balanced representation of the country's biota. For example, in the U.S., most national parks have been chosen for their scenic beauty, not because they preserve the richest habitat for wildlife.

When designing nature reserves, countries should also consider how to finance their management. It is interesting that the amount of money spent to protect nature reserves may better determine species extinction rates than reserve size. According to island biogeography theory, large areas minimize the risk of extinctions because they contain sizable populations. In Africa, several parks, such as Serengeti and Selous in Tanzania, Tsavo in Kenya, and Luangwa in Zambia, are large enough to fulfill this theoretical ideal. However, in the 1980s, populations of black rhinoceros and elephants declined dramatically within these areas because of poaching, showing that a wide gap may exist between theory and reality. In reality, the rates of decline of rhinos and elephants, largely a result of poaching, have been related directly to conservation efforts and spending (**Figure 21.15**). The remaining black rhinos, lowland gorillas, and pygmy chimpanzees in Africa and the vicuna, a llama-like animal in South America, have all shown the greatest stabilization of numbers in areas that have been heavily patrolled and where resources have been concentrated.

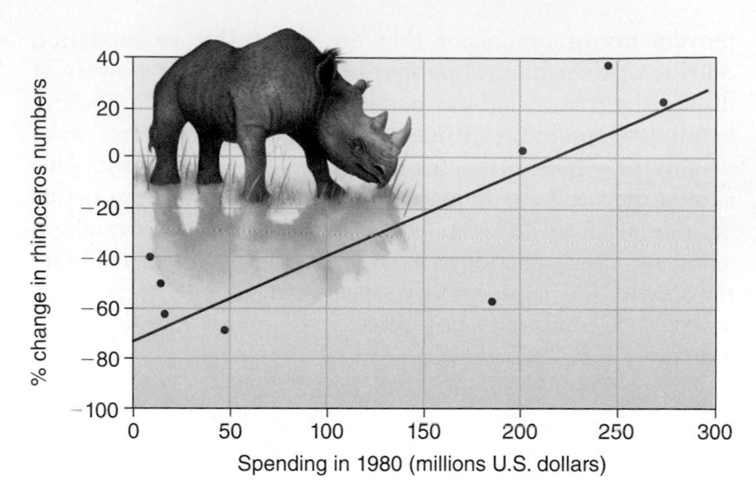


**Figure 21.15** **The economics of conservation.** A positive relationship is seen between change in black rhinoceros numbers between 1980 and 1984, and conservation spending in various African countries. (After Leader-Williams and Albon, 1988.)

### Check Your Understanding

**21.2** An area of habitat is reduced from 1,000 to 500 hectares. If $c = 10$ and $z = 0.3$, what percent of species is lost?

## SUMMARY

**21.1 The Theory of Island Biogeography Considers Succession on Islands**

- Species succession on islands after a disturbance is influenced by mode of dispersal (Figure 21.1).
- In the theory of island biogeography, the number of species on an island tends toward an equilibrium number determined by the balance between immigration rates and extinction rates (Figure 21.2).
- Some mainland areas can be viewed as habitat islands in a sea of unsuitable habitat (Figure 21.3).
- The theory of island biogeography predicts that the number of species increases with increasing island size, that the number of species decreases with distance from the source pool, and that there is frequent turnover of species (Figure 21.4).
- Much evidence exists to support the species-area effect (Figures 21.5, 21.6, 21.8).
- The nature of the relationship between species richness and area can be simultaneously affected by many factors, including habitat productivity and the taxon of interest (Figure 21.7).
- Experimental reduction of island area in the Florida Keys caused a reduction in species richness, supporting the species-area effect (Figure 21.9).
- Species loss during deforestation may be predicted using species-area relationships (Figure 21.A).
- Much evidence, from birds on islands to plants on volcanic soils, supports the species-distance effect (Figures 21.10–21.12).
- Evidence suggests that species turnover is minimal (Figure 21.B).

**21.2 Nature Reserve Designs Incorporate Principles of Island Biogeography and Landscape Ecology**

- The principles of island biogeography can be applied to the design of nature reserves (Figure 21.13), though in practice, collections of small parks often have more species than single larger parks.
- The species richness of some areas is increased by the presence of corridors (Figure 21.14).
- In terms of park management, practicalities and economics often override ecological considerations (Figure 21.15).

## TEST YOURSELF

1. On which type of island would you expect species richness to be greatest?
   a. Small, near mainland
   b. Small, distant from mainland
   c. Large, near mainland
   d. Large, distant from mainland
2. The relationship between species richness and island area is described by the equation:
   a. $S = A^{cz}$
   b. $A = cS^z$
   c. $S = cA^z$
   d. $c = AS^z$
   e. $S = AC^z$
3. Which is part of the original MacArthur-Wilson theory of island biogeography?
   a. $\hat{S}$ is increased by distance from the source pool
   b. $\hat{S}$ is decreased by island size
   c. $\hat{S}$ is a balance between immigration and extinction
   d. Island size influences immigration rates
   e. Distance from source pool influences extinction rates
4. According to MacArthur and Wilson's equilibrium model of island biogeography, which of the following nature reserve designs is favored?
   a. Small
   b. Fragmented
   c. Oval shaped
   d. With corridors
   e. Large
5. The acronym SLOSS stands for:
   a. Several large or single small
   b. Single large or several small
   c. Single large or single small
   d. Several large or several small
6. In the MacArthur-Wilson theory of island biogeography, $z$ represents:
   a. The number of species
   b. A constant measuring number of species per unit area
   c. A constant measuring slope
   d. The intercept
   e. Area
7. Why are the immigration and extinction lines in the MacArthur-Wilson model both curved?
   a. Species arrive at different rates
   b. Some organisms are more mobile than others
   c. Competition increases as more species arrive
   d. Later-arriving species tend to be better competitors
   e. All of the above
8. Because of the wide range of areas and species richness, species-area relationships are traditionally plotted using a:
   a. Bar graph
   b. Log-log plot
   c. Semi-log plot
   d. Line graph
   e. Pie chart
9. Habitat islands on the mainland tend to have lower $z$ values than true islands.
   a. True
   b. False
10. In the Great Basin region of the U.S., the species-area relationship for resident boreal birds of mountain tops has a steeper slope than that for boreal mammals.
    a. True
    b. False

## CONCEPTUAL QUESTIONS

1. What are the main tenets of the MacArthur-Wilson theory of island biogeography?
2. What evidence is there to support the MacArthur-Wilson theory of island biogeography, and what evidence is there to contradict it?
3. Discuss the application of island biogeography to nature reserve design.

## DATA ANALYSIS

William Tonn and John Magnuson investigated the number of fish species, lake area, and other habitat variables, including vegetation diversity, measured as $H_S$, from 18 shallow lakes in Wisconsin. Plot the number of fish species against the $\log_{10}$ lake area and against vegetational diversity. What can you conclude?

| Lake | Area (ha) | Vegetation diversity, $H_S$ | Species richness |
|---|---|---|---|
| 1 | 76.1 | 1.69 | 9 |
| 2 | 38.0 | 1.62 | 8 |
| 3 | 4.9 | 0.37 | 4 |
| 4 | 5.7 | 1.08 | 4 |
| 5 | 3.2 | 0.93 | 6 |
| 6 | 42.9 | 1.47 | 10 |
| 7 | 9.7 | 0.58 | 10 |
| 8 | 89.0 | 0.78 | 6 |
| 9 | 23.9 | 0.91 | 5 |
| 10 | 19.0 | 0.50 | 7 |
| 11 | 53.0 | 0.96 | 7 |
| 12 | 8.1 | 0.65 | 7 |
| 13 | 44.5 | 1.55 | 8 |
| 14 | 6.1 | 0.00 | 3 |
| 15 | 89.8 | 0.91 | 6 |
| 16 | 3.2 | 0.35 | 2 |
| 17 | 2.4 | 0.40 | 2 |
| 18 | 2.8 | 0.32 | 1 |

SECTION SIX

# Biomes

In the early 1960s, British chemist James Lovelock was working for NASA, designing instruments that might detect the presence of life on Mars. Lovelock noted that life might leave characteristic chemical signatures in the Martian atmosphere. He began to compare the atmospheres of Mars and Earth. The Martian atmosphere has 95% carbon dioxide, with a little oxygen and methane. However, Lovelock noted that the chemical signature of the Earth's atmosphere has changed over large time spans. About 3 billion years ago, it contained a little oxygen and 30% carbon dioxide spewed out from volcanic eruptions. Over time, bacteria and photosynthetic algae started to use the carbon dioxide and produce oxygen as a by-product. Eventually, organisms with aerobic respiration evolved and began to use the oxygen. The composition of the Earth's atmosphere is currently 77% nitrogen, 21% oxygen, and a little methane and carbon dioxide.

Viewed from outer space, the Earth might appear as a living entity that regulates its atmosphere. From this distance, Earth appears to be a closely knit community of organisms, each type of which is necessary to maintain the existing conditions. Lovelock's neighbor at the time, William Golding, who wrote the novel *Lord of the Flies,* suggested the name Gaia to describe this whole Earth assemblage, after the Greek goddess of the Earth. Lovelock's Gaia hypothesis proposed that the physical and chemical conditions of the Earth form a complex interacting system that is maintained by life itself. As further evidence of the maintenance of stable physical and chemical conditions, Lovelock noted the constancy of the Earth's temperature and the salinity of its oceans. The heat of the sun has increased 25% since life on Earth began, yet the temperature on Earth has remained fairly constant. Likewise, the ocean salinity has been constant at around 3.4% for a long time, despite the constant addition of river salts. Implicit in the notion of Gaia is a close association and interdependency of different types of organisms in one supercommunity. The Gaia hypothesis was not without criticism. Prominent scientists of the day, such as Stephen J. Gould and Richard Dawkins, argued vehemently against the Gaia idea, noting that organisms generally respond to environmental change rather than act in concert to keep environmental conditions constant. But the Gaia hypothesis has not gone away. The fourth conference on the Gaia hypothesis was held in 2006, and Lovelock's 2009 book *The Vanishing Face of Gaia: A Final Warning* argues that we are unlikely to be able to reverse climate change and discusses the negative effects this will have. We will not examine Gaia further in this book, but we will examine the notion of large-scale ecological communities called **biomes**. Originally, a biome was a large-scale terrestrial community defined by abiotic conditions and the type of vegetation present. Currently, less emphasis is placed on vegetation type, and distinct marine and freshwater biomes are recognized.

In this section we consider the influence of variation in solar radiation in creating different temperature and rainfall patterns on Earth. In Chapter 22, we describe the influence of climate on biome type, from the poles to the Tropics. Next, we describe large-scale terrestrial biomes, such as tropical rain forests and prairies. In Chapters 23 and 24 we discuss aquatic biomes, first in marine systems then in freshwater.

Deep Canyon transect, Santa Rosa Mountains, California.

CHAPTER 22

# Terrestrial Biomes

## Outline and Concepts



Anne Kelly and Michael Goulden (2008) compared surveys of plant cover that were made in Southern California's Santa Rosa Mountains in 1977 and 30 years later in 2007. The Deep Canyon transect spanned an altitudinal range from 244 m to 2,560 m over 16 km, climbing through desert and woodland up to coniferous forest. Over this 30-year period the Southern California climate warmed, precipitation variability increased, and the amount of snow decreased. The average elevation of the dominant plant species rose by about 65 m. The authors ruled out air pollution or changes in fire frequency as causes and attributed the elevated shift to climate change. Species at lower elevations can gradually disperse upward as the climate warms, but species at the highest elevations have nowhere to go and may be threatened with extinction.

Temperature, wind, water, and light are components of **climate**, the prevailing weather pattern in a given region. As we saw in Chapters 5–7, the distribution and abundance of organisms are often influenced by the abiotic environment, so to understand the patterns of abundance of life on Earth, ecologists need to study the global climate. We begin this chapter by examining global climate patterns, focusing on how temperature variation and atmospheric circulation drive climate and how features such as elevation and proximity of landmasses to water can alter these patterns. Knowing how and why climate changes around the world enables us to understand and predict the occurrence of biomes, large-scale terrestrial communities, such as tropical rain forests and hot deserts.

## 22.1 Variation in Solar Radiation Determines the Climate in Different Areas of the World

Substantial differences in temperature occur over the Earth, mainly due to variations in the incoming solar radiation. In higher latitudes, such as northern Canada and Russia, the sun's rays hit the Earth obliquely and are spread out over more of the planet's surface than they are in equatorial areas (**Figure 22.1**). More heat is also lost in the atmosphere of higher latitudes because the sun's rays travel a greater distance through the atmosphere, allowing more energy to be dissipated by cloud cover. The result is that a much smaller amount of solar energy (40% less) strikes polar latitudes than equatorial areas. Generally, temperatures increase as the amount of solar radiation increases (**Figure 22.2**). However, at the Tropics both cloudiness and rain reduce average temperature, so that temperatures do not continue to increase toward the equator.

In 1735 English meteorologist George Hadley made the initial contribution to a model of general atmospheric circulation. Hadley proposed that solar energy drives winds, which in turn influence the global circulation of the atmosphere. In his model, the warmth at the equator causes the surface equatorial air to heat up and rise vertically into the atmosphere. As the warm air rises away from its source of heat, it cools and becomes less buoyant, but because of the warm air behind it the cool air does not sink back to the surface. Instead, the rising air spreads north and south away from the equator, eventually returning to the surface at the poles. From there it flows back toward the equator to close the circulation loop. Hadley suggested that on a nonrotating Earth, this air movement would take the form of one large convection cell in each hemisphere, as shown in **Figure 22.3**.

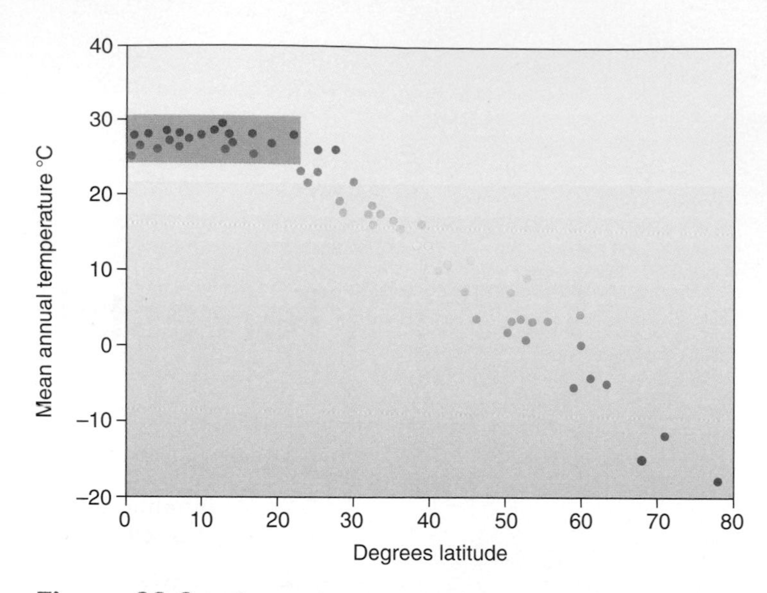


**Figure 22.2 The Earth's temperature generally varies with latitude.** The temperatures shown in this figure were measured at moderately moist continental locations of low elevation. (Reproduced from Terborgh, 1973.)

**ECOLOGICAL INQUIRY**

Why is there a wide band of similar temperatures at the Tropics?

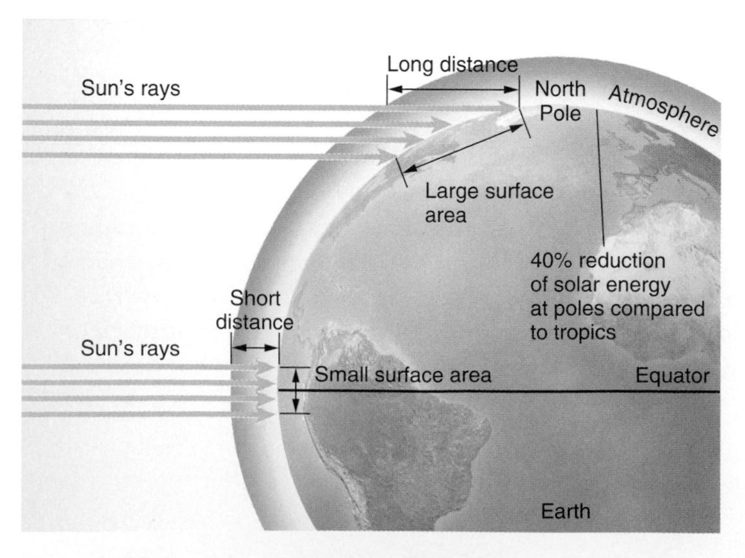


**Figure 22.1 The intensity of solar radiation varies with latitude.** In polar areas the sun's rays strike the Earth at an oblique angle and deliver less energy than at tropical locations. In tropical locations the energy is concentrated over a smaller surface and travels a shorter distance through the atmosphere.

Hadley's one-cell circulation has since been modified to account for additional factors. Once air is moving, any surface flow is deflected toward the west as a result of the Earth's rotation, a phenomenon called the **Coriolis force**, after the French scientist Gaspard Gustave de Coriolis. Imagine a rocket fired from the North or South Pole toward the equator. By the time the rocket reached the equator, the target would have moved to the east. The rocket always travels due south or north, but to an observer at the equator it would bend some distance to the west before landing (**Figure 22.4**). A similar phenomenon occurs with winds and aircraft, though pilots compensate for Coriolis forces when flying. The opposite phenomenon happens when a rocket is fired from the equator either north or south. The rocket's strong eastward velocity, before it was fired, causes its path to appear to veer to the west. The Coriolis force also gives spin to storm systems and is a reason hurricanes spin counterclockwise in the Northern Hemisphere and clockwise in the Southern Hemisphere.

In the 1920s a three-cell circulation in each hemisphere was proposed to fit the Earth's heat balance (**Figure 22.5**). The contribution of George Hadley is still recognized, in that the cell nearest the equator is called the **Hadley cell**. In the Hadley cell, the warm air rising near the equator forms towers of cumulus clouds that provide rainfall, which in turn maintains the lush vegetation of the tropical rain forests. As the upper flow in this cell moves toward the poles, it begins

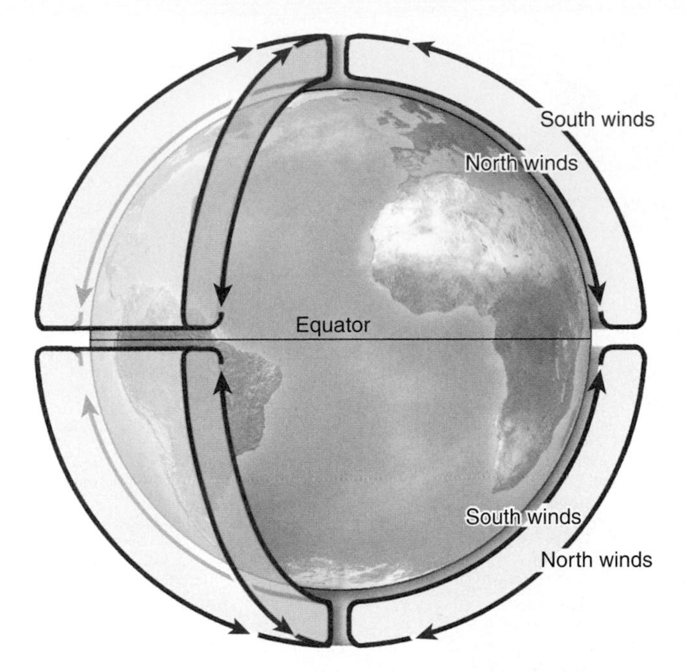


**Figure 22.3 George Hadley's 1735 model of atmospheric circulation.** In Hadley's model, simple convective circulation of air on a uniform, nonrotating Earth, heated at the equator and cooled at the poles, took the form of one large convection cell in each hemisphere. Winds are named according to the direction from which they blow, so the south wind blows from south to north.

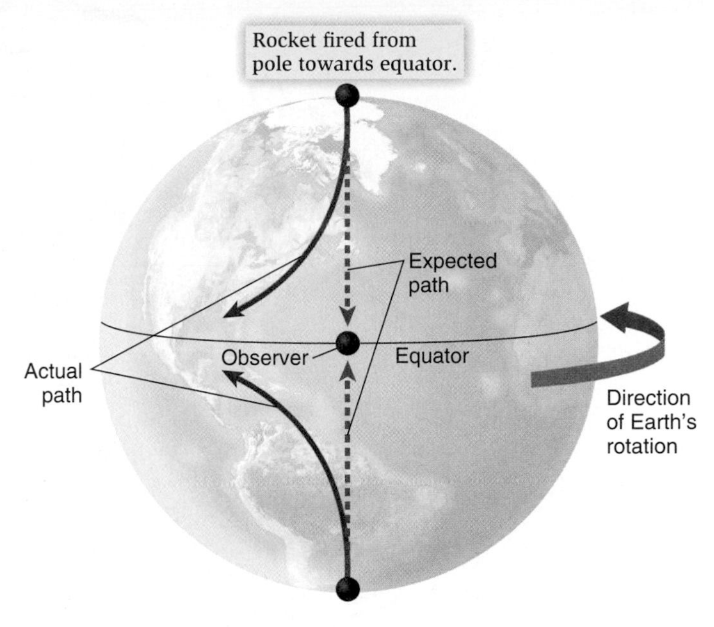


**Figure 22.4 Diagrammatic representation of the Coriolis force.** Even though a rocket fired from the North Pole flies due south, by the time it reaches the equator its intended target would have moved and the landing point would be at a more westward spot.

to subside, or fall back to Earth, at about 20–30° north and south of the equator. These **subsidence zones** are areas of high pressure and are the sites of the world's hot deserts, because the subsiding air is relatively dry, having released all of its moisture over the equator. Winds are generally weak and variable near the center of this zone of descending air. Subsidence zones have popularly been called the horse latitudes. The name is said to have been coined by Spanish sailors crossing the Atlantic, whose ships were sometimes rendered motionless in these waters and who reportedly were forced to throw horses overboard, as they could no longer water or feed them.

From the center of the subsidence zones, the surface flow splits into a poleward branch and an equatorial branch. The equatorial flow is deflected by the Coriolis force and forms the reliable trade winds. In the Northern Hemisphere, the trades are from the northeast, the direction from which they provided the sail power to explore the New World; in the Southern Hemisphere, the trades are from the southeast. The trade winds from both hemispheres meet near the equator in a region called the intertropical convergence zone (ITCZ), also known as the doldrums. Here the light winds and humid conditions provide the monotonous weather that may be the basis for the expression "in the doldrums."

In the three-cell model, the circulation between 20–30° and 40–60° latitude, called the **Ferrel cell**, is opposite that of the Hadley cell. William Ferrel was an American meteorologist who realized that Coriolis forces affect wind flow. In the Ferrel cell the net surface flow is poleward, and because of the Coriolis force, the winds have a strong westerly component

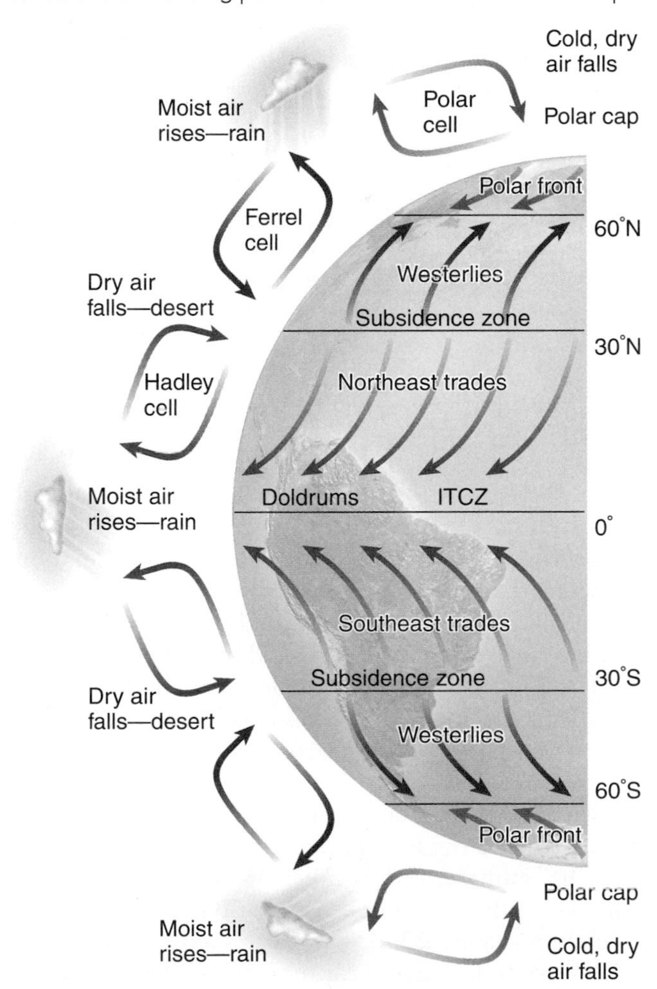


**Figure 22.5 Patterns of atmospheric circulation.** The three-cell model of atmospheric circulation on a uniform, rotating Earth that is heated more at the equator than at the poles. The ITCZ is the intertropical convergence zone.

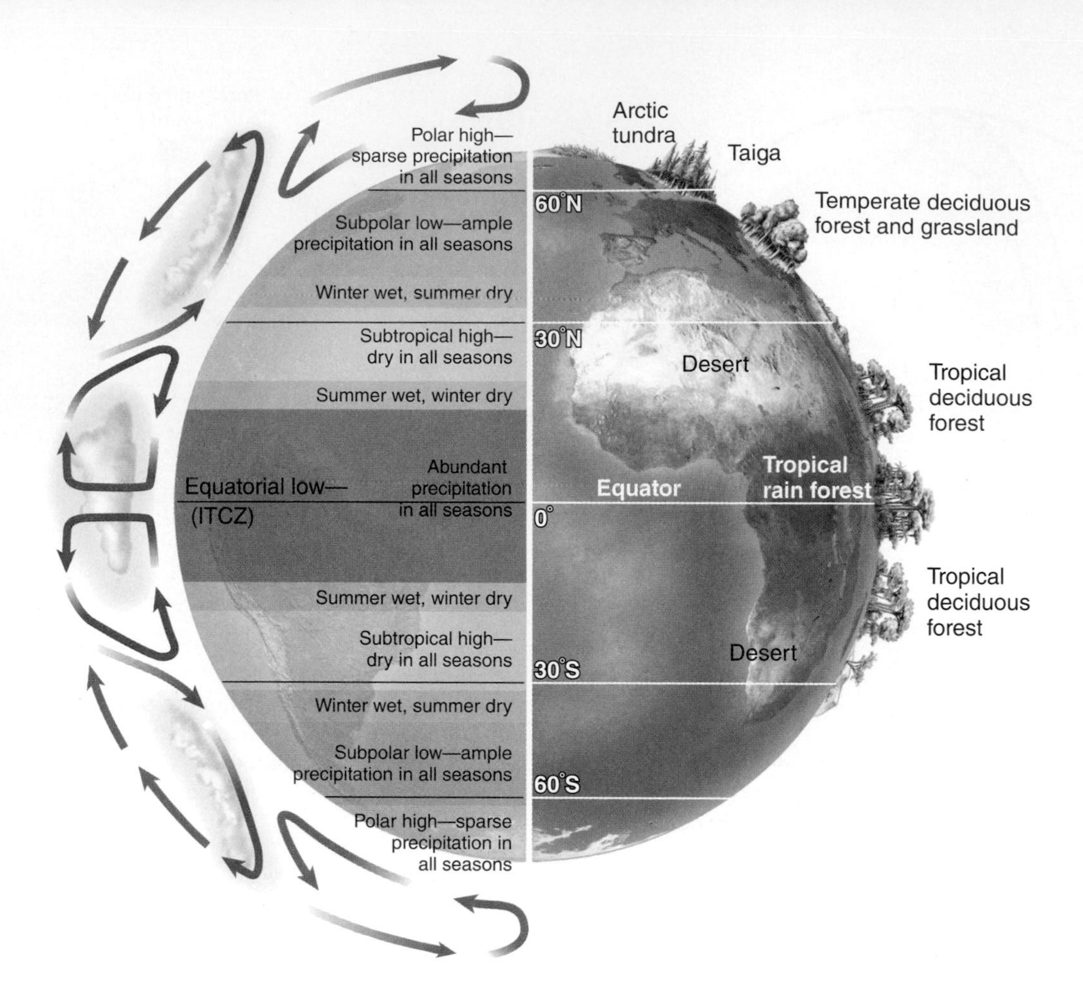


**Figure 22.6 Global circulation based on a modified three-cell model.** Tropical forests exist mainly in a band around the equator, where it is hot and rainy. At around 20–30° north and south, the air is hot and dry, and deserts exist. A secondary zone of precipitation exists at around 45–55° north and south, where temperate forests are located. The polar regions are generally cold and dry.

(flowing from west to east). These prevailing westerlies were known to Benjamin Franklin, who noted that storms migrated eastward across the colonies. In the Southern Hemisphere there is little landmass to impede these winds, and they can become intense. This has given rise to these latitudes being termed the "roaring forties." The final circulation cell is known as the **polar cell**. At the poles, the air has cooled and descends, but it has little moisture left, explaining why many high-latitude regions are actually desert-like in condition.

The three-cell model provides a good understanding of global circulation, but it is still oversimplified. In reality, the Hadley and polar cells are strong but the Ferrel cell is weak, with passing high- and low-pressure systems, depending on the seasons. In fact, the secondary zones of high precipitation associated with the Ferrel cell can come anywhere from about 35° to 65° latitude, with between 45° and 55° being most common. This effect modifies the positions of the three cells of circulation (**Figure 22.6**). The Earth's axis of rotation is tilted at 23.5° from the vertical for the full 365 days it takes to orbit the sun (**Figure 22.7**). The **solar equator**, the area receiving the most solar energy, varies seasonally and reaches 23.5° north on June 21 and 23.5° south on December 21. In the Northern Hemisphere these dates are called the summer and winter solstices, from the Latin *sol,* sun, and *sistit,* stands. For several days before and after each solstice, the noontime elevation of the sun appears in the same place. On March 21 and September 22, the so-called spring and autumn equinoxes, all locations in the Northern and Southern Hemisphere have approximately equal amounts of solar radiation. This means that for half of the year the Northern Hemisphere receives more solar energy, and for the other half of the year the Southern Hemisphere receives more solar energy. At 60° north, during the northern winter, temperatures in Siberia may only average −12°C, whereas in the summer they may average 16°C, a difference of 28°C. In contrast, tropical temperatures vary relatively little, perhaps 2–3°C, year round. Southern Hemisphere temperatures also vary seasonally, but the large expanses of open water moderate the temperature extremes.

Associated with these seasonal changes in temperature are changes in day length and precipitation. The ITCZ follows the solar equator (**Figure 22.8**). Thus, as the solar equator moves seasonally, the band of rainfall associated with it drifts north or south of the equator, depending on the season. The result is a seasonality of tropical rainfall between about 20° north and 20° south of the equator. Two rainy seasons of equal length exist at the equator, both of about 3–4 months

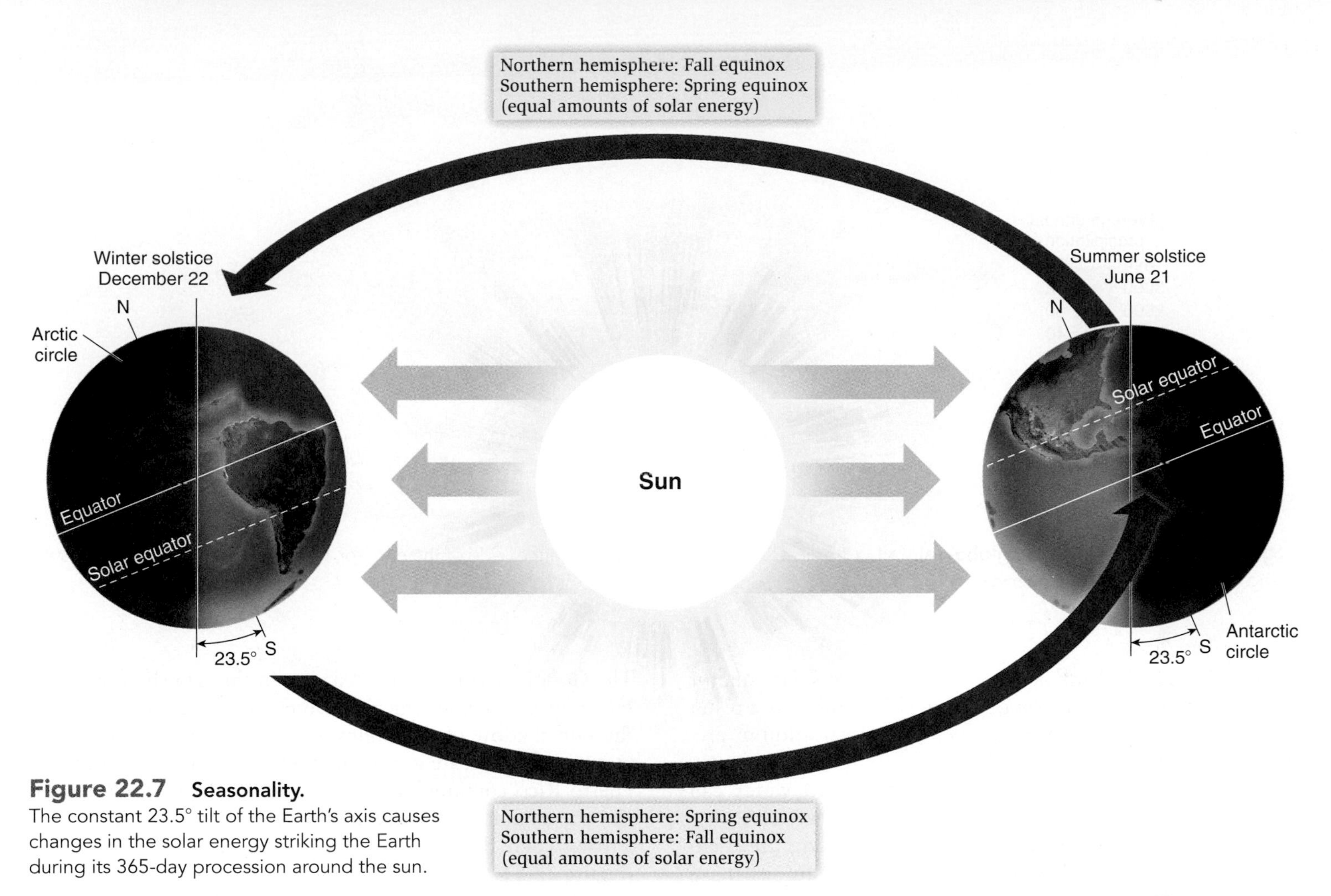


**Figure 22.7** Seasonality.
The constant 23.5° tilt of the Earth's axis causes changes in the solar energy striking the Earth during its 365-day procession around the sun.

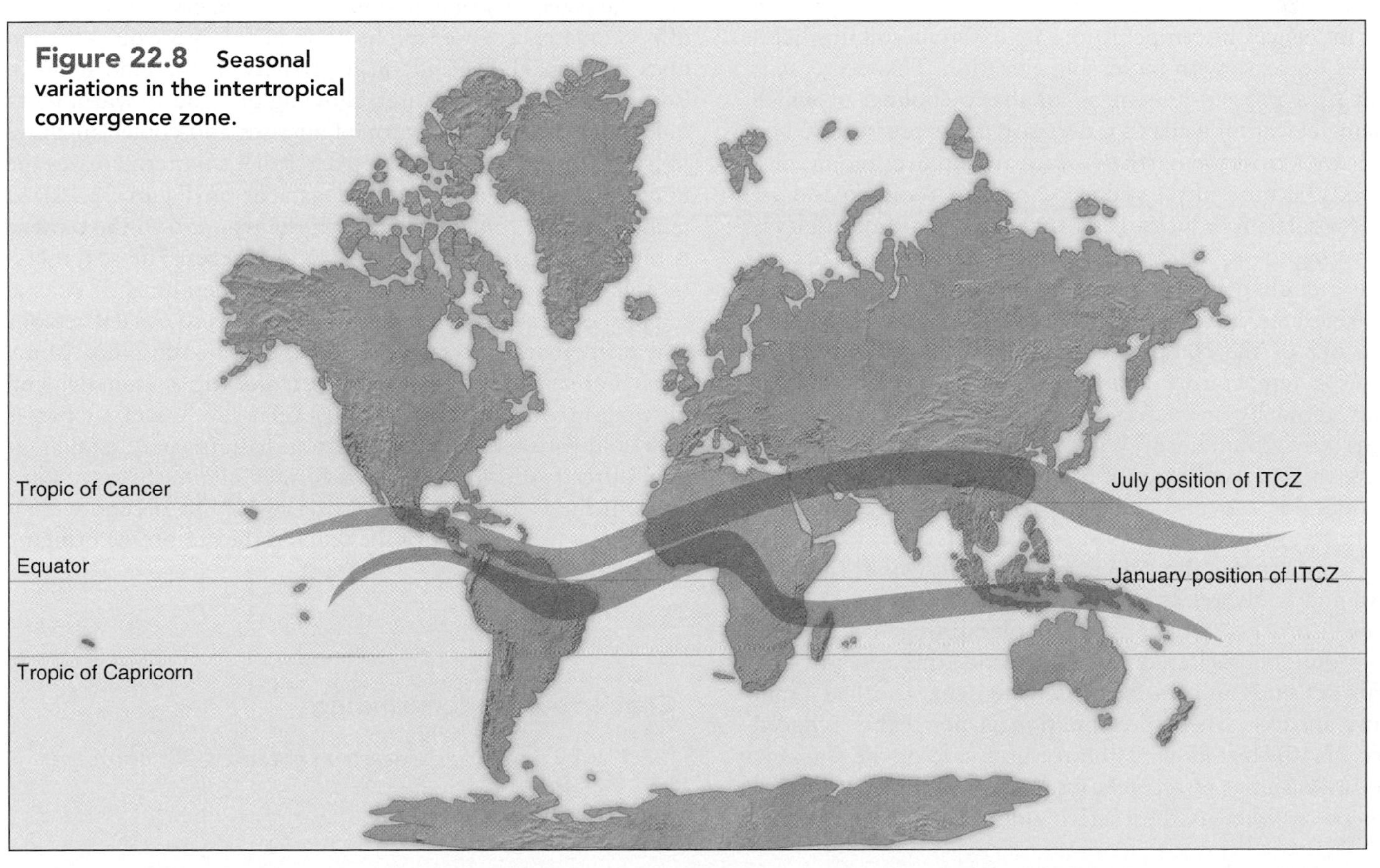


**Figure 22.8** Seasonal variations in the intertropical convergence zone.

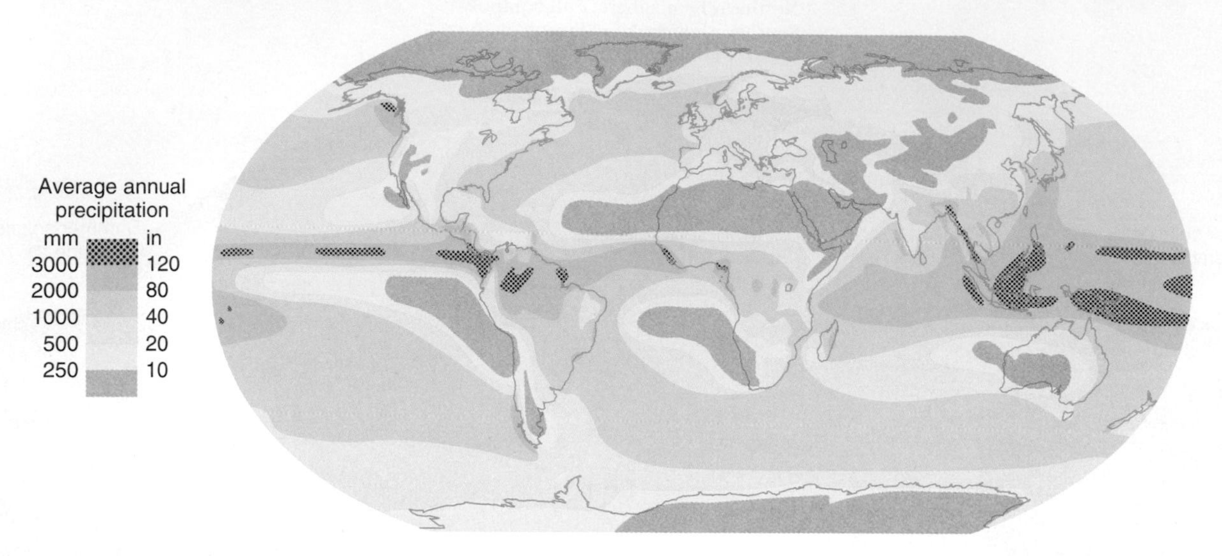


**Figure 22.9** **Variations in global rainfall.** Tropical rain forests are found on either side of the equator, whereas temperate rain forests are found mainly in the Pacific Northwest of North America.

long, as the ITCZ drifts north or south. At the solar equator there is just one rainy season of about 4–5 months. As a result of changes in latitude, there is much global variation in precipitation patterns (**Figure 22.9**).

Elevation and proximity of a landmass to water can affect temperature and precipitation. Thus far we have considered how global temperatures and wind patterns affect climate. The geographic features of a landmass can also have an important impact. For example, the elevation of a region greatly influences its temperature range. On mountains, temperatures decrease with increasing elevation. This decrease is a result of a process known as **adiabatic cooling,** in which increasing elevation leads to a decrease in air pressure. When air is blown across the Earth's surface and up over mountains, it expands because of the reduced pressure. As it expands, it cools at a rate of about 5–10°C for every 1,000 m in elevation, as long as no water vapor or cloud formation occurs. Adiabatic cooling is also the principle behind the function of a refrigerator, in which refrigerant gas cools as it expands coming out of the compressor. A vertical ascent of 600 m produces a temperature change roughly equivalent to that brought about by an increase in latitude of 1,000 km. This explains why mountaintop vegetation, even in tropical areas, can have the characteristics of tundra.

Mountains can also influence patterns of precipitation. For example, when warm, moist air encounters the windward side of a mountain (the side exposed to the wind), it flows upward and cools, releasing precipitation in the form of rain or snow, a process known as **orographic lifting.** On the side of the mountain sheltered from the wind, the leeward side, drier air descends and warms, producing what is called a **rain shadow,** an area where precipitation is noticeably reduced (**Figure 22.10**). In this way, the western side of the Cascade Range in Washington receives more than 500 cm of annual precipitation, whereas the eastern side receives only 50 cm. The Gobi Desert in central Asia lies in the rain shadow of the Himalayas, and the Atacama Desert in Chile and Peru is in the rain shadow of the Andes. Even on small islands, a large mountain can create a rain shadow. The Caribbean island of Puerto Rico contains lush rain forest on the Cordillera Central mountains, but these mountains create a rain shadow so that cacti thrive in the dry conditions of the southwest (**Figure 22.11**).

The proximity of a landmass to a large body of water can affect climate, because land heats and cools more quickly than does the sea. The specific heat capacity of the land is much lower than that of the water, allowing the land to warm more quickly in the day. The warmed air rises and cooler air flows in from above the ocean to replace it. This pattern creates the familiar onshore sea breezes in coastal areas (**Figure 22.12**). At night, the land cools quicker than the sea, and so the pattern is reversed, creating offshore or land breezes. The sea, therefore, has a moderating effect on the temperatures of coastal regions and especially islands. The climates of coastal regions may differ markedly from those of their climatic zones. Many coastal areas never experience frost, and fog is often evident, blowing in with the sea breeze as relatively warm air passes over cold water. Thus, the vegetation patterns in coastal areas may differ from those in areas farther inland. In fact, some areas of the U.S., including Florida, would be deserts were it not for the warm water of the sea and the sea breeze bringing rain inland.

### Check Your Understanding

**22.1** Why do we find deserts at about 20–30° north and south of the equator?

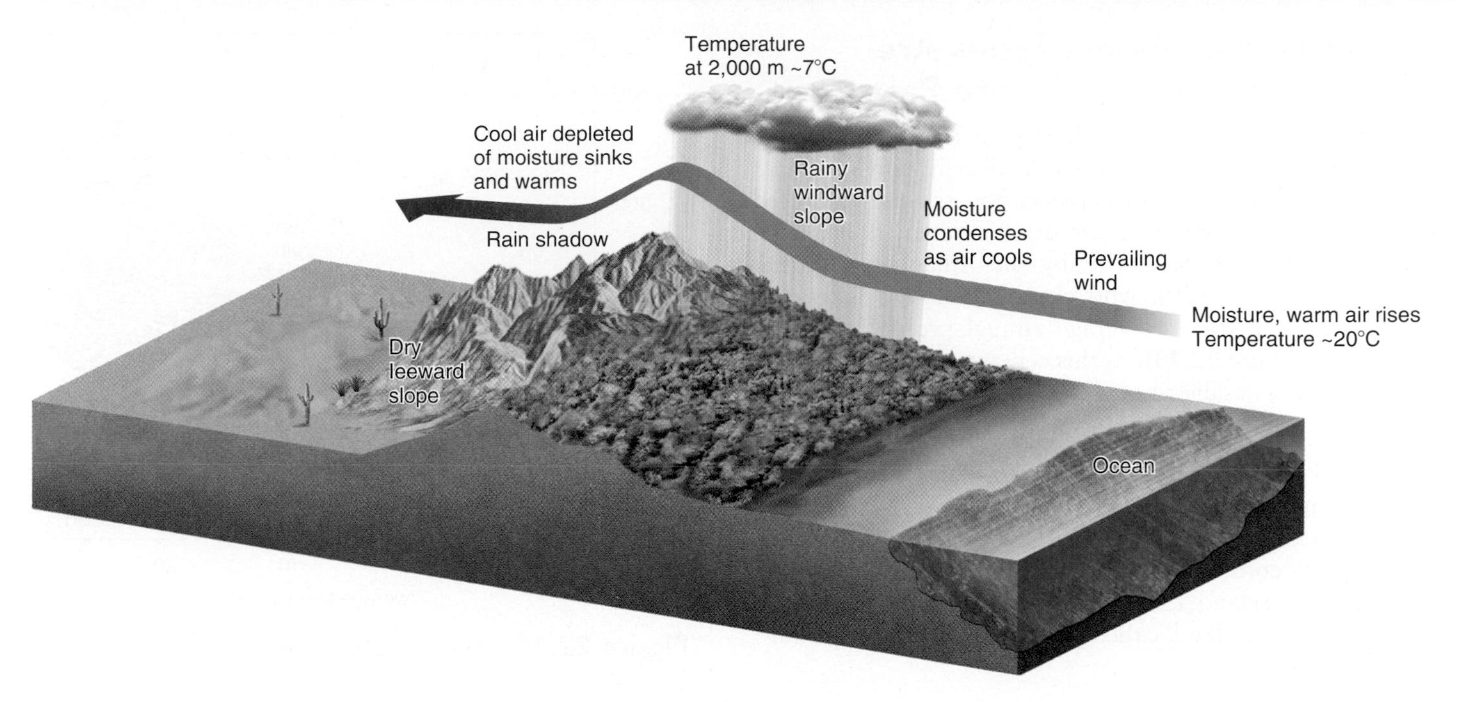


**Figure 22.10** **The influence of elevation on climate.** Warm, moist air cools as it rises up the windward surface of a mountain, creating rain. On the leeward side, dry air sinks and warms.

**Figure 22.11** **Rain shadows.** **(a)** Even on a relatively small island such as Puerto Rico, the mountains cause orographic lifting and precipitation, resulting in rain forests. **(b)** On the leeward side, desert-like vegetation may be found.

**Figure 22.12** **Proximity to water affects climate.** **(a)** When the landmass is warmer than the ocean, a sea breeze is created, bringing in cooler, moisture-laden air and, sometimes, fog. **(b)** When the sea is warmer than the land, a land breeze results.

## 22.2 Terrestrial Biome Types Are Determined by Climate Patterns

Differences in climate help to determine the Earth's different terrestrial biomes, the large-scale distribution of plants and animals. Many types of classification schemes are used for mapping the geographic extent of biomes, but one of the most useful was developed by the American ecologist Robert Whittaker (1970), who classified biomes according to the physical factors of average annual precipitation and temperature (**Figure 22.13**). In this scheme, we can recognize 10 terrestrial biomes distributed over the globe (**Figure 22.14**):

1. Tropical rain forest
2. Tropical deciduous forest
3. Temperate rain forest
4. Temperate deciduous forest
5. Temperate coniferous forest, called taiga or boreal forest
6. Tropical grassland, called savanna
7. Temperate grassland, called prairie, including chaparral
8. Hot desert
9. Cold desert
10. Tundra

While these broad terrestrial biomes are a useful way of defining the main types of communities on Earth, ecologists acknowledge that not all communities fit neatly into

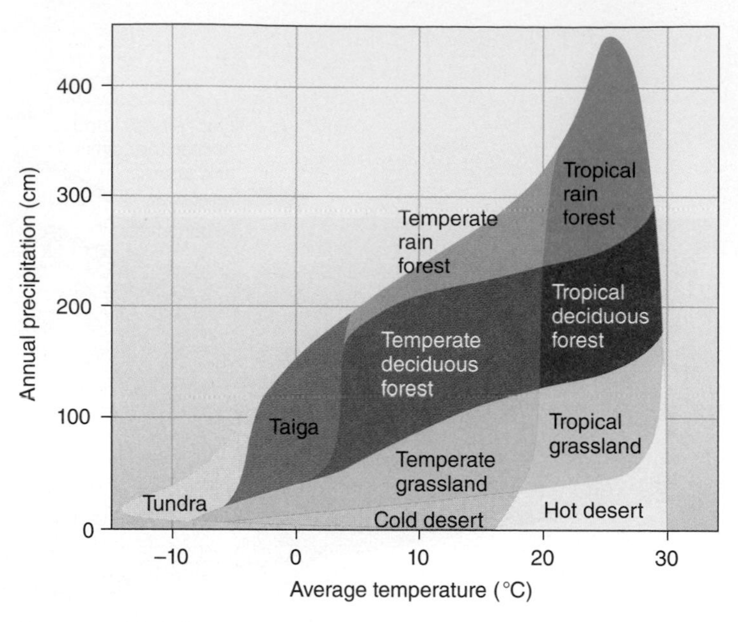


**Figure 22.13** **The world's biome types.**
(From Whittaker, 1970.)

**ECOLOGICAL INQUIRY**

What other factors may influence the vegetation growing in different biome types?

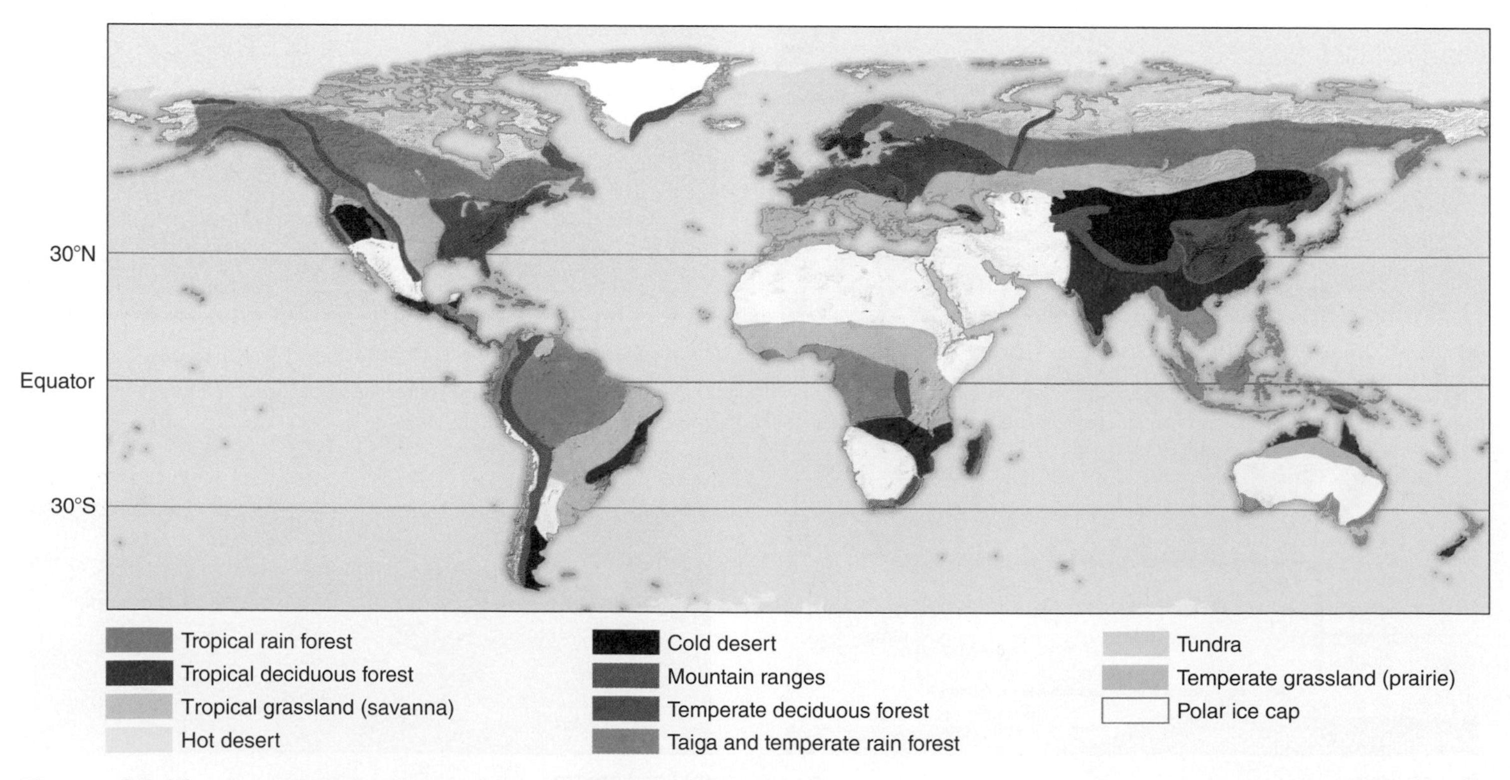


**Figure 22.14** **The influence of geographic location on the distribution of biomes.** The distribution patterns of taiga and temperate rain forest are combined because of their similarity in tree species and because temperate rain forest is actually limited to a very small area.

one of these 10 major biome types. One biome type often grades into another, as seen on mountain ranges. Furthermore, habitat change, such as that brought about by agriculture and forestry, has radically altered the landscape, converting natural biomes to crop fields, tree plantations, and urban settlements. Ecologists are beginning to grapple with these changes. Soil conditions can also influence biome vegetation. In California, serpentine soils, which are dry and nutrient-poor, support only sparse vegetation, whereas adjacent soils may support lush vegetation (refer back to Figure 7.2). In the eastern U.S., most of New Jersey's coastal plain, called the Pine Barrens, consists of sandy, nutrient-poor soil that cannot support the surrounding deciduous forest and instead contains grasses and low shrubs growing among open stands of pygmy pine and oak trees. Some biome classification schemes recognize these finer divisions. For example, the World Wildlife Federation has classified 867 terrestrial ecoregions. This classification is used to define a Global 200 list of the most imperiled ecoregions as priorities for classification. Clearly, we do not have enough space to examine this detailed system. It is valuable to note, however, that even within our broader classification scheme there is useful information for conservation biologists. The numbers of species of mammals, birds, and amphibians is by far the greatest in tropical rain forests (**Figure 22.15**). This area also contains the greatest number of species threatened by extinction.

As we discuss each biome, we will note details of its physical environment, location, plant and animal life, and the more significant effects of humans. Climate characteristics are provided in the form of Walter diagrams, a schematic technique pioneered by the plant ecologist Heinrich Walter (**Figure 22.16**). Walter diagrams summarize valuable information on variations in temperature and precipitation.

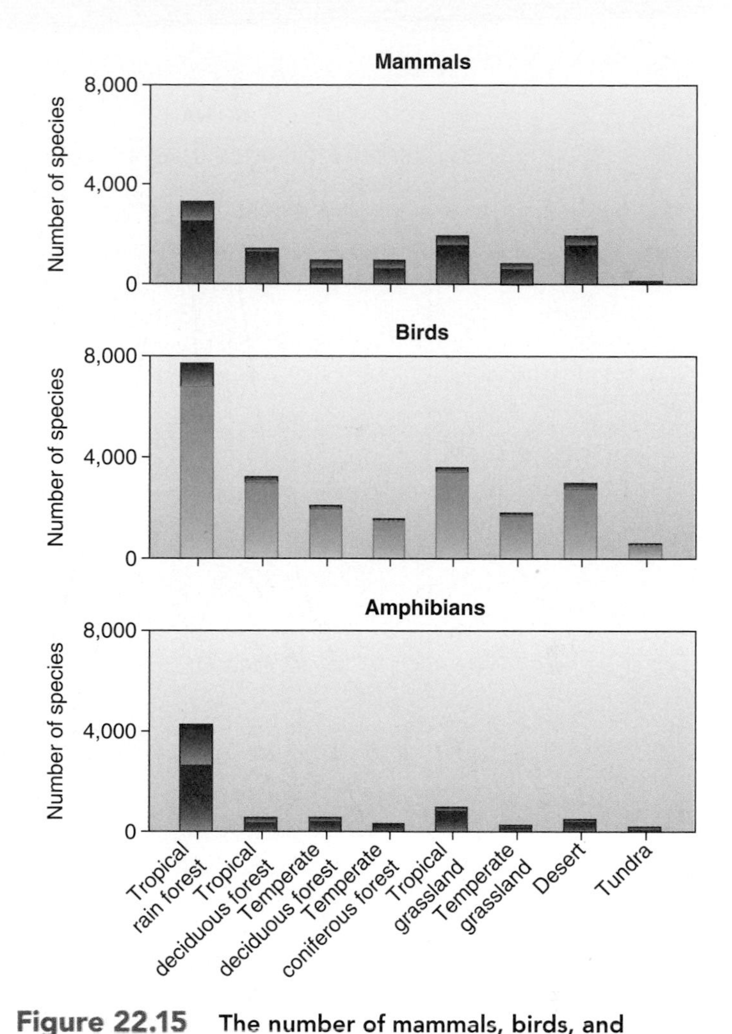


**Figure 22.15 The number of mammals, birds, and amphibians occurring in the major biomes on Earth.** The numbers of species threatened with extinction is given in red. (From Baille et al., 2004.)

## Tropical Rain Forest (Figure 22.17)

**Physical Environment:** Rainfall exceeds 230 cm per year, and temperature is hot year round, averaging 25–29°C. Soils are commonly shallow and nutrient-poor, because many of the nutrients are leached out by heavy rainfall. There is no rich leaf litter layer as there is in temperate systems; fallen leaves are quickly decomposed and nutrients returned to the vegetation, where most of the minerals are locked up. The continual movement of the water table often leaves iron and aluminum oxides in hard, red layers.

**Location:** This biome is equatorial, located between the Tropics of Cancer and Capricorn (see Figure 22.14). Tropical forests cover much of northern South America, Central America, western and central Africa, Southeast Asia, and various islands in the Indian and Pacific Oceans.

**Plant Life:** The numbers of plant species found in tropical forests can be staggering, often reaching as many as 100 tree species per square kilometer. Alwyn Gentry (1988) recorded 283 tree species in one hectare of Peruvian rain forest; 63% of the species were represented by a single individual. Leaves often narrow to "drip-tips" at the apex so that rainwater drains quickly. Many trees have large buttresses that help support their shallow root systems. Little light penetrates the canopy, the uppermost layer of tree foliage, and the ground cover is often sparse. Epiphytes, plants that live perched on trees and are not rooted in the ground, are common. Orchids and bromeliads are common epiphytes in Central and South American forests. Lianas, or climbing vines, are also common.

**Animal Life:** Animal life in the tropical rain forests is diverse; insects, reptiles, amphibians, and mammals are well represented. Large mammals, however, are not common, although monkeys may be important herbivores. Many of the plant species are widely scattered in tropical forests, and animals are important in pollinating flowers and dispersing fruits and seeds. Mimicry and bright protective coloration, warning of bad taste or the existence of toxins, are common. Tropical rain forests are great reservoirs of the Earth's species, with as many as half the animal species on Earth living in them.

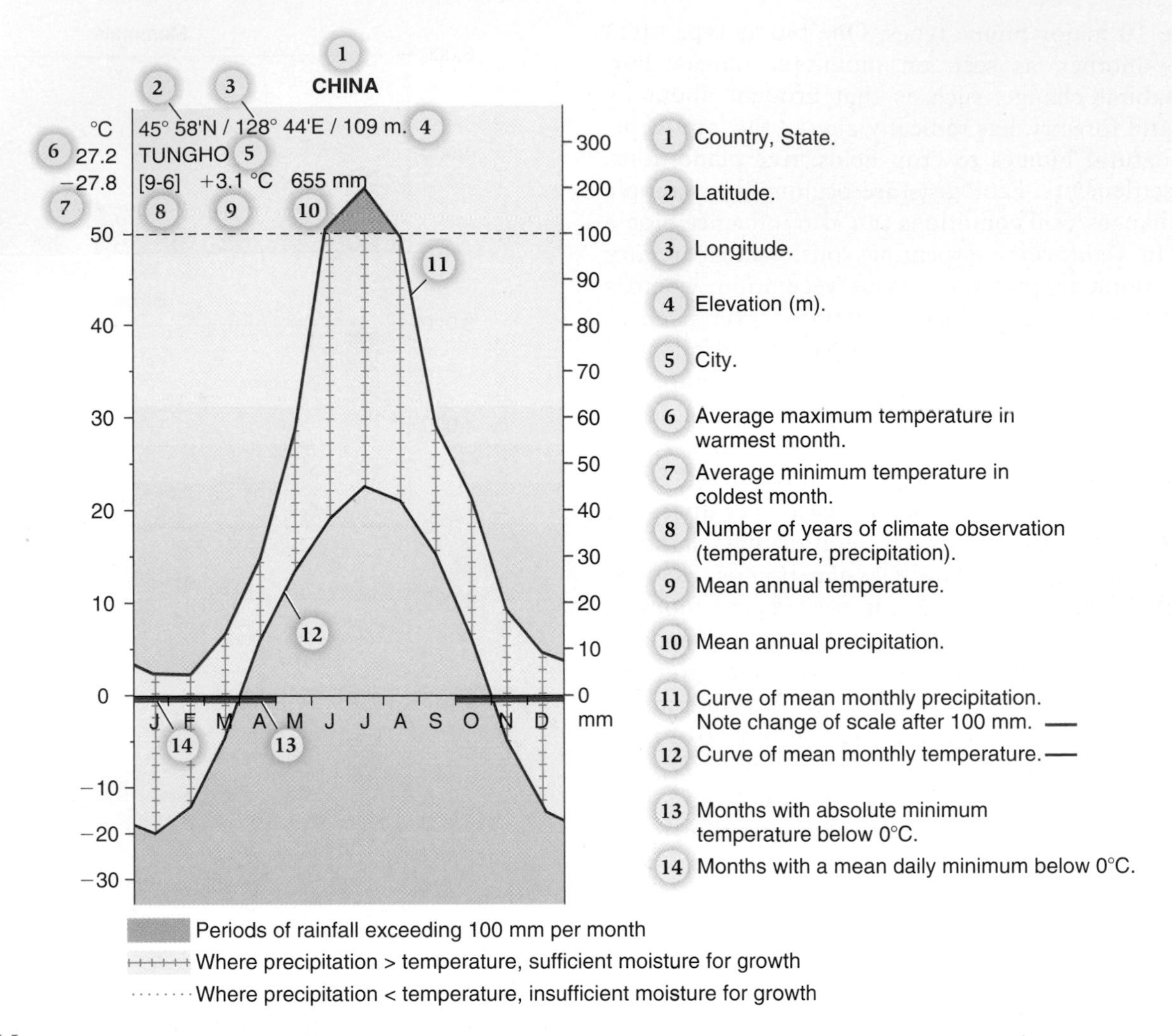


**Figure 22.16** **A Walter climate diagram.** On average, about 20 mm of precipitation per month is needed for plant growth for every 10°C in temperature. The two scales, temperature and precipitation, are therefore aligned in this manner. Not all data are recorded for all stations. Months are arranged January to December for locations in the Northern Hemisphere, and July to June for locations in the Southern Hemisphere.

**Effects of Humans:** Humans are impacting tropical forests greatly by logging and by clearing the land for agriculture (see **Global Insight**). However, because of the low nutrient levels of the soils, cleared tropical forest land does not support agricultural practices well for long. Many South American tropical forests are cleared to create grasslands for cattle.

**Tropical Deciduous Forest (Figure 22.18)** **Physical Environment:** Temperatures are hot year round, averaging 25–39°C. Rainfall is substantial, at around 130–280 cm a year, but the dry season is distinct, often 2–3 months or longer. Soil water shortages can occur in the dry season.

**Location:** Closer to the Tropics of Cancer and Capricorn, where rainfall is more seasonal, than to the equator. Much of India consists of tropical deciduous forest, containing teak trees. Southern Africa, Madagascar, coastal Brazil, Thailand, Northern Australia, and Mexico also contain tropical deciduous forest. The wet edges of this biome may grade into tropical rain forests; the dry edges may grade into tropical grasslands or savannas.

**Plant Life:** Because of the distinct dry season, many of the trees in tropical deciduous forests shed their leaves, and an understory of herbs and grasses may grow during this time. Where the dry season is 6–7 months long, tropical deciduous forests may contain shorter, thorny plants such as acacia trees, and the forest is referred to as tropical woodland. The litter layer of dead, decaying leaves is much thicker than in tropical rain forests.

**Animal Life:** The diversity of animal life is high, and species such as monkeys, antelopes, wild pigs, and tigers are present. However, as with plant diversity, animal diversity is less than in tropical rain forests. Tropical woodland may contain more browsing mammals; many plants employ thorns as a defense.

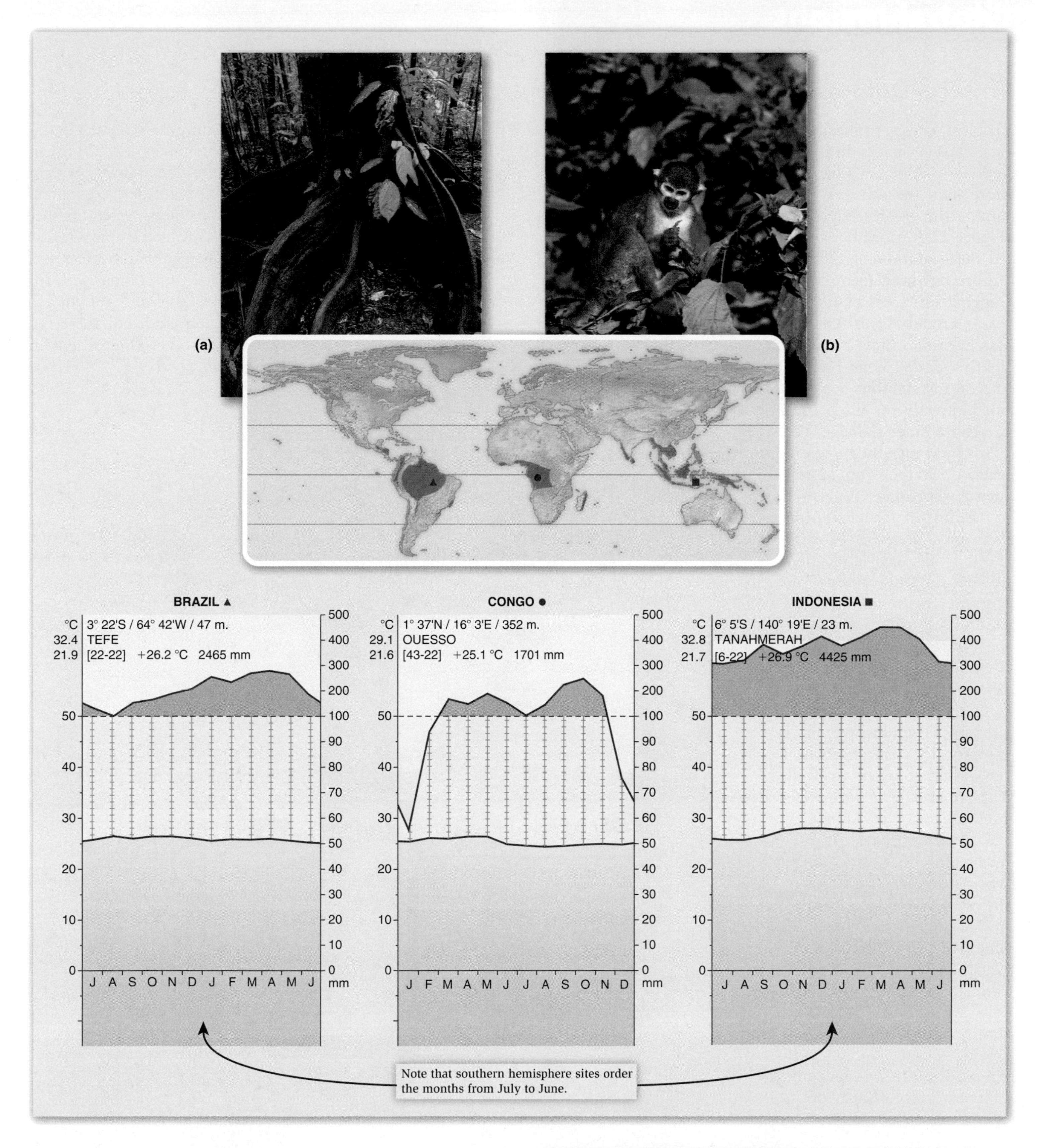


**Figure 22.17** **Tropical rain forest.** **(a)** Buttressed tree roots, Daintree National Park, Queensland, Australia. **(b)** Squirrel monkey, South America.

**ECOLOGICAL INQUIRY**

Why do many species of tropical trees have buttress roots?

## Tropical Deforestation Threatens Many Species with Extinction

Tropical forests, primarily rain forests but also deciduous forests, exist primarily in three areas of the globe: Africa, Asia, and Latin America (**Figure 22.A-i**). Latin America contains more than the other two areas combined, with Brazil containing the greatest amount of tropical forest of any nation (**Figure 22.A-ii**). Although there are different causes for tropical deforestation in different areas, clearing land for agriculture has been identified as the prime cause of forest loss. Logging in excess of regrowth is also a significant cause of loss, particularly in Asian forests. Fuelwood collection can also be important, but generally is more of a problem in lightly wooded areas. Finally, the construction of mines, dams, and oil installations is a minor cause of direct deforestation, but indirectly the discharge of chemicals and silt into rivers can cause much damage. The roads built into these regions often open them up to further development.

Although tropical forests once covered 14% of the Earth's dry land, they now cover only about 7%. Current rates of tropical deforestation increased from 0.67% per year in the 1990s to 0.84% annually between 2000 and 2005.

Conservation of tropical forests would save many rare species from extinction but has other benefits as well. The Amazon rain forest has been termed "the lungs of the planet" because it produces more than 20% of the world's oxygen. Many of the world's crops, including oranges, lemons, bananas, pineapples, tomatoes, cacao (chocolate), coffee, and vanilla, evolved in tropical rain forests. Rain forests remain a depository of genetic variation that could be used in future breeding of these crops. Rain forests are particularly rich in plants with unusual chemicals that they use in defense (refer back to Chapter 14). Many of these chemicals also have medicinal value. Many prescription drugs sold worldwide come from plant-derived sources, yet less than 1% of tropical plants have been tested for their medicinal properties. One estimate has valued tropical forests at $2,400 an acre if renewable and sustainable resources such as fruit, rubber, and nuts are harvested. In comparison, forests are worth only $400 an acre for timber and $60 an acre when converted to agricultural grazing land. For these reasons, it seems to make good economic and ecological sense to slow the rate of tropical deforestation.

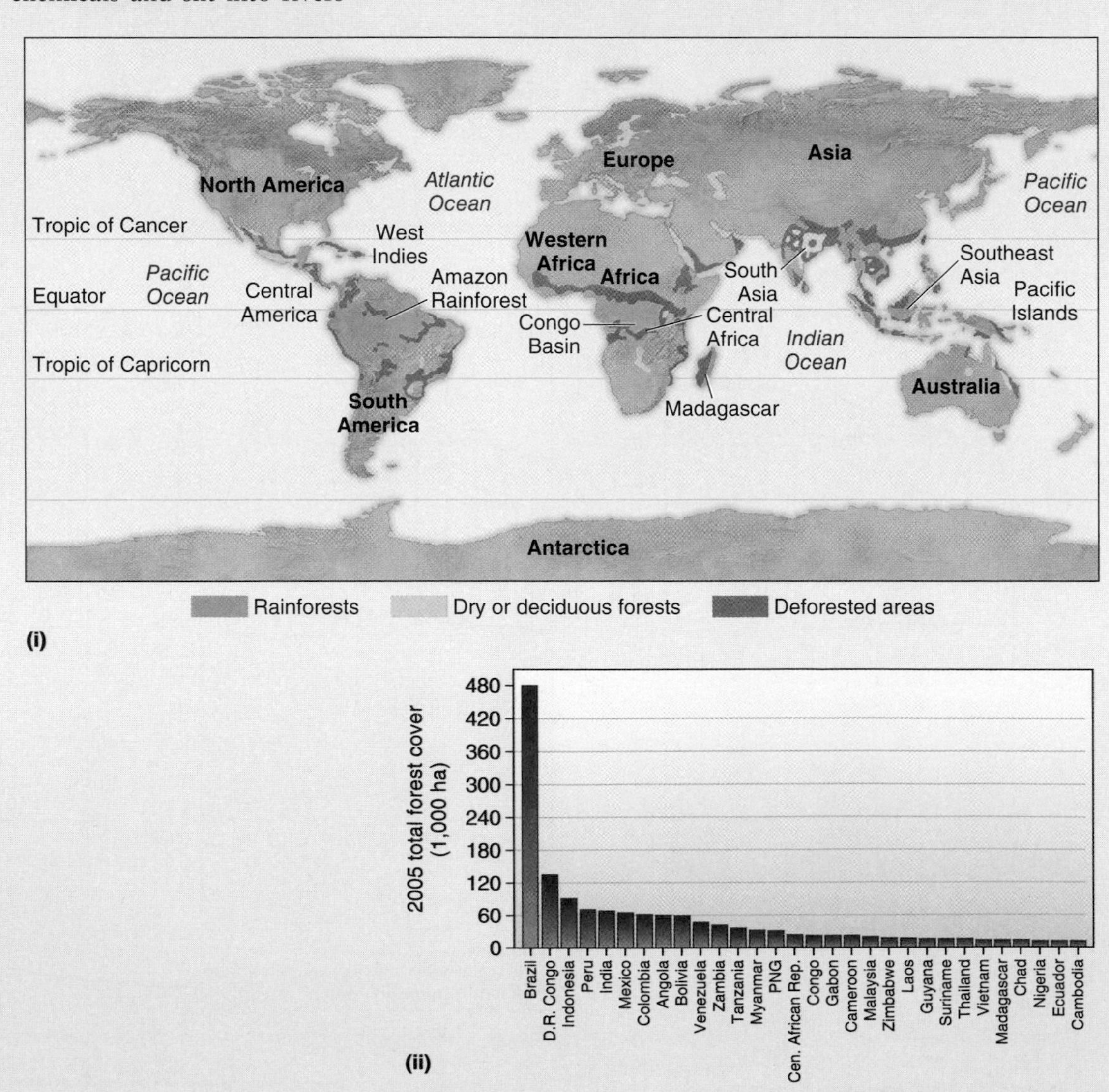


**Figure 22.A Tropical deforestation.** **(i)** Tropical forests consist mainly of rain forest but also of deciduous forest. Deforestation is rampant in both biomes. **(ii)** Brazil contains by far the largest areas of tropical forest (2005 data).

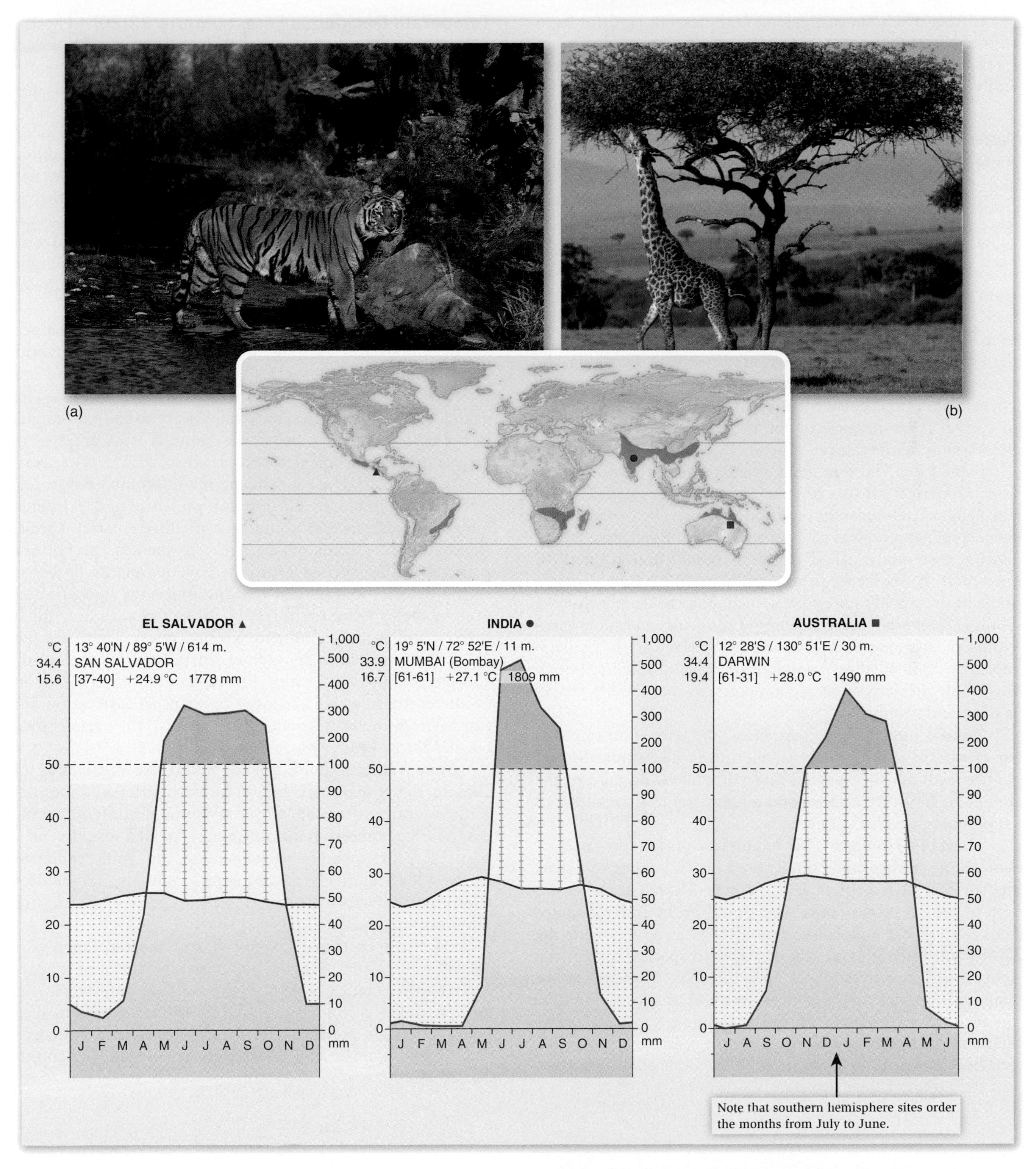


**Figure 22.18** **Tropical deciduous forest.** **(a)** Tiger in the Bandhavgarh National Park, India. **(b)** Masai Mara National Reserve, Kenya, with giraffe.

**ECOLOGICAL INQUIRY**

What factors affect the abundance of acacia trees in the tropical woodland?

**Effects of Humans:** Logging and clearing land for agriculture have had an even larger impact on tropical deciduous forests than on tropical rain forests, because the soils are generally of better quality.

### Temperate Rain Forest (**Figure 22.19**)

**Physical Environment:** Temperatures are a little cooler, 5–25°C on average, and winters are mild, but there is abundant rainfall, usually exceeding 200 cm a year. The condensation of water from dense coastal fogs augments the normal rainfall.

**Location:** The area of this biome type is small, consisting of a thin strip along the northwest coast of North America from northern California through Washington State, British Columbia, and into southeast Alaska, where it is called tongass. It also exists in southwestern South America along the Chilean coast. Smaller areas are present in Tasmania, the western coast of the south island of New Zealand, and in pockets along the Norwegian coast. Indeed, it is found only in coastal areas because of the moderating influence of the ocean on air temperature.

**Plant Life:** The dominant vegetation type, especially in North America, consists of large evergreen trees such as western hemlock, Douglas fir, and Sitka spruce. In the Southern Hemisphere, broad-leaved trees, especially of the genus *Nothofagus,* commonly called southern beech, and conifers of the family Podocarpaceae are common. This biome contains some of the world's tallest trees, including the giant redwood, *Sequoia sempervirens,* in northern California, and some eucalyptus in Australia, both with heights of over 100 m. The high moisture content allows epiphytes to thrive. Cool temperatures slow the activity of decomposers, so that the litter layer is thick and spongy.

**Animal Life:** In North America, the temperate rain forest is rich in species such as mule deer, elk, squirrels, and numerous birds such as jays and nuthatches. Because of the abundant moisture and moderate temperatures, reptiles and amphibians are also common.

**Effects of Humans:** This biome is a prolific producer of wood and supplies much timber. As a result, logging threatens the survival of the forest in some areas. Rates of deforestation match or surpass those in tropical areas. Bitter disputes between loggers and conservationists often erupt, with the flash point often a particular endangered species, such as the spotted owl or redwoods. In New Zealand, where no native mammals apart from bats exist, introduced species threaten the forest. The imported European red deer, *Cervus elaphus*, hinders regeneration of *Nothofagus* trees, and the Australian brushtail possum, *Trichosurus vulpecula,* imported for its fur, can strip trees of their leaves.

### Temperate Deciduous Forest (**Figure 22.20**)

**Physical Environment:** Temperatures fall below freezing each winter but not usually below −12°C, and annual rainfall is generally between 75 and 200 cm. Soils are rich in organic and inorganic nutrients, and quite fertile.

**Location:** Located between 30° and 60°, with most forests between 40° and 55°. Large tracts of temperate deciduous forest are evident in the eastern U.S., western Europe, and eastern Asia. In the Southern Hemisphere, the coverage is much sparser because at these latitudes most landmasses are small and the climate is mediated by the ocean. Eucalyptus forests occur in Australia, and deciduous forests are found in smaller patches in central Chile, northern New Zealand, and southeast Australia.

**Plant Life:** Species diversity is much lower in temperate deciduous forests than in tropical forests, with about only three to four tree species per square kilometer, and several tree genera may be dominant in a given locality—for example, oaks, hickories, and maples are usually dominant in the eastern U.S. However, the biomass is as great as or greater than in tropical forests (**Figure 22.21**). The exact forest composition is dependent on different abiotic features. For example, in Europe, soil moisture and pH determine the location and abundance of different tree species (**Figure 22.22**). Commonly, leaves are shed in the fall and reappear in the spring. Many herbaceous plants flower in spring before the trees leaf out and block the light, though even in the summer the forest is not as dense as in tropical forests, and there is abundant ground cover. There are few epiphytes and lianas. In sandier areas, such as the southeast United States, soils have low nutrient levels. This favors evergreen trees, which are more resistant to desiccation and can tolerate lower nutrient levels because they retain their needles for several years.

**Animal Life:** Animals are adapted to the vagaries of the climate; many mammals hibernate during the cold months, birds migrate to warmer areas in winter, and insects enter diapause, a condition of dormancy passed usually as a pupa. Reptiles, which are dependent on solar radiation for heat, are relatively uncommon. Mammals include deer, squirrels, wolves, wolverines, bobcats, foxes, bears, and mountain lions.

**Effects of Humans:** Logging has eliminated much temperate forest from heavily populated portions of Europe and North America. Little of the original forests remain, though much secondary forest, which has regrown since being logged, is present. With careful agricultural practices, soil richness can be conserved, and as a result, agriculture can flourish.

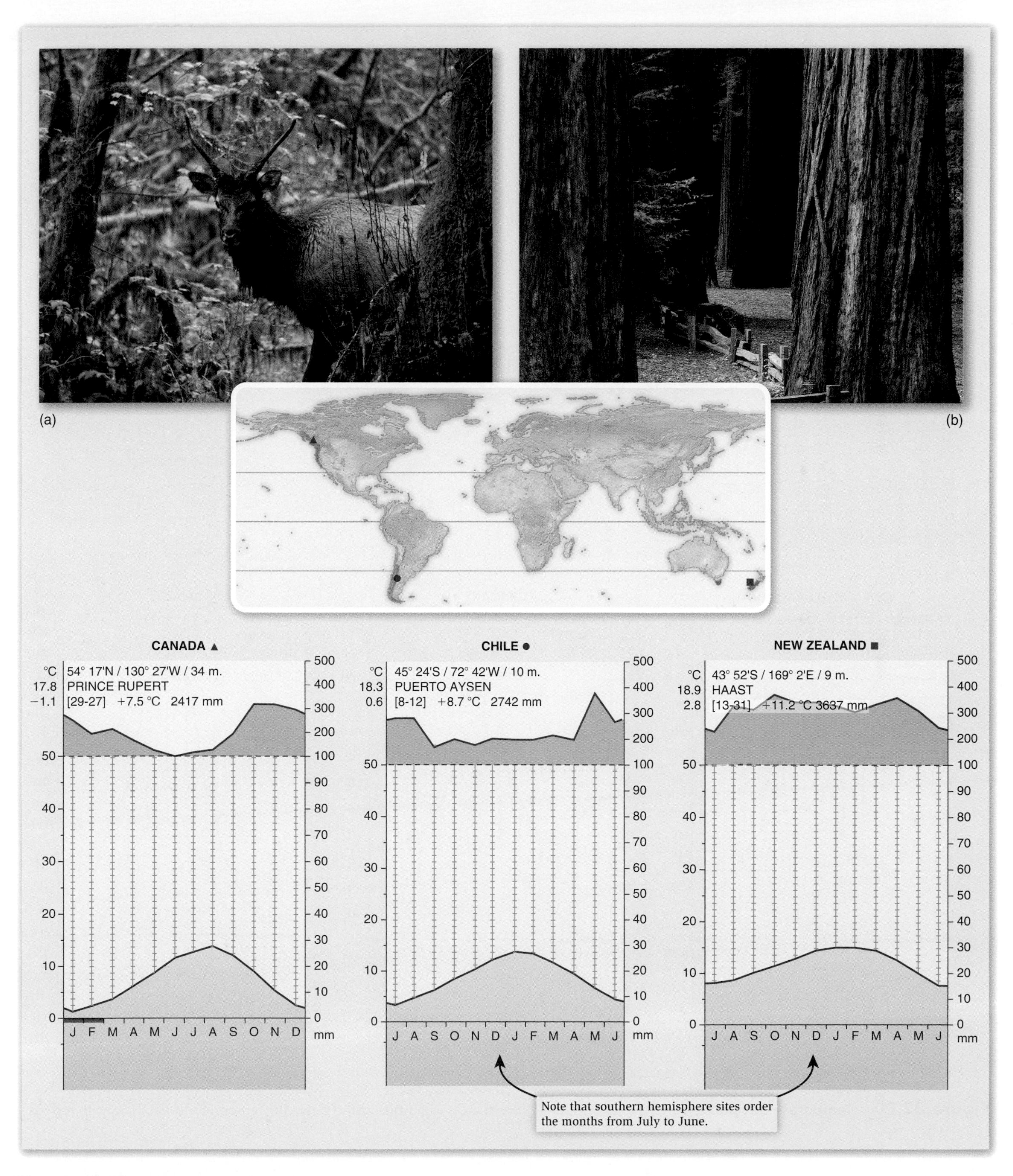


**Figure 22.19 Temperate rain forest.** **(a)** Hoh Rainforest, Olympic National Park, Washington, with elk. The cool temperatures and high rainfall allow lichens and ferns to thrive. **(b)** Redwood trees, Jebediah Smith Redwoods State Park, California.

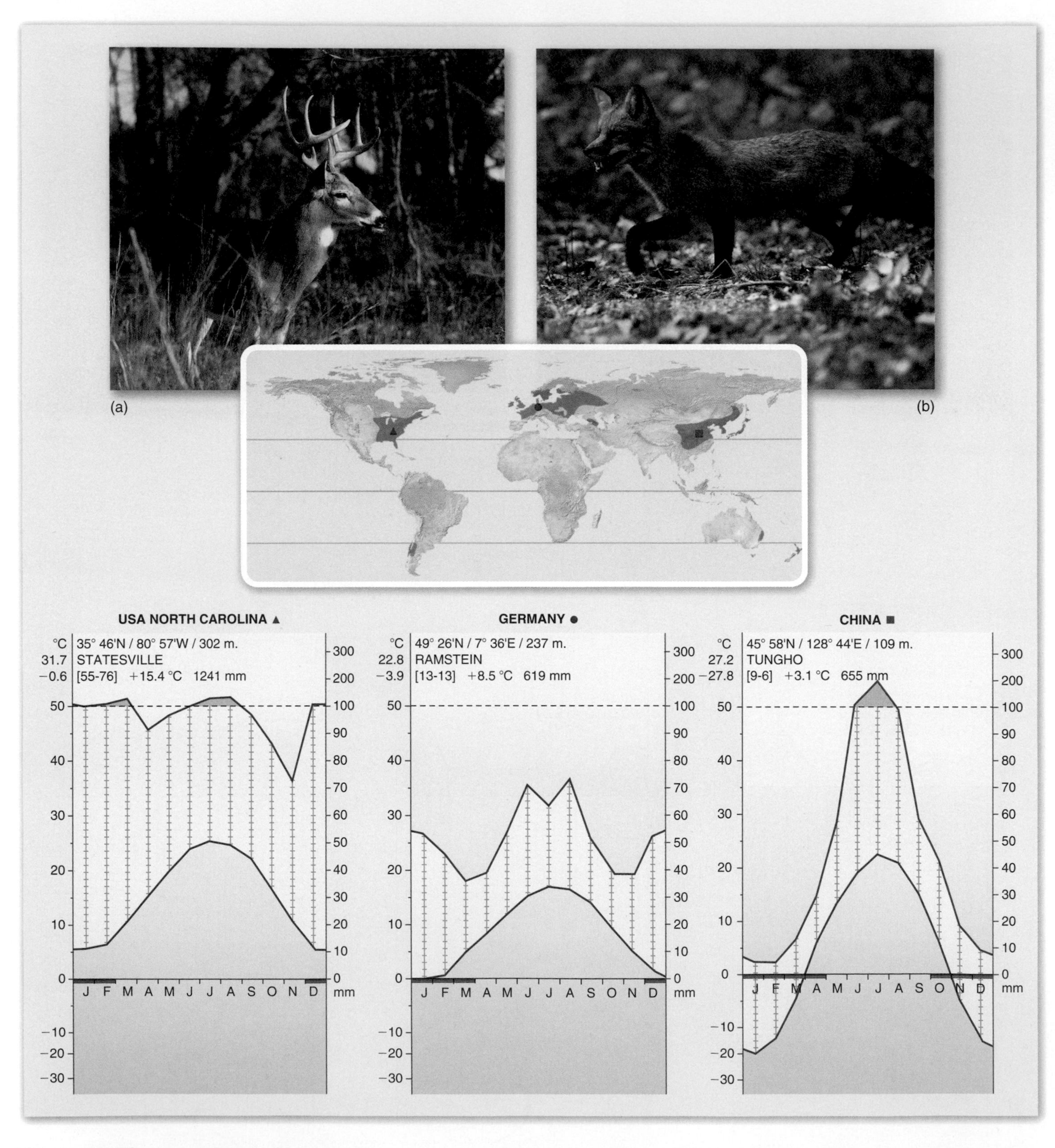


**Figure 22.20 Temperate deciduous forest.** **(a)** Temperate forest, U.S., with white-tailed deer. **(b)** Temperate forest, U.S., with red fox.

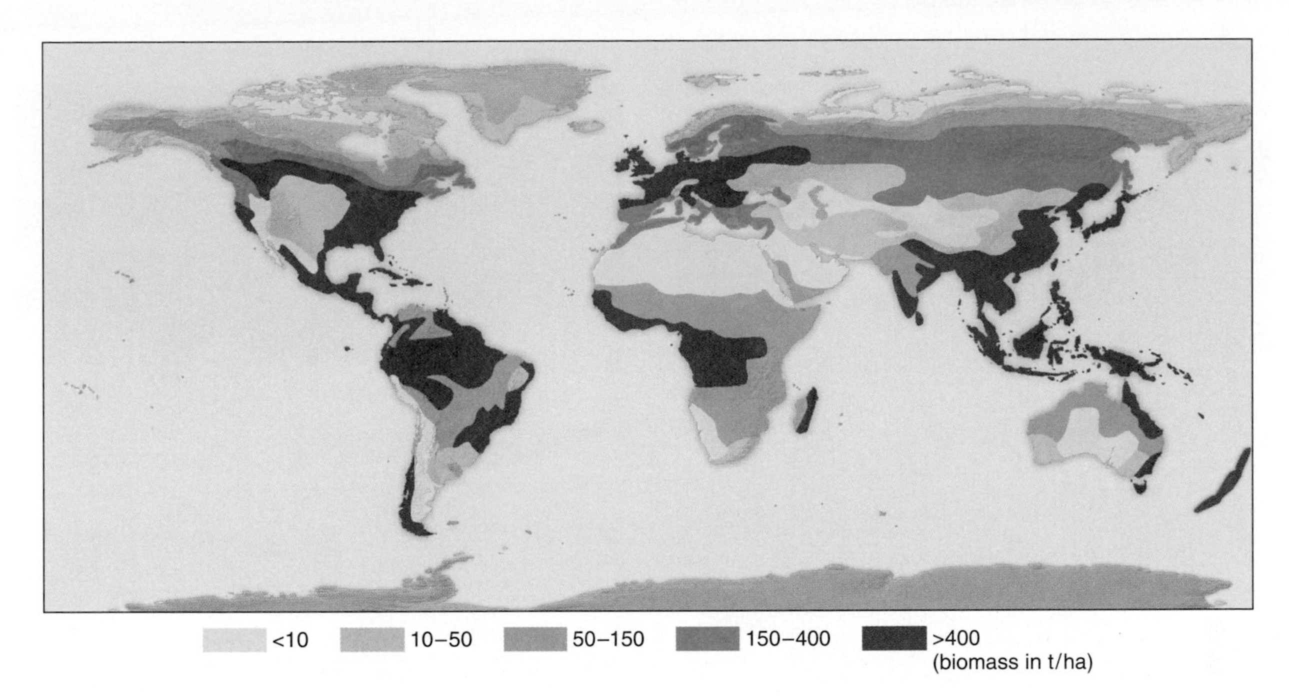


**Figure 22.21** **Distribution of phytomass or standing biomass of plant life on Earth.** Noteworthy is that the forested areas of the world have the largest biomass, and that the biomass of temperate deciduous forests is equal to or greater than that of tropical forests. (After Breckle, 2002.)

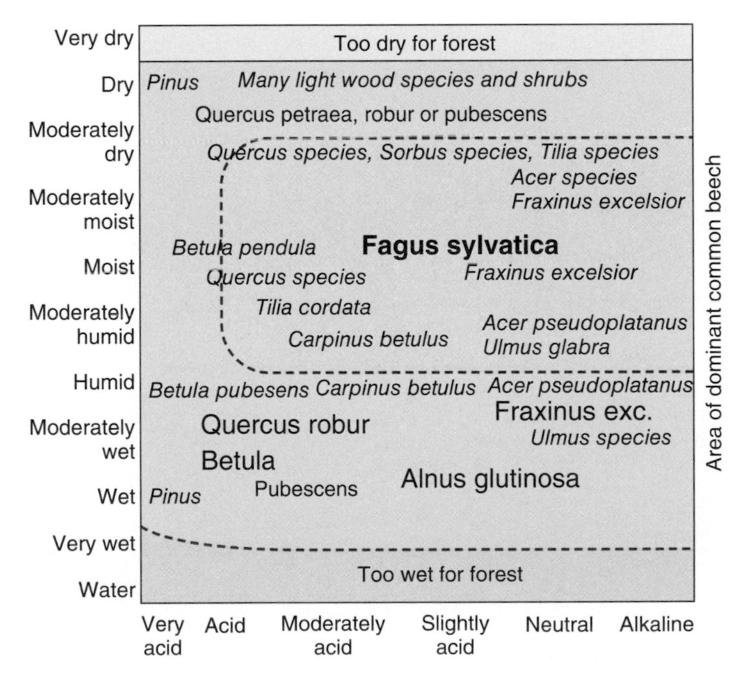


**Figure 22.22** **The species composition of central European forests.** Forest composition can be predicted by soil moisture and pH. The sizes of the letters correspond to the degree of coverage of the tree genera and species. The area occupied by the most dominant species, common beech, *Fagus sylvatica,* is shown as a dotted line. (After Ellenberg, 1988.) Key: Acer = maple, Alnus = alder, Betula = birch, Carpinus = hornbeam, Fraxinus = ash, Quercus = oak, Pinus = pine, Sorbush = mountain ash, Tilia = lime, Ulmus = elm.

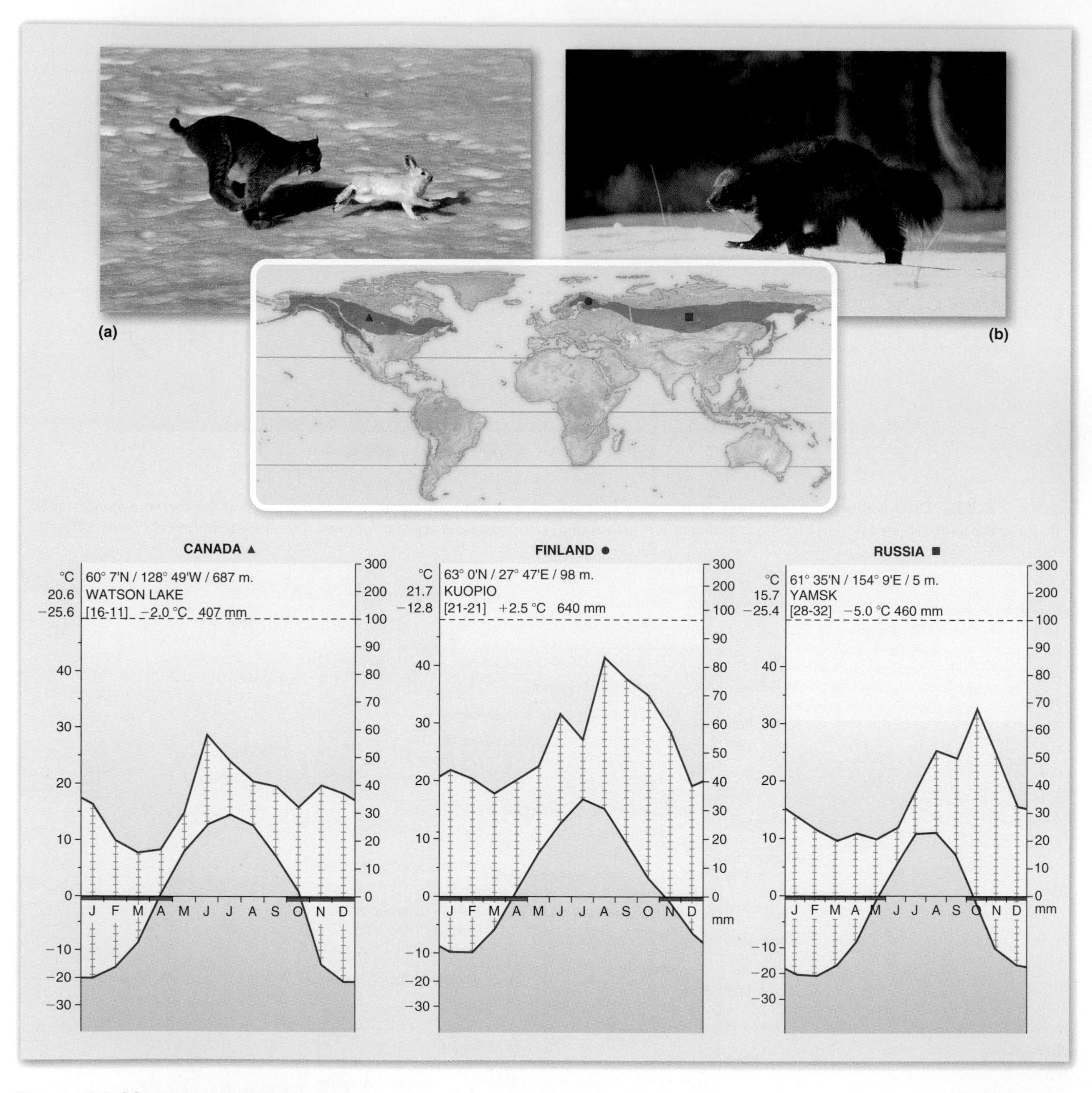


**Figure 22.23** **Temperate coniferous forest or taiga.** **(a)** Canadian taiga with snowshoe hare and lynx. Snow cover may be extensive in winter. **(b)** Wolverine walking on snow.

## Temperate Coniferous Forest or Taiga (Figure 22.23)

**Physical Environment:** Temperatures are very cold, usually averaging below 5°C and below freezing for long periods of time. Precipitation is generally between 40 and 70 cm, often occurring in the form of snow. Growing seasons are usually less than 100 days per year. Soils are poor because fallen needles decay so slowly in the cold temperatures that a layer of needles builds up. This layer acidifies the soil, further reducing the numbers of understory species.

**Location:** The biome of coniferous forests, known commonly by its Russian name, **taiga**, lies north of the temperate-zone forests and grasslands. Vast tracts of taiga exist in North America and Russia, and mountain taiga exists on mountainous areas. As much as 11% of the Earth's land area is covered with taiga. In the Southern Hemisphere, little land area occurs at latitudes at which one would expect extensive taiga to exist.

**Plant Life:** Most of the trees are evergreens or conifers with tough needles, hence its similarity to temperate rain forest. In this biome, spruces, firs, and pines generally dominate, and the number of tree species is relatively low. Some broad-leaved species such as alder, *Alnus* spp., and aspen, *Populus* spp., may occur in light gaps as the community recovers from disturbances. Many of the conifers have conical shapes that reduce bough breakage from heavy loads of snow. As in tropical forests, the understory is sparse because the dense year-round canopies prevent sunlight from penetrating.

**Animal Life:** Reptiles and amphibians are rare because of the low temperatures. Insects are strongly periodic but may often reach outbreak proportions in times of warm temperatures. Mammals that inhabit this biome, such as bears, lynxes, moose, wolverines, beavers, and squirrels, are heavily furred.

**Effects of Humans:** Humans have not extensively settled these areas, although the areas have been quite heavily logged. Trapping of animals for fur was once very common and still occurs today.

### Tropical Grassland or Savanna (Figure 22.24)

**Physical Environment:** These are hot, tropical areas, averaging 24–29°C with a low or seasonal rainfall of between 70 and 130 cm per year. There is often an extensive dry season of 4–5 months with little rainfall. This limited rainfall is insufficient to support large forest trees. On the other hand, the savanna of Colombia and Venezuela, called the Llanos, is too wet for trees to survive. Long periods of standing water prevent their establishment, except on patches of higher ground. A similar phenomenon occurs in Florida's Everglades, where trees exist on slightly elevated areas called hardwood hammocks. In all areas, soils are generally poor in nutrients because they have been leached out just as in tropical forests. Fire may be frequent.

**Location:** Extensive tropical grasslands called savannas occur beyond the tropical deciduous forests in Africa, South America (called Cerrado in Brazil), and northern Australia.

**Plant Life:** Wide expanses of grasses dominate savannas but occasional thorny trees, such as acacias and palms, may occur. Fire is prevalent in this biome, and most plants have well-developed root systems that enable them to resprout quickly. In fact, the frequency of fire and the degree of herbivory affects the resprouting of trees. A higher frequency of either phenomenon retards tree development. When fire and herbivory are controlled, tropical woodland may develop.

**Animal Life:** The world's greatest assemblages of large mammals occur in the savanna biome. In Africa, herds of antelope, zebras, and wildebeest are found, together with their associated predators: lion, cheetah, leopard, and hyena. Many herbivores move considerable distances to follow the rainfall and, with it, the resultant luxuriant vegetation. The extensive herbivory of large grazers, such as elephants, giraffes, and rhinoceros, limit tree establishment and help maintain savannas and prevent their development into forests. In the Llanos, capybaras, large semiaquatic rodents, are common. Termite mounds dot the landscape in some areas, and in South America anteaters are common. Burrowing animals, such as African mole rats, are also common.

**Effects of Humans:** Conversion of this biome to agricultural land is rampant, especially in Africa. Overstocking of land for domestic animals can greatly reduce grass coverage through overgrazing, turning the area more desert-like in a process known as **desertification.** Poaching of large animals is also a problem.

### Temperate Grassland or Prairie (Figure 22.25)

**Physical Environment:** Temperatures in the winter often fall below –10°C, while summers may be very hot, approaching 30°C, and summer droughts are not uncommon. Annual rainfall is generally between 25 and 100 cm, too low to support a forest but higher than that in deserts. Lightning-initiated fire is also frequent and limits the development of trees. The length of the growing season is limited by cold in high latitudes to about 120 days, but in lower latitudes it may approach 300 days. In addition to the limiting amounts of rain, fire and grazing animals may also prevent the establishment of trees in the temperate grasslands. Prairie soils are among the richest in the world, having a nutrient content much greater than that of a typical forest soil. Drier, browner soils of more arid grasslands are less fertile.

**Location:** Temperate grasslands include the prairies of North America, the steppes of Russia, the pampas of Argentina, and the veldt of South Africa. Smaller patches exist in California, southwest Australia, South Africa, and the Mediterranean.

**Plant Life:** From east to west in North America and from north to south in Asia, grasslands show differentiation along moisture gradients. In Illinois, with an annual rainfall of 80 cm, tall prairie grasses such as Indian grass, *Sorghastrum nutans,* big bluestem, *Andropogon gerardii,* and switchgrass, *Panicum virgatum,* grow to about 2 m high. Along the eastern base of the Rockies, 1,300 km to the west, where rainfall is only 40 cm, prairie grasses such as buffalo grass, *Buchloe dactyloides,* and blue grama, *Bouteloua oracilis,* rarely exceed 0.5 m in height. Similar gradients occur in South Africa and Argentina. Where temperatures rarely fall below freezing and most of the rain falls in the winter, chaparral, a fire-adapted community featuring grasses, shrubs, and small trees, occurs. Such conditions are most commonly seen at around 30° latitude, where cool ocean waters moderate the climate.

**Animal Life:** Where the grasslands remain, large mammals are the most prominent members of the fauna: buffalo and pronghorn antelope in North America, wild horses and saiga antelope in Eurasia, and large kangaroos in Australia. Burrowing animals such as North American gophers are also common.

**Effects of Humans:** Worldwide, most prairies have been converted to agricultural cropland, and original grassland habitats are among the rarest biomes in the world. At one time, 35% of the United States consisted of prairie. After the steel plow was invented by John Deere in the 1830s, the rate of conversion of prairie to cropland accelerated enormously. In states such as Iowa and Illinois, less than 1% of the original prairie remains. Hunting has diminished populations of once-common grazers. In Russia the tarpan, *Equus gmelini,*

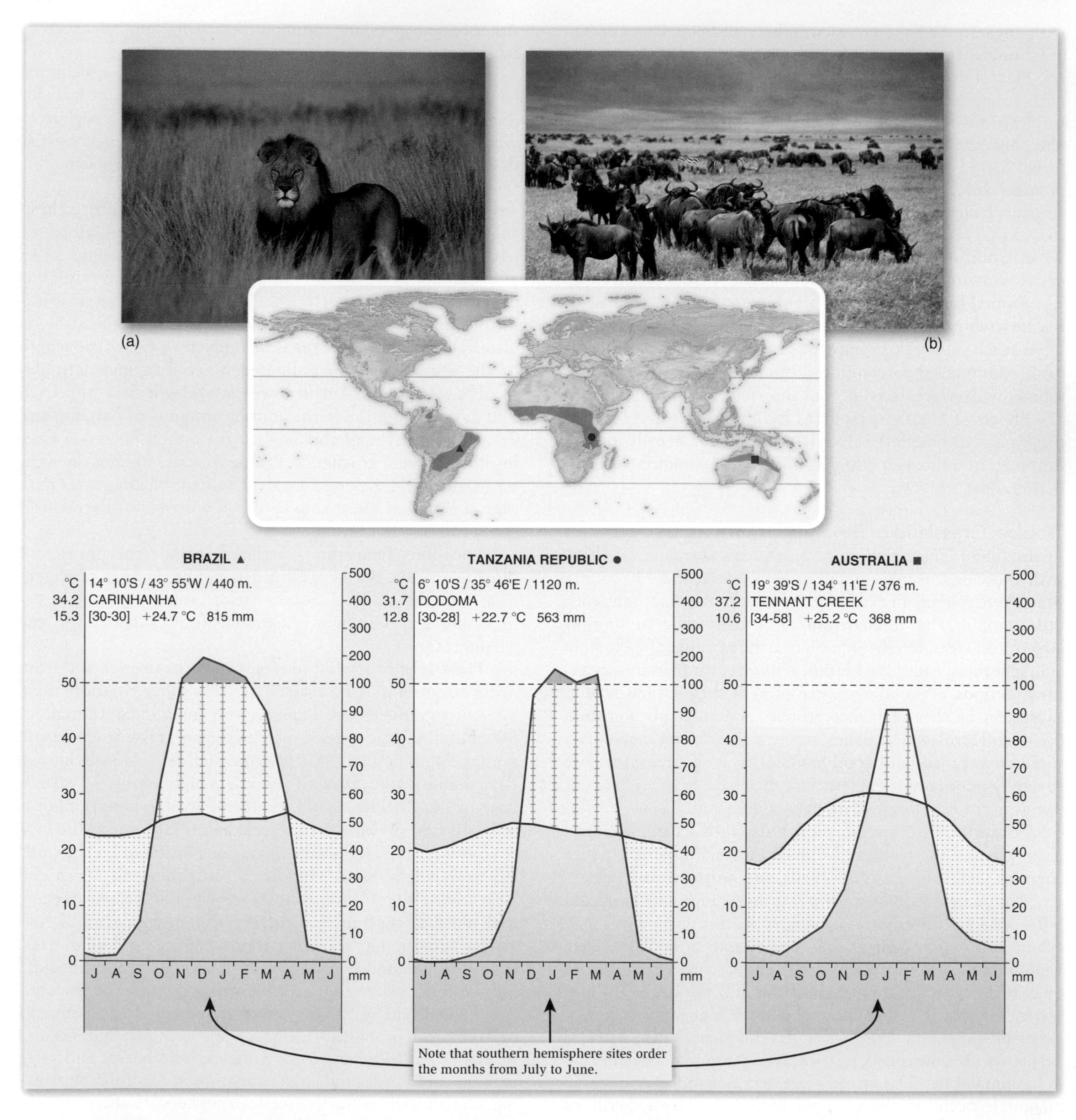


**Figure 22.24** **Tropical grassland or savanna.** **(a)** African savannah with lion. **(b)** Ngorongoro Crater with wildebeest and zebra.

was hunted to extinction by the 1860s. In North America, the area occupied by buffalo has shrunk dramatically as the herds were hunted down from 65 to 75 million in the 18th century to 1,150 by 1899 (**Figure 22.26**). Since then, buffalo numbers have rebounded to about 350,000.

Hot Desert (**Figure 22.27**) **Physical Environment:** Temperatures are variable from below freezing at night to as much as 50°C in the day. Rainfall is less than 30 cm per year, much of which occurs in a short period of winter or summer rainfall. The Sonoran Desert is unusual because it receives rainfall

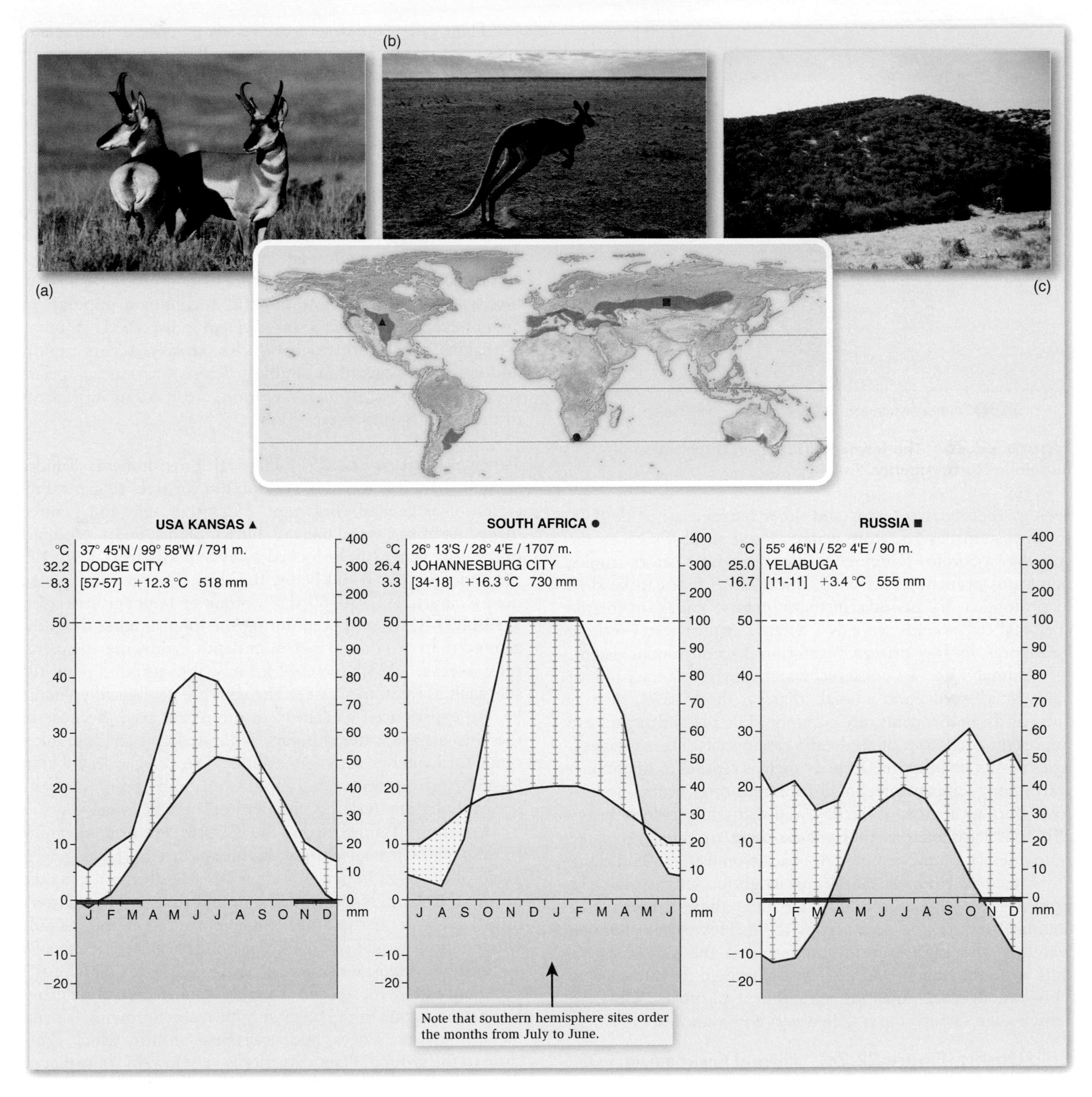


**Figure 22.25 Temperate grassland prairie.** **(a)** Prairie with pronghorn antelope, Wyoming. **(b)** Australian outback with red kangaroo. **(c)** This chaparral habitat in California has numerous small bushes and occasional trees as well as extensive grasslands.

in both these periods and is therefore one of the greenest deserts in the world. Soils are poor and nutrient-limited.

**Location:** Hot deserts are found around latitudes of 20–30° north and south of the equator, and occupy about 20% of the Earth's land surface. Prominent deserts include the Sahara of North Africa; the Kalahari of southern Africa; the Atacama of Chile; the Sonoran, Mojave, and Chihuahuan of northern Mexico and the southwest U.S.; and the Simpson of Australia.

**Plant Life:** Three forms of plant life are adapted to moister deserts: annuals, succulents, and desert shrubs. Many species cannot tolerate freezing temperatures. Annuals circumvent drought by growing only when there is rain. Succulents,

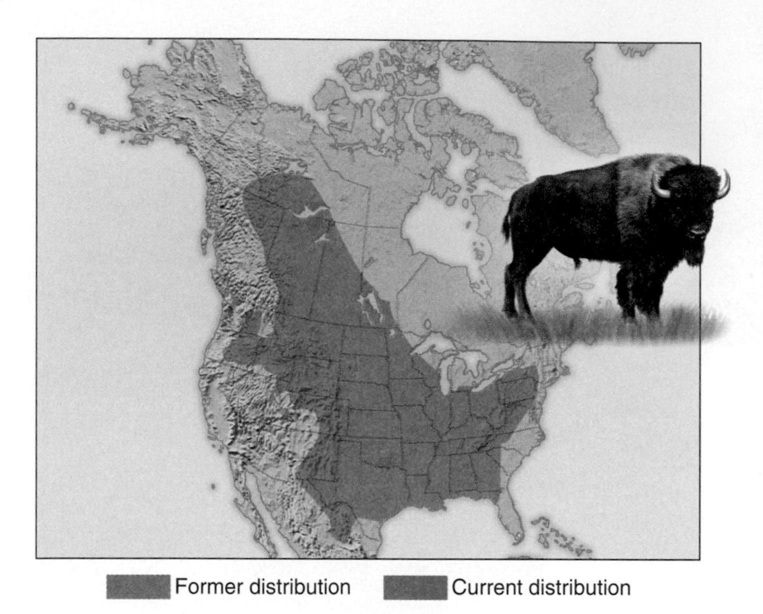


**Figure 22.26** The former and present distribution of buffalo in North America.

such as the saguaro cactus and other barrel cacti of the U.S. southwestern deserts, store water. Desert shrubs, such as the spraylike ocotillo, *Fonquieria splendens,* have short trunks, numerous branches, and small, thick leaves that can be shed in prolonged dry periods. In many plants, spines or volatile chemical compounds serve as a defense against water-seeking herbivores. In drier deserts, vegetation is scant to nonexistent.

**Animal Life:** To conserve water, desert plants produce many small seeds, and animals that eat those seeds, such as ants, birds, and rodents, are common. Lizards and snakes are important predators of seed-eating mammals. High temperatures permit ectothermic animals such as reptiles to maintain a warm body temperature. Small mammals often remain in burrows during the heat of the day, and many insects have heavy chitinous exoskeletons that enable them to conserve water. Many desert animals employ dry excretion using uric acid.

**Effects of Humans:** Ambitious irrigation schemes and the prolific use of underground water have allowed humans to colonize deserts and grow crops there. However, continued watering and high evaporation result in the deposition of salts near the soil surface, a process known as salinization. This can degrade soils for agricultural purposes. Off-road vehicles can disturb the fragile desert communities.

Cold Desert (**Figure 22.28**) **Physical Environment:** In the daytime, temperatures can be high in the summer, 21–26°C, but average around freezing, –2–4°C, in the winter. In the Gobi Desert in Mongolia, winter temperatures are bitterly cold, sometimes below –20°C. Precipitation is less than 25 cm a year, often in the form of snow. Rainfall usually comes in the spring.

**Location:** Cold deserts are found in dry regions at middle to high latitudes, especially in the interiors of continents and in the rain shadows of mountains. Examples of cold deserts are North America's Great Basin Desert, eastern Argentina's Patagonian Desert, and Central Asia's Gobi Desert.

**Plant Life:** Cold deserts are relatively poor in numbers of species of plants. Most plants are small in stature, being only 15–120 cm tall. Succulents are absent, as these are susceptible to freezing. Many species are deciduous and spiny. The Great Basin Desert in Nevada, Utah, and bordering states is a cold desert dominated by sagebrush, *Artemeisa* spp.

**Animal Life:** As in hot deserts, large numbers of plants produce small seeds on which ants, birds, and rodents feed. Many species live in burrows to escape cold and to keep warm. In the Great Basin Desert, pocket mice, jackrabbits, kit fox, and coyote are common.

**Effects of Humans:** Agriculture is hampered because of low temperatures and low rainfall, and human populations are not extensive. If the top layer of soil is disturbed by human intrusions, such as off-road vehicles, erosion occurs rapidly and even less vegetation is able to exist. Introduced species such as cheatgrass, *Bromus tectorum,* are invading large areas of the Great Basin Desert.

Tundra (**Figure 22.29**) **Physical Environment:** Tundra occurs where it is too cold and too dry for trees to grow. Precipitation is generally less than 25 cm per year and is often locked up as snow and unavailable for plants. For a large part of the year, water can be locked away in **permafrost**, a layer of permanently frozen soil below 0.5–1 m. The growing season here is short, only 50–60 days. Summer temperatures average only 3–12°C, and even during the long summer days, the ground thaws to only 0.5–1 m in depth. Midwinter temperatures average –32°C, too cold for any aboveground plant tissue, such as trees, to survive. The soils are geologically young, having experienced glaciation in the recent past. Because of the permafrost, water drainage is limited and shallow lakes often form in the summer months. The oxygen-less waterlogged soil prevents decomposition, and peat (partially decayed plant material) accumulations may be large.

**Location:** Tundra, from the Finnish *tunturia,* meaning treeless plain, exists mainly in the Northern Hemisphere, north of the taiga in places such as Alaska, northern Canada, northern Iceland, Norway, Finland, and Russia. There is very little land area in the Southern Hemisphere at the latitude where tundra would occur, except on the Antarctic Peninsula and neighboring islands such as South Georgia Island.

**Plant Life:** With so little available water, trees cannot grow. Vegetation occurs in the form of fragile, slow-growing lichens, mosses, grasses, sedges, and occasional shrubs, which grow close to the ground. Plant diversity is very low. In some places, desert conditions prevail because so little moisture falls.

**Animal Life:** Animals of the arctic tundra have adapted to the cold with insulating features such as thick fur, layers of fat, and densely packed feathers. Many birds, especially shorebirds and waterfowl, migrate. The fauna is much richer in summer than in winter. Many insects spend the winter at immature stages of growth, which are more resistant to cold than the adult forms. The larger animals include such herbivores as musk oxen and caribou in North America, and in Europe and Asia where they are called reindeer, as well as

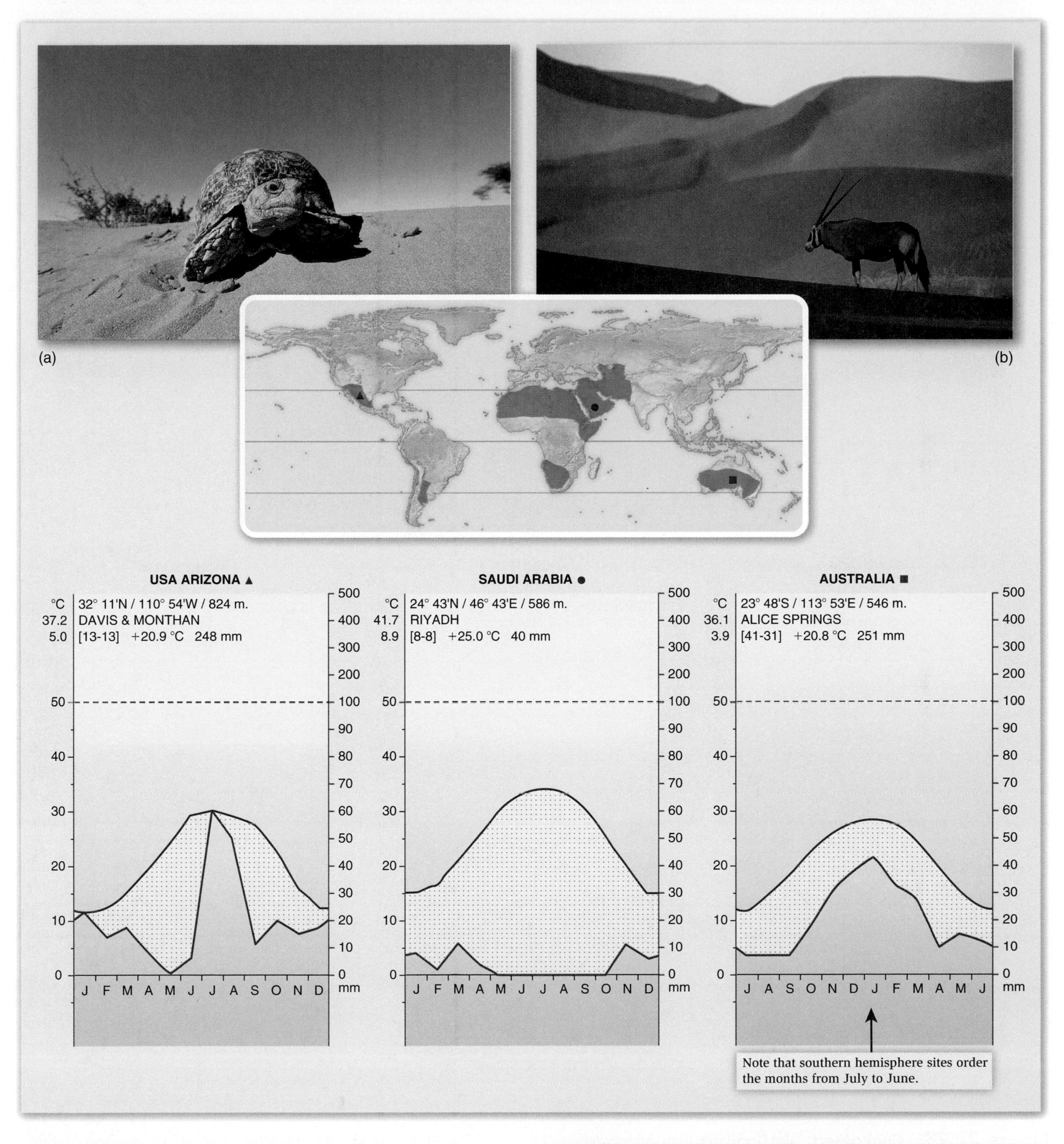


**Figure 22.27** **Hot desert.** **(a)** Kalahari Desert, southern Africa, with leopard tortoise. **(b)** Namib Desert, Africa, with oryx antelope.

**ECOLOGICAL INQUIRY**

Why are reptiles such as lizards and snakes common in the hot desert?

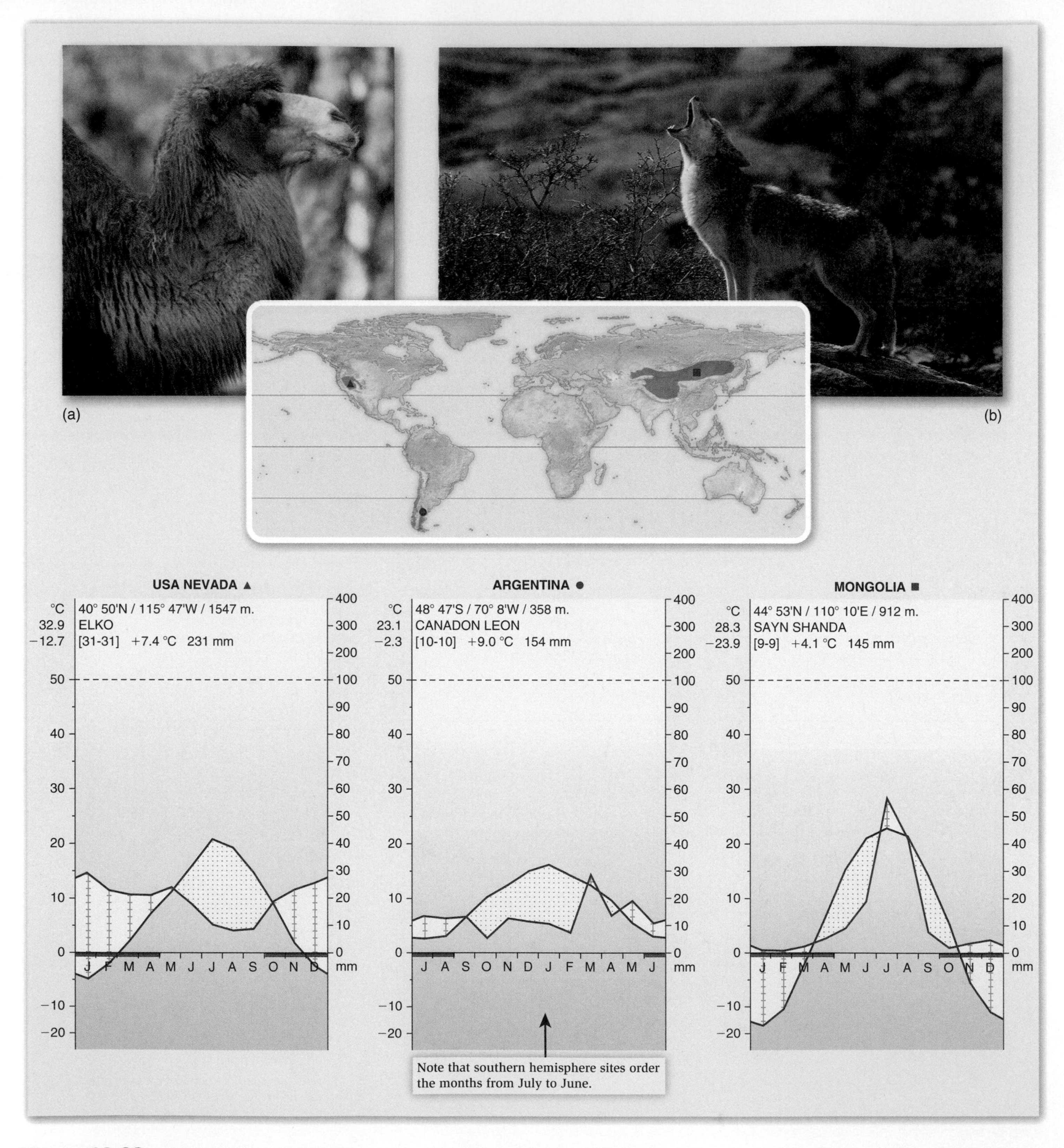


**Figure 22.28** **Cold desert.** **(a)** Bactrian camel, an inhabitant of the Gobi Desert, Mongolia. **(b)** Great Basin Desert with howling coyote.

the smaller hares and lemmings. Common predators include arctic fox, wolves, snowy owls, and polar bears.

**Effects of Humans:** Though this area is sparsely populated, extraction of oil and minerals has the potential to significantly impact this biome. Because vegetation grows very slowly, ecosystem recovery from damage such as an oil spill would be very slow. Global warming threatens to severely decrease the extent of this biome.

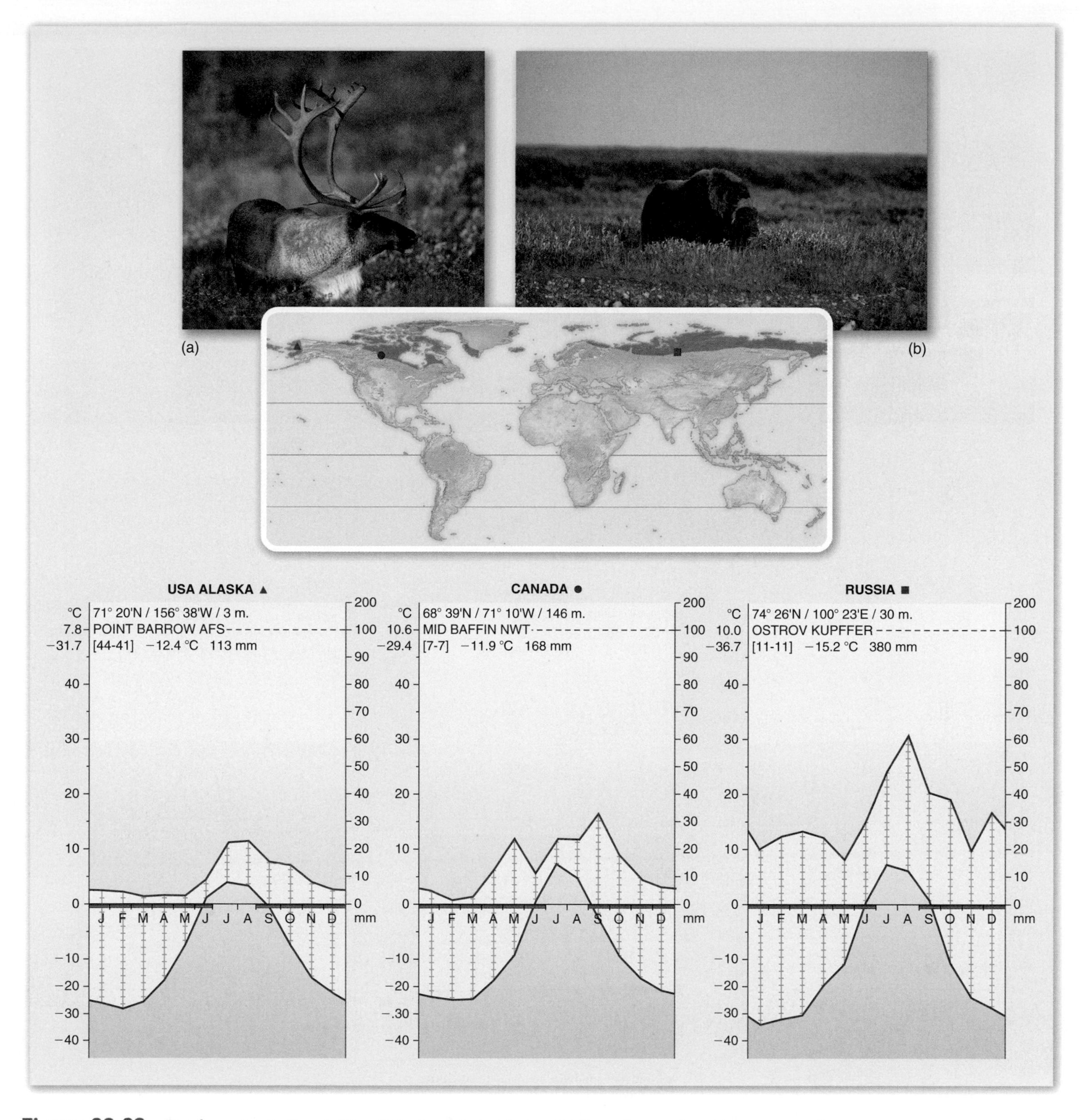


**Figure 22.29** **Tundra.** **(a)** Caribou in Denali National Park, Alaska. **(b)** Canadian tundra with musk ox bull.

Mountain Ranges (**Figure 22.30**) **Physical Environment:** Mountain ranges must be viewed differently than other biomes. On mountains, temperature decreases with increasing elevation through adiabatic cooling, as discussed previously. Precipitation and temperature may change dramatically, depending on elevation and whether the mountainside is on the windward or leeward side.

**Location:** Mountain ranges exist in many areas of the world, but among the largest are the Himalayas in Asia, the Rockies in North America, and the Andes in South America (**Figure 22.31**). Fully one-quarter of the world's landscape is mountainous. Most mountain ranges in the Americas and Australia run north–south, whereas those in Asia and Europe run east–west.

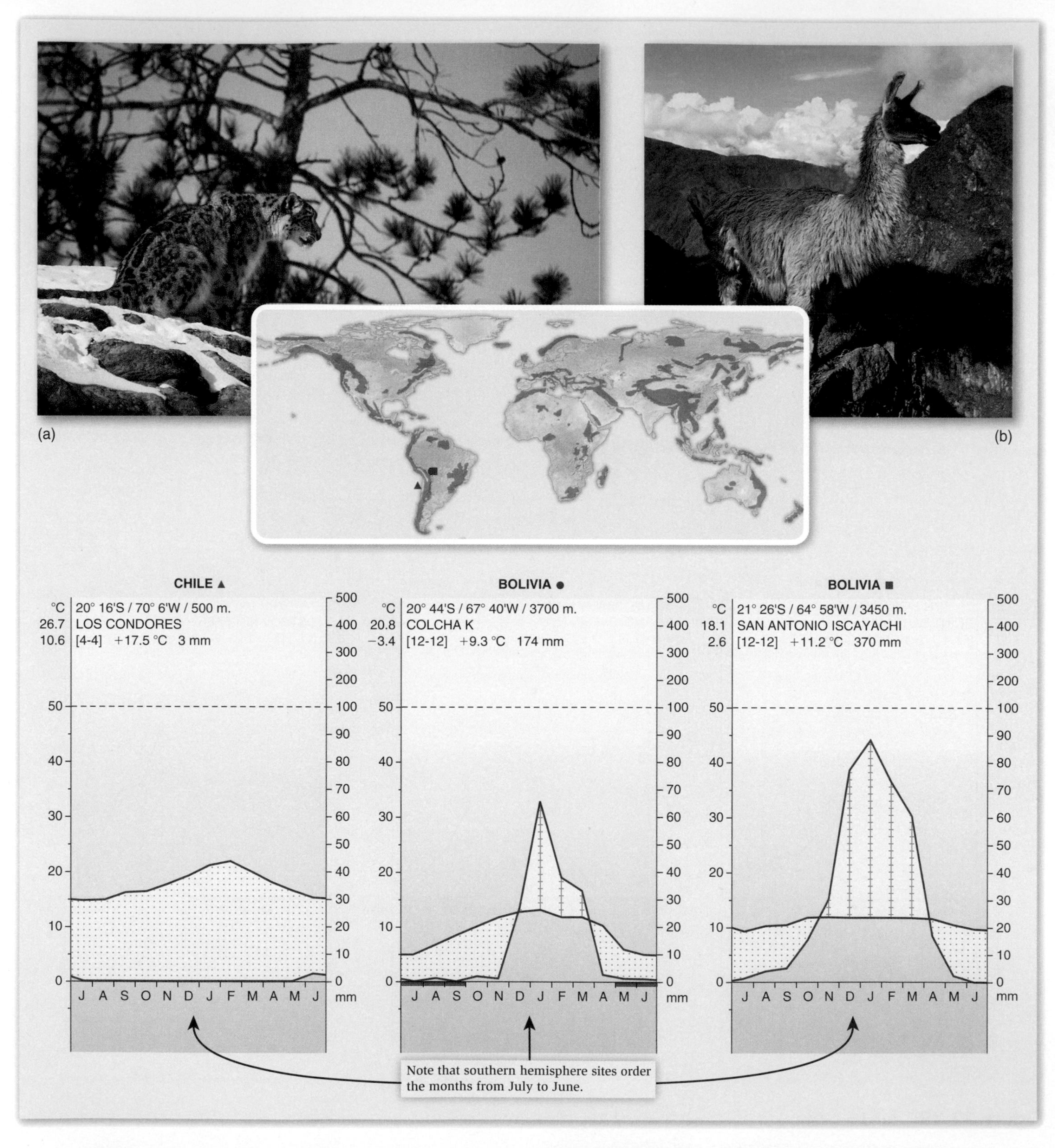


**Figure 22.30** **Mountain ranges.** **(a)** Snow leopard, Central Asia. **(b)** Llama in the Andes, Machu Picchu.

## ECOLOGICAL INQUIRY

Why are the precipitation regimes shown in Chile and Bolivia, areas of similar latitude, so different?

**Plant Life:** Biome type may change from temperate forest through taiga and into tundra on an elevation gradient in the Rocky Mountains, and even from tropical forest to tundra on the highest peaks of the Andes in tropical South America (**Figure 22.32**). In North America, many of the tree species are similar to those of the more northerly taiga, and include firs, pines, hemlock, and other species. In tropical regions, daylight varies little from the 12 hours per day throughout the year. Instead of an intense period of productivity, vegetation in the tropical alpine tundra exhibits slow but steady rates of photosynthesis and growth all year.

**Animal Life:** The animals of this biome are as varied as the number of habitats they contain. Generally, more species of plants and animals are found at lower elevations than at higher ones. At higher elevations in North America, animals such as bighorn sheep and mountain goats have to be very sure-footed to climb the craggy slopes and have skid-proof pads on their hooves. Other large herbivores include llamas in the Andes and

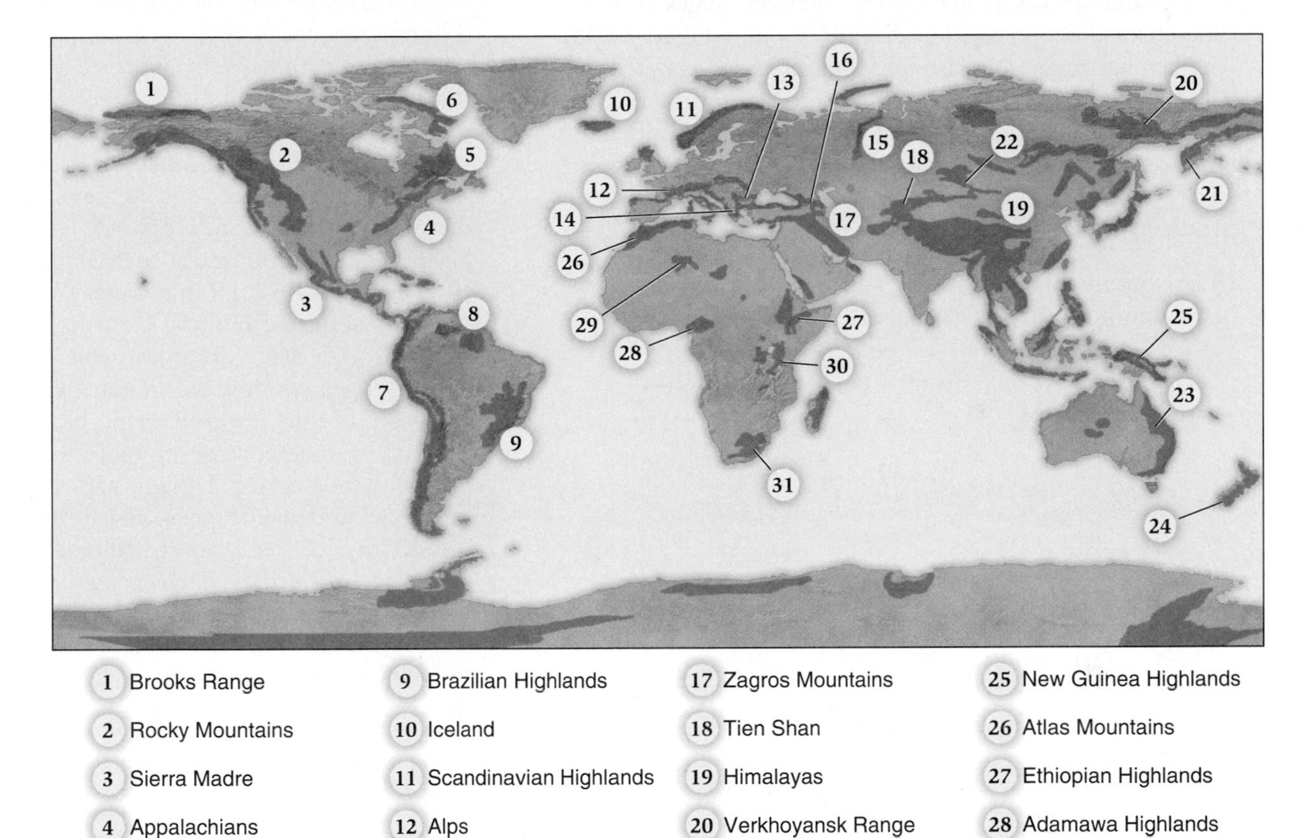


**Figure 22.31** **Location of the world's major mountain ranges.**

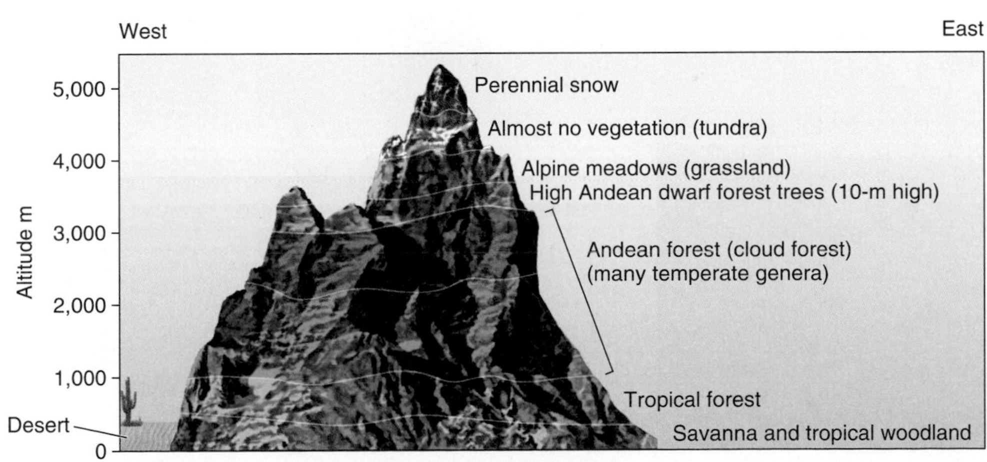


**Figure 22.32** **Diagrammatic representation of the Andes vegetation.** Vegetation is shown along a west–east cross section from the Atacama Desert to the savannas and woodlands of the eastern foothills.

yaks in the Himalayas. Despite the often-strong winds, birds of prey, such as eagles, are frequent predators of the furry rodents found at higher elevations, including guinea pigs and marmots.

**Effects of Humans:** Logging and agriculture at lower elevations can cause habitat degradation. Because of the steep slopes, mountain soils are often well drained, thin, and especially susceptible to erosion. In North America, climate change from regional warming is thought to have increased the mortality rates of all tree species in all western mountain ranges, from the Rockies through the Sierra Nevada and Cascades (**Figure 22.33**). A lengthening of summer drought and increased water deficit are thought to be to blame for this increased mortality. Such anthropogenically induced habitat changes are radically altering the appearance of many natural biomes (see **Global Insight**).

### Check Your Understanding

**22.2** How can global warming cause more extinctions in mountain biomes than in other biomes?

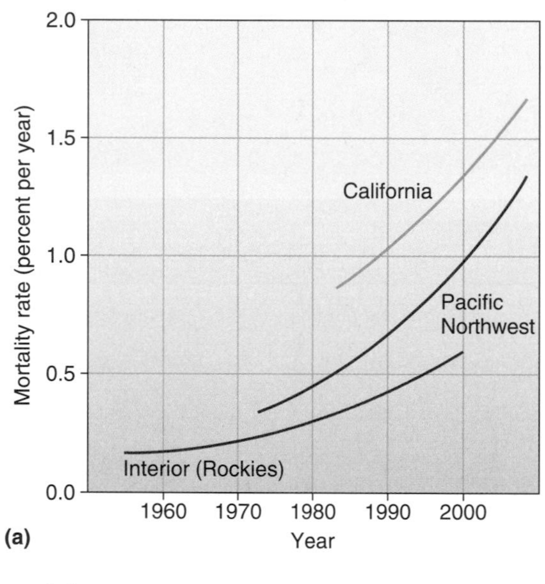


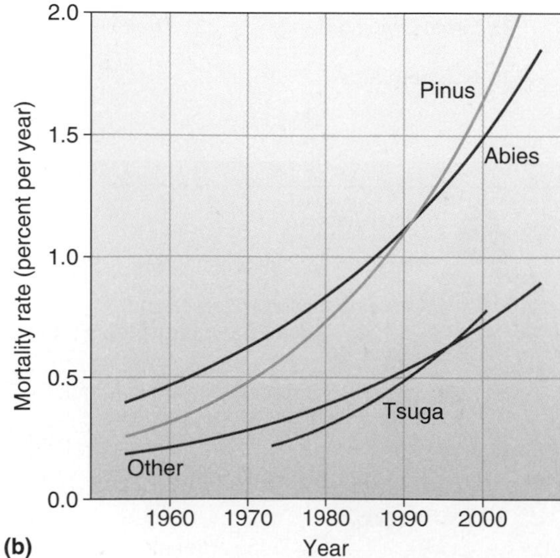


**Figure 22.33 Tree mortality rates.** **(a)** For different regions of the western U.S.; **(b)** for different tree genera. (After van Mantgem et al., 2009.)

## Global Insight

### Anthropogenic Influences Change the Appearance of Many Biomes

Humans are fundamentally changing the appearance of traditional biomes on Earth as they plant crops, cut down old-growth forests and install tree plantations, drain swamps, create irrigated rice fields, and urbanize areas. Some recent studies have suggested that human-dominated biomes now cover more of the Earth's land surface than natural biomes. Because of this, there have been attempts to construct maps of anthropogenic biomes of the world. Early attempts used only a few anthropogenic biome classes such as urban, cropland, pasture, and one or two cropland/natural mosaics such as mixed cropland/forests. Erle Ellis and Navin Ramankutty (2008) described a range of anthropogenic biomes, from lightly impacted pastureland through dense urban settlements. They created a new anthropogenic biome world map (**Figure 22.B**). According to their scheme, anthropogenic biomes dominate the terrestrial biosphere, covering more than 75% of the Earth's ice-free land surface. Though half the people on Earth live in dense settlements, these urban areas cover just 7% of the Earth's ice-free land. Rangeland biomes were the most extensive, covering nearly a third of the global ice-free land, followed by cropland biomes at 20%, and forested biomes at about 18%. Wild lands, without any evidence of human occupation, cover just 22%, generally in the least productive areas of the world, such as deserts.

| Group | Biome | Description |
|---|---|---|
| *Dense settlements* | | *Dense settlements with substantial urban area* |
| | 11 Urban | Dense built environments with very high populations |
| | 12 Dense settlements | Dense mix of rural and urban populations, including both suburbs and villages |
| *Villages* | | *Dense agricultural settlements* |
| | 21 Rice villages | Villages dominated by paddy rice |
| | 22 Irrigated villages | Villages dominated by irrigated crops |
| | 23 Cropped and pastoral villages | Villages with a mix of crops and pasture |
| | 24 Pastoral villages | Villages dominated by rangeland |
| | 25 Rainfed villages | Villages dominated by rainfed agriculture |
| | 26 Rainfed mosaic villages | Villages with a mix of trees and crops |
| *Croplands* | | *Annual crops mixed with other land uses and land covers* |
| | 31 Residential irrigated cropland | Irrigated cropland with substantial human populations |
| | 32 Residential rainfed mosaic | Mix of trees and rainfed cropland with substantial human populations |
| | 33 Populated irrigated cropland | Irrigated cropland with minor human populations |
| | 34 Populated rainfed cropland | Rainfed cropland with minor human populations |
| | 35 Remote croplands | Cropland with inconsequential human populations |
| *Rangeland* | | *Livestock grazing; minimal crops and forests* |
| | 41 Residential rangelands | Rangelands with substantial human populations |
| | 42 Populated rangelands | Rangelands with minor human populations |
| | 43 Remote rangelands | Rangelands with inconsequential human populations |
| *Forested* | | *Forests with human populations and agriculture* |
| | 51 Populated forests | Forests with minor human populations |
| | 52 Remote forests | Forests with inconsequential human populations |
| *Wildlands* | | *Land without human populations or agriculture* |
| | 61 Wild forests | High tree cover, mostly boreal and tropical forests |
| | 62 Sparse trees | Low tree cover, mostly cold and arid lands |
| | 63 Barren | No tree cover, mostly deserts and frozen land |

*Numbers correspond to those used in Figure 22.B(ii).

*Source:* After Ellis and Ramankutty, 2008.

**(i)** Biome types.

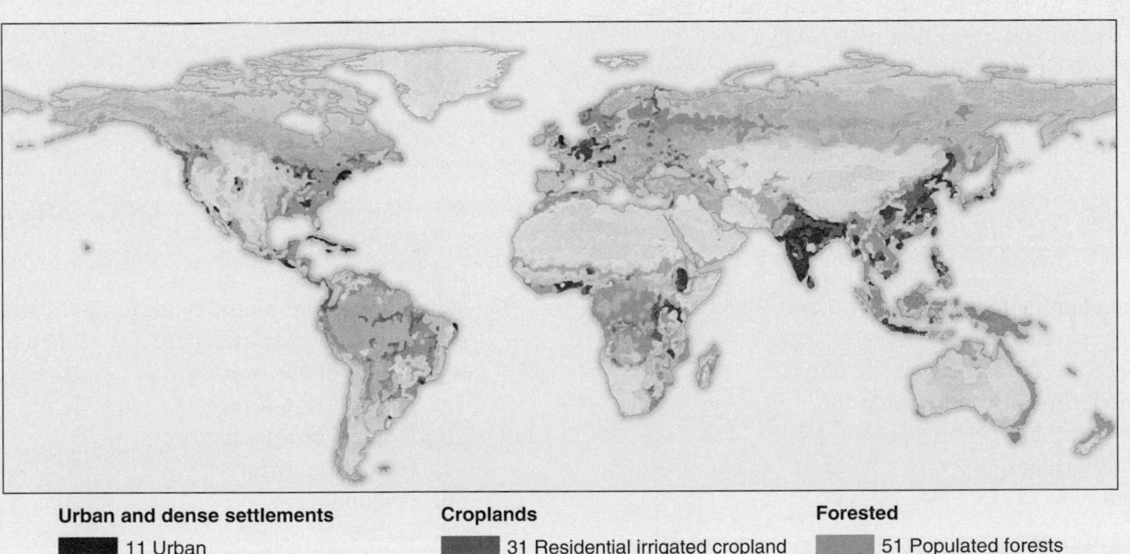

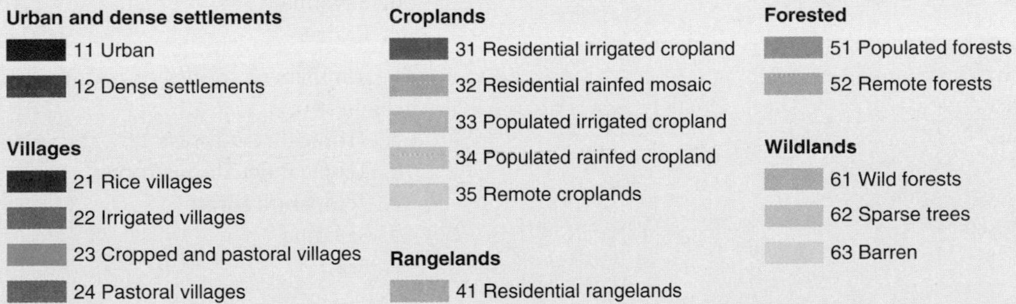


**Figure 22.B Anthropogenic biomes of the world.** **(i)** Biome types. **(ii)** Worldwide distributions. (After Ellis and Ramankutty, 2008.)

**Mosaic*: > 25% tree cover mixed with > 25% pasture and/or cropland

**(ii)** Worldwide distributions.

## SUMMARY

**22.1 Variation in Solar Radiation Determines the Climate in Different Areas of the World**

- Global temperature differentials are caused by variations in incoming solar radiation and patterns of atmospheric circulation (Figures 22.1, 22.2).
- Global models of atmospheric circulation and precipitation are driven by temperature variations (Figures 22.3–22.6).
- The tilt of the Earth's axis causes seasonality, including variations in temperature and rainfall (Figures 22.7–22.9).
- Elevation and the proximity between a landmass and large bodies of water can also affect climate (Figures 22.10–22.12).

**22.2 Terrestrial Biome Types Are Determined by Climate Patterns**

- Temperature and precipitation patterns have a large effect on biomes, the major types of habitats characterized by distinctive plant and animal life (Figures 22.13, 22.14).
- The numbers of species of most taxa are greatest in the Tropics (Figure 22.15).
- Walter diagrams provide climate characteristics of different terrestrial biomes (Figure 22.16).
- Tropical rain forests occur in hot areas with heavy rainfall (Figure 22.17).
- Tropical deforestation threatens many rainforest species with extinction (Figure 22.A).
- Tropical deciduous forest occurs in hot areas with a distinct dry season (Figure 22.18).
- Temperate rain forests have cool temperatures but abundant rainfall and occur mainly in the North American Pacific Northwest (Figure 22.19).
- In temperate deciduous forests, temperatures fall below freezing but rainfall is still substantial (Figures 22.20–22.22).
- Taiga forest occurs in cold temperatures and abundant snowfall, where conical tree shapes help reduce bough breakage (Figure 22.23).
- Savannas are tropical grasslands that are found in hot areas where rainfall is insufficient to permit tree growth (Figure 22.24).
- Prairies are temperate grasslands where summers are hot but winter temperatures often fall below $-10°$ C. Conversion to agriculture has severely reduced the extent of this biome (Figures 22.25, 22.26).
- Hot deserts are areas with extremely low precipitation of less than 30 cm a year (Figure 22.27).
- Cold deserts occur at higher latitudes, often in the interiors of continents (Figure 22.28).
- Tundra occurs where it is too cold and dry for trees to grow (Figure 22.29).
- On mountain ranges, temperature and precipitation change with altitude, as does biome type (Figures 22.30–22.32).
- Climatic changes are increasing mortality rates of species adapted to specific mountain climate zones (Figure 22.33).
- Humans are also changing the distribution of traditional biomes by deforestation, agriculture, and urbanization (Figure 22.B).

## TEST YOURSELF

1. What is the main factor that determines the circulation of the atmospheric air?
   a. Temperature differences of the Earth
   b. The intertropical convergence zone
   c. Ocean currents
   d. Mountain ridges
   e. Rainfall
2. What characteristics are commonly used to identify the biomes of the Earth?
   a. Temperature
   b. Precipitation
   c. Vegetation
   d. a and b only
   e. All of the above
3. Which terrestrial biome type is present where rainfall is substantial, temperatures are hot year round, but there is a distinct dry season?
   a. Tropical rain forest
   b. Tropical deciduous forest
   c. Temperate forest
   d. Savanna
   e. Prairie
4. The numbers of species of mammals, birds, and amphibians is highest in:
   a. Tropical rain forest
   b. Tropical deciduous forest
   c. Temperate forest
   d. Savanna
   e. Prairie

5. The biomass of tropical forests is equaled only by:
   a. Taiga
   b. Temperate deciduous forest
   c. Mangrove forests
   d. Kelp forests
   e. The open ocean
6. The World's major subsidence zones occur at:
   a. 0° and 60°
   b. 30° and the poles
   c. 0° and the poles
   d. 30° and 60°
   e. 0° and 30°
7. The northeast trade winds blow from:
   a. 60°N to 30°N
   b. 0°S to 30°S
   c. 30°N to 0°
   d. 30°S to 60°N
   e. The North Pole to 60°N
8. The equinoxes occur on:
   a. Dec 21 and Sept 22
   b. Mar 21 and Dec 21
   c. June 21 and Dec 21
   d. June 21 and Mar 21
   e. Mar 21 and Sept 22
9. On the leeward of a mountain, air:
   a. Warms and ascends
   b. Warms and descends
   c. Cools and descends
   d. Cools and ascends
10. A rocket fired from the South Pole toward the equator would appear to bend to the East because of the Coriolis force.
   a. True
   b. False

## CONCEPTUAL QUESTIONS

1. Describe the roles of the Hadley cell, Ferrel cell, and polar cell in influencing atmospheric circulation.
2. How do temperature differentials on Earth affect biome type?
3. Based on your knowledge of biomes, identify the biome in which you live. In your discussion, list and describe the plants and animals that you have observed in your area.

Polar bear on thin ice.

CHAPTER 23

# Marine Biomes

## Outline and Concepts



Polar bears, *Ursus maritimus,* are thought to be a good indicator for global climate change, in that their well-being (or lack of) reflects the general health of their environment. Ecologists believe that global warming is causing the ice in the Arctic Ocean to melt earlier in the spring than in the past. Because polar bears rely on the ice to hunt for seals, the earlier break up of the ice is leaving the bears less time to feed and build the fat that enables them to sustain themselves and their young. Ian Stirling and Andrew Derocher (2012) showed that less time to access seals leads to poorer body condition, fewer and smaller cubs, lower survival rate of cubs, and population decline. A U.S. Geological Survey study concluded that future reduction in Arctic ice could result in a loss of two-thirds of the world's polar bear population within 50 years. In addition, as bears' body condition declines, more seek alternate food supplies, resulting in more frequent contact between bears and humans. In May 2008 polar bears were listed as a threatened species under the U.S. Endangered Species Act.

Although the biome concept was originally defined using terrestrial habitats, within marine environments many different biome types are recognized, including the open ocean, shallow-water biomes such as coral reefs, kelp forests and sea grass beds, and the intertidal zone. In addition, some near-shore biomes such as sandy beaches and sand dunes are influenced by tidal regimes and salt spray. These coastal biomes are distinguished by differences in current strength, wave action, and tidal range. Just as with terrestrial biomes, global climate patterns such as temperature variation and atmospheric circulation affect marine biomes. In this chapter we discuss the influence of temperature, salinity, and ocean currents on marine biomes, and describe the variety of marine biomes that exist on Earth.

## 23.1 Winds Strongly Affect Ocean Currents

In this section we see that there are many kinds of currents in marine biomes, from large-scale currents to smaller-scale waves and circulation patterns. All of these features can affect biological processes and help determine marine biome type.

### 23.1.1 Ocean currents are created by winds and the Coriolis force

Surface currents, in the upper 400 m of the oceans, are mostly caused by friction with the wind as it moves over the water. Because of the long distances currents travel, the Coriolis force deflects them, further aiding in the creation of a circular pattern. The major ocean currents act as gyres or "pinwheels" between continents, running clockwise in the ocean basins of the Northern Hemisphere and counterclockwise in those of the Southern Hemisphere (**Figure 23.1**). The Gulf Stream, equivalent in flow to 50 times the world's major rivers combined, brings warm water from the Caribbean and the U.S. coasts across to Europe, the climate of which is correspondingly moderated. At about 30° west, 40° north it splits into two, with the northern part becoming the North Atlantic Drift, warming northern Europe, and the southern part becoming the Canary Current. The Gulf Stream has an energy flow of about 1.4 petawatts of heat (a petawatt is equal to 1 quadrillion watts), which is equivalent to 100 times the world's energy supply. Surprisingly, during November 2004 the Gulf Stream stopped for 10 days, worrying some scientists that this could again occur in the future, possibly for longer. The Humboldt Current brings cool conditions almost to the equator along the western coast of South America. Near the poles, the gyres flow in the opposite direction. Because there are no major landmasses in the Southern Hemisphere at these latitudes to contain the gyres, such currents flow only in the northern polar regions.

When the surface currents move water away from the continents, it is replaced by water from greater depths, a process known as a **coastal upwelling.** This deeper water is usually rich in nutrients from the ocean floor, and thus areas of upwelling are very productive and support coastal fisheries (**Figure 23.2a**). Major areas of upwelling occur along the Pacific coasts of North and South America and close to Antarctica. This process can be greatly modified by a phenomenon known as the El Niño Southern Oscillation, commonly called El Niño, a climate pattern that occurs across the tropical Pacific Ocean every 3–7 years. During an El Niño year, a much greater depth of surface water is warmed in the tropical eastern Pacific Ocean (refer back to Figure 6.9). Upwelled water that comes from beneath this warmer layer contains

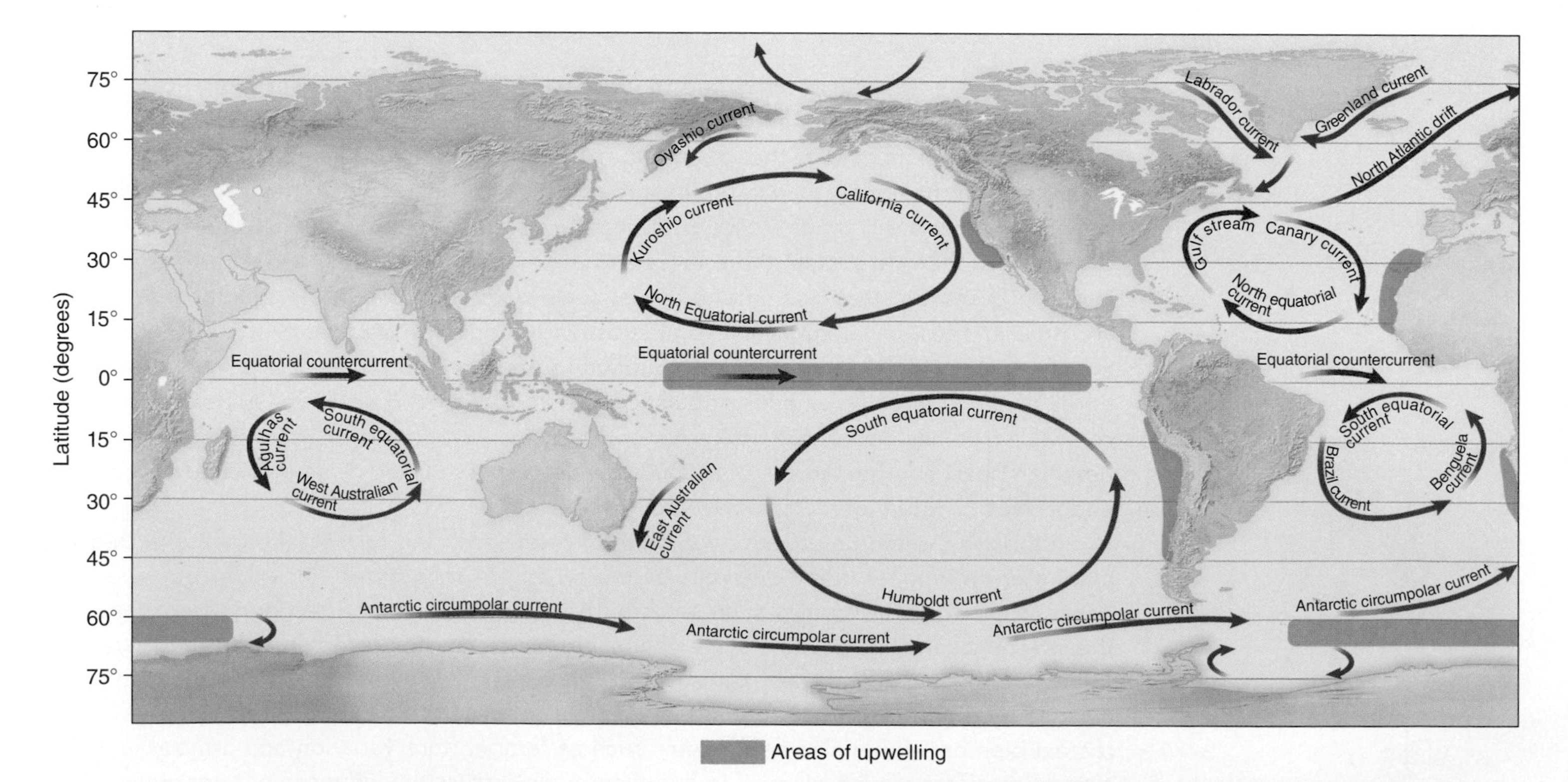


**Figure 23.1 Ocean currents of the world.** The red arrows represent warm water; the blue arrows, cold water. Upwelling zones are shaded.

**ECOLOGICAL INQUIRY**

Based on this figure and our discussion of fog in Chapter 22, where might fog be an important source of moisture for terrestrial biomes?

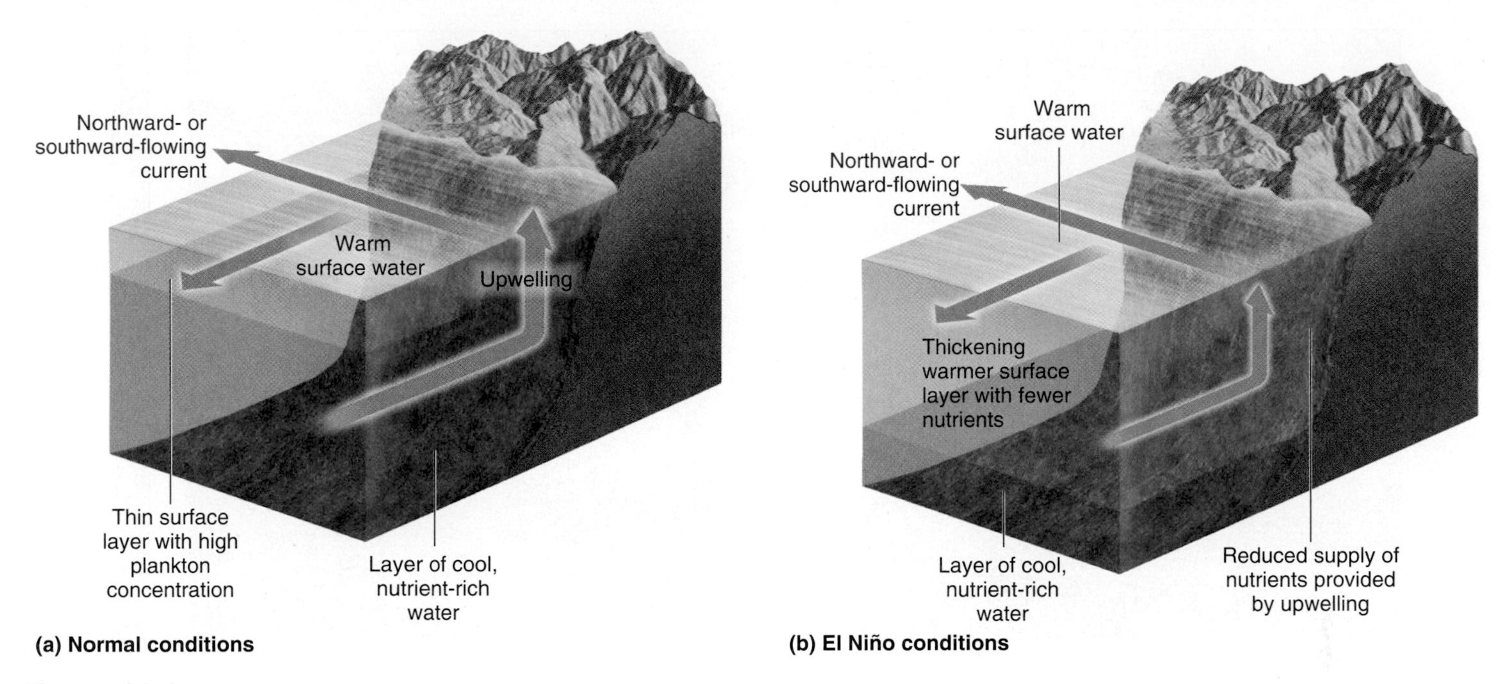


**Figure 23.2** **Coastal upwellings.** **(a)** Under normal conditions, when strong coastal currents move north or southward along a coastline, cooler nutrient-rich water is brought to the surface. **(b)** Under El Niño conditions, this upwelling is much reduced.

fewer nutrients, leading to a broader, relatively nutrient-depleted surface layer. This in turn reduces the amount of phytoplankton and density of fish that are dependent on them (**Figure 23.2b**). The most famous El Niño–caused depression of upwelling occurs along the western South American coast, offshore from Chile, Peru, and Ecuador.

## 23.1.2 Waves are also created by wind

In addition to currents, wind also creates waves, which range in size from small ripples to huge swells (**Figure 23.3**). As the wind blows, friction between the air and the water surface creates small ripples. Once the ripples have formed, the wind has something to push against and the waves can increase in size. The lowest part of the wave is called the trough, and the highest part is the crest. The difference between the trough and the crest is the wave height. The wavelength is the distance from one crest to the next. Four factors influence wave formation: wind speed, fetch, duration time of wind, and water depth. The fetch is the distance of open water over which the wind has blown. Increases in each of these factors increases wave sizes. Waves continue to roll through the ocean even after the wind has stopped blowing. Shorter fetches in lakes tend to reduce wave size. In contrast, the long fetches of the open ocean can regularly create 4- to 5-m-high waves in the

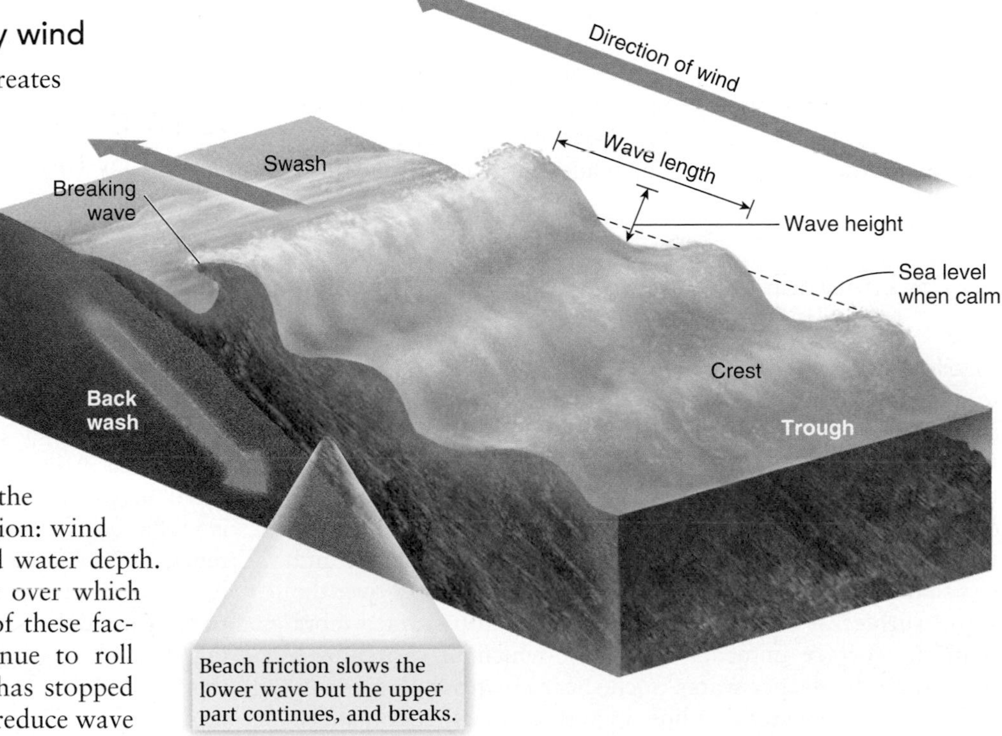


**Figure 23.3** **Wave formation in the ocean.**

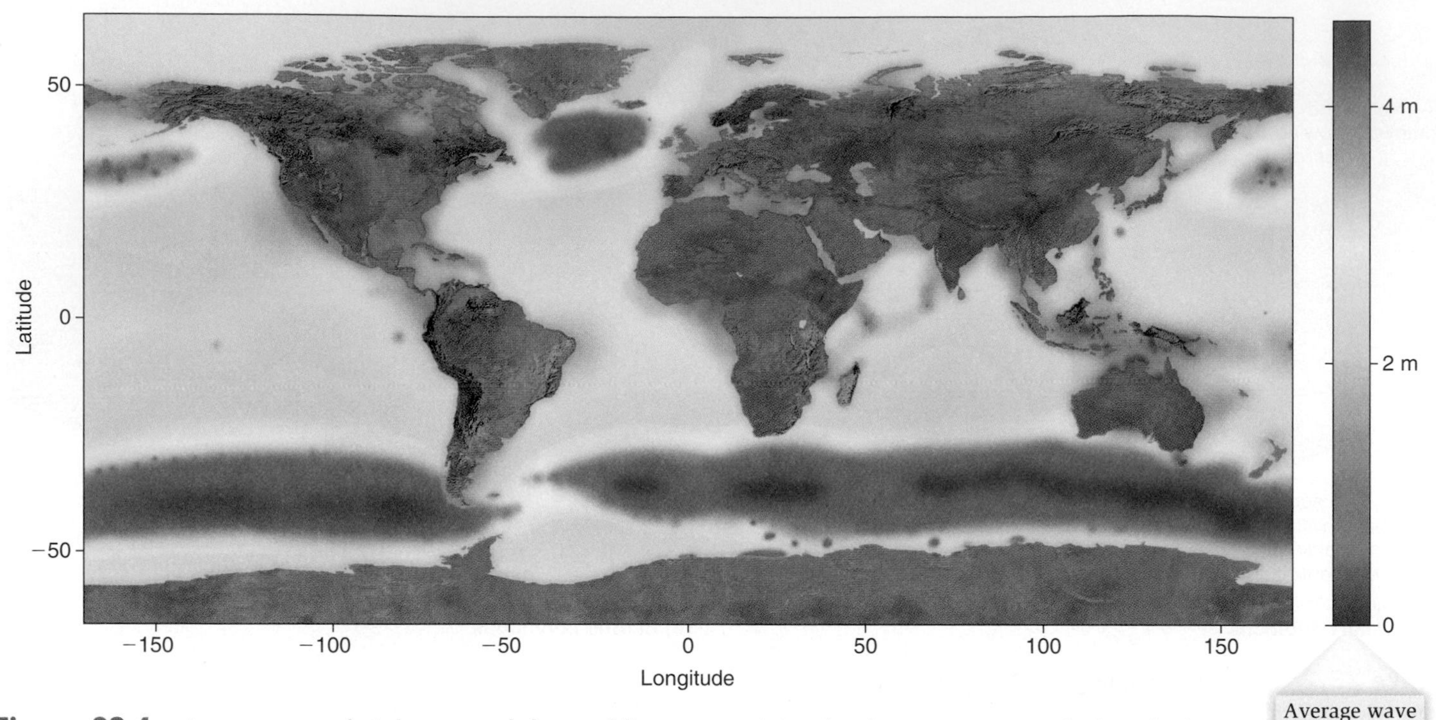


**Figure 23.4 Average wave heights around the world's oceans.** Wave heights are greatest in the long fetches of the Southern Ocean.

North Atlantic and the South Pacific Oceans (**Figure 23.4**), and 10- to 15-m-high waves are not uncommon. Waves in the protected areas of the Mediterranean Sea and the Gulf of Mexico are small, as are waves in equatorial areas of the world, where winds are light. As waves reach the shore, friction between the wave and the substrate slows the bottom of the wave, while the top keeps rolling and cascades forward, breaking against the shore. Larger waves can create devastating effects, dislodging organisms from the rocky shore and eroding or depositing sand along beaches. The breaking wave travels up the shore as swash, while the backwash returns water to the sea.

## 23.1.3 Langmuir circulation may carry material deep below the water surface

In addition to continent-scale surface ocean circulation patterns and waves, smaller-scale circulation patterns, called **Langmuir circulation**, also exist (**Figure 23.5a**). In 1938 Irvin Langmuir noted floating streaks of seaweed while crossing the Atlantic Ocean. He explained these streaks by showing experimentally that they occurred on any fairly flat body of water, sea or lake, in wind speeds of greater than 3 m/sec and less than 13 m/sec (**Figure 23.5b**). As the wind passes over the water surface, it creates shear, moving the top surface more than the surface immediately below, which in turn moves more than the deeper water. Such shear creates long tubes or cells of rotating water that line up in the same direction as the wind. Langmuir circulation commonly carries nonbuoyant material 4–6 m below the surface, but some circulation has been noted at depths of up to 200 m. In turn, other Langmuir cells bring water and nutrients up from the depths. In this way, lake nutrients may be cycled within the water column. Buoyant material, which does not sink, is trapped between the cells and aligned with the wind direction. At high wind speeds of greater than 13 m/sec, the water below becomes turbulent and the cells break down.

## 23.1.4 Deep-ocean currents are caused by thermohaline circulation

Thermohaline circulation, sometimes known as the "global conveyor belt," is an ocean current that is driven by temperature and salinity (haline) gradients (**Figure 23.6**). At the Earth's poles salt water freezes to form a layer of floating pack ice. In this process the salt from the water doesn't freeze in the ice but is left in the ocean. This dense, salty water sinks and more water moves in to replace it, causing a deep-water current. Water near the North Pole in the North Atlantic heads south toward Antarctica, funneled between South America and Africa. At Antarctica it is recharged with more cold water and splits into two currents, one heading toward India and one into the Pacific. As these currents reach the equator, the water warms and rises. Turned back by continental landmasses, the upwelling water loops back into the Indian Ocean and hence to the South Atlantic, then the North Atlantic where the loop begins again. The thermohaline circulation moves slowly, taking around 1,000 years to complete a circuit, but vast volumes of water, more than 100 times the flow of the Amazon,

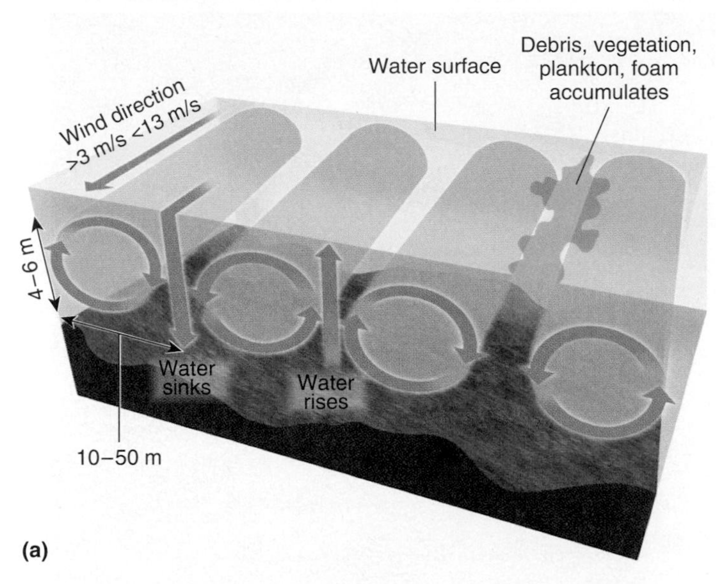

**Figure 23.5 Langmuir circulation.** **(a)** Wind speeds from >3 m/sec to <13 m/sec create small-scale circulation patterns. **(b)** This type of circulation can create streaks on the surface of open water, separated by distances of 10–50 m.

**ECOLOGICAL INQUIRY**

What happens to this circulation at wind speeds of >13 m/sec?

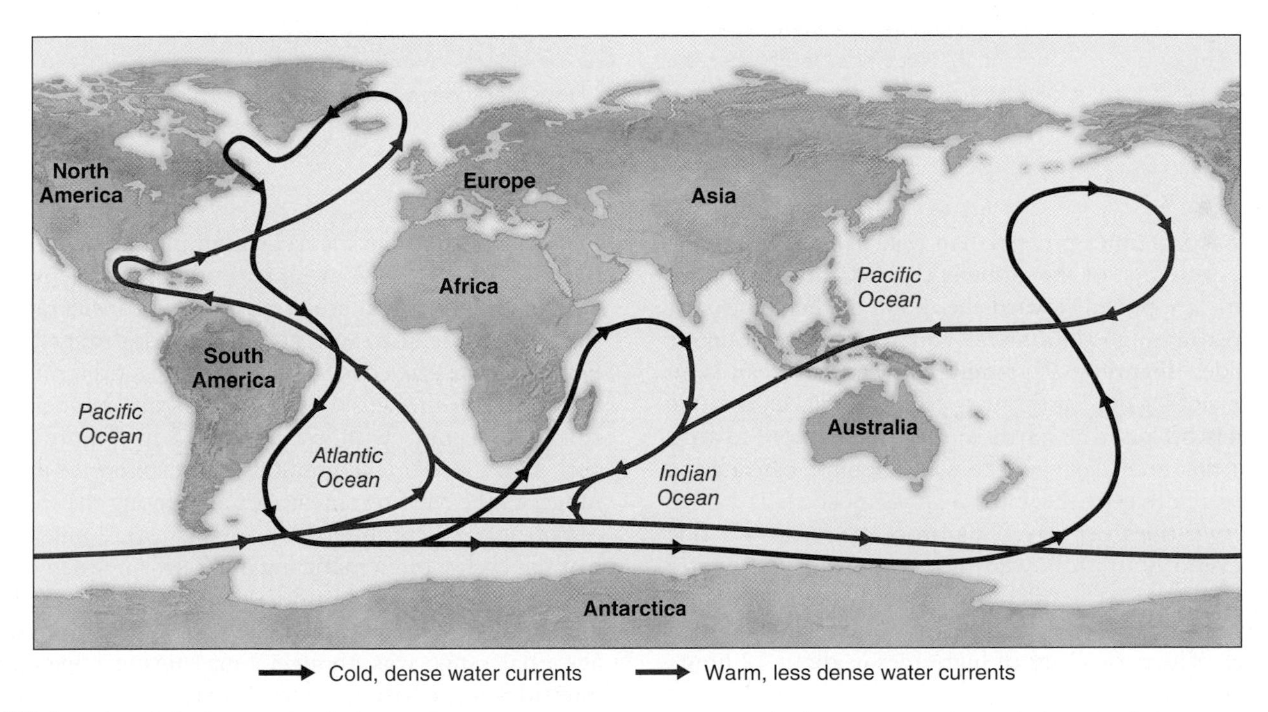


**Figure 23.6 The global conveyor belt.** Differences in water temperature and salinity create a thermohaline circulation pattern around the world.

are moved. Nutrients and carbon dioxide from deeper layers are transported to the surface, helping plankton and algae to grow.

**Check Your Understanding**

**23.1** What causes coastal upwelling close to continents?

## 23.2 Tides Are Caused by the Gravitational Pull of the Moon and the Sun

Marine systems are also influenced by tides, the rise and fall of sea levels. The gravitational pull of the moon and the sun causes tides. The gravitational force of the sun is 179 times stronger than that of the moon, but because the sun is 389 times more distant, its effect is weaker. The tidal force

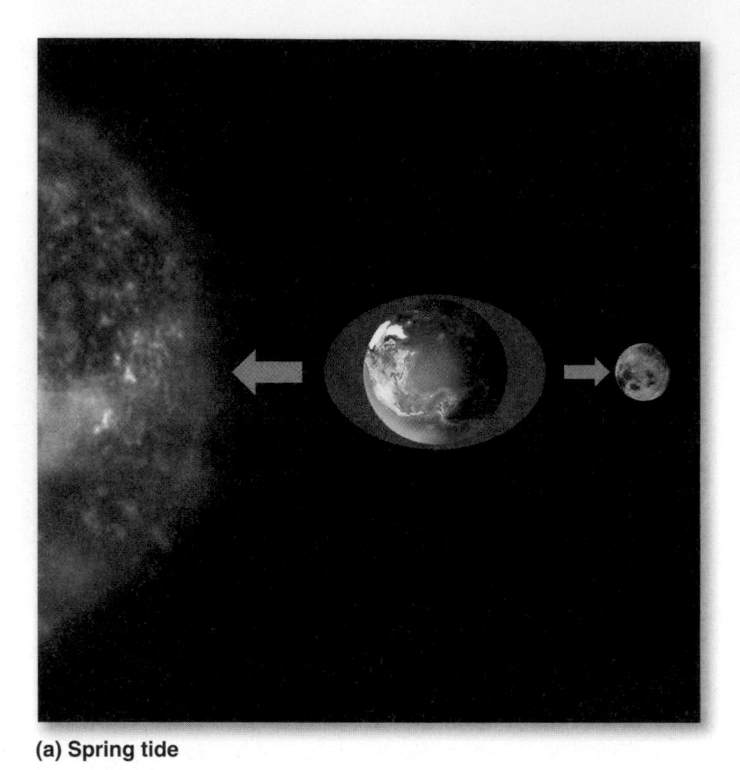

(a) Spring tide

(b) Neap tide

**Figure 23.7 Tide formation.** **(a)** Spring tides form when the sun, Earth, and moon are in alignment. **(b)** Neap tides result when the sun, moon, and Earth are at a 90° angle.

**ECOLOGICAL INQUIRY**

How many days are there between successive spring tides?

of the moon is 2.2 times greater than that of the sun. As the Earth turns, each area of the globe is close to the moon once a day. Oceans are pulled toward the moon at the equator at this time, creating high tides at the equator and low tides at higher latitudes (**Figure 23.7**). Similarly, when an ocean is on the opposite side of the Earth away from the moon, the tide is high. This is because the Earth is itself pulled more toward the moon at this point, leaving the water behind, causing the water to rise relative to the Earth. Thus, most areas of the Earth have two tides per day. This process results in a tide rising for several hours followed by a period of high water known as a slack tide. The tide begins to fall and is known as an ebb tide. The tide stops falling at low water, when the tide is again slack. The periodicity of high tides is about 12 hours and 25.2 minutes.

The tidal range, the difference between the high and low tides, varies over a two-week cycle. When the sun, moon, and Earth form a line, a condition known as syzygy, the gravitational pull of the sun reinforces that of the moon and the tidal range is maximal (**Figure 23.7a**). We term this event a spring tide, because the ocean's surface appears to spring upward, and it occurs at around the time of a new or full moon. When the sun, Earth, and moon are at a 90° angle, the gravitational forces of the sun partially cancel those of the moon, and the tidal range is minimal (**Figure 23.7b**). This is termed a neap tide, and it occurs when the moon is in its first or third quarter. The theoretical amplitude of the spring tide is about 79 cm, while at a neap tide this value is reduced to 29 cm. However, real amplitudes differ considerably. Funnel-shaped bays tend to exaggerate the tidal range. The Bay of Fundy, on the east coast of Canada, has a tidal range of 16 m, the largest in the world, and the Bristol Channel in England is a close second, with a range of 15 m (**Figure 23.8a,b**). In such areas, intertidal communities are extensive in area. Tidal ranges in the Mediterranean Sea are among the world's smallest, at only 29 cm. Intertidal organisms are much influenced by tidal cycles, often maturing and releasing their eggs at periods of high tide. At the time of the Cambrian explosion, the moon was positioned much closer to the Earth and the average tidal range was about 15 m, causing huge expanses of intertidal areas. Rising sea levels may exacerbate the effects of high tides (see **Global Insight**).

**Check Your Understanding**

**23.2** Why are tides affected more by the gravitational effects of the moon, when the sun is so much bigger?

(a)

(b)

**Figure 23.8** **Tidal range.** Tidal range in the Bristol Channel, England, is the second greatest in the world, at 15 m. **(a)** High tide, **(b)** low tide, Birnbeck Pier, Weston-Super-Mare.

**ECOLOGICAL INQUIRY**

How long does it take for the moon to orbit the Earth?

## 23.3 Marine Biomes Are Determined by Water Temperature, Depth, and Wave Action

Differences in water depth and temperature primarily determine the characteristics of marine biomes. These biomes include the open ocean, shallow-water biomes, and biomes of the intertidal zone. The open ocean includes hydrothermal vent communities, locations on the seafloor where two tectonic plates are diverging. Shallow-water biomes include three different types of marine communities. Coral reefs occur in clear tropical waters, with warm waters and hard bottoms. Kelp forests grow in cool, clear temperate waters, with a hard substrate. Sea grasses are present in areas with clear water and sandy bottoms, in both cool and warm water. The **intertidal zone**, sometimes called the littoral zone, the area where the land meets the sea, is alternately submerged and exposed by the daily cycle of tides. The resident organisms are subject to huge daily variations in temperature, light intensity, and availability of seawater. There are several types of intertidal habitats: rocky shore, sandy beach, mangrove forests, and salt marshes.

### The Open Ocean

**Physical Environment:** The open ocean is divided up into different zones according to proximity to shore, depth, and light availability (**Figure 23.9**). The **neritic zone** extends from the intertidal zone to the edge of the

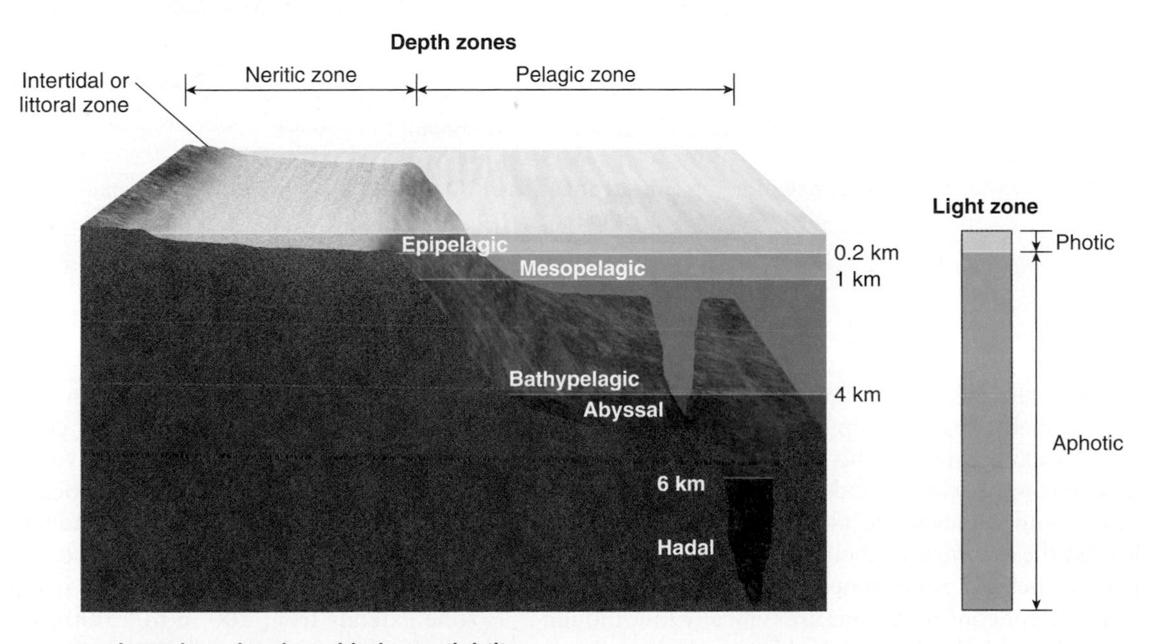


**Figure 23.9** **Ocean zones based on depth and light availability.**

## Global Insight

### The Effects of Sea Level Rise on Coastal Regions

Globally, sea level has risen 10–25 cm over the past century, primarily because of ice melting at the poles and water volume expanding due to warmer temperatures. This rise is an order of magnitude greater than recorded rises of several previous millennia. Data from North America show relatively large sea level rises around Louisiana and other Gulf Coast regions to smaller rises on the Atlantic and Pacific coasts. There are even relative decreases along the Alaskan coastline, brought about by tectonic uplift, but these are comparatively rare (**Figure 23.A**).

The locations of coastal communities such as mangrove forests or salt marshes are set by sea levels. Even though such communities can tolerate salt water, the plants will drown if under water for too long. Normally these coastal communities would gradually retreat inland as the sea level rises, colonizing new areas of freshly salted soil. However, in areas with a steep topography, such as the West Coast of the U.S., or areas that are highly built up, such as Charleston, South Carolina, there will be little opportunity for coastal wetlands to retreat further inland because such movement will be restricted. As sea levels rise, coastal vegetation recedes, resulting in a net loss of coastal wetlands. As wetland areas shrink, services such as food-web support for fish, waste assimilation, and wildlife habitat are also likely to be reduced. Furthermore, protection from tropical cyclones and extratropical storms by coastal wetlands is reduced. Because more than 53% of the U.S. population lives in coastal counties, this is a very real threat to humanity and to natural inland communities such as forests, which cannot tolerate saltwater inundation.

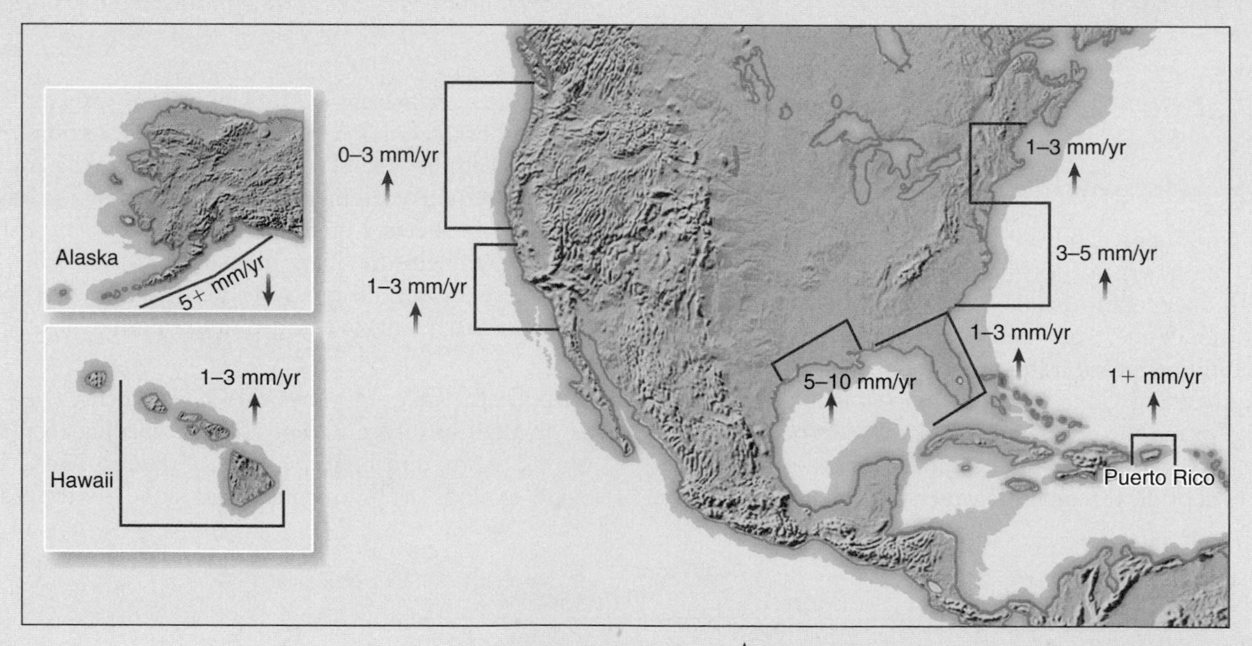


**Figure 23.A Long-term sea-level rises for the U.S. over the past 30 years.** Data are from 117 locations using at least 30 years of sea level data.

continental shelf, where the water averages about 200 m deep. The continental shelf consists of an underwater coastal plain extending from each continent. This coastal plain was part of the continent during glacial periods when much water was locked up as ice and sea level was about 200 m lower than at present. Beyond the continental shelf is the open ocean, sometimes called the **pelagic zone**, where water depth averages 4,000 m and nutrient concentrations are typically low, though the waters may be periodically enriched by ocean upwellings, which carry mineral nutrients from the deep waters to the surface. Pelagic waters are mostly cold, warming only near the surface. The surface extremes vary from about −1°C in the Southern Ocean around Antarctica to 27°C at the equator. At greater depths, below 100 m, the temperature hardly ever varies. The surface waters of the ocean are called the **epipelagic zone** and extend about 200 m in depth. The **mesopelagic zone** extends from 200 m to 1,000 m, and the **bathypelagic zone** from 1,000 m to 4,000 m. Beyond this is the **abyssal**

Oceans and seas of the world

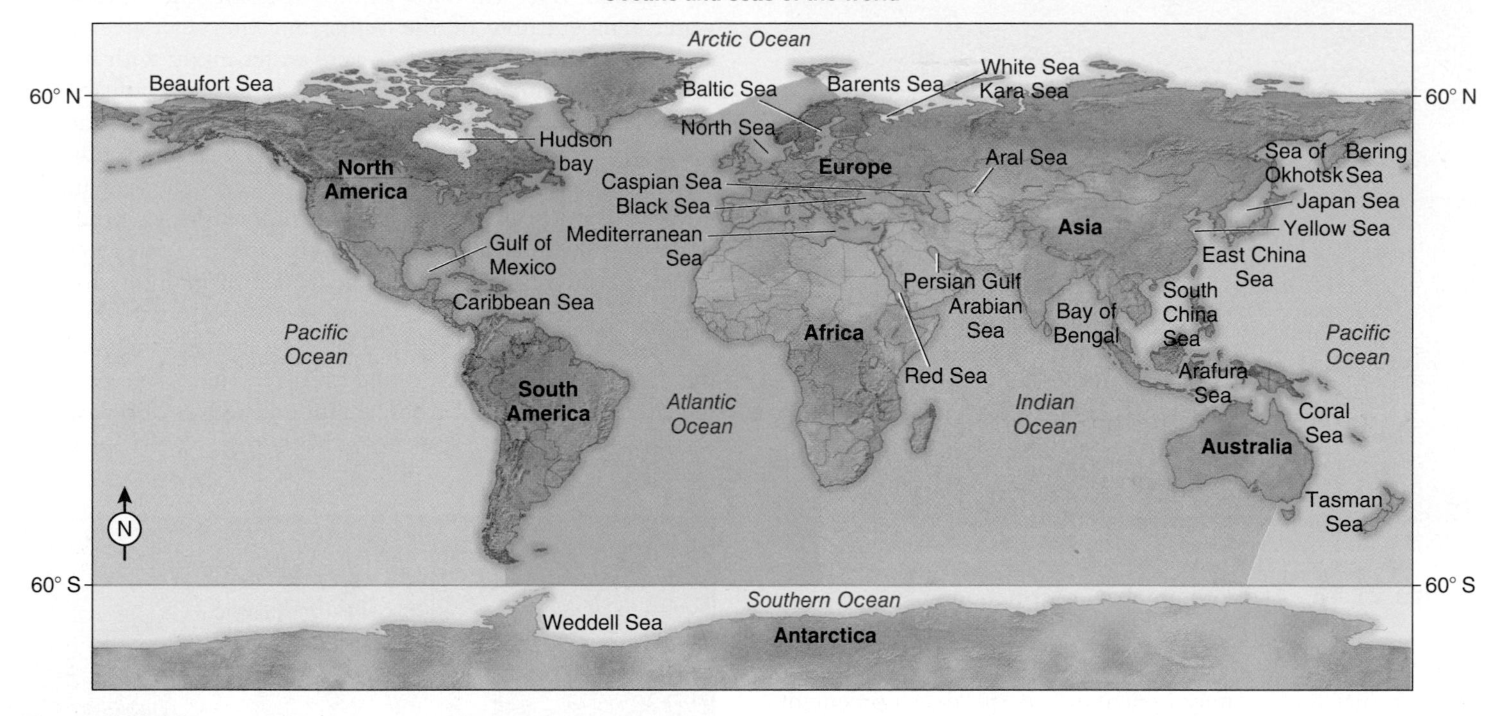


**Figure 23.10** The Earth's oceans and seas.

zone, from 4,000 m to 6,000 m, and finally the **hadal zone**, below 6,000 m. The ocean may also be categorized by light availability. As noted in Chapter 7, water absorbs light readily. Photosynthesis usually occurs only in the first 0–200 m of the ocean's surface, an area known as the **photic zone**. Below is the relatively dark **aphotic zone**. Below 600 m, even in what appears to be the clearest ocean, it is pitch black.

**Location:** Across the globe, the oceans cover 70% of the Earth's surface (**Figure 23.10**). The largest of the world's oceans is the Pacific, with an area of nearly 160 million km$^2$. The second biggest, the Atlantic Ocean, is about 76 million km$^2$, followed by the Indian Ocean at 68 million km$^2$. The Southern Ocean is about 20 million km$^2$, and the Arctic Ocean, 14 million km$^2$.

**Plant Life:** Where light levels are high at the surface, many microscopic, photosynthetic organisms called **phytoplankton** grow and reproduce. Phytoplankton account for nearly half the photosynthetic activity on Earth and produce much of the world's oxygen.

**Animal Life:** Open-ocean organisms include **zooplankton**, such as tiny shrimplike creatures, tiny jellyfish, and the small larvae of invertebrates and fish that graze on the phytoplankton. The open ocean also includes free-swimming animals collectively called **nekton**, which can swim against the currents (**Figure 23.11**). The nekton include large squids, fish, sea turtles, and marine mammals that feed on phytoplankton, zooplankton, or each other. Only a few of these organisms live at any great depth. Bottom-dwelling or **benthic** organisms feed on the rain of dead material from above or eat other benthic organisms.

**Figure 23.11** A manta ray, *Manta birostris*, part of the nekton, in the Bay of Bengal, off the coast of Burma.

**Effects of Humans:** Oil spills and a long history of garbage disposal have polluted the ocean floors of many areas. Overfishing has caused many fish populations to crash, and the whaling industry has greatly reduced the numbers of most species of whales. Pollution, in the form of

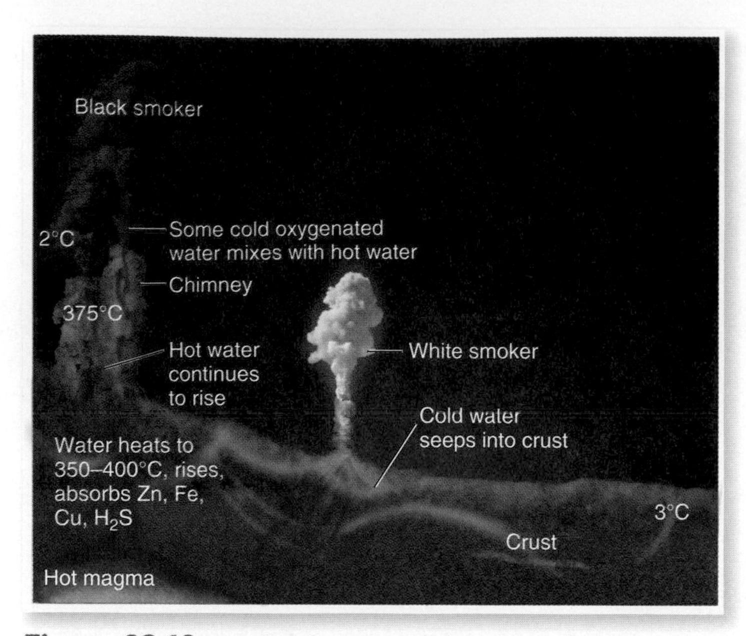


**Figure 23.12** **Hydrothermal vent formation.**

nutrient enrichment, may contribute to the development of algal blooms called red tides, especially in ocean areas closer to the shore. Red tides are caused by dinoflagellate algal species that are red or brown in color and tint the water. They produce neurotoxins that paralyze fish, causing them to stop breathing and die. Red tide dinoflagellates can also be taken in by filter-feeding mollusks such as oysters and mussels; consuming these can result in paralytic shellfish poisoning.

Hydrothermal Vents (Figure 23.12) **Physical Environment:** In 1977 geologists using submersibles discovered a unique assemblage of animals associated with deep-sea **hydrothermal vents** that spew out hot water up to 400°C and rich in hydrogen sulfide. Here, cold seawater finds its way through cracks in the Earth's surface, often in volcanically active areas, where tectonic plates are moving apart. As it moves downward toward hot magma, oxygen, sulfate, and magnesium are removed. Later, barium, calcium, silicon, copper, zinc, iron, and sulfur are added from the crust. As the water returns to the ocean, it has been heated to around 350–375°C but does not boil. Huge pressure from the surrounding ocean prevents boiling. On contacting the cold oxygenated ocean water, the minerals precipitate out to form cylindrical structures, called smokers, reaching to a height of 20–60 m. Vent growths of 30 cm per day have been recorded. Black smokers are rich in sulfides; white smokers emit lighter-hued minerals such as barium, calcium, and silicon and have lower temperatures. Such structures may have been an ideal place for the origin of life.

**Location:** Hydrothermal vents are located worldwide, but only along mid-oceanic ridges such as the East Pacific Rise and the Mid-Atlantic Ridge, at an average depth of 2,100 m (7,000 feet).

**Plant Life:** Absent.

**Animal Life:** Animals cannot tolerate the 350–400°C water temperatures of the vents, but they can live in the cooler water that results when vent water mixes with the surrounding seawater. Among the most common organisms that can tolerate such conditions are chemoautotrophic bacteria that use hydrogen sulfide to produce organic compounds. The thick mats of bacteria support small grazers such as amphipods and copepods. These, in turn, support larger predatory organisms such as snails, shrimp, crabs, fish, and octopuses. Large tube worms are a prominent part of many hydrothermal vent communities and live mutualistically with bacteria inside them (**Figure 23.13**). The tube worm's hemoglobin binds to hydrogen sulfide and transfers it to the bacteria, which in turn provide carbon compounds. Over 300 new species have been discovered at hydrothermal vents, including

**Figure 23.13** **A black smoker on the ocean floor.** Organisms living around deep-sea hydrothermal vents include chemoautotrophic tube worms and clams, and crabs, all of which are relatively large compared to their shallow-water relatives.

**Figure 23.14** Coral reef.

**Figure 23.15** Kelp forest, with harbor seal, *Phoca vitulina.*

the Pompeii worm, *Alvinella pompejana,* which can withstand temperatures up to 80°C. The communities are entirely dependent on the nutrient-rich, warm water emanating from the hydrothermal vents, and any tectonic movements that may cause vent closure result in rapid extinctions.

**Effects of Humans:** Mining damage is possible from companies searching for zinc, copper, and lead sulfides. In 2005 a company was granted 35,000 $km^2$ of exploration rights in the Kermadec Arc, a 1,200-km-long chain of underwater volcanoes stretching northeast from New Zealand, an area with abundant hydrothermal vent activity.

### Coral Reefs (Figure 23.14)

**Physical Environment:** Corals need warm water of at least 20°C but less than 30°C (refer back to Figure 5.1). They are also limited to the photic zone, where light penetrates. Sunlight is important, because many corals harbor mutualistic algae, or dinoflagellates, that contribute photosynthate to the corals but that require light to live. Nutrient levels are fairly low, otherwise algae grow.

**Location:** Coral reefs exist in warm tropical waters where there are solid substrates for attachment and water clarity is good. The largest coral reef in the world is the Great Barrier Reef off the Australian coastline, but other coral reefs are found throughout the Pacific and Atlantic Oceans and Caribbean Sea. The highest numbers of coral species, over 600, live in the western Pacific, compared to only 100 species in the Atlantic Ocean and Caribbean Sea.

**Plant Life:** Dinoflagellate algae live within the coral tissue, and a variety of red and green algae live on the coral reef surface.

**Animal Life:** An immense variety of microorganisms, invertebrates, and fish live among the coral, making the coral reef one of the most interesting and richest biomes on Earth. Probably 30–40% of all fish species on Earth are found on coral reefs, including parrotfish, angelfish, butterfly fish, damselfish, groupers, grunts, and wrasses. Prominent herbivores include snails and sea urchins, as well as fish. These are in turn consumed by octopuses, sea stars, and carnivorous fish. Many invertebrate species are brightly colored, warning predators of their toxic nature.

**Effects of Humans:** Collectors have removed many corals and fish for the aquarium trade, and marine pollution threatens water clarity in some areas. One of the greatest threats is from global warming. Water temperatures that are too high, over 30°C, can cause coral bleaching (refer back to Section 5.2). Increased atmospheric $CO_2$ has lowered ocean pH by nearly 0.1 over the past 250 years. Glenn De'ath and colleagues (2009) showed that both calcification and growth of Pacific coral on the Great Barrier Reef declined 13–14% between 1990 and 2005. A doubling of $CO_2$, as is projected to occur by the 22nd century, may result in oceans too acidic for corals to calcify. Another human-induced change is eutrophication via increased nutrient inputs, which allows overgrowth of corals by algae. Siltation, from increased silt input from nearby rivers, and overfishing are also threatening coral systems.

### Kelp Forests (Figure 23.15)

**Physical Environment:** Kelp forests grow on hard substrates in cool, clear temperate waters.

**Location:** Kelp forests are located along the western coasts of North and South America; northeastern Canada;

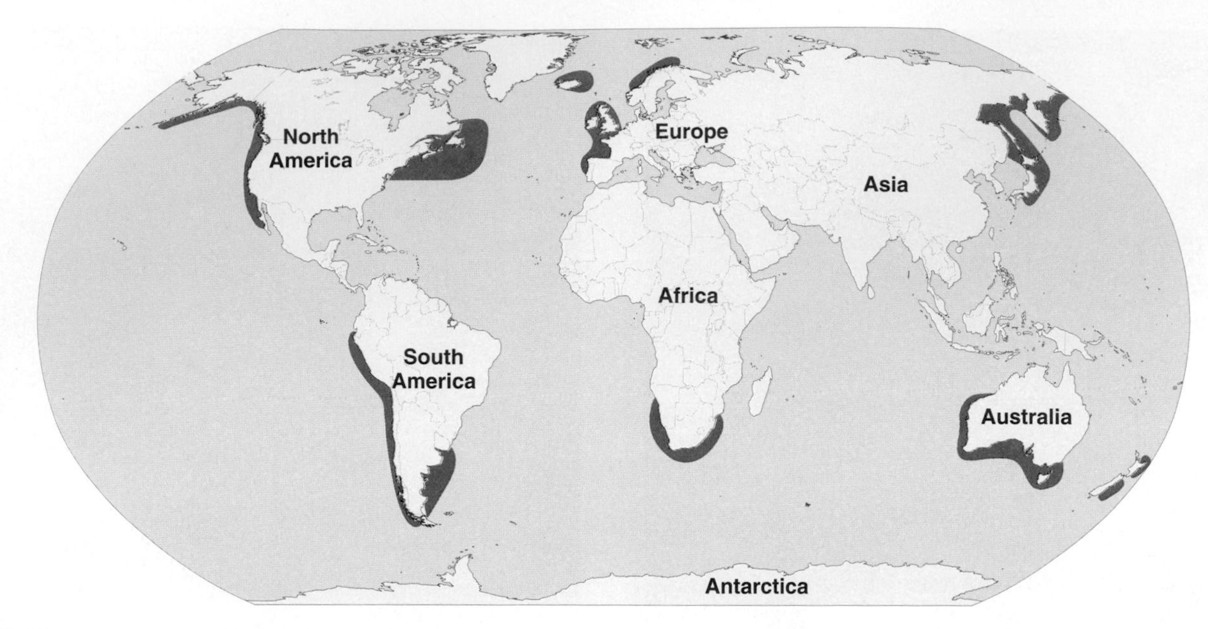


**Figure 23.16** The distribution of the world's kelp forests.

western Europe; South Africa; South Australia and New Zealand; and Japan and Eastern Russia (**Figure 23.16**).

**Plant Life:** Like all algal species, kelp are not true plants but are their ecological equivalents in this community. Most kelp species are in the genera *Macrosystis, Nereocystis, Laminaria,* and *Ecklonia.* Giant kelp, *Macrocystis pyrifera,* the largest submarine algae in the world, is the main constituent of kelp forests in North America. Some individuals reach lengths of more than 30 m and can grow at rates of up to 50 cm per day!

**Animal Life:** The large, three-dimensional structure of kelp forests provides a home for many invertebrates, fish, birds, and marine mammals such as seals. Excess grazing by sea urchins can damage the forest.

**Effects of Humans:** Overharvesting has threatened kelp forests. Algin, a hydrocolloid, is used as a binding agent in the pharmaceutical industry, in cosmetics, and in the food industry (for example, in ice cream). Pollution and runoff from shores can reduce water quality and prevent kelp from growing close to shore.

## Sea Grasses (Figure 23.17)

**Physical Environment:** Sea grasses are found on the bottom of protected bays and other shallow coastal waters with clear water and soft, sandy, or muddy substrates. They are often present in areas where kelp are absent and occur in warm and cool waters worldwide (**Figure 23.18**).

**Plant Life:** Sea grasses are not true grasses but are more closely related to lilies. There are 50 species worldwide. Their thick roots allow them to withstand quite harsh wave action. They may form large underwater meadows. There is a distinct zonation of plants with water depth. In Florida, shoal grass, *Halodule wrightii,* is found in the shallowest waters. This is followed by turtle grass, *Thalassia testudinum,* and, at depths greater than 12 m, manatee grass, *Syringodium filiforme.* At deeper depths other *Halodule* species may grow, but rarely in water deeper than 40 m.

**Figure 23.17** Sea grass bed with green turtle, *Chelonia mydas,* Red Sea, Egypt.

**Animal Life:** Large mammals such as manatees, *Trichechus manatus,* graze on sea grasses, as do green turtles, *Chelonia mydas.* In addition, many smaller animals settle on the grass blades and filter feed. Many species of fish spawn in sea grasses and the young fish use the area as a nursery ground, as do spiny lobsters, *Panulirus guttatus,* and stone crabs, *Menippe mercenaria.*

**Effects of Humans:** Sea grasses stabilize bottom sediments and provide erosion control for coastlines. They trap fine sediments and improve water clarity. However, offshore runoff

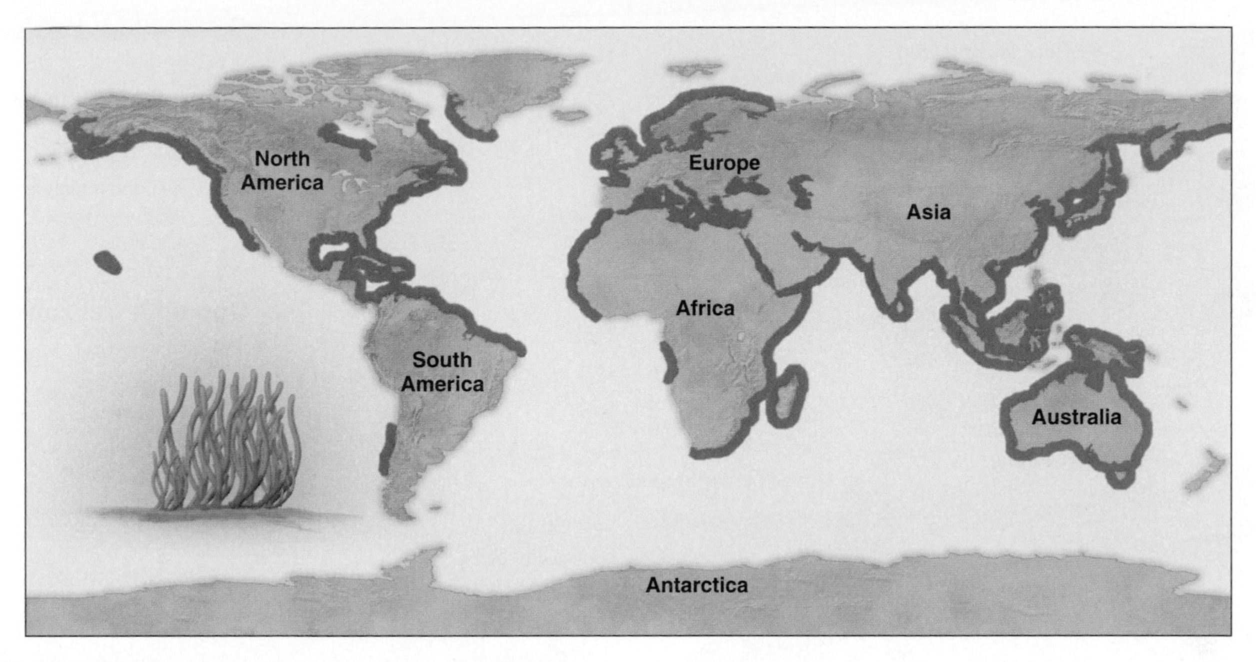


**Figure 23.18** Distribution of the world's sea grass beds.

**Figure 23.19** **The rocky intertidal.** Typical rocky intertidal zonation along the Olympic Coast National Marine Sanctuary, Washington State. Some animals, such as the sea star *Pisaster*, can wander through multiple zones.

and pollution in bays can kill sea grasses. Propeller scars from boats also destroy sea grasses, which has prompted authorities to close off some areas from powerboats.

Rocky Intertidal (Figure 23.19) **Physical Environment:** Rocky intertidal occurs on hard substrates where the land meets the sea. Commonly there is a vertical zonation consisting of four broad zones (**Figure 23.20**). The highest of these zones is the splash zone or **supralittoral**, which is never covered by the tide and receives only splashes or spray from waves. Intertidal organisms are rare in this zone. The **upper littoral zone** is submerged only during the highest tides. The **mid-littoral zone** is submerged during the highest regular tide and exposed during the lowest tide each day. The **lower littoral zone** is exposed only during the lowest tide. Conditions in the intertidal are variable and hostile, ranging from freezing cold when exposed in winter, to hot and dry when exposed in summer. Rocky shores also have high wave action that batters intertidal life and does not allow sand or mud to settle. Salinity can vary dramatically in isolated tide pools when rainfall dilutes the pool or when the hot sun dries it up.

**Location:** Patchily distributed throughout the world. In the U.S., common on the West Coast and New England but absent along the Gulf Coast and most of the East Coast.

**Plant Life:** There is little plant life in the upper littoral zone, because this zone is relatively dry. However, green, red, and brown algae, commonly called seaweeds, abound in the mid- and lower littoral zones. Most have holdfasts to allow them to cling to the rock (refer back to Figure 5.17).

**Animal Life:** Animal life may be quite diverse. In the upper littoral zone, barnacles, limpets, mussels, chitons, and snails occur. Sea anemones, snails, hermit crabs, and small fishes live in tide pools at low tides, escaping harsh conditions. Most have a strong stress response that produces proteins that aid in recovery from temperature stress. On the mid-littoral rock face, there may be a variety of limpets, mussels, sea stars, sea urchins, snails, sponges, tube worms, whelks, isopods, and chitons. Some organisms have hard shells and a sealing plate to minimize water loss and protect them against predators. Limpets occupy a home scar on the rock that is developed by movement of the shell as it grinds to match the contour of a rock it occupies. Many species have secretions that act like

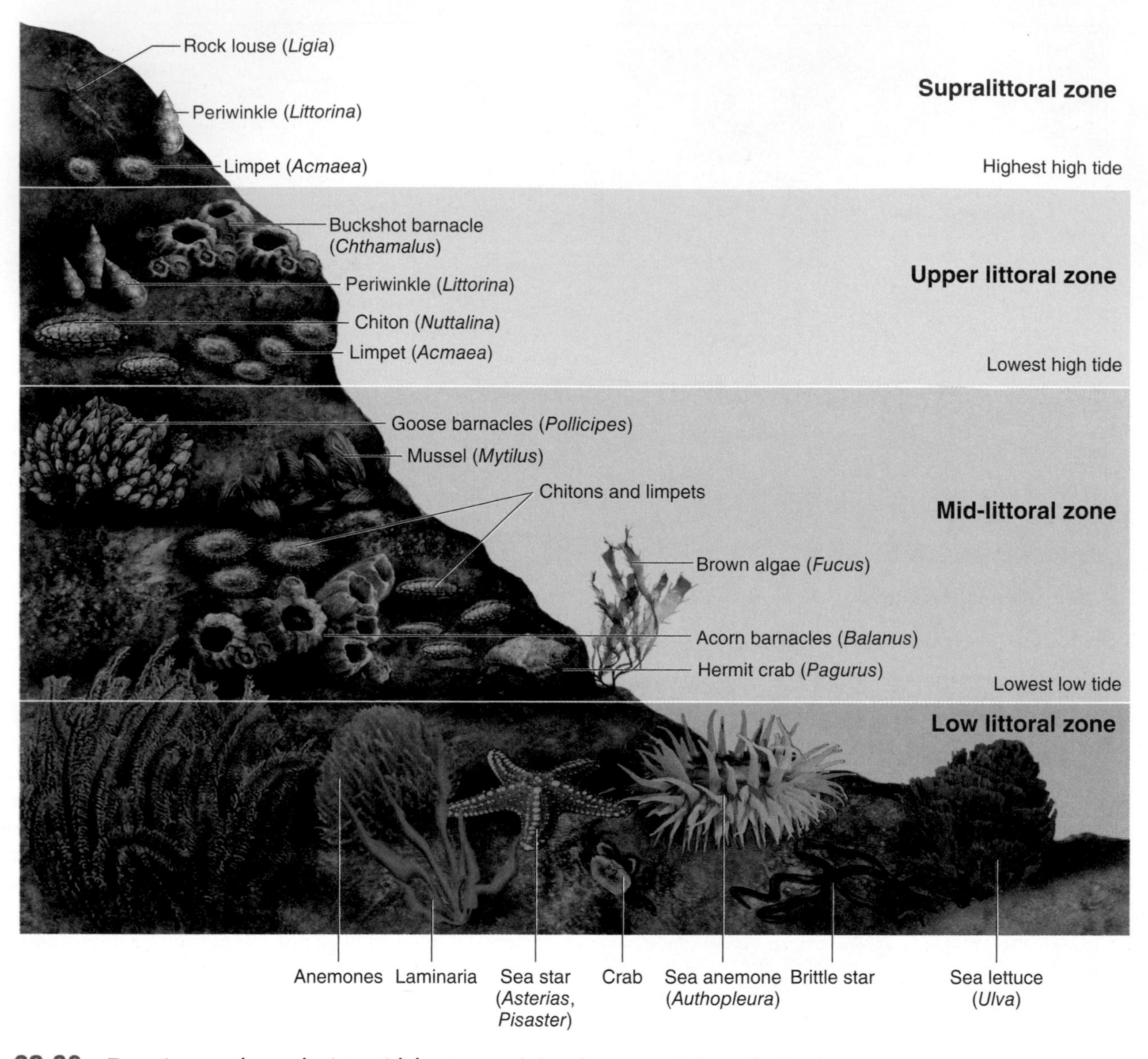


**Figure 23.20 Zonation on the rocky intertidal.** A typical shoreline in the U.S. Pacific Northwest.

glue to hold them tight to the rock face when immersed. Mussels produce strong, silky fibers called byssal threads to attach to rocks and other hard surfaces (refer back to Figure 5.17). In all areas, most species spawn at the same time, releasing their gametes into the ocean. The lower littoral teems with animal life, including many herbivores of seaweeds in addition to anemones, crabs, sea stars, whelks, and other organisms.

**Effects of Humans:** Collection of shells and animals for food has greatly depleted the abundance of rocky intertidal organisms in many parts of the world. Oil spills have greatly impacted some rocky intertidal areas.

Sandy Shores (Figure 23.21) **Physical Environment:** Sand is produced by waves wearing down coastal cliffs or coral reefs to shingle and sand, and by grinding shells from the seafloor. Some white sand beaches, including those in North Florida, are the result of quartz erosion in the nearby mountains and particle deposition by rivers. Although the shifting sand particles of a sandy beach are constantly being rearranged, sand can be very deep. Wave action can be considerable. The beach drains and dries quickly because of its steep slope. There is usually a vertical oxygen gradient within deeper sands. Muds, containing little oxygen, often contain high amounts of hydrogen sulfide.

**Location:** Patchily distributed throughout the world.

**Plant Life:** At the shoreline, little obvious plant life exists because of the constantly shifting substrate, although invisible microalgae called diatoms occur in the beach sands. However, in the supralittoral zone and beyond, the sand is more stable and is not constantly inundated by the tide. Here extensive sand dunes may be present that can be well vegetated. For dune development, there needs to be abundant sand and strong, onshore winds. Small obstacles, such as washed-up sea grass, trap the sand, which begins to build up.

**Figure 23.21 Lamphere dunes, Humboldt Bay National Wildlife Refuge, California.** Dunes form parallel to the prevailing wind.

**Figure 23.22 *Spartina alterniflora* in a New England saltmarsh, Cape Cod, Massachusetts.**

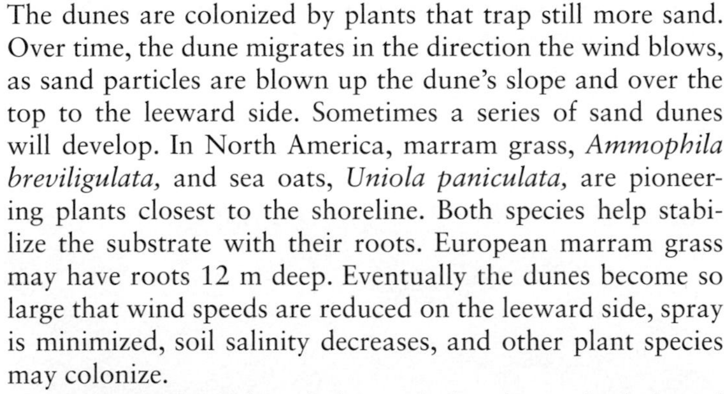

The dunes are colonized by plants that trap still more sand. Over time, the dune migrates in the direction the wind blows, as sand particles are blown up the dune's slope and over the top to the leeward side. Sometimes a series of sand dunes will develop. In North America, marram grass, *Ammophila breviligulata,* and sea oats, *Uniola paniculata,* are pioneering plants closest to the shoreline. Both species help stabilize the substrate with their roots. European marram grass may have roots 12 m deep. Eventually the dunes become so large that wind speeds are reduced on the leeward side, spray is minimized, soil salinity decreases, and other plant species may colonize.

**Animal Life:** Although the sandy beach may look devoid of animal life, many invertebrate species live between the sand grains. There are numerous tube worms that filter feed when submerged by the tide. Worms are abundant and are often dug for bait by anglers. Clams and numerous other bivalves also inhabit the sandy shore, as do sand dollars, types of echinoderms. Many worm species do not produce planktonic larvae but are brooders that take care of their offspring. Although some beaches contain sand to a considerable depth, on average, at about half a meter in depth the animal-rich sandy habitat usually gives way to an anoxic black mud where only microbes can live.

**Effects of Humans:** Urban development has greatly reduced the supralittoral beach area available to breeding turtles and shorebirds. Dune habitat is replaced by hotels and condominiums. Their powerful lights can disorient turtle hatchlings, who normally navigate toward moonlight reflected off the surface of the ocean. In many areas hotels are asked to reduce their outdoor lighting in turtle hatching season to minimize turtle disorientation. Many other beaches, such as those along Cape Cod, restrict access to dune and beach areas to minimize disruption to nesting birds.

### Mangrove Forests and Salt Marshes (Figure 23.22)

**Physical Environment:** These biomes develop in protected intertidal areas including coves, bays, and estuaries, where wave action is limited and fine sands and mud can accumulate. Plant roots slow down tidal flow so that its sediment is deposited, not resuspended as the tide ebbs. Mangrove forests are located along the world's tropical coastlines between about 25° north and 25° south (**Figure 23.23**). Beyond these latitudes, salt marshes are found. Some exceptions occur where warm water extends beyond these latitudes, as in Japan and Bermuda.

**Plant Life:** Worldwide, there are approximately 50 species of mangrove trees. The center of diversity is the Indo-West Pacific region, which has 30 species. Florida has only 3 species. All mangrove species are highly susceptible to freezing. Many mangroves have extensive prop roots descending from the trunk and branches that provide support in the muddy, shifting sediments (**Figure 23.24a**). The limited tidal flow in some areas and the evaporative action of the sun can result in salinities of 60–90 parts per thousand (ppt). Mangroves and *Spartina* grass have salt exclusion or salt excretion adaptations to allow survival in these environments (refer back to Figure 6.13). The oxygen-poor muds make root respiration difficult. Aerial roots allow for atmospheric gas transport to the roots. In addition, black mangroves, *Avicennia germinans,* have special aerial roots called **pneumatophores**, which extend upward above the soil surface (**Figure 23.24b**). Unlike most plants, whose seeds germinate in the soil, many mangroves are viviparous; that is, their seeds germinate while still attached to the parent tree, to form a propagule. The mature propagule drops from the tree and can float horizontally with the tide or change its density so that it floats vertically and is more likely to become lodged in the mud.

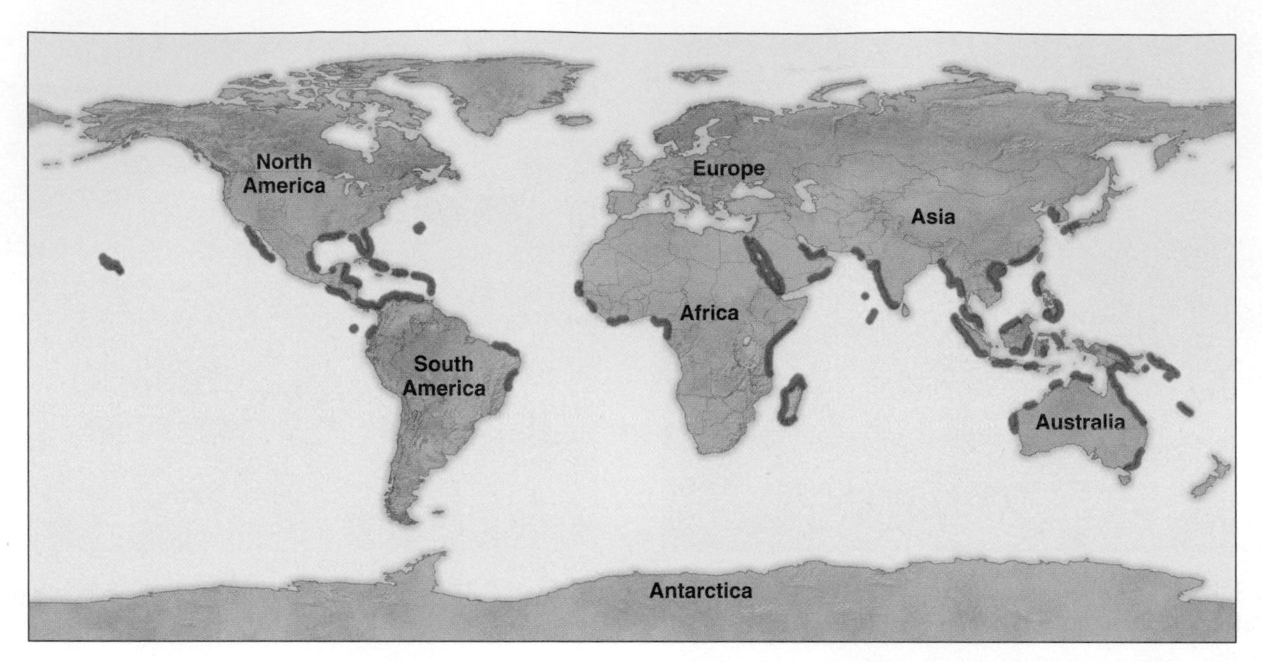


**Figure 23.23 Worldwide distribution of mangroves.** About 50 species exist between approximately 25° north and 25° south.

**ECOLOGICAL INQUIRY**

Why are mangroves generally found between about 25° north and 25° south of the equator?

(a)

(b)

**Figure 23.24 Adaptations of mangroves.** **(a)** Prop roots help trees remain stable in shifting substrates, Okinawa, Japan. **(b)** Pneumatophores, or aerial roots, improve oxygen transport to roots in black mangroves, *Avicennia germinans*, Florida.

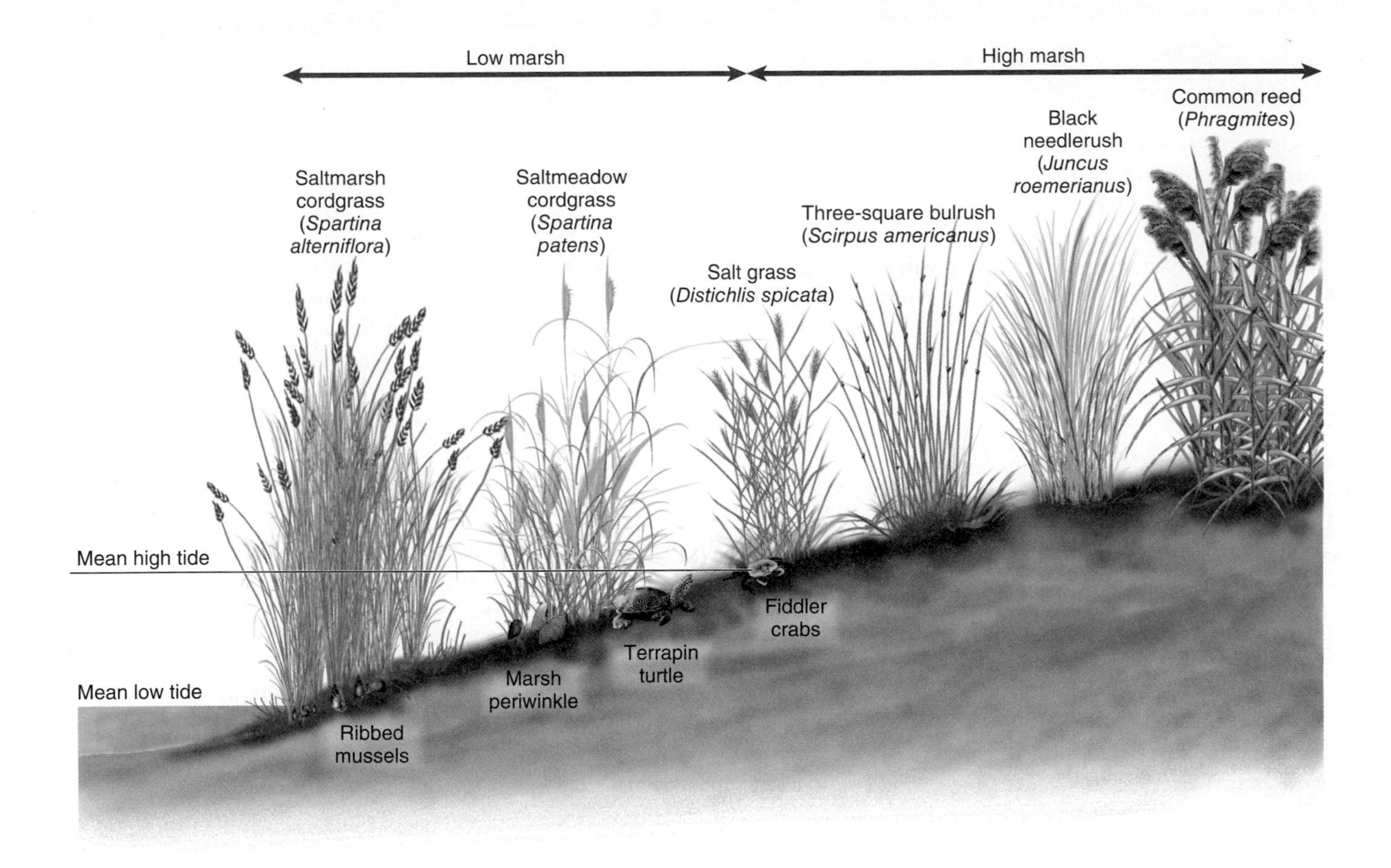


**Figure 23.25** Salt marsh zonation as seen on the U.S. East Coast.

Salt marshes have a distinct zonation, with more salt-tolerant species in the low marsh closer to the shore (**Figure 23.25**). In North America, low-marsh plants such as saltmarsh cordgrass, *Spartina alterniflora,* are inundated by the tide on a daily basis, often for up to 12 hours a day. High-marsh plants are inundated only by higher than average tides, such as would occur during a spring tide. Salt marshes and mangrove forests often occur along the banks of estuaries, where tidal flow influences the salinity of river mouths as they reach the ocean. Growth rates of both saltmarsh plants and mangroves are prodigious, and together with the growth of associated algae, these systems are among the most productive biomes on Earth.

**Animal Life:** A variety of insect herbivores feed on both saltmarsh and mangrove vegetation, but the saline environment restricts many large grazers. Geese may feed on saltmarsh vegetation, and herbivorous crabs feed on mangroves. In Southeast Asia proboscis monkeys, *Nasalis larvatus,* often feed on mangroves. However, much vegetation dies and rots in place to be used by decomposers. Both salt marshes and mangroves are important nursery grounds for fish, and both provide habitat for birds. Mangrove roots are often the only hard substrate for marine algae, sponges, bryozoans, oysters, and barnacles.

**Effects of Humans:** Clearing of land for housing developments impacts both salt marshes and mangroves. Use of the wood for fires and boats can reduce the extent of mangrove forests. In Southeast Asia, mangroves have frequently been cleared to establish shrimp farms. Pollution from the world's many coastal cities reduces water quality in many areas. Coastal oil spills can also greatly impact these habitats, and restoration is difficult because the oil seeps into the soil and cannot easily be removed, as it could on the rocky intertidal. Following the *Deepwater Horizon* oil spill off the coast of Louisiana, which began on April 20, 2010, and lasted until September 19, 2010, oil impacted 800 km (500 miles) of coastline in Louisiana, Mississippi, Alabama, and Florida.

### Check Your Understanding

**23.3** Coral reefs, kelp forests, and sea grass beds all occur in shallow water. What abiotic factors influence their distribution?

## SUMMARY

**23.1 Winds Strongly Affect Ocean Currents and Tidal Range**

- Surface winds and the Coriolis force help create ocean currents (Figure 23.1).
- Where surface currents move water away from continents, coastal upwelling may result (Figure 23.2).
- Winds create waves, which may reach large sizes over long fetches (Figures 23.3, 23.4).
- Langmuir circulation can create small-scale circulation patterns (Figure 23.5).
- Deep-ocean currents are driven by thermohaline circulation and move water very slowly around the globe (Figure 23.6).

**23.2 Tides Are Caused by the Gravitational Pull of the Moon and the Sun**

- Normal tidal ranges are <1 m but may be greatly increased in funnel-shaped bays such as the Bay of Fundy and the Bristol Channel (Figures 23.7, 23.8).
- Global changes in sea level may affect the distribution of emergent coastal wetlands (Figure 23.A).

**23.3 Marine Biomes Are Determined by Water Temperature, Depth, and Wave Action**

- Oceans cover 70% of the Earth's surface and can be divided into different zones according to proximity to shore, depth, and light availability (Figures 23.9, 23.10).
- Free-swimming ocean organisms are known as nekton (Figure 23.11).
- On ocean bottoms some organisms live around hydrothermal vents (Figures 23.12, 23.13).
- Coral reefs grow in shallow, clear water at depths of 1–100 m where temperatures are at least 20°C (Figure 23.14).
- In cooler, temperate waters kelp forests occur where substrates are hard (Figures 23.15, 23.16).
- On soft, sandy, or muddy substrates, sea grasses are present in both cool and warm water (Figures 23.17, 23.18).
- The rocky intertidal can be divided into the supralittoral, upper littoral, mid-littoral, and lower littoral zones (Figures 23.19, 23.20).
- Little plant life exists on sandy shores until the supralittoral zone, where sand dunes may occur (Figure 23.21).
- Mangrove forests and salt marshes occur in secluded bays and estuaries of tropical and temperate intertidal areas, respectively (Figures 23.22–23.25).

## TEST YOURSELF

**1.** What is not one of the four factors that influence wave size?
  **a.** Duration time of wind
  **b.** Water temperature
  **c.** Water depth
  **d.** Length of fetch
  **e.** Wind speed

**2.** When the sun, moon, and Earth form a straight line, the condition is known as a:
  **a.** Low tide
  **b.** High tide
  **c.** Neap tide
  **d.** Eclipse
  **e.** Syzygy

**3.** The greatest long-term sea-level changes in the U.S. over the past 30 years have been seen in:
  **a.** The Northeast
  **b.** The Pacific Northwest
  **c.** California
  **d.** Hawaii
  **e.** Louisiana

**4.** The deepest zone in the open ocean is the:
  **a.** Epipelagic
  **b.** Mesopelagic
  **c.** Bathypelagic
  **d.** Hadal
  **e.** Abyssal

**5.** In aquatic environments, plants are usually found in the _____ zone near the surface of the water, where light is able to penetrate.
  **a.** Aphotic
  **b.** Littoral
  **c.** Photic
  **d.** Limnetic
  **e.** Photosynthetic

**6.** Around hydrothermal vents, chemoautotrophic bacteria use _____ to produce organic compounds.
  **a.** Hydrogen sulfide
  **b.** Iron ore
  **c.** Hot magma
  **d.** Magnesium
  **e.** Copper sulfate

7. The highest zone of the rocky intertidal is the:
   a. Upper littoral
   b. Supralittoral
   c. Mid-littoral
   d. Lower littoral
   e. Euphotic

8. Pneumatophores are:
   a. Mangrove prop roots
   b. Mangrove aerial roots
   c. Marram grass roots
   d. Polychaete tubes
   e. Mangrove propagules

9. Plant life is absent at hydrothermal vents.
   a. True
   b. False

10. There is little plant life in the upper littoral zone.
    a. True
    b. False

## CONCEPTUAL QUESTIONS

1. What are some of the different factors that affect water circulation in the oceans?
2. Explain how tides are generated and the distinguishing features between a neap tide and a spring tide.
3. Describe the processes that allow life to thrive around hydrothermal vents in the absence of light.

Vancouver, British Columbia.

CHAPTER 24

# Freshwater Biomes

## Outline and Concepts



Bodies of fresh water or estuaries are particularly susceptible to global change, especially in urban areas where pollutants can accumulate. Where sewage treatment and storm water runoff share the same pipes, increased runoff may overwhelm the system, causing raw sewage to spill into local waterways. For example, metropolitan Vancouver, in British Columbia, Canada, is serviced by an aging sewage-handling and treatment plant that combines storm water and sewage in the same pipes. The Vancouver sewage system currently overflows 30% of the time, spilling raw sewage into Burrard Inlet and the Strait of Georgia. As far back as 2002, government reports noted that the highest concentrations of fecal coliform bacteria and other pathogens occurred in local water bodies following rain. Vancouver's goal is to change from a combined to a separate system, but this is not projected to be complete until 2050. Taking into account climate change, the volume of rain in this area is expected to increase 18% by 2050, placing even more strain on the system.

Freshwater biomes are traditionally divided into standing-water habitats, termed **lentic**, from the Latin *lenis,* meaning calm, and running-water habitats, called **lotic**, from the Latin *lotus,* meaning washed. Lentic habitats include lakes and ponds. Lotic habitats include fast-moving streams and rivers. Freshwater habitats may also occur in shallow wetlands, which can be found at the margins of both lentic and lotic biomes, and in estuaries, which may have considerable freshwater and marine inputs. In this chapter we will discuss each of these four freshwater biomes; but first we discuss the properties of water, which greatly influences life in freshwater biomes.

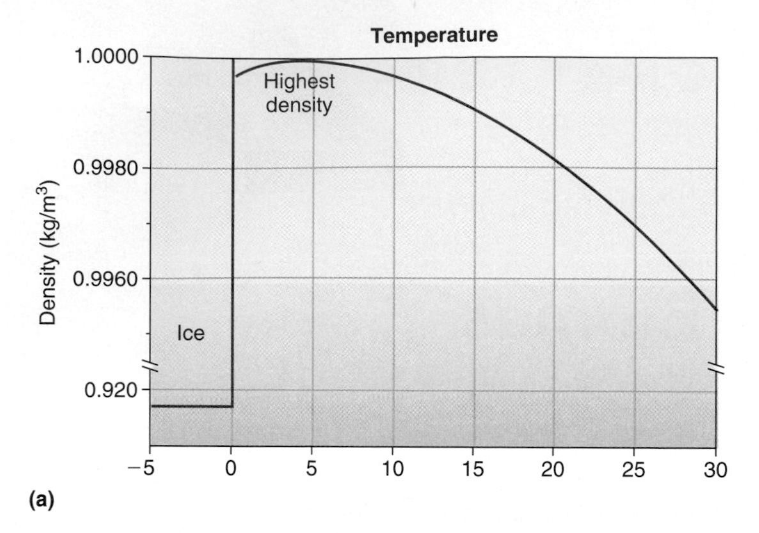


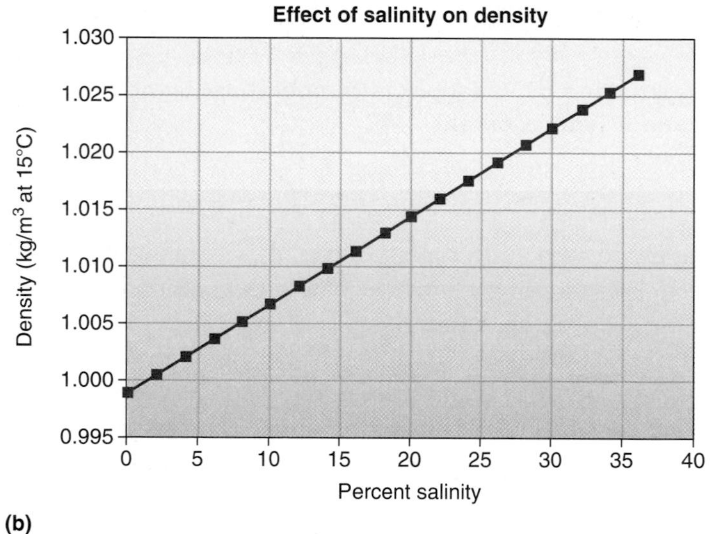


**Figure 24.1** **The physical properties of water.** **(a)** Water is at its most dense at 4°C and its least dense as ice at 0°C. **(b)** The density of water increases as salinity increases.

## 24.1 The Properties of Fresh Water Vary Dramatically with Temperature

The ecology of freshwater habitats is governed largely by the unusual properties of water. First, water is at its most dense at 4°C (**Figure 24.1a**). As water warms or cools from 4°C, it becomes less dense. At 0°C water freezes and is in its least dense state, so ice floats on unfrozen water. A second property of water is that it becomes more dense with increasing salinity (**Figure 24.1b**). This means that where fresh water and salt water mix, as in estuaries, the fresh water stays on top of the salt water.

We can examine how the physical properties of temperate freshwater systems vary seasonally. The relationship between density and temperature explains why bodies of water like lakes and rivers freeze from the top down and why free-flowing water is at the bottom of a frozen lake or pond (**Figure 24.2a**). From a fish's point of view this property is advantageous, because a frozen surface insulates the rest of the lake from freezing. If ice sank, all temperate lakes would freeze solid in winter, and no fish would exist in lakes outside the Tropics. In winter, water temperature usually increases with depth. Oxygen content is depleted toward the lake bottom by the respiration of benthic organisms.

In the spring, ice melts, water warms from 0°C to 1–4°C and sinks, mixing the water. Spring storms also mix the water layers, creating uniform conditions of temperature and oxygen (**Figure 24.2b**). This mixing is termed the **spring overturn.** In deeper temperate lakes, in the summer, three layers are present (**Figure 24.2c**). An upper layer, called the **epilimnion,** is warmed by the sun and mixed well by the wind. Below this lies a transition zone known as the **thermocline,** where the temperature declines rapidly. Lower still is the **hypolimnion,** a cool layer too far below the surface to be much warmed and with low light levels. Without much light in the hypolimnion, photosynthesis is absent and oxygen supply is low. Furthermore, what oxygen exists is used by bacteria decomposing material on the lake bottom. Organisms that need high oxygen levels usually cannot live in the hypolimnion. Algal blooms frequently occur in the epilimnion, fueled by high temperatures, high oxygen, and nutrients brought up from the bottom during spring overturn.

In the fall the upper layers cool, and as their density increases they sink, forcing the bottom layers upward (**Figure 24.2d**). The water in the lake is thoroughly mixed by this fall overturn and by storms, and the thermocline disappears. Later, the water molecules closest to the surface become cooled below 4°C, and eventually ice forms and prevents wind from mixing the layers, causing winter stratification. Not all temperate lakes undergo stratification. Shallow ponds, $<3$ m (10 feet), often do not stratify, because even moderate winds can mix all layers. Bodies of water deeper than 4 m (13 feet) almost always stratify, and you may notice the difference in temperatures when swimming and diving in these lakes.

In contrast to temperate lakes, tropical lakes are more often isothermal; that is, all the water is at the same temperature. At most, only a weak temperature gradient exists from top to bottom. Some mixing occurs based on seasonality and wind, but deep lakes are generally unproductive, with oxygen-poor, fishless lower depths. Because of a lack of mixing, tropical lakes may accumulate large amounts of $CO_2$ or $SO_2$, which can be released when the lower layers of the lake are suddenly brought up to the surface. This occurred in 1986 in Cameroon, when between 1,700 and 1,800 people, many livestock, and much wildlife were asphyxiated from a mixture of $CO_2$ and water droplets that rose violently from Lake Nyos, perhaps as the result of volcanic activity.

Like the ocean, lakes have zonations based on the availability of light. The upper layer, to which sufficient light penetrates to allow photosynthesis, is the photic zone. Below, in

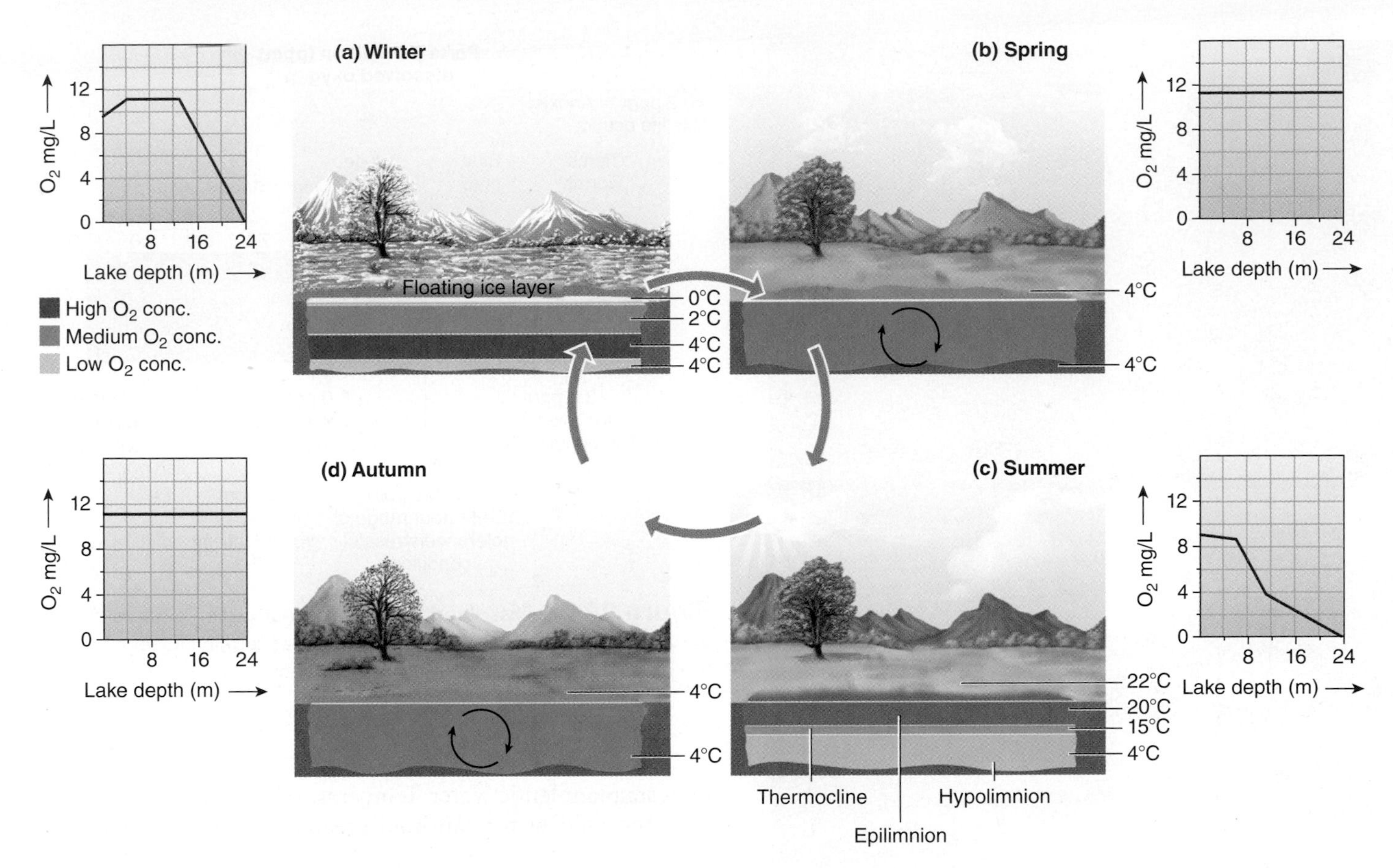


**Figure 24.2 Annual cycle of a temperate lake.** Cross section of a temperate lake with oxygen and temperature profiles with depth for each season. **(a)** The lake surface freezes in winter. **(b)** When the ice melts in the spring, the cold water again sinks and mixes the lake. **(c)** In the summer, the warmest water occurs at the surface, and water temperature decreases with depth. **(d)** Cold air temperatures in the fall cool the upper layers and this dense cold water sinks, thoroughly mixing the lake. Oxygen levels do not vary with depth in the spring and fall when mixing occurs, but generally decrease with depth in summer and winter.

**ECOLOGICAL INQUIRY**

How do tropical lakes differ from temperate lakes?

the relative darkness, is the aphotic zone, where organisms subsist on the rain of material from above. The depth of the photic zone depends on light availability and water clarity. The level at which photosynthate production equals the energy used up by respiration is the lower limit of the photic zone and is known as the **compensation point**. Below the compensation point vegetation cannot survive. In the summer, in temperate lakes, the compensation point is usually above the thermocline.

The degree of productivity in lakes affects their plant and animal life. The least productive lakes are termed **oligotrophic** (**Figure 24.3a**). Such lakes generally have a low nutrient content, largely as a result of their underlying substrate and young geologic age. Young lakes have not had a chance to accumulate as many dissolved nutrients as have older ones. Oligotrophic lakes are relatively clear, and their compensation levels may lie below the thermocline. If so, photosynthesis can take place in the hypolimnion, adding oxygen, so that oxygen levels in oligotrophic lakes are generally high. Low nutrient concentrations keep the algae and rooted plants in the epilimnion sparse, and little debris rains down upon the inhabitants of the hypolimnion. As a result, oligotrophic lakes are clear and many contain few fish.

Eventually nutrients begin to accumulate in oligotrophic lakes. Sediments are deposited, and both algae and rooted vegetation begin to thrive. Organic matter accumulates more rapidly on the lake bottom, the respiration of bacteria involved in decomposition increases, and the oxygen levels of the water decrease. Fish such as trout are excluded by bass and sunfish, which thrive at lower oxygen levels. This process

(a)

(b)

**Figure 24.3** **Lakes vary in their nutrient levels.** **(a)** Oligotrophic lakes are relatively young, clear, nutrient-poor, and well oxygenated. Moraine Lake, Banff National Park, Canada. **(b)** Eutrophic lakes are older, richer in nutrients, and have poorer water clarity. A lake in Wisconsin.

of aging and degradation is natural and is termed **eutrophication**; its end result is a **eutrophic** lake (**Figure 24.3b**).

Another important property of freshwater biomes is **dissolved oxygen** (DO), the amount of molecular oxygen that occurs in the spaces between water molecules and that supports aquatic life. Oxygen enters the water directly via diffusion from the atmosphere, from aquatic plants or algae that release it via photosynthesis, or via waterfalls and water tumbling over rocks, which traps air. Lotic water, such as fast-flowing streams, usually contains higher levels of DO than stagnant lentic water. Temperature affects dissolved oxygen, and cold water can hold greater amounts of DO than warm water.

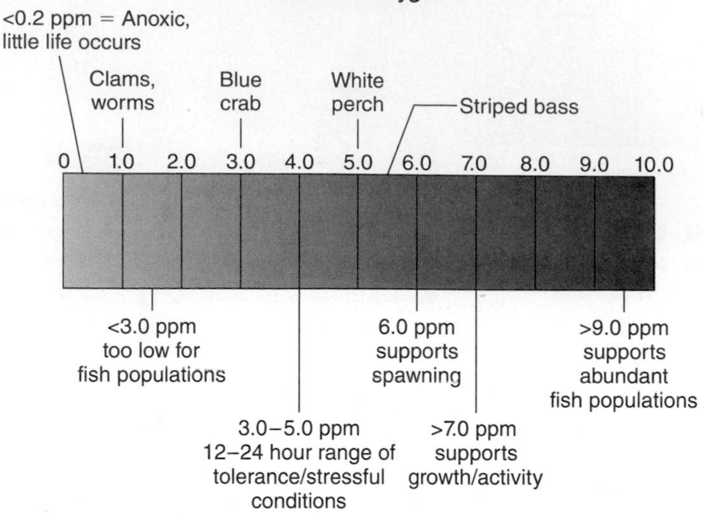


**Figure 24.4** **Dissolved oxygen in freshwater biomes can vary widely.** Some species that can exist at certain DO levels, and the processes that can be supported, are shown.

Bacteria and fungi use oxygen as they decompose dead organic matter in lake and river bottoms. Large amounts of dead organic matter can therefore result in very low DO, too low to support fish populations. Fish populations around sewage discharge areas in Vancouver are small for this reason. Dissolved oxygen is frequently measured as parts per million (ppm) (**Figure 24.4**). Areas with dissolved oxygen of $<0.2$ ppm are called **anoxic** and are unable to support most life. Dissolved oxygen of $<3.0$ ppm is too low for many fish populations, though invertebrates such as clams may survive. Many adult fish species can survive in 3.0–5.0 ppm DO, but spawning and growth of young fish may require 6.0 ppm and above. In lakes, oxygen levels are usually lowest in deeper waters, except in the spring and fall overturns, when water mixing creates uniform dissolved oxygen levels throughout the lake (see Figure 24.2a–d). Low DO levels and eutrophication are just some of the many problems plaguing freshwater biomes (see **Global Insight**).

### Check Your Understanding

**24.1** What would happen to aquatic life in temperate lakes if ice were denser than water?

Global Insight

## Aquatic Fauna in the U.S. Are Threatened by Global Change

Around the world, and throughout the United States, aquatic organisms are particularly threatened by environmental change. The restriction of most aquatic species to specific, relatively small areas has made them easily accessible to humans and readily exploited. Fish and mammal populations have frequently been overharvested. Riversides have been seen as prime locations for housing and industry, both of which discharge pollutants. Some pollutants come from a specific point source such as an industrial discharge outlet; others come from nonpoint sources such as runoff of fertilizer or pesticides from agricultural areas. Point and nonpoint pollution sources concentrate in rivers and lakes, where there is no escape for aquatic organisms. In addition, more than half the wetlands in the U.S. have been drained and filled. Thousands of miles of rivers have been channelized or straightened, and in the U.S. more than 75,000 high dams block 600,000 miles of rivers, equivalent to about 17% of all U.S. river miles. Introduced species also threaten native species, and nearly 30,000 species of exotic water plants, mollusks, fish, or disease-carrying microbes have been brought to the U.S. Approximately 300 invasive species now inhabit the San Francisco Bay and 100 occur in the Chesapeake Bay. Some of these, such as the Asian snakehead fish, *Channa argus*, the African catfish, *Clarias gariepinus*, the Asian clam, *Potamocorbula amurensis*, and the Chinese mitten crab, *Eriocheir sinensis*, were imported for the pet trade or for use in aquaculture programs, but were released in the wild (**Figure 24.A**). Freshwater animals in the U.S. are disappearing at a rate 2–5 times faster than native land animals. Concern is greatest for native species of fish, snails, mussels, crayfish, salamanders, and frogs and toads. At present, 20–50% of these taxa are threatened with extinction.

**Figure 24.A Freshwater habitats are particularly threatened by invasive species.** **(i)** Asian snakehead fish, *Channa argus*, can grow more than a meter in length. **(ii)** African catfish, *Clarias gariepinus*. **(iii)** Asian clam, *Potamocorbula amurensis*. **(iv)** Chinese mitten crab, *Eriocheir sinensis*, with dense patches of hair on the claws that look like mittens.

## 24.2 Freshwater Biomes Are Determined by Variations in Temperature, Light Availability, Productivity, and Oxygen Content

Fresh water occurs in lakes and ponds, rivers and streams, wetlands, and estuaries. In this section we discuss the unique properties of each of these biomes.

### Lakes (Figure 24.5)

**Physical Environment:** Lakes are landlocked standing-water biomes that vary in properties according to depth and location. Deep-water temperate lakes may exhibit seasonal stratifications, but tropical lakes and shallow temperate lakes may have fairly uniform physical properties. Most lakes have a natural outflow in terms of a river or stream. In the absence of such an outlet, evaporation removes water, often creating saline lakes such as the Caspian Sea, the Great Salt Lake, the Aral Sea, and the Dead Sea.

**Location:** Lakes occur throughout all the continents of the world. The locations of the world's largest lakes are shown in **Figure 24.6a,** and their size is appreciated when shown alongside the landmass of Great Britain (**Figure 24.6b**). **Table 24.1** gives more details of the sizes and features of these lakes. Lake Baikal has the largest volume of fresh water, containing more than all the Great Lakes combined.

Most lakes are found in mountainous areas, areas of tectonic activity, or areas of recent glaciation. Glaciation removes topsoil, leaving pockmarked bedrock where water readily accumulates. Because of this, more than 60% of the world's lakes occur in Canada, which was recently glaciated. Over time, most lakes will fill in with sediments or drain from their basins. The exceptions are lakes such as Lake Baikal and Lake Tanganyika, which are formed from rift valleys, areas created by the action of a geologic rift. These lakes are still deepening as the Earth's plates pull apart. Eventually these lakes may join with the oceans, as the Red Sea is thought to have done.

Large lakes are usually referred to with the term "Lake" used before the name, as in Lake Superior. Some large lakes are referred to as seas, as in the Caspian Sea, even though, if they are landlocked, they are technically lakes. Smaller lakes are referred to with the term "Lake" used after the name, as in Emerald Lake. Lakes are generally regarded as being greater than 2 ha in area, and ponds are less than 2 ha. Although lakes are bigger than ponds, it is estimated that there are nine times as many ponds as lakes.

**Plant Life:** In addition to free-floating phytoplankton, lentic habitats may have rooted vegetation, termed emergent vegetation, which often emerges above the water surface, such as cattails, *Typha* spp., plus deeper-dwelling aquatic plants and algae.

**Animal Life:** Animals include fish, frogs, turtles, crayfish, snails, and many species of insects. In tropical and subtropical lakes, alligators and crocodiles commonly are seen. Tropical lakes contain huge numbers of fish species. The rift valley lakes in Africa—Lake Tanganyika and Lake Malawi—together with Lake Victoria, contain over 700 species of fish, including vast numbers of cichlid species (**Figure 24.7**).

**Effects of Humans:** Eutrophication can be greatly speeded up by human activities, which increase nutrient concentrations, especially of phosphorous, through the introduction of sewage and fertilizers from agricultural runoff. We will

**Figure 24.5** Two Harbors, Minnesota, Lake Superior, U.S.

**Table 24.1** The 12 largest lakes in the world by surface area.*

| Lake | Size ($km^2$) | Features |
|---|---|---|
| 1. Caspian Sea | 436,000 | One-third as salty as the ocean |
| 2. Lake Superior | 82,100 | Largest freshwater lake in the world |
| 3. Lake Victoria | 68,800 | Largest freshwater lake in Africa |
| 4. Lake Huron | 59,600 | |
| 5. Lake Michigan | 57,800 | |
| 6. Lake Tanganyika | 32,900 | Second deepest, 1,470 m, largest in Africa |
| 7. Lake Baikal | 31,500 | Oldest and deepest lake in the world, 1,637 m |
| 8. Great Bear Lake | 31,200 | |
| 9. Lake Malawi | 29,600 | |
| 10. Great Slave Lake | 28,400 | Deepest in North America, 614 m |
| 11. Lake Erie | 25,700 | |
| 12. Lake Winnipeg | 24,200 | |

*Numbers refer to locations in Figure 24.6.

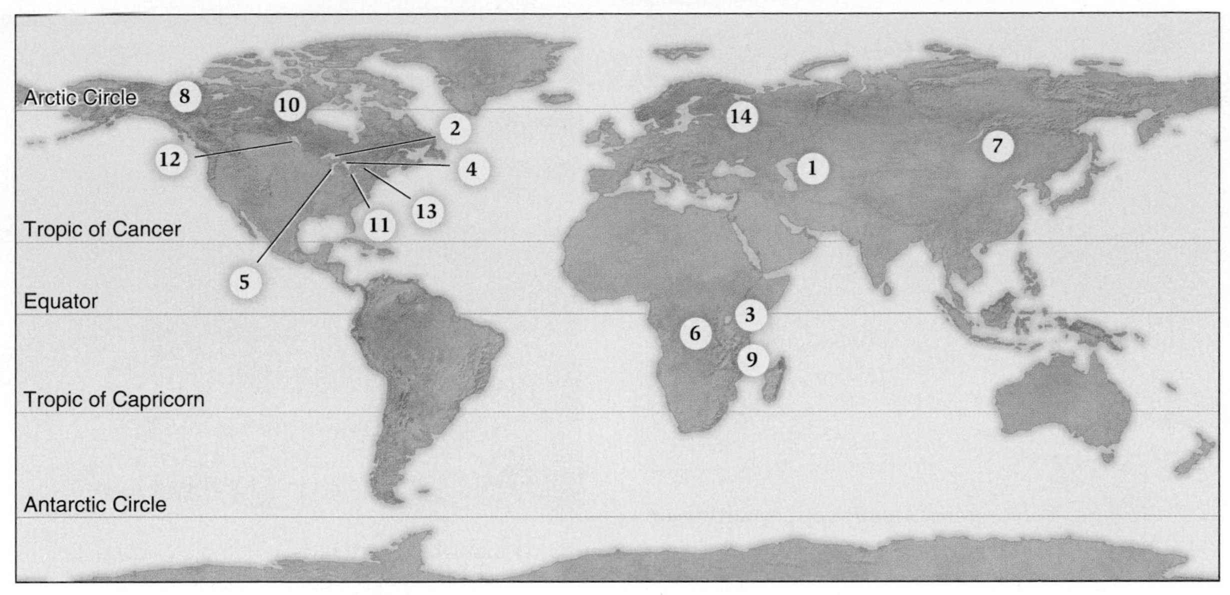


(a)

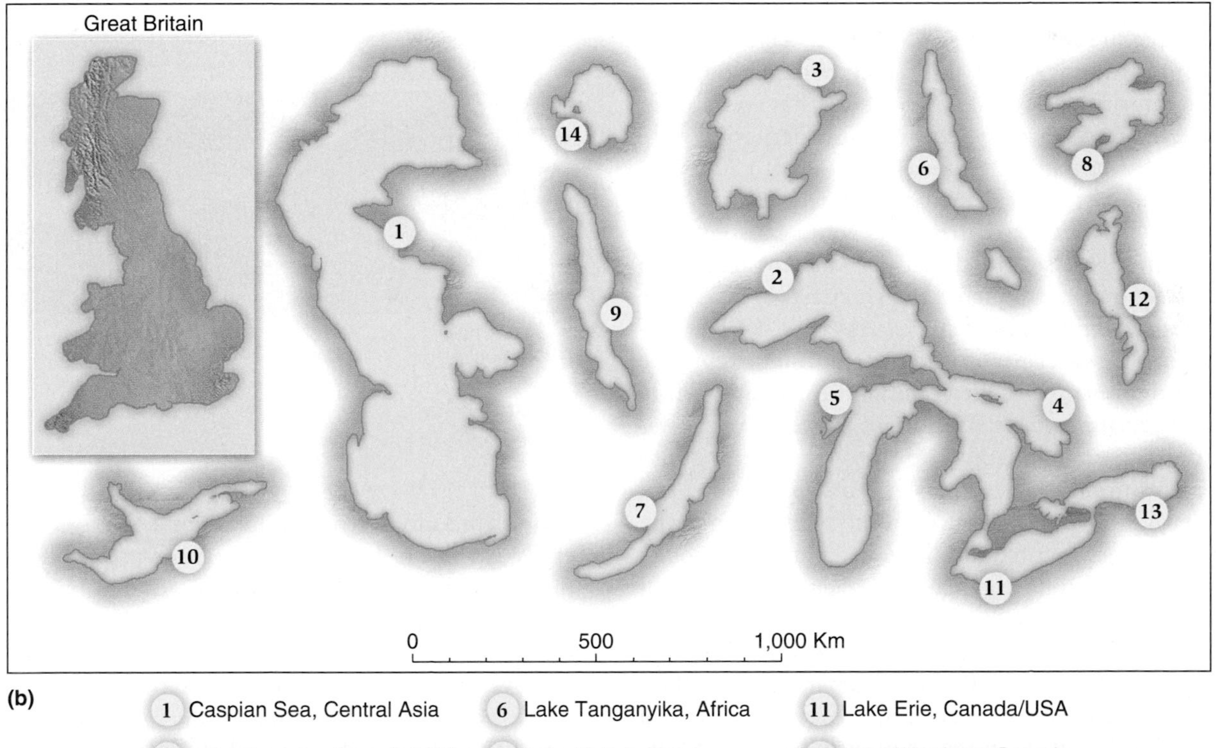


(b)

1 Caspian Sea, Central Asia
2 Lake Superior, Canada/USA
3 Lake Victoria, Africa
4 Lake Huron, Canada/USA
5 Lake Michigan, USA
6 Lake Tanganyika, Africa
7 Lake Baikal, Russia
8 Great Bear Lake, Canada
9 Lake Malawi, Africa
10 Great Slave Lake, Canada
11 Lake Erie, Canada/USA
12 Lake Winnipeg, Canada
13 Lake Ontario, Canada/USA
14 Lake Ladoga, Russia

**Figure 24.6 Major lakes of the world.** **(a)** Location and **(b)** size. (Largest to smallest.)

**ECOLOGICAL INQUIRY**

Why are 60% of the world's lakes in Canada?

**Figure 24.7** **Cichlids, *Boulengerochromis* spp., guard their nest from a terrapin, Lake Tanganyika, Tanzania.**

**Figure 24.8** **Wetlands.** Pothole wetland, South Dakota, with trumpeter swan, *Cygnus buccinator,* exhibiting neck band.

examine this topic in more detail in Chapter 27. This human-induced eutrophication is called **cultural eutrophication.**

In the African rift valley lakes, Nile perch that were introduced in the 1950s are threatening many of the native cichlid species with extinction. In some areas of North America, exotic species of invertebrates, such as crayfish, are outcompeting native crayfish species. Introduced zebra mussels have outcompeted native mussels in the Great Lakes. Changes in lake water pH have been frequent over the past 50 years due to acid rain, precipitation with a pH of less than 5.6. As discussed in Chapter 6, water pH is important in influencing the distribution of organisms in lake communities. Young fish in particular are very susceptible to changes in water pH (refer back to Figure 6.16). Finally, diversion of water feeding into some lakes can diminish their size. The Aral Sea in Central Asia was once the fourth largest lake in the world. Beginning in the 1960s the rivers that fed into it were diverted for Soviet irrigation projects. By 2007 the lake had declined to 10% of its original size.

Wetlands (**Figure 24.8**) **Physical Environment:** Wetlands are areas covered by shallow water or with saturated soil. They may develop at the margins of both lentic and lotic habitats. They act as an interface between terrestrial and aquatic biomes. Wetlands are areas regularly saturated by surface water or groundwater. They can range from marshes and swamps to bogs and more temporary vernal pools and wet meadows (**Figure 24.9**). Many are seasonally flooded when rivers overflow their banks, lake levels rise, snow melts, or heavy rainfall occurs. Because of generally high nutrient levels, and low levels of water, oxygen levels are fairly low. Temperatures vary substantially with location.

Marshes are areas frequently or continually inundated by water and contain soft-stemmed, grassy vegetation (**Figure 24.10a**). They often occur in poorly drained

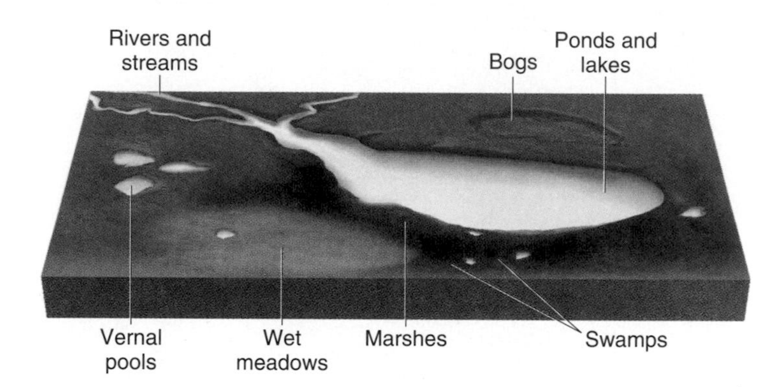


**Figure 24.9** **Various types of wetlands.** Wetlands may border lentic or lotic biomes.

depressions along streams, lakes, and rivers. Water levels generally vary from a few inches to a few feet. The soil is rich in nutrients. African swamps are dominated by papyrus, *Cyperus papyrus.* Wet meadows are a little drier than marshes and for most of the year are without standing water, though a high water table leaves the soil saturated. Vernal pools are depressions that are seasonally filled with shallow water, and range in size from small ponds to large shallow lakes. In the Midwest prairie, vernal pools are known as prairie potholes.

Swamps are wetlands dominated by woody plants (**Figure 24.10b**). They have saturated soils and much standing water at certain times of the year. *Melaleuca* trees are present in Australian swamps. Bogs are characterized by spongy peat deposits and acid waters. They receive most of the water from rainfall and are nutrient-poor. *Sphagnum* mosses predominate in northern bogs.

**Location:** Wetlands occur worldwide, except in Antarctica. Bogs are common in the taiga and tundra biomes. The largest wetlands in the world are shown in **Figure 24.11.** The largest of these are the west Siberian lowlands and the

**Figure 24.10** **Wetlands.** **(a)** Marshes with grassy vegetation, Baxter State Park, Maine. **(b)** Swamps dominated by woody vegetation, cypress trees, Tallahassee, Florida.

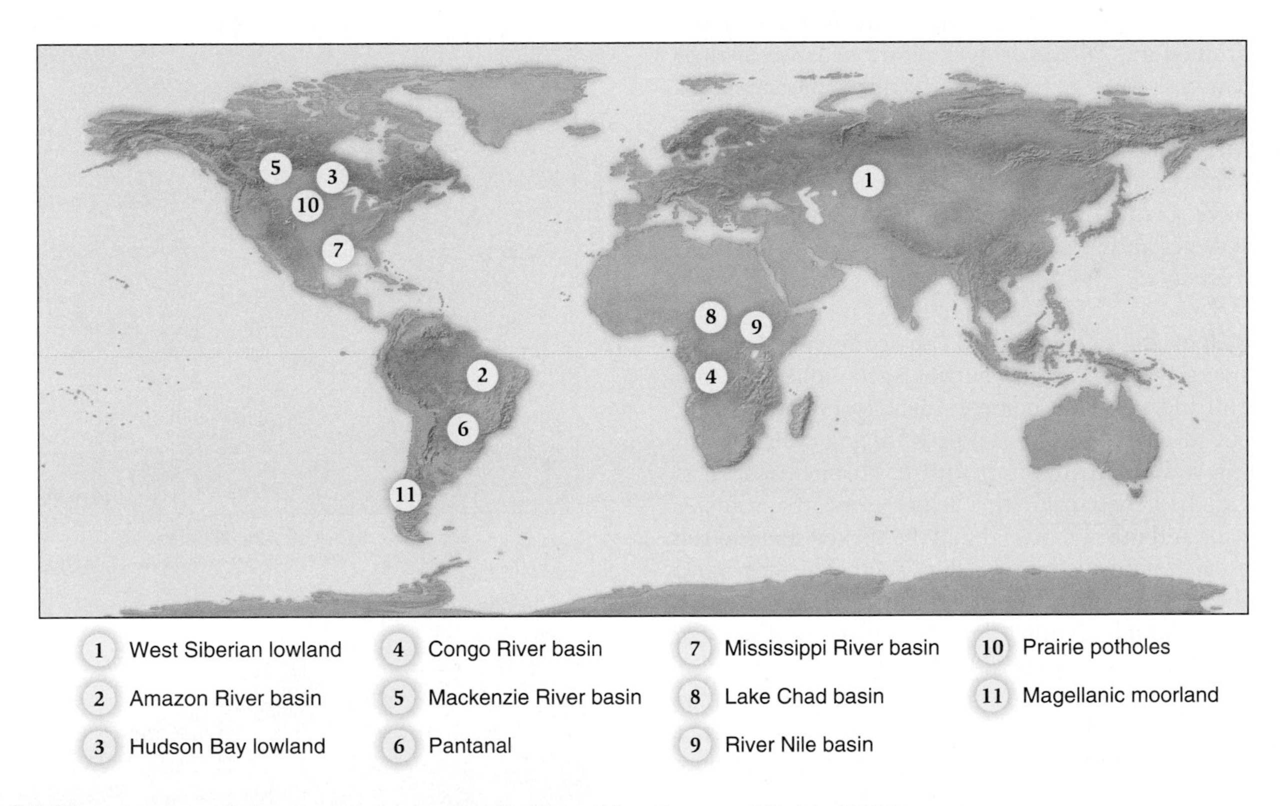


**Figure 24.11** **Location of the world's largest wetlands.** (After Fraser and Keddy, 2005.)

seasonally flooded Amazon basin. The Pantanal, another large South American wetland, is 180,000 km$^2$ (72,000 mi$^2$) in area, rich in species, and designated a UNESCO Biosphere Reserve. Other famous, smaller wetlands include the Florida Everglades, the Okavango Delta, and the Danube Delta. The Everglades is North America's most extensive flooded grassland, 20,000 km$^2$ (8,000 mi$^2$) in size, one-tenth the size of the Pantanal. The Okavango Delta in central Africa supports many large mammals, which move seasonally with the rise and fall of floodwaters. The Danube Delta is one of the largest wetlands in Europe.

**Plant Life:** Wetlands are among the most productive and species-rich areas in the world. In North American marshes, floating plants include natives such as duckweed, *Lemnaceae* spp. Invasives, such as water hyacinth, *Eichhornia crassipes,* are also common. Rooted vegetation includes bulrushes, lilies, sedges, and cattails. In southern U.S. swamps, cypress, *Taxodium* spp., and tupelo, *Nyssa aquatica*, are common. Willows, *Salix* spp., are common in the U.S. Northwest. Dry vernal pools often bloom with a profusion of flowers around the shoreline in the spring. In all cases, species are adapted to low soil oxygen. The nutrient-poor conditions of bogs encourages the growth of carnivorous plants such as pitcher plants, *Sarracenia* spp., and sundews, *Drosera* spp., which supplement their nitrogen intake with insect material (refer back to Figure 7.4).

**Animal Life:** Most wetlands are rich in animal species. Marshes are a prime habitat for wading and diving birds. In addition, they are home to a profusion of invertebrates, from mosquitoes and dragonflies to freshwater shrimp and crayfish. In marshes and swamps, vertebrate predators include many frogs; turtles; snakes such as the cottonmouth, *Agkistrodon piscivorus;* otters; muskrats; and alligators. The endangered American crocodile, *Crocodylus acutus,* is found in extreme south Florida, while different crocodile species thrive in other areas of the world, especially Africa and Australia.

**Effects of Humans:** Long mistakenly regarded as wasteland by humans, many wetlands, especially wet meadows, have been drained and filled for agriculture because of their rich soils. Peat collection for fuel and fertilizer has impacted many northern bogs. Invasive species can also be prevalent. Australian *Melaleuca* trees cover huge areas of Florida's Everglades. This conversion of wetlands to other uses has slowed as people have gradually gained an understanding of the importance of wetlands to flood control, water quality, and biological diversity.

Rivers (**Figure 24.12**) **Physical Environment:** In lotic habitats, flowing water prevents nutrient accumulations and phytoplankton blooms. The currents usually mix the water thoroughly in all areas of a river, providing a well-aerated habitat of relatively uniform temperature. Three broad areas of a river can be recognized: the headwaters; the transfer zone, where silt and other material may be picked up; and the deposition zone, where silt may be deposited. The slope and grain size of the material on the riverbed are the greatest in the headwaters, though channel depth, width, and discharge are lowest in this area (**Figure 24.13**). Mean velocity also increases with distance from the source. It often appears that high mountain streams flow quickly, but velocity is affected by friction with the river bed and banks. As the river gets deeper, relatively less water is in contact with the bottom and sides. Also, the roughness of the river bed is high in the headwaters but decreases further downstream where smoother stones or mud may be present. Oxygen level and water clarity are greater in headwaters than in the lower reaches of rivers, where more silt or alluvium is stored. Nutrient levels and temperatures are generally lower in headwaters. The floodplain is the flatland adjacent to a river that experiences occasional flooding.

There are a variety of river and stream classifications. Headwaters are often referred to as first-order streams (**Figure 24.14**). Where two or more first-order streams join

**Figure 24.12** South Fork of the Holston River, Virginia.

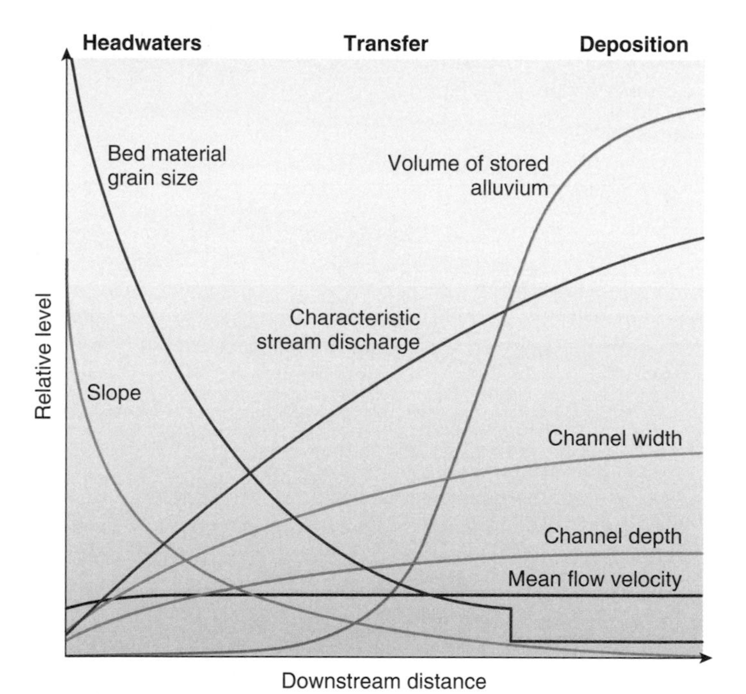

**Figure 24.13** **River zonation.** Rivers are often divided into three main areas: headwaters, transfer zone, and deposition zone. The physical features of rivers differ in these areas.

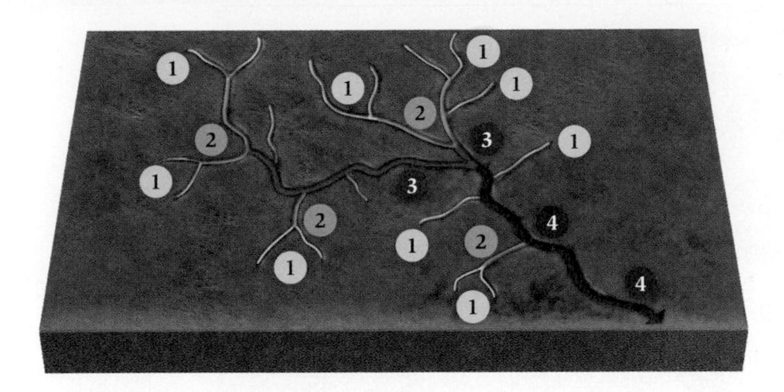


**Figure 24.14** **Stream classification.** Headwaters are referred to as first-order streams. Two or more first-order streams merge to form a second-order stream, and so on. Major rivers are generally fourth-order streams.

together, they form a second-order stream. Where two second-order streams meet, they form a third-order stream, and so on. By the time a fourth-order stream occurs, it is usually a major river. In the headwaters, or first-order streams, the stream gradient is usually quite steep and the stream cuts a narrow V-shaped channel (**Figure 24.15a**). The floodplain is minimal. Lower down in the transfer zone, the grade becomes shallower and the stream meanders more, often spilling its banks over a larger floodplain (**Figure 24.15b**). An exceptionally large river, with a high sediment load, may take on a braided appearance as it forms a delta of many smaller rivers before it discharges into a lake or ocean (**Figure 24.15c**). Here the channel is large and U-shaped and the floodplain is wide. The flow of water through the channel, known as the stream discharge, generally increases from headwaters to mouth. The discharge, $Q$, is given by the equation

$$Q = V \times W \times D$$

where $V$ is the velocity, $W$ is the average width, and $D$ is the average depth. Discharge usually varies seasonally with the rains. Because of frictional drag on the banks, river speed is generally greatest at the river center.

**Location:** Rivers are located on all continents. The major rivers of the world are shown in **Figure 24.16.** The longest river in the world, the Nile, at 6,484 km, discharges only 1,594 $m^3$ per second at its mouth. The Amazon, the second longest river, discharges 180,000 $m^3$ per second, more than the Nile and the next ten rivers combined. The Congo discharges 42,000 $m^3$ per second and is the second largest river in volume.

**Plant Life:** Nearly all rivers have narrow first-order headwaters where fallen leaves from surrounding forests are the primary food source for animals. This material is shredded by invertebrate animals and attacked by fungi and bacteria. In slow-moving streams and rivers, algae and phytoplankton may be present. Rooted vegetation may be concentrated

**(a) V-shaped valley, little floodplain**

**(b) Meandering profile, wider floodplain**

**(c) Braided appearance, large floodplain**

**Figure 24.15** **River valleys.** **(a)** Headwaters often flow over small rapids, creating areas of high flow and pools of relatively slower flow. **(b)** In the middle and lower reaches of a river, the riverbed may meander more. **(c)** Toward its mouth, a river may split into multiple streams, taking on a braided appearance before it discharges into a lake or ocean, though this is uncommon and typically occurs only in very large rivers carrying a high sediment load.

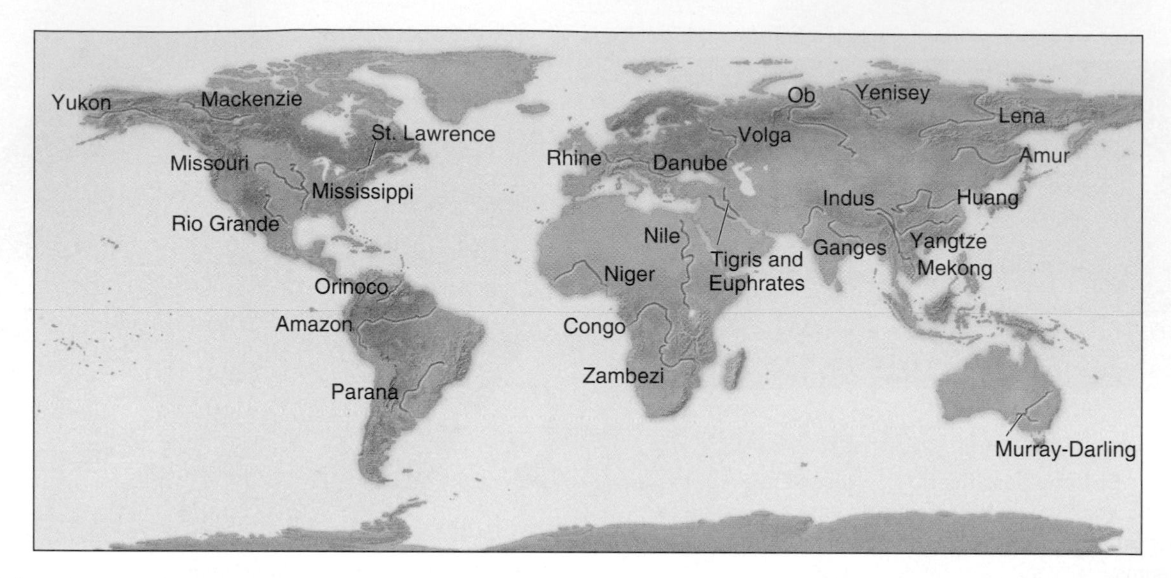


**Figure 24.16** **The world's major rivers.**

around the banks of rivers and streams, where the roots can tap into the groundwater. Such areas are called **riparian zones.**

**Animal Life:** Lotic habitats have a fauna different from that of lentic waters. In the headwaters especially, animals are adapted to stay in place despite an often strong current (**Figure 24.17**). Many of the smaller organisms are flat and attach themselves to rocks to avoid being swept away. Others live on the underside of large boulders, where the current is much reduced. Other invertebrates, such as caddis flies, spin nets on boulders and catch organic matter passing downstream. These species are known as collectors. Fish such as trout may be present in rivers with cool temperatures, high oxygen, and clear water. In warmer, murkier waters, catfish and carp may be abundant.

**Effects of Humans:** Animals of lotic systems are not well adapted for low-oxygen environments and are particularly susceptible to oxygen-reducing pollutants such as sewage. However, rivers are very resilient. Because of the constant flow of water, rivers are cleansed of pollutants relatively quickly and organisms can rebuild their numbers. Thermal pollution at power plants, where warm water is discharged, can increase the incidence of fish diseases, but in Florida warm water also provides a refuge for manatees in periods of cold weather. Diking and channelization reduces floods but also reduces floodplain area and increases water velocity, resulting in the river bottom being scoured of nutrients and organisms. Dams across rivers have prevented the passage of migratory species such as salmon.

(a)

(b)

**Figure 24.17** **River-adapted animals.** **(a)** This Chinese giant salamander, *Andrias davidianus*, is dorsoventrally flattened to minimize the chances of being washed away. It often occurs under boulders. **(b)** Caddis flies may spin small nets to collect fine particulate organic matter that drifts downstream.

### Estuaries (Figure 24.18)

**Physical Environment:** Estuaries are usually semi-enclosed bodies of water that have at least one source of freshwater feeding into them but also have a connection to the open sea. The word estuary is derived from *aestus,* the Latin word for tide, underscoring the fact that estuaries are typically at the mouths of rivers. The water salinity is often intermediate between fresh and salty and is referred to as brackish, but much depends on the relative

**Figure 24.18** **An estuary, Gower Peninsula, Wales.**

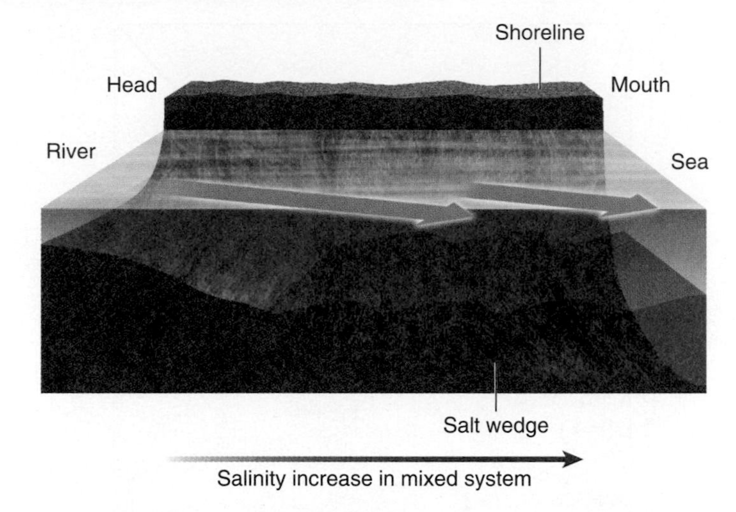


**Figure 24.19** **A salt wedge estuary.** Salt water, being denser than fresh water, usually occurs as a wedge beneath the fresh water at the mouths of estuaries.

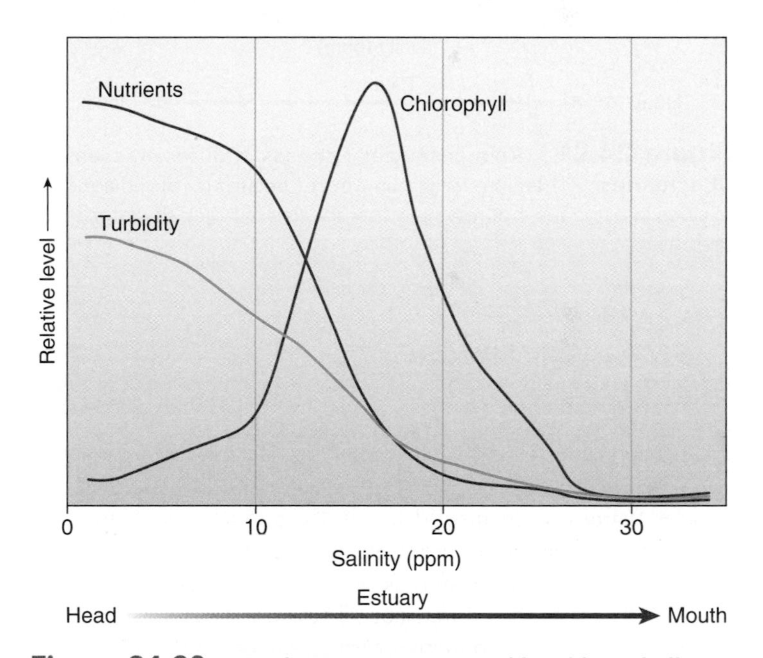


**Figure 24.20** **Production, as measured by chlorophyll, of phytoplankton and algae in estuaries.** Production is usually greatest at intermediate salinities in estuaries.

strengths of both the river feeding the estuary and the tide. Where the river feeding the estuary is large, estuarine conditions can extend well beyond the coast. Where the influence of the tide is strong, saline conditions may extend far up the estuary. Seasonal rains or snowmelt may lower the salinity of estuaries in spring and summer. Often the denser, cooler seawater may form a salt wedge under the lighter, warmer fresh water (**Figure 24.19**). The boundary between fresh and salty water is known as a pycnocline. Dissolved oxygen may be considerably lower below the pycnocline. However, turbulent currents can mix the fresh and saline components to form a well-mixed body of water.

**Location:** Estuaries are located worldwide, at the mouths of rivers.

**Plant Life:** Nutrients are imported from the land and the sea. As a result, estuaries are among the most productive biomes on Earth, with a high biomass of algae, phytoplankton, and some sea grasses. Production may be highest at intermediate salinities (**Figure 24.20**). High turbidity from river sediments at the head of the estuary limits light availability and phytoplankton, and lower nutrient levels at the mouth also limit productivity.

**Animal Life:** There is a mixture of freshwater and marine organisms in estuaries, with only a few brackish-water specialists, including the east Asian catfish, *Mystus gulio,* and the banjo catfish, *Bunocephalus coracoideus*. Although many species are adapted to either fresh water or salt water, few can tolerate the alternating cycles of fresh and salt water that occur in the middle areas of the estuary (**Figure 24.21**). However, for those species that can tolerate these conditions, nutrients are abundant and population densities of these species may be high. The reduced wave action of estuaries provides a refuge for weaker swimmers such as flounders and killifish, while the rich sediment and nutrient loads provide food for

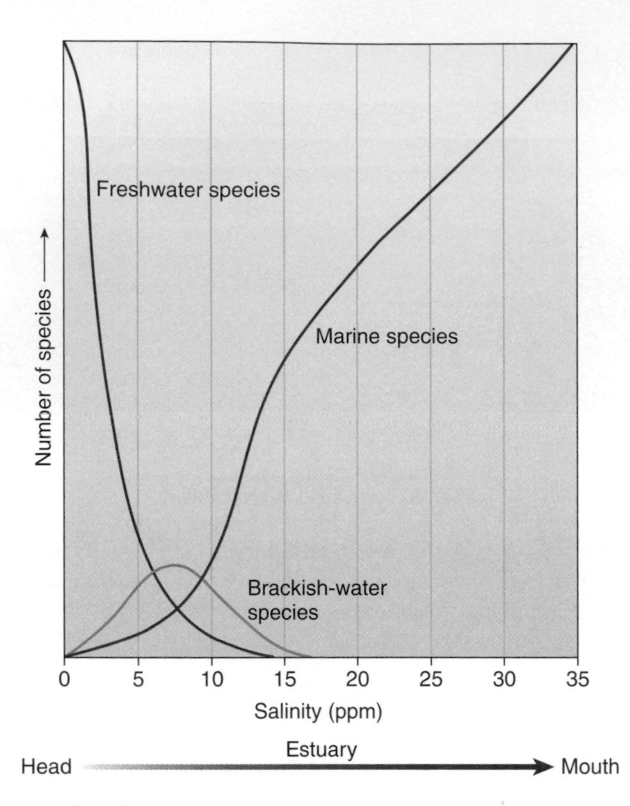


**Figure 24.21 Animal species richness in different areas of estuaries.** There are large numbers of freshwater and marine species in estuaries, but relatively few species of brackish-water specialists.

a variety of filter feeders such as clams, mussels, and oysters. Estuaries are safe places for crabs and other crustaceans to lay their eggs, away from the zooplankton-rich community of the open sea.

**Effects of Humans:** Of the 32 largest cities in the world, 22 are located on estuaries. As a result, many estuaries are under threat from pollution, direct exploitation, introduced species, and habitat destruction. Pollution events far upstream can also impact estuaries. For example, fertilizer runoff can cause eutrophication. Overfishing is also common. For example, oysters in the Chesapeake Bay were almost destroyed by overharvesting; as a result, the water column was not filtered and became more turbid. The release of invasive species from ships' ballast water is also rampant. Farther out at sea, the phosphate- and nitrate-rich runoff from estuaries can be so rich in plankton and bacteria that it is virtually devoid of dissolved oxygen. This oxygen-depleted water is toxic to animals. Discharge from the Mississippi River has caused a large dead zone in the Gulf of Mexico (look ahead to Figure 26.9). All benthic invertebrates have been killed, and fish cannot exist in such areas.

### Check Your Understanding

**24.2** What are the main differences between swamps and marshes?

## SUMMARY

**24.1 The Properties of Fresh Water Vary Dramatically with Temperature**

- Water is at its most dense at 4°C and its least dense as ice. Density also increases with salinity (Figure 24.1).
- Freshwater lakes may be seasonally stratified in temperature and nutrients. They may be clear and relatively unproductive, termed oligotrophic, or less clear and more productive, termed eutrophic (Figures 24.2, 24.3).
- Dissolved oxygen (DO) levels are critical to freshwater aquatic life (Figure 24.4).
- Cultural eutrophication and exotic species make freshwater biomes among the most threatened habitats on Earth (Figure 24.A).

**24.2 Freshwater Biomes Are Determined by Variations in Temperature, Light Availability, Productivity, and Oxygen Content**

- Lakes are lentic biomes varying in size and depth from the deepest, Lake Baikal, to the largest, the Caspian Sea. Tropical lakes contain vast numbers of fish species (Figures 24.5–24.7, Table 24.1).
- Wetlands are areas covered with relatively shallow water and include marshes, swamps, bogs, vernal pools, and wet meadows. Often viewed as wasteland, they are among the most productive and species-rich areas in the world (Figures 24.8–24.11).
- Rivers are lotic habitats of varying depth that occur worldwide. The physical properties of rivers vary greatly according to position in headwaters, transfer zone, or deposition zone (Figures 24.12–24.16).
- Plants and animals in lotic habitats may have adaptations that enable them to remain in place in swift currents (Figure 24.17).
- Estuaries are semi-enclosed bodies of water that have at least one source of fresh water but also have a connection to the open sea (Figures 24.18, 24.19).
- In estuaries, the production of phytoplankton and algae is often greatest at intermediate salinities. Species richness of resident animals is often low in these harsh environments, though the species that can cope with these conditions are often very common (Figures 24.20, 24.21)

## TEST YOURSELF

1. In the fall, in temperate lakes, the greatest DO levels are found in which layer?
   a. Epilimnion
   b. Thermocline
   c. Hypolimnion
   d. DO is equal at all depths of the lake
2. In winter, in cold-temperate lakes, the warmest water usually occurs:
   a. At the surface
   b. Just below the surface
   c. At intermediate depths
   d. At the lowest part of the lake
   e. Temperature is equal at all lake depths
3. In aquatic environments, plants are usually found in the _____ zone near the surface of the water, where light is able to penetrate.
   a. Aphotic
   b. Littoral
   c. Photic
   d. Riparian
   e. Photosynthetic
4. The upper layer of a temperate lake is called the:
   a. Hypolimnion
   b. Epilimnion
   c. Compensation level
   d. Thermocline
   e. Eutrophic zone
5. Most freshwater fishes spawn at DO levels greater than 6.0 ppm.
   a. True
   b. False
6. An example of a temporary wetland is a:
   a. Marsh
   b. Swamp
   c. Bog
   d. Wet meadow
   e. All of the above
7. The largest river in the world, by volume of discharge, is the:
   a. Nile
   b. Amazon
   c. Congo
   d. Mississippi
   e. Yangtze
8. The areas around rivers and lakes, where rooted vegetation taps into groundwater, are called:
   a. Lentic habitats
   b. Lotic habitats
   c. Wet meadows
   d. Vernal pools
   e. Riparian zones
9. Production of phytoplankton and algae is greatest at which areas of an estuary?
   a. Nearest the river feeding it
   b. At areas of intermediate salinity
   c. Nearest the ocean
   d. Production is equal in all areas
10. In lakes, the level at which photosynthate production equals the energy used up by respiration is known as the:
    a. Epilimnion
    b. Compensation point
    c. Hypocline
    d. Pycnocline
    e. Thermocline

## CONCEPTUAL QUESTIONS

1. Describe how seasonal turnover occurs in lakes.
2. Describe the change in river shape, discharge, bed width, and other variables as the river changes from headwaters through transfer to deposition zones.
3. Why are there so few brackish-water animal species in estuaries, but such high densities of those species there?

**Connect Ecology** helps you stay a step ahead in your studies with animations and videos that bring concepts to life and practice tests to assess your understanding of key ecological concepts. Your instructor may also recommend the interactive ebook.

Visit **www.mhhe.com/stilingecology** to learn more.

SECTION SEVEN

# Ecosystems Ecology

The term **ecosystem** was coined in 1935 by the British plant ecologist A. G. Tansley to include both the biotic community of organisms in an area and the abiotic environment affecting that community. **Ecosystem ecology** is concerned with the movement of energy and materials through organisms and their communities. Just like the concept of a community, the ecosystem concept can be applied at any scale: a small pond inhabited by protozoa and insect larvae is an ecosystem, and an oasis with its vegetation, frogs, fish, and birds constitutes another. Most ecosystems cannot be regarded as having definite boundaries. Even in what seems like a clearly defined lake ecosystem, species such as birds may be moving in and out. Thus, an ecosystem is more loosely defined than concepts such as "species" or "population," but it is still a useful ecological construct. Studying ecosystem ecology allows us to use the common currency of energy, biomass, and nutrients to compare the functions between and within ecosystems.

In investigating the different processes of an ecosystem, at least three major constituents can be measured: energy flow, biomass production, and biogeochemical cycling. We begin the section by exploring **energy flow**, the movement of energy through the ecosystem. In Chapter 25 we will document the complex networks of feeding relationships, called food webs, and measure the efficiency of energy transfer between levels in an ecosystem. We look at the distribution of numbers, biomass, and energy between these levels and determine whether some species have an effect that is disproportionate to their abundance.

In Chapter 26 we focus on the measurement of **biomass**, a quantitative estimate of the total mass of living matter in a given area, usually measured in grams or kilograms per square meter. We will examine what limits the amount of biomass produced through photosynthesis, termed primary production, and what limits the amount of biomass produced by the organisms that are the consumers of primary production. We also recognize that the functioning of an ecosystem can sometimes be most limited by the availability of a scarce chemical or mineral. In Chapter 27, the concluding chapter of this text, we examine the movement of important chemicals through ecosystems, called **biogeochemical cycles**. We explore the cycling of chemical elements such as nitrogen, carbon, sulfur, and phosphorus, and the effect that human activities are having on these ecosystem-wide processes.

Gopher tortoises are important members of southeastern U.S. ecosystems. They construct lengthy burrows, over 10 m long in some areas, in which many other species take refuge.

CHAPTER 25

# Food Webs and Energy Flow

## Outline and Concepts



Years ago, especially during the Great Depression of the 1930s, when there was not much else to eat, gopher tortoises, *Gopherus polyphemus,* were a popular food item in Florida. Later, the tortoises were buried alive in their burrows during construction of roads or buildings. Such practices wiped out entire colonies of tortoises in some areas. Only in July 2007 did the state legislature give the gopher tortoise threatened-species status and make it illegal to destroy their habitat. At first glance, the loss of tortoises and their burrows may not seem very critical to local ecosystems. Adult gopher tortoises have few enemies, although raccoons, foxes, and skunks prey on their eggs and hatchlings. In turn, tortoises have minimal impacts on local vegetation. The bulk of their diet consists of grasses, such as wire grass, and some succulents, such as *Opuntia* cacti. But the destruction of the tortoise and its habitats has a significant impact on other members of the community. The gopher tortoise is named for its ability to dig extensive burrows, over 10 m long. Many other animals, such as frogs, burrowing owls, snakes, and invertebrates, take refuge in the burrows. Some species depend on the burrows and would face extinction without them. It has been estimated that 362 different animal species utilize the burrows. Gopher tortoises are keystone species, critical members of communities in the sandy areas of the southeastern United States where they live.

In this chapter we will examine trophic relationships and the flow of energy in food chains, which are representations of feeding relationships in a community, and food webs, which are more complex models of interconnected food chains. We will then explore three of the most important features of food webs: chain length, the pyramid of numbers, and connectance. We will discuss how ecologists have identified keystone species within food webs and other types of species that may have special prominence, such as ecosystem engineers, indicator species, umbrella species, and flagship species.

## 25.1 The Main Trophic Levels Within Food Chains Consist of Primary Producers, Primary Consumers, and Secondary Consumers

Simple feeding relationships between organisms can be characterized by an unbranched **food chain**, a linear depiction of energy flow, with each organism feeding on and deriving energy from the preceding organism. Each feeding level in the chain is called a **trophic level**, from the Greek *trophos*, feeder, and different species feed at different trophic levels. In a food chain diagram, an arrow connects each trophic level with the one above it (**Figure 25.1**).

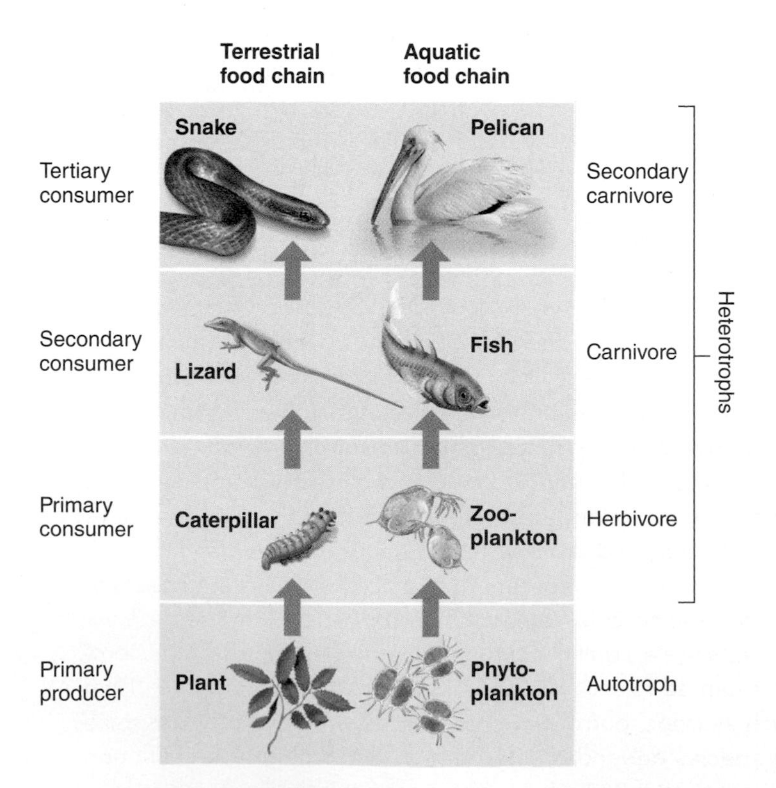


**Figure 25.1 Food chains.** Two examples of the flow of food energy up the trophic levels: a terrestrial food chain, and an aquatic food chain. Arrows flow from prey to predator.

Food chains typically consist of organisms that obtain energy in a few different ways. **Autotrophs** harvest light or chemical energy and store that energy in carbon compounds. Most such organisms, including plants, algae, and photosynthetic prokaryotes, use sunlight for this process. These organisms, called **primary producers**, form the base of the food chain. They produce the energy-rich tissue upon which nearly all other organisms depend. Note that not all primary producers utilize sunlight; some organisms, called chemoautotrophs, obtain their energy by oxidizing inorganic compounds such as sulfides (refer back to **Figure 23.13**).

Organisms in trophic levels above the primary producers are termed **heterotrophs**. These organisms receive their nutrition by eating other organisms. Organisms that obtain their food by consuming primary producers are **primary consumers**; these include many animals, most protists, and even some plants, such as mistletoe, which is parasitic on other plants. Most are also **herbivores**, organisms that eat plants and algae. However, some primary consumers, such as those that eat chemosynthetic organisms around black smokers, are not herbivores.

Organisms that eat primary consumers are **secondary consumers**, also called **carnivores**, from the Latin, *carn*, flesh, because they feed on the tissues of other animals. Organisms that feed on secondary consumers are **tertiary consumers**, also called secondary carnivores; and so on. Thus, energy enters a food chain through primary producers, via photosynthesis, and is passed up the food chain to primary, secondary, and tertiary consumers.

Much energy from the first trophic level, the plants, goes unconsumed by herbivores. Instead, unconsumed plants die and decompose in place. This material, along with animal waste products and remains, is called **detritus.** Consumers that get their energy from detritus, called **detritivores**, break down dead organisms and fecal matter from all trophic levels; these include earthworms, termites, pillbugs, carrion beetles, and dung beetles (**Figure 25.2**). Some detritivores, which do not ingest their food but live by absorbing it on a molecular scale are known as decomposers. Prominent here are fungi and bacteria, which break down dead plant and animal material and release the nutrients back into the soil. Detritivores, in turn, support a variety of nematodes, arthropods, and protozoa, which are themselves preyed upon by larger nematodes and arthropods such as centipedes. These are prey for moles and birds. In terrestrial systems, detritivores probably carry out 80–90% of the consumption of plant matter, with different species working in concert to extract most of the energy.

Usually, many herbivore species feed on the same plant species. For example, dozens of insect species and vertebrate grazers may feed on one type of plant. On the other hand, many species of vertebrate herbivores eat several different plant species. Such branching of food chains also occurs at other trophic levels. For instance, lions eat many different species of prey. It is more accurate, then, to draw relationships between these plants and animals not as a simple chain but as a more elaborate interwoven **food web**, in which there are

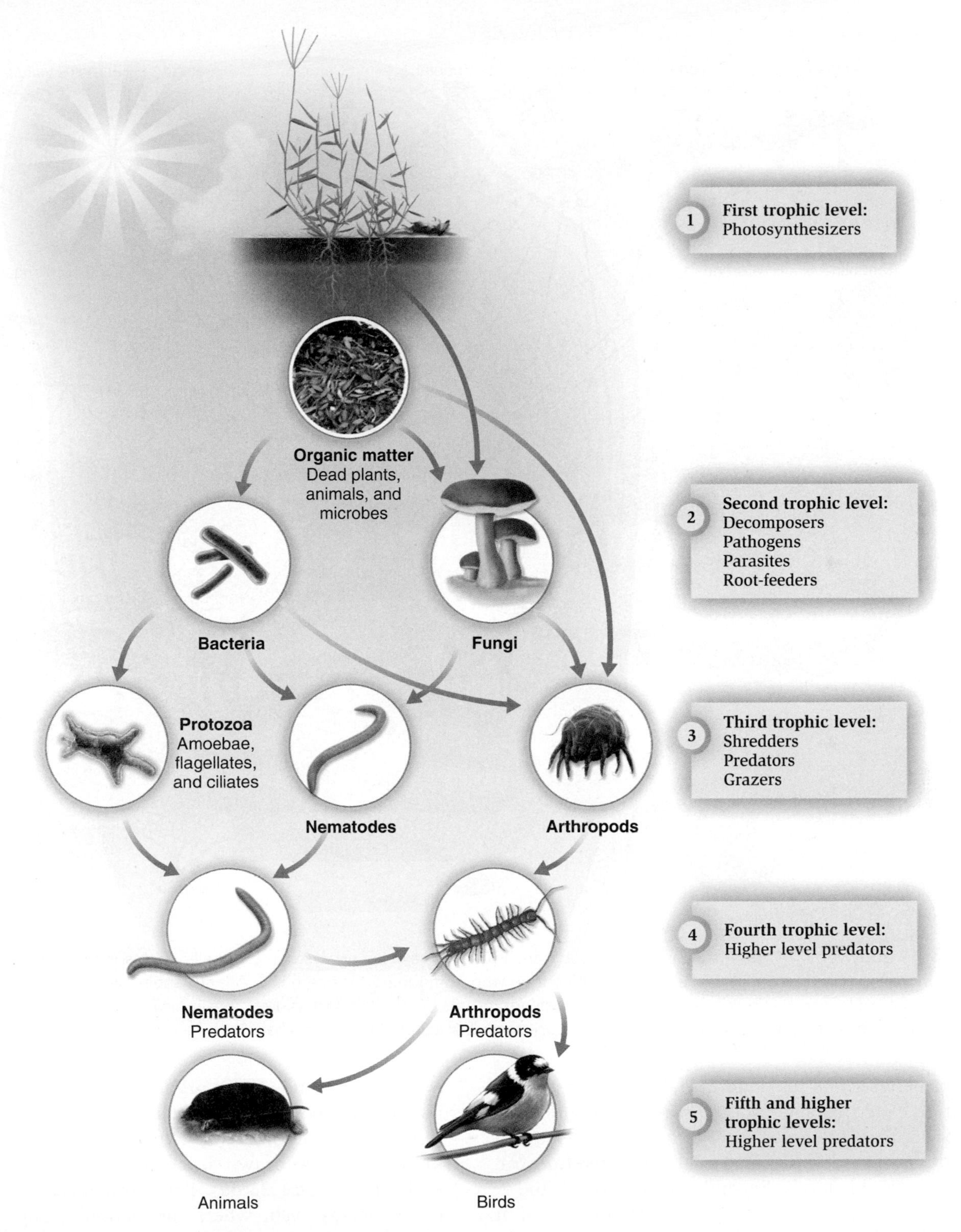


**Figure 25.2 Detritivores are organisms that feed off of dead plant and animal matter.** Many dead plants and animals are eaten by a variety of organisms. Here, fungi and bacteria feed on rotting plant and animal material. These are fed on by protozoa, nematodes, and arthropods. Dead animals, or carrion, may also be fed upon by blowfly larvae and carrion beetles. Many of these species may, in turn, support a variety of predators, including centipedes, moles, and birds.

**ECOLOGICAL INQUIRY**

At which trophic level do decomposers feed?

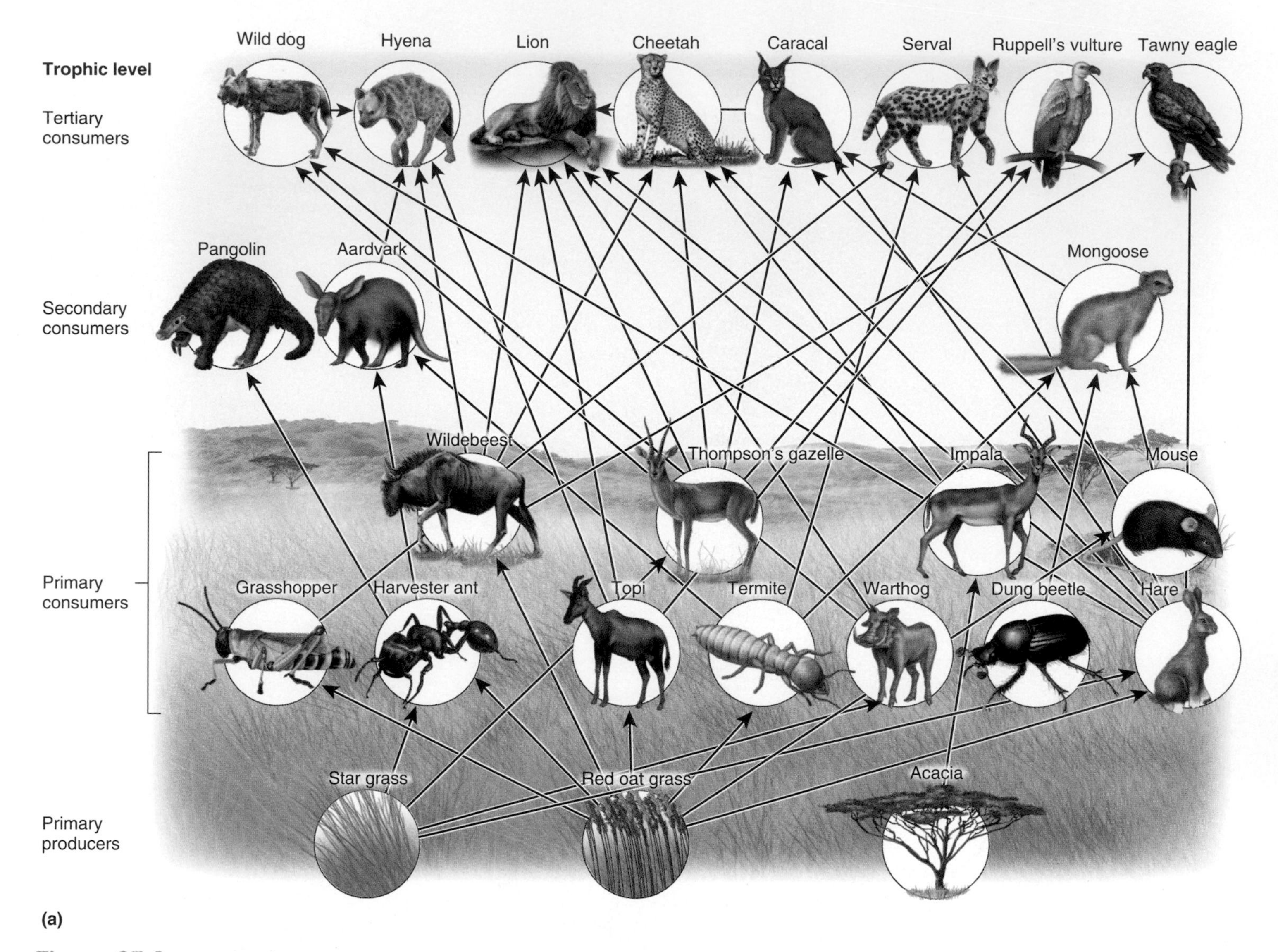


**Figure 25.3 Food webs.** **(a)** A simplified food web from an African savanna ecosystem. Generally, each species feeds on, or is fed upon by, more than one species. **(b)** A real-world marine food web from the northeastern U.S. Atlantic shelf. The left side of the web generally typifies pelagic organisms, and the right to middle represents benthic organisms. Red lines indicate predation on fish. 1 = detritus, 2 = phytoplankton, 3 = *Calanus* sp., 4 = other copepods, 5 = ctenophores, 6 = chaetognatha (such as arrow worms),

multiple links between species (**Figure 25.3a**). Also, it is often hard to ascribe exact trophic levels to many carnivores. For example, hawks may eat herbivores such as field mice and predatory organisms such as snakes. Hawks therefore can feed at multiple trophic levels. Beyond trophic levels 1 and 2, producers and primary consumers, many organisms feed at multiple positions in food webs. Data from real food webs show a more highly tangled bird's nest of feeding relationships (**Figure 25.3b**). Jason Link (2002) constructed a food web of the northeastern U.S. shelf marine community based on stomach content evaluation. The guts of over 3 million individual fish from over 120 species, ranging from Cape Hatteras, North Carolina, to the Gulf of Maine, were examined from 1973 to 2003. The results show a species-rich, highly interconnected food web.

Robert Paine (1980) argued that we can actually recognize three types of food webs: connectedness webs, energy webs, and functional webs. In **connectedness webs**, all the known links are drawn and equal importance is attached to each link (**Figure 25.4a**). Paine provided an early example of an **energy web**, where interaction strengths—based on quantities of food consumed and indicated by the thickness of connecting links—were calculated (**Figure 25.4b**). Paine compared connectedness and energy webs using the rocky intertidal ecosystem in Washington State with which he was familiar. Here, five grazing herbivores, two species of limpets (*Acmaea pelta* and *A. mitra*), two species of chitons (*Katharina tunicata* and *Tonicella lineata*), and one sea urchin (*Strongylocentrotus purpuratus*) fed on numerous species of marine algae and diatoms that adhered to the rock or drifted in.

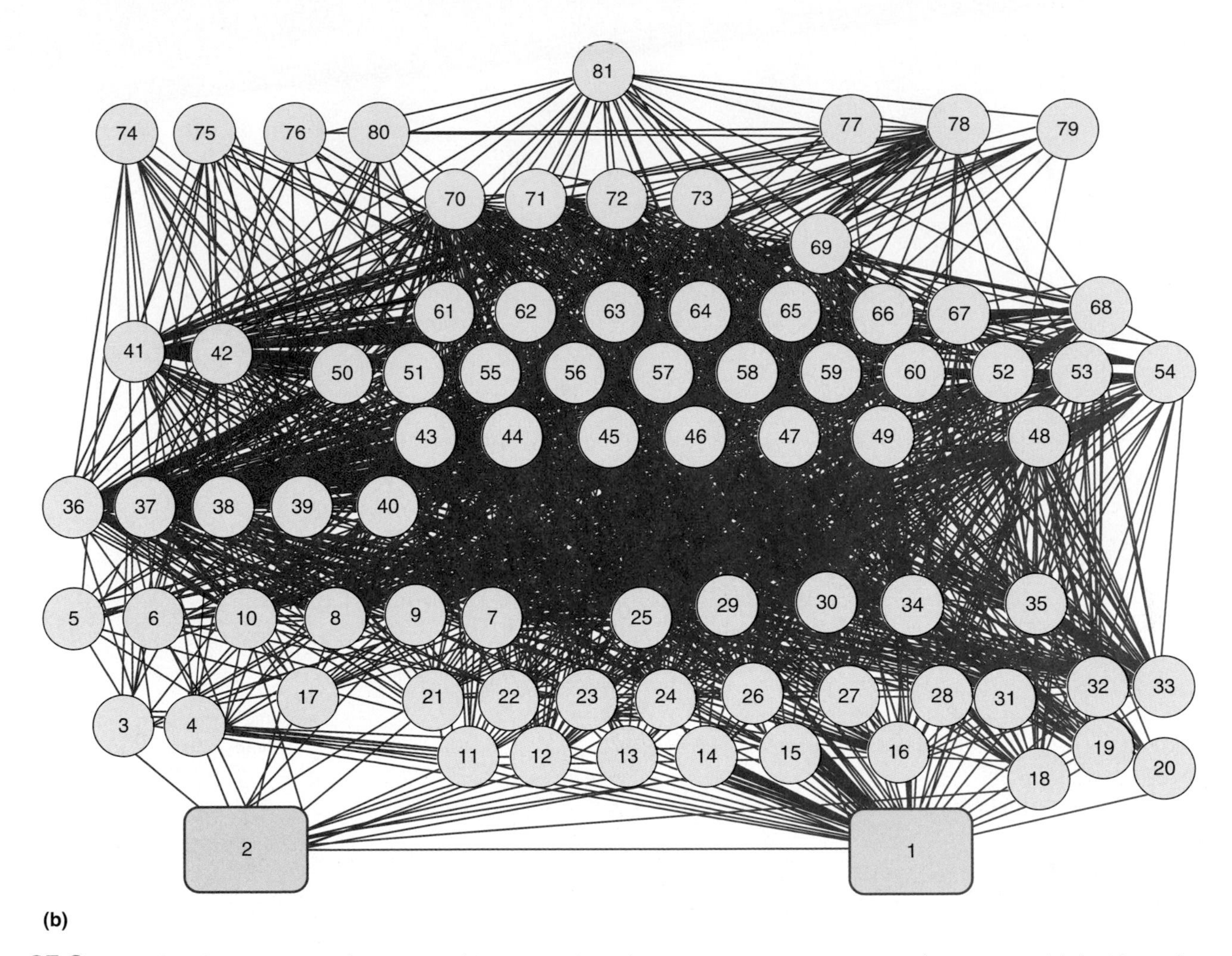


**Figure 25.3** **Food webs. (*continued*)** 7 = jellyfish, 8 = euphasiids, 9 = *Crangon* sp., 10 = mysids, 11 = pandalids, 12 = other decapods, 13 = gammarids, 14 = hyperiids, 15 = caprellids, 16 = isopods, 17 = pteropods, 18 = cumaceans, 19 = mantis shrimps, 20 = tunicates, 21 = porifera, 22 = cancer crabs, 23 = other crabs, 24 = lobster, 25 = hydroids, 26 = corals and anemones, 27 = polychaetes, 28 = other worms, 29 = starfish, 30 = brittlestars, 31 = sea cucumbers, 32 = scallops, 33 = clams and mussels, 34 = snails, 35 = urchins, 36 = sand lance, 37 = Atlantic herring, 38 = alewife, 39 = Atlantic mackerel, 40 = butterfish, 41 = loligo, 42 = illex, 43 = pollock, 44 = silver hake, 45 = spotted hake, 46 = white hake, 47 = red hake, 48 = Atlantic cod, 49 = haddock, 50 = sea raven, 51 = longhorn sculpin, 52 = little skate, 53 = winter skate, 54 = thorny skate, 55 = ocean pout, 56 = cusk, 57 = wolfish, 58 = cunner, 59 = sea robins, 60 = redfish, 61 = yellowtail flounder, 62 = windowpane flounder, 63 = summer flounder, 64 = witch flounder, 65 = four-spot flounder, 66 = winter flounder, 67 = American plaice, 68 = American halibut, 69 = smooth dogfish, 70 = spiny dogfish, 71 = goosefish, 72 = weakfish, 73 = bluefish, 74 = baleen whales, 75 = toothed whales and porpoises, 76 = seals, 77 = migratory scombroids, 78 = migratory sharks, 79 = migratory billfish, 80 = birds, 81 = humans. (From Link, 2002.)

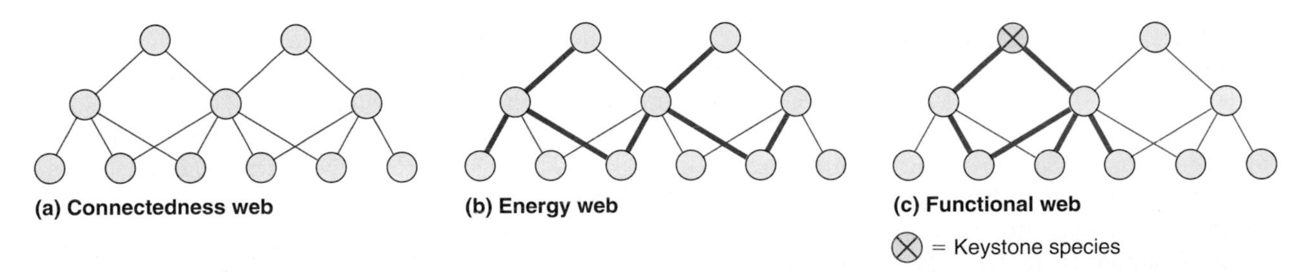


**Figure 25.4** **Different types of food webs.** **(a)** Connectedness web detailing all possible links, regardless of importance. **(b)** Energy web emphasizing the links that process the most energy. Line thickness indicates the relative amount of energy flowing along a particular link. **(c)** Functional webs emphasize the most important links in the food web. Line thickness relates to the importance of a link, which may or may not denote energy flow.

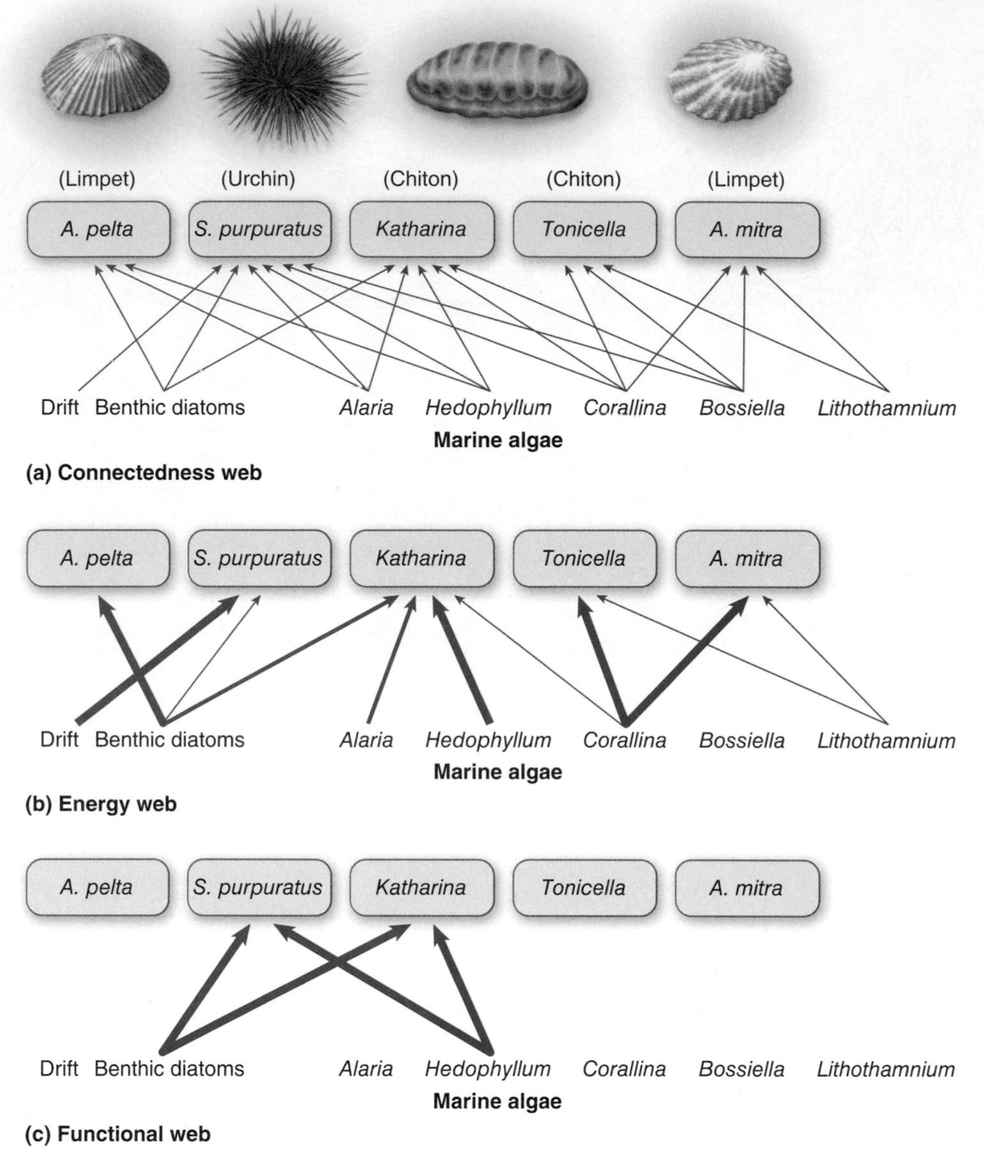


**Figure 25.5** **Connectedness, energy, and functional food webs in a marine intertidal ecosystem in the Pacific Northwest.** Line thickness indicates the amount of energy transferred in that link. Some links conduct so little energy that they have been omitted. (After Paine, 1980.)

Energetically, only five to seven of the observed links supported much energy flow, meaning that most herbivores fed on certain preferred algae, even though they could eat many different species (**Figure 25.5a,b**).

Paine described a third type of food web, a **functional web**, as one that identifies the most important feeding relationships (**Figure 25.4c**). This is not always the same as an energy web. Some species are more important than others in food webs, even if they are not abundant and do not always process a lot of energy. Recall from Chapter 16 that top predators often control populations of herbivores, such as deer, and without such predators those herbivores become superabundant, severely reducing the abundance of local vegetation. Therefore, top predators are clearly very important in influencing energy flow in food webs despite their relatively low biomass. By excluding the various herbivore species from his rocky intertidal study sites and examining the subsequent densities of algae, Paine determined that only four links in the food web were functionally important and were of sufficient strength to affect algal densities (**Figure 25.5c**). Paine used the term "keystone species" to refer to these types of species, and we will discuss them more fully in Section 25.3.

### Check Your Understanding

**25.1** At what trophic level does a carrion beetle feed?

## 25.2 In Most Food Webs, Chain Lengths Are Short and a Pyramid of Numbers Exists

Let's examine some of the characteristics of food webs in more detail. In food webs, the concept of chain length refers to the number of links between the trophic levels involved. For example, if a lion feeds on a zebra, and a zebra feeds on grass, the chain length would be two. Chain lengths are dependent on the efficiencies of energy transfer. The efficiency of energy transfer is influenced by the established laws of thermodynamics. The first law of thermodynamics states that energy cannot be created or destroyed, only transformed. The second law of thermodynamics states that energy conversions are not 100% efficient and that, in any transfer process, some energy is lost (**Figure 25.6**). We can compare the efficiency of energy transfer through trophic levels in different types of food webs. The main measure of the efficiency of consumers as energy transformers is trophic-level transfer efficiency, which examines energy transfer between trophic levels. Trophic-level transfer efficiency can be broken down into three ecological efficiencies: consumption efficiency, assimilation efficiency, and production efficiency (**Figure 25.7**).

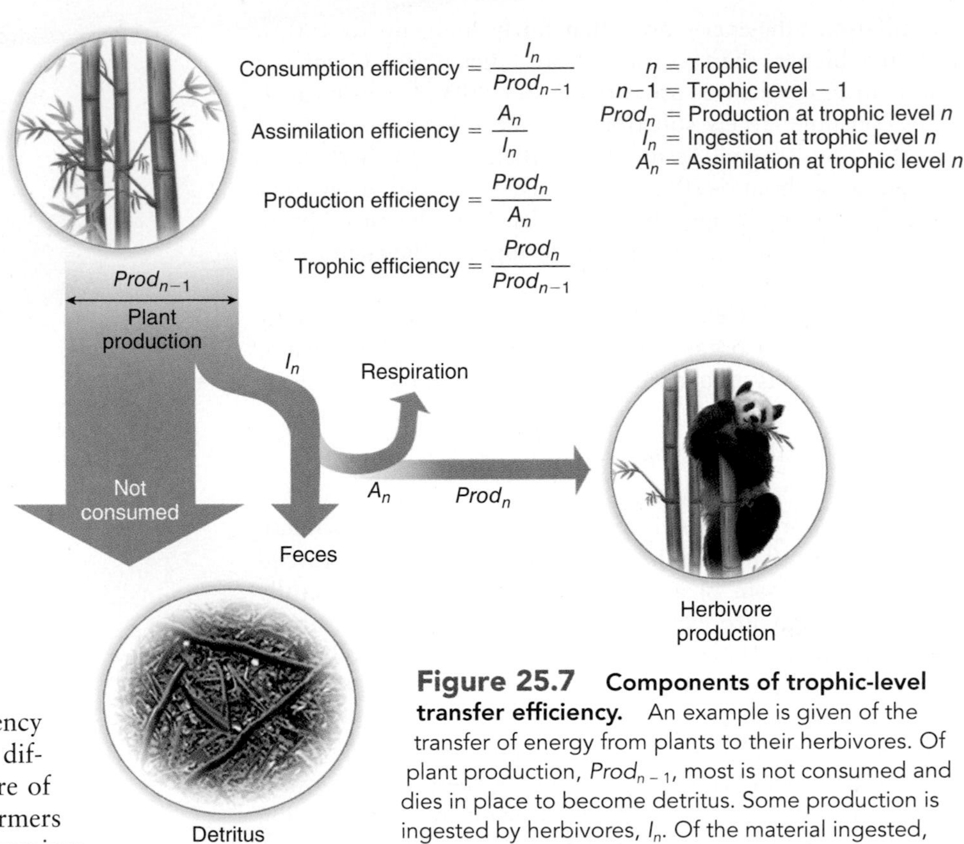


**Figure 25.7 Components of trophic-level transfer efficiency.** An example is given of the transfer of energy from plants to their herbivores. Of plant production, $Prod_{n-1}$, most is not consumed and dies in place to become detritus. Some production is ingested by herbivores, $I_n$. Of the material ingested, most is transferred to the soil as feces. Of the material assimilated to the bloodstream, $A_n$, most is used in maintenance and a small proportion goes to production, $Prod_n$. The same phenomena occur at every trophic level. (From Chapin et al., 2002.)

### 25.2.1 Consumption, assimilation, and production efficiencies are measures of ecological efficiency

Consumption efficiency is the proportion of production at one trophic level ($Prod_{n-1}$) that is eaten by, or ingested, by the next trophic level ($I_n$)

$$\text{Consumption efficiency} = \frac{I_n}{Prod_{n-1}} \times 100$$

Here production is the biomass produced per unit of time. Consumption efficiencies vary dramatically between different ecosystems. Much of this variation is tied to differences in allocations to plant structure. In terrestrial forests, much allocation is to wood, which is largely unconsumed by herbivores. Consumption efficiencies in such forests range from 1 to 2%. In grasslands, most plants are nonwoody, and consumption efficiencies range from 30 to 60%, with only roots left in the ground. In aquatic ecosystems most plant and algal biomass can be consumed by herbivores. Here, consumption efficiencies of 60–99% have been recorded. Consumption efficiencies for carnivores are typically higher than those of herbivores and vary from 5 to 100%. Vertebrate predators commonly consume more than 50% of their prey.

Assimilation efficiency is the proportion of ingested energy that is assimilated into the bloodstream ($A_n$):

$$\text{Assimilation efficiency} = \frac{A_n}{I_n} \times 100$$

Assimilation efficiency depends on the quality of food eaten and the physiological efficiency of the consumer.

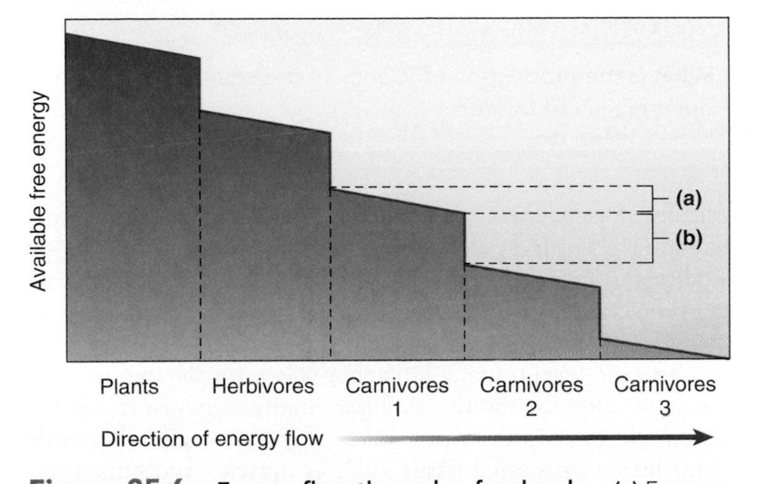


**Figure 25.6 Energy flow through a food web.** **(a)** Energy lost as heat in a single trophic level over time. **(b)** Energy lost as heat in the conversion between one trophic level and another.

Assimilation efficiencies are often fairly high, up to 100%, and are higher than consumption efficiencies. Carnivore assimilation efficiencies are often about 80%, because carnivores consume largely soft tissue, which is easily assimilated into the bloodstream. Assimilation efficiencies for herbivores average only about 5–20% and are lower because even plant leaves have many indigestible cell walls. As with consumption efficiencies, aquatic herbivores have higher efficiencies than terrestrial herbivores.

**Production efficiency** is defined as the percentage of energy assimilated by an organism that becomes incorporated into new biomass. It is influenced largely by animal metabolism:

$$\text{Production efficiency} = \frac{Prod_n}{A_n} \times 100$$

Here, net productivity is measured as energy, stored in biomass, and assimilation is the total amount of energy absorbed by the bloodstream of an organism. Invertebrates generally have fairly high production efficiencies that average about 10–50% (**Figure 25.8a**). Microorganisms also have relatively high production efficiencies. Vertebrates tend to have lower production efficiencies than invertebrates, because they devote more energy to sustaining their metabolism than to new biomass production. Even within vertebrates there is much variation. Fish, which are ectotherms, typically have production efficiencies of around 10%, and birds and mammals, which are endotherms, have production efficiencies in the range of 1–2% (**Figure 25.8b**). In large part the difference reflects the energy cost of maintaining a constant body temperature.

One consequence of these differences is that sparsely vegetated deserts can support healthy populations of snakes and lizards, which have relatively high production efficiencies, whereas mammals might easily starve there. The largest living lizard known, the Komodo dragon, eats the equivalent of its own weight every 2 months, whereas a cheetah consumes approximately four times its own weight in the same period. Production efficiencies are higher in young animals, which are rapidly accruing biomass, than in older animals, which are not.

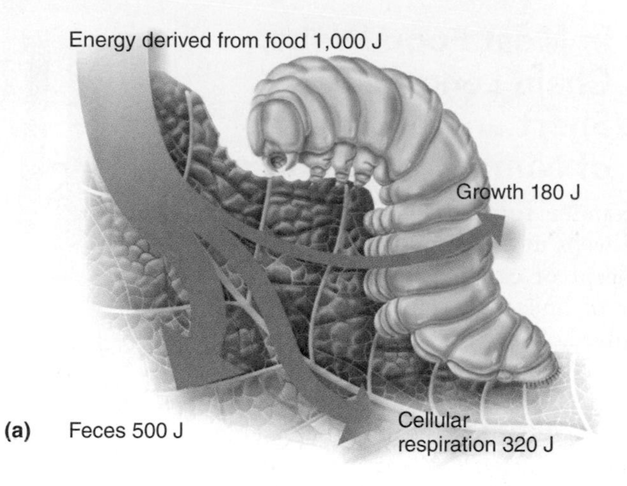


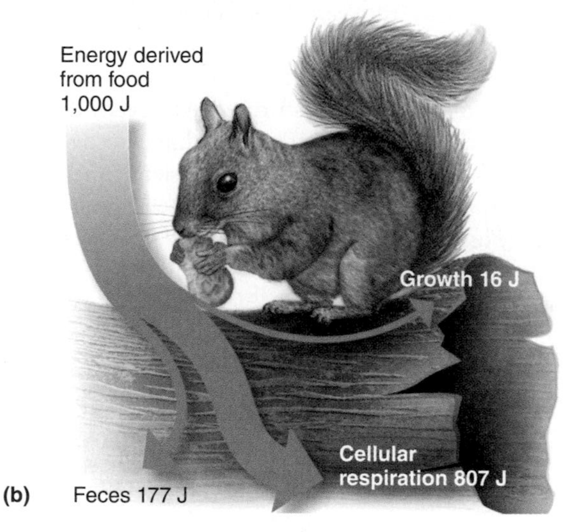


**Figure 25.8 Production efficiency.** **(a)** This caterpillar, an invertebrate, chews leaves to obtain its energy. If a mouthful of food contains 1,000 joules (J) of energy, about 320 J is used to fuel metabolic processes (32%) and 500 J (50%) is lost in feces. This leaves about 180 J to be converted into insect biomass, a production efficiency of 36%. **(b)** The production efficiency of this squirrel, a mammal, is much lower because more energy (807 J) is used to fuel its metabolism.

**ECOLOGICAL INQUIRY**

What is the production efficiency of the squirrel, using the numbers in the figure?

## 25.2.2 Trophic-level transfer efficiency measures energy flow between trophic levels

The final measure of efficiency of consumers as energy transformers is **trophic-level transfer efficiency**, which is the amount of energy at one trophic level that is acquired by the trophic level above and incorporated into biomass. Although this measure of efficiency is affected by changes in the three other measures of efficiency, it provides a way to examine energy flow between trophic levels, not just between individual species. Trophic-level transfer efficiency is calculated as:

$$\text{Trophic-level transfer efficiency} = \frac{Prod_n}{Prod_{n-1}} \times 100$$

For example, if there were 14 g/m$^2$ of zooplankton produced in a lake over a year, at trophic level $n$, and 100 g/m$^2$ of phytoplankton production, at trophic level $n - 1$, the trophic-level transfer efficiency would be 14%. Daniel Pauly and Villi Christensen (1995) surveyed trophic-level transfer efficiencies in 49 studies and found that trophic-level transfer efficiency appears to average around 10%, though there is much variation. In some marine food chains, for example, it can exceed 30%.

Trophic-level transfer efficiency is low for the two reasons we have already discussed. First, many organisms cannot digest all their prey. They take only the easily digestible plant leaves or animal tissue such as muscles and guts, leaving behind the hard wood or energy-rich bones. Second, much of the energy assimilated by animals is used in maintenance, so most energy is lost from the system as heat.

The 10% average transfer rate of energy from one trophic level to another necessitates short food webs of no more than four or five levels. At that point, relatively little energy is available to another trophic level.

### 25.2.3 Ecological pyramids describe the distribution of numbers, biomass, or energy between trophic levels

The effects of trophic-level transfer efficiencies on species abundance can be expressed in a graphical form called an ecological pyramid, first described by the British ecologist Charles Elton. The best-known pyramid, and the one described by Elton in 1927, is the **pyramid of numbers,** in which the number of individuals decreases at each trophic level, with a large number of individuals at the base and fewer individuals at the top. Elton's example was that of a small pond where the numbers of protozoa may run into the millions and the numbers of *Daphnia* and *Cyclops,* their predators, number in the hundreds of thousands. Hundreds of insect larvae may feed on *Daphnia,* and still fewer fish feed on the insects. Many other examples of this type of pyramid are known. For example, in a grassland there may be millions of individual plants per acre, hundreds of thousands of insects that feed on the plants, tens of thousands of insect predators, and a few birds or mice feeding on the predators (**Figure 25.9a**).

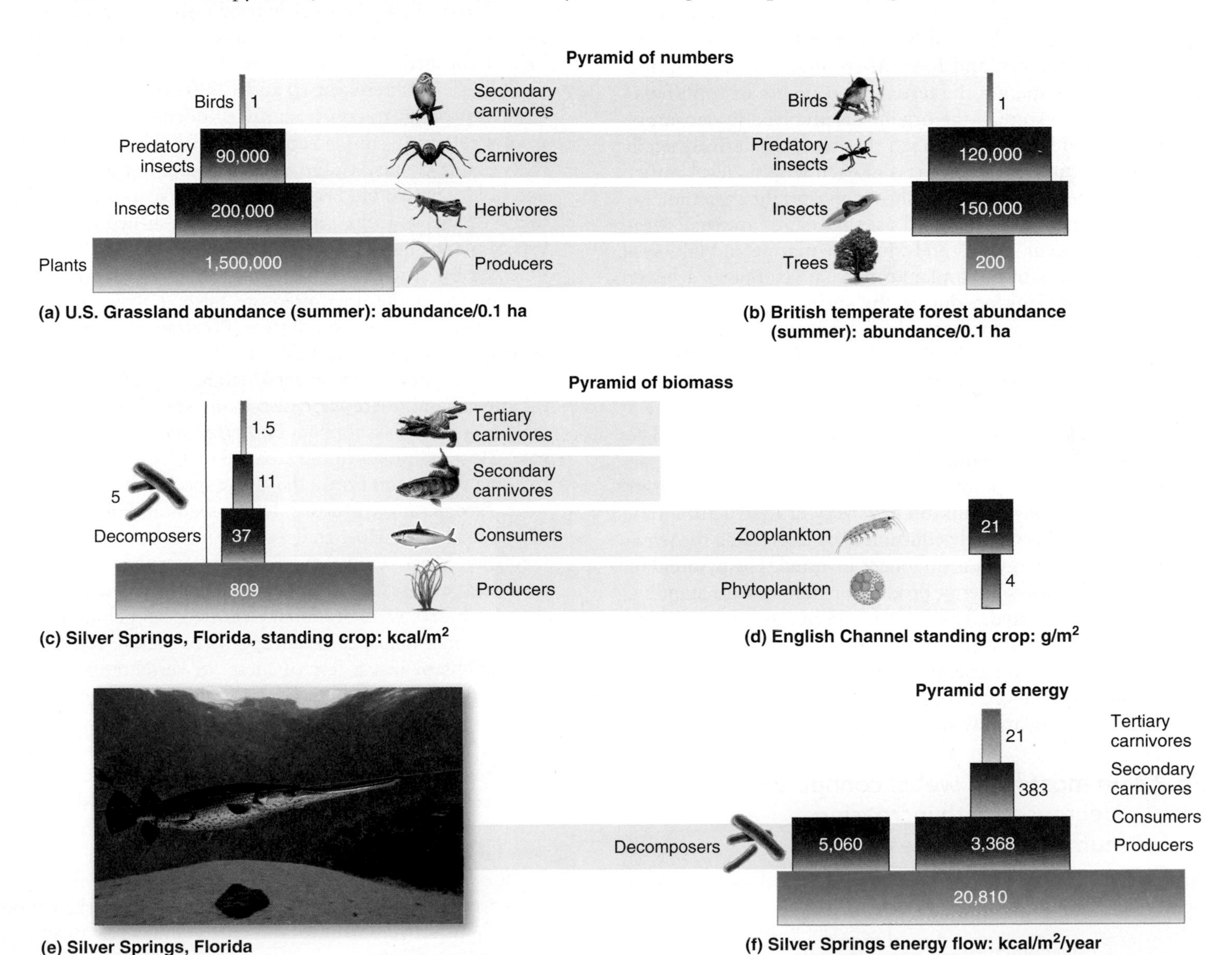


**Figure 25.9 Ecological pyramids in food webs.** **(a)** In this pyramid of numbers, the abundance of species in an American grassland decreases with increasing trophic level. Data are numbers per 0.1 hectare. **(b)** An inverted pyramid of numbers based on organisms living in a British temperate forest. Data are numbers per 0.1 hectare. **(c)** When the amount of biological material is used instead of numbers of individuals, the pyramid is termed a pyramid of biomass. Data from Silver Springs, Florida, are in kcal/m$^2$. Note the presence of decomposers, which decompose material from all trophic levels. **(d)** An inverted pyramid of biomass in the English Channel. Data are in g/m$^2$. **(e)** Silver Springs, Florida. **(f)** A pyramid of energy for Silver Springs, Florida, kcal/m$^2$/year. Note the large flow of energy through the decomposers, despite their small biomass. (After Odum, 1971.)

One can think of several exceptions to the pyramid of numbers. An oak tree, one single producer, supports thousands of herbivorous caterpillars, beetles, and other primary consumers, which in turn may support hundreds of predators and parasites (**Figure 25.9b**). This is called an inverted pyramid of numbers. One way to reconcile such exceptions is to weigh the organisms in each trophic level, creating a **pyramid of biomass,** usually measured as dry biomass. For example, the oak tree weighs much more than all its herbivores and their predators combined. Howard Odum (1957) measured the pyramid of biomass for a freshwater ecosystem in Silver Springs, Florida (**Figure 25.9c,e**). Here, beds of eelgrass, *Sagittara,* and attached algae make up most of the producers. Numerous insects, snails, herbivorous fish, and turtles eat the producers. Other fish and invertebrates form the secondary consumers, and bass, *Micropterus salmoides*, and gar, *Lepisosteus* sp., are the tertiary consumers or top predators. Odum also studied the organisms involved in decomposition, fungi and bacteria, though these had a relatively small biomass. Looking at the biomass at each trophic level rather than at numbers of organisms shows an upright pyramid.

Even when biomass is used as a measure, inverted pyramids can still occur, albeit rarely. In some marine and lake systems, the biomass of phytoplankton usually supports a lower biomass of zooplankton during the spring and summer, the periods of highest productivity. However, in winter the pyramid may be inverted, as in the English Channel (**Figure 25.9d**). This is possible because the rate of production of phytoplankton is much higher than that of zooplankton. Phytoplankton have a relatively short life span and they can reproduce quickly. The small phytoplankton **standing crop**, the total biomass in an ecosystem at any one point in time, processes large amounts of energy. Also, the phytoplankton have very little structural biomass and can be completely consumed. By expressing the pyramid in terms of energy, it is no longer inverted. The **pyramid of energy**, which shows energy production rather than standing crop, is never inverted. The laws of thermodynamics ensure that the highest amounts of free energy are found at the lowest trophic levels. Odum's energy pyramid for Silver Springs also shows that large amounts of energy pass through decomposers, despite their relatively small biomass (**Figure 25.9f**).

### 25.2.4 In most food webs, connectance decreases with increasing numbers of species

Another measure of the relative complexity of a food web can be described by a measure known as connectance, where:

$$\text{Connectance} = \frac{\text{Actual number of links}}{\text{Potential number of links}}$$

In an ecosystem of $n$ species, assuming that links cannot go in both directions, the number of potential links, $N$, is given by:

$$N = \frac{n\,(n-1)}{2}$$

## Feature Investigation

### Gary Polis Showed That Real-World Food Webs Are Complex

Gary Polis (1991) pointed out that a major problem in documenting the properties of food webs is that most food webs reported in the literature are far simpler than real-world food webs. Many authors simply ignore unfamiliar species or aggregate others into higher taxonomic categories such as "insects," "plankton," "spiders," or "parasites." This is problematic because many species in such groups are central to the flow of energy in food webs. As a result, many published food webs do not possess an accurate number of links. Published analyses of diets of carnivores suggest that most species eat between 10 and 1,000 other species.

Polis sought to construct a well-documented food web from a relatively simple ecosystem in the Coachella Valley in Riverside County, California, an area of hot dry summers and mild winters. He began long-term censuses of the area in 1973, and in 1990, after 17 years, published taxonomic lists of plants, insects, arachnids, and vertebrates. Polis documented 60 species of insects feeding on one of the common plants in the area, creosote bush, *Larrea tridentata;* more than 200 on mesquite trees, *Prosopis glandulosa;* and at least 89 on ragweed, *Ambrosia dumosa.* Harvester ants, *Messor pergandei,* fed on seeds from 97 plant species. On the other hand, there were also many specialist herbivores, including the grasshopper, *Bootettix punctatus,* on creosote. Most mammals in the area, 16 of 18 species, ate plants to some degree, and only the 2 bat species did not. Pocket mice, *Perognathus formosus,* fed on seeds and plant parts of 27 plant species. Most rodents also included arthropods in their diet. Most vertebrates, 83% of 95 species, ate arthropods such as insects, spiders, and scorpions. Twenty-four of the vertebrates were secondary carnivores, including some birds and mammals, and eight species of snakes. The trophic interactions of just a few of these 96 vertebrates are very complex and are shown in **Figure 25.A**. Even the top predators are attacked by parasites. For example, coyote parasites include mange mites, ticks, lice, fleas, and tapeworms.

The main point of Polis's work was not merely that food webs are complex, but rather that previously calculated food web statistics from other published food webs were likely to be wrong. When Polis calculated some of the food web properties from the Coachella Valley and compared them to a summary of 40 other published food webs, he discovered quite different results. For example, the number of links per species in the Coachella Valley was 9.6 compared to about an average of 2.0 in other published food webs. The number of prey species per predator was 10.7, compared to 2.5 from other webs, and the number of predator species per prey species was 9.6 compared to only 3.2 in other webs. Thorough, intense surveys are clearly needed to construct food webs accurately.

**HYPOTHESIS** The food web of the Coachella Valley is more complex than previously described food webs.

**STARTING LOCATION** Coachella Valley, Riverside County, California, USA.

| | Conceptual level | Experimental level |
|---|---|---|
| 1 | Document species present in Coachella Valley. |  Conduct extensive surveys of plants, herbivores, and carnivores in area by visual surveys and trapping. Censuses were conducted over 4,000 hours of fieldwork and 4,300 trap days. |
| 2 | Document feeding relationships between species. | Visually observe feeding relationships between species. For example, the scorpion, *Paruroctonus mesaensis,* was observed for more than 2,000 person hours of field time over 5 years. Over 100 species of prey were recorded. |
| 3 | Examine reported plant-herbivore-predator relationships from the area. | Over 820 published papers were examined from the literature. |
| 4 | **THE DATA** | Use data to create a vertebrate food web. |

Great horned owl
Golden eagle
Coyote
Kit fox
Screech and Barn owls
Red-tailed hawk
King snake
Sidewinder rattlesnake
Gopher snake
Burrowing owl
Roadrunner
Loggerhead shrike
Antelope ground squirrel
Grasshopper mouse
Leopard lizard
Arthropod eating snakes
Arthropod eating lizards
Other birds and rodents
Arachnids
Insects
Plants

**Figure 25.A Partial food web involving a few of the 96 vertebrates from the Coachella Valley food web.** More-complex food webs were documented for arachnids, insects, and plants.

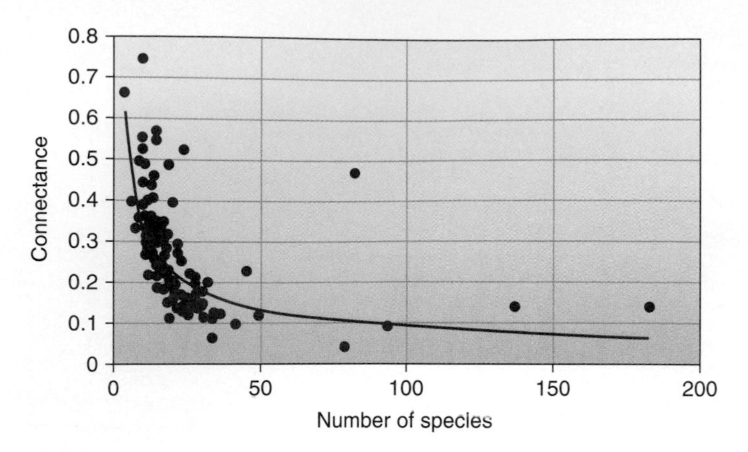


**Figure 25.10** **Food web connectance and species richness.**

This means that each species can eat any other species, but prey species cannot turn around and eat their predators, which seems reasonable. In the simplified food web from the tropical savanna in **Figure 25.3a**, there are 25 species, 49 actual links, and 25 × 24/2 = 300 potential links. Therefore, the connectance of this community is 49/300 = 0.163.

Neo Martinez (1992) argued that connectance values in food webs should remain constant, regardless of species richness. A constant level of connectance can be understood by imagining an insectivorous bird that feeds in two communities, A and B, with A having twice the number of insect species as B. Martinez argued that it is reasonable to suppose that the bird would eat twice as many species of insect in community A as in community B. Because this trend would likely apply to all species in the community, the connectance would remain unchanged as the species richness increased. However, Jason Link (2002) compared connectance in a variety of published food webs, from marine to freshwater and terrestrial (**Figure 25.10**). He showed that connectance decreased as species richness increased. This suggests that each species typically interacts with a few species in its community and that increases in richness merely serve to effectively reduce connectance. With high numbers of species, connectance tends to level out, so that connectance does not reach zero in highly species-rich communities.

Another measure of food web complexity is **linkage density**, which is the number of links per species. For the tropical savanna community the linkage density is 49/25 = 1.96. These three measures of chain length, connectance, and linkage density are widely used to describe the complexity of food webs.

When determining food web structure, great care must be exercised in determining all the links. Often, authors of published works on food webs are not sufficiently knowledgeable about organisms outside of their areas of specialty, so that some links in the web are never drawn and predation on "minor" species is frequently omitted. Gary Polis (1991) argued that food webs in the real world are often very complex. In a relatively simple desert ecosystem, Polis conducted an intensive 17-year study of the local food web and found 174 species of plants, 138 vertebrates, 55 species of spiders and scorpions, 2,000–3,000 species of insects, and an unknown number of microorganisms and nematodes all in a single food web (see **Feature Investigation**).

**Check Your Understanding**

**25.2** What are the connectance and linkage density for Figure 25.5a?

## 25.3 Within Food Webs, Some Species Have Disproportionately Large Effects

Ecologists have found that within food webs, not all species are equally important. As a result, there is a strong impetus to determine which species are the most important. Conservation biologists in particular are keen to know which species they should strive to preserve to maximize the integrity of a functioning ecosystem. Such categories include keystone species and ecosystem engineers. In addition, conservation biologists recognize other categories of species such as indicator species, umbrella species, and flagship species.

A **keystone species** is a species within a community that has a role out of proportion to its abundance (**Figure 25.11a**). The term was introduced by Robert Paine (1966) to describe the importance of the predatory starfish, *Pisaster,* on the intertidal community of Washington State (refer back to Figure 18.10). *Pisaster* preyed on a variety of intertidal organisms, preventing any one species from becoming very common and outcompeting the others, thus promoting species richness. A rare rhinovirus that makes a wildebeest sneeze would not qualify as a keystone species, because of its relatively low impact, but a rare distemper virus that kills lions or wild dogs would. Sea otters have been labeled as keystone predators because they limit the densities of sea urchins, which in turn eat kelp and can destroy kelp forests in the absence of sea otters. We can contrast the keystone species pattern with a random pattern where the importance of each community member varies at random (**Figure 25.11b**) or with a uniform pattern where all species are equally important (**Figure 25.11c**), which is probably unlikely in nature.

John Terborgh (1986) considered palm nuts and figs to be keystone species because they produce fruit during otherwise fruitless times of the year and are thus critical resources for tropical forest fruit-eating animals, including primates, rodents, and many birds (**Figure 25.12**). Together, these fruit eaters account for as much as three-quarters of the tropical forest animal biomass. Without the fruit trees, wholesale extinction of these animals could occur. In many areas, invasive species are becoming what is known as cultural keystone species (see **Global Insight**). Note that a keystone species is not the same as a **dominant species**, one that has a large effect

Global Insight

## Some Invasive Species May Be Viewed as Cultural Keystone Species

Ann Garibaldi and Nancy Turner (2004) noted that certain plant and animal species have become so common and valued in society that they have been termed cultural keystone species. Cultural keystone species can serve as a staple food or a crucial emergency food, or serve a purpose in technology or as an important medicine, and may feature prominently in the language, ceremonies, and stories of native peoples. For example, western red cedar, *Thuja plicata,* is a cultural keystone for the first peoples of coastal British Columbia, who regard themselves as "peoples of the cedar." It is prized for making dugout canoes and totem poles. However, among cultural keystone species are many invasive species. For example, over 100 species of *Eucalyptus* trees have been brought to California from Australia since the late 19th century. Such species have quickly spread and currently constitute a large threat to native species, not just through competition but also because they both promote fires and are able to tolerate fire far better than native species. Removal of *Eucalyptus* has been hindered by groups and people who believe these species are part of the natural landscape and should not be removed. In Florida, Brazilian pepper, *Schinus terebinthifolius,* is an exotic from Brazil that forms dense monocultures, particularly in coastal areas. Nothing else grows underneath large stands of Brazilian pepper. Attempts at removal have been hampered by the public, who use its foliage and bright red fruit around Christmas (**Figure 25.B**). The European periwinkle snail, *Littorina littorea,* was introduced into Nova Scotia in 1840 and has spread into New England, grazing algae off rocks, eating shoots and rhizomes of marsh plants, and reducing the abundance of many intertidal species. Many local people, however, view the current landscape as natural and are unaware of the history of local species. For example, horses are so ingrained into the culture of the American West that it is difficult to remove wild herds, even though they were introduced by the Spanish and can damage local vegetation.

Martin Nuñez and Daniel Simberloff (2005) note that in many cases it is not known whether the invasive species are truly keystone species in the sense that Paine envisaged. Some may better fit the definition of dominant species. However, in the eyes of local societies, many of these species are valuable and have been regarded as cultural keystone species. In such cases, removal will likely prove to be quite problematic.

**Figure 25.B Brazilian pepper, a cultural keystone species.** In Florida, this Brazilian invasive sets bright red fruits in the winter and is known by many as Florida holly.

in a community because of its abundance or large biomass. For example, saltmarsh cordgrass, *Spartina alterniflora,* is a dominant species in a salt marsh because of its large biomass, but it is not a keystone species. Other common dominant species in different ecosystems include trees, prairie grass, corals, and giant kelp (**Figure 25.13**).

Many keystone species have dramatic effects on their environment. The beaver, a relatively small animal, can completely alter a community by building a dam and flooding an entire river valley (**Figure 25.14**). The resultant lake may become habitat for aquatic vegetation, fish species, and wildfowl. A decline in the number of beavers could have serious ramifications for the remaining community members, promoting fish die-offs, waterfowl loss, and the death of vegetation adapted to waterlogged soil.

Because some species such as gopher tortoises and beavers create, modify, and maintain habitats that are used by other organisms, John Lawton and Clive Jones (1995) termed them **ecosystem engineers.** African elephants act as ecosystem engineers through their browsing activity, destroying small trees and shrubs and changing woodland habitats into grasslands. Mollusks and corals add physical structure (shells, coral reefs) to the oceans. Other examples of ecosystem engineers are given in **Table 25.1**. Note that while some ecosystem engineers are keystone species and others are dominant species, some might not be either. Some invasive species of plants may act as ecosystem engineers, changing the local soil conditions.

While keystone species and ecosystem engineers are clearly vital in maintaining the health of ecosystems, conservation biologists have identified other potentially valuable

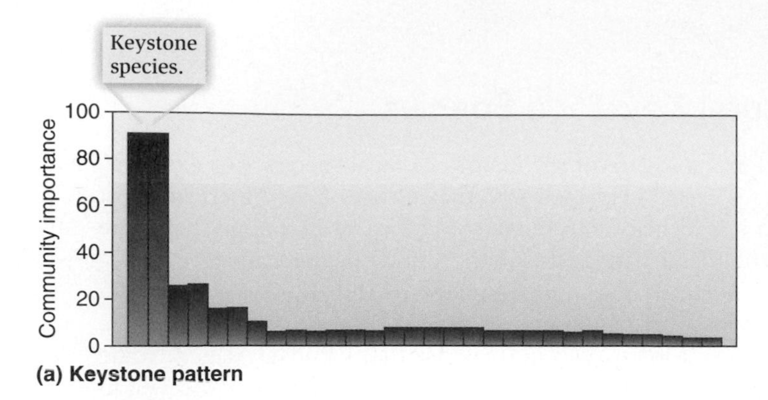


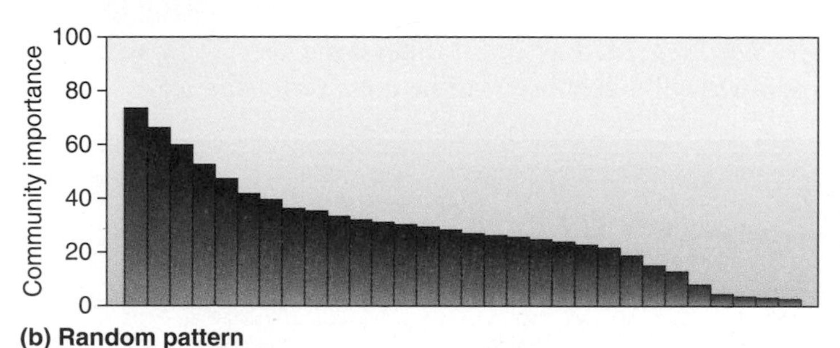


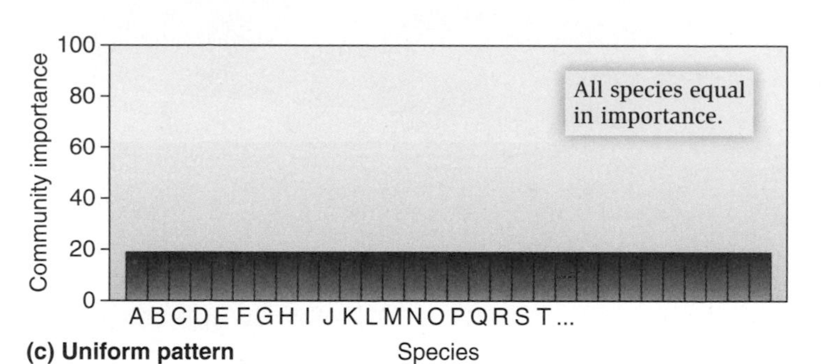


**Figure 25.11 Keystone species.** Community importance values, that is, whether or not a species has a large impact on community diversity or nitrogen uptake or some other function, for species in a hypothetical community based on **(a)** the keystone-species model, **(b)** a random pattern, and **(c)** with all species equally important. (After Mills et al., 1993.)

**Figure 25.12 Tropical fruit trees are keystone species.** These fruits and nuts provide a vital resource to maintain a whole array of fruit eaters, including this channel-billed mountain toucan, *Ramphastos vitelinus.*

types of species within food webs. Such species include indicator species, flagship species, and umbrella species.

Some conservation biologists have suggested that certain organisms can be used as **indicator species**, those species whose status provides information on the overall health of a food web or ecosystem. Corals are good indicators of the degree of marine processes such as siltation, the accumulation of sediments transported by water. Because siltation reduces the availability of light, the abundance of many marine organisms decreases in such situations, with corals among the first to display a decline in health. Coral bleaching is also an indicator of climate change. As we noted in Chapter 2, a proliferation of the dark variety of the peppered moth, *Biston betularia,* has been shown to be a good indicator of air pollution. The darker-colored moths flourish because they are better able to blend in on trees darkened by soot (refer back to Figure 2.A). Polar bears, *Ursus maritimus,* are thought to be an indicator species for global climate change. Scientists believe that global warming is causing the ice in the Arctic to melt earlier in the spring than in the past. Because polar bears rely on the ice to hunt for seals, the earlier breakup of the ice is leaving the bears less time to feed and build the fat that enables them to sustain themselves and their young.

**Umbrella species** are species whose habitat requirements are so large that protecting them would protect many other species existing in the same food web or habitat. The Northern spotted owl, *Strix occidentalis,* of the Pacific Northwest is considered to be an important umbrella species (**Figure 25.15a**). A pair of birds needs at least 800 hectares of old-growth forest for survival and reproduction, so maintaining healthy owl

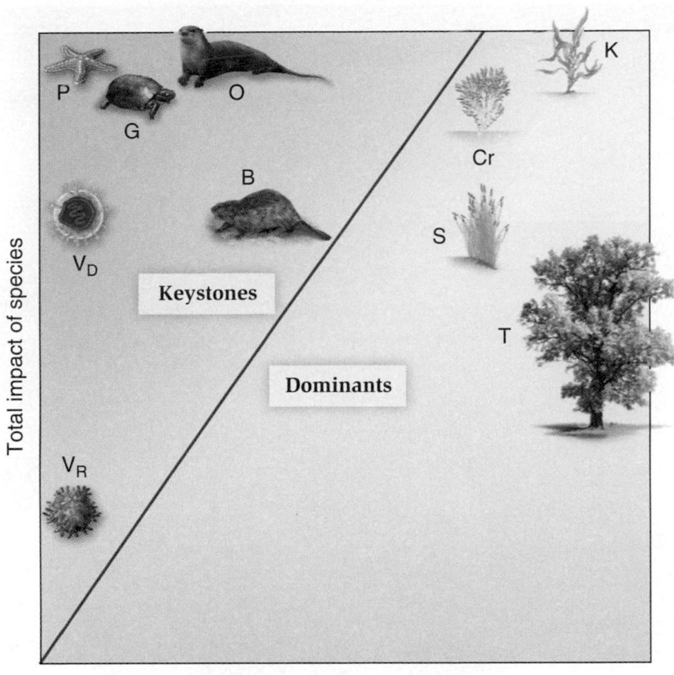


**Figure 25.13 Keystone and dominant species.** Keystone species are those whose effects are out of all proportion to their biomass. Dominants are species that dominate community biomass and whose ecosystem impacts are large, but not disproportionate to their biomass. Letters represent examples of particular species described in the text. (After Power and Mills, 1995.)

**ECOLOGICAL INQUIRY**

Can you think of more examples of keystone species?

**Table 25.1 Examples of organisms as ecosystem engineers.**

| Engineer | Effect |
|---|---|
| Green plants | Oxygen production |
| Soil organisms | Increase and reduction of soil nitrogen and other nutrients |
| Beavers | Altered hydrology |
| Coral reef | Physical structure creates habitat, protection for other species |
| Trees | Physical structure creates habitat for other species |
| Gopher tortoise | Burrow used by other species |
| Woodpecker | Nest cavity used by other species |
| Termite mounds | Homes for other arthropods |
| Vertebrate feces | Homes for dung beetles |

**Figure 25.14 Ecosystem engineers.** The American beaver creates large dams across streams, and the resultant lakes provide habitats for a great diversity of species.

populations is thought to help ensure the survival of many other forest-dwelling species. In the U.S. Southeast, the red-cockaded woodpecker, *Picoides borealis,* is often seen as the equivalent of the spotted owl, because it requires large tracts of old-growth long-leaf pine, *Pinus palustris,* including old diseased trees in which it can excavate its nests. Gopher tortoises can also be considered an umbrella species, in that protecting them would also aid the species that use their burrows. This shows that species can have more than one designation, here keystone species as well as ecological engineer and umbrella species.

In the past, conservation resources were often allocated to a **flagship species**, a single large or instantly recognizable species. Such species were typically chosen because they were attractive and thus more readily engendered support from the public for their conservation. The concept of a flagship species, typically charismatic vertebrates such as the American buffalo, *Bison bison,* has often been used to raise awareness for conservation in general. The giant panda, *Ailuropoda melanoleuca,* is the World Wildlife Fund's emblem for endangered species, and the Florida panther, *Puma concolor,* has become a symbol of that state's conservation campaign (**Figure 25.15b**). Both umbrella species and flagship species are used within the context of conservation goals and have no strict ecological properties.

**Check Your Understanding**

**25.3** Consider the concept of ecosystem engineers and provide more examples and their effects.

**Figure 25.15 Umbrella, and flagship species.** **(a)** The Northern spotted owl is considered an umbrella species for the old-growth forest in the Pacific Northwest. **(b)** The Florida panther has become a flagship species for Florida.

## SUMMARY

**25.1 The Main Trophic Levels Within Food Chains Consist of Primary Producers, Primary Consumers, and Secondary Consumers**

- Ecosystems ecology concerns the movement of energy and materials through organisms and their communities. Some organisms have relatively stronger effects in ecosystems than others.
- Simple feeding relationships between organisms can be characterized by an unbranched food chain, and each feeding level in the chain is called a trophic level (Figure 25.1).
- Organisms that obtain energy from light or chemicals are called autotrophs and are termed primary producers. Organisms that feed on other organisms are called heterotrophs. Those organisms that feed on primary producers are called primary consumers or herbivores. Organisms that feed on primary consumers are called secondary consumers or carnivores. Consumers that get their energy from the remains and waste products of organisms are called detritivores (Figure 25.2).
- Food webs are a complex model of interconnected food chains in which there are multiple links between species (Figure 25.3).
- Food webs may be categorized as connectedness webs, energy webs, or functional webs (Figures 25.4, 25.5).

**25.2 In Most Food Webs, Chain Lengths Are Short and a Pyramid of Numbers Exists**

- Energy conversions are not 100% efficient, and energy is lost within each trophic level and from one trophic level to the next (Figure 25.6).
- There are a variety of ways to measure ecological efficiencies at each trophic level in food webs (Figure 25.7).
- Production efficiency measures the percentage of energy assimilated that becomes incorporated into new biomass. Trophic-level transfer efficiency measures the energy available at one trophic level that is acquired by the level above (Figure 25.8).
- Trophic-level transfer efficiencies can be graphically represented in the form of ecological pyramids, the best known of which is the pyramid of numbers (Figure 25.9).
- Food web connectance decreases with increasing number of species (Figure 25.10).
- Much effort is required in the accurate depiction of a food web, otherwise many links can be overlooked (Figure 25.A).

**25.3 Within Food Webs, Some Species Have Disproportionately Large Effects**

- Keystone species have effects in food webs that are out of proportion to their abundance or biomass (Figures 25.11–25.13, Figure 25.B).
- Other important species in food webs include ecosystem engineers, indicator species, umbrella species, and flagship species (Figures 25.14, 25.15, Table 25.1).

## TEST YOURSELF

1. Which of these organisms is a heterotroph?
   a. Diatom in the phytoplankton
   b. Moss
   c. Oak tree
   d. Corn plant
   e. Fish
2. Lions normally feed at which trophic level?
   a. 1
   b. 2
   c. 3
   d. 4
   e. 5
3. If there are 20 species in a food web, the potential number of links is:
   a. 19
   b. 20
   c. 190
   d. 380
   e. 760
4. If there are 20 species and 20 links in a food web, the actual connectance is:
   a. 0.026
   b. 0.053
   c. 0.105
   d. 1.0
   e. 1.05
5. If there are 20 species and 40 links in a food web, the linkage density is:
   a. 20
   b. 40
   c. 0.5
   d. 2.0
   e. 800
6. Chemoautotrophic bacteria are:
   a. Primary consumers
   b. Secondary consumers
   c. Tertiary consumers
   d. Primary producers
   e. Decomposers
7. A species that has an effect out of all proportion to its commonness is called a:
   a. Dominant species
   b. Keystone species
   c. Indicator species
   d. Umbrella species
   e. Flagship species
8. If production at trophic level $n$ is 50 kcal/m$^2$ and production at trophic level $n - 1$ is 400 kcal/m$^2$, then trophic-level transfer efficiency is:
   a. 10%
   b. 12.5%
   c. 20%
   d. 62.5%
   e. 80%
9. Which of the following organisms is not a keystone species?
   a. *Pisaster* starfish
   b. Sea otter
   c. Distemper virus
   d. Fig tree
   e. *Spartina* grass
10. Polar bears are often thought of as a(n) _____ species.
    a. Umbrella
    b. Keystone
    c. Invasive
    d. Indicator
    e. Ecosystem engineer

## CONCEPTUAL QUESTIONS

1. Why are there generally few links in food chains?
2. Explain the pyramid of numbers and pyramid of biomass concepts. How is it possible to get inverted pyramids? What is the pyramid of energy?
3. What are the advantages and disadvantages of connectedness, energy, and functional webs?
4. Distinguish between flagship, indicator, and umbrella species.
5. Are ecosystem engineers all flagship species?

## DATA ANALYSIS

Invasive beavers, *Castor canadensis,* are having strong effects in Chile, where they were introduced in the 1940s. The table gives data on unmanipulated forest streams, beaver ponds that impact streams, and areas downstream from the ponds. Explain what is happening. Why are beaver ponds compared to two different areas on the streams?

| Category | Forested | Beaver Pond | Downstream |
|---|---|---|---|
| Species richness per $m^2$ | 15.3 | 10.0 | 15.8 |
| Species diversity, $H_S$ | 2.0 | 1.4 | 1.9 |
| Species abundance per $m^2$ | 2611 | 14,350 | 5,086 |
| Biomass, mg per $m^2$ | 257.9 | 864.1 | 443.3 |
| Canopy cover, % | 69.4 | 21.5 | 30.2 |
| Substrate particle type, % | | | |
| Organic | 0.5 | 88.8 | 5.3 |
| Sand | 11.0 | 8.0 | 11.3 |
| Gravel | 37.0 | 2.8 | 46.8 |
| Cobbles | 51.6 | 0.5 | 36.8 |

**Connect Ecology** helps you stay a step ahead in your studies with animations and videos that bring concepts to life and practice tests to assess your understanding of key ecological concepts. Your instructor may also recommend the interactive ebook.

Visit **www.mhhe.com/stilingecology** to learn more.

Gypsy moth caterpillar.

CHAPTER 26

# Biomass Production

## Outline and Concepts



The gypsy moth, *Lymantria dispar,* was introduced into North America around 1868 by Leopold Trouvelot, who was experimenting with them for use in the silk-spinning industry. Native silk moths were susceptible to diseases and Trouvelot was trying to make a resistant caterpillar hybrid. Some moths escaped from his laboratory in Massachusetts and since that time have spread south and west throughout most of the U.S. Northeast. Gypsy moth larvae are voracious eaters that prefer oak and aspen leaves. About every 10 years there is a serious outbreak and virtually anything green is eaten. The first outbreak occurred in Massachusetts in 1889. During outbreaks, entire forest canopies can be defoliated. Nitrogen in the foliage is turned into caterpillar biomass, insect feces (called frass), and fallen leaves, all of which fall to the forest floor. Most of this nitrogen is eventually used by soil microorganisms or is taken up by growing plants, which may increase in biomass. However, some is leached into streams and rivers, and water nitrate concentrations rise dramatically in areas that previously had low nitrate concentrations. Studies of the effects of gypsy moths underscore the interconnectedness of terrestrial and aquatic systems and how energy flows from one area to another, such as from forest canopy to understory or stream, and affects nutrient availability and biomass.

In this chapter we will take a closer look at biomass production and energy flow through ecosystems. Because the bulk of the Earth's biosphere consists of primary producers, when we measure ecosystem biomass production, we are mainly interested in **primary production**, the production of organic compounds through the process of photosynthesis. Plants, algae, and cyanobacteria are the organisms responsible for primary production. Production in heterotrophs and decomposers, termed secondary production, involves far less energy. In this chapter we begin by outlining the effects of water, temperature, nutrients, and light. Understanding how these factors limit primary production is of vital importance if we are to examine ecosystems as energy transformers. Furthermore, by determining which of these factors is important, we can understand how primary production varies globally. We can also examine the effects of primary production on secondary production.

## 26.1 Production Is Influenced by Water, Temperature, Nutrients, and Light Availability

Because plants represent the first, or primary, trophic level, we measure plant production as gross primary production. **Gross primary production** (**GPP**) is equivalent to the carbon fixed during a given time period of photosynthesis. The efficiency of gross primary production can be calculated by comparing the energy produced by photosynthesis to the energy available in incident sunlight. The efficiency of utilization of sunlight is given by the formula:

$$\text{Efficiency of gross primary production} = \frac{\text{Energy fixed by gross primary production}}{\text{Energy in incident sunlight}} \times 100$$

Deserts are not very efficient at converting huge amounts of incident energy from sunlight into plant biomass, because water availability is limited. Most ecosystems based on phytoplankton have very low efficiencies, usually less than 0.5%, because the availability of nutrients is severely limited. Herbaceous plants and deciduous trees are more efficient but have to grow new sets of leaves every year. The highest efficiencies of sunlight utilization occur in coniferous forests, because coniferous trees' numerous needles are always present and create a large surface area for photosynthesis (**Figure 26.1**).

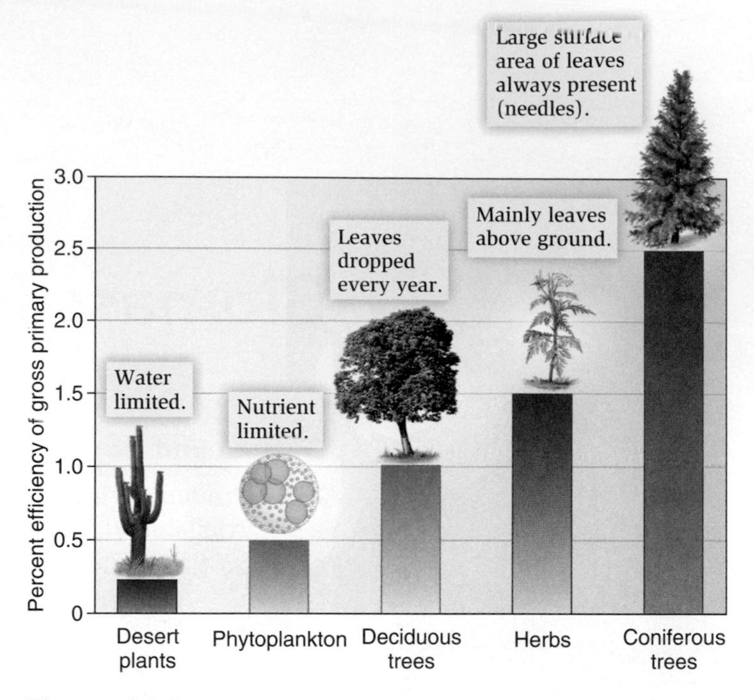


**Figure 26.1 Relative efficiencies of primary production for various types of plants.**

**Net primary production** (**NPP**) is gross primary production minus the energy lost in plant cellular respiration (R).

$$NPP = GPP - R$$

Net primary production is the amount of energy that is available to primary consumers. Energy content is generally measured using dry biomass. Dry weight is used because the bulk of living matter in most species is water, and water content fluctuates widely, often according to wet or dry seasons. Of the dry weight, 95% is made up of carbon compounds, so that measuring energy flow in ecosystems is in many ways equivalent to examining the carbon cycle (look ahead to Section 27.3), and ecologists often measure NPP in terms of carbon fixed per square meter or per hectare per year.

In order to examine ecosystems as energy transformers, we must have some way of determining the energy content of plants. There are many ways to measure energy content, including five that we will describe here: calorimetry, harvesting, $CO_2$ uptake, $O_2$ output, and chlorophyll concentration.

In calorimetry, a sample of dry material is burned in a small chamber so that it completely oxidizes to carbon dioxide and water. High-pressure oxygen is fed into the chamber to ensure complete combustion. A water jacket surrounds the chamber, and the rise in temperature of the water provides a measure of the heat generated by combustion of the sample. The greater the energy in a sample, the more heat will be generated. In harvesting, samples of plant biomass are obtained at the beginning of a growing season, then dried and weighed. At the end of the season, another sample is obtained, dried, and weighed. The difference in weights represents growth in plants in the area studied and provides a measure of productivity in a known unit of time. In the $CO_2$ uptake method, plants are grown inside sealed chambers for short periods. Because we know air contains 0.04% $CO_2$, any decrease in $CO_2$ inside the chambers is the result of $CO_2$ uptake by plants during photosynthesis. The higher the $CO_2$ uptake, the greater the production of plant biomass. The exact plant biomass required to take up a given volume of $CO_2$ is then calculated. A variation of this technique is to add a known amount of the radioactive isotope $^{14}C$. The amount of $^{14}C$ assimilated by the plant is then determined. The rate of carbon fixation is calculated by dividing the amount of $^{14}C$ found in the plant by the amount in the chamber air at the beginning of the experiment.

For many aquatic plants we can estimate NPP by $O_2$ output instead of $CO_2$ intake. For example, to measure phytoplankton production, samples are suspended in two bottles in a lake or ocean. One of the bottles is dark in color and oxygen is used during respiration. The other bottle is light and photosynthesis occurs, giving off oxygen. The difference in oxygen concentration between the light and dark bottles yields the oxygen produced by photosynthesis. The greater the amount of $O_2$ produced, the greater the plant biomass.

Finally, NPP can be estimated by measuring plant chlorophyll concentration. We know how much carbon is assimilated per gram of chlorophyll. By taking a sample of the plant under question and measuring its chlorophyll, we can calculate its carbon production. More modern methods take advantage of the reflectance of plants to measure the chlorophyll concentration. Thus, orbiting satellites can estimate reflectance colors of different habitats and thus estimate their production (**Figure 26.2**). When we look at the oceans, bright greens, yellows, and reds indicate high chlorophyll concentrations. Some of the highest marine chlorophyll concentrations occur at continental margins where river nutrients pour into the oceans. Upwellings along coasts also bring nutrient-rich water to the surface. Northern oceans, and to a lesser extent southern oceans, are also very productive, because cold temperatures in the winter allow vertical mixing of all layers, bringing nutrient-rich water to the surface. In the spring, the high amounts of light and nutrients permit rapid phytoplankton growth until the nutrients are all used up. Many other areas of the oceans are highly unproductive. The only other area of relatively high marine production occurs along the equator as trade winds and Coriolis forces cause water to move north of the equator in the Northern Hemisphere and south of the equator in the Southern Hemisphere. This causes an upwelling as cold water from below rises to replace the wind-blown surface water. Nutrients are brought to the surface and productivity increases.

Over land, productivity is measured as the Normalized Difference Vegetation Index (NDVI), which is an estimate of the photosynthetically absorbed radiation over land surfaces. Plants absorb much visible light but reflect light at near-infrared wavelengths. Other surfaces such as bare earth, snow, clouds, or water reflect more visible light but absorb more infrared. This difference in absorption allows scientists to estimate photosynthetic rates and hence primary production. Plant productivity on land appears large in the broad band of the Tropics, but is equally great in the temperate zones, especially in the north. Productivity is much reduced in the hot deserts of the world.

### 26.1.1 Net primary production in terrestrial ecosystems is limited mainly by water, temperature, and nutrient availability

In terrestrial systems, water is a major determinant of primary production, and net primary production often shows an almost linear increase with annual precipitation, at least in somewhat arid regions (**Figure 26.3**). However, more recent analyses by Edward Schuur (2003), which included data from

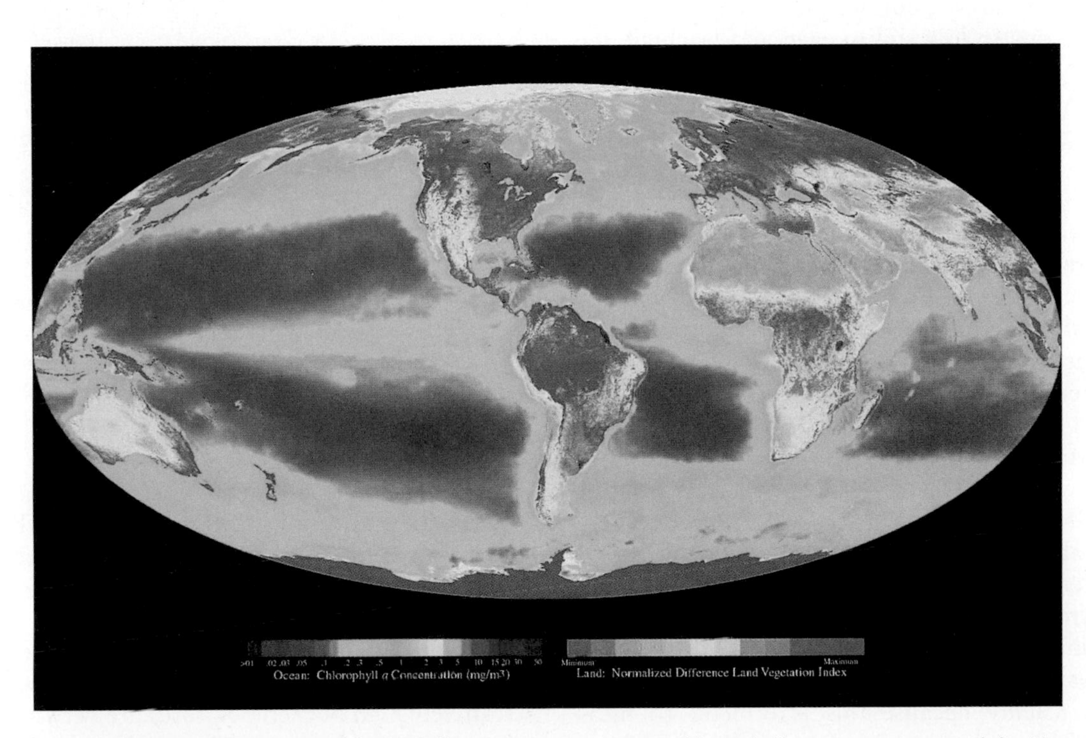


**Figure 26.2 Primary productivity measured by satellite imagery.** Ocean chlorophyll concentrations and the Normalized Difference Vegetation Index on land provide good data on marine and terrestrial productivity, respectively.

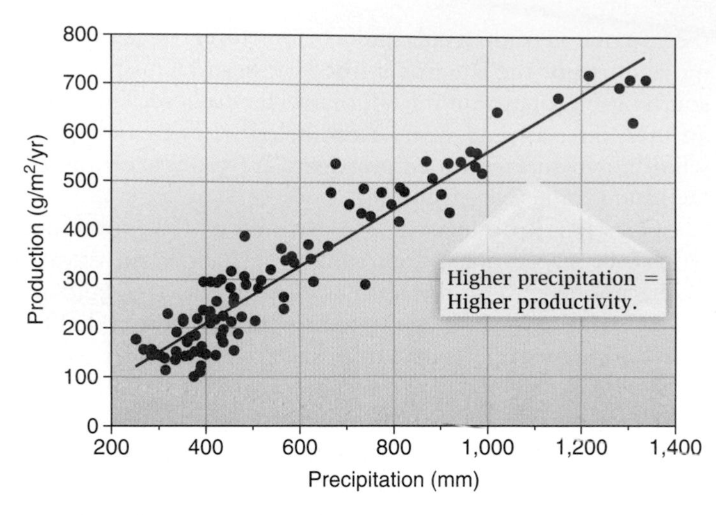


**Figure 26.3 Relationship between mean annual precipitation and mean aboveground net primary production in North American prairies.** Data come from 100 areas across the central grassland region, the Great Plains, of the United States. (Redrawn from Sala et al., 1988.)

**ECOLOGICAL INQUIRY**

What feature might generate the differences in the prairie data?

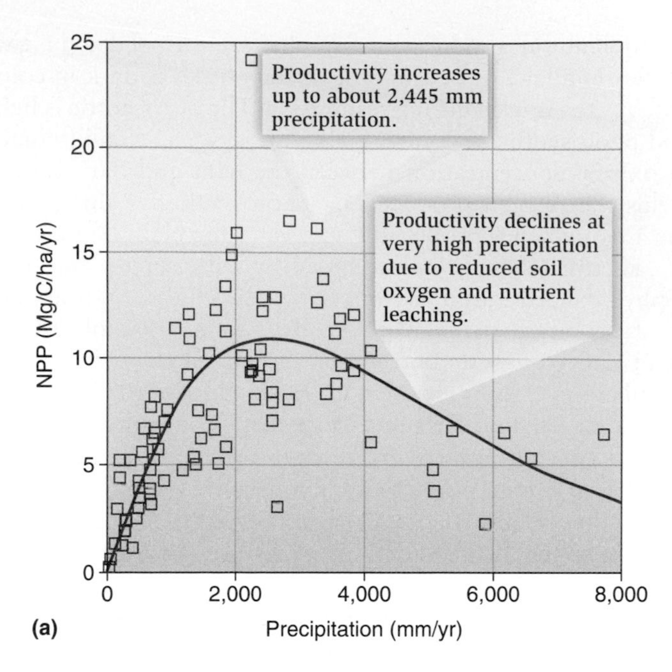


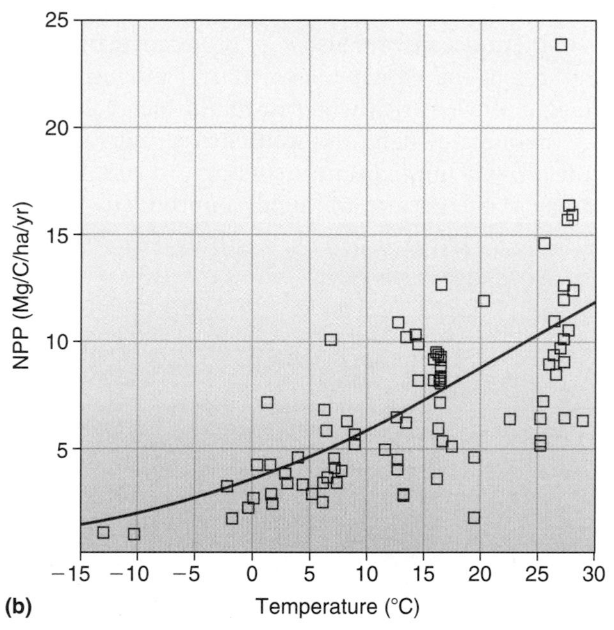


**Figure 26.4 Relationship between mean annual precipitation, mean annual temperature, and net primary productivity across the globe.** **(a)** Precipitation. **(b)** Temperature. Units are megagrams ($10^6$ grams or 1 metric ton) per hectare per year. (After Schuur, 2003.)

a wide variety of biomes including many tropical forests, showed NPP declines at very high precipitation (**Figure 26.4a**). NPP is highest at 2,000–3,000 mm/yr of precipitation, which is typical of many rain forests, but declines at extremely high precipitation levels due to low soil oxygen and leaching, the loss of water-soluble plants nutrients from the soil.

Temperature also affects production, primarily by slowing or accelerating plant metabolic rates. For example, Jim Raich and colleagues (1997) documented a linear relationship between total net primary productivity and mean annual temperature as measured on an elevation gradient in Hawaii. For each 1°C increase in mean annual temperature, total net primary production increased by 54 g/m$^2$/yr. Edward Schuur's (2003) synthesis of data from a wide variety of biomes shows a continual increase in NPP with temperature (**Figure 26.4b**). However, at extremely high temperatures, above 30°C and beyond, NPP can decline as leaf stomata close and plants conserve water. Michael Rosenzweig (1968) noted that both temperature and water availability were often important at the same time. He showed that the evapotranspiration rate could predict the aboveground primary production with good accuracy for different biomes in North America, with two additional sites from African tropical forest (**Figure 26.5**). Aboveground primary production is a measure of the productivity of leaves, twigs, and branches; it does not include root production, usually because this is difficult to measure. However, aboveground and belowground productivity are tightly correlated. Recall from Section 18.3 that the evapotranspiration rate is a measure of the amount of water entering the atmosphere from the ground through the process of evaporation from the soil and transpiration of plants, so it is a measure of both temperature and available water. For example, a desert has a low evapotranspiration rate because water availability is low despite high temperature. Tundra has a low evapotranspiration rate because water availability is relatively high but temperatures are very low. Rates of evapotranspiration are maximized when both temperature and moisture are at high levels, as in tropical forests.

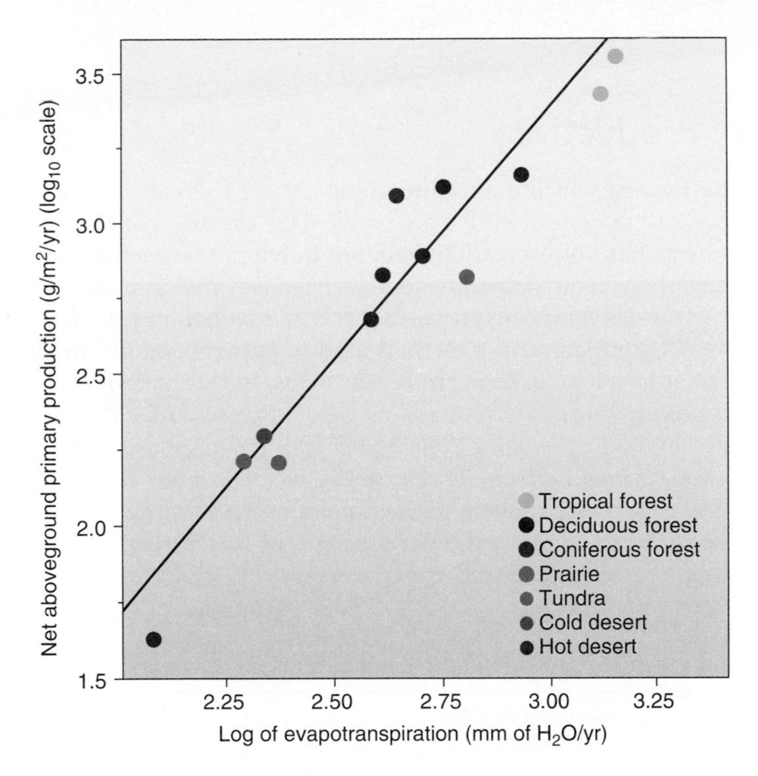


**Figure 26.5 Primary production is positively correlated with the evapotranspiration rate.** Warm, humid environments are ideal for plant growth. Dots represent different ecosystems. (After Rosenzweig, 1968.)

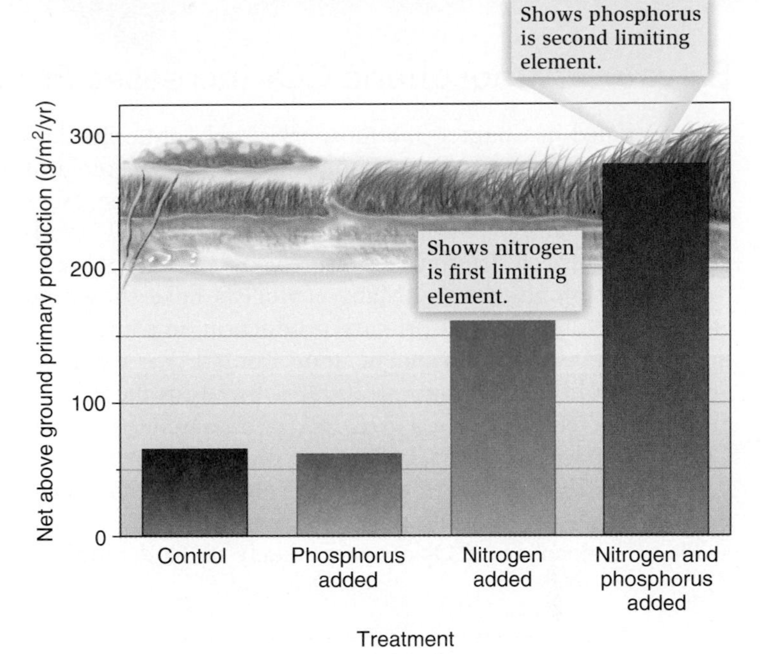


**Figure 26.6 Primary production is constrained by the most limiting factor.** Net aboveground primary production of saltmarsh sedge, *Carex subspathacea*, in response to nutrient addition. Nitrogen is more limiting than phosphorus alone, but once nitrogen becomes available, phosphorus becomes the limiting factor. (From data in Cargill and Jefferies, 1984.)

A lack of **nutrients**, key elements in usable form, particularly nitrogen and phosphorus, can also limit primary production in terrestrial ecosystems, as agricultural practitioners know only too well. Fertilizers are commonly used to boost the production of annual crops. Stewart Cargill and Rob Jefferies (1984) showed that a lack of both nitrogen and phosphorus was limiting to saltmarsh sedges and grasses in subarctic conditions in Hudson Bay, Canada (**Figure 26.6**). Of the two nutrients, nitrogen was the most limiting; without it, the addition of phosphorus did not increase production. However, once nitrogen was added, phosphorus became the **limiting factor**, that is, the scarcest in relation to need. Once nitrogen was added and was no longer limiting, the addition of phosphorus increased production. The addition of nitrogen and phosphorus together increased production the most. This result supports a principle known as **Liebig's law of the minimum**, named for Justus von Liebig, a 19th-century German chemist, which states that species biomass or abundance is limited by the scarcest factor. This factor can change, as the Hudson Bay experiment showed: When sufficient nitrogen is available, phosphorus becomes the limiting factor. Once phosphorus becomes abundant, then productivity will be limited by another nutrient. Another nutrient that limits primary productivity, carbon dioxide, is currently increasing in the atmosphere, and this increase is likely to continue unabated for a long period of time. As a result, net primary productivity is also likely to show significant long-term increases (see **Global Insight**).

## 26.1.2 Net primary production in aquatic ecosystems is limited mainly by light and nutrient availability

Of the factors limiting primary production in aquatic ecosystems, the most important are available light and nutrients. As we saw in Chapter 7, light is particularly likely to be in short supply because water readily absorbs light. At a depth of 1 m, more than half the solar radiation has been absorbed. By 20 m, only 5–10% of the radiation is left. The decrease in light is what limits the depth of water to which plants are restricted. However, if light were of prime importance in limiting ocean productivity, we should expect a gradient of increasing productivity from the poles to the equator. Satellite imagery shows quite the opposite (see Figure 26.2). Oceanic chlorophyll concentrations are low in tropical waters, despite high light regimes. Here, nutrients appear to be a limiting factor.

The most important nutrients affecting primary production in aquatic systems are nitrogen and phosphorus, because they occur in very low concentrations. While soil contains about 0.5% nitrogen, seawater contains only 0.00005% nitrogen. Enrichment of the aquatic environment by the addition of nitrogen and phosphorus can result in large, unchecked growths of algae called algal blooms. Such enrichment occurs naturally in areas of upwellings, where cold, deep, nutrient-rich water containing sediment from the ocean

Global Insight

## Elevated Atmospheric $CO_2$ Increases Primary Productivity

As mentioned in Chapter 3, atmospheric carbon dioxide levels have been linked to the rise of land plants in geological time (refer back to Figure 3.11). Understanding the exact nature of the link between plant production and $CO_2$ availability is critical in determining future ecosystem response to this type of global change. Many ecologists have shown an effect of elevated $CO_2$ on primary production. In some cases, growth increases are dependent upon rainfall. For example, Stanley Smith and colleagues (2000) showed an increase in growth of creosote bush, *Larrea tridentata,* under elevated $CO_2$ in the Mojave Desert, but only in years of plentiful rainfall (**Figure 26.A**). Deserts have relatively low primary productivity, and the world's ecosystem responses to elevated levels of atmospheric $CO_2$ are more likely to be determined by forests, which contribute about 50% of global NPP and 80% of terrestrial NPP. Large-scale $CO_2$ elevation in tropical forests has not been attempted, but in temperate forests there have been many experiments that have elevated atmospheric $CO_2$ levels by means of so-called FACE technology, or free-air $CO_2$ enrichment, a method used to experimentally enrich the atmosphere in terrestrial ecosystems. In this method, $CO_2$ is elevated through freestanding pipes (**Figure 26.B**). Richard Norby and colleagues (2005) analyzed the response of temperate forest NPP to elevated $CO_2$, which was increased by FACE from the ambient 370 ppm to about 550 ppm, the level expected to be reached before the end of the current century. In four separate FACE experiments, NPP was elevated by 23% on average.

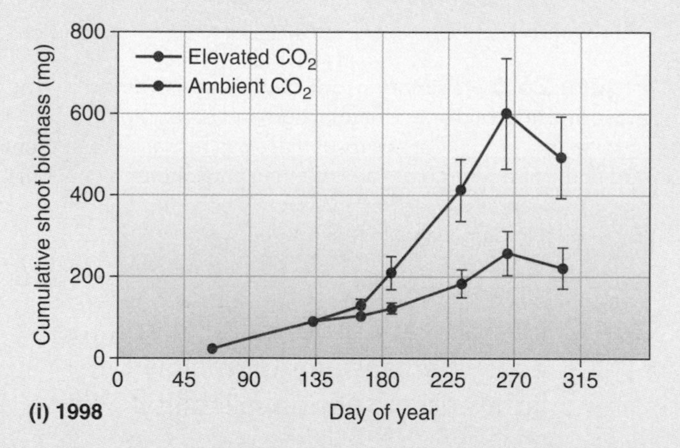


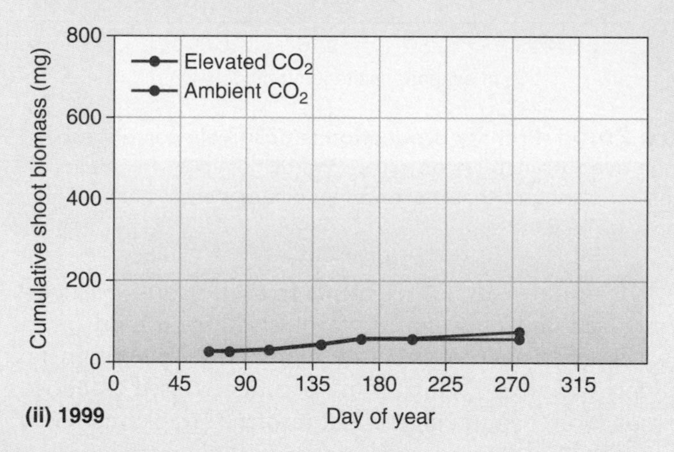


**Figure 26.A Increased net primary production under elevated $CO_2$.** Net primary production was measured as cumulative shoot biomass in ambient and elevated $CO_2$ in the Mojave Desert in **(i)** a wet El Niño year, 1998, and **(ii)** a dry La Niña year, 1999. (After Smith et al., 2000.)

**Figure 26.B Free-air $CO_2$ enrichment (FACE) technology.** Carbon dioxide is increased via a circle of pipes. Here four replicates in a forest are shown.

floor is brought to the surface by strong currents, resulting in very productive ecosystems and plentiful fish. Some of the largest areas of upwelling, as measured by ocean nitrate concentrations, occur off the coast of Antarctica, in the North Pacific and North Atlantic Oceans, and along the coasts of Peru, in South America (**Figure 26.7**).

In marine systems, early experiments showed that iron also limits production in addition to nitrogen and phosphorus. However, iron is only limiting once the nitrogen and phosphorus requirements have been met. Fertilization experiments on water in the Sargasso Sea in the early 1960s showed that once nitrogen and phosphorus were no longer limiting, the addition of iron to surface water boosted production, as revealed by an increase in carbon uptake (**Figure 26.8a**). Since then, many more nutrient enrichment experiments on marine systems have been performed. John Downing and colleagues (1999) summarized the results of 303 of them (**Figure 26.8b**). They examined studies lasting from 1 to 7 days to avoid the effects of primary production reductions by grazers such as zooplankton. In studies of greater time periods, some of the increases in primary production might have been greatly reduced by grazing. Growth rates were measured as changes in weight, carbon accumulation, or chlorophyll concentrations. Most experiments showed a strong effect of nitrogen and iron addition, but, surprisingly, no effects of phosphorus enrichment. However, results were site-dependent. For example, phosphorous and iron limitation was strong in pristine tropical waters, but weak in coastal waters where pollution had probably already increased phosphorus levels. Such experiments are particularly timely given the fact that fertilizing ocean waters with iron and/or nitrogen is one of the prominent proposed strategies to stimulate the growth of phytoplankton and reduce global atmospheric carbon dioxide. Since 1993, 12 large-scale ocean studies have seeded large areas of international waters with iron-rich fertilizer and triggered an increase in phytoplankton blooms.

Too much nutrient supply can also be harmful to aquatic ecosystems. In marine systems, high nutrient levels cause **dead zones**, where fish and other marine life can no longer survive. The problem is caused by fertilizer runoff into rivers, which drain into the oceans. The heavy doses of nitrogen and phosphorus cause blooms of microscopic algae near the water surface. When these algae die, they sink and are decomposed by masses of bacteria. As the bacteria respire, they deplete the oxygen around them, leaving little to support other marine life. A study by Robert Diaz and Rutger Rosenberg (2008) documented 405 dead zones in the world's oceans, totaling more than 245,000 km$^2$. The largest of these is a 22,126-km$^2$ (8,543-mi$^2$) area in the Gulf of Mexico where the Mississippi River dumps runoff high in nutrients from a vast drainage basin. The area of the Gulf dead zone is equivalent to the size of New Jersey (**Figure 26.9**). Dead zones can recover if fertilizer use is reduced.

Daniel Gruner and colleagues (2008) recently compared the effects of nutrient addition and herbivore removal on producer biomass. They examined 191 studies in which

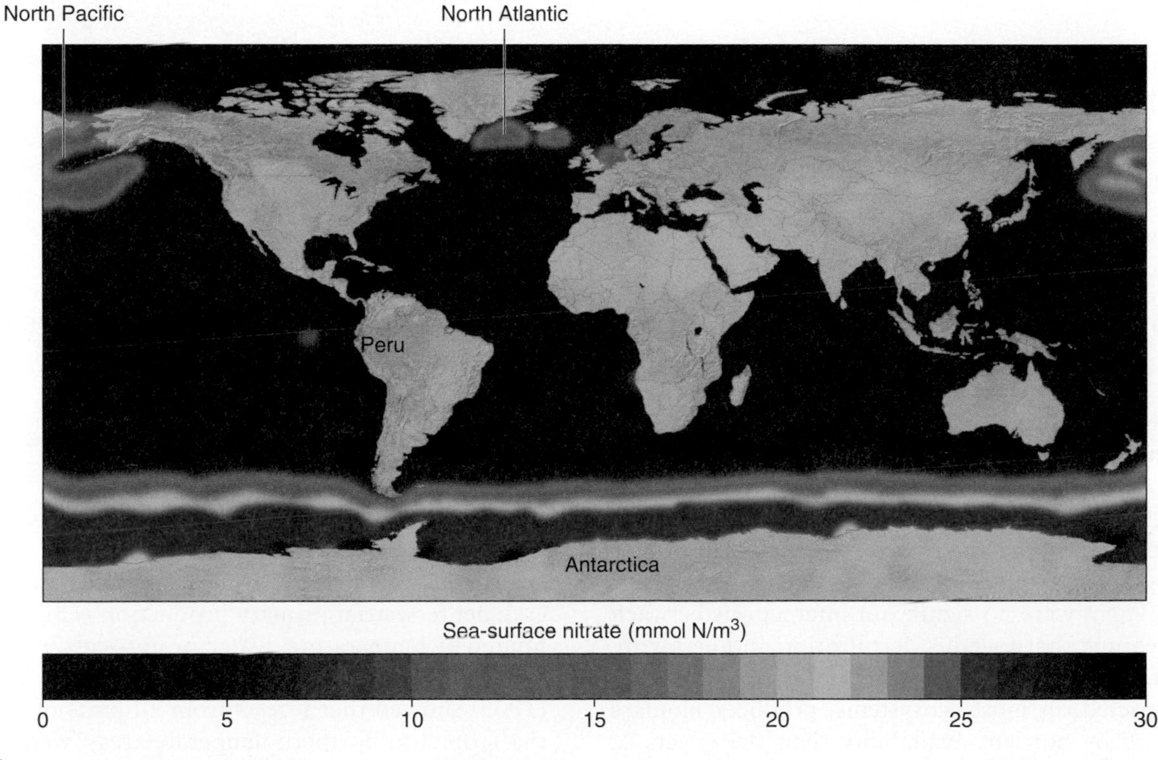


**Figure 26.7 Annual mean sea-surface nitrate concentrations.** Units are mmol N/m$^3$.

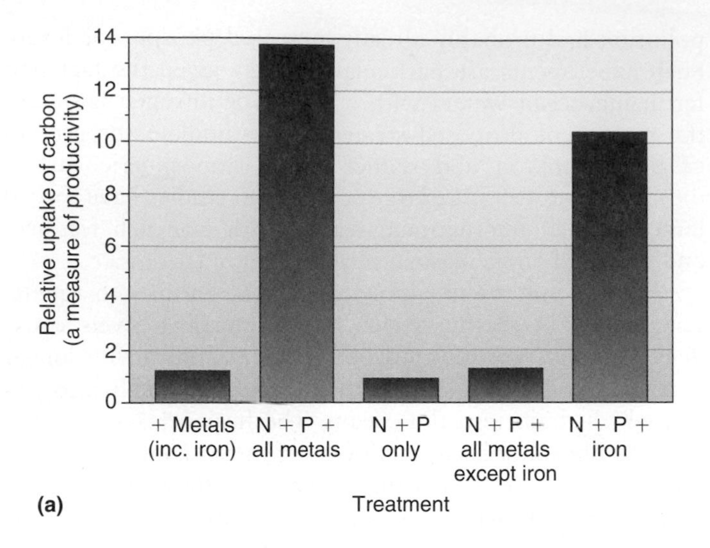


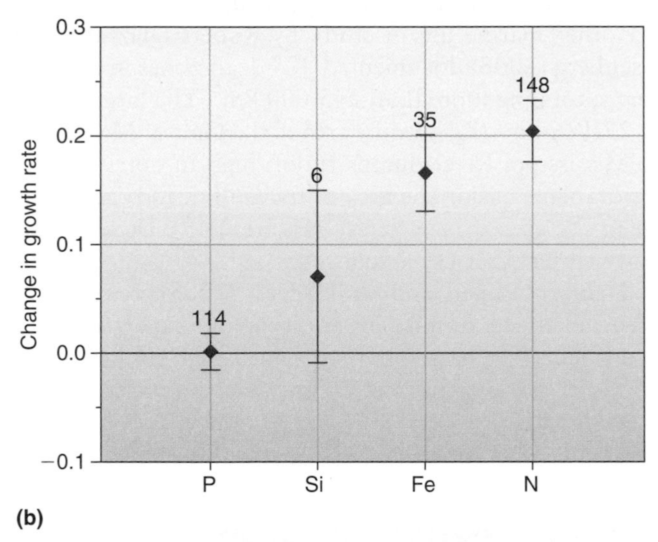


**Figure 26.8 Effects of nutrient enrichment on marine productivity.** **(a)** Results of experiments in which bottled seawater from the Sargasso Sea, south of Bermuda, was enriched with various nutrients. Productivity in experimental bottles was compared relative to that in controls where no nutrients were added. (From data in Menzel and Ryther, 1961.) **(b)** Meta-analysis of marine enrichment studies showing the effects of nitrogen, iron, silicon, and phosphorus addition on growth rate of phytoplankton. The number of studies is on top of the bars. (After Downing et al., 1999.)

**Figure 26.9 A dead zone in the Gulf of Mexico.** Nutrients from the Mississippi River cause algal blooms in the Gulf of Mexico. When these algae die and decompose, the oxygen levels in the ocean become too low to support much marine life.

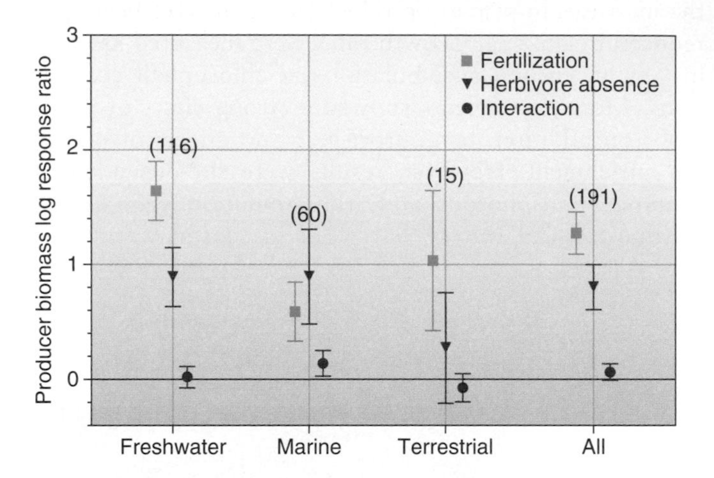


**Figure 26.10 Meta-analysis of responses of producer biomass to nutrient addition and herbivore removal.** Producer biomass was measured as a log response ratio. Sample sizes are given in parentheses. (After Gruner et al., 2008.)

nutrients were added and consumers were removed across freshwater, marine, and terrestrial ecosystems (**Figure 26.10**). Producer biomass increased with fertilization across all biomes, though increases were greatest in freshwater habitats. Herbivore removal generally increased biomass in freshwater and marine systems; the results were less consistent in terrestrial areas. There were no significant interactions between treatments, meaning that together, fertilization and herbivore removal did not cause any extra increase in producer biomass than expected. In most ecosystems, producer biomass is affected more by nutrient availability than the effects of herbivores. This reinforces the idea that strong abiotic effects from the bottom of the food web are more important than effects from higher trophic levels.

## 26.1.3 Net primary production varies in different biomes

Knowing which factors limit primary production helps ecologists understand why mean net primary production varies across the different biomes on Earth (see Figure 26.2). In general, net terrestrial primary production is highest in areas not limited by temperature, water, or nutrients. As a result, forests are very productive. Michael Huston and Steve Wolverton (2009) showed that forests from all parts of the world, from the Tropics to northern temperate areas, were very similar in productivity (**Figure 26.11a**). Indeed, if one looks at the area

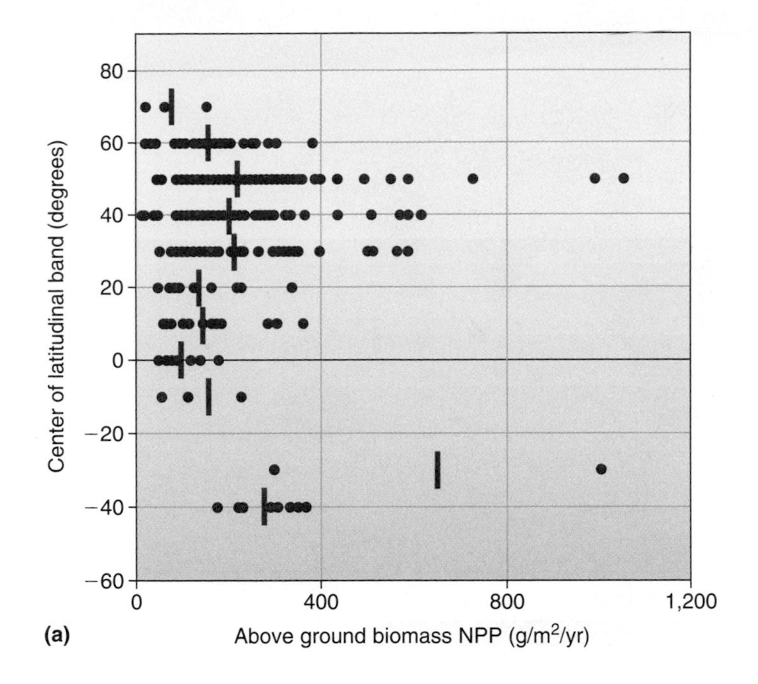


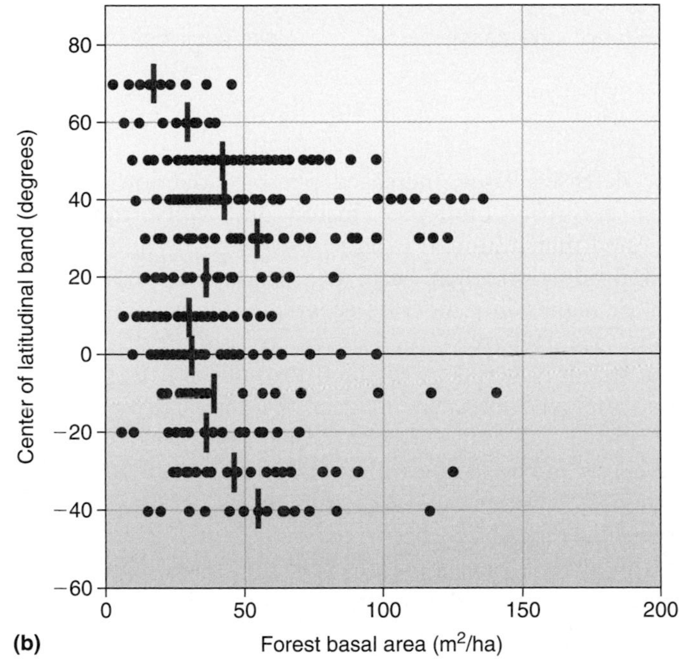


**Figure 26.11 Global variation in forest production.** Forest production is similar in most areas of the globe and may be a little higher in temperate forests than in tropical forests. **(a)** Global distribution of annual aboveground NPP for 362 sites. Means are given by vertical bars. **(b)** Forest basal areas in 800 mature forest sites. (After Huston and Wolverton, 2009.)

of the forest floor covered by trees, called the basal area, production is even higher in temperate than in tropical habitats (**Figure 26.11b**). This matches the pattern of productivity in the oceans. Although tropical forests enjoy warm temperatures and abundant rainfall, soils are rapidly weathered by such conditions. Tropical soils are low in available forms of most plant nutrients because of loss through leaching. In contrast, temperate soils tend to have much greater concentrations of essential nutrients because of lower rates of nutrient loss and more frequent grinding of fresh minerals by the cycles of continental glaciation over the past 3 million years (**Figure 26.12**).

Prairies and savannas are also highly productive, because their plant biomass usually dies and decomposes each year, returning nutrients to the soil, and temperatures and rainfall are not limiting. Deserts and tundra have low productivity because of a lack of water and low temperatures, respectively. Wetlands tend to be extremely productive, primarily because water is not limiting and nutrient levels are high. In aquatic biomes, productivity of the open ocean is very low because of a lack of nutrients, falling somewhere between the productivity of deserts and that of the Arctic tundra. Marine production is high on coastal shelves, particularly in zones where upwelling brings up nutrients from the ocean floor. However, the greatest marine production occurs on coral reefs and algal beds, where temperatures are high and water levels are not so deep that light becomes limiting.

## 26.1.4 Secondary production is generally limited by available primary production

Secondary production, the productivity of consumers, herbivores, carnivores, and detritivores, is dependent on primary production for energy. In order to examine secondary production, we need to be able to measure energy consumption by herbivores and carnivores. Measuring energy consumption is fairly easy: An herbivore is confined to a given area, and the mass of vegetation eaten per unit time is recorded. For carnivores, the intake of prey items of known biomass and energy content can be measured. Measuring how much energy an animal assimilates from its prey is more difficult. Usually we measure the nutrient content of feces and urine excreted to determine how much of the food was assimilated by the organism. Of the energy assimilated, some goes to production (growth) but much goes to maintenance (respiration). For production we can measure the weight of an organism before and after a meal to get its increase in weight, but we have to wait a sufficient period, at least 24 hours, for the meal to have been completely absorbed into the body.

Secondary production is generally thought to be limited largely by available primary production. A strong relationship exists between primary production in a variety of terrestrial ecosystems and the biomass of herbivores (**Figure 26.13**). The correlation is not as obvious as one might think, because it implies that secondary metabolites that make plants poisonous or distasteful to herbivores (as discussed in Chapter 14) are much less important influences on consumption at the ecosystem level than they are at the population level. This means that although individual plant species may successfully defend themselves against herbivores by producing secondary metabolites, on a large scale, involving scores or even hundreds of plant species, many herbivores can overcome

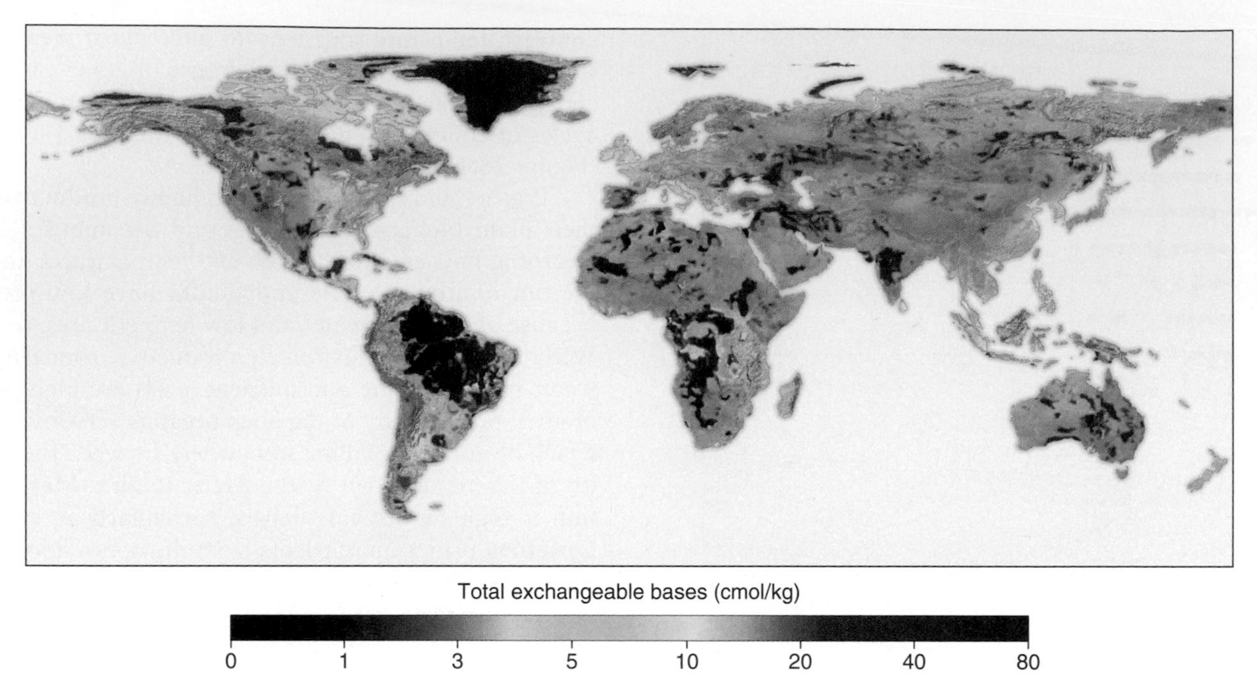


**Figure 26.12 Global variation in soil fertility.** Soil fertility is measured as total exchangeable bases in centimoles per kilogram (cmol/kg). (After Huston and Wolverton, 2009.)

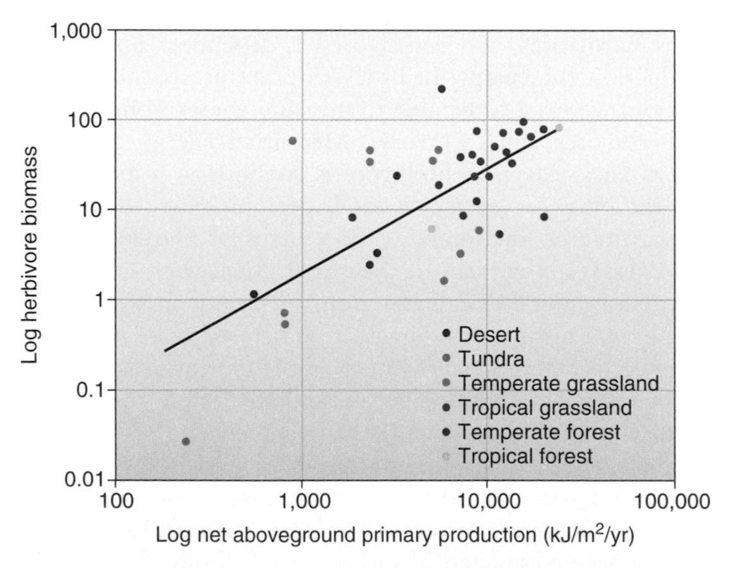


**Figure 26.13 Herbivore biomass is positively correlated with net aboveground primary production.** These data are taken from a variety of case studies from different biomes. Herbivore biomass can be considered a surrogate for secondary production. (After Moen and Oksanen, 1991.)

**ECOLOGICAL INQUIRY**

What does this relationship imply about the effects of plant secondary metabolites, many of which taste bad and some of which are toxic, on secondary production?

these defenses. Thus, increased plant production results in larger herbivore biomass. Huston and Wolverton (2009) suggested that if animal production is heavily influenced by plant production, then herbivore production should be as high, or higher, in temperate ecosystems as in tropical ones. Though the data are scant, they tend to support this idea. The mean monthly aerial insect biomass collected from suction traps employed in parkland near the University of Sterling (57° north) was much greater than that near the University of Malaya in the Tropics (3° north) (**Figure 26.14**). Insects provide food for many other animals, especially birds. Not surprisingly, the clutch size in temperate birds is much larger than in their tropical counterparts. This led Huston and Wolverton to argue that many birds migrate from the Tropics to temperate areas to take advantage of high seasonal productivity as much as leaving temperate areas to avoid winter. This highlights the importance of bottom-up factors such as plant quantity in controlling the numbers of herbivores that feed on them. Just Cebrian and Julien Lartigue (2004) showed that the percent of primary production consumed by herbivores in a variety of different ecosystems was related to nitrogen and phosphorus content of the primary producers (**Figure 26.15**). The greater the plant nutrient content, the higher the amount consumed. Some nutrient enrichment can be harmful to secondary production. Iron-rich dust from the Sahara that is blown into the Caribbean Sea and Gulf of Mexico can stimulate an algal bloom called red tide, in which algae become so numerous that they discolor the coastal

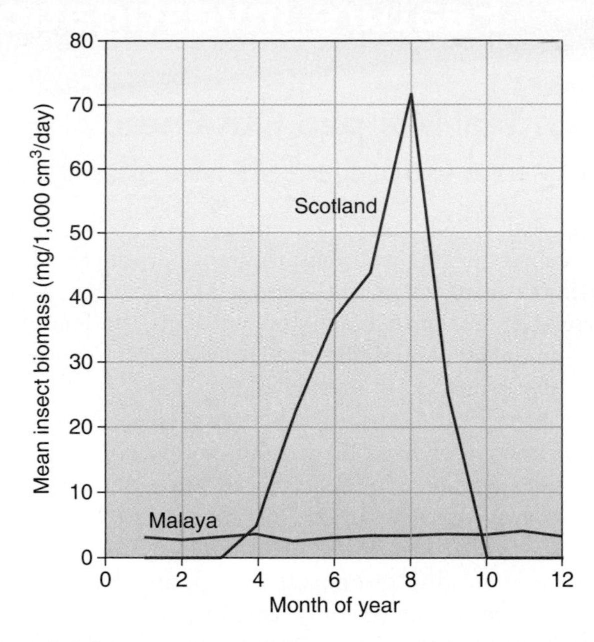


**Figure 26.14 Global variation in insect biomass.** Insect biomass at a tropical (Malaya) and temperate (Scotland) site over a year-long period. Summed over the year, insect biomass is greater in Scotland than Malaya. (After Hails, 1982.)

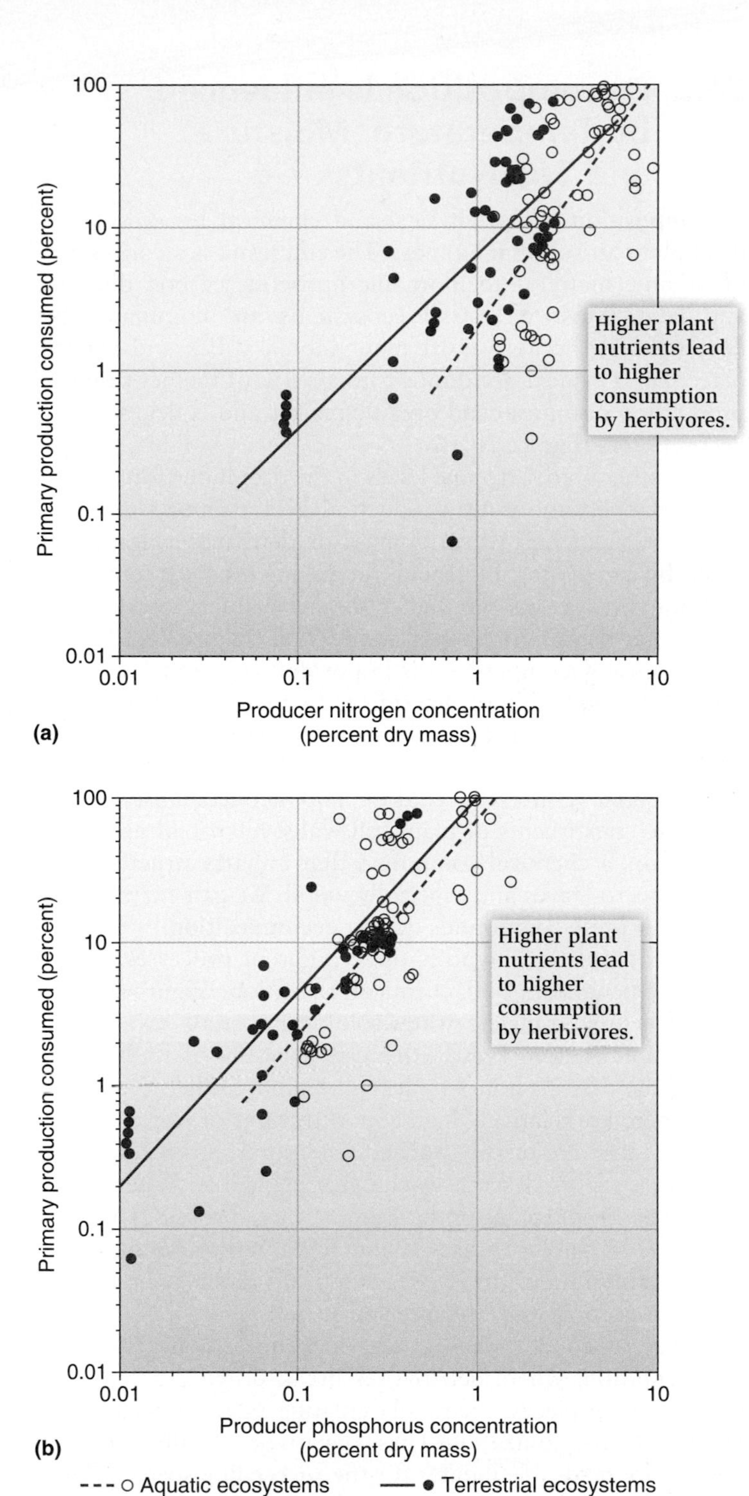


**Figure 26.15 Consumption of primary production by herbivores in relation to nutrients.** In many different ecosystems, consumption is related to **(a)** nitrogen concentration and **(b)** phosphorus concentration. (After Cebrian and Lartigue, 2004.)

waters. Nitrogen and potassium from sewage and runoff from agricultural fields can also cause red tide. The algae produce and release toxins that kill fish and shellfish and can also cause illness in the aquatic mammals and humans that eat affected shellfish.

As discussed in Chapter 25, trophic-level transfer efficiency, the amount of energy at one trophic level that is acquired by the trophic level above and incorporated into biomass, averages about 10%. Thus, after one link in the food web, only one-tenth of the energy captured by plants is transferred to herbivores, and after two links in the food web, only one-hundredth of the energy fixed by plants goes to carnivores. Thus, secondary production is much smaller than primary production (see **Feature Investigation**). Decomposition becomes an important avenue of energy transfer in ecosystems, as we discuss next. In general, decomposition is more important in terrestrial biomes than in freshwater or marine biomes. Only 10–20% of terrestrial primary productivity goes to herbivores, whereas about 30% of freshwater primary productivity and up to 60–80% of oceanic primary productivity is consumed by animals.

## Check Your Understanding

**26.1** Why is it that temperate forests or temperate marine waters may be more productive than, or as productive as, their tropical counterparts?

## 26.2 Decomposition Is Influenced by Temperature, Moisture, and Soil Nutrients

**Decomposition** is the physical and chemical breakdown of dead plant and animal biomass. The end result is a conversion of organic matter into inorganic nutrients, carbon dioxide, and heat. Most terrestrial ecosystems are dominated by decomposers, because most plant biomass dies and rots in place. For a typical deciduous forest, 96% of the net primary production becomes dead organic matter and is degraded by decomposers (**Figure 26.16**).

At other higher trophic levels in the deciduous forest food web, relatively more net production is consumed, such that nearly 90% of herbivore biomass or detritivore biomass is taken by carnivores. In general, in a wide range of terrestrial and aquatic systems the higher the net primary production, the greater the rate of detrital production (**Figure 26.17**).

Because decomposition is important as a transformer of energy, it is valuable to determine the factors that influence it. Much organic matter results from the periodic shedding of plant leaves. These leaves contain four different components: water-soluble material; cellulose and hemicellulose, which are large constituents of plant cell walls; microbial products; and lignin, a chemical compound that imparts structure and toughness to leaves and especially wood. We can examine the fate of all these compounds during decomposition.

Decomposition involves three different processes: leaching, fragmentation, and chemical alteration (**Figure 26.18**). First, **leaching** transfers water-soluble materials away from the organic matter. Fresh litter can lose 5% of its biomass during the first 24 hours by leaching alone. When the leaf is shed, there are relatively high concentrations of water-soluble products that are readily leached. Of course, in very dry ecosystems, such as deserts, leaching is negligible. Where temperatures are warm, as in the Tropics, these processes are fast and 75% of the litter mass decomposes within 3 months. In the frozen tundra of the Arctic, over 100 years may be needed to decompose 75% of the litter mass.

The second phase of decomposition involves **fragmentation**, where detritivores break down organic matter into small pieces, eventually creating a large surface area for microbial colonization. These detritivores facilitate or create a more favorable habitat for the succeeding species. Fragmentation occurs in a sequential fashion. The tough cuticle of leaves and the lignin-impregnated cell walls are designed to protect tissues from microbial attack. Larger animals such as earthworms pierce these protective barriers and increase the ratio of litter surface area to mass, facilitating the entry of microbes. A now classic experiment by Clive Edwards and G.W. Heath (1963) demonstrated this phenomenon. These researchers put oak leaves in nylon bags in the soil and examined decomposition rates. By varying the mesh size of the bags, they could vary the sizes of the organisms entering them. Large-meshed bags permitted access by all soil

## Feature Investigation

### John Teal Mapped Out Energy Flow in a Georgia Salt Marsh

John Teal (1962) examined energy flow in a salt marsh, an ecosystem that is among the most productive habitats on Earth in terms of the amount of vegetation produced (**Figure 26.C**). In salt marshes, most of the energy from the sun, incident sunlight, goes to two types of organisms: *Spartina* plants and marine algae. The *Spartina* plants are rooted in the ground, whereas the algae float on the water surface or live on the mud or on *Spartina* leaves at low tide. Of the two, *Spartina* is much more important. These photosynthetic organisms fix about 6% of the incident sunlight. Most of the plant energy, 77.6%, is used in plant cellular respiration. Of the energy that is accumulated in plant biomass, most dies in place and rots on the muddy ground, to be consumed by decomposers and other detritivores: bacteria, nematodes, and crabs. Bacteria are the major detritivores in this system, followed distantly by nematodes and crabs, which feed on tiny food particles as they sift through the mud. Some of this dead material is also removed from the system by the tide. The herbivores take very little of the plant production, eating only a small proportion of the *Spartina* and none of the algae. Overall, if we view the species in ecosystems as transformers of energy, then plants and algae are by far the most important organisms on the planet, bacteria are next, and animals are a distant third.

**HYPOTHESIS** To examine the energy flow in a natural ecosystem.

**STARTING LOCATION** Sapelo Island, Georgia, USA.

| | Conceptual level | Experimental level |
|---|---|---|
| 1 | Measure energy input as light. | Use light meter. Incident energy was 600,000 kcal/$m^2$/yr. |
| 2 | Measure net primary productivity, respiration, and gross primary productivity. | Measure net primary productivity of *Spartina* by harvesting, and respiration in air, at a variety of sites. |
| 3 | Measure bacterial production on *Spartina* detritus. | Place air-dried grass into flasks with sea water, inoculate with marsh mud, and measure $O_2$ consumption. |
| 4 | Measure primary consumer respiration. | Count number of grasshoppers and leafhoppers per $m^2$ and determine their respiration. Count number of nematodes and crabs per $m^2$ of mud and determine their respiration. |
| 5 | Measure secondary consumer respiration. | Count numbers of spiders per $m^2$ and determine their respiration. |
| 6 | Determine energetic loss of material washed out to sea. | Subtract energy consumed by bacteria, primary and secondary consumers from net production. |

7 **THE DATA** Numbers reflect percentages of gross primary production that flows into different trophic levels or is used in plant respiration. The biggest consumers of energy in a saltmarsh are the primary producers. Most primary production dies and rots in place to be decomposed by bacteria.

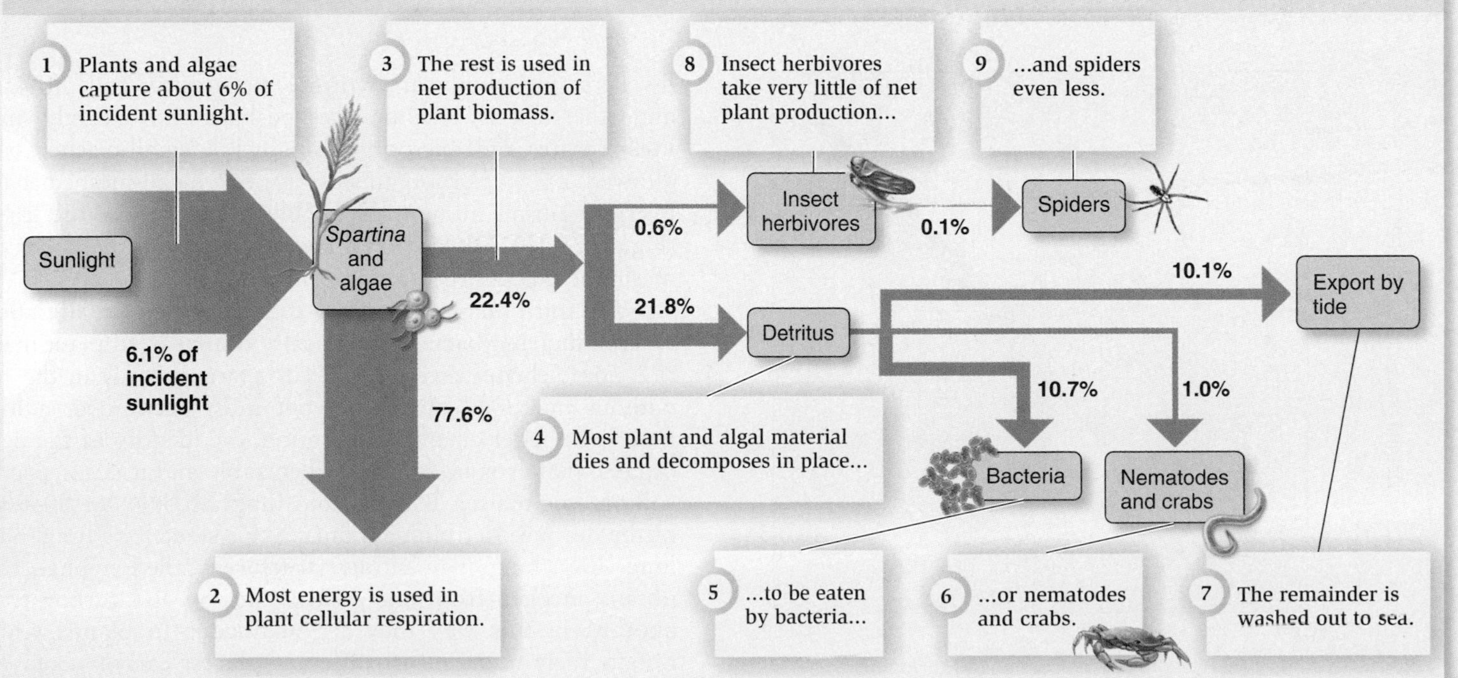

1 Plants and algae capture about 6% of incident sunlight.
3 The rest is used in net production of plant biomass.
8 Insect herbivores take very little of net plant production...
9 ...and spiders even less.
Insect herbivores
Spiders
Sunlight
*Spartina* and algae
0.6%
0.1%
10.1%
Export by tide
22.4%
21.8%
Detritus
6.1% of incident sunlight
77.6%
10.7%
1.0%
4 Most plant and algal material dies and decomposes in place...
Bacteria
Nematodes and crabs
2 Most energy is used in plant cellular respiration.
5 ...to be eaten by bacteria...
6 ...or nematodes and crabs.
7 The remainder is washed out to sea.

**Figure 26.C** Energy flow diagram for a Georgia salt marsh.

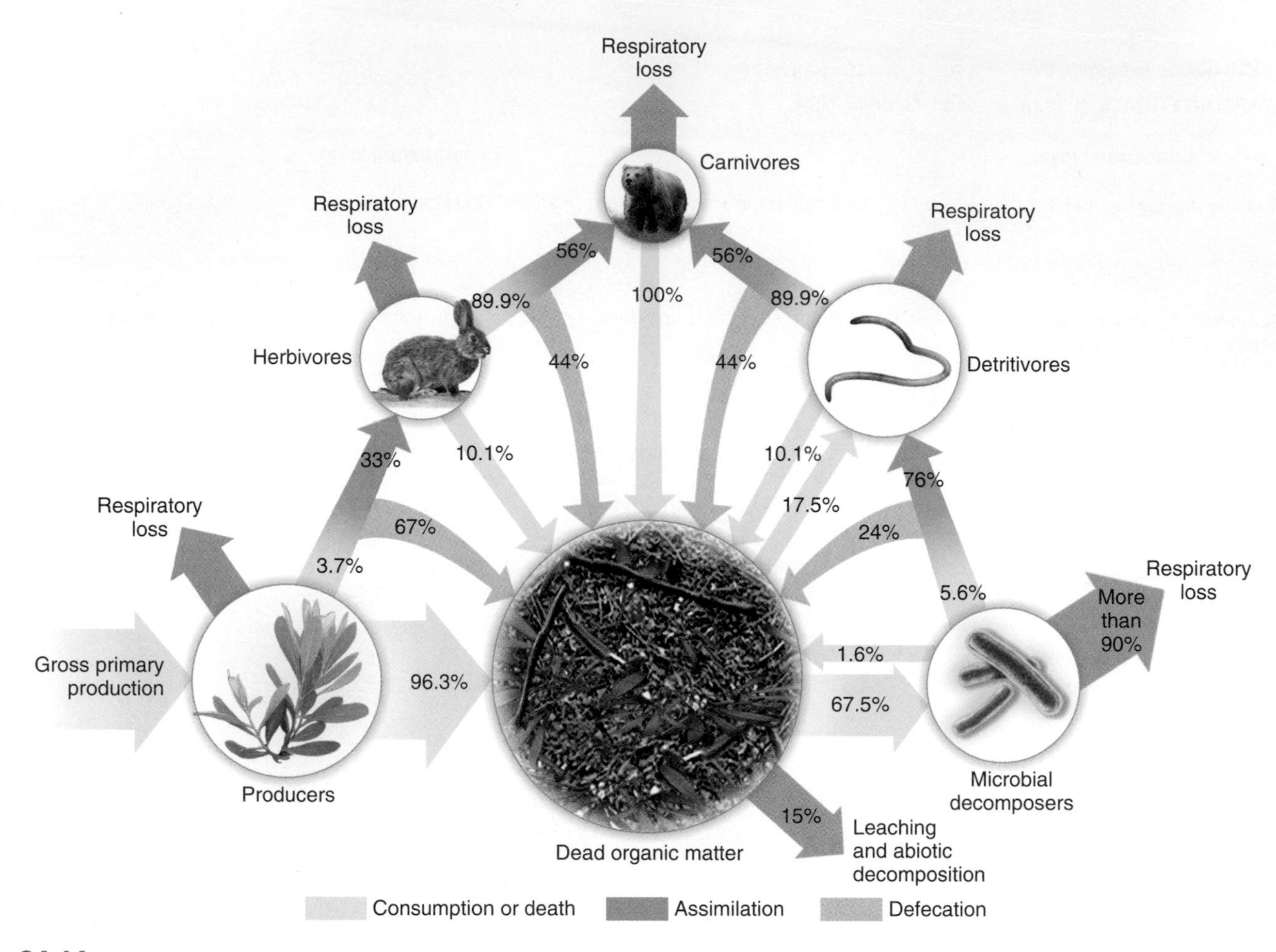


**Figure 26.16 Energy flow through a temperate deciduous forest.** Most energy from plants goes to decomposers. More energy from herbivores goes toward the next trophic level. (After Hairston and Hairston, 1993.)

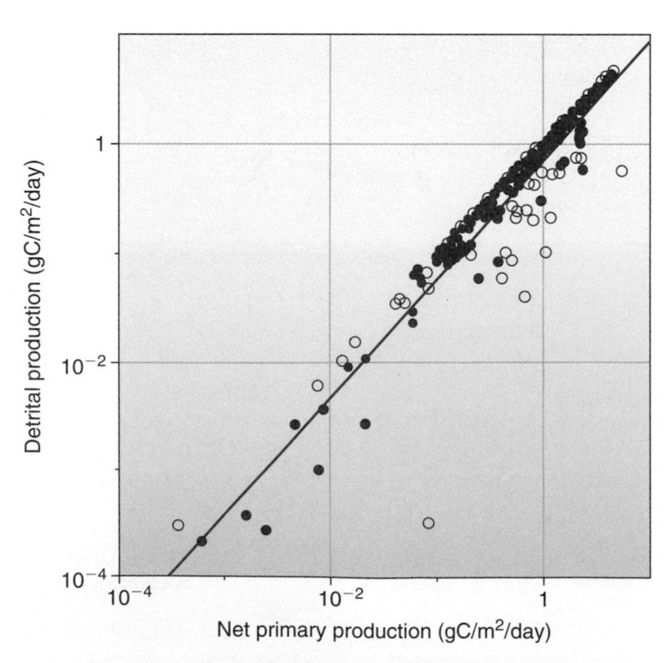


**Figure 26.17 The relationship between plant detrital production and net primary productivity.** There is a strong increase in detrital production as NPP increases, and this is evident across a wide range of biomes. Open symbols are aquatic ecosystems, closed symbols are terrestrial. (After Cebrian, 1999.)

dwellers, including earthworms, nematodes, collembolans, and other insects. Medium-meshed bags allowed only small invertebrates and microorganisms, while small-meshed bags allowed only microorganisms. In the small-meshed bags, microorganisms alone were unable to decompose the leaves (**Figure 26.19**). Edwards and Heath found that the larger the mesh, the quicker the decomposition.

The third phase of decomposition is **chemical alteration**, where fungi and bacteria chemically change dead plant material. Some chemical changes occur spontaneously in the soil without microbial mediation, but most occur over a long time span. After chemical alteration, the identity of the dead organic matter is no longer recognizable and it is considered **soil organic matter.** Bacteria and fungi account for 80–90% of the decomposer biomass during this stage. The fungi have long, branching filamentous structures called hyphae that obtain nitrogen from one area of the soil and carbon from another. In this way they are analogous to plants, which obtain their nutrients in different places: carbon from the atmosphere and nitrogen from the soil. Furthermore, the fungal hyphae are able to penetrate deep within the decaying litter or wood in a way that bacteria cannot.

Temperature and moisture conditions can greatly affect decomposition speed. Microbial activity is favored by warm,

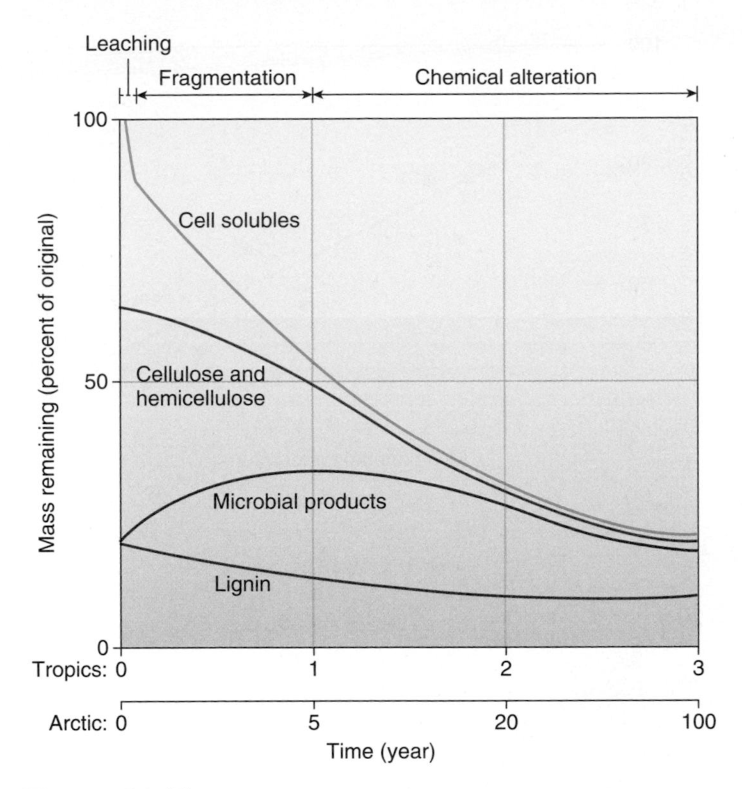


**Figure 26.18 Representative time course of leaf-litter decomposition.** The major chemical constituents are cell solubles, cellulose and hemicellulose, microbial products, and lignin. The three major phases of litter decomposition are leaching, fragmentation, and chemical alteration. Also shown are the time scales commonly found in tropical and arctic environments. (After Chapin et al., 2002.)

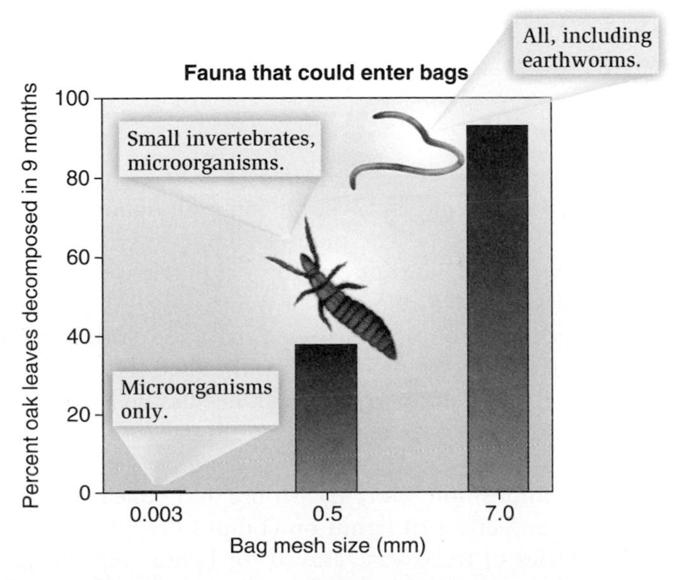


**Figure 26.19 The rate of litter decomposition is affected by the type of organisms involved.** Fragmentation by larger organisms such as earthworms facilitates decay. (After data in Edwards and Heath, 1963.)

moist conditions and is greater during the summer than the winter. Decomposition times in tropical ecosystems are on the order of months, whereas in arctic ecosystems decomposition occurs over decades. Because decomposition occurs much more quickly in tropical environments than in temperate ones, many tropical forests have a smaller litter pool than temperate forests. In a tropical forest, the leaf-litter layer is so thin that one can often see the soil below. In a temperate forest, the litter layer is much thicker and usually obscures the soil. T. R. Moore and colleagues (1999) examined the rate of decomposition of leaf litter at 18 forest sites across a spectrum of Canadian ecosystem types that they called ecoclimatic provinces, which varied in temperature and moisture (**Figure 26.20a**). Litter from 10 different tree species was placed into 20 × 20–cm polypropylene bags with openings of 0.25 × 0.5 mm. After 3 years the bags were reopened, the litter remaining in the bag was weighed, and the proportion of the original litter mass remaining was calculated. There was a direct relationship between mean annual temperature at each site and percent mass remaining averaged across all tree species (**Figure 26.20b**).

In most temperate ecosystems, leaf litter frequently loses 30–70% of its mass in the first year and another 20–30% in the next 5–10 years. As shown in Figure 26.18, in most ecosystems there is an exponential decline in litter mass, implying that a constant proportion of litter is decomposed each year. This rate can be quantified as

$$L_t = L_0 e^{-kt}$$

Where $L_0$ is the litter mass at time zero and $L_t$ is the litter mass at time $t$. The **decomposition constant,** $k$, is an exponent that characterizes the decomposition rate under specified conditions. The mean **residence time,** or time required for litter to decompose under these conditions, is $1/k$. Alternatively, residence time can be calculated by dividing the average pool size of litter by the average annual litter input. Thus,

$$\text{Residence time} = \frac{\text{Litterpool}}{\text{Litterfall}}$$

and

$$k = \frac{\text{Litterfall}}{\text{Litterpool}}$$

Residence times for easily leached substances like sugars are hours to days, whereas residence times for lignin are months to decades. Decomposition constants tend to be much higher in warm, moist tropical forests than in cooler, temperate deciduous forests or, especially, temperate coniferous forests (**Figure 26.21**).

Soil properties also affect litter accumulation and decomposition. Generally, litter accumulation is greater in very wet soils, because decomposition is more restricted by high soil moisture than is net primary productivity. Oxygen levels are reduced in water-logged soils and microbial activity is reduced. Acid soils also tend to reduce decomposition, again because of reduced bacterial abundance. On the other hand, drier soils do not hinder decomposition as much as they reduce net primary productivity, so litter pools are reduced.

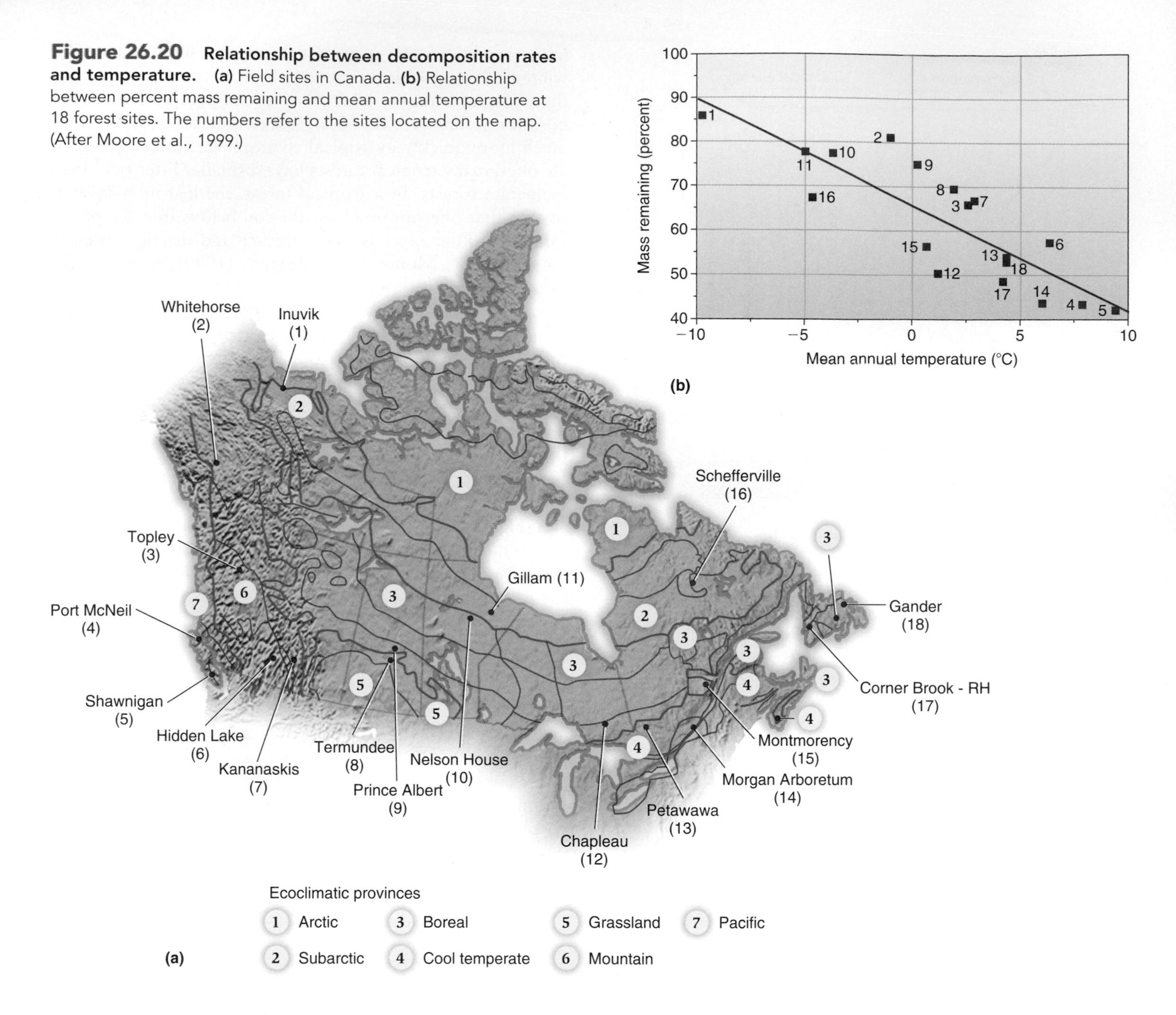


**Figure 26.20 Relationship between decomposition rates and temperature.** **(a)** Field sites in Canada. **(b)** Relationship between percent mass remaining and mean annual temperature at 18 forest sites. The numbers refer to the sites located on the map. (After Moore et al., 1999.)

Finally, the quality of the organic matter itself may affect decomposition rates. Animal carcasses, being nutrient-rich, decompose much more rapidly than plant material. Though carcasses are relatively rare in the environment, the pulse of nutrients that they provide greatly stimulates plant growth around them (**Figure 26.22**). Plant compounds such as lignin have an irregular structure that does not fit the active sites of most enzymes. As such they are broken down slowly by the extracellular enzymes of microbes. Cellulose, which consists of regularly repeating glucose units, is quicker to decompose. Organic nitrogen and phosphorus support microbial growth, and plant material rich in these nutrients decomposes relatively rapidly. We can see the differences in decomposition of a pine branch, pine needle, and deciduous leaf of pin cherry, *Prunus pensylvanica*, in a Canadian temperate forest (**Figure 26.23**). The pine branch, having high lignin content, decomposes very slowly. The pine needle decomposes faster than the branch, but pines often grow on nutrient-poor soils and their needles have relatively low nitrogen. As a result, the pine needle decomposes slower than the softer, more nutrient-rich deciduous pin-cherry leaf. This also explains why there is a bigger litter layer of needles in temperate coniferous forests than there is of leaves in temperate deciduous forests.

The decomposition study by Moore and colleagues also showed a strong effect of lignin on composition rates. Litter from 10 species of trees was assayed for lignin and nitrogen content, and the results were expressed as a ratio of lignin to nitrogen (**Figure 26.24**). Tree species with high lignin : nitrogen ratios were more resistant to decomposition, and more mass

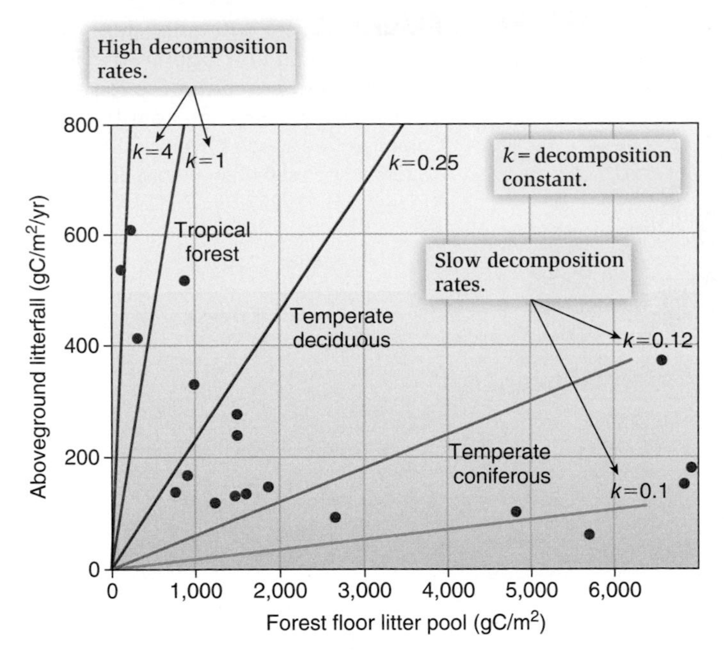


**Figure 26.21 Litterfall and leaf-litter layers.** Fast leaf decomposition in tropical forests results in a thin leaf-litter layer compared to temperate ecosystems, where decomposition is slower. The lines represent the relationship between aboveground litterfall and forest floor biomass for selected decomposition constants. (After Olsen, 1963.)

**Figure 26.22 A musk ox carcass with plant growth around it.** The nutrients released via decomposition have allowed tundra vegetation to grow.

remained at the end of 3 years of decomposition in these species, compared to species with low lignin : nitrogen ratios.

Lignin concentration also impacts decomposition in aquatic habitats. Leaf litter falling into streams provides much of the nutrients used by aquatic organisms. In Germany, Markus Schindler and Mark Gessner (2009) put leaves of nine local deciduous tree species into small mesh bags and examined decomposition in streams over time. Once again,

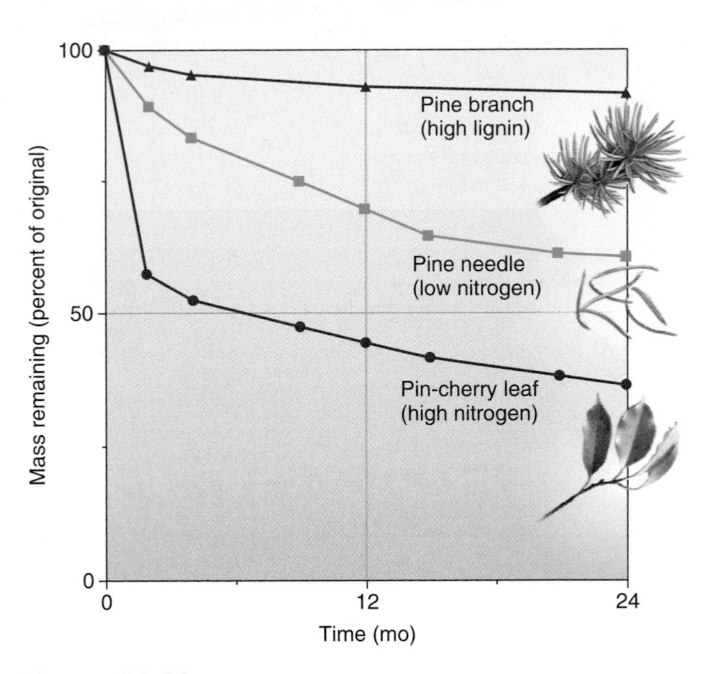


**Figure 26.23 Time course of decomposition of a pine branch, pine needle, and pin-cherry leaf in a Canadian forest.** (After MacLean and Wein, 1978.)

the decay rate was linked to the leaf-litter lignin content (**Figure 26.25**).

Just Cebrian and Julien Lartigue (2004) surveyed decomposition rates and detritus nutrient concentration in a wide variety of biomes, from freshwater and marine to tundra and temperate and tropical grasslands and forests. They also found that litter nutrient concentrations affected decomposition rates (**Figure 26.26**). Both aquatic and terrestrial detritus with higher nitrogen and higher phosphorus concentrations tended to have a larger proportion of mass decomposed per day. Decomposition rates were higher in aquatic than in terrestrial biomes.

**Check Your Understanding**

**26.2** Why isn't there a thick litter layer in tropical rain forests, given that primary productivity is high in moist environments?

## 26.3 Living Organisms Can Affect Nutrient Availability

So far, much of our discussion has focused on how organismal biomass can be affected by temperature, moisture, and nutrient availability.

However, some studies have shown that organisms themselves can affect the availability of nutrients. For example, Mark Ritchie and colleagues (1998) experimentally excluded herbivores from an oak savanna. In the savanna the herbivores preferentially eat nitrogen-fixing legumes and nitrogen-rich woody plants, leaving prairie grasses that have lower nitrogen content

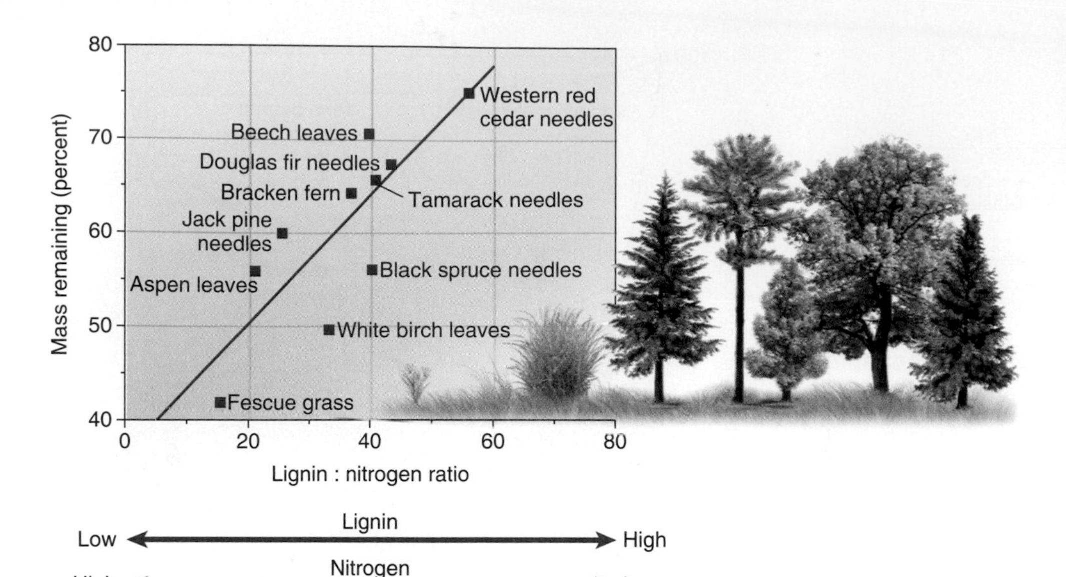


**Figure 26.24 Relationship between lignin : nitrogen ratios in leaf litter and percent mass remaining after 3 years.** Data are from 10 different Canadian tree species averaged over the 18 forest sites shown in Figure 26.20. (After Moore et al., 1999.)

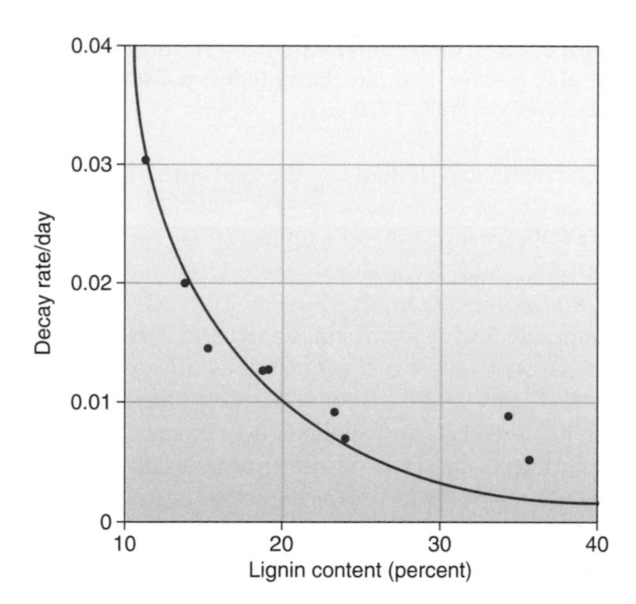


**Figure 26.25 Decay rates of litter from nine species of trees in German streams.** (After Schindler and Gessner, 2009.)

(**Figure 26.27**). The effect of this preference for high-nitrogen plants is the removal of nitrogen from the local system.

Invasive species can also change soil nutrient availabilities. At the northern end of the Great Plains in Canada, 20 miles north of the Montana state line, crested wheat grass, *Agropyron cristatum*, an introduced $C_3$ perennial grass from Russia, has been planted widely since the 1930s and dominates millions of acres of grassland. It has outcompeted native prairie grass over large areas. In these areas, soil carbon and available soil nitrogen are lower than in native prairie. Janice Christian and Scott Wilson (1999) compared soil nitrogen levels of normal undisturbed prairie, disturbed prairie that had been ploughed but had returned to native prairie, and areas invaded by the exotic *Agropyron* (**Figure 26.28**). Disturbance and revegetation by native species decreased soil nitrogen but not significantly, whereas disturbance and colonization by *Agropyron* depleted soil nitrogen greatly. The low soil nitrogen in turn reduced the diversity of plant species growing in the area.

Invasive animals may also have substantial effects on nutrient availability. John Maron and colleagues (2006) showed that the introduction of arctic foxes, *Alopex lagopus* (**Figure 26.29a**), onto the Aleutian Islands, a chain of 450 islands off the coast of Alaska, reduced abundant seabird populations, thereby disrupting nutrient subsidies vectored by seabirds from sea to land. The islands sit at the confluence of the highly productive North Pacific Ocean and Bering Sea, and historically they supported millions of seabirds. The seabird droppings called guano, derived from ocean fish, were extensive. This added nutrient input, particularly rich in phosphorus and nitrogen, supported dense swards of waist-high grasses, graminoids, the most abundant of which was *Leymus mollis*. In the 1910s–1920s, arctic foxes were introduced onto a hundred or so of these islands to maintain the North Pacific fur trade, fur seals and sea otters having been overexploited in the late 19th century. Together with Norway rats, *Rattus norvegicus*, the foxes extirpated many seabird species, particularly the ground nesters. From 1949 through 1981, the U.S. Fish and Wildlife Service eradicated foxes from most of the Aleutian Islands and seabird populations began to recover. However, many seabird species such as the tufted puffin, *Fratercula cirrhata* (**Figure 26.29b**), were slow to recolonize and their intrinsic rate of increase was also slow. Even by 2001, the seabird abundance on fox-free islands was still an order of magnitude higher than on islands where foxes had been introduced but eradicated.

Maron and colleagues surveyed nine historically fox-free islands and 9 islands where foxes were once present but had been eradicated (**Figure 26.29c**). They noted a greatly decreased guano input, and decreased soil nitrogen, soil phosphorus, and grass nitrogen, on previously fox-infested islands

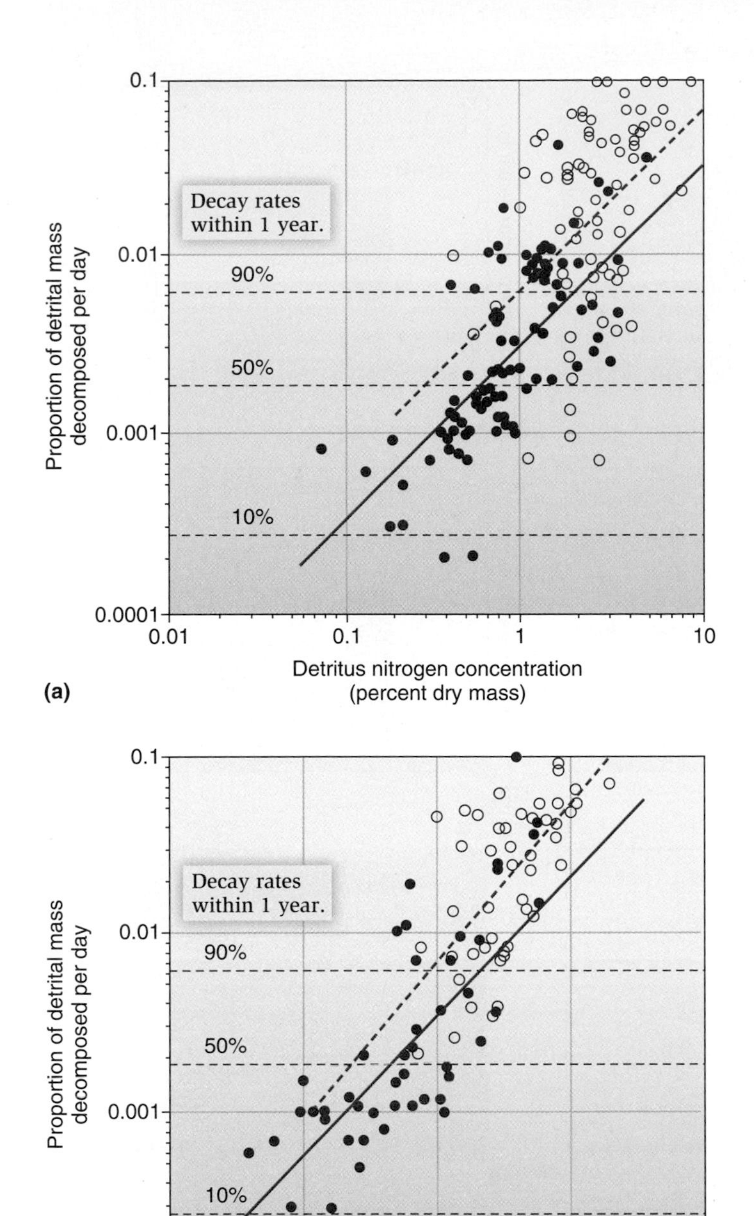


**Figure 26.26 Relationship between detrital decomposition and nutrient concentration for (a) nitrogen and (b) phosphorus.** Dashed lines represent the relationship for aquatic biomes, solid lines for terrestrial biomes. (After Cebrian and Lartigue, 2004.)

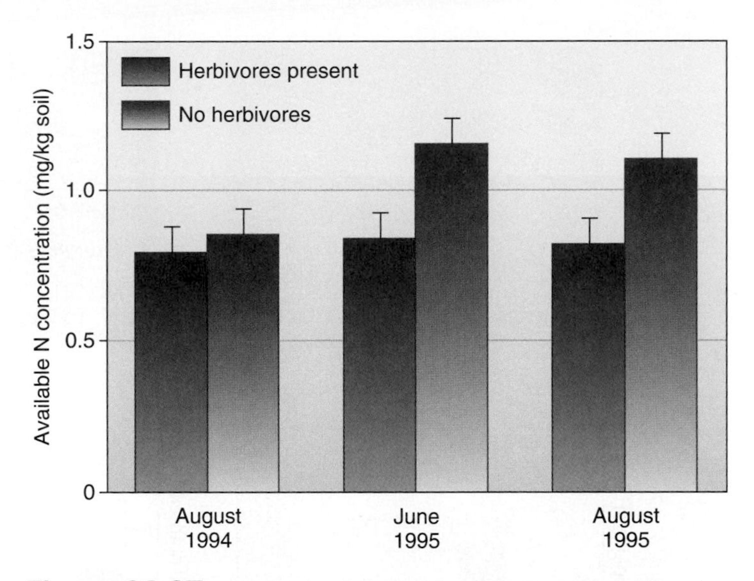


**Figure 26.27 Increase in available soil nitrogen at three different time periods following the exclusion of herbivores from plots in Minnesota.** Herbivores eat the nitrogen-rich plants that absorb a lot of the soil nitrogen. (After Ritchie et al., 1998.)

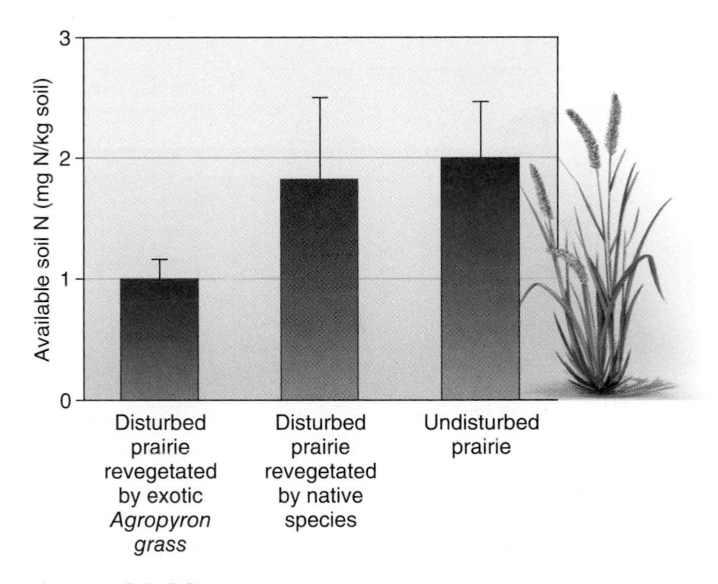


**Figure 26.28 Soil nitrogen in Canadian prairies changed by exotic species.** The decrease in soil nitrogen in ploughed soils in Canadian prairies that have been recolonized by native prairie soils is slight compared to areas that have been disturbed and planted with the exotic *Agropyron*. (After Christian and Wilson, 1999.)

(**Figure 26.29d,e,f**). As a result, the vegetation on previously fox-infested islands changed to become less dominated by graminoid grasses and *L. mollis,* to greater cover by low-lying dwarf shrubs such as *Empetrum nigrum,* and more mosses and lichens (**Figure 26.29g,h,i**). These new plant assemblages were low-productivity maritime tundra communities, with increased plant species richness, resulting from a competitive release from the dominant graminoid species.

### Check Your Understanding

**26.3** In many areas of the world, forests have been logged and planted with crops. In the Southern Hemisphere, such crops fail more often than in the Northern Hemisphere. Explain this phenomenon.

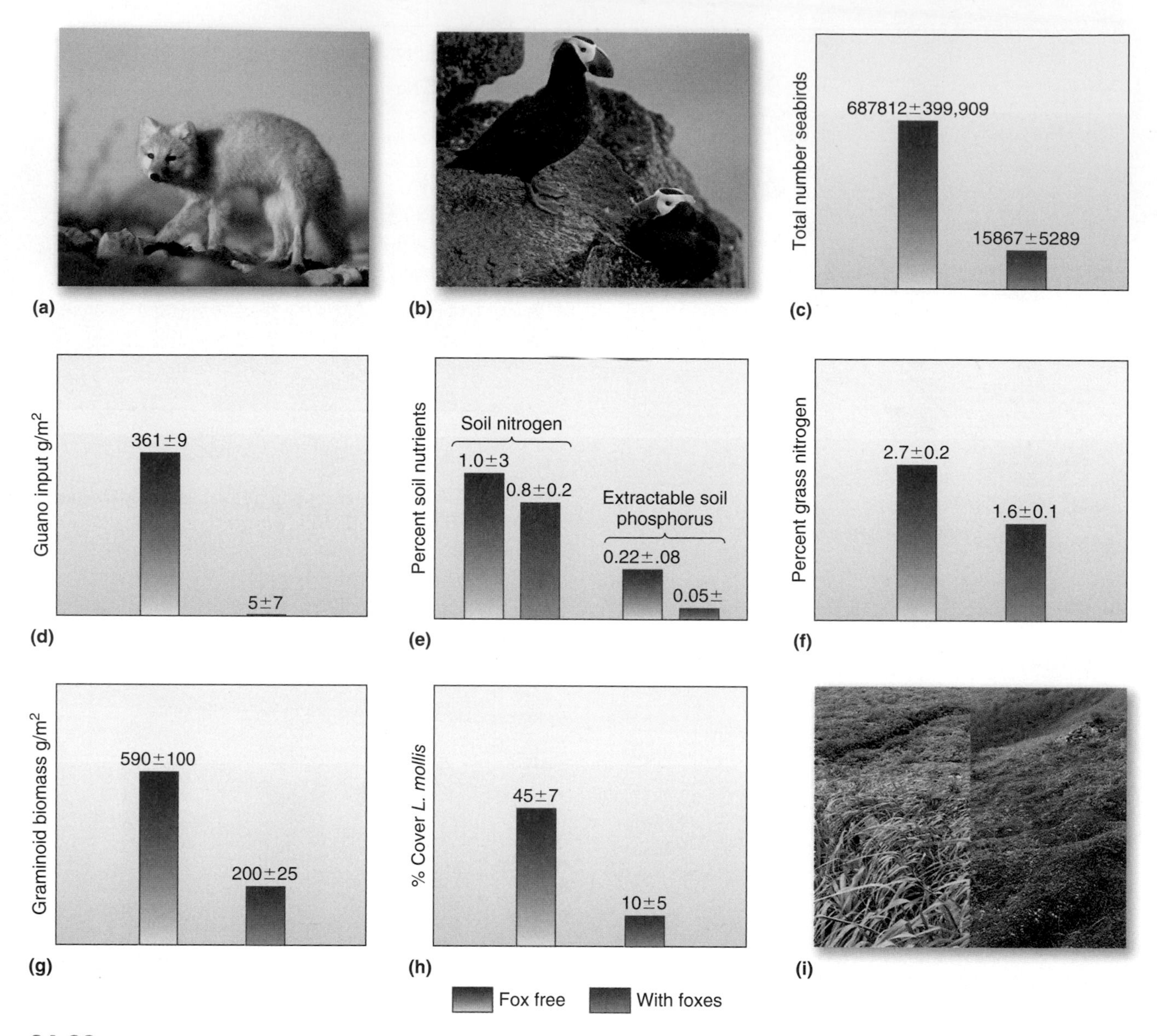


**Figure 26.29 The effects of introduced arctic foxes on Aleutian Islands flora.** **(a)** Arctic fox, **(b)** tufted puffins. Compared to islands which never had foxes, previously fox-infested islands show **(c)** lower densities of seabirds, **(d)** lower guano input, **(e)** decreased soil nitrogen, **(f)** decreased plant nitrogen, and **(g)** decreased graminoid grasses, especially of **(h)** *L. mollis.* **(i)** Appearance of vegetation without foxes (left) and with foxes (right). (From data in Maron et al., 2006.)

## SUMMARY


**26.1 Production Is Influenced by Water, Temperature, Nutrients, and Light Availability**

- Plant production can be measured as gross primary production. The efficiency of GPP varies between different types of plants (Figure 26.1).
- Net primary production is gross primary production minus the energy lost in plant respiration. NPP can be measured using a variety of techniques, including satellite imagery (Figure 26.2).
- Primary production in terrestrial ecosystems is limited primarily by availability of water and nutrients (Figures 26.3–26.6).
- Increasing atmospheric $CO_2$ is increasing primary production globally (Figures 26.A, 26.B).
- Primary production in aquatic ecosystems is limited mainly by availability of light and nutrients (Figures 26.7, 26.8).
- Excess nutrients in marine ecosystems may produce oxygen-poor dead zones (Figure 26.9).

- Producer biomass is generally increased more by nutrient addition than by herbivore removal (Figure 26.10).
- Biomes differ in their net primary production. However, production in temperate forests may equal or even surpass that of tropical forests because of poor soil quality in the Tropics (Figures 26.11, 26.12).
- Secondary production is limited by available primary production. Primary production consumption is increased where primary production is high in nutrients such as nitrogen and phosphorus (Figures 26.13–26.15).
- Most primary production dies and rots in place, to be broken down by decomposers. Detrital production increases as primary production increases (Figures 26.C, 26.16, 26.17).

**26.2 Decomposition Is Influenced by Temperature, Moisture, and Soil Nutrients**

- There are three recognizable phases of decomposition: leaching, fragmentation, and chemical alteration (Figure 26.18).
- Decomposition is facilitated by larger soil organisms and may be affected by temperature, soil conditions, litter composition, litter C : N ratios, litter lignin content, and litter lignin : nitrogen ratios of plant material (Figures 26.19–26.26).

**26.3 Living Organisms Can Affect Nutrient Availability**

- Plants and animals, including exotic species, may cause changes in nutrient availability (Figures 26.27–26.29).

## TEST YOURSELF

1. The amount of energy that is fixed during photosynthesis is:
   a. Net primary production
   b. Biomagnification
   c. Trophic-level transfer efficiency
   d. Gross primary production
   e. Production efficiency
2. Which type of producers have the highest efficiencies of gross primary production?
   a. Desert plants
   b. Phytoplankton
   c. Herbs
   d. Deciduous trees
   e. Coniferous trees
3. Primary production in aquatic ecosystems is limited mainly by:
   a. Temperature and moisture
   b. Temperature and light
   c. Temperature and nutrients
   d. Light and nutrients
   e. Light and moisture
4. The most highly productive terrestrial communities on Earth are:
   a. Forests
   b. Grasslands
   c. Tundra
   d. Deserts
5. In unpolluted areas of tropical oceans, primary productivity is limited most often by:
   a. Nitrogen
   b. Iron
   c. Phosphorus
   d. Nitrogen and iron
   e. Nitrogen, iron, and phosphorus
6. Which of the following leaf constituents are the first to decline during decomposition?
   a. Cell solubles
   b. Cellulose
   c. Hemicellulose
   d. Microbial products
   e. Lignin
7. Decomposition rates are affected by:
   a. Temperature
   b. Nutrient availability
   c. Lignin : carbon ratios
   d. a and b
   e. All of the above
8. Which organisms are the most important consumers of energy in a Georgia salt marsh?
   a. *Spartina* grass and algae
   b. Insects
   c. Spiders
   d. Crabs
   e. Bacteria
9. As measured by satellite imagery, ocean chlorophyll concentrations are highest in the Tropics.
   a. True
   b. False.
10. Across the globe, increasing precipitation always increases net primary productivity.
    a. True
    b. False.

## CONCEPTUAL QUESTIONS

1. What are the main limiting factors for terrestrial primary production?
2. Describe the sequential steps of decomposition and the main organisms involved.
3. What effects might the elimination of tens of millions of buffalo have had on ecosystem processes in the North American Great Plains?

## DATA ANALYSIS

This figure shows the relative consumption of leaf litter of three species of temperate forest trees—beech, *Fagus sylvatica;* oak, *Quercus robur;* and sycamore, *Acer pseudoplatanus*—by a common litter-dwelling isopod, the pillbug, *Oniscus asellus.* The litter was offered to the pillbug as each species alone or as all three species together, and when grown under ambient or elevated $CO_2$ conditions. What can we conclude?

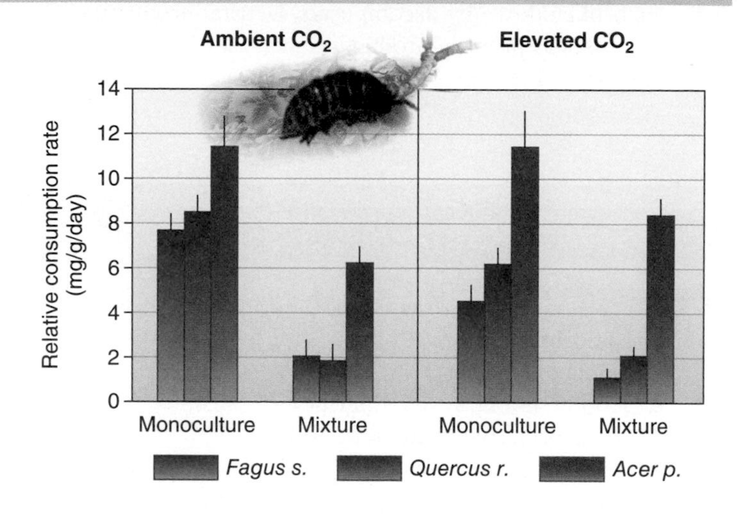

The reintroduction of tortoises on Round Island, off the coast of Mauritius, in 2007.

CHAPTER 27

# Biogeochemical Cycles

## Outline and Concepts



Many northern forests in North America lacked earthworm populations before European settlement. Though there are about 120 native species of earthworms in North America, all died out in northern areas due to glaciations. After the glaciers retreated, 12,000 to 15,000 years ago, many forests remained without earthworms because the northward expansion of the surviving earthworms to the south was very slow. Forests in Minnesota and Wisconsin evolved in the absence of earthworms. Thick litter layers were present, and the cycling of nutrients was relatively slow. Invasion of exotic earthworm species from Europe and Asia, especially of *Lumbricus* spp., began during the colonial period in the 1600s and 1700s, most likely facilitated by relocation of horticultural materials and by releases of fishing bait. Dramatic changes in understory vegetation and herb layers appeared in areas invaded by non-native earthworms. Earthworms eat much carbon-rich organic matter, especially dead leaves, and quickly incorporate it into the soil, mixing the soil layers via their burrowing activities. Earthworms also eat the roots of some plant species and thus diminish the fungal mycorrhizae—fungi that grow in association with plant roots—on which many plant species depend. They shift the soil ecosystem from a slower nutrient-cycling, fungal-dominated system to a faster cycling, bacterial-dominated system. This, in turn, can lead to a net loss of carbon from the soil. In response to these changes, plant species richness in earthworm-invaded areas has dropped dramatically. The forest floor has literally been eaten away, and relatively few, usually nonmycorrhizal, plant species remain (**Figure 27.1**).

(a)

(b)

**Figure 27.1 Changes in understory vegetation brought about by invasive earthworms.** **(a)** Noninvaded and **(b)** invaded plots in northern Minnesota. Note the lack of herb layer, exposed soil surface, and reduced litter layer in the invaded plot.

In Chapter 26 we explored energy flow through ecosystems. A unit of energy passes through a food web only once. In contrast, chemical nutrients required by life, such as nitrogen or carbon, cycle through ecosystems continuously. In this chapter we examine the cycles of phosphorus, carbon, nitrogen, sulfur, and water.

## 27.1 Biogeochemical Cycles Transfer Elements Among the Biotic and Abiotic Components of Ecosystems

In addition to the basic building blocks—hydrogen, oxygen, and carbon—the elements required in the greatest amounts by living organisms are nitrogen, phosphorus, and sulfur. In this section we take a detailed look at the cycles of these nutrients. Because these cycles involve biological, geological, and chemical transport mechanisms, they are termed **biogeochemical cycles.** Chemical transport mechanisms include dissolved matter in rain and snow, atmospheric gases, and dust blown by the wind. Geological mechanisms include weathering and erosion of rocks, and elements transported by surface and subsurface drainage. Biological mechanisms involve the absorption of chemicals by living organisms and their subsequent release back into soil or water via decomposition. Biogeochemical cycles can be divided into two broad types: local cycles, such as the phosphorus cycle, which involve elements with no atmospheric mechanism for long-distance transfer; and global cycles, which involve an interchange between the atmosphere and the ecosystem. Global cycles, such as the carbon, nitrogen, sulfur, and water cycles, unite the Earth and its living organisms into one giant interconnected ecosystem called the **biosphere.** Most biogeochemical cycles involve assimilation of nutrients from the soil by plants, then animals, and decomposition of plants and animals, releasing nutrients back into the soil. A generalized and simplified biogeochemical cycle involves both biotic and abiotic components and is shown in **Figure 27.2.** In our discussion of biogeochemical cycles, we will take a particular interest in the alteration of these cycles through human activities that increase nutrient inputs, such as the burning of fossil fuels. Although energy dissipates as heat at each trophic level, chemical elements often become more concentrated in organisms at higher trophic levels (**Figure 27.3**). This can have disastrous results in the case of chemical pollutants (see **Global Insight**).

**Check Your Understanding**

**27.1** What is a fundamental difference between the passage of energy and the passage of nutrients through ecosystems?

## 27.2 Phosphorus Cycles Locally Between Geological and Biological Components of Ecosystems

All living organisms require phosphorus, which becomes incorporated into ATP, the compound that provides energy for most metabolic processes. Phosphorus is also a key component of other biological molecules such as DNA and RNA and the phospholipid bilayer of cell membranes, and it is an essential mineral that in many animals helps maintain a strong, healthy skeleton.

The phosphorus cycle is a relatively simple cycle (**Figure 27.4**). Phosphorus has no gaseous phase and no

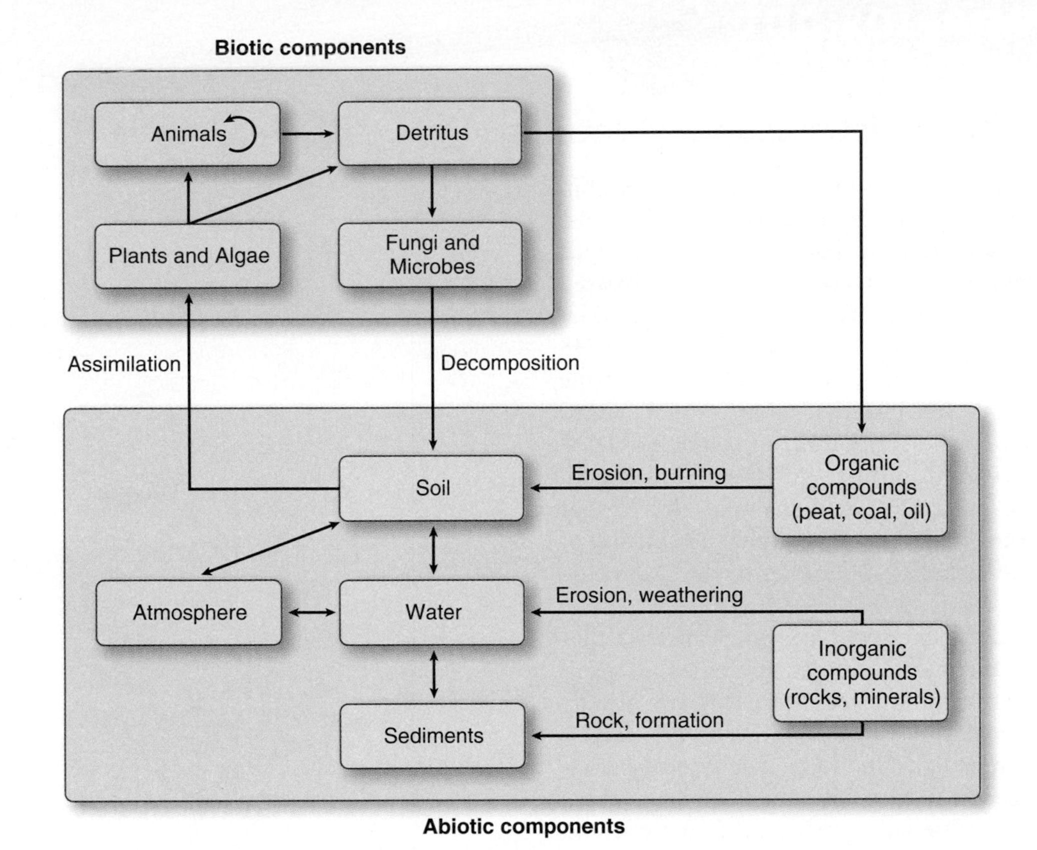


**Figure 27.2** **Generalized and simplified model of a biogeochemical cycle.**

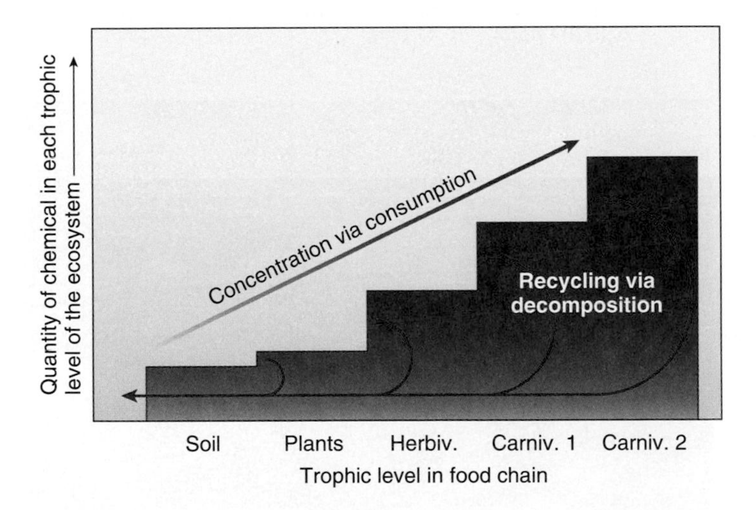


**Figure 27.3** **Chemical cycling.** Chemicals concentrate in higher trophic levels and return to the soil via decomposition.

substantial atmospheric component. It is moved a little as dust by the wind. As a result, phosphorus tends to cycle only locally. The Earth's crust is the main storehouse for this element. Weathering and erosion of sedimentary rocks release phosphorus into the soil. Plants have the metabolic means to absorb dissolved ionized forms of phosphorus, the most important of which occurs as phosphate, $HPO_4^{2-}$ or $H_2PO_4^-$. Herbivores obtain their phosphorus only from eating plants, and carnivores obtain it by eating other animals. When plants and animals excrete wastes or die, the phosphorus becomes available to decomposers, which release it back to the soil.

Leaching and runoff eventually wash much phosphate into aquatic systems, where plants and algae utilize it. In addition, rivers transport phosphorus to lakes or oceans, where it is often quickly taken up by phytoplankton. Phosphate that is not taken up into the food chain settles to the ocean floor or lake bottom, eventually forming sedimentary rock. Phosphorus can remain locked in sedimentary rock for millions of years, becoming available again through the geological process of uplift, which exposes the element to weathering.

Plants can take up phosphate rapidly and efficiently. In fact, they can do this so quickly that they often reduce soil or water concentrations of phosphorus to extremely low levels, so that phosphorus becomes limiting (refer back to Figure 26.6). The more phosphorus that is added to an aquatic system, the more the production of algae and aquatic plants increases (**Figure 27.5**). However, in a series of pivotal studies, David Schindler (1974, 1977) showed that an overabundance of phosphorus caused the rapid growth of algae and other plants in an experimental lake in Canada. Not only does phytoplankton biomass increase with increased phosphorus, but the composition of the algal community also changes, from eukaryotic green algae to prokaryotic blue-green algae. Zooplankton prefer the green algae as opposed

Global Insight

## Biomagnification of Pesticides Can Occur in Higher Trophic Levels

Certain chemicals concentrate at high trophic levels within food chains, a phenomenon called **biomagnification.** The passage of DDT in food chains provides a dramatic example. Dichlorodiphenyltrichloroethane (DDT) was first synthesized in 1874. In 1939 its insecticidal properties were recognized by Paul Müller, a Swiss scientist who won a Nobel Prize in 1948 for his discovery and subsequent research on the uses of the chemical. DDT was an effective insecticide that was believed to be harmless to humans. The first important application of DDT was in public health programs during and after World War II, particularly to control mosquito-borne malaria, and at that time its use in agriculture also began. The global production of DDT peaked in 1970, when 175 million kg of the insecticide were manufactured.

DDT has several chemical and physical properties that profoundly influence the nature of its ecological impact. First, DDT is persistent in the environment; it is not rapidly degraded to other, less toxic chemicals by microorganisms or by physical agents such as light and heat. The typical persistence in soil of DDT is about 10 years, which is two to three times longer than the persistence of many other insecticides. Another important characteristic of DDT is its low solubility in water and its high solubility in fats or lipids. In the environment, most lipids are present in living tissue. Therefore, because of its high lipid solubility, DDT tends to concentrate in biological tissues.

Biomagnification occurs at each step of the food chain, which means that organisms at higher trophic levels can amass especially large concentrations of DDT in their lipids. A typical pattern of the biomagnification is illustrated in **Figure 27.A-i**, which shows the relative amounts of DDT found in a Lake Michigan food chain. The largest concentration of the insecticide was found in gulls, tertiary consumers that feed on fish that, in turn, eat small insects. An unanticipated effect of DDT on bird species was its interference with the metabolic process of eggshell formation. The result was thin-shelled eggs that often broke under the weight of incubating birds (**Figure 27.A-ii**). DDT was responsible for a dramatic decrease in the populations of many birds of prey due to failed reproduction. Relatively high levels of the chemical were also found to be present in some fish, which became unfit for human consumption.

Because of growing awareness of the adverse effects of DDT, most industrialized countries, including the U.S., had banned the use of the chemical by the early 1970s. The good news is that following the outlawing of DDT, populations of the most severely affected bird species have recovered. Had scientists initially understood how DDT accumulated in food chains, however, some of the damage to the bird populations might have been prevented. DDT is now used in only a few developing countries to control malaria, although even in these locations there has been significant movement toward the use of alternative pest control technologies.

Cl Cl H C $CCl_3$

**DDT (dichlorodiphenyltrichloroethane)**
- Persists in environment
- High solubility in lipids
- Found in high concentrations at higher trophic levels

(ii)

**Figure 27.A Biomagnification in a Lake Michigan food chain.** **(i)** The DDT tissue concentration in gulls, a tertiary consumer, was about 240 times that in the small insects sharing the same environment. The biomagnification of DDT in lipids causes its concentration to increase at each successive link in the food chain. **(ii)** DDT causes thinning of eggshells. These ibis eggs are thin-shelled and show signs of having been crushed by the incubating adult.

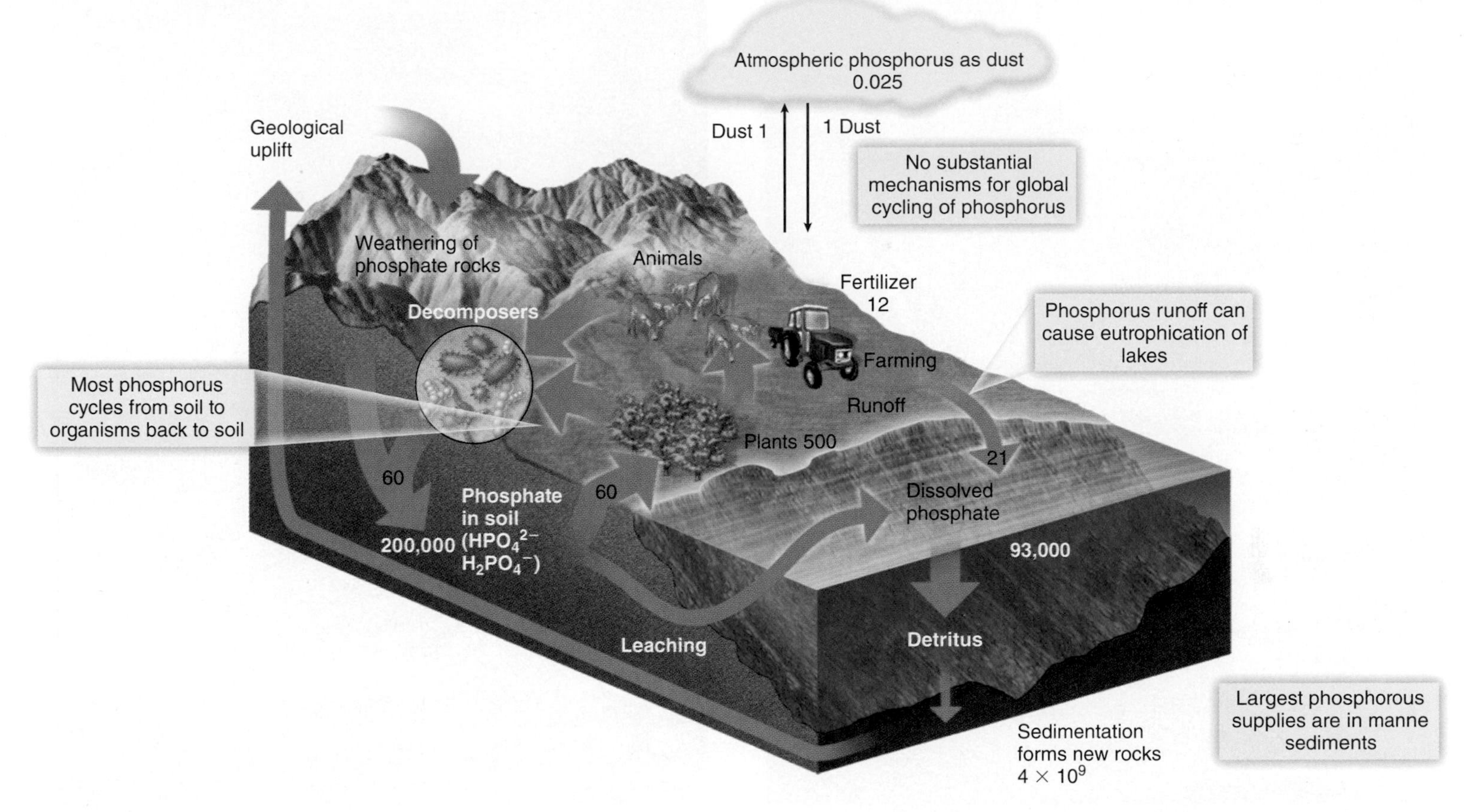


**Figure 27.4 The phosphorus cycle.** Unlike other major biogeochemical cycles, the phosphorus cycle does not have an atmospheric component and thus cycles locally. Storage units are teragrams = $10^{12}$ g and flux rates are $10^{12}$ g per year. The width of the arrows indicates the relative importance of each process.

to the blue-green algae, which can produce some secondary metabolites. As a result, the phosphorus-laden lakes become clogged with a scum of blue-green algae. When the algae die, they sink to the bottom where bacteria decompose them, using the dissolved oxygen. Oxygen levels drop, resulting in fish kills. The process by which elevated nutrient levels leads to an overgrowth of algae and the subsequent depletion of water oxygen levels is known as **eutrophication**. Cultural eutrophication refers to the enrichment of water with nutrients derived from human activities such as fertilizer use and sewage dumping. The annual application of phosphorus to

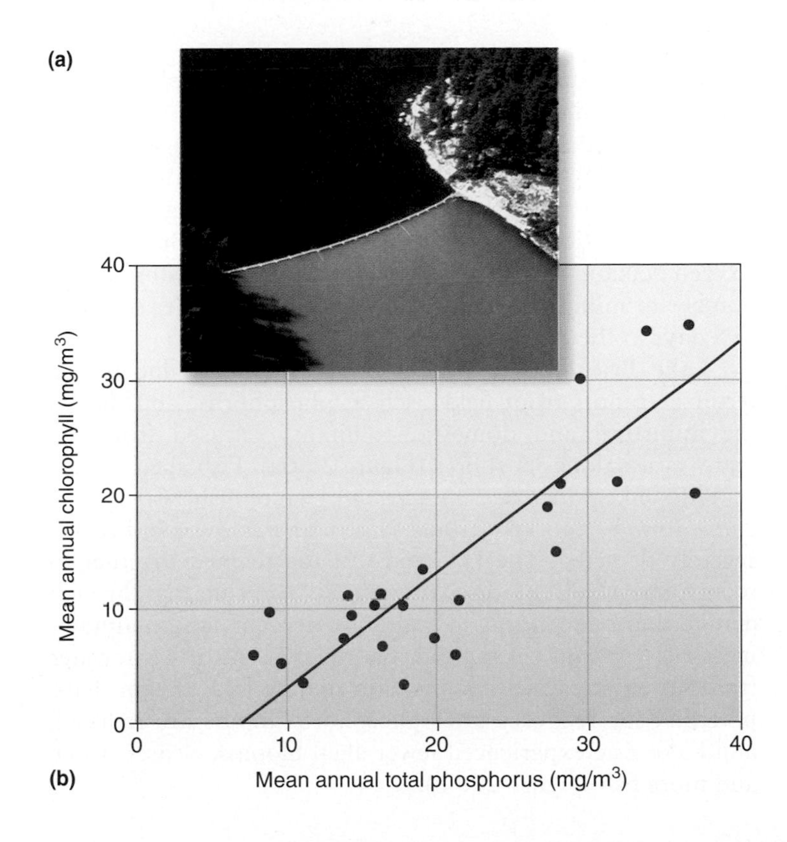


**Figure 27.5 Primary production increases with an increase in total phosphorus concentration.** **(a)** This aerial view shows the contrast in water quality of two basins of an experimental lake in Canada, separated by a plastic curtain. The lower basin received additions of carbon, nitrogen, and phosphorus, while the upper basin received only carbon and nitrogen. The bright green color is from a surface film of algae that resulted from the added phosphorus. **(b)** As phosphorus increases in different North American lakes, production increases. The increase in primary production is measured as an increase in chlorophyll concentration. A higher chlorophyll concentration in the water means more algae are present. (After Schindler, 1977.)

(a)

(b)

**Figure 27.6** **Recovery of a eutrophic lake.** Lake Erie **(a)** in the 1960s when it was polluted by fertilizer runoff and industrial effluent; and **(b)** as it appeared in 2007, after eutrophication was reversed by pollution control.

ecosystems via fertilizers is about 20–30% of the phosphorus that cycles naturally through all ecosystems.

A measure of eutrophication is the **biochemical oxygen demand (BOD)**, which is the difference between the production of oxygen by plants and the amount of oxygen needed for the respiration of the organisms in the water. Biochemical oxygen demand is normally measured in the laboratory as the number of milligrams of oxygen consumed per liter of water in 5 days in the dark at 20°C.

Lake Erie became eutrophic in the 1960s due to the fertilizer runoff from fields rich in phosphorus, and due to the industrial and domestic pollutants released from the many cities along its shores (**Figure 27.6a**). Fish species such as blue pike, *Stizostedion vitreum,* white fish, *Coregonus clupeaformis,* and lake trout, *Salvelinus namaycush,* became severely depleted. The U.S. and Canada teamed together to reduce the levels of discharge by 80%, primarily through eliminating phosphorus in laundry detergents and maintaining strict controls on the phosphorus content of wastewater from sewage treatment plants. Fortunately, lake systems have potential for recovery after phosphorus inputs are reduced, and Lake Erie experienced fewer algal blooms, clearer water, and more fish (**Figure 27.6b**).

### Check Your Understanding

**27.2** Why is it that many terrestrial and, especially, aquatic plants are phosphorus-limited?

## 27.3 Carbon Cycles Among Biological, Geological, and Atmospheric Pools

The movement of carbon from the atmosphere into organisms and back again is known as the carbon cycle (**Figure 27.7**). Terrestrial and aquatic autotrophs, primarily plants and algae, acquire carbon dioxide from the atmosphere, incorporate it into the organic matter of their own biomass via photosynthesis, and release nearly as much through respiration. At the same time, the decomposition of plant material and subsequent bacterial respiration releases a similar amount of carbon back into the atmosphere as $CO_2$. Herbivores also contribute to the release of carbon dioxide back to the atmosphere, but this is a relatively small amount. Under certain conditions organic matter may also form fossil fuels. This occurs primarily when organic matter becomes buried

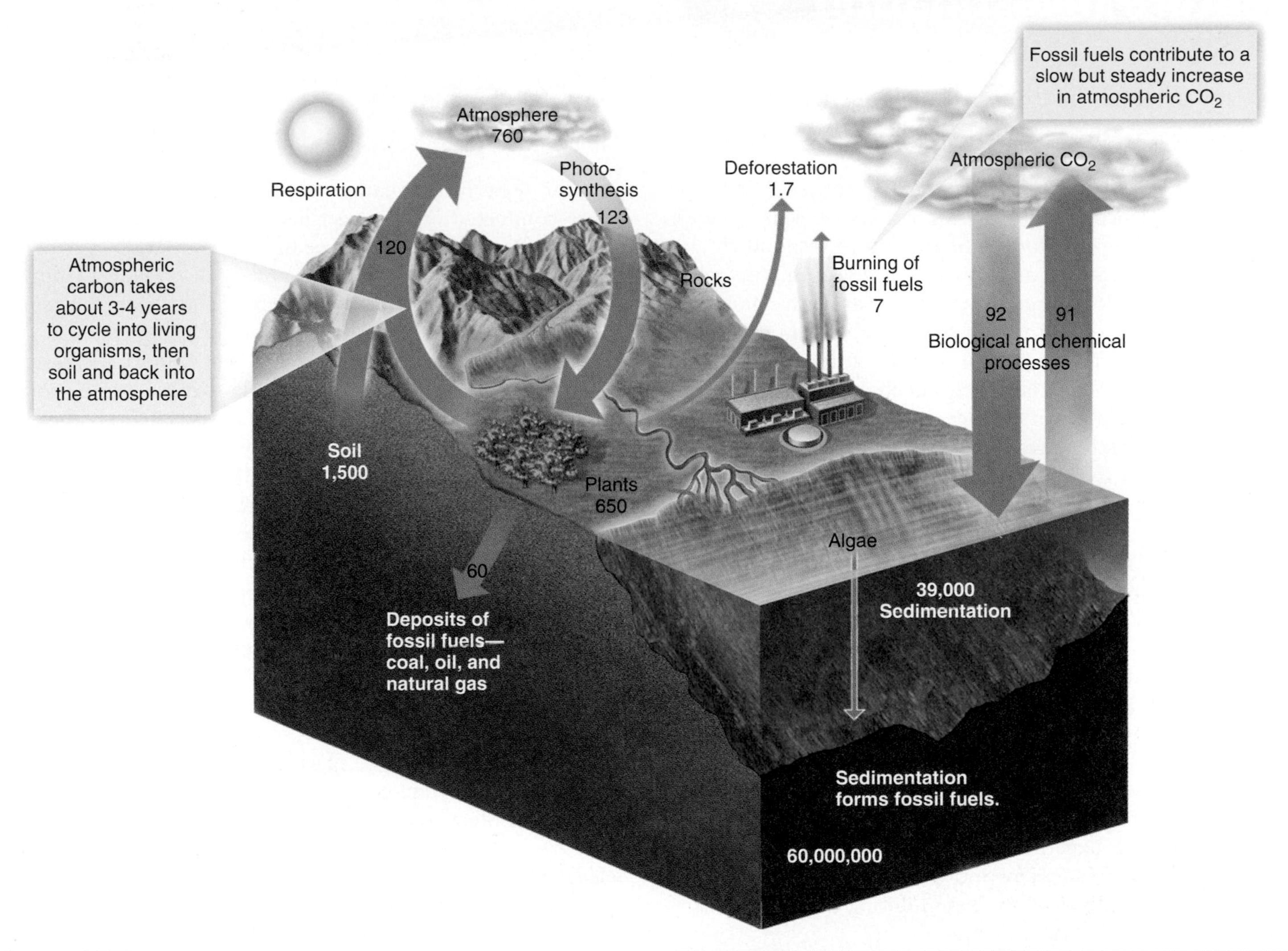


**Figure 27.7 The carbon cycle.** Each year, plants and algae remove about one-seventh of the $CO_2$ in the atmosphere. Animal respiration is so small, it is not represented. Storage units are petagrams = $10^{15}$ g and flux rates are $10^{15}$ g per year. The width of the arrows indicates the relative importance of each process.

**ECOLOGICAL INQUIRY**

Where are the largest stores of global carbon?

under layers of sediment. High pressure and heat convert the organic matter to fossil fuels. In aquatic environments, organisms usually created oil and gas deposits, whereas coal formed from the dead remains of trees and ferns that lived 300–400 million years ago on land. Atmospheric carbon is the most dynamic of the carbon pools, exhibiting a **turnover** that completely replenishes in 3–4 years. In marine environments, much carbon is also incorporated into the shells of marine organisms. When these organisms die, their shells sink and accumulate on the ocean floor, eventually forming huge limestone deposits. Natural sources of $CO_2$ such as volcanoes, hot springs, and fires release large amounts of $CO_2$ into the atmosphere.

Human activities, primarily the burning of **fossil fuels**, are increasingly causing large amounts of $CO_2$ to enter the atmosphere together with large volumes of particulate matter. Deforestation also increases atmospheric $CO_2$ by reducing the carbon dioxide fixed by vegetation. Direct measurements over the past nearly 50 years show a steady rise in atmospheric $CO_2$ (refer back to Figure 5.19), a pattern that shows no sign of slowing. Because of its high concentration in the atmosphere, $CO_2$ is the most significant of the greenhouse gases, which are a primary source of global warming (described in Chapter 5). However, elevated atmospheric $CO_2$ has other dramatic environmental effects, boosting plant growth and lowering herbivory (see **Feature Investigation**).

**Check Your Understanding**

**27.3** Why is elevated $CO_2$ likely to increase crop growth in the future?

## Feature Investigation

### Stiling and Drake's Experiments with Elevated $CO_2$ Show an Increase in Plant Growth but a Decrease in Herbivory

Ecologists are working to determine the climatic and biological consequences of the ongoing rise in the atmosphere's $CO_2$ content. How will forests of the future respond to elevated $CO_2$? To begin to answer such a question, ecologists ideally would enclose large areas of forests with chambers, increase the $CO_2$ content within the chambers, and measure the responses. This has proven to be difficult for two reasons. First, it is hard to enclose large trees in chambers, and second, it is expensive to increase $CO_2$ levels over such a large area. Some ecologists have overcome this problem by simply pumping $CO_2$ through a circle of freestanding pipes to enrich the forests within the circles (refer back to Figure 26.B). However, ecologists Peter Stiling, Bert Drake, and colleagues (Stiling et al., 2003) were able to increase $CO_2$ levels around small patches of forest at the Kennedy Space Center in Cape Canaveral, Florida, by the use of open-top chambers. In much of Florida's forests, trees are small, only 3–5 m when mature, because frequent lightning-initiated fires prevent the growth of larger trees. Drake and Stiling teamed up with NASA engineers to create 16 circular, open-top chambers (**Figure 27.B**), and in 8 of these they increased atmospheric $CO_2$ from the ambient level 360 ppm to double that amount—720 ppm, which is the atmospheric level predicted by the end of the 21st century. The experiments were initiated in 1996 and lasted until 2007. Plants grew more in elevated $CO_2$, because normal levels of carbon dioxide are limiting to plant growth, but the data revealed much more.

Because the chambers were open-topped, insect herbivores could come and go. Insect herbivores cause the largest amount of herbivory in North American forests, because vertebrate herbivores cannot access the high foliage. Censuses were conducted of all insect herbivores but focused on leaf miners, the most common type of herbivore at this site, which are small moths whose larvae are small enough to burrow between the surfaces of plant leaves and create blister-like "mines" on leaves.

Densities of leaves damaged by leaf miners decreased in elevated $CO_2$ in every year studied. Part of the reason for the decline was that even though plants increased in mass, the existing soil nitrogen was diluted over a greater volume of plant material, so that the nitrogen level in leaves decreased. Also, elevated $CO_2$ tends to inhibit the uptake of nitrogen by plants. This increased insect mortality by two means. First, poorer leaf quality directly increased insect death, because plant nitrogen levels may have been too low to support the normal development of the leaf miners. Second, lower leaf quality increased the length of time insects had to feed to gain sufficient nitrogen. Increased feeding times in turn led to increased exposure to natural enemies, such as parasitoids and predators like spiders and ants, and top-down mortality also increased (see the data of Figure 27.B). Thus, in a world of elevated $CO_2$, plant growth may increase but herbivory could decrease.

Peter Stiling and Tatiana Cornelissen (2007) conducted a meta-analysis of 75 studies that examined the effects of elevated $CO_2$ on plants and their herbivores. Plants exhibited a 38.4% increase in biomass, a 16.4% decrease in foliar nitrogen, and a 30% increase in tannins and other phenolics, secondary metabolites that deter insect feeding (refer back to Chapter 14). As a result, herbivore abundance decreased by 21.6% and the herbivore growth rate decreased by 8%.

**GOAL** To determine the effects of elevated $CO_2$ on a forest ecosystem; effects on herbivory are highlighted here.

**STUDY LOCATION** Patches of forest at the Kennedy Space Center in Cape Canaveral, Florida.

| | Conceptual level | Experimental level |
|---|---|---|
| 1 | Expected atmospheric $CO_2$ level is 720 ppm by end of the 21st century. Open-top chambers allow movement of herbivores in and out of chambers. | Erect 16 open-top chambers around native vegetation. Increase $CO_2$ levels from 360 ppm to 720 ppm in half of them. |

| | Conceptual level | Experimental level |
|---|---|---|
| 2 | | Conduct a yearly count of numbers of leaf mines per 200 leaves in each chamber. |

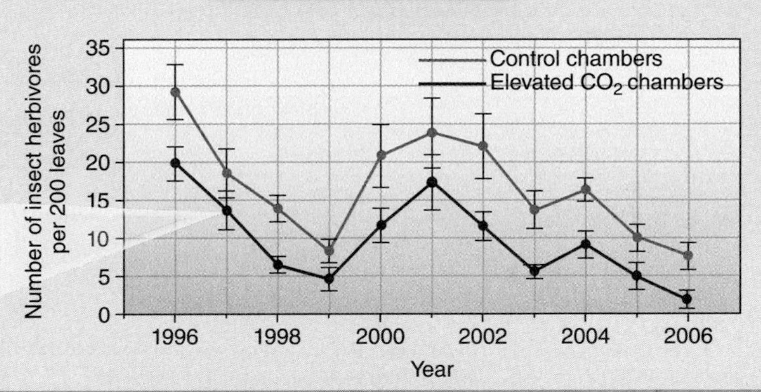


| | Conceptual level | Experimental level |
|---|---|---|
| 3 | Elevated $CO_2$ reduces foliar nitrogen, inhibits normal insect development, and prolongs the feeding time of herbivores, allowing natural enemies greater opportunities to attack them. | Count number of herbivores that died due to nutritional inadequacy. Monitor attack rates on insect herbivores by natural enemies such as predators and parasitoids. |

**4 THE DATA**

| Source of mortality* | Elevated $CO_2$ (% mortality) | Control (% mortality) |
|---|---|---|
| Nutritional inadequacy | 10.2 | 5.0 |
| Predators | 2.4 | 2.0 |
| Parasitoids | 10.0 | 3.2 |

*Data refer only to mortality of larvae within leaves and do not sum to 100%. Mortality of eggs on leaves, pupae in the soil, and flying adults is unknown.

**5 CONCLUSION** Elevated $CO_2$ decreases insect herbivory in a Florida forest.

**Figure 27.B** The effects of elevated atmospheric $CO_2$ on insect herbivory.

## 27.4 The Nitrogen Cycle Is Strongly Influenced by Biological Processes That Transform Nitrogen into Usable Forms

Nitrogen is a limiting nutrient because it is not always available in readily usable forms and because it is an essential component of proteins, nucleic acids, and chlorophyll. Because 78% of the Earth's atmosphere consists of nitrogen gas ($N_2$), it may seem that nitrogen should not be in short supply for organisms. However, $N_2$ molecules must be broken apart before nitrogen atoms are available to combine with other elements. Because of its triple bond, nitrogen gas is very stable and only certain bacteria can break it apart into usable forms. This process, called nitrogen fixation, is a critical component of the five-part nitrogen cycle, which also includes nitrification, assimilation, ammonification, and denitrification (**Figure 27.8**):

1. *Nitrogen fixation.* Only certain bacteria and a few cyanobacteria can accomplish nitrogen fixation, that is,

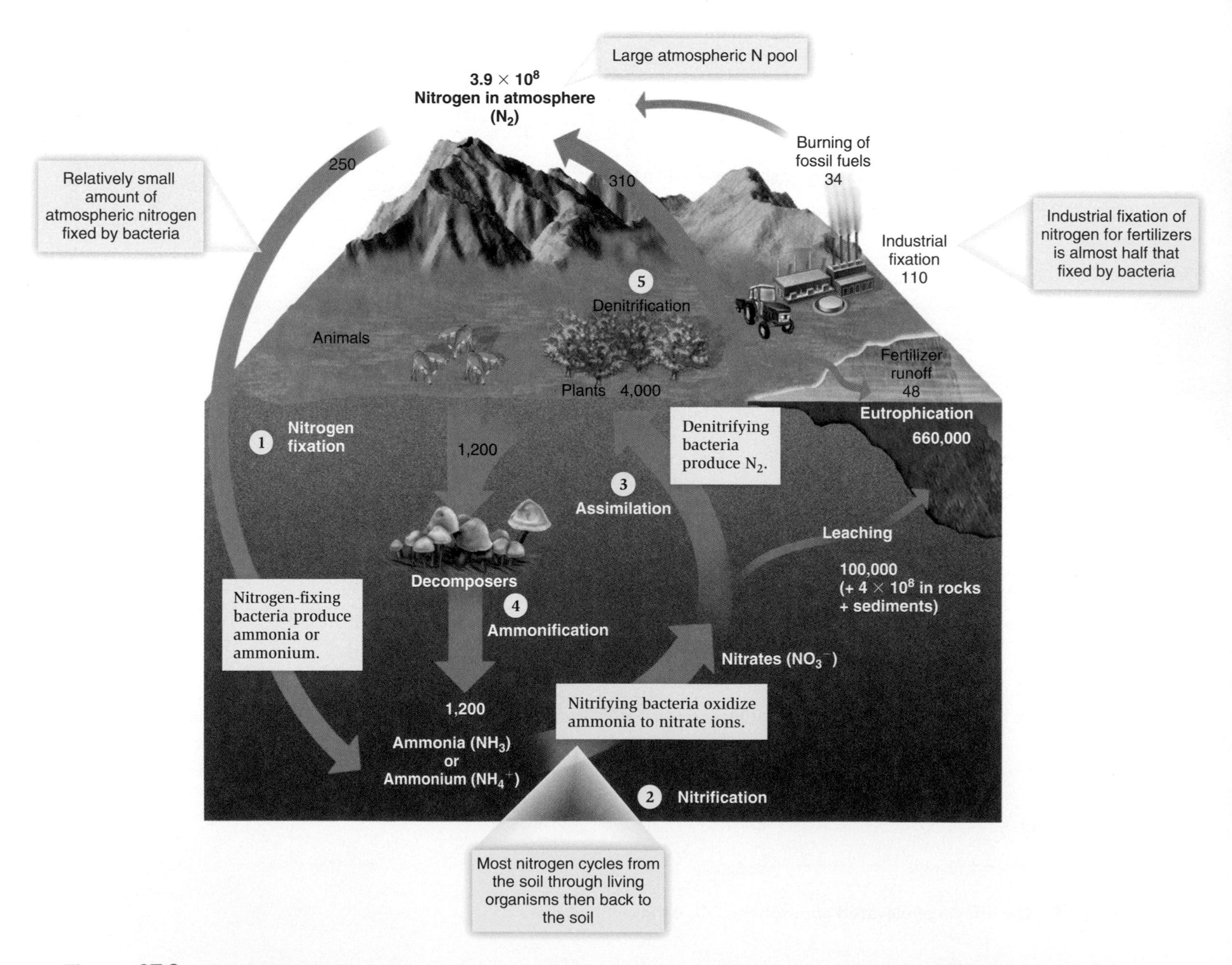


**Figure 27.8** **The nitrogen cycle.** There are five main parts of the nitrogen cycle: **(1)** nitrogen fixation, **(2)** nitrification, **(3)** assimilation, **(4)** ammonification, and **(5)** denitrification. The recycling of nitrogen from dead plants and animals into the soil and then back into plants is of paramount importance. Units are teragrams or teragrams per year. The width of the arrows indicates the relative importance of each process.

convert atmospheric nitrogen to forms usable by other organisms. These organisms use the enzyme nitrogenase to break the nitrogen triple bond. The bacteria that fix nitrogen are fulfilling their own metabolic needs, but in the process they release excess ammonia ($NH_3$) or ammonium ($NH_4^+$), which can be used by some plants. The ammonium ion is water-soluble, and root absorption is the means by which most plants obtain their nitrogen. The most important of these nitrogen-fixing bacteria in the soil are called *Rhizobium,* which live in nodules on the roots of legumes, including peas, beans, lentils, and peanuts, and some woody plants (refer back to Figure 7.3). In other ecosystems such as forests and savannas, nitrogen-fixing bacteria such as *Frankia* spp. form a similar mutualism with plants. In both cases, the bacteria supply ammonia to the plant and receive organic material as an energy source. Cyanobacteria are important nitrogen fixers in aquatic systems. A much smaller amount of nitrogen is fixed by lightning, which breaks the stable nitrogen triple bond and enables the nitrogen atoms to combine with oxygen.

2. *Nitrification.* In the process of nitrification, soil bacteria convert $NH_3$ or $NH_4^+$ to nitrate ($NO_3^-$), a form of nitrogen also commonly used by plants. Bacteria in the genera *Nitrosomonas* and *Nitrococcus* first oxidize the forms of ammonia to nitrite ($NO_2^-$), after which bacteria in the genus *Nitrobacter* convert $NO_2^-$ to $NO_3^-$. These bacteria commonly compete with plants for ammonia.
3. *Assimilation.* Assimilation is the process by which inorganic substances are incorporated into organic molecules. In the nitrogen cycle, organisms assimilate nitrogen by taking up ammonia and $NO_3^-$ formed through nitrogen fixation and nitrification and incorporating them into other molecules. Plant roots take up these forms of nitrogen through their roots, and animals assimilate nitrogen from plant tissue.
4. *Ammonification.* Ammonia can also be formed in the soil through the decomposition of plants and animals and the release of animal waste. Ammonification is the conversion of organic nitrogen to $NH_3$ and $NH_4^+$. This process is carried out by bacteria and fungi. Most soils are slightly acidic and, because of an excess of $H^+$, the $NH_3$ rapidly gains an additional proton to form $NH_4^+$. Because many soils lack nitrifying bacteria, ammonification is the most common pathway for nitrogen to enter the soil.
5. *Denitrification.* Denitrification is the reduction of nitrate ($NO_3^-$) to gaseous nitrogen ($N_2$). Denitrifying bacteria, in the genus *Pseudomonas,* which are anaerobic and use $NO_3^-$ in their metabolism instead of oxygen, perform the reverse of their nitrogen-fixing counterparts by delivering nitrogen to the atmosphere. This process delivers only a relatively small amount of nitrogen to the atmosphere.

In terms of the global nitrogen budget, industrial fixation of nitrogen for the production of fertilizer makes a significant contribution to the pool of nitrogen-containing material in the soils and waters of agricultural regions. In many agricultural areas of Europe, Southeast Asia, and the United States, the amount of nitrogen deposition is three to five times the normal amount.

One problem with increased nitrogen deposition is that fertilizer runoff can cause eutrophication of rivers and lakes; as the resultant algae die, decomposition increases and the increased bacterial activity depletes the oxygen level of the water, resulting in fish die-offs. Excess nitrates in surface or groundwater systems used for drinking water are also a health hazard, particularly for infants. In the body, nitrate is converted to nitrite, which then combines with hemoglobin to form methemoglobin, a type of hemoglobin that does not carry oxygen. In infants, the production of large amounts of nitrites can cause methemoglobinemia, a dangerous condition in which the level of oxygen carried through the body decreases.

The dramatic effects of human activities on nutrient levels in watersheds were illustrated by a famous long-term study by Gene Likens, Herbert Bormann, and their colleagues (1970) at Hubbard Brook Experimental Forest in New Hampshire in the 1960s. Hubbard Brook is a 3,160-hectare reserve that consists of six catchments along a mountain ridge. A catchment is an area of land where all water eventually drains to a single outlet. In Hubbard Brook, each outlet is fitted with a permanent concrete dam that enables researchers to monitor the outflow of water and nutrients (**Figure 27.9a**). In this large-scale experiment, researchers felled all of the trees in one of the Hubbard Brook catchments (**Figure 27.9b**). The catchment was then sprayed with herbicides for 3 years to prevent regrowth of vegetation. An untreated catchment was used as a control.

Researchers monitored the concentrations of key nutrients in the streams exiting the two catchments for over 3 years. Their results revealed that the overall export of dissolved nutrients from the disturbed catchment rose to many times the normal rate (**Figure 27.9c**). The researchers determined that two phenomena were responsible. First, the enormous reduction in plants reduced water uptake by vegetation and led to 40% more runoff to be discharged to the streams. This increased outflow caused greater rates of chemical leaching and rock and soil weathering. Second, and more significant, in the absence of nutrient uptake in spring, when the deciduous trees would have started production, the inorganic nutrients released by decomposer activity were simply leached in the drainage water. Similar processes operate in the majority of terrestrial ecosystems where deforestation is significant.

### Check Your Understanding

**27.4** If the atmosphere consists of 78% nitrogen, why is nitrogen such a limiting nutrient?

(a) Hubbard Brook dam and weir

(b) Hubbard Brook Experimental Forest, New Hampshire

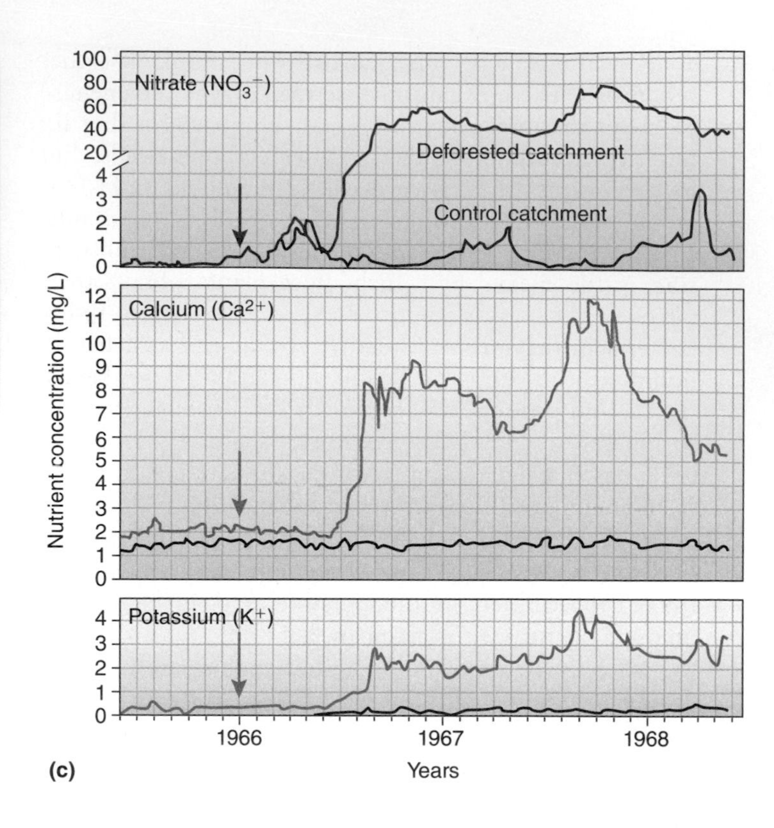


**Figure 27.9 The effects of deforestation on nutrient concentrations.** **(a)** Concrete dam and weir used to monitor nutrient flow from a Hubbard Brook catchment. **(b)** Clear-cut catchment at Hubbard Brook. **(c)** Nutrient concentrations in stream water from the experimentally deforested catchment and a control catchment at Hubbard Brook. The timing of deforestation is indicated by arrows. (After Likens et al., 1970.)

## 27.5 The Sulfur Cycle Is Heavily Influenced by Anthropogenic Effects

Sulfur is an important macronutrient because it is an essential component of proteins, which in turn form enzymes and the nucleic acids DNA and RNA. Most naturally produced sulfur in the atmosphere comes from the gas hydrogen sulfide, $H_2S$, which is released from volcanic eruptions and deep-sea hydrothermal vents and during decomposition (**Figure 27.10**). The $H_2S$ quickly oxidizes into sulfur dioxide, $SO_2$. Because $SO_2$ is soluble in water, it returns to the Earth in precipitation as weak sulfuric acid, $H_2SO_4$. Sulfuric acid contributes to making the pH of natural rainwater slightly acidic, about 5.6 (refer back to Chapter 6). The sulfate ions, $SO_4^-$, enter the soil, where sulfate-reducing bacteria may release sulfur as $H_2S$, or the sulfate may be incorporated by plants into their tissue.

In the presence of iron, sulfur can precipitate as ferrous sulfide, $FeS_2$, and be incorporated in pyritic rocks. The weathering of rocks and the decomposition of organic matter release sulfur to solution, which runs through rivers to the sea. Because iron-rich rocks commonly overlay coal deposits, mining exposes them to the air and water, resulting in a discharge of sulfuric acid and other sulfur-containing compounds into aquatic ecosystems. Mining has polluted hundreds of kilometers of streams and rivers in this way in mid-Atlantic states such as West Virginia, Kentucky, and Pennsylvania.

Certain marine algae and a few saltmarsh plants produce relatively large amounts of the sulfurous gas dimethyl sulfide, $CH_3SCH_3$, commonly abbreviated as DMS. Small particles of DMS that diffuse into the atmosphere often form the nuclei around which water vapor condenses and form the water droplets making up clouds. Because of the sheer global extent of the oceans, changes in algal abundance and thus global DMS levels have the potential to alter cloud cover and thus climate. Because of the potential ability of DMS to cool the climate, some researchers are investigating how DMS production might potentially offset global warming.

Human activity involving the combustion of fossil fuels has altered the sulfur cycle proportionately more than any of the other biogeochemical cycles. The burning of coal and oil to provide energy for heating or to fuel electric power stations produces huge amounts of sulfur dioxide. This reacts with rain or snow to make human-produced acid rain. In North America, natural rainwater has a pH of about 5.6, while measurements of rain falling in southern Ontario, Canada, in the 1980s showed pH values in the range of 4.1–4.5. Such acidity is enough to kill much aquatic life and also impact terrestrial systems (refer back to Section 6.3). Huge areas of the industrial northeastern U.S. and Europe were affected by acid rain in the 1950s through the 1990s, but a reduction in the use of high-sulfur coal and the use of scrubbers to prevent sulfur dioxide from passing through smokestacks has reduced the problem in more recent times (**Figure 27.11**).

### Check Your Understanding

**27.5** Why is acid rain damaging to terrestrial systems?

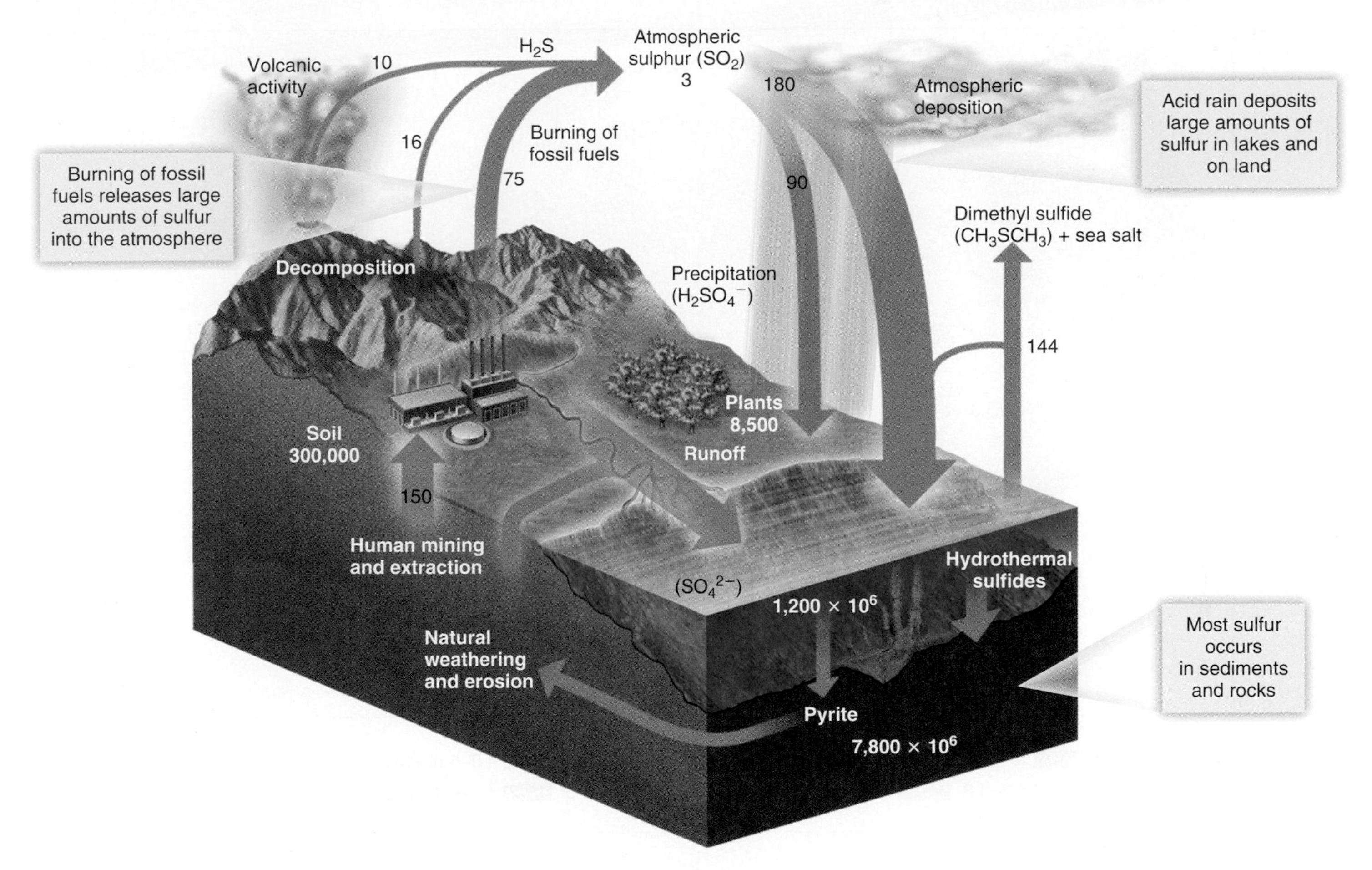


**Figure 27.10** **The sulfur cycle.** This cycle can be dramatically affected by human outputs through the burning of fossil fuels. Units are teragrams or teragrams/year. The width of the arrows indicates the relative importance of each process.

## 27.6 The Water Cycle Is Largely a Physical Process of Evaporation and Precipitation

The water cycle, also called the hydrological cycle, differs from the cycles of other nutrients in that very little of the water that cycles through ecosystems is chemically changed by any of the cycle's components (**Figure 27.12**). It is a physical process, fueled by the sun's energy, rather than a chemical one, because it consists of essentially two phenomena: evaporation and precipitation. However, the water cycle has a huge influence on other biogeochemical cycles because so many other nutrients are water-soluble. Over land, 90% of the water that reaches the atmosphere is moisture that has passed through plants and exited from their leaves via transpiration. The total evapotranspiration from land is only about 15% of the Earth's total, despite the fact that land is about 30% of the Earth's surface. This indicates that evaporation rates are double on the oceans compared to those on land. One-third of the precipitation on land comes from water evaporated from the oceans.

The mean residence time for water vapor in the atmosphere is only about 10 days. Precipitation falling on land follows one of several routes. In colder locations it may be stored as snow, either temporarily or permanently. In warmer locations rain may percolate into groundwater or runoff via streams and rivers to the oceans.

As we noted in Chapter 6, water is limiting to the abundance of many organisms, including humans. Approximately 228 L of water are needed to produce a pound of dry wheat, and 9,500 L of water are needed to support the necessary vegetation to produce a pound of meat. Industry is also a heavy user of water, with goods such as oil, iron, and steel requiring up to 20,000 L of water per ton of product. Humans have therefore interrupted the hydrological cycle in many ways to increase the amount available to them. Prominent among these activities is the use of dams to create reservoirs. Such dams, including those on the Columbia River in Washington State, can greatly interfere with the migration of fish such as salmon and affect their ability to reproduce and survive. The Three Gorges Dam in China, completed in 2006, is a hydroelectric dam on the Yangtze River that is the world's largest

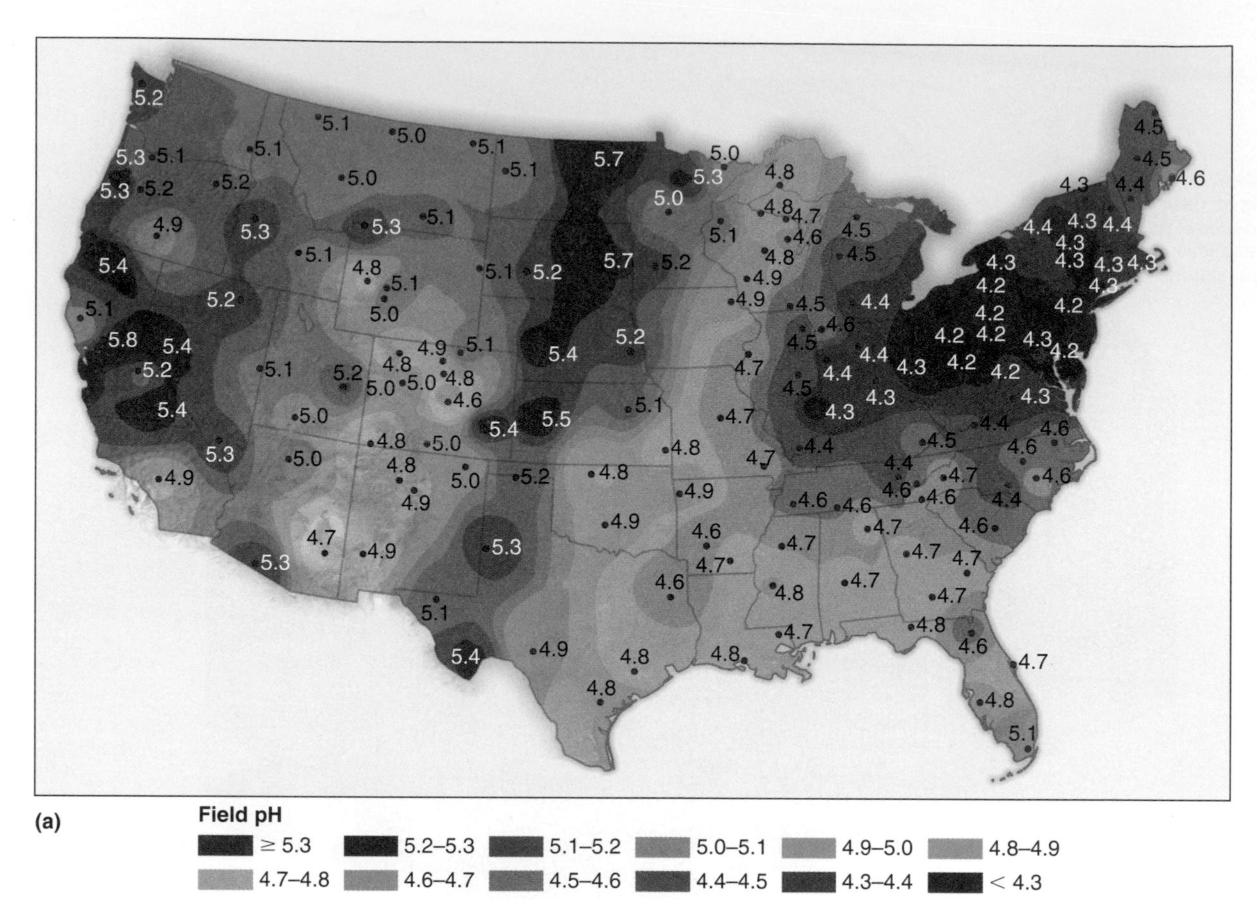


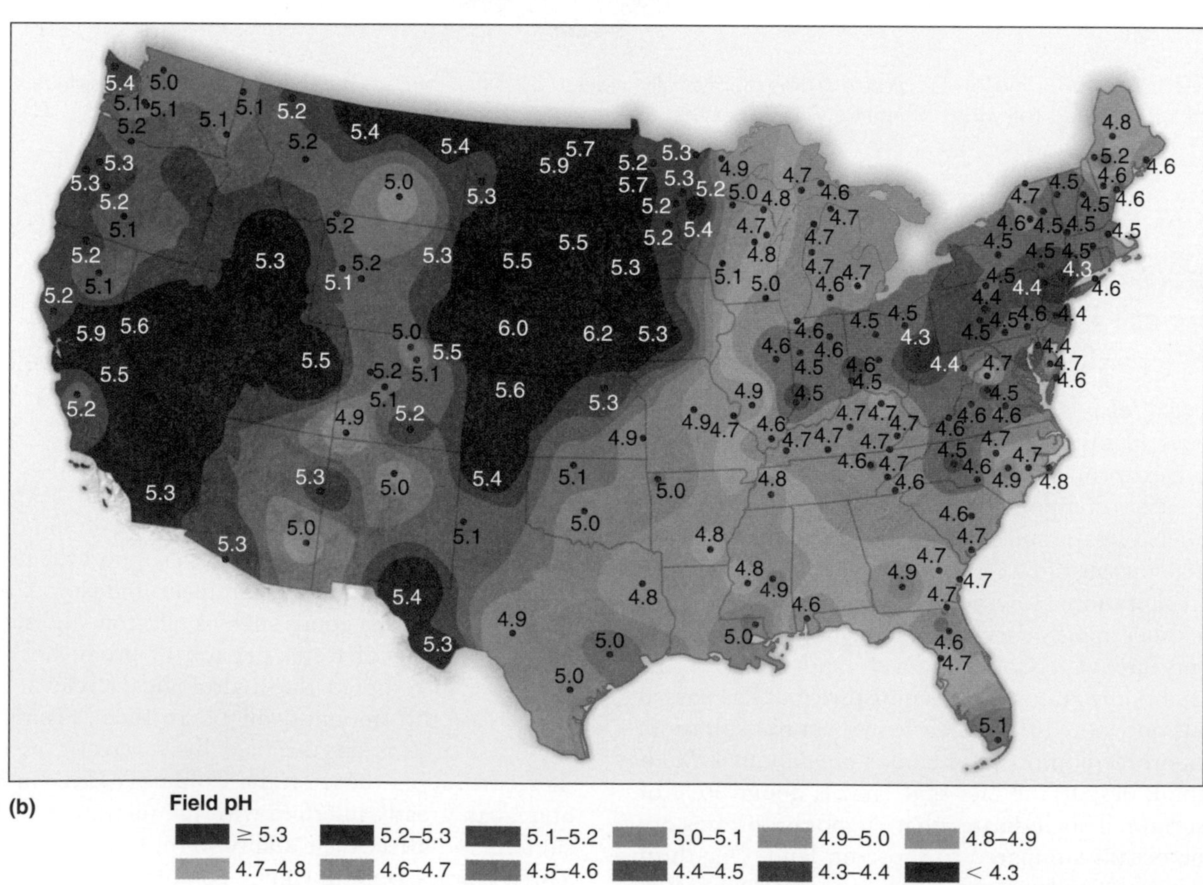


**Figure 27.11** **The extent of acid rain in the U.S. has decreased.** Data show pH measurements taken in the field **(a)** in 1994 and **(b)** in 2004.

### ECOLOGICAL INQUIRY

Why did relatively high levels of wet sulfate deposition occur in the 1980s in upstate New York, which was not an industrial area?

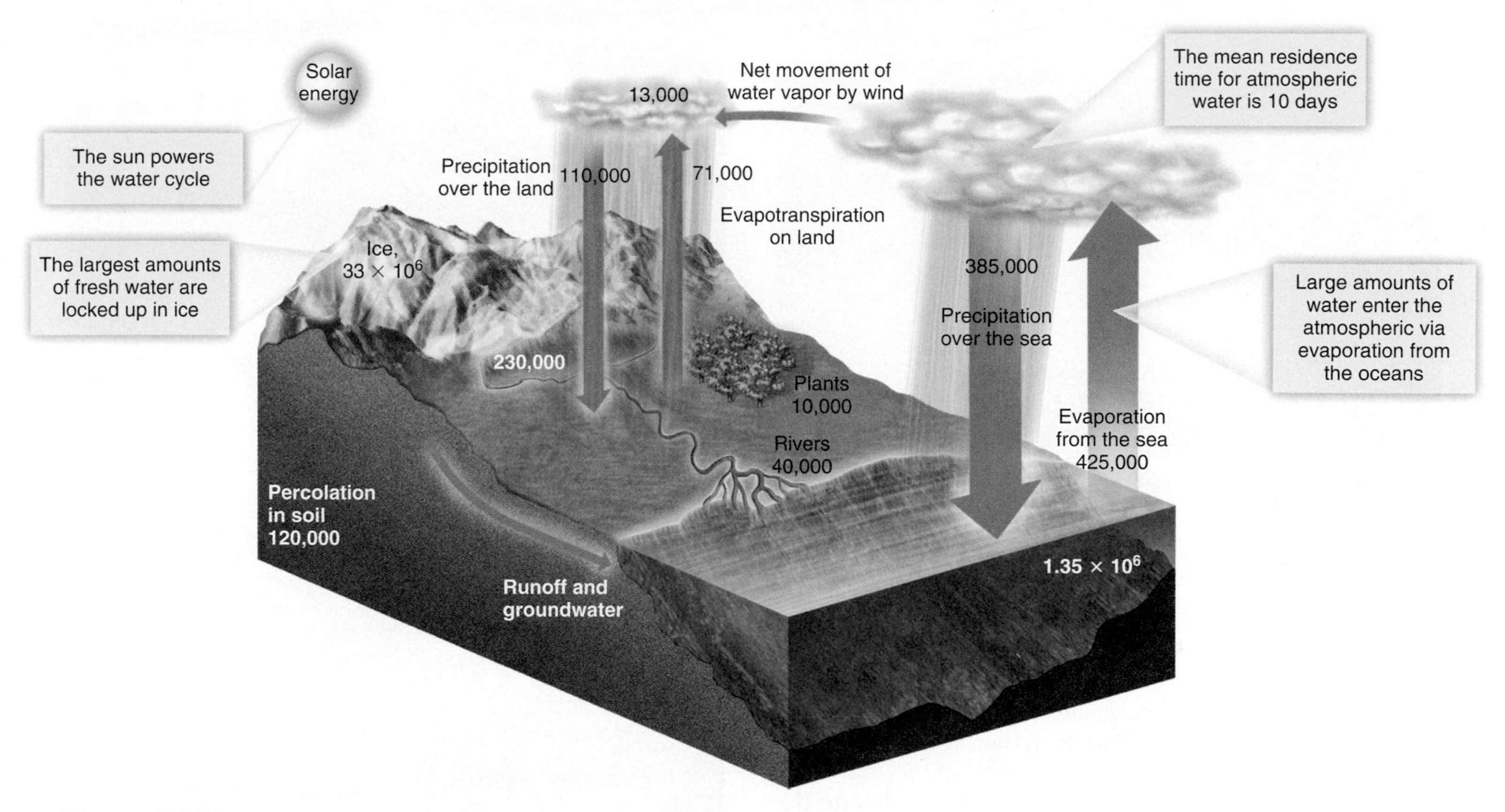


**Figure 27.12** **The water cycle is primarily a physical process, not a chemical one.** Solar energy drives the water cycle, causing evaporation of water from the ocean and evapotranspiration from the land. This is followed by condensation of water vapor into clouds and precipitation. Units are $km^3$ or $km^3$/ year. The width of the arrows indicates the relative importance of each process.

electricity generating plant (**Figure 27.13**). Although it has proved to be an economic success for China, some scientists worry that it could have contributed to the decline of endangered species in the area, reducing the available habitat for Siberian cranes, changing water flow and temperature, and reducing river habitat for the endangered Yangtze sturgeon and the Baiji or Chinese river dolphin. Other activities, such as tapping into underground water supplies, or **aquifers**, for irrigation water, remove more water than is put back by rainfall and can cause shallow ponds and lakes to dry up and sinkholes to develop. Global warming is also altering the water cycle via increased melting of glaciers and polar ice, leading to rising sea levels.

Deforestation can also significantly impact the water cycle. When forests are cut down, less moisture is transpired into the atmosphere. This reduces cloud cover and diminishes precipitation, subjecting the area to drought. Reestablishing the forests, which calls for increased water, then becomes very difficult. Such a problem has occurred on the island of Madagascar, located off the east coast of Africa. In this country, clearing of the forests for cash crops such as cotton, coffee, and tobacco has been so rapid and extensive that areas have become devoid of vegetation (**Figure 27.14**). In Madagascar, as in so many other areas on Earth, deforestation and other environmental degradations are having an impact on much of the natural habitat.

We have seen that humans have impacted the biogeochemical cycles of all the major nutrients, carbon, nitrogen, phosphorus, and sulfur, affecting their deposition rates in terrestrial and aquatic systems. However, as we have learned throughout this book, the effects of humans don't stop there. Increases in atmospheric carbon dioxide levels are causing global warming and impacting the distributions of organisms around the world. In Key West, Florida, dengue fever, a mosquito-borne disease, reappeared in 2009 after an absence of more than 60 years. Surveys in 2010 showed at least 5% of the residents had been exposed to the virus. Humans are also destroying habitat for native species and at the same time are introducing invasive species. These phenomena are often linked. For example, as we discussed in Chapter 7, Burmese pythons are now invasive in the Florida Everglades and are predicted to spread. Global warming would increase the habitat available. Research on the effects of abiotic conditions and prey availability are vital in order to predict the impact of this snake. Even now, ecologists are debating its predicted range. A recent study by Michael Avery and colleagues (2010)

**Figure 27.13** **The Three Gorges Dam on the Yangtze River, China.**

showed that 7 of 9 captive Burmese pythons held in outdoor pens in Gainesville, Florida, died or would have died had not the researchers intervened in a period of cold weather in January 2010. Avery and colleagues cast doubt on the ability of the pythons to spread much farther north than South Florida. And yet, as we have seen, not all species introductions have negative impacts. Biological control relies on the introduction of predators and parasites to control invasive pest species. In many cases this involves reuniting natural enemies from the pest's native area with pest populations in invaded areas. Other positive effects of non-native species are possible. Christine Griffiths and colleagues (2010) have suggested introducing nonindigenous tortoises to the Mascarene Islands, a series of islands off the coast of Madagascar, including Mauritius and Reunion. Here, the endemic giant *Cylindraspis* tortoises became extinct in the early 19th century. As we learned in Chapter 25, tortoises are often ecosystem engineers and their burrows support many other species. However, in this case the extinct tortoises were the most common herbivores and seed dispersers, and in their absence low-lying grazing-intolerant species outcompete and replace grazing-tolerant species. Also, the regeneration of many tree species is hampered by a lack of seed dispersal. In the 1970s and 1980s, non-native herbivores such as goats were removed from these islands and the vegetation changed dramatically. While goats were seed predators and did not aid in seed dispersal, their feeding did allow grazing-tolerant plant species to thrive. A small-scale introduction of Madagascan tortoises was performed in 2007 on Round Island, a tiny 1.7-$km^2$ island off the coast of Mauritius. Tortoise survival has been good, with 14 of 23 tortoises still alive after 5 years, in 2012. In the coming years, ecologists will be keen to observe the effects on native plant species. In a globally changing world, there is much work for ecologists to do.

**Figure 27.14** **Severe erosion following deforestation in Madagascar.** After clearing and a few years of farming, the shallow soil can no longer support crops and is susceptible to erosion by rainfall.

### Check Your Understanding

**27.6** How is the water cycle different from the cycles of other nutrients?

## SUMMARY

**27.1 Biogeochemical Cycles Transfer Elements Among the Biotic and Abiotic Components of Ecosystems**

- Invasive species may change rates of nutrient cycling (Figure 27.1).
- Elements such as phosphorus, carbon, and nitrogen cycle continuously, moving from the physical environment to organisms and back in what are called biogeochemical cycles. Biogeochemical cycles involve both abiotic and biotic components (Figure 27.2).
- The increase in the concentration of a substance in living organisms, called biomagnification, can occur at each step of the food chain (Figures 27.3, 27.A).

**27.2 Phosphorus Cycles Locally Between Geological and Biological Components of Ecosystems**

- The phosphorus cycle lacks an atmospheric component and thus is a local cycle (Figure 27.4).
- An overabundance of phosphorus can cause the overgrowth of algae and subsequent depletion of oxygen levels, called eutrophication (Figures 27.5, 27.6).

**27.3 Carbon Cycles Among Biological, Geological, and Atmospheric Pools**

- In the carbon cycle, autotrophs incorporate carbon dioxide from the atmosphere into their biomass; decomposition of plants and respiration recycles most of this $CO_2$ back to the atmosphere. Human activities, primarily the burning of fossil fuels, are causing increased amounts of $CO_2$ to enter the atmosphere (Figure 27.7).
- Experiments have shown that elevated levels of carbon dioxide result in an increase in plant growth but a decrease in herbivory (Figure 27.B).

**27.4 The Nitrogen Cycle Is Strongly Influenced by Biological Processes That Transform Nitrogen into Usable Forms**

- The nitrogen cycle has five parts: nitrogen fixation, nitrification, assimilation, ammonification, and denitrification. In the nitrogen cycle, atmospheric nitrogen is unavailable for use by most organisms and must be converted to usable forms by bacteria and fungi (Figure 27.8).
- The activities of humans, including fertilizer use, fossil fuel use, and deforestation, have dramatically altered the nitrogen cycle. Deforestation can result in nitrogen runoff into rivers and lakes (Figure 27.9).

**27.5 The Sulfur Cycle Is Heavily Influenced by Anthropogenic Effects**

- Sulfur enters the atmosphere through both natural sources, such as volcanoes and decomposition, and human sources, including the combustion of fossil fuels. Atmospheric sulfur dioxide returns to Earth as weak sulfuric acid, $H_2SO_4$ (Figure 27.10).
- Human-produced acid rain results primarily from the combustion of fossil fuels (Figure 27.11).

**27.6 The Water Cycle Is Largely a Physical Process of Evaporation and Precipitation**

- The water cycle is a physical rather than a chemical process, because it consists of essentially two phenomena: evaporation and precipitation (Figure 27.12).
- Alteration of the water cycle by dams or deforestation can result in reduction of available habitat for native species (Figures 27.13, 27.14).
- Careful ecological research can be used to help reduce human effects on biogeochemical cycles, to reduce the impacts of invasive species, and to promote the restoration of degraded areas.

## TEST YOURSELF

1. The concentration of certain chemicals, such as DDT, in higher trophic levels is known as:
   a. Eutrophication
   b. Biomagnification
   c. Biogeochemical cycling
   d. Energy transfer
   e. Turnover

2. Which of the following nutrients does not cycle globally?
   a. Carbon
   b. Nitrogen
   c. Phosphorus
   d. Sulfur
   e. Water

3. Eutrophication is highest where biological oxygen demand is:
   a. Low
   b. Intermediate
   c. High

4. Eutrophication is:
   a. Caused by an overabundance of nitrogen, which leads to an increase in bacteria populations
   b. Caused by an overabundance of nutrients, which leads to an increase in algal populations
   c. The normal breakdown of algal plants following a pollution event
   d. Normally seen in dry, hot regions of the world
   e. None of the above

5. Primary producers acquire the carbon necessary for photosynthesis from:
   a. Decomposing plant material
   b. Carbon monoxide released from the burning of fossil fuels
   c. Carbon dioxide in the atmosphere
   d. Carbon sources in the soil
   e. Both a and d
6. In the immediate future, atmospheric carbon dioxide levels are likely to:
   a. Increase  b. Decrease  c. Stay the same
7. Nitrogen fixation is the process:
   a. That converts organic nitrogen to ammonia
   b. By which plants and animals take up nitrates
   c. By which bacteria convert nitrite to gaseous nitrogen
   d. By which atmospheric nitrogen is converted to ammonia or ammonium ions
   e. All of the above
8. Which is the correct order of processes in the nitrogen cycle?
   a. Nitrogen fixation, nitrification, denitrification, assimilation, ammonification
   b. Denitrification, nitrification, nitrogen fixation, assimilation, ammonification
   c. Ammonification, denitrification, nitrification, assimilation, nitrogen fixation
   d. Assimilation, ammonification, denitrification, nitrogen fixation, nitrification
   e. Nitrogen fixation, nitrification, assimilation, ammonification, denitrification
9. What organisms are important nitrogen fixers in aquatic ecosystems?
   a. Diatoms
   b. Cyanobacteria
   c. Fungi
   d. Insects
   e. Fish
10. More rainfall occurs over land than at sea.
   a. True  b. False

## CONCEPTUAL QUESTIONS

1. What is biomagnification and how does it work?
2. Explain the causes and consequences of eutrophication.
3. Describe the carbon cycle.

## DATA ANALYSIS

The figure shows proportional legume abundance over 8 years of elevated $CO_2$ and/or N enrichment from fertilizer in a Minnesota grassland. Shown are the proportion of four species of legumes in a series of plots containing 16 total grassland species. Explain the results.

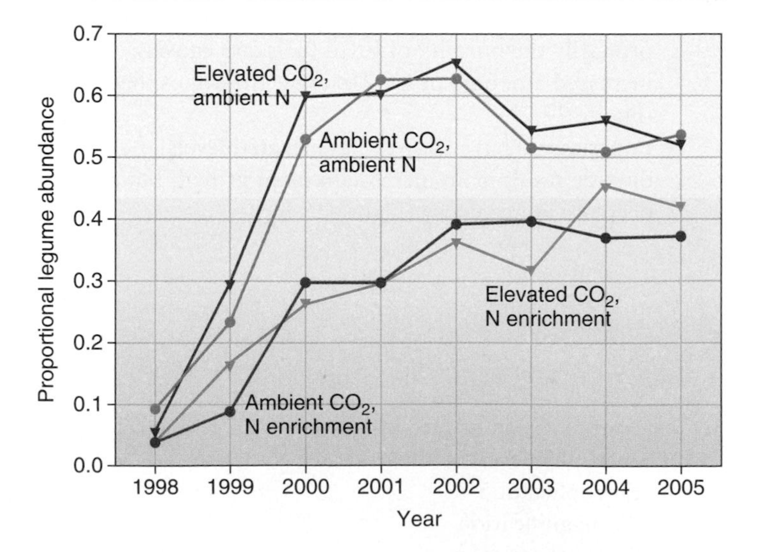

# Appendix A

# Answers to Questions

## Chapter 1

### Check Your Understanding

**1.1** Abiotic factors concern the physical environment, whereas biotic factors concern interactions among living organisms, so in this example the organism would be limited by an abiotic factor.

**1.2** A disadvantage of group living for many species is that competition between individuals for resources or space is increased. In addition, the spread of diseases among individuals may be facilitated.

**1.3** Populations of some mouse predators, such as raptors, have also increased, and these may negatively impact populations of other small mammals. The increased deer mice densities may also reduce recruitment of native plants because of seed predation.

**1.4** You might set up cages around the crops to exclude the pests and compare crop yields inside the cages to uncaged controls. However, cages may prevent pollinators from visiting crops, so timing of cage installation may be important. You might also spray some plants with pesticides and compare yields to unsprayed plants. However, spraying might also increase yields due to an increase in water from the sprays, so an appropriate control may involve spraying plants simply with water.

### Test Yourself

**1.** b, **2.** b, **3.** e, **4.** e, **5.** e, **6.** e, **7.** b, **8.** c, **9.** e, **10.** c

### Data Analysis

The average nest success in trapped areas was 42% for all years and sites combined, compared to only 23% for untrapped sites. Predators are affecting nesting success of ducks in North Dakota.

### Ecological Inquiry

**Figure 1.6** Pollution is more important to freshwater species such as fish, mollusks, and crayfish because pollutants often concentrate in freshwater systems due to runoff from neighboring terrestrial habitats. Many mammals are selectively killed for sport or fur. Some plants, such as orchids and cacti, are removed by collectors.

**Figure 1.9** Amphibians' ability to fight infection depends on temperature, which is gradually changing. Also, amphibians' skin is relatively porous, so these species are more susceptible to water pollution.

**Figure 1.12** Higher winter moth abundance occurs where predation levels are highest. In this case, predation might have little impact on winter moth populations.

**Figure 1.14** Because the 95% confidence interval (the vertical bar) of insects does not overlap that of mammals, it means the effect of insects is significantly greater than that of mammals.

## Chapter 2

### Check Your Understanding

**2.1** Through the discovery of Mendel's work in the early part of the 20th century. Between 1856 and 1863 Gregor Mendel's experiments with peas provided evidence that inherited factors can be passed down from generation to generation.

**2.2** Gene mutations involve changes in the sequence of nucleotide bases that make up the DNA. Chromosomal mutations do not add to the variability of the gene pool, they merely rearrange it into different combinations. Chromosomal mutations are the most common source of new genes.

**2.3** If $R = 0.4$, then $r = 0.6$. The frequency of red flowers $= p^2 = 0.16$. The frequency of the pink flowers $= 2pq = 0.48$. The frequency of the white flowers $= q^2 = 0.36$.

**2.4** To reduce the risks associated with inbreeding in especially small populations.

### Test Yourself

**1.** b, **2.** b, **3.** c, **4.** a, **5.** a, **6.** c, **7.** d, **8.** c, **9.** c, **10.** b

### Ecological Inquiry

**Figure 2.B** No, some melanics will always occur at low frequencies. This is the situation that endured before industrial pollution. Such natural variation allows natural selection to act.

**Figure 2.10** If $R = 0.5$ then $r = 0.5$. The frequency of white flowers is then $q^2$ or 0.25 or 25%.

**Figure 2.18** 400

## Chapter 3

### Check Your Understanding

**3.1** The drought caused a large reduction in small seeds on the island Daphne Major, and only larger seeds were available. Only birds with larger beaks were able to eat these seeds and survive. They passed this trait on to their offspring, and the next generation of finches had larger beaks, supporting the idea of natural selection.

**3.2** Because plants commonly exhibit polyploidy while animals do not. Polyploidy changes species chromosome numbers, effectively isolating them reproductively.

**3.3** Convergent evolution. Even though both species have very different ancestors, a common environment and food source promote the evolution of similar morphologies in different species.

3.4 Because the environment may be changing rapidly. According to the Red Queen hypothesis, species must continually evolve and change just to keep pace with environmental change. There is no evidence that older lineages are more likely to be successful in the future than newer lineages.

3.5 Rare species may be locally abundant, albeit limited to a small area. The more widespread species may be uncommon across all its range and may be particularly susceptible to climate change or introduced species.

### Test Yourself

**1.** d, **2.** b, **3.** a, **4.** b, **5.** b, **6.** e, **7.** d, **8.** e, **9.** d, **10.** a

### Data Analysis

The frequency of melanics before clean air legislation was 89.1%. Melanics were favored because bird predation was reduced on dark-colored and lichen-less trees. After the clean air legislation, the frequency of melanics was 18.33%. Melanics were susceptible to bird predators because they were more obvious on lighter-colored and lichen-encrusted trees.

### Ecological Inquiry

**Figure 3.A** The red morphs would suffer both higher rates of parasitism and higher rates of predation, compared to green morphs, and would likely become rare or go extinct.

**Figure 3.7** In theory, fertile offspring of intermediate color would result, showing that the races were not separate biological species.

**Figure 3.13** Climate change, possibly caused by a meteor impact that reduced global temperatures. Competition from mammals, as well as disease, may also have contributed.

**Figure 3.24** Because of small population sizes, limited distributions, lack of dispersal ability, lower reproductive rates, and tameness.

**Figure 3.26** Because mollusks have a distribution limited to freshwater lakes and rivers and these habitats often receive much pollution from runoff and point sources.

**Figure 3.28** Rare species may be abundant but limited to a small area, or be at low population densities over a wide area.

## Chapter 4

### Check Your Understanding

4.1 It lends more weight to Alexander's ideas about lifestyle promoting eusociality. In ant colonies fathered by multiple males, sisters are less likely to share the same father. This lowers the chance of being related by $r = 0.75$, and thus lowers the likelihood of cooperation between sisters.

4.2 Living in a tighter group is a defensive tactic of guppies. The geometry of the selfish herd gives them some protection. When guppies are moved from predator-rich to predator-poor areas of Trinidad streams, they grow bigger, live longer, and have bigger offspring.

4.3 Unlike whelks, walnuts are damaged by each drop, and so crows adjust for the increased likelihood that a repeatedly dropped nut will break on a subsequent drop by decreasing drop height. The probability of food stealing by neighboring crows also resulted in a decrease in drop height to lessen the chance of food stealing.

4.4 Males with tails cut and re-glued to their original length would be the best controls. This controls for any effects of tail clipping itself, which might alter reproductive success.

### Test Yourself

**1.** d, **2.** a, **3.** b, **4.** c, **5.** a, **6.** b, **7.** d, **8.** c, **9.** c, **10.** c

### Data Analysis

1. Small mussels are easy to crack open but contain little nutritive value. Large mussels contain high nutritive value but are difficult to crack open. Subtracting the energy used to crack open the shells from the energy obtained from the meal yields the net energy gained. The net energy gained is relatively small for small and large mussels but is greatest for medium-sized mussels.
2. Camouflaged species are less likely to be toxic to predators. Their survival is based on their ability to avoid detection, which is increased by their solitary nature. Warning coloration and toxicity are enhanced by aggregation into kin groups. Even some solitary species are warningly colored, suggesting some direct benefits as well. If a solitary brightly colored larva is lucky enough to survive an attack by predator, it probably won't be attacked again.
3. Alarm calling calls attention to the caller, so females often bolt into their warren and escape if no relatives are present. If daughters are present, then females call to alert them. However, females may also pass on copies of their own genes via their kin, if their sisters or mothers have their own pups, so alarm calls are made when these relatives are present also.

### Ecological Inquiry

**Figure 4.3** No, according to Hamilton's theory you would sacrifice your life only for more than 8 cousins.

**Figure 4.5** Females, because females are more likely to stay in the same burrow system and are more related to the other group members. Males are more likely to disperse to other areas.

**Figure 4.18** There are at least two different hypotheses for the existence of leks. The first is the "hotshot" hypothesis. Hotshot males attract a lot of males, and subordinate males are drawn to these leks to have a chance to mate with females that they might otherwise never encounter. The second is the "hotspot" hypothesis. Leks occur because there are simply few good areas to display, such as open areas in a forest. Hotspots might also be close to the travel routes for females, or they may occur in areas where predation risks are low.

**Figure 4.22** The males are often not the father of the pups they crush, so there often is no genetic cost for them.

## Chapter 5

### Check Your Understanding

5.1 In theory, "warm-blooded" species such as birds, bats, and terrestrial mammals should be relatively immune to cold temperatures. In practice, they often cannot feed quickly enough to generate sufficient heat to stay warm. For this reason, cold temperatures often limit the distribution of endotherms.

5.2 Microclimate concerns local variations in climate. Organisms can alter their microclimate by growing or living in shady conditions or sunny areas, on windward or leeward sides of mountains, or northerly versus southerly facing mountain slopes. Animals can dig burrows or rest on hot sunny surfaces.

**5.3** Global warming would advance the flowering time of plants and increase the vegetative growing season. The time of emergence from hibernation of mammals, the egg-laying and hatching times of birds, the time of ice melt, the calling (mating) time of frogs, the appearance of butterflies, and the appearance of phytoplankton blooms in lakes would all occur earlier in the year.

### Test Yourself

**1.** e, **2.** b, **3.** d, **4.** b, **5.** d, **6.** c, **7.** c, **8.** b, **9.** d, **10.** c

### Data Analysis

Temperature and moisture are the main factors limiting the distribution of this species. The westward edge of the range is almost certainly set by water availability. In the north, cold winter temperatures limit distribution through availability of soil moisture. In cold weather the water uptake by the roots is severely limited. In south Florida, conditions may be too wet in the fall to allow ripening and dispersal of seed.

### Ecological Inquiry

**Figure 5.1** This is due to cold, southerly ocean currents along the coasts of both continents (look ahead to Figure 23.1).

**Figure 5.2** No, some ectotherms that rely on external heat sources can be homeotherms if their environment is very stable. For example, the temperature of deep-sea fish does not change.

**Figure 5.6** Yes, in theory vampire bats could survive in south Florida, and geological fossils of such bats have been found there.

**Figure 5.13** Mainly by convection, because the movement creates air currents.

**Figure 5.17** Many species, such as barnacles, limpets, chitons, whelks, sea anemones, and starfish.

## Chapter 6

### Check Your Understanding

**6.1** Large trees, especially those as large as redwoods, require substantial water. Their distribution is limited to coastal areas where winter rainfall is greatest and summer moisture is increased by coastal fog.

**6.2** Hypo-osmotic fish have a lower concentration of ions than their environment, so marine bony fish are hypo-osmotic. This means they lose water from their mouth and gills and have to drink continually to replace it.

**6.3** Because at low pH, nitrifying bacteria are less able to decompose leaf litter and other organic material. This decreases soil nitrogen content. Because most plant species grow better in nitrogen-rich soils, more species are found in alkaline than in acid soils.

### Test Yourself

**1.** c, **2.** a, **3.** e, **4.** a, **5.** b, **6.** a, **7.** b, **8.** a, **9.** b, **10.** b

### Data Analysis

In the watering treatments, the root mass of both species continued to grow. However, when watering ceased, the root mass of *Eleusine* grew only slowly, and toward the end of the experiment grew very little. The root mass of *Digitaria,* however, continued to grow even when unwatered. The weight increased and so did the length (not shown). Park suggested that the ability of the *Digitaria* to produce a large root mass and to deeply penetrate the substrate would permit it to tap deeper soil moisture and allow a high rate of water intake, even in the driest environments.

### Ecological Inquiry

**Figure 6.3** Both. Trees cannot grow where there is insufficient moisture. On mountainsides this point is set by low temperatures and high winds, which cause transpiration to exceed water uptake.

**Figure 6.5** Ammonia is toxic and needs to be excreted immediately. This requires large volumes of water.

**Figure 6.14** Close to that of beer, which is also carbonated. The pH of carbonated drinks is 3.5.

## Chapter 7

### Check Your Understanding

**7.1** Minerals are commonly leached from the A horizon and deposited in the B horizon. Other material, such as clay, may also be deposited in this layer, forming an impermeable pan.

**7.2** The Venus flytrap would waste energy closing its trap around organic debris, such as a leaf, that happened to blow into the trap. For this reason, the traps shut when more than one trigger is touched or one hair is touched twice or more. This increases the likelihood that the trap will close around live prey.

**7.3** Many rodents gnaw on old bones or on shed antlers to gain their supply of sodium, so bones don't last very long in nature. In addition, herbivores, such as cows, have been seen eating animals, such as dead rabbits, to supplement their intake of essential nutrients such as sodium.

**7.4** Red algae appear red only at the surface, where there is white light, or when photographed at depth with a flash, which also supplies white light.

**7.5** $C_3$ plants absorb $CO_2$ less effectively than $C_4$, so in a world of increasing $CO_2$, $C_3$ plants would fare better. $C_4$ plants contain less nitrogen than $C_3$ plants and so increased numbers of $C_3$ plants would provide more nutrients to herbivores.

**7.6** At the surface of cold, freshwater lakes at sea level.

### Test Yourself

**1.** e, **2.** a, **3.** c, **4.** a, **5.** b, **6.** c, **7.** c, **8.** c, **9.** b, **10.** b

### Data Analysis

Leaf nitrogen content decreased as leaf life span increased. Leaf nitrogen is low in low-nutrient environments. In such environments, it is energetically costly to replace leaves. In higher-quality habitats, nitrogen is less limiting and leaves can be more readily replaced. Pine trees grow in low-quality soils and needles have low nitrogen levels; they may last for years on the tree.

### Ecological Inquiry

**Figure 7.3** Because oxygen also binds to the enzyme nitrogenase, such bacteria perform best in a low-oxygen habitat.

**Figure 7.5** It may be that in nitrogen-limited conditions, *Spartina* puts more effort into aboveground biomass and possibly sexual reproduction than into belowground biomass.

## Chapter 8

### Check Your Understanding

**8.1** (a) Bats are most commonly caught in mist nets placed along flyways. (b) Butterflies are often caught on the wing with butterfly nets, or caterpillars can be counted on vegetation. (c) Ants fall into pitfall traps or may be attracted to baited traps, such as peanut butter, and the number seen over a given time interval is recorded.

**8.2** Generally clumped, because some people sit next to their friends. This might change to a uniform pattern at test time as the instructor asks for an equal space between students.

**8.3** Road construction, power lines, canals, deforestation, urbanization, agriculture.

**8.4** Because habitat corridors promote the spread of individuals among patches, they might also promote the spread of invasive species or parasites. Also, fire can spread more easily between connected patches.

**8.5** The increased ratio of habitat edge to habitat area allows greater access of prey to predators in patchy habitats. This is one reason fragmented habitats are not as good as unfragmented ones. For example, percent loss of juvenile scallops in North Carolina sea grass beds was >90% in patchy sea grass and 50% in continuous areas.

### Test Yourself

**1.** c, **2.** b, **3.** b, **4.** b, **5.** b, **6.** e, **7.** b, **8.** b, **9.** d, **10.** a

### Ecological Inquiry

**Figure 8.5** $110 \times 100/20 = 550$

**Figure 8.7** Clumped, because resources are often clumped in nature.

**Figure 8.8** Fires, floods, rivers, mountain ranges, soils (especially for plants), soil moisture levels.

**Figure 8.A** Yes, in general. Over time, habitats are continually divided to create more numerous, but smaller, fragments.

## Chapter 9

### Check Your Understanding

**9.1**

| $x$ | $l_x$ | $q_x$ |
|---|---|---|
| 1 | 1.000 | 0 |
| 2 | 1.000 | 0.061 |
| 3 | 0.939 | 0.197 |
| 4 | 0.754 | 0.330 |
| 5 | 0.505 | 0.396 |
| 6 | 0.305 | 0.290 |
| 7 | 0.186 | 0.290 |
| 8 | 0.132 | 0.810 |
| 9 | 0.025 | 1.000 |

**9.2**

| $x$ | $l_x m_x$ | $\Sigma l_x m_x$ |
|---|---|---|
| 0 | 0 | 0 |
| 1 | 0 | 0 |
| 2 | 0.077 | 0.154 |
| 3 | 0.290 | 0.870 |
| 4 | 0.262 | 1.048 |
| 5 | 0.189 | 0.945 |
| 6 | 0.143 | 0.858 |
| 7 | 0.106 | 0.742 |
| 8 | 0.078 | 0.624 |
| 9 | 0.064 | 0.576 |
| 10 | 0.064 | 0.640 |
| 11 | 0.055 | 0.605 |
| 12 | 0.041 | 0.492 |
| 13 | 0.028 | 0.364 |
| 14 | 0.023 | 0.322 |
| 15 | 0.018 | 0.270 |
| 16 | 0.009 | 0.144 |

$R_0 = \Sigma l_x m_x = 1.447$

$T = \frac{\Sigma l_x m_x}{R_0} = \frac{8.654}{1.447} = 5.981$

### Test Yourself

**1.** c, **2.** d, **3.** c, **4.** b, **5.** e, **6.** a, **7.** d, **8.** d, **9.** a, **10.** c

### Data Analysis

**1.**

| $x$ | $l_x$ | $m_x$ | $l_x m_x$ |
|---|---|---|---|
| 1 | 1.0 | 0.05 | 0.05 |
| 2 | 0.253 | 1.28 | 0.324 |
| 3 | 0.116 | 2.28 | 0.264 |
| 4 | 0.089 | 2.28 | 0.203 |
| 5 | 0.058 | 2.28 | 0.132 |
| 6 | 0.039 | 2.28 | 0.089 |
| 7 | 0.025 | 2.28 | 0.057 |
| 8 | 0.022 | 2.28 | 0.050 |

$R_0 = 1.169$, so the population is increasing.

**2.**

| Age, $x$ | $d_x$ | $n_x$ | $L_x$ | $e_x$ |
|---|---|---|---|---|
| 0 | 160 | 1,000 | 920 | 8.40 |
| 1 | 141 | 840 | 770 | 8.90 |
| 2 | 93 | 699 | 652 | 9.60 |
| 3 | 68 | 606 | 572 | 10.00 |
| 4 | 50 | 538 | 513 | 10.19 |
| 5 | 33 | 488 | 471 | 10.18 |
| 6 | 31 | 455 | 439 | 9.89 |
| 7 | 31 | 424 | 408 | 9.58 |
| 8 | 31 | 393 | 377 | 9.30 |
| 9 | 25 | 362 | 350 | 9.04 |
| 10 | 25 | 337 | 324 | 8.68 |
| 11 | 26 | 312 | 299 | 8.33 |
| 12 | 25 | 286 | 273 | 8.06 |
| 13 | 25 | 261 | 248 | 7.78 |
| 14 | 25 | 236 | 223 | 7.55 |
| 15 | 25 | 211 | 198 | 7.38 |
| 16 | 21 | 186 | 175 | 7.32 |
| 17 | 17 | 165 | 156 | 7.20 |
| 18 | 16 | 148 | 140 | 6.97 |
| 19 | 14 | 132 | 125 | 6.74 |
| 20 | 12 | 118 | 112 | 6.49 |
| 21 | 11 | 106 | 100 | 6.16 |
| 22 | 11 | 95 | 90 | 5.82 |
| 23 | 10 | 84 | 79 | 5.51 |
| 24 | 10 | 74 | 69 | 5.19 |
| 25 | 9 | 64 | 59 | 4.92 |
| 26 | 7 | 55 | 51 | 4.65 |
| 27 | 6 | 48 | 45 | 4.28 |
| 28 | 6 | 42 | 39 | 3.81 |
| 29 | 6 | 36 | 33 | 3.36 |

| | | | | |
|---|---|---|---|---|
| 30 | 6 | 30 | 27 | 2.93 |
| 31 | 6 | 24 | 21 | 2.54 |
| 32 | 5 | 18 | 15 | 2.22 |
| 33 | 4 | 13 | 11 | 1.94 |
| 34 | 4 | 9 | 7 | 1.56 |
| 35 | 2 | 5 | 4 | 1.40 |
| 36 | 2 | 3 | 2 | 1.00 |
| 37 | 1 | 1 | 1 | 1.00 |
| 38 | 0 | 0 | 0 | — |

Young skulls may be eaten by rodents. This would result in fewer young skulls and lead to underestimation of mortality. Older skulls are tougher and are not completely eaten.

3. Survivorship of both African and Asian elephants in zoos is reduced compared to wild-born individuals experiencing natural mortality. This phenomenon is true across all age classes. Inbreeding can reduce survivorship, but wild-born elephants exhibit the same trends as captive-born individuals. The median age of zoo-born females was only 16.9 years compared to 56.0 years for wild-born females in Amboseli National Park, Africa, and for zoo-born Asian elephants, 18.9 years compared to 41.7 years in the Burmese logging industry. Stress and/or obesity have been suggested as likely causes of accelerated zoo mortality rates.

### Ecological Inquiry

**Figure 9.4** Butterfly, type I, much mortality in egg and caterpillar stages; turtle, type II, fairly uniform death rates throughout life; human, type III, strong parental care of young, more death in older individuals.

## Chapter 10

### Check Your Understanding

**10.1**

1. $t = \log(N_t/N_0)/\log(\lambda)$
   $= \log(10\text{ billion}/6.5\text{ billion})/\log 1.0123$
   $= 0.187/0.005$
   $= 37.4$ years

2. $\lambda = (N_t/N_0)^{(1/t)} - 1$
   $= (1{,}800/1{,}000)^{(1/5)} - 1$
   $= 1.125$ or 12.5% per year

**10.2** Laboratory populations often have uniform conditions of temperature, humidity, and food availability. In the field, many of these conditions vary, especially resources, which strongly affect population growth. The existence of time lags also disrupts population growth and may prevent populations from reaching an upper asymptote.

**10.3** The data reveal a pattern of inverse density dependence. In this case, the introduced parasites failed to limit the population growth of the codling moth.

**10.4** (a) Competitors, (b) stress tolerators, (c) ruderals.

**10.5** The net reproductive rate measures the number of daughters a woman would have in her lifetime if she were subject to prevailing age-specific fertility and mortality rates in all given years. The total fertility rate, TFR, is the average number of children a woman would have in a given year, were she able to fast-forward through all her childbearing years in a single year. TFR is not as useful as net reproductive rate, because the fertility rate of young women now could change from those of older women now. The difference is somewhat analogous to the difference between cohort and static life tables introduced in Chapter 5. Surprisingly, the United Nations stopped reporting $R_0$ in 1999.

### Test Yourself

**1.** a, **2.** c, **3.** e, **4.** e, **5.** c, **6.** b, **7.** d, **8.** c, **9.** b, **10.** c

### Data Analysis

**1.** (a) $\lambda = 1{,}200 / 1{,}000 = 1.2$

(b) After 5 years, $N_5 = N_0\lambda^5$
$= (1{,}000)(1.2)^5$
$= 2{,}488$

(c) $10{,}000 = (1{,}000)(1.2)^t$
or $10{,}000 / 1{,}000 = 1.2^t$
therefore $10 = 1.2^t$
taking logs of both sides,
$\ln 10 = (\ln 1.2)^t$
or $2.302 = 0.182 \times t$
so $t = 2.302 / 0.182$
$= 12.65$ years

**2.** No evidence for density dependence exists in this data set, therefore parasitism is density independent.

**3.** Logistic. Plotting the numbers of willows against time yields an S-shaped curve.

### Ecological Inquiry

**Figure 10.1** 54

**Figure 10.6** 12

**Figure 10.13** Semelparity, because an organism could devote less energy to maintenance and more to reproduction.

**Figure 10.B** Go to one of the many websites that enable you to create a personal ecological footprint.

## Chapter 11

### Check Your Understanding

**11.1** The competition is interspecific, between different species. It is also exploitative, caused through the consumption of a common resource, acorns and pinecones. The competition also appears to be amensalism, because the gray squirrels have a negative effect on the reds but the reds do not appear to be affecting the spread of the grays, although this has not been investigated in great detail.

**11.2** As we have noted in earlier chapters, much genetic variation exists between members of a population. In later experiments, Park noted that particular genetic strains of both *T. confusum* and *T. castaneum* could cause differences in the outcomes of competition experiments.

**11.3** The data support the propagule pressure hypothesis, because the longer the time period, the greater the likelihood of more individuals being "released" into the environment via seed, and the greater the likelihood of invasiveness.

**11.4** The per capita effect of species 1 on species 2 is the same as the per capita effect of species 2 on species 1. Both species are equivalent competitors. Predicting the winner of a competitive interaction requires the knowledge of the carrying capacity of both competitors.

11.5 According to the data, the size ratio between each bird species, 1–2, 2–3, 3–4 and 4–5 is 2.0, 1.895, 1.397 and 1.133. According to theory, the pair least likely to coexist is the last pair, *L. minimus* and *T. glareola*, so one of these two species is the most likely to go extinct.

### Test Yourself

**1.** c, **2.** e, **3.** b, **4.** c, **5.** c, **6.** d **7.** c, **8.** d, **9.** b, **10.** d

### Data Analysis

**a.** Both species grow best in the absence of competition.

**b.** The effects of interspecific competition were asymmetric. The growth of *D. nudiflorum* was depressed much more by *D. glutinosum* than vice versa.

**c.** Intraspecific competition was more important than interspecific competition for *D. glutinosum* but not for *D. nudiflorum*.

### Ecological Inquiry

**Figure 11.3** Interspecific interference competition.

**Figure 11.4** Four of ten species pairs equals 40%

**Figure 11.5** Leaf feeders, sap suckers, root feeders, seed borers, and flower feeders are some possibilities.

**Figure 11.10** Species 1 wins, see part (a).

**Figure 11.15** Yes, because $d/w = 2.0$, so species would compete but coexist.

**Figure 11.17** No, theoretically the PS value would have to less than $0.7 \times 0.7 \times 0.7 = 0.343$.

## Chapter 12

### Check Your Understanding

**12.1** (a) Because in theory each mutualist can increase the carrying capacity for its fellow mutualist, which can lead to runaway increases in population sizes. Reducing the mutualism coefficients, $\alpha$ and $\beta$, as the populations grow, can result in population stability.

(b) Janzen suggested that fallen fruit is desirable to a variety of organisms, including mammals and microbes. Microbes that colonize fallen fruit manufacture ethanol, and give the fruit its "rotten" appearance, so that it is distasteful to mammals.

**12.2** Inquilines. An inquiline uses a second species or something made by the second species, in this case the gall created by the first species, for housing.

**12.3** Facilitation may be more common in stressful habitats such as deserts, the arctic, mountaintops, cold temperate areas, salt marshes, or the intertidal exposed to pounding surf. It might be less common in more benign habitats such as the Tropics, freshwater wetlands, and lowland terrestrial temperate areas.

### Test Yourself

**1.** c, **2.** a, **3.** c, **4.** a, **5.** e, **6.** d, **7.** b, **8.** e, **9.** b, **10.** b

### Data Analysis

As the intensity of poaching increases, the density of many mammals, including white-faced and howler monkeys, decreases. As seed dispersers become rare in areas with high poaching, the proportion of seeds dispersed away from the parent tree decreases (panel c). Without dispersal, the density of subsequent seedlings under the parent tree increases (panel d). Because of the mutualism between seed dispersers and seed density, any change in the density of one mutualist strongly affects the density of the other. (Data from Wright et al., 2000.)

### Ecological Inquiry

**Figure 12.3** Cages might shade the plants or deter the pollinators by their mere presence. To control for such effects, there were always cages present differing only in mesh size.

**Figure 12.5** Facultative; both ants and aphids can live without the other.

**Table 12.2** From top to bottom: I, O, P & I, M, O, I, I, I, I, I, O, O, O, O, P.

## Chapter 13

### Check Your Understanding

**13.1** If the predator keys in on one color form, the prey could proliferate via the other morph. Once this morph becomes common, the predator might switch. The prey are able to proliferate via the less-preferred morph.

**13.2** Prey refuges tend to stabilize predator-prey dynamics within the context of predator-prey oscillations. At low prey densities, all prey would be safe within refuges. Predator density would be low. As prey density increases and prey individuals become more common, more individuals live outside of refuges or preferred habitats. Predators gain access to these individuals and predator numbers increase. This drives prey densities down and the cycle begins again. However, the interaction is stable in that neither prey nor predator is likely to go extinct.

**13.3** Most predators are polyphagous, that is, they feed on more than one prey species. Predator populations may be sustained by populations of common prey. Predators can then become common enough to threaten populations of rare prey.

**13.4** Because changes in habitat use and increased availability of food, such as agricultural crops, may also cause large increases.

**13.5** Maximum sustainable yield represents the number of individuals that can be removed from a population without affecting population growth. This is rather like removing the interest from a bank account and not touching the principal. Maximum sustainable yield occurs at the steepest point of the growth curve, which is at the midpoint of the logistic curve.

### Test Yourself

**1.** d, **2.** a, **3.** c, **4.** a, **5.** c, **6.** c, **7.** c, **8.** b, **9.** a, **10.** a

### Data Analysis

Attacks by predators more often fail than succeed. Where prey are easily caught, substandard individuals are not often taken. Where prey are more difficult to catch, weak, sick, or injured animals are more often taken.

### Ecological Inquiry

**Figure 13.2** Yes, for example, skunks are able to squirt foul-smelling liquids.

**Figure 13.4** Batesian mimicry, since the fly is completely harmless.

**Figure 13.10** Because the birds evolved in the absence of snakes and have few defenses against them.

**Figure 13.14** It is possible that the contiguous continents of Eurasia, Africa, and the Americas shared predator types, which may render prey less naive to introduced predators. Australia never had placental carnivores, and these predators may use different tracking and hunting techniques than native marsupial carnivores. Because many marsupial carnivores are now extinct, this is difficult to test.

## Chapter 14

### Check Your Understanding

**14.1** Roots and leaves, which are of equal importance to plants, are equally chemically defended, supporting the optimal defense hypothesis.

**14.2** Think back to Chapter 13. Monarch caterpillars sequester poisonous cardiac glycosides from their milkweed hosts and advertise their toxicity to predators with striking colors.

**14.3** Short-lived leaves are less tough than long-lived leaves and tend to have lesser amounts of secondary metabolites. Together with higher N levels, this leads to high herbivory. Long-lived leaves have to be well defended against herbivores, and they have high levels of secondary metabolites.

### Test Yourself

**1.** a, **2.** b, **3.** d, **4.** d, **5.** d, **6.** b, **7.** a, **8.** a, **9.** a, **10.** b

### Data Analysis

1. Induced defenses. Tannin concentration was greatest in grazed algae.
2. Support for the optimal defense hypothesis. In this case basal shoots support all the vegetative and reproductive tissues, and tannins increased 55% in grazed basal tissues over ungrazed basal tissues. Apical shoots are less valuable, and grazing increases tannin content less in this tissue.

### Ecological Inquiry

**Figure 14.4** Qualitative, they are present only in small amounts in the plant.

**Figure 14.11** Flies and aphids.

## Chapter 15

### Check Your Understanding

**15.1** Cats are the next host in the parasite's life cycle. The parasite acts an enslaver, changing the behavior of the rats, to make them easier prey for the cats.

**15.2** It is possible that by removing parasites from a neighbor, an individual may be reducing the likelihood of the parasite spreading.

**15.3** Vaccinate local dogs. A comprehensive program of free vaccination for pet dogs has been offered to local residents, and collars designate whether a dog has been vaccinated.

**15.4** Mosquitoes feed only on live rabbits, so killing the host quickly means there was only a short time when the mosquito could transmit the virus. Rabbit fleas were introduced later as another vector of the disease.

**15.5** Habitat fragmentation reduced the number of fruit trees available for fruit bats to roost in. Eventually the bats began sharing trees with horses, which used them for shade. The virus leapt first from bats to horses, then from horses to humans.

### Test Yourself

**1.** d, **2.** c **3.** e, **4.** a, **5.** d, **6.** c, **7.** d, **8.** c, **9.** e, **10.** a

### Data Analysis

Brucellosis prevalence increased with increasing herd size, hence the positive slope to the data. The researchers were able to calculate the threshold density, the minimum herd size needed to maintain the disease, at about 200 individuals, and they drew a vertical line to represent this point. The point to the left of this line is somewhat of an anomaly. The herd size was less than 200 but individuals exhibited high disease prevalence. The researchers explained this point by associational susceptibility. A large neighboring elk population supported brucellosis and continually infected the bison, despite their relatively small herd size.

### Ecological Inquiry

**Figure 15.1** The bont ticks are ectoparasitic macroparasites; the parasitoids are endoparasitic macroparasites that only emerge from their host's body to pupate.

**Figure 15.8** Cooler weather reduces the development rate of the flies and they are exposed longer to searching parasitoids. Parasitism increases to high levels on both treatments by the end of the season and gall numbers are reduced.

**Figure 15.B** Geometric growth.

## Chapter 16

### Check Your Understanding

**16.1** If you take away the plants, the bottom-up effect is so strong that nothing of the system will remain. If you take away natural enemies, herbivores and plants might still exist.

**16.2** This research supports Oksanen's ecosystem exploitation hypothesis. Removing secondary carnivores permits primary carnivores to increase; this increases herbivore densities and decreases plant biomass. Such examples remind us that great care is needed in adding predators, such as biological control agents to agricultural systems. If the added predator also feeds on existing predators, causing what is known as intraguild predation, herbivore numbers might actually increase.

**16.3** The 500 hatchlings now develop into 500 juveniles. Ninety percent are still lost to fishing nets, leaving only 50 turtles that survive to become adults. This equates to 95 percent mortality. Indispensable mortality at the hatchling stage is $99 - 95 = 4\%$.

### Test Yourself

**1.** e, **2.** d, **3.** c, **4.** b, **5.** b, **6.** a, **7.** d, **8.** e, **9.** b, **10.** b

### Data Analysis

1. See figure. The key factor is $k_{oa}$, mortality caused from reduction in the numbers of eggs laid. $k_4$ and $k_1$ are substantial, but not key factors. $k_{ob}$, $k_2$, and $k_3$ are not important and are grouped together at the bottom of the figure.

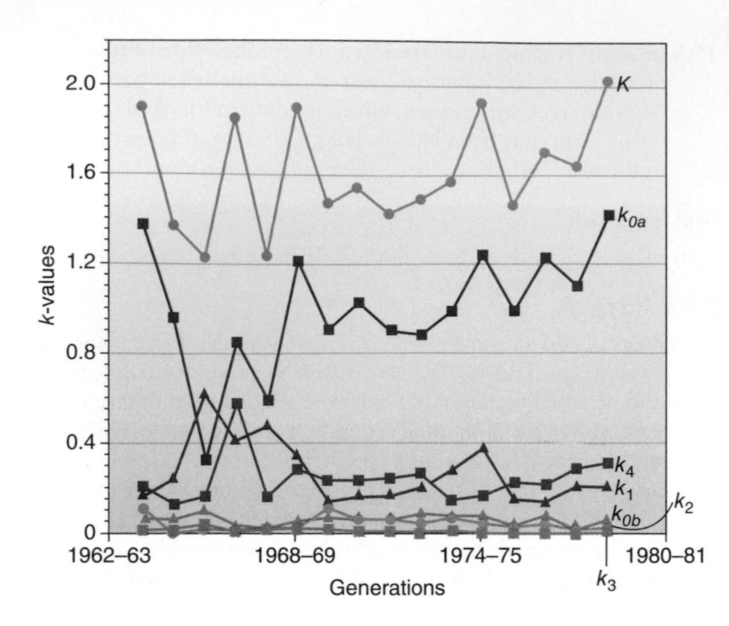


2. The results showed that, despite lowered egg parasitism, insect densities were not significantly changed by parasitoid removal and so top-down control was minimal. However, fertilization increased insect densities, showing that bottom-up control in this system was strong. The fertilizer effects were stronger in *Spartina,* perhaps indicating that nitrogen limitation was more pronounced in this system.

### Ecological Inquiry

**Figure 16.1** An indirect effect, because the natural enemies act directly on the herbivores, not on the plants.

## Chapter 17

### Check Your Understanding

**17.1** Each species is distributed according to its individual needs, and there are few, tightly knit communities with mutually dependent species.

**17.2** If we reconsider Stuart Marsden's data from the logged forests of Indonesia, we find that the unlogged areas have an effective number of species of 9.816 compared to 7.668 for the logged areas, a difference of 28.0%. This is greater than the difference for the values of the Shannon diversity index.

**17.3** Because an abundance of a single environmental factor, such as oil, high pH, or a soil toxin, is likely to kill a large number of species. Only a relatively few species can tolerate such conditions, and these species tend to dominate the community, leading to a good fit to the dominance-preemption model.

**17.4** The Sorenson index weights matches in species composition twice as much as the Jaccard index.

### Test Yourself

1. e, **2.** b, **3.** a, **4.** d, **5.** a, **6.** b, **7.** c, **8.** b, **9.** b, **10.** e

### Data Analysis

Shannon indices = 2.303 and 1.667; effective number of species = 10.00 and 5.296.

### Ecological Inquiry

**Figure 17.1** Scenario 3, since the physical boundaries in soil conditions are sharp and different groups of species would be adapted to different soil types.

**Table 17.1** 1.677

**Figure 17.8** The random fraction model, since the rare species on the right-hand portion of Figure 17.5b would not be sampled.

## Chapter 18

### Check Your Understanding

**18.1** Species richness is affected by evolutionary speed of species formation and rate of extinction. If species formation is high, but extinction rates are also high, then richness levels will not be increased.

**18.2** No, the Great Lakes have a huge combined area yet support fewer fish species than smaller tropical lakes.

**18.3** Yes, Brazil has the largest number of ant species and has a huge area; Alaska also has a huge area but only has 7 ant species. However, Alaska has a much lower productivity than Brazil, which is mainly tropical.

**18.4** Yes. At low levels of grazing, some algal species outcompete the others and dominate. At moderate levels of grazing, these competitive dominants are eaten, permitting other species to survive. At the highest grazing levels, few species can survive. (After Lubchenko, 1978.)

**18.5** There is a general increase in the richness of all species from the poles to the equator. However, there are other trends occurring simultaneously. First, trees need large amounts of rainfall, so species richness is greatest in the wettest areas, the Southeast. Second, richness increases with habitat diversity. The Southwest is mountainous and provides many different habitats for organisms which aren't as water-limited, mammals and birds. This explains their high richness in the Southwest.

**18.6** Because leaves are the softest part of the plant and are easier to chew or bore into. Alternatively, they offer the biggest area to feed on compared to other parts of the plant.

**18.7** Tropical rain forests, because these are the richest in species.

### Test Yourself

1. d, **2.** a, **3.** b, **4.** c, **5.** c, **6.** b, **7.** b, **8.** b, **9.** e, **10.** b

### Data Analysis

Areas highest in coral species have the highest ocean temperatures and highest coral biomass. This supports the species-energy hypothesis. However, species richness is much lower in the Atlantic Ocean and Caribbean Sea than in the Pacific Ocean, suggesting that evolutionary processes are also important. Caribbean coral reefs were affected by Northern Hemisphere glaciations that reduced water temperatures; Pacific coral reefs were not affected. Finally, disturbances in the form of tropical cyclones are more frequent in the Pacific than in the Atlantic and Caribbean.

### Ecological Inquiry

**Figure 18.1** Bird species diversity is usually high in mountainous areas because mountain areas contain a variety of habitats, from low valleys to high mountains, each occupied by different bird species.

**Figure 18.5** Much of what we learned in Chapter 14 related to how defensive chemicals influenced herbivory. This graph shows that a tree's range influences species richness, regardless of defensive chemicals. However, population abundance of individual herbivore species may be influenced by the presence of secondary metabolites.

## Chapter 19

### Check Your Understanding

**19.1** First, the results could have been caused by a sampling effect, not species complementarity. Community performance was not compared to performance of monocultures. Second, it is possible, though perhaps unlikely, that a particular combination of species caused the high species richness treatment to perform well. This differs from Tilman's studies, where many different random draws of species were used to create the different species richness treatments.

**19.2** Greater native species richness reduced the biomass of invading *C. tectorum*. High native species richness increased crowding, decreased available nutrients, and decreased available light, all of which reduced *C. tectorum* biomass. The results support the biotic resistance hypothesis.

### Test Yourself

**1.** b, **2.** e, **3.** e, **4.** b, **5.** b, **6.** e, **7.** b, **8.** d, **9.** b, **10.** a

### Data Analysis

First, there is a species richness effect. Plant biomass increases as the number of species increases. Second, this effect is not simply a sampling effect. The highest species number, 16, outperforms the best monoculture, and numerous replicates in lower species richness treatments also outperform it. Therefore, the effect is likely due to species complementarity.

### Ecological Inquiry

**Figure 19.2** The redundancy hypothesis.

**Figure 19.3** As we increase species richness, we also increase the likelihood of including a "superspecies" in our treatment, which by itself greatly increases community function. We have to ensure that species richness itself, and not the inclusion of a "superspecies," is responsible for increases in community function.

**Figure 19.7** No, it is possible, though unlikely, that some bird species go extinct and are replaced by exactly the same number of new species in different years.

## Chapter 20

### Check Your Understanding

**20.1** Facilitation. *Calluna* litter enriches the soil with nitrogen, facilitating the growth of the grasses. Adding fertilizer also increases soil nitrogen.

**20.2** Humans can suppress fire, which may promote the expansion of one or two tree species in place of a higher number of grass species. The introduction of large grazing mammals may great increase species richness, as competitive dominants may be preferentially eaten. Alternatively, grazers may decrease richness by promoting the germination of tree species after their seeds pass through the animals' guts. Finally, humans may directly increase species richness by inadvertently transporting seeds.

**20.3** In the early stages of succession, nutrient levels are low and competition for nutrients is intense, which selects for large root systems. Later on in succession, nutrient levels are higher but competition for light increases, favoring species that grow tall to intercept light. David Tilman (1985) termed this the resource ratio hypothesis.

### Test Yourself

**1.** d, **2.** a, **3.** b, **4.** c, **5.** a, **6.** e, **7.** e, **8.** a, **9.** b, **10.** b

### Data Analysis

Animal-dispersed plants accumulate more slowly than wind- and sea-dispersed plants. This is because successional change has to occur to provide habitat for the animal transporters, namely birds.

### Ecological Inquiry

**Figure 20.2** Many different types of periodic disturbances could reset succession, from fires or drought on land, to floods or pollution in aquatic environments.

**Figure 20.3** Because William Cooper established permanent study plots there in 1916, almost a 100 years ago.

**Figure 20.7** Competition and facilitation feature equally. Early species facilitate the entry of later species, but late-arriving species outcompete early species.

**Figure 20.9** Quite appropriate but not 100% accurate. A normal jigsaw only has one complete picture or end point, whereas in community succession, various different pictures or end points may result, depending upon the order of species arrival.

## Chapter 21

### Check Your Understanding

**21.1** New Guinea, Java and Sri Lanka have more bird species than expected based on area. Sri Lanka has more bird species than expected because it is close to mainland India and New Guinea may be enriched by the Australian mainland.

**21.2** If $A = 1000$, then $S = 10 \times 1000^{0.3} = 79.43$
If $A = 500$, then $S = 10 \times 500^{0.3} = 64.51$

$$\text{Species loss} = \frac{79.43 - 64.51}{79.43} \times 100 = 18.78\%$$

### Test Yourself

**1.** c, **2.** c, **3.** c, **4.** e, **5.** b, **6.** c, **7.** e, **8.** b, **9.** a, **10.** b

### Data Analysis

The log area of a lake predicts its number of species, but habitat diversity, in this case vegetational diversity, predicts it equally as well. (Data from Tonn and Magnuson, 1982.)

### Ecological Inquiry

**Figure 21.5** From approximately log 1.1 to log 1.6, or 12 to 40 species, a difference of 28 species.

**Figure 21.6** Because birds are better dispersers than mammals, and many species can be found even on relatively small "sky islands" as they pass from one larger mountaintop to another.

## Chapter 22

### Check Your Understanding

**22.1** Because of the effects of the strong Hadley cell and weaker Ferrell cell. Together their actions causes a subsidence zone where hot, dry air falls back to Earth.

**22.2** As temperatures rise, species distributions move up mountainsides. Species that occur at the top of the mountain have nowhere to go, and if they cannot tolerate the new environmental conditions, they may go extinct. Populations of butterflies and frogs living on mountainsides have shown range retractions and even extinctions in recent years.

### Test Yourself

**1.** a, **2.** d, **3.** b, **4.** a, **5.** b, **6.** b, **7.** c, **8.** e, **9.** b, **10.** b

### Ecological Inquiry

**Figure 22.2** Temperature increases linearly with latitude but only up to the Tropics. After that, cloud cover and rainfall result in similar temperatures throughout the Tropics.

**Figure 22.13** Soil conditions.

**Figure 22.17** Because shallow tropical soils do not permit deep roots. Therefore, many large trees rely on buttress roots to stabilize them during storms.

**Figure 22.18** As well as the abundance of water, the frequency of fire and degree of herbivory can affect the abundance of acacia trees.

**Figure 22.27** Because warm temperatures allow these ectotherms to maintain a warm body temperature. Being "cold-blooded," they also require less food where a mammal of comparable size would starve.

**Figure 22.30** The Walter diagrams illustrate how precipitation increases across a west–east gradient in the Andes. The gradient occurs from the leeward, rain-shadow areas of Chile, through the high mountains of Bolivia and into the windward, lower mountains of Bolivia.

## Chapter 23

### Check Your Understanding

**23.1** Winds that move water away from the continental land mass cause upwellings close to continents. This means that water is replaced by deeper water, which moves upward, often bringing nutrients from the ocean floor and increasing productivity.

**23.2** The sun may be much bigger than the moon, but it is also much farther away from the Earth, so its effects on tides are much less than those of the smaller but closer moon.

**23.3** Water temperature and substrate type. Where the substrate is hard, coral reefs grow in warm water and kelp forests in cooler water. Where the substrate is sandy, sea grasses occur in both tropical and temperate conditions.

### Test Yourself

**1.** b, **2.** e, **3.** e, **4.** d, **5.** c, **6.** a, **7.** b, **8.** b, **9.** a, **10.** a

### Ecological Inquiry

**Figure 23.1** Where cold currents abut warm terrestrial areas, such as along the western shorelines of North and South America.

**Figure 23.5** The water becomes turbulent and the Langmuir cells disappear.

**Figure 23.7** 14.

**Figure 23.8** 24 hours and 50 minutes.

**Figure 23.23** Mangroves are susceptible to freezing temperatures, which often occur north or south of this zone.

## Chapter 24

### Check Your Understanding

**24.1** Ice would sink and the lake would eventually freeze solid. All aquatic life would freeze, and biodiversity in temperate freshwater biomes would plummet.

**24.2** Swamps are dominated by woody plants, whereas marshes support mainly soft-stemmed grassy vegetation.

### Test Yourself

**1.** d, **2.** d, **3.** c, **4.** b, **5.** a, **6.** d, **7.** b, **8.** e, **9.** b, **10.** b

### Ecological Inquiry

**Figure 24.2** There is less seasonal mixing of water in tropical lakes. While the upper layers may be productive, the lower layers are often oxygen-poor and fishless.

**Figure 24.6** Lakes are common in previously glaciated areas that have depressions scoured into the rocks. Canada is the largest previously glaciated landmass on Earth.

## Chapter 25

### Check Your Understanding

**25.1** Carrion beetles are detritivores. They feed on dead animals, such as mice, at trophic level 3 or 4. Mice generally feed on vegetative material (trophic level 1) or crawling arthropods (trophic level 2), so mice themselves feed at trophic level 2 or 3.

**25.2** There are 12 "species," so the number of potential links is 66. The number of links is 20. The connectance is 20/66 = 0.303. The linkage density is 20/12 = 1.67.

**25.3** Humans, multiple effects; large tropical ungulates, mud wallows where rainwater collects and mosquitoes breed; tropical vines, connect trees providing walkways for monkeys; prairie dogs, burrows are used by burrowing owls and plant species grow on the entrance mounds; *Spartina* grass, reduces velocity of tidal flow, allowing other saltmarsh plants to colonize.

### Test Yourself

**1.** e, **2.** c, **3.** c, **4.** c, **5.** d, **6.** d, **7.** b, **8.** b, **9.** e, **10.** d

### Data Analysis

Beavers are ecosystem engineers. They create deep pools in Chilean streams by cutting down surrounding trees, thus reducing canopy cover. The beaver dams reduce water flow and permit increased retention of organic matter. This increased organic matter permits large increases in the number of benthic macroinvertebrates. At the same time, the reduced flow permits large volumes of small organic matter to cover the stream bottom so that relative coverage by sand, gravel, and cobbles decreases. A reduced diversity of substrate types in turn reduces the richness and diversity of benthic macroinvertebrates. Some changes may be expected as we progress from forested areas of streams to lower areas. This is why beaver dam areas are compared to both upstream and downstream sites. (Data from Anderson and Rosemond, 2007.)

### Ecological Inquiry

**Figure 25.2** It depends on the trophic level of their food, whether dead vegetation or dead animals. Many decomposers feed at multiple trophic levels.

**Figure 25.8** 2%

**Figure 25.13** We have discussed many keystone natural enemies throughout this text, where absence results in dramatic community changes. Examples include wolves, sharks, and many invasive species such as *Cactoblastic cactorum* and diseases such as rinderpest and chestnut blight.

## Chapter 26

### Check Your Understanding

**26.1** Nutrient availability. Though tropical biomes have warm temperatures and abundant water, they are often nutrient-depleted. High rainfall leaches out soil nutrients. Temperate forests occur on relatively young soils with higher nutrient concentrations. Temperate marine systems occur in well-mixed water, with high nutrient concentrations. Tropical waters usually have low nutrient levels, except in areas of coastal upwellings.

**26.2** Because high temperature and moisture levels increase arthropod and microbial activity. Arthropods and microbes are prominent in litter decomposition. A reduction of such activity in temperate winters allows a thicker leaf layer to accumulate in temperate forests than in tropical rain forests, where decomposition occurs year round.

**26.3** Soil nutrient levels are fairly high in the young, often glaciated soils of temperate North America, Europe, and Northern Asia; this promotes crop growth. However, the older, unglaciated soils of the Southern Hemisphere are nutrient-poor, especially in tropical areas where heavy rainfall leaches out nutrients. As a result, crop growth in the Southern Hemisphere is often low.

### Test Yourself

**1.** d, **2.** e, **3.** d, **4.** a, **5.** e, **6.** a, **7.** e, **8.** a, **9.** b, **10.** b

### Data Analysis

First, isopods change their feeding rate depending on whether they did or did not have a choice among litter of the three species. They ate similar amounts of all species when grown in monocultures but showed distinct preferences in mixtures. Second, relative preferences changed when the litter was grown under elevated $CO_2$, with *Acer* more strongly preferred. This behavior likely has consequences for decomposition rates of litter and nutrient cycling. Litter-feeding fauna have strong and important effects on ecosystems because of the huge amounts of litter they process. (Data from Hättenschwiler and Bretscher, 2001.)

### Ecological Inquiry

**Figure 26.3** The prairies stretch from a rain shadow just east of the Rocky Mountains through relatively wet areas such as eastern Illinois. Though these sites may be of relatively similar latitude, the effects of the Rocky Mountains on rainfall generates much difference in water availability.

**Figure 26.13** On a population level, plant secondary metabolites can deter generalist herbivores from feeding on individual plant species. However, on an ecosystem level these effects are not as important, because higher primary production tends to result in higher secondary production.

## Chapter 27

### Check Your Understanding

**27.1** A unit of energy passes through a food web only once and energy is lost at each transfer between trophic levels. In contrast, chemicals cycle repeatedly through food webs and may become more concentrated at higher trophic levels.

**27.2** Plants take up phosphorus rapidly and efficiently, often reducing soil or water phosphorus concentrations to very low levels, so that it becomes limiting to production.

**27.3** First, carbon dioxide is a limiting nutrient, and increasing atmospheric carbon dioxide will directly increase crop growth. Second, elevated carbon dioxide decreases plant nitrogen and depresses herbivory, which again increases crop growth. Third, in many areas of the world, global warming will increase rainfall, which will also boost crop yields.

**27.4** Nitrogen molecules have a triple bond, making them hard to break apart. Only a few species of bacteria can break apart atmospheric nitrogen or fix nitrogen. The most common of these bacteria, in terrestrial systems, are *Rhizobium,* which live in the roots of legumes and some other plants. The excess ammonia, $NH_3$, or ammonium, $NH_4^+$, gradually accumulates and can be used by plants.

**27.5** Acid rain is damaging to terrestrial systems because it damages plant roots, stops nitrifying bacteria from functioning properly, and prevents organic matter from decomposing.

**27.6** The water cycle is different from other nutrient cycles because very little of the water is chemically changed as it cycles through ecosytems. Also, the water cycle is a physical process fueled by the sun's energy, rather than a chemical process.

### Test Yourself

**1.** b, **2.** c, **3.** c, **4.** b, **5.** c, **6.** a, **7.** d, **8.** e, **9.** b, **10.** b

### Data Analysis

Legumes fix nitrogen, so in nitrogen-enriched plots they have no competitive advantage and do not increase in abundance. In ambient N plots, legumes are at an advantage and they increase in relative abundance. At this site, legumes show no response to elevated $CO_2$ and they make up similar fractions of the community at both ambient and elevated $CO_2$.

### Ecological Inquiry

**Figure 27.7** In rocks and fossil fuels.

**Figure 27.11** Because prevailing westerly winds carried acid rain from the industrial areas of the Midwest, where it was produced, to areas of the U.S. Northeast.

# Glossary

## A

**abiotic interactions** Interactions between living organisms and their physical environment.
**abyssal zone** The zone of the open ocean that reaches from 4,000 m to 6,000 m.
**acclimation** Changes by an organism subjected to new environmental conditions that enable it to withstand those conditions.
**acidic** A solution that has a pH below 7.
**acid rain** Rainfall acidified by contact with sulfur dioxide and nitrogen oxide (a by-product of the burning of fossil fuels) in the atmosphere; precipitation with a pH of less than 5.6.
**acids** Molecules that release hydrogen ions in solution.
**acquired characteristics** Nonhereditary changes of function or structure made in response to the environment.
**adaptation** The process and structures by which organisms adjust to changes in their environment.
**adaptation hypothesis** The idea that invasive species are already preadapted to conditions in their new environment.
**additive mortality** Mortality that occurs in addition to other existing mortality; for example, hunting could act in an additive way.
**adiabatic cooling** The cooling of air that results as air is blown up over mountains and pressure is lessened.
**age class** Individuals in a population of a particular age.
**age-specific fertility rate** The rate of offspring production for females of a certain age.
**age structure** The relative numbers of individuals in each age class.
**aggressive mimicry** The mimicry of harmless models by predators, which enables them to get close to prey.
**alkaline** A solution that has a pH above 7.
**allele** One of two or more alternative forms of a gene located at a single point (locus) on a chromosome.
**allele frequencies** The number of copies of a particular allele in a population divided by the total number of alleles in that population.
**allelochemicals** A substance produced by one organism that affects the growth and behavior of another species.
**allelopathy** The negative chemical influence of plants upon one another.
**Allen's rule** The hypothesis that vertebrates living in cold environments tend to have shorter appendages than those living in warmer environments.
**allopatric** Occurring in different geographic areas.
**allopatric speciation** A form of speciation whereby a population becomes separated into two or more evolutionary units as a result of geographic separation.
**altruism** Enhancement of the fitness of a recipient individual by acts that reduce the evolutionary fitness of the donor individual.
**amensalism** Asymmetric competition where one species has a large effect on the other, but the other species has little effect on the first.
**ammonification** The conversion of organic nitrogen to $NH_3$ and $NH_4^+$.
**annual** An organism, usually a plant, that completes its life cycle, from birth through reproduction to death, in a year.
**anoxic** Low in dissolved oxygen and unable to support most life.
**aphotic zone** The zone of a body of water where sunlight fails to penetrate, generally below 200 m.
**aposematic coloration** Coloration that warns of a toxic nature.
**apparent competition** The reduction of one species caused by a natural enemy of a different species.
**apparent plants** Usually long-lived plants that are easy for herbivores to locate; for example, trees.
**aquifers** Underground water supplies.
**assimilation** The process by which inorganic substances are incorporated into organic molecules (in living things).
**assimilation efficiency** The percentage of energy ingested in food that is assimilated into the protoplasm of an organism.
**assisted migration** Human-assisted movement of species outside of their native ranges in order to promote their survival in a globally changed world.
**associational resistance** The protection of one species by its close association with unpalatable neighbors.
**associational susceptibility** Where herbivores attacking one host plant species spill over and attack neighboring species.
**autotroph** An organism that obtains energy from the sun and materials from inorganic sources; contrast with "heterotroph."

## B

**balanced polymorphism** The phenomenon in which two or more alleles are kept in balance, and therefore are maintained in a population over the course of many generations.
**balancing selection** The pattern of natural selection that maintains genetic diversity in a population.
**base** A molecule that when dissolved in water lowers the hydrogen ion concentration.
**Batesian mimicry** Resemblance of an edible (mimic) species to an unpalatable (model) species to deceive predators.
**bathypelagic zone** The zone of the open ocean that reaches from 1,000 m to 4,000 m.
**behavior** The observable response of organisms to stimuli.
**behavioral ecology** The study of how the behavior of an individual contributes to its survival and reproductive success.
**benthic** Pertaining to aquatic bottom or sediment habitats.
**benthic invertebrates** Invertebrate animals that live on the bottom of lakes, rivers, or marine environments.

**Bergmann's rule** (1847) Among homeotherms, the tendency for organisms in colder climates to have larger body size (and thus smaller surface-to-volume ratio) than those in warm climates.

**biochemical oxygen demand** (BOD) The amount of oxygen that would be consumed if all the organic substances in a given volume of water were oxidized by bacteria and other organisms; reported in milligrams per liter.

**biodegradable** Capable of being decomposed quickly by the action of microorganisms.

**biodiversity** Biological diversity, including genetic diversity, species diversity, and ecosystem diversity.

**biodiversity crisis** The elevated extinction rates of species caused by human activities.

**biodiversity hot spots** Those areas with greatest number of endemic species.

**biogeochemical cycle** The passage of a chemical element (such as nitrogen, carbon, or sulfur) from the environment into organic substances and back into the environment.

**biogeographic realms** Large-scale divisions of the Earth's surface based on amount and historic distribution patterns of organisms.

**biogeography** The branch of ecology that deals with the geographic distribution of plants and animals.

**biological control** Use of natural enemies (diseases, parasites, predators) to regulate populations of pest species.

**biological species concept** The concept that a species consists of groups of populations that can interbreed with each other but that are reproductively isolated from other such groups.

**biomagnification** The concentration of a substance as it "moves up" the food chain from consumer to consumer.

**biomass** Dry weight of living material in all or part of an organism, population, or community; commonly expressed as weight per unit area.

**biome** Originally defined as a major terrestrial climax community, for example, coniferous forest, tundra. Now broadened to include freshwater and marine systems.

**biophilia** The idea that humans have an innate love of life; a term coined by E. O. Wilson.

**bioremediation** The use of living organisms to detoxify polluted habitats.

**biosphere** The whole Earth ecosystem.

**biotic interactions** Interactions between living organisms.

**biotic resistance hypothesis** The idea that species-rich communities are more resistant to invasion that species-poor communities.

**boreal** Occurring in the temperate and subtemperate zones of the Northern Hemisphere.

# C

**canopy** The uppermost layer of tree foliage.

**carbon-nitrogen balance hypothesis** The idea that the allocation of carbon and nitrogen to plant defenses are dependent on their availability in the environment.

**carnivore** An animal (or plant) that eats other animals; contrast with "herbivore."

**carrying capacity** *(K)* The number of individuals that the resources of a habitat can support.

**character** A visible characteristic, such as a pea plant's seed color or pod texture.

**character displacement** Divergence in the characteristics of two otherwise similar species where their ranges overlap; caused by competition between the species in the area of overlap.

**chemical alteration** The first phase of decomposition where fungi and bacteria chemically change dead organic matter.

**chronosequence** The change in community structure primarily influenced by time such that older communities appear different to younger communities.

**climate** The prevailing weather pattern in a given area.

**climax community** The community capable of indefinite self-perpetuation under given climatic conditions.

**clumped** A spatial dispersion pattern where individuals are clustered in certain areas, especially around resources.

**coastal upwelling** The upwelling of deeper water near the coasts of continents.

**coefficient of relatedness** The probability that any two individuals will share a copy of a particular gene.

**cohort** Those members of a population that are of the same age, usually in years or generations.

**cohort life table** A life table that follows a cohort of individuals from birth to death.

**colonization hypothesis** The idea that seed dispersal is advantageous to plants because parental locations are not always ideal for seed germination.

**commensalism** An association between two organisms in which one benefits and the other is not affected.

**community** An assemblage of microbes, plants, and animals in a given place; used in a broad sense to refer to ecological units of various sizes and degrees of integration.

**community ecology** The branch of ecology that investigates why communities contain different numbers of species.

**compensation point** The depth within a lake or other body of water at which the photosynthate production equals the energy used up by respiration.

**compensatory mortality** Mortality that increases or decreases following changes in other mortality factors operating on a population; for example, hunting could act in a compensatory way.

**competition** The interaction that occurs when organisms of the same or different species use a common resource that is in short supply ("exploitation" competition) or when they harm one another in seeking a common resource ("interference" competition).

**competition avoidance hypothesis** The idea that seed dispersal is advantageous to plants because competition between seedlings and parent plants is avoided.

**competitive exclusion principle** The hypothesis that two or more species cannot coexist and use a single resource.

**conduction** The process in which the body surface loses or gains heat through direct contact with cooler or warmer substances.

**connectance** In a food web, the actual number of links divided by the potential number of links.

**connectedness webs** Food webs detailing all the possible known feeding relationships between organisms in a community.

**conservation biology** The study that uses principles and knowledge from molecular biology, genetics, and ecology to protect the biological diversity of life at all levels.

**conspecific** Belonging to the same species.

**constitutive defenses** Plant defenses that are always present.

**consumption efficiency** The percentage of energy at one trophic level that is eaten by the trophic level higher up.

**continental drift** The movement of the continents, by tectonic processes operating over millions of years, from their original positions as parts of a common landmass to their present locations.

**convection** The transfer of heat by the movement of air or water next to the body.

**convergent evolution** The development of similar adaptations by genetically unrelated species, usually under the influence of similar environmental conditions.

**core-satellite** Metapopulations based on the existence of one or more large extinction-resistant populations that supply colonists to peripheral satellite patches.

**Coriolis force** The effect of the Earth's rotation on the surface flow of wind.

**correlation** The strength and direction of a linear relationship between two variables.

**countercurrent heat exchange** A method of regulating heat loss to the environment by keeping the body core much warmer than the extremities.

**coupled oscillations** An endless cycle of predator and prey densities, driven by a delay in the response of the predators to the prey.

**cross-fertilization** The fusion of gametes from different individuals.

**cryptic coloration** Coloration or appearance that tends to prevent detection of an organism by predators.

**cultural eutrophication** Human-induced eutrophication of water bodies.

# D

**dead zone** A hypxic (low-oxygen) area in an ocean or large lake caused by excessive nutrient pollution and subsequent algal growth and decomposition, which depletes oxygen levels.

**decomposition** The physical and chemical breakdown of detritus.

**decomposition constant, *k*** An exponent that characterizes the decomposition rate under specified conditions.

**defensive mutualism** Mutualism involving defense of one species by another.

**definitive host** The host in which macroparasites exhibit sexual reproduction.

**deforestation** Removal of trees from natural forests and woodlands.

**degree-days** A combination of time at a certain temperature; determines organisms' development rates.

**demographic transition** The shift in birth and death rates accompanying human societal development.

**demography** The study of birth rates, death rates, age distributions, and the sizes of populations.

**denitrification** Enzymatic reduction by bacteria of nitrates to nitrogen gas.

**density-dependent factor** A mortality factor whose influence on a population varies with the number of individuals per unit area in the population.

**density-independent factor** A mortality factor whose influence on a population does not vary with the number of individuals per unit area in the population.

**desertification** The overstocking of land with domestic animals that can greatly reduce grass coverage through overgrazing, turning the area more desert-like.

**detritivores** Organisms that eat detritus.

**detritus** Dead plant and animal material and animal waste products.

**diploid** Having two copies of each gene in a genotype.

**directed dispersal hypothesis** The idea that seed dispersal is advantageous to plants because dispersers are likely to distribute seeds into optimal sites.

**directional selection** The pattern of natural selection that favors individuals of a particular phenotype that have the greatest reproductive success in a particular environment.

**dispersive mutualism** Mutualism between plants and their pollinators or seed dispersers.

**disruptive selection** The pattern of selection that favors the survival of two or more genotypes that produce different phenotypes.

**dissolved oxygen** The amount of oxygen that occurs in microscopic bubbles of gas mixed in with the water and that supports aquatic life.

**diversity index** An index that measures the relative number of species in an area and the distribution of individuals among them.

**diversity-stability hypothesis** Elton's idea that species-rich communities are more likely to be stable than species-poor communities.

**dominant** A term that denotes the displayed trait.

**dominant species** A species that has a large effect in a community because of its high abundance or biomass.

# E

**ecological footprint** The amount of productive land needed to support each person on Earth.

**ecological species concept** The concept that a species is distinct from other species if it occupies a distinct portion of habitat or niche.

**ecology** The study of interactions among organisms and between organisms and their environment.

**ecosystem** A biotic community and its abiotic environment.

**ecosystem ecology** The study of the flow of energy and cycling of nutrients among organisms in a community.

**ecosystem engineer** A keystone species that has a dramatic effect on an ecosystem by modifying habitat; for example, a beaver.

**ecosystem exploitation hypothesis** The idea that the strength of mortality factors in communities varies with plant productivity.

**ectoparasites** Parasites that live on the outside of the host's body.

**ectotherm** An animal that depends on an external heat source to warm itself.

**edge effects** Special physical conditions that exist at the boundary or edge of a habitat.

**effective number of species** The conversion of a species diversity index to an equivalent number of species.

**effective population size** The number of individuals that contribute genes to future populations; often smaller than the number of individuals in a population.

**emigration** The movement of organisms out of a population.

**endangered species** A species with so few living members that it will soon become extinct unless measures are taken to slow its loss.

**endemic** An organism that is native to a particular region.

**endoparasites** Parasites that live inside the host's body.

**endosymbiosis** A close association between species where the smaller species, the symbiont, lives inside the body of the larger species.

**endosymbiosis theory** The idea that eukaryotic mitochondria evolved from aerobic bacteria and that chloroplasts evolved from cyanobacteria, and both took up residence within a primordial eukaryotic cell.

**endotherm** An animal that metabolically generates its own heat.

**enemy release hypothesis** The idea that invasive species are successful because they have been released from their natural enemies.

**energy flow** The movement of energy through an ecosystem.

**energy web** Food webs where the links are drawn in various thicknesses to represent the flow of energy between the species.

**environmental science** The application of ecology to real-world problems; the study of human interactions with the environment.

**environmental stress hypothesis** The idea that the strength of mortality factors in communities is governed by the degree of environmental stress.

**epilimnion** The upper layer of water in a lake, usually warm and containing high levels of dissolved oxygen.

**epipelagic zone** The surface waters of the open ocean, to about 200 m in depth.

**epiphyte** A plant that lives on another plant but uses it only for support, drawing its water and nutrients from natural runoff and the air.

**equilibrium** A condition of balance, such as that between immigration and emigration or birth rates and death rates in a population of fixed size.

**ethology** The scientific study of animal behavior.

**eusociality** Relating to species that possess nonreproducing castes that assist the reproductive individuals.

**eutrophic** Freshwater habitats rich in nutrients and organisms; high productivity; contrast with "oligotrophic."

**eutrophication** The normally slow aging process by which an oligotrophic lake accumulates nutrients, fills with organic matter, becomes more turbid, and turns eutrophic.

**evaporation** The transformation of water from the liquid to the gaseous state.

**evapotranspiration** The sum of the water lost from the land by evaporation and plant transpiration. Potential evapotranspiration is the evapotranspiration that would occur if water were unlimited.

**evenness** The species diversity of a community divided by the maximum possible diversity; a measure of relative diversity.

**evolution** Changes in gene frequencies in a population over time; descent with modification.

**evolution of increased competitive ability hypothesis (EICA)** The idea that loss of natural enemies allows invasive species to devote more resources to competition and less to defense.

**evolutionary ecology** The branch of ecology that examines the environmental factors that drive species adaptation and therefore distribution.

**evolutionary species concept** The concept that a species is distinct from other species if it is derived from a single lineage that is distinct from other lineages.

**evolutionarily stable strategy** A behavioral strategy that cannot be replaced by another behavioral strategy.

**exploitation competition** Competition mediated by use of a common resource.

**exponential growth** Growth of populations with overlapping generations in unlimited resources, which often yields a J-shaped curve.

**extinction** The process by which species die out.

# F

**$F_1$ generation** The first filial generation; the first generation offspring.

**$F_2$ generation** The second filial generation, in genetic crosses, often the offspring of two $F_1$ parents.

**facilitation** Enhancement of a population of one species by another, often during succession, a type of one-way mutualism.

**facultative anaerobes** Species that may or may not use oxygen, depending on its availability.

**facultative mutualism** An interaction between two species that is beneficial to both, but not essential to either.

**fecundity** The potential of an organism to produce living offspring.

**female-enforced monogamy hypothesis** The idea that females actively prevent their male partners from mating with other females.

**Ferrel cell** The middle cell in the three-cell circulation of wind in each hemisphere.

**fertility** The actual reproductive output of a living organism.

**finite rate of increase** The ratio of population size from one time interval to the next.

**Fisher's principle** The principle that explains why a 1:1 sex ratio of males : females occurs in many species.

**fixed nitrogen** Nitrogen combined with another element, for example, ammonia ($NH_3$) or nitrite ($NO_3^-$).

**flagship species** A large or instantly recognizable species, often used to gain support for conservation purposes.

**food chain** A linear depiction of energy flow showing the feeding links of organisms on others, usually in a series beginning with plants and ending with the largest carnivores.

**food web** A complex model of interconnected food chains in which there are multiple links between species.

**fossil fuels** Coal, oil, and natural gas, so-called because they are derived from the fossil remains of ancient plant and animal life.

**fragmentation** (a) Habitat that is broken up into isolated patches. (b) The second phase of decomposition where soil animals break down organic matter into small pieces.

**frameshift mutation** A mutation that involves the addition or deletion of nucleotides that are not in multiples of three.

**functional web** Food webs where the links are drawn in various thicknesses to represent the strength of interactions between species.

**fundamental niche** The optimal range in which a particular species best functions.

# G

**game theory** In behavioral ecology, mathematical formulations of interactions between individuals.

**gene** A unit of genetic information.

**generation time** The time between the birth of a parent and the birth of its offspring.

**genetic drift** Change in gene frequency caused solely by chance, usually unidirectional and more important in small populations.

**genome** The entire genetic complement of an individual.

**genotype** The genetic constitution of an organism, in contrast to its observable characteristics.

**genotype frequencies** In a population, the number of individuals within a given genotype divided by the total number of individuals.

**geometric growth** Growth of populations with discrete generations in unlimited resources; yields a J-shaped curve.

**global warming** A gradual elevation of the Earth's surface temperature caused by the greenhouse effect.

**gradualism** The idea that species evolve continuously over long periods of time; contrast "punctuated equilibrium."

**grassland** A region with sufficient average annual precipitation (25–75 cm [10–30 inches]) to support grass but not trees.

**green Earth hypothesis** The idea that since the Earth appears green, herbivores must have little impact on plant abundance.

**greenhouse effect** The heating effect of the atmosphere upon the Earth, particularly as $CO_2$ concentration rises, caused by its ready admission of light waves but its slower release of the heat they generate on striking the ground.

**gross primary production** Production before respiration losses are subtracted; photosynthetic production for plants.

**group selection** The idea that natural selection produces outcomes that benefit groups rather than individuals.

**guild** A group of organisms that feed on a similar resource in the same manner; for example, leaf chewers, sap suckers, or grain feeders.

## H

**habitat** The sum of the environmental conditions in which an organism, population, or community lives; the place where an organism normally lives; the environment in which the life needs of an organism are supplied.
**hadal zone** The zone of the open ocean that reaches below 6,000 m.
**Hadley cell** In climatology, the convection cell nearest to the equator, named in honor of George Hadley.
**halophytes** Plants adapted to live in saline conditions.
**Hamilton's rule** The idea that an altruistic gene will be favored if $r > C/B$, where $r$ is the coefficient of relatedness, $B$ is the benefit to the recipient of the altruism, and $C$ is the cost incurred to the donor.
**handicap principle** The hypothesis that excessive ornamentation signals high genetic quality because the bearer must be able to afford this energetically costly trait.
**haplodiploidy** The presence of haploid males, which develop from unfertilized eggs, and diploid females, which develop from fertilized eggs, in the same species; for example, in the Hymenoptera.
**haploid** Having one set of genes in a genotype, as do sperm and eggs.
**Hardy-Weinberg equation** An equation ($p^2 + 2pq + q^2 = 1$) that relates allele and genotype frequencies.
**harem** A group of females controlled by one male.
**hemiparasite** A parasitic plant that is partly dependent on its host, for example for water, e.g., mistletoe.
**herbivore** An organism that eats plants; contrast "carnivore."
**heterotherm** An animal that has a body temperature that varies with environmental conditions.
**heterotroph** An organism that obtains energy and materials from other organisms; contrast "autotroph."
**heterozygous** An individual with two different alleles of the same gene.
**holoparasite** A parasitic plant that is wholly dependent on its host; for example, *Rafflesia.*
**homeotherm** An animal that maintains its body temperature within a narrow range.
**homozygous** An individual with two identical copies of an allele.
**host** The organism that furnishes food, shelter, or other benefits to an organism of another species.
**humus** The finely ground organic matter in soil.
**hybridization** Breeding (crossing) of individuals from genetically different strains, populations, or, sometimes, species.
**hydrothermal vents** Deep-water oceanic vents in the sea floor that spew out superheated water and around which distinct animal communities form.
**hypolimnion** The layer of cold, dense water at the bottom of a lake, often with low levels of dissolved oxygen.
**hypothesis** An explanation for a phenomenon.

## I

**ideal weed hypothesis** The idea that invasive species are successful because they are preadapted to existing environmental conditions.
**idiosyncratic hypothesis** The idea that community function and species richness are not linked in a predictable way.
**immigration** The movement of individuals into a population.
**inbreeding** A mating system in which adults mate with relatives more often than would be expected by chance.
**inclusive fitness** The total genetic contribution of an individual to future generations by way of its sons, daughters, and all other relatives such as nieces, nephews, and cousins.
**indicator species** Species whose status provides information on the overall health of an ecosystem; a canary in a coal mine.
**indirect effect** An effect of one species on another that is mediated by a third species; for example, spiders benefit plants by eating insect herbivores.
**individual selection** The idea that traits are selected for because they benefit individuals rather than groups of individuals.
**individualistic model** A view of the nature of a community that considers it to be an assemblage of species coexisting primarily because of similarities in their physiological requirements and tolerances.
**induced defenses** Plant defenses that are only switched on following herbivore attack.
**industrial melanism** Increased amount of black or nearly black pigmentation in response to pollution of the environment.
**inhibition** Type of succession whereby early colonists inhibit establishment of later arriving species.
**inquilinism** The use of one species by a second species for housing or support.
**interference competition** The physical interaction of one species with another, usually in competition over territory or resources.
**intermediate disturbance hypothesis** The proposal that moderately disturbed communities are more diverse than undisturbed or highly disturbed communities.
**intermediate host** One or more species of host in which macroparasites develop but do not undergo sexual reproduction.
**internal fragmentation** Habitat bisected by thin barriers such as fences, power lines, or roads.
**interspecific** Between species; between individuals of different species.
**intertidal zone** The shallow zone where the land meets the sea; usually refers to rocky intertidal, but may be sandy shore, mangrove, or salt marsh also.
**intraspecific** Within species; between individuals of the same species.
**intrinsic rate of increase, $r_{max}$** The maximum rate of increase of a population under ideal conditions.
**invasional meltdown** The idea that invasion of a community by exotic species predisposes the community to further invasion by more exotic species.
**invasive species** Introduced species that are spreading in their new range and often cause harm to native species.
**inverse density-dependent factor** A mortality factor whose influence on a population decreases with increasing population density.
**iteroparity** Being able to breed continuously throughout a lifetime.

## K

***K*-selected species** Species that have a relatively low rate of per capita population growth, *r,* but that exist near the carrying capacity, *K,* of the environment.
**key factor** A mortality factor that mirrors most closely the overall population mortality.
**keystone hypothesis** The idea that most species are vital to the functioning of ecosystems and that function decreases immediately as species richness declines.
**keystone species** A species having a huge effect on a community, out of all proportion to its biomass.
**kin selection** Selection for behavior that lowers an individual's own fitness but promotes the survival of kin who carry the same alleles.
**kleptoparasitism** A type of parasitism involving theft of resources from one individual by another.

## L

**landscape connectivity** In landscape ecology, the extent to which different patches are connected to one another.
**landscape ecology** The study of the influence of large-scale spatial patterns of land use or habitat type on ecological processes.
**Langmuir circulation** Small-scale circulation patterns in aquatic environments caused by wind speeds of between 3 m/s and 13 m/s.
**latitudinal species richness gradient** The change in species richness from the poles to the equator.
**law of segregation** States that two copies of a gene segregate from each other during gamete formation and during transmission from parent to offspring.
**leaching** The first process of decomposition during which soluble materials in organic matter in the soil, such as nutrients, are washed into a lower layer of soil or are dissolved and carried away by water.
**lek** A communal courtship area on which several males display to attract and mate with females.
**lentic** Pertaining to standing freshwater habitats (ponds and lakes).
**Liebig's law of the minimum** The principle that species biomass or abundance is limited by the scarcest factor.
**life history strategies** Sets of physiological and behavioral features that incorporate reproductive and survivorship traits.
**life table** Tabulation presenting complete data on the numbers of individuals of various ages alive in a population.
**limiting factor** The nutrient or substance that is in shortest supply in relation to organisms' demand for it.
**line transect** A long length of string, rope, or tape, along which the number of organisms touching are counted.
**linkage density** In a food web, the average number of links per species.
**logistic equation** An equation governing population growth with limited resources, $dN/dt = rN(K - N)/K$.
**logistic growth** Population growth limited by resources; described by a symmetrical S-shaped curve with an upper asymptote.
**lotic** Pertaining to running freshwater habitats (streams and rivers).
**lower littoral zone** The area of the rocky intertidal zone exposed only during the lowest tide.

## M

**macroparasites** Parasites that live inside the host but do not cause disease.
**mafia hypothesis** The idea that parasitic birds, such as cuckoos or cowbirds, destroy all the eggs in a nest if their own egg has been removed.
**male assistance hypothesis** The idea that females need males to help them raise their offspring.
**Malthusian theory of population** The idea that the Earth is overrun by humans because of food shortage, disease, war, or conscious population control.
**many eyes hypothesis** The idea that living in groups increases the detection of predators while maximizing the time available for the prey to feed.
**mark-recapture** A method of estimating population density by catching, tagging, and recatching animals.
**mate-guarding hypothesis** The idea that monogamy is enforced on females by males who guard them against mating with other males.
**matrix** In landscape ecology, the most extensive element of an area.
**maximum sustainable yield** The largest number of individuals that can be harvested from a population without causing long-term decreases to the population.
**megadiversity country** Those countries with the greatest numbers of species; used in targeting areas for conservation.
**mesopelagic zone** The zone of the open ocean that extends from 200 m to 1,000 m.
**metabiosis** The use of something produced by one organism, usually after its death, by another species.
**metapopulation** A series of small, separate populations that mutually affect one another.
**microclimate** Local variations of the climate within a given area.
**microparasites** Parasites that cause diseases.
**mid-littoral zone** The area of the rocky intertidal zone submerged during the highest regular tide and exposed during the lowest tide each day.
**mimicry** The resemblance of one species (the mimic) to another species (the model).
**monogamy** A mating system in which one male mates with only one female, and one female mates with only one male.
**monohybrids** The $F_1$ generation of two different, true-breeding parents that differ in regard to a single trait.
**monophagous** Feeding on one species or two or three closely related species.
**movement corridors** Thin strips of habitat that permit the movement of species between larger habitat patches.
**Müllerian mimicry** Mutual resemblance of two or more conspicuously marked, distasteful species to reinforce predator avoidance.
**mutant** An organism with a changed characteristic resulting from a genetic change.
**mutation** A heritable change in the genetic material of an organism resulting from a change in its DNA.
**mutualism** An interaction between two species in which both benefit.

## N

**natural selection** The natural process by which the organisms best adapted to their environment survive and reproduce and those less well-adapted are eliminated.
**nekton** Free-swimming marine animals that can swim against the currents.
**neritic zone** Pertaining to the shallow, coastal marine zone.
**net primary production** Production after respiration losses are subtracted; the amount of energy available to herbivores.
**net reproductive rate, $R_0$** The number of offspring a female can be expected to bear during her lifetime; for species with clearly defined discrete generations.
**neutralism** The occurrence together of two species with no interaction between them.
**niche** The role of an organism in an ecosystem; the range of conditions within the environment within which an organism can exist.
**niche overlap** The overlap in resource use by two or more species; sharing of niche space.
**nitrification** The conversion of $NH_3$ or $NH_4^+$ to nitrite, $NO_3^-$, by bacteria.
**nitrogen fixation** A specialized metabolic process in which some prokaryotes use the enzyme nitrogenase to convert atmospheric nitrogen gas into ammonia.
**nitrogen-limitation hypothesis** The idea that organisms select their food based on its nitrogen content.
**nonequilibrium metapopulation** A metapopulation in decline, where local extinctions are occurring within patches.
**nutrient turnover time** The time taken for a given nutrient, for example nitrogen, to pass through one complete cycle of the nitrogen cycle.

**nutrients** Chemicals required for the growth, maintenance, or reproduction of an organism.

## O

**obligate aerobes** Organisms that require oxygen to live.

**obligate anaerobes** Organisms that live only in the complete absence of oxygen.

**obligate mutualism** Where one species cannot live without the other.

**oligotrophic** Freshwater habitats low in nutrients and organisms; low in productivity; contrast with "eutrophic."

**optimal defense hypothesis** The idea that certain plant parts, for example, flowers and seeds, are more heavily defended against herbivores than other plant parts, such as leaves or twigs.

**optimality modeling** The idea that organisms should behave in a way that maximizes benefits minus costs.

**organic** Of biological origin; in chemistry, containing carbon.

**organismal ecology** The study of how adaptations and choices by individuals affect their fitness.

**organismic model** A view of the nature of a community that considers it to be a tightly knit, interdependent association of species in much the same way as an organism is an interdependent association of organs.

**orographic lifting** The release of precipitation as wind flows upward over mountains.

## P

**P generation** The parent generation; in genetics, often true-breeding parents.

**parasite** The organism that benefits in an interspecific interaction in which two species live symbiotically, and one organism benefits and the other is harmed. Lives in intimate association with its host.

**parasitoid** A specialized insect parasite that is usually fatal to its host and therefore might be considered a predator rather than a classical parasite.

**patchy population** A metapopulation consisting of many patches that exchange individuals at such a high rate that no patch becomes extinct.

**pelagic zone** Pertaining to the upper layers of the open ocean.

**per capita rate of increase, *r*** Rate of population growth per individual; used for species with overlapping, nondiscrete generations.

**perforated** Habitat with small clear areas within it.

**permafrost** A permanently frozen layer of soil underlying the Arctic and Antarctic tundra biome.

**pH** The negative logarithm of the hydrogen ion concentration; a measure of acidity.

**phenology** Study of the periodic (seasonal) phenomena of animal and plant life (for example, flowering time in plants) and their relations to weather and climate.

**phenotype** The physical expression in an organism of the genotype; the outward appearance of an organism.

**pheromones** Chemicals that act as sex attractants between males and females.

**phoresy** The transport of one organism by another of a different species.

**photic zone** The zone of a body of water that is penetrated by sunlight, usually 0–100 m.

**photosynthesis** Synthesis of carbohydrates from carbon dioxide and water, with oxygen as a by-product.

**phylogenetic species concept** The concept that species are defined by having unique physical or genetic characteristics.

**physiological ecology** The study of how organisms are physiologically adapted to their environment and how this limits their distribution patterns.

**phytoplankton** The plant community in marine and freshwater habitats, containing many species of algae and diatoms, that floats free in the water.

**phytoremediation** The reduction of environmental contaminants via the use of plants.

**Pleistocene re-wilding** The advocacy of reintroducing the descendants of Pleistocene megafauna, especially large herbivore and predator species, or their close ecological equivalents, to areas where they once existed but were driven to extinction by humans.

**pneumatophores** Aerial roots of mangroves that extend above the soil surface.

**point mutation** A mutation that affects only a single base pair within DNA or that induces the addition or deletion of a single base pair to the DNA sequence.

**polar cell** The highest latitude cell in the three-cell circulation of wind in each hemisphere.

**pollination syndromes** The pattern of co-evolved traits between particular types of flowers and their specific pollinators.

**polyandry** A mating system in which one female mates with several males in a single breeding season, but males mate only with one female.

**polygyny** A mating system in which one male mates with several females in a single breeding season, but females mate only with one male.

**polygamy** A mating system in which one individual of one sex mates with more than one individual of the opposite sex, such as where a male pairs with more than one female at a time (polygyny) or a female pairs with more than one male (polyandry).

**polyphagous** Feeding on many species; with a wide diet.

**population** A group of potentially interbreeding individuals of a single species.

**population density** The number of organisms of one species in a given unit area.

**population ecology** The study of how populations grow and interact with other species.

**predator** An organism that benefits in an interspecific interaction where it feeds on prey and always kills the prey. Lives in loose association with prey.

**predator escape hypothesis** The idea that seed dispersal is advantageous to plants because seedlings escape the seed predators that tend to congregate under the parent tree.

**predator satiation** The synchronous production of many progeny by all individuals in a population to satiate predators.

**primary consumer** Organisms that consume primary producers; usually herbivores.

**primary producer** A green plant or chemosynthetic bacteria that converts light or chemical energy into organismal tissue.

**primary production** Production by autotrophs, normally green plants.

**primary succession** Succession on completely sterile ground or water.

**principle of species individuality** A view of the nature of a community that regards species distributions according to physiological needs and that most communities therefore integrate continuously.

**production efficiency** The percentage of assimilated energy that becomes incorporated into new biomass.

**promiscuous** A mating system where females mate with a different male every year or breeding season.

**propagule pressure hypothesis** The idea that invasive species are successful because they produce more progeny than native species.

**proportional similarity analysis** A measure of how much overlap exists between species in their use of resources.

**proximate causes** Specific genetic and physiological mechanisms of behavior.

**punctuated equilibrium** The idea that species evolve very quickly and then spend long periods of time without changing; contrast with "gradualism."

**pyramid of biomass** A graphical representation of how biomass decreases with increasing trophic level.

**pyramid of energy** A graphical representation of how energy production decreases with increasing trophic level.

**pyramid of numbers** A graphical representation of how population sizes decrease with increasing trophic level.

## Q

**quadrat** A square frame, of known area, used to count the population density of organisms, usually in the field.

**qualitative defenses** Plant defenses that are effective against herbivores in small doses; for example, highly toxic substances.

**quantitative defenses** Plant defenses that are more effective as more are digested by herbivores; for example, tannins.

## R

***r*-selected species** Species that have a high rate of per capita population growth, $r$, but poor competitive ability.

**rain shadow** An area where precipitation is reduced, often on the leeward sides of mountains.

**random** A spatial dispersion pattern where there is no obvious pattern to the distribution of individuals.

**rank abundance diagram** A graphical representation of the number of individuals per species plotted against the rank of species commonness.

**realized niche** The actual range of an organism in nature. Contrast with fundamental niche.

**recessive** A term that denotes the trait masked by the dominant trait.

**reciprocal altruism** The idea that altruism to other organisms will later be repaid, as in "you scratch my back, I'll scratch yours."

**Red Queen hypothesis** The hypothesis that species have to evolve just to keep pace with environmental change and to avoid extinction.

**redundancy hypothesis** The idea that most species are not vital to the functioning of ecosystems; in the same way only a few people, the crew, are needed in the functioning of an airplane, the passengers are redundant.

**relative abundance** The frequency of occurrence of species in a community.

**replication** The repetition level of an experiment.

**residence time** The time required for litter to decompose under specified conditions.

**resilience** The ability of the community to return to equilibrium following disturbance; usually measured by the speed of the return, or the degree of disturbance from which the community can recover.

**resistance** The size of a force needed to change community structure.

**resource partitioning** The differentiation of niches among species that permits similar species to coexist in a community.

**resource-based mutualism** Mutualism involving the increased acquisition of resources by both species.

**restoration ecology** The process of rehabilitating damaged communities and ecosystems; returning human-impacted communities to a more natural condition.

**riparian** Living in, or located on, the bank of a natural watercourse, usually a river, sometimes a lake or tidewater.

**runaway selection** Selection of males by females based on plumage color or courtship display rather than parenting skills or material benefits such as nests or territories.

## S

**sampling effect** The idea that species-rich communities perform better than species-poor communities because they have a better chance of containing a "superspecies."

**scientific method** A series of steps to test the validity of a hypothesis. The steps often involve a comparison between control and experimental samples.

**secondary consumers** Organisms that eat primary consumers.

**secondary metabolites** Chemicals that are produced by plants that are not essential for cell function but are useful deterrents against herbivores.

**secondary production** Production by herbivores, carnivores, or detritus feeders; contrast "primary production."

**secondary succession** Succession on partially cleared land.

**seed rain** The deposition of seeds by wind and animals.

**segregate** In genetics, to separate, as in the separation of chromosomes during meiosis.

**selfish herd** The concept of individuals banding together to use some members as protection against predators.

**self-fertilization** Fertilization of a male and female gamete from the same individual.

**semelparity** Having one reproductive episode per lifetime.

**semiochemicals** Behavior-altering chemical messengers.

**seral stage** A distinct phase of succession, also called a sere.

**sere** The series of successional communities leading from bare substrate to the climax community, also called a seral stage.

**sessile** Attached to an object or fixed in place; for example, barnacles.

**sexual dimorphism** A condition where individuals of one sex are substantially bigger than individuals of the other sex.

**sexual selection** Selection that promotes traits that will increase an organism's mating success.

**siblings** Brothers or sisters.

**similarity indices** Indices that directly compare how many species are in common to two communities.

**SLOSS debate** In conservation biology, the debate over whether it is preferable to protect one single, large reserve or several smaller ones.

**soil organic matter (SOM)** Dead organic matter (DOM) that has been so chemically changed it is no longer recognizable.

**soil profile** The vertical layers in soil.

**solar equator** The area of Earth directly under the sun and receiving the most solar energy; varies seasonally from 23.5° north on June 21 and 23.5° south on December 21.

**source pool** The pool of species available to colonize an area.

**spatial dispersion** The physical distribution patterns of organisms in a given area; clumped, random, or uniform.

**speciation** The formation of new species.

**species-area hypothesis** The idea that communities diversify with area so that larger areas contain the highest numbers of species.

**species complementarity** The idea that species are complementary in their use of resources in a community; for example, some species tap into deep soil moisture while others use water near the soil surface.

**species-distance effect** The idea that immigration rates are highest on islands nearest the source pool because species do not have so far to travel.

**species-energy hypothesis** The idea that communities diversify with energy so that communities rich in energy, from the sun and with abundant water, as in moist tropical forests, contain the highest numbers of species.

**species interactions** Interactions between species, including competition, predation, mutualism, commensalism, herbivory, and parasitism.

**species richness** The number of species in a community.

**species-time hypothesis** The idea that communities diversify with time so that older communities' areas contain the highest numbers of species.

**species turnover** The change in species composition in an area as some species go extinct but new species arrive.

**spring overturn** The mixing of lake water as ice melts and storms churn up water from the bottom.

**stability** Absence of fluctuations in populations; ability to withstand perturbations without large changes in community species composition.

**stabilizing selection** The pattern of natural selection that favors the survival of individuals with intermediate phenotypes.

**standing crop** The total biomass in an ecosystem at any one point in time.

**static life table** A life table of a population taken at one moment in time.

**strong acid** An acid that completely ionizes in solution.

**subsidence zone** Areas where cool air from the upper atmosphere falls toward the Earth, creating areas of high pressure.

**succession** Replacement of one kind of community by another; the progressive changes in vegetation and animal life that tend toward climax.

**supercooling** The ability of some organisms to prevent their body fluids from freezing despite temperatures lower than 0°C.

**superior competitor hypothesis** The idea that invasive species are successful because they are more efficient users of natural resources than native species.

**supralittoral zone** The area of the rocky intertidal zone that is never submerged by the tide and receives only spray from waves.

**survival of the fittest** A phrase used by Darwin as a metaphor for natural selection, the latter of which is more commonly used.

**survivorship curve** A graphical representation of the number of individuals of various ages alive in a population.

**symbiosis** The living together of two or more organisms of different species in close proximity.

**sympatric** Occurring in the same place.

**sympatric speciation** A form of speciation without geographic isolation, whereby one species divides into two or more species within the same habitat.

## T

**taiga** The northern boreal forest zone; a broad band of coniferous forest south of the Arctic tundra.

**territory** Any area defended by one or more individuals and protected against intrusion by others of the same or different species.

**tertiary consumers** Organisms that eat secondary consumers.

**theory of island biogeography** A theory that explains the process of succession on islands; states that the number of species on an island tends toward an equilibrium number that is determined by the balance between immigration and extinction.

**thermocline** The thin transitional zone in a lake that separates the epilimnion from the hypolimnion.

**thermohaline circulation** A type of long-distance oceanic circulation driven by variations in water temperature and salinity.

**time lag** Delay in response to a change.

**tolerance** Succession that is not affected by previous colonists.

**total fertility rates (TFR)** The average number of live births a female has during her lifetime, assuming life to a maximum age.

**tree line** The point at which trees stop growing on a mountainside.

**trophic cascade** The idea that in a food web, each trophic level strongly influences the one below it so that in the end, top predators influence the density of primary producers.

**trophic level** The functional classification of an organism in a community according to its feeding relationships.

**trophic-level transfer efficiency** The percentage of energy of one trophic level that is incorporated into the bodies of individuals at the next trophic level.

**true-breeding lines** Strains of organisms that exhibit the same trait after several generations of self-fertilization.

**tundra** Level or undulating treeless land, characteristic of arctic regions and high altitudes, having permanently frozen subsoil.

**turnover** Rate of replacement of resident species by new, immigrant species.

## U

**ultimate causes** Reasons why behavior evolved, in terms of their affects on fitness.

**umbrella species** Species whose habitat requirements are so large that protecting them would protect many other species existing in the same habitat.

**unapparent plants** Usually short-lived plants that are difficult for herbivores to locate; for example, weedy species.

**uniform** A spatial dispersion pattern where there is an identical distance between individuals.

**upper littoral zone** The area of the rocky intertidal zone submerged only during the highest tides.

**upwelling** The process whereby, as a result of wind patterns, nutrient-rich bottom waters rise to the surface of the ocean.

**urohydrosis** The behavior of urinating on the legs to cool the body by evaporation.

## W

**watershed** The land area that drains into a particular lake, river, or reservoir.

**weak acid** An acid that only partially ionizes in solution.

## Z

**zero population growth** The situation in which no changes in population size occur.

**zooplankton** The animal community, predominantly single-celled animals, that floats free in marine and freshwater environments, moving passively with the currents.

# References


Abbott, I., and R. Black. 1980. Changes in species compositions of floras on islets near Perth, Western Australia. Journal of Biogeography 7: 399–410.

Alexander, R. D. 1974. The evolution of social behavior. Annual Review of Ecology and Systematics 5: 325–83.

Alexander, R. D., J. L. Hoogland, R. D. Howard, K. M. Noonan, and P. W. Sherman. 1979. Sexual dimorphisms and breeding systems in pinnipeds, ungulates, primates and humans, pp. 402–435 in Evolutionary Biology and Human Social Behavior: An Anthropological Perspective, N. A. Chagnon and W. Irons, eds., Duxbury Press, Scituate, MA.

Allendorf, F. W. 1994. Genetically effective sizes of grizzly bear populations, pp. 155–156 in Principles of Conservation Biology, G. K. Meffe and C. R. Carrol, eds., Sinauer Associates, MA.

Alvarez, L. W., W. Alvarez, F. Asaro, and H. V. Michel. 1980. Extraterrestrial cause of the Cretaceous-Tertiary extinction. Science 208: 1095–1108.

Anderson, C. B., and A. D. Rosemond. 2007. Ecosystem engineering by invasive exotic beavers reduces in-stream diversity and enhances ecosystem function in Cape Horn, Chile. Oecologia 154: 141–153.

Andersson, M. 1982. Sexual selection, natural selection and quality advertisement. Biological Journal of the Linnean Society 17: 375–393.

Antonovics, J., and A. D. Bradshaw. 1970. Evolution in closely adjacent plant populations. VIII. Clinal patterns at a mine boundary. Heredity 25: 349–362.

Avery, M., R. Engeman, K. Keacher, J. Humphrey, W. Bruce, T. Mathies, and R. Mauldin. 2010. Cold weather and the potential range of invasive Burmese pythons. Biological Invasions 12: 3649–3652.

Baille, J. E. M., C. Hilton-Taylor, and S. N. Stuart. 2004. 2004 IUCN Red List of Threatened Species. IUCN, Gland, Switzerland.

Baird, R. W., and L. M. Dill. 1996. Ecological and social determinants of group size in transient killer whales. Behavioral Ecology 7: 408–416.

Baker, B. W., T. R. Stanley, and J. A. Sedgwick. 1999. Predation of artificial ground nests on white-tailed prairie dog colonies. Journal of Wildlife Management 63: 270–277.

Barkalow, F. S., Jr., R. B. Hamilton, and R. F. Soots Jr. 1970. The vital statistics of an unexploited gray squirrel population. Journal of Wildlife Management 34: 489–500.

Bazely, D. R., M. Vicari, S. Emmerich, L. Filip, D. Lin, and A. Inman. 1998. Interactions between herbivores and endophyte-infected *Festuca rubra* from the Scottish Islands of St. Kilda, Berbecula and Rum. Journal of Applied Ecology 34: 847–860.

Bazzaz, F. A. 1975. Plant species diversity in old-field successional ecosystems in southern Illinois. Ecology 56: 485–488.

Beamish, R., W. Lockhardt, J. Van Loon, and H. Harvey. 1975. Long term acidification of a lake and resulting effects on fishes. Ambio 4: 98–101.

Beauchamp, G. 2008. What is the magnitude of the group-size effect on vigilance? Behavioral Ecology 19: 1361–1368.

Begon, M., J. L. Harper, and C. R. Townsend. 1996. Ecology: Individuals, Populations, and Communities, 3rd ed. Blackwell Science, Cambridge, MA.

Belt T. 1874. The Naturalist in Nicaragua. J. Murray, London.

Benayas, J. R., A. C. Newton, A. Diaz, and J. M. Bullock. 2009. Enhancement of biodiversity and ecosystem services by ecological restoration: a meta-analysis. Science 325: 1121–1124.

Berenbaum, M. 1981. Patterns of furanocoumarin distribution and insect herbivory in the Umbelliferae: plant chemistry and community structure. Ecology 62: 1254–1266.

Berger, J. 1990. Persistence of different-sized populations: an empirical assessment of rapid extinctions in bighorn sheep. Conservation Biology 4: 91–98.

Biesmeijer J. C., S. P. M. Roberts, M. Reemer, R. Ohlemüller, M. Edwards, T. Peeters, A. P. Schaffers, S. G. Potts, R. Kleukers, C. D. Thomas, J. Settele, and W. E. Kunin. 2006. Parallel declines in pollinators and insect-pollinated plants in Britain and the Netherlands. Science 313: 351–354.

Bigger, P. S., and M. A. Marvier. 1998. How different would a world without herbivory be? A search for generality in ecology. Integrative Biology 1: 60–67.

Birks, H. J. B. 1980. British trees and insects: a test of the time hypothesis over the last 13,000 years. American Naturalist 115: 600–605.

Blueweiss, L. H., H. Fox, V. Kudzma, D. Nakashima, R. Peters, and S. Sams. 1978. Relationships between body size and some life history parameters. Oecologia 37: 257–272.

Bonner, J. T. 1965. Size and Cycle: An Essay on the Structure of Biology. Princeton University Press, Princeton, NJ.

Breckle, S. W. 2002. Walter's Vegetation of the Earth: The Ecological Systems of the Geo-biosphere. Springer, Berlin.

Brooker, R. J., E. P. Widmaier, L. E. Graham, and P. D. Stiling. 2008. Biology. McGraw-Hill, Dubuque, IA.

Brooks, T. M., R. A. Mittermeier, G. A. B. da Fonseca, J. Gerlach, M. Hoffman, J. F. Lamoreux, C. G. Mittermeier, J. D. Pilgrim, and A. S. L. Rodrigues. 2006. Global biodiversity conservation priorities. Science 313: 58–61.

Brown, J. H. 1978. The theory of insular biogeography and the distribution of boreal birds and mammals. Great Basin Naturalist Memoirs 2: 209–227.

Brown, J. H. 1989. Patterns, modes, and extents of invasions by vertebrates, pp. 85–109 in Biological Invasions: A Global Perspective, J. A. Drake, H. A. Mooney, F. di Castri, R. H. Groves, F. J. Kruger, M. Rejmanek, and M. Williamson, eds., Wiley, Chichester, England.

Brown, J. H., J. F. Gillooly, A. P. Allen, V. M. Savage, and G. B. West. 2004. Towards a metabolic theory on ecology. Ecology 85: 1770–1789.

Bruno, J. F., J. J. Stachowicz, and M. D. Bertness. 2003. Inclusion of facilitation into ecological theory. Trends in Ecology and Evolution 18: 119–125.

Bush, A. O., J. M. Aho, and C. R. Kennedy. 1990. Ecological versus phylogenetic determinants of helminth parasite community richness. Evolutionary Ecology 4: 1–20.

Callaway, R. M. 1995. Positive interactions among plants. Botanical Review 61: 306–349.
Callaway, R. M. 1998. Are positive interactions species-specific? Oikos 82: 202–207.
Callaway, R. M., and E. T. Aschehoug. 2000. Invasive plants versus their new and old neighbors: a mechanism for exotic invasion. Science 290: 521–523.
Callaway, R. M., R. W. Brooker, P. Choler, Z. Kikvidze, C. J. Lortie, R. Michalet, L. Paolini, F. I. Pugnaire, B. Newingham, E. T. Aschehoug, C. Armas, D. Kikodze, and B. J. Cook. 2002. Positive interactions among alpine plants increase with stress. Nature 417: 844–848.
Cardinale, B. J., D. S. Srivastava, J. E. Duffy, J. P. Wright, A. L. Downing, M. Sankaran, and C. Jousseau. 2006. Effects of biodiversity on the functioning of trophic groups and ecosystems. Nature 443: 989–992.
Cargill, S. M., and R. L. Jeffries. 1984. Nutrient limitation of primary production in a sub-arctic salt marsh. Journal of Applied Ecology 21: 657–668.
Case, T. J. 2000. An Illustrated Guide to Theoretical Ecology. Oxford University Press, New York.
Caughley, G. 1966. Mortality patterns in mammals. Ecology 47: 906–918.
Cebrian J. 1999. Patterns in the fate of production in plant communities. American Naturalist 154: 449–468.
Cebrian, J., and J. Lartigue. 2004. Patterns of herbivory and decomposition in aquatic and terrestrial ecosystems. Ecological Monographs 74: 237–259.
Center, T. D., and F. A. Dray Jr. 2010. Bottom-up control of water hyacinth weevil populations: do the plants regulate the insects? Journal of Applied Ecology, 47: 329–337.
Cézilly, G., A. Grégoire, and A. Bertin. 2000. Conflict between co-occurring manipulative parasites? An experimental study of the joint influence of two acanthrocephalan parasites on the behaviour of *Gammarus pulex*. Parasitology 120: 625–630.
Chapin, F. C., III, P. A. Matson, and H. A. Mooney. 2002. Principles of Terrestrial Ecosystem Ecology, Springer, New York.
Cheng, T. L., S. M. Rovito, D. B. Wake, and V. T. Vredenburg. 2011. Coincident mass extirpation of neotropical amphibians with the emergence of the infectious fungal pathogen *Batrachochytrium dendrobatidis*. Proceedings of the National Academy of Sciences of the United States of America 108: 9502–9507.
Chittibabu, C. V., and N. Parthasarathy. 2000. Attenuated tree species diversity in human-impacted tropical evergreen forest sites at Kolli hills, Eastern Ghats, India. Biodiversity and Conservation 9: 1493–1519.
Christian, J. M., and D. Wilson. 1999. Long-term ecosystem impacts of an introduced grass in the northern Great Plains. Ecology 80: 2397–2407.
Clark, C. M., and D. Tilman. 2008. Loss of plant species after chronic low-level nitrogen deposition to prairie grasslands. Nature 451: 712–715.
Clarke, C. A., F. M. M. Clarke, and H. C. Dawkins. 1990. *Biston betularia* (the peppered moth) in West Kirby, Wirral, 1959–1989: updating the decline in *F. cabonaria*. Biological Journal of the Linnean Society 39: 323–326.
Clements, F. E. 1905. Research Methods in Ecology. University Publishing Co., Lincoln. Reprinted Arno Press, New York, 1977.
Clubb R., P. Lee, K. U. Mar, C. Moss, M. Rowcliffe, and G. J. Mason. 2008. Compromised survivorship in zoo elephants. Science 322: 1649.
Coleman, F. C., W. F. Figueira, J. S. Ueland, and L. B. Crowder. 2004. The impact of United States recreational fisheries on marine fish populations. Science 305: 1958–1960.
Compton, S. G., J. H. Lawton, and V. K. Rashbrook. 1989. Regional diversity, local community structure, and vacant niches: the herbivorous insects of bracken in South Africa. Ecological Entomology 14: 365–373.
Conant, R. 1975. A Field Guide to the Reptiles and Amphibians. Riverside Press, Cambridge, MA.
Connell, J. H. 1970. A predator-prey system in the marine intertidal region I. *Balanus glandular* and several species of *Thais*. Ecological Monographs 40: 49–78.
Connell, J. H. 1978. Diversity in tropical rain forests and coral reefs. Science 199: 1302–1310.
Connell, J. H. 1983. On the prevalence and relative importance of interspecific competition: evidence from field experiments. American Naturalist 122: 661–696.
Connell, J. H., and R. O. Slatyer. 1977. Mechanisms of succession in natural communities and their role in community stability and organization. American Naturalist 111: 1119–134.
Costanza, R., R. d'Arge, R. de Groot, S. Farber, M. Grasso, B. Hannon, K. Limburg, S. Naeem, R. V. O'Neill, J. Paruelo, R. G. Raskin, P. Sutton, and M. van den Belt. 1997. The value of the world's ecosystem services and natural capital. Nature 387: 253–260.
Cook, R. E. 1969. Variation in species diversity of North American birds. Systematic Zoology 18: 63–84.
Côté, S. D. 2005. Extirpation of a large black bear population by introduced white-tailed deer. Conservation Biology 19: 1668–1671.
Cowles, H. C. 1899. The ecological relations of the vegetation on the sand dunes of Lake Michigan. Botanical Gazette 27: 95–117, 167–202, 361–391.
Crissey, W. F. 1969. Prairie pot holes from a continental viewpoint, pp. 161–171 in Transactions of the Saskatoon Wetlands Seminar. Department of Indian Affairs and Northern Development, Canada. Wildlife Service Report Series No. 6, 262 pp.
Crombie, A. C. 1946. Further experiments on insect competition. Proceedings of the Royal Society, series B, 133: 76–109.
Crouse, D. T., L. B. Crowder, and H. Caswell. 1987. A stage-based population model for loggerhead sea turtles and implication for conservation. Ecology 68: 1412–1423.
Crow, J. F., and M. Kimura. 1970. An Introduction to Population Genetics Theory. Harper and Row, New York.
Culver, M., P. W. Hedrick, K. Murphy, S. O'Brien, and M. G. Hornocker. 2008. Estimation of the bottleneck size in Florida panthers. Animal Conservation 11: 104–110.
Currie, D. J., and V. Paquin. 1987. Large-scale biogeographical patterns of species richness of trees. Nature 329: 326–327.
Curtis, J. T. 1956. The modifications of mid-latitude grasslands and forests by man, pp. 721–736 in Man's Role in Changing the Face of the Earth, W. L. Thomas, ed., University of Chicago Press, Chicago.
Cyr, H., and M. L. Pace. 1993. Magnitude and patterns of herbivory in aquatic and terrestrial ecosystems. Nature 361: 148–150.
Daday, H. 1954. Gene frequencies in wild populations of *Trifolium repens* L. I. Distribution by latitude. Heredity 8: 61–78.
Damschen, E. I., N. M. Haddad, J. L. Orrock, J. J. Tewksbury, and D. J. Levey. 2006. Corridors increase plant species richness at large scales. Science 313: 1284–1286.
Darwin, C. 1859. On the Origin of Species by Means of Natural Selection. John Murray, London.
Davidson, J., and H. G. Andrewartha. 1948. The influence of rainfall, evaporation and atmospheric temperature on fluctuations in the size of a natural population of *Thrips imaginis*. Journal of Animal Ecology 17: 200–222.
Davies, N. B. 1992. Dunnock behaviour and social evolution. Oxford University Press, Oxford.
Davis, M. G., and A. Zabinski. 1992. Changes in geographical range resulting from greenhouse warming: effects on biodiversity in

forests, pp. 297–308 in Global Warming and Biodiversity, R. Peters and T. Lovejoy, eds., Yale University Press, New Haven, CT.

Dean, T. A., and L. E. Hurd. 1980. Development in an estuarine fouling community: the influence of early colonists on later arrivals. Oecologia 46: 295–301.

De'ath, G., J. M. Lough, and K. E. Fabricius. 2009. Declining coral calcification on the Great Barrier Reef. Science 323: 116–119.

del Cerro, S., S. Merino, J. Martínez-de la Puente, E. Lobato, R. Ruiz-de-Castañeda, J. Rivero-de Aguilar, J. Morales, G. Tomás, and J. Moreno. 2010. Carotenoid-based plumage colouration is associated with blood parasite richness and stress protein levels in blue tits (*Cyanistes caeruleus*). Oecologia 162: 825–835.

Del Moral, R. 2000. Succession and species turnover on Mount St. Helens, Washington. Acta Phytogeographica Suecica 85: 53–62.

De Moraes, C. M., W. J. Lewis, P. W. Paré, H. T. Alborn, and J. H. Tumlinson. 1998. Herbivore-infested plants selectively attract parasitoids. Nature 393: 570–573.

DeSteven, D. 1991. Experiments on mechanisms of tree establishment in old-field succession: seedling survival and growth. Ecology 72: 1075–1088.

Diamond, J. M. 1969. Avifaunal equilibria and species turnover rates on the Channel Islands of California. Proceedings of the National Academy of Science of the United States of America 64: 57–63.

Diamond, J. M. 1986. Overview: laboratory experiments, field experiments, and natural experiments, pp. 3–22 in Community Ecology, J. Diamond and T. J. Case, eds., Harper and Row, New York.

Diaz, R. J., and R. Rosenberg. 2008. Spreading dead zones and consequences for marine ecosystems. Science 21: 926–929.

Dixon, J. D., M. K. Oli, M. C. Wooten, T. H. Eason, J. W. McCown, and D. Paetkau. 2006. Effectiveness of a regional corridor in connecting Florida black bear populations. Conservation Biology 20: 155–162.

Dobson, A. P., K. D. Lafferty, A. M. Kuris, R. F. Hechinger, and W. Jetz. 2008. Homage to Linnaneus: how many parasites? How many hosts? Proceedings of the Natural Academy of the United States of America 105: 11482–11489.

Dobson, A. P., J. P. Rodriguez, W. M. Roberts, and D. S. Wilcove. 1997. Geographic distribution of endangered species in the United States. Science 275: 550–553.

Dorcas, M. E., J. D. Willson, R. N. Reed, R. W. Snow, M. R. Rochford, M. A. Miller, W. E. Meshaka Jr., P. T. Andreadish, F. J. Mazzottie, C. M. Romagosai, and K. M. Hart. 2012. Severe mammal declines coincide with proliferation of invasive Burmese pythons in Everglades National Park. Proceedings of the National Academy of Sciences of the United States of America 109: 2418–2422.

Doty S. L., C. A. James, A. L. Moore, A. Vajzovic, G. L. Singleton, C. Ma, Z. Khan, G. Xin, J. W. Kang, J. Y. Park, R. Meilan, S. H. Strauss, J. Wilkerson, F. Farin, and S. E. Strand. 2007. Enhanced phytoremediation of volatile environmental pollutants with transgenic trees. Proceedings of the National Academy of Sciences of the United States of America 104: 16816–16821.

Downing, J. A., C. W. Osenberg, and O. Sarnelle. 1999. Meta-analysis of marine nutrient enrichment experiments: variation in the magnitude of nutrient limitation. Ecology 80: 1157–1167.

Drake, J. A. 1990. Communities as assembled structures: Do rules govern patterns? Trends in Ecology and Evolution 5: 159–164.

Edwards, C. A., and G. W. Heath. 1963. The role of soil animals in breakdown of leaf material, pp. 76–84 in Soil Organisms, D. Doiksen and J. van der Pritt, eds., North-Holland, Amsterdam.

Egler, F. E. 1954. Vegetation science concepts: initial floristic composition—a factor in old-field development. Vegetatio 4: 412–417.

Ehrlich, P. R., and P. H. Raven. 1964. Butterflies and plants: a study in coevolution. Evolution 18: 586–608.

Eisner, T., and D. J. Aneshansley. 1982. Spray aiming in bombardier beetles: jet deflection by the Coanda effect. Science 215: 83–85.

Ellenberg, H. 1988. Vegetation Ecology of Central Europe. Cambridge University Press, Cambridge.

Ellis, E. C., and N. Ramankutty. 2008. Putting people in the map: anthropogenic biomes of the world. Frontiers in Ecology and the Environment 6: 439–447.

Elner, R. W., and R. N. Hughes. 1978. Energy maximization in the diet of the shore crab, *Carcinus meanas*. Journal of Animal Ecology 47: 103–116.

Elton, C. 1927. Animal Ecology. Sidgwick and Jackson, London.

Elton, C. 1958. The Ecology of Invasions by Animals and Plants. Methuen, London.

Elton, C., and M. Nicholson. 1942. The ten-year cycle in numbers of the lynx in Canada. Journal of Animal Ecology 11: 215–244.

Faeth, S., P. S. Warren, E. Shochat, and W. A. Marussich. 2005. Trophic dynamics in urban communities. BioScience 55: 399–407.

Farner, D. S. 1945. Age groups and longevity in the American robin. Wilson Bulletin 57: 56–74.

Feeny, P. 1970. Seasonal changes in the oak leaf tannins and nutrients as a cause of spring feeding by winter moth caterpillars. Ecology 51: 565–581.

Fitzpatrick, J. W., M. Lammertink, M. D. Luneau Jr., T. W. Gallagher, B. R. Harrison, G. M. Sparling, K. V. Rosenberg, R. W. Rohrbaugh, E. C. H. Swarthout, P. H. Wrege, S. B. Swarthout, M. S. Dantzker, R. A. Charif, T. R. Barksdale, J. V. Remsen Jr., S. D. Simon, and D. Zollner. 2005. Ivory-billed woodpecker (*Campephilus principalis*) persists in continental North America. Science 308: 1460–1462.

Francis, W. J. 1970. The influence of weather on population fluctuations in California quail. Journal of Wildlife Management 34: 249–266.

Fraser, L. H., and P. A. Keddy (eds.). 2005. The World's Largest Wetlands: Ecology and Conservation. Cambridge University Press, Cambridge.

Frazier, M. R., R. B. Huey, and D. Berrigan. 2006. Thermodynamics constrains the evolution of insect population growth rates: "warmer is better." American Naturalist 168: 512–520.

Garibaldi, A., and N. Turner. 2004. Cultural keystone species: implications for ecological conservation and restoration. Ecology and Society 9 (3): 1. http://www.ecologyandsociety.org/vol9/iss3/art1/.

Garrettson, P. R., and F. C. Rohwer. 2001. Effects of mammalian predator removal on production of upland-nesting ducks in North Dakota. Journal of Wildlife Management 65: 398–405.

Gates, D. M. 1993. Climate Change and Its Biological Consequences. Sinauer Associates, Sunderland, MA.

Gentry, A. H. 1988. Tree species of upper Amazonian forests. Proceedings of the National Academy of Sciences of the United States of America 85: 156.

Gilbert, F. S. 1980. The equilibrium theory of island biogeography: fact or fiction. Journal of Biogeography 7: 209–235.

Gill, G. B., and L. L. Wolf. 1975. Economies of feeding territoriality in the golden-winged sunbird. Ecology 56: 333–345.

Gillespie, T. R., and C. A. Chapman. 2006. Prediction of parasite infection dynamics in primate metapopulations based on attributes of forest fragmentation. Conservation Biology 20: 444–448.

Giraldeau, L. A. 2008. Social foraging, pp. 257–283 in Behavioural Ecology, E. Danchin, L. A. Giraldeau, and F. Cezilly, eds., Oxford University Press, Oxford.

Gleason, H. A. 1926. The individualistic concept of the plant association. Torrey Botanical Club Bulletin 53: 7–26.

Gómez, J. M., and R. Zamora. 2002. Thorns as induced mechanical defense in a long-lived shrub (*Hormathophylla spinosa,* Cruciferae). Ecology 83: 885–890.

Gould, S. J., and N. Eldredge. 1977. Punctuated equilibria: the tempo and mode of evolution reconsidered. Paleobiology 3: 115–151.

Grange, S., P. Duncan, J. M. Gaillard, A. R. E. Sinclaire, P. J. P Gogan, C. Packer, H. Hofer, and M. East. 1984. What limits the Serengeti zebra population? Oecologia, 140: 523–532.

Grenier, M. B., D. B. McDonald, and S. W. Buskirk. 2007. Rapid population growth of a critically endangered carnivore. Science 317: 779.

Griffen, B. D., T. Guy, and J. C. Buck. 2008. Inhibition between invasives: a newly introduced predator moderates the impacts of a previously established invasive predator. Journal of Animal Ecology 77: 32–40.

Griffiths C. J., C. G. Jones, D. M. Hansen, M. Puttoo, R. V. Tatayah, C. B. Müller, and S. J. Harris. 2010. The use of extant non-indigenous tortoises as a restoration tool to replace extinct ecosystem engineers. Restoration Ecology 18: 1–7.

Grime, J. P. 1977. Evidence for the existence of three primary strategies in plants and its relevance to ecological and evolutionary theory. American Naturalist 111: 1169–1194.

Grime, J. P. 1979. Plant Strategies and Vegetation Process. John Wiley, New York.

Groom, M. J., G. K. Meffe, and C. R. Carroll. 2006. Principles of Conservation Biology, 3rd ed. Sinauer Associates, Sunderland, MA.

Gruner, D. S., J. E. Smith, E. W. Seabloom, S. A. Sandin, J. T. Ngai, H. Hillebrand, W. S. Harpole, J. J. Elser, E. E. Cleland, M. E. S. Bracken, E. T. Borer, and B. M. Bolker. 2008. A cross-system synthesis of herbivore and nutrient resource control on producer biomass. Ecology Letters 11: 740–755.

Gurrevitch, J., L. L. Morrow, A. Wallace, and J. S. Walsh. 1992. A meta-analysis of field experiments on competition. American Naturalist 140: 539–572.

Hails, C. J. 1982. A comparison of tropical and temperate aerial insect abundance. Biotropica 14: 310–313.

Hairston, N. G., Jr., and N. G. Hairston Sr. 1993. Cause-effect relationships in energy flow, trophic structure, and interspecific interactions. American Naturalist 142, 379–411.

Hairston, N. G., F. E. Smith, and L. B. Slobodkin. 1960. Community structure, population control, and competition. American Naturalist 44: 421–425.

Halaj, J., and D. H. Wise. 2001. Terrestrial trophic cascades: how much do they trickle? American Naturalist 157: 262–281.

Hamback, P. A., J. A. Agren, and L. Ericson. 2000. Associational resistance: insect damage to purple loosestrife reduced in thickets of sweet gale. Ecology 81: 1784–1794.

Hames, R. S., K. V. Rosenberg, J. D. Lowe, S. E. Barker, and A. A. Dhondt. 2002. Adverse effects of acid rain on the distribution of the wood thrush *Hylocichla mustelina* in North America. Proceedings of the National Academy of Science 99: 11235–11240.

Hamilton, W. D. 1964. The genetical evolution of social behaviour. I, II. Journal of Theoretical Biology 7: 1–52.

Hamilton, W. D. 1971. Geometry for the selfish herd. Journal of Theoretical Biology 31: 295–311.

Hanssen, S. A., I. Folstad, K. E. Erikstad, and A. Oksanen. 2003. Costs of parasites in common eiders: effects of anti-parasite treatment. Oikos 100: 105–111.

Hardin, G. 1960. The competitive exclusion principle. Science 162: 1243–1248.

Harris, S., and G. C. Smith. 1987. Demography of two urban fox (*Vulpes vulpes*) populations. Journal of Animal Ecology 24: 75–86.

Harrison, S. 1991. Local extinction in a metapopulation context: an empirical evaluation. Biological Journal of the Linnean Society 42: 73–88.

Harrison, S., D. D. Murphy, and P. R. Ehrlich. 1988. Distribution of the bay checkerspot butterfly, *Euphydryas editha bayensis:* evidence for a metapopulation model. American Naturalist 132: 360–382.

Harvell, C. D., S. Altizer, I. M. Cattadori, L. Harrington, and E. Weil. 2009. Climate change and wildlife diseases: When does the host matter the most? Ecology 90: 912–920.

Harvey, P. H., and J. W. Bradbury. 1991. Sexual selection, pp. 203–231 in Behavioural Ecology: An Evolutionary Approach, J. R. Krebs and N. B. Davies, eds., Blackwell Scientific, Oxford.

Harvey, P. H., J. J. Bull, and R. J. Paxton. 1983. Looks pretty nasty. New Scientist 97: 26–27.

Hättenschwiler, S., and D. Bretscher. 2001. Isopod effects on decomposition of litter produced under elevated $CO_2$, N deposition and different soil types. Global Change Biology 7: 565–579.

Hattersley, P. W. 1983. The distribution of C3 and C4 grasses in Australia in relation to climate. Oecologia 57: 113–128.

Hawkes, C. V., and J. J. Sullivan. 2001. The impact of herbivory on plants in different resource conditions: a meta-analysis. Ecology 82: 2045–2058.

Hayes, K. R., and S. C. Barry. 2008. Are there any consistent predictors of invasion success? Biological Invasions 10: 483–506.

Heads, M. 2006. Panbiogeography of *Nothofagus* (Nothofagaceae): analysis of the main species massings. Journal of Biogeography 33: 1066–1075.

Heatwole, H. 1965. Some aspects of the association of cattle egrets with cattle. Animal Behavior 13: 79–83.

Hebblewhite, M., C. White, C. Nietvelt, J. McKenzie, and T. Hurd. 2005. Human activity mediates a trophic cascade caused by wolves. Ecology 86: 2135–2144.

Heil, L., E. Fernández-Juricic, D. Renison, A. Cingolani, and D. T. Blumstein. 2007. Avian responses to tourism in the biogeographically isolated high Córdoba mountains, Argentina. Biodiversity and Conservation 16: 1009–1026.

Heithaus, M. R., A. Frid, A. J. Wirsing, and B. Worm. 2008. Predicting ecological consequences of marine top predator declines. Trends in Ecology and Evolution 23: 202–210.

Heron, A. C. 1972. Population ecology of a colonizing species: the pelagic tunicate *Thalia democratica.* Oecologia 10: 269–293, 294–312.

Hickling, R., D. B. Roy, J. K. Hill, R. Fox, and C. D. Thomas. 2006. The distributions of a wide range of taxonomic groups are expanding polewards. Global Change Biology 12: 450–455.

Hilderbrand, R. H., A. C. Watts, and A. N. Randle. 2005. The myths of restoration ecology. Ecology and Society 10 (1): 19.

Hiura, T. 1995. Gap formation and species diversity in Japanese beech forests: a test of the intermediate disturbance hypothesis on a geographic scale. Oecologia 104: 265–271.

Hocker, H. W., Jr. 1956. Certain aspects of climate as related to the distribution of loblolly pine. Ecology 37: 824–834.

Hoegh-Guldberg, O., P. J. Mumby, A. J. Hooten, R. S. Steneck, P. Greenfield, E. Gomez, C. D. Harvell, P. F. Sale, A. J. Edwards, K. Caldeira, N. Knowlton, C. M. Eakin, R. Iglesias-Prieto, N. Muthiga, R. H. Bradbury, A. Dubi, and M. E. Hatziolos. 2007. Coral reefs under rapid climate change and ocean acidification. Science 318: 1737–1742.

Hoekstra, J. M., T. M. Boucher, T. H. Ricketts, and C. Roberts. 2005. Confronting a biome crisis: global disparities of habitat loss and protection. Ecology Letters 8: 23–29.

Holt, B. G., L. Jean-Philippe, M. K. Borregaard, S. A. Fritz, M. B. Araújo, D. Dimitrov, P. Fabre, C. H. Graham, G. R. Graves, K. A. Jønsson, D. Nogués-Bravo, Z. Wang, R. J. Whittaker, J. Fjeldså, and C. Rahbek. 2013. An update of Wallace's zoogeographic regions of the world. Science 339: 74–78.

Hoover, J. P., and S. K. Robinson. 2007. Retaliatory mafia behavior by a parasitic cowbird favors host acceptance of parasitic eggs. Proceedings of the National Academy of Sciences of the United States of America 104: 4479–4483.

Huffaker, C. B., and C. E. Kennett. 1969. Some aspects of assessing efficiency of natural enemies. Canadian Entomologist 101: 425–440.

Hurtrez-Boussess, S., P. Perret, F. Renaud, and J. Blondel. 1997. High blowfly parasitic loads affect breeding success in a Mediterranean population of blue tits. Oecologia 112: 514–517.

Huston, M. A. 1997. Hidden treatments in ecological experiments: re-evaluating the ecosystem function of biodiversity. Oecologia 110: 449–460.

Huston, M. A., and S. Wolverton. 2009. The global distribution of net primary production: resolving the paradox. Ecological Monographs 79: 343–377.

Hutchinson, G. E. 1959. Homage to Santa Rosalia, or why are there so many kinds of animals? American Naturalist 93: 145–159.

Hutchinson, G. E. 1978. An Introduction to Population Ecology. Yale University Press, New Haven, CT.

Inouye, R. S., N. J. Huntly, D. Tilman, J. R. Tester, M. Stillwell, and K. C. Zinnel. 1987. Old-field succession on a Minnesota sand plain. Ecology 68: 12–26.

Jaccard, P. 1912. The distribution of the flora of the alpine zone. New Phytologist 11: 37–50.

Janzen, D. H. 1966. Coevolution of mutualism between ants and acacias in Central America. Evolution 20: 249–275.

Janzen, D. H. 1968. Host plants as islands in evolutionary and contemporary time. American Naturalist 102: 592–595.

Janzen, D. H. 1979. Why fruit rots. Natural History Magazine 88(6): 60–64.

Johnson W. E., D. P. Onorato, M. E. Roelke, E. D. Land, M. Cunningham, R. C. Belden, R. McBride, D. Jansen, M. Lotz, D. Shindle, J. Howard, D. E. Wildt, L. M. Penfold, J. A. Hostetler, M. K. Oli, and S. J. O'Brien. 2010. Genetic restoration of the Florida panther. Science 329: 1641–1645.

Johnston, D. W., and E. P. Odum. 1956. Breeding bird populations in relation to plant succession on the Piedmont of Georgia. Ecology 37: 50–62.

Jost, L. 2006. Entropy and diversity. Oikos 113: 363–375.

Jost, L. 2007. Partitioning diversity into independent alpha and beta components. Ecology 88: 2427–2439.

Keever, C. 1950. Causes of succession on old fields of the Piedmont, North Carolina. Ecological Monographs 20: 230–250.

Kelly, A. E., and M. L. Goulden. 2008. Rapid shifts in plant distribution with recent climate change. Proceedings of the National Academy of Sciences of the United States of America 105: 11823–11827.

Kempton, R. A. 1979. The structure of species abundance and measurement of diversity. Biometrics 35: 307–321.

Kennedy, T. A., S. Naeem, K. M. Howe, J. M. H. Knops, D. Tilman, and P. Reich. 2002. Biodiversity as a barrier to ecological invasion. Nature 417: 636–638.

Kenward, R. E. 1978. Hawks and doves: factors affecting success and selection in goshawk attacks on wood-pigeons. Journal of Animal Ecology 47: 449–460.

Kerr, R. A. 2007. How urgent is climate change? Science 318: 1230–1231.

Kettlewell, H. B. D. 1955. Selection experiments on industrial melanism in the Lepidoptera. Heredity 10: 287–301.

Knops, J., D. Tilman, N. M. Haddad, S. Naeem, C. Mitchell, J. Haarstad, M. E. Ritchie, K. M. Howe, P. B. Reich, E. Siemann, and J. Groth. 1999. Effects of plant species richness on invasion dynamics, disease outbreaks, insect abundances and diversity. Ecology Letters 2: 286–293.

Koella, J. C., F. L. Sørensen, and R. Anderson. 1998. The malaria parasite *Plasmodium falciparum* increases the frequency of multiple feeding of its mosquito vector *Anopheles gambiae*. Proceedings of the Royal Society of London B 265: 763–768.

Komdeur, J. 1992. Importance of habitat saturation and territory quality for evolution of cooperative breeding in the Seychelles warbler. Nature 358: 493–495.

Krakauer, A. H. 2005. Kin selection and cooperative courtship in turkeys. Nature 434: 69–72.

Kump, L. R. 2008. The rise of atmospheric oxygen. Nature 451: 277–278.

Lacey, R. C. 1987. Loss of genetic diversity from unmanaged populations: interacting effects of drift, mutation, immigration, selection, and population subdivision. Conservation Biology 1: 143–158.

Lack, D. 1947. The significance of clutch size. Ibis 89: 302–352.

Lahaye, W. S., R. J. Gutiérrez, and H. R. Ankçakaya. 1994. Spotted owl metapopulation dynamics in southern California. Journal of Animal Ecology 63: 775–785.

Lambeck, R. J., and Hobbs, R. J. 2002. Landscape planning for conservation, pp. 360–380 in Applying Landscape Ecology in Biological Conservation, K. Gutzwiller, ed., Springer Verlag, New York.

Latham, R. E., and R. E. Ricklefs. 1993. Continental comparisons of temperate-zone tree species diversity, pp. 294–314 in Species Diversity in Ecological Communities, R. E. Ricklefs and P. Schluter, eds., University of Chicago Press, Chicago.

Lavers, C. P., R. H. Haines-Young, and M. I. Avery. 1996. The habitat associations of dunlin (*Calidris alpina*) in the flow country of northern Scotland and an improved model for predicting habitat quality. Journal of Applied Ecology 33: 279–290.

Lawton, J. H., and C. G. Jones. 1995. Linking species and ecosystems: organisms as ecosystem engineers, pp. 141–150 in Linking Species and Ecosystems, C. G. Jones and J. H. Lawton, eds., Chapman and Hill, New York.

Lawton, J. H., T. M. Lewinsohn, and S. G. Compton. 1993. Patterns of diversity for the insect herbivores on bracken, pp. 178–184 in Species Diversity in Ecological Communities, R. E. Ricklefs and D. Schluter, eds., University of Chicago Press, Chicago.

Leader-Williams, N., and S. D. Albon. 1988. Allocation of resources for conservation. Nature 336: 533–535.

Levine, J. M., P. B. Adler, and S. G. Yelenik. 2004. A meta-analysis of biotic resistance to exotic plant invasions. Ecology Letters 7: 975–989.

Levine, J. M., and C. M. D'Antonio. 1999. Elton revisited: a review of evidence linking diversity and invasibility. Oikos 87: 15–26.

Levins, R. 1969. Some demographic and genetic consequences of environmental heterogeneity for biological control. Bulletin of the Entomological Society 15: 237–240.

Likens, F. H., N. Bormann, N. M. Johnson, D. W. Fisher, and R. S. Pierce. 1970. The effects of forest cutting and herbivore treatment on nutrient budgets in the Hubbard Brook Watershed-Ecosystem. Ecological Monographs 40: 23–47.

Link, J. 2002. Does food web theory work for marine ecosystems? Marine Ecology Progress series 230: 1–9.

Liu, H., and R. W. Pemberton. 2009. Solitary invasive orchid bee outperforms co-occurring native bees to promote fruit set of an invasive *Solanum*. Oecoogia 159: 515–525.

Liu, Z., B. J. Richmond, E. A. Murray, R. C. Saunders, S. Steenrod, B. K. Stubblefield, D. M. Montague, and E. I. Ginns. 2004. DNA targeting of rhinal cortex D2 receptor protein reversibly blocks learning of cues that predict reward. Proceedings of the National Academy of Sciences of the United States of America 101: 12336–12341.

Lomolino, M. V. 1989. Interpretations and comparisons of constants in the species–area relationship: an additional caution. American Naturalist 133: 277–280.

Losey, J. E., A. R. Ives, J. Harman, F. Ballantyre, and C. Brown. 1997. A polymorphism maintained by opposite patterns of parasitism and predation. Nature 388: 269–272.

Lotka, A. J. 1925. Elements of Physical Biology. Reprinted, Dover, New York, 1956.
Lubchenko, J. 1978. Plant species diversity in a marine intertidal community: importance of herbivore food preference and algal competitive abilities. American Naturalist 112: 23–39.
MacArthur, R. H. 1965. Patterns of species diversity. Biological Review 40: 510–533.
MacArthur, R. H., and E. O. Wilson. 1963. An equilibrium theory of insular biogeography. Evolution 17: 373–387.
MacArthur, R. H., and E. O. Wilson. 1967. The Theory of Island Biogeography. Princeton University Press, Princeton, NJ.
MacLean, D. A., and R. W. Wein. 1978. Weight loss and nutrient changes in decomposing litter and forest floor material in New Brunswick forest stands. Canadian Journal of Botany 56: 2730–2749.
Maron, J. L., J. A. Estes, D. A. Croll, E. M. Danner, S. C. Elmendorf, and S. Buckalew. 2006. An introduced predator transforms Aleutian Island plant communities by disrupting spatial subsidies. Ecological Monographs 76: 3–34.
Marsden, S. J. 1998. Changes in bird abundance following selective logging on Seram, Indonesia. Conservation Biology 12: 605–611.
Martinez, N. D. 1992. Constant connectance in community food webs. American Naturalist 139: 1208–1218.
Marzal, A., F. de Lope, C. Navarro, and A. P. Møller. 2005. Malarial parasites decrease reproductive success: an experimental study in a passerine bird. Oecologia 142: 541–545.
May, R. M. 1976. Simple mathematical models with very complicated dynamics. Nature 261: 459–467.
Maynard Smith, J. 1976. Evolution and the theory of games. American Scientist 64: 41–45.
Maynard Smith, J. 1982. Evolution and the Theory of Games, Cambridge University Press, Cambridge.
Mayr, E. 1942. Systematics and the Origin of Species. Columbia University Press, New York.
McClenachan, L. 2009. Documenting loss of large trophy fish from the Florida Keys with historical photographs. Conservation Biology 23: 636–643.
McNab, B. K. 1973. Energetics and the distribution of vampires. Journal of Mammalogy 54: 131–144.
Menge, B. A., and J. P. Sutherland. 1976. Species diversity gradients: synthesis of the roles of predation, competition, and temporal heterogeneity. American Naturalist 110: 351–369.
Menzel, D. W., and J. H. Ryther. 1961. Nutrients limiting to the production of phytoplankton in the Sargasso sea, with special reference to iron. Deep Sea Research 7: 276–281.
Mills, L. S., and F. W. Allendorf. 1996. The one-migrant-per-generation rule in conservation and management. Conservation Biology 10: 1509–1518.
Mills, L. S., M. E. Soulé, and D. F. Doak. 1993. The keystone-species concept in ecology and conservation. BioScience 43: 219–224.
Mittelbach, G. G., et al. 2007. Evolution and the latitudinal diversity gradient: speciation, extinction and biogeography. Ecology Letters 10: 315–331.
Mittermeier, R. A., P. Robles Gil, M. Hoffmann, J. Pilgrim, T. Brooks, C. G. Mittermeier, J. Lamoreux, and G. A. B. da Fonseca. 2004. Hotspots Revisited. CEMEX, Mexico City.
Mittermeier, R. A., P. Robles-Gil, and C. G. Mittermeier (eds.). 1997. Megadiversity: Earth's Biologically Wealthiest Nations. CEMEX/ Agrupación Sierra Madre, Mexico City.
Moen, J., and L. Oksanen. 1991. Ecosystem trends. Nature 355: 510.
Moon, D. C., and P. D. Stiling. 2002. The influence of species identity and herbivore feeding mode on top-down and bottom-up effects in a salt marsh system. Oecologia 133: 243–253.
Moore, J. 1984. Altered behavioral responses in intermediate hosts: an acanthocephalan parasite strategy. American Naturalist 123: 572–577.
Moore T. R., J. A. Trofymow, B. Taylor, B. C. Prescott, C. Camiré, L. Duschene, J. Fyle, L. Kozak, M. Kranabette, I. Morrison, M. Siltanen, S. Smith, B. Titus, S. Visser, R. Wein, and S. Zoltai. 1999. Litter decomposition rates in Canadian forests. Global Change Biology 5: 75–82.
Morris, R. F. 1957. The interpretation of mortality data in studies on population dynamics. Canadian Entomologist 89: 49–69.
Morrison, L. W. 1997. The insular biogeography of small Bahamian cays. Journal of Ecology 85: 441–454.
Moulton, M. P., and S. L. Pimm. 1986. The extent of competition in shaping an introduced avifauna, pp. 80–97 in Community Ecology, J. Diamond and T. J. Case, eds., Harper and Row, New York.
Munir, B., and R. I. Sailer. 1985. Population dynamics of the tea scale, *Fiorinia theae* (Homoptera: Diaspididae), with biology and life tables. Environmental Entomology 14: 742–748.
Myers, R., J. K. Baum, T. D. Shepherd, S. P. Powers, and C. H. Peterson. 2007. Cascading effects of the loss of predatory sharks from a coastal ocean. Science 315: 1846–1850.
Naeem, S., J. M. H. Knops, D. Tilman, K. M. Howe, T. Kennedy, and S. Gale. 2000. Plant diversity increases resistance to invasion in the absence of covarying extrinsic factors. Oikos 91: 97–108.
Naeem, S., M. Loreau, P. Inchausti. 2002. Biodiversity and ecosystem functioning: the emergence of a synthetic ecological framework, pp. 3–11 in Biodiversity and Ecosystem Functioning: Synthesis and Perspectives, M. Loreau, S. Naeem, and P. Inchausti, eds., Oxford University Press, Oxford.
Naeem, S., L. J. Tompson, S. P. Lawler, J. H. Lawton, and R. M. Woodfin. 1994. Declining biodiversity can alter the performance of ecosystems. Nature 368: 734–737.
Nelleman, C., I. Vistnes, P. Jordhoy, and O. Strand. 2001. Winter distribution of wild reindeer in relation to powerlines, roads and resorts. Biological Conservation 101: 351–360.
Nelson, T. C. 1955. Chestnut replacement in the southern Highlands. Ecology 36: 352–353.
Norby, R. J., E. H. DeLucia, B. Gielen, C. Calfapietra, C. P. Giardina, J. S. King, J. Ledford, H. R. McCarthy, D. J. P. Moore, R. Ceulemans, P. De Angelis, A. C. Finzi, D. F. Karnosky, M. E. Kubiske, M. Lukac, K. S. Pregitzer, G. E. Scarascia-Mugnozza, W. H. Schlesinger, and R. Oren. 2005. Forest response to elevated $CO_2$ is conserved across a broad range of productivity. Proceedings of the National Academy of Sciences 102: 18052–18056.
Nores, M. 1995. Insular biogeography of birds on mountaintops in northwestern Argentina. Journal of Biogeography 22: 61–70.
Nuñez, M. A., and D. Simberloff. 2005. Invasive species and the cultural keystone species concept. Ecology and Society 10 (1): r4. http://www.ecologyandsociety.org/vol10/iss1/resp4/.
O'Dowd, D. J., P. T. Green, and P. S. Lake. 2003. Invasional "meltdown" on an oceanic island. Ecology Letters 6: 812–817.
Odum, E. P. 1971. Fundamentals of Ecology, 3rd ed. W. B. Saunders, Philadelphia.
Odum, H. T. 1957. Trophic structure and productivity of Silver Springs, Florida. Ecological Monographs 27: 55–112.
Oksanen, L., S. D. Fretwell, J. Arruda, and P. Niemela. 1981. Exploitation ecosystems in gradients of primary productivity. American Naturalist 118: 240–261.
Olden, J. D., J. M. McCarthy, J. T. Maxted, W. W. Fetzer, and M. J. Vander Zanden. 2006. The rapid spread of rusty crayfish (*Orconectes rusticus*) with observations on native crayfish declines in Wisconsin (U.S.A.) over the past 130 years. Biological Invasions 8: 1621–1628.

Olsen, J. S. 1963. Energy storage and the balance of producers and decomposers in ecological systems. Ecology 44: 322–331.

Oprea, M., P. Mendes, T. B. Vieira, and A. D. Ditchfield. 2009. Do wooded streets provide connectivity for bats in an urban landscape? Biodiversity and Conservation 18: 2361–2371.

Orr, M. R. 1992. Parasitic flies (Diptera: Phoridae) influence foraging rhythms and cast division of labor in the leaf cutter ant, *Atta cephalotes* (Hymenoptera: Formicidae). Behavioral Ecology and Sociobiology 30: 395–402.

Paine, R. T. 1966. Food web complexity and species diversity. American Naturalist 100: 65–75.

Paine, R. T. 1980. Food webs: linkage, interaction strength and community infrastructure. Journal of Animal Ecology 49: 667–695.

Park, T. 1948. Experimental studies of interspecies competition. I. Competition between populations of the flour beetles *Tribolium confusum* Duval and *Tribolium castaneum* Herbst. Ecological Monographs 18: 265–307.

Park, T. 1954. Experimental studies of interspecies competition. II. Temperature, humidity, and competition in two species of *Tribolium*. Physiological Zoology 27: 177–238.

Park, Y.-M. 1990. Effects of drought on two grass species with a different distribution around coastal dunes. Functional Ecology 4: 735–741.

Pascual, M., and M. J. Bouma. 2009. Do rising temperatures matter? Ecology 90: 906–912.

Pauly, D., and V. Christensen. 1995. Primary production required to sustain global fisheries. Nature 374: 255–257.

Payne, N. F. 1984. Mortality rates of beaver in Newfoundland. Journal of Wildlife Management 48: 117–126.

Pearl, R. 1927. The growth of populations. Quarterly Review of Biology 2: 532–548.

Pearson, D. E., and R. M. Callaway. 2006. Biological control agents elevate hantavirus by subsidizing deer mouse populations. Ecology Letters 9: 443–450.

Pemberton, R. W., and H. Liu. 2009. Marketing time predicts naturalization of horticultural plants. Ecology 90: 69–80.

Peterson, R. K., R. S. Davis, L. G. Higley, and O. A. Fernandes. 2009. Mortality risk in insects. Environmental Entomology 38: 2–10.

Pianka, E. R. 1970. On *r*- and *K*-selection. American Naturalist 104: 592–597.

Polis, G. A. 1991. Complex trophic interactions in deserts: an empirical critique of food-web theory. American Naturalist 138: 123–155.

Pollard, A. J. 1992. The importance of deterrence. Responses of grazing animals to plant variation, pp. 216–239 in Plant Resistance to Herbivores and Pathogens: Ecology, Revolution, and Genetics, R. S. Fritz and E. L. Simms, eds., University of Chicago Press, Chicago.

Post, E., R. O. Peterson, N. C. Stenseth, and B. E. McLaren. 1999. Ecosystem consequences of wolf behavioural response to climate. Nature 401: 905–907.

Pounds, J. A., M. R. Bustamante, L. A. Coloma, J. A. Consuegra, M. P. L. Fogden, P. N. Foster, E. La Marca, K. L. Masters, A. Merino-Viteri, R. Puschendorf, S. R. Ron, G. A. Sánchez-Azofeifa, C. J. Still, and B. E. Young. 2006. Widespread amphibian extinctions from epidemic disease driven by global warming. Nature 439: 161–167.

Power, M. E., and L. S. Mills. 1995. The Keystone cops meet in Hilo. Trends in Ecology and Evolution 10: 182–184.

Preston, C. D., D. A. Pearman, and A. R. Hall. 2004. Archaeophytes in Britain. Botanical Journal of the Linnean Society 145: 257–294.

Preston, F. W. 1948. The commonness and rarity of species. Ecology 29: 254–283.

Preston, F. W. 1962. The canonical distribution of commonness and rarity. Ecology 43: 185–215, 410–432.

Preszler, R. W., and W. J. Boecklen. 1996. The influence of elevation on tri-trophic interactions: opposing gradients of top-down and bottom-up effects on a leaf-mining moth. Ecoscience 3: 75–80.

Pugnaire, F., P. Haase, and J. Puigdefabregas. 1996. Facilitation between higher plant species in a semiarid environment. Ecology 77: 1420–1426.

Quinn, J. F., and S. P. Harrison. 1988. Effects of habitat fragmentation and isolation on species richness: evidence from biogeographic patterns. Oecologia 75: 132–140.

Rabinowitz, D. 1981. Seven forms of rarity, pp. 205–217 in The Biological Aspects of Rare Plant Conservation, H. Synge, ed., John Wiley, London.

Raich, J. W., A. E. Russell, and P. M. Vitusek. 1997. Primary productivity and ecosystem development along an elevational gradient on Mauna-Loa, Hawaii. Ecology 78: 707–721.

Ralls, K., and J. Ballou. 1983. Extinction: lessons from zoos, pp. 164–184 in Genetics and Conservation: A Reference for Managing Wild Animal and Plant Populations, C. M. Schonewald-Cox, S. M. Chambers, B. MacBryde, and L. Thomas, eds., Benjamin/Cummings, Menlo Park, CA.

Raymundo, L. J., A. R. Halford, A. P. Maypa, and A. M. Kerr. 2009. Functionally diverse reef-fish communities ameliorate coral disease. Proceedings of the National Academy of Sciences of the United States of America 106: 17067–17070.

Reed, D. H., J. J. O'Grady, B. W. Brook, J. D. Ballou, and R. Frankham. 2003. Estimates of minimum sizes for vertebrates and factors influencing those estimates. Biological Conservation 113: 23–34.

Reich, P. B., M. B. Walters, and D. S. Ellsworth. 1992. Leaf life-span in relation to leaf, plant and stand characteristics among diverse ecosystems. Ecological Monographs 62: 365–392.

Rey, J. R. 1983. Insular ecology of salt marsh arthropods: species-level patterns. Journal of Biogeography 12: 96–107.

Ricciardi, A. 2001. Facilitative interactions among aquatic invaders: is an "invasional meltdown" occurring in the Great Lakes? Canadian Journal of Fisheries and Aquatic Sciences 58: 2512–2525.

Ricklefs, R. E., and I. J. Lovette. 1999. The roles of island area per se and habitat diversity in the species-area relationships of four Lesser Antillean faunal groups. Journal of Animal Ecology 68: 1132–1160.

Ritchie, M. E., D. Tilman, and J. M. H. Knops. 1998. Herbivore effects on plant and nitrogen dynamics in oak savanna. Ecology 79: 165–177.

Robbins, C. S., D. K. Dawson, and B. A. Dowell. 1989. Habitat area requirements of breeding forest birds of the middle Atlantic states. Wildlife Monographs 103: 1–34.

Rocky Mountain Wolf Recovery 2005 Interagency Annual Report. http://www.r6.fws.gov/wolf/annualrpt05/.

Rodda, G. H., C. S. Jarnevich, and R. N. Reed. 2009. What parts of the US mainland are climatically suitable for invasive alien pythons spreading from Everglades National Park? Biological Invasions 11: 241–252.

Rodriquez de la Fuente, F. 1975. Animals of South America (World of Wildlife series), English-language version by John Gilbert. Orbis, London.

Roland, J. 1986a. Parasitism of winter moth in British Columbia during build-up of its parasitoid *Cyzenis albicans:* attack rate on oak vs. apple. Journal of Animal Ecology 55: 215–234.

Roland, J. 1986b. Success and failure of *Cyzenis albicans* in controlling its host the winter moth. Ph.D. thesis. University of British Columbia, Vancouver.

Roman, J., and S. R. Palumbi. 2003. Whales before whaling in the North Atlantic. Science 301: 508–510.

Root, R. 1967. The niche exploitation pattern of the blue-gray gnatcatcher. Ecological Monographs 37: 317–350.

Root, T. L. 1988. Energy constraints in avian distributions and abundances. Ecology 69: 330–339.
Root, T. L., J. T. Price, K. R. Hall, S. H. Schneider, C. Rosenzweig, and J. A. Pounds. 2003. Fingerprints of global warming on wild animals and plants. Nature 421: 57–60.
Rorison, I. H. 1969. Ecological Aspects of the Mineral Nutrition of Plants. Blackwell, Oxford.
Rosenzweig, M. L. 1968. Net primary productivity of terrestrial communities: prediction from climatological data. American Naturalist 102: 67–74.
Rosenzweig, M. L. 1992. Species diversity gradients: we know more and less than we thought. Journal of Mammalogy 73: 715–730.
Rosumek, F. B., F. A. Silveira, F. de S Neves, N. P. de U Barbosa, L. Diniz, Y. Oki, F. Pezzini, G. W. Fernandes, and T. Cornelissen. 2009. Ants on plants: a meta-analysis of the role of ants as plant biotic defenses. Oecologia 160: 537–549.
Royama, T. 1996. A fundamental problem in key factor analysis. Ecology 77: 87–93.
Ryan, P. G., P. Blommer, C. L. Moloney, T. J. Grant, and W. Delport. 2007. Ecological speciation in south Atlantic island finches. Science 315: 1420–1422.
Saccheri, I., M. Kuussaari, M. Kankare, P. Vikman, W. Forteluis, and I. Hanski. 1998. Inbreeding and extinction in a butterfly metapopulation. Nature 392: 491–494.
Sala, O. S., W. J. Parton, L. A. Joyce, and W. K. Lauenroth. 1988. Primary production of the central grassland region of the United States. Ecology 69: 40–45.
Salisbury, E. J. 1926. The geographical distribution of plants in relation to climate factors. Geographical Journal 67: 312–335.
Salo, P., E. Korpimäki, P. B. Banks, M. Nordström, and C. R. Dickman. 2007. Alien predators are more dangerous than native predators to prey populations. Proceedings of the Royal Society B 274: 1237–1243.
Sand, H., G. Cederlund, and K. Danell. 1995. Geographical and latitudinal variation in growth patterns and adult body size of Swedish moose (*Alces alces*). Oecologia 102: 433–442.
Sanderson, E., M. Jaiteh, M. A. Levy, K. H. Redford, A. V. Wannebo, and G. Woolmer. 2002. The human footprint and the last of the wild. BioScience 52: 891–904.
Sarukhán, J., and J. L. Harper. 1973. Studies on plant demography: *Ranunculus repens* L., *R. bulbosus* L., and *R. acris* L. I. Population flux and survivorship. Journal of Ecology 61: 675–716.
Schaffer, M. L. 1981. Minimum population sizes for species conservation. BioScience 31: 131–134.
Schindler, D. W. 1974. Eutrophication and recovery in experimental lakes: implications for lake management. Science 184: 397–399.
Schindler, D. W. 1977. Evolution of phosphorus limitation in lakes. Science 195: 260–262.
Schindler, M. H., and M. O. Gessner. 2009. Functional leaf traits and biodiversity effects on litter decomposition in a stream. Ecology 90: 1641–1649.
Schluter, D., and R. E. Ricklefs. 1993. Convergence and the regional component of species diversity, pp. 230–240 in Species Diversity in Ecological Communities, R. Ricklefs and D. Schluter, eds., University of Chicago Press, Chicago.
Schoener, T. W. 1983. Field experiments on interspecific competition. American Naturalist 122: 240–285.
Schuur, E. A. G. 2003. Productivity and global climate revisited: the sensitivity of tropical forest growth to precipitation. Ecology 84: 1165–1170.
Schwartz, M. W., C. A. Brigham, J. D. Hoeksema, K. G. Lyons, M. H. Mills, and P. J. van Mantgem. 2000. Linking biodiversity to ecosystem function: implications for conservation ecology. Oecologia 122: 297–305.
Silvertown, J. W., M. Franco, I. Pisanty, and A. Mendoza. 1993. Comparative planned demography: relative importance of life-cycle components to the finite rate of increase in woody and herbaceous perennials. Journal of Ecology 81: 465–476.
Simberloff, D. S. 1976. Species turnover and equilibrium island biogeography. Science 194: 572–578.
Simberloff, D. S. 1978. Colonization of islands by insects: immigration, extinction and diversity, pp. 139–153 in Diversity of Insect Faunas, L. A. Mound and N. Waloff, eds., Blackwell Scientific, Oxford.
Simberloff, D. 2003. How much information on population biology is needed to manage introduced species? Conservation Biology 17: 83–92.
Simberloff, D. 2006. Invasional meltdown 6 years later: important phenomenon, unfortunate metaphor, or both? Ecology Letters 9: 912–919.
Simberloff, D., and B. Von Holle. 1999. Positive interactions of nonindigenous species: invasional meltdown? Biological Invasions 1: 21–32.
Simberloff, D., and E. O. Wilson. 1969. Experimental zoogeography of islands: the colonization of empty islands. Ecology 50: 278–296.
Simberloff, D. S., and E. O. Wilson. 1970. Experimental zoogeography of islands: a two year record of recolonization. Ecology 51: 934–937.
Simpson, E. H. 1949. Measurement of diversity. Nature 163: 688.
Simpson, G. G. 1961. Principles of Animal Taxonomy, Columbia University Press, New York.
Simpson, G. G. 1964. Species density of North American recent mammals. Systematic Zoology 13: 57–73.
Sinclair, A. R. E. 1977. The African Buffalo. University of Chicago Press, Chicago.
Sinclair, A. R. E. 1989. Population regulation in animals, pp. 197–242 in Ecological Concepts, J. M. Cherret, ed., Blackwell Scientific, Oxford.
Smith, S. D., T. E. Huxman, S. F. Zitzer, T. N. Charlet, D. C. Housman, J. S. Coleman, L. K. Fenstermaker, J. R. Seemann, and R. S. Nowak. 2000. Elevated $CO_2$ increases productivity and invasive species success in an arid ecosystem. Nature 408: 79–82.
Snyder, W. E., G. B. Snyder, D. L. Finke, and C. S. Straub. 2006. Predator biodiversity strengthens herbivore suppression. Ecology Letters 9: 789–796.
Sorensen, T. 1948. A method of establishing groups of equal amplitude in plant sociology based on similarity of species content. Det. Kong. Danske Vidensk. Selsk. Biol. Skr (Copenhagen) 5 (4): 1–34.
Sousa, W. P. 1979. Disturbance in marine intertidal boulder fields: the nonequilibrium maintenance of species diversity. Ecology 60: 1225–1239.
Southwood, T. R. E. 1978. Ecological Methods with Particular Reference to the Study of Insect Populations, 2nd ed. Methuen, London.
Sterling, I., and D. E. Derocher. 2012. Effects of climate warming on polar bears: a review of the evidence. Global Change Biology 18: 2694–2706.
Stiling, P. D. 1990. Calculating the establishment rates of parasitoids in classical biological control. Bulletin of the Entomological Society of America 36: 225–230.
Stiling, P. D. 1993. Why do natural enemies fail in classical biological control programs? American Entomologist 39: 31–37.
Stiling, P. D., and T. Cornelissen. 2007. How does elevated carbon dioxide ($CO_2$) affect plant-herbivore interactions? A field experimental and meta-analysis of $CO_2$-mediated changes on plant chemistry and herbivore performance. Global Change Biology 13: 1823–1842.
Stiling, P. D., D. C. Moon, M. D. Hunter, A. M. Rossi, G. J. Hymus, and B. G. Drake. 2003. Elevated $CO_2$ lowers relative and absolute

herbivore density across all species of a scrub oak forest. Oecologia 134: 82–87.
Stiling, P. D., and A. M. Rossi. 1997. Experimental manipulations of top-down and bottom-up factors in an in-trophic system. Ecology 78: 1602–1606.
Stiling, P. D., and D. R. Strong. 1983. Weak competition among *Spartina* stem borers by means of murder. Ecology 64: 770–778.
Stohlgren, T. J., D. Barnett, and J. Kartesz. 2003. The rich get richer: patterns of plant invasions in the United States. Frontiers in Ecology and the Environment 1: 11–14.
Strickberger, M. W. 1986. Genetics, 3rd ed. Macmillan, New York.
Strong, D. R. 1974. Nonasymptotic species richness models and the insects of British trees. Proceedings of the National Academy of Sciences of the United States of America 71: 2766–2769.
Suttle, K. B., M. Thomsen, and M. Power. 2007. Species interactions reverse grassland responses to changing climate. Science 315: 640–642.
Swenson, J. E., F. Sandegren, A. Söderberg, A. Bjärvall, R. Franzén, and P. Wabakken. 1997. Infanticide caused by hunting of male bears. Nature 386: 450–451.
Teal, J. 1962. Energy flow in the salt marsh ecosystem. Ecology 43: 614–624.
Teer, J. G., J. W. Thomas, and E. A. Walker. 1965. Ecology and management of white-tailed deer in the Llano Basin of Texas. Wildlife Monographs No. 15, 62 pp.
Teeri, J. A., and L. G. Stowe. 1976. Climate patterns and the distribution of C4 grasses in North America. Oecologia 28: 1–12.
Terborgh, J. 1973. On the notion of favorableness in plant ecology. American Naturalist 107: 481–501.
Terborgh, J. 1986. Keystone plant resources in the tropical forest, pp. 330–344 in Conservation Biology: The Science of Scarcity and Diversity, M. E. Soulé, ed., Sinauer Associates, Sunderland, MA.
Terborgh, J., L. Lopez, P. Nuñez, M. Rao, G. Shahabuddin, G. Orihuela, M. Riveros, R. Ascanio, G. H. Adler, T. D. Lambert, and L. Balbas. 2001. Ecological meltdown in predator-free forest fragments. Science 294: 1923–1926.
Tewksbury, J. J., D. J. Levey, N. M. Haddad, S. Sargent, J. L. Orrock, A. Weldon, B. J. Danielson, J. Brinkerhoff, E. I. Damschen, and P. Townsend. 2002. Corridors affect plants, animals, and their interactions in fragmented landscapes. Proceedings of the National Academy of Sciences 99: 12923–12926.
Thomas, C. D., A. Cameron, R. E. Green, M. Bakkenes, L. J. Beaumont, Y. C. Collingham, B. F. N. Erasmus, M. Ferreira de Siqueira, A. Grainger, L. Hannah, L. Hughes, B. Huntley, A. S. Van Jaarsveld, G. E. Midgely, L. Miles, M. A. Ortega-Huerta, A. T. Peterson, O. L. Phillips, and S. E. Williams. 2004. Extinction risk from climate change. Nature 427: 145–148.
Tilman, D. 1982. Resource Competition and Community Structure. Princeton University Press, Princeton, NJ.
Tilman, D. 1985. The resource ratio hypothesis of plant succession. American Naturalist 125: 827–852.
Tilman, D. 1996. Biodiversity: population versus ecosystem stability. Ecology 77: 350–363.
Tilman, D. 1997. Mechanisms of plant competition, pp. 239–261 in Plant Ecology, M. J. Crawley, ed., 2nd ed., Blackwell Scientific Oxford, England.
Tilman, D., M. Mattson, and S. Langer. 1981. Competition and nutrient kinetics along a temperature gradient: an experimental test of a mechanistic approach to niche theory. Limnology Oceanography 26: 1020–1033.
Tilman, D., D. Wedin, and J. Knops. 1996. Productivity and sustainability influenced by biodiversity in grassland ecosystems. Nature 379: 718–720.
Tokeshi, M. 1999. Species coexistence: ecological and evolutionary perspectives. Blackwell Scientific, Oxford.
Tonn, W. M., and J. J. Magnusson. 1982. Patterns in the species composition and richness of fish assemblages in northern Wisconsin lakes. Ecology 63: 1139–1166.
Toth, G. B., O. Langhamer, and H. Pavia. 2005. Inducible and constitutive defenses of valuable seaweed tissues: consequences for herbivore fitness. Ecology 86: 612–618.
Tyler, A. C., J. G. Lambrinos, and E. D. Grossholz. 2007. Nitrogen inputs promote the spread of an invasive marsh grass. Ecological Applications 17: 1886–1898.
Vahed, K. 1998. The function of nuptual feeding in insects: a review of empirical studies. Biological Reviews 73: 43–78.
Van der Veken, S., M. Hermy, M. Vellend, A. Knapen, and K. Verheyen. 2008. Garden plants get a head start on climate change. Frontiers in Ecology and the Environment 6: 212–216.
van Mantgem, P. J., N. L. Stephenson, J. C. Byrne, L. D. Daniels, J. F. Franklin, P. Z. Fule, M. E. Harmon, A. J. Larson, J. M. Smith, A. H. Taylor, and T. T. Veblen. 2009. Widespread increase of tree mortality rates in the western United States. Science 323: 521–524.
Van Valen, L. M. 1973. A new evolutionary law. Evolutionary Theory 1: 1–30.
Van Valen, L. M. 1976. Ecological species, multi-species, and oaks. Taxon 25: 233–239.
Varley, G. C. 1941. On the search for hosts and the egg distribution of some chalcid parasites of the knapweed gall-fly. Parasitology 33: 47–66.
Varley, G. C. 1971. The effects of natural predators and parasites on winter moth populations in England, pp. 103–116 in Proceedings, Tall Timbers Conference on Ecological Animal Control by Habitat Management No. 2, Tall Timbers Research Station, Tallahassee, FL.
Varley, G. C., G. R. Gradwell, and M. P. Hassell. 1973. Insect Population Ecology: An Analytical Approach. Blackwell Scientific, Oxford.
Visser, M. E., and C. Both. 2005. Shifts in phenology due to global climate change: the need for a yardstick. Proceedings of the Royal Society B 272: 2561–2569.
Vitousek, P. M., L. R. Walker, L. D. Whiteaker, D. Muellar-Dombois, and P. A. Matson. 1987. Biological invasion by *Myrica faya* alters ecosystem development in Hawaii. Science 238: 802–804.
Vogel, M., H. Remmert, and R. I. Lewis Smith. 1984. Introduced reindeer and their effects on the vegetation and the epigeic invertebrate fauna of South Georgia (subantarctic). Oecologia 62: 102–109.
Volterra, V. 1926. Fluctuations in the abundance of a species considered mathematically. Nature 118: 558–560.
Vredenburg, V. T. 2004. Reversing introduced species effects: experimental removal of introduced fish leads to rapid recovery of a declining frog. Proceedings of the National Academy of Sciences 101: 7646–7650.
Wackernagel, M., L. Onisto, P. Bello, A. Callegjas Linares, I. S. Lopez Falfan, J. Mendez Garcia, A. I. Suarez Guerrero, and M. G. Suarez Cuerrero. 1999. National natural capital accounting with the ecological footprint concept. Ecological Economics 29: 375–390.
Wackernagel, M., and W. E. Rees. 1996. Our Ecological Footprint: Reducing Human Impact on the Earth. New Society, Gabriola Island, B.C.
Walker, B. H. 1992. Biodiversity and ecological redundancy. Conservation Biology 6: 18–23.
Ward, P. D. 2006. Out of Thin Air: Dinosaurs, birds, and Earth's ancient atmosphere. Joseph Henry Press, Washington, DC.
Waring, G. L., and N. S. Cobb. 1992. The impact of plant stress on herbivore population dynamics, pp. 168–226 in Insect-Plant Interactions, vol. 4, E. Bernays, ed., CRC Press, Boca Raton, FL.
Weeks, P. 2000. Red-billed oxpeckers: vampires or tickbirds? Behavioral Ecology 11: 154–160.

Weir, J. T., and D. Schluter. 2007. The latitudinal gradient in recent speciation and extinction rates of birds and mammals. Science 315: 1574–1576.

Westemeier, R. L., J. D. Brown, S. A. Simpson, T. L. Esker, R. W. Jansen, and J. W. Walk. 1998. Tracing the long-term decline and recovery of an isolated population. Science 282: 1695–1698.

White, J. A., and T. G. Whitham. 2000. Associational susceptibility of cottonwood to a box elder herbivore. Ecology 81: 1975–1883.

Whitehorn, P. R., S. O'Connor, F. L. Wackers, and D. Goulson. 2012. Neonicotinoid pesticide reduces bumble bee colony growth and queen production. Science 336: 351–352.

Whittaker, R. H. 1970. Communities and Ecosystems. MacMillan, London.

Whittaker, R. H. 1975. Communities and Ecosystems, 2nd ed. Macmillan, New York.

Whittaker, R. J., M. B. Bush, T. Partomihardjo, N. M. Asquith, and K. Richards. 1992. Ecological aspects of plant colonization of the Krakatau Islands. Geojournal 28: 81–302.

Wilcove, D. S., D. Rothstein, J. Dubow, A. Phillips, and E. Losos. 1998. Quantifying threats to imperiled species in the United States. BioScience 48: 607–615.

Wiles, G. J., J. Bart, R. E. Beck Jr., and C. Aguon. F. 2003. Impacts of the brown tree snake: patterns of decline and species persistence in Guam's avifauna. Conservation Biology 17: 1350–1360.

Wilkinson, G. S. 1990. Food sharing in vampire bats. Scientific American 262: 76–82.

Williams, C. B. 1964. Patterns in the Balance of Nature and Related Problems in Quantitative Ecology. Academic Press, New York.

Williams, G. C. 1966. Adaptation and Natural Selection. Princeton University Press, Princeton, NJ.

Williamson, M. 1987. Are communities ever stable? pp. 353–371 in Colonization, Succession and Stability, A. J. Gray, M. J. Crawley, and P. J. Edwards, eds., Blackwell Scientific, Oxford.

Wilson, Edward O. 1984. Biophilia. Harvard University Press, Cambridge.

Wilson, E. O. 2002. The Future of Life. Knopf, New York.

Wilson, J. B., E. Spijkerman, and J. Huisman. 2007. Is there really insufficient support for Tilman's R* concept? A comment on Miller et al. American Naturalist 169: 700–706.

Wilson, J. D., M. S. Dorcas, and R. W. Snow. 2011. Identifying plausible scenarios of the establishment of invasive Burmese pythons (*Python malorus*) in southern Florida. Biological Invasions. 13: 1493–1504.

Witz, B. W. 1989. Antipredator mechanisms in arthropods: a twenty-year literature survey. Florida Entomologist 73: 71–99.

Woodell, S. R., J. H. A. Mooney, and A. J. Hill. 1969. The behavior of *Larrea divaricata* (creosote bush) in response to rainfall in California. Journal of Ecology 57: 37–44.

Wright, S., J. Keeling, and L. Gillman. 2006. The road from Santa Rosalia: a faster temper of evolution in tropical climates. Proceedings of the National Academy of Sciences 103: 7718–7722.

Wynne-Edwards, V. C. 1962. Animal Dispersion in Relation to Social Behaviour. Oliver and Boyd, Edinburgh.

Yom-Tov, Y., S. Yom-Tov, J. Wright, C. J. R. Thorne, and R. Du Feu. 2006. Recent changes in body weight and wing length among some British passerine birds. Oikos 112: 91–101.

Zach, R. 1979. Shell-dropping: decision making and optimal foraging in Northwestern crows. Behaviour 68: 106–117.

Zelitch, I. 1971. Photosynthesis, Photorespiration and Plant Productivity. Academic Press, New York.

Zuk, K. M., R. Thornhill, J. D. Ligen, and K. Johnson. 1990. Parasites and mate choice in red jungle fowl. American Zoologist 30: 235–244.

# Photo Credits

Image Research by Danny Meldung/Photo Affairs, Inc.

## Section Openers

1: © Dave Watts/Visuals Unlimited; 2: © David Chapman/Alamy; 3: Pedro Ramirez, Jr./U.S. Fish & Wildlife Service; 4: © Image Source/Alamy RF; 5: © Dr. Morley Read/Science Source; 6: © Getty Images/Digital Vision RF; 7: © Marc Anderson.

## Chapter 1

Opener: © Thomas Marent/Visuals Unlimited; 1.2a: © Hulton Archive/Getty Images; 1.2b: © Science Source; 1.3a: © Brand X Pictures/PunchStock RF; 1.3b: © Digital Vision/Getty Images RF; 1.3c: © Paul Springett/Alamy RF; 1.3d: © DLILLC/Corbis RF; 1.4: © Charles V. Angelo/Science Source; 1.7: Library of Congress; 1.Ba: Norman E. Rees, USDA Agricultural Research Service - Retired, Bugwood.org; 1.Ba(inset): © John W. Bova/Science Source; 1.Bb: Robert D. Richard, USDA APHIS PPQ, Bugwood.org; 1.Bb(inset): Jim Story, Montana State University, Bugwood.org; 1.Bc: © Nicholas Bergkessel, Jr./Science Source; 1.Bd: © Albuquerque Journal, Paul Bearce/AP Photo; 1.8a: © James T. Tanner/Science Source; 1.8b: © Jack Jeffrey; 1.8c: © Topham/The Image Works; 1.8d: © Frans Lanting/Corbis.

## Chapter 2

Opener: © Brand X Pictures/PunchStock RF; 2.2a–b: © Diane Nelson; 2.A(i–ii): © David Fox/Getty Images; 2.3: © Geoff Renner/Getty Images; 2.8a–b: © Stan Flegler/Visuals Unlimited; 2.15: © Charles G. Summers Jr./Photoshot; 2C(bottom right): © blickwinkel/Alamy; 2.16: © Royalty-Free/Corbis.

## Chapter 3

Opener: © Peter Ryan; 3.2a: © D. Parer & E. Parer-Cook/ardea.com; 3.2b: © Gerald & Buff Corsi/Visuals Unlimited; 3.6a: © Gary Meszaros/Visuals Unlimited; 3.6b: © R. Andrew Odum/Getty Images; 3.B(i): © Chris Hepburn/Getty Images RF; 3.B(ii): © Rob Walls/Alamy; 3.B(iii): © Bob Gibbons/Science Source; 3.9(*A. taeniatus*): © Hal Beral/V&W/imagequestmarine.com; 3.9(*A. virginicus*): © Amar and Isabelle Guillen/Guillen Photo LLC/Alamy; 3.12: © altrendo nature/Getty Images; 3.13: © Jim Zuckerman/Corbis; 3.14: © Dorling Kindersley/Getty Images.

## Chapter 4

Opener: © Dr. Jeremy Burgess/Science Source; 4.1: © Cyril Ruoso/Biosphoto; 4.4: © Peter Stiling; 4.5: © Richard H. Hansen/Science Source; 4.6: © Raymond Mendez/Animals Animals; 4.10: © Biosphoto/SuperStock; 4.12: Photo by Scott Bauer/USDA; 4.13a: © Digital Vision/PunchStock RF; 4.13b: © Gregory G. Dimijian/Science Source; 4.13c: © PhotoAlto/James Hardy/PunchStock RF; 4.14: © Stuart Crump/Alamy; 4.16a: © Masahiro Iijima/ardea.com; 4.16b: © WILDLIFE GmbH/Alamy; 4.16c: © Millard H. Sharp/Science Source; 4.17: © John Devries/Science Source; 4.18: © Chris Knights/ardea.com; 4.19: © Stephen A. Marshall; 4.22: © Ben Osborne/Getty Images.

## Chapter 5

Opener: © Norbert Rosing/Getty Images; 5.1b: © Comstock Images/PictureQuest RF; 5.4b(both): Courtesy of Uffe Midtgård, University of Copenhagen; 5.6b: © Stephen Dalton/Science Source; 5.10: © Getty Images RF; 5.11a: © Rick & Nora Bowers/Alamy; 5.11b: © Royalty-Free/Corbis; 5.11c: © Image Source/PunchStock RF; 5.11d: © Image Plan/Corbis RF; 5.12a: © John Kapriclian/Science Source; 5.12b: © John Glover/Gap Photo/Visuals Unlimited; 5.13: © Frank Oberle/Getty Images; 5.14: © Arno Vlooswijk/TService/Science Source; 5.16a: © Jeffrey Lepore/Science Source; 5.16b: © Creatas/PunchStock RF; 5.17a–b: © Heather Angel/Natural Visions; 5.21c: © Steffen Hauser/agefotostock; 5.21d: © Organica/Alamy.

## Chapter 6

Opener: © NHPA/Photoshot; 6.1b: © Aaron Haupt/Science Source; 6.1d: © Chris McGrath/Getty Images; 6.1e: © Dana Tezarr/Getty Images; 6.1f: © Anthony Bannister/Gallo Images/Corbis; 6.1g: © Hermann Eisenbeiss/Science Source; 6.3: © Michael Diggin/Alamy; 6.6(inset): © DLILLC/Corbis RF; 6.7b(inset): © Getty Images RF; 6.8: © Rolf Nussbaumer Photography/Alamy; 6.10a: © Nigel Cattlin/Visuals Unlimited; 6.12: © Frans Lanting/Frans Lanting Stock; 6.13: © Virginia P. Weinland/Science Source; 6.15a: © dbphots/Alamy; 6.15b: © G.A. Matthews/Science Source; 6.17: © Frederica Georgia/Science Source; 6.C: © William Leaman/Alamy RF.

## Chapter 7

Opener: © Tom Campbell/Purdue Agricultural Communication file photo; 7.2: © Catherine Koehler; 7.3: © Wally Eberhart/Visuals Unlimited; 7.4a: © Royalty-Free/Corbis; 7.4b, 7.10a–b: © Steven P. Lynch; 7.12c: Photo courtesy of National Park Service, Everglades National Park.

## Chapter 8

Opener: © Digital Vision/Getty Images RF; 8.1: © Paul Glendell/Alamy; 8.2a: © Marlin E. Rice; 8.2b: © Rincon-Vitova Insectaries; 8.4a: © Inga Spence/Visuals Unlimited; 8.4b: © AfriPics.com/Alamy; 8.4c: U.S. Air Force; 8.5: © W. Wayne Lockwood, M.D./Corbis; 8.7a, c: © Digital Vision/Getty Images RF; 8.7b: © Getty Images RF; 8.8a: © Comstock Images/Alamy RF; 8.8b: © PhotoLink/Getty Images RF; 8.8c: © Digital Vision/PunchStock RF; 8.B: © Royalty-Free/Corbis; 8.11(inset): © Stockbyte RF; 8.13b: © Kent Knudson/PhotoLink/Getty Images RF; 8.14(inset): © Richard Wear/Design Pics/Corbis RF; 8.C(i): © Ellen Damschen; 8.17(inset): J&K Hollingsworth/U.S. Fish & Wildlife Service.

## Chapter 9

Opener: © Roy Toft/National Geographic Creative/Getty Images; 9.1(inset): © Peter Stiling; 9.3(inset): © Creatas/PunchStock RF; 9.A: © Artostock.com/Alamy RF; 9.5(inset): © Photodisc Collection/Getty Images RF; 9.6(inset): © MJ Photography/Alamy RF.

## Chapter 10

Opener: © Mike Lockhart; 10.10a: © David Sieren/Visuals Unlimited; 10.10b: © Dr. Phil Gates/Biological Photo Service; 10.13a: © Doug Sherman/Geofile; 10.13b: © Digital Vision/PunchStock RF; 10.13c: © Creatas/PunchStock RF.

## Chapter 11

Opener: © FLPA/Phil McLean/agefotostock; 11.3(left): Courtesy CSIRO Publishing; 11.3(right): © Nigel Cattlin/Visuals Unlimited; 11.6a: © Kenneth Lilly/Getty Images RF; 11.6b: © Royalty-Free/Corbis; 11.6c: © John Sohlden/Visuals Unlimited; 11.6d: © Dr. Barry Rice; 11.7b: © James H. Robinson/Science Source; 11.B(inset): © Gary Meszaros/Visuals Unlimited.

## Chapter 12

Opener: © ARCO/J. Meul/agefotostock; 12.A(i): USDA; 12.A(ii): © Martial Colomb/Getty Images RF; 12.2a: © Steven P. Lynch; 12.2b: © Doug Sherman/Geofile; 12.2c: © Chris Raper; 12.2d: Courtesy Joaquim Alves Gaspar, Wikimedia Commons; 12.4a: © Mike Wilkes/Nature Picture Library; 12.4b: © Adrian Warren/ardea.com; 12.4c: © Rodger Jackman/Getty Images; 12.5a: © Creatas Images/PictureQuest RF; 12.5b: © WILDLIFE GmbH/Alamy; 12.6a: © sharky/

Alamy; 12.6b: © Chris Gomersall/Alamy; 12.7a: © Dr. Merton Brown/Visuals Unlimited; 12.7b: © Bryan Mullennix/Getty Images; 12.7c: © Alex Wild/Visuals Unlimited; 12.12: © Bill Banaszewski/Visuals Unlimited.

### Chapter 13

Opener: Whitney Cranshaw, Colorado State University, Bugwood.org; 13.2a: © Dr. Thomas Eisner/Visuals Unlimited; 13.2b–c: © Creatas/PunchStock RF; 13.2d: © Doug Sherman/Geofile; 13.3a: © Andrew Hewitt/Alamy RF; 13.3b: © Creatas Images/PictureQuest RF; 13.3c: © Diane Nelson; 13.4a: © Suzanne L. & Joseph T. Collins/Science Source; 13.4b: © Jason Ondreicka/Alamy RF; 13.4c: © Purestock/Alamy RF; 13.4d: © Creatas/PunchStock RF; 13.4e: © Doug Sherman/Geofile; 13.5a: © Design Pics Inc./agefotostock RF; 13.5b: © Digital Vision/Getty Images RF; 13.6: © Charles Volkland/agefotostock; 13.9: © Thomas Kitchin & Victoria/agefotostock; 13.11a: U.S. Fish and Wildlife Service; 13.19a–b: From the archives of Monroe County Public Library, Key West, Florida. Photo a by Wil-Art Studio/Art Stickel. Photo b by Dale McDonald; 13.19c: Photo courtesy Loren McClenachan, Ph.D.

### Chapter 14

Opener: © Paul E. Tessier/Getty Images RF; 14.2a: © Steven P. Lynch; 14.2b: © FLPA/Richard Becker/agefotostock; 14.2c: © Digital Vision/PunchStock RF; 14.3: © EggImages/Alamy RF; 14.4a: © CuboImages srl/Alamy; 14.4b: © Author's Image/PunchStock RF; 14.4c: © Brand X Pictures/PunchStock RF; 14.7: © Dr. David Dussourd; 14.11a–b: © Hawaii Department of Agriculture; 14.11c: Courtesy Forest and Kim Starr; 14.13a–b: Courtesy Peter Coyne, www.petaurus.com; 14.14a–b: State Library of Queensland, neg no.'s API-101-01-0001r & API-101-01-0002r.

### Chapter 15

Opener: © Colin Chapman, McGill University; 15.1a: © Roger De LaHarpe/Gallo Images/Corbis; 15.1b: © Gerry Bishop/Visuals Unlimited; 15.3a: © A & J Visage/Alamy; 15.3b: © Malcolm Schuyl/Alamy; 15.6: © Stuart Wilson/Science Source; 15.8a: © Peter Stiling; 15.10: © Lisa Norman-Hudson/AP Photo; 15.11: © inga spence/Alamy; 15.B(i): Photo courtesy of Texas A & M University College of Veterinary Medicine & Biomedical Sciences.

### Chapter 16

Opener: © Piers Calvert Photography; 16.5: © Nigel Cattlin/Alamy.

### Chapter 17

Opener: © Daryl & Sharna Balfour/Okapia/Science Source; 17.A(i): © Danita Delimont/Alamy; 17.A(ii): © Wayne Lawler/Ecoscene/Corbis; 17.A(iii): © Hugh Lansdown/agefotostock; 17.6d: © The Natural History Museum/Alamy.

### Chapter 18

Opener: © Martin Harvey/Corbis RF; 18.9a: © amana images inc./Alamy RF; 18.10: © David Wrobel/Visuals Unlimited; 18.12: © Steven P. Lynch.

### Chapter 19

Opener: © Michael Patrick O'Neill/Science Source; 19.1a: © Florapix/Alamy; 19.1b: © Michael J. Tyler/Science Source; 19.A(3): © Pete Manning, Ecotron Facility, NERC Centre for Population Biology; 19.4a: © 2009 Elizabeth Livermore/Getty Images RF; 19.8b: © David Tilman, University of Minnesota, 2000; 19.13a: © Philip Stewart; 19.13b, e, f: © Peter T. Green.

### Chapter 20

Opener: © Steve Terrill/Corbis; 20.1a–b: © Dr. Juerg Alean/www.swisseduc.ch; 20.3a: © Fred Hirschmann; 20.4a: © Grant Dixon/Lonely Planet Images/Getty Images; 20.4b: © Robert Harding World Imagery/Alamy; 20.4c: © Accent Alaska.com/Alamy; 20.4d: © Harald Sund/Getty Images RF; 20.5: © John Lemker/Animals Animals; 20.6a: © Wayne Sousa/University of California, Berkeley; 20.B(i): 1899, W.H. Jackson/U.S. Geological Survey; 20.B(ii): 1977, H.E. Malde/U.S. Geological Survey; 20.C: Courtesy Quivira Coalition/www.quiviracoalition.org; 20.D(i): J.D. Wright/U.S. Geological Survey; 20.D(ii): R.M. Turner/U.S. Geological Survey; 20.14a: © The Board of Regents of the University of Wisconsin System; 20.14b: © D.L. Rockwood, School of Forest Resources and Conservation, University of Florida, Gainesville, FL; 20.14c: © Sally A. Morgan/Ecoscene/Corbis; 20.15a–b: © Wild Earth Guardians; 20.17: © Visual&Written SL/Alamy.

### Chapter 21

Opener: © Dr. Richard Roscoe/Visuals Unlimited; 21.3: Courtesy National Park Service; 21.9, 21.B(1–3): Courtesy Dr. D. Simberloff, University of Tennessee.

### Chapter 22

Opener: © Anne E. Kelly; 22.11a: © Nick Hanna/Alamy; 22.11b: © A.E. Amador, PhotosPR.com; 22.17a: © Digital Vision/PunchStock RF; 22.17b: © Creatas/PunchStock RF; 22.18a: © Purestock/PunchStock RF; 22.18b: © Anup Shah/Photodisc/Getty Images RF; 22.19a: © Alex L. Fradkin/Getty Images RF; 22.19b: © Royalty-Free/Corbis; 22.20a: © Purestock/PunchStock RF; 22.20b, 22.23a: © Creatas/PunchStock RF; 22.23b: © Photodisc Collection/Getty Images RF; 22.24a: © Getty Images RF; 22.24b: © Brand X Pictures/PunchStock RF; 22.25a–b: © Royalty-Free/Corbis; 22.25c: © Steven P. Lynch; 22.27a–b: © Digital Vision/Getty Images RF; 22.28a: © image 100/PunchStock RF; 22.28b: © Jeremy Woodhouse/Getty Images RF; 22.29a: © Royalty-Free/Corbis; 22.29b: © Photodisc Collection/Getty Images RF; 22.30a: © Royalty-Free/Corbis; 22.30b: © Brand X Pictures/PunchStock RF.

### Chapter 23

Opener: © Ralph Lee Hopkins/Getty Images; 23.5b: © Andreas M. Thurnherr; 23.8a: © LH Images/Alamy; 23.8b: © Paul Glendell/Alamy; 23.11: © Martin Strmiska/Alamy; 23.13: © Kenneth L. Smith, Jr./Oxford Scientific/Getty Images; 23.14: © Darryl Leniuk/Getty Images RF; 23.15: © Steven Trainoff Ph.D./Getty Images RF; 23.17: © WaterFrame/Alamy; 23.19: © Nature Picture Library/Alamy; 23.21: © Andrea Pickart; 23.22: © Tim Laman/National Geographic Creative; 23.24a: © Kazuo Ogawa/Getty Images; 23.24b: © Gary Meszaros/Visuals Unlimited/Corbis.

### Chapter 24

Opener: © José Fuste Raga/agefotostock; 24.3a: © Doug Sherman/Geofile; 24.3b: Photograph by Belinda Rain, courtesy of EPA/National Archives; 24.A(i): Brett Billings/U.S. Fish and Wildlife Service; 24.A(ii): © Ken Lucas/Visuals Unlimited; 24.A(iii): © Andrew N. Cohen, Center for Research on Aquatic Bioinvasions; 24.A(iv): © blickwinkel/Alamy; 24.5: © Morey Milbradt/Brand X Pictures/PunchStock RF; 24.7: © Mark Deeble and Victoria Stone/Getty Images; 24.8: Photo by Dennis Larson, USDA Natural Resources Conservation Service; 24.10a: © Robert Cable/Getty Images RF; 24.10b: © Comstock/PunchStock RF; 24.12: Photo by Jeff Vanuga, USDA Natural Resources Conservation Service; 24.17a: © Tom McHugh/Science Source; 24.17b: © Garold W. Sneegas/www.gwsphotos.com; 24.18: © 1999 Copyright IMS Communications Ltd./Capstone Design. All Rights Reserved RF.

### Chapter 25

Opener: © Zigmund Leszczynski/Animals Animals; 25.9e: © Paul Sutherland/National Geographic Creative; 25.A(1): © Fotosearch/SuperStock RF; 25.12: © George Grall/National Geographic Creative/Getty Images; 25.14: © Robert Glusic/Getty Images RF; 25.15a: J & K Hollingsworth/U.S Fish & Wildlife Service; 25.15b: © Nature Picture Library/Alamy RF; 25.B: George Gentry/U.S. Fish and Wildlife Service.

### Chapter 26

Opener: © George Grall/Getty Images; 26.2: Courtesy of NASA and GeoEye Inc. Copyright 2010. All rights reserved; 26.B: © Oak Ridge National Laboratory/photograph by Curtis Boles; 26.9: Image courtesy Liam Gumley, Space Science and Engineering Center, University of Wisconsin-Madison and the MODIS Science Team; 26.22: © Simon Fraser/Science Source; 26.29a: © PhotoAlto/PunchStock RF; 26.29b: © Galen Rowell/Corbis; 26.29i: © Don Croll.

### Chapter 27

Opener: © Vikash Tatayah/Mauritian Wildlife Foundation; 27.1a–b: © Patrick J. Bohlen; 27.A(ii): © Robert T. Smith/ardea.com; 27.5a: Courtesy of Experimental Lakes Area, Fisheries and Oceans Canada, photograph by E. Dubruyn. Reproduced with the permission of Her Majesty the Queen in Right of Canada, 2010; 27.6a: © JK Enright/Alamy RF; 27.6b: © RGB Ventures LLC dba SuperStock/Alamy; 27.B(1, 3): © The McGraw-Hill Companies, Inc./Peter Stiling, photographer; 27.9a: © Science VU/Visuals Unlimited; 27.9b: Provided by the Northern Research Station, Forest Service, USDA; 27.13: © Wen Zhenxiao/XinHua/Xinhua Press/Corbis; 27.14: © Frans Lanting/Frans Lanting Stock.

# Index

Note: Page numbers followed by *f* denotes additional figures; *t* denotes additional tables; G denotes glossary

# B

## C

# E

# F



# G

# H

## J



## K



## L

## N



## O

# P

# Q



# R

# S

# T